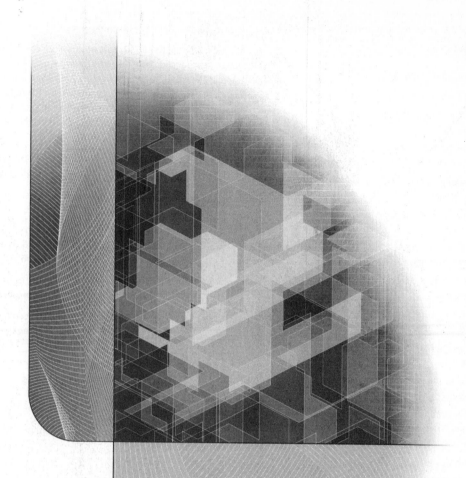

AutoCAD 2012: A Problem Solving Approach

autodesk Press

SHAM TICKOO

Professor
Department of Manufacturing Engineering Technology
Purdue University Calumet
Hammond, Indiana, USA

CADCIM TECHNOLOGIES

(www.cadcim.com)
USA

CENGAGE
Learning™

D1254063

Australia • Brazil • Japan • Korea • Mexico • Singapore • Spain • United Kingdom • Unite

AutoCAD 2012: A Problem Solving Approach
Sham Tickoo

Vice President, Editorial:
Dave Garza

Director of Learning Solutions:
Sandy Clark

Acquisitions Editor:
Stacy Masucci

Managing Editor:
Larry Main

Senior Product Manager:
John Fisher

Editorial Assistant:
Andrea Timpano

Vice President, Marketing:
Jennifer Baker

Marketing Director:
Deborah Yarnell

Marketing Manager:
Katie Hall

Associate Marketing Manager:
Jillian Borden

Senior Production Director:
Wendy Troeger

Senior Content Project Manager:
Glenn Castle

Senior Art Director:
David Arsenault

Technology Project Manager:
Joe Pliss

Technical Editors:
Gaurav Verma
D Swapna
Surinder Raina
(CADCIM Technologies)

Copy Editors:
Anju Jethwani
Rajendra Badola (CADCIM Technologies)

Cover image:
Courtesy of Shutterstock © Yury Zap.
Used by permission.

For product information and technology assistance, contact us at
Cengage Learning Customer & Sales Support,
1-800-354-9706

For permission to use material from this text or product, submit all requests online at **www.cengage.com/permissions**.

Further permissions questions can be e-mailed to
permissionrequest@cengage.com.

Library of Congress Control Number: 2010931966
ISBN-13: 978-1-1116-4850-3
ISBN-10: 1-1116-4850-6

Delmar
5 Maxwell Drive
Clifton Park, NY 12065-2919
USA

Cengage Learning is a leading provider of customized learning solutions with office locations around the globe, including Singapore, the United Kingdom, Australia, Mexico, Brazil, and Japan. Locate your local office at:
international.cengage.com/region

Cengage Learning products are represented in Canada by
Nelson Education, Ltd.

To learn more about Delmar, visit www.cengage.com/delmar
Purchase any of our products at your local college store or at our preferred online store www.cengagebrain.com

Notice to the Reader
Publisher does not warrant or guarantee any of the products described herein or perform any independent analysis in connection with any of the product information contained herein. Publisher does not assume, and expressly disclaims, any obligation to obtain and include information other than that provided to it by the manufacturer. The reader is expressly warned to consider and adopt all safety precautions that might be indicated by the activities described herein and to avoid all potential hazards. By following the instructions contained herein, the reader willingly assumes all risks in connection with such instructions. The publisher makes no representations or warranties of any kind, including but not limited to, the warranties of fitness for particular purpose or merchantability, nor are any such representations implied with respect to the material set forth herein, and the publisher takes no responsibility with respect to such material. The publisher shall not be liable for any special, consequential, or exemplary damages resulting, in whole or part, from the readers' use of, or reliance upon, this material.

Printed in the United States of America
1 2 3 4 5 6 7 14 13 12 11

DEDICATION

*To teachers, who make it possible to disseminate knowledge
to enlighten the young and curious minds
of our future generations*

*To students, who are dedicated to learning new technologies
and making the world a better place to live in*

THANKS

*To the faculty and students of the MET Department of
Purdue University Calumet for their cooperation*

To Employees of CADCIM Technologies for their valuable help

CONTENTS

Chapter 2: Getting Started with AutoCAD

Chapter 3: Starting with Advanced Sketching

Chapter 4: Working with Drawing Aids

Chapter 5: Editing Sketched Objects-I

Chapter 6: Editing Sketched Objects-II

Chapter 7: Creating Text and Tables

Chapter 8: Basic Dimensioning, Geometric Dimensioning, and Tolerancing

Chapter 9: Editing Dimensions

Chapter 10: Dimension Styles, Multileader Styles, and System Variables

Chapter 11: Adding Constraints to Sketches

Chapter 12: Technical Drawing with AutoCAD

Chapter 13: Isometric Drawings

Chapter 16: Plotting Drawings

Chapter 17: Template Drawings

AutoCAD Part II

Chapter 18: Working with Blocks

Chapter 19: Defining Block Attributes

Chapter 20: Understanding External References

Chapter 21: Working with Advanced Drawing Options

Chapter 22: Grouping and Advanced Editing of Sketched Objects

Chapter 23: Working with Data Exchange & Object Linking and Embedding

Chapter 24: The User Coordinate System

Chapter 25: Getting Started with 3D

Chapter 26: Creating Solid Models

Chapter 30: Mesh Modeling

AutoCAD Part III

Chapter 33: Accessing External Database

Chapter 34: Script Files and Slide Shows

Chapter 35: Creating Linetypes and Hatch Patterns

Chapter 36: Customizing the acad.pgp File

INTRODUCTION

AutoCAD, developed by Autodesk Inc., is the most popular PC-CAD system available in the market. Today, over 7 million people use AutoCAD and other AutoCAD based design products. 100% of the Fortune 100 firms are Autodesk customers and 98% of the Fortune 500 firms are Autodesk customers. AutoCAD's open architecture has allowed third-party developers to write an application software that has significantly added to its popularity. For example, the author of this book has developed a software package "**SMLayout**" for sheet metal products that generates a flat layout of various geometrical shapes such as transitions, intersections, cones, elbows, and tank heads. Several companies in Canada and the United States are using this software package with AutoCAD to design and manufacture various products. AutoCAD also facilitates customization that enables the users to increase their efficiency and improve their productivity.

The **AutoCAD 2012: A Problem Solving Approach** textbook contains a detailed explanation of AutoCAD 2012 commands and their applications to solve drafting and design problems. Every AutoCAD command is thoroughly explained with the help of examples and illustrations. This makes it easy for the users to understand its function and application in the drawing. After reading this textbook, you will be able to use AutoCAD commands to make a drawing, dimension a drawing, apply constraints to sketches, insert symbols, apply materials, render a scene as well as create text, blocks and dynamic blocks, 3D objects, drafting views of a model, surface objects, mesh objects, and solid models.

The book also covers basic drafting and design concepts that provide you with the essential drafting skills to solve the drawing problems in AutoCAD. These include orthographic projections, dimensioning principles, sectioning, auxiliary views, and assembly drawings. While going through this textbook, you will discover some new unique applications of AutoCAD that will have a significant effect on your drawings. In addition, you will be able to understand why AutoCAD has become such a popular software package and an international standard in PC-CAD.

Formatting Conventions Used in the Text

Please refer to the following list for the formatting conventions used in this textbook.

Convention

Example

- AutoCAD 2012 features in the text are designated by an asterisk symbol at the end of the feature.

  SURFNETWORK*, **SURFPATCH***

- All the exercises that are designated by a double asterisk symbol at the end of the question are courtesy of CADCIM Technologies.

  Draw a detail drawing whose top, side, and section views are given in Figure 15-53. Then, hatch the section view. **

- Command names are capitalized and bold.

  Example: The **MOVE** command

- A key icon appears when you have to respond by pressing the ENTER or the RETURN key.

  [Enter]

- Command sequences are indented. The responses are indicated in a boldface. The directions are indicated in italics and the comments are enclosed in parentheses.

  Command: **MOVE**
  Select object: **G**
  Enter group name: *Enter a group name (the group name is group1)*

- The methods of invoking a tool/option from the **Ribbon, Menu Bar, Quick Access toolbar, Tool Palettes, Application menu**, toolbars, Status Bar, and Command prompt are enclosed in a shaded box.

Ribbon:	Draw > Line
Menu Bar:	Draw > Line
Tool Palettes:	Draw > Line
Toolbars:	Draw > Line
Command:	LINE or L

- Icons are placed near the topics that are relevant for the *Certified Associate Exam* and *Certified Professional Exam*

  Object Properties

Naming Conventions Used in the Text

Tool

If you click on an item in a toolbar or a panel of the **Ribbon** and a command is invoked to create/edit an object or perform some action, then that item is termed as **tool**.

For example:
To Create: Line tool, **Circle** tool, **Extrude** tool
To Edit: Fillet tool, **Array** tool, **Stretch** tool
Action: Zoom tool, **Move** tool, **Copy** tool

If you click on an item in a toolbar or a panel of the **Ribbon** and a dialog box is invoked wherein you can set the properties to create/edit an object, then that item is also termed as **tool**, refer to Figure 1.

For example:
To Create: Define Attributes tool, **Create** tool, **Insert** tool
To Edit: Edit Attributes tool, **Block Editor** tool

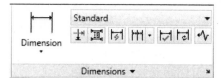

Figure 1 *Various tools in the Ribbon*

Button

If you click on an item in a toolbar or a panel of the Ribbon and the display of the corresponding object is toggled on/off, then that item is termed as **Button**. For example, **Grid** button, **Snap** button, **Ortho** button, **Properties** button, **Tool Palettes** button, and so on; refer to Figure 2.

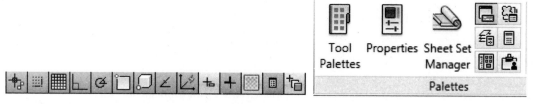

Figure 2 *Various buttons displayed in the Status Bar and Ribbon*

The item in a dialog box that has a 3d shape like a button is also termed as **Button**. For example, **OK** button, **Cancel** button, **Apply** button, and so on. Refer to Figure 3 given below for the terminologies used for the components in a dialog box.

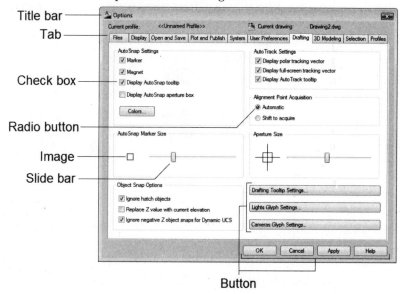

Figure 3 *The components in a dialog box*

Drop-down

A drop-down is one in which a set of common tools are grouped together for creating an object. You can identify a drop-down with a down arrow on it. These drop-downs are given a name based on the tools grouped in them. For example, **Circle** drop-down, **Fillet/Chamfer** drop-down, **Create Light** drop-down, and so on; refer to Figure 4.

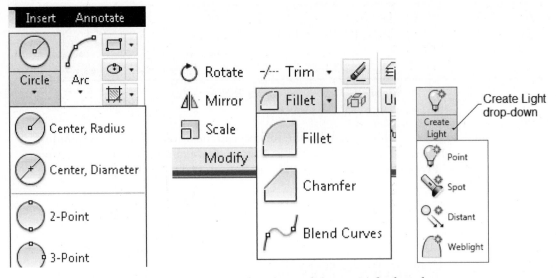

*Figure 4 The **Circle**, **Fillet/Chamfer**, and **Create Light** drop-downs*

Drop-down List

A drop-down list is one in which a set of options are grouped together. You can set various parameters using these options. You can identify a drop-down list with a down arrow on it. To know the name of a drop-down list, move the cursor over it; its name will be displayed as a tool tip. For example, **Lineweight** drop-down list, **Linetype** drop-down list, **Object Color** drop-down list, **Visual Styles** drop-down list, and so on; refer to Figure 5.

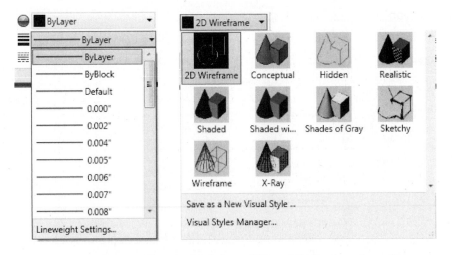

*Figure 5 The **LineWeight** and **Visual Styles** drop-down lists*

Options

Options are the items that are available in shortcut menu, drop-down list, Command prompt, **Properties** panel, and so on. For example, choose the **Properties** option from the shortcut menu displayed on right-clicking in the drawing area, refer to Figure 6.

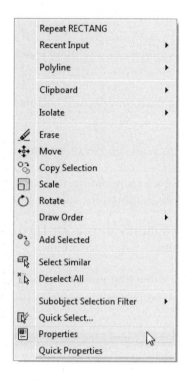
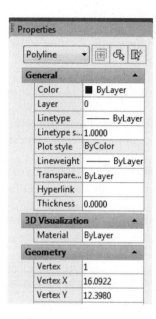

Figure 6 Options in the shortcut menu and the **Properties** *palette*

Tools and Options in Menu Bar

A menu bar consists of both tools and options. As mentioned earlier, the term **tool** is used to create/edit something or perform some action. For example, in Figure 7, the item Box has been used to create a box shaped surface, therefore it will be referred as **Box** tool.

Similarly, an option in the menu bar is one that is used to set some parameters. For example, in Figure 7, the item Linetype has been used to set/load the linetype, therefore it will be referred as an option.

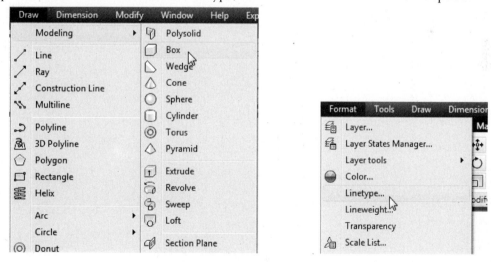

Figure 7 Tools and options in the menu bar

e-resource

This is an educational resource that creates a truly electronic classroom. It is a CD-ROM containing tools and instructional resources that enrich your classroom and reduce your preparation time. The features of **e-resource** link directly to the text and tie together to provide a unified instructional system. With **e-resource**, you can spend quality time teaching, rather than preparing to teach.

Features contained in an e-resource include:

Syllabus: Lesson plans created for each chapter. You have the option of using these lesson plans with your own course information.

Chapter Hints: Objectives and teaching hints that provide the basis for a lecture outline and help you present concepts and material.

Answers to Review Questions: These solutions enable you to grade and evaluate the end-of-chapter tests.

PowerPoint® Presentations: These slides provide the basis for a lecture outline and help you present concepts and material. The key points and concepts can be graphically highlighted for student retention.

Exam View Test Bank: Over 800 questions of varying levels of difficulty are provided in true/false and multiple-choice formats, so that you can assess the knowledge of students.

AVI Files: AVI files, listed by topic, enable you to view a quick video that illustrates and explains key concepts.

Drawing Files
The drawing files have been used in examples and exercises.

Publisher's Website
You can also get learning materials from the publisher's website:
www.cengage.com/cad/autodeskpress

AutoCAD
Part I

Author's Website

For Faculty: Please contact the author at **Stickoo@purduecal.edu** or **tickoo525@gmail.com** to access the website that contains the following:

1. PowerPoint presentations, program listings, and drawings used in this textbook.
2. Syllabus, chapter objectives and hints, and questions with answers for every chapter.

For Students: To download drawings, exercises, tutorials, and programs, please visit the author's website: *http://www.cadcim.com*. Alternatively, you can visit: *www.cengage.com/cad/autodeskpress*, choose the *Online Companions* option, and then click on the ***AutoCAD 2012: A Problem Solving Approach*** link to download student resources.

Chapter 1

Introduction to AutoCAD

CHAPTER OBJECTIVES

In this chapter, you will learn:
- *To start AutoCAD.*
- *About the components of the initial AutoCAD screen.*
- *To invoke AutoCAD commands from the keyboard, menu, toolbar, shortcut menu, Tool Palettes, and Ribbon.*
- *About the components of dialog boxes in AutoCAD.*
- *To start a new drawing using the New tool and the Startup dialog box.*
- *To save a work using various file-saving commands.*
- *To close a drawing.*
- *To open an existing drawing.*
- *To exit AutoCAD.*
- *Various options of AutoCAD's help.*
- *About the use of Active Assistance, Learning Assistance, and other interactive help topics.*

KEY TERMS

- *Initial Setup*
- *AutoCAD Screen Components*
- *Ribbon*
- *Application Menu*
- *Tool Palettes*

- *Menu Bar*
- *Toolbar*
- *New*
- *Save*
- *Save As*
- *Close*

- *STARTUP*
- *Open*
- *Partial open*
- *Drawing Recovery Manager*
- *Workspaces*

- *Help*
- *Autodesk Exchange*

STARTING AutoCAD

After you have installed AutoCAD 2012, an AutoCAD 2012 icon is displayed on the desktop. You can start AutoCAD by double-clicking on it. You can also load AutoCAD from the Windows taskbar by choosing the **Start** button at the bottom left corner of the screen (default position). On doing so, a menu will be displayed. From this menu, choose **Programs** to display program folders. Now, choose **Autodesk > AutoCAD 2012** folder to display AutoCAD programs and then choose **AutoCAD 2012-English** to start AutoCAD, see Figure 1-1.

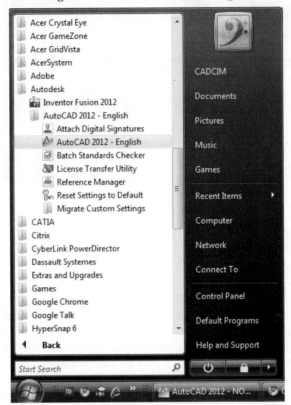

Figure 1-1 Starting AutoCAD 2012 using the ***Start*** *Menu*

When you start AutoCAD 2012, the **Autodesk Exchange** window will be displayed, as shown in Figure 1-2. In this window, various links are available providing information on enhancements, new features, products and services, subscription, and so on. You can click on a link to find information contained in that link. For example, in the **New in AutoCAD 2012** area, you can learn about new features and enhancements of AutoCAD 2012.

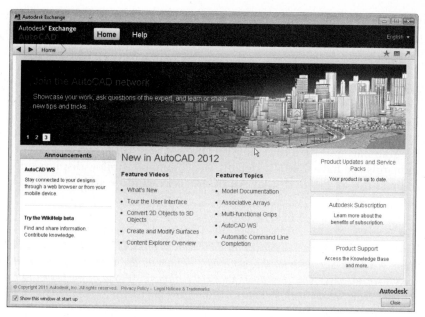

*Figure 1-2 The **Autodesk Exchange** window of AutoCAD 2012*

AutoCAD SCREEN COMPONENTS*

Various components of the initial AutoCAD screen are drawing area, command window, menu bar, several toolbars, model and layout tabs, and status bar (Figure 1-3). A title bar that has AutoCAD symbol and the current drawing name is displayed on top of the screen.

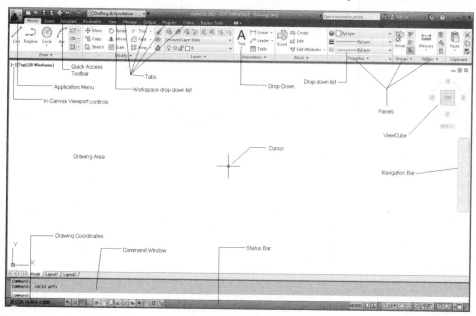

*Figure 1-3 AutoCAD screen components in AutoCAD **Drafting** & **Annotation** Workspace*

Drawing Area

The drawing area covers the major portion of the screen. In this area, you can draw the objects and use the commands. To draw the objects, you need to define the coordinate points, which can be selected by using your pointing device. The position of the pointing device is represented on the screen by the cursor. There is a coordinate system icon at the lower left corner of the drawing area. The window also has the standard Windows buttons such as close, minimize, scroll bar, and so on, on the top right corner. These buttons have the same functions as for any other standard window.

Command Window

The command window at the bottom of the drawing area has the Command prompt where you can enter the commands. It also displays the subsequent prompt sequences and the messages. You can change the size of the window by placing the cursor on the top edge (double line bar known as the grab bar) and then dragging it. This way you can increase its size to see all the previous commands you have used. By default, the command window displays only three lines. You can also press the F2 key to display **AutoCAD Text window**, which displays the previous commands and prompts.

Tip
You can hide all toolbars displayed on the screen by pressing the CTRL+0 keys or by choosing **View > Clean Screen** *from the menu bar. To turn on the display of the toolbars again, press the CTRL+0 keys. Note that the 0 key on the numeric keypad of the keyboard cannot be used for the* **Clean Screen** *option. You can also choose the* **Clean Screen** *button in the Status Bar to hide all toolbars.*

Navigation Bar

The **Navigation Bar** is displayed in the drawing area and contains navigation tools. These tools are grouped together, refer to Figure 1-4 and are discussed next.

SteeringWheels

The SteeringWheels has a set of navigation tools such as pan, zoom, and so on. You will learn more about the SteeringWheel in the later chapters.

Pan

This tool allows you to view the portion of the drawing that is outside the current display area. To do so, choose this tool, press and hold the left mouse button, and then drag the drawing area. Press ESC to exit this command.

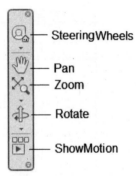

Figure 1-4 Tools in the Navigation Bar

Zoom

The tools to enlarge the view of the drawing on the screen without affecting the actual size of the objects are grouped together. You will learn more about zoom in later chapters.

Rotate

The tools to rotate the view in the 3D space are grouped together.

ShowMotion

Choose this button to capture different views in a sequence and animate them when required.

ViewCube

ViewCube is available on the top right corner of the drawing area and is used to switch between the standard and isometric views or roll the current view. The ViewCube and its options are discussed in later chapters.

In-canvas Viewport Controls

In-canvas Viewport Controls is available on the top left corner of the drawing screen. It enables you to change the view, visual style as well as the viewport.

Status Bar

The Status Bar is displayed at the bottom of the screen and is called Application Status Bar. It contains some useful information and buttons (see Figure 1-5) that make it easy to change the status of some AutoCAD functions. You can toggle between the on and off states of most of these functions by choosing them.

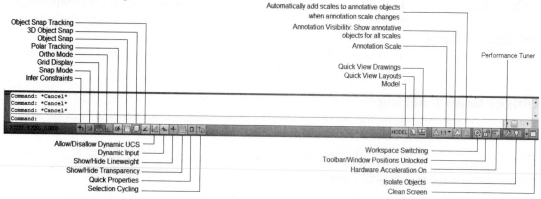

Figure 1-5 *The Status Bar displayed in the* **Drafting & Annotation** *workspace*

Drawing Coordinates

The information about the coordinates is displayed on the left-corner of the Status Bar. You can select this coordinate button to toggle between the on and off states. The **COORDS** system variable controls the type of display of the coordinates. If the value of the **COORDS** variable is set to 0, the coordinate display is static, that is, the coordinate values displayed in the Status Bar change only when you specify a point. If the value of the **COORDS** variable is set to 1 or 2, the coordinate display is dynamic. When the variable is set to 1, AutoCAD constantly displays the absolute coordinates of the graphics cursor with respect to the UCS origin. The polar coordinates (length<angle) are displayed if you are in an AutoCAD command and the **COORDS** variable is set to 2. You can also click on the **Drawing Coordinates** area to change the coordinate status from on to off and vice versa.

Infer Constraints

If this button is chosen then some of the geometric constraints will be automatically applied to sketch while it is drawn.

Snap Mode

If the **Snap Mode** button is chosen, the snap mode is on. So, you can move the cursor in fixed increments. The F9 key acts as a toggle key to turn the snap off or on.

Grid Display

In AutoCAD, the grid lines are used as reference lines to draw objects. If the **Grid Display** button is chosen, the grid display is on and the grid lines are displayed on the screen. The F7 function key can be used to turn the grid display on or off.

Ortho Mode

If the **Ortho Mode** button is chosen, you can draw lines at right angles only. You can use the F8 function key to turn ortho on or off.

Polar Tracking

If you turn the polar tracking on, the movement of the cursor is restricted along a path based on the angle set as the polar angle. Choose the **Polar Tracking** button to turn the polar tracking on. You can also use the F10 function key to turn on this option. Note that turning the polar tracking on, automatically turns off the ortho mode.

Object Snap

When the **Object Snap** button is chosen, you can use the running object snaps to snap on to a point. You can also use the F3 function key to turn the object snap on or off. The status of **OSNAP** (off or on) does not prevent you from using the immediate mode object snaps.

3D Object Snap

When this button is chosen, you can snap the key point on a solid or a surface. You can also use the F4 function key to turn on or off the 3D object snap.

Object Snap Tracking

When you choose this button, the inferencing lines will be displayed. Inferencing lines are dashed lines that are displayed automatically when you select a sketching tool and track a particular keypoint on the screen. Choosing this button turns the object snap tracking on or off.

Allow/Disallow Dynamic UCS

Choosing this button allows or disallows the use of dynamic UCS. Allowing the dynamic UCS ensures that the XY plane of the UCS got dynamically aligned with the selected face of the model. You can also use the F6 function key to turn the **Dynamic UCS** button on or off.

Dynamic Input

The **Dynamic Input** button is used to turn the **Dynamic Input** on or off. Turning it on facilitates the heads-up design approach because all commands, prompts, and dimensional inputs will now be displayed in the drawing area and you do not need to look at the Command prompt all the time. This saves the design time and also increases the efficiency of the user. If the **Dynamic Input** mode is turned on, you will be allowed to enter the commands through the **Pointer Input** boxes, and the numerical values through the **Dimensional Input** boxes. You will also be allowed to select the command options through the **Dynamic Prompt** options in the graphics window. To turn the **Dynamic Input** on or off, use the F12 key.

Show/Hide Lineweight

Choosing this button in the Status Bar allows you to turn on or off the display of lineweights in the drawing. If this button is not chosen, the display of lineweight will be turned off.

Show/Hide Transparency

This button is available in the Status Bar and is chosen to turn on or off the transparency set for a drawing. You can set the transparency in the **Properties** panel or in the layer in which the sketch is drawn.

Quick Properties

If you select a sketched entity when this button is chosen in the Status Bar, the properties of the selected entity will be displayed in a panel.

Selection Cycling

When this button is chosen, you can cycle through the objects to be selected, if they are overlapping or close to other entities. On selecting an entity when this button is chosen, the **Selection** list box with a list of the entities that can be selected will be displayed.

Model

The **Model** button is chosen by default because you are working in the model space to create drawings. You will learn more about the model space in later chapters.

Quick View Layouts

Choose this button to display a panel from which you can choose the layout you need to invoke.

Quick View Drawings

Choose this button to display a panel from which you can choose the drawings you need to invoke.

Annotation Scale

The annotation scale controls the size and display of the annotative objects in the model space. The **Annotation Scale** button has a flyout that displays all the annotation scales available for the current drawing.

Annotation Visibility

This button is used to control the visibility of the annotative objects that do not support the current annotation scale in the drawing area.

Automatically Add Scale

This button, if chosen, automatically adds all the annotation scales that are set current to all the annotative objects present in the drawing.

Toolbar/Window Positions Unlocked

The **Toolbar/Window Positions Unlocked** button is used to lock and unlock the positions of toolbars and windows. When you click on this icon, a shortcut menu is displayed. Choosing the **Floating Toolbars/Panels** option allows you to lock the current position of the floating toolbars. A checkmark will be displayed in the shortcut menu on the type of toolbars that are currently locked. Choosing the **Docked Toolbars/Panels** option from the shortcut menu allows you to lock the current position of all the docked toolbars. Similarly, you can lock or unlock the position of floating and docked windows, such as the **Properties** window or the **Tool Palettes**. If you move the cursor on the **All** option, a cascading menu is displayed that provides the option to lock and unlock all the toolbars and windows.

Note
*The **LOCKUI** system variable is responsible for the locking and unlocking of the toolbars and windows. The following are the values of the system variable:*

Lockui<0> No toolbar or window locked
Lockui<1> Locks all docked toolbars
Lockui<2> Locks all docked windows
Lockui<4> Locks all floating toolbars
Lockui<8> Locks all floating windows

Hardware Acceleration On

This button is used to set the performance of the software to an acceptable level.

Isolate Objects

This button is used to hide or isolate objects from the drawing area. On choosing this button, a flyout will be displayed with two options. Choose the required option from this flyout and then select the objects to hide or isolate. To end isolation or display a hidden object, click this button again and choose the **End Object Isolation** option.

Drawing Status Bar

The **Drawing Status Bar** is displayed in between the drawing area and the command window. Choose the **Application Status Bar Menu** arrow and choose the **Drawing Status Bar** option from the flyout; the **Drawing Status Bar** will be displayed, as shown in Figure 1-6. Turn on the **Drawing Status Bar**; the **Annotation Scale**, **Annotation Visibility**, and **Automatically Add Scale** buttons will move automatically to the **Drawing Status Bar**. If you turn off the **Drawing Status Bar**, these buttons will move back to the **Application Status Bar**.

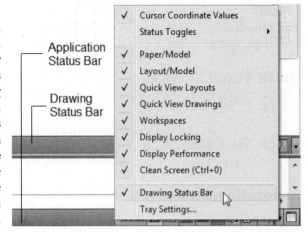

Figure 1-6 The Drawing Status Bar

Tray Settings

Choose the **Tray Settings** option from the flyout displayed on clicking the arrow in the **Application Status Bar**; the **Tray Settings** dialog box will be displayed. You can control the display of icons and notifications in the tray at the right end of the status bar by selecting appropriate options.

Clean Screen

The **Clean Screen** button is at the lower right corner of the screen. This button, when chosen, displays an expanded view of the drawing area by hiding all the toolbars except the command window, Status Bar, and menu bar. The expanded view of the drawing area can also be displayed by choosing **View > Clean Screen** from the menu bar or by using the CTRL+0 keys. Choose the **Clean Screen** button again to restore the previous display state.

Status Toggles

You can hide the display of some of the buttons in the Status Bar. To do so, right-click on the **Application Status Bar**; a shortcut menu will be displayed. Move the cursor on the **Status Toggles** option in the shortcut menu; a cascading menu will be displayed. Clear the check mark near the names of the corresponding buttons in the cascading menu.

Plot/Publish Details Report Available

This icon is displayed when some plotting or a publishing activity was performed in the background. When you click on this icon, the **Plot and Publish Details** dialog box, which provides the details about the plotting and publishing activity, will be displayed. You can copy this report to the clipboard by choosing the **Copy to Clipboard** button from the dialog box.

Manage Xrefs

The **Manage Xrefs** icon is displayed whenever an external reference drawing is attached to the selected drawing. This icon displays a message and an alert whenever the Xreffed drawing needs to be reloaded. To find detailed information regarding the status of each Xref in the drawing and the relation between the various Xrefs, click on the **Manage Xrefs** icon; the **External References Palette** will be displayed. The Xrefs are discussed in detail in Chapter 20, Understanding External References.

INVOKING COMMANDS IN AutoCAD*

On starting AutoCAD, when you are in the drawing area, you need to invoke AutoCAD commands to perform any operation. For example, to draw a line, first you need to invoke the **LINE** command and then define the start point and the endpoint of the line. Similarly, if you want to erase objects, you must invoke the **ERASE** command and then select the objects for erasing. AutoCAD has provided the following methods to invoke the commands:

Keyboard	Ribbon	Application Menu	Tool Palettes
Tool Palettes	Menu bar	Shortcut menu	Toolbar

Keyboard

You can invoke any AutoCAD command from the keyboard by typing the command name and then pressing the ENTER key. As you type the first letter of command, AutoCAD displays all available commands starting with the letter typed. If the **Dynamic Input** is on and the cursor is in the drawing area, by default, the command will be entered through the **Pointer Input** box. The **Pointer Input** box is a small box displayed on the right of the cursor, as shown in Figure 1-7. However, if the cursor is currently placed on any toolbar or menu bar, or if the **Dynamic Input** is turned off, the command will be entered through the Command prompt. Before you enter a command, the Command prompt is displayed as the last line in the command window area. If it is not displayed, you must cancel the existing command by pressing the ESC (Escape) key. The following example shows how to invoke the **LINE** command using the keyboard:

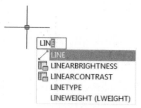

*Figure 1-7 The **Pointer Input** box displayed when the **Dynamic Input** is on*

Command: **LINE** or **L** [Enter]   (L is command alias)

Ribbon

In AutoCAD, you can also invoke a tool from the Ribbon. The tools for creating, modifying, and annotating the 2D & 3D designs are available in the panels instead of being spread out in the entire drawing area in different toolbars and menus, see Figure 1-8.

*Figure 1-8 The **Ribbon** for the **Drafting & Annotation** workspace*

When you start the AutoCAD session for the first time, by default the **Ribbon** is displayed horizontally below the **Quick Access Toolbar**. The **Ribbon** consists of various tabs. The tabs have different panels, which in turn, have tools arranged in rows. Some of the tools have small black down arrow. This indicates that the tools having similar functions are grouped together. To choose a tool, click on the down arrow; a drop-down will be displayed. Choose the required tool from the drop-down displayed. Note that if you choose a tool from the drop-down, the corresponding command will be invoked and the tool that you have chosen will be displayed in the panel. For example, to draw a circle using the **2-Point** option, click on the down arrow next to the **Center, Radius** tool in the **Draw** panel of the **Home** tab; a flyout will be displayed. Choose the **2-Point** tool from the flyout and then draw the circle. You will notice that the **2-Point** tool is displayed in place of the **Center, Radius** tool. In this textbook, the tool selection sequence will be written as, choose the **2-Point** tool from **Home > Draw > Circle** drop-down.

Choose the down arrow to expand the panel. You will notice that a push pin is available at the left end of the panel. Click on the push pin to keep the panel in the expanded state. Also, some of the panels have an inclined arrow at the lower-right corner. When you left click on an inclined arrow, a dialog box is displayed. You can define the setting of the corresponding panel in the dialog box.

You can reorder the panels in the tab. To do so, press and hold the left mouse button on the panel to be moved and drag it to the required position. To undock the **Ribbon**, right-click on the blank space in the **Ribbon** and choose the **Undock** option. You can move, resize, anchor, and auto-hide the **Ribbon** using the shortcut menu that will be displayed when you right-click on the heading strip. To anchor the floating **Ribbon** to the left or right of the drawing area in the vertical position, right-click on the heading strip of the floating **Ribbon**; the shortcut menu is displayed. Choose the corresponding option from this shortcut menu. The **Auto-hide** option will hide the **Ribbon** into the heading strip and will display it only when you move the cursor over this strip.

You can customize the display of tabs and panels in the **Ribbon**. To customize the **Ribbon**, right-click on any one of the tools in it; a shortcut menu will be displayed. On moving the cursor over one of the options, a flyout will be displayed with a tick mark before all options and the corresponding tab or panel will be displayed in the **Ribbon**. Select/clear appropriate option to display/hide a particular tab or panel.

Application Menu

The **Application Menu** is available at the top-left of the AutoCAD window. It contains some of the tools that are available in the **Standard** toolbar. Click the down arrow on the **Application Menu** to display the tools, as shown in Figure 1-9. You can search a command using the search field on the top of the **Application Menu**. To search a tool, enter the complete or partial name of the command in the search field; the possible tool list will be listed. If you click on a tool from the list, the corresponding command will get activated.

By default, the **Recent Document** button is chosen in the **Application Menu**. Therefore, the recently opened drawings will be listed. If you have opened multiple drawing files, choose the **Open Documents** button; the documents that are opened will be listed in the **Application Menu**. To set the preferences of the file, choose the **Options** button available at the bottom-right of the **Application Menu**. To exit AutoCAD, choose the **Exit** button next to the **Options** button.

Tool Palettes

AutoCAD has provided **Tool Palettes** as an easy and convenient way of placing and sharing hatch patterns and blocks in the current drawing. By default, the **Tool Palettes** are not displayed. Choose the **Tool Palettes** button from the **Palettes** panel in the **View** tab or choose the CTRL+3 keys to display the **Tool Palettes** as a window on the right of the drawing area. You can resize the **Tool Palettes** using the resizing cursor that is displayed when you place the cursor on the top or bottom extremity of the **Tool Palettes**. The **Tool Palettes** are discussed in detail in Chapter 14, *Hatching Drawings*.

Menu Bar

You can also select commands from the menu bar. Menu Bar is not displayed by default. To display the menu bar, choose the down arrow in the **Quick Access Toolbar**; a flyout is displayed. Choose the **Show Menu Bar** option from it; the menu bar will be displayed. As you move the cursor over the menu bar, different titles are highlighted. You can choose the desired item by left-clicking on it; the corresponding menu is displayed directly under the title. You can invoke a command by left-clicking on a menu. Some of the menu items display an arrow on the right side, which indicates that they have a cascading menu. The cascading menu provides various options to execute the same AutoCAD command. You can display the cascading menu by choosing the menu item or by moving the arrow pointer to the right of that item. You can then choose any item from the cascading menu by highlighting the item or command and pressing the pick button of your pointing device. For example, to draw an ellipse using the **Center** option, choose the **Draw** menu and then choose the **Ellipse** option; a cascading menu will be displayed. From the cascading menu, choose the **Center** option. In this text, this command selection sequence will be referenced as choosing **Draw > Ellipse > Center** from the menu bar.

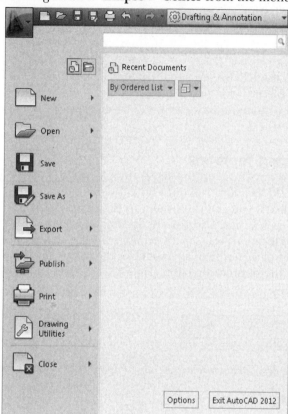

Figure 1-9 The Application Menu

Toolbar

Toolbars are not displayed by default. To display a toolbar, choose the **View** tab in the **Ribbon** and click on **Toolbars** in the **Windows** panel; a flyout will be displayed. Move the cursor over the **AutoCAD** option; a list of toolbars will be displayed. Select the required toolbar. Alternatively, display the menu bar and then choose **Tools > Toolbars > AutoCAD** from it; a list of toolbars will be displayed. Select the required toolbar.

In a toolbar, the similar tools representing various AutoCAD commands are grouped together. When you move the cursor over the button of a toolbar, the button gets lifted and a three-dimensional (3D) box encloses it. The tooltip (name of the tool and related information) is also displayed below the tool. Once you have located the desired tool, left-click on it to invoke the corresponding command. For example, you can invoke the **LINE** command by choosing the **Line** tool from the **Draw** toolbar, see Figure 1-10.

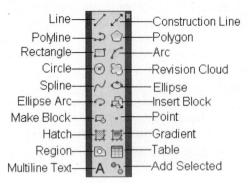

*Figure 1-10 The **Draw** toolbar*

Some of the tools in a toolbar have a small triangular arrow at the lower-right corner. This indicates that the tool has a flyout attached to it. If you press and hold the left mouse button on those tools, a flyout containing more tools will be displayed. Choose the required tool from this flyout.

Moving and Resizing Toolbars

Toolbars can be moved anywhere on the screen by placing the cursor on the strip and then dragging it to the desired location. You must hold the left mouse button down while dragging. While moving the toolbars, you can dock them to the top or sides of the screen by dropping them in the docking area. You may also prevent docking by holding the CTRL key when moving the toolbar to a desired location. You can also change the size of a toolbar by placing the cursor anywhere on the border of the toolbar where it takes the shape of a double arrow (Figure 1-11), and then pulling it in the desired direction (Figure 1-12). You can also customize toolbars to meet your requirements.

*Figure 1-11 The **Draw** toolbar*   *Figure 1-12 The **Draw** toolbar reshaped*

Shortcut Menu

AutoCAD has provided shortcut menus as an easy and convenient way of invoking the recently used tools. These shortcut menus are context-sensitive, which means that the tools present in them are dependent on the place/object for which they are displayed. A shortcut menu is invoked by right-clicking and is displayed at the cursor location. You can right-click anywhere in the drawing area to display the general shortcut menu. It generally contains an option to select the previously invoked tool again, apart from the common tools for Windows, refer to Figure 1-13.

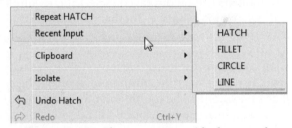

Figure 1-13 Shortcut menu with the recently used commands

If you right-click in the drawing area while a command is active, a shortcut menu is displayed, containing the options of that particular command. Figure 1-14 shows the shortcut menu when the **Polyline** tool is active.

If you right-click on the **Layout** tab, a shortcut menu wil be displayed, containing the options for layouts (Figure 1-15).

You can also right-click on the command window to display the shortcut menu. This menu displays the six most recently used commands and some of the window options like **Copy** and **Paste** (Figure 1-16). The commands and their prompt entries are displayed in the History window (previous command lines not visible) and can be selected, copied, and pasted in the command line using the shortcut menu. As you press the up arrow key, the previously entered commands are displayed in the command window. Once the desired command is displayed at the Command prompt, you can execute it by simply pressing the ENTER key. You can also copy and edit any previously invoked command by locating it in the History window and then selecting the lines.After selecting the desired command lines from the History window, right-click to display a shortcut menu. Choose Copy from the menu and then paste the selected lines at the end of the command line.

You can right-click on the coordinate display area of the Status Bar to display the shortcut menu. This menu contains the options to modify the display of coordinates, as shown in Figure 1-17. You can also right-click on any of the toolbars to display the shortcut menu from where you can choose any toolbar to be displayed.

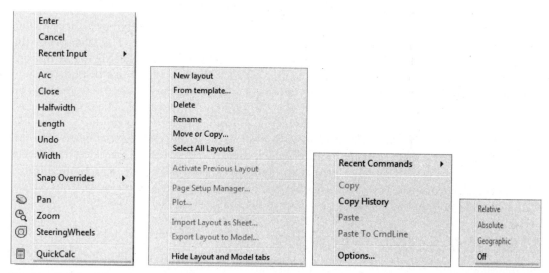

Figure 1-14 *Shortcut menu with the POLYLINE command active* **Figure 1-15** *Shortcut menu for the Layout tab* **Figure 1-16** *Command line window shortcut menu* **Figure 1-17** *The Status Bar shortcut menu*

AutoCAD DIALOG BOXES

There are certain commands, which when invoked, display a dialog box. A dialog box is a convenient method of a user interface. When you choose an item in the menu bar with the ellipses [...], it displays the dialog box. For example, **Options** in the **Tools** menu displays the **Options** dialog box. A dialog box contains a number of parts like the dialog label, radio buttons, text or edit boxes, check boxes, slider bars, image boxes, and command buttons. These components are also referred to as tiles. Some of the components of a dialog box are shown in Figure 1-18.

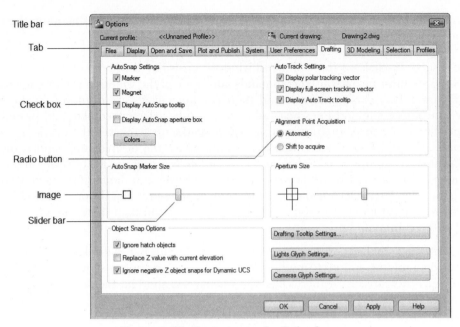

Title bar —
Tab —
Check box —
Radio button —
Image —
Slider bar —

Figure 1-18 *Components of a dialog box*

You can select the desired tile using the pointing device, which is represented by an arrow when a dialog box is invoked. The titlebar displays the name of the dialog box. The tabs specify the various sections with a group of related options under them. The check boxes are toggle options for making the particular option available or unavailable. The drop-down list displays an item and an arrow on the right which when selected displays a list of items to choose from. You can make a selection in the radio buttons. Only one can be selected at a time. The image displays the preview image of the item selected. The text box is an area where you can enter a text like a file name. It is also called an edit box, because you can make any change to the text entered. In some dialog boxes, there is the [...] button, which displays another related dialog box. There are certain buttons (**OK**, **Cancel**, **Help**) at the bottom of the dialog box. The name implies their functions. The button with a dark border is the default button. The dialog box has a Help button for getting help on the various features of the dialog box.

STARTING A NEW DRAWING

Application Menu: New › Drawing	**Command:** NEW or QNEW
Quick Access Toolbar: New	**Menu Bar:** New › Drawing

You can open a new drawing using the **New** tool in the **Quick Access Toolbar**. When you invoke the **New** tool, by default AutoCAD will display the **Select template** dialog box, as shown in Figure 1-19. This dialog box displays a list of default templates available in AutoCAD 2012. The default template is *acad.dwt*, which starts the 2D drawing environment. You can select the *acad3D.dwt* template to start the 3D modeling environment. Alternatively, you can select any other template to start a new drawing that will use the settings of the selected template. You can also open any drawing without using any template either in metric or imperial system. To do so, choose the down arrow on the right of the **Open** button and select the **Open with no Template-Metric** option or the **Open with no Template-Imperial** option from the flyout.

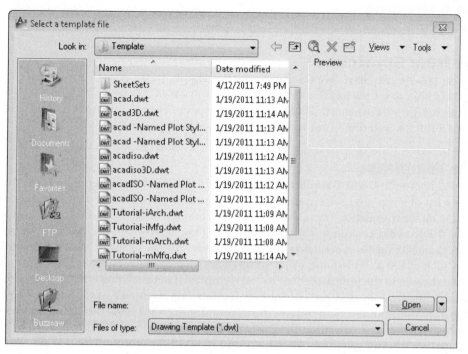

Figure 1-19 *The Select template dialog box*

You can also open a new drawing using the **Use a Wizard** and **Start from Scratch** options from the **Create New Drawing** dialog box. By default, this dialog box is not invoked. To invoke the **Create New Drawing** dialog box, enter **STARTUP** in the command window and then enter **1** as the new value for this system variable. After setting 1 as the new value for the system variable, whenever you invoke the **New** tool, the **Create New Drawing** dialog box will be displayed, as shown in Figure 1-20. The options in this dialog box are discussed next.

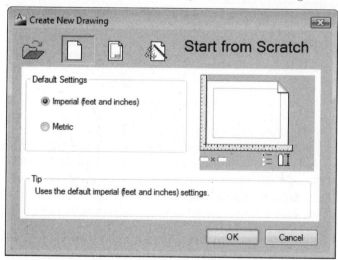

Figure 1-20 *The Create New Drawing dialog box*

Note

If you have started a new AutoCAD session with the STARTUP variable set to 1, then the Startup dialog box is displayed instead of the Create New Drawing dialog box.

Open a Drawing

By default, this option is not available.

Start from Scratch

When you choose the **Start from Scratch** button (Figure 1-20), AutoCAD provides you with options to start a new drawing that contains the default AutoCAD setup for Imperial (*Acad.dwt*) or Metric drawings (*Acadiso.dwt*). If you select the Imperial default setting, the limits are 12X9, text height is 0.20, and dimensions and linetype scale factors are 1.

Use a Template

When you choose the **Use a Template** button in the **Create New Drawing** dialog box, AutoCAD displays a list of templates, see Figure 1-21. The default template file is *acad.dwt* or *acadiso.dwt*, depending on the installation. You can directly start a new file in the 2D sketching environment by selecting the *acad.dwt* or *acadiso.dwt* template. If you use a template file, the new drawing will have the same settings as specified in the template file. All the drawing parameters of the new drawing such as units, limits, and other settings are already set according to the template file used. The preview of the template file selected is displayed in the dialog box. You can also define your own template files that are customized to your requirements (see Chapter 17, *Template Drawings*). To differentiate the template files from the drawing files, the template files have a *.dwt* extension whereas the drawing files have a *.dwg* extension. Any drawing file can be saved as a template file. You can use the **Browse** button to select other template files. When you choose the **Browse** button, the **Select a template file** dialog box is displayed with the **Template** folder open, displaying all the template files.

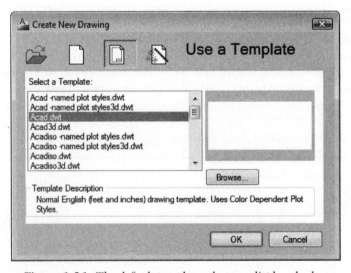

Figure 1-21 *The default templates that are displayed when you choose the **Use a Template** button*

Use a Wizard

The **Use a Wizard** option allows you to set the initial drawing settings before actually starting a new drawing. When you choose the **Use a Wizard** button, AutoCAD provides you with the option for using the **Quick Setup** or **Advanced Setup**, see Figure 1-22. In the **Quick Setup**, you can specify the units and the limits of the work area. In the **Advanced Setup**, you can set the units, limits, and the different types of settings for a drawing.

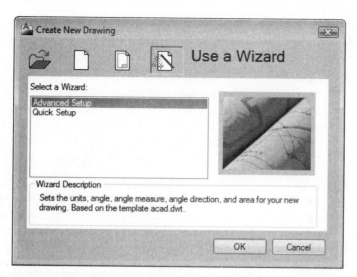

Figure 1-22 *The wizard options displayed when you choose the* **Use a Wizard** *button*

Advanced Setup

This option allows you to preselect the parameters of a new drawing such as the units of linear and angular measurements, type and direction of angular measurements, approximate area desired for the drawing, precision for displaying the units after decimal, and so on. When you select the **Advanced Setup** wizard option from the **Create New Drawing** dialog box and choose the **OK** button, the **Advanced Setup** wizard is displayed. The **Units** page is displayed by default, as shown in Figure 1-23.

This page is used to set the units for measurement in the current drawing. You can select the required unit of measurement by selecting the respective radio button. You will notice that the preview image is modified accordingly. The different units of measurement you can choose from are Decimal, Engineering, Architectural, Fractional, and Scientific. You can also set the precision for the measurement units by selecting it from the **Precision** drop-down list.

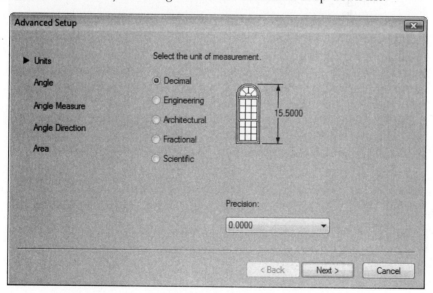

Figure 1-23 *The* **Units** *page of the* **Advanced Setup** *wizard*

Choose the **Next** button to open the **Angle** page, as shown in Figure 1-24. You will notice that an arrow appears on the left of **Angle** in the **Advanced Setup** wizard. This suggests that this page is current.

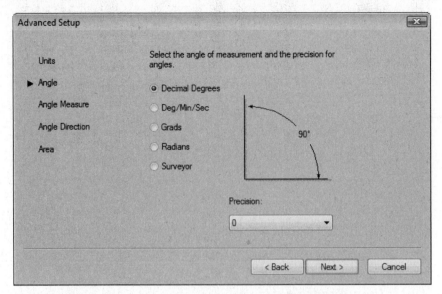

*Figure 1-24 The **Angle** page of the **Advanced Setup** wizard*

This page is used to set the units for angular measurements and its precision. The units for angle measurement are Decimal Degrees, Deg/Min/Sec, Grads, Radians, and Surveyor. The units for angle measurement can be set by selecting any one of these radio buttons as required. The preview of the selected angular unit is displayed on the right of the radio buttons. The precision format changes automatically in the **Precision** drop-down list depending on the angle measuring system selected. You can then select the precision from the drop-down list.

The next page is the **Angle Measure** page, as shown in Figure 1-25. This page is used to select the direction of the baseline from which the angles will be measured. You can also set your own direction by selecting the **Other** radio button and then entering the value in its edit box. This edit box is available when you select the **Other** radio button.

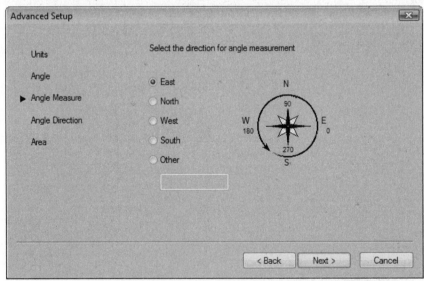

*Figure 1-25 The **Angle Measure** page of the **Advanced Setup** wizard*

Choose **Next** to display the **Angle Direction** page (Figure 1-26) to set the orientation for the angle measurement. By default the angles are positive, if measured in a counterclockwise direction. This is because the **Counter-Clockwise** radio button is selected. If you select the **Clockwise** radio button, the angles will be considered positive when measured in the clockwise direction.

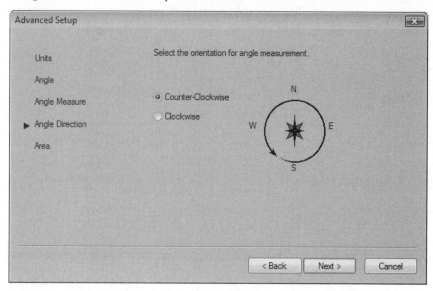

*Figure 1-26 The **Angle Direction** page of the **Advanced Setup** wizard*

To set the limits of the drawing, choose the **Next** button; the **Area** page will be displayed, as shown in Figure 1-27. You can enter the width and length of the drawing area in the respective edit boxes.

 Note
*Even after you increase the limits of the drawing, the drawing display area is not increased. You need to invoke the **Zoom All** tool from the Navigation Bar to increase the drawing display area.*

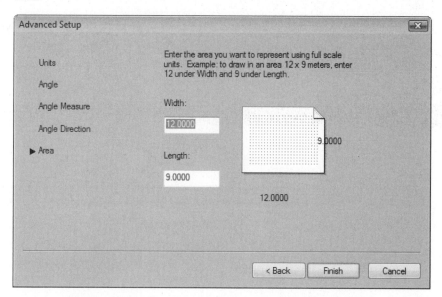

*Figure 1-27 The **Area** page of the **Advanced Setup** wizard*

Quick Setup

When you select the **Quick Setup** option and choose the **OK** button, the **QuickSetup** wizard is displayed. This wizard has two pages: **Units** and **Area**. The **Units** page is opened by default, as shown in Figure 1-28. The options in the **Units** page are similar to those in the **Units** page of the **Advanced Setup** wizard. The only difference is that you cannot set the precision for the units in this wizard.

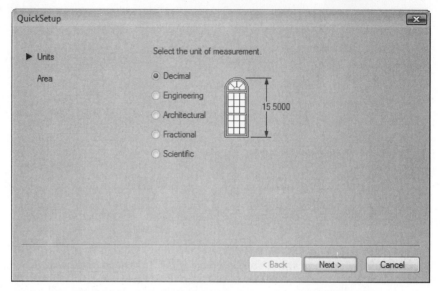

Figure 1-28 *The **Units** page of the **QuickSetup** wizard*

Choose **Next** to display the **Area** page, as shown in Figure 1-29. The **Area** page of the **QuickSetup** is similar to that of the **Advanced Setup** wizard. In this page, you can set the drawing limits.

Tip
*By default, when you open an AutoCAD session, a drawing opens automatically. But you can open a new drawing using options such as **Start from Scratch** and **Use a Wizard** before entering into AutoCAD environment using the **Startup** dialog box. As mentioned earlier, the display of the **Startup** dialog box is turned off by default. Refer to the section of **Starting a New Drawing** to know how to turn on the display of this dialog box.*

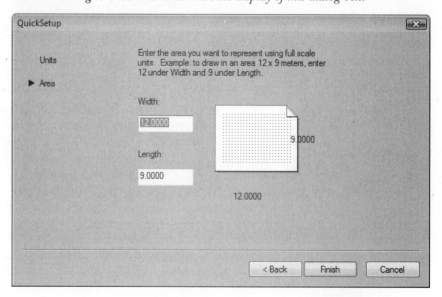

Figure 1-29 *The **Area** page of the **QuickSetup** wizard*

SAVING YOUR WORK

Application Menu: SAVEAS, SAVE	**Command:** QSAVE, SAVEAS, SAVE
Quick Access Toolbar: Save or Save As	**Menu Bar:** File > Save or Save As

You must save your work before you exit from the drawing editor or turn off your system. Also, it is recommended that you save your drawings after regular intervals, so that in the event of a power failure or an editing error, all works saved before the problem started will be retained.

AutoCAD has provided the **QSAVE**, **SAVEAS**, and **SAVE** commands that allow you to save your work. These commands allow you to save your drawing by writing it to a permanent storage device, such as a hard drive or in any removable drive.

When you choose the **Save** tool from the **Quick Access toolbar** or the **Application Menu**, the **QSAVE** command is invoked. If the current drawing is unnamed and you save the drawing for the first time in the present session, the **SAVEAS** command will be invoked and you will be prompted to enter the file name in the **Save Drawing As** dialog box, as shown in Figure 1-30. You can enter the name for the drawing and then choose the **Save** button. If you have saved a drawing file once and then edited it, you can use the **Save** tool to save it, without the system prompting you to enter a file name. This allows you to do a quick save.

When you choose **SaveAs** from the **Application Menu** or choose the **Save As** tool from the **Quick Access Toolbar**, the **Save Drawing As** dialog box will be displayed, similar to that shown in Figure 1-30. Even if the drawing has been saved with a file name, this tool gives you an option to save it with a different file name. In addition to saving the drawing, it sets the name of the current drawing to the file name you specify, which is displayed in the title bar. This tool is used when you want to save a previously saved drawing under a different file name. You can also use this tool when you make certain changes to a template and want to save the changed template drawing but leave the original template unchanged.

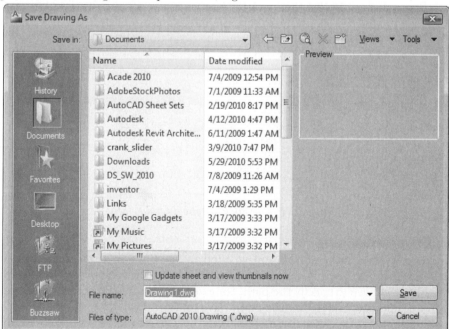

Figure 1-30 *The **Save Drawing As** dialog box*

Save Drawing As Dialog Box

The **Save Drawing As** dialog box displays the information related to the drawing files on your system. The various components of the dialog box are described next.

Places List

A column of icons is displayed on the left side of the dialog box. These icons contain the shortcuts to the folders that are frequently used. You can quickly save your drawings in one of these folders. The **History** folder displays the list of the most recently saved drawings. You can save your personal drawings in the **Documents** or the **Favorites** folder. The **FTP** folder displays the list of the various FTP sites that are available for saving the drawing. By default, no FTP sites are shown in the dialog box. To add a FTP site to the dialog box, choose the **Tools** button on the upper-right corner of the dialog box to display a shortcut menu and select **Add/Modify FTP Locations**. The **Desktop** folder displays the list of contents on the desktop. The **Buzzsaw** icons connect you to their respective pages on the Web. You can add a new folder in this list for an easy access by simply dragging the folder on to the **Places** list area. You can rearrange all these folders by dragging them and then placing them at the desired locations. It is also possible to remove the folders, which are not in frequent use. Right-click on the particular folder and then select **Remove** from the shortcut menu.

File name Edit Box

To save your work, enter the name of the drawing in the **File name** edit box by typing the file name or selecting it from the drop-down list. If you select the file name, it automatically appears in the **File name** edit box. If you have already assigned a name to the drawing, the current drawing name is taken as the default name. If the drawing is unnamed, the default name *Drawing1* is displayed in the **File Name** edit box. You can also choose the down arrow at the right of the edit box to display the names of the previously saved drawings and choose a name here.

Files of type Drop-Down List

The **Files of type** drop-down list (Figure 1-31) is used to specify the drawing format in which you want to save the file. For example, to save the file as an AutoCAD 2004 drawing file, select **AutoCAD 2004/LT 2004 Drawing (*.dwg)** from the drop-down list.

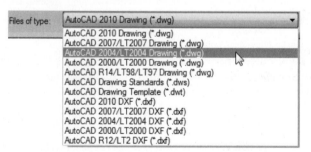

*Figure 1-31 The **Files of type** drop-down list*

Save in Drop-Down List

The current drive and path information is listed in the **Save in** drop-down list. AutoCAD will initially save the drawing in the default folder, but if you want to save the drawing in a different folder, you have to specify the path. For example, to save the present drawing as *house* in the *C1* folder, choose the arrow button in the **Save in** drop-down list to display the drop-down list. Select **C:** from the drop-down list; all folders in the C drive will be listed in the **File** list box. Double-click on the **C1** folder, if it is already listed there or create a folder C1 by choosing the **Create New Folder** button. Select *house* from the list, if it is already listed there, or enter it in the **File name** edit box and then choose the **Save** button. Your drawing (*house*) will be saved in the *C1* folder (*C:\C1\house.dwg*). Similarly, to save the drawing in the D drive, select **D:** in the **Save in** drop-down list.

Tip
The file name you enter to save a drawing should match its contents. This helps you to remember the drawing details and makes it easier to refer to them later. Also, the file name can be 255 characters long and can contain spaces and punctuation marks.

Views drop-down list

The **Views** drop-down list has the options for the type of listing of files and displaying the preview images (Figure 1-32).

List, Details, Thumbnails, and Preview Options

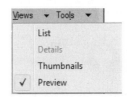

*Figure 1-32 The **Views** drop-down list*

If you choose the **Details** option, it will display the detailed information about the files (size, type, date, and time of modification) in the **Files** list box. In the detailed information, if you click on the **Name** label, the files are listed with the names in alphabetical order. If you double-click on the **Name** label, the files will be listed in reverse order. Similarly, if you click on the **Size** label, the files are listed according to their size in ascending order. Double-clicking on the **Size** label will list the files in descending order of size. Similarly, you can click on the **Type** label or the **Modified** label to list the files accordingly. If you choose the **List** option, all files present in the current folder will be listed in the **File** list box. If you select the **Preview** option, the list box displays the **Preview** image box wherein the bitmap image of the file chosen is displayed. If cleared, the **Preview** image box is not displayed. If you select the **Thumbnails** option, the list box displays the preview of all the drawings, along with their names displayed at the bottom of the drawing preview.

Create New Folder Button

If you choose the **Create New Folder** button, AutoCAD creates a new folder under the name **New Folder**. The new folder is displayed in the **File** list box. You can accept the name or change it to your requirement.

Up one level Button

The **Up one level** button displays the folders that are up by one level. For example, if you are in the *Sample* subfolder of the *AutoCAD 2012* folder, then choosing the **Up one level** button will take you to the *AutoCAD 2012* folder.

Search the Web

It displays the **Browse the Web** dialog box that enables you to access and store AutoCAD files on the Internet. You can also use the ALT+3 keys to browse the Web when this dialog box is available on the screen.

Tools drop-down list

The **Tools** drop-down (Figure 1-33) has an option for adding or modifying the FTP sites. These sites can then be browsed from the FTP shortcut in the **Places** list. The **Add Current Folder to Places** and **Add to Favorites** options add the folder displayed in the **Save in** edit box to the **Places** list or to the Favorites folder. The **Options** button displays the **Saveas Options** dialog box where you can save the proxy images of

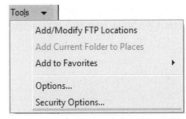

*Figure 1-33 The **Tools** drop-down list*

the custom objects. It has the **DWG Options** and **DXF Options** tabs. The **Security Options** button displays the **Security Options** dialog box, which is used to configure the security options of the drawing.

AUTOMATIC TIMED SAVE

AutoCAD allows you to save your work automatically at specific intervals. To change the time intervals, you can enter the intervals duration in minutes in the **Minutes between saves** text box in the **File Safety Precautions** area in the **Options** dialog box (**Open and Save** tab). This dialog box can be invoked by choosing the **Options** button from the **Application Menu**. Depending on the power supply, hardware, and type of drawings, you should decide on an appropriate time and assign it to this variable. AutoCAD saves the drawing with the file extension *.sv$*. You can also change the time interval by using the **SAVETIME** system variable.

Tip
Although the automatic save feature saves your drawing after a certain time interval, you should not completely depend on it because the procedure for converting the ac$ file into a drawing file is cumbersome. Therefore, it is recommended that you save your files regularly using the **QSAVE** *or* **SAVEAS** *commands.*

CREATING BACKUP FILES

If the drawing file already exists and you use **Save** or **Save As** tools to update the current drawing, AutoCAD creates a backup file. AutoCAD takes the previous copy of the drawing and changes it from a file type *.dwg* to *.bak*, and the updated drawing is saved as a drawing file with the *.dwg* extension. For example, if the name of the drawing is *myproj.dwg*, AutoCAD will change it to *myproj.bak* and save the current drawing as *myproj.dwg*.

Changing Automatic Timed Saved and Backup Files into AutoCAD Format

Sometimes, you may need to change the automatic timed saved and backup files into AutoCAD format. To change the backup file into an AutoCAD format, open the folder, in which you have saved the backup or the automatic timed saved drawing using **Computer** or **Windows Explorer**. Choose the **Organize > Folder and Search Options** from the menu bar to invoke the **Folder Options** dialog box. Choose the **View** tab and under the **Advanced settings** area, and clear the **Hide extensions for known file types** text box, if selected. Exit the dialog box. Rename the automatic saved drawing or the backup file with a different name and also change the extension of the drawing from *.sv$* or *.bak* to *.dwg*. After you rename the drawing, you will notice that the icon of the automatic saved drawing or the backup file is replaced by the AutoCAD icon. This indicates that the automatic saved drawing or the backup file is changed to an AutoCAD drawing.

Using the Drawing Recovery Manager to Recover Files

The files that are saved automatically can also be retrieved by using the **Drawing Recovery Manager**. You can open the **Drawing Recovery Manager** again by choosing **Drawing Utilities > Open the Drawing Recovery Manager** from the **Application Menu** or by entering **DRAWINGRECOVERY** at the Command prompt.

If the automatic save operation is performed in a drawing and the system crashes accidentally, the next time you run AutoCAD, the **Drawing Recovery** message box will be displayed, as shown in Figure 1-34. The message box informs you that the program unexpectedly failed and you can open the most suitable among the backup files created by AutoCAD. Choose the **Close** button from the **Drawing Recovery** message box; the **Drawing Recovery Manager** is displayed on the left of the drawing area, as shown in Figure 1-35.

The **Backup Files** rollout lists the original files, the backup files, and the automatically saved files. Select a file; its preview will be displayed in the **Preview** rollout. Also, the information corresponding to the selected file will be displayed in the **Details** rollout. To open a backup file, double-click on its name in the **Backup Files** rollout. Alternatively, right-click on the file name and then choose **Open** from the shortcut menu. It is recommended that you save the backup file at the desired location before you start working on it.

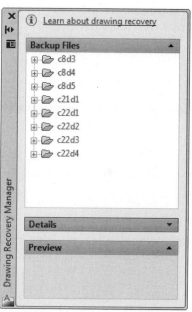

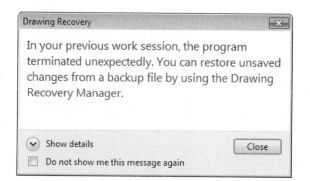

*Figure 1-34 The **Drawing Recovery** message box*

*Figure 1-35 The **Drawing Recovery Manager***

CLOSING A DRAWING

You can use the **CLOSE** command to close the current drawing file without actually quitting AutoCAD. If you choose **Close > Current Drawing** from the **Application Menu** or enter **CLOSE** at the Command prompt, the current drawing file will be closed. If multiple drawing files are opened, choose **Close > All Drawings** from the **Application Menu**. If you have not saved the drawing after making the last changes to it and you invoke the **CLOSE** command, AutoCAD displays a dialog box that allows you to save the drawing before closing. This box gives you an option to discard the current drawing or the changes made to it. It also gives you an option to cancel the command. After closing the drawing, you are still in AutoCAD from where you can open a new or an already saved drawing file. You can also use the close button (**X**) of the drawing area to close the drawing.

Note
You can close a drawing even if a command is active.

OPENING AN EXISTING DRAWING

Application Menu: Open > Drawing	**Quick Access Toolbar:** Open
Menu Bar: File > Open	**Command:** OPEN

You can open an existing drawing file that has been saved previously. There are three methods that can be used to open a drawing file: by using the **Select File** dialog box, by using the **Create New Drawing** dialog box, and by dragging and dropping.

Opening an Existing Drawing Using the Select File Dialog Box

If you are already in the drawing editor and you want to open a drawing file, choose the **Open** tool from the **Quick Access Toolbar**; the **Select File** dialog box will be displayed. Alternatively, invoke the **OPEN** command to display the **Select File** dialog box, see Figure 1-36. You can select the drawing to be opened using this dialog box. This dialog

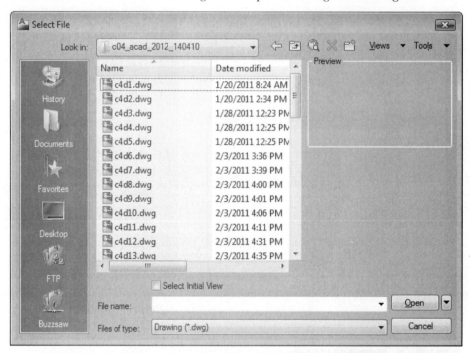

*Figure 1-36 The **Select File** dialog box*

box is similar to the standard dialog boxes. You can choose the file you want to open from the folder in which it is stored. You can change the folder from the **Look in** drop-down list. You can then select the name of the drawing from the list box or you can enter the name of the drawing file you want to open in the **File name** edit box. After selecting the drawing file, you can select the **Open** button to open the file. Here, you can choose *Drawing1* from the list and then choose the **Open** button to open the drawing.

When you select a file name, its image is displayed in the **Preview** box. If you are not sure about the file name of a particular drawing but know the contents, you can select the file names and look for the particular drawing in the **Preview** box. You can also change the file type by selecting it in the **Files of type** drop-down list. Apart from the *dwg* files, you can open the *dwt* (template) files or the *dxf* files. You have all the standard icons in the **Places** list that can be used to open drawing files from different locations. The **Open** button has a drop-down list, as shown in

Figure 1-37. You can choose a method for opening the file using this drop-down list. These methods are discussed next.

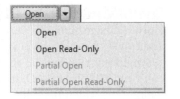

*Figure 1-37 The **Open** drop-down list*

Open Read-Only

To view a drawing without altering it, you must select the **Open Read-Only** option from the drop-down list. In other words, read only protects the drawing file from changes. AutoCAD does not prevent you from editing the drawing, but if you try to save the opened drawing with the original file name, AutoCAD warns you that the drawing file is write protected. However, you can save the edited drawing to a file with a different file name using the **SAVEAS** command. This way you can preserve your drawing.

Partial Open

The **Partial Open** option enables you to open only a selected view or a selected layer of a selected drawing. This option can be used to edit small portions of a complicated drawing and then save it with the complete drawing. When you select the **Partial Open** option from the **Open** drop-down list, the **Partial Open** dialog box (Figure 1-38) is displayed, which contains

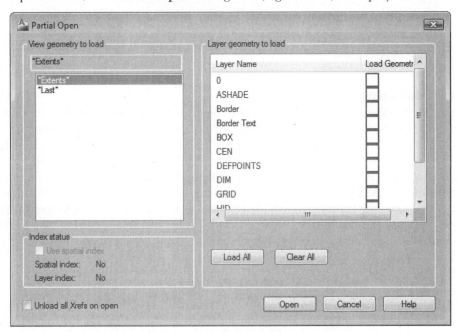

*Figure 1-38 The **Partial Open** dialog box*

different views and layers of the selected drawing. When you select a check box for a layer and then choose the **Open** button, only the objects drawn in that particular layer for the drawing are displayed in the new drawing window. You can make the changes and then save it. For example, in the *C:/Program Files/AutoCAD 2012* folder, double-click on the **Sample** folder and then select *Blocks and Tables - Metric.dwg* from the list. Now, choose the down arrow on the right of the **Open** button to display the drop-down list and choose **Partial Open**. All the views and layers of this drawing are displayed in the **Partial Open** dialog box. Select the check box on the right of the layer that you want to open. When you choose the **Open** button, after selecting the layers, only the selected layers of the drawing will be opened.

Note
The concept of layers is discussed in Chapter 4, Working with Drawing Aids.

Loading Additional Objects to Partially Opened Drawing

Once you have opened a part of a drawing and made the necessary changes, you may want to load additional objects or layers on the existing ones. This can be done by using the **PARTIALOAD** command, which can be invoked by choosing **File > Partial Load** from the menu bar or by entering **PARTIALOAD** at the Command prompt. This command displays the **Partial Load** dialog box, which is similar to the **Partial Open** dialog box. You can choose another layer and the objects drawn in it will be added to the partially loaded drawing.

Note

1. *The **Partial Load** option is not enabled in the **File** menu unless a drawing is partially opened.*

2. *Loading a drawing partially is a good practice when you are working with objects on a specific layer in a large complicated drawing.*

3. *In the **Select File** dialog box, the preview of a drawing which was partially opened and then saved is not displayed.*

Tip

If a drawing was partially opened and saved previously, it is possible to open it again with the same layers and views. AutoCAD remembers the settings so that while opening a previously partially opened drawing, a dialog box is displayed asking for an option to fully open it or restore the partially opened drawing.

Select Initial View

A view is defined as the way you look at an object. Select the **Select Initial View** check box if you want to load a specific view initially when AutoCAD loads the drawing. This option will work, if the drawing has saved views. This is generally used while working on a large complicated drawing, in which you want to work on a particular portion of the drawing. You can save that particular portion as a view and then select it to open the drawing next time. You can save a desired view, by using AutoCAD's **VIEW** command (see "**Creating Views**", Chapter 6). If the drawing has no saved views, selecting this option will load the last view. If you select the **Select Initial View** check box and then the **OK** button, AutoCAD will display the **Select Initial View** dialog box. You can select the view name from this dialog box, and AutoCAD will load the drawing with the selected view displayed.

Tip

*Apart from opening a drawing from the **Startup** dialog box or the **Select File** dialog box, you can also open a drawing from the **Application Menu**. By default, the **Recent Documents** option is chosen in the **Application Menu**, so the most recently opened drawings will be displayed and you can open the required file from it.*

It is possible to open an AutoCAD 2000 drawing in AutoCAD 2012. When you save this drawing, it is automatically converted and saved as an AutoCAD 2012 drawing file.

Opening an Existing Drawing Using the Startup Dialog Box

If you have configured the settings to show the **Startup** dialog box by setting the **STARTUP** system variable value as **1**, the **Startup** dialog box will be displayed every time you start a new AutoCAD session. The first button in this dialog box is the **Open a Drawing** button. When you choose this button, a list of the most recently opened drawings will be displayed for you to select from, see Figure 1-39. The **Browse** button displays the **Select File** dialog box, which allows you to browse to another file.

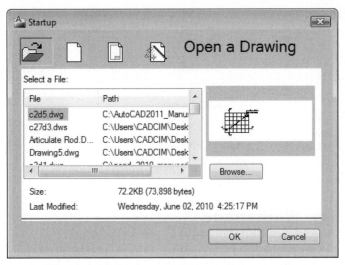

Figure 1-39 List of the recently opened drawings

Note
The display of the dialog boxes related to opening and saving drawings will be disabled, if the
STARTUP *and the* ***FILEDIA*** *system variables are set to 0. The initial value of these variables*
is 1.

Opening an Existing Drawing Using the Drag and Drop Method

You can also open an existing drawing in AutoCAD by dragging it from the Window Explorer
and dropping it into AutoCAD. If you drop the selected drawing in the drawing area, the drawing
will be inserted as a block and as a result you cannot modify it. But, if you drag the drawing
from the Window Explorer and drop it anywhere other than the drawing area, AutoCAD opens
the selected drawing.

QUITTING AutoCAD

You can exit the AutoCAD program by using the **EXIT** or **QUIT** commands. Even if you have
an active command, you can choose **Exit AutoCAD 2012** from the **Application Menu** to quit
the AutoCAD program. In case the drawing has not been saved, it allows you to save the work
first through a dialog box. Note that if you choose **No** in this dialog box, all the changes made
in the current list till the last save will be lost. You can also use the close button (**X**) of the main
AutoCAD window (present in the title bar) to end the AutoCAD session.

CREATING AND MANAGING WORKSPACES

A workspace is defined as a customized arrangement of **Ribbon**, toolbars, menus, and
window palettes in the AutoCAD environment. You can create your own workspaces, in which only
specified toolbars, menus, and palettes are available. When you start AutoCAD, by default, the
Drafting & Annotation workspace is the current workspace. You can select any other predefined
workspace from the **Workspace** drop-down list available in the title bar, next to the **Quick Access
Toolbar**, see Figure 1-40. You can also set the workspace from the flyout that will be displayed
on choosing the **Workspace Switching** button on the Status Bar or by choosing the required
Workspace from the menu bar. You can also choose the workspace using the toolbar.

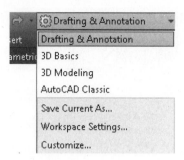

Figure 1-40   The predefined workspaces

Creating a New Workspace

To create a new workspace, customize the **Ribbon** and invoke the palettes to be displayed in the new workspace. Next, select the **Save Current As** option from the **Workspace** drop-down list in the titlebar; the **Save Workspace** dialog box will be displayed, as shown in Figure 1-41. Enter the name of the new workspace in the **Name** edit box and choose the **Save** button.

*Figure 1-41   The **Save Workspace** dialog box*

The new workspace is now the current workspace and is added to the drop-down list in the title bar. Likewise, you can create workspaces based on your requirement and switch from one workspace to the other by selecting the name from the drop-down list in the **Workspaces** toolbar or the drop-down list in the title bar.

Modifying the Workspace Settings

AutoCAD allows you to modify the workspace settings. To do so, select the **Workspace Settings** option in the **Workspace** drop-down list in the title bar; the **Workspace Settings** dialog box will be displayed, as shown in Figure 1-42. All workspaces are listed in the **My Workspace** drop-down list. You can make any of the workspaces as My Workspace by selecting it in the **My Workspace** drop-down list. You can also choose the **My Workspace** button from the **Workspaces** toolbar to change the current workspace to the one that was set as My Workspace in the **Workspace Settings** dialog box. The other options in this toolbar are discussed next.

Menu Display and Order Area

The options in this area are used to control the display and the order of display of workspaces in the **Workspace** drop-down list. By default, workspaces are listed in the sequence of their creation. To change the order, select a workspace and choose the **Move Up** or **Move Down** button. To control the display of the workspaces, you can select or clear the check boxes. You can also add a separator between workspaces by choosing the **Add Separator** button. A separator is a line that is placed between two workspaces in the **Workspace** drop-down list in the title bar, as shown in Figure 1-43.

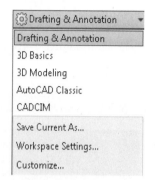

Figure 1-42 *The **Workspace Settings** dialog box*

Figure 1-43 *The **Workspace** drop-down list after adding separators*

When Switching Workspaces Area

By default, the **Do not save changes to workspace** radio button is selected in this area. This ensures that while switching the workspaces, the changes made in the current workspace will not be saved. If you select the **Automatically save workspace changes** radio button, the changes made in the current workspace will be automatically saved when you switch to the other workspace.

AutoCAD'S HELP

Titlebar: ? > Help	**Shortcut Key:** F1	**Command:** HELP or ?

You can get the on-line help and documentation about the working of AutoCAD 2012 commands from the **Help** menu in the title bar, see Figure 1-44. You can also access the **Help** menu by pressing the F1 function key. An **InfoCenter** bar is displayed at the top right corner in the title bar that will help you sign into the Autodesk Online services, see Figure 1-45. You can also access AutoCAD community by using certain keywords. Some important options in the **Help** menu are discussed next.

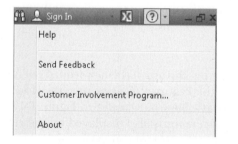

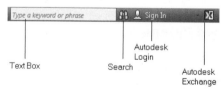

Figure 1-44 *The **Help** menu*

Figure 1-45 *The **InfoCenter** bar*

Customer Involvement Program

This option is used to share information about your system configuration and uses of Autodesk products with Autodesk. The collected information is used by Autodesk for the improvement of Autodesk software.

About

This option gives you information about the Release, Serial number, Licensed to, and also the legal description about AutoCAD.

Autodesk Exchange*

Autodesk Exchange enables you to learn the new features in AutoCAD 2012 through videos and text, get connected to the AutoCAD network, share information and designs, and so on. On choosing the **Exchange** button from the title bar, the **Autodesk Exchange** window will be displayed, as shown in Figure 1-46. In this window, there are two tabs, **Home** and **Help**. These tabs are discussed next.

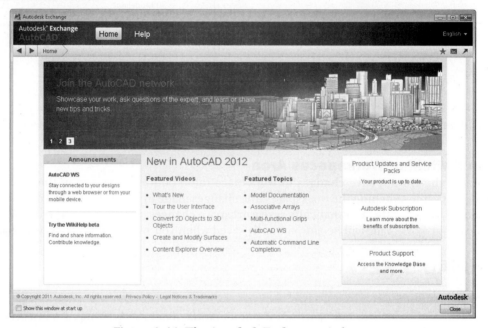

Figure 1-46 The Autodesk Exchange window

Home

This tab is chosen by default. Using this tab, you can overview the videos and topics on new features of AutoCAD 2012, tour the user interface, get connected to AutoCAD WS, and so on. The **Featured Videos** and **Featured Topics** areas are displayed in this tab, and these areas are discussed next.

Featured Videos

In the **Featured Videos** area, links to the videos showing the use of the new features of AutoCAD are displayed. When you choose **What's New** from the **Featured Videos** area, the video related to all latest enhancements starts playing in the **What's New** page, refer to Figure 1-47. In this window, a list of videos related to each latest enhancement is also displayed on the right. You can play any of these videos. Also, you can take a tour of the user interface of AutoCAD 2012 by clicking on the **Tour the User Interface** link. Note that when you click on any of the links in the **Home** tab, the **Help** tab is automatically activated.

Featured Topics

In the **Featured Topics** area, the text links are provided to explain new features of AutoCAD. There are five links available in this area. By using these links, you can find information about many important topics like **Associative Arrays**, **Multi-functional Grips**, **AutoCAD WS**, and so on.

*Figure 1-47 The **What's New** page*

Help

On choosing the **Help** tab in the **Autodesk Exchange** window, the **Browse Help** page will be displayed, as shown in Figure 1-48. The entire help documentation on AutoCAD 2012 is available in this page. You can search information about any command or tool on this page. As the feature is provided online, you cannot access this page without an active internet connection. If you press the F1 key while offline, an internet explorer window will be displayed with help topics, as shown in Figure 1-49. This window has all the information that is available in the online help window, except the videos and other internet-linked topics.

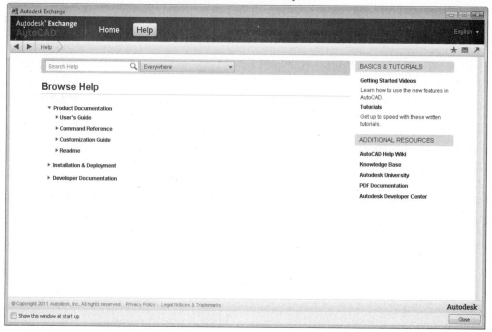

*Figure 1-48 The **Browse Help** page in the **Help** tab*

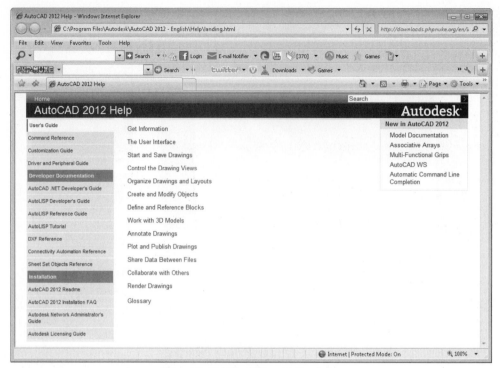

Figure 1-49 *The Offline Help window*

ADDITIONAL HELP RESOURCES

1. You can get help for a command while working by pressing the F1 key. The help html containing information about the command is displayed. You can exit the dialog box and continue with the command.

2. You can get help about a dialog box by choosing the **Help** button in that dialog box.

3. Autodesk has provided several resources that you can use to get assistance with your AutoCAD questions. The following is a list of some of the resources:

    a. Autodesk website *http://www.autodesk.com*
    b. AutoCAD Technical Assistance website *http://www.autodesk.com/support*
    c. AutoCAD Discussion Groups website *http://discussion.autodesk.com/index.jspa*
    d. Autodesk Press website *http://www.cengage.com/cad/autodeskpress*

4. You can also get help by contacting the author, Sham Tickoo, at *Stickoo@purduecal.edu* and *tickoo525@gmail.com*.

5. You can download AutoCAD drawings, programs, and special topics by registering yourself at the faculty's website by visiting: *http://cadcim.com/New/FacultyBooks_page.aspx*

Self-Evaluation Test

Answer the following questions and then compare them to those given at the end of this chapter:

1. You can press the F3 key to display the **AutoCAD** text window, which displays the previous commands and prompts. (T/F)

2. If you do not have internet connection, you cannot access the Help files. (T/F)

3. If a drawing was partially opened and saved previously, it is not possible to open it again with the same layers and views. (T/F)

4. If the current drawing is unnamed and you save the drawing for the first time, the **Save** tool will prompt you to enter the file name in the **Save Drawing As** dialog box. (T/F)

5. The _____ button is used to set the performance of the software at an acceptable level.

6. If the _____ variable is set to 1 and you invoke the **New** tool, the **Create New Drawing** dialog box will be displayed.

7. If you want to work on a drawing without altering the original drawing, you must select the _____ option from the **Open** drop-down list in the **Select File** dialog box.

8. The _____ option enables you to open only a selected view or a selected layer of the current drawing.

9. You can use the _____ command to close the current drawing file without actually quitting AutoCAD.

10. The _____ system variable can be used to change the time interval for automatic save.

Review Questions

Answer the following questions:

1. The shortcut menu invoked by right-clicking in the command window displays the most recently used commands and some of the window options such as **Copy**, **Paste**, and so on. (T/F)

2. It is possible to open an AutoCAD 2002 drawing in AutoCAD 2012. (T/F)

3. The file name that you enter to save a drawing in the **Save Drawing As** dialog box can be 255 characters long, but cannot contain spaces and punctuation marks. (T/F)

4. You can close a drawing in AutoCAD 2012 even if a command is active. (T/F)

5. Which of the following combination of keys should be pressed to hide all toolbars displayed on the screen?

    (a) CTRL+3   (b) CTRL+0
    (c) CTRL+5   (d) CTRL+2

6. Which of the following combination of keys should be pressed to turn on or off the display of the **Tool Palettes** window?

    (a) CTRL+3   (b) CTRL+0
    (c) CTRL+5   (d) CTRL+2

7. Which of the following commands is used to exit from the AutoCAD program?

    (a) **QUIT**   (b) **END**
    (c) **CLOSE**   (d) None of these

8. Which of the following options in the **Startup** dialog box is used to set the initial drawing settings before actually starting a new drawing?

    (a) **Start from Scratch**   (b) **Use a Template**
    (c) **Use a Wizard**   (d) None of these

9. When you choose **Save** from the **File** menu or choose the **Save** tool in the **Quick Access** toolbar, which of the following commands is invoked?

    (a) **SAVE**   (b) **LSAVE**
    (c) **QSAVE**   (d) **SAVEAS**

10. AutoCAD has provided _____ as an easy and convenient way of placing and sharing hatch patterns and blocks in the current drawing.

11. By default, the angles are positive if measured in the _____ direction.

12. You can change the size of toolbars by placing the cursor anywhere on the _____ of the toolbar where it takes the shape of a double-sided arrow.

13. To differentiate the template files from the drawing files, the template files have the _____ extension, whereas the drawing files have the _____ extension.

14. You can also use _____ and _____ instead of dragging and dropping the objects from one drawing to another while multiple drawings are opened.

15. The _____ page of the offline **AutoCAD 2012 Help** window displays the help topics that are organized by categories pertaining to different sections of AutoCAD.

Answers to Self-Evaluation Test

1. F, **2.** F, **3.** F, **4.** T, **5. Hardware Acceleration**, **6. STARTUP**, **7. Open Read-Only**, **8. Partial Open**, **9. CLOSE**, **10. SAVETIME**

Chapter 2

Getting Started with AutoCAD

CHAPTER OBJECTIVES

In this chapter, you will learn:

- *To draw lines by using the Line tool.*
- *About various coordinate systems used in AutoCAD.*
- *To clear the drawing area by using the Erase tool.*
- *About the two basic object selection methods: Window and Window Crossing.*
- *To draw circles by using various tools.*
- *To use the Zoom and Pan tools.*
- *To set up units by using the UNITS command.*
- *To set up and determine limits for a given drawing.*
- *To plot drawings by using the basic plotting options.*
- *To use the Options dialog box and specify settings.*

KEY TERMS

- *Dynamic Input*
- *Line*
- *Coordinate Systems*
- *Absolute Coordinate System*
- *Relative Coordinate System*
- *Direct Distance Entry*
- *Erase*
- *Object Selection*
- *Circle*
- *Zoom*
- *Pan*
- *Units Format*
- *Options*
- *Plot*
- *Limits*

DYNAMIC INPUT MODE

In AutoCAD, the **Dynamic Input** mode allows you to enter the commands through the pointer input and the dimensions using the dimensional input. When this mode is turned on, all prompts are available at the tooltip as dynamic prompts and you can select the command options through the dynamic prompt. The settings for the **Dynamic Input** mode are done through the **Dynamic Input** tab of the **Drafting Settings** dialog box. To invoke the **Drafting Settings** dialog box, right-click on the **Dynamic Input** button in the Status Bar; a shortcut menu will be displayed. Choose the **Settings** option from the shortcut menu; the **Drafting Settings** dialog box will be displayed, as shown in Figure 2-1. The options in this tab are discussed next.

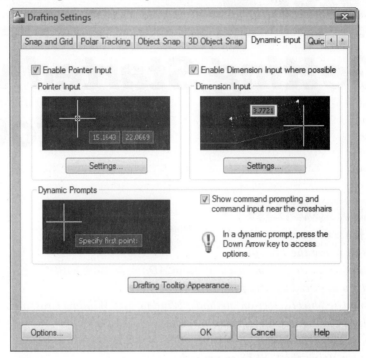

Figure 2-1 The **Dynamic Input** tab of the **Drafting Settings** dialog box

Enable Pointer Input

If the **Enable Pointer Input** check box selected, you can enter the commands through the pointer input. Figure 2-2 shows the **CIRCLE** command entered through the pointer input. If this check box is cleared, the **Dynamic Input** will be turned off and commands have to be entered through the Command prompt, in a way similar to the old releases of AutoCAD. With this release of AutoCAD, if you enter any alphabet at the **Dynamic Input**, all tools whose names start with the entered alphabet will be displayed in a list at the **Dynamic Input**, see Figure 2-2.

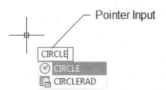

Figure 2-2 Entering a command using the pointer input

Choosing the **Settings** button from the **Pointer Input** area displays the **Pointer Input Settings** dialog box, as shown in Figure 2-3. The radio buttons in the **Format** area of this dialog box are used to set the default settings for specifying the other points, after specifying the first point. By default, the **Polar format** and **Relative coordinates** radio buttons are selected. As a result, the coordinates will be specified in the polar form and with respect to the relative coordinates system. You can select the **Cartesian format** radio button to enter the coordinates in cartesian form. Likewise, if you select the **Absolute coordinates** radio button, the numerical entries will be measured with respect to the absolute coordinate system.

The **Visibility** area in the **Pointer Input Settings** dialog box is used to set the visibility of the coordinates tool tips. By default, the **When a command asks for a point** radio button is selected. You can select the other radio buttons to modify this display.

Enable Dimension Input where possible

This check box is selected by default. As a result, the dimension input field is displayed in the graphics area showing the preview of that dimension. Figure 2-4 displays the dimension input fields. The options under the Dynamic prompt will be available when you press the down Arrow key from the keyboard. The dotted lines shows the geometric parameters like length, radius, or diameter corresponding to that dimension. Figure 2-4 shows a line being drawn using the **Pline** command. The two dimension inputs that are shown are for the length of the line and the angle with a positive direction of the X axis.

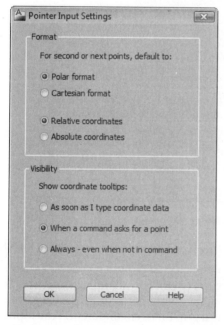

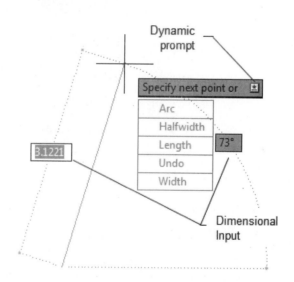

Figure 2-3 *The **Pointer Input Settings** dialog box*

Figure 2-4 *Input fields displayed when the **Enable Dimension Input where possible** check box is selected*

Using the TAB key, you can toggle between the dimension input fields. As soon as you have specified one dimension and moved to the other, the previous dimension will be locked. If the **Enable Dimensional Input where possible** check box is cleared, the preview of dimensions will not be displayed. You can only enter the dimensions in the dimension input fields below the cursor, as shown in Figure 2-5. Choose the **Settings** button from the **Dimension Input** area to display the **Dimension Input Settings** dialog box, as shown in Figure 2-6.

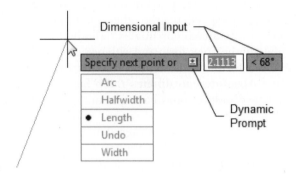

Figure 2-5 Input fields displayed when the
Enable Dimension Input where possible
check box is cleared

*Figure 2-6 The **Dimension Input Settings** dialog box*

By default, the **Show 2 dimension input fields at a time** radio button is selected. As a result, two dimension input fields will be displayed in the drawing area while stretching a sketched entity. The two input fields will depend on the entity that is being stretched. For example, if you stretch a line using one of its endpoints, the input field will show the total length of the line and the change in its length. Similarly, while stretching a circle using a grip on its circumference, the input fields will show the total radius and the change in the radius. You can set the priority to display only one input field or various input fields, simultaneously, by selecting their respective check boxes.

Tip
If multiple dimension input fields are available, use the TAB key to switch between the dimension input fields

Show command prompting and command input near the crosshairs

If this check box is selected, the prompt sequences will be dynamically displayed near the crosshairs. Whenever a blue arrow appears at the pointer input, it suggests that the access options are available. To access these options, press the down arrow key to see the dynamic prompt listing all options. In the dynamic prompt, you can use the cursor or the down arrow key to jog through the options. A black dot will appear before the option that is currently active. In Figure 2-5, the **Length** option is currently active. Press ENTER to confirm the polyline creation with the **Length** option.

Drafting Tooltip Appearance

When you choose the **Drafting Tooltip Appearance** button, the **Tooltip Appearance** dialog box will be displayed, as shown in Figure 2-7. This dialog box contains the options to customize the tooltip appearance. The **Colors** button is chosen to change the color of the tooltip in the model space or layouts.

The edit box in the **Size** area is used to specify the size of the tooltip. You can also use the slider to control the size of the tool tip. The preview is displayed in the **Model Preview** area and the **Layout Preview** area, as soon as the value is changed in the **Size** edit box. Likewise, the transparency of the tooltip can be controlled using the edit box or the slider in the **Transparency** area.

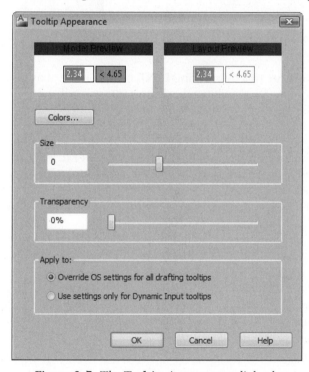

*Figure 2-7 The **Tooltip Appearance** dialog box*

Selecting the **Override OS settings for all drafting tooltips** radio button in the **Apply to** area ensures that changes made in the **Tooltip Appearance** dialog box will be applied to all drafting tooltips. If you select the **Use settings only for Dynamic Input tooltips** radio button, the changes will be applied only to the **Dynamic Input** tooltips. For example, if you change any of the parameters using the **Tooltip Appearance** dialog box and select the **Use settings only for Dynamic Input tooltips** radio button, the tooltips for the dynamic input will be modified, but for the polar tracking it will consider the original values. On the other hand, if you select the **Override OS settings for all drafting tooltips** radio button, the tooltips displayed for the polar tracking will also be modified based on the values in the **Tooltip Appearance** dialog box.

DRAWING LINES IN AutoCAD

Ribbon: Home > Draw > Line	**Toolbar:** Draw > Line	**Menu Bar:** Draw > Line
Tool Palettes: Draw > Line	**Command:** LINE or L	

The most commonly used fundamental object in a drawing is line. In AutoCAD, a line is drawn between two points by using the **LINE** command. You can invoke the **LINE** command by choosing the **Line** tool from the **Draw** panel of the **Home** tab in the **Ribbon**, as shown in Figure 2-8. Besides this, you can choose the **Line** tool from the **Draw** tab of the **Tool Palettes**. To invoke the **Tool Palettes**, choose the **Tool Palettes** button from the **Palettes** panel in the **View** tab, as shown in Figure 2-9. Alternatively, you can invoke the **LINE** command by choosing the **Line** tool from the **Draw** toolbar, as shown in Figure 2-10. However, the **Draw** toolbar is not displayed by default. To invoke this toolbar, choose **View > Windows > Toolbars > AutoCAD** from the **Ribbon**.

*Figure 2-8 The **Line** tool in the **Draw** panel*

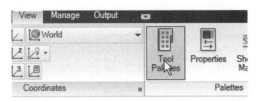

*Figure 2-9 Invoking the **Tool Palettes** from the **Palettes** panel.*

*Figure 2-10 The **Line** tool in the **Draw** toolbar*

You can also invoke the **Line** tool by entering **LINE** or **L** (L is the alias for the LINE command) at the Command prompt. On invoking the **Line** tool, you will be prompted to specify the starting point of the line. Select a point by using the mouse or by entering its coordinates at the command prompt. After specifying the first point, you will be prompted to specify the second point. Specify the second point; a line will be drawn. You may continue specifying points and draw lines or terminate the **LINE** command by pressing ENTER, ESC, or SPACEBAR. You can also right-click to display the shortcut menu and then choose the **Enter** or **Cancel** option from it to exit the **Line** tool. After terminating the **LINE** command, AutoCAD will again display the Command prompt. The prompt sequence for the drawing shown in Figure 2-11 is given next.

Start a new file with the *Acad.dwt* template in the **2D Drafting & Annotation** or **AutoCAD Classic** workspace.

Command: *Choose the **Line** tool*
Specify first point: *Move the cursor (mouse) and left-click to specify the first point.*
Specify next point or [Undo]: *Move the cursor and left-click to specify the second point.*
Specify next point or [Undo]: *Specify the third point.*
Specify next point or [Close/Undo]: [Enter] *(Press ENTER to exit the **Line** tool.)*

Note
When you specify the start point of the line by pressing the left mouse button, a rubber band line stretches between the selected point and the current position of the cursor. This line is sensitive to the movement of the cursor and helps you select the direction and the placement of the next point for the line.

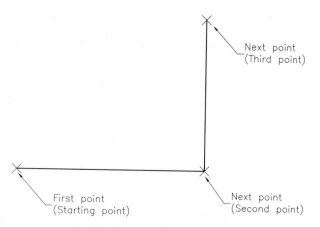

Figure 2-11 *Drawing lines using the* **Line** *tool*

Tip
To clear the drawing area and draw new drawings, choose the **Erase** *tool from the* **Modify** *panel in the* **Home** *tab or type* **ERASE** *at the Command prompt and press ENTER; the cross hairs will change into a box called pick box and you will be prompted to select objects. You can select an object by positioning the pick box on it and then by pressing the left mouse button. After selecting the objects, press ENTER to terminate the* **ERASE** *command; the selected objects will be erased. If you enter* **All** *at the* **Select objects** *prompt and press ENTER, all objects in the screen will be erased. You can use the* **U (undo)** *command to undo the last command. Alternatively, choose the* **Undo** *button from the* **Quick Access Toolbar** *to undo the last command. (See "Erasing Objects" discussed later in this chapter.)*

The **LINE** command has the following two options. They are discussed next.

Close   Undo

The Close Option

After drawing two continuous lines by using the **Line** tool, you will notice that the **Close** option is displayed at the Command prompt. The **Close** option is used to join the current point to the start point of the first line when two or more continuous lines are drawn. If you are specifying the endpoint by using the mouse, then click at the start point of the first line or enter **C** at the Command prompt, as given in the Command prompt below.

> Command: *Choose the* **Line** *tool*
> _line Specify first point: *Pick the first point.*
> Specify next point or [Undo]: *Pick the second point.*
> Specify next point or [Undo]: *Pick the third point.*
> Specify next point or [Close/Undo]: *Pick the fourth point.*
> Specify next point or [Close/Undo]: **C** Enter *(The fifth point joins with the first point). See Figure 2-12.*

You can also choose the **Close** option from the shortcut menu, which appears when you right-click in the drawing area.

Tip
After exiting the **Line** *tool, you may want to draw another line starting from the endpoint of the previous line. In such cases, press ENTER twice; a new line will start from the endpoint of the previous line. You can also type the @ symbol to start the line from the last point. For example, if you have drawn a circle and then immediately start the* **Line** *tool, the @ symbol will snap to the center point of the circle.*

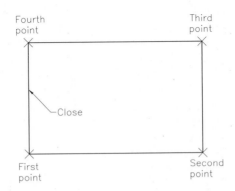

Figure 2-12 Using the **Close** option with the **Line** tool

The Undo Option

While drawing a line, if you have specified a wrong endpoint by mistake, then you can remove that line by using the **Undo** option of the **Line** tool. You can use this option multiple times and remove as many lines as you want. To use this option, type **Undo** (or just **U**) at the **Specify next point or [Undo]** prompt. You can also right-click to display the shortcut menu and then choose the **Undo** option from it.

Note

*By default, whenever you open a new drawing, you need to modify the drawing display area. To modify the display area, type **ZOOM** at the Command prompt and press ENTER. Then, type **ALL** and press ENTER; the drawing display is modified. You will learn more about the **ZOOM** command later in this chapter.*

INVOKING TOOLS USING DYNAMIC INPUT/COMMAND PROMPT

With this release of AutoCAD, if you enter any alphabet at the Command prompt or **Dynamic Input**, all tools whose names start with the entered alphabet will be displayed in a list at the Command prompt or **Dynamic Input**. For example, if you enter **L** at the Command prompt or **Dynamic Input**, all tools whose names start with the alphabet L will be displayed, refer to Figure 2-13. In this way, you can view all the tool names starting with a particular alphabet and select the required tool. In Figure 2-13, you can select the **Line** option, the **Layer** option, and so on.

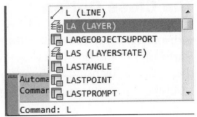

Figure 2-13 List displayed after typing L at the Command prompt

COORDINATE SYSTEMS

In AutoCAD, the location of a point is specified in terms of Cartesian coordinates. In this system, each point in a plane is specified by a pair of numerical coordinates. To specify a point in a plane, take two mutually perpendicular lines as references. The horizontal line is called the X axis, and the vertical line is called the Y axis. The X and Y axes divide the XY plane into four parts, generally known as quadrants. The point of intersection of these two axes is called the origin and the plane is called the XY plane. The origin has the coordinate values of $X = 0$, $Y = 0$. The origin is taken as the reference for locating a point on the XY plane. Now, to locate a point, say P, draw a vertical line intersecting the X axis. The horizontal distance between the origin and the intersection point will be called the X coordinate of P. It will be denoted as P(x). The X coordinate specifies how far the point is to the left or right from the origin along the X axis. Now, draw a horizontal line intersecting the Y axis. The vertical distance between the origin and the intersection point will be the Y coordinate of P. It will be denoted as P(y). The *Y* coordinate specifies how far the point is to the top or bottom from the origin along the *Y* axis. The intersection point of the horizontal and vertical lines is the coordinate of the point and is denoted as P(x,y). The *X* coordinate is positive, if measured to the right of the origin and is negative, if measured to the left of the origin. The *Y* coordinate is positive, if measured above the origin and is negative, if measured below the origin, see Figure 2-14.

In AutoCAD, the default origin is located at the lower left corner of the drawing area. AutoCAD uses the following coordinate systems to locate a point in an XY plane.

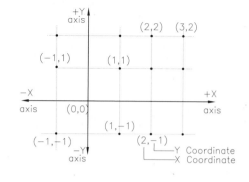

Figure 2-14 Cartesian coordinate system

1. Absolute coordinates
2. Relative coordinates
      a. Relative rectangular coordinates
      b. Relative polar coordinates
3. Direct distance entry

If you are specifying a point by entering its location at the Command prompt then you need to use any one of the coordinate systems.

Absolute Coordinate System

In the absolute coordinate system, points are located with respect to the origin (0,0). For example, a point with X = 4 and Y = 3 is measured 4 units horizontally (distance along the *X* axis) and 3 units vertically (distance along the *Y* axis) from the origin, as shown in Figure 2-15. In AutoCAD, the absolute coordinates are specified at the Command prompt by entering *X* and *Y* coordinates, separated by a comma. However, remember that if you are specifying the coordinates by using the **Dynamic Input** mode, you need to add # as the prefix to the X coordinate value. For example, enter #1,1 in the dynamic input boxes to use the absolute coordinate system. The following example illustrates the use of absolute coordinates at the Command prompt to draw the rectangle shown in Figure 2-16.

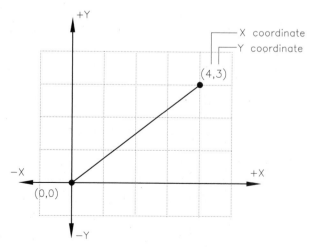

Figure 2-15 Absolute coordinate system

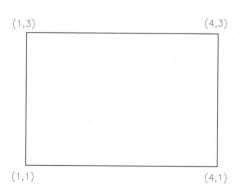

Figure 2-16 Lines created by using absolute coordinates

Command: *Choose the **Line** tool (Ensure that the **Dynamic Input** button is not chosen)*
_line Specify first point: **1,1** `Enter`   *(X = 1 and Y = 1.)*
Specify next point or [Undo]: **4,1** `Enter`   *(X = 4 and Y = 1.)*
Specify next point or [Undo]: **4,3** `Enter`
Specify next point or [Close /Undo]: **1,3** `Enter`
Specify next point or [Close/Undo]: **C** `Enter`

EXAMPLE 1   *Absolute Coordinate System*

Draw the profile shown in Figure 2-17 by using the Absolute Coordinate system. The absolute coordinates of the points are given in the following table. Save the drawing with the name *Exam1.dwg*.

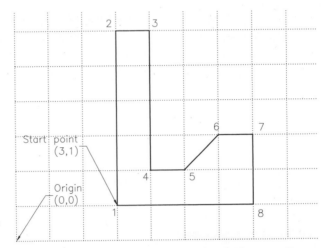

Figure 2-17 *Drawing a figure using the absolute coordinates*

Point	Coordinates	Point	Coordinates
1	3,1	5	5,2
2	3,6	6	6,3
3	4,6	7	7,3
4	4,2	8	7,1

Start a new file with the *Acad.dwt* template in the **Drafting & Annotation** or **AutoCAD Classic** workspace. Once you know the coordinates of the points, you can draw the sketch by using the **Line** tool. But, before you proceed with drawing the object, you need to turn off the **Dynamic Input** mode, if it is on by default. By doing so, you will be able to enter the command in the Command prompt. The prompt sequence is given next.

Choose the **Dynamic Input** button from the Status Bar, if it has been already chosen.

Command: **ZOOM** `Enter`
Specify corner of window, enter a scale factor (nX or nXP), or [All/Center/Dynamic/Extents/Previous/Scale/Window/Object] <real time>: **ALL** `Enter`
Command: *Choose the **Line** tool* `Enter`
_ Specify first point: **3,1** `Enter`   *(Start point.)*

Specify next point or [Undo]: **3,6** `Enter`
Specify next point or [Undo]: **4,6** `Enter`
Specify next point or [Close/Undo]: **4,2** `Enter`
Specify next point or [Close/Undo]: **5,2** `Enter`
Specify next point or [Close/Undo]: **6,3** `Enter`
Specify next point or [Close/Undo]: **7,3** `Enter`
Specify next point or [Close/Undo]: **7,1** `Enter`
Specify next point or [Close/Undo]: **C** `Enter`

Choose the **Save** tool from the **Quick Access Toolbar** to display the **Save Drawing As** dialog box. Enter **Exam1** in the **File name** edit box and then choose the **Save** tool. The drawing will be saved with the specified name in the default *Documents* folder.

EXERCISE 1   *Absolute Coordinate System*

Draw the profile shown in Figure 2-18. The distance between the dotted lines is 1 unit. Enter the absolute coordinates of the points given in the following table. Then, use these coordinates to draw the same figure.

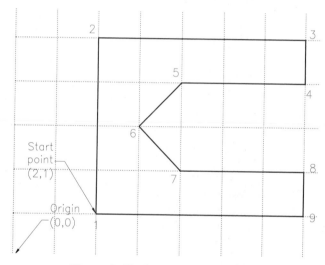

Figure 2-18 Drawing for Exercise 1

Point	Coordinates	Point	Coordinates
1	2, 1	6	_____
2	_____	7	_____
3	_____	8	_____
4	_____	9	_____
5	_____		

Relative Coordinate System

There are two types of relative coordinates: relative rectangular and relative polar.

Relative Rectangular Coordinates

In the relative rectangular coordinate system, the location of a point is specified with respect to the previous point and not with respect to the origin. To enter coordinate values in terms of the Relative Rectangular Coordinate system, check whether the **Dynamic Input** is on or not. If the **Dynamic Input** is turned on, then by default the profile will be drawn using the Relative Rectangular Coordinate system. Therefore, in this case, enter the X coordinate, type comma (,), and then enter the Y coordinate. However, if the **Dynamic Input** is turned off, the coordinate values have to be prefixed by the @ symbol, so that the profile will be drawn using the Relative Rectangular Coordinate system. For example, to draw a rectangle (see Figure 2-19) of length 4 units and width 3 units and the lower left corner at the point (1,1) using the Relative Rectangular Coordinate system, you need to use the following prompt sequence:

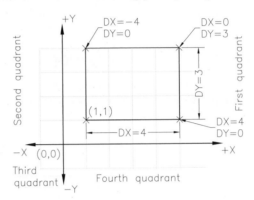

Figure 2-19 *Drawing lines using the relative rectangular coordinates*

Command: *Choose the* **Line** *tool*
_line Specify first point: **1,1** Enter *(Start point)*
Specify next point or [Undo]: **@4,0** Enter
Specify next point or [Undo]: **@0,3** Enter
Specify next point or [Close/Undo]:**@-4,0** Enter
Specify next point or [Close/Undo]: **@0,-3** Enter
Specify next point or [Close/Undo]: Enter

Remember that if the **Dynamic Input** is on, you need to use a comma (,) after entering the first value in the Dynamic Input boxes. Else, AutoCAD will take coordinates in relative polar form.

Sign Convention. As just mentioned, in the relative rectangular coordinate system, the distance along the X and Y axes is measured with respect to the previous point. To understand the sign convention, imagine a horizontal line and a vertical line passing through the previous point so that you get four quadrants. If the new point is located in the first quadrant, then both the distances (DX and DY) will be specified as positive values. If the new point is located in the third quadrant, then both the distances (DX and DY) will be specified as negative values. In other words, the point will have a positive coordinate values, if it is located above or right of an axis. Similarly, the point will have a negative coordinate values, if it is located below or left of an axis.

EXAMPLE 2   *Relative Rectangular Coordinates*

Draw the profile shown in Figure 2-20 using relative rectangular coordinates. The coordinates of the points are given in the table below.

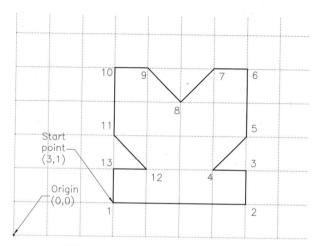

Figure 2-20 *Profile for Example 2*

Point	Coordinates	Point	Coordinates
1	3,1	8	@-1,-1
2	@4,0	9	@-1,1
3	@0,1	10	@-1,0
4	@-1,0	11	@0,-2
5	@1,1	12	@1,-1
6	@0,2	13	@-1,0
7	@-1,0	14	@0,-1

Start a new file with the *Acad.dwt* template in the **Drafting & Annotation** or **AutoCAD Classic** workspace. Before you proceed, you need to make sure that the **Dynamic Input** is turned on.

> Command: **ZOOM** Enter
> Specify corner of window, enter a scale factor (nX or nXP), or
> [All/Center/Dynamic/Extents/Previous/Scale/Window/Object] <real time>: **ALL** Enter
> Command: *Choose the **Line** tool*
> _line Specify first point: *Type* **3,1** *in the dynamic input boxes and press* Enter *(Start point)*
> Specify next point or [Undo]: *Type* **4,0** *in the dynamic input boxes and press* Enter
> Specify next point or [Undo]: *Type* **0,1** *in the dynamic input boxes and press* Enter
> Specify next point or [Close/Undo]: *Type* **-1,0** *in the dynamic input boxes and press* Enter
> Specify next point or [Close/Undo]: *Type* **1,1** *in the dynamic input boxes and press* Enter
> Specify next point or [Close/Undo]: *Type* **0,2** *in the dynamic input boxes and press* Enter
> Specify next point or [Close/Undo]: **-1,0** *and press* Enter
> Specify next point or [Close/Undo]: **-1,-1** *and press* Enter
> Specify next point or [Close/Undo]: **-1,1** *and press* Enter

Specify next point or [Close/Undo]: **-1,0** *and press* Enter
Specify next point or [Close/Undo]: **0,-2** *and press* Enter
Specify next point or [Close/Undo]: **1,-1** *and press* Enter
Specify next point or [Close/Undo]: **-1,0** *and press* Enter
Specify next point or [Close/Undo]: **0,-1** *and press* Enter
Specify next point or [Close/Undo]: Enter

EXERCISE 2   *Absolute Coordinates*

For Figure 2-21, enter the relative rectangular coordinates of the points given in the following table. Then, use these coordinates to draw the figure. The distance between the dotted lines is 1 unit.

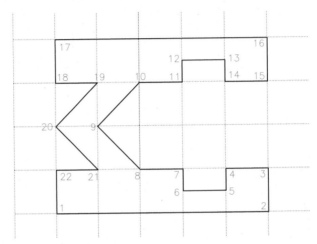

Figure 2-21 Drawing for Exercise 2

Point	Coordinates	Point	Coordinates
1	2, 1	12	_____
2	_____	13	_____
3	_____	14	_____
4	_____	15	_____
5	_____	16	_____
6	_____	17	_____
7	_____	18	_____
8	_____	19	_____
9	_____	20	_____
10	_____	21	_____
11	_____	22	_____

Relative Polar Coordinates

In the relative polar coordinate system, the location of a point is specified by defining the distance of the point from the current point and the angle between the two points with respect to the positive X axis. The prompt sequence to draw a line of length 5 units whose start point is at 1,1 and inclined at an angle of 30 degrees to the X axis, as shown in Figure 2-22, is given next.

Command: *Choose the **Line** tool*
Specify first point: **1,1** [Enter]
Specify next point or [Undo]: **@5<30** [Enter]

If the **Dynamic Input** is on, by default the relative polar coordinate mode will be activated. Therefore, when you invoke the **Line** tool and specify the start point, two input boxes will be displayed. The second input box shows the angle value, preceded by the **<** symbol. Now, enter the distance value, press the TAB key to shift to the second input box, and then enter the angle value.

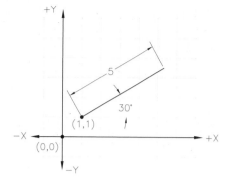

Sign Convention. By default, in the relative polar coordinate system, the angle is measured from the horizontal axis as the zero degree. Also, the angle

Figure 2-22 Drawing a line by using relative polar coordinates

is positive, if measured in counterclockwise direction and is negative, if measured in clockwise direction. Here, it is assumed that the default setup of the angle measurement has not been changed.

Note
*You can modify the default settings of the angle measurement direction by using the **UNITS** command, which is discussed later.*

EXAMPLE 3   *Relative Polar Coordinates*

Draw the profile shown in Figure 2-23 by using the relative polar coordinates. The relative coordinate values of each point are given in the table. The start point is located at 1.5, 1.75. Save this drawing with the name *Exam3.dwg*. The dimensions and the numbering are for reference only.

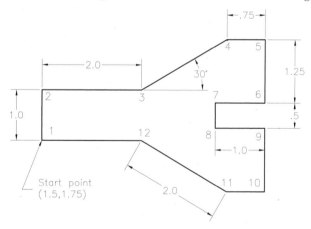

Figure 2-23 Drawing for Example 3

Point	Coordinates	Point	Coordinates
1	1.5,1.75	7	@1.0<180
2	@1.0<90	8	@0.5<270
3	@2.0<0	9	@1.0<0
4	@2.0<30	10	@1.25<270
5	@0.75<0	11	@0.75<180
6	@1.25<-90 (or <270)	12	@2.0<150

Start a new file with the *Acad.dwt* template in the **Drafting & Annotation** or **AutoCAD Classic** workspace. Next, you need to modify the drawing display area. To do so, choose the **Zoom Extents** tool from the Navigation bar. Next, turn off the **Dynamic Input** option by choosing the **Dynamic Input** button from the Status Bar.

> Command: *Choose the* **Line** *tool*
> _line Specify first point: **1.5,1.75** `Enter`   *(Start point)*
> Specify next point or [Undo]: **@1<90** `Enter`
> Specify next point or [Undo]: **@2.0<0** `Enter`
> Specify next point or [Close/Undo]: **@2<30** `Enter`
> Specify next point or [Close/Undo]: **@0.75<0** `Enter`
> Specify next point or [Close/Undo]: **@1.25<-90** `Enter`
> Specify next point or [Close/Undo]: **@1.0<180** `Enter`
> Specify next point or [Close/Undo]: **@0.5<270** `Enter`
> Specify next point or [Close/Undo]: **@1.0<0** `Enter`
> Specify next point or [Close/Undo]: **@1.25<270** `Enter`
> Specify next point or [Close/Undo]: **@0.75<180** `Enter`
> Specify next point or [Close/Undo]: **@2.0<150** `Enter`
> Specify next point or [Close/Undo]: **C** `Enter` *(The last point joins with the first point.)*

To save this drawing, choose the **Save** tool from the **Quick Access Toolbar**; the **Save Drawing As** dialog box will be displayed. Enter **Exam3** in the **File name** edit box and then choose the **Save** tool; the drawing will be saved with the specified name in the **My Documents** folder.

EXERCISE 3

Draw the profile shown in Figure 2-24 by specifying points using the absolute, relative rectangular, and relative polar coordinate systems. Do not dimension the profile. They are given for reference only.

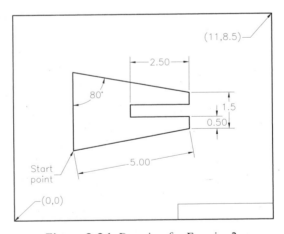

Figure 2-24 Drawing for Exercise 3

Direct Distance Entry

The easiest way to draw a line in AutoCAD is by using the Direct Distance Entry method. Before drawing a line by using this method, ensure that the **Dynamic Input** button is chosen in the Status Bar. Next, choose the **Line** tool; you will be prompted to specify the start point. Enter the coordinate values in the text box and press ENTER; you will be prompted to specify the next point. Now, enter the absolute length of the line and its angle with respect to the current position of the cursor in the corresponding text boxes, as shown in Figure 2-25. Note that you can use the TAB key to toggle between the text boxes. If the **Ortho** mode is on while drawing lines using this method, you can position the cursor only along the *X* or *Y* axis. If the **Dynamic Input** button is not chosen, then you need to enter the length of the line at the Command prompt. Therefore, position the cursor at the desired angle, type the length at the Command prompt, and then press ENTER, as shown in Figure 2-25.

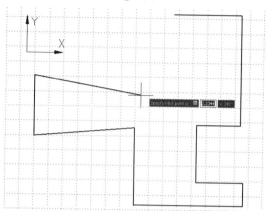

Figure 2-25 *Drawing lines using the Direct Distance Entry method*

Command: *Choose the **Line** tool*
_line Specify first point: *Start point.*
Specify next point or [Undo]: *Position the cursor and then enter distance.*
Specify next point or [Undo]: *Position the cursor and then enter distance.*

EXAMPLE 4   *Direct Distance Entry*

In this example, you will draw the profile shown in Figure 2-26, by using the Direct Distance Entry method. The start point is 2,2.

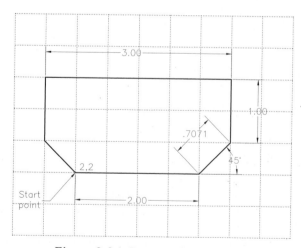

Figure 2-26 *Drawing for Example 4*

Also, you will use the polar tracking option to draw lines. The polar tracking option allows you to track the lines that are drawn at specified angles. The default angle specified for polar tracking is 90-degree. Therefore, by default, you can track lines at an angle that is multiple of 90 degrees, such as 90, 180, 270, and 360. In this example, you need to draw lines at the angles that are multiples of 45-degree such as 45, 90, 135, and so on. Therefore, first you need to set the polar tracking angle as 45-degree.

Note
You will learn more about polar tracking in Chapter 4, Working with Drawing Aids.

1. Start a new file with the *Acad.dwt* template in the **Drafting & Annotation** or **AutoCAD Classic** workspace.

2. To add a 45-degree angle to polar tracking, right-click on the **Polar Tracking** button on the Status Bar and then choose **45** from the shortcut menu. Again, choose the **Polar Tracking** button in the Status Bar to turn polar tracking on.

3. Choose the **Line** tool from the **Draw** panel of the **Home** tab; you are prompted to specify the start point.

4. Enter **2,1** at the Command prompt and press ENTER; you are prompted to specify the next point.

5. Move the cursor horizontally toward the right and when the tooltip displays 0 as polar angle, type **2** and press ENTER; you are prompted to specify the next point.

6. Move the cursor at an angle close to 45-degree and when the tooltip displays 45 as polar angle, type **0.7071** and press ENTER; you are prompted to specify next point.

7. Move the cursor vertically upward and when the tooltip displays 90 as polar angle, type **1** and press ENTER; you are prompted to specify the next point.

8. Move the cursor horizontally toward the left and when the tooltip displays 180 as polar angle, type **3** and press ENTER; you are prompted to specify the next point.

9. Move the cursor vertically downward and when the tooltip displays 90 polar angle, type **1** and press ENTER; you are prompted to specify the next point.

10. Type **C** and press ENTER.

Tip
*To add angular values other than those displayed in the shortcut menu of polar tracking, choose **Settings** from the shortcut menu. Next, select the **Additional angles** check box in the **Polar Angle Settings** area and choose the **New** button. Enter a new value in the field that appears and then press ENTER. Choose **OK** to close the dialog box.*

EXERCISE 4   *Direct Distance Entry*

Use the Direct Distance Entry method to draw a parallelogram. The base of the parallelogram equals 4 units, the side equals 2.25 units, and the angle equals 45-degree. Draw the same parallelogram using the absolute, relative, and polar coordinates. Note the differences and the advantages of using this method over relative and absolute coordinate methods.

ERASING OBJECTS

Ribbon: Home > Modify > Erase	**Toolbar:** Modify > Erase
Menu Bar: Modify > Erase	**Tool Palettes:** Modify > Erase
Command: ERASE or E	

Sometimes, you may need to erase the unwanted objects from the objects drawn. You can do so by using the **Erase** tool. This tool is used exactly the same way as an eraser is used in manual drafting to delete the unwanted lines. To erase an object, choose the **Erase** tool from the **Modify** panel, see Figure 2-27. You can also choose the **Erase** button from the **Modify** toolbar, as shown in Figure 2-28. To invoke the **Modify** toolbar, choose **View > Windows > Toolbars > AutoCAD > Modify** from the **Ribbon**. On invoking the **Erase** tool, a small box, known as pick box, replaces the screen cursor. To erase the object, select it by

*Figure 2-27 The **Erase** tool in the **Modify** panel*

using the pick box (see Figure 2-29); the selected object will be displayed in dashed lines and the **Select objects** prompt will be displayed again. You can either continue selecting the objects or press ENTER to terminate the object selection process and erase the selected objects. The prompt sequence is given next.

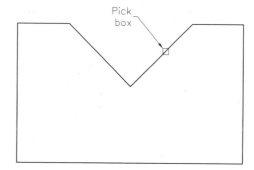

*Figure 2-28 The **Erase** tool in the **Modify** toolbar*

Figure 2-29 Selecting the object by positioning the pick box at the top of the object

Command: *Choose the **Erase** tool*
Select objects: *Select the first object.*
Select objects: *Select the second object.*
Select objects:

If you enter **ALL** at the **Select objects** prompt, all objects in the drawing area will be selected, even if they are outside the display area. Now, if you press ENTER, all the selected objects will be erased.

To erase objects, you can also first select the objects to be erased from the drawing and then choose the **Erase** option from this shortcut menu that is displayed on right-click in the drawing area.

CANCELING AND UNDOING A COMMAND

If you have erased an object by mistake, then to restore the erased object, enter the **OOPS** or **UNDO** command. The **OOPS** command is used to restore the objects erased by the previous **ERASE** command. The **U** (Undo) command is used to undo the action of the previously performed command.

To cancel or exit a command, press the ESC (Escape) key on the keyboard.

OBJECT SELECTION METHODS

The usual method to select objects is by selecting them individually. But it will be time-consuming, if you have a number of objects to select. This problem can be solved by creating a selection set that enables you to select several objects at a time. The selection set options can be used with those tools that require object selection, such as **Erase** and **Move**. There are many object selection methods, such as **Last**, **Add**, **Window**, **Crossing**, and so on. In this chapter, you will learn two methods: **Window** and **Crossing**. The remaining options are discussed in Chapter 5.

Window Selection

The window selection is one of the selection methods in which an object or group of objects are selected by drawing a window. The objects that are completely enclosed within the window are selected and the objects that lie partially inside the boundaries of the window are not selected. To select the objects by using the **Window** option after invoking a tool, type **W** at the **Select objects** prompt and press ENTER; you will prompted to specify the first corner of the window. Select the first corner and then move the cursor to specify the opposite corner. As you move the cursor, a blue color window of continuous line will be displayed. The size of this window changes as you move the cursor. Specify the opposite corner of the window; the objects that are enclosed in this window are displayed as dashed objects. Figure 2-30 shows the window drawn to select objects by using the **Window** option. The objects that will be selected are shown in dashed lines.

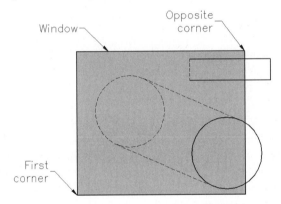

Figure 2-30 *Selecting objects using the **Window** option*

You can also invoke the **Window** option without entering **W** at the Command prompt. To do so, specify a point on the screen at the **Select objects** prompt. This is considered as the first corner of the window. Moving the cursor to the right will display a blue-shaded window. After enclosing the required objects, specify the other corner of the window. The objects that are completely enclosed within the window will be selected and displayed in dashed lines. The following is the prompt sequence for automatic window after invoking the **Erase** tool:

> Command: *Choose the **Erase** tool*
> Select objects: *Select a blank point as the first corner of the window.*
> Specify opposite corner: *Drag the cursor to the right to select the other corner of the window.*
> Select objects: Enter

Tip
*In AutoCAD, the entities are highlighted when you move the cursor over them. This feature is known as the selection preview. To set the selection preview, invoke the **Options** dialog box, choose the **Selection** tab, and then the **Visual Effects Settings** button in the **Selection Preview** area; the **Visual Effect Settings** dialog box will be displayed. Set the type of selection preview by selecting the appropriate radio button in the **Line highlighting** area of this dialog box.*

Window Crossing Method

The window crossing selection is one of the selection methods in which an object or group of objects that are completely or partially enclosed by the selection window are selected. The objects to be selected should touch the window boundaries or completely enclosed within it. To select the objects by using the window crossing method after invoking a tool, type **C** at the **Select objects** prompt and press ENTER; you will be prompted to select the first corner of the window. Select the first corner and then move the cursor to specify the opposite corner. As you move the cursor, a green color window with dashed outline is displayed. Specify the opposite corner of the window; the objects that touch the window boundaries and the objects that are enclosed by the window are selected and displayed as dashed objects. Figure 2-31 shows a window drawn to select objects by using the **Window Crossing** method. The objects that will be selected are shown in dashed lines.

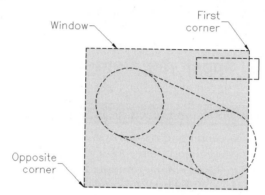

Figure 2-31 Selecting objects using the **Window Crossing** option

You can also invoke the **Window Crossing** method without entering **C** at the Command prompt. To do so, specify a point in the drawing area at the **Select objects** prompt and move the cursor to the left. As you move the cursor, a green color window with dashed outline will be displayed. Specify the opposite corner of the window; the objects touching the window boundary and that are enclosed within this window are selected and displayed as dashed objects. The prompt sequence for the automatic window crossing method when you choose the **Erase** tool is given next.

Select objects: *Select a blank point as the first corner of the crossing window.*
Specify opposite corner: *Drag the cursor to the left to select the other corner of the crossing window.*
Select objects: [Enter]

Tip
You can also select the objects by using the window or window crossing methods before invoking a command. To do so, specify the start point of the selection window and then drag the cursor to enclose the objects in a window. If you move the cursor to the left of the start point, the window crossing method will be activated. But, if you move the cursor to the right of the start point, the window option will be activated.

If you do not invoke any tool and click to specify the first corner of the window for window selection or window crossing, the Command prompt provides you with three selection options: **Fence**, **WPolygon**, and **CPolygon**. If you enter **FENCE** or **F** at the Command prompt, you can select objects by drawing a fence around them. If you enter **WP** at the Command prompt, you can select objects by drawing a polygon around them. If you enter **CP** at the Command prompt, you can select objects by drawing a polygon around them. These options will be discussed in detail in Chapter 5.

DRAWING A CIRCLE

Command: CIRCLE or C	**Toolbar:** Draw > Circle
Menu Bar: Draw > Circle	**Tool Palettes:** Draw > Circle
Ribbon: Home > Draw > Circle drop-down > Center, Radius	

A circle is drawn by using the **CIRCLE** command. In AutoCAD, you can draw a circle by using six different tools. All these tools are grouped together in the **Draw** panel of the **Ribbon**. To view these tools, choose the down arrow next to the **Center, Radius** tool in the **Draw** panel, as shown in Figure 2-32; all tools will be listed in a drop-down. Note that the tool chosen last will be displayed in the **Draw** panel. You can also invoke the **CIRCLE** command by choosing the **Circle** tool from the **Draw** toolbar or the **Tool Palettes**. The different methods to draw a circle are discussed next.

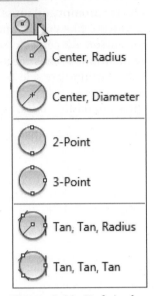

Figure 2-32 Tools in the Circle drop-down

Drawing a Circle by Specifying Center and Radius

Ribbon: Home > Draw > Circle drop-down > Center, Radius

To draw a circle by specifying its center and radius, first ensure that the **Dynamic Input** button is chosen, and then choose the **Center, Radius** tool from the **Draw** panel; you will be prompted to specify the center of the circle. Type the coordinates and press ENTER or specify the center by using the left mouse button. After specifying the center of the circle, move the cursor to define its radius; the current radius of the circle will be displayed in the dimension input box, as shown in Figure 2-33. This radius value will change as you move the cursor. Type a radius value in the dimension input box or click to define the radius; a circle of the specified radius value will be drawn.

Drawing a Circle by Specifying Center and Diameter

Ribbon: Home > Draw > Circle drop-down > Center, Diameter

To draw a circle by specifying its center and diameter, first ensure that the **Dynamic Input** button is chosen, and then choose the **Center, Diameter** tool from the **Draw** panel; you will be prompted to specify the center. Type the coordinates and press ENTER or specify the center by using the left mouse button. After specifying the center of the circle, move the cursor to define its diameter; the current diameter of the circle will be displayed in the dimension input box, as shown in Figure 2-34. This diameter value will change as you move the cursor. Type a diameter value in the dimension input box or click to define the diameter; a circle of the specified diameter value will be drawn.

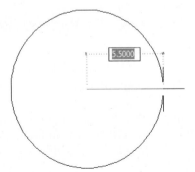

Figure 2-33 Drawing a circle by specifying the center and the radius

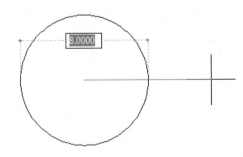

Figure 2-34 Drawing a circle by specifying the center and the diameter

Drawing a Circle by Specifying Two Diametrical Ends

Ribbon: Home > Draw > Circle drop-down > *2*-Point
Command: C > 2P

You can also draw a circle by specifying its two diametrical ends, see Figure 2-35. To do so, first ensure that the **Dynamic Input** button is chosen, and then choose the **2-Point** tool from the **Draw** panel; you will be prompted to specify the first end of the diameter. Type the coordinates and press ENTER or specify the center by using the left mouse button. After specifying the center of the circle, move the cursor to define its diameter. Now, you can type the coordinates or diameter in the dimension input box.

Drawing a Circle by Specifying Three Points on a Circle

Ribbon: Home > Draw > Circle drop-down > *3*-Point
Command: C > 3P

To draw a circle by specifying three points on its periphery, choose the **3-Point** tool from the **Draw** panel and specify the three points in succession. You can type the coordinates of the points or specify them by using the left mouse button. The prompt sequence to type the three coordinates on choosing the **3-Point** tool is given below.

Specify center point for circle or [3P/2P/Ttr(tan tan radius)]: _3p Specify first point on circle: **3,3**
Specify second point on circle: **#3,1**
Specify third point on circle: **#4,2** *(see Figure 2-36)*

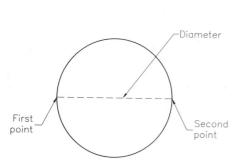

Figure 2-35 *A circle drawn by using the 2-Point option*

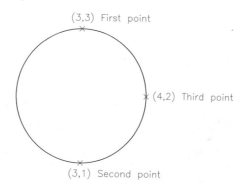

Figure 2-36 *A circle drawn by using the 3-Point option*

You can also use the relative rectangular coordinates to define the points.

Drawing a Circle Tangent to Two Objects

Ribbon: Home > Draw > Circle drop-down > Tan, Tan, Radius   **Command:** C > Ttr

An object (line, circle, or arc) is said to be tangent to a circle or an arc, if it touches the circumference of the circle or the arc at only one point. To draw a circle that has specified radius and is tangent to two objects, first ensure that the **Dynamic Input** button is chosen, and then choose the **Tan, Tan, Radius** tool from the **Draw** panel; you will be prompted to specify a point on the first object to be tangent to the circle. Move the cursor near the object to be made tangent to the circle; a tangent symbol will be displayed. Specify the first point; you will be prompted to specify a point on the second object to be made tangent to the circle. Move the cursor near the second object that is to be tangent to the circle; a tangent symbol will be displayed. Specify the second point; you will be prompted to specify the radius. Type the radius value in the dimension input box and press ENTER; a circle of the specified radius and tangent to two specified objects will be drawn.

In Figures 2-37 through 2-40, the dotted circle represents the circle that is tangent to two objects. The circle actually drawn depends on how you select the objects to be made tangent to the new circle. The figures show the effect of selecting different points on the objects. If you specify too small or large radius, you may get unexpected results or the "**Circle does not exist**" prompt.

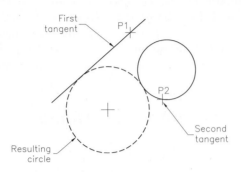

Figure 2-37 *Drawing a circle tangent to two objects*

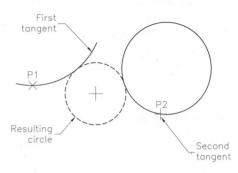

Figure 2-38 *Drawing a circle tangent to two objects*

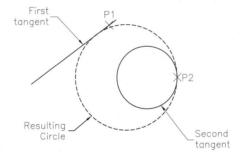

Figure 2-39 *Drawing a circle tangent to two objects*

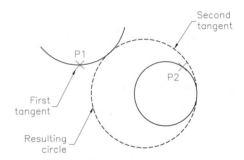

Figure 2-40 *Drawing a circle tangent to two objects*

Drawing a Circle Tangent to Three Objects

| Ribbon: | Home > Draw > Circle drop-down > Tan, Tan, Tan |

You can also draw a circle that is tangent to three objects. To do so, choose the **Tan, Tan, Tan** tool from the **Draw** panel and select the three objects in succession to which the resulting circle is to be tangent; the circle will be drawn, as shown in Figure 2-41.

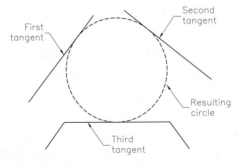

Figure 2-41 *Drawing a circle tangent to three objects*

EXERCISE 5   *Line and Circle*

Draw the profile shown in Figure 2-42 using various options of the **LINE** and **CIRCLE** commands. Use the absolute, relative rectangular, or relative polar coordinates for drawing the triangle. The vertices of the triangle will be used as the center of the circles. The circles can be drawn by using the **Center, Radius**, or **Center, Diameter**, or **Tan, Tan, Tan** tools. Do not apply dimensions; they are for reference only.

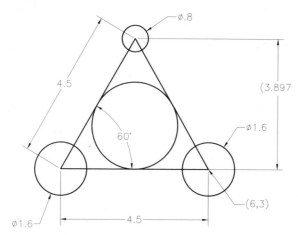

Figure 2-42 Drawing for Exercise 5

BASIC DISPLAY COMMANDS

Sometimes while drawing a sketch, it may be very difficult to view and alter minute details. You can overcome this problem by viewing only a specific portion of the drawing. This is done by using the **ZOOM** command. This command lets you enlarge or reduce the size of the drawing displayed on the screen. Similarly, you may need to slide the drawing view. This can be done by using the **Pan** tool. These are called display commands and are discussed next.

Zooming Drawings

The **ZOOM** command is used to enlarge or reduce the view of a drawing on the screen, without affecting the actual size of entities. In AutoCAD 2012, the ZOOM tools are grouped together and are available in the Navigator Bar. To invoke different **ZOOM** tools, press the down arrow next to the **Zoom Extents** tool in the Navigator Bar; all options of the **ZOOM** command will be displayed in the drop-down, as shown in Figure 2-43. You can also invoke the **ZOOM** command by choosing **View > Zoom** from the menu bar. To display the menu bar, left-click on the arrow in the **Quick Access Toolbar** and then select **Show Menu Bar**. Some **ZOOM** commands are also available in the **Standard** toolbar.

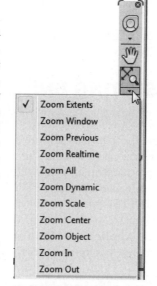

*Figure 2-43 The Zoom tools in the **Navigator Bar***

Zoom Extents

 Choose the **Zoom Extents** tool to increase or decrease the drawing display area so that all sketched entities or dimensions fit inside the current view.

Zoom Window

 This is the most commonly used option of the **ZOOM** command. On choosing this tool, you need to draw a window by specifying its two opposite corners. The center of the zoom window becomes the center of the new display area and the objects in this window are magnified.

Zoom Realtime

The **Zoom Realtime** tool is used to dynamically zoom in or out a drawing. When you choose this option, the cursor will be replaced by the zoom cursor. To zoom out a drawing, press and hold the left mouse button and drag the cursor downward. Similarly, to zoom in a drawing, press and hold the left mouse button and drag the cursor upward. As you drag the cursor, the drawing display changes dynamically. After you get the desired view, exit this tool by right-clicking and then choosing **Exit** from the shortcut menu. On exiting this tool, the zoom cursor will change into cross hairs. Next, press the ESC key. You can also exit the **Zoom Realtime** tool by pressing the ESC key twice. If you have a mouse with scroll wheel, then scroll the wheel to zoom in/out the drawing.

Zoom Previous

While working on a complex drawing, you may need to zoom in a drawing multiple times to edit some minute details. After completing the editing, if you want to view the previous views, choose the **Zoom Previous** tool. You can view up to the last ten views by using the **Zoom Previous** tool.

Zoom In / Zoom Out

Choose the **Zoom In / Zoom Out** tool to increase/decrease the size of the drawing view twice/half of the original drawing size, respectively.

Note
*You will learn about the other options of the **ZOOM** command in detail in Chapter 6.*

Moving the View

You can use the **Pan** tool to move a view by sliding and placing it at the required position. To pan a drawing view, invoke the **Pan** tool from the Navigator Bar; a hand cursor will be displayed. Drag the cursor in any direction to move the drawing. To exit the **Pan** tool, right-click and then choose **Exit** from the shortcut menu. You can also press the ESC or ENTER key to exit the tool.

SETTING UNITS TYPE AND PRECISION

Application Menu: Drawing Utilities > Units	**Command:** UNITS

In the previous chapter, you learned to set units while starting a drawing by using the **Use a Wizard** option in the **Startup** dialog box. But, if you are drawing a sketch in an existing template or in a new template, you need to change the format of the units for distance and angle measurements. This is done by using the **UNITS** command. To set the units format, enter **UNITS** at the Command prompt; the **Drawing Units** dialog box will be displayed, as shown in Figure 2-44. You can also invoke this dialog box by choosing **Drawing Utilities > Units** from the **Application Menu** or by choosing **Format > Units** from the menu bar. The procedure to change the units format is discussed next.

Specifying the Format

In the **Drawing Units** dialog box, you can select the desired format of units from the **Type** drop-down list. You can select any one of the five formats given next.

Architectural (0'-01/16")	Decimal (0.00)	Engineering (0'-0.00")
Fractional (0 1/16)	Scientific (0.00E+01)	

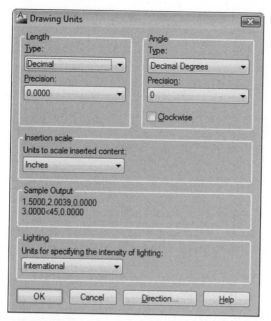

*Figure 2-44 The **Drawing Units** dialog box*

If you select the scientific, decimal, or fractional format, you can enter distance or coordinate values in any of these three formats, but not in engineering or architectural units. If you select the engineering or architectural format, you can enter distances or coordinates in any of the five formats.

Note
The inch symbol (") is optional. For example, 1'1-3/4" is the same as 1'1-3/4, and 3/4" is the same as 3/4.

Specifying the Angle Format

You can select any one of the following five angle measuring formats:

1. Decimal Degrees (0.00)   2. Deg/min/sec (0d00'00")

3. Grads (0.00g)   4. Radians (0.00r)

5. Surveyor's Units (N 0d00'00" E)

If you select any one of the first four measuring formats, you can specify the angle in the Decimal, Degrees/minutes/seconds, Grads, or Radians system, but you cannot enter the angle in Surveyor's Units system. However, if you select **Surveyor's Units**, you can enter angle values in any of the five systems. To enter a value in another system, use the appropriate suffixes and symbols, such as r (Radians), d (Degrees), or g (Grads). If you enter an angle value without indicating the symbol of a measuring system, it is taken in the current system.

In Surveyor's units, you must specify the angle that the line makes with respect to the north-south direction, as shown in Figure 2-45. For example, if you want to define an angle of 60-degree with north, in the Surveyor's units the angle will be specified as N 60d E. Similarly, you can specify angles such as S 50d E, S 50d W, and N 75d W, refer to Figure 2-45. You cannot specify an angle that exceeds 90-degree (N 120 E). Angles can also be specified in radians or grads; for example, 180-degree is equal to π (3.14159) radians. You can convert degrees into radians, or radians into degrees by using the equations given below.

$$radians = degrees \times 3.14159/180;$$
$$degrees = radians \times 180/3.14159$$

Grads are generally used in land surveys. There are 400 grads or 360 degree in a circle. 90 degree angle is equal to 100 grads.

In AutoCAD, if an angle is measured in the counterclockwise direction, then it is positive. Also, the angles are measured about the positive X axis, see Figure 2-46. If you want the angles to be measured as positive in the clockwise direction, select the **Clockwise** check box from the **Angle** area. You can specify the precision for the length and angle in the respective **Precision** drop-down lists in this dialog box.

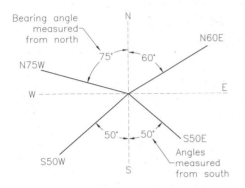

Figure 2-45 *Specifying angles in Surveyor's Units*

Figure 2-46 *Measuring angles*

Setting the Direction for Angle Measurement

As mentioned above, angles are measured about the positive X axis. This means the base angle (0-degree) is set along the east direction, see Figure 2-47. To change this base angle, choose the **Direction** button in the **Drawing Units** dialog box; the **Direction Control** dialog box will be displayed, as shown in Figure 2-48. Select the appropriate radio button to specify the direction for the base angle (0-degree).

If you select the **Other** option, you can set the direction of your choice for the base angle (0-degree) by entering a value in the **Angle** edit box or by choosing the **Pick an angle** button and picking two points to specify the angle. After specifying the base angle direction, choose the **OK** button to apply the settings.

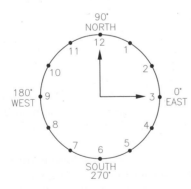

Figure 2-47 *North, South, East, and West directions*

Figure 2-48 *The **Direction Control** dialog box*

Tip
*If you are entering values by using the dimension input boxes that are displayed when the **Dynamic Input** option is on, then you are not required to set the base angle.*

Specifying Units for the Drawing or Block to be Inserted

To set units for a block or a drawing to be inserted, select a unit from the **Units to scale inserted content** drop-down list. Now, if you insert a block or a drawing from the **DesignCenter**, the specified unit will be applied to the block. Even if the block was created using a different measuring unit, AutoCAD scales it and inserts it using the specified measuring unit. If you select **Unitless** from the drop-down list, then the units specified in the **Insertion Scale** area of the **User Preferences** tab in the **Options** dialog box will be used.

Note

Inserting blocks in a drawing is discussed in detail in Chapter 18.

Sample Output

The **Sample Output** area in this dialog box shows the example of the current format used for specifying units and angles. When you change the type of length and angle measure in the **Length** and **Angle** areas of the **Drawing Units** dialog box, the corresponding example is displayed in the **Sample Output** area.

Specifying Units for Lighting

You can also specify the units to be used for the measurement of intensity of photometric lights. Photometric lights are used for rendering the objects. Select the desired unit system from the **Units for specifying the intensity of lighting** drop-down list in the **Lighting** area of the **Drawing Units** dialog box. While using photometric lights in rendering, you cannot use the **Generic** unit for the measurement of intensity of light. The concept of rendering and lights is discussed in Chapter 31.

EXAMPLE 5   *Setting Units*

In this example, you will set the units of a drawing based on the specifications given below and then draw Figure 2-49.

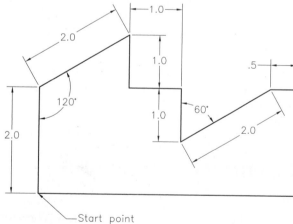

Figure 2-49 *Drawing for Example 5*

a. Set the units of length to fractional, with the denominator of the smallest fraction equal to 32.
b. Set the angular measurement to Surveyor's Units, with the number of fractional places for display of angles equal to zero.
c. Set the base angle (0-degree) to North and the direction of measurement of angles to clockwise.

The following steps are required to complete this example:

1. Start a new file with the *Acad.dwt* template in the **Drafting & Annotation** or **AutoCAD Classic** workspace and invoke the **Drawing Units** dialog box by choosing **Drawing Utilities > Units** from the **Application Menu**. You can also invoke this dialog box by entering **UNITS** at the Command prompt.

2. In the **Length** area of this dialog box, select **Fractional** from the **Type** drop-down list. Select **0 1/32** from the **Precision** drop-down list.

3. In the **Angle** area of this dialog box, select **Surveyor's Units** from the **Type** drop-down list. From the **Precision** drop-down list, select **N 0d E**, if it has not been already selected. Also, select the **Clockwise** check box to set the clockwise angle measurement to positive.

4. Choose the **Direction** button to display the **Direction Control** dialog box. Next, select the **North** radio button. Choose the **OK** button to exit the **Direction Control** dialog box.

5. Choose the **OK** button to exit the **Drawing Units** dialog box.

6. With the units set, you need to draw Figure 2-49 by using the relative polar coordinates. Turn off the dynamic input. The prompt sequence to complete the sketch is as follows:

    > Command: *Choose the* **Line** *tool*
    > _line Specify first point: **2,2** `Enter`
    > Specify next point or [Undo]: **@2.0<0** `Enter`
    > Specify next point or [Undo]: **@2.0<60** `Enter`
    > Specify next point or [Close/Undo]: **@1<180** `Enter`
    > Specify next point or [Close/Undo]: **@1<90** `Enter`
    > Specify next point or [Close/Undo]: **@1<180** `Enter`
    > Specify next point or [Close/Undo]: **@2.0<60** `Enter`
    > Specify next point or [Close/Undo]: **@0.5<90** `Enter`
    > Specify next point or [Close/Undo]: **@2.0<180** `Enter`
    > Specify next point or [Close/Undo]: **C** `Enter`

    Here, the units are fractional and the angles are measured from north (90-degree axis). Also, the angles are measured as positive in the clockwise direction and negative in the counterclockwise direction.

7. To modify the drawing display area, choose the **Zoom All** tool from the Navigator Bar.

SETTING THE LIMITS OF A DRAWING

Command: LIMITS

In AutoCAD, limits represent the drawing area and it is endless. Therefore, you need to define the drawing area before starting the drawing. In the previous chapter, you learned to set limits while starting a drawing by using the **Use a Wizard** option in the **Startup** dialog box. If you are working in a drawing by using the default template, you need to change limits. For example, the template *Acad.dwt* has the default limits set to 12,9. To draw a rectangle of dimension 150x75 in this template, you need to change its limits to 200x100. This can be done by using the **LIMITS** command. The following is the prompt sequence of the **LIMITS** command for setting the limits to 24,18 for the *Acad.dwt* template, which has the default limits 12,9.

Command: **LIMITS** Enter
Reset Model space limits:
Specify lower left corner or [ON/OFF]<0.0000,0.0000>: **0,0** Enter
Specify upper right corner <12.0000,9.0000>: **24,18** Enter

Tip
*Whenever you reset the drawing limits, the display area does not change automatically. You need to use the **All** option of the **ZOOM** command to display the complete drawing area.*

The limits of the drawing area are usually determined by the following factors:

1. The actual size of drawing.
2. The space needed for adding dimensions, notes, bill of materials, and other necessary details.
3. The space between various views so that the drawing does not look cluttered.
4. The space for the border and title block, if any.

Setting Limits

To get a good idea of how to set up limits, it is better to draw a rough sketch of a drawing. This will help in calculating the required drawing area. For example, if an object has a front view size of 5 X 5, a side view size of 3 X 5, and a top view size of 5 X 3, the limits should be set so that the drawing and everything associated with it can be easily accommodated within the set limit. In Figure 2-50, the space between the front and side views is 4 units and between the front and top views is 3 units. Also, the space between the border and the drawing is 5 units on the left, 5 units on the right, 3 units at the bottom, and 2 units at the top. (The space between the views and between the borderline and the drawing depends on the drawing.)

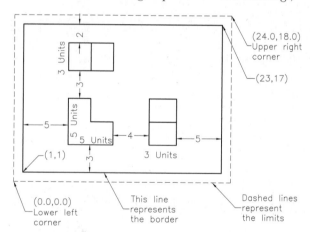

Figure 2-50 *Setting limits in a drawing*

After knowing the size of different views, the space required between views, the space between the border and the drawing, and the space required between the borderline and the edges of the paper, you can calculate the space in the following way:

Space along (*X* axis) = 1 + 5 + 5 + 4 + 3 + 5 + 1 = 24
Space along (*Y* axis) = 1 + 3 + 5 + 3 + 3 + 2 + 1 = 18

This shows that the limits you need to set for this drawing is 24 X 18. Once you have determined the space, select the sheet size that can accommodate your drawing. In the case just explained, you will select a D size (34 X 22) sheet. Therefore, the actual drawing limits will be 34,22.

Tip

*To display the grid, choose the **Grid Display** button in the Status Bar. By default, the grid will be displayed beyond the limits. To display the grid up to the limits, use the **Limits** option of the **GRID** command and set the **Display grid beyond Limits [Yes/No] <Yes>**: option to **No**; the grids will be displayed only up to the limits set.*

Limits for Architectural Drawings

Most architectural drawings are drawn at the scale of 1/4" = 1', 1/8" = 1', or 1/16" = 1'. You must set the limits accordingly. The following example illustrates how to calculate the limits in architectural drawings.

Given

        Sheet size = 24 X 18
        Scale is 1/4" = 1'

Calculate limits

        Scale is 1/4" = 1'
          or 1/4" = 12"
          or 1" = 48"
     X limit = 24 X 48
           = 1152" or 1152 Units
           = 96'
     Y limit = 18 X 48
           = 864" or 864 Units
           = 72'

Thus, the scale factor is 48 and the limits are 1152",864", or 96',72'.

EXAMPLE 6   *Setting Limits*

In this example, you will calculate the limits and determine an appropriate drawing scale factor for the drawing shown in Figure 2-51. You will plot the drawing on a 12" X 9" sheet. Assume the missing dimensions.

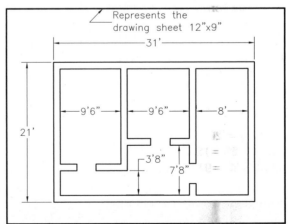

Figure 2-51 *Drawing for Example 6*

The calculation for the scale factor is given next.

Given or known

        Overall length of the drawing = 31'
        Length of the sheet = 12"

Approximate space between the drawing and the edges of the paper = 2"

Calculate the scale factor

To calculate the scale factor, you have to try various scales until you find the one that satisfies the given conditions. After some experience, you will find this fairly easy to do. For this example, assume a scale factor of 1/4" = 1'.

Scale factor 1/4" = 1'   or 1" = 4'

Thus, a line 31' long will be = 31'/4' = 7.75" on paper. Similarly, a line 21' long = 21'/4' = 5.25". Approximate space between the drawing and the edges of paper = 2".

Therefore, the total length of the sheet = 7.75 + 2 + 2 = 11.75"

Similarly, the total width of the sheet = 5.25 + 2 + 2 = 9.25"

Because you selected the scale 1/4" = 1', the drawing will definitely fit in the given sheet of paper (12" x 9"). Therefore, the scale for this drawing is 1/4" = 1'.

Calculate limits

Scale factor = 1" = 48" or 1" = 4'
The length of the sheet is 12"
Therefore, X limit = 12 X 4' = 48' and Y limit = 9 X 4' = 36'

Limits for Metric Drawings

When the drawing units are in metric, you must use **standard metric size sheets** or calculate limits in millimeters (mm). For example, if the sheet size is 24 X 18, the limits, after conversion to the metric system, will be 609.6,457.2 (multiplying length and width by 25.4). You can round these numbers to the nearest whole numbers 610,457. Note that metric drawings do not require any special setup, except for the limits. Metric drawings are like any other drawings that use decimal units. Similar to architectural drawings, you can draw metric drawings to a scale. For example, if the scale is 1:20, you must calculate the limits accordingly. The following example illustrates how to calculate limits for metric drawings:

Given

Sheet size = 24" X 18"
Scale = 1:20

Calculate limits

Scale is 1:20
Therefore, scale factor = 20
X limit = 24 X 25.4 X 20 = 12192 units
Y limit = 18 X 25.4 X 20 = 9144 units

EXERCISE 6   *Setting Units and Limits*

Set the units of the drawing according to the specifications given below and then make the drawing shown in Figure 2-52 (leave a space of 3 to 5 units around the drawing for dimensioning and title block). The space between the dotted lines is 1 unit.

1. Set UNITS to decimal units, with two digits to the right of the decimal point.
2. Set the angular measurement to decimal degrees, with the number of fractional places for display of angles equal to 1.

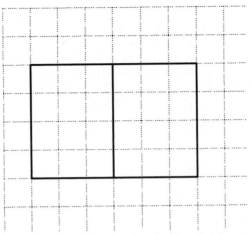

Figure 2-52 Drawing for Exercise 6

3. Set the direction to 0-degree (east) and the direction of measurement of angles to counterclockwise (angles measured positive in a counterclockwise direction).
4. Set the limits leaving a space of 3 to 5 units around the drawing for dimensioning and title block.

INTRODUCTION TO PLOTTING DRAWINGS

Ribbon: Output > Plot > Plot   **Quick Access Toolbar:** Plot
Application Menu: Print > Plot   **Command:** PLOT or PRINT

After creating a drawing of an architectural plan or a mechanical component, you may need to send it to the client or have a hard copy for reference. To do so, you need to plot the drawing. To plot the drawing, choose **Plot** from the **Quick Access Toolbar**; the **Plot** dialog box will be displayed. If the dialog box is not expanded by default, choose the **More Options** button at the lower right corner of the dialog box to expand it, as shown in Figure 2-53.

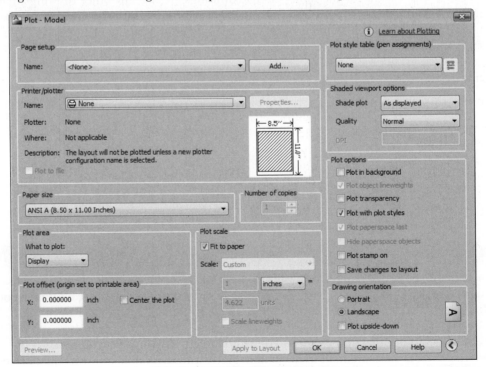

*Figure 2-53 The **Plot** dialog box*

Basic Plotting

In this section, you will learn to set up the basic plotting parameters. Later, you will learn about the advance options to plot your drawings according to plot drawing specifications. Basic plotting involves selecting the correct output device (plotter), specifying the area to plot, selecting the paper size, specifying the plot origin, orientation, and the plot scale.

To learn the basic plotting, you will plot the drawing drawn in Example 3 of this chapter. It is assumed that AutoCAD is configured for two output devices: Default System Printer and Linotronic 300v52.3.

1. Invoke the **Plot** dialog box by choosing the **Plot** tool in the **Quick Access Toolbar**.

2. The name of the default system printer is displayed in the **Name** drop-down list in the **Printer/plotter** area. Select **Linotronic 300v52.3** in the **Name** drop-down list.

3. Select the **Window** option from the **What to plot** drop-down list in **Plot area**. The dialog box will close temporarily and the drawing area will appear. Next, select two opposite corners to define a window that can enclose the entire area you want to plot. Note that the complete drawing, along with the dimensions should be enclosed in the window. Once you have defined the two corners, the **Plot** dialog box will reappear.

4. To set the size of the plot, you need to select a paper size from the drop-down list in the **Paper size** area. After selecting the paper size, you need to set the orientation of the paper. To set the orientation, expand the **Plot** dialog box by choosing the **More Options** button at the lower right corner of the dialog box. You can set the orientation as **Landscape** or **Portrait** by selecting the appropriate radio button from the **Drawing orientation** area. The sections in the **Plot** dialog box related to the paper size and orientation are automatically revised to reflect the new paper size and orientation. For this example, you will specify the **A4** paper size and the **Portrait** orientation.

5. You can also modify values for the plot offset in the **Plot offset** area; the default value for X and Y is 0. For this example, you can select the **Center the plot** check box to get the drawing at the center of the paper.

6. In AutoCAD, you can enter values for the plot scale in the **Plot scale** area. Clear the **Fit to paper** check box if it is selected and then click the **Scale** drop-down list in the **Plot scale** area to display various scale factors. From this list, you can select a scale factor based on your requirement. For example, if you select the scale factor **1/4" = 1'-0"**, the edit boxes below the drop-down list will show 0.25 inches = 12 units. If you want the drawing to be plotted so that it fits on the specified sheet of paper, select the **Fit to paper** check box. On selecting this check box, AutoCAD will determine the scale factor and display it in the edit boxes. For this example, you will plot the drawing so that it scales to fit the paper. Therefore, select the **Fit to paper** check box and notice the change in the edit boxes. You can also enter arbitrary values in the edit boxes.

7. To preview a plot, choose the **Preview** button. You can preview the plot on the specified paper size. In the preview window, the realtime zoom icon will be displayed. If needed, you can zoom in or zoom out the preview image for better visualization.

8. If the plot preview is satisfactory, you can plot your drawing by right-clicking and then choosing **Plot** from the shortcut menu. If you want to make some changes in the settings, choose **Exit** in the shortcut menu or press the ESC or ENTER key to get back to the dialog box. Change the parameters and choose the **OK** button in the dialog box to plot the drawing.

MODIFYING AutoCAD SETTINGS BY USING THE OPTIONS DIALOG BOX

Application Menu: Options	**Command:** OPTIONS

You can use the **Options** dialog box to change the default settings and customize them to your requirements. For example, you can use this dialog box to turn off the settings that is used to display the shortcut menu, change the display color of the objects, or specify the support directories containing the files you need.

To invoke the **Options** dialog box, right-click at the Command prompt or in the drawing area when no command is active or no object is selected and then choose **Options** from the shortcut menu; the dialog box will be displayed, as shown in Figure 2-54. The name of the current profile and the current drawing names will be displayed below the title bar. You can save a set of custom settings in a profile to be used later for other drawings.

Files

This tab stores the directories in which AutoCAD looks for the driver, support, menu, project, template, and other files. It uses three icons: folder, paper stack, and file cabinet. The folder icon is for a search path, the paper stack icon is for files, and the file cabinet icon is for a specific folder. Suppose you want to know the path of the font mapping file. To do so, you need to click on the + symbol before the **Text Editor**, **Dictionary**, and **Font File Names** folders and then select the **Font Mapping Files** icon, see Figure 2-54. Similarly, you can define a custom hatch pattern file and then add its search path.

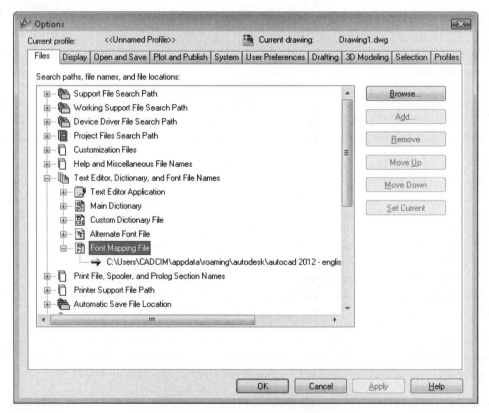

*Figure 2-54 The **Options** dialog box*

Display

This tab is used to control the drawing and window settings like screen menu display and scroll bar. For example, if you want to display the drawing status bar, select the **Display Drawing status bar** check box in the **Window Elements** area. You can change the background color of the graphics window, layout window, and command line as well as the color of the command line text by using the **Drawing Window Colors** dialog box that is displayed on choosing the **Colors** button. In the **Display** tab, you can specify the parameters to set the display resolution and display performance. You can also set the smoothness and resolutions of certain objects such as circle, arc, rendered object, and polyline curve. You can also toggle on and off the display performance such as pan and zoom with raster images, apply the solid fills, and so on. In this tab, you can toggle on and off the various layout elements such as the layout tabs on the screen, margins, paper background, and so on.

Open and Save

This tab is used to control the parameters related to the opening and saving of files in AutoCAD. You can specify the file type while saving using the **SAVEAS** command. Various formats are **AutoCAD 2010 Drawing (*.dwg)**, **AutoCAD 2007 Drawing (*.dwg)**, **AutoCAD 2004/LT 2004 Drawing (*.dwg)**, **AutoCAD 2000/LT2000 Drawing (*.dwg)**, **AutoCAD R14/LT98/LT97 Drawing (*.dwg)**, **AutoCAD Drawing Template(*.dwt)**, **AutoCAD 2007 DXF (*.dxf)**, **AutoCAD 2004/ LT2004 DXF (*.dxf)**, **AutoCAD 2000/LT2000 DXF(*.dxf)**, and **AutoCAD R12/LT2 DXF (*.dxf)**. You can also set various file safety precautions such as the Automatic Save feature or the creation of a backup copy. To add a password and digital signatures to your drawing while saving, choose the **Security Options** button in the **File Safety Precautions** area and set the password. On selecting the **Display digital signature information** check box, you can view the digital signature information when a file with a valid digital signature is opened. You can change the number of the recently saved files to be displayed in the **File Open** area. You can also set the various parameters for external references and the ObjectARX applications.

Plot and Publish

The options in this tab are used to control the parameters related to the plotting and publishing of the drawings in AutoCAD. You can set the default output device and also add a new plotter. You can set the general parameters such as the layout or plot device paper size and the background processing options while plotting or publishing. It is possible to select the spool alert for the system printer and also the OLE plot quality. You can also set the parameters for the plot style such as the color-dependent plot styles or the named plot styles.

System

This tab contains AutoCAD system settings options such as the 3D graphics display and pointing device settings options where you can choose the pointing device driver. Here you can also set the system parameters such as the single drawing mode instead of MDE, the display of the Startup option while opening a new session of AutoCAD and the **OLE Properties** dialog box, and beep for wrong user input. You also have options to set the parameters for database connectivity.

User Preferences

The parameters in this tab are used to control the settings such as the right-click customization to change the shortcut menus according to the user's preferences. You can set the units parameters for the blocks or drawings that are inserted as well as the priorities for various coordinate data entry methods. You can also set the order of object sorting methods and also set the lineweight options.

Drafting

The options in this tab are used to control the settings such as autosnap settings and aperture size. Here you can also set the toggles on and off for the various autotracking settings. Using this tab, you can also set the tool tip appearance of **Dynamic Input** mode in the Model and Paper space.

3D Modeling

This tab is used to control the settings related to 3D modeling, such as the display of cursor in the 3D modeling environment, visual styles, 3D navigation, and so on.

Selection

This tab is used to control the set the methods of object selection, grips, grip colors, and the grip size. You can also set the toggles on or off for the various selection modes.

Profiles

This tab is used to save and restore the system settings. To save a profile and have different settings, choose the **Add to List** button; the **Add Profile** dialog box will be displayed. Enter the name and description of the profile and then choose the **Apply & Close** button. Next, make the new profile current and then change the settings in each tab. The settings thus applied will be saved in the new profile and can be restored anytime by making the profile current.

Note
*The options in various tabs of the **Options** dialog box have been discussed throughout the book wherever applicable.*

Tip
*Some options in the **Options** dialog box have drawing file icon in front of them. For example, the options in the **Display resolution** area of the **Display** tab have drawing file icons. This specifies that these parameters are saved with the current drawing only; therefore, it affects only that drawing. The options without the drawing file icon are saved with the current profile and affect all drawings present in that AutoCAD session or future sessions.*

EXAMPLE 7	*Modifying the Default Options*

In this example, you will create a profile with specific settings by using the **Options** dialog box.

1. Choose **Options** from the **Application Menu** to invoke the **Options** dialog box. Alternatively, right-click in the drawing area to display the shortcut menu, and then choose **Options** to invoke this dialog box.

2. Choose the **Profiles** tab, and then choose the **Add to List** button to display the **Add Profile** dialog box. Enter **CADCIM** as the name of the new profile and **Profile created for chapter 2, Tutorial 8** as the description of the new profile, and then choose the **Apply & Close** button to exit.

3. Select the **CADCIM** profile and then choose the **Set Current** button to make this profile current. You will notice that the **Current Profile** name above the tabs displays *CADCIM*.

4. Choose the **Display** tab and then choose the **Colors** button; the **Drawing Window Colors** dialog box is displayed. Select the **2D model space** in the **Context** area and **Uniform**

background in the **Interface element** area. Select **White** from the **Color** drop-down list; the background color of the model tab will change into white. Choose the **Apply & Close** button to return to the **Options** dialog box.

5. Choose the **Drafting** tab and then change **AutoSnap Marker Size** to the maximum using the slider bar. Choose the **Apply** button and then the **OK** button to exit the dialog box.

6. Draw a line and then choose the **Object Snap** button from the Status Bar. Again, invoke the **LINE** command and move the cursor on the previously drawn line; a marker will be displayed at the endpoint. Notice the size of the marker now.

7. Invoke the **Options** dialog box again and choose the **Profiles** tab. Double-click on the default profile to reload the default settings. The screen settings will change as specified in the default profile.

Self-Evaluation Test

Answer the following questions and then compare them to those given at the end of this chapter:

1. You can draw a line by specifying the length and direction of the line, using the Direct Distance Entry method. (T/F)

2. While using the **Window Crossing** method of object selection, only those objects that are completely enclosed within the boundaries of the crossing box are selected. (T/F)

3. Choose the **3-Point** tool from the **Circle** drop-down to draw a circle by specifying the two endpoints of the circle's diameter. (T/F)

4. If you choose the engineering or architectural format for units in the **Drawing Units** dialog box, you can enter distances or coordinates in any of the five formats. (T/F)

5. You can erase a previously drawn line by using the _____ option of the **Line** tool.

6. Choose the _____ tool from the **Circle** drop-down to draw a circle that is tangent to the two previously drawn objects.

7. The _____ tool is used to enlarge or reduce the view of a drawing without affecting the actual size of entities.

8. After increasing the drawing limits, you need to choose the _____ tool from the **Navigator Bar** to display the complete area inside the drawing area.

9. In _____ units, you must specify the bearing angle that a line makes with the north-south direction.

10. You can preview a plot before actually plotting it by using the _____ button in the **Plot** dialog box.

Review Questions

Answer the following questions:

1. In the relative rectangular coordinate system, the displacements along the X and Y axes (DX and DY) are measured with respect to the previous point and not with respect to the origin. (T/F)

2. In AutoCAD, by default, angles are measured along the positive X axis and it will be positive if measured in the counterclockwise direction. (T/F)

3. You can also invoke the **PLOT** command by choosing **Plot** from the shortcut menu displayed on right-clicking in the Command window. (T/F)

4. The **Files** tab of the **Options** dialog box is used to store the directories in which AutoCAD looks for the driver, support, menu, project, template, and other files. (T/F)

5. Which of the following keys is used to terminate the **Line** tool at the **Specify next point or [Close/Undo]:** prompt?

    (a) SPACEBAR   (b) BACKSPACE
    (c) ENTER   (d) ESC

6. Which of the following **ZOOM** commands is used to zoom a drawing up to the limits or the extents, whichever is greater?

    (a) **Zoom Previous**   (b) **Zoom Window**
    (c) **Zoom All**   (d) **Zoom Realtime**

7. How many formats of units can be chosen from the **Drawing Units** dialog box?

    (a) Three   (b) Five
    (c) Six   (d) Seven

8. Which of the following input methods cannot be used to invoke the **OPTIONS** command for displaying the **Options** dialog box?

    (a) Menu   (b) Toolbar
    (c) Shortcut menu   (d) Command prompt

9. When you define direction by specifying angle, the output of the angle does not depend on which one of the following factors?

    (a) Angular units   (b) Angle value
    (c) Angle direction   (d) Angle base

10. The _____ option of the **Line** tool can be used to join the current point with the initial point of the first line when two or more lines are drawn in succession.

11. The _____ option of drawing a circle cannot be invoked by entering the command at the Command prompt.

12. When you select any type of unit and angle in the **Length** or **Angle** area of the **Drawing Units** dialog box, the corresponding example is displayed in the _____ area of the dialog box.

13. If you want a drawing to be plotted so that it fits on the specified sheet of paper, select the _____ option in the **Plot** dialog box.

14. The _____ tab in the **Options** dialog box is used to store the details of all profiles available in the current drawing.

15. You can use the _____ command to change the settings that affect the drawing environment or the AutoCAD interface.

EXERCISE 7   *Relative Rectangular & Absolute Coordinates*

Invoke the **Line** tool and use the following relative rectangular and absolute coordinate values to draw the object.

Point	Coordinates	Point	Coordinates
1	3.0, 3.0	5	@3.0,5.0
2	@3,0	6	@3,0
3	@-1.5,3.0	7	@-1.5,-3
4	@-1.5,-3.0	8	@-1.5,3

EXERCISE 8   *Relative Rectangular & Polar Coordinates*

Draw the profile shown in Figure 2-55 by using the relative rectangular and relative polar coordinates of the points given in the following table. The distance between the dotted lines is 1 unit. Save this drawing with the name *C02_Exer8.dwg*.

Point	Coordinates	Point	Coordinates	Point	Coordinates
1	3.0, 1.0	7	_____	13	_____
2	_____	8	_____	14	_____
3	_____	9	_____	15	_____
4	_____	10	_____	16	_____
5	_____	11	_____		
6	_____	12	_____		

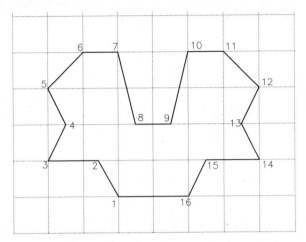

Figure 2-55 *Drawing for Exercise 8*

EXERCISE 9   *Relative Polar Coordinates*

For the drawing shown in Figure 2-56, enter the relative polar coordinates of the points in the following table. Then, use these coordinates to create the drawing. Do not dimension the drawing.

Point	Coordinates		Point	Coordinates
1	1.0, 1.0		6	_____
2	_____		7	_____
3	_____		8	_____
4	_____		9	_____
5	_____			

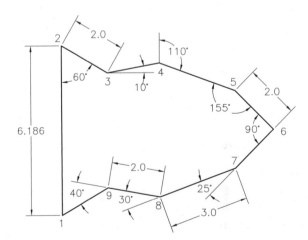

Figure 2-56 *Drawing for Exercise 9*

EXERCISE 10   *Line and Circle*

Draw the sketch shown in Figure 2-57 by using the **Line** and **Center,Radius** tools. The distance between the dotted lines is 1.0 unit.

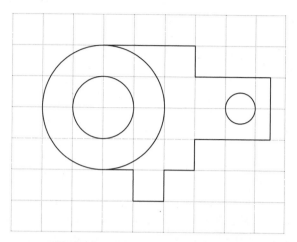

Figure 2-57 *Drawing for Exercise 10*

EXERCISE 11   *Line and Circle Tangent to Two objects*

Draw the sketch shown in Figure 2-58 using the **Line** and **Tan,Tan,Radius** tools.

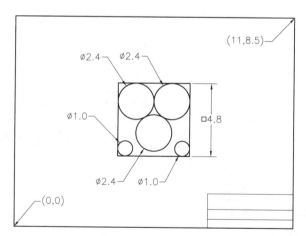

Figure 2-58 *Drawing for Exercise 11*

EXERCISE 12   *Setting Units*

Set the units for a drawing based on the following specifications.

1. Set the UNITS to architectural, with the denominator of the smallest fraction equal to 16.
2. Set the angular measurement to degrees/minutes/seconds, with the number of fractional places for the display of angles equal to 0d00'.
3. Set the direction to 0-degree (east)·and the direction of measurement of angles to counterclockwise (angles measured positive in counterclockwise direction).

Chapter 2

Based on Figure 2-59, determine and set the limits for the drawing. The scale for this drawing is 1/4" = 1'. Leave enough space around the drawing for dimensioning and title block. (HINT: Scale factor = 48 sheet size required is 12 x 9; therefore, the limits are 12 X 48, 9 X 48 = 576, 432. Use the **ZOOM** command and then select the **All** option to display the new limits.)

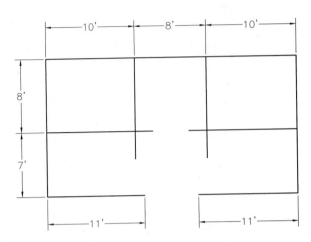

Figure 2-59 *Drawing for Exercise 12*

EXERCISE 13

Draw the sketch shown in Figure 2-60. The distance between the dotted lines is 10 feet. Determine the limits for this drawing and use the Architectural units with 0'-01/32" precision.

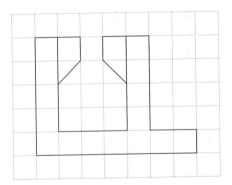

Figure 2-60 *Drawing for Exercise 13*

EXERCISE 14

Draw the object shown in Figure 2-61. The distance between the dotted lines is 5 inches. Determine the limits for this drawing and use the Fractional units with 1 1/16 precision.

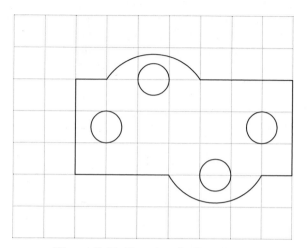

Figure 2-61 *Drawing for Exercise 14*

EXERCISE 15

Draw the object shown in Figure 2-62. The distance between the dotted lines is 1 unit. Determine the limits for this drawing and use the Decimal units with 0.00 precision.

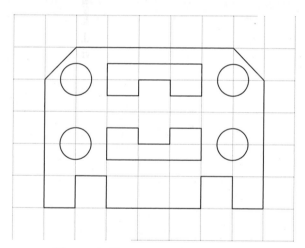

Figure 2-62 *Drawing for Exercise 15*

EXERCISE 16

Draw the object shown in Figure 2-63. The distance between the dotted lines is 10 feet. Determine the limits for this drawing and use the Engineering units with 0'0.00" precision.

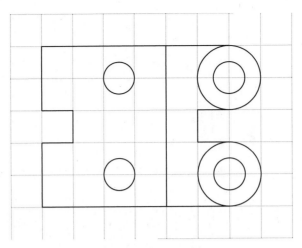

Figure 2-63 *Drawing for Exercise 16*

Problem-Solving Exercise 1

Draw the object shown in Figure 2-64, using the **Line** and **Center,Diameter** tools. In this exercise only the diameters of the circles are given. To draw the lines and small circles (Dia 0.6), you need to find the coordinate points for the lines and the center points of the circles. For example, if the center of concentric circles is at 5,3.5, then the X coordinate of the lower left corner of the rectangle will be 5.0 - 2.4 = 2.6.

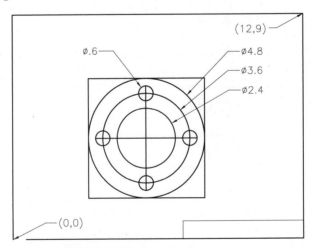

Figure 2-64 *Drawing for Problem-Solving Exercise 1*

Answers to Self-Evaluation Test

1. T, **2.** F, **3.** F, **4.** T, **5.** Undo, **6. Tan, Tan, Radius**, **7. ZOOM**, **8. Zoom All**, **9.** Surveyor's, **10. Preview**

Chapter 3

Starting with Advanced Sketching

CHAPTER OBJECTIVES

In this chapter, you will learn to:
* *Draw arcs using various options.*
* *Draw rectangles, ellipses, elliptical arcs, and polygons.*
* *Draw polylines and donuts.*
* *Place points and change their style and size.*
* *Create simple text.*
* *Draw infinite lines.*
* *Write simple text.*

KEY TERMS

* *Arc*
* *Rectangles*
* *Explode*
* *Ellipse*

* *Elliptical Arc*
* *Polygon*
* *Polylines*
* *Donut*

* *FILLMODE*
* *Points*
* *DDPTYPE*
* *PDMODE*

* *XLINE*
* *Text*
* *Ray*

DRAWING ARCS

Ribbon: Home > Draw > Arc drop-down	**Toolbar: Draw > Arc**
Menu Bar: Draw > Arc	**Command:** ARC or A

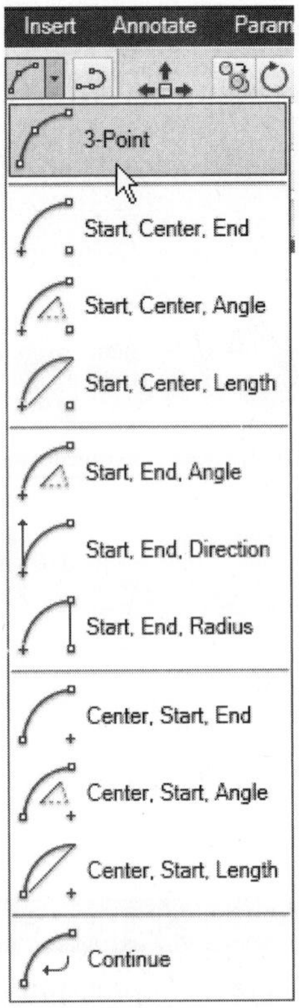

An arc is defined as a segment of a circle. In AutoCAD, an arc is drawn by using the **Arc** tool. There are eleven different tools to draw an arc. The tools to draw an arc are grouped together in the **Arc** drop-down of the **Draw** panel in the Ribbon, see Figure 3-1. You can choose the appropriate tool depending upon the parameters known and then draw the arc. Remember that, the tool that was used last to create an arc will be displayed in the **Draw** panel. The different methods to draw an arc are discussed next.

Drawing an Arc by Specifying Three Points

To draw an arc by specifying the start point, endpoint, and another point on its periphery, choose the **3-Point** tool from the **Draw** panel (see Figure 3-1). On doing so, you will be prompted to specify the start point. Specify the first point or enter coordinates. Then, specify the second point and endpoint of the arc, see Figure 3-2.

Following is the prompt sequence to draw an arc by specifying three points (You can also specify the points by using the mouse).

> Command: ***ARC*** 🔲 *(Ensure that dynamic input is off)*
> Specify start point of arc or [Center]: **2,2** 🔲
> Specify second point of arc or [Center/End]: **3,3** 🔲
> Specify end point of arc: **3,4** 🔲

Figure 3-1 *The tools in the **Arc** drop-down*

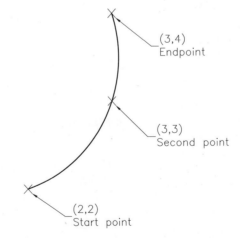

Figure 3-2 *Drawing an arc using the **3-Point** tool*

EXERCISE 1   *3-Point*

Draw several arcs by using the **3-Point** tool. The points can be selected by entering coordinates or by specifying points on the screen. Also, try to create a circle by drawing two separate arcs and a single arc and notice the limitations of the **ARC** command.

Drawing an Arc by Specifying its Start Point, Center Point, and Endpoint

If you know the start point, endpoint, and center point of an arc, choose the **Start, Center, End** tool from the **Draw** panel and then specify the start, center, and end points in succession; the arc will be drawn. The radius of the arc is determined by the distance between the center point and the start point. Therefore, the endpoint is used to calculate the angle at which the arc ends. Note that in this case, the arc will be drawn in counterclockwise direction from the start point to the endpoint around the specified center, as shown in Figure 3-3.

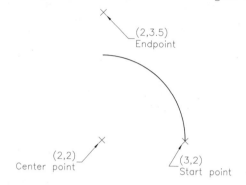

Figure 3-3 *Drawing an arc using the* **Start, Center, End** *tool*

Drawing an Arc by Specifying its Start Point, Center Point, and Included Angle

Included angle is the angle between the start and end points of an arc about the specified center. If you know the location of the start point, center point, and included angle of an arc, choose the **Start, Center, Angle** tool from the **Draw** panel and specify the start point, center point, and included angle; the arc will be drawn in counterclockwise direction with respect to the specified center and start point, see Figure 3-4.

If you enter a negative angle value, the arc will be drawn in clockwise direction, see Figure 3-5. Following is the prompt sequence to draw an arc by specifying the center point at (2,2), start point at (3,2), and an included angle of -60 degrees:

Command: *Choose* **Start, Center, Angle** *from the* **Draw** *panel (Ensure that dynamic input is off)*
_Specify start point of arc or [Center]: **3,2** Enter
Specify second point of arc or [Center/End]: _c Specify center point of arc: **C** Enter
Specify center point of arc: **2,2** Enter
Specify end point of the arc or [Angle/chord Length]: _a Specify included angle: **A** Enter
Specify included angle: **-60** Enter *see Figure 3-5.*

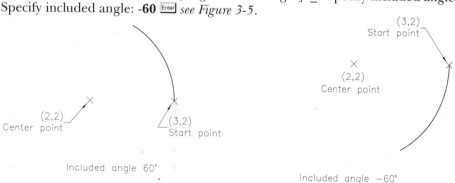

Figure 3-4 *Drawing an arc using the* **Start, Center, Angle** *tool*

Figure 3-5 *Drawing an arc by specifying a negative angle in the* **Start, Center, Angle** *tool*

Chapter 3

EXERCISE 2   *Start, Center, Angle*

a. Draw an arc whose start point is at 6,3, center point is at 3,3, and the included angle is 240 degrees.

b. Draw the profile shown in Figure 3-6. The distance between the dotted lines is 1.0 unit. Create the arcs by using different arc command options as indicated in the figure.

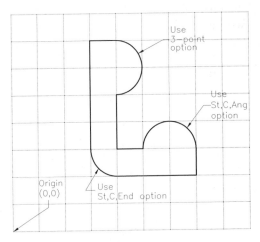

Figure 3-6 Drawing for Exercise 2

Drawing an Arc by Specifying the Start Point, Center Point, and Chord Length

A chord is defined as a straight line connecting the start point and endpoint of an arc. To draw an arc by specifying its chord length, choose the **Start, Center, Length** tool from the **Draw** panel and specify the start point, center point, and length of the chord in succession. On specifying the chord length, AutoCAD will calculate the included angle and an arc will be drawn in counterclockwise direction from the start point. A positive chord length gives the smallest possible arc with that length, as shown in Figure 3-7. This arc is known as minor arc. The included angle in a minor arc is less than 180 degrees. A negative value for the chord length results in the largest possible arc, also known as major arc, as shown in Figure 3-8.

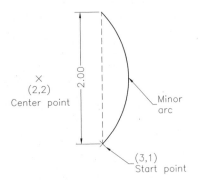

Figure 3-7 Drawing an arc by specifying a positive chord length in the **Start, Center, Length** tool

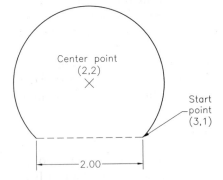

Figure 3-8 Drawing an arc by specifying a negative chord length in the **Start, Center, Length** tool

EXERCISE 3   *Center, Start, Length*

Draw a minor arc with the center point at (3,4), start point at (4,2), and chord length of 4 units.

Drawing an Arc by Specifying its Start Point, Endpoint, and Included Angle

To draw an arc by specifying its start point, endpoint, and the included angle, choose the **Start, End, Angle** tool from the **Draw** panel and specify the start point, endpoint, and the included angle in succession; the arc will be drawn. A positive included angle value draws an arc in the counterclockwise direction from the start point to the endpoint, as shown in Figure 3-9. Similarly, a negative included angle value draws the arc in clockwise direction.

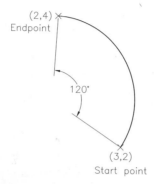

Figure 3-9 *Drawing an arc using the* **Start, End, Angle** *option*

Drawing an Arc by Specifying its Start Point, Endpoint, and Direction

This option is used to draw a major or minor arc, whose size and position are determined by the distance between the start point and endpoint and the direction specified. You can specify the direction by selecting a point on a line that is tangent to the start point or by entering an angle between the start point of the arc and the end point of the tangent line.

To draw an arc by specifying its direction, choose the **Start, End, Direction** tool from the **Draw** panel and specify the start and end points in succession; you will be prompted to specify the direction. Specify a point on the line that is tangent to the start point or enter an angle between the start point of the arc and the end point of the tangent line; an arc will be drawn.

In other words, on using this option, the arc will start in the direction you specify (the start of the arc is established tangent to the direction you specify). The prompt sequence to draw an arc, whose start point is at 3,6, endpoint is at 4.5,5, and direction of -40 degrees is given next.

Command: *Choose* **Start, End, Direction** *from the* **Draw** *panel (Ensure that dynamic input is off)*
_arc Specify start point of arc or [Center]: **3,6** [Enter]
Specify second point of arc or [Center/End]: _e Specify end point of arc: **4.5,3** [Enter]
Specify center point of arc or [Angle/Direction/Radius]: _d Specify tangent direction for the start point of arc: **-40** [Enter], *see Figure 3-10.*

The prompt sequence to draw an arc, whose start point is at 4,3, endpoint is at 3,5, and direction of 90 degrees, is given next.

Command: *Choose **Start, End, Direction** from the **Draw** panel (Ensure that dynamic input is off)*
_arc Specify start point of arc or [Center]: **4,3** [Enter]
Specify second point of arc or [Center/End]: _e Specify end point of arc: **3,5** [Enter]
Specify center point of arc or [Angle/Direction/Radius]: _d Specify tangent direction for the start point of arc: **90** [Enter] See Figure 3-11.

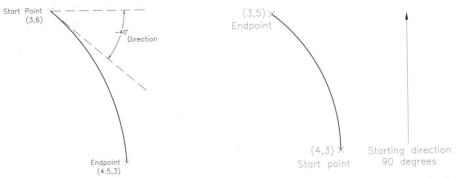

Figure 3-10 *Drawing an arc in the negative direction using the **Start, End, Direction** tool*

Figure 3-11 *Drawing an arc using the **Start, End, Direction** tool*

EXERCISE 4

a. Specify the directions and coordinates of two arcs in such a way that they form a circular figure.
b. Draw the profile shown in Figure 3-12. Create the curves by using the **ARC** command. The distance between the dotted lines is 1.0 unit and the diameter of the circles is 1 unit.

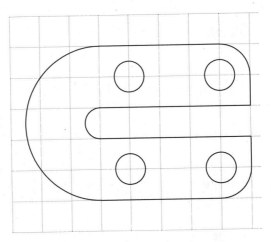

Figure 3-12 *Drawing for Exercise 4*

Drawing an Arc by Specifying its Start Point, Endpoint, and Radius

If you know the location of the start point, endpoint, and radius of an arc, choose the **Start, End, Radius** tool from the **Draw** panel and specify the start and end points; you will be prompted to specify the radius. Enter the radius value; the arc will be drawn. In this case, the arc will be drawn in counterclockwise direction from the start point. This means that a negative radius value results in a major arc (the largest arc between the two endpoints), see Figure 3-13(a). Whereas a positive radius value results in a minor arc (smallest arc between the start point and the endpoint), see Figure 3-13(b).

Drawing an Arc by Specifying its Center Point, Start Point, and Endpoint

The **Center, Start, End** tool is the modification of the **Start, Center, End** tool. Use this tool, whenever it is easier to start drawing an arc by establishing the center first. Here, the arc is always drawn in the counterclockwise direction from the start point to the endpoint, around the specified center. The prompt sequence for drawing the arc shown in Figure 3-14, which has a center point at (3,3), start point at (5,3), and endpoint at (3,5), is given next.

Command: *Choose* **Center, Start, End** *from the* **Draw** *panel.*
_arc Specify start point of arc or [Center]: _c Specify center point of arc: **3,3** Enter
Specify start point of arc: **5,3** Enter
Specify endpoint of arc or [Angle/chord Length]: **3,5** Enter

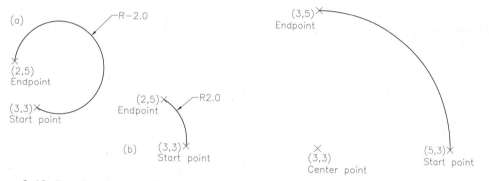

Figure 3-13 *Drawing an arc using the* **Start, End, Radius** *tool*

Figure 3-14 *Drawing an arc using the* **Center, Start, End** *tool*

Drawing an Arc by Specifying its Center Point, Start Point, and Angle

You can use the **Center, Start, Angle** tool if you need to draw an arc by specifying the center first. The prompt sequence for drawing the arc shown in Figure 3-16, that has a center point at (4,5), start point at (5,4), and included angle of 120 degrees is given next.

Command: *Choose* **Center, Start, Angle** *from the* **Draw** *panel (Ensure that dynamic input is off)*
_arc Specify start point of arc or [Center]: _c Specify center point of arc: **4,5** Enter
Specify start point of arc: **5,4** Enter
Specify end point of arc or [Angle/chord Length]: _a Specify included angle: **120** Enter, see *Figure 3-15.*

Drawing an Arc by Specifying the Center Point, Start Point, and Chord Length

The **Center, Start, Length** tool is used whenever it is easier to draw an arc by establishing the center first. The prompt sequence for drawing the arc shown in Figure 3-16, that has a center point at (2,2), start point at (4,3), and length of chord as 3 is given next.

Command: *Choose* **Center, Start, Length** *from the* **Draw** *panel (Ensure that dynamic input is off)*
_arc Specify start point of arc or [Center]: **2,2** Enter

Specify start point of arc: **4,3** [Enter]
Specify end point of arc or [Angle/chord Length]: _l Specify length of chord: **3** [Enter], see Figure 3-16.

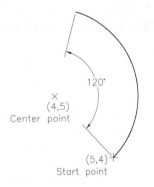

Figure 3-15 *Drawing an arc using the* **Center, Start, Angle** *tool*

Figure 3-16 *Drawing an arc using the* **Center, Start, Length** *tool*

Continue Option

To continue drawing an arc from a previously drawn arc or line, choose the **Continue** tool from the **Draw** panel; the start point and the direction of the arc will be taken from the endpoint and the ending direction of the previous line or arc. Specify the endpoint to draw an arc. If this option is used to draw arcs, each successive arc will be tangent to the previous one. This option is used to draw arcs that are tangent to a previously drawn line.

Tip
You can also invoke the **Continue** *option automatically. To do so, first draw a line or an arc and choose a tool from the* **Arc** *drop-down in the* **Draw** *panel. Now, press ENTER at the* **Specify start point of arc or [Center]** *prompt; the* **Continue** *option will be invoked automatically. The endpoint of the line or the arc drawn previously will be selected as the start point of the arc. Now, you will be prompted to specify the endpoint to complete the arc.*

EXERCISE 5

a. Use the **Center, Start, Angle**, and **Continue** tools to draw the profiles shown in Figure 3-17.

b. Draw the profile shown in Figure 3-18. The distance between the dotted lines is 1.0 unit. Create the radii as indicated in the drawing by using the **ARC** command options.

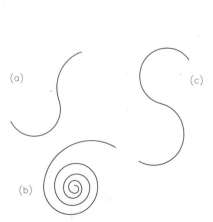

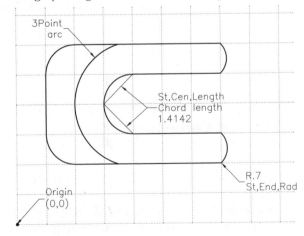

Figure 3-17 *Drawing for Exercise 5(a)*

Figure 3-18 *Drawing for Exercise 5(b)*

DRAWING RECTANGLES

Ribbon: Home > Draw > Rectangle	**Toolbar:** Draw > Rectangle
Tool Palettes: Draw > Rectangle	**Command:** RECTANG

A rectangle is drawn by choosing the **Rectangle** tool (see Figure 3-19) from the **Draw** panel. In AutoCAD, you can draw rectangles by specifying two opposite corners of the rectangle, by specifying the area and the size of one of the sides, or by specifying the dimensions of the rectangle. All these methods of drawing rectangles are discussed next.

Drawing Rectangles by Specifying Two Opposite Corners

On invoking the rectangle command, you will be prompted to specify the first corner of the rectangle. Enter the coordinates of the first corner or specify the start point by using the mouse. The first corner can be any one of the four corners. Next, you will be prompted to specify the other corner. Specify the diagonally opposite corner by entering the coordinates or by using the left mouse button, as shown in Figure 3-20.

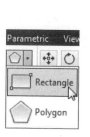

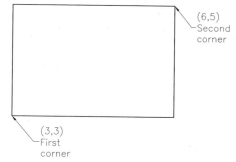

Figure 3-19 *Invoking the **Rectangle** tool from the **Draw** panel*

Figure 3-20 *Drawing a rectangle by specifying two opposite corners*

Drawing Rectangles by Specifying the Area and One Side

To draw a rectangle by specifying its area and the length of one of the sides, first specify the start point. Next, invoke the shortcut menu by right-clicking and then choose the **Area** option. Next, specify the parameters; the rectangle is drawn. Following is the prompt sequence to draw a rectangle whose start point is at 3,3, has area 15 units and length 5 units:

Command: **RECTANG**
Specify first corner point or [Chamfer/Elevation/Fillet/Thickness/Width]: **3,3** Enter
Specify other corner point or [Area/Dimensions/Rotation]: **A** Enter
Enter area of rectangle in current units <100.000>: **15** Enter
Calculate rectangle dimensions based on [Length/Width] <Length>: **L** Enter
Enter rectangle length <10.0000>: **5** Enter

In the above case, the area and length of the rectangle were entered. The system automatically calculates the width of the rectangle by using the following formula:

Area of rectangle = Length x Width
Width = Area of rectangle/Length
Width = 15/5
Width = 3 units

Drawing Rectangles by Specifying their Dimensions

You can also draw a rectangle by specifying its dimensions. This can be done by choosing the **Dimensions** option from the shortcut menu at the **Specify other corner point or [Area/Dimensions/Rotation]** prompt and entering the length and width of the rectangle. The prompt sequence for drawing a rectangle at 3,3 with a length of **5** units and width of **3** units is given next.

Command: **RECTANG**
Specify first corner point or [Chamfer/Elevation/Fillet/Thickness/Width]: **3,3** [Enter]
Specify other corner point or [Area/Dimensions/Rotation]: **D** [Enter]
Specify length for rectangles <0.0000>: **5** [Enter]
Specify width for rectangles <0.0000>: **3** [Enter]
Specify other corner point or [Area/Dimensions/Rotation]: *Click on the screen to specify the orientation of rectangle.*

Here, you are allowed to choose any one of the four locations for placing the rectangle. You can move the cursor to see the four quadrants. Depending on the location of the cursor, the corner point that is specified first holds the position of either the lower left corner, the lower right corner, the upper right corner, or the upper left corner. After deciding the position, you can click to place the rectangle.

Drawing Rectangle at an Angle

You can also draw a rectangle at an angle. This can be done by choosing the **Rotate** option from the shortcut menu, at the **Specify other corner point or [Area/Dimensions/Rotation]** prompt and entering the rotation angle. After entering the rotation angle, you can continue sizing the rectangle using any one of the above discussed methods. The prompt sequence for drawing a rectangle at an angle of 45-degree is:

Command: **RECTANG**
Specify first corner point or [Chamfer/Elevation/Fillet/Thickness/Width]: *Select a point as lower left corner location.*
Specify other corner point or [Area/Dimensions/Rotation]: **R** [Enter]
Specify rotation angle or [Pick points] <current>: **45** [Enter]
Specify other corner point or [Area/Dimensions/Rotation]: *Select a diagonally opposite point.*

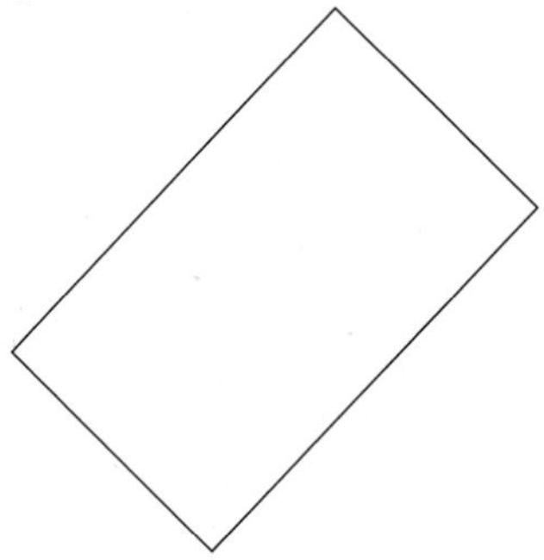

While specifying the other corner point, you can place the rectangle in any of the four quadrants. Move the cursor in different quadrants and then select a point in the quadrant in which you need to draw the rectangle. Figure 3-21 shows a rectangle drawn at an angle of 45-degree.

Figure 3-21 Rectangle drawn at an angle

Note
Once the rotation angle had been specified, the subsequent rectangles will be drawn at an angle as specified. If you do not want to draw the subsequent rectangles at an angle, you need to set the rotation angle to zero.

You can also set some of the parameters of a rectangle before specifying the start point. These parameters are the options in the Command Prompt and are discussed next.

Chamfer

The **Chamfer** option is used to create a chamfer, which is an angled corner, by specifying the chamfer distances, see Figure 3-22. The chamfer is created at all four corners. You can give two different chamfer values to create an unequal chamfer.

Command: **RECTANG**
Specify first corner point or [Chamfer/Elevation/Fillet/Thickness/Width]: **C** [Enter]

Specify first chamfer distance for rectangles <0.0000>: *Enter a value, d1.*
Specify second chamfer distance for rectangles <0.0000>: *Enter a value, d2.*
Specify first corner point or [Chamfer/Elevation/Fillet/Thickness/Width]: *Select a point as lower left corner.*
Specify other corner point or [Area/Dimensions/Rotation]: *Select a point as upper right corner.*

Fillet

The **Fillet** option is used to create a filleted rectangle, see Figure 3-23. You can specify the required fillet radius. The following is the prompt sequence for specifying the fillet:

Specify first corner point or [Chamfer/Elevation/Fillet/Thickness/Width]: **F** ⏎
Specify fillet radius for rectangles <0.0000>: *Enter a value.*

Note that the rectangle will be filleted only if the length and width of the rectangle are equal to or greater than twice the value of the specified fillet. Otherwise, AutoCAD will draw a rectangle without fillets.

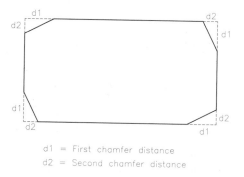

d1 = First chamfer distance
d2 = Second chamfer distance

Figure 3-22 *Drawing a rectangle with chamfers*   **Figure 3-23** *Drawing a rectangle with fillets*

Note
*You can draw a rectangle with chamfers or fillets. If you specify the chamfer distances first and then specify the fillet radius in the same **RECTANG** command, the rectangle will be drawn with fillets only.*

Width

The **Width** option is used to create a rectangle whose line segments have some specified width, as shown in Figure 3-24.

Specify first corner point or [Chamfer/Elevation/Fillet/Thickness/Width]: **W** ⏎
Specify line width for rectangles <0.0000>: *Enter a value.*

Thickness

The **Thickness** option is used to draw a rectangle that is extruded in the Z direction by a specified value of thickness. For example, if you draw a rectangle with thickness of 2 units, you will get a cuboid whose height is 2 units, see Figure 3-25. To view the cuboid, choose the **Home** option of the **ViewCube** from the drawing area or choose **SE Isometric** from **View > Views** panel. To restore the view to the plan view, choose **Top** from **View > Views** panel or choose **Top** from the **ViewCube** in the drawing area.

Specify first corner point or [Chamfer/Elevation/Fillet/Thickness/Width]: **T** ⏎
Specify thickness for rectangles <0.0000>: *Enter a value.*

Chapter 3

Elevation

The **Elevation** option is used to draw a rectangle at a specified distance along the *Z* axis and from the *XY* plane. For example, if the elevation is 2 units, the rectangle will be drawn two units above the *XY* plane. If the thickness of the rectangle is 1 unit, you will get a rectangular box of 1 unit height, located 2 units above the *XY* plane, see Figure 3-25.

> Chamfer/Elevation/Fillet/Thickness/Width/<First corner>: **E** [Enter]
> Specify elevation for rectangles <0.0000>: *Enter a value.*

To view the rectangle in 3D space, choose the **Home** option of the **ViewCube** from the drawing area or choose **SE Isometric** from **View > Views** panel. To restore the view to the plan view, choose **Top** from **View > Views** panel or choose **Top** from the **ViewCube**.

Figure 3-24 *Drawing a rectangle of specified width*

Figure 3-25 *Drawing rectangles with thickness and elevation specified*

Note
The values you enter for fillet, width, elevation, and thickness become the current values for the rectangles drawn subsequently. Therefore, you need to reset the values based on the requirement.

*A rectangle generated by using the **RECTANG** command is treated as a single object. To edit the individual lines of the rectangle, you need to explode the rectangle by using the **EXPLODE** command and then edit them.*

EXERCISE 6   *Rectangle*

Draw a rectangle of length 4 units, width 3 units, and start point at (1,1). Draw another rectangle of length 2 units and width 1 units, with its first corner at 1.5,1.5, and which is at an angle of 65-degree.

DRAWING ELLIPSES

Ribbon: Home > Draw > Ellipse drop-down	**Toolbar:** Draw > Ellipse
Tool Palettes: Draw > Ellipse	**Command:** ELLIPSE

If you cut a cone by a cutting plane at an angle and view the cone perpendicular to the cutting plane, the shape created is called an ellipse. An ellipse is created by using the **ELLIPSE** command. An ellipse can be created by using different methods and the tools to invoke these methods are grouped together in the **Draw** panel, refer to Figure 3-26. In AutoCAD, you can create a true ellipse, also known as a NURBS-based (Non-Uniform Rational Bezier Spline) ellipse. A true ellipse has center and quadrant points. If you select it, grips (small blue squares) will be displayed at the center and quadrant points of the ellipse. If you move one of the grips located on the perimeter of the ellipse, the size of the ellipse will be changed.

Once you invoke the **ELLIPSE** command, the **Specify axis endpoint of ellipse or [Arc/Center]** or **Specify axis endpoint of ellipse or [Arc/Center/Isocircle]** (if isometric snap is on) prompt will be displayed. The response to this prompt depends on the option you choose. The various options are explained next.

Note
*By default, the **Isocircle** option is not available in the **ELLIPSE** command. To display this option, you have to select the **Isometric snap** radio button in the **Snap and Grid** tab of the **Drafting Settings** dialog box. The **Isocircle** option will be discussed in Chapter 13.*

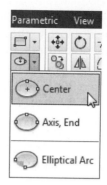

*Figure 3-26 Tools in the **Ellipse** drop-down in the **Draw** panel*

Drawing Ellipse Using the Center Option

To draw an ellipse by specifying its center point, endpoint of one axis, and length of other axis, choose the **Center** tool from the **Draw** panel; you will be prompted to specify the center of the ellipse. The center of an ellipse is defined as the point of intersection of the major and minor axes. Specify the center point or enter coordinates; you will be prompted to specify the endpoint. Specify the endpoint of the major or minor axis; you will be prompted to specify the distance of the other axis. Specify the distance; the ellipse will be drawn.

After specifying the endpoint of one axis, you can enter **R** at the **Specify distance to other axis or [Rotation]** prompt to specify the rotation angle around the major axis. In this case, the first axis specified is considered as the major axis. On specifying the rotation angle, the ellipse will be drawn at an angle with respect to the major axis. Note that the rotation angle should range between 0 and 89.4-degree. Figure 3-27 shows the ellipse created at different rotation angles.

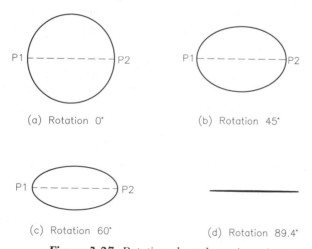

(a) Rotation 0°   (b) Rotation 45°

(c) Rotation 60°   (d) Rotation 89.4°

Figure 3-27 Rotation about the major axis

Drawing an Ellipse by Specifying its Axis and Endpoint

To draw an ellipse by specifying one of its axes and the endpoint of the other axis, choose the **Axis, End** tool from the **Draw** panel; you will be prompted to specify the axis end point. Specify the first endpoint of one axis of the ellipse; you will be prompted to specify the other endpoint of the axis. Specify the other endpoint of the axis. Now, you can specify the distance to other axis from the center or specify the rotation angle around the specified axis.

Figure 3-28 shows an ellipse with one endpoint of the axis located at (3,3), the other at (6,3), and the distance to the other axis as 1 unit. Figure 3-29 shows an ellipse with one endpoint of the axis located at (3,3), the other at (4,2), and the distance to the other axis as 2 units.

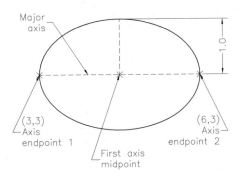

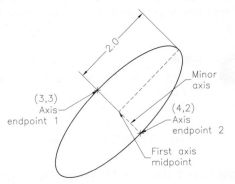

Figure 3-28 *Drawing an ellipse using the Axis and Endpoint option*

Figure 3-29 *Drawing an ellipse using the Axis and Endpoint option*

EXERCISE 7   *Ellipse*

Draw an ellipse whose major axis is 4 units and the rotation around this axis is 60 degrees. Draw another ellipse whose rotation around the major axis is 15 degrees.

Drawing Elliptical Arcs

Ribbon: Home > Draw > Ellipse drop-down > Elliptical Arc	**Toolbar:** Draw > Ellipse Arc
Tool Palettes: Draw > Ellipse Arc	**Command:** ELLIPSE > Arc

 In AutoCAD, you can draw an elliptical arc by choosing the **Elliptical Arc** tool from the **Draw** panel. On choosing this tool, you need to specify the endpoints of one of the axes, the distance to other axis from the center, and any one of the following information:

1. Start and End angles of the arc.
2. Start and Included angles of the arc.
3. Start and End parameters.

Remember that in case of elliptical arcs, angles are measured from the first point and in counterclockwise direction.

In this section, you will draw an elliptical arc by using the information given below.

a. Start angle = -45, end angle = 135
b. Start angle = -45, included angle = 225
c. Start parameter = @1,0, end parameter = @1<225

Specifying the Start and End Angles of the Elliptical Arc [Figure 3-30(a)]

> Command: *Choose the **Elliptical Arc** tool from the **Draw** panel*
> Specify axis endpoint of ellipse or [Arc/Center]: _a
> Specify axis endpoint of elliptical arc or [Center]: *Select the first endpoint*
> *Switch on the Ortho mode by pressing F8, if it is not already chosen*
> Specify other endpoint of axis: *Select the second point to the left of the first point*
> Specify distance to other axis or [Rotation]: *Select a point or enter a distance*
> Specify start angle or [Parameter]: **-45** Enter
> Specify end angle or [Parameter/Included angle]: **135** Enter *(Angle where arc ends.)*

Specifying the Start and Included Angles of the Elliptical Arc [Figure 3-30(b)]

> Command: *Choose the* **Elliptical Arc** *tool from the* **Draw** *panel*
> Specify axis endpoint of ellipse or [Arc/Center]: _a
> Specify axis endpoint of elliptical arc or [Center]: *Select the first endpoint*
> Specify other endpoint of axis: *Select the second point*
> Specify distance to other axis or [Rotation]: *Select a point or enter a distance*
> Specify start angle or [Parameter]: **-45** [Enter]
> Specify end angle or [Parameter/Included angle]: **I** [Enter]
> Specify included angle for arc<current>: **225** [Enter] *(Included angle.)*

Specifying the Start and End Parameters (Figure 3-30(c))

> Command: *Choose the* **Elliptical Arc** *tool from the* **Draw** *panel*
> Specify axis endpoint of ellipse or [Arc/Center]: _a
> Specify axis endpoint of elliptical arc or [Center]: *Select the first endpoint*
> Specify other endpoint of axis: *Select the second endpoint*
> Specify distance to other axis or [Rotation]: *Select a point or enter a distance*
> Specify start angle or [Parameter]: **P**
> Specify start parameter or [Angle]: **@1,0**
> Specify end parameter or [Angle/ Included angle]: **@1<225**

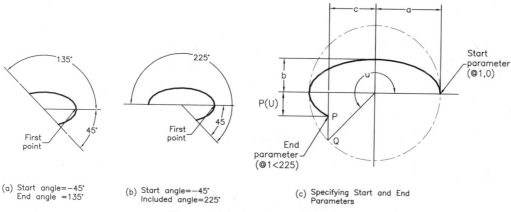

(a) Start angle=−45°
    End angle =135°

(b) Start angle=−45°
    Included angle=225°

(c) Specifying Start and End
    Parameters

Figure 3-30 *Drawing elliptical arcs*

EXERCISE 8 — *Elliptical Arc*

a. Construct an ellipse with center at (2,3), axis endpoint at (4,6), and the other axis endpoint at a distance of 0.75 unit from the midpoint of the first axis.

b. Draw the profile shown in Figure 3-31. The distance between the dotted lines is 1.0 unit.

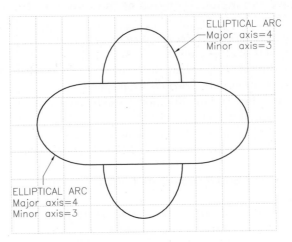

Figure 3-31 *Drawing for Exercise 8*

DRAWING REGULAR POLYGONS

Ribbon: Home > Draw > Polygon	**Toolbar:** Draw > Polygon
Tool Palettes: Draw > Polygon	**Command:** POLYGON

A regular polygon is a closed geometric entity with equal sides. The number of sides of a polygon varies from 3 to 1024. For example, a triangle is a three-sided polygon and a pentagon is a five-sided polygon. To draw a regular 2D polygon, choose the **Polygon** tool from the **Draw** panel; you will be prompted to specify the number of sides. Type the number of sides and press ENTER. Now, you can draw the polygon by specifying the length of an edge or by specifying the center of the polygon. Both these methods are discussed next.

Drawing a Polygon by Specifying the Center of Polygon

After you specify the number of sides and press ENTER, you will be prompted to specify the center of polygon. Specify the center point; you will be prompted to specify whether the polygon to be drawn is inscribed in a circle or circumscribed about an imaginary circle. A polygon is said to be inscribed when it is drawn inside an imaginary circle such that the vertices of the polygon touch the circle, see Figure 3-32. Whereas, a polygon is said to be circumscribed when it is drawn outside the imaginary circle such that the sides of the polygon are tangent to the circle, see Figure 3-33. Type **I** or **C** to draw an inscribed or a circumscribed polygon respectively, and press ENTER; you will be prompted to specify the radius. Specify the radius and press ENTER; a polygon will be created.

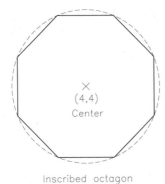

Figure 3-32 *Drawing an inscribed polygon using the **Center of Polygon** option*

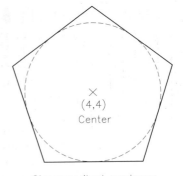

Figure 3-33 *Drawing a circumscribed polygon using the **Center of Polygon** option*

Drawing a Polygon by Specifying an Edge

To draw a polygon by specifying the length of an edge, you need to type **E** at the **Specify center of polygon or [Edge]** Command prompt and press ENTER. Next, specify the first and second endpoints of the edge in succession; the polygon will be drawn in counterclockwise direction, as shown in Figure 3-34.

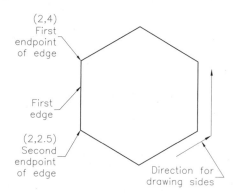

Figure 3-34 Drawing a polygon (hexagon) using the Edge option

EXERCISE 9

Draw a circumscribed polygon of eight sides by using the **Center of Polygon** method.

EXERCISE 10

Draw a polygon of ten sides by using the **Edge** option and an elliptical arc, as shown in Figure 3-35. Let the first endpoint of the edge be at (7,1) and the second endpoint be at (8,2).

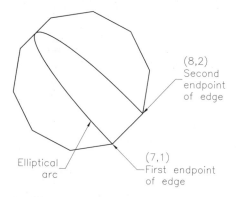

Figure 3-35 Polygon and elliptical arc for Exercise 10

DRAWING POLYLINES

Ribbon: Home > Draw > Polyline	**Toolbar:** Draw > Polyline
Tool Palettes: Draw > Polyline	**Command:** PLINE or PL

Etymologically, polylines means many lines. Some of the features of a polyline are listed below.

1. Polylines are thick lines with desired width. They are very flexible and can be used to draw any shape, such as a filled circle or a donut.
2. Polylines can be used to draw objects in any linetype (for example, hidden linetype).
3. Polylines are edited by using the advanced editing commands such as the **PEDIT** command.

4. A single polyline object can be formed by joining polylines and polyarcs of different thicknesses.

5. It is easy to determine the area or perimeter of a polyline feature. Also, it is easy to offset a polyline when drawing walls.

To draw a polyline, you need to invoke the **PLINE** command. To do so, choose the **Polyline** tool from the **Draw** panel, see Figure 3-36. The **PLINE** command functions fundamentally like the **LINE** command, except that some additional options are provided and all segments of the polyline form a single object. After invoking the **PLINE** command and specifying the start point, the following prompt is displayed.

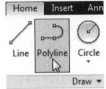

*Figure 3-36 Choosing the **Polyline** tool from the **Draw** panel*

> Specify start point: *Specify the starting point or enter its coordinates.*
> Current line-width is nn.nnnn
> Specify next point or [Arc/Halfwidth/Length/Undo/Width]:

Note that the message **Current line-width is nn.nnnn** is displayed automatically indicating that the polyline drawn will have nn.nnnn width.

When you are prompted to specify the next point, you can continue specifying the next point and draw a polyline, or depending on your requirements, the other options can be invoked. All these options are discussed next.

Next Point of Line

This is the default option that is displayed after specifying the start point and is used to specify the next point of the current polyline segment. If additional polyline segments are added to the first polyline, AutoCAD automatically makes the endpoint of the previous polyline segment as the start point of the next polyline segment. The prompt sequence is given next.

> Command: **PLINE** [Enter]
> Specify start point: *Specify the start point of the polyline.*
> Current line-width is 0.0000.
> Specify next point or [Arc/Halfwidth/Length/Undo/Width]: *Specify the endpoint of the first polyline segment.*
> Specify next point or [Arc/Close/Halfwidth/Length/Undo/Width]: *Specify the endpoint of the second polyline segment, or press ENTER to exit the command.*

Width

To change the current width of a polyline, enter **W** (width option) at the last prompt. You can also right-click and choose the **Width** option from the shortcut menu. On doing so, you will be prompted to specify the width at the start and end of the polyline. Specify the widths of the polyline and press ENTER; the polyline will be drawn with the specified width.

The starting width value is taken as the ending width value by default. Therefore, to have a uniform polyline, you need to press ENTER at the **Specify ending width < >** prompt. However, if you specify different value for the ending width, the resulting polyline will be tapered.

For example, to draw a polyline of uniform width 0.25 unit, start point at (4,5), endpoint at (5,5), and the next endpoint at (3,3), use the following prompt sequence:

> Command: **PLINE** [Enter]
> Specify start point: **4,5** [Enter]

Current line-width is 0.0000
Specify next point or [Arc/Halfwidth/Length/Undo/Width]: **W** Enter
Specify starting width <current>: **0.25** Enter
Specify ending width <0.25>: Enter
Specify next point or [Arc/Halfwidth/Length/Undo/Width]: **5,5** Enter
Specify next point or [Arc/Close/Halfwidth/Length/Undo/Width]: **3,3** Enter
Specify next point or [Arc/Close/Halfwidth/Length/Undo/Width]: Enter, *See Figure 3-37.*

Similarly, to draw a tapered polyline of starting width 0.5 units and an ending width 0.15 units, a start point at (2,4), and an endpoint at (5,4), use the prompt sequence given next.

Command: **PLINE** Enter
Specify start point: **2,4** Enter
Current line-width is 0.0000
Specify next point or [Arc/Halfwidth/Length/Undo/Width]: **W** Enter
Specify starting width <0.0000>: **0.50** Enter, *See Figure 3-38.*
Specify ending width <0.50>: **0.15** Enter
Specify next point or [Arc/Halfwidth/Length/Undo/Width]: **5,4** Enter
Specify next point or [Arc/Close/Halfwidth/Length/Undo/Width]: Enter

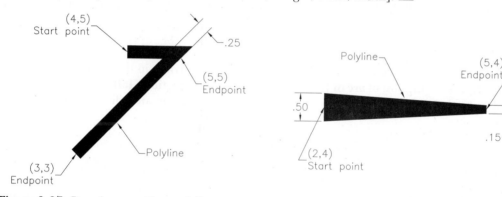

Figure 3-37 *Drawing a uniform polyline using the **Polyline** tool*

Figure 3-38 *Drawing a tapered polyline using the **Polyline** tool*

Halfwidth

The halfwidth distance is equal to the half of the actual width of a polyline. This option is invoked by entering **H** or choosing **Halfwidth** from the shortcut menu. The prompt sequence to specify the starting and ending halfwidths of a polyline is given next.

Specify next point or [Arc/Halfwidth/Length/Undo/Width]: **H** Enter
Specify starting half-width <0.0000>: **0.12** Enter *(Specify the desired starting halfwidth)*
Specify ending half-width <0.1200>: **0.05** Enter *(Specify the desired ending halfwidth)*

Length

The **Length** option is used to draw a new polyline segment of specified length and at the same angle as the last polyline segment or tangent to the previous polyarc segment. This option is invoked by entering **L** at the following prompt or by choosing **Length** from the shortcut menu.

Specify next point or [Arc/Close/Halfwidth/Length/Undo/Width]: **L** Enter
Specify length of line: *Specify the desired length of the Pline*

Undo

This option erases the most recently drawn polyline segment. It can be invoked by entering **U** at the **Specify next point or [Arc/Close/Halfwidth/Length/Undo/Width]** prompt. You can use this option repeatedly until you reach the start point of the first polyline segment. If you use this option again, the message **All segments already undone** will be displayed.

Close

This option will be available only when at least one segment of a polyline is drawn. It closes the polyline by drawing a polyline segment from the most recent endpoint to the initial start point and exits from the **PLINE** command.

Arc

This option is used to switch from polylines to polyarcs. You can also set the parameters associated with polyarcs. By default, the arc segment is drawn tangent to the previous segment of the polyline. The direction of the previous line, arc, or polyline segment is the default direction for the polyarc. On invoking the arc option, you need to choose the sub-options to draw the polyarc. Some of the sub-options to draw a polyarc are similar to that of drawing an arc. The sub-options that are different are discussed next.

Close

This option will be available only when you specify two or more than two points for creating the polyline segments. To join the start and last points of a polyline in the form of an arc, type **CL** at the **Specify endpoint of arc or [Angle/CEnter/CLose/Direction/Halfwidth/Line/Radius/Second pt/Undo/Width]** Command prompt and press ENTER; an arc will be created that will close the loop.

Direction

Usually, the arc drawn by using the **Arc** option in the **PLINE** command is tangent to the previously drawn polyline segment. In other words, the starting direction of the arc depends upon the ending direction of the previous segment. The **Direction** option is used to specify the direction of the tangent for the arc segment to be drawn. You need to specify the direction by specifying a point. The prompts are given next.

> Specify tangent direction for the start point of arc: *Specify the direction.*
> Specify endpoint of arc: *Specify the endpoint of arc.*

Halfwidth

The use of this option is similar to the option used for the polyline segment. It is used to specify the starting and ending halfwidths of an arc segment.

Line

This option is used to invoke the **Line** mode again.

Radius

This option is used to specify the radius of the arc segment. The prompt sequence for specifying the radius is given next.

> Specify endpoint of arc or [Angle/CEnter/CLose/Direction/Halfwidth/Line/Radius/Second pt/Undo/Width]: **R**
> Specify radius of arc: *Specify the radius of the arc segment.*
> Specify endpoint of arc or [Angle]: *Specify the endpoint of arc or choose an option.*

If you specify a point, the arc segment will be drawn. If you enter an angle, you will have to specify the angle and the direction of the chord at the **Specify included angle** and **Specify direction of chord for arc<current>** prompts respectively.

Second pt

This option is used to select the second point of an arc in the three-point arc option. The prompt sequence is given next.

> Specify second point of arc: *Specify the second point on the arc*
> Specify endpoint of arc: *Specify the third point on the arc*

Width

This option is used to enter the width of the arc segment. The prompt sequence for specifying the width is the same as that of the polyline. To draw a tapered arc segment, as shown in Figure 3-39, you need to enter different values at the starting width and ending width prompts.

The prompt sequence to draw an arc, whose start point is at (3,3), endpoint is at (3,5), starting width is 0.50 unit, and ending width is 0.15 unit, is given next.

> Command: **PLINE** [Enter]
> Specify start point: **3,3** [Enter]
> Current line-width is 0.0000
> Specify next point or [Arc/Halfwidth/Length/Undo/Width]: **A** [Enter]
> Specify endpoint of arc or [Angle/CEnter/CLose/Direction/Halfwidth/Line/Radius/Second pt/ Undo/Width]: **W** [Enter]
> Specify starting width <current>: **0.50** [Enter]
> Specify ending width <0.50>: **0.15** [Enter]
> Specify endpoint of arc or [Angle/CEnter/CLose/Direction/Halfwidth/Line/Radius/Second pt/ Undo/Width]: **3,5** [Enter]
> Specify endpoint of arc or [Angle/CEnter/CLose/Direction/Halfwidth/Line/Radius/Second pt/ Undo/Width]: [Enter]

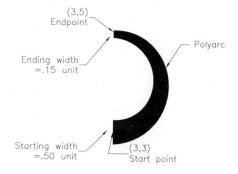

Figure 3-39 *Drawing a polyarc*

EXERCISE 11

Draw the objects shown in Figures 3-40 and 3-41 by using the polylines of different width.

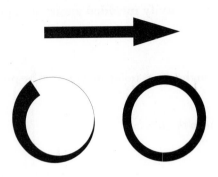

Figure 3-40 *Drawing for Exercise 11*

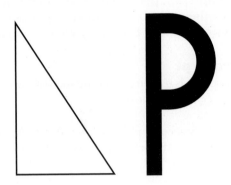

Figure 3-41 *Drawing for Exercise 11*

DRAWING DONUTS

Ribbon: Home > Draw > Donut   **Command:** DONUT

In AutoCAD, the **DONUT** or **DOUGHNUT** command is used to draw an object that looks like a filled circle ring called donut. AutoCAD's donuts are made of two semicircular polyarcs with a certain width. Therefore, the **DONUT** command allows you to draw a thick circle. The donuts can have any inside and outside diameters. If **FILLMODE** is off, a donut will look like a circle (if the inside diameter is zero) or a concentric circle (if the inside diameter is not zero). On invoking the **DONUT** command, you will be prompted to specify the diameters. After specifying the two diameters, the donut gets attached to the crosshairs. Specify a point for the center of the donut in the drawing area to place the donut. In this way, you can place as many donuts as required without exiting the command. Press ENTER to exit the command.

The default values for the inside and outside diameters of donuts are saved in the **DONUTID** and **DONUTOD** system variables. A solid-filled circle is drawn by specifying the inside diameter as zero and if **FILLMODE** is on.

EXAMPLE 1   *Donut*

You will draw an unfilled donut shown in Figure 3-43 with an inside diameter of 0.75 unit, an outside diameter of 2.0 units, and centered at (2,2). You will also draw a filled donut and a solid-filled donut with the given specifications.

The following is the prompt sequence to draw an unfilled donut shown in Figure 3-42.

> Command: **FILLMODE** [Enter]
> New value for FILLMODE <1>: **0** [Enter]
> Command: *Choose the **Donut** tool from the **Draw** panel*
> Specify inside diameter of donut<0.5000>: **0.75** [Enter]
> Specify outside diameter of donut <1.000>: **2** [Enter]
> Specify center of donut or <exit>: **2,2** [Enter]
> Specify center of donut or <exit>: [Enter]

The prompt sequence for drawing a filled donut with an inside diameter of 0.5 unit, outside diameter of 2.0 units, centered at a specified point.

> Command: **FILLMODE** [Enter]
> Enter new value for FILLMODE <0>: **1** [Enter]
> Command: *Choose the **Donut** tool from the **Draw** panel*

Specify inside diameter of donut<0.5000>: **0.50** Enter
Specify outside diameter of donut <1.000>: **2** Enter
Specify center of donut or <exit>: *Specify a point*
Specify center of donut or <exit>: Enter, *See Figure 3-43*

To draw a solid-filled donut with an outside diameter of 2.0 units, use the following prompt sequence:

Command: *Choose the* **Donut** *tool from the* **Draw** *panel*
Specify inside diameter of donut <0.50>: **0** Enter
Specify outside diameter of donut <1.0>: **2** Enter
Specify center of donut or <exit>: *Specify a point*
Specify center of donut or <exit>: Enter, *See Figure 3-44*

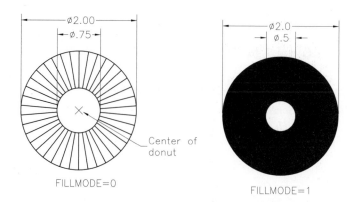

Figure 3-42 *Unfilled donut*　　**Figure 3-43** *Filled donut*　　**Figure 3-44** *Solid-filled donut*

PLACING POINTS

A point is one of the basic drawing objects and is specified as a dot (a period). A point is defined as a geometric object that has no dimension and properties, except location. However, in AutoCAD, you can control the size and appearance (style) of a point. You will first learn to change the point style and size and then about various methods to place a point.

Changing the Point Style and Size

Ribbon: Home > Utilities > Point Style　　**Command:** DDPTYPE

To change the style and size of a point, choose the **Point Style** tool from the **Utilities** panel in the **Draw** tab; the **Point Style** dialog box will be displayed, as shown in Figure 3-45. Alternatively, use the **DDPTYPE** command to invoke this dialog box. Select any one of the twenty point styles in the **Point Style** dialog box. You can specify the point size as a specified percentage of the drawing area or as an absolute size by selecting the **Set Size Relative to Screen** or **Set Size in Absolute Units** radio button respectively. After selecting the appropriate radio button, enter the point size in the **Point Size** edit box. Choose the **OK** button after specifying the parameters. Now, the points will be drawn according to the selected style and size, until you change the style and size. You can also change the point style and the point size by using the **PDMODE** and **PDSIZE** system variables respectively. The different values of **PDMODE** and the resulting style are shown in Figure 3-46.

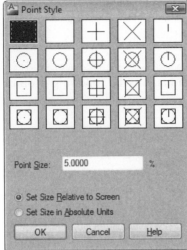

Figure 3-45 The Point Style dialog box

Pdmode Value	Point Style	Pdmode Value	Point Style
0		64+0=64	□
1		64+1=65	□
2	+	64+2=66	⊞
3	×	64+3=67	⊠
4	ǀ	64+4=68	⊓
32+0=32	○	96+0=96	○
32+1=33	○	96+1=97	○
32+2=34	⊕	96+2=98	⊕
32+3=35	⊗	96+3=99	⊗
32+4=36	⊙	96+4=100	⊙

*Figure 3-46 Different point styles for **PDMODE** values*

Note

*On selecting the **Set Size Relative to Screen** radio button, the point size will not change when you zoom in or zoom out the drawing. However, on selecting the **Set Size in Absolute Units** radio button, the size of the point will change when you zoom in or zoom out the drawing.*

Placing Multiple Points

Ribbon: Home > Draw > Multiple Points   **Tool Palettes:** Draw > Point
Menu Bar: Draw > Point > Multiple Points

 To place points, choose the **Multiple Points** tool from the **Draw** panel (see Figure 3-47); the current **PDMODE** and **PDSIZE** values will be displayed and you will be prompted to specify a point. Left-click to place a point. Next, continue placing as many points as needed and then press ESC to exit the command. You can also enter the coordinate value and place a point at a particular location.

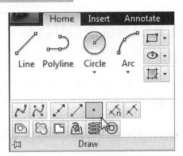

*Figure 3-47 Choosing the **Multiple Points** tool from the **Draw** panel*

Placing Points at Equal Distance

Ribbon: Home > Draw > Divide

To place points at an equal distance on an object, choose the **Divide** tool from the **Draw** panel; you will be prompted to select an object. Select an object; you will be prompted to specify the number of segments. Enter the number of segments; the points will be created on the object.

Placing Points at Specified Intervals

Ribbon: Home > Draw > Measure

You can place points at specified intervals on an object by selecting the object and specifying the length of the segment between two points. To do so, choose the **Measure** tool from the **Draw** panel; you will be prompted to select the object to measure. Select the object; you will be prompted to specify the length of the segment. Specify the length of the segment; the points will be placed at specified intervals.

EXERCISE 12   *PDMODE and PDSIZE*

a.   Try various combinations of the **PDMODE** and **PDSIZE** variables.

b.   Check the difference between the points generated by using the negative values of **PDSIZE** and the points generated by using positive values of **PDSIZE**.

DRAWING INFINITE LINES

In AutoCAD, you can draw a construction line or a ray that aids in construction or projection. A construction line (xline) is a 3D line that extends to infinity at both ends. As the line is infinite in length, it does not have any endpoint. Whereas, a ray is a 3D line that extends to infinity at only one end. The other end of the ray has a finite endpoint. The xlines and rays have zero extents. This means that the extents of the drawing will not change, if you use the **Zoom All** tool. Most of the object snap modes work with both xlines and rays, with some limitations. You cannot use the **Endpoint** object snap with the xline because by definition an xline does not have any endpoints. However for rays you can use the **Endpoint** snap on one end only. Also, xlines and rays take the properties of the layer, in which they are drawn.

Tip

Sometimes, Xlines and rays when plotted like any other object may create confusion. Therefore, it is recommended to create the construction lines in a different layer altogether, such that you can recognize them easily. You will learn about layers in later chapters.

Drawing Construction Lines

Ribbon: Home > Draw > Construction Line	**Toolbar:** Draw > Construction Line
Tool Palettes: Draw > Construction Line	**Command:** XLINE

To draw a construction line that extends to infinity at both sides, choose the **Construction Line** tool from the **Draw** panel; the **XLINE** command will be invoked. The prompt sequence that is displayed on choosing this tool is as follows:

Command: **XLINE**
Specify a point or [Hor/Ver/Ang/Bsect/Offset]: *Specify an option or select a point through which the xline will pass.*

The various options in the **XLINE** command are discussed next.

Point

If you use the default option, you need to specify two points through which the xline will pass. After specifying the first point, move the cursor; a line will be attached to the cursor. On specifying the second point, an xline will be created that passes through the first and second points (Figure 3-48).

> Specify a point or [Hor/Ver/Ang/Bisect/Offset]: *Specify a point.*
> Specify through point: *Specify the second point.*

You can continue to select more points to create more xlines. All these xlines will pass through the first point you had selected at the **Specify a point** prompt. This point is also called the root point. Right-click or press ENTER to end the command.

Horizontal

This option is used to create horizontal xlines of infinite length that pass through the selected points. The xlines will be parallel to the *X* axis of the current UCS, see Figure 3-49. On invoking this option, a horizontal xline will be attached to the cursor and you will be prompted to select a point through which the horizontal xline will pass. Specify a point. You can continue specifying more points to draw more horizontal xlines. To exit the command, right-click or press ENTER.

Vertical

This option is used to create vertical xlines of infinite length that pass through the selected points. The xlines will be parallel to the *Y* axis of the current UCS, see Figure 3-49.

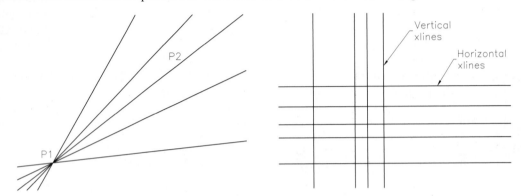

Figure 3-48 *Drawing the xlines*   *Figure 3-49* *Horizontal and vertical xlines*

Angular

This option is used to create xlines of infinite length that pass through the selected point at a specified angle (see Figure 3-50). The angle can be specified by entering a value. You can also invoke the **Reference** option by selecting an object and then specifying an angle relative to it. The **Reference** option is useful when the actual angle is not known, but the angle relative to an existing object can be specified.

> Command: _xline
> _xline Specify a point or [Hor/Ver/Ang/Bisect/Offset]: **A** [Enter]
> Enter angle of xline (0) or [Reference]: **R** [Enter]   (*Use the* **Reference** *method to specify the angle*)
> Select a line object: *Select a line.*
> Enter angle of xline <0>: *Enter angle (the angle will be measured counterclockwise with respect to the selected line)*
> Specify through point: *Specify the second point.*

Bisect

This option is used to create an xline that bisects an angle. In this case, you need to specify the vertex, start point, and endpoint of the angle. The xline will pass through the vertex and bisect the angle specified by selecting two points. The xline created using this option will lie on the plane defined by the selected points. The following is the prompt sequence for this option, refer to Figure 3-51.

> Command: _xline
> _xline Specify a point or [Hor/Ver/Ang/Bisect/Offset]: **B** [Enter]
> Specify angle vertex point: *Enter a point (P1).*

Specify angle start point: *Enter a point (P2).*
Specify angle end point: *Enter a point (P3).*
Specify angle end point: *Select more points or press ENTER or right-click to end the command.*

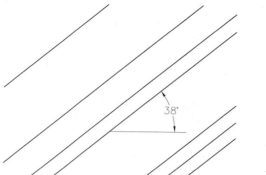

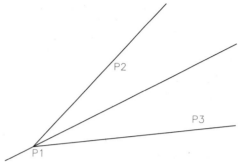

Figure 3-50 *The angular xlines* **Figure 3-51** *Using the **Bisect** option to draw xlines*

Offset

The **Offset** option is used to create xlines that are parallel to a selected line/xline at a specified offset distance. You can specify the offset distance by entering a numerical value or by selecting two points on the screen. If you choose the **Through** option, the offset line will pass through the selected point. The following is the prompt sequence for this option:

Command: _xline
_xline Specify a point or [Hor/Ver/Ang/Bisect/Offset]: **O** [Enter]
Specify offset distance or [Through] <Through>: *Press ENTER to accept the **Through** option or specify a distance from the selected line object at which the xline shall be drawn.*
Select a line object: *Select the object to which the xline is drawn parallel and at a specified distance.*
Specify through point: *Select a point through which the xline should pass.*

After specifying the offset distance and selecting a line object, you need to specify the direction in which the xline has to be offset. You can continue drawing xlines or right-click or press ENTER to exit the command.

Drawing Ray

Ribbon: Home > Draw > Ray   **Command:** RAY

A ray is a 3D line similar to the xline construction line with the difference being that it extends to infinity only in one direction. It starts from a specified point and extends to infinity through the specified point. The prompt sequence is given next.

Specify start point: *Select the start point for the ray.*
Specify through point: *Specify the second point.*

Press ENTER or right-click to exit the command.

Note
When you trim an xline, it gets converted into a ray, and when a ray is trimmed, it gets converted into a line object.

WRITING A SINGLE LINE TEXT

Ribbon: Home > Annotation > Text drop-down > Single Line   **Command:** TEXT

 The **TEXT** command is used to write a single line text. Although you can write more than one line of text using this command, but each line will be a separate text entity. To write text, choose the **Single Line** tool from **Home > Annotation > Text** drop-down. This tool is also available in the **Text** drop-down available in the **Text** panel of the **Annotate** tab. After invoking this command, you need to specify the start point for the text. Next, you need to specify the text height and the rotation angle. As you enter the characters, they start appearing on the screen. After typing a line if you press ENTER, the cursor will be automatically placed at the start of the next line and the prompt for entering another line will be repeated. Type another line or press ENTER again to exit the command. You can use the BACKSPACE key to edit the text on the screen while writing it. The prompt sequence is given next.

Command: **TEXT**
Current text style: "Standard" Text height: 0.2000 Annotative: No
Specify start point of text or [Justify/Style]: *Specify the starting point of the text.*
Specify height<current>: *Enter the text height.*
Specify rotation angle of text <0>: `Enter`
Enter the first line of the text in the text box displayed in the drawing window. `Enter`
Enter the second line of the text in the text box displayed in the drawing window.
Press ENTER twice to exit.

Note
The other commands to enter text are discussed in detail in Chapter 7.

Self-Evaluation Test

Answer the following questions and then compare them to those given at the end of this chapter:

1. While drawing an arc by choosing the **Start, Center, Angle** tool, if you enter the included angle with a negative value, an arc will be drawn in clockwise direction. (T/F)

2. After choosing the **ARC** command, if you press ENTER instead of specifying the start point, the start point and direction of the arc will be taken from the endpoint and ending direction of the previous line or the arc drawn. (T/F)

3. By using the **Single Line Text** tool, you can write more than one line of text. (T/F)

4. The start and end parameters of an elliptical arc are determined by specifying a point on the circle whose diameter is equal to the minor diameter of an ellipse. (T/F)

5. The _____ option of the **RECTANG** command is used to draw a rectangle at a specified distance from the *XY* plane along the *Z* axis.

6. If the **FILLMODE** is set to _____, only the outlines for the new polyline will be drawn.

7. You can get a _____ polyline by entering two different values at the starting width and ending width prompts.

8. Choose the _____ tool from the **Draw** panel to place points at specified intervals on an object.

9. Choose the _____ tool from the **Draw** panel to draw as many points as you want in a single command.

10. The size of a point will be taken as a percentage of the viewport size, if you enter a _____ value for the **PDSIZE** variable.

Review Questions

Answer the following questions:

1. While drawing an arc by choosing the **Start, End, Angle** tool, a negative included angle value draws the arc in clockwise direction. (T/F)

2. When the **Continue** option of the **ARC** command is used to draw arcs, each successive arc is perpendicular to the previous one. (T/F)

3. If you specify the chamfer distances first and then specify the fillet radius in the same **RECTANG** command, the rectangle will be drawn with chamfers only. (T/F)

4. A rectangle drawn using the **Rectangle** tool is treated as a combination of different objects; therefore, the individual sides can be edited independently. (T/F)

5. On drawing an arc by choosing the **Start, Center, Length** tool, a positive chord length generates the smallest possible arc (minor arc), and the arc is always less than

    (a) 90 degree   (b) 180 degree
    (c) 270 degree   (d) 360 degree

6. Which one of the following options of the **Rectangle** tool is used to draw a rectangle that is extruded in the Z direction by a specified value?

    (a) **Elevation**   (b) **Thickness**
    (c) **Extrude**   (d) **Width**

7. Which of the following commands should be used to draw a line in 3D space that starts from a specified start point and the other end extends to infinity?

    (a) **PLINE**   (b) **RAY**
    (c) **XLINE**   (d) **MLINE**

8. A polygon is said to be _____, when it is drawn inside an imaginary circle and its vertices touch the circle.

9. If additional polyline segments are added to the first polyline, the _____ of the first polyline segment becomes the start point of the next polyline segment.

10. To create a solid-filled circle by the **DONUT** tool, the value of the inside diameter of the circle should be _____.

11. With the **DONUT** tool, you can draw a solid-filled circle by specifying the inside diameter as _____ and keeping the **FILLMODE** on.

12. The _____ option of the **XLINE** tool is used to create xlines of infinite length that are parallel to the *Y* axis of the current UCS.

13. You can use the _____ key to edit text on the screen, while writing it using the **Single Line Text** tool.

EXERCISE 13   Arc

Draw the sketch shown in Figure 3-52. The distance between the dotted lines is 1.0 unit. Create radii by choosing appropriate tools from the **Arc** drop-down.

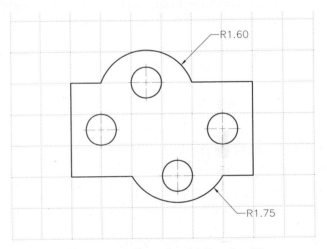

Figure 3-52 Drawing for Exercise 13

EXERCISE 14   Arc

Draw the sketch shown in Figure 3-53. The distance between the dotted lines is 1.0 unit. Create arcs by choosing appropriate tools from the **Arc** drop-down.

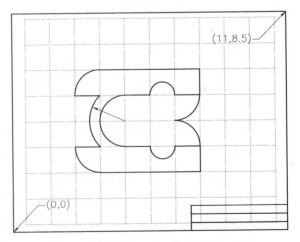

Figure 3-53 Drawing for Exercise 14

EXERCISE 15   *Ellipse*

Draw the sketch shown in Figure 3-54. The distance between the dotted lines is 0.5 unit. Create ellipses using the tools in the **Ellipse** drop-down.

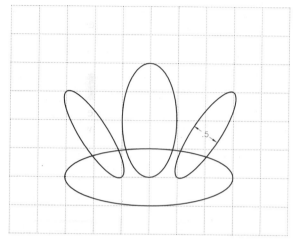

Figure 3-54 *Drawing for Exercise 15*

EXERCISE 16

Draw the sketch shown in Figure 3-55 using the **LINE**, **CIRCLE**, and **ARC** commands. The distance between the dotted lines is 1.0 unit and the diameter of the circles is 1.0 unit.

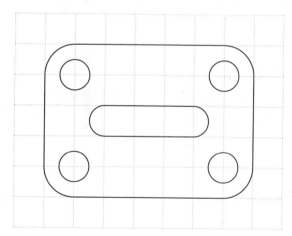

Figure 3-55 *Drawing for Exercise 16*

Chapter 3

EXERCISE 17

Draw the sketch shown in Figure 3-56 using the **LINE, CIRCLE,** and **ARC** commands or their options. The distance between the grid lines is 1.0 unit and the diameter of the circle is 1.0 unit.

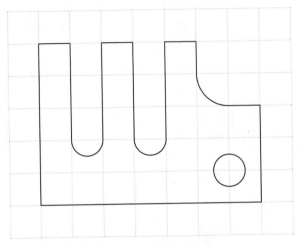

Figure 3-56 *Drawing for Exercise 17*

Problem-Solving Exercise 1   *Arc*

Draw the sketch shown in Figure 3-57. Create arcs by using the **ARC** command options indicated in the drawing. (Use the @ symbol to snap to the previous point. Example: Specify start point of arc or [Center]: @)

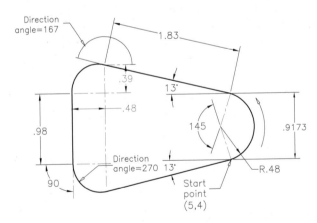

Figure 3-57 *Drawing for Problem-Solving Exercise 1*

Problem-Solving Exercise 2   *Arc*

Draw the sketch shown in Figure 3-58. Create arcs by using the **ARC** command options. The distance between the dotted lines is 0.5 unit.

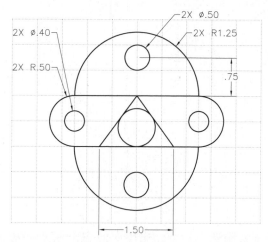

Figure 3-58 *Drawing for Problem-Solving Exercise 2*

Problem-Solving Exercise 3   *Arc*

Draw the sketch shown in Figure 3-59. Create arcs by using the **ARC** command options. The distance between the dotted lines is 1.0 unit.

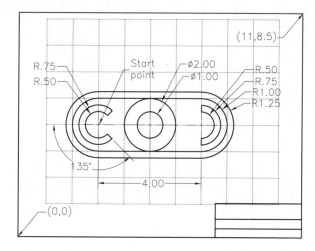

Figure 3-59 *Drawing for Problem-Solving Exercise 3*

Problem-Solving Exercise 4

Draw the sketch shown in Figure 3-60 using the **POLYGON**, **CIRCLE**, and **LINE** commands.

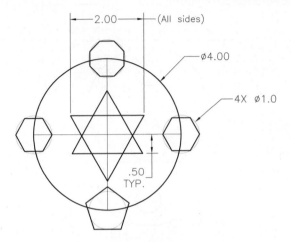

Figure 3-60 *Drawing for Problem-Solving Exercise 4*

Problem-Solving Exercise 5

Draw the sketch shown in Figure 3-61 by using different tools in the **Draw** panel. Note, Sin30=0.5, Sin60=0.866. The distance between the dotted lines is 1 unit.

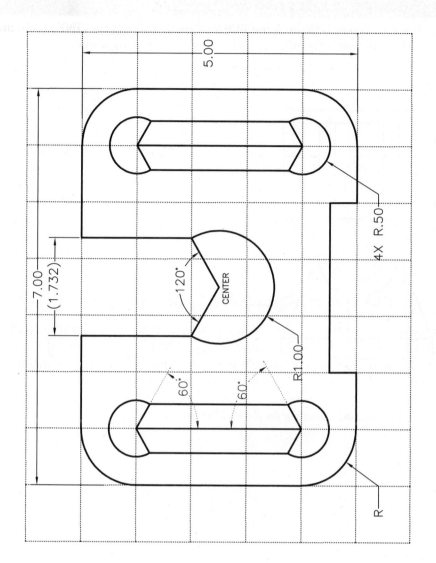

Figure 3-61 *Drawing for Problem-Solving Exercise 5*

Problem-Solving Exercise 6

Draw the sketch shown in Figure 3-62 by using different tools in the **Draw** panel. Also, draw the hidden lines and centerlines as continuous line. Note that the dimensions are given only for your reference. **

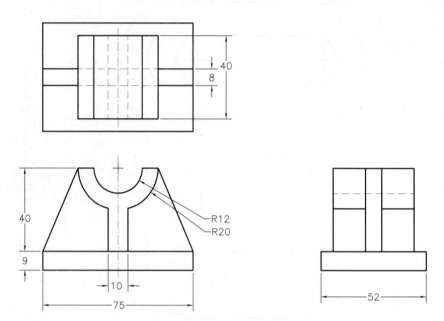

Figure 3-62 *Drawing for the Problem-Solving Exercise 6*

Answers to Self-Evaluation Test

1. T, **2.** T, **3.** T, **4.** F, **5. Elevation**, **6.** 0, **7.** tapered, **8. Measure**, **9. Multiple Point**, **10.** negative

Chapter 4

Working with Drawing Aids

CHAPTER OBJECTIVES

In this chapter, you will learn:
* *To set up layers, and assign colors and line type to layers.*
* *To change general object properties using the Properties toolbar.*
* *To change object properties using the PROPERTIES command.*
* *To determine the line type scaling and the LTSCALE factor for plotting.*
* *To set up Grid, Snap, and Ortho modes.*
* *To use the object snaps and running Object Snap modes.*
* *To use AutoTracking to locate keypoints in a drawing.*

KEY TERMS

* *Layer*
* *Freeze*
* *Thaw*
* *Plot Style*

* *Reconciling Layers*
* *Isolating Layers*
* *Quick Properties*

* *Global Linetype scaling*
* *Current Linetype scaling*

* *DesignCenter*
* *DSETTINGS*
* *ONSAP*

INTRODUCTION

In this chapter, you will learn about the drawing setup and the factors that affect the quality and accuracy of a drawing. This chapter contains a detailed description of how to set up layers. You will also learn about some other drawing aids, such as Grid, Snap, and Ortho. These aids will help you create drawings accurately and quickly.

UNDERSTANDING THE CONCEPT AND USE OF LAYERS

The concept of layers can be best explained by using the concept of overlays in manual drafting. In manual drafting, different details of a drawing can be drawn on different sheets of paper or overlays. Each overlay is perfectly aligned with the others. Once all of them are placed on top of each other, you can reproduce the entire drawing. For example, in Figure 4-1, the object lines are drawn in the first overlay and the dimensions in the second overlay. You can place these overlays on top of each other and get a combined look of the drawing.

In AutoCAD, instead of using overlays, you can use layers. Each layer is assigned a name. You can also assign a color and a line type to a layer. For example, in Figure 4-2, the object lines are drawn in the OBJECT layer and the dimensions are drawn in the DIM layer. The object lines will be red because red has been assigned to the OBJECT layer. Similarly, the dimension lines will be green because green has been assigned to the DIM layer. You can display the layers together, individually, or in any combination.

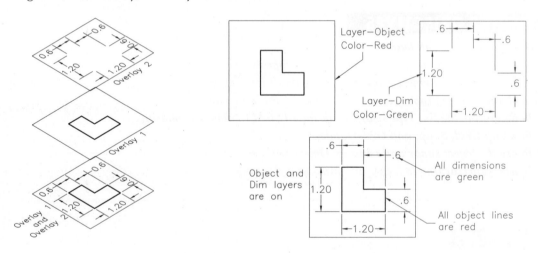

Figure 4-1 *Drawing lines and dimensions in different overlays*

Figure 4-2 *Drawing lines and dimensions in different layers*

Advantages of Using Layers

1. Each layer can be assigned a different color. Assigning a particular color to a group of objects is very important for plotting. For example, if all object lines are red, then at the time of plotting, you can assign the red color to a slot (pen) that has the desired tip width (e.g., medium). Similarly, if the dimensions are green, you can assign the green color to another slot (pen) that has a thin tip. By assigning different colors to different layers, you can control the width of lines while plotting the drawing. You can also make a layer plottable or non-plottable.
2. Layers are also useful for performing some editing operations. For example, to erase all dimensions in a drawing, you can freeze or lock all layers except the dimension layer, then select all objects by using the Window Crossing option, and erase all dimensions.
3. You can turn off a layer or freeze a layer that you do not want to be displayed or plotted.
4. You can lock a layer to prevent the user from accidentally editing the objects in it.

5. Colors also help you distinguish different groups of objects. For example, in architectural drafting, the plans for foundation, floors, plumbing, electrical work, and heating systems may be made in different layers. In electronic drafting and in PCB (printed circuit board), the design of each level of a multilevel circuit board can be drawn on a separate layer. Similarly, in mechanical engineering, the main components of an assembly can be made in one layer, other components such as nuts, bolts, keys, and washers can be made in another layer, and the annotations such as datum symbols and identifiers, texture symbols, Balloons, and Bill of Materials can be made in yet another layer.

WORKING WITH LAYERS

| Ribbon: | Home > Layers > Layer Properties | Command: | LAYER or LA |
| Toolbar: | Layers > Layer Properties Manager | | |

You can freeze, thaw, lock, unlock, and so on by using the **Layers** panel in the **Ribbon** (see Figure 4-3). However, to add new layers, to delete the existing layers, or to assign colors and linetypes to layers, you need to invoke the **Layer Properties Manager**. To invoke the **Layer Properties Manager**, choose the **Layer Properties** button from the **Layers** panel in the **Home** tab, or invoke it from the **Layers** toolbar, (see Figure 4-4).

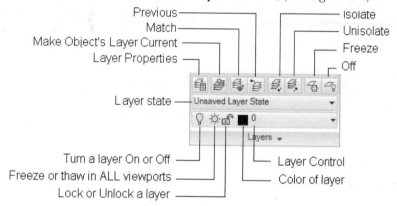

Figure 4-3 The **Layers** panel

Figure 4-4 The **Layers** toolbar

On invoking the **Layer Properties Manager**, a default layer with the name 0 is displayed. It is the current layer and any object you draw is created in it. The current layer can be recognized by the green colored tick mark in the **Status** column. There are certain features such as color, linetype, and lineweight are associated with each layer. The **0** layer has the default color as white, linetype as continuous, and lineweight as default.

Creating New Layers

To create new layers, choose the **New Layer** button in the **Layer Properties Manager**; a new layer, named Layer1, with the same properties as that of the current layer will be created and listed, as shown in Figure 4-5. Alternatively, right-click anywhere in the **Layers** list area of the **Layer Properties Manager** and then choose the **New Layer** option from the shortcut menu to create a new layer. If there are more layers and you right-click on a layer other than the current layer and then choose **New Layer** from the shortcut menu, a new layer will be created with the properties similar to the layer on which you right-clicked. To create a new layer that is automatically frozen in all viewports, choose the **New Layer VP Frozen in All Viewports** button.

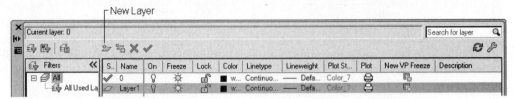

*Figure 4-5 Partial view of the **Layer Properties Manager** with the new layer created*

Naming a New Layer

New layers are created with the name layer 1, layer 2, and so on. To change or edit the name of a layer, select it, then click once in the field corresponding to the **Name** column and enter a new name. Some of the points to be remembered while naming a layer are given below.

1. A layer name can be up to 255 characters long, including letters (a-z), numbers (0-9), special characters ($ _ -), and spaces. Any combination of lower and uppercase letters can be used while naming a layer. However, characters such as <>;:,'?*|"=, and so on are not valid characters while naming a layer.

2. Layers should be named to help the user identify the contents of the layer. For example, if a layer name is HATCH, a user can easily identify it and its contents. On the other hand, if a layer's name is X261, it is hard to identify its contents.

3. Layer names should be short, but should also convey the meaning.

Tip
If you exchange drawings or provide drawings to consultants or others, it is very important that you standardize and coordinate layer names and other layer settings.

Making a Layer Current

To draw an object in a particular layer, you need to make it the current layer. There are different methods to make a layer current and are listed below.

1. Double-click on the name of a layer in the list box in the **Layer Properties Manager**; the selected layer is made current.
2. Select the name of the layer in the **Layer Properties Manager** and then choose the **Set Current** button in the **Layer Properties Manager**.
3. Right-click on a layer in the **Layer** list box in the **Layer Properties Manager** and choose the **Set current** option from the shortcut menu displayed, see Figure 4-6.
4. Select a layer from the **Layer** drop-down list in the **Layers** toolbar or from the **Layers** panel.
5. Choose the **Make Object's Layer Current** button from the **Layers** panel; you will be prompted to select the object whose layer you want to make current. Select the required object; the layer associated with that object will become current.

On making a layer as the current layer, a green color tick mark will be displayed in the **Status** column of that layer. The name and properties of the current layer will be displayed in the **Layers** toolbar and the name of the current layer will be displayed at the upper-left corner of the **Layer Properties Manager**.

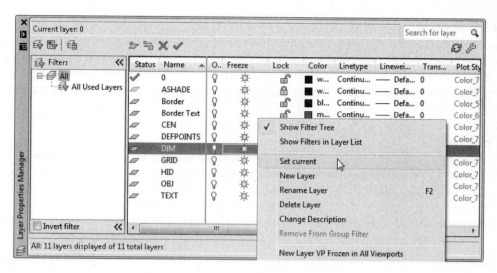

Figure 4-6 The **Layer Properties Manager** *with the shortcut menu*

Note

If you select more than one layer at a time using the SHIFT or CTRL key, the **Make Current** *option does not appear in the shortcut menu in the* **Layer Properties Manager**. *This is because only one layer can be made current at a time.*

Controlling the Display of Layers

You can control the display of layers by choosing the **Turn a layer On or Off**, **Freeze or thaw in ALL viewports**, and **Lock or Unlock a layer** toggle buttons in the list box of any particular layer.

Turn a layer On or Off

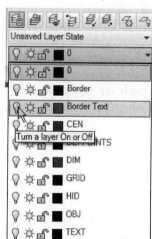

Choose the **Turn a layer On or Off** toggle icon (bulb) to turn a layer on or off. You can also turn the layer on or off by clicking on the **On/Off** toggle icon from the **Layer** drop-down list in the **Layers** toolbar or the **Layers** panel, as shown in Figure 4-7. The objects in the layers that are turned on are displayed and can be plotted, while the objects in the layers that are turned off are not displayed and cannot be plotted. However, you can perform the operations such as drawing and editing of the objects in the layer that has been turned off. You can turn the current layer off, but AutoCAD will display a message box informing you that the current drawing layer has been turned off.

Freeze or thaw in ALL Viewports

Figure 4-7 Turning off the **Border Text** layer

Sometimes in an architectural drawing, you may not need the door tag, window tag, or surveyor data to be displayed, so that they are not changed. These types of information or entities can be placed in a particular layer and that layer can be frozen. You cannot edit the entities in the frozen layer. Also, frozen layers are invisible and cannot be plotted. To freeze a layer, select it and choose the **Freeze or thaw in ALL viewports** toggle icon (sun/snowflakes) in the **Layer Properties Manager**. Note that the current layer cannot be frozen. You can also select the **Freeze or thaw in ALL viewports** toggle icon in the **Layers** panel of the **Ribbon** or in the **Layers** toolbar to freeze or thaw a layer. The **Thaw** option negates the effect of the **Freeze** option, and the frozen layers are restored to normal. The difference between the usage of the **Off** option and the **Freeze** option is that in using the **Freeze** option, the layer is frozen and the entities in the frozen layer are not regenerated thereby saving time.

New VP Freeze

On choosing the **Layout** tab, you will observe that a default viewport is displayed. You can also create new viewports anytime during the design. If you want to freeze some layers in all the subsequent new viewports, select the layers, and then choose the **New VP Freeze** toggle icon. The selected layers will be frozen in all the subsequently created viewports, without affecting the existing viewports.

VP Freeze

This icon will be available only if you invoke the **Layer Properties Manager** in the **Layout** tab, see Figure 4-8. The **Freeze or thaw in current viewport** icon will also be available the **Layer** drop-down list in the **Layers** panel of the **Ribbon** or in the **Layers** toolbar. If there are multiple viewports, make a viewport as the current viewport by double-clicking in it and freeze or thaw the selected layer in it by choosing the **VP Freeze** icon in the **Layout Properties Manager**. However, a layer that is frozen in the model space cannot be thawed in the current viewport.

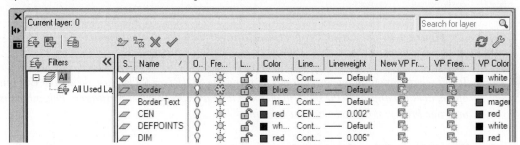

*Figure 4-8 Partial view of the **Layer Properties Manager** with the **VP Freeze** icon*

Tip
*To increase or decrease the width of column headings in the **Layer Properties Manager**, place the cursor on the separator that is between column headings; the cursor turns into a two-sided arrow. Now, press and hold the left mouse button and then drag the cursor to the right or left. This way you can change the width of the column headings.*

Lock or Unlock a Layer

While working on a drawing, if you want to avoid editing some objects on a particular layer but still need to have them visible, use the **Lock/Unlock** toggle icon to lock the layer. However, you can still use the objects in the locked layer for **Object Snaps** and inquiry commands such as **LIST**. You can also make the locked layer as the current layer and draw objects on it. Note that you can also plot a locked layer. The **Unlock** option negates the **Lock** option and allows you to edit objects on the layers that were previously locked.

Make a Layer Plottable or Nonplottable

By default, you can plot all layers, except the layers that are turned off or frozen. If you do not want to plot a layer that is not turned off or frozen, select the layer and choose the **Plot** icon (printer). This is a toggle icon, therefore you can choose this icon again to plot the layer.

Tip
*The faster and convenient way to make a layer current and control the display features of the layer (**On/Off, Freeze/Thaw, Lock/Unlock**) is by using the **Layer** drop-down list in the **Layers** panel.*

Assigning Linetype to a Layer

By default, continuous linetypes are assigned to a layer, if no layer is selected while creating a new layer. Otherwise, new layer takes the properties of the selected layer. To assign a new linetype to a layer, click on the field under the **Linetype** column of that layer in the **Layer Properties**

Manager; the **Select Linetype** dialog box showing the linetypes loaded on your computer will be displayed. Select the new linetype and then choose the **OK** button; the selected linetype will be assigned to the layer.

If you are opening the **Select Linetype** dialog box for the first time, only the **Continuous** linetype will be displayed, as shown in Figure 4-9.

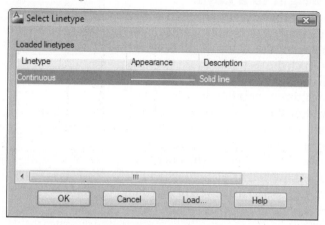

*Figure 4-9 The **Select Linetype** dialog box*

You need to load linetypes and then assign them to layers. To load linetypes, choose the **Load** button in the **Select Linetype** dialog box; the **Load or Reload Linetypes** dialog box will be displayed, see Figure 4-10. This dialog box displays all linetypes in the *acad.lin* or *acadiso.lin* file. In this dialog box, you can select a single linetype, or a number of linetypes by pressing and holding the SHIFT or CTRL key and then selecting the linetypes. After selecting the linetypes, choose the **OK** button; the selected linetypes are loaded in the **Select Linetype** dialog box. Now, select the desired linetype and choose **OK**; the selected linetype is assigned to the selected layer.

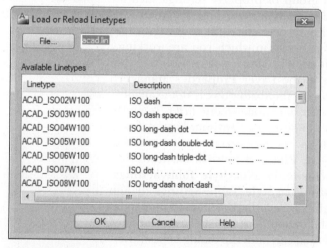

*Figure 4-10 The **Load or Reload Linetypes** dialog box*

Assigning Transparency to a Layer

By default, no transparency is assigned to a layer, if **Layer 0** is selected while creating that layer. Otherwise, new layer takes the transparency of the selected layer. To assign transparency to a layer, click on the **Transparency** field of that layer; the **Layer Transparency** dialog box will be displayed. Select the required transparency level from the **Transparency Value (0-90)** drop-down list and choose **OK**. You can also enter a value in this drop-down list. Now, if you place an object on this layer, the object will have a faded color according to the specified transparency value.

Assigning Color to a Layer

To assign a color to a layer, select the color swatch in that layer in the **Layer Properties Manager**; the **Select Color** dialog box will be displayed. Select the desired color and then choose the **OK** button; the selected color will be assigned to the layer.

Assigning Lineweight to a Layer

Lineweight is used to give thickness to objects in a layer. For example, if you create a sectional plan, you can assign a layer with a larger value of lineweight to create the objects through which the section is made. Another layer with a lesser lineweight can be used to show the objects through which the section does not pass. This thickness is displayed on the screen if the display of the lineweight is on. The lineweight assigned to an object can also be plotted. To assign a lineweight to a layer, select the layer and then click on the lineweight associated with it; the **Lineweight** dialog box will be displayed, see Figure 4-11. Select a lineweight and then choose **OK** from this dialog box to return to the **Layer Properties Manager**.

Figure 4-11 The Lineweight dialog box

Tip

*Remember that the **Linetype Control**, **Lineweight Control**, **Color Control**, and **Plot Style Control** list boxes in the **Properties** panel should display **ByLayer** as the current property of objects. This is to ensure that the objects drawn in a layer take the properties assigned to the layer in which they are drawn.*

Assigning Plot Style to a Layer

The plot style is a group of property settings such as color, linetype, and lineweight that can be assigned to a layer. The assigned plot style affects the drawings while plotting only. The drawing, in which you are working, should be in a named plot style mode (*.stb*) to make the plot style available in the **Layer Properties Manager**. If the **Plot Style** icon in the dialog box is not available, then you are in a color-dependent mode (*.ctb*). To make this icon available, choose the **Options** button from the **Application Menu**; the **Options** dialog box will be displayed. Choose the **Plot and Publish** tab from this dialog box. Next, choose the **Plot Style Table Settings** button to invoke the **Plot Style Table Settings** dialog box. In this dialog box, select the **Use named plot style** radio button from the **Default plot style behavior for new drawings** area and then exit both the dialog boxes. After changing the plot style to named plot style dependent, you have to start a new AutoCAD session to apply this setting. Start a new AutoCAD session with templates like *acad -Named Plot Styles* and invoke the **Layer Properties Manager**. You will notice that the default plot style is **Normal**, in which the color, linetype, and lineweight are BYLAYER. To assign a plot style to a layer, select the layer and then click on its plot style; the **Select Plot Style** dialog box will be displayed, as shown in Figure 4-12. In this dialog box, you can select a specific plot style from the **Plot styles** list of available plot styles. Plot styles need to

Figure 4-12 The Select Plot Style dialog box

be created before you can use them (see Chapter 16, *Plotting Drawings*). Choose **OK** to return to the **Layer Properties Manager**.

Note
*When you are in a viewport, you can override the color, linetype, lineweight, and plot style settings assigned to a selected layer for the particular viewport (current viewport). These settings for all other viewports and model space will remain unaffected. All these overrides are highlighted in different background colors. Also, the icon of the selected layer under the **Status** column will change indicating that some of the properties of the selected layer have been overridden by the new ones (see Chapter 15, Model Space Viewports, Paper Space Viewports, and Layouts).*

Tip
*You can also change the plot style mode from the command line by using the **PSTYLEPOLICY** system variable. A value of 0 sets it to the color-dependent mode and a value of 1 sets it to the named mode.*

Deleting Layers

You can delete a layer by selecting it and then choosing the **Delete Layer** button in the **Layer Properties Manager**. Remember that you cannot a delete a layer that contains any object. Additionally, you cannot delete layers 0, Defpoints (created while dimensioning), Ashade (created while rendering), current layer, and an Xref-dependent layer.

Managing the Display of Columns

You can change the display order of columns in the **Layer Properties Manager**. To do so, drag and drop the desired column head to the desired location. You can also change the display order by right-clicking on the column head and choosing the **Customize** option from the shortcut menu, as shown in Figure 4-13. On choosing this option, the **Customize Layer Columns** dialog box will be displayed, as shown in Figure 4-14. Select the desired column from the list and then choose the **Move Up** or **Move Down** button to change the display order.

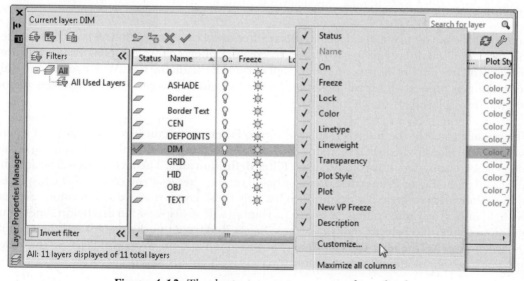

Figure 4-13 *The shortcut menu to manage column heads*

You can also control the column heads to be displayed in the **Layer Properties Manager** by using the shortcut menu shown in Figure 4-13. The tick marked entries in the shortcut menu will be displayed in the list. Choose the entry that you want to display in the **Layer Properties Manager** from the shortcut menu.

Figure 4-14 The **Customize Layer Columns**
dialog box

Selective Display of Layers

If a drawing has a limited number of layers, it is easy to scan through them. However, if the drawing has a large number of layers, it is sometimes difficult to search through the layers. To solve this problem, you can use layer filters. By defining filters, you can specify the properties and display only those layers in the **Layer Properties Manager** dialog box that match the properties in the filter. By default, the **All** and **All Used Layers** filters are created. The **All** filter is selected by default, and therefore, all layers are displayed. If you select the **All Used Layers** filter, only the layers that are used in the drawing will be displayed.

AutoCAD allows you to create a layer property filter or a layer group filter. A layer group filter can have additional layer property filters. To create a filter, choose the **New Property Filter** button, which is the first button on the top left corner of the dialog box. When you choose this button, a new property filter will be added to the list and the **Layer Filter Properties** dialog box will be displayed, as shown in Figure 4-15.

The default name of the filter is displayed in the **Filter name** edit box. You can enter a new name for the filter in this edit box. Using the **Layer Filter Properties** dialog box, you can create filters based on any property column in the **Filter definition** area. The layers that will be actually displayed in the **Layer Properties Manager** based on the filter that you create will be displayed in the **Filter preview** area. For example, to list only those layers that are red in color, click on the field under the **Color** column; a swatch [**...**] button will be displayed in this field. Choose this button to display the **Select Color** dialog box. Select **Red** from this dialog box and then exit it. You will notice that the filter row color has changed to red and the display of layers in the **Filter preview** has been modified such that only the red layers are displayed.

After creating the filter, exit the **Layer Filter Properties** dialog box. The layer filter is selected automatically in the **Layer Properties Manager** and only the layers that satisfy the filter properties are displayed. You can restore the display of all layers again by clicking on the **All** filter.

Choose the **New Group Filter** button (second button on the top left corner of the dialog box) to group all filters under a common group name. You can create groups and subgroups of filters to categorize the filters with similar criteria.

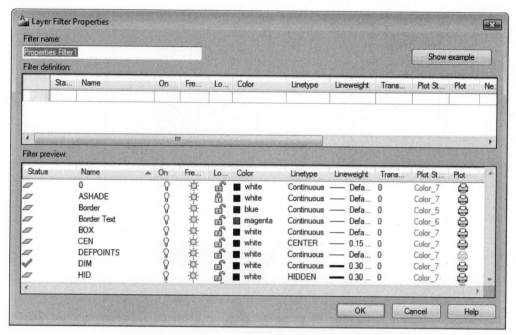

Figure 4-15 The *Layer Filter Properties* dialog box

Note
To modify a layer filter, double-click on it; the **Layer Filter Properties** *dialog box will be displayed. Modify the filter and then exit the dialog box.*

In the **Layer Properties Manager**, when you select the **Invert filter** check box, the selected filter parameters are inverted. For example, if you have selected a filter to show all layers, none of the layers will be displayed. Note that in the **Layer Properties Manager**, the current layer is not displayed in the list box if it is not among the filtered layers.

Layer States

While working in a drawing, you can save the properties of all layers under one name and restore the properties later. The properties thus saved can also be imported to other drawings. This saves time in setting the layer properties again in a new drawing. To save the properties of all layers in a drawing, choose the **Layer States Manager** button in the **Layer Properties Manager**; the **Layer States Manager** dialog box will be displayed, as shown in Figure 4-16. Choose the **New** button to invoke the **New Layer State to Save** dialog box. Specify the name and description related to the state of a layer in the **New layer state name** and **Description** edit boxes respectively, and then choose the **OK** button. Choose the **More Restore Options (>>)** button next to the **Help** button; the dialog box will expand. Specify different states and properties of the layers by selecting the corresponding check boxes in the **Layers properties to restore** area. Choose the **OK** button to save the layer state. However, this layer state is saved only for the current file. If you want to use the current layer state in other files also, you need to save them. To do so, choose the **Export** button; the **Export layer state** dialog box will be invoked. Enter a name for the layer state. The layer state will be exported with the *.las* extension.

You can also import a layer state later in any other file by using the **Layer States Manager** dialog box. On importing a layer state, you can edit the states and properties of the imported layer state. You can also rename and delete a layer state.

Select the **Don't list layer states in Xrefs** check box; the layer states that are defined in an externally referenced drawing will be referenced from the **Layer states** list box. Select the **Turn off layers not found in layer state** check box to turn off the newly created layers (that were

Chapter 4

*Figure 4-16 The **Layer States Manager** dialog box*

created after defining the layer state) so that the drawing appears the same as it was when the layer state was saved. The **Apply properties as viewport overrides** check box is used to apply the layer property overrides to the current viewport. This check box is selected by default and is available only when a viewport is active.

Note
*Select the **New Layer State** option from the **Layer State** drop-down list in the **Layers** panel of the **Home** tab to invoke the **New Layer State to Save** dialog box and save the new layer state. Similarly, select the **Manage Layer States** option from the **Layer State** drop-down list to invoke the **Layer States Manager** dialog box.*

*The linetypes will be restored with the layer state in a new drawing file only if the linetypes are already loaded using the **Select Linetype** dialog box.*

Reconciling New Layers

When you save or plot a drawing for the first time, AutoCAD compiles a list of existing layers. These layers are known as reconciled layers. When new layers are inserted or an external reference is attached to the current drawing, AutoCAD compares these new layers (unreconciled layers) with the previously created list. This helps prevent any new layers from being added to the current drawing without user's permission.

If you need AutoCAD to notify the unreconciled layer, choose the **Settings** button from the **Layer Properties Manager**; the **Layer Settings** dialog box will be displayed. Select the **Evaluate all new layers** and **Notify when new layers are present** check boxes from the **Layer Settings** dialog box. You also need to select the options under these check boxes based on your requirement. After selecting the **Notify when new layers are present** check box and the corresponding options, AutoCAD, displays the **Unreconciled New Layers** information bubble by default, see Figure 4-17, whenever new unreconciled layers are found in the current drawing or in the externally referenced drawings. As soon as the unreconciled layers are found in the drawing, an

Unreconciled New Layers filter is automatically created in the Layer Properties Manager. Click on the blue link in the information bubble or open the Layer Properties Manager to view the unreconciled layers. To reconcile the layers, select all the unreconciled layers in the Layer Properties Manager, right-click on them and choose the Reconcile Layer option from the shortcut menu.

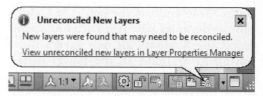

Figure 4-17 The information bubble for unreconciled layers

Isolating and Unisolating Layers

| Ribbon: | Home > Layers > Isolate/Unisolate | Command: | LAYISO/LAYUNISO |
| Toolbar: | Layers II > Layer Isolate/Layer Unisolate | | |

 This option is used to lock or turn off all layers except those for the selected objects. To isolate a layer, choose the **Isolate** tool from the Layers panel; you will be prompted to select the objects. Select one or more objects and press the ENTER key; the layers corresponding to the selected objects will get isolated and unlocked. All other layers will be turned off or frozen in the current viewport or locked depending on the current settings. The unisolated layers (locked) will appear faded. This will help you to differentiate between the isolated and unisolated layers. You can change the settings by choosing the **Settings** option from the shortcut menu at the **Select objects on the layer(s) to be isolated or [Settings]** prompt. The options at the **Settings** prompt are discussed next.

Off

This option is used to turn off or freeze all layers except the selected layers. Choose the **Off** suboption to turn off all layers, except the selected one in all viewports, and choose the **Vpfreeze** suboption to freeze all layers, except the selected one only in the current floating viewport.

Lock and fade

This option is used to lock all layers except the selected one and prompts you to enter the fade value to specify the fading intensity. You can also control the fading intensity by dragging the **Locked layer fading** sliding bar in the **Layers** panel. You can also turn off or on the fading effect by choosing the **Locked Layer Fading** button from the **Layers** panel.

To unisolate the previously isolated layer, choose the **Unisolate** tool from the **Layers** panel.

Controlling the Layer Settings

The **Settings** button in the **Layer Properties Manager** is used to control the new layer notification settings. You can also change the background color used to highlight the layer properties that are overridden in the current viewport. Choose the **Settings** button in the **Layer Properties Manager**; the **Layer Settings** dialog box will be displayed (see Figure 4-18). The options in the **Layer Settings** dialog box are discussed next.

New Layer Notification Area

By default, the **Evaluate new layers added to drawing** check box is cleared. Therefore, the new layers will not be evaluated. Also, the user will not be informed about the unreconciled layers. Select the **Evaluate new layers added to drawing** check box; all other options in this area will be available. The **Evaluate new xref layers only** radio button is selected by default and allows

AutoCAD to evaluate the new layers added only to the externally referenced drawings. Select the **Evaluate all new layers** radio button to evaluate all the layers in the drawing, including the unreconciled layers of the externally referenced drawings. This setting can also be controlled by setting the **LAYEREVAL** system variable to 2.

The **Notify when new layers are present** check box is selected by default and is used to specify the operations after which AutoCAD will evaluate the unreconciled files. These operations include open, save, attach/reload xrefs, insert, restore layer states, and plot. Select the respective check boxes to specify your preferences. These settings can also be specified by using the **LAYERNOTIFY** system variable. The values for this system variable are 0, 1, 2, 4, 8, 16, 32, and 64. The respective operations after which an information bubble will be displayed are Off, Plot, Open, Load/Reload/Attach for xrefs, Restore layer state, Save, and Insert.

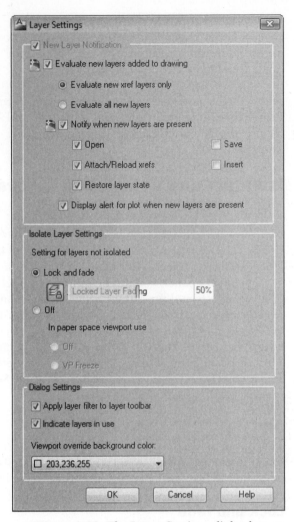

*Figure 4-18 The **Layer Settings** dialog box*

Isolate Layer Settings Area

The options in this area are used to set the display state of the unisolated layers. If you need to display a locked layer in a faded mode, select the **Lock and fade** radio button and set the intensity of fading by using the slider. Similarly, you can make the unisolated layers turned off or VP Freeze in the viewport by selecting the **Off** radio button and the corresponding option.

Dialog Settings Area

You can apply the current layer filter to the **Filter applied** drop-down list in the **Layers** toolbar. To do so, invoke the **Layer Filter Properties** dialog box by choosing the **New Property Filter** button from the **Layer Properties Manager**. Select the layer state or property from the **Filter definition** list box and choose the **OK** button. Only the filtered layers will be displayed in the **Filter applied** drop-down list of the **Layers** toolbar.

EXAMPLE 1   *Layers*

Set up three layers with the following linetypes and colors. Then create the drawing shown in Figure 4-19 (without dimensions).

Layer Name	Color	Linetype	Lineweight
Obj	Red	Continuous	0.012"
Hid	Yellow	Hidden	0.008"
Cen	Green	Center	0.006"

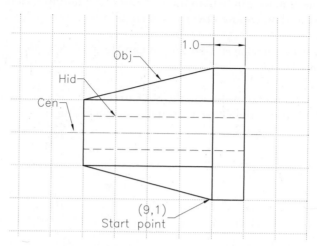

Figure 4-19 *Drawing for Example 1*

In this example, assume that the limits and units are already set. Before drawing the lines, you need to create layers and assign colors, linetypes, and lineweights to them. Also, depending on the objects that you want to draw, you need to set a layer as the current layer. In this example, you will create layers by using the **Layer Properties Manager**. You will use the **Layer** drop-down list in the **Layers** panel to set a layer as the current layer and then draw the figure.

1. As the lineweights specified are given in inches, first you need to change their units, if they are in millimeters. To do so, invoke the **Lineweight Settings** dialog box by right-clicking on the **Show/Hide Lineweight** button on the Status Bar and then choose the **Settings** option from the shortcut menu. Select the **Inches [in]** radio button in the **Units for Listing** area of the dialog box and then choose the **OK** button.

2. Choose the **Layer Properties** button from the **Layers** panel in the **Home** tab of the **Ribbon** to display the **Layer Properties Manager**. The layer **0** with default properties is displayed in the list box.

3. Choose the **New Layer** button; a new layer (Layer1) with the default properties is displayed in the list box. Change the default name, **Layer1** to **Obj**.

4. Left-click on the **Color** field of this layer; the **Select Color** dialog box is displayed. Select the **Red** color and then choose **OK**; red color is assigned to the **Obj** layer.

5. Left-click on the **Lineweight** field of the **Obj** layer; the **Lineweight** dialog box is displayed. Select **0.012"** and then choose **OK**; the selected lineweight is assigned to the **Obj** layer.

6. Again, choose the **New Layer** button; a new layer (Layer1) with properties similar to that of the **Obj** layer is created. Change the default name to **Hid**.

7. Left-click on the **Color** field of this layer; the **Select Color** dialog box is displayed. Select the **Yellow** color and then choose **OK**.

8. Left-click on the **Linetype** field of the **Hid** layer; the **Select Linetype** dialog box is displayed. If the **HIDDEN** linetype is not displayed in the dialog box, choose the **Load** button; the **Load and Reload Linetypes** dialog box is displayed. Select **HIDDEN** from the list and choose the **OK** button from the **Load and Reload Linetypes** dialog box. Next, select **HIDDEN** from the **Select Linetype** dialog box and then choose **OK**.

9. Left-click on the **Lineweight** field of the **Hid** layer; the **Lineweight** dialog box is displayed. Select **0.008"** and then choose **OK**; the selected lineweight is assigned to the **Hid** layer.

10. Similarly, create the new layer **Cen** and assign the color **Green**, linetype **CENTER**, and lineweight **0.006"** to it.

11. Select the **Obj** layer and then choose the **Set Current** button to make the **Obj** layer current, see Figure 4-20. Choose the **Close** button to exit the dialog box.

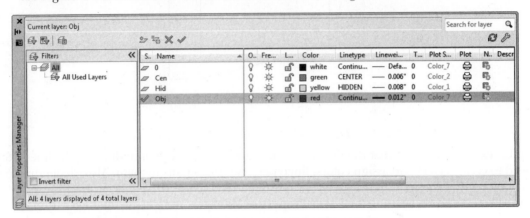

Figure 4-20 Layers created for Example 1

12. Choose the **Show/Hide Lineweight** button from the Status Bar to turn on the display of the lineweights of the lines to be drawn.

13. Choose the **Dynamic Input** button from the Status Bar to turn it on.

14. Choose the **Line** button from the **Draw** panel and draw solid lines in the drawing shown in Figure 4-19. Make sure that the start point of the line is at 9,1. You will notice that a continuous line is drawn in red color. This is because the **Obj** layer is the current layer and red color is assigned to it.

15. Click the down arrow in the **Layer** drop-down list available in the **Layers** panel to display the list of layers and then select the **Hid** layer from the list to make it current, as shown in Figure 4-21.

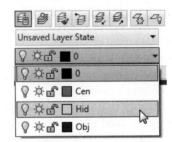

16. Draw two hidden lines; the lines are displayed in yellow color with hidden linetype.

17. Draw the center line; the centerline is displayed in yellow color with hidden linetype. This is because the **Hid** layer is the current layer.

*Figure 4-21 Selecting the **Hid** layer from the **Layers** panel to set it as the current layer*

18. Now, select the centerline and then select the **Cen** layer from the **Layer** drop-down list available in the **Layers** panel; the color and linetype of the centerline is changed.

EXERCISE 1   *Layers*

Set up layers with the following linetypes and colors. Then make the drawing (without dimensions) as shown in Figure 4-22. The distance between the dotted lines is 1 unit.

Layer Name	Color	Linetype
Object	Red	Continuous
Hidden	Yellow	Hidden
Center	Green	Center
Dimension	Blue	Continuous

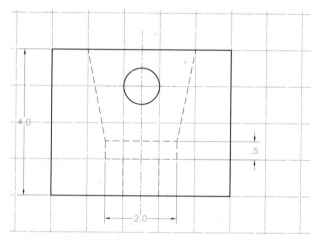

Figure 4-22 Drawing for Exercise 1

OBJECT PROPERTIES

Drawing entities have properties such as color, linetype, lineweight, and plot style based on the layer it is associated with. However, you can change these properties by using the **Properties** palette or the **Properties** panel in the **Home** tab of the **Ribbon**. By default, the **By Layer** option will be selected in the **Object Color**, **Lineweight**, and **Linetype** drop-down lists of the **Properties** panel in the **Ribbon**. Therefore, the properties set in the current layer will be applied to the entities in it. The procedure to change the properties of the selected objects is discussed next.

Changing the Color

Select the object whose color you want to change; the current color of the object will be displayed in the **Object Color** drop-down list. To change this color, select the down arrow in the **Object Color** drop-down list, as shown in Figure 4-23, and select a new color; the selected color will be applied to the object. If you want to assign a color that is not displayed in the list, select the **Select Colors** option; the **Select Color** dialog box will be displayed. You can select the desired color in this dialog box, and then choose the **OK** button.

To set a color other than the one set in the current layer as the current color, choose a color from the **Object Color** drop-down list without selecting an object. Now, all new objects will be drawn in this color.

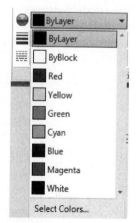

Figure 4-23 Different colors in the **Object Color** drop-down list

Changing the Linetype

To set a linetype other than the one set in the current layer as the current linetype, select a linetype from the **Linetype** drop-down list in the **Properties** panel from the **Home** tab, as shown in Figure 4-24. Now, all objects will be drawn in this linetype.

Chapter 4

To assign a linetype that is not displayed in the list, choose the **Other** option; the **Linetype Manager** dialog box will be displayed, as shown in Figure 4-25. By default, the **ByLayer**, **ByBlock**, and **Continuous** linetypes will be listed in this dialog box. To load other linetypes, choose the **Load** button; the **Load or Reload Linetypes** dialog box will be displayed. Select a linetype in this dialog box; they will be displayed in the **Linetype Manager** dialog box. Next, select the required linetype from this dialog box and choose **OK**. You can also make a linetype current by using the **Current** button in the **Linetype Manager** dialog box.

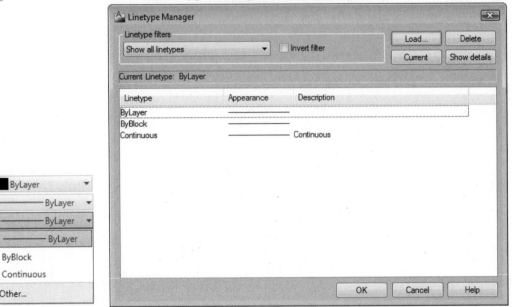

*Figure 4-24 The **Line type** drop-down list*

*Figure 4-25 The **Linetype Manager** dialog box*

Changing the Lineweight

To change the lineweight of a selected object, select a different lineweight value from the **Lineweight** drop-down list in the **Properties** panel. Also, you can set a new lineweight as current by selecting it from the list. To assign a lineweight value that is not listed in the drop-down list, select the **Lineweight Settings** option from the **Lineweight** drop-down list; the **Lineweight Settings** dialog box will be displayed, as shown in Figure 4-26. You can also invoke this dialog box by choosing the **Settings** option from the shortcut menu that is displayed on right-clicking on the **Hide/Show Lineweight** button on the Status Bar. In this dialog box, set the current lineweight for the objects. You can change the units for the lineweight and also the

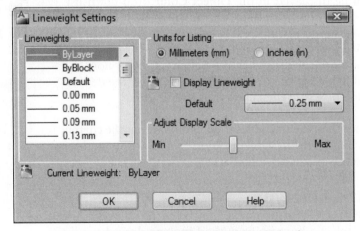

*Figure 4-26 The **Lineweight Settings** dialog box*

display of lineweights for the current drawing. Lineweights are displayed in pixel widths and depending on the lineweight value chosen, the lineweights are displayed if the **Display Lineweight** check box is selected. For a large drawing, displaying lineweight increases the regeneration time, and therefore the corresponding check box should be cleared. As mentioned earlier, you can also turn the display of lineweights to on or off directly from the Status Bar by choosing the **Hide/Show Lineweight** button. The value set in the **Adjust Display Scale** slider bar also affects the regeneration time. You can keep the slider bar at **Max** for getting a good display of different lineweights on the screen in the Model space; otherwise, keep it at **Min** for a faster regeneration.

Changing the Plot Style

The options in this drop-down list will be available only if you are using the named plot style modes. To change the plot style, select a plot style from the **Plot Style** drop-down list in the **Properties** panel. Also, you can set a new plot style as current by selecting it from the drop-down list.

Properties Palette

Ribbon: View > Palettes > Properties	**Command:** PROPERTIES, CH, MO
Toolbar: Standard > Properties	

The **Properties** palette is used to set the current properties and to change the general properties of the selected objects. The **Properties** palette (see Figure 4-27) is displayed on choosing the **Properties** button from the **Palettes** panel. It can also be invoked by right-clicking on an object and then choosing the **Properties** option from the shortcut menu. Right-clicking in the **Properties** palette displays a shortcut menu from where you can choose **Allow Docking** or **Hide** to dock or hide the palette.

When you select an object, the **Properties** palette displays the properties of the selected object. Depending on the object selected, the properties differ. If you change the properties in the **Properties** palette without selecting any object, the current properties will be changed in the **Properties** panel in the **Home** tab. Now, if you draw an object, it will have the properties set in the **Properties** palette.

When you select **Color** from the **General** list, the color of the selected object is displayed, along with a down arrow. Click the down arrow and select a color from the list to assign to the selected object. If you want to assign a color that is not displayed in the list, choose the **Select Color** option to display the **Select Color** dialog box. You can select the desired color in this dialog box, and then choose the **OK** button.

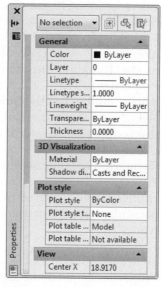

*Figure 4-27 The **Properties** palette*

Similarly, you can set current the linetype, lineweight, linetype scale, and other properties for the other objects individually. You can also assign a hyperlink to an object. The **Hyperlink** field in this window displays the name and description of the hyperlink, if any, assigned to the object. If there is no hyperlink attached to the object, this field will be blank. To add a hyperlink to a selected object, click on this field; the **[...]** button will be displayed in this field. Choose the **[...]** button; the **Insert Hyperlink** dialog box will be displayed where you can enter the path of the URL or the file that you want to link to the selected object.

Note

*You can enter the **CHPROP** command at the command line to change the properties of an object. You can also change the general properties of an object by using the **Properties** option of the **CHANGE** command.*

*The system variables **CECOLOR**, **CELTYPE**, and **CELWEIGHT** control the current color, linetype, and lineweight of an object. You can set the properties of objects current by using these variables from the command line.*

When you set the property of an object current, it does not consider the property of the layer, in which the object will be drawn.

EXERCISE 2   *Object Properties*

Draw a hexagon on the **Obj** layer in red. Keep the linetype hidden. Now, use the **Properties** palette to change the layer, the color to yellow, and the linetype to continuous.

Quick Properties Palette

Status Bar: Quick Properties

In AutoCAD, you can view some of the properties of the selected entity in the **Quick Properties** palette or in the **Properties** palette. To view the properties, keep the **Quick Properties** button chosen in the Status Bar and then select an entity; some of the properties of the selected entity will be displayed in the **Quick Properties** palette. Alternatively, enter **QP** at the Command prompt or the **Dynamic Input** to invoke the **Quick Properties** palette. You can change the properties that are displayed in the **Quick Properties** palette. Figure 4-28 shows the **Quick Properties** palette that is displayed on selecting a line.

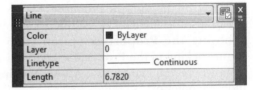

Figure 4-28 *The **Quick Properties** palette*

If you select multiple objects of different types, then in the drop-down list available in the **Quick Properties** palette, the name of the objects will be defined as **All** and the number of selected entities will be displayed in parenthesis. After selecting multiple entities, you can also select a particular type of object from the drop-down list to view the properties of the particular object type.

To set the properties displayed in the **Quick Properties** palette, right-click on the **Quick Properties** button in the Status Bar and choose the **Settings** option; the **Drafting Settings** dialog box will be displayed with the **Quick Properties** tab chosen. You can specify the display of the object type, the position of palette on invoking, and the height of palette by selecting a suitable option in this dialog box.

You can customize the **Quick Properties** palette such that it displays the required properties of a selected entity. To do so, choose the **Customize** button available next to the drop-down list; the **Customize User Interface** dialog box will be displayed. In the dialog box, different object types and their corresponding properties will be displayed on the right. If you select an object type, you will notice that the check boxes of some of the properties are selected. The properties, whose check boxes are selected, will be listed in the **Quick Properties** palette. You can also select the check boxes of the properties that you want to be displayed in the **Quick Properties** palette. If you place the cursor over an entity and pause, a tooltip, called the rollover tooltip, will be displayed. You can synchronize the properties displayed in the rollover tooltip and the **Quick Properties** panel. To do so, invoke the **Customize User Interface** dialog box, right-click on the **Rollover Tooltips** subnode in it, and choose the **Synchronize with Quick Properties** option.

GLOBAL AND CURRENT LINETYPE SCALING

The **LTSCALE** system variable controls the global scale factor of the lines in a drawing. For example, if **LTSCALE** is set to 2, all lines in the drawing will be affected by a factor of 2. Like **LTSCALE**, the **CELTSCALE** system variable controls the linetype scaling with the only difference that **CELTSCALE** determines the current linetype scaling. For example, if you set **CELTSCALE** to 0.5, all lines drawn after setting the new value for **CELTSCALE** will have the linetype scaling factor of 0.5. The value is retained in the **CELTSCALE** system variable. For example, line (a) in Figure 4-29 is drawn with a **CELTSCALE** factor of 1 and line (b) is drawn with a **CELTSCALE** factor of 0.5. The length of the dash is reduced by a factor of 0.5 when **CELTSCALE** is 0.5. The net scale factor is equal to the product of **CELTSCALE** and **LTSCALE**.

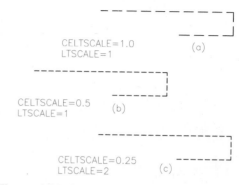

Figure 4-29 Using CELTSCALE to control current linetype scaling

Figure 4-29(c) shows a line that is drawn with **LTSCALE** of 2 and **CELTSCALE** of 0.25. The net scale factor = **LTSCALE** X **CELTSCALE** = 2 X 0.25 = 0.5. You can also change the global and current scale factors by entering a desired value in the **Linetype Manager** dialog box. If you choose the **Show details** button, the properties associated with the selected linetype will be displayed, as shown in Figure 4-30. You can change the values according to your drawing requirements.

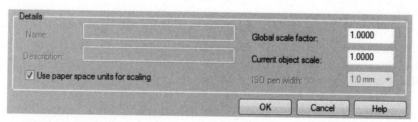

*Figure 4-30 The **Details** area of the **Linetype Manager** dialog box*

LTSCALE FACTOR FOR PLOTTING

The **LTSCALE** factor for plotting depends on the size of the sheet you use to plot the drawing. For example, if the limits are 48 by 36, the drawing scale is 1:1, and to plot the drawing on a 48" by 36" size sheet, then the **LTSCALE** factor is 1. If you check the specification of the **Hidden** linetype in the *acad.lin* file, the length of each dash is 0.25. Hence, when you plot a drawing with 1:1 scale, the length of each dash in a hidden line will be 0.25.

However, if the drawing scale is 1/8" = 1' and you want to plot the drawing on 48" by 36" paper, the **LTSCALE** factor must be 8 X 12 = 96. The length of each dash in the hidden line will increase by a factor of 96, because the **LTSCALE** factor is 96. Therefore, the length of each dash will be (0.25 X 96 = 24) units. At the time of plotting, the scale factor must be 1:96 to plot the 384' by 288' drawing on 48" by 36" paper. Each dash of the hidden line that was 24" long on the drawing will be 24/96 = 0.25" long when plotted. Similarly, if the desired text size on the paper is 1/8", the text height in the drawing must be 1/8 X 96 = 12".

LTSCALE factor for PLOTTING = Drawing Scale

Sometimes your plotter may not be able to plot a 48" by 36" drawing, or you might like to decrease the size of the plot so that the drawing fits within the specified area. To get the correct dash length for hidden, center, or other lines, you must adjust the **LTSCALE** factor. For example, if you want to plot the previously mentioned drawing in a 45" by 34" area, the correction factor is:

$$
\begin{aligned}
\text{Correction factor} \quad &= 48/45 \\
&= 1.0666
\end{aligned}
$$

$$
\begin{aligned}
\text{New } \textbf{LTSCALE} \text{ factor} \quad &= \textbf{LTSCALE} \text{ factor x Correction factor} \\
&= 96 \times 1.0666 \\
&= 102.4
\end{aligned}
$$

New **LTSCALE** factor for PLOTTING = Drawing Scale x Correction Factor

Note

*If you change the **LTSCALE** factor, all lines in the drawing will be affected by the new ratio.*

Changing the Linetype Scale Using the Properties Palette

You can also change the linetype scale of an object by using the **Properties** palette. To do so, select the object and then invoke the **PROPERTIES** command to display the **Properties** palette. Next, locate the **Linetype scale** edit box in the **General** list and enter a new linetype scale in the edit box. The linetype scale of the selected objects is changed to the new value entered.

WORKING WITH THE DesignCenter

Ribbon:	View > Palettes > DesignCenter	**Command:**	ADCENTER
Toolbar:	Standard > DesignCenter		

 The **DesignCenter** allows you to reuse and share contents in different drawings. You can use this palette to locate the drawing data with the help of search tools and then use it in your drawing. You can insert layers, linetypes, blocks, layouts, external references, and other drawing content in any number of drawings. Hence, if a layer is created once, it can be used repeatedly any number of times.

Open the **DesignCenter** palette and then choose the **Tree View Toggle** button to display the **Tree pane** and the **Palette** side by side (if they are not already displayed). Open the folder that contains the drawing having the required layers and linetypes to be inserted. You can use the **Load** and **Up** buttons to open the folder. For example, to insert some layers and linetypes from the drawing *c4d1* saved in the *c:\AutoCAD 2012\c04_acad_2012* location, browse to the folder and open *c4d1* to display its contents. Select **Layers**; the layers created in *c4d1* will be displayed in the **Palette**. Press and hold the CTRL key and select the **Border** and **CEN** layers. Right-click to display the shortcut menu and choose **Add Layer(s)**, as shown in Figure 4-31; the two layers are added to the current drawing. You can also drag and drop the desired layers into the current drawing. If you open the drop-down list in the **Layers** toolbar, you will notice that the two layers are listed there. Similarly, you can insert the other elements such as linetypes, blocks, dimension styles, and so on. You will learn more about **DesignCenter** in later chapters.

DRAFTING SETTINGS DIALOG BOX

Command:	DESETTINGS

You can use the **Drafting Settings** dialog box to set the drawing modes such as Grid, Snap, Object Snap, Polar, Object Snap tracking, and Dynamic Input. All these modes help you draw accurately and also increase the drawing speed. You can right-click on the **Snap Mode**, **Grid Display**, **Ortho**

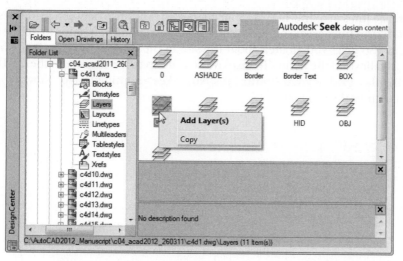

Figure 4-31 The *DesignCenter* with the shortcut menu

Mode, **Polar Tracking**, **Object Snap**, **3D Object Snap**, **Object Snap Tracking**, **Dynamic Input**, **Quick Properties**, or **Selection Cycling** button in the Status Bar to display a shortcut menu. In this shortcut menu, choose **Settings** to display the **Drafting Settings** dialog box, as shown in Figure 4-32. This dialog box has seven tabs: **Snap and Grid**, **Polar Tracking**, **Object Snap**, **3D Object Snap**, **Dynamic Input**, **Quick Properties**, and **Selection Cycling**. On starting AutoCAD, these tabs have default settings. You can change them according to your requirements.

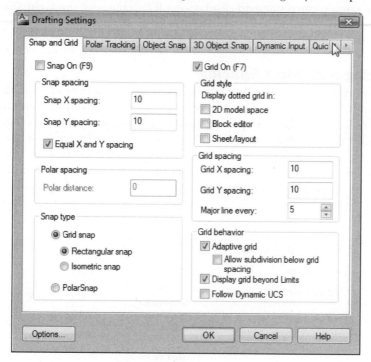

Figure 4-32 The *Drafting Settings* dialog box

Setting Grid

Grid lines are the checked lines on the screen at predefined spacing, see Figure 4-33. In AutoCAD 2012, by default, the grids are displayed as checked lines. In the earlier versions, the grids were displayed as dotted lines, see Figure 4-34. To display the grids as dotted lines, set the **2D model space** check box in the **Grid style** area. These dotted lines act as a graph that can be used as reference lines in a drawing. You can change the distance between grid lines as per your requirement. If grid lines are displayed within the drawing limits, it helps to define the working area. The grid also gives you an idea about the size of the drawing objects. By default, the grid will be displayed beyond the limits. To display the grid up to the limits, use the **GRID** command and set the **Display grid beyond Limits [Yes/No] <Yes>:** option to **No**. Now, the grids will be displayed only up to the limits set.

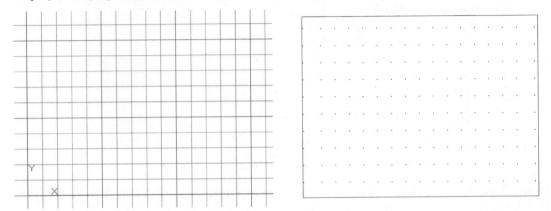

Figure 4-33 *Grid as checked lines*   **Figure 4-34** *Grid as dots*

Grid On (F7): Turning the Grid On or Off

You can turn the grid display on/off by using the **Grid On** check box in the **Drafting Settings** dialog box. You can also turn the grid display on/off by choosing the **Grid Display** button in the Status Bar or by using the F7 key. **Grid Display** is a toggle button. When the grid display is turned on after it has been turned off, the grid is set to the previous grid spacing.

Grid X Spacing and Grid Y Spacing

The **Grid X spacing** and **Grid Y spacing** edit boxes in the **Drafting Settings** dialog box are used to define the desired grid spacing along the X and Y axes. For example, to set the grid spacing to 0.5 units, enter **0.5** in the **Grid X spacing** and **Grid Y spacing** edit boxes. You can also enter different values for the horizontal and vertical grid spacing, see Figure 4-35. If you specify only the grid X spacing value and then choose the **OK** button in the dialog box, the corresponding Y spacing value will automatically be set to match the X spacing value. Therefore, if you want different X and Y spacing values as shown in Figure 4-36, clear the **Equal X and Y spacing** check box in the **Snap spacing** area, and enter different X and Y spacing values.

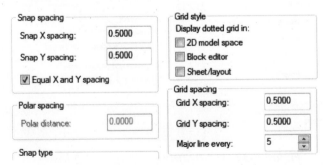

Figure 4-35 *The **Grid Spacing** area of the **Drafting Settings** dialog box*

Note
Grids are especially effective in drawing when the objects in the drawing are placed at regular intervals.

Setting Snap

After displaying the grid, you need to switch on the snap mode so that the cursor snaps to the snap point. The snap points are invisible points (see Figure 4-37) that are created at the intersection of invisible horizontal and vertical lines. A snap point is independent of the grid spacing and the two can have equal or different values. Therefore, when the snap mode is on, the cursor moves in specified intervals from one point to another. You can turn on the snap mode even when the grid lines are invisible. To set snap spacing, enter the X and Y values in the corresponding edit boxes in

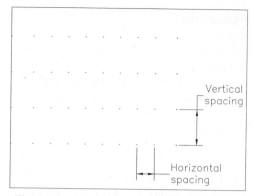

Figure 4-36 Creating unequal grid spacing

the **Snap spacing** area of the **Snap and Grid** tab in the **Drafting Settings** dialog box, as shown in Figure 4-35.

You can turn the snap mode on/off by selecting the **Snap On** check box from the **Drafting Settings** dialog box. You can also turn the snap on or off by choosing the **Snap Mode** button in the Status Bar, or from the shortcut menu displayed by right-clicking on the **Snap Mode** button in the Status Bar, or by using the function key F9 as a toggle key.

Snap Type

There are two snap types, **PolarSnap** and **Grid snap**. On selecting **Grid snap**, the cursor snaps along the grid. The **Grid snap** is either of the **Rectangular snap** type or of the **Isometric snap** type. Rectangular snap is the default snap and has been discussed earlier. The other types are discussed next.

Isometric Snap

The isometric mode is used to make isometric drawings. In isometric drawings, the isometric axes are at angles 30, 90, and 150 degrees. The isometric snap/grid enables you to display the grid lines along these axes, see Figure 4-38. Select the **Isometric snap** radio button in the **Snap type** area of the **Drafting Settings** dialog box to set the snap grid to the isometric mode. The default mode is off (standard). Once you select the **Isometric snap** radio button and choose **OK** in the dialog box, the cursor aligns with the isometric axis. You will notice that the X spacing is not available for this option. You can change the vertical snap and grid spacing by entering values in the **Snap Y spacing** and **Grid Y spacing** edit boxes. While drawing in isometric mode, you can adjust the cursor orientation to the left, top, or right plane of the drawing by using the F5 key. The procedure to create drawings using the **Isometric snap** option is discussed in later chapters.

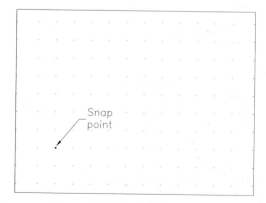

Figure 4-37 Invisible Snap grid

Figure 4-38 Isometric Snap grid

PolarSnap

The polar snap is used to snap points at a specified distance along the polar alignment angles. To set the snap mode as polar, select the **PolarSnap** radio button in the **Snap type** area; the **Polar distance** edit box will be enabled and the snap spacing options will be disabled. Enter a value for the distance in the **Polar distance** edit box and choose the **OK** button. If this value is zero, AutoCAD takes the same value as specified for Snap X spacing earlier. Now, if you draw a line, the cursor snaps along an imaginary line according to the Polar tracking angles that are relative to the last point selected. The polar tracking angle can be set in the **Polar Tracking** tab of the **Drafting Settings** dialog box (discussed later in this chapter).

For example, select the **PolarSnap** radio button and enter **0.5** in the **Polar distance** edit box. Choose the **Polar Tracking** tab and enter **30** in the **Increment angle** drop-down list and then choose the **OK** button. Choose the **Snap Mode** and **Polar Tracking** buttons in the Status Bar. Invoke the **Line** tool and select the start point anywhere on the screen. Now if you move the cursor at angles in multiples of 30-degree, a dotted line will be displayed and if you move the cursor along this dotted line, a small cross mark will be displayed at a distance of 0.5 (polar snap).

DRAWING STRAIGHT LINES USING THE ORTHO MODE

The **Ortho** mode is used to draw lines at right angles only. You can turn the Ortho mode on or off by choosing the **Ortho Mode** button in the Status Bar, by using the function key F8, or by using the **ORTHO** command. When the **Ortho** mode is on and you move the cursor to specify the next point, a rubber-band line is connected to the cursor in horizontal (parallel to the *X* axis) or vertical (parallel to the *Y* axis) direction. To draw a line in the Ortho mode, specify the start point at the **Specify first point** prompt. To specify the second point, move the cursor and specify a desired point. The line drawn will be either vertical or horizontal, depending on the direction in which you move the cursor, see Figures 4-39 and 4-40.

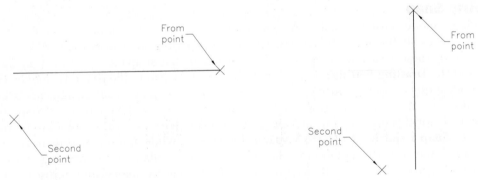

Figure 4-39 *Drawing a horizontal line using the ORTHO mode*

Figure 4-40 *Drawing a vertical line using the ORTHO mode*

Tip
*You can use the buttons in the Status Bar (at the bottom of the graphics area) to toggle between on or off for the different drafting functions like **Snap**, **Grid**, and **Ortho**.*

WORKING WITH OBJECT SNAPS

Toolbar: Object Snap

Object snaps are one of the most useful features of AutoCAD. They improve your performance and the accuracy of the drawing. Also, drafting is much simpler than it normally would be. The term object snap refers to the cursor's ability to snap exactly to a geometric point on an object.

The advantage of using object snaps is that you do not have to specify an exact point. For example, to snap the midpoint of a line, use the **Midpoint** object snap and move the cursor closer to the object; a marker (in the form of a geometric shape, a triangle for Midpoint) will automatically be displayed at the mid point. Click to snap the mid point.

Object snaps recognize only the objects that are visible on the screen, which include the objects on the locked layers. The objects on the layers that are turned off or frozen are not visible, and they cannot be used for object snaps.

Object snapping can be done by using the **Object Snap modes** shortcut menu. To invoke this shortcut menu, choose a tool, press and hold the SHIFT key, and then right-click, see Figure 4-41. You can also do object snapping by using the **Object Snap** toolbar shown in Figure 4-42. Alternatively, you can invoke the **Object Snap modes** shortcut menu when you are inside any other sketching command. For example, invoke the **Center,Radius** tool from the **Draw** panel and then right-click to invoke the shortcut menu. From the shortcut menu, choose the **Snap Overrides** option to display the **Object Snaps** shortcut menu. The following are the object snap modes in AutoCAD.

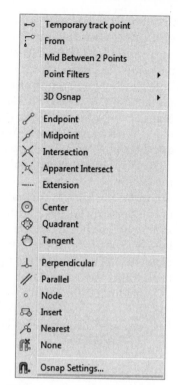

Figure 4-41 The **Object snap modes** shortcut menu

Endpoint	Center
Perpendicular	Nearest
Midpoint	Quadrant
Parallel	None
Intersection	Tangent
Insert	From
Apparent Intersection	Extension
Node	
Midpoint Between two points	

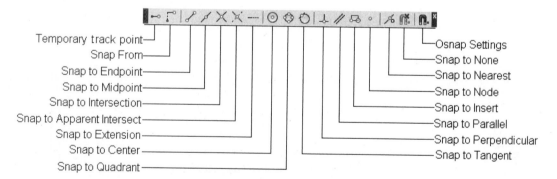

Figure 4-42 The **Object Snap** toolbar

AutoSnap

The AutoSnap feature controls various characteristics for the object snap. As you move the target box over the object, AutoCAD displays the geometric marker corresponding to the shapes shown in the **Object Snap** tab of the **Drafting Settings** dialog box. You can change different AutoSnap settings such as attaching a target box to the cursor when you invoke any object snap, or change the size and color of the marker. These settings can be changed from the **Options** dialog box (**Drafting** tab). You can invoke the **Options** dialog box from the **Application Menu**, or by entering **OPTIONS** at the Command prompt. You can also invoke the dialog box by choosing the **Options** button in the **Drafting Settings** dialog box. After invoking the **Options** dialog box, choose the **Drafting** tab in this dialog box; the **AutoSnap Settings** will be displayed, see Figure 4-43. Select the **Marker** check box to toggle the display of the marker. Select the **Magnet** check box to toggle the magnet that snaps the crosshairs to the particular point of the object for that object snap. You can use the **Display AutoSnap tooltip** and **Display AutoSnap aperture box** check boxes to toggle the display of the tooltip and the aperture box. Note that if the **Display AutoSnap tooltip** check box is selected in the **Options** dialog box and you move the cursor over an entity, after invoking a tool, the autosnap marker will be displayed. Now, do not move the cursor; the name of the object snap will be displayed. You can change the size of the marker and the aperture box by moving the **AutoSnap Marker Size** and **Aperture Size** slider bars, respectively. You can also change the color of the markers by choosing the **Colors** button. The size of the aperture is measured in pixels (short form of picture elements). Picture elements are dots that make up the screen picture. The aperture size can also be changed using the **APERTURE** command. The default value for the aperture size is 10 pixels. The display of the marker and the tooltip is controlled by the **AUTOSNAP** system variable. The following are the bit values for **AUTOSNAP**.

Bit Values	Functions
0	Turns off the Marker, AutoSnap Tooltip, and Magnet
1	Turns on the Marker
2	Turns on the AutoSnap Tooltip
4	Turns on the Magnet
47	Turns on the Marker, AutoSnap Tooltip, and Magnet

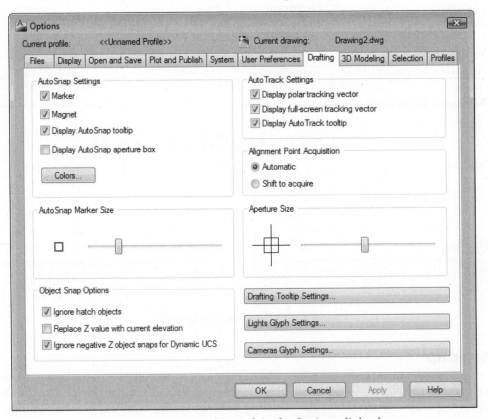

Figure 4-43 The **Drafting** tab in the **Options** dialog box

The basic functionality of Object Snap modes is discussed next.

Endpoint

The **Endpoint** Object Snap mode is used to snap the cursor to the closest endpoint of a line or an arc. To use this Object Snap mode, invoke a draw tool such as **Line** and select the **Endpoint** button and move the cursor (crosshairs) anywhere close to the endpoint of the object; a marker will be displayed at the endpoint. Click to select that point; the endpoint of the object will be snapped. If there are several objects near the cursor crosshairs, the endpoint of the object that is closest to the crosshairs will be snapped. However, if the **Magnet** is on, you can move to grab the desired endpoint. Figure 4-44 shows the cursor snapping the end point of a circle. The dotted line is the proposed line to be drawn.

Midpoint

The **Midpoint** Object Snap mode is used to snap the midpoint of a line or an arc. To use this Object Snap mode, select **Midpoint** osnap and select an object anywhere; the midpoint of the object will be snapped. Figure 4-45 shows the cursor snapping to the midpoint of a line. The dotted line is the proposed line to be drawn.

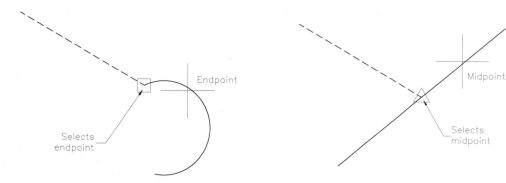

Figure 4-44 The **Endpoint** Object Snap mode   **Figure 4-45** The **Midpoint** Object Snap mode

Nearest

The **Nearest** Object Snap mode is used to select a point on an object (line, arc, circle, or ellipse) that is visually closest to the crosshairs. To use this mode, invoke a tool, choose the **Nearest** object snap, and move the crosshairs near the intended point on the object; a marker will be displayed. Left-click to snap that point. Figure 4-46 shows the cursor snapping the point in a line. The dotted line is the proposed line to be drawn.

Center

The **Center** Object Snap mode is used to snap the cursor to the center point of an ellipse, circle, or arc. After choosing this option, move the cursor on the circumference of a circle or an arc; a marker will be displayed at the center of the circle. Left-click to snap the center of the circle. Figure 4-47 shows the cursor snapping the center point of a circle. The dotted line is the proposed line to be drawn.

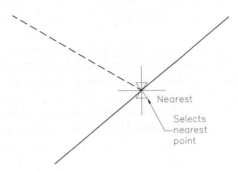

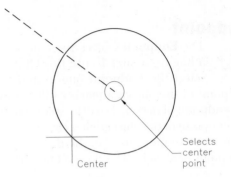

Figure 4-46 *The **Nearest** Object Snap mode*   Figure 4-47 *The **Center** Object Snap mode*

Tangent

The **Tangent** Object Snap mode is used to snap the cursor to the tangent point of an existing ellipse, circle, or arc. To use this object snap mode, place the cursor on the circumference of a circle or an arc and select it. Figure 4-48 shows the cursor selecting the circle. The dashed line is the proposed tangential line.

Note

*If the start point of a line is defined by using the **Tangent** Object Snap, the tip shows **Deferred Tangent**. However, if you end the line using this Object Snap, the tip shows **Tangent**.*

Figure 4-49 shows the use of the **Nearest**, **Endpoint**, **Midpoint**, and **Tangent** Object Snap modes.

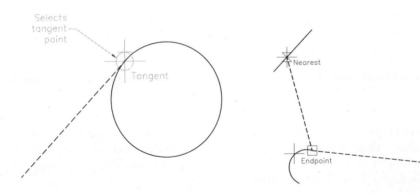

Figure 4-48 *The **Tangent** Object Snap mode*   Figure 4-49 *Using the **Nearest**, **Endpoint**, **Midpoint**, and **Tangent** Object Snap modes*

Quadrant

The **Quadrant** Object Snap mode is used to snap the cursor to the quadrant point of an ellipse, an arc, or a circle. A circle has four quadrants, and each quadrant subtend an angle of 90-degree. Therefore, the quadrant points are located at 0, 90, 180, and 270-degree positions. If the circle is inserted as a block (see Chapter 16) and rotated, the quadrant points will also be rotated by the same degree, see Figures 4-50 and 4-51.

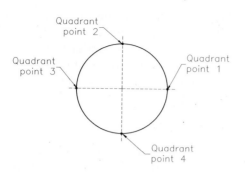

Figure 4-50 *Location of the circle quadrants*

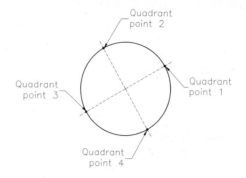

Figure 4-51 *Quadrants in a rotated circle*

To use this object snap, position the cursor on the circle or arc closest to the desired quadrant and click when the marker is displayed. Figure 4-52 shows the cursor selecting the third quadrant and the proposed line.

Intersection

The **Intersection** Object Snap mode is used to snap the cursor to a point where two or more lines, circles, ellipses, or arcs intersect. To use this object snap, move the cursor to the desired intersection point such that the intersection is within the target box and click when the intersection marker is displayed. Figure 4-53 shows the cursor selecting the intersection point and the proposed line is in dotted line.

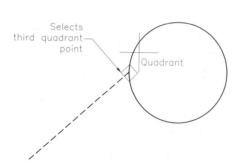

Figure 4-52 *The **Quadrant** Object Snap mode*

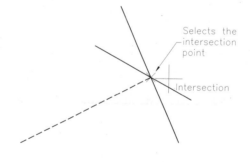

Figure 4-53 *The **Intersection** Object Snap mode*

After setting the **Intersection** Object Snap mode, if the cursor is close to an object and not close to actual intersection, the **Extended Intersection** tooltip will be displayed. If you select this object, you need to select another object. Move the cursor close to another object; the intersection point will be snapped. Left-click to select the intersection point. But, if the object selected second does not intersect with the object selected first, then the cursor will snap to the extended intersection point, as shown in Figure 4-54. The extended intersections are the intersections that do not exist, but are imaginary and formed if a line or an arc is extended.

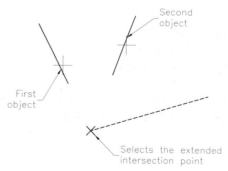

Figure 4-54 *The **Extended Intersection** Object Snap mode*

Chapter 4

Apparent Intersection

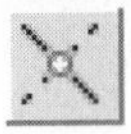

The **Apparent Intersection** Object Snap mode is used to select the projected intersections of two objects in 3D space. Sometimes, two objects appear to intersect one another in the current view, but in 3D space they do not actually intersect. The **Apparent Intersection** snap mode is used to select such intersections. This mode works on wireframes in 3D space (See Chapter 25 for wireframe models). If you use this object snap mode in 2D, it works like an extended intersection.

Perpendicular

The **Perpendicular** Object Snap mode is used to draw a line perpendicular to or from another line, or normal to or from an arc or a circle, or to an ellipse. After invoking the **Line** tool, if you choose this mode and then select an object, a perpendicular line will be attached to the cursor. Move the cursor and place the line, as shown in Figure 4-55. But, after invoking the **Line** tool, if you select the start point of the line first and then choose the **Perpendicular** snap mode, you need to select an object. On doing so, AutoCAD selects a point such that the resulting line is perpendicular to the selected object, as shown in Figure 4-56.

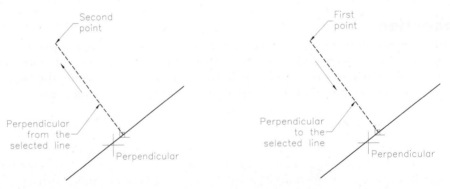

Figure 4-55 Selecting the perpendicular snap first

Figure 4-56 Selecting the start point and then the perpendicular snap

Figure 4-57 shows the use of the various object snap modes.

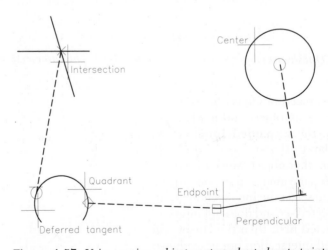

Figure 4-57 Using various object snap modes to locate points

EXERCISE 3   Quadrant & Tangent

Draw the sketch shown in Figure 4-58. P1 and P2 are the center points of the top and bottom arcs. The space between the dotted lines is 1 unit.

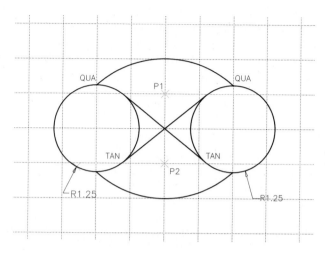

Figure 4-58 *Drawing for Exercise 3*

Node

You can use the **Node** Object Snap mode to snap the points placed by using the **Multiple Points**, **Divide**, or **Measure** command. Figure 4-59 shows the three points placed by using the **Multiple Points** command, the dotted line indicates the proposed line to be drawn, and the **Node** tooltip displayed on moving the cursor close to a point.

Insertion

The **Insertion** Object Snap mode is used to snap to the insertion point of a text, shape, block, attribute, or attribute definition. The insertion point is the point with respect to which a text, shape, or a block is inserted. Figure 4-60 shows the insertion point of the text and a block. The dotted line is the proposed line to be drawn.

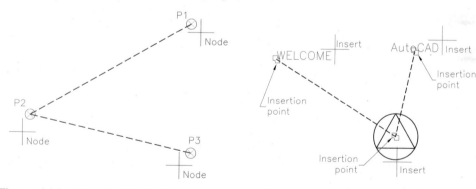

Figure 4-59 *Using the Node object snap*   *Figure 4-60* *The Insertion object snap*

Snap to None

The **Snap to None** Object Snap mode is used to turn off any running object snap (see the section "Running Object Snap Mode" that follows) for one point only. The following example illustrates the use of this Object Snap mode.

Invoke the **Drafting Settings** dialog box and select the **Object Snap** tab. Select the **Midpoint** and **Center** check boxes. This sets the object snap to midpoint and center. Now, to draw a line whose start point is closer to the endpoint of another line, invoke the **Line** tool and move the cursor to the desired position on the previous line. The cursor automatically snaps to the midpoint of the line. You can disable this automatic snap by choosing the **Snap to None** object snap from the **Object Snap** toolbar.

Chapter 4

Parallel

 When you need to draw a line parallel to a line or a polyline, use the **Parallel** Object Snap. For example, when you are in the middle of the **Line** tool, and you have to draw a line parallel to the one already drawn, you can use the **Parallel** object snap mode as discussed next.

1. Choose **Line** from the **Draw** toolbar.
2. Select a point on the screen.
3. Choose the **Parallel** button from the **Object Snap** toolbar.
4. Move the cursor close to the reference object and pause for a while; the parallel tooltip appears on it, indicating that the line has been selected.
5. Now, move the cursor at an angle parallel to the line; an imaginary parallel line (construction line) appears, on which you can select the next point. As you move the cursor on the construction line, a tooltip with relative polar coordinates is displayed near the cursor. This helps you to select the next point. The line thus drawn is parallel to the selected object, as shown in Figure 4-61.

Extension

The **Extension** Object Snap mode is used to locate a point on the proposed extension path of a line or an arc (Figure 4-62). It can also be used with intersection to determine the extended intersection point. To snap a point by using this snap mode, choose the **Line** tool from the **Draw** toolbar, and then choose the **Snap to Extension** button from the **Object Snap** toolbar. Briefly pause at the end of the line or the arc; a small plus sign (+) or a tooltip will appear at the end of the line or the arc, indicating that it has been selected. If you move the cursor along the line, a temporary extension path will be displayed and the tooltip displays the relative polar coordinates from the end of the line. Select a point or enter a distance to begin a line and then select another point to finish the line.

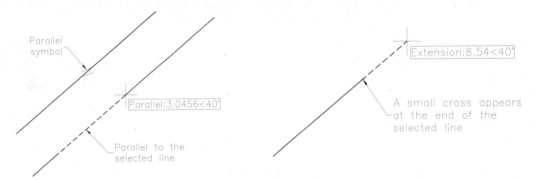

Figure 4-61 *Using the **Parallel** Object Snap mode*

Figure 4-62 *Using the **Extension** Object Snap mode*

Tip

*While using the **Parallel** object snap mode, you can choose more than one line as reference lines that are indicated by a plus sign. Depending upon the direction in which you move the cursor, AutoCAD will choose any one of these reference lines and the chosen line will be marked with the parallel symbol in place of the plus sign.*

*Similarly, while using the **Extension** object snap mode, you can choose more than one endpoint as the reference.*

From

 The **From** Object Snap mode is used to locate a point relative to a given point (Figure 4-63). For example, to locate a point that is 2.5 units up and 1.5 units right from the endpoint of a given line, you can use the **From** object snap as follows:

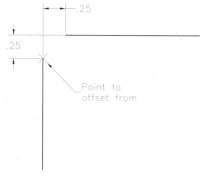

Command: *Choose the **Line** tool.*
Specify first point: *Choose the **Snap From** button from the **Object Snap** toolbar.*
_from Base point: *Choose the **Snap to Endpoint** button from the **Object Snap** toolbar.*
_endp of *Specify the endpoint of the given line*
<Offset>: **@1.5,2.5**

Note
*The **From** Object Snap mode cannot be used as the running object snap.*

*Figure 4-63 Using the **From** Object Snap mode to locate a point*

Midpoint Between 2 Points

This Object Snap mode is used to select the midpoint of an imaginary line drawn between two selected points. Note that this Object Snap mode can only be invoked from the shortcut menu. To understand the working of this Object Snap mode, refer to the sketch shown in Figure 4-64. In this sketch, there are two circles and you need to draw another circle with the center point at the midpoint of an imaginary line drawn between the center points of two existing circles. The following is the prompt sequence:

Command: *Choose the **Center,Radius** tool*
Specify center point for circle or [3P/2P/Ttr (tan tan radius)]: *Right-click and choose **Snap Overrides > Midpoint Between 2 Points** from the shortcut menu*
_m2p First point of mid: *Snap the center of the left circle*
Second point of mid: *Snap the center of the right circle*
Specify radius of circle or [Diameter] <current>: *Specify the radius of the new circle*

Temporary Tracking Point

The **Temporary track point** is used to locate a point with respect to two different points. If you choose this option and then select a point and move the cursor, an orthogonal imaginary line will be displayed either horizontally or vertically. Then, you need to select another point to display another orthogonal imaginary line in the other direction. The required point will be located where these two imaginary lines intersect. You can also use this option along with the other Object Snap modes. To locate a point, invoke the **Line** tool from the **Draw** toolbar and then use the temporary tracking as follows.

Specify first point: *Choose the **Temporary track point** button from the **Object Snap** toolbar.*
Specify temporary OTRACK point: *Choose the **Snap to Midpoint** button from the **Object Snap** toolbar.*
Select the midpoint of the line and move the cursor horizontally toward the right.
Specify first point: *Choose the **Temporary Tracking Point** button from the **Object Snap** toolbar.*
Specify temporary OTRACK point: *Choose the **Snap to Endpoint** button from the **Object Snap** toolbar.*
Select the upper endpoint of the line and move the cursor vertically down.

As you move the cursor down, both the horizontal and vertical imaginary lines will be displayed, as shown in Figure 4-65. Select their intersection point and then draw the line.

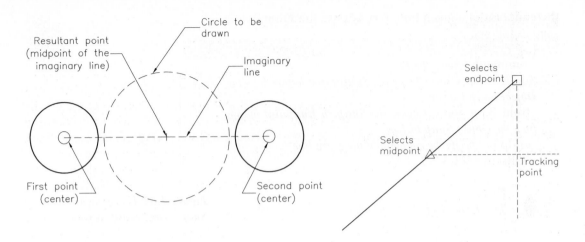

Figure 4-64 *Using the Midpoint Between 2 Points Object Snap mode to locate a point*

Figure 4-65 *Using temporary tracking point*

Combining Object Snap Modes

You can also combine the snaps from the command line by separating the snap modes with a comma. In this case, AutoCAD searches for the specified modes and grabs the point on the object that is closest to the point where the object is selected. The prompt sequence for using the **Midpoint** and **Endpoint** object snaps is given next.

Command: *Choose the* **Line** *tool*
_line Specify first point: **MID, END** [Enter]   *(MIDpoint or ENDpoint object snap.)*
of: *Select the object*

Note
In reference to object snaps, "line" generally includes xlines, rays, and polyline segments, and "arc" generally includes polyarc segments.

EXERCISE 4

Draw the sketch shown in Figure 4-66. The space between the dotted lines is 1 unit.

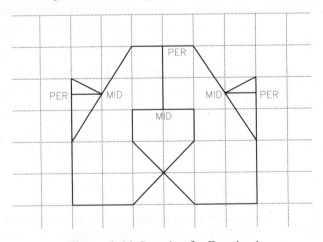

Figure 4-66 *Drawing for Exercise 4*

RUNNING OBJECT SNAP MODE

Toolbar: Object Snap > Osnap Settings   **Command:** OSNAP

In the previous sections, you have learned to use object snaps to snap to different points of an object. The object snaps are to be selected from the toolbar or from the shortcut menu. One of the drawbacks of this method is that you have to select them every time, even if you are using the same snap mode again. This problem can be resolved by using **running object snaps**. The Running Osnap can be invoked from the **Object Snap** tab of the **Drafting Settings** dialog box (Figure 4-67). If you choose the **Osnap Settings** button from the toolbar, or enter **OSNAP** at the Command line, or choose **Settings** from the shortcut menu displayed on right-clicking on the **OSNAP** button in the Status Bar, the **Object Snap** tab will be displayed in the **Drafting Settings** dialog box. In this tab, you can set the running object snap modes by selecting the check boxes next to the snap modes. For example, to set **Endpoint** as the running Object Snap mode, select the **Endpoint** check box and then the **OK** button in the dialog box.

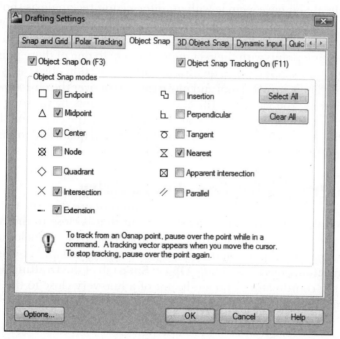

*Figure 4-67 The **Drafting Settings** dialog box (**Object Snap** tab)*

Once you set the running Object Snap mode, a marker will be displayed when you move the crosshairs over the key points. If you have selected a combination of modes, AutoCAD selects the mode that is closest to the crosshairs. For example, if you have selected the **Endpoint**, **Midpoint**, and **Center** check boxes in the dialog box and move the cursor over a line, the endpoint or the midpoint will be displayed, depending upon the position of the crosshair. Similarly, if you place the cursor over the circumference of a circle, the center will be displayed. The running Object Snap mode can be turned on or off without losing the object snap settings by choosing the **OSNAP** button in the Status Bar. You can also accomplish this by pressing the function key F3 or CTRL+F keys.

Overriding the Running Snap

When you select the running object snaps, all other Object Snap modes are ignored unless you select another Object Snap mode. Once you select a different osnap mode, the running OSNAP mode is temporarily overruled. After the operation has been performed, the running OSNAP mode becomes active again. If you want to discontinue the current running Object Snap mode completely, choose the **Clear all** button in the **Drafting Settings** dialog box. If you want to temporarily disable the running object snap, choose the **OSNAP** button (off position) in the Status Bar.

If you override the running object snap modes for a point selection and do not find a point to satisfy the override Object Snap mode, AutoCAD displays a message to this effect. For example, if you specify an override Object Snap mode of Center and no circle, ellipse, or arc is found at that location, AutoCAD will display the message "**No center found for specified point. Point or option keyword required**".

Cycling through Snaps

AutoCAD displays the geometric marker corresponding to the shapes shown in the **Object snap settings** tab of the **Drafting Settings** dialog box. You can use the TAB key to cycle through the snaps. For example, if you have a circle with an intersecting rectangle as shown in Figure 4-68 and you want to snap to one of the geometric points on the circle, you can use the TAB key to cycle through geometric points. The geometric points for a circle are the center point, quadrant points, and intersecting points with the rectangle. To snap to one of these points, first you need to set the running object snaps (center, quadrant, and intersection object snaps) in the **Drafting Settings** dialog box. Then, invoke a tool and move the cursor over the objects; AutoSnap displays a marker and a tooltip. You can cycle through the snap points by pressing the TAB key. For example, if you press the TAB key while the aperture box is on the circle and the rectangle (near the lower left intersection point), the intersection, center, and quadrant points will be displayed one by one. Left-click to select the required key point

Setting the Priority for Coordinate Entry

Sometimes you may want the keyboard entry to take precedence over the running Object Snap modes. This is useful when you want to locate a point that is close to the running osnap. By default, when you specify the coordinates (by using the keyboard) of a point located near a running osnap, AutoCAD ignores the point and snaps to the running osnap. For example, if you have selected the **Intersection** object snap in the **Object Snap** tab of the **Drafting Settings** dialog box and if you enter the coordinates of the endpoint of a line very close to the intersection point (Figure 4-69a), the line will snap to the intersection point and ignore the keyboard entry point.

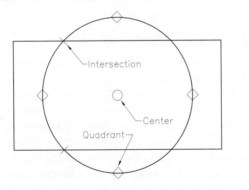

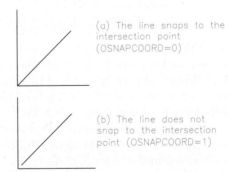

Figure 4-68 Using the TAB key to cycle through the snaps

Figure 4-69 Setting the coordinate entry priority

Therefore, to set the priority between the keyboard entry and the object snap, select the **Keyboard entry** radio button in the **User Preferences** tab of the **Options** dialog box, see Figure 4-70, and then choose **OK**. Now, if you enter the coordinates of the endpoint of a line very close to the intersection point (Figure 4-69b), the start point will snap to the coordinates specified and the running osnap (intersection) is ignored. This setting is stored in the **OSNAPCORD** system variable and you can also set the priority through this variable. A value **0** gives priority to running osnap, value **1** gives priority to the keyboard entry, and value **2** gives priority to the keyboard entry except scripts.

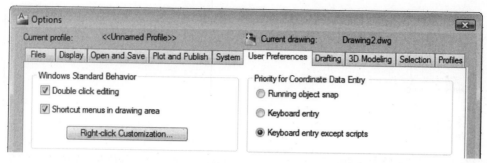

Figure 4-70 Setting priority for coordinate data entry

Tip

*If you do not want the running osnap to take precedence over the keyboard entry, you can simply disable the running osnap temporarily by choosing the **OSNAP** button in the Status Bar to disable it. This lets you specify a point close to a running osnap. This way you do not have to change the settings for coordinate entry in the **Options** dialog box.*

USING AUTOTRACKING

When using AutoTracking, the cursor moves along temporary paths to locate key points in a drawing. It can be used to locate points with respect to other points or objects in the drawing. There are two types of AutoTracking options: **Object Snap Tracking** and **Polar Tracking**.

Object Snap Tracking

Object Snap Tracking is used to track the movement of the cursor along the alignment paths based on the Object Snap points (running osnaps) are selected in the **Object Snap** tab of the **Drafting Settings** dialog box. You can set the object snap tracking on by selecting the **Object Snap Tracking On (F11)** check box in the **Object Snap** tab of the **Drafting Settings** dialog box or by choosing the **Object Snap Tracking** button in the Status Bar, or by using the function key F11.

The direction of a path is determined by the motion of the cursor or the point you select on an object. For example, to draw a circle whose center is located at the intersection of the imaginary lines passing through the center of two existing circles (Figure 4-71), you can use AutoTracking. To do so, select **Center** in the **Object Snap** tab of the **Drafting Settings** dialog box and turn on the **Object Snap** button in the Status Bar. Now, activate the object tracking by choosing the **Object Snap Tracking** button in the Status Bar. Choose the **Center,Radius** button from the **Draw** toolbar and pause the cursor at the center of the first circle till the marker or a plus sign is displayed. Then, move it horizontally to get the imaginary horizontal line. Move the cursor toward the second circle; the horizontal path disappears. Now, place the cursor for a while at the center of the second circle and move it vertically to get the imaginary vertical line. Move the cursor along the vertical alignment path and when you are in line with the other circle, the horizontal path will also be displayed. Select the intersection of the two alignment paths. The selected point becomes the center of the circle. Then, enter radius or specify a point.

Similarly, you can use AutoTracking in combination with the **Midpoint** object snap to locate the center of the rectangle and then draw a circle (Figure 4-72).

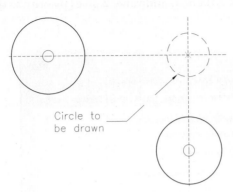

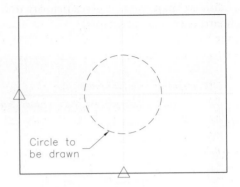

Figure 4-71 Using **AutoTracking** to locate a point (center of circle)

Figure 4-72 Using **AutoTracking** to locate a point (midpoint of rectangle)

 Note
*Object tracking works only when **OSNAP** is on and some running object snaps have been set.*

Polar Tracking

Polar Tracking is used to locate points on an angular alignment path. Polar tracking can be selected by choosing the **Polar Tracking** button in the Status Bar, by using the function key F10, or by selecting the **Polar Tracking On (F10)** check box in the **Polar Tracking** tab of the **Drafting Settings** dialog box (Figure 4-73). Polar Tracking constrains the movement of the cursor along a path that is based on the polar angle settings. For example, if the **Increment angle** list box value is set to 15-degree in the **Polar Angle Settings** area, the cursor will move along the alignment paths that are multiples of 15-degree (0, 15, 30, 45, 60, and so on) and a tooltip will also display the distance and angle. Selecting the **Additional angles** check box and choosing

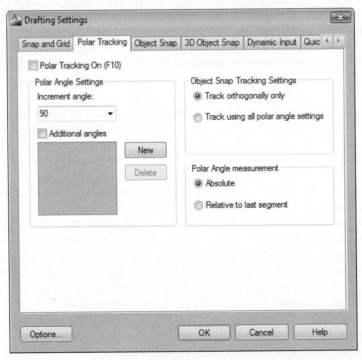

Figure 4-73 The **Polar Tracking** tab in the **Drafting Settings** dialog box

the **New** button allow you to add an additional angle value. The imaginary path will also be displayed at these new angles, apart from the increments of the increment angle selected. For example, if the increment angle is set to **15** and you add an additional angle of **22**, the imaginary path will be displayed at 0, 15, 22, 30, 45, and the increments of 15. Polar tracking is on only when the Ortho mode is off.

In the **Drafting Settings** dialog box (**Polar Tracking** tab), you can set the polar tracking to absolute or relative to the last segment. If you select the **Absolute** radio button, the base angle is taken from 0. If you select the **Relative to last segment** radio button, the base angle for the increments is set to the last segment drawn. You can also use Polar tracking together with Object tracking (Otrack). You can select the **Track using all polar angle settings** radio button in the dialog box.

AutoTrack Settings

You have different settings while working with autotracking. These settings can be specified in the **Drafting** tab of the **Options** dialog box (Figure 4-74). If you choose the **Options** button in the **Polar Tracking** tab of the **Drafting Settings** dialog box, the **Options** dialog box with the **Drafting** tab chosen will be displayed. You can use the **Display polar tracking vector** check box to toggle the display of the angle alignment path for

Figure 4-74 *The AutoTrack Settings area in the **Options** dialog box (**Drafting** tab)*

Polar tracking. You can also use the **Display full-screen tracking vector** check box to toggle the display of a full-screen construction line for Otrack. You can use the **Display AutoTrack tooltip** check box to toggle the display of tooltips with the paths. You can also use the **TRACKPATH** system variable to set the path display settings.

Tip
*You need the **Options** dialog box quite frequently to change different drafting settings. When you choose the **Options** button from the **Drafting Settings** dialog box, it directly opens the required tab. After making changes in the dialog box, choose the **OK** button to get back to the **Drafting Settings** dialog box. The **Options** button is available in all the three tabs of the **Drafting Settings** dialog box.*

FUNCTION AND CONTROL KEYS

You can also use the function and control keys to change the status of the coordinate display, Snap, Ortho, Osnap, tablet, screen, isometric planes, running Object Snap, Grid, Polar, and Object tracking. The following is a list of functions and their control keys.

F1	Help	F7	Grid On/Off (CTRL+G)
F2	Graphics Screen/AutoCAD Text Window	F8	Ortho On/Off (CTRL+L)
F3	Osnap On/Off (CTRL+F)	F9	Snap On/Off (CTRL+B)
F4	3DOsnap On/Off	F10	Polar tracking On/Off
F5	Isoplane top/right/left (CTRL+E)	F11	Object Snap tracking
F6	Dynamic UCS On/Off (CTRL+D)	F12	Dynamic Input On/Off

Self-Evaluation Test

Answer the following questions and then compare them to those given at the end of this chapter:

1. The layers that are turned off are displayed on the screen but cannot be plotted. (T/F)

2. The drawing, in which you are working, should be in the named plot style mode (*.stb*) to make the plot style available in the **Layer Properties Manager**. (T/F)

3. The grid pattern appears within the drawing limits, which helps define the working area. (T/F)

4. If a circle is inserted as a rotated block, the quadrant points are not rotated by the same amount. (T/F)

5. You can change the plot style mode from the command prompt by using the _____ system variable.

6. The _____ command enables you to set up an invisible grid that allows the cursor to move in fixed increments from one snap point to another.

7. The _____ snap works along with polar and object tracking only.

8. The _____ Object Snap mode is used to select the projected or visual intersections of two objects in 3D space.

9. The _____ can be used to locate a point with respect to two different points.

10. You can set the priority between the keyboard entry and the object snap through the _____ tab of the **Options** dialog box.

Review Questions

Answer the following questions:

1. You cannot enter different values for the horizontal and vertical grid spacing. (T/F)

2. You can lock a layer to prevent a user from accidentally editing the objects in it. (T/F)

3. When a layer is locked, you cannot use the objects in it for Osnaps. (T/F)

4. The thickness given to the objects in a layer, by using the **Lineweight** option, is displayed on the screen and can be plotted. (T/F)

5. Which of the following options is not displayed in the **Layer** shortcut menu in the **Layer Properties Manager** when you select more than one layer by using the SHIFT key?

    (a) **New Layer**   (b) **Select All**
    (c) **Make Current**   (d) **Clear All**

6. Which of the following function keys acts as a toggle key for turning the grid display on or off?

    (a) F5   (b) F6
    (c) F7   (d) F8

7. Which one of the following object snap modes is used to turn off any running object snap for one point only?

    (a) **NODe**   (b) **NONe**
    (c) **From**   (d) **NEArest**

8. Which of the following object snap modes cannot be used as the running object snap?

    (a) **EXTension**   (b) **PARallel**
    (c) **From**   (d) **NODe**

9. Which of the following keys can be used to cycle through different running object snaps?

    (a) ENTER   (b) SHIFT
    (c) CTRL   (d) TAB

10. While working on a drawing, you can save all layers with their current properties' settings anytime under one name and then restore them later using the _____ button in the **Layer Properties Manager**.

11. You can use the _____ window to locate the drawing data with the help of the search tools and then use it in your drawing.

12. The _____ function key is used to turn on/off the Osnap.

13. The difference between the usage of the **Off** option and the **Freeze** option is that in the **Freeze** option the frozen layers are not _____ by the computer while regenerating the drawing.

14. The size of the aperture is measured in _____.

15. In the **Extension** object snap mode, when you move the cursor along a path, a temporary extension path and the tooltip is displayed with the cursor. The tooltip displays _____ coordinates from the end of the line.

EXERCISE 5   *Line type and Object Color*

Set up layers with the following linetypes and colors. Then make the drawing, as shown in Figure 4-75. The distance between the dotted lines is 1.0 unit.

Layer name	Color	Linetype
Object	Red	Continuous
Hidden	Yellow	Hidden
Center	Green	Center

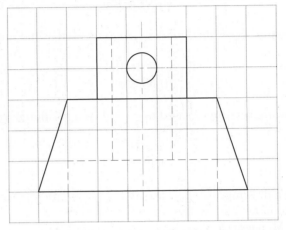

Figure 4-75 Drawing for Exercise 5

EXERCISE 6   *Line type and Object Color*

Set up layers, linetypes, and colors, as given in Exercise 5. Then make the drawing shown in Figure 4-76. The distance between the dotted lines is 1.0 unit.

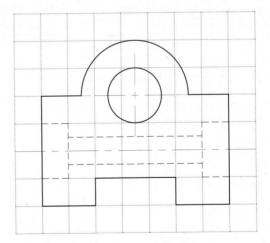

Figure 4-76 Drawing for Exercise 6

EXERCISE 7   *Line type and Object Color*

Set up layers, linetypes, and colors and then make the drawing shown in Figure 4-77. The distance between the dotted lines is 1.0 unit.

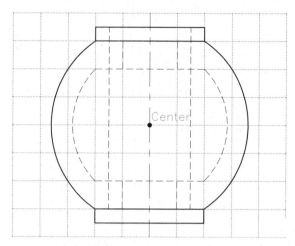

Figure 4-77 Drawing for Exercise 7

EXERCISE 8   *Line type and Object Color*

Set up layers, linetypes, and colors and then make the drawing shown in Figure 4-78. Use the object snaps as indicated.

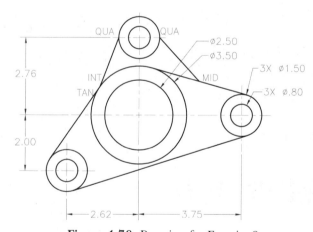

Figure 4-78 Drawing for Exercise 8

Chapter 4

Problem-Solving Exercise 1

Draw the object shown in Figure 4-79. First draw the lines and then draw the arcs using the appropriate **ARC** command options.

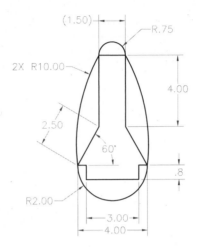

Figure 4-79 *Drawing for Problem-Solving Exercise 1*

Problem-Solving Exercise 2

Draw the object shown in Figure 4-80. First draw the front view (bottom left) and then the side and top views. Assume the missing dimensions.

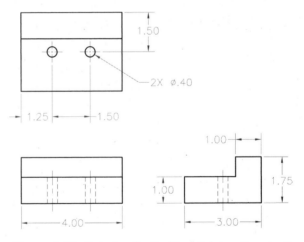

Figure 4-80 *Drawing for Problem-Solving Exercise 2*

Answers to Self-Evaluation Test

1. F, **2.** T, **3.** T, **4.** F, **5. PSTYLEPOLICY**, **6.** SNAP, **7.** Polar , **8. Apparent Intersection**, **9. Temporary Tracking**, **10. User Preferences**

Chapter 5

Editing Sketched Objects-I

CHAPTER OBJECTIVES

In this chapter, you will learn:
- *To create selection sets using various object selection options.*
- *To use the Move and Copy tools.*
- *To copy objects using the ARRAY command.*
- *Various editing and measuring tools.*

KEY TERMS

- *Move*
- *Copy*
- *Break*
- *Offset*

- *Chamfer*
- *Trim*
- *Extend*
- *Stretch*

- *Array*
- *Rotate*
- *Mirror*
- *Scale*

- *Lengthen*
- *Measure*
- *Divide*
- *Blend*

CREATING A SELECTION SET

For most of the commands, the default object selection method is to use the pick box (cursor) and select one entity at a time. You can also click in a blank area by using a pick box and select objects by using the **Window** option or the **Window Crossing** option. In Chapter 2, these two options were discussed. The other options used for selection are listed below and are discussed in this chapter.

Last	CPolygon	Add	Undo	Previous	SUbobject
Fence	BOX	SIngle	ALL	Group	AUto
WPolygon	Remove	Multiple	Object		

Last

This option is used to select the last drawn object, which is partially or fully visible in the current display. This is a convenient option to select the last drawn object while editing it. Keep in mind that if the last drawn object is not in the current display, the last drawn object in the current display will be selected. Although a selection can be formed using the **Last** option, only one object is selected at a time. However, you can use the **Last** option a number of times. You can use the **Last** selection option with any tool that requires the selection of multiple objects (e.g., **Copy**, **Move**, and **Erase**). To select object by using this option, invoke a particular tool, and then enter **LAST** or **L** at the **Select objects** prompt to use this object selection method.

EXERCISE 1   *Erase*

Draw the profile shown in Figure 5-1(a) by using the **Line** tool. Then, use the **Last** option of the **Erase** tool to erase the three most recently drawn lines to obtain Figure 5-1(d).

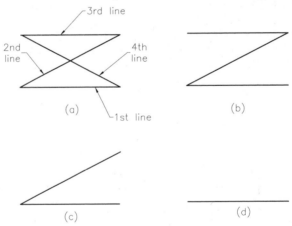

Figure 5-1 *The initial and final profiles for Exercise 1*

Previous

The **Previous** option is used to select the objects that were selected in the previous selection set. Generally, AutoCAD saves the previous selection set; therefore, you can select those objects again by using the **Previous** option, instead of invoking the same selection method. In other words, with the help of the **Previous** option, you can select the previous set without selecting the objects individually. Another advantage of using the **Previous** option is that you do not need to remember the objects if more than one editing operation has to be carried out on the same set of objects. To invoke this option, enter **P** at the **Select objects** prompt. For example, to copy objects that are selected in the previous step and then move them, you can use the **Previous** option to select the same group of objects with the **Move** tool. The prompt sequence for doing so is given next.

Command: **Copy**
Select objects: *Select the objects.*
Select objects: `Enter`
Specify base point or [displacement] <Displacement>: *Specify the base point.*
Specify second point of displacement or <use first point as displacement>: *Specify the point for displacement.*
Command: **Move**
Select objects: **P** `Enter`
found

The previous selection set is cleared by various deletion operations and the commands associated with them, like **UNDO**. You cannot select objects in the model space and then use the same selection set in the paper space, or vice versa. This is because AutoCAD keeps the record of the space (paper space or model space), in which the individual selection set is created.

WPolygon

This option is similar to the **Window** option with the only difference that by using this option, you can define a window consisting of an irregular polygon. You can specify the selection area by specifying points around the object that you want to select (Figure 5-2). Similar to the window method, the objects to be selected by using this method should be completely enclosed within the polygon. The polygon is formed as you specify points and can take any shape except a self-intersecting profile. The last segment of the polygon is automatically drawn to close the polygon. The polygon can be created by specifying the coordinates of the points or by specifying the points with the help of a pointing device. With the **Undo** option, the most recently specified WPolygon point can be undone. To use the **WPolygon** option with the object selection tools such as **Erase**, **Move**, **Copy**, invoke a tool, enter **WP** at the **Select objects** prompt, and then select the vertices of the polygon.

EXERCISE 2   *WPolygon*

Draw several objects, select some of them using the **WPolygon** option, and then erase them.

CPolygon

This option is similar to the **Window Crossing** option with the only difference that by using this option you can define a window consisting of an irregular polygon. The procedure to create a polygon is the same as discussed above. In this option, in addition to the objects that are completely enclosed within the polygon, an object lying partially inside the polygon or even touching it is also selected (Figure 5-3). CPolygon is formed as you specify the points. The points

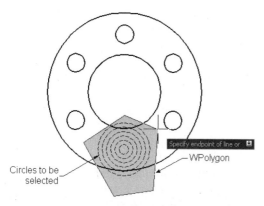

*Figure 5-2 Selecting objects using the **WPolygon** option*

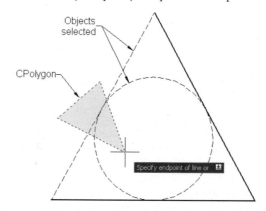

*Figure 5-3 Selecting objects using the **CPolygon** option*

can be specified at the Command line or by specifying points using the mouse. Just as in the **WPolygon** option, the crossing polygon can take any shape except a self-intersecting profile. As the last segment of the polygon is drawn automatically, the CPolygon is closed at all times.

Remove

The **Remove** option is used to remove the objects from a selection set, but not from the drawing. After selecting a number of objects in a drawing by any selection method, you may need to remove some of the objects from the selection set. The following prompt sequence displays the use of the **Remove** option for removing the objects from the selection set shown in Figure 5-4.

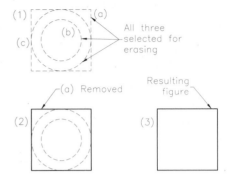

Figure 5-4 *Removing objects using the* **Remove** *option*

> Command: *Choose the* **Erase** *tool.*
> Select objects: *Select objects (a), (b), and (c).*
> Select objects: **R**
> Remove objects: *Select object (a).*
> 1 found, 1 removed, 2 total.
> Remove objects: [Enter]

Tip
Objects can also be removed from a selection set by using the SHIFT key. For example, pressing the SHIFT key and selecting an object (a) with the pointing device will remove the object from the selection set. On doing so, the message: **1 found, 1 removed, 2 total** *will be displayed.*

Add

You can use the **Add** option to add objects to a selection set. When you start creating a selection set, you are in the **Add** mode. Once you create a selection set by using any selection method, you can add more objects by simply selecting them with the pointing device, if the system variable **PICKADD** is set to 2 (default). If it is set to 0, you need to press the SHIFT key and then select the objects to add objects to the selection set.

ALL

The **ALL** selection option is used to select all objects in the current working environment of the current drawing. Note that the objects in the "OFF" layers can also be selected by using the **ALL** option. However, you cannot select the objects that are in the frozen layers by using this method. You can use this selection option with any command that requires object selection. After invoking a command, enter **ALL** at the **Select objects** prompt. Once you enter this option, all the objects drawn on the screen will be highlighted (dashed). For example, if there are four objects on the screen and you want to erase all of them, follow the prompt sequence given next.

> Command: *Choose the* **Erase** *tool.*
> Select objects: **ALL** [Enter]
> 4 found
> Select objects: [Enter]

You can also use this option in combination with the other selection options. For example, consider that there are five objects on the drawing screen and you want to erase three of them. To do so, invoke the **Erase** tool, enter **ALL** at the **Select objects** prompt. Then, press the SHIFT key and select the two objects that you want to remove from the selection set. Next, press ENTER; the remaining three objects will be erased.

Fence

In the **Fence** option, a selection set is created by drawing an open polyline through the objects to be selected. In such a case, any object touched by the polyline is selected (Figure 5-5). The selection fence can be created by entering the coordinates at the Command line or by specifying the points with the pointing device. With this option, more flexibility of selection is provided because a fence can intersect itself. The **Undo** option can be used to undo the most recently selected fence point. Like the other selection options, this option is also used with the commands that need an object selection. The prompt sequence to use the **Fence** option is given next.

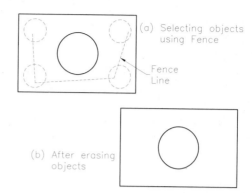

Figure 5-5 Erasing objects using the Fence option

Select objects: **F** [Enter]
Specify first fence point: *Specify the first point.*
Specify next fence point of line or [Undo]: *Specify the second point.*
Specify next fence point of line or [Undo]: *Specify the third point.*
Specify next fence point of line or [Undo]: *Specify the fourth point.*
Specify next fence point of line or [Undo]: [Enter]

Group

The **Group** option enables you to select a group of objects by their group name. You can create a group and assign a name to it with the help of the **GROUP** command (see Chapter 22). Once a group has been created, you can select it by using the **Group** option for editing. This makes the object selection process easier and faster, since a set of objects can be selected by just entering the name of their group. The prompt sequence to use the **Group** option is given next.

Command: **MOVE** [Enter]
Select objects: **G** [Enter]
Enter group name: *Enter the name of the predefined group you want to select.*
4 found
Select objects: [Enter]

EXERCISE 3   *Erase*

Draw six circles and then erase all of them by using the **ALL** option of the **Erase** tool. Next, change the contents of the selection set by removing alternate circles from the selection set by using the fence, so that alternate circles are erased.

Box

If the system variable **PICKAUTO** is set to 1 (default), the **BOX** selection option is used to select the objects inside a rectangle. After you have entered **BOX** at the **Select objects** prompt, you need to specify the two corners of a rectangle at the **Specify first corner** and **Specify opposite corner** prompts. If you define the box from right to left, it will be equivalent to the **Window Crossing** option. Therefore, it also selects the objects that touch the rectangle boundaries, in addition to the ones that are completely enclosed within the rectangle. If you define the box from left to right, this option will be equivalent to the **Window** option and select only the objects that are completely enclosed within the rectangle. The prompt sequence to use the **Box** option is given next.

Select objects: **BOX** ⏎
Specify first corner: *Specify a point.*
Specify opposite corner: *Specify the opposite corner point of the box.*
Select objects: ⏎

AUto

The **AUto** option is used to establish an automatic selection. Using this option, you can select either a single object or a number of objects by creating a window or a crossing. If you select a single object, it will be selected. If you specify a point in the blank area, you are automatically in the **BOX** selection option, and the point you have specified becomes the first corner of the box. **AUto** and **Add** are the default selection methods in such cases.

Multiple

On entering **M** (Multiple) at the **Select objects** prompt, you can select multiple objects at a single **Select objects** prompt. In this case, the selected objects will not be highlighted. Once you give a null response to the **Select objects** prompt, all selected objects will be highlighted together.

Undo

This option removes the most recently selected object from the selection set.

SIngle

When you enter **SI** (SIngle) at the **Select objects** prompt, the selection takes place in the **SIngle** selection mode. Once you select an object or a number of objects using the **Window** or **Window Crossing** option, the **Select objects** prompt does not repeat. AutoCAD proceeds with the command for which the selection has been made.

Command: *Choose the **Erase** tool.*
Select objects: **SI** ⏎
Select objects: *Select the object to erase; the selected object is erased.*

SUbobject

This option is used to select sub-entities such as the edges and faces of 3D solids. This option can also be used to select the individual members of a composite solid created by applying Boolean operations such as union, intersection, and so on. You can use this option on solid models with modify operations such as erase, move, rotate, and so on.

Tip
Sub-objects can also be selected by pressing and holding the CTRL key when you are prompted to select objects for a selection set.

Object

This option is used to exit the **SUbobject** option and it enables you to select objects as a whole again.

You can also create a selection set by using the **SELECT** command. The prompt sequence to do so is given next.

Command: **SELECT** ⏎
Select objects: *Use any selection method.*

EDITING SKETCHES

To use AutoCAD efficiently, you need to know the editing commands and how to use them. In this section, you will learn about the editing commands. These commands can be invoked from the **Ribbon**, or toolbar, or by entering them at the Command prompt. Some of the editing commands such as **ERASE** and **OOPS** have been discussed in Chapter 2 (Getting Started with AutoCAD). The rest of them will be discussed in this chapter.

MOVING THE SKETCHED OBJECTS

Ribbon:	Home > Modify > Move	**Command:** MOVE
Toolbar:	Modify > Move	

The **Move** tool is used to move one or more objects from their current location to a new location without changing their size or orientation. On invoking this tool from the **Modify** panel, as shown in Figure 5-6, you will be prompted to select the objects to be moved. Select the object by using any one of the selection techniques discussed earlier; you will be prompted to specify the base point. The base point is the reference point with respect to which the object will be picked and moved. It is recommended to select the base point on the object selected to be moved. On specifying the base point, you will be prompted to specify the second point of displacement. This is the new location where you want to move the object. On specifying this point, the selected objects will move to this point. Figure 5-7 shows objects moved by using the **Move** tool. The prompt sequence that will be followed on choosing the **Move** tool from the **Modify** panel is given next.

Select objects: *Select the objects to be moved.*
Select objects: Enter
Specify base point or [Displacement] <Displacement>: *Specify the base point to move the selected object(s).*
Specify second point or <use first point as displacement>: *Specify the second point or press ENTER to use the first point.*

If you press ENTER at the **Specify second point of displacement or <use first point as displacement>** prompt, AutoCAD interprets the first point as the relative value of the displacement in the *X* axis and *Y* axis directions. This value will be added in the *X* and *Y* axis coordinates and the object will be automatically moved to the resultant location. For example, draw a circle with its center at (3,3) and then select the center point of the circle as the base point. Now, at the **Specify second point of displacement or <use first point as displacement>** prompt, press ENTER. You will notice that the circle is moved such that its center is now placed at 6,6. This is because 3 units (initial coordinates) are added along both the X and Y directions.

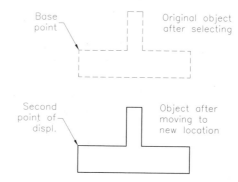

Figure 5-6 *Invoking the **Move** tool from the **Modify** panel*

Figure 5-7 *Moving the objects to a new location*

COPYING THE SKETCHED OBJECTS*

Ribbon: Home > Modify > Copy   **Toolbar:**   Modify > Copy
Command: COPY

The **Copy** tool is used to copy an existing object. This tool is used to make the copies of the selected objects and place them at the specified location. On invoking this tool, you need to select the objects and then specify the base point. Next, you need to specify the second point where the copied objects have to be placed. Figure 5-8 shows the objects copied by using this tool. The prompt sequence that will be followed when you choose the **Copy** tool from the **Modify** panel is given next.

Select objects: *Select the objects to copy.*
Select objects: [Enter]
Specify base point or [Displacement/mOde]<Displacement>: *Specify the base point.*
Specify second point or Array <use first point as displacement>: *Specify a new position on the screen using the pointing device or by entering coordinates.*
Specify second point or [Array/Exit/Undo] <Exit>: [Enter]

Creating Multiple Copies

On specifying the second point, the copy of the selected object will be created and you will be prompted again to specify the second point. You can continue specifying the second point for creating multiple copies of the selected entities, as shown in Figure 5-9. You can use **U (Undo)** to undo the last copied instance at any stage of the **COPY** command. After entering **U**, you can again specify the position of the last instance. The prompt sequence for creating multiple copies of an object by using the **Copy** tool is given next.

Specify base point or displacement: *Specify the base point.*
Specify second point or <use first point as displacement>: *Specify a point for placement.*
Specify second point or <Exit/Undo>: *Specify another point for placement.*
Specify second point or <Exit/Undo>: *Specify another point for placement.*
Specify second point or <Exit/Undo>: [Enter]

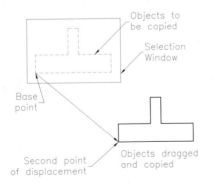

Figure 5-8 *The objects copied by using the COPY tool*

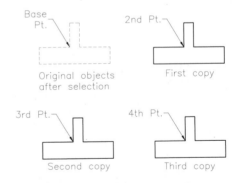

Figure 5-9 *Making multiple copies*

Creating an Array of Selected Objects

After specifying the base point, you will be prompted to specify the second point (where the copied object will be placed). You can create an array (multiple copies arranged in linear manner) of the selected objects by entering **ARRAY** or **A** at the **Specify second point or [Array] <use first point as displacement>:** Command prompt. On entering the required command at the Command prompt, you will be prompted to specify the number of items. Enter the number of items and press ENTER; you will be prompted to specify the second point and the preview of the array of selected items will be displayed attached to the cursor. Click to specify the second point; the linear array of selected object will be created, refer to Figure 5-10. In this array, the distance between any two consecutive items will be equal to the distance specified between the first and second points. If you enter **FIT** at the **Specify second point or [Fit]** Command prompt, all array items will fit inside the first and second points, refer to Figure 5-11.

Figure 5-10 *Linear array created by using the **Copy** tool*

Figure 5-11 *Array created by using the **Fit** option of the **Copy** tool*

The prompt sequence for creating a linear array of a rectangle with the second point option of the **Copy** tool is given next, refer to Figure 5-10.

    Command: **COPY**
    Select objects: 1 found
    Select objects: [Enter]
    Current settings: Copy mode = Multiple
    Specify base point or [Displacement/mOde] <Displacement>: *Specify the base point.*
    Specify second point or [Array] <use first point as displacement>: **ARRAY**
    Enter number of items to array: **3**
    Specify second point or [Fit]: *Specify another point for placement.*

The prompt sequence for creating the linear array of a rectangle by using the **Fit** option of the **Copy** tool is given next, refer to Figure 5-11.

    Command: **COPY**
    Select objects: 1 found
    Select objects: [Enter]
    Current settings: Copy mode = Multiple
    Specify base point or [Displacement/mOde] <Displacement>: *Specify the base point.*
    Specify second point or [Array] <use first point as displacement>: **ARRAY**
    Enter number of items to array: **3**
    Specify second point or [Fit]: **FIT**
    Specify second point or [Array]: *Specify another point for placement.*

Creating a Single Copy

By default, AutoCAD creates multiple copies of the selected objects. However, you can also create a single copy of the selected object. To create a single copy of the selected object, choose the **mOde** option from the shortcut menu at the **Specify base point or [Displacement/mOde] <current>** prompt. Next, choose the **Single** option from the shortcut menu at the **Enter a copy mode option [Single/Multiple] <current>** prompt. On specifying the second point of displacement, a copy will be placed at that point. Now, you can exit the **Copy** tool.

Chapter 5

EXERCISE 4   *Copy*

In this exercise, you will create the drawing shown in Figure 5-12. Use the **COPY** command for creating the drawing.

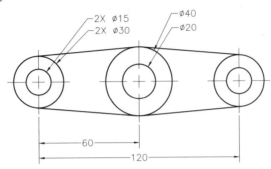

Figure 5-12 Drawing for Exercise 4

COPYING OBJECTS USING THE BASE POINT

Command: COPYBASE

Sometimes you may need to paste the copied object very precisely in the same diagram or in another diagram. This is done by using the **COPYBASE** command. Unlike the **COPY** command where the objects are dragged and placed at the desired location, this command is used to copy the selected object on to the Clipboard. Then from the Clipboard, objects are placed at the specified location. The Clipboard is defined as a medium of storing data while transferring data from one place to the other. Note that the contents of the Clipboard are invisible. The objects copied by using this command can be copied from one drawing file to the other or from one working environment to the other.

You can invoke the **COPYBASE** command from the shortcut menu. To do so, select an object and right-click. Then, choose **Clipboard > Copy with Base Point**, see Figure 5-13. On invoking this command, you will be prompted to specify the base point for

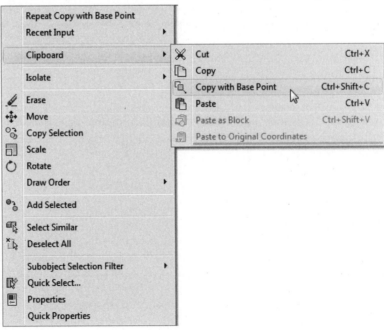

*Figure 5-13 Invoking the **COPYBASE** command from the shortcut menu*

selected the objects. Specify the base point; the objects will be copied. On copying objects using the **COPYBASE** command, you can paste the objects at the desired place with a greater accuracy.

 Note
You will learn more about the working environments and the Clipboard in the following chapters.

PASTING CONTENTS FROM THE CLIPBOARD

Ribbon: Home > Clipboard > Paste > Paste as Block   **Command:** PASTEBLOCK

The **Paste as Block** tool is used to paste the contents of the Clipboard into a new drawing or in the same drawing at a new location. You can also invoke the **PASTEBLOCK** command from the shortcut menu by right-clicking in the drawing area and then choosing **Clipboard > Paste as Block**.

PASTING CONTENTS USING THE ORIGINAL COORDINATES

Ribbon: Home > Clipboard > Paste > Paste to Original Coordinates   **Command:** PASTEORIG

The **Paste to Original Coordinates** tool is used to paste the contents of the Clipboard into a new drawing by using coordinates from the original drawing. You can invoke the **PASTEORIG** command from the shortcut menu by right-clicking in the drawing area and then choosing **Clipboard > Paste to Original Coordinates**. Note that this command will be available only when the Clipboard contains AutoCAD data from a drawing other than the current drawing.

OFFSETTING SKETCHED OBJECTS

Ribbon: Home > Modify > Offset   **Toolbar:** Modify > Offset
Command: OFFSET

You can use the **Offset** tool to draw parallel lines, polylines, concentric circles, arcs, curves, and so on (Figure 5-14). However, you can offset only one entity at a time. While offsetting an object, you need to specify the offset distance and the side to offset, or specify the distance through which the selected object has to be offset. Depending on the side to offset, the resulting object will be smaller or larger than the original object. For example, while offsetting a circle if the offset side is toward the inner side of the perimeter, the resulting circle will be smaller than the original one. The prompt sequence that will follow when you choose the **Offset** tool from the **Modify** panel is given next.

Current settings: Erase source=No Layer=Source OFFSETGAPTYPE=0
Specify offset distance or [Through/Erase/Layer] <Through>: *Specify the offset distance*
Select object to offset or [Exit/Undo]<Exit>: *Select the object to offset*
Specify point on side to offset or <Exit/Multiple/Undo>: *Specify a point on the side to offset*
Select object to offset or [Exit/Undo]<Exit>*Select another object to offset or press* [Enter].

Through Option

While offsetting the entities, the offset distance can be specified by entering a value or by specifying two points. The distance between these two points will be used as the offset distance. The **Through** option is generally used to create orthographic views. In this case, you do not need to specify a distance; you need to specify an offset point, see Figure 5-15. The offset distance is stored in the **OFFSETDIST** system variable. A negative value indicates that the offset value is set to the **Through** option. Using this option, you can offset lines, arcs, 2D polylines, xlines, circles, ellipses, elliptical arcs, rays, and planar splines. If you try to offset objects other than these, the message **Cannot offset that object** is displayed.

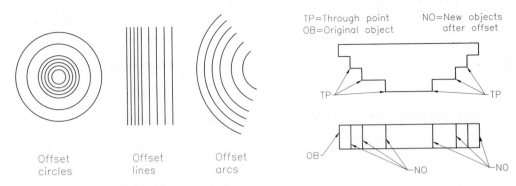

Figure 5-14 *Using the **OFFSET** command multiple times to create multiple offset entities*

Figure 5-15 *Using the **Through** option*

Erase Option

The **Erase** option is used to specify whether the source object has to be deleted or not. Enter **Yes** at the **Erase source object after offsetting** prompt to delete the source object after creating the offset. The prompt sequence that will follow when you choose the **Erase** option is given next.

Current settings: Erase source=No Layer=Source OFFSETGAPTYPE=0
Specify offset distance or [Through/Erase/Layer] <current>: **E**
Erase source object after offsetting? [Yes/No] <No>: **Y**
Specify offset distance or [Through/Erase/Layer] <current>:

Layer Option

The **Layer** option is used to specify whether the offset entity will be placed in the current layer or the layer of the source object. The prompt sequence to offset the entity in the **Source** layer is given next.

Current settings: Erase source=Yes Layer=Current OFFSETGAPTYPE=0
Specify offset distance or [Through/Erase/Layer] <current>: **L**
Enter layer option for offset objects [Current/Source] <Current>: **S**
Specify offset distance or [Through/Erase/Layer] <current>:

EXERCISE 5   *Offset*

Use the **Offset** tool to draw entities shown in Figures 5-16 and 5-17.

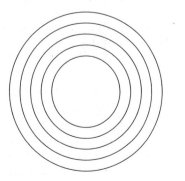

Figure 5-16 *Drawing for Exercise 5 (a)*

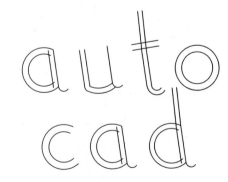

Figure 5-17 *Drawing for Exercise 5 (b)*

ROTATING SKETCHED OBJECTS

Ribbon: Home > Modify > Rotate   **Toolbar:** Modify > Rotate
Command: ROTATE

While creating designs, sometimes you have to rotate an object or a group of objects. You can accomplish this by using the **Rotate** tool. On invoking this tool, you will be prompted to select the objects and the base point about which the selected objects will be rotated. You should be careful in selecting the base point, if the base point is not located on the known object. After you specify the base point, you need to enter the rotation angle. By default, a positive angle results in a counterclockwise rotation, whereas a negative angle results in a clockwise rotation (Figure 5-18). The **Rotate** tool can also be invoked from the shortcut menu by selecting an object and right-clicking in the drawing area and choosing **Rotate** from the shortcut menu.

If you need to rotate objects with respect to a known angle, you can do so by using the **Reference** option in two different ways. The first way is to specify the known angle as the reference angle, followed by the proposed angle to which the objects will be rotated (Figure 5-19). Here the object is first rotated clockwise from the *X* axis, through the reference angle. Then the object is rotated through the new angle from this reference position in a counterclockwise direction. The prompt sequence is given next.

Current positive angle in UCS: ANGDIR=*current* ANGBASE=*current*
Select objects: *Select the objects for rotation.*
Select objects: Enter
Specify base point: *Specify the base point.*
Specify rotation angle or [Copy/Reference]<current>: **R** Enter
Specify the reference angle <0>: *Enter reference angle.*
Specify the new angle or [Points]: *Enter new angle or enter **P** to select two points for specifying the angle value.*

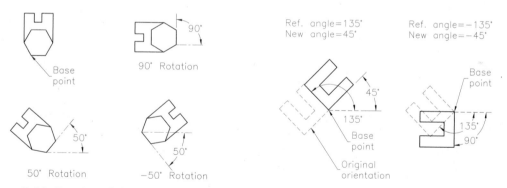

Figure 5-18 Rotation of objects in different rotation angles

*Figure 5-19 Rotation using the **Reference** option*

The other method is used when the reference angle and the new angle are not known. In this case, you can use the edges of the original object and the reference object to specify the original object and the reference angle, respectively. Figure 5-20 shows a model and a line created at an unknown angle. In this case, this line will be used as a reference object for rotating the object. In such cases, remember that the base point should be taken on the reference object. This is because you cannot define two points for specifying the new angle. You have to directly enter the angle value or specify only one point. Therefore, the base point will be taken as the first point and the second point can be defined for a new angle. Figure 5-21 shows the model after rotating it with reference to the line such that the line and the model are inclined at similar angles. The prompt sequence to rotate the model shown in Figure 5-20 is given next.

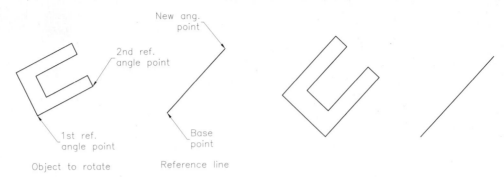

Figure 5-20 *Rotating the model using a reference line*

Figure 5-21 *The model after rotating with reference to the line*

Current positive angle in UCS: ANGDIR=*current* ANGBASE=*current*
Select objects: *Select the object for rotation.*
Select objects: Enter
Specify base point: *Specify the base point as the lower endpoint of the line, see Figure 5-20.*
Specify rotation angle or [Copy/Reference]<current>: **R** Enter
Specify the reference angle <0>: *Specify the first point on the edge of the model, see Figure 5-20.*
Specify second point: *Specify the second point on the same edge of the model, see Figure 5-20.*
Specify the new angle or [Points]<0>: *Select the other endpoint of the reference line, see Figure 5-20.*

You can also specify a new angle by entering a numeric value at the **Specify the new angle** prompt.

If you want to retain the original object and create a copy while rotating, use the **Copy** option. The source entity is retained in its original orientation and a new instance is created and rotated through the specified angle. The prompt sequence for the **Copy** option is given next.

Current positive angle in UCS: ANGDIR=counterclockwise ANGBASE=0
Select objects: *Select the object for rotation.*
Select objects: Enter
Specify base point: *Specify a base point about which the selected objects will be rotated.*
Specify rotation angle or [Copy/Reference] <current>: **C** Enter
Rotating a copy of the selected objects.
Specify rotation angle or [Copy/Reference] <current>: *Enter a positive or negative rotation angle, or specify a point.*

SCALING THE SKETCHED OBJECTS

Ribbon: Home > Modify > Scale	**Toolbar:** Modify > Scale
Command: SCALE	

Sometimes you need to change the size of objects in a drawing. You can do so by using the **Scale** tool. This tool dynamically enlarges or shrinks a selected object about a base point, keeping the aspect ratio of the object constant. This means that the size of the object will be increased or reduced equally in the X, Y, and Z directions. The dynamic scaling property allows you to view the object as it is being scaled. This is a useful and time saving editing tool because instead of redrawing objects to the required size, you can scale the objects. Another advantage of this tool is that if you have dimensioned the drawing, they will also change accordingly. You can also invoke the **Scale** tool from the shortcut menu by right-clicking in the drawing area and choosing the **Scale** tool. The prompt sequence that will follow when you choose the **Scale** tool is given next.

Select objects: *Select the objects to be scaled.*
Select objects: [Enter]
Specify base point: *Specify the base point, preferably a known point.*
Specify scale factor or [Copy/Reference]<current>: *Specify the scale factor.*

The base point will not move from its position and the selected object(s) will be scaled with respect to the base point, as shown in Figures 5-22 and 5-23. To reduce the size of an object, the scale factor should be less than 1 and to increase its size, the scale factor should be greater than 1. You can enter a scale factor or select two points to specify a distance as a factor. When you select two points to specify a distance as a factor, the first point should be on the referenced object.

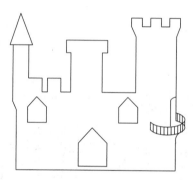

Figure 5-22 *Original object*   **Figure 5-23** *Objects after scaling to 0.5 of the actual size*

Sometimes, it is time-consuming to calculate the relative scale factor. In such cases, you can scale the object by specifying a desired size in relation to the existing size (a known dimension). In other words, you can use a reference length. This can be done by entering **R** at the **Specify scale factor or [Copy/Reference]** prompt. Then, you can either specify two points to specify the length or enter a length. At the next prompt, enter the length relative to the reference length. For example, if a line is 2.5 units long and you want its length to be 1.00 unit, then instead of calculating the relative scale factor, you can use the **Reference** option. The prompt sequence for using the **Reference** option is given next.

Select objects: *Select the object to scale.*
Select objects: [Enter]
Specify base point: *Specify the base point.*
Specify scale factor or [Copy/Reference]<current>: **R** [Enter]
Specify reference length <1>: *Specify the reference length.*
Specify new length [Point]: *Specify the new length or enter **P** to select two points for specifying the angle value.*

Similar to the **Rotate** tool, you can also scale one object by using the reference of another object. Again, in this case also, the base point has to be taken on the reference object as you can define only one point for the new length.

You can use the **Copy** option to retain the source object and scale the copied instance of the source object. The prompt sequence for using the **Copy** option is given next.

Select objects: *Select the object to scale.*
Select objects: [Enter]
Specify base point: *Specify the base point.*

Specify scale factor or [Copy/Reference] <current>: **C** Enter
Scaling a copy of the selected objects.
Specify scale factor or [Copy/Reference] <current>: *Specify the scale factor.*

Tip
If you need to change the dimensions of all objects with respect to other units, you can use the ***Reference*** *option of the* ***SCALE*** *command to correct the error. To do so, select the entire drawing by using the* ***ALL*** *selection option. Next, specify the* ***Reference*** *option, and then select the endpoints of the object whose desired length is known. Specify the new length; all objects in the drawing will be scaled automatically to the desired size.*

FILLETING THE SKETCHES

Ribbon: Home > Modify > Fillet/Chamfer drop-down	**Toolbar:** Modify > Fillet
Command: FILLET	

The edges in a model are generally filleted to reduce the area of stress concentration. The **Fillet** tool helps you form round corners between any two entities that form a sharp vertex. As a result, a smooth round arc is created that connects the two objects. A fillet can also be created between two intersecting or parallel lines as well as non-intersecting and nonparallel lines, arcs, polylines, xlines, rays, splines, circles, and true ellipses. The fillet arc created will be tangent to both the selected entities. The default fillet radius is 0.0000. Therefore, after invoking this tool, you first need to specify the radius value. The prompt sequence displayed on choosing the **Fillet** tool from the **Fillet/Chamfer** drop-down in the **Modify** panel (Figure 5-24) is given next.

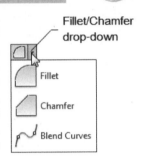

*Figure 5-24 Tools in the **Fillet/Chamfer** drop-down*

Current Settings: Mode= TRIM, Radius= 0.0000
Select first object or [Undo/Polyline/Radius/Trim/Multiple]:

Creating Fillets Using the Radius Option

The fillet you create depends on the radius distance specified. The default radius is 0.0000. You can enter a distance or specify two points. The new radius you enter becomes the default radius and remains in effect until changed. Note that the **FILLETRAD** system variable controls and stores the current fillet radius. The default value of this variable is 0.0000. The prompt sequence that is displayed on invoking the **Fillet** tool is given next.

Select first object or [Undo/Polyline/Radius/Trim/Multiple]: **R** Enter
Specify fillet radius <current>: *Enter a value or press ENTER to accept the current value.*

Tip
A fillet with a zero radius has sharp corners and is used to clean up lines at corners if they overlap or have a gap.

Creating Fillets Using the Select First Object Option

This is the default method to fillet two objects. As the name implies, it prompts for the first object required for filleting. The prompt sequence to use this option is given next.

Current Settings: Mode= TRIM, Radius= modified value
Select first object or [Undo/Polyline/Radius/Trim/Multiple]: *Specify the first object.*
Select second object or shift-select to apply corner: *Select the second object or press the SHIFT key while selecting the object to create sharp corner.*

The **Fillet** tool can also be used to cap the ends of two parallel as well as non-parallel lines, see Figure 5-25. The cap is a semicircle whose radius is equal to half the distance between the two parallel lines. The cap distance is calculated automatically when you select the two parallel lines for filleting. You can select lines by using the **Window**, **Window Crossing**, or **Last** option, but to avoid unexpected results, select the objects by picking them individually. Also, selection by picking objects is necessary in the case of arcs and circles that have the possibility of more than one fillet. They are filleted closest to the selected points, as shown in Figure 5-26.

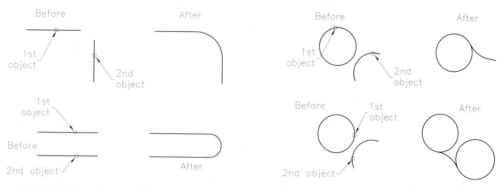

Figure 5-25 *Filleting the parallel and non-parallel lines*

Figure 5-26 *Using the **FILLET** command on circles and arcs*

Creating Fillets Using the Trim Option

When you create a fillet, an arc is created and the selected objects are either trimmed or extended at the fillet endpoint. This is because the **Trim** mode is set to **Trim**. If it is set to **No Trim**, they are left intact. Figure 5-27 shows a model filleted with the **Trim** mode set to **Trim** and **No Trim**. The prompt sequence is given next.

> Select first object or [Undo/Polyline/Radius/Trim/Multiple]: **T** [Enter]
> Specify Trim mode option [Trim/No trim] <current>: *Enter **T** to trim edges, **N** to leave them intact. You can also choose the required option from the dynamic preview.*

Creating Fillets Using the Polyline Option

If an object is created by using the **Polyline** or **Rectangle** tools, then the object will be a polyline. You can fillet all sharp corners in a polyline by using the **Polyline** option of the **Fillet** tool, as shown in Figure 5-28. On selecting this option after specifying the fillet radius,

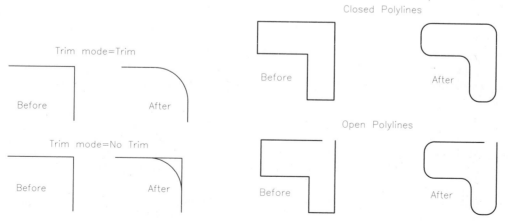

Figure 5-27 *Filleting the lines with the **Trim** mode set to **Trim** and **No Trim***

Figure 5-28 *Filleting closed and open polylines*

you will be prompted to select a polyline. Select the polyline; all vertices of the polyline will be filleted with the same fillet radius. The prompt sequence for using this option is given next.

Current Settings: Mode= *current*, Radius= *current*
Select first object or [Undo/Polyline/Radius/Trim/Multiple]: **P** Enter
Select 2D polyline: *Select the polyline.*

Creating Fillets Using the Multiple Option

When you invoke the **Fillet** tool, by default, a fillet is created between a pair of entities only. However, with the help of the **Multiple** option, you can add fillets to more than a pair of entities. On selecting this option, you will be prompted to select the first object and then the second object. On selecting two objects, a fillet will be created between the two entities and again you will be prompted to select the first object and then the second object. This prompt will continue until you press ENTER to terminate the **Fillet** tool. The prompt sequence to use this option is given next.

Current Settings: Mode= TRIM, Radius= current
Select first object or [Undo/Polyline/Radius/Trim/Multiple]: **M** Enter
Select first object or [Undo/Polyline/Radius/Trim/Multiple]: *Specify the first object of one set.*
Select second object or shift-select to apply corner: *Select the second object or press down the SHIFT key while selecting the object to create sharp corner.*
Select first object or [Undo/Polyline/Radius/Trim/Multiple]: *Specify the first object of the other set.*
Select second object or shift-select to apply corner: *Select the second object or press down the SHIFT key while selecting the object to create sharp corner.*
Select first object or [Undo/Polyline/Radius/Trim/Multiple]: *Specify the first object of the other set or press* Enter

Note
*Use the **Undo** option to undo the fillet created.*

Filleting Objects with a Different UCS

The **Fillet** tool also fillets the objects that are not in the current UCS plane. To create a fillet for these objects, AutoCAD automatically changes the UCS transparently so that it can generate a fillet between the selected objects.

Note
Filleting objects in different planes is discussed in detail in Chapter 25 (Getting Started with 3D).

Setting the TRIMMODE System Variable

The **TRIMMODE** system variable eliminates any size restriction on the **Fillet** tool. By setting **TRIMMODE** to **0**, you can create a fillet of any size without actually cutting the existing geometry. Also, there is no restriction on the fillet radius. This means that the fillet radius can be larger than one or both objects that are being filleted. The default value of this variable is 1.

Note
TRIMMODE = 0   *Fillet or chamfer without cutting the existing geometry.*
TRIMMODE = 1   *Extend or trim the geometry.*

CHAMFERING THE SKETCHES

Ribbon: Home > Modify > Fillet/Chamfer drop-down > Chamfer   **Toolbar:** Modify > Chamfer
Command: CHAMFER

 Chamfering the sharp corners is another method of reducing the areas of stress concentration in a model. Chamfering is defined as the process by which the sharp edges or corners are beveled. The size of a chamfer depends on its distance from the corner. If a chamfer is equidistant from the corner in both directions, it is a 45-degree chamfer. A chamfer can be drawn between two lines that may or may not intersect. This command also works on a single polyline. In AutoCAD, the chamfers are created by defining two distances or by defining one distance and the chamfer angle.

The prompt sequence that will follow when you choose the **Chamfer** tool from the **Fillet/Chamfer** drop-down in the **Modify** panel is given next.

> (TRIM mode) Current chamfer Dist1 = 0.0000, Dist2 = 0.0000
> Select first line or [Undo/Polyline/Distance/Angle/Trim/mEthod/Multiple]:

The different options to create chamfers are discussed next.

Creating Chamfer Using the Distance Option

To create chamfer by entering the distances, enter **D** at the **Select first line or [Undo/Polyline/Distance/Angle/Trim/Method/mUltiple]** prompt. Next, enter the first and second chamfer distances. The first distance is the distance of the corner calculated along the edge selected first. Similarly, the second distance is calculated along the edge that is selected last. The new chamfer distances remain in effect until you change them. Instead of entering the distance values, you can specify two points to indicate each distance, see Figure 5-29. The prompt sequence is given next.

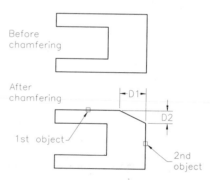

Figure 5-29 *Chamfering the model using the* **Distance** *option*

> Select first line or [Undo/Polyline/Distance/Angle/Trim/mEthod/Multiple]: **D** [Enter]
> Specify first chamfer distance <0.0000>: *Enter a distance value or specify two points.*
> Specify second chamfer distance <0.0000>: *Enter a distance value or specify two points.*

Note
If Dist 1 and Dist 2 are set to zero, the **Chamfer** *tool will extend or trim the selected lines so that they end at the same point.*

The first and second chamfer distances are stored in the **CHAMFERA** *and* **CHAMFERB** *system variables. The default value of these variables is 0.0000.*

Creating Chamfer Using the Select First Line Option

In this option, you need to select two nonparallel objects so that they are joined with a beveled line. The size of a chamfer depends on the values of the two distances. The prompt sequence to do so is given next.

> (TRIM mode) Current chamfer Dist1 = current, Dist2 = current
> Select first line or [Undo/Polyline/Distance/Angle/Trim/mEthod/Multiple]: *Specify the first line.*

Select second object or shift-select to apply corner: *Select the second object or press down the SHIFT key while selecting the object to create sharp corner.*

Creating Chamfer Using the Polyline Option

Similar to the **Fillet** tool, you can chamfer all sharp corners in a polyline by using the **Polyline** option of the **Chamfer** tool, as shown in Figure 5-30. The prompt sequence to chamfer the polylines is given next.

(TRIM mode) Current chamfer Dist1 = current, Dist2 = current
Select first line or [Undo/Polyline/Distance/Angle/Trim/mEthod/Multiple]: **P** Enter
Select 2D polyline or [Distance/Angle/Method]: *Select the polyline.*

Creating Chamfer Using the Angle Option

The other method of creating a chamfer is by specifying the distance and the chamfer angle, as shown in Figure 5-31. The prompt sequence to do so is given next.

Select first line or [Undo/Polyline/Distance/Angle/Trim/mEthod/Multiple]: **A** Enter
Specify chamfer length on the first line <current>: *Specify a length.*
Specify chamfer angle from the first line <current>: *Specify an angle.*

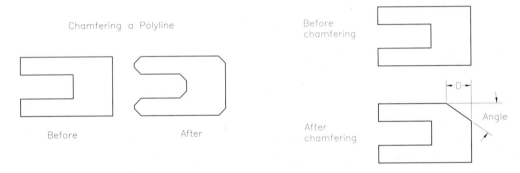

Figure 5-30 Chamfering a polyline   *Figure 5-31 Chamfering using the **Angle** option*

Creating Chamfer Using the Trim Option

On selecting the **Trim** option, the selected objects are either trimmed or extended to the endpoints of the chamfer line or left intact. The prompt sequence to invoke this option is given next.

Select first line or [Undo/Polyline/Distance/Angle/Trim/mEthod/Multiple]: **T** Enter
Enter Trim mode option [Trim/No Trim] <current>:

Creating Chamfer Using the Method Option

On using this option, you can toggle between the **Distance** method and the **Angle** method for creating a chamfer. The current settings of the selected method will be used for creating the chamfer. The prompt sequence to create the chamfer is given next.

Select first line or [Undo/Polyline/Distance/Angle/Trim/mEthod/Multiple]: **E** Enter
Enter trim method [Distance/Angle] <current>: *Enter **D** for the **Distance** option, **A** for the **Angle** option.*

Note
*If you set the value of the **TRIMMODE** system variable to **1** (default value), the objects will be trimmed or extended after they are chamfered and filleted. If **TRIMMODE** is set to zero, the objects will be left untrimmed.*

Creating Chamfers Using the Multiple Option

When you invoke the **Chamfer** tool, by default, a chamfer is created between a pair of entities only. However, with the help of the **Multiple** option, you can add chamfers to multiple pairs. On selecting this option, you will be prompted to select the first line and then the second line. On selecting the two lines, the chamfer will be created. Next, you will again be prompted to select the first line and then the second line. This prompt will continue until you press ENTER to terminate the **Chamfer** tool. The prompt sequence that follows to create multiple chamfers is given next.

(TRIM mode) Current chamfer Dist1 = current, Dist2 = current
Select first line or [Undo/Polyline/Distance/Angle/Trim/mEthod/Multiple]: **M** Enter
Select first line or [Undo/Polyline/Distance/Angle/Trim/mEthod/Multiple]: *Specify the first line of one set.*
Select second object or shift-select to apply corner: *Select the second object or press the SHIFT key while selecting the object to create sharp corner.*
Select first line or [Undo/Polyline/Distance/Angle/Trim/mEthod/Multiple]: *Specify the first line of the other set.*
Select second object or shift-select to apply corner: *Select the second object or press the SHIFT key while selecting the object to create sharp corner.*
Select first line or [Undo/Polyline/Distance/Angle/Trim/mEthod/Multiple]: *Specify the first line of the other set or press **ENTER** to terminate the **Chamfer** tool.*

Setting the Chamfering System Variables

The chamfer modes, distances, length, and angle can also be set by using the following variables:

CHAMMODE = 0	Distance/Distance (default)
CHAMMODE = 1	Length/Angle
CHAMFERA	Sets first chamfer distance on the first selected line (default = 0.0000)
CHAMFERB	Sets second chamfer distance on the second selected line (default = 0.0000)
CHAMFERC	Sets the chamfer length (default = 0.0000)
CHAMFERD	Sets the chamfer angle from the first line (default = 0)

BLENDING THE CURVES*

Ribbon: Home > Modify > Fillet/Chamfer drop-down > Blend Curves
Command: BLEND

In AutoCAD 2012, you can create a smooth or tangent continuous spline between the endpoints of two existing curves. The existing curves can be arcs, lines, helices, open polylines, or open splines. To create a smooth curve between two open curves, choose the **Blend** tool from the **Modify** panel (see Figure 5-24); you will be prompted to select the curves one after the other. Select the two curves; a blend curve will be created between the selected curves. The length of the selected curves do not change. The prompt sequence to use this tool is given next.

Command: **BLEND**
Continuity = Tangent
Select first object or [CONtinuity]: *Select the first object*
Select second object: *Select the second object*

This tool provides two options to specify the continuity of the blend curve with the two selected curves: **Tangent** and **Smooth**. By default, the **Tangent** option is chosen. As a result, a tangent continuous curve will be created between the two selected curves. The prompt sequence to use this option has been given earlier in this topic.

You can also blend two curves with a smooth continuity between them. To do so, first enter **CON** at the **Select first object or [CONtinuity]:** Command prompt. Next, enter **S** at the Command prompt and then select the two curves; a smooth curve will be created between the two selected curves. The prompt sequence to use this option is given next.

Command: **BLEND**
Continuity = Tangent
Select first object or [CONtinuity]: **CON** Enter
Enter continuity [Tangent/Smooth] <Smooth>: **S** Enter
Select first object or [CONtinuity]: *Select the first object*
Select second object: *Select the second object*

Figure 5-32 shows the two curves selected and Figure 5-33 shows the resulting tangent and smooth curves as the blend curves.

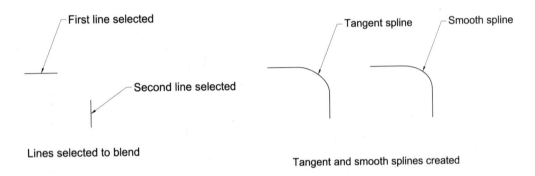

Figure 5-32 Curves selected to be blended *Figure 5-33 Tangent and smooth blend curves created*

To view the continuity of the spline (blend curve) with the selected curves, select the spline; the control points will be displayed in the form of control vertices. You can change the display of control points from control vertices to the fit points. To do so, click on the down-arrow on the spline; a shortcut menu will be displayed. Choose the **Show Fit Points** option from the shortcut menu to change the control vertices to fit points. Figure 5-34 shows the control vertices of a tangent continuous spline and a smooth spline created between two open lines. You can change the shape of the splines by dragging its control vertices.

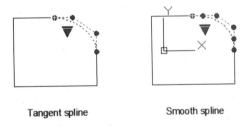

Figure 5-34 Control vertices of tangent and smooth splines

TRIMMING THE SKETCHED OBJECTS

Ribbon: Home › Modify › Trim   **Toolbar:** Modify › Trim
Command: TRIM

When creating a design, you may need to remove the unwanted and extended edges. Breaking individual objects takes time if you are working on a complex design with many objects. In such cases, you can use the **Trim** tool. This command is used to trim the objects that extend beyond a required point of intersection. When you invoke this tool from the **Trim/Extend** drop-down in the **Modify** panel (Figure 5-35), you will be prompted to select the cutting edges or boundaries. These edges can be lines, polylines, circles, arcs, ellipses, xlines, rays, splines, text, blocks, or even viewports. There can be more than one cutting edge and you can use any selection method to select them. After the cutting edge or edges are selected, you must select each object to be trimmed. An object can be both a cutting edge and an object to be trimmed. You can trim lines, circles, arcs, polylines, splines, ellipses, xlines, and rays. The prompt sequence to trim objects is given next.

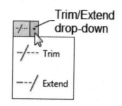

Figure 5-35 *Tools in the* **Trim/Extend** *drop-down*

Current settings:Projection=UCS Edge=Extend
Select cutting edges...
Select objects or <select all>: *Select the cutting edges.*
Select objects: Enter
Select object to trim or shift-select to extend or [Fence/Crossing/Project/Edge/eRase/Undo]:

Various options used to trim objects are discussed next.

Select object to trim Option

On selecting this option, you have to specify the objects you want to trim and the side from which the objects will be trimmed. This prompt is repeated until you press ENTER. This way you can trim several objects on invoking this tool once. The prompt sequence to use this option is given next.

Current settings: Projection= UCS Edge= Extend
Select cutting edges...
Select objects or <select all>: *Select the first cutting edge.*
Select objects: *Select the second cutting edge.*
Select objects: Enter
Select object to trim or shift-select to extend or [Fence/Crossing/Project/Edge/eRase/Undo]: *Select the first object.*
Select object to trim or shift-select to extend or [Fence/Crossing/Project/Edge/eRase/Undo]: *Select the second object. (Figure 5-36)*
Select object to trim or shift-select to extend or [Fence/Crossing/Project/Edge/eRase/Undo]: Enter

Shift-select to extend Option

This option is used to switch from trim mode to the extend mode. It is used to extend an object instead of trimming. In case the object to be extended does not intersect with the cutting edge, you can press the SHIFT key and then select the object to be extended; the selected edge will be extended taking the cutting edge as the boundary for extension. Note that you need to click near the endpoint that is closest to the cutting edge.

Chapter 5

Edge Option

This option is used to trim those objects that do not intersect the cutting edges, but they intersect if the cutting edges were extended, see Figure 5-37. The prompt sequence is given next.

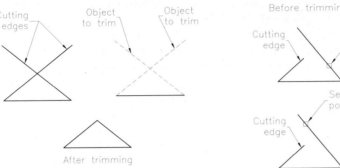

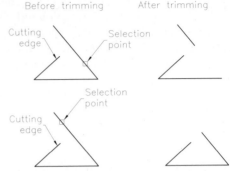

Figure 5-36 Using the **TRIM** command

Figure 5-37 Trimming an object using the **Edge** option (**Extend**)

Current settings: Projection= UCS Edge= Extend
Select cutting edges...
Select objects: *Select the cutting edge.*
Select objects: Enter
Select object to trim or shift-select to extend or [Fence/Crossing/Project/Edge/eRase/Undo]: **E** Enter
Enter an implied edge extension mode [Extend/No extend] <current>: **E** Enter
Select object to trim or shift-select to extend or [Fence/Crossing/Project/Edge/eRase/Undo]: *Select object to trim.*

Project Option

The **Project** option is used to trim those objects that do not intersect the cutting edges in 3D space, but do visually appear to intersect in a particular UCS or the current view. The prompt sequence to invoke this option is given next.

Select object to trim or shift-select to extend or [Fence/Crossing/Project/Edge/eRase/Undo]: **P** Enter
Enter a projection option [None/Ucs/View] <current>:

The **None** option is used whenever the objects to be trimmed intersect the cutting edges in 3D space. The **UCS** option is used to project the objects to the *XY* plane of the current UCS, while the **View** option is used to project the objects to the current view direction (trims to their apparent visual intersections).

Fence Option

As discussed in earlier chapters, the **Fence** option is used for the selection purpose. Using the **Fence** option, all the objects crossing the selection fence are selected. The prompt sequence for the **Fence** option is given next.

Select object to trim or shift-select to extend or [Fence/Crossing/Project/Edge/eRase/Undo]: **F** Enter
Specify first fence point: Specify the first point of the fence
Specify next fence point or [Undo]: Specify the second point of the fence

Specify next fence point or [Undo]: *Specify the third point of the fence*
Specify next fence point or [Undo]: Enter

Crossing Option

The **Crossing** option is used to select the entities by using a crossing window. On using this option, the objects touching the window boundaries or completely enclosing them are selected. The prompt sequence to trim objects using this option is given next.

Select object to trim or shift-select to extend or [Fence/Crossing/Project/Edge/eRase/Undo]: **C** Enter
Specify first corner: *Select the first corner of the crossing window.*
Specify opposite corner: *Select the opposite corner of the crossing window.*

Erase Option

The **Erase** option in the **Trim** tool is used to erase the entities without canceling the **Trim** tool.

Select object to trim or shift-select to extend or [Fence/Crossing/Project/Edge/eRase/Undo]: **R** Enter
Select objects to erase or <exit>: *Select object to erase.*
Select objects to erase: *Select object to erase or* Enter
Select objects to erase: Enter

Undo Option

If you want to remove the previous change made by the **Trim** tool, enter **U** at the **Select object to trim or [Fence/Crossing/Project/Edge/eRase/Undo]** prompt.

EXERCISE 6   *Fillet, Chamfer, and Trim*

Draw the top illustration in Figure 5-38 and then use the **Fillet, Chamfer,** and **Trim** tools to obtain the next illustration given in the same figure. Assume the missing dimensions.

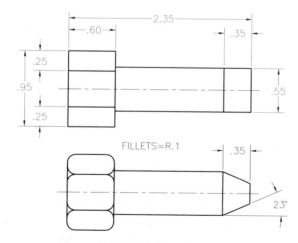

Figure 5-38 *Drawing for Exercise 6*

EXTENDING THE SKETCHED OBJECTS

Ribbon: Home > Modify > Trim/Extend drop-down > Extend
Toolbar: Modify > Extend   **Command:** EXTEND

The **Extend** tool may be considered as the opposite of the **Trim** tool. You can trim objects using the **Trim** tool. Whereas, you can extend lines, polylines, rays, and arcs to connect to other objects by using the **Extend** tool. However, you cannot extend closed loops. The command format of the **Extend** tool is similar to that of the **Trim** tool. You are required to select the boundary edges first. The boundary edges are those objects that the selected lines or arcs extend to meet. These edges can be lines, polylines, circles, arcs, ellipses, xlines, rays, splines, text, blocks, or even viewports. The prompt sequence that will follow when you choose the **Extend** tool is given next.

Current settings: Projection= UCS, Edge=Extend
Select boundary edges...
Select objects or <select all>: *Select boundary edges.*
Select objects: [Enter]
Select object to extend or shift-select to trim or [Fence/Crossing/Project/Edge/Undo]:

Various options used to extend the objects are discussed next.

Select object to extend Option

In this option, you have to specify the object that you want to extend to the selected boundary (Figure 5-39). This prompt is repeated until you press ENTER. Note that you can select a number of objects in a single **EXTEND** command.

Shift-select to trim Option

This option is used to switch to the trim mode in the **Extend** tool, if two entities are intersecting. You can press the SHIFT key and then select the object to be trimmed. In this case, the boundary edges are taken as the cutting edges.

Project Option

The **Project** option is used to extend objects in the 3d space. The prompt sequence that will follow when you choose the **Extend** tool is given next.

Current settings: Projection= UCS Edge= Extend
Select boundary edges...
Select objects or <select all>: *Select boundary edges.*
Select objects: [Enter]
Select object to extend or shift-select to trim or [Fence/Crossing/Project/Edge/Undo]: **P**
[Enter]
Enter a projection option [None/UCS/View] <current>:

The **None** option is used whenever the objects to be extended intersect with the boundary edge in 3D space. If you want to extend those objects that do not intersect the boundary edge in 3D space, use the **UCS** or **View** option. The **UCS** option is used to project the objects to the *XY* plane of the current UCS, while the **View** option is used to project the objects to the current view.

Edge Option

You can use this option whenever you want to extend the objects that do not actually intersect the boundary edge, but would intersect its edge if the boundary edges were extended (Figure 5-40).

If you enter **E** at the prompt, the selected object is extended to the implied boundary edge. If you enter **N** at the prompt, only those objects that would actually intersect the real boundary edge are extended (the default). The prompt sequence to use this option is given next.

Current settings: Projection= UCS Edge= Extend
Select boundary edges...
Select objects or <select all>: *Select the boundary edge.*
Select objects: Enter
Select object to extend or shift-select to trim or [Fence/Crossing/Project/Edge/Undo]: **E** Enter

Enter an implied extension mode [Extend/No extend] <current>: **E** Enter
Select object to extend or shift-select to trim or [Project/Edge/Undo]: *Select the line to extend.*

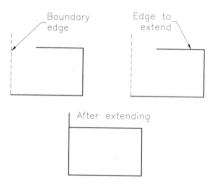

Figure 5-39 *Extending an edge*

Figure 5-40 *Extending an edge using the **Edge** option (**Extend**)*

Note
*The function of the **Fence** and **Crossing** options in the **Extend** tool is the same as that in the **Trim** tool.*

Undo Option
If you want to remove the previous change created by the **Extend** tool, enter **U** at the **Select object to extend or [Fence/Crossing/Project/Edge/Undo]:** prompt.

Trimming and Extending with Text, Region, or Spline
The **Trim** and **Extend** tools can be used with text, regions, or splines as edges (Figure 5-41). This makes the **Trim** and **Extend** tools the most useful editing tools. The **Trim** and **Extend** tools can also be used with arcs, elliptical arcs, splines, ellipses, 3D Pline, rays, and lines. The system variables **PROJMODE** and **EDGEMODE** determine how the **Trim** and **Extend** tools are executed, see Figure 5-42. The values that can be assigned to these variables are discussed next.

Value	PROJMODE	EDGEMODE
0	True 3D mode	Use regular edge without extension (default)
1	Project to current UCS *XY* plane (default)	Extend the edge to the natural boundary
2	Project to current view plane	

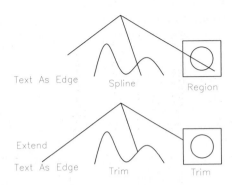

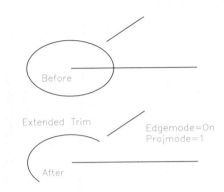

Figure 5-41 *Using the* **TRIM** *and* **EXTEND** *commands with text, spline, and region*

Figure 5-42 *Using the* **PROJMODE** *and* **EDGEMODE** *option to trim*

STRETCHING THE SKETCHED OBJECTS

Ribbon: Home > Modify > Stretch   **Toolbar:** Modify > Stretch
Command: STRETCH

This tool can be used to lengthen objects, shorten them, and alter their shapes, see Figure 5-43. To invoke this tool, choose the **Stretch** tool from the **Modify** panel; you will be prompted to select objects. Use the **Crossing** or **CPolygon** selection method to select the objects to be stretched. After selecting the objects, you will be prompted to specify the base point of displacement. Select the portion of the object that needs to be stretched; you will be prompted to specify the second point of displacement. Specify the new location; the object will lengthen or shorten.

It is recommended to use Crossing or CPolygon and select the objects to stretch, because if you use a Window, those objects that cross the window are not selected, and the selected objects will be moved but not stretched. The object selection and stretch specification process of **STRETCH** is a little unusual. You are really specifying two things: first, you are selecting objects. Second, you are specifying the portions of those selected objects to be stretched. You can use a **Crossing** or **CPolygon** selection to simultaneously specify both, or you can select objects by any method, and then use any window or crossing specification to specify what parts of those objects to stretch. Objects or portions of the selected objects completely within the window or crossing specification are moved. If the selected objects cross the window or crossing specification, their defining points within the window or crossing specification are moved, their defining points outside the window or crossing specification remain fixed, and the parts crossing the window or the crossing specification are stretched. Only the last window or the crossing specification determines what is stretched or moved. Figure 5-43 illustrates the usage of a crossing selection to simultaneously select two angled lines and specify that their right ends will be stretched. Alternatively, you can select the lines by any method and then use a Crossing selection (which will not actually select anything) to specify that the right ends of the lines will be stretched.

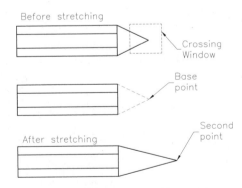

Figure 5-43 *Stretching the entities*

Note

The regions and solids cannot be stretched. If you select them, they will move instead of getting stretched.

LENGTHENING THE SKETCHED OBJECTS

Ribbon: Home > Modify > Lengthen   **Command:** LENGTHEN

Like the **Trim** and **Extend** tools, the **LENGTHEN** command can also be used to extend or shorten lines, polylines, elliptical arcs, and arcs. The **LENGTHEN** command has several options that allow you to change the length of objects by dynamically dragging the object endpoint, entering the delta value, entering the percentage value, or entering the total length of the object. This command also allows the repeated selection of the objects for editing. The prompt that will follow when you invoke the **LENGTHEN** command is given next.

Select an object or [DElta/Percent/Total/DYnamic]:

Select an object Option

This is the default option that returns the current length or the included angle of the selected object. If the object is a line, AutoCAD returns only the length. However, if the selected object is an arc, AutoCAD returns the length and the angle. The same prompt sequence will be displayed, after you select the object.

DElta Option

The **DElta** option is used to increase or decrease the length or angle of an object by defining the distance or angle by which the object will be extended. The delta value can be entered by entering a numerical value or by specifying two points. A positive value will increase (Extend) the length of the selected object; whereas a negative value will decrease it (Trim), see Figure 5-44. The prompt sequence to change the length of the object is given next.

Select an object or [DElta/Percent/Total/DYnamic]: **DE** ⏎
Enter delta length or [Angle] <current>: *Enter A for angle.*
Enter delta angle <current>: *Specify the delta angle.*
Select object to change or [Undo]: *Select the object to be extended.*
Select object to change or [Undo]: ⏎

Percent Option

The **Percent** option is used to extend or trim an object by defining the change as a percentage of the original length or the angle, see Figure 5-44. The current length of the line is taken as 100 percent. If you enter a value more than 100, the length will increase by that amount. Similarly, if you enter a value less than 100, then the length will decrease by that amount. For example, the numerical value 150 will increase the length by 50 percent and the numerical value 75 will decrease the length by 25 percent of the original value (negative values are not allowed).

Total Option

The **Total** option is used to extend or trim an object by defining the new total length or angle, see Figure 5-44. For example, if you enter a total length of 1.25, AutoCAD will automatically increase or decrease the length of the object so that the new length is 1.25. The value can be entered by entering a numerical value or by specifying two points. The object is shortened or lengthened with respect to the endpoint that is closest to the selection point. The selection point is determined from where the object was selected.

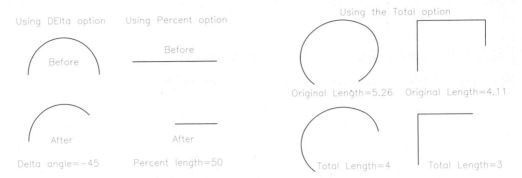

Figure 5-44 *Using the **DElta**, **Percent**, and **Total** options*

Tip
*By default, the **Lengthen** tool is not available in the **Modify** toolbar. To add the **Lengthen** tool in the **Modify** toolbar, right-click on any toolbar to display the shortcut menu. Next, choose **Customize** from the shortcut menu to display the **Customize User Interface** dialog box. Select the **Modify** option in the **Filter the command list by category** drop-down list; all tools that can be added or are currently available in the **Modify** toolbar will be displayed. Drag the **Lengthen** button from the **Command list** area to the **Modify** toolbar in the drawing area. Choose the **OK** button from the **Customize User Interface** dialog box.*

DYnamic

The **DYnamic** option allows you to dynamically change the length or angle of an object by specifying one of the endpoints and dragging it to a new location. The other end of the object remains fixed and unaffected by dragging. The angle of lines, radius of arcs, and shape of elliptical arcs remain unaffected on using this option.

ARRAYING THE SKETCHED OBJECTS*

Command:	ARRAY

In some drawings, you may need to create an object multiple times in a rectangular or circular arrangement. This type of arrangement can be obtained by creating an array of objects. An array is defined as the method of creating multiple copies of a selected object and arranging them in a rectangular or circular fashion. For example, to draw six chairs around a table, you can draw each chair separately or use the **Copy** tool to make multiple copies of the chair. But this is a very tedious process and also in this case, the alignment of the chairs will have to be adjusted. On the other hand, you can obtain the circular or rectangular arrangement of chairs easily by using the **Array** tool. All you have to do is to create one chair, and the remaining five will be created and arranged automatically around the table by using the **Array** tool. This method is more efficient and less time-consuming. In this type of arrangement, each resulting element of the array can be controlled separately. The arrays can be created by using the **Array** drop-down, refer to Figure 5-45. You can use this drop-down to create a rectangular, polar, or path array. The prompt sequence to be followed to create an array of an object is given next.

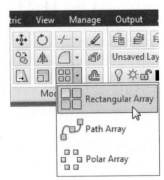

Figure 5-45 *Tools in the **Array** drop-down*

Command: **ARRAY**
Select objects: *Select the object*
Specify opposite corner: *Click to specify the opposite corner*
Select objects: [Enter]
Enter array type [Rectangular/PAth/POlar] <Rectangular>:

At the last prompt, you can specify the type of array to be created. The different types of arrays that can be created using the **ARRAY** tool are discussed next.

Rectangular Array

Ribbon:	Home > Modify > Array drop-down > Rectangular Array
Toolbar:	Modify > Rectangular Array

A rectangular array is formed by making copies of the selected object along the *X* and *Y* directions of an imaginary rectangle (along rows and columns). To create a rectangular array, choose the **Rectangular Array** tool from the **Modify** panel, see Figure 5-46; you will be prompted to select objects. Select the objects to be arrayed and press ENTER; you will be prompted to specify the opposite corner to specify the number of items in the array. Move the cursor and click in the drawing window to specify the opposite corner; you will be prompted to specify the spacing. Enter the value of spacing or click to specify the spacing for the array. Next, press ENTER or X; the array of the selected object will be created, refer to Figure 5-46. This figure also shows the location of the cursor while specifying the opposite corner. The prompt sequence to create the rectangular array of an object is given next.

Command: **ARRAYRECT**
Select objects: *Select the object*
Specify opposite corner: 1 found
Select objects: [Enter]
Type = Rectangular Associative = Yes
Specify opposite corner for number of items or [Base point/Angle/Count] <Count>: *Click to specify the opposite corner*
Specify opposite corner to space items or [Spacing] <Spacing>: *Enter the spacing or click to specify the spacing*
Press ENTER to accept or [ASsociative/Base point/Rows/Columns/Levels/eXit]<eXit>: *Press* [Enter] *or* **X**

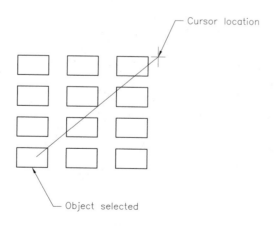

Figure 5-46 *Reactnagular array created*

While specifying the opposite corner, if you move the cursor along the vertical or horizontal direction, the array will be created in the specified direction. Figures 5-47 and 5-48 shows the array created in the vertical and horizontal directions, respectively.

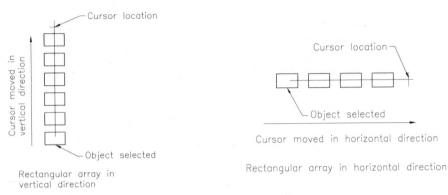

Figure 5-47 *A vertical rectangular array created*

Figure 5-48 *A horizontal rectangular array created*

Rectangular Array Options

There are three options to create a rectangular array. These options are displayed at the **Specify opposite corner for number of items or [Base point/Angle/Count] <Count>:** Command prompt. These options are discussed next.

Base point. This option is used to specify the base point of the array to be created. You can specify the base point by selecting a keypoint on the selected object. By default, the centroid is used as the base point for the array. However, you can specify any other point than the centroid of the object as the base point of the array. The prompt sequence that will follow when you invoke the **Base point** command is given next. In this prompt, the **Key point** option has been used.

> Command: **ARRAYRECT**
> Select objects: *Select the object*
> Specify opposite corner: 1 found
> Select objects: Enter
> Type = Rectangular Associative = Yes
> Specify opposite corner for number of items or [Base point/Angle/Count] <Count>: **B** Enter
> Specify base point or [Key point] <centroid> : **K** Enter
> Specify a key point on a source object as the base point: *Select any keypoint on the object*

Angle. This option is used to specify the rotation angle of the array, refer to Figure 5-49. If you enter **A** at the **Specify opposite corner for number of items or [Base point/Angle/Count] <Count>:** Command prompt, you will be prompted to specify the row axis angle. Enter the required angle value and then specify the opposite corner to specify the number of items. The prompt sequence that will follow when you invoke the **Angle** command is given next.

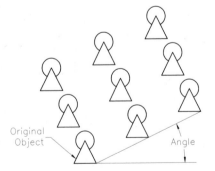

Figure 5-49 *Rotated rectangular array*

> Specify opposite corner for number of items or [Base point/Angle/Count] <Count>: **A** Enter

Specify row axis angle <current>: *Enter the angle value* [Enter]
Specify opposite corner for number of items or [Base point/Angle/Count] <Count>: *Click to specify the opposite corner*

Count. This option is used to specify the number of rows and columns in the array. If you enter **C** at the **Specify opposite corner for number of items or [Base point/Angle/Count] <Count>:** Command prompt, you will be prompted to specify the number of rows. Enter the number of rows in the array and press ENTER; you will be prompted to specify the number of columns in the array. Enter the number of columns and press ENTER; you will be prompted to specify the opposite corner or the spacing between the components of the array. Click to specify the opposite corner or enter the spacing at the Command prompt to create the array.

Specify opposite corner for number of items or [Base point/Angle/Count] <Count>: **C** [Enter]
Enter number of rows or [Expression] <current>: *Specify the number of rows*
Enter number of columns or [Expression] <current>: *Specify the number of columns*
Specify opposite corner to space items or [Spacing] <Spacing>: *Click to specify the opposite corner or specify the spacing between the components of the array*

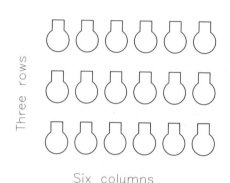

Figure 5-50 Rectangular array of three rows and six columns

You can also specify the number of columns and rows by using expressions or mathematical formulae. Figure 5-50 shows a rectangular array of three rows and six columns.

You can also specify the associativity of the objects in the array. Associativity determines whether the objects of the array are dependent upon each other or not. To specify the associativity, enter **ASSOCIATIVE** at the **Press Enter to accept or [ASsociative/Base point/Rows/Columns/Levels/eXit]<eXit>:** Command prompt; you will be prompted to specify whether the objects of the resultant array will be associative or non-associative. Enter **Yes** at the Command prompt to make the objects of the array associative. The associative array behaves as a single block. If you enter **No** at the Command prompt, the objects of the resultant array will be non-associative. If you edit one of the objects in the non-associative array, the other objects in the array will not be affected. After creating the array, press ENTER or X to exit the **ARRAYRECT** command.

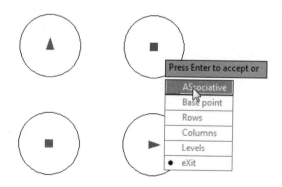

Figure 5-51 Dynamic input options for rectangular array

All aforementioned options for creating a rectangular array are also available at the **Dynamic Input**, see Figure 5-51, provided you have chosen the **Dynamic Input** button from the status bar.

Editing the Associative Rectangular Array with Grips

You can change the number of rows, columns, and the distance between the objects in the associative array dynamically. To do so, select an object from the array; grips will be displayed on it, refer to Figure 5-52. If you hover the cursor over the parent grip of a multiple row and column array, a tooltip with two options, **Move** and **Level Count** will be displayed, as shown in Figure 5-53. Choose the **Move** option from the tooltip to move the array. If you choose the **Level Count** option from the tooltip, you will be prompted to specify the number of levels. Levels allow you to create a 3D array. Specify the number of levels at the Command prompt or in the **Dynamic Input**; the 3D array will be created. You can view the 3D array in any of the isometric views. The isometric views can be invoked by using the ViewCube or by using the options in the **Views** panel of the **View** tab in the **Ribbon**.

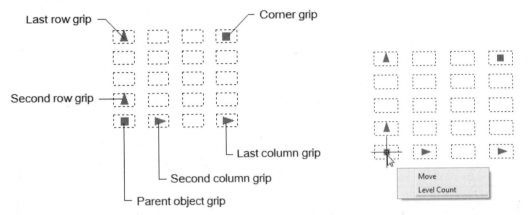

Figure 5-52 *Grips of a rectangular array*   **Figure 5-53** *The parent grip tooltip*

If you hover the cursor over the corner grip of the rectangular array, a tooltip with two options, **Row and Column Count** and **Total Row and Column Spacing**, will be displayed, as shown in Figure 5-54. Choose the **Row and Column Count** option from the tooltip to change the number of rows and columns in the array. Note that the array created using this option will have equal number of rows and columns. If you choose the **Total Row and Column Spacing** option, you will be prompted to specify the spacing between the rows and columns. Specify the distance value; the array with equal spacing between the objects in the rows and columns will be created.

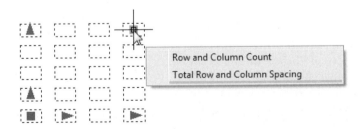

Figure 5-54 *The corner grip tooltip*

Hover the cursor over the second column grip of the rectangular array. If the **Dynamic Input** option is turned on, the value between the two columns will be displayed in the graphics window, refer to Figure 5-55. Click on the grip; an edit box will be displayed in the graphics window. Enter a distance in this edit box to specify the spacing between the columns of the array. If the **Dynamic Input** is turned off, you can specify the spacing between the columns by entering values at the Command prompt or by clicking in the graphics window.

Hover the cursor over the grip on the last column; a tooltip with three options, **Column Count**, **Total Column Spacing**, and **Axis Angle** will be displayed, as shown in Figure 5-56. Choose the **Column Count** option from the tooltip; an edit box showing the total number of columns will be displayed. If you choose the **Total Column Spacing** option, you will be prompted to specify the spacing between the columns. If the **Dynamic Input** is turned off, you can specify the spacing between the columns by entering a value at the Command prompt or by clicking in the graphics window. If you choose the **Axis Angle** option, you will be prompted to specify the axis angle. Specify the required angle at the **Dynamic Input** or the Command prompt and press ENTER.

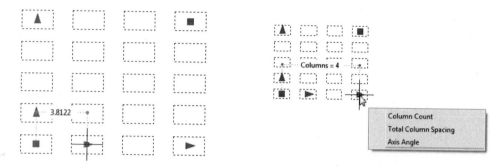

Figure 5-55 *Distance between two columns*   **Figure 5-56** *The last column grip tooltip*

Similarly, if you hover the cursor over the grip on the second row or the last row of the rectangular array, the options related to row count, row spacing, and axis angle will be displayed. These options are used to specify the parameters for rows of the array. Figure 5-57 shows a rectangular array with specific distances between rows and columns.

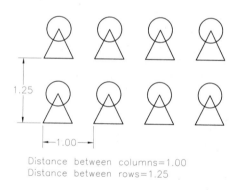

Figure 5-57 *A rectangular array with rows and columns at specific distance*

Editing the Associative Rectangular Array using the Array Contextual Tab

You can also edit the associative rectangular array by using the **Array** contextual tab that will be displayed when you select the rectangular array. The **Array** (Rectangular) contextual tab is shown in Figure 5-58.

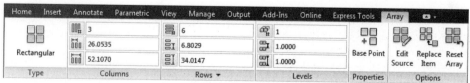

Figure 5-58 *The **Array** (Rectangular) contextual tab*

The different panels and their corresponding options in this tab are discussed next.

Type
This panel displays the type of array to be modified.

Columns
This panel displays three options which are used for changing the number of columns, the distance between two columns, and the total distance between the first and last columns. The description of these options are the same as those discussed in the previous topic.

Rows
This panel displays the options for changing the number of rows, the distance between two rows, and the total distance between the first and last rows. You can also specify the incremental elevation between the rows.

Levels
This panel displays three options for changing the number of levels, the distance between two levels, and the total distance between the first and last levels. The levels allow you to create a 3D array.

Properties
The **Base Point** tool in the **Properties** panel allows you to change the base point of the array.

Options
There are three tools in the **Options** panel: **Edit Source**, **Replace Item**, and **Reset Array**. These options are discussed next.

Edit Source. The **Edit Source** tool is used to edit the source of the array. On choosing this tool, you will be prompted to select an item in the array. Select any object in the array that will act as the source of the array. On selecting an item, the **Array Editing State** message box will be displayed, as shown in Figure 5-59, prompting you either to edit the source object or exit the editing state. Choose **OK** in this message box; the **Edit Array** panel will be added to the **Home** tab of the **Ribbon**, indicating that you are now in the editing mode. Next, modify the source object geometry. On modifying the source object geometry, you will notice that the other items of the array are also modified. Next, choose **Save Array** from the **Edit Array** panel; the changes made in the array will be saved.

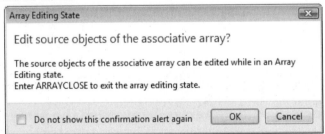

*Figure 5-59 The **Array Editing State** message box*

Figure 5-60 shows the rectangular array of a triangle and Figure 5-61 shows the source of the array being edited.

Replace Item. This tool is used to replace an item in the array with another item. On choosing this tool, you will be prompted to select the replacement object. Select the replacement object and press ENTER; you will be prompted to specify the base point of the replacement object.

Click to specify the base point of the replacement object. You can also specify a keypoint as the base point of the replacement object. To do so, enter **K** at the command prompt and then select any keypoint from the array. After specifying the base point, you will be prompted to specify the item to be replaced. Select an item from the array; the selected item will be replaced with the replacement item. You can continue replacing rest of the items in the array by selecting them. Figure 5-62 shows a rectangular array and the replacement object. Figure 5-63 shows the objects of the array after replacing with the selected object.

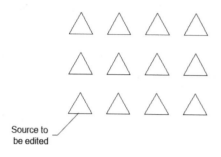

Figure 5-60 A rectangular array

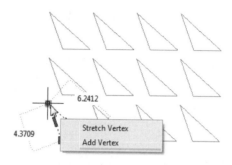

Figure 5-61 Source of the array being edited

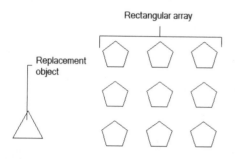

Figure 5-62 Array with the replacement object

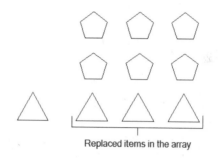

Figure 5-63 Replaced items in the first row of the array

Reset Array. This tool is used to restore the objects that have been deleted from the array. If you have made any replacements in the array, those replacements can also be reverted by using this tool.

Polar Array

Ribbon:	Home > Modify > Array drop-down > Polar Array
Toolbar:	Modify > Rectangular Array > Polar Array

Polar array is an arrangement of objects around a point in circular pattern. A polar array can be created by choosing the **Polar Array** tool from the **Modify** panel, refer to Figure 5-45. When you choose this tool, you will be prompted to select objects. Select objects to be arrayed; you will be prompted to specify the center point of the array. Select the center point of the array; you will be prompted to specify the number of items. Specify the number of items and press ENTER; you will be prompted to specify the fill angle. Enter an angle value at the Command prompt and then press ENTER; the polar array will be created, and you will prompted to exit the **Polar Array** tool. Press ENTER or X to exit the tool. Figure 5-64 shows a polar array with its center.

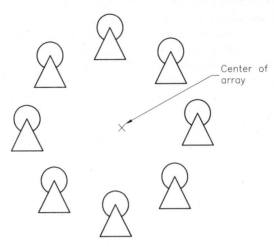

Figure 5-64 *Polar array created*

The prompt sequence to create the polar array of an object is given next.

Command: **ARRAYPOLAR**
Select objects: *Select the object*
Select objects: [Enter]
Type = Polar Associative = Yes
Specify center point of array or [Base point/Axis of rotation]: *Click to specify the center point of array*
Enter number of items or [Angle between/Expression] <current>: *Enter the number of items to be created*
Specify the angle to fill (+=ccw, -=cw) or [EXpression] <current>: *Enter the fill angle*
Press Enter to accept or [ASsociative/Base point/Items/Angle between/Fill angle/ROWs/Levels/ROTate items/eXit]<eXit>: **X** [Enter]

Polar Array Options

There are two options used for creating a polar array. These options are displayed at the **Specify center point of array or [Base point/Axis of rotation]:** Command prompt. These options are discussed next.

Base point. This option is used to specify the base point of the polar array. You can specify the base point by clicking on the graphics window or by selecting a keypoint on the selected object. By default, the centroid is used as the base point for the array. You can also click in the drawing window to specify the base point of the array. The prompt sequence followed while using the **Base point** command is given next. In this prompt, the **Key point** option has been used.

Command: **ARRAYPOLAR**
Select objects: *Select the object*
Specify opposite corner: 1 found
Select objects: [Enter]
Type = Polar Associative = Yes
Specify center point of array or [Base point/Axis of rotation]: **B** [Enter]
Specify base point or [Key point] <centroid> : **K** [Enter]
Specify a key point on the source object as the base point: *Select any keypoint on the object*

Axis of rotation. This option is used to specify the rotation angle of the array, refer to Figure 5-65. If you enter **A** at the **Specify center point of array or [Base point/Axis of rotation]:** Command prompt, you will be prompted to specify the first point on the axis of rotation. Click to specify the first point; you will be prompted to specify the second point on the axis of rotation. Click to specify the second point; the preview of the polar array depending upon the axis of rotation will be displayed. You can also specify the first and second points of the axis of rotation by using the **Dynamic Input**.

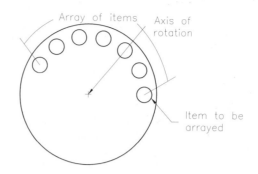

Figure 5-65 Polar array created around the axis of rotation

The prompt sequence followed while using the **Axis of rotation** option is given next.

> Specify center point of array or [Base point/Axis of rotation]: **A** [Enter]
> Specify first point on axis of rotation: *Click to specify the first point*
> Specify second point on axis of rotation: *Click to specify the second point*

After specifying the required polar array option, you will be prompted to enter the number of items to be created. The prompt sequence to enter the number of items in the polar array is given next.

> Enter number of items or [Angle between/Expression] <current>:

As evident from the Command prompt, you can specify the number of items in various ways. The first way is that you can enter the number of items directly at the Command prompt. The second way to specify the number of items is that you can enter the angle between the items in the polar array. If you enter **ANGLE** or **A** at the Command prompt, you will be prompted to enter the angle between the items. Specify the angle value at the Command prompt and press ENTER. You can also specify the angle by using expressions or mathematical formulae. Another way of specifying the number of items of the polar array is by specifying the fill angle at the **Specify number of items or [Fill angle/Expression] <current>:** Command prompt. This command prompt is displayed when you enter the angle between the items in the polar array. The third method to specify the number of items is to use expressions. The prompt sequence to specify the number of items in the polar array of an object is given next. In this prompt sequence, the **Angle between** option is used.

> Command: **ARRAYPOLAR**
> Select objects: Select the objects to be arrayed
> Select objects: [Enter]
> Type = Polar Associative = Yes
> Specify center point of array or [Base point/Axis of rotation]: *Click to specify the center point of the array*
> Enter number of items or [Angle between/Expression] <current>: **A** [Enter]
> Specify angle between items or [EXpression] <current>: *Enter the angle value*
> Specify number of items or [Fill angle/Expression] <current>: *Enter the number of items of the polar array*
> Press ENTER to accept or [ASsociative/Base point/Items/Angle between/Fill angle/ROWs/Levels/ROTate items/eXit]<eXit>: **X** [Enter]

Chapter 5

If you enter **F** at the **Specify number of items or [Fill angle/Expression] <current>:** Command prompt, you will be prompted to specify the fill angle. Enter the fill angle; all items of the polar array will be adjusted within the specified fill angle value. You can specify the direction of the polar array either clockwise or counterclockwise. If you enter a positive angle value at the Command prompt, the polar array will be created in the counterclockwise direction. If you enter a negative angle value, the polar array will be created in the clockwise direction.

After specifying all options for the polar array, enter X at the **Press Enter to accept or [ASsociative/Base point/Items/Angle between/Fill angle/ROWs/Levels/ROTate items/eXit]<eXit>:** Command prompt to exit the **Polar Array** tool. All options in the Command prompt, except the **ROWs** and **Levels**, have already been discussed. The **ROWs** option is used to specify the number of items in the radially outward direction, whereas the **Levels** option is used to create a 3D polar array. You can view the 3D polar array in any of the isometric views.

Figure 5-66 shows a polar array created by specifying the number or items and the angle between the items. Figure 5-67 shows a polar array created by specifying the number or items and the fill angle.

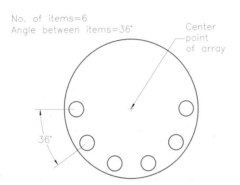

Figure 5-66 Array created by specifying the number of items and the angle between items

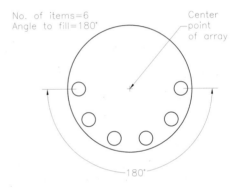

Figure 5-67 Array created by specifying the number of items and the fill angle

In an array, you can also make the objects associative. Associativity determines whether the objects of the array will be dependent upon each other or not. To specify the associativity, enter **ASSOCIATIVE** or **AS** at the **Press Enter to accept or [ASsociative/Base point/Items/Angle between/Fill angle/ROWs/Levels/ROTate items/eXit]<eXit>:** Command prompt. On doing so, you will be prompted to specify whether the objects of the array created will be associative or non-associative. Enter **Yes** at the Command prompt to make the objects of the array associative. An associative array behaves as a single block. If you enter **No** at the Command prompt, the objects of the array created will be non-associative. If you edit one of the objects in the non-associative array, the other objects in the array will not be affected. After creating the array, press ENTER or X to exit the **ARRAYPOLAR** command.

All aforementioned options for creating the polar array will also be available at the **Dynamic Input**, see Figure 5-68, provided the **Dynamic Input** button is chosen in the status bar.

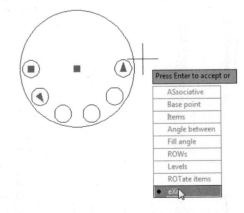

Figure 5-68 Dynamic input options for polar array

Editing the Associative Polar Array with Grips

You can change the number of rows, levels, and the fill angle between the objects in the associative array dynamically. To do so, select an object from the array; the grips will be displayed on the array, as shown in Figure 5-69. If you hover the cursor over the parent grip of a multiple row array, a tooltip with two options, **Stretch Radius** and **Level Count**, will be displayed, as shown in Figure 5-70. Choose the **Stretch Radius** option from the tooltip to change the radius of the array. If you choose the **Level Count** option from the tooltip, you will be prompted to specify the number of levels. Specify the number of levels at the Command prompt or at the **Dynamic Input**; the 3D array will be created. You can view the 3D array in any of the isometric views. The isometric views can be invoked by using the ViewCube or by using the options in the **Views** panel of the **View** tab in the **Ribbon**.

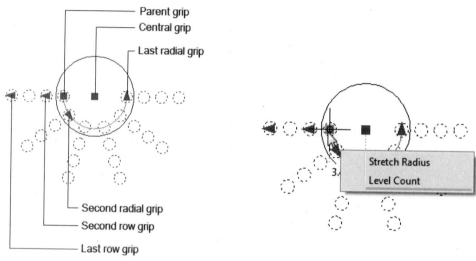

Figure 5-69 *Grips of a polar array*   *Figure 5-70* *The parent grip tooltip*

If you hover the cursor over the second radial grip of the polar array, the angle between the parent item and the second items of the array will be displayed, as shown in Figure 5-71. If you click at this time, an edit box will be displayed and you will be prompted to specify the angle between the items. Enter the desired angle value in this edit box.

If you hover the cursor over the last grip of the polar array, a tooltip with two options, **Item Count** and **Fill Angle**, will be displayed, as shown in Figure 5-72. If you choose **Item Count** from the tooltip, an edit box will be displayed and you will be prompted to specify the number of items. Enter the number of items in this edit box. If you choose **Fill Angle** from the tooltip, an edit box will be displayed and you will be prompted to specify the fill angle. Enter the fill angle in this edit box.

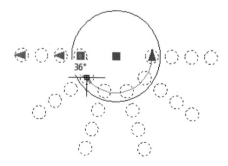

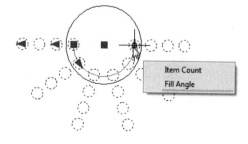

Figure 5-71 *Angle displayed between the parent and second radial grips*   *Figure 5-72* *Tooltip displayed on the last radial grip*

If you hover the cursor over the second row grip of the polar array, the current distance between two rows will be displayed, as shown in Figure 5-73. Click on this grip; an edit box will be displayed and you will be prompted to specify the distance between the two rows. Enter the new value in this edit box to specify the distance between the two rows.

If you hover the cursor over the last row grip of the polar array, a tooltip with two options, **Row Count** and **Total Row Spacing**, will be displayed, as shown in Figure 5-74. If you choose **Row Count** from the tooltip, an edit box will be displayed and you will be prompted to specify the number of rows. Specify the number of rows in the edit box to determine the number of rows in the array. If you choose **Total Row Spacing** from the tooltip, an edit box will be displayed and you will be prompted to specify the distance between the first and last row. Enter the desired value in this edit box.

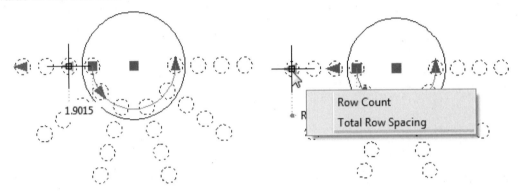

Figure 5-73 *Distance displayed between parent and second row grips*

Figure 5-74 *Tooltip displayed on the last row grip*

In case of single row polar array, if you hover the cursor over the parent object, a tooltip with three options, **Stretch Radius, Row Count**, and **Level Count**, will be displayed. You can specify the radius, number of rows, and the number of levels with these options.

Editing the Associative Polar Array using the Array Contextual Tab

You can also edit the associative polar array by using the **Array** contextual tab that will be displayed when you select the polar array. The **Array** contextual tab is shown in Figure 5-75.

Figure 5-75 *The **Array** (Polar) contextual tab*

The different panels and their corresponding options in this tab are discussed next.

Type

This panel displays the type of array to be modified.

Items

This panel displays three options that are used for changing the number of items, angle between two items, and total fill angle between the first and last items. These options are the same as those discussed in the previous topic.

Rows

This panel displays the options that are used for changing the number of rows, distance between two rows, and total distance between the first and last rows of the polar array. You can also specify the incremental elevation between the rows.

Levels

This panel displays three options that are used for changing the number of levels, distance between two levels, and total distance between the first and last levels. The levels allow you to create a 3D array.

Properties

By default, the **Rotate Items** button is chosen in this panel. As a result, the polar array will be created on rotating the items. Choose the **Base Point** tool if you want to change the base point of the array.

Options

There are three tools in the **Options** panel: **Edit Source**, **Replace Item**, and **Reset Array**. The options in this panel function in a way similar to the options in **Options** panel of the rectangular array.

Path Array

Ribbon:	Home > Modify > Array drop-down > Path Array
Toolbar:	Modify > Rectangular Array > Path Array

In AutoCAD, you can create an array of objects along a path called path arrays. The path can be a line, polyline, circle, helix, and so on. To create a path array, choose the **Path Array** tool from the **Modify** panel, see Figure 5-45. On doing so, you will be prompted to select objects. Select the objects to be arrayed and press ENTER; you will be prompted to select the path curve along which the object will be arrayed. Select the path curve; the preview of the path array will be displayed. You will notice that as you move the cursor, the number of items get arranged on the path and you will be prompted to enter the number of items in the array. Enter the desired number at the Command prompt and press ENTER; you will be prompted to specify the distance between the items along the path. You will notice that the spacing between the objects gets adjusted as you move the cursor. Specify the distance at the Command prompt and press ENTER; the array of the selected item will be created along the path and you will be prompted to exit the **Path Array** tool. Press ENTER to accept or X to exit the tool. Figure 5-76 shows the item and the path curve and Figure 5-77 shows the path array created.

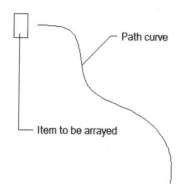

Figure 5-76 *Item and the path curve*

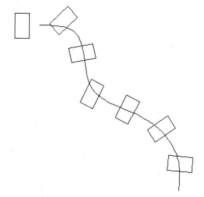

Figure 5-77 *Path array created*

The prompt sequence to create the path array of an object is given next.

Command: **ARRAYPATH**
Select objects: *Select the object*
Select objects: [Enter]
Type = Path Associative = Yes
Select path curve: *Select the path*
Enter number of items along path or [Orientation/Expression] <Orientation>: *Specify the number of items of the path array*
Specify the distance between items along path or [Divide/Total/Expression] <Divide evenly along path>: *Enter the distance value between the items*
Press Enter to accept or [ASsociative/Base point/Items/Rows/Levels/Align items/Z direction/ eXit]<eXit>: **X** [Enter]

For creating a path array, you need to specify the orientation of the items along the path curve. If you enter **O** or **Orientation** at the **Enter number of items along path or [Orientation/ Expression] <Orientation>:** Command prompt, you will be prompted to specify the base point. You can click at the desired point on the graphics window to specify the base point by the using a keypoint on the source object. You can also use the endpoint of the path curve as the base point. On specifying the base point, you will be prompted to specify the direction along which you have to align the object in the path. You can specify the alignment direction at the **Specify direction to align with path or [2Points/NORmal] <current>:** prompt. Click to specify the alignment direction with the path; the preview of the array will be displayed. You can also specify the alignment direction by specifying two points. This can be done by entering **2P** at the **Specify direction to align with path or [2Points/NORmal] <current>:** prompt. If you enter **NOR** at the **Specify direction to align with path or [2Points/NORmal] <current>:** prompt, the items will be oriented normal to the path and you will be prompted to specify the number of items. Specify the number of items to be created and press ENTER. Figures 5-78 and 5-79 show the path arrays created with the aligned direction and normal orientation of items, respectively.

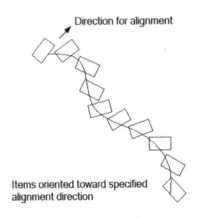

Figure 5-78 Path array created with aligned direction orientation

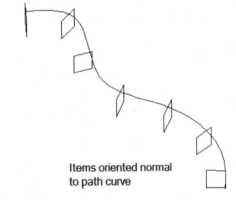

Figure 5-79 Path array created with normal orientation of items

After you have specified the number of items to be created, you need to specify the distance between the items at the **Specify the distance between items along path or [Divide/Total/ Expression] <Divide evenly along path>:** prompt. You can either enter a value or use other options at this prompt to specify the distance between items. If you enter **D** at the command prompt, the specified number of items will be placed in equally-spaced arrangement along the path, as shown in Figure 5-80. If you enter **T** at the Command prompt, you will be prompted to specify the total distance between the first and last items. Enter the required value at the command prompt and press ENTER; the path array will be created within the specified distance, as shown in Figure 5-81. You can also specify the distance between the items by using mathematical expressions or formulae.

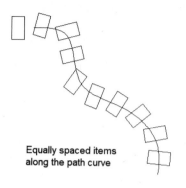

Equally spaced items along the path curve

*Figure 5-80 Path array created using the **Divide** option*

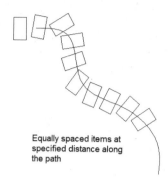

Equally spaced items at specified distance along the path

*Figure 5-81 Path array created using the **Total** option*

The prompt sequence to create the path array of an object by specifying the distance between the items by using the **Divide** option is given next.

Command: **ARRAYPATH**
Select objects: *Select the object*
Select objects: [Enter]
Type = Path Associative = Yes
Select path curve: *Select the path*
Enter number of items along path or [Orientation/Expression] <Orientation>: Specify the number of items to be created
Specify the distance between items along path or [Divide/Total/Expression] <Divide evenly along path>: **D** [Enter]
Press Enter to accept or [ASsociative/Base point/ Items/Rows/Levels/Align items/Z direction/ eXit]<eXit>: **X**

All options discussed here for creating the path array are also available at the **Dynamic Input**, see Figure 5-82, which will be available only if the **Dynamic Input** button is chosen from the status bar.

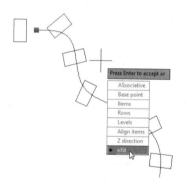

Figure 5-82 Dynamic input options for path array

Editing the Associative Path Array

Like rectangular and polar arrays, you can edit the associative path array by using either the grips or the **Array** contextual tab. The **Array** (Path) contextual tab, as shown in Figure 5-83, will be displayed when you select the path array. Most of the options in the **Array** contextual tab function the same way as those in the rectangular and polar arrays. The rest of the options are discussed next.

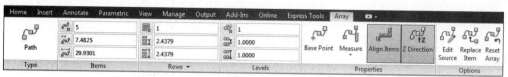

*Figure 5-83 The **Array** (Path) contextual tab*

Properties
By default, the **Align Items** and **Z Direction** tools are chosen in this panel. As a result, the path array will be aligned tangent to the path. Also, the original Z direction of the objects will be maintained in case the path is 3-dimensional. You can also use the **Measure** tool to specify the distance between the items of the path array.

MIRRORING THE SKETCHED OBJECTS

Ribbon: Home > Modify > Mirror	**Toolbar:** Modify > Mirror
Command: MIRROR	

The **Mirror** tool is used to create a mirror copy of the selected objects. The objects can be mirrored at any angle. This tool is helpful in drawing symmetrical figures. On invoking this tool, you will be prompted to select objects. On selecting the objects to be mirrored, you will be prompted to enter the first point of the mirror line and the second point of the mirror line. A mirror line is an imaginary line about which the objects are mirrored. You can specify the endpoints of the mirror line by specifying the points in the drawing area or by entering their coordinates. The mirror line can be specified at any angle. On selecting the first point of the mirror line, the preview of the mirrored objects will be displayed. Next, you need to specify the second endpoint of the mirror line, as shown in Figure 5-84. On selecting the second endpoint, you will be prompted to specify whether you want to delete the source object or not. Enter **Yes** to delete the source object and **No** to retain the source object, see Figure 5-85.

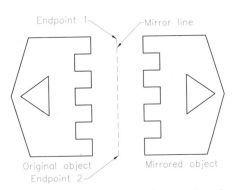

*Figure 5-84 Creating a mirror image of an object by using the **Mirror** tool*

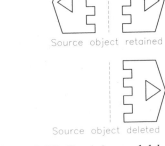

Figure 5-85 Retaining and deleting old objects after mirroring

The prompt sequence that will follow when you choose this tool is given next.

Select objects: *Select the objects to be mirrored.*
Select objects: Enter
Specify first point of mirror line: *Specify the first endpoint.*
Specify second point of mirror line: *Specify the second endpoint.*
Erase source objects? [Yes/No] <N>: *Enter **Y** for deletion, **N** for retaining previous objects.*

To mirror the objects at some angle, define the mirror line accordingly. For example, to mirror an object such that the mirrored object is placed at an angle of 90-degree from the original object, define the mirror line at an angle of 45-degree, see Figure 5-86.

Text Mirroring
By default, the **Mirror** tool reverses all objects, except the text. But, if you want the text to be mirrored (written backward) then you need to modify the value of the **MIRRTEXT** system variable. This variable has two values that are given next (Figure 5-87).

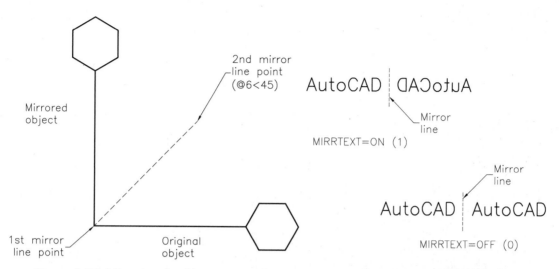

Figure 5-86 *Mirroring the object at an angle*

Figure 5-87 *Using the **MIRRTEXT** system variable for mirroring the text*

1 = Text is reversed in relation to the original object.
0 = Restricts the text from being reversed with respect to the original object. This is the default value.

BREAKING THE SKETCHED OBJECTS

Ribbon: Home > Modify > Break at Point or Break   **Toolbar:** Modify > Break at Point, Break
Command: BREAK

The **BREAK** command breaks an existing object into two parts or erases a portion of an object. This command can be used to remove a part of the selected objects or to break objects such as lines, arcs, circles, ellipses, xlines, rays, splines, and polylines. You can break the objects using the methods discussed next.

1 Point Option

 Choose the **Break at Point** tool from the **Modify** panel to break an object into two parts by specifying a breakpoint. On choosing this tool, you will be prompted to select the object to be broken. Once you have selected the object, you will be prompted to specify the first break point. Specify the first break point; the object will be broken. You can select the object to view the broken object.

2 Points Option

 To break an object by removing a portion of the object between two selected points, choose the **Break** tool from the **Modify** panel; you will be prompted to select the object. Select the object; you will be prompted to select the second break point. Select a point; the portion of the object between the two selected points will be removed. Note that the point at which you select an object becomes the first break point

The prompt sequence that will follow when you choose this tool is given next.

Select object: *Select the object to be broken.*
Specify second break point or [First point]: *Specify the second break point on the object.*

The object is broken and the portion in between the broken objects is removed, see Figure 5-88.

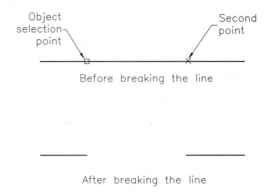

Figure 5-88 *Using the **2 Points** option for breaking the line*

2 Points Select Option

This option is similar to the **2 Points** option; the only difference being that, in this case, instead of making the selection point as the first break point, you need to specify a new first point, see Figure 5-89. The prompt sequence that will follow when you choose the **Break** tool is given next.

Select object: *Select the object to be broken*.
Enter second break point or [First point]: **F** Enter
Specify first break point: *Specify a new break point*.
Specify second break point: *Specify the second break point on the object*.

If you need to work on arcs or circles, make sure that you work in the counterclockwise direction, or you may end up cutting the wrong part. In this case, the second point should be selected in the counterclockwise direction with respect to the first one (Figure 5-90).

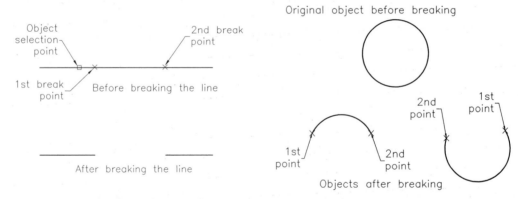

Figure 5-89 *Re-specifying the first break point for breaking the line*

Figure 5-90 *Breaking a circle*

You can use the **2 Points Select** option to break an object into two without removing a portion in between. This can be done by specifying the same point on the object as the first and second break points. If you specify the first break point on a line and the second break point beyond the end of the line, one complete end starting from the first break point will be removed.

EXERCISE 7   *Break*

Break a line at five different places and then erase the alternate segments. Next, draw a circle and break it into four equal parts.

PLACING POINTS AT SPECIFIED INTERVALS

Ribbon: Home > Draw > Measure   **Command:** MEASURE

While drawing, you may need to segment an object at fixed distances without actually dividing it. You can use the **Measure** tool to accomplish this task. This tool places points or blocks (nodes) on a given object at a specified distance. The shape of these points is determined by the **PDMODE** system variable and the size is determined by the **PDSIZE** variable.

To measure an object, invoke the **Measure** tool from the **Draw** panel and select the object to be measured; you will be prompted to specify the length. Enter the length; the points will be placed at regular intervals of the specified length. Instead of entering a value, you can also select two points that will be taken as distance. The **Measure** tool starts measuring the object from the endpoint that is closest to the object selection point. Note that markers are placed at equal intervals of the specified distance without considering whether the last segment is at the same distance or not.

When a circle is to be measured, an angle from the center is formed that is equal to the Snap rotation angle. This angle becomes the start point of measurement, and markers are placed at equal intervals in the counterclockwise direction.

In Figure 5-91, a line and a circle are measured. The Snap rotation angle is 0-degree. The **PDMODE** variable is set to 3 so that X marks are placed as markers. The prompt sequence that will follow when you invoke this option from the **Draw** menu is given next.

> Select object to measure: *Select the object to be measured.*
> Specify length of segment or [Block]: *Specify the length for measuring the object.*

You can also place blocks as markers (Figure 5-92), but the block must already be defined within the drawing. You can align these blocks with the object to be measured. The prompt sequence is given next.

> Select object to measure: *Select the object to be measured.*
> Specify length of segment or [Block]: **B** [Enter]
> Enter name of block to insert: *Enter the name of the block.*
> Align block with object? [Yes/No] <Y>: *Enter **Y** to align, **N** to not align.*
> Specify length of segment: *Enter the measuring distance.*

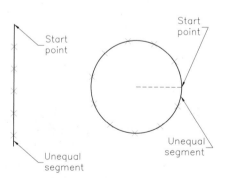

Figure 5-91 *Measuring a line and a circle*

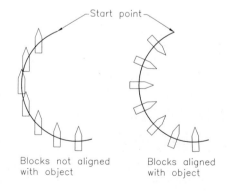

Figure 5-92 *Blocks placed as markers*

Note
You will learn more about creating and inserting blocks in Chapter 18 (Working with Blocks).

Chapter 5

DIVIDING THE SKETCHED OBJECTS

Ribbon: Home > Draw > Divide   **Command:** DIVIDE

The **Divide** tool is used to divide an object into a specified number of equal length segments without actually breaking it. This tool is similar to the **Measure** tool except that here you do not need to specify the distance. The **Divide** tool is used to calculate the total length of an object and place markers at equal intervals. This makes the last interval equal to the rest of the intervals. If you want to divide a line, invoke this tool and select the line. Then, enter the number of divisions or segments. The number of divisions entered can range from 2 to 32,767. The prompt sequence that will follow when you invoke this tool is given next.

> Select object to divide: *Select the object you want to divide.*
> Enter number of segments or [Block]: *Specify the number of segments.*

You can also place blocks as markers, but the blocks must be defined within the drawing. You can then align these blocks with the object to be measured. The prompt sequence to place a block as a marker is given next.

> Select object to divide: *Select the object to be measured.*
> Enter number of segments or [Block]: **B** Enter
> Enter name of block to insert: *Enter the name of the block.*
> Align block with object? [Yes/No] <Y>: *Enter **Y** to align, **N** to not align.*
> Enter number of segments: *Enter the number of segments.*

Figure 5-93 shows a line and a circle divided by using this tool and Figure 5-94 shows the use of blocks for dividing the selected segment.

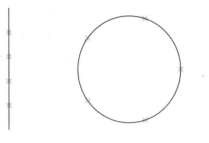

Dividing into five equal parts

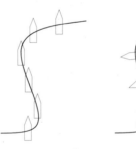

Blocks not aligned with object

Blocks aligned with object

Figure 5-93 *Dividing a line and a circle*

Figure 5-94 *Using blocks for dividing the segments*

> **Note**
> *The size and shape of the points placed by using the **Divide** and **Measure** tools are controlled by the **PDSIZE** and **PDMODE** system variables.*

JOINING THE SKETCHED OBJECTS

Ribbon: Home > Modify > Join   **Command:** JOIN

The **Join** tool is used to join two or more collinear lines or arcs lying on the same imaginary circle. The lines or arcs can have gap between them. This tool can also be used to join two or more polylines or splines, but they should be on the same plane and should not have any gap between them. Note that this tool can also be used to join the two endpoints of an arc to convert it into a circle. On invoking this tool, you will be prompted to select the source object. The source object is the one with which other objects will be joined.

The prompt sequence depends on the type of source object selected. The following prompt sequence is displayed when you invoke this tool:

Select source object or multiple objects to join at once: *Select a sketched entity (line, polyline, arc, elliptical arc, or spline)*

Joining Collinear Lines

If you have selected a line as the source object, you will be prompted to select lines to join to source. Remember that the lines will be joined only if they are collinear. If there is any gap between the endpoints of lines, the gap will be filled. The prompt sequence for joining two lines with a source line is given next.

Select lines to join to source: *Select the first line to join to the source.*
Select lines to join to source: *Select the second line to join to the source.*
Select lines to join to source: Enter
2 lines joined to source

Joining Arcs

Disjointed arcs can be joined by using the **Join** tool only if all of them lie on the same imaginary circle. The prompt sequence that follows when you select an arc as the source object is given next.

Select arcs to join to source or [cLose]: *Select the first arc to join to the source.*
Select arcs to join to source: *Select the second arc to join to the source or* Enter

After selecting the first arc, enter **L** at the **Select arcs to join to source or [cLose]** prompt to close the arc and convert it into a circle.

Joining Elliptical Arcs

Elliptical arcs can be joined by using the **Join** tool only if they lie on the same imaginary ellipse on which the source elliptical arc lies. The prompt sequence that follows when you select an elliptical arc as the source object is given next.

Select elliptical arcs to join to source or [cLose]: *Select the first elliptical arc to join to the source.*
Select elliptical arcs to join to source or: *Select the second elliptical arc to join to the source or* Enter

After selecting an elliptical arc, enter **L** at the **Select arcs to join to source or [cLose]** prompt to close the arc and convert it into an ellipse.

Joining Splines

In this case, you need to select a spline as the source object. On doing so, you will be prompted to select splines to join to the source. The splines will be joined only if they are coplanar and are connected at their ends. The prompt sequence that follows when you select **Spline** as the source object is given next.

Select any open curves to join to source: *Select the first spline to join to the source.*
Select any open curves to join to source: *Select the second spline to join to the source or* Enter

Chapter 5

Joining Polylines

Only the end-connected polylines can be joined using this tool. The prompt sequence when you select a polyline as the source object is given next.

Select objects to join to source: *Select a polyline/arc/line to join to the source.*
Select objects to join to source: *Select a polyline/arc/line to join to the source or* Enter

EXAMPLE 1   *Rectangular and Polar Arrays*

In this example, you will create the drawing of an end plate. The dimensions of the end plate are shown in Figure 5-95.

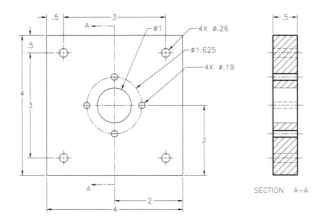

Figure 5-95 *Drawing for Example 1*

First you need to create a rectangle by using the **Rectangle** tool.

1. Start a new drawing file in **Drafting & Annotation** workspace.

2. Choose the **Rectangle** tool from **Home > Draw > Rectangle** drop-down and draw a rectangle of 4 units in length and breadth each. The prompt sequence to draw the rectangle is:

    Command: *Choose the **Rectangle** tool from the **Draw** panel*
    Specify first corner point or [Chamfer/Elevation/Fillet/Thickness/Width]: **0,0** Enter
    Specify other corner point or [Area/Dimensions/Rotation]: **@4,4** Enter

After creating the rectangle, you need to create three circles of different radii. These circles can be created by using the **Center, Radius** tool. You need to use the **From** object snap mode to create the third circle.

3. Choose the **Center, Radius** tool from **Home > Draw > Circle** drop-down and draw a circle of diameter **1** unit, as shown in Figure 5-96. The prompt sequence to draw the circle is:

    Command: *Choose the **Center, Radius** tool from the **Draw** panel*
    Specify center point for circle or [3P/2P/Ttr (tan tan radius)]: **2,2** Enter
    Specify radius of circle or [Diameter] <current>: **D** Enter
    Specify diameter of circle <current>: **1** Enter

4. Press ENTER to invoke the **Center, Radius** tool again and draw a circle of diameter **0.26** unit, as shown in Figure 5-97. You can also invoke the **Center, Radius** tool from the **Ribbon**. The prompt sequence to draw the circle is:

   Command: *Choose the **Center, Radius** tool from the **Draw** panel*
   Specify center point for circle or [3P/2P/Ttr (tan tan radius)]: **0.5, 0.5**
   Specify radius of circle or [Diameter] <current>: **D**
   Specify diameter of circle <current>: **0.26**

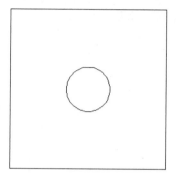

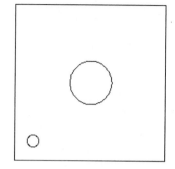

Figure 5-96 *Rectangle and circle created*   *Figure 5-97* *Circle of diameter 0.26 unit created*

5. Again, press ENTER to invoke the **Center, Radius** tool. You can also invoke this tool from the **Draw** panel. Next, press SHIFT and then right-click to invoke a shortcut menu.

6. Choose the **From** option from the shortcut menu and then draw the circle, as shown in Figure 5-98 by following the prompt given next.

   Command: *Choose the **Center, Radius** tool from the **Draw** panel*
   Specify center point for circle or [3P/2P/Ttr (tan tan radius)]: *Click the center of the circle of diameter 1 unit*
   _from Base point: <Offset>: **0.8125** ⏎
   Specify radius of circle or [Diameter] <current>: **D** ⏎
   Specify diameter of circle <0.2600>: **0.19** ⏎

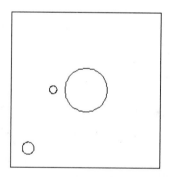

Figure 5-98 *Top view of the model*

After creating all circles, you need to create the pattern of two circles created last. The first array will be a rectangular array and second array will be a polar array. These arrays will be created by using the **Rectangular Array** and **Polar Array** tools.

7. Choose the **Rectangular Array** tool from **Draw > Modify > Array** drop-down and follow the Command prompt given next to create the rectangular array of the circle of diameter 0.26 unit shown in Figure 5-99.

   Command: *Choose the **Rectangular Array** tool from the **Modify** panel*
   Select objects: *Select the circle of diameter 0.26 unit and press ENTER*
   Select objects:
   Type = Rectangular Associative = Yes
   Specify opposite corner for number of items or [Base point/Angle/Count]<Count>: **C** ⏎

Enter number of rows or [Expression] <current>: **2** [Enter]
Enter number of columns or [Expression] <current>: **2** [Enter]
Specify opposite corner to space items or [Spacing] <Spacing>: **3,3** [Enter]
Press Enter to accept or [ASsociative/Base point/Rows/Columns/Levels/eXit]<eXit>: **X**

8. Next, choose the **Polar Array** tool from **Draw > Modify > Array** drop-down and follow the prompt given next to create the polar array of the circle of diameter **0.19** unit shown in Figure 5-100.

Command: *Choose the* **Polar Array** *tool from the* **Modify** *panel*
Select objects: *Select the circle of diameter* **0.19** *unit and press ENTER*
Select objects:
Type = Polar Associative = Yes
Specify center point of array or [Base point/Axis of rotation]: *Click at the center point of circle of diameter 1 unit*
Enter number of items or [Angle between/Expression] <current>: **4** [Enter]
Specify the angle to fill (+=ccw, -=cw) or [EXpression] <current>: **360** [Enter]
Press Enter to accept or [ASsociative/Base point/Items/Angle between/Fill angle/ROWs/Levels/ROTate items/eXit]<eXit>: **X**

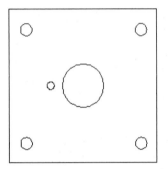

Figure 5-99 *Rectangular array created*

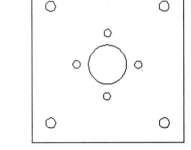

Figure 5-100 *Polar array created*

Self-Evaluation Test

Answer the following questions and then compare them to those given at the end of this chapter:

1. When you shift a group of objects by using the **Move** tool, the size and orientation of those objects will get changed. (T/F)

2. The **Copy** tool is used to make copies of the selected object, leaving the original object intact. (T/F)

3. A fillet cannot be created between two parallel and non-intersecting lines. (T/F)

4. On selecting an object after invoking the **Break** tool, the selection point becomes the first break point. (T/F)

5. You can create small or large circles, ellipses, and arcs by using the _____ tool, depending on the side to be offset.

6. The _____ tool prunes the objects that extend beyond the required point of intersection.

7. The offset distance is stored in the _____ system variable.

8. Instead of specifying the scale factor, you can use the _____ option to scale an object with reference to another object.

9. If the _____ system variable is set to 1, the mirrored text is not reversed with respect to the original object.

10. There are three types of arrays: _____ , _____ , and _____ .

Review Questions

Answer the following questions:

1. In the case of the **Through** option of the **Offset** tool, you do not need to specify a distance; you simply have to specify an offset point. (T/F)

2. Using the **Break** tool, you cannot break an object in two parts without removing a portion in between two selected points. (T/F)

3. While creating a fillet by using the **Fillet** tool, the extrusion direction of the selected object must be parallel to the Z axis of the UCS. (T/F)

4. In AutoCAD, you can rotate a rectangular array. (T/F)

5. Which of the following object selection methods is used to create a polygon that selects all the objects, touching or lying inside it?

    (a) **Last**   (b) **WPolygon**
    (c) **CPolygon**   (d) **Fence**

6. Which of the following commands is used to copy an existing object using a base point to another drawing?

    (a) **COPY**   (b) **COPYBASE**
    (c) **MOVE**   (d) None of these

7. Which of the following tools is used to change the size of an existing object with respect to an existing entity?

    (a) **Rotate**   (b) **Scale**
    (c) **Move**   (d) None of these

8. Which of these options of the **Lengthen** tool is used to modify the length of the selected entity such that irrespective of the original length, the entity acquires the specified length?

    (a) **DElta**   (b) **DYnamic**
    (c) **Percent**   (d) **Total**

Chapter 5

9. Which selection option is used to select the most recently drawn entity?

   (a) **Last**   (b) **Previous**
   (c) **Add**   (d) None of these

10. If a selected object is within a window or crossing specification, the _____ tool works like the **Move** tool.

11. The _____ option of the **Extend** tool is used to extend objects to the implied boundary.

12. If a polyline is not closed and while creating a fillet by using the **Fillet** tool with the **Polyline** option selected, the _____ corner is not filleted.

13. When the chamfer distance is zero, the chamfer created is in the form of a _____.

14. AutoCAD saves the previous selection set and lets you select it again by using the _____ selection option.

15. The _____ selection option is used to select objects by creating a section line touching the objects to be selected.

EXERCISE 8   *Divide*

Create the drawing shown in Figure 5-101. Use the **Divide** tool to divide the circle and use the **NODE** object snap to select the points. Assume the dimensions of the drawing.

Figure 5-101 *Drawing for Exercise 8*

EXERCISE 9

Create the drawing shown in Figure 5-102 and save it. Assume the missing dimensions.

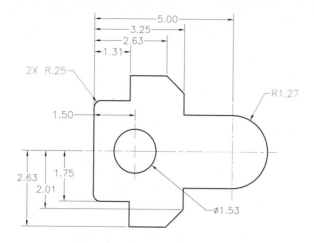

Figure 5-102 *Drawing for Exercise 9*

EXERCISE 10

Create the drawing shown in Figure 5-103 and save it.

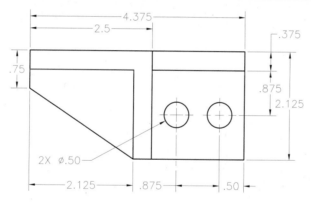

Figure 5-103 *Drawing for Exercise 10*

EXERCISE 11

Create the drawing shown in Figure 5-104 and save it.

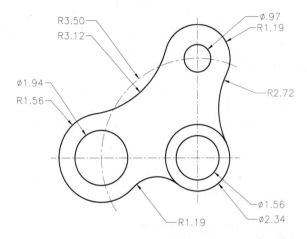

Figure 5-104 *Drawing for Exercise 11*

EXERCISE 12

Create the drawing shown in Figure 5-105 and save it. Assume the missing dimensions.

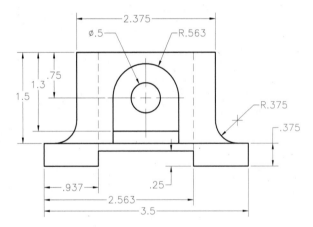

Figure 5-105 *Drawing for Exercise 12*

Problem-Solving Exercise 1

Create the drawing shown in Figure 5-106 and save it. Assume the missing dimensions.

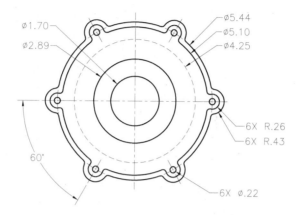

Figure 5-106 *Drawing for Problem-Solving Exercise 1*

Problem-Solving Exercise 2

Create the drawing shown in Figure 5-107 and save it. Assume the missing dimensions.

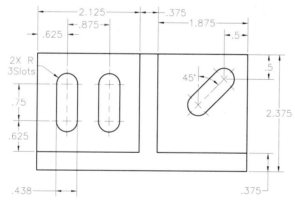

Figure 5-107 *Drawing for Problem-Solving Exercise 2*

Problem-Solving Exercise 3

Draw the dining table with chairs, as shown in Figure 5-108 and save the drawing. Assume the missing dimensions.

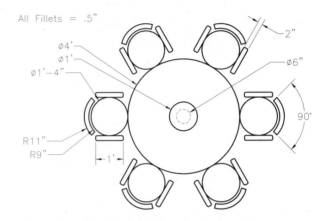

Figure 5-108 Drawing for Problem-Solving Exercise 3

Problem-Solving Exercise 4

Draw a reception table with chairs, as shown in Figure 5-109 and save the drawing. The dimensions of the chairs are the same as those used in Problem-Solving Exercise 3.

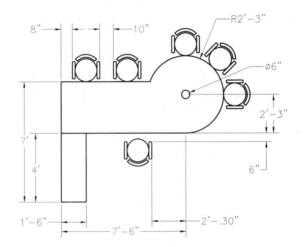

Figure 5-109 Drawing for Problem-Solving Exercise 4

Problem-Solving Exercise 5

Draw a center table with chairs, as shown in Figure 5-110, and save the drawing. The dimensions of the chairs are the same as those used in Problem-Solving Exercise 3.

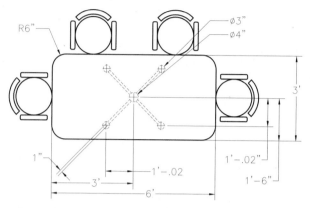

Figure 5-110 *Drawing for Problem-Solving Exercise 5*

Problem-Solving Exercise 6

Create the drawing shown in Figure 5-111 and save it. Refer to the note given in the drawing to create an arc of radius 30.

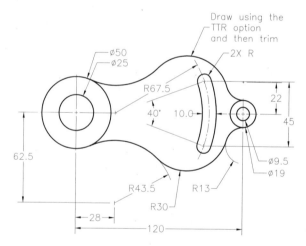

Figure 5-111 *Drawing for Problem-Solving Exercise 6*

Problem-Solving Exercise 7

Create the drawing shown in Figure 5-112 and save it.

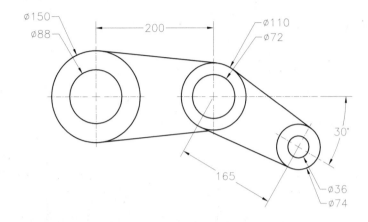

Figure 5-112 *Drawing for Problem-Solving Exercise 7*

Problem-Solving Exercise 8

Create the drawing shown in Figure 5-113 and save it.

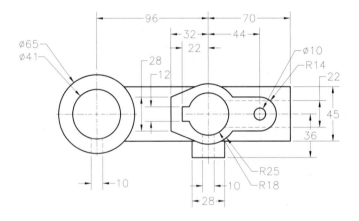

Figure 5-113 *Drawing for Problem-Solving Exercise 8*

Answers to Self-Evaluation Test

1. F, **2.** T, **3.** F, **4.** T, **5. OFFSET**, **6. TRIM**, **7. OFFSETDIST**, **8. Reference**, **9. MIRRTEXT**, **10. Rectangular**, **Polar**, **Path**

Chapter 6

Editing Sketched Objects-II

CHAPTER OBJECTIVES

In this chapter, you will learn:
- *To use grips and adjust their settings.*
- *To stretch, move, rotate, scale, and mirror objects with grips.*
- *To use the Match Properties tool to match properties of the selected objects.*
- *To use the Quick Select tool to select objects.*
- *To manage contents using the DesignCenter.*
- *To use the Inquiry tools.*
- *To use the REDRAW and REGEN commands.*
- *To use the ZOOM command and its options.*
- *To use the PAN and VIEW commands.*
- *To use Sheet Set Manager.*

KEY TERMS

- *Grips*
- *Properties Panel*
- *Quick Select*
- *DesignCenter*

- *Area*
- *Distance*
- *Locate Point*
- *List*

- *Time*
- *Status*
- *Regen*
- *Redraw*

- *Zoom*
- *Sheet Set Manager*

INTRODUCTION TO GRIPS

Grips provide a convenient and quick means of editing objects. Grips are small squares that are displayed on the key points of an object on selecting the object, as shown in Figure 6-1. If the grips are not displayed on selecting an object, invoke the Options dialog box, choose the Selection tab, and select the **Enable grips** check box. Using grips you can stretch, move, rotate, scale, and mirror objects, change properties, and load the Web browser. The number of grips depends on the selected object. For example, a line has three grip points and an arc has three triangular grips along with the small square grips. The triangular grips point towards the direction, in which the arc can be edited dynamically. Similarly, a circle has five grip points and a dimension (vertical, horizontal, or inclined) has five grips. Note that **Noun/verb selection** check box will be selected in the **Selection modes are** of the **Selection** tab of the **Options** dialog box. On clearing this selection box the small square (aperture box) will not be displayed at the intersection of crosshairs.

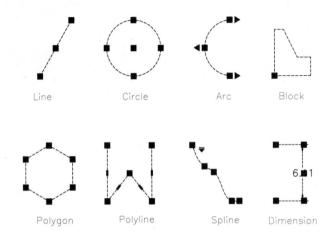

Line   Circle   Arc   Block

Polygon   Polyline   Spline   Dimension

Figure 6-1 *Grips displayed on various objects*

TYPES OF GRIPS

Grips are classified into three types: unselected grips, hover grips, and selected grips. The selected grips are also called hot grips. When you select an object, the grips are displayed at the definition points of the object and object is displayed as dashed line. These grips are called unselected grips and displayed in blue. Now, move the cursor over an unselected grip, and pause for a second, the grip will be displayed in orange. These grips are called hover grips. Dimensions corresponding to a hover grip are displayed when you place the cursor on the grip, as shown in Figure 6-2.

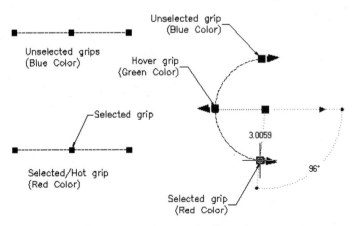

Unselected grip
(Blue Color)

Unselected grips
(Blue Color)

Hover grip
(Green Color)

Selected grip

Selected/Hot grip
(Red Color)

3.0059

96°

Selected grip
(Red Color)

Figure 6-2 *Hover grip dimensions*

If you select a grip, it is displayed in red color and called as hot grips. Once the grip is hot, the object can be edited. To cancel the grip, press ESC twice. If you press ESC once, the hot grip changes to an unselected grip.

ADJUSTING GRIP SETTINGS

Application Menu: Options   **Command:** OPTIONS

The grip settings can be adjusted by using the options in the **Selection** tab of the **Options** dialog box. This dialog box can also be invoked by choosing **Options** from the shortcut menu displayed upon right-clicking in the drawing area or by choosing the **Options** button from the **Application Menu**.

The options in the **Selection** tab of the **Options** dialog box (Figure 6-3) are discussed next.

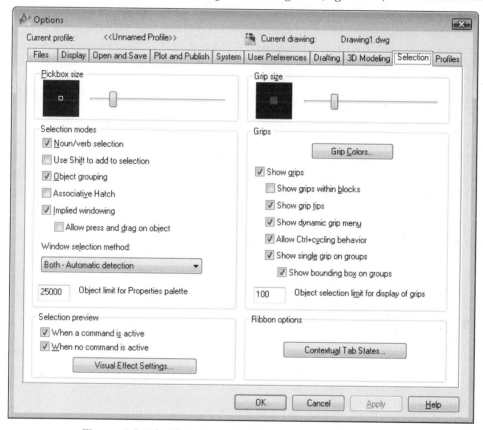

Figure 6-3   *The **Selection** tab of the **Options** dialog box*

Grip Size Area

The **Grip Size** area of the **Selection** tab of the **Options** dialog box consists of a slider bar and a rectangular box that displays the size of the grip. To adjust the size, move the slider box left or right. The size can also be adjusted by using the **GRIPSIZE** system variable. The **GRIPSIZE** variable is defined in pixels, and its value can range from 1 to 255 pixels.

Grips Area

To set the color of the unselected, hot, hover, and grip contour, choose the **Grip Colors** button; the **Grip Colors** dialog box will be displayed. Specify the colors in this dialog box. Grip contour is the color applied on the outline of the grip outline. You can also set the grip colors by using the following system variables:

Unselected grip color: GRIPCOLOR
Selected grip color: GRIPHOT
Hover grip color: GRIPHOVER
Grip contour color: GRIPCONTOUR

The **Grips** area has three check boxes: **Enable grips**, **Enable grips within blocks**, and **Enable grip tips**. Grips are displayed on selecting on object only if the **Enable Grips** check box is selected. They can also be enabled by setting the **GRIPS** system variable to 1. On selecting the **Enable grips within blocks** check box, every grip in the blocks is displayed. You can also enable the grips within a block by setting the value of the **GRIPBLOCK** system variable to 1 (On). If you clear the **Enable grips within blocks** check box, only one grip will be displayed at the insertion point of the block. Alternatively, set the **GRIPBLOCK** system variable to 0 (Off), see Figure 6-4. To display the grip tips when the cursor moves over the

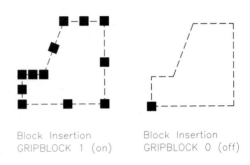

Block Insertion
GRIPBLOCK 1 (on)

Block Insertion
GRIPBLOCK 0 (off)

*Figure 6-4 Block insertion with **GRIPBLOCK** set to 1 and 0*

custom object that supports grip tips, select the **Enable grip tips** check box. If you disable this check box, the grip tips are not displayed when the cursor moves over the custom object. You can also enable the grip tips by setting the value of the **GRIPTIPS** system variable to 0 (Off). If **GRIPTIPS** is set to 1 (On), the grip tips for the custom object will be displayed.

Note

*If the block has a large number of objects, and if **GRIPBLOCK** is set to 1 (On), AutoCAD will display grips for every object in the block. Therefore, it is recommended that you set the system variable **GRIPBLOCK** to 0 or clear the **Enable grips within blocks** check box in the **Selection** tab of the **Options** dialog box.*

Object selection limit for display of grips

This text box is used to specify the maximum number of objects that can be selected at a single attempt for the display of grips. If you select objects more than that specified in the text box using a single selection method, the grips will not be displayed. Note that this limit is set only for those objects that are selected at a single attempt using any of the **Crossing**, **Window**, **Fence**, **All, or Multiple objects selection** options.

EDITING OBJECTS BY USING GRIPS

As mentioned earlier, you can perform different kinds of editing operations using the selected grip. The editing operations are discussed next.

Stretching the Objects by Using Grips (Stretch Mode)

If you select an object, the unselected grips are displayed at the definition points of the object. If you select a grip for editing, you are automatically in the **Stretch** mode. The **Stretch** mode has a function similar to that of the **Stretch** tool. When you select a grip, it acts as a base point and is called a base grip. To select several grips, press and hold the SHIFT key and then select the grips. Now, release the SHIFT key and select one of the hot grips and stretch it; all selected grips will be stretched. The geometry between the selected base grips will not be altered. You can also make copies of the selected objects or define a new base point. When selecting grips on text objects, blocks, midpoints of lines, centers of circles and ellipses, and point objects in the stretch mode, the selected objects are moved to a new location. The following example illustrates the use of the **Stretch** mode and other associated options.

1. Use the **Polyline** tool to draw a W-shaped figure and select the object; grips will be displayed at the endpoints and midpoints of each line, as shown in Figure 6-5.

2. Hold the SHIFT key and select the grips on the lower endpoints of the two vertical lines, refer to Figure 6-5(b). The selected grips will become hot grips and their color will change from blue to red.

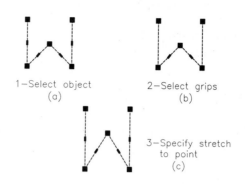

Figure 6-5 *Using the* ***Stretch*** *mode to stretch the lines*

Note

Make sure that you hold the SHIFT key before selecting even the first grip. You cannot hold the SHIFT key and select more grips if the first grip is selected without holding it.

3. Select one of the selected grips, and drag downwards; the profile will be stretched. When you select a grip, the following prompt is displayed in the Command prompt area.

   STRETCH
   Specify stretch point or [Base point/Copy/Undo/eXit]:

Tip

To stretch a line to desired length, enter a numerical value in the dimensional input box that will be displayed on selecting a grip of the line.

The Stretch mode has several options: **Base point**, **Copy**, **Undo**, and **eXit**. The **Base point** and **Copy** options can also be invoked from the shortcut menu, as shown in Figure 6-6. The **Base point** option is used to move an object with respect to a base point and the **Copy** option is used to make copies.

4. Select the grip where the two inclined lines intersect, as shown in Figure 6-7(a), and right-click to display a shortcut menu. Choose the **Copy** option and specify a point; the two inclined lines will be copied, as shown in Figure 6-7(b). You can also select a grip, press the CTRL key and specify the point to make multiple copies of the selected object.

5. To move a grip with respect to a base point, select the grip and choose the **Base Point** option from the shortcut menu. Next, select the base point and then specify the displacement point, as shown in Figure 6-7(d).

6. To terminate the grip editing mode, right-click when the grip is hot to display the shortcut menu and then choose **Exit**. You can also enter X at the Command prompt or press ESC to exit.

Note

*You can choose an option (**Copy** or **Base Point**) from the shortcut menu that can be invoked by right-clicking your pointing device after selecting a grip. The* *different modes can also be selected from the shortcut menu. You can also cycle through all the different modes by selecting a grip and pressing the ENTER key or the SPACEBAR.*

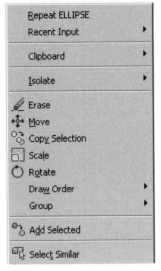

Figure 6-6 *Partial view of shortcut menu with editing options*

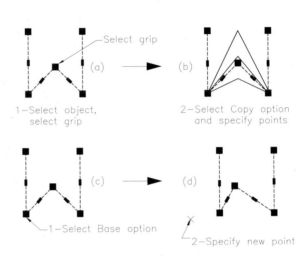

Figure 6-7 *Using the* **Copy** *and* **Base point** *options of the* **Stretch** *mode*

Moving the Objects by Using Grips (Move Mode)

The **Move** mode lets you move the selected objects to a new location. When you move objects, their size and angles do not change. You can also use this mode to make the copies of the selected objects or to redefine the base point. The following example illustrates the use of the **Move** mode:

1. Invoke the **Line** tool and draw the shape, as shown in Figure 6-8(a). When you select the objects, grips will be displayed at the definition points and the objects will be highlighted.

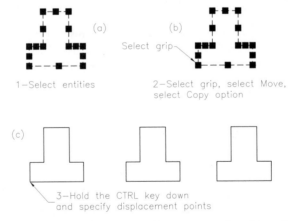

Figure 6-8 *Using the* **Move** *mode to move and make copies of the selected objects*

2. Select the grip located at the lower left corner and then choose **Move** from the shortcut menu. You can also select a grip and give a null response by pressing the SPACEBAR or ENTER key to invoke the **Move** mode; the following prompt will be displayed at the Command prompt area:

MOVE
Specify move point or [Base point/Copy/Undo/eXit]:

3. Specify the displacement point. You can also enter coordinates to specify the displacement. If you hold down the CTRL key and then specify the displacement point, the objects will be copied. Also, the distance between the first and the second object defines the snap offset for subsequent copies.

Rotating the Objects by Using Grips (Rotate Mode)

The **Rotate** mode allows you to rotate objects around the base point without changing their size. The options of the **Rotate** mode can be used to redefine the base point, specify a reference angle, or make multiple copies that are rotated about the specified base point. You can access the **Rotate** mode by selecting the grip and then choosing **Rotate** from the shortcut menu, or by giving a null response twice by pressing the SPACEBAR or the ENTER key. The following example illustrates the use of the **Rotate** mode:

1. Use the **Line** tool to draw the shape, as shown in Figure 6-9(a). When you select the objects, grips will be displayed at the definition points and the shape will be highlighted.

2. Select the grip located at the lower left corner and then invoke the **Rotate** mode. AutoCAD will display the following prompt.

    ROTATE
    Specify a rotation angle or [Base point/Copy/Undo/Reference/eXit]:

3. At this prompt, enter the rotation angle. By default, the rotation angle will be entered through the dimensional input below the cursor in the drawing area. AutoCAD will rotate the selected objects by the specified angle [Figure 6-9(b)].

4. To make a copy of the original drawing, as shown in Figure 6-9(c), select the objects, and then select the grip located at its lower left corner. Invoke the **Rotate** mode and then choose the **Copy** option from the shortcut menu or enter **C** (Copy) at the Command prompt or the **Copy** option from the dynamic preview. Enter the rotation angle. AutoCAD will rotate the copy of the object through the specified angle [Figure 6-9(d)].

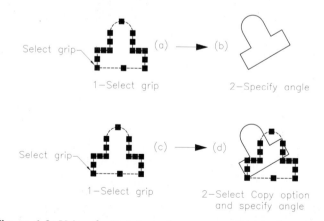

Figure 6-9 *Using the **Rotate** mode to rotate and make the copies of the selected objects*

5. To make another copy of the object, as shown in Figure 6-10(a), select it, and then select the grip at point (P0). Access the **Rotate** mode and the copy option as described earlier. Choose

the **Reference** option from the shortcut menu, enter **R** at the prompt sequence, or select the **Reference** option from the dynamic preview. The prompt sequence is given next.

**ROTATE (multiple) **
Specify rotation angle or [Base point/Copy/Undo/Reference/eXit]: **R**
Specify reference angle <0>: *Select the grip at (P1).*
Specify second point: *Select the grip at (P2).*
Specify new angle or [Base point/Copy/Undo/Reference/eXit]: **45**

In response to the **Specify reference angle <0>:** prompt, select the grips at points (P1) and (P2) to define the reference angle. When you enter the new angle, AutoCAD will rotate and insert a copy at the specified angle [Figure 6-10(c)]. For example, if the new angle is 45°, the selected objects will be rotated about the base point (P0) so that the line P1P2 makes a 45° angle with respect to the positive *X* axis.

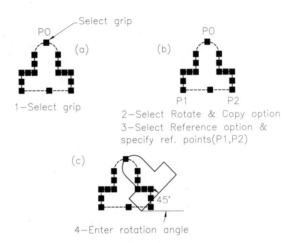

*Figure 6-10 Using the **Rotate** mode to rotate the selected objects by giving a reference angle*

Scaling the Objects by Using Grips (Scale Mode)

The **Scale** mode allows you to scale objects with respect to the base point without changing their orientation. The options of **Scale** mode can be used to redefine the base point, specify a reference length, or make multiple copies that are scaled with respect to the specified base point. You can access the **Scale** mode by selecting the grip and then choosing **Scale** from the shortcut menu, or giving a null response three times by pressing the SPACEBAR or the ENTER key. The following example illustrates the use of the **Scale** mode:

1. Use the **Polyline** tool to draw the shape shown in Figure 6-11(a). Then, select the objects; the grips will be displayed at the definition points.

2. Select the grip located at the lower left corner as the base grip, and then invoke the **Scale** mode. AutoCAD will display the following prompt in the Command prompt area.

   SCALE
   Specify scale factor or [Base point/Copy/Undo/Reference/eXit]:

3. At this prompt, enter the scale factor or move the cursor and select a point to specify a new size. AutoCAD will scale the selected objects by the specified scale factor [Figure 6-11(b)]. If the scale factor is less than 1 (<1), the objects will be scaled down by the specified factor.

If the scale factor is greater than 1 (>1), the objects will be scaled up.

4. To make a copy of the original drawing, as shown in Figure 6-11(c), select the objects, and then select the grip located at their lower left corner. Invoke the **Scale** mode. At the following prompt, enter **C** (Copy), and then enter **B** for the base point.

**SCALE (multiple) **
Specify scale factor or [Base point/Copy/Undo/Reference/eXit]: **B**

5. At the **Specify base point** prompt, select the point (P0) as the new base point, and then enter **R** at the following prompt:

**SCALE (multiple) **
Specify scale factor or [Base point/Copy/Undo/Reference/eXit]: **R**
Specify reference length <1.000>: *Select grips at (P1) and (P2).*

After specifying the reference length at the **Specify new length or [Base point/Copy/Reference/ eXit]** prompt, enter the actual length of the line. AutoCAD will scale the objects so that the length of the bottom edge is equal to the specified value [Figure 6-11(c)].

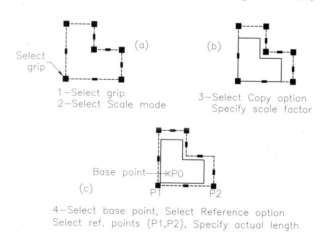

Figure 6-11 *Using the **Scale** mode to scale and make the copies of the selected objects*

Mirroring the Objects by Using Grips (Mirror Mode)

The **Mirror** mode allows you to mirror the objects across the mirror axis without changing their size. The mirror axis is defined by specifying two points. The first point is the base point, and the second point is the point that you select when AutoCAD prompts for the second point. The options of the **Mirror** mode can be used to redefine the base point and make a mirror copy of the objects. You can access the **Mirror** mode by selecting a grip and then choosing **Mirror** from the shortcut menu, or giving a null response four times by pressing the SPACEBAR or the ENTER key. The following is the example for the **Mirror** mode.

1. Use the **Polyline** tool to draw the shape shown in Figure 6-12(a). Select the objects; grips will be displayed at the definition points.

2. Select the grip located at the lower right corner (P1), and then invoke the **Mirror** mode. The following prompt will be displayed:

MIRROR
Specify second point or [Base point/Copy/Undo/eXit]:

3. At this prompt, specify the second point (P2); the selected objects will be mirrored, with line P1P2 as the mirror axis, as shown in Figure 6-12(b).

4. To make a copy of the original figure, as shown in Figure 6-12(c), select the object, and then select the grip located at its lower right corner (P1). Invoke the **Mirror** mode and then choose the **Copy** option to make a mirror image while retaining the original object. Alternatively, you can also hold down the SHIFT key and make several mirror copies by specifying the second point.

5. Select point (P2) in response to the prompt **Specify second point or [Base point/Copy/ Undo/eXit]**. AutoCAD will create a mirror image, and the original object will be retained.

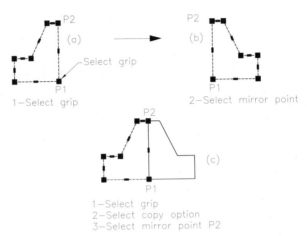

*Figure 6-12 Using the **Mirror** mode to create a mirror image of the selected objects*

Note
*You can use some editing commands such as **ERASE**, **MOVE**, **ROTATE**, **SCALE**, **MIRROR**, and **COPY** on an object with unselected grips. However, this is possible only if the **PICKFIRST** system variable is set to 1 (On).*

You cannot select an object when a grip is hot.

To remove an object from the selection set displaying grips, press the SHIFT key and then select the particular object. This object, which is removed from the selection set, will no longer be highlighted.

Editing a Polyline by Using Grips

In AutoCAD, when you select a polyline, three grips are displayed in each segment of the polyline. You can edit the polyline by using these grips. To do so, move the cursor on one of the grips and pause for a while; a tooltip will be displayed, as shown in Figure 6-13. To stretch the polyline, choose the **Stretch Vertex** option and specify a new point; the polyline will be stretched. After choosing the **Stretch Vertex** option, you can also invoke the **Base Point** or **Copy** option from the shortcut menu as discussed earlier.

To add a new vertex, choose the **Add Vertex** option from the tooltip, as shown in Figure 6-14; you will be prompted to specify a new vertex point. Specify a point; a new vertex will be added between the selected vertex and the next vertex, as shown in Figure 6-15. However, if the selected vertex is the last vertex, then a new segment will be added to the last vertex.

To remove a vertex, move the cursor over the vertex to be removed and choose the **Remove Vertex** option from the tooltip displayed.

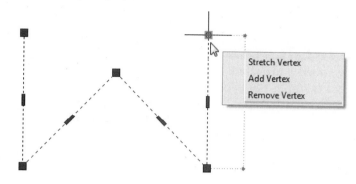

Figure 6-13 Tooltip displayed near the grip

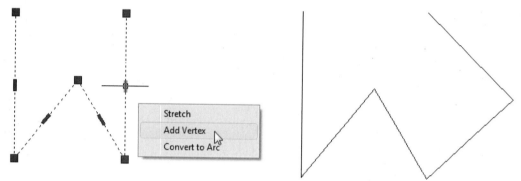

*Figure 6-14 Choosing the **Add Vertex** option from the tooltip*

Figure 6-15 Polyline after adding a new vertex

If you place the cursor on the middle grip, the tooltip will display the **Convert to Arc** option instead of the **Remove Vertex** option. Choose this option to convert (see Figure 6-16) the selected line to an arc. On choosing this option, you will be prompted to specify the midpoint of the arc. Specify a point on the periphery of the arc; the line will be converted to an arc, as shown in Figure 6-17. If you have drawn an arc using the **Polyline** tool or converted a polyline to an arc, you can convert the arc back to the line by choosing the **Convert to Line** option from the tool tip that is displayed on placing the cursor on the middle grip of the arc.

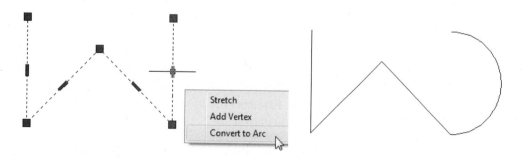

*Figure 6-16 Selecting the **Convert to Arc** option from the tooltip*

Figure 6-17 Polyline after converting a line to an arc

Chapter 6

Tip
After choosing an option from the tooltip, you can cycle through the options in the tooltip by pressing the CTRL key .

LOADING HYPERLINKS

If you have already added a hyperlink to the object, you can also use the grips to open a file associated with it. For example, the hyperlink could start a word processor, or activate the Web browser and load a Web page that is embedded in the selected object. To launch the Web browser that provides hyperlinks to other Web pages, select the URL-embedded object and then right-click to display the shortcut menu. In the shortcut menu, choose the **Hyperlink > Open** option and AutoCAD will automatically load the Web browser. When you move the cursor over or near the object that contains a hyperlink, AutoCAD displays the hyperlink information with the cursor.

Note
Working with hyperlinks is discussed in later chapters.

EDITING GRIPPED OBJECTS

You can also edit the properties of the gripped objects by using the **Properties** panel in the **Home** tab, see Figure 6-18. When you select objects without invoking a tool, the grips (rectangular boxes) will be displayed on the selected objects. The gripped objects are highlighted and will display grips (rectangular boxes) at their grip points. For example, to change the

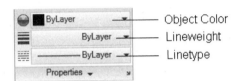

*Figure 6-18 The **Properties** panel*

color of the gripped objects, select the **Object Color** drop-down list in the **Properties** panel and then select a color from it. The color of the gripped objects will change to the selected color. Similarly, to change the layer, lineweight, or linetype of the gripped objects, select the required linetype, lineweight, or layer from the corresponding drop-down lists. If the gripped objects have different colors, linetypes, or lineweights, the **Object Color**, **Linetype**, and **Lineweight** drop-down lists will appear blank. You can also change the plot style of the selected objects using this panel.

CHANGING THE PROPERTIES USING THE PROPERTIES PALETTE

Ribbon:	View > Palettes > Properties
Command:	PROPERTIES
Toolbar:	Quick Access > Properties

As mentioned earlier, each object has properties associated with it such as the color, layer, linetype, line weight, and so on. You can modify these properties by using the **Properties** palette. To view this palette, choose the **Properties** button from the **Palettes** tab in the **View** tab; the **Properties** palette will be displayed, as shown in Figure 6-19. The **Properties** palette can also be displayed when you double-click on the object to be edited. The contents of the **Properties** palette change depending upon the objects selected. For example, if you select a text entity, the related properties such as its height, justification, style, rotation angle, obliquing factor, and so on, will be displayed.

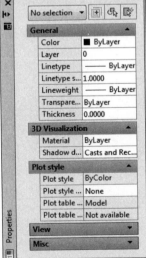

*Figure 6-19 The **Properties** palette for editing the properties of an entity*

The **Properties** palette can also be invoked from the shortcut menu. To do so, select an entity and right-click in the drawing area. Choose the **Properties** option from the shortcut menu. If you select more than one entity, the common properties of the selected entities will be displayed in the **Properties** palette. To change the properties of the selected entities, click in the cell next to the name of the property and change the values manually. Alternatively, you can choose from the available options in the drop-down list, if one is available. You can cycle through the options by double-clicking in the property cell.

Note
*Some of the options of the **Properties** palette have been explained in Chapter 4. Other options are explained in detail in later chapters.*

CHANGING THE PROPERTIES BY USING GRIPS

You can also use the grips to change the properties of a single or multiple objects. To change the properties, select the object to display the grips and then right-click to display the shortcut menu. Choose the **Properties** option to display the **Properties** palette. If you select a circle, AutoCAD will display **Circle** in the **No selection** drop-down list on the upper left corner of the **Properties** palette. Similarly, if you select text, **Text** is displayed in the drop-down list. If you select several objects, AutoCAD will display all objects in the selection drop-down list of the **Properties** palette. You can use this palette to change the properties (color, layer, linetype, linetypes scale, lineweight, thickness, and so on) of the gripped objects.

MATCHING THE PROPERTIES OF SKETCHED OBJECTS

Ribbon: Home > Clipboard > Match Properties   **Command:** MATCHPROP
Quick Access Toolbar: Match Properties (*Customize to Add*)

The **Match Properties** tool is used to apply properties like color, layer, linetype, and linetype scale of a source object to the selected objects. On invoking this tool, you will be prompted to select the source object and then the destination objects. The properties of the destination objects will be replaced with the properties of the source object. This is a transparent tool and can be used when another tool is active. The prompt sequence that will follow when you choose the **Match Properties** tool is given next.

Select source object: *Select the source object*.
Current active settings: Color Layer Ltype Ltscale Lineweight TransparencyThickness PlotStyle Dim Text Hatch Polyline Viewport Table Material Shadow display Multileader
Select destination object(s) or [Settings]:

If you select the destination object in the **Select destination object(s) or [Settings]** prompt, the properties of the source object will be forced on it. If you select the **Settings** option, the **Property Settings** dialog box will be displayed, as shown in Figure 6-20. The properties displayed are those of the source object. You can use this dialog box to edit the properties that are copied from the source to destination objects.

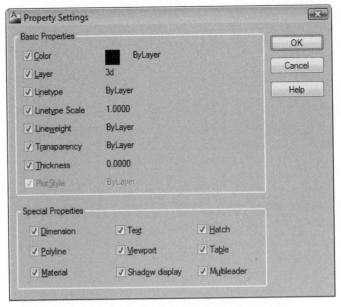

Figure 6-20 The **Property Settings** *dialog box*

QUICK SELECTION OF SKETCHED OBJECTS

Ribbon: Home > Utilities > Quick Select   **Command:** QSELECT

The **Quick Select** tool is used to create a new selection set that will either include or exclude all objects whose object type and property criteria matches as specified for the selection set. The **Quick Select** tool can be used to select the entities in the entire drawing or in the existing selection set. If a drawing is partially opened, the **Quick Select** tool does not consider the objects that are not loaded. The **Quick Select** tool can be invoked from the **Utilities** panel. You can also select the **Quick Select** option from the shortcut menu displayed by right clicking in the drawing area. On invoking this tool, the **Quick Select** dialog box will be displayed, see Figure 6-21. The **Quick Select** dialog box is used to specify the object filtering criteria and creates a selection set from it.

Apply to

The **Apply to** drop-down list specifies whether to apply the filtering criteria to the entire drawing or to the current selection set. If there is an existing selection set, the **Current selection** is the default value. Otherwise, the entire drawing

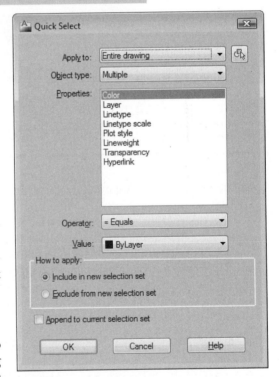

Figure 6-21 The **Quick Select** *dialog box*

is the default value. You can select the objects to create a selection set by choosing the **Select objects** button on the right side of this drop-down list. The **Quick Select** dialog box is temporarily closed when you choose this button and you will be prompted to select the objects. The dialog box will be displayed again once the selection set is made.

Object type

This drop-down list specifies the type of object to be filtered. It lists all the available object types and if some objects are selected, it lists all the selected object types. **Multiple** is the default setting.

Properties

This list box displays the properties to be filtered. All the properties related to the object type will be displayed in this list box. The property selected from this list box will define the options that will be available in the **Operator** and **Value** drop-down list.

Operator

This drop-down list is used to specify the range of the filter for the chosen property. The filters that are available are given next.

- Equals   =   • Not Equal   <>
- Greater than   >   • Less than   <
- Select All
- Wildcard Match (Available only for text objects that can be edited and if you select **Hyperlink** option from the **Properties** list.)

Note
*The **Value** drop-down list will not be available when you select **Select All** from the **Operator** drop-down list.*

Value

This drop-down list is used to specify the property value of the filter. If the values are known, it becomes a list of the available values from which you can select a value. Otherwise, you can enter a value.

How to apply Area

The options under this area are used to specify whether the filtered entities will be included or excluded from the new selection set. This area provides the following two radio buttons.

Include in new selection set

If this radio button is selected, the filtered entities will be included in the new selection set. If selected, this radio button creates a new selection set composed only of those objects that conform to the filtering criteria.

Exclude from new selection set

If this radio button is selected, the filtered entities will be excluded from the new selection set. This radio button creates a new selection set of objects that do not conform to the filtering criteria.

Append to current selection set

This creates a cumulative selection set by using multiple uses of Quick Select. It specifies whether the objects selected using the **Quick Select** tool replace the current selection set or append the current selection set.

Tip
*Quick Select supports objects and their properties that are created by some other applications. If the objects have properties other than AutoCAD, then the source application of the object should be running for the properties to be available in the **QSELECT**.*

CYCLING THROUGH SELECTION

In AutoCAD, you can cycle through the objects to be selected if they are overlapping or close to other entities. This new feature helps in selecting the entities easily and quickly. To enable this feature, choose the **Selection Cycling** button in the Status Bar. Now, if you move the cursor near an entity that has other entities nearby its, then the selection cycling symbol will be displayed, as shown in Figure 6-22. If you select anyone of the entities when the selection cycling symbol is displayed, the **Selection** list box with a list of the entities that can be selected will be displayed, as shown in Figure 6-23. You can select the entities from the **Selection** list box.

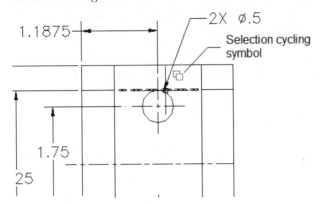

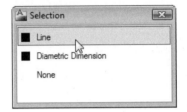

Figure 6-22 *The selection cycling symbol displayed*   *Figure 6-23* *The* **Selection** *list box*

MANAGING CONTENTS USING THE DesignCenter

| **Ribbon:** View > Palettes > DesignCenter | **Command:** ADCENTER |
| **Menu bar:** Tools > Palettes > DesignCenter | |

The **DesignCenter** window is used to locate and organize drawing data, and to insert blocks, layers, external references, and other customized drawing content. These contents can be selected from either your own files, local drives, a network, or the Internet. You can even access and use the contents between the files or from the Internet. You can use the **DesignCenter** to conveniently drag and drop any information that has been previously created into the current drawing. This powerful tool reduces the repetitive tasks of creating information that already exists. To invoke the **DesignCenter** window, choose the **DesignCenter** tool from the **Palettes** panel; the **DesignCenter** window will be displayed, see Figure 6-24.

To move the **DesignCenter**, drag the title bar located on the left of the window. To resize it, click on the borders and drag them to the right or left. Right-clicking on the title bar of the window displays a shortcut menu to move, resize, close, dock, and hide the **DesignCenter** window. The **Auto-Hide** button on the title bar acts as a toggle for hiding and displaying the **DesignCenter**. Also, double-clicking on the title bar of the window docks the **DesignCenter** window. To use this option, make sure that the **Allow Docking** option is selected from the shortcut menu that is displayed on right-clicking on the title bar.

Note
The **DesignCenter** *can be turned on and off by pressing the CTRL+2 keys.*

Figure 6-25 shows the buttons in the **DesignCenter** toolbar. When you choose the **Tree View Toggle** button on the **DesignCenter** toolbar, it displays the **Tree View** (Left Pane) with a tree view of the contents of the drives. If the tree view is not displayed, you can also right-click in the window and choose **Tree** from the shortcut menu that is displayed. Now, the window is divided into two parts, the **Tree View** (left pane) and the **Palette** (right pane). The **Palette** displays folders, files, objects in a drawing, images, Web-based content, and custom content. You can also resize both the **Tree View** and the **Palette** by clicking and dragging the bar between them to the right or the left.

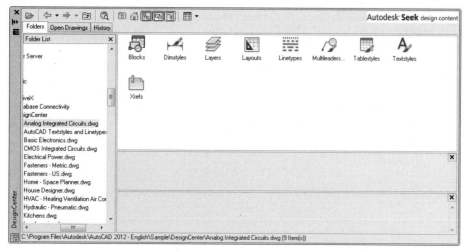

Figure 6-24 The **DesignCenter** *window*

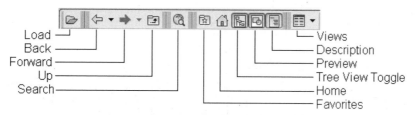

Figure 6-25 The **DesignCenter** *toolbar*

The **DesignCenter** has four tabs below the **DesignCenter** toolbar buttons. They are **Folders**, **Open drawings**, **History**, and **DC Online**. The description of these tabs is given next.

Folders Tab

The **Folders** tab lists all folders and files in the local and network drives. When this tab is chosen, the **Tree View** displays the tree view of the contents of the drives and the **Palette** displays various folders, and files in a drawing, images, and the Web-based content in the selected drive.

In the **Tree View**, you can browse the contents of any folder by clicking on the plus sign (+) adjacent to it to expand the view. Further, expanding the contents of a file displays the categories such as **Blocks**, **Dimstyles**, **Layers**, **Layouts**, **Linetypes**, **Textstyles**, and **Xrefs**. Clicking on any one of these categories in the **Tree View** displays the listing under the selected category in the **Palette** (Figure 6-26). Alternately, right-clicking on a particular folder, file, or category of the file contents displays a shortcut menu.

Choose the **Explore** option in this shortcut menu to further expand the selected folder, file, or category of contents to display the listing of the contents respectively. Choosing the **Preview** button from the toolbar displays an image of the selected object or file in a **Preview pane** below the Palette. Choosing the **Description** button displays a brief text description of the selected

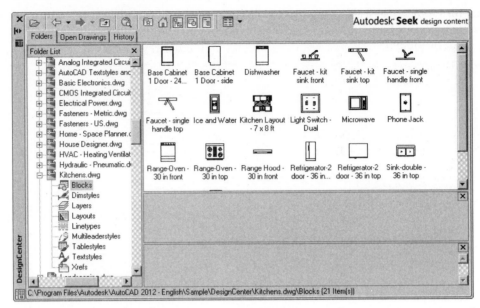

Figure 6-26 *The **DesignCenter** displaying the **Tree pane**, **Palette**, **Preview pane**, and **Description box***

item, if it has one in the **Description** box. When you click on a specific block name in the palette, its preview image and description that was defined earlier when creating the block are displayed in the **Preview pane** and the **Description box**, respectively.

You can drag and drop any of the contents into the current drawing, or add them by double-clicking on them. These are then reused as part of the current drawing. When you double-click on specific Xrefs and blocks, AutoCAD displays the **External Reference** dialog box and the **Insert** dialog box, respectively, to help in attaching the external reference and inserting the block, respectively. Right-clicking a block displays the options of **Insert Block**, **Copy**, or **Create Tool Palette** and right-clicking an Xref displays the options of **Attach Xref** or **Copy** in the shortcut menus. Similarly, when you double-click a layer, text style, dimstyle, layout, or linetype style, they also get added to the current drawing. If any of these named objects already exist in the current drawing, duplicate definition is ignored and it is not added again. When you right-click on a specific linetype, layer, textstyle, layout, or dimstyle in the palette, a shortcut menu is displayed that gives you an option to **Add** or **Copy**. The **Add** option directly adds the selected named object to the current drawing. The **Copy** option copies the specific named object to the clipboard from where you can paste it into a particular drawing.

Note
You will learn more about inserting blocks in Chapter 18, Working with Blocks.

Right-clicking a particular folder or file in the **Tree View** displays a shortcut menu. The various options in the shortcut menu, besides those discussed earlier, are **Add to Favorites**, **Organize Favorites**, **Create Tool Palette**, and **Set as Home**. **Add to Favorites** adds the selected file or folder to the **Favorites** folder, which contains the most often accessed files and folders. **Organize Favorites** allows you to reorganize the contents of the **Favorites** folder. When you select **Organize Favorites** from the shortcut menu, the **Autodesk** folder is opened in a window. **Create Tool Palette** adds the blocks of the selected file or folder to the **Tool Palettes** window, which contains the predefined blocks. **Set as Home** sets the selected file or folder as the **Home** folder. You will notice that when the **Design Center** tool is invoked the next time, the file that was last set as the **Home** folder is displayed selected in the **DesignCenter**.

Open Drawings Tab

The **Open Drawings** tab lists all the drawings that are open, including the current drawing which is being worked on. When you select this tab, the **Tree View** (left pane) displays the tree view of all the drawings that are currently open and the **Palette** (right pane) displays the various contents in the selected drawing.

History Tab

The **History** tab lists the locations of the most recently accessed files through the **DesignCenter**. When you select this tab, the **Tree View** (left pane) and the **Palette** (right pane) are replaced by a list box. Right-clicking a particular file displays a shortcut menu. The various options in the shortcut menu are **Explore, Folders, Open Drawings, Delete,** and **Search** etc. The **Explore** option invokes the **Folders** tab of the **DesignCenter** with the file selected in the **Tree View** and the contents in the selected file displayed in the **Palette View**. The **Folders** option invokes the **Folders** tab of the **DesignCenter**. The **Open Drawings** option invokes the **Open Drawings** tab of the **DesignCenter**. The **Delete** option deletes the selected drawing from the History list. The **Search** option allows you to search for drawings or named objects such as blocks, textstyles, dimstyles, layers, layouts, external references, or linetypes.

Autodesk Seek design content Link

The **Autodesk Seek design content** link allows you to download the *.dwg, .dwf, .pdf, .dgn* files of various products from the Autodesk online source for product specification and design files. To access the online source, click on the **Autodesk seek design content** in the **DesignCenter**; a new web page will be displayed. In this page, you can search for the required file type using the **Search** option. Once you get the required files, you can download them to your system.

Choosing the **Back** button in the **DesignCenter** toolbar displays the last item selected in the **DesignCenter**. If you pick the down arrow on the **Back** button, a list of the recently visited items is displayed. You can view the desired item in the **DesignCenter** by selecting it from the list. The **Forward** button is available only if you have chosen the **Back** button once. This button displays the same page as the current page before you choose the **Back** button. The **Up** button moves one level up in the tree structure from the current location. Choosing the **Favorites** button displays shortcuts to files and folders that are accessed frequently by you and are stored in the **Favorites** folder. This reduces the time you take to access these files or folders from their normal location. Choosing the **Tree View Toggle** button in the **DesignCenter** toolbar displays or hides the tree pane with the tree view of the contents in a hierarchical form. Choosing the **Load** button displays the **Load** dialog box, whose options are similar to those of the standard **Select file** dialog box. When you select a file here and choose the **Open** button, AutoCAD displays the selected file and its contents in the **DesignCenter**.

The **Views** button gives four display format options for the contents of the palette: **Large icons, Small icons, List,** and **Details**. The **List** option lists the contents in the palette, while the **Details** option gives a detailed list of the contents in the palette with the name, file size, and type.

Right-clicking in the palette displays a shortcut menu with all the options provided in the **DesignCenter** in addition to the **Add to Favorites, Organize favorites, Refresh,** and **Create Tool Palette of Blocks** options. The **Refresh** option refreshes the palette display if you have made any changes to it. The **Create Tool Palette of Blocks** option adds the drawings of the selected file or folder to the **Tool Palettes**, which contains the predefined blocks. The following example will illustrate how to use the **DesignCenter** to locate a drawing and then use its contents in a current drawing.

EXAMPLE 1   *DesignCenter*

Use the **DesignCenter** to locate and view the contents of the drawing *Kitchens.dwg*. Also, use the **DesignCenter** to insert a block from this drawing and import a layer and a textstyle from the *Blocks and Tables - Imperial.dwg* file located in the **Sample** folder. Use these to make a drawing of a Kitchen plan (*MyKitchen.dwg*) and then add text to it, as shown in Figure 6-27.

Figure 6-27 Drawing for Example 1

1. Open a new drawing using the **Start from Scratch** option. Make sure the **Imperial (feet and inches)** radio button is selected in the **Create New Drawing** dialog box.

2. Change the units to **Architectural** using the **Drawing Units** dialog box. Increase the limits to 10',10'. Use the **Zoom All** tool to increase the drawing display area.

3. Choose the **DesignCenter** tool from the **Palettes** panel in the **View** tab; the **DesignCenter** window is displayed at its default location.

4. In the **DesignCenter** toolbar, choose the **Tree View Toggle** button to display the **Tree View** and the **Palette** (if is not already displayed). Also, choose the **Preview** button, if it is not displayed already. You can resize the window, if needed, to view both the **Tree View** and the **Palette**, conveniently.

5. Choose the **Search** button in the **DesignCenter** to display the **Search** dialog box. Here, select **Drawings** from the **Look for** drop-down list and **C:** (or the drive in which AutoCAD 2012 is installed) from the **In** drop-down list. Select the **Search subfolders** check box. In the **Drawings** tab, type **Kitchens** in the **Search for the word(s)** edit box and select **File Name** from the **In the field(s)** drop-down list. Now, choose the **Search Now** button to commence the search. After the drawing has been located, its details and path are displayed in a list box at the bottom of the dialog box.

6. Now, right-click on *Kitchens.dwg* in the list box of the **Search** dialog box and choose **Load into Content Area** from the shortcut menu. You will notice that the drawing and its contents are displayed in the Tree view.

7. Close the **Search** dialog box, if it is still open.

8. Double-click on *Kitchens.dwg* in the Tree View to expand the tree view and display its contents,

in case they are not displayed. You can also expand the contents by clicking on the + sign located on the left of the file name in the Tree view.

9. Select **Blocks** in the Tree View to display the list of blocks in the drawing in the **Palette**. Using the left mouse button, drag and drop the block **Kitchen Layout-7x8 ft** in the current drawing.

10. Now, double-click on the *AutoCAD Textstyles and Linetypes.dwg* file located in the **DesignCenter** folder in the same directory to display its contents in the **Palette**.

11. Double-click on **Textstyles** to display the list of text styles in the **Palette**. Select **Dutch Bold Italic** in the **Palette** and drag and drop it in the current drawing. You can use this textstyle for adding text to the current drawing.

12. Invoke the **Multiline Text** tool and use the imported textstyle to add the text to the current drawing and complete it refer to Figure 6-20.

Note
1. To create the text, refer to Figure 6-20, you need to change the text height.
2. You can import Blocks, Dimstyles, Layers, and so on to the current drawing from any existing drawing.

13. Save the current drawing with the name *MyKitchen.dwg*.

MAKING INQUIRIES ABOUT OBJECTS AND DRAWINGS

When you create a drawing or examine an existing one, you often need some information about the entities in your drawing. The information can be about the distance from one location to another, the area of an object like a polygon or circle, coordinates of a location on the drawing, and so on. In manual drafting, you need to measure and calculate manually to get the required information. In AutoCAD, you need to use the inquiry commands to obtain information about the selected objects and they do not affect the drawings in any way. The following is the list of Inquiry commands:

AREA	DIST	ID	LIST	DBLIST
STATUS	TIME	DWGPROPS	MASSPROP	

Note
*The **MASSPROP** command will be discussed in Chapter 28.*

For most of the Inquiry commands, you are prompted to select objects. Once the selection is complete, AutoCAD switches from the graphics mode to the text mode, and all relevant information about the selected objects is displayed. For some commands, information is displayed in the AutoCAD Text Window. The display of the text screen can be tailored to your requirements using a pointing device. Therefore, by moving the text screen to one side, you can view the drawing screen and the text screen, simultaneously. If you choose the **minimize** button or the **Close** button, you will return to the graphics screen. You can also return to the graphics screen by entering the **GRAPHSCR** command at the Command prompt. Similarly, you can return to the AutoCAD Text Window by entering **TEXTSCR** at the Command prompt.

Measuring Area of Objects

Ribbon: Home > Utilities > Measure drop-down > Area   **Command:** AREA
Menu bar: Tools > Inquiry > Area   **Toolbar:** Inquiry > Distance > Area

Calculating the area of an object manually is time consuming. In AutoCAD, the **Area** tool is used to automatically calculate the area of an object in square units. This tool saves time when calculating the area of complicated or irregular shapes. You can use the default option of the **Area** tool to calculate the area and perimeter or circumference of the space enclosed by the sequence of specified points.

For example, to find the area of a pentagon created with the help of the **Line** tool, invoke the **Area** tool (Figure 6-28) and select all vertices of the polygon to define the shape whose area is to be found. This is the default method for determining the area of an object. Note that all the points you specify should be in a plane parallel to the *XY* plane of the current UCS. You can use object snaps such as the ENDpoint, INTersect, and TANgent, or running Osnaps, to select the vertices quickly and accurately. The prompt sequence that will be displayed on choosing the **Area** tool from the **Utilities** panel is given next.

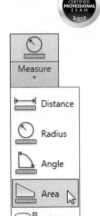

Figure 6-28 Tools in the Measure drop-down

Specify first corner point or [Object/Add area/Subtract area] <current>: *Specify first point.*
Specify next corner point or press ENTER for total: *Specify the second point.*
Specify next corner point or press ENTER for total: *Continue selecting until all points enclosing the area have been selected.*
Specify next corner point or press ENTER for total: Enter
Area = X, Perimeter = Y

Where, X represents the numerical value of the area and Y represents the circumference/perimeter.

It is not possible to determine the area of a curved object accurately with the default option of the **Area** tool. However, the approximate area of a curved object can be calculated by specifying several points on the curved entity. If the object whose area you want to find is not closed (formed of independent segments) and has curved lines, you should use the following steps to determine the accurate area of such an object:

1. Convert all the segments in that object into polylines using the **Edit Polyline** tool.
2. Join all the individual polylines into a single polyline. Once you have performed these operations, the object becomes closed and you can then use the **Object** option of the **Area** tool to determine the area.

Object Option

You can use the **Object** option to find the area of objects such as polygons, circles, polylines, regions, solids, and splines. If the selected object is a polyline or polygon, the area and perimeter of the polyline will be displayed. In case of open polylines, the area is calculated assuming that the last point is joined to the first point also the length of this segment is calculated not the perimeter. If the selected object is a circle, ellipse, or planar closed spline curve, information about its area and circumference will be displayed. For a solid, the surface area is displayed. For a 3D polyline, all vertices must lie in a plane parallel to the *XY* plane of the current UCS. The extrusion direction of a 2D polyline whose area you want to determine should be parallel to the *Z* axis of the current UCS. In case of polylines that have a width, the area and length of the polyline are calculated using the centerline. If any of these conditions is violated, an error message is displayed. The prompt sequence for using the Object option of the **Area** tool is given below.

Specify first corner point or [Object/Add/Subtract]: **O** Enter
Select objects: *Select an object.* Enter
Area = (X), Perimeter = (Y)

X represents the numerical value of the area, and Y represents the circumference/perimeter.

Tip
*The easiest and most accurate way to find the area of a region enclosed by multiple objects is to use the **Boundary** tool to create a polyline, and then use the **Object** option.*

Add area Option

Sometimes you want to add areas of different objects to determine a total area. For example, while drawing the plan of a house, you need to add the areas of all the rooms to get the total floor area. In such cases, you can use the **Add area** option of the **Area** tool. The **Object** option adds the areas and perimeters of selected objects. After choosing this option specify the first corner point at the **Specify first corner point or [Object/Subtract area/eXit]** prompt. Then continue specifying other corner points and press ENTER; the selected region will be displayed in green color and its area and perimeter will be displayed at the Command prompt. As the **Add area** option is on, you will be again prompted to select next corner point. Keep selecting all corner points to define the area and press ENTER twice to exit the command; the total area will be displayed.

After invoking the **Add area** option, you can also invoke the **Object** option, at the **Specify first corner point or [Object/Subtract area/eXit]** prompt, to select the objects. On selecting the objects, the regions are displayed in green color, and the area and length of the selected object as well as the combined area will be displayed at the Command prompt. In this manner, you can add areas of different objects. Until the **Add area** mode is active, the string **ADD mode** is displayed along with all subsequent object selection prompts to remind you that the **Add** mode is active.

If the polygon whose area is to be added is not closed, the area and perimeter are calculated assuming that there exists a line that connects the first point to the last point. The length of this imaginary is added in the perimeter.

Subtract Option

The **Subtract area** option is used along with the **Add area** option to subtract an area from the previously added area. For example, to calculate the area of the closed profile after subtracting the areas of the circles in Figure 6-29, you need to use the **Subtract area** option. First, invoke the **Area** tool and calculate the area of the closed profile using the **Add area** option; the region will be enclosed in green color area and the total area and the perimeter will be displayed. Give a single null response to exit the **Add area** option. Then, choose the **Subtract area** option and invoke the **Object** option. Now, select both two circles; the circles will be displayed in different colors and their areas will be displayed. Also, the total area will be displayed.

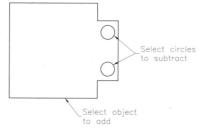

***Figure 6-29** Measuring the area of a sketch using the **Add** and **Subtract** options*

The prompt sequence for these two modes is given next.

Specify first corner point or [Object/Add area/Subtract area/eXit]: **A** Enter
Specify first corner point or [Object/Subtract area/eXit]: **O** Enter
(ADD mode) Select objects: *Select the polyline.*
Area = 2.4438, Perimeter = 6.4999
Total area = 2.4438
(ADD mode) Select objects: Enter
Specify first corner point or [Object/Subtract area/eXit]: **S** Enter
Specify first corner point or [Object/Add area/eXit]: **O** Enter
(SUBTRACT mode) Select object: *Select the circle.*
Area = 0.0495, Circumference = 0.7890
Total area = 2.3943
(SUBTRACT mode) Select objects: *Select the second circle.*
Area = 0.0495, Circumference = 0.7890
Total area = 2.3448
(SUBTRACT mode) Select object: Enter
Specify first corner point or [Object/Add area/eXit]: Enter

The **AREA** and **PERIMETER** system variables hold the area and perimeter (or circumference in the case of circles) of the previously selected polyline (or circle). Whenever you use the **AREA** command, the **AREA** variable is reset to zero.

Tip
*If an architect wants to calculate the area of flooring and skirting in a room, the **AREA** command provides you with its area and perimeter. You can use these parameters to calculate the skirting.*

Measuring the Distance between Two Points

Ribbon: Home > Utilities > Measure drop-down > Distance	**Command:** DIST
Menu bar: Tools > Inquiry > Distance	**Toolbar:** Inquiry > Distance

The **Distance** tool is used to measure the distance between two selected points, as shown in Figure 6-30. The angles that the selected points make with the *X* axis and the *XY* plane are also displayed. The measurements are displayed in the current units. Delta X (horizontal displacement), delta Y (vertical displacement), and delta Z are also displayed. The distance computed by the **Distance** tool is saved in the **DISTANCE** variable.

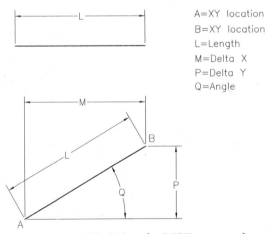

*Figure 6-30 Using the **DIST** command*

The prompt sequence that will follow when you choose the **Distance** tool is given next.

Specify first point: *Specify a point.*
Specify second point: *Specify a point.*

AutoCAD returns the following information.

Distance = *Calculated distance between the two points.*
Angle in XY plane = *Angle between the two points in the XY plane.*
Angle from XY plane = *Angle the specified points make with the XY plane.*
Delta X = *Change in X*, Delta Y = *Change in Y*, Delta Z = *Change in Z.*

If you give a null response at the **Specify first point** prompt and enter a single number or fraction at the **Specify second point** prompt, AutoCAD will convert it into the current unit of measurement and display in the command line. Also, the other values will be calculated with respect to the last used point.

Command: _MEASUREGEOM
Enter an option [Distance/Radius/Angle/ARea/Volume] <Distance>: _distance
Specify first point: `Enter`
Specify second point: 3-3/4 *(Enter a number or a fraction.)*
Distance = 3.7500

Note
The Z coordinate is used in 3D distances. If you do not specify the Z coordinates of the two points between which you want to know the distance, AutoCAD takes the current elevation as the Z coordinate value.

Identifying the Location of a Point

Ribbon: Home > Utilities > ID Point	**Command:** ID
Menu bar: Tools > Inquiry > ID Point	**Toolbar:** Inquiry > Locate Point

The **ID Point** tool is used to identify the position of a point in terms of its X, Y, and Z coordinates. The prompt sequence that will follow when you choose the **ID Point** tool from the **Utilities** panel is given next.

Specify point: *Specify the point to be identified.*
X = X coordinate   Y = Y coordinate   Z = Z coordinate

The current elevation is taken as the Z coordinate value. If an **Osnap** mode is used to snap to a 3D object in response to the **Specify point** prompt, the Z coordinate displayed will be that of the selected feature of the 3D object. You can also use the **ID** command to identify a location in the drawing area by entering the coordinate values. For example, the following is the prompt sequence to find where the position X = 2.345, Y = 3.674, and Z = 1.0000 is located on the screen.

Specify point: 2.345,3.674,1.00 `Enter`
X = 2.345 Y = 3.674 Z = 1.0000

The coordinates of the point specified in the **ID Point** tool are saved in the **LASTPOINT** system variable. You can locate a point with respect to the **ID** point by using the relative or polar coordinate system. You can also snap to the last point by typing @ at the **Specify point** prompt.

Listing Information about Objects

Ribbon: Home > Properties > List	**Command:** LIST
Menu bar: Tools > Inquiry > List	**Toolbar:** Inquiry > List

 The **List** tool displays all the information pertaining to the selected objects. The information are displayed in the **AutoCAD Text Window**. The prompt sequence that follows, when you invoke the **List** tool from the **Properties** panel is given next.

Select objects: *Select objects whose data you want to list.*
Select objects: Enter

Once you select the objects to be listed, a new AutoCAD Text Window will be displayed. The information displayed (listed) varies from object to object. The information on an object's type, its coordinate position with respect to the current UCS (user coordinate system), the name of the layer on which it is drawn, and whether the object is in model space or paper space is listed for all types of objects. If the color, lineweight, and the linetype are not BYLAYER, they are also listed. Also, if the thickness of the object is greater than 0, that is also displayed. The elevation value is displayed in the form of a Z coordinate (in the case of 3D objects). If an object has an extrusion direction different from the Z axis of the current UCS, the object's extrusion direction is also provided.

More information based on the objects in the drawing is also provided. For example, the following information is displayed for a line.

1. The coordinates of the endpoints of the line.
2. Its length (in 3D).
3. The angle made by the line with respect to the X axis of the current UCS.
4. The angle made by the line with respect to the XY plane of the current UCS.
5. Delta X, Delta Y, and Delta Z: this is the change in each of the three coordinates from the start point to the endpoint.
6. The name of the layer in which the line was created.
7. Whether the line is drawn in Paper space or Model space.

For a circle, the center point, radius, true area, and circumference are displayed. For polylines, this command displays the coordinates. In addition, for a closed polyline, its true area and perimeter are also given. If the polyline is open, AutoCAD lists its length and also calculates the area by assuming a segment connecting the start point and endpoint of the polyline. In the case of wide polylines, all computation is done based on the centerlines of the wide segments. For a selected viewport, the **LIST** command displays whether the viewport is on and active, on and inactive, or off. Information is also displayed about the status of Hideplot and the scale relative to paper space. If you use the **LIST** command on a polygon mesh, the size of the mesh (in terms of M, X, N), the coordinate values of all the vertices in the mesh, and whether the mesh is closed or open in M and N directions, are all displayed. As mentioned before, if all the information does not fit on a single screen, AutoCAD pauses to allow you to press ENTER to continue the listing.

Listing Information about all Objects in a Drawing

Command: DBLIST

The **DBLIST** command displays information pertaining to all objects in a drawing. Once you invoke this command, information is displayed at the Command prompt. If you want to display the drawing information in the AutoCAD Text Window, press the F2 key. If the information does not fit on a single screen, AutoCAD pauses to allow you to press ENTER to continue the listing.

To terminate the command, press ESC. To return to the graphics screen, close the AutoCAD Text Window.

Checking Time-Related Information

Menu Bar: Tools > Inquiry > Time | **Command:** TIME

The time and date set in your system are used by AutoCAD to provide information about several time factors related to the drawings. Hence, you should be careful about setting the current date and time in your computer. The **TIME** command is used to display information pertaining to time related to a drawing and the drawing session. The display obtained by invoking the **TIME** command is similar to the following:

```
Command: TIME
Current time:                Tuesday, March 05, 2009 at 6:59:41:157 PM
Times for this drawing:
Created:                     Tuesday, March 05, 2009 at 3:51:19:396 PM
Last updated:                Tuesday, March 05, 2009 at 3:51:19:396 PM
Total editing time:          0 days 03:08:22.522
Elapsed timer (on):          0 days 03:08:21.961
Next automatic save in:      0 days 00:37:11.432
```

Enter option [Display/ON/OFF/Reset]: *Enter the required option.*

Obtaining Drawing Status Information

Menu Bar: Tools > Inquiry > Status | **Command:** STATUS

The **STATUS** command displays information about the prevalent settings of various drawing parameters, such as snap spacing, grid spacing, limits, current space, current layer, current color, and various memory parameters. Once you enter this command, AutoCAD displays information similar to the following.

```
106 objects in Drawing.dwg
Model space limits are     X: 0.0000      Y: 0.0000(On)
                           X: 6.0000      Y: 4.4000
Model space uses           X:0.6335       Y:-0.2459      **Over
                           X:8.0497       Y: 4.9710      **Over
Display shows              X: 0.0000      Y:-0.2459
                           X: 8.0088      Y: 5.5266
Insertion base is          X: 0.0000      Y: 0.0000      Z: 0.0000
Snap resolution is         X: 0.2500      Y: 0.2500
Grid spacing is            X: 0.2500      Y: 0.2500
Current space:        Model space
Current layout:       Model
Current layer:        OBJ
Current color:        BYLAYER .... .... 7  (white)
Current linetype:     BYLAYER .... .... CONTINUOUS
Current lineweight:   BYLAYER
Current elevation:    0.0000   thickness:   0.0000
Fill on  Grid on  Ortho off  Qtext off  Snap off  Tablet  on
Object snap modes: None
Free dwg disk space: 2047.7 MBytes
Free temp disk: 2047.7 MBytes
Free physical memory: 15.0 MBytes
Free swap file space: 1987.5 MBytes
```

All the values (coordinates and distances) on this screen are given in the format declared in the **UNITS** command. You will also notice **Over in the **Model space uses** or **Paper space uses** line. This signifies that the drawing is not confined within the drawing limits. The amount of memory on the disk is given in the **Free dwg disk space** line. Information on the name of the current layer, current color, current space, current linetype, current lineweight, current elevation, snap spacing (snap resolution), grid spacing, various tools that are on or off (such as Ortho, Snap, Fill, Tablet, Qtext), and which object snap modes are active is also provided by the display obtained by invoking the **STATUS** command.

Displaying Drawing Properties

Application Menu: Drawing Utilities > Drawing Properties	**Command:** DWGPROPS

The **DWGPROPS** command is used to display information about drawing properties. Choose **Drawing Utilities > Drawing Properties** from the **Application Menu**; the **Drawing Properties** dialog box will be displayed, as shown in Figure 6-31. This dialog box has four tabs under which information about the drawing is displayed. This information helps you look for the drawing more easily. These tabs are discussed next.

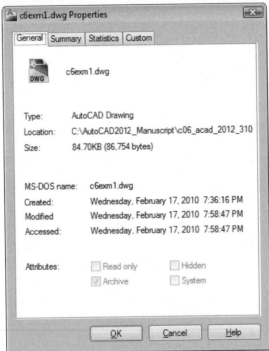

*Figure 6-31 The **Drawing Properties** dialog box*

General
This tab displays general properties about a drawing like the **Type**, **Size**, and **Location**.

Summary
The **Summary** tab displays predefined properties like the Author, title, and subject. Also this tab is displayed at the last that allows you to specify the information related to the drawing.

Statistics
This tab stores and displays data such as the file size and data such as the dates when the drawing was last saved or modified.

Custom
This tab displays custom file properties including values assigned by you.

BASIC DISPLAY OPTIONS

Drawing in AutoCAD is much simpler than manual drafting in many ways. Sometimes while drawing manually, it is very difficult to see and alter minute details. In AutoCAD, you can overcome this problem by viewing only a specific portion of the drawing. For example, to display a part of the drawing on a larger area use the **ZOOM** command to enlarge or reduce the size of the drawing displayed. Similarly, use the **REGEN** command to regenerate the drawing and **REDRAW** to refresh the screen. In this chapter, you will learn some of the drawing display commands, such as **REDRAW**, **REGEN**, **PAN**, **ZOOM**, and **VIEW**. These commands can also be used in the transparent mode. It means you can use these commands while another command is in progress.

REDRAWING THE SCREEN

Menu Bar: View > Redraw   **Command:** REDRAW

The **REDRAW** command is used to redraw the drawing area. It is used to removes the temporary graphics from the drawing area which are left after using the **VSLIDE** command and some other operations. The **REDRAW** command also redraws the objects that do not display on the screen as a result of editing some other object. In AutoCAD, several commands redraw the screen automatically (for example, when a grid is turned off), but sometimes it is useful to redraw the screen explicitly.

To use the **REDRAW** command in transparent mode, add apostrophe as a prefix. For example, if you are drawing a line after specifying few points, type **'Redraw** at the Command prompt as given below.

Specify next point or [Close/Undo]: *Specify a point.*
Specify next point or [Close/Undo]: **'REDRAW**
Resuming LINE command.
Specify next point or [Close/Undo]: *Specify a point.*

In this case, the redrawing process takes place without any prompting for information.

The **REDRAW** command affects only the current viewport. If you have more than one viewport, use the **REDRAWALL** command to redraw all the viewports. **Redraw** in the **View** menu is the **REDRAWALL** command.

REGENERATING DRAWINGS

Menu Bar: View > Regen   **Command:** REGEN

The **REGEN** command is used to regenerate the entire drawing to update it. The need for regeneration usually occurs when you change certain aspects of the drawing. All objects in the drawing are recalculated and redrawn in the current viewport. One of the advantages of this command is that the drawing is refined by smoothing out circles and arcs. To use this command, type **REGEN** at the Command prompt and press enter; the message **Regenerating model** will be displayed while the system regenerates the drawing. The **REGEN** command affects only the current viewport. If you have more than one viewport, use the **REGENALL** command to regenerate all of them. The **REGEN** command can be aborted by pressing ESC. This saves time if you are going to use another command that causes automatic regeneration.

Note
*Under certain conditions, the **ZOOM** and **PAN** commands automatically regenerate the drawing. Some other commands also perform regenerations under certain conditions.*

ZOOMING DRAWINGS

Ribbon: View > Navigate 2D > Extents	**Command:** ZOOM
Menu bar: View > Zoom	**Toolbar:** Navigation Bar > Zoom Extents

Creating drawings in AutoCAD would not be of much use, if you cannot magnify the drawing view to work on the minute details. The ability to zoom in, or magnify, has been helpful in creating the minuscule circuits used in the electronics and computer industries. This is performed using the **ZOOM** command. This is one of the most frequently used commands. Getting close to or away from the drawing is the function of the **ZOOM** command. In other words, this command enlarges or reduces the view of the drawing on the screen, but it does not affect the actual size of the objects. In this way, the **ZOOM** command functions like the zoom lens of a camera. When you magnify the apparent size of a section of the drawing, you see that area in greater detail. On the other hand, if you reduce the apparent size of the drawing, you see a larger area.

The **ZOOM** command can also be used in transparent mode. This means that this command can be used while working with other commands. Various options of the **ZOOM** command are grouped as tools in the **Zoom** drop-down, as shown in Figure 6-32. You can also right-click in the drawing area and choose the **Zoom** option from the shortcut menu even when you are working with some other command, refer to Figure 6-33. Various options of the **ZOOM** command have been made available in the **Navigation bar** in the drawing area, as shown in Figure 6-34.

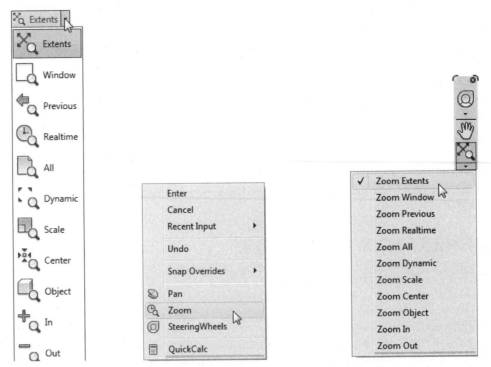

*Figure 6-32 Tools in the **Zoom** drop-down*

*Figure 6-33 The **Zoom** option in the shortcut menu*

*Figure 6-34 The **Zoom** tools in the **Navigation Bar***

This command has several options and can be used in a number of ways. The following is the prompt sequence that is displayed when you invoke this command.

Command: **ZOOM** [Enter]
Specify corner of window, enter a scale factor (nX or nXP), or
[All/Center/Dynamic/Extents/Previous/Scale/Window/Object] <real time>:

Realtime Zooming

To zoom in and zoom out interactively, invoke the **Realtime Zoom** option by choosing the **Zoom Realtime** tool from the **Navigation Bar** in the drawing area. To zoom in, invoke the command, then hold the pick button down and move the cursor up. If you want to zoom in further, release the pick button and bring the cursor down. Specify a point and move the cursor up again. Similarly, to zoom out, hold the pick button down and move the cursor down. If you move the cursor vertically up from the midpoint of the screen to the top of the window, the drawing will be magnified by 100% (zoom in 2x magnification). Similarly, if you move the cursor vertically down from the midpoint of the screen to the bottom of the window, the drawing display will be reduced to 100% (zoom out 0.5x magnification). Realtime zoom is the default setting for the **ZOOM** command. Pressing ENTER after entering the **ZOOM** command automatically invokes the realtime zoom.

When you use the realtime zoom, the cursor becomes a magnifying glass and displays a plus sign (+) and a minus sign (–). When you reach the zoom out limit, AutoCAD does not display the minus sign (–) while dragging the cursor. Similarly, when you reach the zoom in limit, AutoCAD does not display the plus sign (+) while dragging the cursor. To exit the realtime zoom, press ENTER or ESC, or select **Exit** from the shortcut menu.

All Option

This option of the **ZOOM** command is used to adjust the display area on the basis of the drawing limits (Figure 6-35) or extents of an object, whichever is greater. Even if the objects are not within the limits, they are still included in the display. Therefore, with the **All** option, you can view the entire drawing in the current viewport (Figure 6-36).

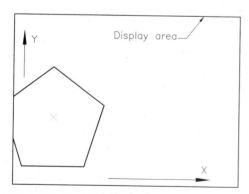

Figure 6-35 *Drawing showing the limits*

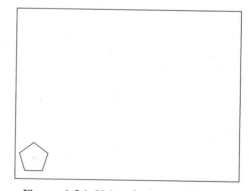

Figure 6-36 *Using the **Zoom All** option*

Center Option

This option is used to define a new display window by specifying its center point (Figures 6-37 and 6-38) and magnification or height. Here, you are required to enter the center and the height of the subsequent screen display. If you press ENTER instead of entering a new center point, the center of the view will remain unchanged. Instead of entering a height, you can enter the **magnification factor** by typing a number. If you press ENTER at the height prompt, or if the height you enter is the same as the current height, magnification does not take place. For example, if the current height is 2.7645 and you press ENTER at the **magnification or height <current>** prompt, magnification will not take place. The smaller the value, the greater is the enlargement of the image. You can also enter a number followed by **X**.

This indicates the change in magnification, not as an absolute value, but as a value relative to the current screen.

The prompt sequence is given next.

Command : **ZOOM** Enter
Specify corner of window, enter a scale factor (nX or nXP), or
[All/Center/Dynamic/Extents/Previous/Scale/Window/Object] <real time>: **C** Enter
Specify center point: *Specify a center point.*
Enter magnification or height <current>: **5X** Enter

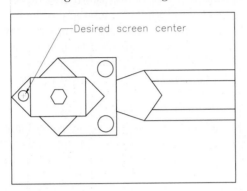

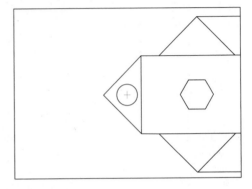

Figure 6-37 *Drawing before using the **ZOOM** Center option*

Figure 6-38 *Drawing after using the **ZOOM** Center option*

In Figure 6-36, the current magnification height is 5X, which magnifies the display five times. If you enter the value **2**, the size (height and width) of the zoom area changes to 2 X 2 around the specified center. In Figure 6-39, if you enter 0.12 as the height after specifying the center point as the circle's center, the circle will zoom to fit in the display area since its diameter is 0.12. The prompt sequence is given next.

Command: **ZOOM** Enter
Specify corner of window, enter a scale factor (nX or nXP), or
[All/Center/Dynamic/Extents/Previous/Scale/Window/Object]<real time>: **C** Enter
Center point: *Select the center of the circle.*
Magnification or Height <5.0>: **0.12** Enter

Extents Option

As the name indicates, this option is used to zoom to the extents of the biggest object in the drawing. The extents of the drawing comprise the area that has the drawings in it. The rest of the empty area is neglected. With this option, all objects in the drawing are magnified to the largest possible display, see Figure 6-40.

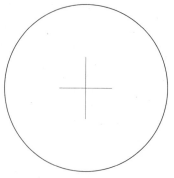
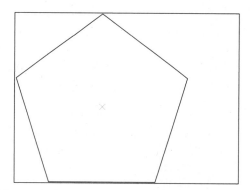

Figure 6-39 *Drawing after using the* **ZOOM** *Center option*

Figure 6-40 *Drawing after using the* **ZOOM** *Extents option*

Dynamic Option

 This option displays the portion of the drawing that you have already specified. The prompt sequence for using this option of the **ZOOM** command is given next.

Command: **ZOOM** `Enter`
Specify corner of window, enter a scale factor (nX or nXP), or
[All/Center/Dynamic/Extents/Previous/Scale/Window/Object]<real time>: **D** `Enter`

You can then specify the area to be displayed by manipulating a view box representing the viewport. This option lets you enlarge or shrink the view box and move it around. When the view box is in the proper position and size, the current viewport is cleared and a special view selection screen will be displayed. This special screen comprises information regarding the current view as well as the available views. In a color display, the different viewing windows are very easy to distinguish because of their different colors, but in a monochrome monitor, they can be distinguished by their shape.

Blue dashed box representing drawing extents

Drawing extents are represented by a dashed blue box (Figure 6-41), which constitutes the drawing limits or the actual area occupied by the drawing.

Green dashed box representing the current view

A green dashed box is formed to represent the area of the current viewport (Figure 6-42).

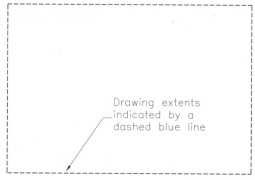
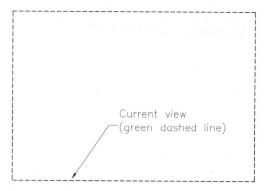

Figure 6-41 *Box representing drawing extents*

Figure 6-42 *Representation of the current view*

Panning view box (X in the center)

A view box initially of the same size as the current view box is displayed with an X in the center

(Figure 6-43). You can move this box with the help of your pointing device. This box is known as the panning view box and it helps you to find the center point of the zoomed display you want. When you have found the center, press the pick button to display the zooming view box.

Zooming view box (arrow on the right side)

On pressing the pick button at the center of the panning view box, the X at the center of the view box is replaced by an arrow pointing to the right edge of the box. This zooming view box (Figure 6-44) indicates the ZOOM mode. You can now increase or decrease the area of this box according to the area you want to zoom into. To shrink the box, move the pointer to the left; to increase it, move the pointer to the right. The top, right, and bottom sides of the zooming view box move as you move the pointer, but the left side remains fixed, with the zoom base point at the midpoint of the left side. You can slide it up or down along the left side. When you have the zooming view box in the desired size for your zoom display, press ENTER to complete the command and zoom into the desired area of the drawing. Before pressing ENTER, if you want to change the position of the zooming view box, click the pick button of your pointing device to redisplay the panning view box. After repositioning, press ENTER.

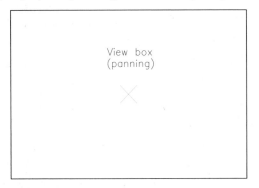

Figure 6-43 *The panning view box*

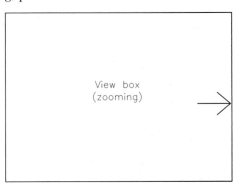

Figure 6-44 *The zooming view box*

Previous Option

While working on a complex drawing, you may need to zoom in on a portion of the drawing to edit some minute details. Once the editing is over you may want to return to the previous view. This can be done by choosing the **Zoom Previous** tool from the **Zoom** drop-down in the **Navigation Bar**. Without this command, it would be very tedious to zoom back to the previous views. AutoCAD saves the view specification of the current viewport whenever it is being altered by any of the ZOOM options or by the **PAN**, **VIEW Restore**, **DVIEW**, or **PLAN** commands (which are discussed later). Up to ten views are saved for each viewport. The prompt sequence for this option is given next.

> Command: **ZOOM** `Enter`
> Specify corner of window, enter a scale factor (nX or nXP), or
> [All/Center/Dynamic/Extents/Previous/Scale/Window/Object]<real time>: **P** `Enter`

Successive **ZOOM > P** commands can restore up to ten previous views. **VIEW** here refers to the area of the drawing defined by its display extents. If you erase some objects and then issue the **ZOOM > Previous** command, the previous view is restored, but the erased objects are not.

Window Option

This is the most commonly used option of the **ZOOM** command. After invoking this command, you need to specify the area you want to zoom in by specifying two opposite corners of a rectangular window. The center of the specified window becomes the center of the zoomed area. The area inside the window is magnified or reduced in size to fill the display as completely as possible. The points can be specified either by selecting them with the help of the pointing device or by entering their coordinates. The prompt sequence is given next.

Command: **ZOOM** Enter
Specify corner of window, enter a scale factor (nX or nXP), or
[All/Center/Dynamic/Extents/Previous/Scale/Window/Object]<real time>: *Specify a point.*
Specify opposite corner: *Specify another point.*

Whenever the **ZOOM** command is invoked, the window method is one of two default options. This is illustrated by the previous prompt sequence where you can specify the two corner points of the window without invoking any option of the **ZOOM** command. The **Window** option can also be used by entering **W**. In this case, the prompt sequence is given next.

Command: **ZOOM** Enter
Specify corner of window, enter a scale factor (nX or nXP), or
[All/Center/Dynamic/Extents/Previous/Scale/Window/Object]<real time>: **W** Enter
Specify first corner: *Specify a point.*
Specify opposite corner: *Specify another point.*

Scale Option

 The **Scale** option of the **ZOOM** command is a very versatile option. It can be used in the following ways:

Scale: Relative to full view

This option of the **ZOOM** command is used to magnify or reduce the size of a drawing according to a scale factor (Figure 6-45). A scale factor equal to 1 displays an area equal in size to the area defined by the established limits. This may not display the entire drawing if the previous view was not centered on the limits or if you have drawn outside the limits. To get a magnification relative to the full view, you can enter any other number. For example, you can type 4 if you want the displayed image to be enlarged four times. If you want to decrease the magnification relative to the full view, you need to enter a number that is less than 1. In Figure 6-46, the image size is decreased because the scale factor is less than 1. In other words, the image size is half of the full view because the scale factor is 0.5.

The prompt sequence is given next.

Command: **ZOOM** Enter
Specify corner of window, enter a scale factor (nX or nXP), or
[All/Center/Dynamic/Extents/Previous/Scale/Window/Object]<real time>: **S** Enter
Enter a scale factor (nX or nXP): **0.5** Enter

Figure 6-45 *Drawing before the* **ZOOM Scale** *option selected*

Figure 6-46 *Drawing after the* **ZOOM** **Scale** *option selected*

Scale: Relative to current view

The second way to scale is with respect to the current view (Figure 6-47). In this case, instead of entering only a number, enter a number followed by an **X**. The scale is calculated with reference to the current view. For example, if you enter **0.25X**, each object in the drawing will be displayed at one-fourth (¼) of its current size. The following example shows how to increase the display magnification by a factor of **0.25** relative to its current value (Figure 6-48).

Command: **ZOOM** [Enter]
Specify corner of window, enter a scale factor (nX or nXP), or
[All/Center/Dynamic/Extents/Previous/Scale/Window/Object]<real time>: **0.25X** [Enter]

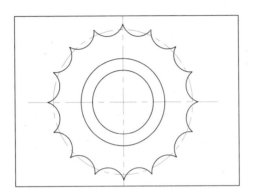

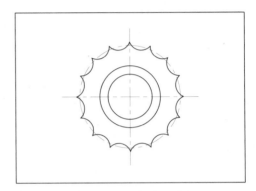

Figure 6-47 Current view of the drawing

Figure 6-48 Drawing after applying magnification factor of 0.25X

Scale: Relative to paper space units

The third method of scaling is with respect to paper space. You can use paper space in a variety of ways and for various reasons. For example, you can array and plot various views of your model in the paper space. To scale each view relative to paper space units, you can use the **ZOOM XP** option. Each view can have an individual scale. The drawing view can be at any scale of your choice in a model space viewport. For example, to display a model space at one-fourth (¼) the size of the paper space units, the prompt sequence is given next.

Command: **ZOOM** [Enter]
Specify corner of window, enter a scale factor (nX or nXP), or
[All/Center/Dynamic/Extents/Previous/Scale/Window/Object]<real time>: **1/4XP** [Enter]

Note
For a better understanding of this topic, refer to "Model Space Viewports, Paper Space Viewports and Layouts" in Chapter 15.

Object Option

 The **Object** option of the **ZOOM** command is used to select one or more than one objects and display them at the center of the screen in the largest possible size.

Zoom In and Out

 You can also zoom into the drawing by using the **In** option, which doubles an image size. Similarly, you can use the **Out** option to decrease the size of the image by half. To invoke these options from the command line, enter **ZOOM 2X** for the **In** option or **ZOOM .5X** for the **Out** option at the Command prompt. The center of the screen is taken as the reference point for enlarging or reducing the view of a drawing.

EXERCISE 1   *Zoom*

Draw the profile shown in Figure 6-49 based on the given dimensions. Use the **ZOOM** command to get a bigger view of the drawing. Do not dimension the drawing.

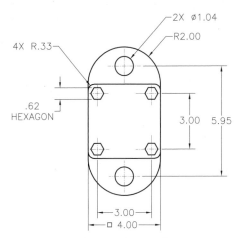

Figure 6-49 Drawing for Exercise 1

PANNING DRAWINGS

Ribbon:	View > Navigate 2D > Pan	**Command:** PAN
Menu Bar:	View > Pan > Realtime	

You may want to view or draw on a particular area outside the current display. You can do this using the **Pan** tool. If done manually, this would be like holding one corner of the drawing and dragging it across the screen. The **Pan** tool allows you to bring into view the portion of the drawing that is outside the current display area. This is done without changing the magnification of the drawing. The effect of this command can be illustrated by imagining that you are looking at a big drawing through a window (**display window**) that allows you to slide the drawing right, left, up, and down to bring the part you want to view inside this window. You can invoke the **Pan** tool from the shortcut menu also.

Panning in Realtime

You can use the **Realtime Pan** to pan the drawing interactively. To pan a drawing, invoke the tool and then hold the pick button down and move the cursor in any direction. When you select the realtime pan, an image of a hand will be displayed indicating that you are in the **Realtime Pan**. This is the default setting for the **Pan** tool. Choosing the **Pan Realtime** tool in the toolbar and entering **PAN** at the Command prompt automatically invokes the realtime pan. To exit the realtime pan, press ENTER or ESC, or choose **Exit** from the shortcut menu.

The **Pan** tool has various options to pan a drawing in a particular direction. These options can be invoked only from the menu bar, as shown in Figure 6-50.

Point

This option is used to specify the actual displacement. To do this, you need to specify in what direction to move the drawing and by what distance. You can give the displacement either by entering the coordinates of the points or by specifying the coordinates by using a pointing device. The coordinates can be entered in two ways. One way is to specify a single coordinate pair. In this case, AutoCAD takes it as a relative displacement of the drawing with respect to the screen. For example, in the following case, the **Pan** tool would shift the displayed portion of the drawing 2 units to the right and 2 units up.

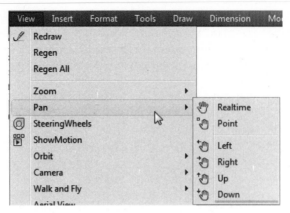

Figure 6-50 *Invoking the **PAN** command options from the **View** menu*

Command: *Select the **Point** option from the **View** menu.*
Specify base point or displacement: **2,2** Enter
Specify second point: Enter

In the second case, you can specify two coordinate pairs. AutoCAD computes the displacement from the first point to the second. Here, the displacement is calculated between point (3,3) and point (5,5).

Command: *Specify the **Point** item from the **View** menu.*
Specify base point or displacement: **3,3** Enter   *(Or specify a point.)*
Specify second point: **5,5** Enter   *(Or specify a point.)*

Left

Moves the drawing left so that some of the right portion of the drawing is brought into view.

Right

Moves the drawing right so that some of the left portion of the drawing is brought into view.

Up

Moves the drawing up so that some of the bottom portion of the drawing is brought into view.

Down

Moves the drawing down so that some of the top portion of the drawing is brought into view.

Tip
*You can use the scroll bars to pan the drawing vertically or horizontally. The scroll bars are located at the right side and the bottom of the drawing area. You can control the display of the scroll bars in the **Display** tab of the **Options** dialog box.*

CREATING VIEWS

Ribbon: View > Views > View Manager	**Command:** VIEW
Menu Bar: View > Named Views	

While working on a drawing, you may frequently be working with the **ZOOM** and **PAN** commands, and you may need to work on a particular drawing view (some portion of the drawing) more often than others. Instead of wasting time by recalling your zooms and pans and selecting the same area from the screen over and over again, you can store the view under a name and restore it using the name you have given. To do so, choose the **View Manager** button from the **View** panel; the **View Manager** dialog box will be displayed, as shown in Figure 6-51. This dialog box is used to save the current view with a name so that you can restore (display) it later. It does not save any drawing object data, only the view parameters needed to redisplay that portion of the drawing.

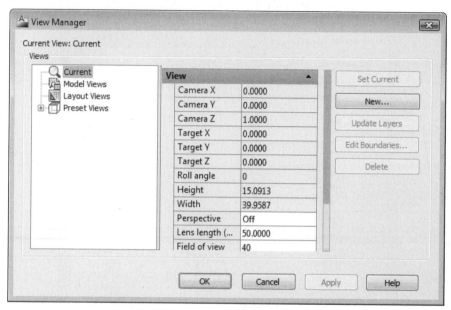

*Figure 6-51 The **View Manager** dialog box*

This dialog box is very useful when you are saving and restoring many views. The **Views** area lists the current view and all the existing named views in the model space and layouts. This area also lists all the preset views such as orthographic and isometric views. As soon as you select a view from the list box, the information about it will be displayed in the **Information** area provided on the right of the list box.

Other options in the **Views** area of the **View Manager** dialog box are discussed next.

Set Current

The **Set Current** button is used to replace the current view by the view you select from the list box in the **Views** area. You can select any predefined view or any preset view to be made current. AutoCAD uses the center point and magnification of each saved view and executes a **ZOOM Center** with this information when a view is restored.

New

The **New** button is used to create and save a new view by giving it a name. When you choose the **New** button, the **New View / Shot Properties** dialog box is displayed, as shown in Figure 6-52. The options in this dialog box are discussed next.

View name

You can enter the name for the view in the **View name** edit box.

View category

You can specify the category of the view from the **View category** edit box. The categories include the front view, top view, and so on. If there are some existing categories, you can select them from this drop-down list also.

View type

You can specify the type of view in this drop-down list.

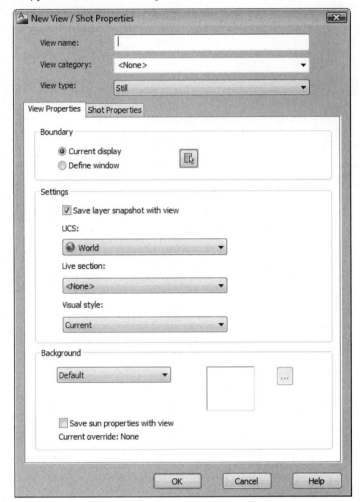

Figure 6-52 The New View / Shot Properties dialog box

Boundary Area

This area is used to specify the boundary of the view. To save the current display as the view, select the **Current display** radio button. To define a window that will specify the new view (without first zooming in on that area), select the **Define window** radio button. As soon as you select this radio button, the dialog boxes will be temporarily closed and you will be prompted to define two corners of the window. You can modify the window by choosing the **Define view window** button. You can also enter the *X* and *Y* coordinates in the **Specify first corner** and **Specify other corner** in the command lines.

Settings Area

This area allows you to save the layer visibility, UCS, live section, and view style in which you want to visualize the object with the new view. To save the settings of the visibility of the current layers with the view, select the **Save layer snapshot with view** check box. You can select the WCS or the UCS to be saved with the view from the **UCS** drop-down list. If you have created a live section in model space for 3D objects, you can select it from the **Live section** drop-down list, in order to save it with the view. You can also select the visual style to view the object from the **Visual style** drop-down list.

Note
The concept of WCS, UCS, live section, and visual style will be discussed in later chapters.

Background Area

This area allows you to save a view in different background types. To do so, select the type of background from the drop-down list in the **Background** area; the **Background** dialog box will be displayed. You can select the background type from the **Type** drop-down list. Alternatively, you can select these options directly from the drop-down list in the **Background** area of the **New View** dialog box. On selecting the **Solid** type background, you can select any one color to be displayed in the background. To change the color, click on the **Color** area and select the type of color from the **Select Color** dialog box. On selecting the **Gradient** type background, you can apply two or three colors to the background. On selecting the **Image** type background, you can insert an image in the background. To select an image file, choose the **Browse** button from the **Image options** area of the **Background** dialog box. You can also set sun and sky in the background of a view. To do so, select the **Sun & Sky** option from the drop-down list in the **Background** area of the **New View / Shot Properties** dialog box; the **Adjust Sun & Sky Background** dialog box will be displayed. Specify the settings of the sun and sky background in this dialog box and choose the **OK** button. To save the specified sun and sky properties with the view, select the **Save sun properties with view** check box from the **New View / Shot Properties** dialog box.

Note
The effect of background can be visualized only if you have saved the view in the model space with a visual style other than the 2D Wireframe.

*The options in the **Adjust Sun & Sky Background** dialog box and the **Adjust Image** button in the **Image options** area of the **Background** dialog box will be explained in detail in Chapter 31 (Rendering and Animating Designs) of this book.*

*The options in the **Shot Properties** tab are used when you create Motion animations. This is discussed in detail in Chapter 28.*

Update Layers

The **Update Layers** button is chosen to update the layer information saved with an existing view.

Edit Boundaries

The **Edit Boundaries** button is chosen to edit the boundary that was defined using the **New View Manager** dialog box while creating the new view.

Tip
You can use the shortcut menu to rename or delete any named view in the dialog box. You can also update the layer information or edit the boundary of a view using this shortcut menu.

*The options of the **VIEW** command can be used in the transparent mode by entering '-VIEW at the Command prompt.*

UNDERSTANDING THE CONCEPT OF SHEET SETS

The sheet sets feature allows you to logically organize a set of multiple drawings as a single unit, called the sheet set. For example, consider a setup, in which there are a number of drawings in different folders in the hard drive of a computer. Organizing or archiving these drawings is tedious and time consuming. However, this can be easily and efficiently done by creating sheet sets. In a sheet set, you can import the layouts from an existing drawing or create a new sheet with a new layout and place the views in the new sheet. You can easily plot and publish all the drawings in the sheet set. You can also manage and create sheet sets using the **Sheet Set Manager**. To do so, choose the **Sheet Set Manager** tool from the **Palettes** panel in the **View** tab or press the CTRL+4 keys.

Creating a Sheet Set

AutoCAD allows you to create two different types of sheet sets. The first one is an example sheet set that uses a well organized structure of settings. The second one is used to organize existing drawings. The procedure for creating both these types of sheet sets are discussed next.

Creating an Example Sheet Set

To create an example sheet set, select **New Sheet Set** from the **Open** drop-down list in the **Sheet Set Manager** or choose **New > Sheet Set** from the **Application Menu**. You can also enter **NEWSHEETSET** at the Command prompt. When you invoke this command, the **Create Sheet Set** wizard will be displayed with the **Begin** page, as shown in Figure 6-53. In this page, select the **An example sheet set** radio button, if it is not selected by default. Choose the **Next** button; the **Sheet Set Example** page will be displayed, as shown in Figure 6-54.

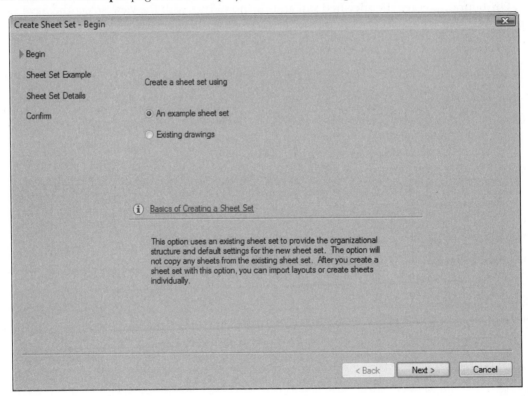

*Figure 6-53 The **Begin** page of the **Create Sheet Set** wizard*

By default, the **Select a sheet set to use as an example** radio button is selected in this page. The list box below this radio button displays the list of sheet sets that you can use as an example. Each of these sheet sets has structurally organized settings for the sheets. You can select the required sheet set from this list box. The title and the description related to the selected sheet set are displayed in the lower portion of the dialog box.

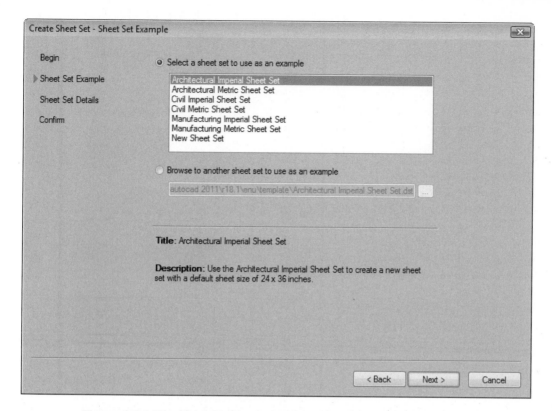

*Figure 6-54 The **Sheet Set Example** page of the **Create Sheet Set** wizard*

You can also select the **Browse to another sheet set to use as an example** radio button to select another sheet set located in a different location. You can enter the location of the sheet set in the edit box below this radio button or choose the [...] button to display the **Browse for Sheet Set** dialog box. Using this dialog box, you can locate the sheet set file, which is saved with the *.dst* extension.

After selecting the sheet set to use as an example, choose the **Next** button; the **Sheet Set Details** page will be displayed, as shown in Figure 6-55.

Enter the name of the new sheet set in the **Name of new sheet set** edit box. By default, some description is added in the **Description (optional)** area. You can enter additional description in this area. The **Store sheet set data file (.dst) here** edit box displays the default location in which the sheet set data file will be stored. You can modify this location by entering the new location or by selecting the folder using the **Browse for Sheet Set Folder** dialog box, which is displayed by choosing the [...] button.

You can modify the sheet set properties such as name, storage location, template, description, and so on by choosing the **Sheet Set Properties** button.

Once all parameters on this page are configured, choose the **Next** button; the **Confirm** page will be displayed, as shown in Figure 6-56. This page shows the detailed structure of the sheet set and also lists its parameters and properties.

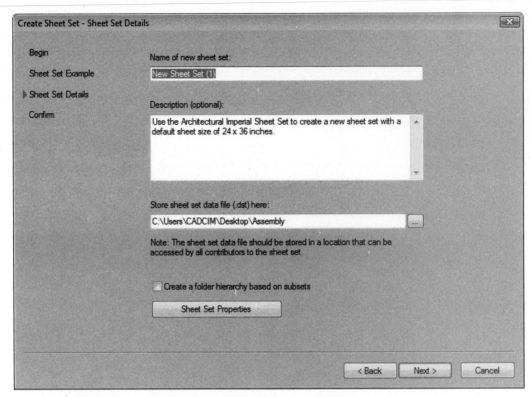

Figure 6-55 *The **Sheet Set Details** page of the **Create Sheet Set** wizard*

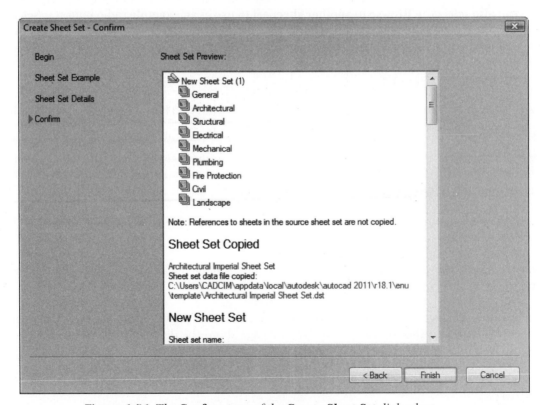

Figure 6-56 *The **Confirm** page of the **Create Sheet Set** dialog box*

After checking all parameters and properties, choose the **Finish** button; the **Sheet Set Manager** displays the sheet structure in the **Sheets** area and the details of that sheet set in the **Details** area, as shown in Figure 6-57. If the **Details** area is not displayed, right-click in the **Sheet Set Manager** and select the **Preview/Details Pane** option from the shortcut menu.

Creating a Sheet Set Using Existing Drawings

As mentioned earlier, this sheet set is used to organize and archive an existing set of drawings. To create this type of sheet set, select the **Existing drawings** radio button from the **Begin** page of the **Create Sheet Set** wizard and choose **Next**. The **Sheet Set Details** page will be displayed, which is similar to the one shown in Figure 6-54. Enter the name of the sheet set and the description on this page. Note that by default, there will be no description given about the new sheet set. After setting the parameters on this page, choose the **Next** button; the **Choose Layouts** page will be displayed, as shown in Figure 6-58.

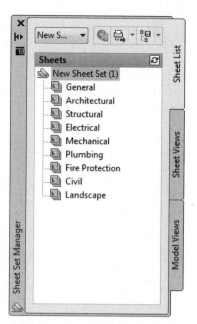

Figure 6-57 The Sheet Set Manager

Choose the **Browse** button from this page and browse for the folder in which the files to be included in the sheet set are saved. All the drawing files, along with their initialized layouts, are displayed in the list box available below the **Browse** button. You can select as many folders as you want by choosing the **Browse** button.

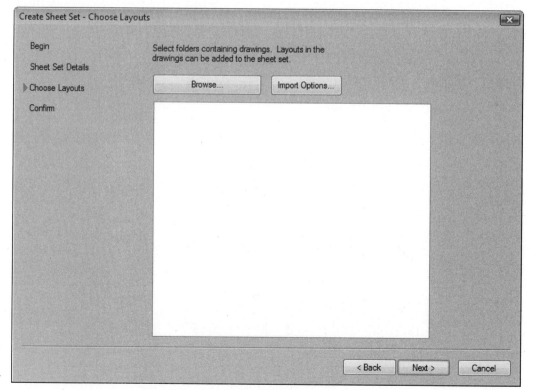

*Figure 6-58 The **Choose Layouts** page of the **Create Sheet Set** dialog box*

Tip
*You can remove the folders from the list box in the **Choose Layouts** page by selecting them and pressing the DELETE key.*

When you select a folder, all the drawings in it and all the initialized layouts in the drawings have a check mark on their left. This suggests that all these drawings and layouts will be included in the sheet set. You can clear the check box of the folder to clear all the check boxes and then select the check boxes of only the required drawings and layouts. You can modify the import options by using the **Import Options** dialog box, which is displayed by choosing the **Import Options** button.

After selecting the layouts to be included, choose the **Next** button to display the **Confirm** page similar to that shown in Figure 6-56. This page lists all the layouts that will be included in the sheet set. Choose **Finish** button to complete the process of creating the sheet set.

Adding a Subset to a Sheet Set

For a better and more efficient organization of a sheet set, it is recommended that you add subsets to the sheet set. For example, consider a case where you have created a sheet set called Mechanical Drawings, in which you want to store all the mechanical drawings. In this sheet set, you can create subsets such as Bolts, Nuts, Washers, and so on and place the sheets of bolts, nuts, and washers for a more logical organization of the sheet set.

To add a subset to a sheet set, right-click on the sheet set or subset and choose **New Subset** from the shortcut menu. The **Subset Properties** dialog box is displayed, see Figure 6-59.

Enter the name of the subset in the **Subset name** field. Also, specify the location for saving the DWG file and the template for creating the sheets using this dialog box. If you specify **Yes** in

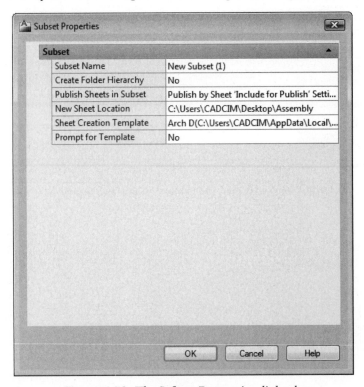

*Figure 6-59 The **Subset Properties** dialog box*

the **Create Folder Hierarchy** field, a new folder will be created under the sheet set. If you specify Yes in the **Prompt for template** check box, which will prompt you to select the template for the drawings. Choose **OK** after configuring all the parameters. A new subset will be added to the sheet set or the subset that you selected.

Adding Sheets to a Sheet Set or a Subset

To add a new sheet to a sheet set or a subset, right-click on it in the **Sheet Set Manager** window and choose **New Sheet** from the shortcut menu. The **New Sheet** dialog box is displayed. In this dialog box, enter the number and the title of the sheet, along with the file name. You can also set the path of the folder and the sheet template to be used, using this dialog box. Choose **OK** button after configuring all the parameters. A new sheet will be added to the sheet set or the subset that you selected.

Archiving a Sheet Set

AutoCAD allows you to archive a sheet set as a zip file, a self-extracting executable file (exe), or a file folder. All files related to the sheet set are automatically included in the zip file. To archive a sheet set, right-click on its name in the **Sheet Set Manager** and choose **Archive** from the shortcut menu. After AutoCAD gathers the archive information, the **Archive a Sheet Set** dialog box will be displayed, as shown in Figure 6-60.

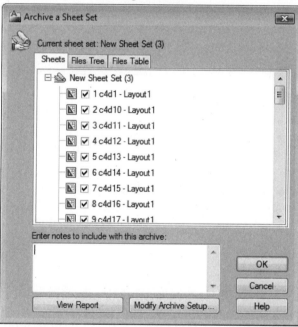

*Figure 6-60 The **Archive a Sheet Set** dialog box*

Before archiving the sheet set, you can modify the archiving options by using the **Modify Archive Setup** dialog box, as shown in Figure 6-61. This dialog box is displayed when you choose the **Modify Archive Setup** button. Using the **Archive package type** drop-down list, you can specify whether the archived file is a zip file, a self-extracting executable file (*exe*), or a file folder. You can also specify the format, in which you want to save the files. You can select the current release format, AutoCAD 2007/LT 2007, AutoCAD 2004/LT 2004, or AutoCAD 2000/LT 2000 formats for archiving the files.

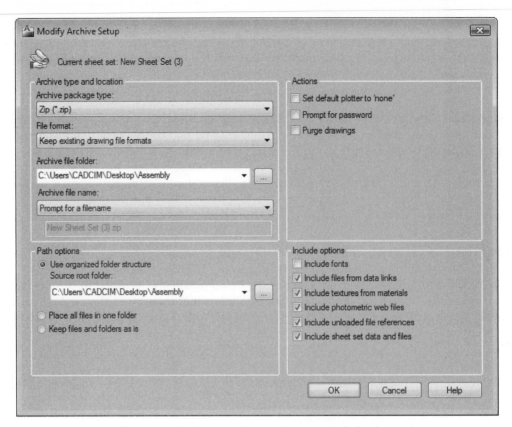

*Figure 6-61 The **Modify Archive Setup** dialog box*

After setting the parameters, choose the **OK** button from the **Archive a Sheet Set** dialog box. The standard save dialog box will be displayed, which can be used to specify the name and the location of the resultant file.

Resaving all Sheets in a Sheet Set

The **Sheet Set Manager** allows you to easily resave all sheets in a sheet set. To save all sheets in a sheet set again, right-click on the name of the sheet set in the **Sheet Set Manager** and choose **Resave All Sheets** from the shortcut menu, as shown in Figure 6-62. All the sheets in the sheet set will be saved once again.

PLACING VIEWS ON A SHEET OF A SHEET SET

You can place views in the sheets of a sheet set. To place a view on a sheet, open the sheet by double-clicking on it in the **Sheet Set Manager**. Next, you need to locate the file from which you want to copy the view. To locate this file, choose the **Model Views** tab. If the location of the file is not already listed in this tab, double-click on **Add New Location** to display the **Browse for Folder** dialog box. Using this dialog box, select the folder in which you have saved the file. All the files of that folder are listed. Click on the plus sign (+) located on the left of the file to display all the views created in that file.

Now, right-click on the view and choose **Place on Sheet** from the shortcut menu, as shown in Figure 6-63; the preview of the view will be attached to the cursor and you will be prompted to specify the insertion point. By default, the view will be placed with 1:1 scale. But, you can change the view scale before placing the view. To change the scale, right-click to display the shortcut menu that has various preset view scales. Choose the desired view scale from this shortcut menu and then place the view. The view will be placed inside the paper space viewport in the layout. Also, the name of the view is displayed below the view in the layout where you placed the view. You can view the list of all the views in the current sheet set by choosing the **View List** tab.

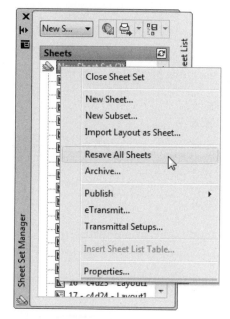

Figure 6-62 *Resaving all sheets in a sheet set using the* **Sheet Set Manager**

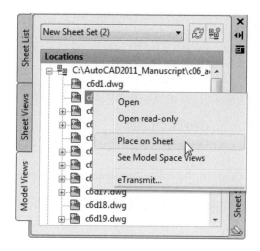

Figure 6-63 *The* **Model Views** *tab of the* **Sheet Set Manager** *displaying the views in a selected drawing*

Self-Evaluation Test

Answer the following questions and then compare them to those given at the end of this chapter:

1. The number of grips that will be displayed on an object depends on the object itself. (T/F)

2. You can use the **Options** dialog box to modify grip parameters. (T/F)

3. You need at least one source object while using the **Match Properties** tool. (T/F)

4. You cannot drag and drop entities from the **DesignCenter** window. (T/F)

5. The **Zoom All** tool displays drawing limits or extents, whichever is greater. (T/F)

6. While using the **Zoom Scale** tool with respect to the current view, you need to enter a number followed by **X**. (T/F)

7. After selecting a polyline, if you place the cursor on the _____ grip, the **Convert to Arc** option will be displayed in the tooltip.

8. The color of the unselected grips can also be changed by using the _____ system variable.

9. You can access the **Mirror** mode by selecting a grip and then entering _____ or _____ or giving a null response by pressing the SPACEBAR four times.

10. The _____ drop-down list will not be available if you select **Select All** from the **Operator** drop-down list of the **Quick Select** dialog box.

Review Questions

Answer the following questions:

1. Which of these system variables is used to modify the color of a selected grip?

   (a) **GRIPCOLOR**   (b) **GRIPHOT**
   (c) **GRIPCOLD**   (d) **GRIPBLOCK**

2. Which of these system variables is used to enable the display of grips inside blocks?

   (a) **GRIPCOLOR**   (b) **GRIPHOT**
   (c) **GRIPCOLD**   (d) **GRIPBLOCK**

3. Which of the following system variables is used to modify the size of grips?

   (a) **GRIPCOLOR**   (b) **GRIPSIZE**
   (c) **GRIPCOLD**   (d) **GRIPBLOCK**

4. How many views are saved with the **Zoom Previous** tool?

   (a) 6   (b) 8
   (c) 10   (d) 12

5. Which of the following commands recalculate all objects in a drawing and redraws the current viewport only?

   (a) **REDRAW**   (b) **REDRAWALL**
   (c) **REGEN**   (d) **REGENALL**

6. If you select a grip of an object, the grip becomes a hot grip. (T/F)

7. The **Rotate** mode is used to rotate objects around the base point without affecting their size. (T/F)

8. If you have already added a hyperlink to an object, you can also use the grips to open a file associated with it. (T/F)

9. You can view the entire drawing (even if it is beyond limits) with the help of the _____ option.

10. The **GRIPSIZE** is defined in pixels, and its value can range from _____ to _____ pixels.

11. The **VIEW** command does not save any drawing object data. Only the _____ parameters required to redisplay that portion of the drawing are saved.

12. The _____ mode is used to move the selected objects to a new location.

13. The _____ mode allows you to scale objects with respect to the base point without changing their orientation.

14. The **Mirror** mode allows you to mirror objects across the _____ without changing the size of the objects.

15. With the **ZOOM** command, the actual size of an object changes. (T/F)

16. The **REDRAW** command can be used as a transparent command. (T/F)

EXERCISE 2

1. Use the **Line** tool to draw the shape shown in Figure 6-64(a).

2. Use grips (**Stretch** mode) to get the shape shown in Figure 6-64(b).

3. Use the **Rotate** and **Stretch** modes to get the copies shown in Figure 6-64(c).

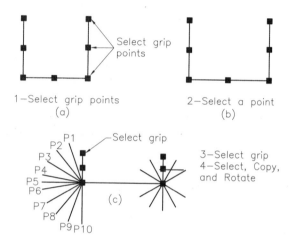

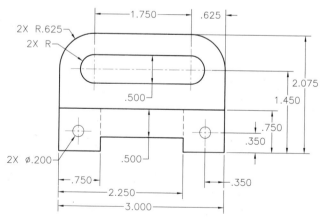

Figure 6-64 *Drawing for Exercise 2*

EXERCISE 3

Use the drawing and editing tools to create the sketch shown in Figure 6-65. Use the display tools to facilitate the process.

Figure 6-65 *Drawing for Exercise 3*

EXERCISE 4

Use the drawing and editing tools to create the sketch shown in Figure 6-66. Use the display tools to facilitate the process.

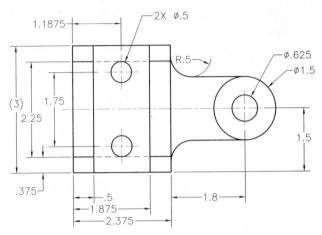

Figure 6-66 Drawing for Exercise 4

Problem-Solving Exercise 1

Draw the sketch shown in Figure 6-67 using the draw and edit tools. Use the **Mirror** tool to mirror the shape 9 units across the Y axis so that the distance between the two center points is 9 units. Mirror the shape across the X axis and then reduce the mirrored shape by 75 percent. Join the two ends to complete the shape of the open end spanner. Save the file. Assume the missing dimensions. Note that this is not a standard size spanner.

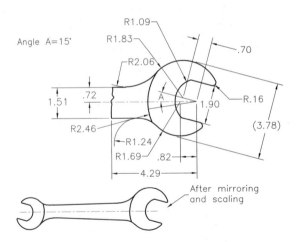

Figure 6-67 Drawing for Problem-Solving Exercise 1

Problem-Solving Exercise 2

Use the drawing and editing tools to create the drawing shown in Figure 6-68. Use the display tools to facilitate the process.

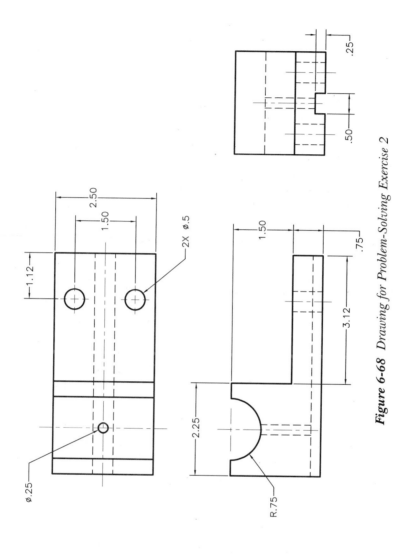

Figure 6-68 *Drawing for Problem-Solving Exercise 2*

Problem-Solving Exercise 3

Use the drawing and editing tools to create the drawing shown in Figure 6-69. Use the display tools to facilitate the process. Assume the missing dimensions.

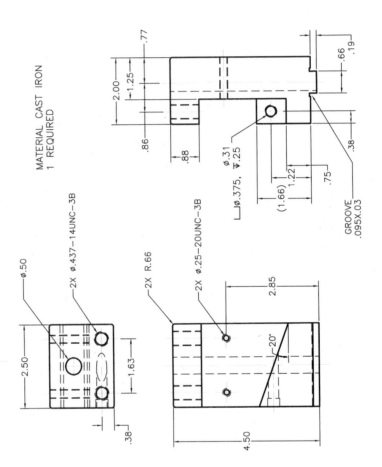

Figure 6-69 Drawing for Problem-Solving Exercise 3

Problem-Solving Exercise 4

Draw the reception desk shown in Figure 6-70. To get the dimensions of the chairs, refer to Problem-Solving Exercise 3 of Chapter 5.

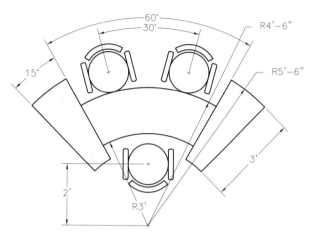

Figure 6-70 *Drawing for Problem-Solving Exercise 4*

Problem-Solving Exercise 5

Draw the views of Vice Body shown in Figure 6-71. The dimensions given in the figure are only for your reference. **

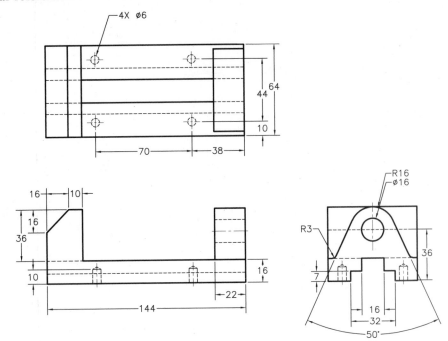

Figure 6-71 *Views and dimensions of the Vice Body*

Answers to Self-Evaluation Test

1. T, **2.** T, **3.** T, **4.** F, **5.** T, **6.** T, **7.** middle, **8. GRIPCOLOR**, **9. MIRROR, MI**, **10. Value**

Chapter 7

Creating Texts and Tables

CHAPTER OBJECTIVES

In this chapter, you will learn:
- *About annotative objects.*
- *To write text using the Single Line and Multi Line Text tools.*
- *To edit text using the Edit tool.*
- *To create text styles using the STYLE command.*
- *To draw tables using the Table tool.*
- *To create and modify a table style.*
- *To use the Properties palette to change properties .*

KEY TERMS

- *Annotative Objects*
- *Annotation Scale*
- *Text*
- *DDEDIT*
- *Table*
- *Data Link*
- *Text Style*
- *Annotative Text*
- *Sheet Set Table*

ANNOTATIVE OBJECTS

One of the recent enhancements in AutoCAD is the improved functionality of the drawing annotation that enables you to annotate your drawing easily and efficiently. Annotations are the notes and objects that are inserted into a drawing to add important information to it. The list of annotative objects is given next.

Text	Mtext	Dimensions	Leaders	Blocks
Multileaders	Hatches	Tolerances	Attributes	

All the above mentioned annotative objects have an annotative property. This annotative property allows you to automate the process of display of the drawings at a correct desired scale on the paper.

ANNOTATION SCALE

Annotation Scale is a feature that allows you to control the size of the annotative objects in model space. In the older releases of AutoCAD, you had to specify the height of the annotative object in model space, keeping in mind the extent up to which the factor would be scaled down while printing the drawing. But the **Annotation scale** option allows you to set the height of the printed text while annotating, and then allows you to control the height of the object in model space.

Generally, the formula used to calculate the height of the annotative objects in model space is:

Height of the annotative object in model space = Annotation scale X Height of the annotative object in paper space.

For example, if you need to create an annotative object with a height of 1/4" on paper, and assume that annotation scale is set to 1/2" = 1' (implies 1" = 2', or 1" = 24"). Then, the height of the object in model space will be 1/4" x 24"= 6" and the object will be displayed in model space with its height equal to 6".

Note
The above calculation is performed automatically by AutoCAD and has been explained to make the concept clear.

Assigning Annotative Property and Annotation Scales

You can apply the annotative properties to objects in many ways. The objects (Mtext, Hatches, Blocks, and Attributes) that are drawn with the help of dialog boxes have the **Annotative** check boxes in their respective dialog box. You can also create the annotative styles, and all the objects created using that style will have the annotative property automatically. The objects that are created using the Command prompt can be converted into annotative after creating them through the **Properties** palette. You can also override the annotative property of an individual object through the **Properties** palette, see Figure 7-1.

All annotative objects are indicated with an annotative symbol attached to the cursor. When you move the cursor over an annotative object, the annotative symbol starts rolling on, see Figure 7-2. All annotative styles are indicated with the same annotative symbol displayed next to the cursor.

To assign the annotation scale while working in model space, choose the **Annotation Scale** button displayed on the right of the status bar; a flyout will be displayed. Select the desired annotation scale from the flyout. This flyout displays 33 commonly used imperial and metric

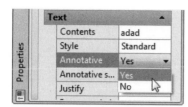

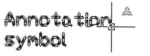

Figure 7-1 *Assigning annotative property through the Properties palette*

Figure 7-2 *Annotation symbol attached to the cursor*

scales, and this list can also be customized. Among these scales, 1:1 annotation scale is selected by default. To change the default annotation scale, enter **CANNOSCALE** in the Command prompt and specify the new annotation scale. To assign the annotation scale while working in viewport, select a viewport first and then assign the annotation scale.

Customizing Annotation Scale

In any drawing, only few annotation scales are used. Therefore, you can delete unwanted scales from the list according to your requirement. To customize the **Annotation Scale** list, choose the **Custom** option from the flyout that is displayed when you click on the **Annotation Scale** button; the **Edit Drawing Scales** dialog box will be displayed, as shown in Figure 7-3. With this dialog box, you can add, modify, delete, change the sequence of scales displayed, and restore the default list of scales.

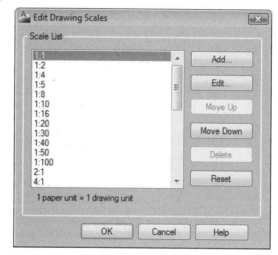

Figure 7-3 *The Edit Drawing Scales dialog box*

Note
You cannot delete the annotation scales that have been used to draw any annotative object in the drawing.

MULTIPLE ANNOTATION SCALES

Ribbon: Annotate > Annotation Scaling > Add/Delete Scales   **Command:** OBJECTSCALE
Menu Bar: Modify > Annotative Object Scale > Add/Delete Scales

You can assign more than one annotation scale to an annotative object. This enables you to display or print the same annotative object in different sizes. The annotative object is displayed or printed based on the current annotation scale. Assigning multiple annotation scales to annotative objects saves a considerable amount of time that is lost while creating a set of objects with different scales in different layers. You can add multiple annotation scales to objects manually and automatically. Both these methods are discussed next.

Assigning Multiple Annotation Scales Manually

To assign multiple annotation scales manually to an annotative object, invoke the **Add/Delete Scales** tool; you will be prompted to select an annotative object to which you want to add another annotation scale. Select one or more annotation objects and then press ENTER; the **Annotation Object Scale** dialog box will be displayed, see Figure 7-4. Alternatively, select an annotative object and right-click to display the shortcut menu. Now, choose **Annotative Object Scale > Add / Delete Scales** from the shortcut menu to display the **Annotation Object Scale** dialog box. Using this dialog box, you can add annotation scale to an annotative object or delete the annotation scale assigned to a selected annotative object. However, you cannot delete the current

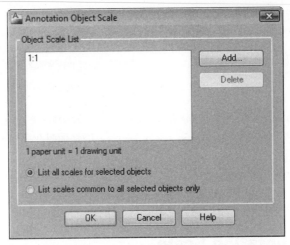

*Figure 7-4 The **Annotation Object Scale** dialog box*

annotation scale by using this dialog box. When you place the cursor over an annotative object, all annotative objects with multiple annotation scales will be indicated with a double-annotation symbol attached to the cursor, see Figure 7-5. You can also control the assignment of multiple annotation scales to annotative objects by using the **Annotation Scaling** panel in the **Annotate** tab, see Figure 7-6.

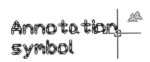

Figure 7-5 Multiple annotation scale symbol attached to the cursor

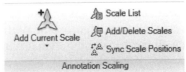

*Figure 7-6 The **Annotation Scaling** panel in the **Ribbon***

Assigning Multiple Annotation Scales Automatically

The annotation scales that have been set as current can be assigned to annotative objects without invoking the **Annotation Object Scale** dialog box. To do so, choose the **Automatically add scales to annotative objects when the annotation scale changes** button located at the right corner of the Status Bar to activate it. Now onward, all annotation scales that are set as current will be automatically added to the annotation objects.

By default, the annotation scale will be applied to all objects irrespective of their layers. However, you can control the assignment of annotation scale depending upon the layer of the object by using the **ANNOAUTOSCALE** system variable. This system variable can have values ranging from -4 to 4 except 0. The description of these values is given next.

1 = This value assigns the current annotation scale to all annotative objects, except the annotative objects that are in the locked, turned off, frozen, and viewport frozen layers.
2 = This value assigns the current annotation scale to all annotative objects, except the annotative objects that are in the turned off, frozen, and viewport frozen layers.
3 = This value assigns the current annotation scale to all annotative objects, except the ones that are in the locked layers.
4 = This value assigns the current annotation scale to all objects.
-1 = On assigning this value, the **ANNOAUTOSCALE** gets turned off, but when turned on again, it regains the value 1.
-2 = On assigning this value, the **ANNOAUTOSCALE** gets turned off, but when turned on again, it regains the value 2.
-3 = On assigning this value, the **ANNOAUTOSCALE** gets turned off, but when turned on again, it regains the value 3.

-4 = On assigning this value, the **ANNOAUTOSCALE** gets turned off, but when turned on again, it regains the value 4.

CONTROLLING THE DISPLAY OF ANNOTATIVE OBJECTS

 After assigning annotation scale to objects, the paper height display scale (in case of plotting) or the viewport scale (in case of viewports) is automatically applied to annotative objects and they are displayed accordingly in the model space or viewport. The annotative objects that do not support the current annotation scale will not be scaled and will be displayed according to the previous annotation scale applied to them.

You can control the display of annotative objects by using the **Annotation Visibility** button on the right of the **Annotation Scale** flyout in the Status Bar. If this button is turned on, all annotative objects will be displayed in the model space or viewport regardless of the current annotation scale. If this button is turned off, only the annotative objects that have annotation scale equal to that of the current annotation scale will be displayed. Note that the **Annotation Visibility** buttons in the model and layout tabs are independent of each other.

When you select an annotative object, you can observe that the selected annotative object is displayed in all annotative scales assigned to it. The annotation scale that is equal to the current scale is displayed in dark dashed lines, whereas the other scales of the object are displayed in faded dashed lines. You can position the annotation object of different scales at different locations with the help of grips. However, if you need to synchronize the start point of annotative objects of different scale factors with current annotative objects, select the annotative object and then right-click in the drawing area; a shortcut menu will be displayed. Choose **Annotative Object Scale > Synchronize Multiple-Scale positions** from the shortcut menu; the start point of annotative objects will be synchronized. Alternatively, you can also use the **ANNORESET** command to synchronize the start point.

SELECTIONANNODISPLAY system variable is used to control the display of the supported scale representations. If you assign 0 to the system variable, it will stop displaying all the supported scale representations selected; only the representations with the current annotation scale will be displayed. If you assign 1 to this system variable, it will display all the supported scale representations selected.

> **Note**
> *Whenever you create a drawing that involves annotative objects in it, use the newly introduced* **Drafting & Annotation** *workspace. This workspace displays the* **Ribbon** *with drafting and annotation panels visible by default and hides many of the tools. This also makes the drawing and annotating work much more easy. For all examples and exercises given onward and involving annotative objects, you can use this workspace.*

CREATING TEXT

In manual drafting, lettering is accomplished by hand using a lettering device, pen, or pencil. This is a very time-consuming and tedious job. Computer-aided drafting has made this process extremely easy. Engineering drawings invoke certain standards to be followed in connection with the placement of a text in a drawing. In this section, you will learn how a text can be added in a drawing by using the **TEXT** and **MTEXT** commands.

Writing Single Line Text

Ribbon:	Home > Annotation > Text drop-down > Single Line Or	
	Annotate > Text > Text drop-down > Single Line	**Command:** TEXT or DTEXT
Menu Bar:	Draw > Text > Single Line Text	**Toolbar:** Text > Single Line Text

The **Single Line** tool is used to write text on a drawing. While writing you can delete what has been typed by using the BACKSPACE key. On invoking the **Single Line** tool, you will be prompted to specify the start point. The default and the most commonly used option in the **Single Line** tool is the **Start Point** option. By specifying a start point, the text is left-justified along its baseline. Baseline refers to the line along which their bases lies. After specifying the start point, you need to set the height and the rotation angle of the text.

The **Specify height** prompt determines the distance by which the text extends above the baseline, measured by the capital letters. This distance is specified in drawing units. You can specify the text height by specifying two points or by entering a value. In the case of a null response, the default height, that is, the height used for the previous text drawn in the same style, will be used.

The **Specify rotation angle of text** prompt determines the angle at which the text line will be drawn. The default value of the rotation angle is 0-degree (along east); and in this case, the text is drawn horizontally from the specified start point. The rotation angle is measured in counterclockwise direction. The last angle specified becomes the current rotation angle, and if you give a null response, the last angle specified will be used as default rotation angle. You can also specify the rotation angle by specifying two points. The text will be drawn upside down if a point is specified at a location to the left of the start point.

You can now enter the text in the **Text Editor**. The characters will be displayed as you type them. After entering the text, click outside the **Text Editor**. The prompt sequence displayed on choosing this tool is given next.

Specify start point of text or [Justify/Style]: *Specify the start point.*
Specify height <0.2000>: **0.15** [Enter]
Specify rotation angle of text <0>: [Enter]; *the textbox will be displayed. Start typing in it.*

After completing a line, press ENTER; the cursor will automatically be placed at the start of the next line and you can enter the next line of the text. However, if you place the cursor at a new location, a new textbox will be displayed, and you can start typing text at this location with new start point and the same text parameters. You can enter multiple lines of text at any desired location in the drawing area on invoking the **Single Line** tool once. By pressing BACKSPACE, you can delete one character to the left of the current position of the cursor box. Even if you have entered several lines of text, you can use BACKSPACE to delete the text, until you reach the start point of the first line. To exit the tool, press the ESC key or click outside the text box.

This tool can be used with most of the text alignment modes, although it is most useful in the case of left-justified texts. In the case of aligned texts, this tool is used to assign a height appropriate for the width of the first line to every line of the text. Even if you select the **Justify** option, the text is first left-aligned at the selected point.

Justify Option

AutoCAD offers various options to align text. Alignment refers to the layout of a text. The main text alignment modes are **left**, **center**, and **right**. You can align a text by using a combination of modes, for example, top/middle/bottom and left/center/right (Figure 7-7). **Top** refers to a line along which lie the top points of the capital letters. Letters with descenders (such as p, g, y) dip below the baseline to the bottom. When the **Justify** option is invoked, the user can place the text in one of the fourteen various alignment types by selecting the desired alignment option using the Command prompt or the selection preview (Figure 7-8). The orientation of the text style determines the command interaction for Text Justify. (Text styles and fonts are discussed later in this chapter). For now, assume that the text style orientation is horizontal.

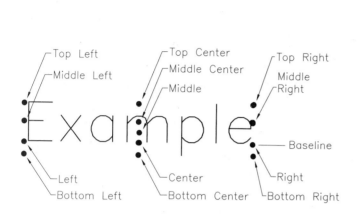

Figure 7-7 *Text alignment positions*

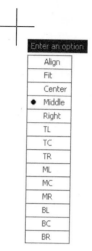

Figure 7-8 *Selection preview for the* ***Justify*** *option*

The prompt sequence that will follow when you choose the **Single Line Text** button to use this option is given next.

Specify start point of text or [Justify/Style]: **J** Enter
Enter an option [Align/Fit/Center/Middle/Right/TL/TC/TR/ML/MC/MR/BL/BC/ BR]: *Select any of these options.*

If the text style is vertically oriented (refer to the "Creating Text Styles" section later in this chapter), only four alignment options are available. The prompt sequence is given next.

Specify start point of text or [Justify/Style]: **J** Enter
Enter an option [Align/Center/Middle/Right]: *You can select the desired alignment option.*

If you know what justification you want, you can enter it directly at the **Specify start point of text or [Justify/Style]** prompt instead of first entering **J** to display the justification prompt. If you need to specify a style as well as a justification, you must specify the style first. The various alignment options are as follows.

Align Option. In this option, the text string is written between two points (Figure 7-9). You must specify the two points that act as the endpoints of the baseline. The two points may be specified horizontally or at an angle. AutoCAD adjusts the text width (compresses or expands) so that it fits between the two points. The text height is also changed, depending on the distance between points and the number of letters.

Specify start point of text or [Justify/Style]: **J** Enter
Enter an option [Align/Fit/Center/Middle/Right/TL/TC/TR/ML/MC/MR/BL/BC/BR]: **A** Enter

Specify first endpoint of text baseline: *Specify a point.*
Specify second endpoint of text baseline: *Specify a point.*

Fit Option. This option is very similar to the previous one. The only difference is that in this case, you select the text height, and it does not vary according to the distance between the two points. AutoCAD adjusts the letter width to fit the text between the two given points, but the height remains constant (Figure 7-9). The **Fit** option is not accessible for vertically oriented text styles. If you try the **Fit** option on the vertical text style, you will notice that the text string does not appear in the prompt. The prompt sequence is given next.

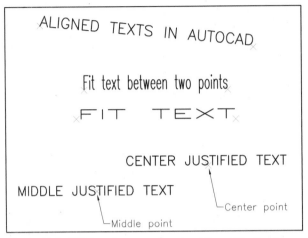

Figure 7-9 *Writing the text using the **Align**, **Fit**, **Center**, and **Middle** options*

Enter an option [Align/Fit/Center/Middle/Right/TL/TC/TR/ML/MC/MR/BL/BC/BR]: **F**
Specify first endpoint of text baseline: *Specify a point.*
Specify second endpoint of text baseline: *Specify a point.*
Specify height<current>: *Enter the height.*

Note
*You do not need to select the **Justify** option (J) for selecting the text justification. You can enter the text justification by directly entering justification when AutoCAD prompts "Specify start point of text or [Justify/Style]:"*

Center Option. You can use this option to select the midpoint of the baseline for the text. This option can be invoked by entering **Justify** and then **Center** or **C**. After you select or specify the center point, you must enter the letter height and the rotation angle (Figure 7-9).

Specify start point of text or [Justify/Style]: **C** `Enter`
Specify center point of text: *Specify a point.*
Specify height<current>: **0.15** `Enter`
Specify rotation angle of text<0>: `Enter`

Middle Option. Using this option, you can center text not only horizontally, as with the previous option, but also vertically. In other words, you can specify the middle point of the text string (Figure 7-9). You can alter the text height and the angle of rotation to meet your requirement. The prompt sequence that follows is given next.

Specify start point of text or [Justify/Style]: **M** `Enter`
Specify middle point of text: *Specify a point.*
Specify height<current>: **0.15** `Enter`
Specify rotation angle of text<0>: `Enter`

Right Option. This option is similar to the default left-justified start point option. The only difference is that the text string is aligned to the lower right corner (the endpoint you specify), that is, the text is **right-justified** (Figure 7-10). The prompt sequence that will follow when you choose this button is given next.

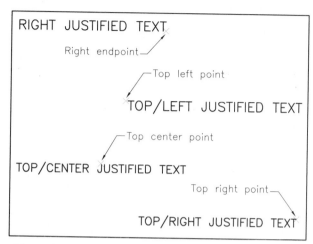

Figure 7-10 *Writing the text using the* **Right**, **Top-Left**, **Top-Center**, *and* **Top-Right** *options*

Specify start point of text or [Justify/Style]: **R** Enter

Specify right endpoint of text baseline: *Specify a point.*
Specify height<current>: **0.15** Enter
Specify rotation angle of text<0>: Enter

TL Option. In this option, the text string is justified from top left (Figure 7-10). The prompt sequence is given next.

Specify start point of text or [Justify/Style]: **TL** Enter
Specify top-left point of text: *Specify a point.*
Specify height<current>: **0.15** Enter
Specify rotation angle of text <0>: Enter

Note
The rest of the text alignment options are similar to those already discussed, and you can try them on your own. The prompt sequence is almost the same as those given for the previous examples.

Style Option

With this option, you can specify another existing text style. Different text styles can have different text fonts, heights, obliquing angles, and other features. This option can be invoked by entering **TEXT** and then **S** at the next prompt. The prompt sequence that will follow when you choose the **Single Line Text** tool for using this option is given next.

Specify start point of text or [Justify/Style]: **S** Enter
Enter style name or [?] <current>: *Specify the desired style or enter ? to list all styles.*

If you want to work in the previous text style, just press ENTER at the last prompt. If you want to activate another text style, enter the name of the style at the last prompt. You can also choose from a list of available text styles, which can be displayed by entering **?**. After you enter **?**, the next prompt is given next.

Enter text style(s) to list <*>: *Specify the text styles to list or enter * to list all styles.*

Press ENTER to display the available text style names and the details of the current styles and commands in the **AutoCAD Text Window**.

Note

*With the help of the **Style** option of the **Text** tool, you can select a text style from an existing list. If you want to create a new style, use the **STYLE** command, which is explained later in this chapter.*

ENTERING SPECIAL CHARACTERS

In almost all drafting applications, you need to use special characters (symbols) in the normal text and in the dimension text. For example, you may want to use the degree symbol (°) or the diameter symbol (ø), or you may want to underscore or overscore some text. This can be achieved with the appropriate sequence of control characters (control code). For each symbol, the control sequence starts with a percent sign written twice (%%). The character immediately following the double percent sign depicts the symbol. The control sequences for some of the symbols are given next.

Control sequence	Special character
%%c	Diameter symbol (ø)
%%d	Degree symbol (°)
%%p	Plus/minus tolerance symbol (±)

For example, if you want to write 25° Celsius, you need to enter **25%%dCelsius**. If you want to write 43.0ø, you need enter **43.0%%c**.

In addition to the control sequences shown earlier, you can use the %%nnn control sequence to draw special characters. The nnn can take a value in the range of 1 to 126. For example, to use the & symbol, enter the text string **%%038**.

To insert an euro symbol, press and hold the ALT key and enter 0128 from the numeric keypad and then release the ALT key. Before entering the numbers, make sure that the NUM LOCK is on.

CREATING MULTILINE TEXT

Ribbon: Home > Annotate > Text drop-down > MultilineText	**Command:** MTEXT
Menu Bar: Draw > Text > Multiline Text	
Tool Palette: Draw > MText Or Annotate > Text > Multiline Text	
Toolbar: Draw, Text > Multiline Text	

The **Multiline Text** tool in the **Text** drop-down in the **Text** panel (see Figure 7-11) is used to write a multiline text whose width is specified by defining two corners of the text boundary or by entering a width, using the coordinate entry. The text created by using the **Multiline Text** tool is a single object regardless of the number of lines it contains. On invoking the **Multiline Text** tool, a sample text "abc" is attached to the cursor and you are prompted to specify the first corner. Specify the first corner and move the pointing device so that

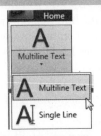

*Figure 7-11 The **Multiline Text** tool in the **Text** drop-down*

a box that shows the location and size of the paragraph text is formed. An arrow is displayed within the boundary indicating the direction of the text flow. Specify the other corner to define the boundary. When you define the text boundary, it does not mean that the text paragraph will fit within the defined boundary. AutoCAD only uses the width of the defined boundary as the width of the text paragraph. The height of the text boundary has no effect on the text paragraph.

Note
*On invoking the **Multiline Text** tool, a sample text "abc" is attached to the cursor. The **MTJIGSTRING** system variable stores the default contents of the sample text. You can specify a string of ten alphanumeric characters as the default sample text.*

The prompt sequence that will be displayed is given next.

Command: **MTEXT** [Enter]
Current text style: "Standard". Text height: 0.2000 Annotative: No
Specify first corner: *Select a point to specify first corner.*
Specify opposite corner or [Height/Justify/Line spacing/Rotation/Style/Width/Columns]: *Select an option or select a point to specify other corner. Now, you can write text in this boundary.*

Note
Although the box boundary that you specify controls the width of the paragraph, a single word is not broken to adjust inside the boundary limits. This means that if you write a single word whose width is more than the box boundary specified, AutoCAD will write the word irrespective of the box width and therefore, will exceed the boundary limits.

Once you have defined the boundary of the paragraph text, AutoCAD displays the **Text Editor** tab along with a **Text Window**, as shown in Figure 7-12. The window & the tab are discussed next.

Figure 7-12 The **Text Editor** *tab and the* **Text** *Window*

Text Window

The **Text** Window is used to enter the multiline text. The width of the active text area is determined by the width of the window that you specify when you invoke the **Multiline Text** tool. You can increase or decrease the size of the text window by dragging the double-headed arrow provided at the top-right and bottom left corner of the text window. You can also move the scroll bar up or down to display the text.

The ruler on the top of the text window is used to specify the indentation of the current paragraph. The top slider of the ruler specifies the indentation of the first line of the paragraph while the bottom slider specifies the indentation of the other lines of the paragraph.

Text Editor Tab

On creating the text window a contextual tab will be added to the **Ribbon**. The options in this tab are discussed next.

Text Style

The list of available text styles are grouped on the left of the **Style** panel in the
Text Editor tab. Click the down arrow to scroll down the list of text styles. If you
choose the double arrow, a flyout containing text styles created in the current
drawing will be displayed. You can select the desired text style from this flyout. You can also
create a new text style by using the **STYLE** command, which is explained later in this chapter.

Annotative

The **Annotative** button is used to write the annotative text. The annotative text scales itself
according to the current annotative setting. The annotative text also keeps on scaling according
to the change in the annotative scale. The annotative texts are defined in the drawing area in
terms of the paper height and the current annotation scale setting.

> **Note**
> *While editing the text, the **Annotative** button is used to convert the non-annotative multiline
> text into annotative multiline text and vice-versa.*
>
> *You can also change the non-annotative text to annotative by setting the text's **Annotative**
> property to **Yes** in the **Properties** palette. This applies to both the single line and the multiline
> text.*

Text Height

The **Text Height** drop-down list is used to specify the text height of the multiline text. The
default value in this drop-down list is 0.2000. Once you modify the height, AutoCAD retains
that value till you change it. Remember that the multiline text height does not affect the height
specified for the **TEXT** command.

Bold, Italic, Underline, Overline

You can use the appropriate buttons in the **Formatting** panel to make the selected text bold,
italics, underlined, or create overlined text. Bold and italics are not supported by SHX fonts
and hence, they will not be available for the particular fonts. These four buttons toggle between
on and off.

Font

The **Font** drop-down list in the **Formatting** panel displays all fonts in AutoCAD. You can select
the desired font from this drop-down list, see Figure 7-13. Irrespective of the font assigned to
a text style, you can assign a different font to that style for the current multiline text by using
the **Font** drop-down list.

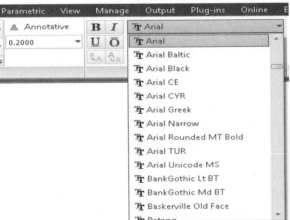

*Figure 7-13 The **Font** drop-down list*

Uppercase

When you select text in the text editor and choose this button, the alphabets written in the lowercase are converted to the uppercase.

Lowercase

Choosing this button converts the alphabets written in the uppercase to the lowercase.

Color

The **Color** drop-down list is used to set the color for the multiline text. You can also select the color from the **Select Color** dialog box that is displayed by selecting **Select Colors** in the drop-down list.

Background Mask

This option is used to define a background color for a multiline text. When you choose this option, the **Background Mask** dialog box will be displayed, as shown in Figure 7-14. To add a background mask, select the **Use background mask** check box and select the required color from the drop-down list in the **Fill Color** area. You can set the size of the colored background behind the text using the **Border offset factor**

*Figure 7-14 The **Background Mask** dialog box*

edit box. The previous value of the offset factor is automatically selected in the edit box. The box that defines the background color will be offset from the text by the value you define in this edit box. The value in this edit box is based on the height of the text. If you enter **1** as the value, the height of the colored background will be equal to the height of the text and will extend through the length of the window defined to write the multiline text. Similarly, if you enter **2** as the value, the height of the colored box will be twice the height of the text and will be equally offset above and below the text. However, the length of the colored box will still be equal to the length of the window defined to write the multiline text. You can select the **Use drawing background color** check box in the **Fill Color** area to use the color of the background of the drawing area to add the background mask. The previous color used is automatically selected.

Oblique Angle

Expand the **Formatting** panel to view this option. This option is used to specify the slant angle for the text. Enter the angle in the **Oblique Angle** edit box or use spinner to specify the slant angle, which is measured from the positive direction of X-axis in the clockwise direction.

Tracking

Expand the **Formatting** panel to view this option. This option is used to control the spacing between the selected characters. Select the characters and enter the value of the spacing in the **Tracking** edit box. You can also use the **Tracking** spinner to specify this value. The default tracking value is 1.000. You can specify values ranging from 0.7500 to 4.0000.

Width Factor

Expand the **Formatting** panel to view this option. This option is used to control the width of characters. By default, the value is set to 1.000. Select the characters and use the **Width Factor** edit box or the **Width Factor** spinner to specify the width of characters.

Justification

In large complicated technical drawings, the **Justification** option is used to fit the text matter with a specified justification and alignment. For example, if the text justification is bottom-right (BR), the text paragraph will spill to the left and above the insertion point, regardless of how you define the width of the paragraph. When you choose the **Justification** button in the **Paragraph** panel of the **Text Editor** in the **Ribbon**, a cascading menu appears that displays the predefined text justifications. By default, the text is **Top Left** justified. You can choose the new justification from the cascading menu. The various justifications are TL, ML, BL, TC, MC, BC, TR, MR, and BR. Figure 7-15 shows various text justifications for multiline text.

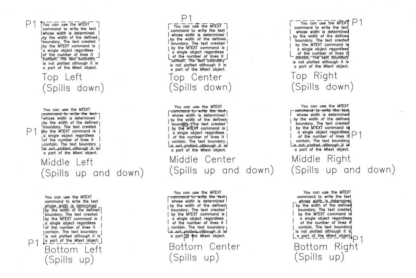

Figure 7-15 Text justifications for multiline text (P1 is the text insertion point)

Paragraph

This button is used to control the setting of a paragraph such as tab, indent, alignment, paragraph spacing, and line spacing, Choose this button to display the **Paragraph** dialog box on the screen, as shown in Figure 7-16. The options in the **Paragraph** dialog box are discussed next.

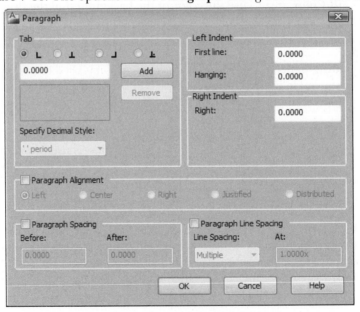

*Figure 7-16 The **Paragraph** dialog box*

Tab. This area is used to specify the tab type by selecting the required radio button. This area is also used to add or remove the additional tab positions up to which the cursor is moved while pressing the TAB key once at the start of a new paragraph for the left, center, right, and decimal justification. This option can also be set using the mouse by clicking on the ruler of the **In-Place Text** editor.

Left Indent. In this area, you can set the indent of the first line and the following lines of multiline text by entering the values in the **First Line** and the **Hanging** edit boxes.

Right Indent. In this area, you can set the right indent of the paragraph of the multiline text by entering the value in the **Right** edit box.

Paragraph Alignment. Select the **Paragraph Alignment** check box to set the alignment for the current or the selected paragraph.

Paragraph Spacing. Select the **Paragraph Spacing** check box to set the gap before the start of the paragraph and after the end of the paragraph. According to the gap set by you, AutoCAD calculates the gap between the two consecutive paragraphs. This gap is equal to the sum of after paragraph gap for the upper paragraph and the before paragraph gap for the lower paragraph.

Paragraph Line Spacing. Select the **Paragraph Line Spacing** check box to set the gap between the two consecutive lines. Select the **Exactly** option from the **Line Spacing** drop-down list to maintain the gap exactly specified by you in the **At** edit box, irrespective of the text height. On selecting the **At least** option, both the user specified arbitrary value as well as the text height will be considered to determine the gap. If the text height is smaller than the specified value, the line space will be determined by the user specified value. If the text height is larger, the line spacing will be equal to the text height value. By selecting the **Multiple** option, you can specify the spacing according to the text height. When the text height is not equal in a line, the line space will be determined by the largest text height value of that line.

Left
This radio button is used to left align the text written in the **Text** Window.

Center
This radio button is used to center align the text written in the **Text** Window.

Right
This radio button is used to right align the text written in the **Text** Window.

Justify
This radio button is used to adjust horizontal spacing so that the text is aligned evenly in between the left and right margins. Justifying the text creates a smooth edge on both sides.

Distribute
This button is used to adjust the horizontal spacing so that the text is evenly distributed throughout the width of the column in the **In-Place Text Editor**.

 Note
The spaces entered at the end of the line are also considered while justifying the text.

Line Spacing
Line spacing is the distance between two consecutive lines in a multiline text. Choose **Paragraph > Line Spacing** from the **Ribbon** to display a flyout containing options to select some predefined line spacing or opening the **Paragraph** dialog box, see Figure 7-17. The **1.0x**, **1.5x**, **2x**, **2.5x**

options set the line spacing in the multiples of the largest text height value in the same line. The **More** option is chosen to open the **Paragraph** dialog box that has been discussed earlier. The **Clear Line Space** option removes the space setting for the lines of the selected or the current paragraph. As a result, the line spacing will retrieve the default mtext line space setting.

Bullets and Numbering

This option is used to control the settings to display the bullets and numbering in a multiline text. Choose the **Bullets and Numbering** option; a flyout will be displayed, as shown in Figure 7-18. The options in this flyout are discussed next.

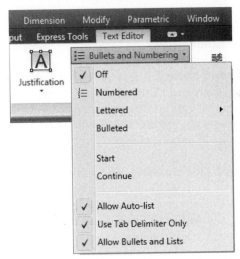

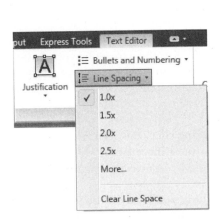

*Figure 7-17 Options in the **Line Spacing** flyout*

*Figure 7-18 Flyout displayed on choosing the **Bullets and Numbering** option*

Off. Select this option to remove letters, numbers, and bullets from the selected text.

Lettered. This option inserts the letters for list formatting. You can choose the uppercase letters and lower case letters to be used for the list formatting.

Note
If the number of items exceed the number of alphabets, the numbering continues again from the start, but by using the double alphabet.

Numbered. This option inserts numbers for the list formatting.

Bulleted. This option inserts bullets for the list formatting.

Start. This option restarts a new letter or a numbering sequence for the list formatting.

Continue. This option adds the selected paragraphs to the last list and continues the sequence.

Allow Auto-list. Select the **Allow Auto-list** option and follow the procedure given below to start the formatting of list automatically.

1. Start a line of text by entering a number.
2. Use period (.), comma (,), close angle bracket (>), close square bracket (]), close parenthesis ()), or close curly bracket (}) as punctuation after the number.
3. Provide space by pressing the TAB key.
4. Enter the required text.

5. Press the ENTER key to move to the next line; the new line will be automatically listed.
6. Press the SHIFT+ENTER key to add a plain paragraph.
7. To end auto-listing, press the ENTER key twice.

Use Tab Delimiter Only. If this option is checked, list formatting will be applied to the text only when the space after the letter and number is created by the TAB key.

Allow Bullets and Lists. If this option is checked, the list formatting will be applied to the whole text lines in the multiline text area. If you clear this option, any list formatting in the multiline text object will be removed and the items will be converted into the plain text. And, all the **Bullets and Lists** options, including the **Allow Auto-list** available on the contextual tab will become unavailable. To make these options available, again right-click in the text window and choose the **Allow Bullets and Lists** sub-option from the **Bullets and Lists** option.

Combine Paragraphs

This option is used to combine the selected paragraphs into a single paragraph. AutoCAD replaces the returns between all the paragraphs by a space. As a result, the lines in the resultant paragraph are in continuation.

Columns

This button is chosen to create columns in the multiline text. On choosing this button from the **Insert** panel, a flyout will be displayed that has the options to control the number, height, and gap between the columns. These options are shown in Figure 7-19, and are discussed next.

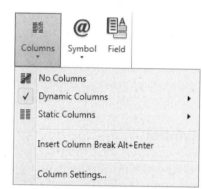

Figure 7-19 *The options in the* **Column** *flyout*

No Columns. This option does not create any column. The whole of the multiline text is written in a single column. When you start a multiline text, this option is selected by default.

Dynamic Columns. The dynamic columns are the text driven columns that change according to the quantity of the text. According to the quantity of text written, the columns are added or removed automatically. There are two options to control the height of the dynamic columns, **Auto height** and **Manual height**. The **Auto height** option automatically starts a new column when the text exceeds the range that the preset column can accommodate. The default height of the first column is decided by the height of the window that you create while starting the multiline text. All the other new columns generated are equal in height to the first column. If you change the height of the first column, the height of the remaining columns changes accordingly, in order to be in level with the first one. The **Manual height** option, on the other hand, provides the flexibility to control the height of each column individually. A new column will not start if the text exceeds the range that the present column can accommodate. The text will keep on adding to the same column by increasing the height of that column. Later on, you can add extra columns by adjusting the height of the present column. When you decrease the height of the column in which you have entered the text that does not fit the height of the new column, a new column will start automatically. Similarly, you can add more columns by adjusting the height of the last column.

Static Columns. The **Static Columns** option allows you to specify the number of columns before writing the text. All the generated columns are of the same height and width. You can also change the number of columns after writing the text.

Insert Column Break Alt + Enter. This option is used to start writing the text in a new column manually. When you insert a column break, the cursor automatically moves to the start of the next column. You can also press the ALT+ENTER key to insert a column break.

Column Settings. This option controls the setting to create a column. Choose this option to invoke the **Column Settings** dialog box, see Figure 7-20. The **Column Settings** dialog box has four areas: **Column Type**, **Column Number**, **Height**, and **Width**. The **Column Type** area specifies the type of column you want to create. All options in this area are similar to the options discussed above. In the **Column Number** area, you can specify the number of columns to be generated when you create the manual type of columns. In the **Height** area, you can specify the height of the columns to be generated for **Auto Height Dynamic Columns**, and **Static Column** type. The **Width** area controls the width of the columns and the gap between two consecutive columns.

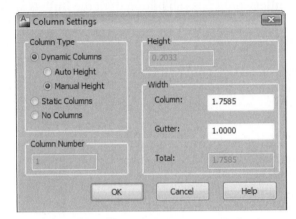

*Figure 7-20 The **Column Settings** dialog box*

Symbol

This option is used to insert the special characters in the text. When you choose the **Symbol** button from the **Insert** panel of the **Text Editor** tab in the **Ribbon**, a flyout will be displayed with some predefined special characters, see Figure 7-21. You can also choose **Other** from the flyout to display the **Character Map** dialog box. This dialog box has a number of other special characters that you can insert in the multiline text. To insert characters from the dialog box, select the character that you want to copy and then choose the **Select** button. Once you have selected all the required special characters, choose the **Copy** button and then close the dialog box. Now in the text window, position the cursor where you want to insert the special characters and right-click to display the shortcut menu. Choose **Paste** to insert the selected special character in the **Text Editor**.

Field

AutoCAD allows you to insert a field in the multiline text. A field contains data that is associated with the property that defines the field. For example, you can insert a field having the author's name of the current drawing. If you have already defined the author of the current drawing in the **Drawing Properties** dialog box, it will automatically be displayed in the field. If you modify the author and update the field, the changes will automatically be made in the text. When you choose this option, the **Field** dialog box will be displayed, as shown in Figure 7-22. You can select the field to be added in the **Field names** list box and define the format of the field using the **Format** list box. Choose **OK** after selecting the field and format. If the data in the selected field is already defined, it will be displayed in the **Text** window. If not, the field will display dashes (----).

Note
*To update a field after the text of that field is modified, double-click on the multiline text to display the **In-Place Text Editor**. Click on the field in the **Text** window to select it and then right-click to display the shortcut menu. Choose **Update Field** from this menu; the field will be updated. You can also edit the field or convert it into text using the same shortcut menu.*

Spell Check

To check spelling errors, choose the **Spell Check** button; the misspelt words will appear with a red underline. You can specify settings to check spelling errors as per your requirement. To do so, left-click on the inclined arrow in the **Spell Check** panel; the **Check Spelling Settings** dialog box will be displayed. This dialog box is used to specify what to include for spell check such as the dimension text, block attributes, sub option, and text in external references.

*Figure 7-21 Flyout displayed when you choose the **Symbol** button from the **Insert** panel*

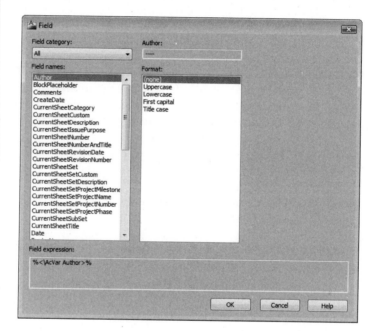

*Figure 7-22 The **Field** dialog box*

You can also specify what not to include like capitalized words, mixed cases, uppercase, and words with numbers or punctuation.

Edit Dictionaries

This option is used to match the word with the words in dictionaries while a text is run on spell check. On choosing this option, the **Dictionaries** dialog box will be displayed. This dialog box has two areas, **Main dictionary** and **Custom dictionary**. In the **Main dictionary** area, you can select the required dictionary from a list of different language dictionaries. And in the **Custom dictionary** area, you can add the commonly used words to a *.cus* file. You can also import the words from other dictionaries to the current customized *.cus* file.

Check Spelling Settings. Choose this sub-option to display the **Check Spelling Settings** dialog box.

Find and Replace

When you choose the **Find & Replace** button from the **Tools** panel, the **Find and Replace** dialog box will be displayed, as shown in Figure 7-23. The options that can be chosen from this dialog box are discussed next.

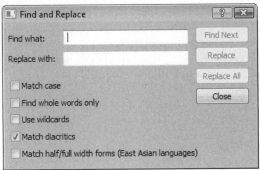

Find what. This edit box is used to enter the text or a part of a word or a complete word that you need to find in the drawing.

Replace with. If you want some text to be replaced, enter new text for replacement in this text box.

Match case. If this check box is selected, AutoCAD will find out the word only if the case of all characters in the word are identical to the word mentioned in the **Find what** edit box.

*Figure 7-23 The **Find and Replace** dialog box*

Find whole words only. If this check box is selected, AutoCAD will match the word in the text only if it is a single word, identical to the one mentioned in the **Find what** text box. For example, if you enter **and** in the **Find what** edit box and select the **Find whole words only** check box, AutoCAD will not match the string **and** in words like sand, land, band, and so on.

Use wildcards. This check box is selected to use the wildcard characters such as * , ?, and so on in the **Find what** edit box and then replace the selected text.

Match diacritics. This check box is selected to match the Latin characters, diacritical marks, or assent in your search.

Match half/full width forms (East Asian languages). If this check box is selected, you can match half or full width characters in your search.

Find Next. Choose this button to continue the search for the text entered in the **Find what** box.

Replace. Choose this button to replace the highlighted text with the text entered in the **Replace with** text box.

Replace All. If you choose this button, all the words in the current multiline text that match the word specified in the **Find what** text box will be replaced with the word entered in the **Replace with** box.

Import Text

Expand the **Tools** panel to view this option. When you choose this option, AutoCAD displays the **Select File** dialog box. In this dialog box, select any text file that you want to import as the multiline text; the imported text will be displayed in the text area. Note that only the ASCII or RTF file is interpreted properly.

AutoCAPS

Expand the **Tools** panel to view this option. If you choose this option, the case of all the text written or imported after choosing this option will be changed to uppercase. However, the case of the text written before choosing this option is not changed.

More

To edit text, select it and then choose the **More** button from the **Options** panel in the **Text Editor** tab; a flyout will be displayed. The options in the **More** flyout are discussed next.

Character Set. This option is used to define the character set for the current font. You can select the desired character set from the cascading menu that is displayed when you choose this option.

Remove Formatting. This option is used to remove the formatting such as bold, italics, or underline from the selected text. To use this option, select the text whose formatting you need to change and then right-click to display the shortcut menu. In the menu, choose the **Remove Formatting** option. The formatting of the selected text will be removed.

Editor Settings. The **Editor Settings** option has the sub-options to control the display of the **Text Formatting** toolbar. These sub-options are shown in Figure 7-24 and are discussed next.

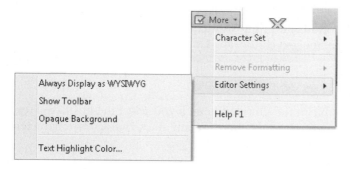

*Figure 7-24 The sub-options of **Editor Settings** option*

Always Display as WYSIWYG (What you see is what you get). This option controls the display size and orientation of the text in the **Text Editor**. If you choose this sub-option, the text in the **Text Editor** will be displayed in the **In-Place Text Editor** with its original size, position, and orientation. But, if the text is very small, or very large or rotated, then you can deselect this sub-option, so that the displayed text is oriented horizontally and is displayed in the readable size. Thus making it easy to read and edit.

Show Toolbar. This sub-option is used to show or hide the **Text Formatting** toolbar.

Opaque Background. Choose this sub-option to make the background of the **Text** window opaque. By default, the background is transparent. Invoke the text window over the existing objects/entities in the drawing window to see the effect of the opaque background sub-option.

Ruler

The **Ruler** button in the **Options** panel is used to turn on or off the display of ruler in the **Text Window**.

Undo

The **Undo** button allows you to undo the actions in the **In-Place Text Editor**. You can also press the CTRL+Z keys to undo the previous actions.

Redo

The **Redo** button allows you to redo the actions in the **In-Place Text Editor**. You can also press the CTRL+Y keys to redo the previous actions.

The options discussed next are not available in the **Text Editor** tab. However, in the **Text Formatting** toolbar, these options will be displayed on choosing **More > Editor Settings > Show Toolbar** from the **Text Editor** tab.

Stack*

To create a fraction text, you must use the stack option with the special characters /, ^, and #. In AutoCAD 2012, this option is available in sleep mode in the **Formatting** panel of the **Text Editor** tab. When you enter two numbers separated by / or ^, the stack option becomes active. You can use this option to stack. Alternatively, press ENTER or SPACEBAR after entering two characters separated by / or ^; the **AutoStack Properties** dialog box will be displayed, as shown in Figure 7-25. You can use this dialog box to control the stacking properties. The character / stacks the text vertically with a line, and the character ^ stacks the text vertically without a line (tolerance stack). The character # stacks the text with a diagonal line. After you enter the text with the required special character between them, select the text, and then choose the **Stack/Unstack** button. The stacked text is displayed equal to 70 percent of the actual height.

Unstack

This option is also available in the shortcut menu that is displayed on right-clicking on the stack characters from the text editor. The **Unstack** option is used to unstack the selected stacked text.

Stack

This option is available in the shortcut menu if you select the unstacked text. The **Stack** option is used to stack the selected text if there are any stack characters (characters separated by /, #, or ^) available in the multiline text.

Stack Properties

This option is available in the shortcut menu only when you select a stacked text. When you choose this option, the **Stack Properties** dialog box will be displayed, as shown in Figure 7-26. This dialog box is used to edit the text and the appearance of the selected stacked text. The options in this dialog box are discussed next.

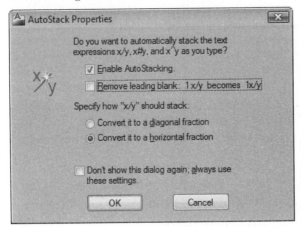
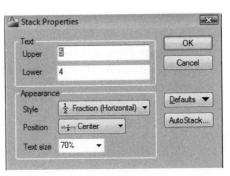

*Figure 7-25 The **AutoStack Properties** dialog box*   *Figure 7-26 The **Stack Properties** dialog box*

Text Area. You can change the upper and the lower values of the stacked text by entering their values in the **Upper** and **Lower** text boxes, respectively.

Appearance Area. You can change the style, position, and size of the stacked text by entering their values in the **Style**, **Position**, and **Text size** text boxes, respectively.

Defaults. This option allows you to restore the default values or save the new settings as the default settings for the selected stacked text.

AutoStack. When you choose this button, the **AutoStack Properties** dialog box will be displayed, in the same way as shown earlier in Figure 7-25. The options in this dialog box were discussed under the **Stack** heading.

Note
*The **Cut** and **Copy** sub-options of the **Edit** option can be used to move or copy the text from the text editor to any other application. Similarly, using the **Paste** sub-option, you can paste text from any windows text-based application to the **Text Editor**.*

EXAMPLE 1   *Multiline Text*

In this example, you will use the **Multiline Text** tool to write the following text.

For long, complex entries, create multiline text using the MTEXT option. The angle is 10°, dia = 1/2", and length = 32 1/2".

The font of the text is **Swis721 BT**, text height is 0.20, color is red, and written at an angle of 10-degree with Middle-Left justification. Make the word "multiline" bold, underline the text "multiline text", and make the word "angle" italic. The line spacing type and line spacing between the lines are **At least** and **1.5x**, respectively. Use the symbol for degrees. After writing the text in the **Text window**, replace the word "option" with "command".

1. Choose the **Multiline Text** tool from the **Annotation** panel in the **Home** tab in the **Drafting & Annotation** workspace. After invoking this tool, specify the first corner on the screen to define the first corner of the paragraph text boundary. You need to specify the rotation angle of the text before specifying the second corner of the paragraph text boundary. The prompt sequence is given next.

   Current text style: STANDARD. Text height: 0.2000. Annotative: No
   Specify first corner: *Select a point to specify the first corner.*
   Specify opposite corner or [Height/Justify/Line spacing/Rotation/Style/Width/Columns]: **R** Enter
   Specify rotation angle <0>: 10.
   Specify opposite corner or [Height/Justify/Line spacing/Rotation/Style/Width/Columns]: **L** Enter
   Enter line spacing type [At least/Exactly] <At least>: Enter
   Enter line spacing factor or distance <1x>: **1.5x**
   Specify opposite corner or [Height/Justify/Line spacing/Rotation/Style/Width/Columns]: *Select another point to specify the other corner.*

   The **Text Editor** and the **Text Editor** tab are displayed.

2. Select the **Swis721 BT** true type font from the **Ribbon Combo Box - Font** drop-down list of the **Formatting** panel in the **Text Editor** tab.

3. Enter **0.20** in the **Ribbon Combo Box - Text height** edit box of the **Style** panel, if the value in this edit box is not 0.2.

4. Select **Red** from the **Text Editor Color Gallery** drop-down list in the **Formatting** panel.

5. Now, enter the text in the **Text Editor**, as shown in Figure 7-27. To add the degree symbol, choose the **Symbol** button from the **Insert** panel of the **Text Editor** tab in the **Ribbon;** a

flyout is displayed. Choose **Degrees** from the flyout. When you type 1/2 after Dia = and then press the " key, AutoCAD displays the **AutoStack Properties** dialog box. Select the **Convert it to a diagonal fraction** radio button, if it is not already selected. Also, make sure the **Enable AutoStacking** check box is selected. Now, close the dialog box.

Similarly, when you type 1/2 after length = 32 1/2 and then press the " key, AutoCAD displays the **AutoStack Properties** dialog box. Select the **Convert it to a diagonal fraction** radio button and then close the dialog box.

6. Double-click on the word "multiline" to select it (or pick and drag to select the text) and then choose the **B** button to make it boldface and underline the words "multiline text" by using the **U** button.

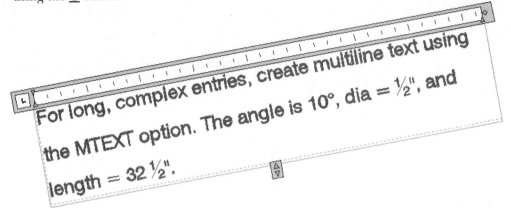

Figure 7-27 Text entered in the **Text Editor**

7. Highlight the word "angle" by double-clicking on it and then choose the **Italic** button.

8. Choose the **Middle Left** option from the **Justification** drop-down in the **Paragraph** panel.

9. Choose the **Find & Replace** button from the **Tools** panel in the **Text Editor** tab. Alternatively, right-click on the Text Window and select **Find and Replace** from the shortcut menu. Alternatively, you can use CTRL+R keys. On doing so, the **Find and Replace** dialog box is displayed.

10. In the **Find what** edit box, enter **option** and in the **Replace with** edit box, enter **command**.

11. Choose the **Find Next** button. AutoCAD finds the word "option" and highlights it. Choose the **Replace** button to replace **option** by **command**. The **AutoCAD** information box is displayed informing you that AutoCAD has finished searching for the word. Choose **OK** to close the information box.

Note
To set the width of the multiline text objects, hold and drag the arrowhead on the right side of the ruler. In this example, the text has been accommodated into two lines.

12. Now, choose **Close** to exit the **Find and Replace** dialog box and return to the **In-Place Text Editor**. To exit the editor mode, click outside the **Text Editor**; a text is displayed on the screen, as shown in Figure 7-28.

For long, complex entries, create **multiline** text using MTEXT command. The *angle* is 10°. dia $=\frac{1}{2}"$, and length $=32\frac{1}{2}"$.

Figure 7-28 *Multiline text for Example 1*

Tip
The text in the **Text Editor** *can be selected by double-clicking on the word, by holding the left mouse button of the pointing device and then dragging the cursor, or by triple-clicking on the text to select the entire line or paragraph.*

EXERCISE 1   *Single Line & Multiline Text*

Write the text using the **Single Line** and **Multiline Text** tools, as shown in Figures 7-29 and 7-30. Use the special characters and the text justification options shown in the drawing. The text height is 0.1 and 0.15 respectively in Figures 7-29 and 7-30.

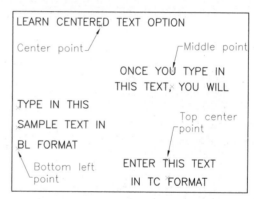

Figure 7-29 *Drawing with special characters*   ***Figure 7-30*** *Drawing for Exercise 1*

EDITING TEXT

The contents of **MTEXT** and **TEXT** object can be edited by using the **DDEDIT** and **Properties** commands. You can also use the AutoCAD editing commands, such as **MOVE**, **ERASE**, **ROTATE**, **COPY**, **MIRROR**, and **GRIPS** with any text object.

In addition to editing, you can also modify the text in AutoCAD. The modification that you can perform on the text include changing its scale and justification. The various editing and modifying operations are discussed next.

Editing Text Using the DDEDIT Command

Menu Bar: Modify > Object > Text > Edit   **Command:** DDEDIT
Toolbar: Text > Edit

You can use the **DDEDIT** command to edit text. The most convenient way of invoking this command is by double-clicking on the text. If you double-click on a single line text written by using the **Single Line** tool, AutoCAD creates an edit box around the text and highlights it. You can modify the text string in this edit box. The size of the bounding box increases or decreases as you add more text or remove the existing text. Note that for the text object, you cannot modify any of its properties in the bounding box. However, if you double-click on a multiline text written by using the **Multiline Text** tool, the text will be displayed in the

Text Editor. You can make changes using various options in the editor. Apart from changing the text string, you can also change the properties of a paragraph text.

You can also select the text for editing and then right-click in the drawing area; a shortcut menu is displayed. Depending on the text object you have selected, the **Edit** or **Mtext Edit** options will be available in the shortcut menu. On choosing the appropriate option, the **Text Editor** and the **Text Editor** tab will be displayed.

Editing Text Using the Properties Palette

Using the **DDEDIT** command with the text object, you can only change the text string, and not its properties such as height, angle, and so on. In this case, you can use the **Properties** palette for changing the properties. Select the text, right-click to invoke the shortcut menu, and choose the **Properties** option from the shortcut menu; the **Properties** palette with all properties of the selected text will be displayed, as shown in Figure 7-31. In this palette, you can change any value and the text string.

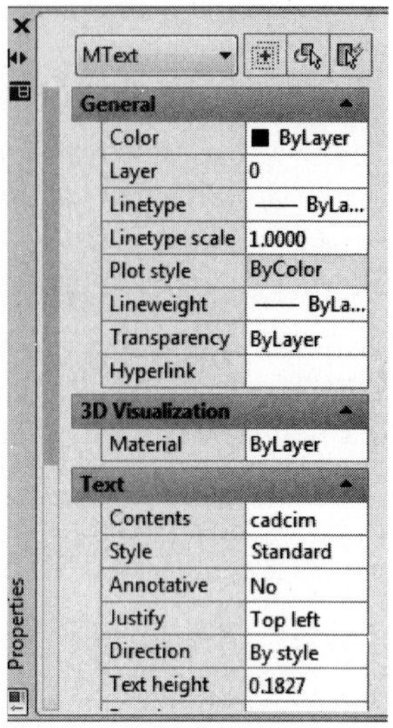

You can edit a single line text in the **Contents** edit box. However, to edit a multiline text, you must choose the **Full editor** button in the **Contents** edit box. On doing so, the **Text Editor** will be displayed and you can make the changes.

*Figure 7-31 The **Properties** palette*

Modifying the Scale of the Text

Ribbon: Annotate > Text > Scale	
Menu Bar: Modify > Object > Text > Scale	
Toolbar: Text > Scale	
Command: SCALETEXT	

You can modify the scale factor of a text by using the **Scale** tool in the extended options of the **Text** panel of the **Annotate** tab. On invoking this tool, you can select the text and specify the base point for scaling the text. AutoCAD lists the justification options as the base point to scale the text. Specify the appropriate base point and press ENTER; you will be prompted to specify the new model height or select one of the options. All these options are discussed next.

Paper Height

It scales the text height depending on the annotative properties of a drawing. The **Paper height** option can only be applied to the annotative objects.

Match object

You can use the **Match object** option to select an existing text whose height is used to scale a selected text.

Scale factor

You can use the **Scale factor** option to specify a scale factor to scale a text. You can also use the **Reference** option to specify the scale factor for the text.

Modifying the Justification of the Text

Ribbon: Annotate > Text > Justify
Menu Bar: Modify > Object > Text > Justify
Toolbar: Text > Justify
Command: JUSTIFYTEXT

You can modify the justification of a text by choosing the **Justify** tool in the extended options of the **Text** panel in the **Annotate** tab; you will be prompted to select the objects. Select the text and press ENTER; AutoCAD lists the justification options and you will be prompted to specify the justification. Specify the justification; the justification will be modified. Note that even after modifying the justification using this tool, the location of the text will not be changed. However, you can notice the changed justification location on selecting the text.

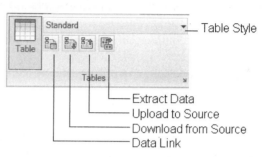

— Table Style
— Extract Data
— Upload to Source
— Download from Source
— Data Link

*Figure 7-32 Tools in the **Tables** panel of the **Annotate** tab*

INSERTING TABLE IN THE DRAWING

Ribbon: Home > Annotation > Table Or Annotate > Tables > Table
Menu Bar: Draw > Table
Tool Palette: Draw > Table
Toolbar: Draw > Table
Command: JUSTIFYTEXT

A number of mechanical, architectural, electric, or civil drawings require a table in which some information about the drawing is displayed. For example, the drawing of an assembly needs the Bill of Material, which is a table that provides details such as the number of parts in the drawing, their names, their material, and so on. To enter these information, AutoCAD allows you to create tables using the **Table** tool, see Figure 7-32. When you invoke this tool, the **Insert Table** dialog box is displayed, as shown in Figure 7-33. The options in this dialog box are discussed next.

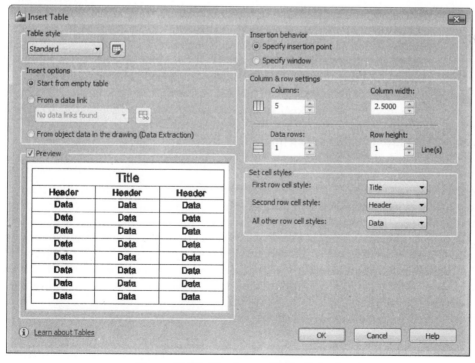

*Figure 7-33 The **Insert Table** dialog box*

Table style Area

The drop-down list in this area displays the names of the various table styles in the current drawing. By default, the **Standard** table style is displayed in the drop-down list. Choose the **Launch the Table Style dialog** button on the right of this drop-down list to display the **Table Style** dialog box. This dialog box can be used to create a new table style, modify, or delete an existing table style. You can also set the selected table style as the current table style. You will learn more about creating a new table style in the next section.

Insert options Area

The options in this area enable you to create tables with different types of data. You can add static data, externally-linked data, or object data to a table. The options in this area are discussed next.

Start from empty table

This option enables you to enter the data manually in the table. An empty table is created in which you have to enter the values manually. This type of table data is known as the static data. This is the default option selected for creating a table.

Tip
*You can copy an existing excel spreadsheet and paste it on the existing drawing as a table with the static data. To do so, use the **PASTESPEC** command and paste the copied data as AutoCAD entity. The resulting table will be similar to the table created using the **Start from empty table** option.*

From a data link

This option enables you to create a table automatically in AutoCAD from an excel sheet. This excel sheet remains linked with the drawing and any changes made in the excel sheet will reflect in the table. From the drop-down list, you can select an already linked excel sheet or you can also attach a new excel sheet to the drawing. To link a new excel file, choose the **Data Link Manager** button; the **Select a Data Link** dialog box will be displayed, see Figure 7-34. Select the **Create a new Excel Data Link** option; the **Enter Data Link Name** dialog box will be invoked. Enter a name for the new data link that you are creating and choose the **OK** button; the **New Excel Data Link** dialog box will be invoked, see Figure 7-35. Choose the **Browse [...]** button; the **Save As** dialog box will be invoked. Specify the location of the excel file to be linked with the current drawing and choose the **Open** button from the **Save As** dialog box. After choosing the **Open** button, more options will be added in the **New Excel Data Link** dialog box, see Figure 7-36. The options in the modified **New Excel Data Link** dialog box are discussed next.

Choose an Excel file. This drop-down list displays the excel file attached to the current drawing. Choose the **Browse [...]** button to change the excel file to be linked with the current drawing.

Link options Area. In this area, you can specify the part of the excel file to be linked with the drawing. Select the **Link entire sheet** radio button to link the entire worksheet in the form of a table to the drawing. Select the **Link to a named range** radio button to link the already defined name ranges from the excel sheet to the drawing. Select the **Link to range** radio button to enter the range of cells to be included in the data link. Choose the **Preview** button; the preview of the range of cells included in the data link will be displayed. The preview of the attached data from the excel sheet will be displayed in the preview area. Clear the **Preview** check box to disable the preview of the table.

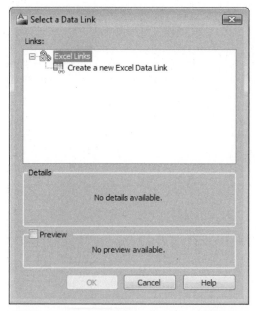

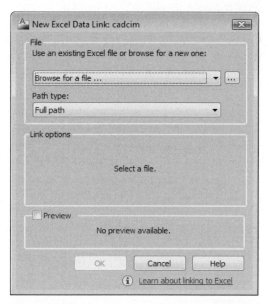

Figure 7-34 *The **Select a Data Link** dialog box*

Figure 7-35 *The **New Excel Data Link** dialog box*

Cell contents Area. The **Keep data formats and formulas** radio button in this area allows you to link the excel file data along with the supported formulas. The **Keep data formats, solve formulas in Excel** radio button allows you to import data from the excel file, but the calculations made using the formulas are done in Excel. The **Convert data formats to text, solve formulas in Excel** radio button in this area is selected by default and lets you import the data of the excel file as text, and the calculations are done in Excel. The **Allow writing to source file** check box is selected by default. This check box enables you to choose the **Download from Source** button from the **Tables** panel of the **Annotate** tab in the **Ribbon** to update the changes made in the drawing table that is linked to the excel sheet.

Cell formatting Area. Select the **Use Excel formatting** check box to import any formatting settings if specified in the excel sheet. Other options in this area will be available only when the **Use Excel formatting** check box is selected. The **Keep table updated to Excel formatting** radio button allows you to update the drawing table according to the changes made in the excel sheet. If you select the **Start with Excel formatting, do not update** radio button, any formatting changes specified in the excel sheet will be imported to the table, but the changes made in the excel sheet after importing will not be updated in the drawing table. Choose the **OK** button twice to display the **Insert Table** dialog box.

Tip
*If you have made any changes in the drawing table, choose the **Upload to Source** tool from the **Tables** panel in the **Annotate** tab; the changes made in the drawing file will reflect in the excel sheet that is linked to the drawing table. Similarly, if you have made any changes in the excel file after uploading, choose the **Download from Source** tool from the **Tables** panel in the **Annotate** tab; the changes made in the excel file will reflect in the drawing file.*

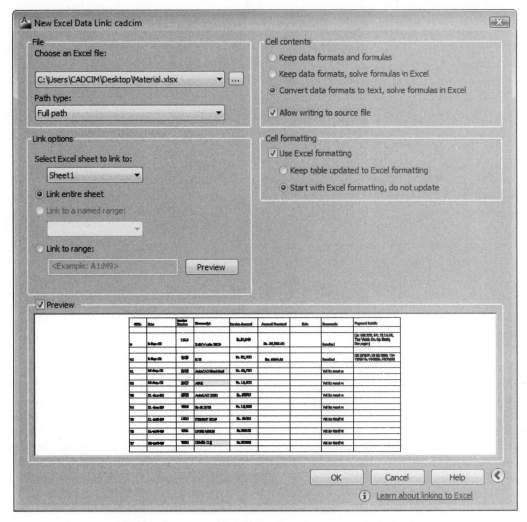

*Figure 7-36 The modified **New Excel Data Link** dialog box*

From object data in the drawing (Data Extraction)

To create a table by automatically extracting the data related to the objects created in the current drawing file, select the **From object data in the drawing (Data Extraction)** radio button from the **Insert Table** dialog box and choose the **OK** button; AutoCAD will start the **Data Extraction** wizard that will guide you step-by-step to insert the desired data into the current drawing.

Note

*For more information regarding the **Data Extraction** wizard, refer to Chapter-19 (Defining Block Attributes).*

Insertion behavior Area

The options in this area are used to specify the method of placing the table in the drawing. These options are discussed next.

Specify insertion point

This radio button is selected to place the table using the upper left corner of the table. If this radio button is selected and you choose **OK** from the **Insert Table** dialog box, you will be prompted to select the insertion point, which is by default the upper left corner of the table. By creating a different table style, you can change the point using which the table is inserted.

Specify window

If this radio button is selected and you choose **OK** from the **Insert Table** dialog box, you will be prompted to specify two corners for placing the table. The number of rows and columns in the table will depend on the size of the window you define.

Column and row settings Area

The options in this area are used to specify the number and size of rows and columns. The availability of these options depend on the option selected from the **Insert options** and **Insertion behavior** area. These options are discussed next.

Columns

This spinner is used to specify the number of columns in the table.

Column width

This spinner is used to specify the width of columns in the table.

Data rows

This spinner is used to specify the number of rows in the table.

Row height

This spinner is used to specify the height of rows in the table. The height is defined in terms of lines and the minimum value is one line.

Set cell styles Area

The options in this area are used to assign different cell styles for the rows in the new table. You can assign different cell styles to all the rows of a table. The options in this area are discussed next.

First row cell style

This drop-down list is used to specify a cell style for the first row of the table. By default, the **Title** cell style is selected for the first row of the table.

Second row cell style

This drop-down list is used to specify a cell style for the second row of the table. By default, the **Header** cell style is selected for the second row of the table.

All other row cell styles

This drop-down list is used to specify a cell style for all the other rows of the table. By default, the **Data** cell style is selected for the remaining rows of the table.

After setting the parameters in the **Insert Table** dialog box, choose the **OK** button. Depending on the type of insertion behavior selected, you will be prompted to insert the table. As soon as you complete the insertion procedure, the **In-Place Text editor** is displayed and you are allowed to enter the parameters in the first row of the table. By default, the first row is the title of the table. After entering the data, press ENTER. The first field of the first column is highlighted, which is the column head, and you are allowed to enter the data in it.

AutoCAD allows you to use the arrow keys on the keyboard to move to the other cells in the table. You can enter the data in the field and then press the arrow key to move to the other cells in the table. After entering the data in all the fields, press ENTER to exit the **Text Formatting** toolbar.

Tip
*You can also right-click while entering the data in the table to display the shortcut menu. This shortcut menu is similar to that shown in the **In-Place Text Editor** and can be used to insert field, symbols, text, and so on.*

CREATING A NEW TABLE STYLE

Ribbon: Home > Annotation > Table Style	**Toolbar:** Styles > Table Style
Menu Bar: Format > Table Style	**Command:** TABLESTYLE

 To create a new table style, choose **Table Style** from the extended options of the **Annotation** panel in the **Home** tab; the **Table Style** dialog box will be displayed, as shown in Figure 7-37. You can also invoke this dialog box by choosing the inclined arrow of the **Tables** panel in the **Annotate** panel.

To create a new table style, choose the **New** button from the **Table Style** dialog box; the **Create New Table Style** dialog box will be displayed, as shown in Figure 7-38.

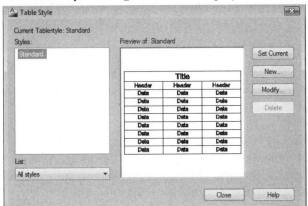

*Figure 7-37 The **Table Style** dialog box*

*Figure 7-38 The **Create New Table Style** dialog box*

Enter the name of the table style in the **New Style Name** edit box. Select the style on which you want to base the new style from the **Start With** drop-down list. By default, this drop-down list shows only the **Standard** table style. After specifying the settings, choose the **Continue** button from the **Create New Table Style** dialog box; the **New Table Style** dialog box will be displayed, see Figure 7-39. The options in this dialog box are discussed next.

Starting table Area

The options in this area enable you to select a table in your drawing to be used as reference for formatting the current table style. Once the table is selected, the table style associated with that table gets copied to the current table style and then you can modify it according to your requirement. The **Remove Table** button allows you to remove the initial table style from the current table style. This button is highlighted only when a table style is attached to the current one.

General Area

The **Table direction** drop-down list from the **General** area is used to specify the direction of the table. By default, this direction is down. As a result, the title and headers will be at the top and the data fields will be below them. If you select **Up** from the **Table direction** drop-down list, the title and headers will be at the bottom of this table and the data fields will be on the top of the table.

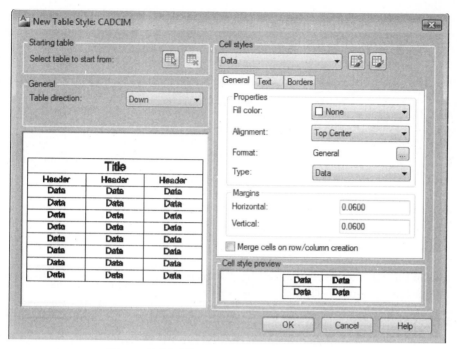

Figure 7-39 *The **New Table Style** dialog box*

Cell styles Area

This area has the options to define a new cell style or to modify the existing ones. The **Cell style** drop-down list displays the existing cell styles within the table. It also displays the options to create a new cell style or to manage the existing ones. These options are discussed next.

Create new cell style

To create a new cell style, select this option from the **Cell styles** drop-down list. Alternatively, you can choose the **Create a new cell style** button on the right of the **Cell styles** drop-down list. On choosing this button, the **Create New Cell Style** dialog box will be invoked. Next, enter a style name for the new cell style. Select the existing cell style from the **Start With** drop-down list; the settings from the existing style will be used as a reference for the new one to be created. Next, choose the **Continue** button; the new cell style will get added to the **Cell style** drop-down list. Next, you can modify this new cell style by modifying the options in the **General**, **Text**, and **Borders** tab of the **New Table Style** dialog box.

General Tab

The options in this tab are used to control the general appearance, alignment, and formatting of the table cells. These options are shown in Figure 7-40 and are discussed next.

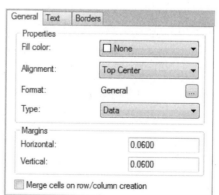

Fill color. This drop-down list is used to specify the fill color for the cells.

Alignment. This drop-down list is used to specify the alignment of the text entered in the cells. The default alignment is top center.

Figure 7-40 *The **General** tab of the New Table Style dialog box*

Format. If you choose the **Browse** [...] button available on the right of **Format**, the **Table Cell Format** dialog box will be displayed. This dialog box is used to specify the data type and format of the data type to be entered in the table. While creating the table, once the data type and format are specified, you cannot enter any other type of data or format without changing or modifying the table style. The default data type is **General**. In this data type, you can enter any alphanumeric characters.

Type. This drop-down list is used to specify the cell style either as a **Label** or **Data**.

Horizontal. This edit box is used to specify the minimum spacing between the data entered in the cells and the left and right border lines of the cells.

Vertical. This edit box is used to specify the minimum spacing between the data entered in the cells and the top and bottom border lines of the cells.

Merge cells on row/column creation. This check box is selected to merge all the new rows and columns created by using this cell style into one cell.

Text Tab

The options in this tab are used to control the display of the text to be written in the cells. These options are shown in Figure 7-41 and are discussed next.

Text style. This drop-down list is used to select the text style that is used for entering the text in the cells. By default, it shows only **Standard**, which is the default text style. You will learn to create more text styles later in this chapter.

Text height. This edit box is used to specify the height of the text to be entered in the cells.

Text color. This drop-down list is used to specify the color of the text that will be entered in the cells. If you select the **Select Color** option, the **Select Color** dialog box will be displayed, which can be used to select from index color, true color, or from the color book.

Text angle. This edit box is used to specify the slant angle of the text to be entered in the cell.

Borders Tab

The options in this area are used to set the properties of the border of the table, see Figure 7-42. The line weight and color settings that you specify using this area will be applied to all borders, outside borders, inside borders, bottom border, left border, top border, right border, or without border depending on which button is chosen from this area. Select the **Double line** check box to display the borders with double lines. You can also control the spacing between the double lines by entering the gap value in the **Spacing** edit box.

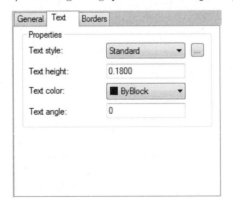

Figure 7-41 *The* ***Text*** *tab of the* ***New Table Style*** *dialog box*

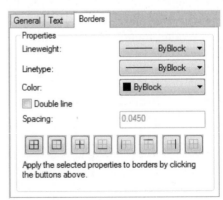

Figure 7-42 *The* ***Borders*** *tab of the* ***New Table Style*** *dialog box*

Manage Cell Styles

To create a new cell style, select this option from the **Cell style** drop-down list. Alternatively, you can choose the **Manage Cell Style dialog** button on the right of the **Cell styles** drop-down list. On choosing this button, the **Manage Cell Styles** dialog box will be invoked, see Figure 7-43. This dialog box displays all cell styles in the current table. You can also create a new cell style, delete, or rename an existing cell style. Note that the cell styles **Title**, **Header**, and **Data** are the default table styles provided in AutoCAD, and they cannot be deleted or renamed.

SETTING A TABLE STYLE AS CURRENT

*Figure 7-43 The **Manage Cell Styles** dialog box*

To set a table style as the current style for creating all the new tables, invoke the **Table Style** dialog box by choosing **Table Style** from the extended options of the **Annotation** panel in the **Home** tab. Next, select the table style from the **Styles** list box in the **Table Style** dialog box and choose the **Set Current** button. You can also set a table style current by selecting it from the **Table Style** drop-down list in the **Tables** panel. This is a convenient method of setting a table style current.

MODIFYING A TABLE STYLE

To modify a table style, invoke the **Table Style** dialog box by choosing **Table Style** in the extended options of the **Annotation** panel in the **Home** tab. Select the table style from the **Styles** list box in the **Table Style** dialog box and choose the **Modify** button; the **Modify Table Style** dialog box is displayed. This dialog box is similar to the **New Table Style** dialog box. Modify the options in the various tabs and areas of this dialog box and then choose **OK** to exit from the **Modify Table Style** dialog box.

MODIFYING TABLES

Select any cell in the table by double-clicking in it; the **Table Cell** tab will be added to the **Ribbon**, as shown in Figure 7-44. The options in the **Table Cell** tab are used to modify the table, insert block, add formulas, and perform other operations.

*Figure 7-44 The **Table Cell** tab added to the **Ribbon***

Modifying Rows

To insert a row above a cell, select a cell and choose the **Insert Above** tool from the **Row** panel in the **Table Cell** tab. To insert a row below a cell, select a cell and choose the **Insert Below** tool from the **Row** panel in the **Table Cell** tab. To delete the selected row, choose the **Delete Row(s)** tool from the **Row** panel in the **Table Cell** tab. You can also add more than one row by selecting more than one row in the table.

Modifying Columns

This option is used to modify columns. To add a column to the left of a cell, select the cell and choose the **Insert Left** tool from the **Columns** panel in the **Table Cell** tab. To add a column to the right of a cell, select a cell and choose the **Insert Right** tool from the **Columns** panel in the **Table Cell** tab. To delete the selected column, choose the **Delete Column(s)** tool from the **Columns** panel in the **Table Cell** tab. You can also add more than one column by selecting more than one row in the table.

Merge Cells

This button is used to merge cells. Choose this button; the **Merge Cells** drop-down is displayed. There are three options available in the drop-down. Select multiple cells using the **SHIFT** key and then choose **Merge All** from the drop-down to merge all the selected cells. To merge all the cells in the row of the selected cell, choose the **Merge By Row** button from the drop-down. Similarly, to merge all cells in the column of the selected cell, select the **Merge By Column** tool from the drop-down. You can also divide the merged cells by choosing the **Unmerge Cells** button from the **Merge** panel.

Match Cells

This button is used to inherit the properties of one cell into the other. For example, if you have specified **Top Left Cell Alignment** in the source cell, then using the **Match Cells** button, you can inherit this property to the destination cell. This option is useful if you have assigned a number of properties to one cell, and you want to inherit these properties in some specified number of cells. Choose **Match Cell** from the **Cell Styles** panel in the **Table Cell** tab; the cursor is changed to the match properties cursor and you are prompted to choose the destination cell. Choose the cells to which you want the properties to be inherited and then press ENTER.

Table Cell Styles

This drop-down list displays the preexisting cell styles or options to modify the existing ones. Select the desired cell style to be assigned to the selected cell. The **Cell Styles** drop-down list also has the options to create a new cell style or manage the existing ones. These options have been discussed earlier in the **Creating a New Table Style** topic.

Cell Borders

Choose the **Cell Borders** button from the **Table Cell** tab; the **Cell Border Properties** dialog box will be displayed, as shown in Figure 7-45. The options in this dialog box are similar to those in the **Border** tab of the **New Table Style** dialog box.

Text Alignment

The down arrow on the right of the **Middle-Center** button of the **Cell Styles** panel in the **Ribbon** is used to align the text written in cells with respect to the cell boundary. Choose this button; the **Text Alignment** drop-down will be displayed. Select the desired text alignment from this drop-down; the text of the selected cell will get aligned accordingly.

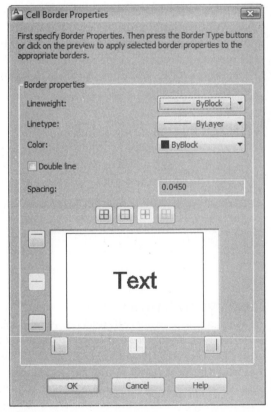

Figure 7-45 The **Cell Border Properties** *dialog box*

Locking

This button is used to lock the cells so that they cannot be edited by accident. Select a cell and choose **Cell Locking** from the **Cell Format** panel in the **Table Cell** tab; the **Cell Locking** drop-down list will be displayed. Four options are available in the drop-down list. The **Unlocked** option is chosen by default. Choose the **Content Locked** option to prevent the modification in the content of the text, but you can modify the formatting of the text. Choose the **Format Locked** option to prevent the modification in the formatting of the text, but in this case, you can modify the content of the text. Select the **Content and Format Locked** option to prevent the modification of both formatting and content of the text.

Data Format

The display of the text in the cell depends on the format type selected. Choose **Cell Format** to change the format of the text in the cell. On doing so, the **Data Format** drop-down list is displayed. Choose the required format from it. You can also select the **Custom Table Cell** option and choose the required format from the **Data type** list box in the **Table Cell Format** dialog box and then choose the **OK** button.

Block

This tool is used to insert a block in the selected cell. Choose the **Block** tool from the **Insert** panel; the **Insert a Block in a Table Cell** dialog box will be displayed, as shown in Figure 7-46. Enter the name of the block in the **Name** edit box or choose the **Browse** button to locate the destination file of the block. If you have browsed the file path, it will be displayed in the **Path** area. The options available in the **Insert a Block in a Table Cell** dialog box are discussed next.

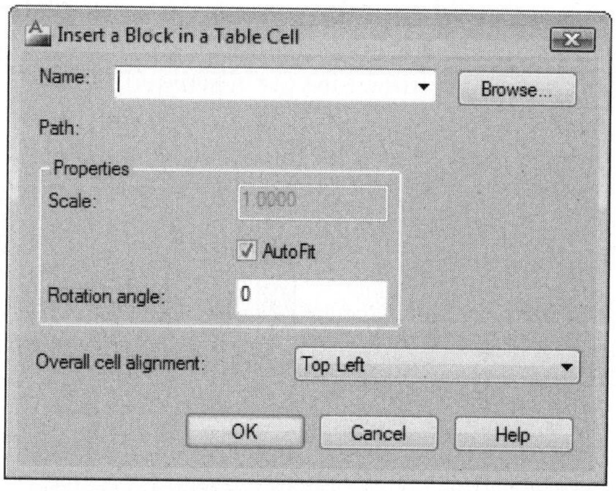

*Figure 7-46 The **Insert a Block in a Table Cell** dialog box*

Properties Area

The options in this area are discussed next.

Scale. This drop-down list is used to specify the scale of the block. By default, this edit box is not available because the **AutoFit** check box is selected below this drop-down list. Selecting the **AutoFit** check box ensures that the block is scaled such that it fits in the selected cell.

Rotation angle. The **Rotation angle** edit box is used to specify the angle by which the block will be rotated before being placed in the cell.

Overall cell alignment. This drop-down list is used to define the block alignment in the selected cell.

Field

You can also insert a field in the cell. The field contains the data that is associative to the property that defines the field. For example, you can insert a field that has the name of the author of the current drawing. If you have already defined the author of the current drawing in the **Drawing Properties** dialog box, it will automatically be displayed in the field. If you modify the author name and update the field, the changes will automatically be made in the text. When you choose the **Field** button from the **Insert** panel, the **Field** dialog box will be displayed. You can select the field to be added from the **Field names** list box and select the format of the field from the **Format** list box. Choose **OK** after selecting the field and format. If the data in the selected field is already defined, it will be displayed in the **Text window**. If not, the field will display dashes (----).

Formula

Choose **Formula** from the **Insert** panel; a drop-down list is displayed. This drop-down list contains the formulas that can be applied to a given cell. The formula calculates the values for that cell using the values of other cell. In a table, the columns are named with letters (like A, B, C, ...) and rows are named with numbers (like 1, 2, 3, ...). The **TABLEINDICATOR** system variable controls the display of column letters and row numbers. By default, the **TABLEINDICATOR** system variable is set to 1, which means the row numbers and column letters will be displayed when **Text Editor** is invoked. Set the system variable to zero to turn off the visibility of row numbers and column letters. The nomenclature of cells is done using the column letters and row numbers. For example, the cell corresponding to column A and row 2 is A2. For a better understanding, some of the cells have been labeled accordingly in Figure 7-47.

Formulas are defined by the range of cells. The range of cells is specified by specifying the name of first and the last cell of the range, separated by a colon (:). The range takes all the cells falling between specified cells. For example, if you write A2 : C3, this means all the cells falling in 2nd and 3rd rows, Column A and B will be taken into account. To insert a formula, double click on the cell; **Text Editor** is invoked. You can now write the syntax of the formula in the cell. The syntax for different formulas are discussed later while explaining different formulas. Formulas can also be inserted by using the **Formula** drop-down list. Different formulas available in the **Formula** drop-down list are discussed next.

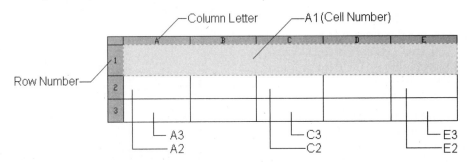

*Figure 7-47 The **Table** showing nomenclature for **Columns**, **Rows**, and **Cells***

Sum

The **Sum** option gives output for a given cell as the sum of the numerical values entered in a specified range of cells. Choose the **Sum** option from the **Table Cell > Insert > Formula** drop-down; you will be prompted to select the first corner of the table cell range and then the second corner. The sum of values of all the cells that fall between the selected range will be displayed as the output. As soon as you specify the second corner, the **Text Editor** is displayed and also the formula is displayed in the cell. In addition to the formula, you can also write multiline text in the cell. Choose **Close Text Editor** from the **Close** panel to exit the editor. When you exit the text editor, the formula is replaced by a hash (#). Now, if you enter numerical values in the cells included in the range, the hash (#) is replaced according to the addition of those numerical values. The prompt sequence, when you select the **Sum** option, is given next.

Select first corner of table cell range: *Specify a point in the first cell of the cell range.*
Select second corner of table cell range: *Specify a point in the last cell of the cell range.*

Note
*The syntax for the **Sum** option is: =Sum{Number of the first cell of cell range (for example: A2): Number of the last cell of the cell range (for example: C5)}*

Average

This option is used to insert a formula that calculates the average of values of the cells falling in the cell range. Prompt sequence is the same as for the **Sum** option.

Note
*The syntax for the **Average** option is: =Average{Number of the first cell of the cell range (for example: A2): Number of the last cell of the cell range (for example: C5)}*

Count

This option is used to insert a formula that calculates the number of cells falling under the cell range. The prompt sequence is the same as for the **Sum** option.

Note
*The syntax for the **Count** option is: =Count{Number of the first cell of cell range (for example: A2): Number of the last cell of cell range (for example: C5)}*

Cell

This option equates the current cell with a selected cell. Whenever there is a change in the value of the selected cell, the change is automatically updated in the other cell. To do so, choose the **Cell** option from **Table Cell > Insert > Formula** drop-down; you will be prompted to select a table cell. Select the cell with which you want to equate the current cell. The prompt sequence for the **Cell** option is given next.

Select table cell: *Select a cell to equate with the current cell.*

Note
*The syntax for the **Cell** option is: =Number of the cell.*

Equation

Using this option, you can manually write equations. The syntax for writing the equations should be the same as explained earlier.

Manage Cell Content

The **Manage Cell Content** button is used to control the sequence of the display of blocks in the cell, if there are more than one block in a cell. Choose **Manage Cell Content** from the **Insert** panel; the **Manage Cell Content** dialog box will be displayed, see Figure 7-48. The options in the **Manage Cell Content** dialog box are discussed next.

Cell content Area

This area lists all blocks entered in the cell according to the order of their insertion sequence.

Move Up

This button is used to move the selected block to one level up in the display order.

Move Down

This button is used to move the selected block to one level down in the display order.

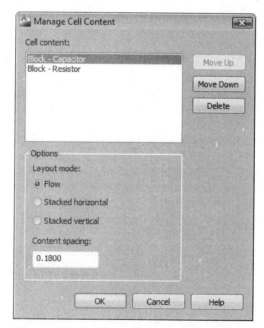

*Figure 7-48   The **Manage Cell Content** dialog box*

Delete

This button is used to delete the selected block from the table cell.

Options Area

The options in this area are used to control the direction in which the inserted block is placed in the cell. If you select the **Flow** radio button, the direction of placement of the blocks in the cell will depend on the width of the cell. Select the **Stacked horizontal** radio button to place the inserted blocks horizontally. Similarly, select the **Stacked vertical** radio button to place the inserted blocks vertically. You can also specify the gap to be maintained between the two consecutive blocks by entering the desired gap value in the **Content spacing** edit box.

Link Cell

To insert data from an excel into the selected table, choose **Data > Link Cell** from the **Table Cell** tab; the **Select a Data Link** dialog box will be displayed. The options in this dialog box are similar to the **Select a Data Link** dialog box that has been discussed earlier in the "**Inserting Table in the Drawing**" topic of this chapter.

Download from source

If the contents of the attached excel spreadsheet are changed after linking it to a cell, you can update these changes in the table by choosing this button. AutoCAD will inform you about the changes in the content of the attached excel sheet by displaying an information bubble at the lower-right corner of the screen.

Tip
*The **Cut** and **Copy** options can be used to move or copy the content from one cell to another. Using the **Paste** option, you can paste the content that you have cut or copied from one cell to another. Place the cursor on the **Recent Input** option; a flyout containing the recent command inputs will be displayed. Choose the **Remove All Property Overrides** option to restore all the default properties of the table.*

Note
*The options in the **Table Cell** tab are also available in the shortcut menu that is displayed when you select any cell in the table and then right-click on it.*

SUBSTITUTING FONTS

AutoCAD provides you the facility to designate the fonts that you want to substitute for the other fonts used in the drawing. The information about font mapping is specified in the font mapping file (*acad.fmp*). The font mapping has the following advantages:

1. You can specify a font to be used when AutoCAD cannot find a font used in the drawing.

2. You can enforce the use of a particular font in the drawings. If you load a drawing that uses different fonts, you can use font mapping to substitute the desired font for the fonts used in the drawing.

3. You can use *.shx* fonts while creating or editing a drawing. When you are done and ready to plot the drawing, you can substitute other fonts for *.shx* fonts.

The font mapping file is an ASCII file with *.fmp* extension containing one font mapping per line. The format on the line is given next.

> **Base name of the font file; Name of the substitute font with extension** (ttf, shx, etc.)

For example, if you want to substitute the ROMANC font for *SWISS.TTF*, the entry is given next.

> **SWISS;ROMANC.SHX**

You can enter this line in the *acad.fmp* file or create a new file. To create a new font mapping file, you need to specify this new file. You can use the **Options** dialog box to specify the new font mapping file. Choose the **Options** button available in the **Application Menu**; the **Options** dialog box will be displayed. Choose the **Files** tab and click on the **plus** sign next to **Text Editor, Dictionary,** and **Font File Names**. Now, click on the plus sign next to **Font Mapping File** to display the path and the name of the font mapping file. Double-click on the file; the **Select a file** dialog box will be displayed. Next, select the new font mapping file and exit the **Options** dialog box. At the Command prompt, enter **REGEN** to convert the existing text font to the font as specified in the new font mapping file. You can also use the **FONTMAP** system variable to specify the new font map file.

Command: **FONTMAP** [Enter]
Enter new value for FONTMAP, or . for none <"path and name of the current font mapping file">: *Enter the name of the new font mapping file.*

The following file is a partial listing of the *acad.fmp* file with the new font mapping line added (swiss;romanc.shx).

swiss;romanc.shx	cibt;CITYB___.TTF	cobt;COUNB___.TTF
eur;EURR____.TTF	euro;EURRO___.TTF	par;PANROMAN.TTF
rom;ROMANTIC.TTF	romb;ROMAB___.TTF	romi;ROMAI___.TTF
sas;SANSS___.TTF	sasb;SANSSB__.TTF	sasbo;SANSSBO_.TTF
saso;SANSSO__.TTF		

Note
The text styles that were created using the PostScript fonts are substituted with an equivalent TrueType font and plotted using the substituted font.

Specifying an Alternate Default Font

When you open a drawing file that specifies a font file that is not on your system or is not specified in the font mapping file, AutoCAD by default, substitutes the *simplex.shx* font file. You can specify a different font file in the **Options** dialog box or do so by changing the **FONTALT** system variable.

Command: **FONTALT** [Enter]
Enter new value for FONTALT, or . for none <"simplex.shx">: *Enter the font file name.*

CREATING TEXT STYLES

Ribbon: Annotate > Text > Text Style (*Inclined arrow*)	
Menu Bar: Format > Text Style	
Toolbar: Text > Text Style or Styles > Text Style	
Command: STYLE	

By default, the text in AutoCAD is written using the default text style which is called **Standard**. This text style is assigned a default text font (*txt.shx*). Another default text style that is available is **Annotative**. This text style is also assigned a default text font (*txt.shx*). If you need to write a text using some other fonts and other parameters, you need to use the **Text Editor**. This is because you can change the formatting and font of the text only by using this command.

However, it is a tedious job to use the **Text Editor** every time to write the text and change its properties. That is why, AutoCAD provides you with an option for modifying the default text style or creating a new text style. After creating a new text, you can make it current. All the texts written after making the new style current will use this style.

To create a new text style or modify the default style, left click on the inclined arrow in the **Text** panel of the **Annotate** tab; the **Text Style** dialog box will be displayed, as shown in Figure 7-49.

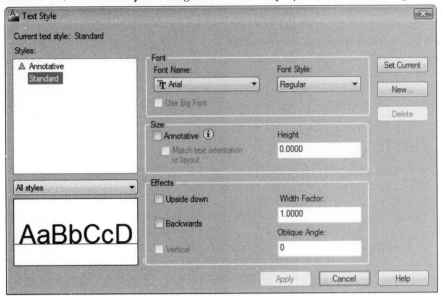

*Figure 7-49 The **Text Style** dialog box*

The **Styles** area displays the styles present in the drawing along with the current style highlighted in blue. An annotative symbol is displayed in front of the annotative text styles. The **Style List Filter** drop-down list below the **Styles** area is used to specify whether all styles will be displayed or only the styles that have been used in the drawing will be displayed. To create a new style, choose the **New** button from the **Text Style** dialog box; the **New Text Style** dialog box will be displayed, as shown in Figure 7-50. Choose the **OK** button from the **Style Name** edit box.

A new style having the entered name and the properties present in the **Text Style** dialog box will be created. To modify this style, select the style name from the list box and then change the different settings by entering new values in the appropriate boxes. You can change the font by selecting a new font from the **Font Name** drop-down list. Similarly, you can change the text height, width, and oblique angle.

*Figure 7-50 The **New Text Style** dialog box*

Remember that if you have already specified the height of the text in the **Text Style** dialog box, AutoCAD will not prompt you to enter the text height while writing the text using the **Single Line** tool. The text will be created using the height specified in the text style. If you want AutoCAD to prompt you for the text height, specify 0 text height in the dialog box. Select the **Annotative** check box to automate the process of scaling the text height. Annotative texts are defined according to the height of the text to be displayed on the paper. According to the annotation scale set for the spaces, the text will be displayed in the viewports and the model space. Select the **Match text orientation to layout** check box to match the orientation of the text in the paper space viewport with the orientation of the layout.

For **Width Factor**, 1 is the default value. If you want the letters expanded, enter a width factor greater than 1. For compressed letters, enter a width factor less than 1. Similarly, for the **Oblique Angle**, 0 is the default value. If you want the slant of the letters toward the right, the value should be greater than 0; to slant the letters toward the left, the value should be less than 0. You can also force the text to be written upside down, backwards, and vertically by checking their respective check boxes. As you make the changes, you can see their effect in the **Preview** box. After making the desired changes, choose the **Apply** button and then the **Close** button to exit the dialog box. Figure 7-51 shows the text objects with all these settings.

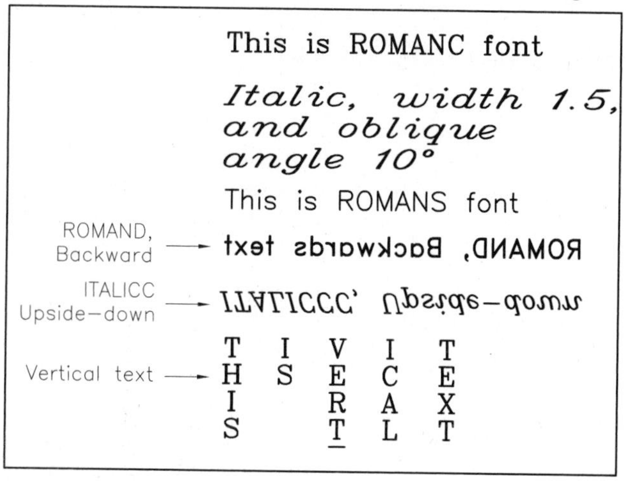

Figure 7-51 *Specifying different features to text style files*

DETERMINING TEXT HEIGHT

The actual text height is equal to the product of the **scale factor** and the **plotted text height**. Therefore, scale factors are important numbers for plotting the text at the correct height. This factor is a reciprocal of the drawing plot scale. For example, if you plot a drawing at a scale of ¼ = 1, you calculate the scale factor for text height given next.

¼" = 1" (i.e., the scale factor is 4)

The scale factor for an architectural drawing that is to be plotted at a scale of ¼" = 1'0" is calculated as given next

¼" = 1'0", or ¼" = 12", or 1 = 48
Therefore, in this case, the scale factor is 48.

For a civil engineering drawing with a scale 1"= 50', the scale factor is shown next.

1" = 50', or 1" = 50X12", or 1 = 600

Therefore, the scale factor is 600.

Next, calculate the height of the AutoCAD text. If it is a full-scale drawing (1=1) and the text is to be plotted at 1/8" (0.125), it should be drawn at that height. However, in a civil engineering drawing, a text drawn 1/8" high will look like a dot. This is because the scale for a civil engineering drawing is 1"= 50', which means that the drawing you are working on is 600 times larger. To draw a normal text height, multiply the text height by 600. Now, the height will be as calculated below.

0.125" x 600 = 75

Similarly, in an architectural drawing, which has a scale factor of 48, a text that is to be 1/8" high on paper must be drawn 6 units high, as shown in the following calculation:

0.125 x 48 = 6.0

It is very important to evaluate scale factors and text heights before you begin a drawing. It would be even better to include the text height in your prototype drawing by assigning the value to the **TEXTSIZE** system variable.

CREATING ANNOTATIVE TEXT

One of the recent inclusions in AutoCAD is now you do not need to calculate the text height in advance. While creating the annotative text, if you decide the text height to be displayed on the paper, the current annotation scale will automatically decide the display size of the text in the model space or the paper space viewport. For example, if you want the text to be displayed at a height of 1/4" on the paper, you can define a text style having a **Paper Text Height** of 1/4". When you add text to a viewport having a scale of 1/4"=1'0", the current annotation scale, which is set as the same scale as of the viewport, automatically scales the text to display appropriately at 12". You can create annotative text by assigning the annotative text style that has been explained earlier.

To write a single line annotative text, choose an annotative type text style from the **Text Style** dialog box and set it as the current text style. The annotative type text style will be displayed with an annotative symbol on its side. Next, enter the text in the drawing using the **Single Line** tool in the Command prompt. To write multiline annotative text, enter the **Multiline Text** tool at the Command prompt and specify the two opposite corners of the box denoting the width of the multiline text; the **Text Editor** will be displayed on the screen. You can also select an existing annotative text style from **Text Style** option in the **Text Editor** tab or choose **Annotative** button in the **Style** panel **Text Editor** tab to create the annotative multiline text.

The existing non-annotative texts whether it is single line or multiline can also be converted into the annotative text. To do so, select the text object and right-click on it to choose the **Properties** option from the shortcut menu. In the **Properties** palette below the **Text** area, click on the **Annotative** edit box and select the **Yes** option from the drop-down list.

CHECKING SPELLING

Ribbon: Annotate > Text > Check Spelling	**Menu Bar:** Tools > Spelling
Toolbar: Text > Spell Check	**Command:** SPELL

You can check the spelling of all texts in a drawing. The text that can be checked for spelling by using the **Check Spelling** tool are: single line or multiline text, dimension text, text in external references, and block attribute text. To check the spelling in a given text, choose the **Check Spelling** tool from the **Text** panel in the **Annotate** tab; the **Check Spelling** dialog box will be displayed, see Figure 7-52. To start the spell check, choose the **Start** button from the **Check Spelling** dialog box. If there is no spelling mistake in the drawing, the **AutoCAD Message** dialog box with the message 'Spelling check complete' will be displayed. Choose the **OK** button from the **AutoCAD Message** dialog box to close it. If any spelling error is found in the drawing, AutoCAD will highlight the word and zoom it according to the size of the window so that it is easily visible. The options in the **Check Spelling** dialog box are discussed next.

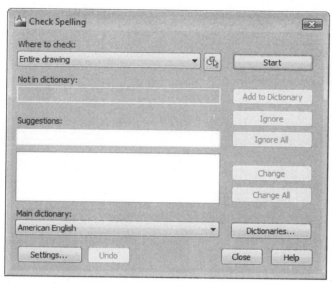

*Figure 7-52 The **Check Spelling** dialog box*

The **Where to check** drop-down list displays three options to specify the portion of the drawing to be Spell checked during the spelling check. These options are: **Entire drawing**, **Current space/layout**, and **Selected objects**. If you select the **Selected objects** option, the **Select text objects** button will be highlighted to enable you to select the text in the drawing to be checked. The **Not in dictionary** box displays the text that is found to be misspelled. The correct spelled alternate words are listed in the **Suggestions** box. The **Main dictionary** drop-down list displays the language options in which you want to check the text.

The **Add to Dictionary** button is used to add the identified misspelled word to the dictionary. The **Ignore** button leaves the identified misspelled words intact. The **Ignore All** button leaves all the identified misspelled words intact. The **Change** button substitutes the present word with the word suggested in the **Suggestions** box. The **Change All** button substitutes all the words that are similar to the currently selected word with the word suggested in the **Suggestions** box. You can also use the MS Word dictionary or any other dictionary for checking spellings. Choose the **Dictionaries** button; the **Dictionaries** dialog box will be displayed, see Figure 7-53. Next, select the **Manage custom dictionaries** option from the **Current custom dictionary** drop-down list; the **Manage Custom Dictionaries** dialog box will be displayed, see Figure 7-54. From this dialog box, you can create a new dictionary, add an existing dictionary, or delete a dictionary from the **Custom dictionaries** list. To add the words listed in the custom dictionary, choose the **Add** button from the **Dictionaries** dialog box. The **Settings** button in the **Check Spelling** dialog box is used to specify the type of text to be checked for spelling in your drawing.

Note
*The dictionaries can also be changed by specifying the name in the **DCTMAIN** or **DCTCUST** system variable.*

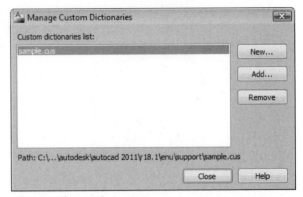

Figure 7-53 The **Dictionaries** dialog box   *Figure 7-54* The **Manage Custom Dictionaries** dialog box

TEXT QUALITY AND TEXT FILL

AutoCAD supports **TrueType fonts**. You can use your own **TrueType fonts** by adding them to the Fonts directory. You can also keep your fonts in a separate directory, in that case you must specify the location of your fonts directory in the AutoCAD search path.

The resolution and text fill of the **TrueType font** text is controlled by the **TEXTFILL** and **TEXTQLTY** system variables. If **TEXTFILL** is set to 1, the text will be filled. If the value is set to 0, the text will not be filled. On the screen the text will appear filled, but when it is plotted the text will not be filled. The **TEXTQLTY** variable controls the quality of the **TrueType font** text. The value of this variable can range from 0 to 100. The default value is 50, which gives a resolution of 300 dpi (dots per inch). If the value is set to 100, the text will be drawn at 600 dpi. The higher the resolution, the more time it takes to regenerate or plot the drawing.

FINDING AND REPLACING TEXT

Ribbon: Annotate > Text > Find Text	**Menu Bar:** Edit > Find
Toolbar: Text > Find	**Command:** FIND

You can use the **Find Text** search box to find and replace a text. The text can be a line text created by the **Single Line** tool, paragraph text created by the **Multiline Text** tool, dimension annotation text, block attribute value, hyperlinks, or hyperlink description. To find a text, enter the text in the **Find Text** search box in the **Text** panel and press ENTER; the **Find and Replace** dialog box will be displayed, as shown in Figure 7-55. You can use this dialog box to perform the following functions:

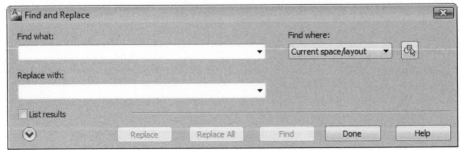

Figure 7-55 The **Find and Replace** dialog box

Finding Text

To find a text, enter the text that you want to find in the **Find what** edit box. You can search the entire drawing or confine your search to the selected text. To select the text, choose the **Select objects** button available to the right of the **Find where** drop-down list. The **Find and Replace** dialog box will temporarily be closed and AutoCAD switches to the drawing window. Once you have selected the text, the **Find and Replace** dialog box is displayed again. In the **Find where** drop-down list, you can specify if you need to search the entire drawing or the current selection. Choose the **Options** button available at the bottom left corner of the **Find and Replace** dialog box; the **Find and Replace** dialog box gets expanded, as shown in Figure 7-56. In this dialog box, you can specify whether to find the whole word and whether to match the case of the specified text. To find the text, choose the **Find** button. The text found along with the surrounding text will be displayed in the **Text String** column of the **List results** area, provided the **List results** check box is selected. To find the next occurrence of the text, choose the **Find** button again.

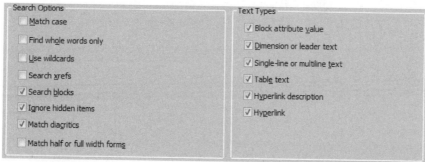

*Figure 7-56 The expanded view of the **Find and Replace** dialog box*

Replacing Text

If you want to replace the specified text with the new text, enter the new text in the **Replace with** edit box. Now, if you choose the **Replace** button, only the found text will be replaced. If you choose the **Replace All** button, all occurrences of the specified text will be replaced with the new text.

CREATING TITLE SHEET TABLE IN A SHEET SET

While working with a sheet set, it is recommended that you create a title sheet that has the details about the sheets in the sheet set. You can enter the details about the sheets in the table. The advantage of using a table is that the information in it can be automatically updated if there is a change in the sheet number or name. Also, if a sheet is removed from the current sheet set, you can easily update the table to reflect the change in the sheet set.

To create a table in a sheet set, double-click on any one of the layouts added to the sheet set. Remember that this layout will be the title sheet.

Tip
*If the title sheet is displayed at the bottom of the list in the **Sheet Set Manager**, you can drag and move it to the top, below the name of the sheet set. Hold the left mouse button down on the title sheet and drag the cursor upward. Next, release the left mouse button below the name of the sheet set.*

When the title sheet is opened, right-click on the name of the sheet set in the **Sheet Set Manager**, and then choose the **Insert Sheet List Table** option from the shortcut menu, as shown in Figure 7-57. Note that this option will not be available, if the model tab is active or if the layout is not from the current sheet set. On choosing the **Insert Sheet List Table** option, the **Sheet List Table** dialog box will be displayed, as shown in Figure 7-58.

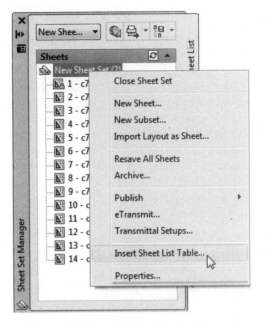

*Figure 7-57 Inserting table in the title sheet using the **Sheet Set Manager***

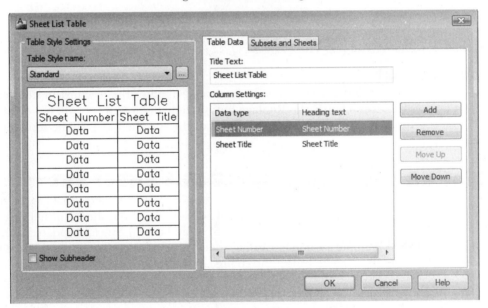

*Figure 7-58 The **Sheet List Table** dialog box*

Enter the title of the table in the **Title Text** text box in the **Table Data Settings** area. By default, two rows will be displayed for each subset in the table. This is because there are only two rows displayed in the **Column Settings** area. You can add additional rows by choosing the **Add** button. To change the data type of a row, click on the field in the **Data type** column. The field will be changed into a drop-down list. Select the required data type from this drop-down list. After specifying all the parameters, choose **OK** from the **Insert Sheet List Table** dialog box. The table will be attached to the cursor and you will be prompted to specify the insertion point. Specify the insertion point for the table. The table will be inserted and based on the parameters selected, the sheets are displayed in the table. Figure 7-59 shows a sheet set table inserted in a title sheet.

Sheet List Table	
1	Title Sheet
Main Parts	
2	Base Part
3	Brasses
4	Cap
Fastners	
5	Bolt
6	Nut
7	Washer

Figure 7-59 *Sheet list table*

If any changes are made in the sheet set numbering or any other property of the sheets in the sheet set, you can easily highlight those changes in the sheet list table by updating it. To update the sheet list table, right-click on it, and then choose **Update Table Data Links** from the shortcut menu. The sheet list table is automatically updated.

Self-Evaluation Test

Answer the following questions and then compare them to those given at the end of this chapter:

1. Tables in AutoCAD are created using the **TABLET** command. (T/F)

2. An annotative text automatically gets scaled according to the viewport's scale. (T/F)

3. You can insert a block into a table cell. (T/F)

4. The **Standard** text style cannot be used for creating annotative text. (T/F)

5. You can control the height of an individual column separately by choosing the **Manual height** sub-option from **Text Editor > Insert > Columns >Dynamic Columns** of the **Ribbon**. (T/F)

6. Multiple lines of text can be entered at any desired location in the drawing area by using the __________ tool.

7. With the __________ justification option of the **Single Line** tool, AutoCAD adjusts the letter width to fit the text between the two given points, but the height remains constant.

8. While writing a text by using the **Multiline Text** tool, the height specified in the **Text Editor** does not affect the __________ system variable.

9. You can change the text string in the edit box by using the __________ tool to edit a single line text. However, to edit a multiline text, you must choose the **Full editor** button in the **Content** edit box of the **Properties** palette.

10. You can use the __________ system variable to specify the new font mapping file.

Review Questions

Answer the following questions:

1. You cannot insert a field by using the **Multiline Text** tool. (T/F)

2. You can add or delete rows and columns from a table by using the **Table** toolbar. (T/F)

3. The **Single Line** tool does not allow you to see the text on the screen as you type it. (T/F)

4. Which of the following buttons in the **Table** toolbar is used to inherit the properties from one cell to another?

   (a) **Cell Styles**   (b) **Match Cells**
   (c) **Link Cells**   (d) **Manage Cell Content**

5. Which of the following text styles is not present in AutoCAD by default?

   (a) **Standard**   (b) **Annotative**
   (c) **Auto text**   (d) None of these

6. If you want the text to be displayed at a height of 1/4" on the paper, you can define a text style having a paper height of 1/4". When you add a text to a viewport having a scale of 1/4"=1'0", the current annotation scale, which is set to the same scale as the viewport, automatically scales the text to display approximately at _____.

   (a) 0.1875"   (b) 0.375"
   (c) 2.66"   (d) 4.8"

7. Which of the following characters in the **Text Editor** tab is used to stack the text with a diagonal line without using the **Autostack Properties** dialog box?

   (a) ^   (b) /
   (c) #   (d) @

8. Which of the following commands can be used to create a new text style and modify the existing ones?

   (a) **TEXT**   (b) **MTEXT**
   (c) **STYLE**   (d) **SPELL**

9. The columns created by the _____ sub-option of the **Dynamic Columns** option of the **Columns** button in the **Insert** panel of the **Text Editor** tab are equal in size. (T/F)

10. The _____ sub-option in the **Formula** flyout of the **Insert** panel in the **Table** tab is used to equate the current cell with the selected cell.

11. The four main text alignment modes are _____, _____, _____, and _____.

12. You can use the _____ tool to write a paragraph text whose width can be specified by defining the _____ of the text boundary.

13. When the **Justify** option is invoked, the user can place the text in one of the _____ various types of alignment by choosing the desired alignment option.

14. A text created by using the _____ tool is a single object irrespective of the number of lines it contains.

15. Using the **Multiline Text** tool, the character _____ stacks a text vertically without a line (tolerance stack).

16. If you want to edit text, select it and then right-click such that various editing options _____ in the menu are available.

17. The _____ tool is used to check the spelling of all the texts written in the current drawing.

EXERCISE 2   *Text*

Write the text, shown in Figure 7-60, on the screen. Use the text justification that will produce the text as shown in the drawing. Assume a value for text height. Use the **Properties** palette to change the text, as shown in Figure 7-61.

Figure 7-60 Drawing for Exercise 2

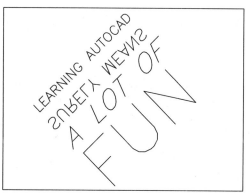

Figure 7-61 Drawing for Exercise 2 (After changing the text)

EXERCISE 3   *Text Style*

Write the text on the screen, as shown in Figure 7-62. First, you must define new text styles by using the **STYLE** command with the attributes, as shown in the figure. The text height is 0.25 units.

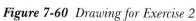

Figure 7-62 Drawing for Exercise 3

EXERCISE 4

Draw the sketch shown in Figure 7-63 using the draw, edit, and display commands. Do not dimension the drawing.

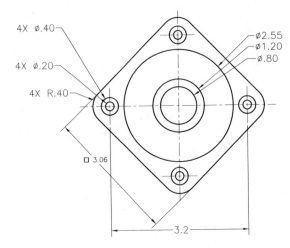

Figure 7-63 Drawing for Exercise 4

EXERCISE 5   *Mirror*

Draw the sketches shown in Figures 7-64 and 7-65. Use the **Mirror** tool to duplicate the features that are identical. Do not dimension the drawing.

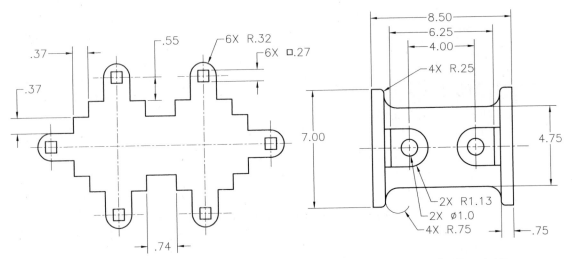

Figure 7-64 Drawing for Exercise 5   *Figure 7-65 Drawing for Exercise 5*

EXERCISE 6

Draw Figure 7-66. Do not dimension the drawing.

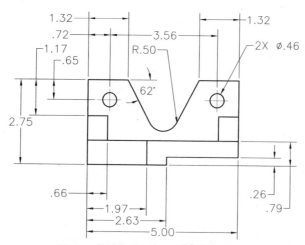

Figure 7-66 *Drawing for Exercise 6*

Problem-Solving Exercise 1

Draw the sketch shown in Figure 7-67 using the draw, edit, and display commands. Assume the missing dimensions. Do not dimension the drawing.

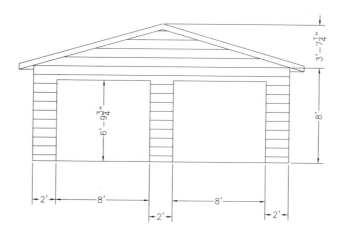

Figure 7-67 *Drawing for Problem-Solving Exercise 1*

Problem-Solving Exercise 2

Draw Figure 7-68 using AutoCAD's draw, edit, and display commands. Also, add text to the drawing. Assume the missing dimensions. Do not dimension the drawing.

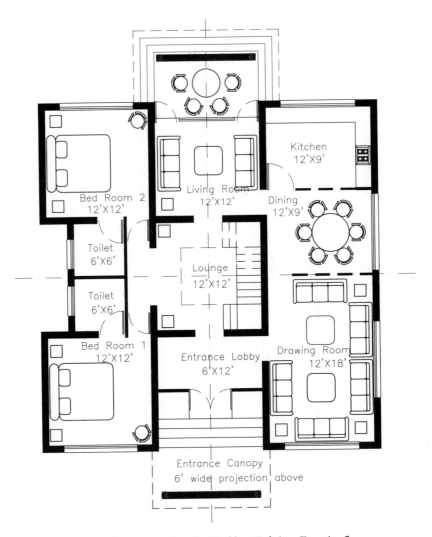

Figure 7-68 Drawing for Problem-Solving Exercise 2

Answers to Self-Evaluation Test

1. F, **2.** T, **3.** T, 4. F, **5.** T, **6. Single Line, 7. Fit, 8. TEXTSIZE, 9. Edit, 10. FONTMAP**

Chapter 8

Basic Dimensioning, Geometric Dimensioning, and Tolerancing

CHAPTER OBJECTIVES

In this chapter, you will learn:
- *The need for dimensioning in drawings.*
- *The Fundamental dimensioning terms.*
- *About the associative and annotative dimensioning.*
- *To use the Quick Dimension option for quick dimensioning.*
- *To create various types of dimensions in a drawing.*
- *To create center marks and centerlines.*
- *To attach leaders to objects.*
- *To attach and modify multileaders.*
- *To use geometric tolerancing, feature control frames, and characteristic symbols.*
- *To combine geometric characteristics and create composite position tolerancing.*
- *To use the projected tolerance zone.*
- *To use feature control frames with leaders.*

KEY TERMS

- *Associative Dimensions*
- *Definition Points*
- *Annotative Dimensions*
- *Dimension Break*
- *Center Marks and Centerlines*
- *Inspection Dimensions*
- *Leaders*
- *Multileaders*
- *Geometric Tolerance*
- *Complex Feature Control Frames*
- *Projected Tolerance Zone*

NEED FOR DIMENSIONING

To make designs more informative and practical, a drawing must convey more than just the graphic picture of a product. To manufacture an object, the drawing of that object must contain size descriptions such as the length, width, height, angle, radius, diameter, and location of features. These informations are added to the drawing by dimensioning. Some drawings also require information about tolerances with the size of features. The information conveyed through dimensioning are vital and often as important as the drawing itself. With the advances in computer-aided design/drafting and computer-aided manufacturing, it has become mandatory to draw part to actual size so that dimensions reflect the actual size of features. At times, it may not be necessary to draw the object of the same size as the actual object would be, but it is absolutely essential that the dimensions be accurate. Incorrect dimensions will lead to manufacturing errors.

By dimensioning, you not only give the size of a part, but also give a series of instructions to a machinist, an engineer, or an architect. The way the part is positioned in a machine, the sequence of machining operations, and the location of various features of the part depend on how you dimension the part. For example, the number of decimal places in a dimension (2.000) determines the type of machine that will be used to do that machining operation. The machining cost of such an operation is significantly higher than for a dimension that has only one digit after the decimal (2.0). Similarly, whether a part is to be forged or cast, the radii of the edges, and the tolerance you provide to these dimensions determine the cost of the product, the number of defective parts, and the number of parts you get from a single die.

DIMENSIONING IN AutoCAD

The objects that can be dimensioned in AutoCAD range from straight lines to arcs. The dimensioning commands provided by AutoCAD can be classified into four categories:

Dimension Drawing Commands   **Dimension Style Commands**
Dimension Editing Commands   **Dimension Utility Commands**

While dimensioning an object, AutoCAD automatically calculates the length of the object or the distance between two specified points. Also, settings such as the gap between the dimension text and the dimension line, the space between two consecutive dimension lines, arrow size, and text size are maintained and used when the dimensions are being generated for a particular drawing. The generation of arrows, lines (dimension lines, extension lines), and other objects that form a dimension is automatically performed by AutoCAD to save the user's time. This also results in uniform drawings. However, you can override the default measurements computed by AutoCAD and change the settings of various standard values. The modification of dimensioning standards can be achieved through the dimension variables.

The dimensioning functions offered by AutoCAD provide you with extreme flexibility in dimensioning by letting you dimension various objects in a variety of ways. This is of great help because different industries, such as architectural, mechanical, civil, or electrical have different standards for the placement of dimensions.

FUNDAMENTAL DIMENSIONING TERMS

Before studying AutoCAD's dimensioning commands, it is important to know and understand various dimensioning terms that are common to linear, angular, radius, diameter, and ordinate dimensioning. Figures 8-1 and 8-2 show various dimensioning parameters.

Dimension Line

The dimension line indicates the distance or the angle being measured. Usually, this line has arrows at both ends, and the dimension text is placed along the dimension line. By default, the dimension line is drawn between the extension lines (Figure 8-1 and Figure 8-2). If the dimension line does not fit inside, two short lines with arrows pointing inward are drawn outside the extension lines. The dimension line for angular dimensions (which are used to dimension angles) is an arc. You can control the positioning and various other features of the dimension lines by setting the parameters in the dimension styles. (The dimension styles are discussed in Chapter 10.)

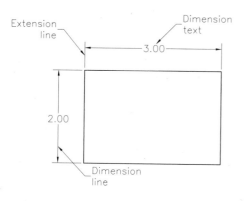

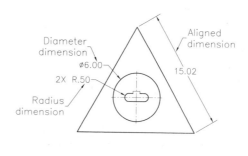

Figure 8-1 *Various dimensioning parameters*   **Figure 8-2** *Various dimensioning parameters*

Dimension Text

The dimension text is a text string that reflects the actual measurement (dimension value) between the selected points as calculated by AutoCAD. You can accept the value that AutoCAD returns or enter your own value. In case you use the default text, AutoCAD can be supplied with instructions to append the tolerances to it. Also, you can attach prefixes or suffixes of your choice to the dimension text.

Arrowheads

An arrowhead is a symbol used at the end of a dimension line (where dimension lines meet the extension lines). Arrowheads are also called terminators because they signify the end of the dimension line. Since drafting standards differ from company to company, AutoCAD allows you to draw arrows, tick marks, closed arrows, open arrows, dots, right angle arrows, or user-defined blocks (Figure 8-3). The user-defined blocks at the two ends of the dimension line can be customized to your requirements. The size of the arrows, tick marks, user blocks, and so on can be regulated by using the dimension variables.

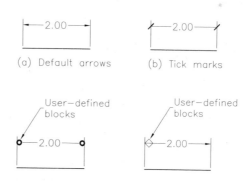

Figure 8-3 *Using arrows, tick marks, and user-defined blocks*

Extension Lines

Extension lines are drawn from the object measured to the dimension line (Figure 8-4). These lines are also called witness lines. Extension lines are used in linear and angular dimensioning. Generally, extension lines are drawn perpendicular to the dimension line. However, you can make extension lines inclined at an angle by choosing the **Oblique** tool from the **Dimensions** panel of the **Annotate** tab. Alternatively, you can choose the **Dimension Edit** tool from the

Dimension toolbar. AutoCAD also allows you to suppress either one or both extension lines in a dimension (Figure 8-5). You can insert breaks in a dimension or an extension line, in case they intersect other geometric objects or dimension entities. You can also control various other features of the extension lines by setting parameters in dimension styles. (Dimension styles are discussed in Chapter 10.)

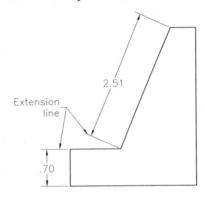

Figure 8-4 Extension lines

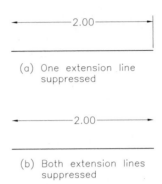

Figure 8-5 Extension line suppressed

Leader

A leader is a line that stretches from the dimension text to the object being dimensioned. Sometimes the text for dimensioning and other annotations do not adjust properly near the object. In such cases, you can use a leader and place the text at the end of the leader line. For example, the circle shown in Figure 8-6 has a keyway slot that is too small to be dimensioned. In this situation, a leader can be drawn from the text to the keyway feature. Also, a leader can be used to attach annotations such as part numbers, notes, and instructions to an object. You can also draw multileaders that are used to connect one note to different places or many notes to one place.

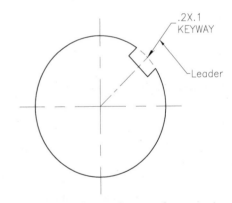

Figure 8-6 Leader used to attach annotation

Center Mark and Centerlines

The center mark is a cross mark that represents the center point of a circle or an arc. Centerlines are mutually perpendicular lines that pass through the center of a circle/arc and intersect the circumference of the circle/arc. A center mark or a centerline is automatically drawn when you dimension a circle or an arc (see Figure 8-7). The length of center mark and the extension of centerline beyond the circumference of circle are determined by the value assigned to the **DIMCEN** dimension variable. You can toggle between the center mark and the centerlines by entering a positive and a negative value respectively for the **DIMCEN** variable. Alternatively you can use the **Dimension Style Manager** dialog box to toggle between the center mark and the centerlines. This will be discussed in detail in Chapter 10.

Alternate Units

With the help of alternate units, you can generate dimensions for two systems of measurement at the same time (Figure 8-8). For example, if the dimensions are in inches, you can use the alternate units dimensioning facility to append metric dimensions to the dimensions (controlling the alternate units through the dimension variables is discussed in Chapter 10).

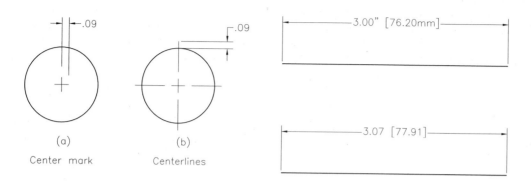

Figure 8-7 *Center mark and centerlines*   **Figure 8-8** *Using alternate units for dimensioning*

Tolerances

Tolerance is the amount by which the actual dimension can vary (Figure 8-9). AutoCAD can attach the plus/minus tolerances to the dimension text (actual measurement computed by AutoCAD). This is also known as deviation tolerance. The plus and minus tolerances that you specify can be same or different. You can use the dimension variables to control the tolerance feature (these variables are discussed in Chapter 10).

Limits

Instead of appending tolerances to dimension text, you can apply tolerances to the measurement itself (Figure 8-10). Once you define tolerances, AutoCAD automatically calculates the upper and lower limit values of the dimension. These values are then displayed as a dimension text.

For example, if the actual dimension as computed by AutoCAD is 2.6105 units and the tolerance values are +0.025 and -0.015, the upper and lower limits will be 2.6355 and 2.5955. After calculating the limits, AutoCAD will display them as a dimension text, as shown in Figure 8-10. The dimension variables that control the limits are discussed in Chapter 10.

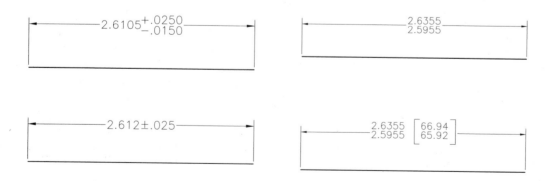

Figure 8-9 *Using tolerances with dimensions*   **Figure 8-10** *Using limits with dimensions*

ASSOCIATIVE DIMENSIONS

The Associative dimensioning is a method of dimensioning, in which the dimension is associated with the object that is dimensioned. In other words, the dimension is influenced by the changes in the size of the object. In the earlier releases of AutoCAD, the dimensions were not truly associative, but were related to the objects being dimensioned by definition points on the DEFPOINTS layer. To cause the dimension to be associatively modified, these definition points had to be adjusted along with the object being changed. If, for example, you use the **Scale** tool to change an object's size and select the object, the dimensions will not be modified. If you select the object and its defpoints (using the Crossing selection method), then the dimension will be modified. If the dimensions are associated to the object and the object changes its size, the dimensions will also change automatically. With the introduction of the true associative dimensions, there is no need to select the definition points along with the object. This eliminates the use of definition points for updating the dimensions.

The values and location of the associative dimensions are updated automatically if the value or location of the object is modified. For example, if you edit an object using simple editing operations such as breaking an object using the **Break** tool, then the true associative dimension will be modified automatically. The dimensions can be converted into the true associative dimensions using the **Reassociate** tool in the **Dimensions** panel of the **Annotate** tab to reassociate the dimension. The association of the dimensions with the objects can be removed using the **DIMDISASSOCIATE** command. Both these commands will be discussed later in this chapter.

The dimensioning variable **DIMASSOC** controls the associativity of dimensions. The default value of this variable is **2**, which means the dimensions are associative. When the value is **1**, the dimensions placed are non-associative. When the **DIMASSOC** is turned off (value of this variable is **0**), then the dimension will be placed in the exploded format. This means that the dimensions will now be placed as a combination of individual arrowheads, dimension lines, extension lines, and text. Also, note that the exploded dimensions cannot be associated to any object.

DEFINITION POINTS

Definition points are the points drawn at the positions used to generate a dimension object. The definition points are used by the dimensions to control their updating and rescaling. AutoCAD draws these points on a special layer called **DEFPOINTS**. These points are not plotted by the plotter because AutoCAD does not plot any object on the **DEFPOINTS** layer. If you explode a dimension (which is as good as turning **DIMASSOC** off), the definition points are converted to point objects on the **DEFPOINTS** layer. In Figure 8-11, the small circles indicate the definition points for different objects.

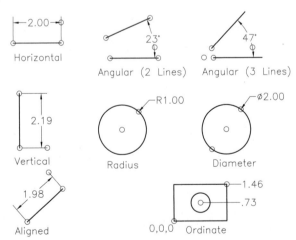

Figure 8-11 *Definition points of linear, radial, angular, and ordinate dimensions*

The definition points for linear dimensions are the points used to specify the extension lines and the point of intersection of the first extension line and the dimension line. The definition points for the angular dimension are the endpoints of the lines used to specify the dimension and the point used to specify the dimension line arc. For example, for a three-point angular dimension, the definition points are the extension line endpoints, angle vertex, and the point used to specify the dimension line arc.

The definition points for the radius dimension are the center point of the circle or arc, and the point where the arrow touches the object. The definition points for the diameter dimension are the points where the arrows touch the circle. The definition points for the ordinate dimension are the UCS origin, feature location, and leader endpoint.

Note
In addition to the definition points just mentioned, the middle point of the dimension text serves as the definition point for all types of dimensions.

ANNOTATIVE DIMENSIONS

When all the elements of a dimension such as text, spacing, and arrows get scaled according to the specified annotation scale, it is known as Annotative Dimension. They are created in the drawing by assigning annotative dimension styles to them. You can also change the non-annotative dimensions to annotative by changing their **Annotative** property to **Yes** in the **Properties** palette.

SELECTING DIMENSIONING COMMANDS

AutoCAD provides the following fundamental dimensioning types:

Quick dimensioning	**Linear dimensioning**	**Diameter dimensioning**
Radius dimensioning	**Angular dimensioning**	**Ordinate dimensioning**
Arc Length dimensioning	**Aligned dimensioning**	**Jogged dimensioning**

Figures 8-12 and 8-13 show various fundamental dimension types.

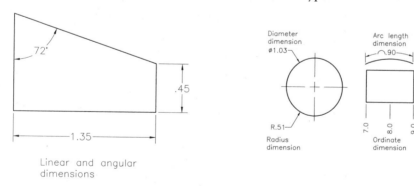

Figure 8-12 *Linear and angular dimensions*   **Figure 8-13** *Radius, diameter, and ordinate dimensions*

You can select the requisite command from the menu bar, toolbar, and Ribbon to apply dimensions. You can also use the command bar to work with the dimensioning. The procedure to select the dimensioning command is discussed next.

Using the Ribbon and the Toolbar

You can select the dimensioning tools from the **Dimensions** panel of the **Annotate** tab (Figure 8-14) or from the **Dimension** toolbar (Figure 8-15). The **Dimension** toolbar can be displayed by choosing **Tools > Toolbars > AutoCAD > Dimension** from the menu bar, if the menu bar is displayed.

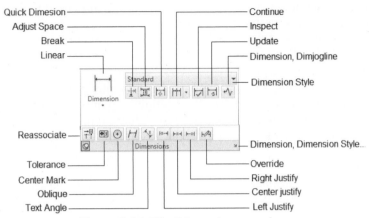

Figure 8-14 The **Dimensions** *panel*

Figure 8-15 The **Dimension** *toolbar*

Using the Command Line

You can directly enter a dimensioning command in the Command line or use the **DIM** or the **DIM1** commands to invoke the dimensioning commands.

> **Note**
> The **DIMDEC** *variable sets the number of decimal places for the value of primary dimension and the **DIMADEC** variable for angular dimensions. For example, if **DIMDEC** is set to **3**, AutoCAD will display the decimal dimension up to three decimal places (2.037).*

DIMENSIONING A NUMBER OF OBJECTS TOGETHER

Ribbon: Annotate > Dimensions > Quick Dimension	**Command:** QDIM
Menu Bar: Dimension > Quick Dimension	**Toolbar:** Dimension > Quick Dimension

The **Quick Dimension** tool is used to dimension a number of objects at the same time. It also helps you to quickly edit dimension arrangements already existing in the drawing and also create new dimension arrangements. It is especially useful when creating a series of baseline or continuous dimensions. It also allows you to dimension multiple arcs and circles at the same time. When you are using the **Quick Dimension** tool, you can relocate the datum base point for baseline and ordinate dimensions. The prompt sequence that will follow when you choose this tool is given next.

Select geometry to dimension: *Select the objects to be dimensioned and press ENTER.*
Select geometry to dimension: ⌞Enter⌟
Specify dimension line position, or [Continuous/Staggered/Baseline/Ordinate/Radius/Diameter/datumPoint/Edit/seTtings] <Continuous>: *Press ENTER to accept the default dimension arrangement and specify dimension line location or enter new dimension arrangement or edit the existing dimension arrangement.*

For example, you can dimension all circles in a drawing (Figure 8-16) by using the quick dimensioning as follows:

Associative dimension priority = Endpoint.
Select geometry to dimension: *Select all circles.*

Select geometry to dimension: `Enter`
Specify dimension line position, or
[Continuous/Staggered/Baseline/Ordinate/Radius/Diameter/datumPoint/Edit/
seTtings] <Continuous>: *Press D for diameter dimensioning and select a point where you want to position the radial dimension.*

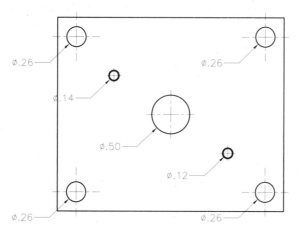

Figure 8-16 *Using the QDIM command to dimension multiple circles*

CREATING LINEAR DIMENSIONS

Ribbon: Annotate > Dimensions > Dimension drop-down > Linear
Menu Bar: Dimension > Linear   **Toolbar:** Dimension > Linear
Command: DIMLIN or DIMLINEAR

Linear dimensioning is used to measure the shortest distance between two points. You can directly select the object to dimension or select two points. The points can be any two points in the space, endpoints of an arc or line, or any set of points that can be identified. To achieve accuracy, points must be selected with the help of object snaps or by selecting an object to dimension. In case, the object selected is aligned, then the linear dimensions will add the **Horizontal** or **Vertical** dimension to the object. The prompt sequence that will follow when you choose the **Linear** tool is given next.

Specify first extension line origin or <select object>: `Enter`
Select object to dimension: *Select the object.*
Specify dimension line location or
[Mtext/Text/Angle/Horizontal/Vertical/Rotated]: *Select a point to locate the position of the dimension.*

Instead of selecting the object, you can also select the two endpoints of the line that you want to dimension (Figure 8-17). Usually the points on the object are selected by using the **object snaps** (endpoints, intersection, center, etc.). The prompt sequence is as follows.

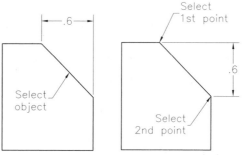

Specify first extension line origin or <select object>: *Select a point.*
Specify second extension line origin: *Select second point.*
Specify dimension line location or

Figure 8-17 *Drawing linear dimensions*

[Mtext/Text/Angle/Horizontal/Vertical/Rotated]: *Select a point to locate the position of the dimension.*

Using the **Linear** tool, you can obtain the horizontal or vertical dimension by simply defining the appropriate dimension location point. If you select a point above or below the dimension, AutoCAD creates a horizontal dimension. If you select a point that is on the left or right of the dimension, AutoCAD creates a vertical dimension through that point.

DIMLINEAR Command Options
The options under this command are discussed next.

Mtext Option
The **Mtext** option is used to override the default dimension text and also change the font, height, and so on, by using the **Text Editor**. When you enter **M** at the **Specify dimension line location or [Mtext/Text/Angle/Horizontal/Vertical/Rotated]** prompt, the **Text Editor** is displayed. You can change the text by entering a new text. You can also use various options of the **Text Editor** (explained in Chapter 7). Choose the **OK** button. However, if you override the default dimensions, the dimensional associativity of the dimension text is lost. This means that if you modify the object using the definition points, AutoCAD will not recalculate the dimension text. Even if the dimension is a true associative dimension, the text will not be recalculated when the object is modified. The prompt sequence to invoke this option is given next.

> Specify first extension line origin or <select object>: *Specify a point.*
> Specify second extension line origin: *Specify the second point.*
> Specify dimension line location or
> [Mtext/Text/Angle/Horizontal/Vertical/Rotated]: **M** *(Enter the dimension text in the **Text Editor** and then click outside the text editor to accept the changes.)*
> Specify dimension line location or
> [Mtext/Text/Angle/Horizontal/Vertical/Rotated]: *Specify the dimension location.*

Text Option
This option also allows you to override the default dimension. However, this option will prompt you to specify the new text value in the Command prompt itself, see Figure 8-18. The prompt sequence to invoke this option is given next.

> Specify first extension line origin or <select object>: *Select a point.*
> Specify second extension line origin: *Select second point.*
> Specify dimension line location or
> [Mtext/Text/Angle/Horizontal/Vertical/Rotated]: **T**
> Enter dimension text <Current>: *Enter new text.*
> Specify dimension line location or
> [Mtext/Text/Angle/Horizontal/Vertical/Rotated]: *Specify the dimension location.*

Angle Option
This option lets you change the angle of the dimension text, see Figure 8-18.

Rotated Option
This option lets you create a dimension that is rotated at a specified angle, see Figure 8-18.

Horizontal Option
This option lets you create a horizontal dimension regardless of where you specify the dimension location, see Figure 8-19.

Vertical Option

This option lets you create a vertical dimension regardless of where you specify the dimension location, see Figure 8-19.

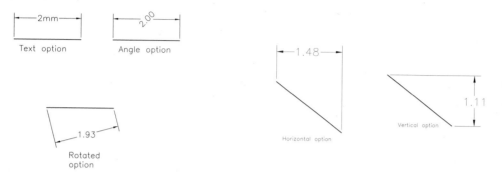

Figure 8-18 *The **Text**, **Angle**, and **Rotated** options*

Figure 8-19 *The **Horizontal** and **Vertical** options*

Note

If you override the default dimensions, the dimensional associativity of the dimension text is lost and AutoCAD will not recalculate the dimension when the object is scaled.

EXAMPLE 1   *Horizontal Dimension*

In this example, you will use linear dimensioning to dimension a horizontal line of 4 units length. The dimensioning will be done first by selecting the object and later on by specifying the first and second extension line origins. Using the **Text Editor**, modify the default text such that the dimension is underlined.

Selecting the Object

1. Start a new file in the **Drafting & Annotation** workspace and draw a line of 4 units length.

2. Choose the **Linear** tool from **Annotate > Dimensions > Dimension** drop-down; you will be prompted to specify the extension line origin or the object. The prompt sequence to apply the linear dimension is as follows:

   Specify first extension line origin or <select object>: `Enter`
   Select object to dimension: *Select the line.*
   Specify dimension line location or
   [Mtext/Text/Angle/Horizontal/Vertical/Rotated]: **M** `Enter`
   *The **Text Editor** will be displayed, as shown in Figure 8-20. Select the default dimension value and then choose the **Underline** button from the **Formatting** panel of the **Text Editor** tab of the **Ribbon** to underline the text. Click anywhere in the drawing area to exit the **Text Editor** tab.*
   Specify dimension line location or
   [Mtext/Text/Angle/Horizontal/Vertical/Rotated]: *Place the dimension.*
   Dimension text = 4.0000

Figure 8-20 *The **Text Editor** tab*

Specifying Extension Line Origins

1. Choose the **Linear** tool from **Annotate** > **Dimensions** > **Dimension** drop-down. The prompt sequence is as follows:

Specify first extension line origin or <select object>: *Select the first endpoint of the line using the* **Endpoint** *object snap, see Figure 8-21.*

Specify second extension line origin: *Select the second endpoint of the line using the* **Endpoint** *object snap, see Figure 8-21.*
Specify dimension line location or
[Mtext/Text/Angle/Horizontal/Vertical/Rotated]: **M**
Select the text and then choose the **Underline** *button from the* **Ribbon** *to underline the text in the* **Text Editor**.
Specify dimension line location or
[Mtext/Text/Angle/Horizontal/Vertical/Rotated]: *Place the dimension.*
Dimension text = 4.00

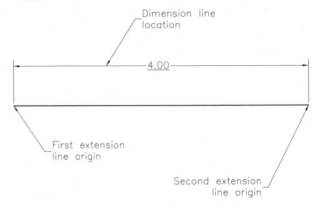

Figure 8-21 *Line for Example 1*

CREATING ALIGNED DIMENSIONS

Ribbon: Annotate > Dimensions > Dimension drop-down > Aligned
Menu Bar: Dimension > Aligned   **Toolbar:** Dimension > Aligned
Command: DIMALIGNED

Generally, the drawing consists of various objects that are neither parallel to the X axis nor to the Y axis. Dimensioning of such objects can be done using aligned dimensioning. In horizontal or vertical dimensioning, you can only measure the shortest distance from the first extension line origin to the second extension line origin along the horizontal or vertical axis, respectively, whereas with the help of aligned dimensioning, you can measure the true aligned distance between the two points. The function of the **Aligned** tool is similar to that of the other linear dimensioning commands. The dimension created with the **Aligned** tool is parallel to the object being dimensioned. The prompt sequence that will follow when you choose this tool is given next.

Specify first extension line origin or <select object>: *Specify the first point or press ENTER.*
Specify second extension line origin: *Specify second point.*
Specify dimension line location or [Mtext/Text/Angle]: *Specify the location for the dimension line.*
Dimension text = Current

The options in this tool are similar to those of the **Linear** tool. Figure 8-22 illustrates the aligned dimensioning.

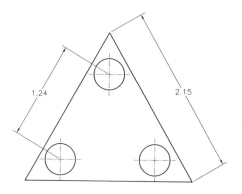

Figure 8-22 *The aligned dimensioning*

EXERCISE 1   *Aligned Dimension*

Draw the object shown in Figure 8-23 and then use linear and aligned dimensioning to dimension the part. The distance between the dotted lines is 0.5 units. The dimensions should be up to 2 decimal places. To get dimensions up to 2 decimal places, enter DIMDEC at the Command prompt and then enter 2. (There will be more information about dimension variable in Chapter 10.)

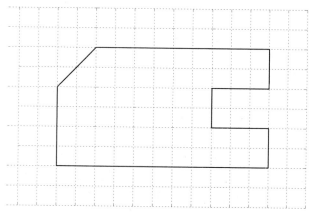

Figure 8-23 *Drawing for Exercise 1*

CREATING ARC LENGTH DIMENSIONS

Ribbon: Annotate > Dimensions > Dimension drop-down > Arc Length
Menu Bar: Dimension > Arc Length   **Toolbar:** Dimension > Arc Length
Command: DIMARC

The Arc Length dimensioning is used to dimension the length of an arc or the polyline arc segment. You are required to select an arc or a polyline arc segment and the dimension location. Figure 8-24 shows the Arc Length dimensioning of an arc. You can invoke this command by choosing the **Arc Length** tool in the **Dimensions** panel. The prompt sequence that will follow is given next.

Select arc or polyline arc segment: *Select arc or polyline arc segment to dimension.*
Specify arc length dimension location, or [Mtext/Text/Angle/Partial/Leader]: *Specify the location for the dimension line.*
Dimension text = *Current.*

Using the **Partial** option, you can dimension a selected portion of the arc, as shown in Figure 8-25. The prompt sequence for the **Partial** option is given next.

Select arc or polyline arc segment: *Select arc or polyline arc segment to dimension.*
Specify arc length dimension location, or [Mtext/Text/Angle/Partial]: **P** [Enter]
Specify first point for arc length dimension: *Specify the first point on arc.*
Specify second point for arc length dimension: *Specify the second point on arc.*
Specify arc length dimension location, or [Mtext/Text/Angle/Partial]: *Specify the location for the dimension line.*
Dimension text = *Current*

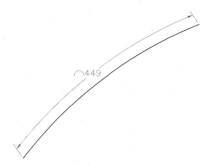

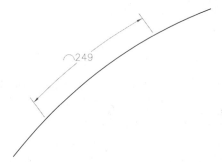

Figure 8-24 *Arc Length dimensioning*   **Figure 8-25** *Partial Arc Length dimensioning*

Using the **Leader** option, you can attach a leader to the dimension text, starting from its circumference. This leader is drawn radial to the arc, as shown in Figure 8-26.

Note
*The **Leader** option is displayed only when the arc subtend an included angle greater than 90 degrees at its centre.*

CREATING ROTATED DIMENSIONS

Rotated dimensioning is used when you want to place the dimension line at an angle (if you do not want to align the dimension line with the selected extension line origins), as shown in Figure 8-27. You can invoke this option by entering **ROTATED** at the command line after choosing the **Linear** tool from the **Dimensions** panel. The **ROTATED** dimension option will prompt you to specify the dimension line angle. The prompt sequence is given next.

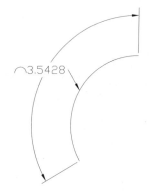

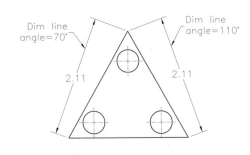

Figure 8-26 *Leader Arc Length dimensioning*   **Figure 8-27** *Rotated dimensioning*

Specify first extension line origin or <select object>: *Select the origin of the first extension line.*
Specify second extension line origin: *Select the origin of the second extension line.*
Non-associative dimension created.
Specify dimension line location or
[Mtext/Text/Angle/Horizontal/Vertical/Rotated]: **R**

Specify angle of dimension line <0>: **110**
Specify dimension line location or [Mtext/Text/Angle/Horizontal/Vertical/Rotated]: *Select the location for the dimension line.*
Dimension text = current

Note
You can draw horizontal and vertical dimensioning by specifying the rotation angle of 0-degree for horizontal dimensioning and 90° for vertical dimensioning.

CREATING BASELINE DIMENSIONS

Ribbon: Annotate > Dimensions > Continue drop-down > Baseline
Menu Bar: Dimension > Baseline   **Toolbar:** Dimension > Baseline
Command: DIMBASE or DIMBASELINE

Sometimes in manufacturing, you may want to locate different points and features of a part with reference to a fixed point (base point or reference point). This can be accomplished by using the Baseline dimensioning (Figure 8-28). To invoke the Baseline dimension, choose the **Baseline** tool from the **Dimensions** panel. Using this tool, you can continue a linear dimension from the first extension line origin of the first dimension to the dimension point. The new dimension line is automatically offset by a fixed amount to avoid overlapping of the dimension lines. This has to be kept in mind that there must already exist a linear, ordinate, or angular associative dimension to use the Baseline dimensions. When you choose the **Baseline** tool, the last linear, ordinate, or angular dimension created will be selected and used as the baseline. The prompt sequence that will follow when you choose this tool is given next.

Specify a second extension line origin or [Undo/Select] <Select>: *Select the origin of the second extension line.*
Dimension text = current
Specify a second extension line origin or [Undo/Select] <Select>: *Select the origin of the second extension line.*
Dimension text = current
Specify a second extension line origin or [Undo/Select] <Select>: *Select the origin of the second extension line or press ENTER.*
Select base dimension: [Enter]

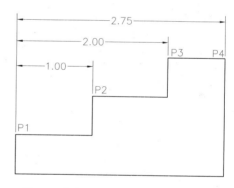

Figure 8-28 *Baseline dimensioning*

When you use the **Baseline** tool, you cannot change the default dimension text. However, the **DIM** command allows you to override the default dimension text.

Command: **DIM**
Dim: **HOR**
Specify first extension line origin or <select object>: *Select left corner (P1, Figure 8-28). (Use Endpoint object snap.)*
Specify second extension line origin: *Select the origin of the second extension line (P2).*
Specify dimension line location or [MText/Text/Angle]: **T**
Enter dimension text <1.0000>: **1.0**
Specify dimension line location or [MText/Text/Angle]: *Select the dimension line location.*
Dim: **BASELINE (or BAS)**

Specify a second extension line origin or [Select] <Select>: *Select the origin of the next extension line (P3).*
Enter dimension text <2.0000>: **2.0**
Dim: **BAS**
Specify a second extension line origin or [Select] <Select>: *Select the origin of the next extension line (P4).*
Enter dimension text <3.000>: **2.75**

The next dimension line is automatically spaced and drawn by AutoCAD.

CREATING CONTINUED DIMENSIONS

Ribbon: Annotate > Dimensions > Continue drop-down > Continue
Menu Bar: Dimension > Continue   **Toolbar:** Dimension > Continue
Command: DIMCONT or DIMCONTINUE

Using the **Continue** tool, you can continue a linear dimension from the second extension line of the previous dimension. This is also called as Chained or Incremental dimensioning. Note that there must exist linear, ordinate, or angular associative dimension to use the Continue dimensions. The prompt sequence that will follow when you choose this tool is given next.

Specify a second extension line origin or [Undo/Select] <Select>: *Specify the point on the origin of the second extension line.*
Dimension text = current
Specify a second extension line origin or [Undo/Select] <Select>: *Specify the point on the origin of the second extension line.*
Dimension text = current
Specify a second extension line origin or [Undo/Select] <Select>: ⏎
Select continued dimension: ⏎

Also, in this case, the **DIM** command should be used if you want to change the default dimension text.

Command: **DIM**
Dim: **HOR**
Specify first extension line origin or <select object>: *Select left corner (P1, see Figure 8-29).* *(Use Endpoint object snap.)*
Specify second extension line origin: *Select the origin of the second extension line (P2, see Figure 8-29).*

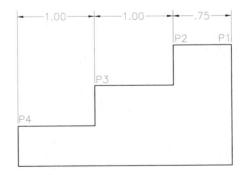

Figure 8-29 *Continue dimensioning*

Specify dimension line location or [MText/Text/Angle]: **T**
Enter dimension text <current>: **0.75**
Specify dimension line location or [MText/Text/Angle]: *Select the dimension line location.*
Dim: **CONTINUE**
Specify a second extension line origin or [Select] <Select>: *Select the origin of the next extension line (P3, see Figure 8-29).*
Enter dimension text <current>: [Enter]
Dim: **CONTINUE**
Specify a second extension line origin or [Select] <Select>: *Select the origin of next extension line (P4, see Figure 8-29).*
Enter dimension text <current>: [Enter]

The default base (first extension line) for the dimensions created with the **CONTINUE** command is the previous dimension's second extension line. You can override the default extension by pressing ENTER at the **Specify a second extension line origin or [Select] <Select>** prompt, and then specifying the other dimension. The extension line origin nearest to the selection point is used as the origin for the first extension line.

Tip
*You can use the **Select** option of the **DIMBASELINE** or the **DIMCONTINUE** command to select any other existing dimension to be used as the baseline or continuous dimension.*

EXERCISE 2   *Baseline Dimension*

Draw the object shown in Figure 8-30 and then use baseline dimensioning to dimension the top half and continue dimensioning to dimension the bottom half. The distance between the dotted lines is 0.5 unit.

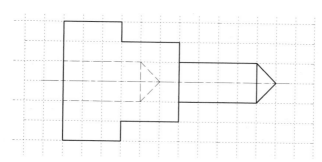

Figure 8-30   *Drawing for Exercise 2*

CREATING ANGULAR DIMENSIONS

Ribbon: Annotate > Dimensions > Dimension drop-down > Angular
Menu Bar: Dimension > Angular   **Toolbar:** Dimension > Angular
Command: DIMANG or DIMANGULAR

The Angular dimensioning is used for applying angular dimension to an entity. The **Angular** tool is used to generate a dimension arc (dimension line in the shape of an arc with arrowheads at both ends) to indicate the angle between two nonparallel lines. This tool can also be used to dimension the vertex and two other points, a circle with another point, or the angle of an arc. For every set of points, there exists one acute angle and one obtuse angle (inner and outer angles). If you specify the dimension arc location between the two points, you will get the acute angle; if you specify it outside the two points, you will get the obtuse angle. Figure 8-31 shows the four ways to dimension two nonparallel lines. The prompt sequence that will follow when you choose this tool is given next:

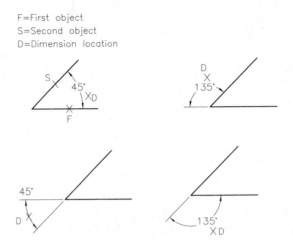

Figure 8-31 *Angular dimensioning between two nonparallel lines*

Select arc, circle, line, or <specify vertex>: *Select the object or press ENTER to select a vertex point where two segments meet.*
Select second line: *Select the second object.*
Specify dimension arc line location or [Mtext/Text/Angle/Quadrant]: *Place the dimension or select an option.*
Dimension text = current

If you want to override the default angular value, use the **Mtext** or **Text** option. Use the %%d control sequence after the number at the text prompt. For example, for 45°, type 45%%d and then press ENTER.

The methods of dimensioning various entities using this command are discussed next.

Dimensioning the Angle between Two Nonparallel Lines

The angle between two nonparallel lines or two straight line segments of a polyline can be dimensioned using the **DIMANGULAR** dimensioning command. The vertex of the angle is taken as the point of intersection of the two lines.

The following example illustrates the dimensioning of two nonparallel lines by using the **DIMANGULAR** command. Alternatively, you can apply the angular dimension by choosing the **Angular** tool from the **Dimension** drop-down in the **Dimensions** panel.

Select arc, circle, line, or <specify vertex>: *Select the first line.*
Select second line: *Select the second line.*
Specify dimension arc line location or [Mtext/Text/Angle/Quadrant]: **M** *(Enter the new value in the **Text Editor**. Specify dimension arc line location or [Mtext/Text/Angle/Quadrant]: Specify the dimension arc location or select an option.*

The location of the extension lines and the dimension arc is determined by the placement of the dimension arc. In AutoCAD, you can place the dimension text outside the quadrant in which you measure the angle by extending the dimension arc. Choose the **Quadrant** option from the shortcut menu or from the **Specify dimension arc line location or [Mtext/Text/Angle/Quadrant]** prompt. Next, you will be prompted to specify the quadrant in which you want to measure the angle. Then, specify the quadrant with mouse. If the dimension arc line is lying outside the quadrant that is being measured, the dimension arc will be extended up to that location with the help of extension line. Figure 8-32(a) shows a dimension created without using the **Quadrant** option and Figure 8-32(b) shows the same dimension created by using the **Quadrant** option.

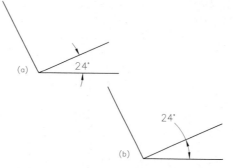

*Figure 8-32 The angular dimension created with and without choosing the **Quadrant** option*

Dimensioning the Angle of an Arc

Angular dimensioning can also be used to dimension the angle of an arc. In this case, the center point of the arc is taken as the vertex and the two endpoints of the arc are used as the extension line origin points for the extension lines (Figure 8-33). The following example illustrates the dimensioning of an arc by using the **Angular** tool:

Select arc, circle, line, or <specify vertex>: *Select the arc.*
Specify dimension arc line location or [Mtext/Text/Angle/Quadrant]: *Specify a location for the arc line or select an option.*

Angular Dimensioning of Circles

The angular feature associated with the circle can be dimensioned by selecting a circle object at the **Select arc, circle, line, or <specify vertex>** prompt. The center of the selected circle is used as the vertex of the angle. The first point selected (when the circle is selected for angular dimensioning) is used as the origin of the first extension line. In the similar manner, the second point selected is taken as the origin of the second extension line (Figure 8-34). The following is the prompt sequence for dimensioning a circle:

Select arc, circle, line, or <specify vertex>: *Select the circle at the point where you want the first extension line.*
Specify second angle endpoint: *Select the second point on or away from the circle.*
Specify dimension arc line location or [Mtext/Text/Angle/Quadrant]: *Select the location for the dimension line.*

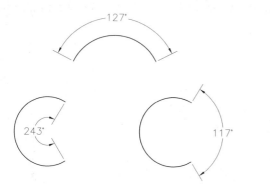

Figure 8-33 Angular dimensioning of arcs

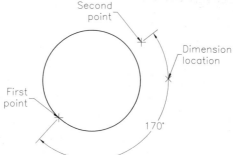

Figure 8-34 Angular dimensioning of a circle

Angular Dimensioning based on Three Points

If you press ENTER at the **Select arc, circle, line, or <specify vertex>** prompt, AutoCAD allows you to select three points to create an angular dimension. The first point is the vertex point, and the other two points are the first and second angle endpoints of the angle (Figure 8-35). The coordinate specifications of the first and the second angle endpoints must not be identical. However, the angle vertex coordinates and one of the angle endpoint coordinates can be identical. The following example illustrates angular dimensioning by defining three points:

Select arc, circle, line, or <specify vertex>: Enter
Specify angle vertex: *Specify the first point, vertex. This is the point where two segments meet. If the two segments do not meet, use the* **Apparent Intersection** *object snap.*
Specify first angle endpoint: *Specify the second point. This point will be the origin of the first extension line.*
Specify second angle endpoint: *Specify the third point. This point will be the origin of the second extension line.*
Specify dimension arc line location or [Mtext/Text/Angle/Quadrant]: *Select the location for the dimension line.*
Dimension text = current

EXERCISE 3   *Angular Dimension*

Draw the profile as shown in Figure 8-36 and then use angular dimensioning to dimension all angles of the part. The distance between the dotted lines is 0.5 unit.

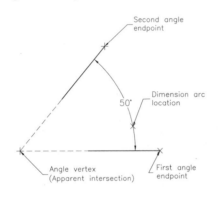

Figure 8-35   *Angular dimensioning using 3 points*

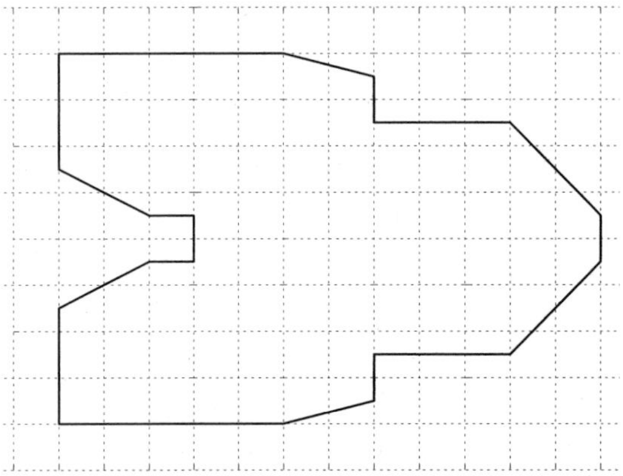

Figure 8-36   *Drawing for Exercise 3*

CREATING DIAMETER DIMENSIONS

Ribbon: Annotate > Dimensions > Dimension drop-down > Diameter
Menu Bar: Dimension > Diameter   **Toolbar:** Dimension > Diameter
Command: DIMDIA

Diameter dimensioning is used to dimension a circle or an arc. Here, the measurement is done between two diametrically opposite points on the circumference of the circle or the arc (Figure 8-37). The dimension text generated by AutoCAD begins with the ø symbol to indicate a diameter dimension. The prompt sequence that will follow when you choose the **Diameter** tool in the **Dimensions** panel is given next.

Select arc or circle: *Select an arc or circle by selecting a point anywhere on its*

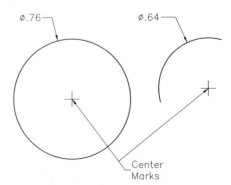

Figure 8-37   *Diameter dimensioning*

circumference.
Dimension text = Current
Specify dimension line location or [Mtext/Text/Angle]: *Specify a point to position the dimension.*

If you want to override the default value of the dimension text, use the **Mtext** or the **Text** option. The control sequence %%C is used to obtain the diameter symbol ø. It is followed by the dimension text that should appear in the diameter dimension. For example, if you want to write a text that displays a value ø20, then enter %%c20 at the text prompt.

CREATING JOGGED DIMENSIONS

Ribbon: Annotate > Dimensions > Dimension drop-down > Jogged
Menu Bar: Dimension > Jogged   **Toolbar:** Dimension > Jogged
Command: DIMJOGGED

The necessity of the Jogged dimension arises because of the space constraint. Also the Jogged dimension is used when you want to avoid the merging of the dimension line with other dimensions. Also, there are instances when it is not possible to show the center of the circle in the sheet. In such situations, the jogged dimensions are used, as shown in Figure 8-38. The jogged dimensions can be added using the **Jogged** tool. Note that with this command, you can add only jogged radius dimensions. To add jogged dimensions, invoke the **Jogged** tool from the **Dimensions** panel and select an arc or a circle from the drawing area; you will be prompted to select the center location override. Specify a new location to override the existing center. This point will also become the start point of the dimension line. Next, specify the location of the dimension line and the jog. The Command prompt that will follow when you invoke the **Jogged** tool is given next.

Select arc or circle: *Select arc or circle to be dimensioned.*
Specify center location override: *Specify a point which will currently override the actual center point location.*
Dimension text = *Current*
Specify dimension line location or [Mtext/Text/Angle]: *Specify a point to position dimension line.*
Specify jog location: *Specify a point for the positioning of jog.*

Figure 8-38 *Jogged dimensioning*

CREATING RADIUS DIMENSIONS

Ribbon: Annotate > Dimensions > Dimension drop-down > Radius
Menu Bar: Dimension > Radius   **Toolbar:** Dimension > Radius
Command: DIMRAD

The Radius dimensioning is used to dimension a circle or an arc (Figure 8-39). Radius and diameter dimensioning are similar; the only difference is that instead of the diameter line, a radius line is drawn (half of the diameter line), which is measured from the center to any point on the circumference. The dimension text generated by AutoCAD is preceded by the letter **R** to indicate a radius dimension. If you want to use the default dimension text (dimension text generated automatically by AutoCAD), simply specify a point to position the dimension at the **Specify dimension line location or [Mtext/Text/Angle]** prompt. You can also enter a new value or specify a prefix or suffix, or suppress the entire text by entering a blank space following the **Enter dimension text <current>** prompt. A center mark for the circle/arc is drawn automatically, provided the center mark value controlled by the **DIMCEN** variable is not 0. The prompt sequence that will follow when you choose this button is given next.

Select arc or circle: *Select the object that you want to dimension.*
Dimension text = Current
Specify dimension line location or [Mtext/Text/Angle]: *Specify the dimension location.*

If you want to override the default value of the dimension text, use the **Text** or the **Mtext** option. You can also enter the required value at the text prompt.

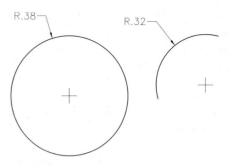

Figure 8-39 Radius dimensioning

Note
*In case of diametric, radial, and jogged radial dimensions, you can extend the dimension line beyond the endpoints of the arc by creating the arc extension line beyond the endpoints. These arc extension lines are similar to the other extension lines. The system variables **DIMSE1** and **DIMSE2** allow the display of the first and the second extension lines, respectively. These variables should be **ON** in order to suppress the respective extension line.*

CREATING JOGGED LINEAR DIMENSIONS

| **Ribbon:** Annotate > Dimensions > Dimension, Dimjogline | **Toolbar:** Dimension > Jogged Linear |
| **Menu Bar:** Dimension > Jogged Linear | **Command:** DIMJOGLINE |

The **Dimension, Dimjogline** tool is used to add or remove a jog in the existing dimensions. This kind of dimensioning technique is generally used to dimension the components that have a high length to width ratio, see Figure 8-40. To add a jog to a linear dimension, choose the **Dimjogline** tool from the **Dimensions** panel and select the dimension to which you want to add a jog. Next, pick a point along the dimension line to specify the location of the jog placement. Alternatively, you can press the ENTER key to place the jog automatically. To remove a jog, invoke the **DIMJOGLINE** command and choose **Remove** from the shortcut menu. Next, specify the dimension line from which you want to remove the jog. You can modify the location of the jog symbol with the help of grips. Note that you can add a jog only to the linear or aligned dimensions with this command.

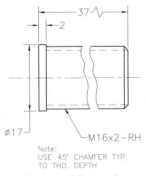

Figure 8-40 Jogged linear dimensioning

Note
*The height of jog symbol can be changed by varying the **Jog height factor** in the **Lines** & **Arrows** rollout of the **Properties** palette.*

GENERATING CENTER MARKS AND CENTERLINES

Ribbon: Annotate > Dimensions > Center Mark	**Toolbar:** Dimension > Center Mark
Menu bar: Dimension > Center Mark	**Command:** DIMCENTER

When circles or arcs are dimensioned with the **DIMRADIUS** or **DIMDIAMETER** command, a small mark known as center mark, or a line known as centerline, may be drawn at the center of the circle/arc. Sometimes, you need to mark the center of a circle or an arc without using these dimensioning commands. This can be achieved with the help of the **Center Mark** tool. You can invoke this tool by choosing the **Center Mark** tool from the **Dimensions** panel or by entering **CENTER** (or **CEN**) at the **Dim:** prompt. When you invoke this tool, you are prompted to select the arc or the circle. The result of this command will depend upon the value of the **DIMCEN** variable. If the value of this variable is positive, center marks are drawn, see Figure 8-41 and if the value is negative, centerlines are drawn, see Figure 8-42.

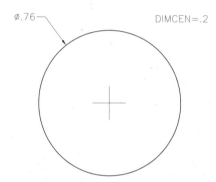

Figure 8-41 *Using a positive value for the* **DIMCEN** *variable*

Figure 8-42 *Using a negative value for the* **DIMCEN** *variable*

Note
The center marks created by **DIMCENTER** *or* **DIM CENTER** *are lines, but not associative dimensioning objects, and they have an explicit linetype.*

CREATING ORDINATE DIMENSIONS

Ribbon: Annotate > Dimensions > Dimension drop-down > Ordinate	
Menu Bar: Dimension > Ordinate	**Toolbar:** Dimension > Ordinate
Command: DIMORD	

Ordinate dimensioning is used to dimension the *X* and *Y* coordinates of the selected point. This type of dimensioning is also known as arrowless dimensioning because no arrowheads are drawn in it. Ordinate dimensioning is also called datum dimensioning because all dimensions are related to a common base point. The current UCS (user coordinate system) origin becomes the reference or the base point for ordinate dimensioning. With ordinate dimensioning, you can determine the X or Y displacement of a selected point from the current UCS origin.

Dimension text (*X* or *Y* coordinate value) and the leader line along the *X* or *Y* axis are automatically placed using Ordinate dimensioning (Figure 8-43). Since ordinate dimensioning pertains to either the *X* coordinate or the *Y* coordinate, you should keep ORTHO on. When ORTHO is off, the leader line is automatically given a bend when you select the second leader line point that is offset from the first point. This allows you to generate offsets and avoid overlapping text on closely spaced dimensions. In ordinate dimensioning, only one extension line (leader line) is drawn.

The leader line for an *X* coordinate value will be drawn perpendicular to the *X* axis, and the leader line for a *Y* coordinate value will be drawn perpendicular to the *Y* axis. Since you cannot override this, the leader line drawn perpendicular to the *X* axis will have the dimension text aligned with the leader line. The dimension text is the *X* datum of the selected point. The leader line drawn perpendicular to the *Y* axis will have the dimension text, which is the *Y* datum of the selected point, aligned with the leader line. Any other alignment specification for the dimension text is nullified. Hence, changes in the Text Alignment in the **Dimension Style Manager** dialog box (**DIMTIH** and **DIMTOH** variables) have no effect on the alignment of the dimension text. You can specify the coordinate value that you want to dimension at the **Specify leader endpoint or [Xdatum/Ydatum/MText/Text/Angle]** prompt.

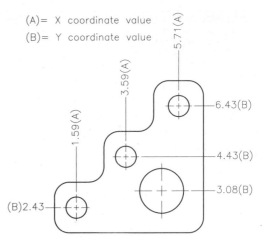

Figure 8-43 Ordinate dimensioning

If you select or enter a point, AutoCAD checks the difference between the feature location and the leader endpoint. If the difference between the *X* coordinates is greater, the dimension measures the *Y* coordinate; otherwise, the *X* coordinate is measured. In this manner, AutoCAD determines whether it is an X or Y type of ordinate dimension. However, if you enter Y instead of specifying a point, AutoCAD will dimension the *Y* coordinate of the selected feature. Similarly, if you enter X, AutoCAD will dimension the *X* coordinate of the selected point. The prompt sequence that follows when you choose this tool is given next.

Specify feature location: *Select a point on an object.*
Specify leader endpoint or [Xdatum/Ydatum/Mtext/Text/Angle]: *Enter the endpoint of the leader.*

You can override the default text with the help of the **Mtext** or **Text** option. If you use the **Mtext** option, the **Text Editor** will be displayed. If you use the **Text** option, you will be prompted to specify the new text in the Command line itself.

EXERCISE 4   *Ordinate Dimension*

Draw the model shown in Figure 8-44 and then use ordinate dimensioning to dimension the part. The distance between the dotted lines is 0.5 unit.

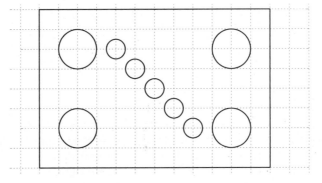

Figure 8-44 Drawing for Exercise 4

MAINTAINING EQUAL SPACING BETWEEN DIMENSIONS

Ribbon: Annotate > Dimensions > Adjust Space	**Toolbar:** Dimension > Dimension Space
Menu Bar: Dimension > Dimension Space	**Command:** DIMSPACE

This command is used to equally space the overlapping or the unequally spaced linear and angular dimensions. Maintaining equal spacing between the dimension lines increases the clarity of the drawing display. Figure 8-45 shows a drawing with unequally spaced dimensions, and Figure 8-46 shows the same drawing with equally spaced dimensions after the use of the **Adjust Space** tool. Figure 8-47 shows a drawing with its angular dimensions equally spaced.

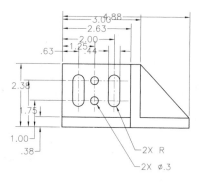

Figure 8-45 Drawing with its linear dimensions unequally spaced

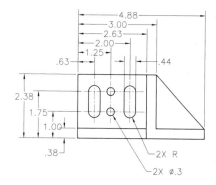

Figure 8-46 Drawing with its linear dimensions equally spaced

The prompt sequence that will follow when you choose the **Adjust Space** tool is given next.

Command: _**DIMSPACE**
Select base dimension: *Select a linear or an angular dimension with respect to which all other dimensions will get spaced.*
Select dimensions to space: *Select the dimensions that you want to be equally spaced from the specified base dimension.*
Select dimensions to space: *Select more similar dimensions or press* Enter
Enter value or [Auto] <Auto>: *Enter a value to specify the spacing between the selected dimensions or enter **A** to calculate automatically the gap between the dimensions.*

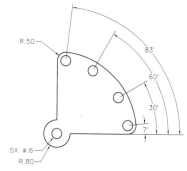

Figure 8-47 Drawing with its angular dimensions equally spaced

Enter **0** in the above prompt to align all the selected dimensions in a single line. With the **Auto** option selected, AutoCAD will calculate the space value on the basis of the dimension text height. The spacing is automatically maintained to a value that is double of the dimension text height.

CREATING DIMENSION BREAKS

Ribbon: Annotate > Dimensions > Break	**Toolbar:** Dimension > Dimension Break
Menu bar: Dimension > Dimension Space	**Command:** DIMBREAK

The **Break** tool is used to create a break in the dimension line, extension line, or multileaders at the point where they intersect with other drawing entities or annotations. Inserting dimension breaks will keep annotative objects separate from the main drawing, which increases the clarity of the drawing reading, see Figure 8-48. The prompt sequence that will follow when you choose this tool is given next.

Command: **_DIMBREAK**
Select a dimension to add/remove break or [Multiple]: *Select a dimension in which you want to insert a break.*
Select object to break dimension or [Auto/Manual/Remove] <Auto>: *Select objects or dimensions that intersect the above selected dimension, enter an option or press* Enter *to choose the* **Auto** *option.*

The options in the **Break** tool are discussed next.

Multiple

This option allows you to select more than one dimension for adding or removing a break. The prompt sequence that will follow is given next.

Select dimensions: *Select the dimensions by any object selection method and press* Enter
Select object to break dimensions or [Auto/Remove] <Auto>: *Enter an option or press* Enter

The **Auto** option, available in the above prompt, creates breaks automatically at the intersections of the selected dimensions with the other drawing and annotation entities. Also, if the selected dimensions are self-intersecting, then the **Auto** option, if chosen, creates a break at the point of intersection. The **Remove** option removes all the dimension breaks that exist in the selected dimensions.

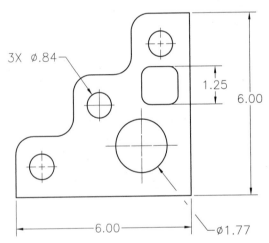

Figure 8-48 *Drawing with dimension breaks*

Auto

This option creates the break automatically at the intersection of the selected dimension with the other drawing and annotation entities.

Note
All dimension breaks created by using the **Multiple** *and* **Auto** *options update automatically when you modify the broken dimension or the intersecting objects. However, the new objects that are created over the broken dimension do not have any effect on the display of the broken dimension. For example, the dimension that has already been broken will not be further broken with respect to newly created objects even if the object(s) interfere with the dimension. To break the newly created dimension, you need to use the* **Break** *tool again.*

Remove

This option removes all the dimension breaks from the selected dimension.

Manual

This option is used to create a break in the dimension manually. You have to specify two points between which you want to break the dimension or the extension line. You can also create the dimension break at only one location with this option.

> **Note**
> *The dimension breaks created by the* **Manual** *option do not update automatically when you modify the broken dimension or the intersecting objects. For updating the dimension break, you need to remove the dimension, and then again add the dimension break.*
>
> *You cannot add dimension breaks to the straight leaders, spline leaders, and spline multileaders.*

CREATING INSPECTION DIMENSIONS

Ribbon: Annotate > Dimensions > Inspect	**Toolbar:** Dimensions > Inspect
Menu Bar: Dimension > Inspection	**Command:** DIMINSPECT

A drawing sheet provides every minute information about the dimension of the product, including the inspection rate. The **Inspect** tool in AutoCAD is used to describe Inspection Rate of critical dimension of the product to ensure its quality. The inspection rate is used to specify how frequently the dimensions are to be checked to ensure the variations in the dimensions are within the range. Select a dimension from the drawing and choose the **Inspect** tool from the **Dimensions** panel; the **Inspection Dimension** dialog box will be displayed, as shown in Figure 8-49. Choose the **OK** button; the selected dimension will change into the inspection dimension that consists of three fields (see Figure 8-50), which are discussed next.

Inspection Label

The inspection label is located at the extreme left of the inspection dimension and is used to specify the particular inspection dimension, see Figure 8-50.

Dimension Value

The dimension value is located in the middle of the inspection dimension and is used to display the value of the dimension to be maintained, see Figure 8-50.

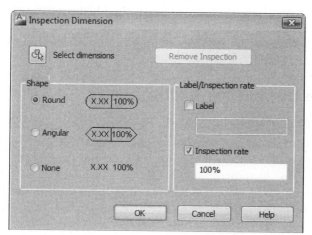

*Figure 8-49 The **Inspection Dimension** dialog box*

Inspection Rate

The inspection rate is located at the right-end of the inspection dimension. It is used to specify the frequency of carrying out inspections, see Figure 8-50. The frequency of inspection is specified in terms of percentage value. The value 100% signifies that you need to check the dimension each time you inspect a component. Similarly, if the value is 50 % then you need to check the dimension for every second component.

If you have not selected the dimension before invoking the **Inspect** tool, choose the **Select dimensions** button from the **Inspection Dimension** dialog box (Refer to Figure 8-49); the dialog box will temporarily disappear from the screen, thereby enabling you to select the dimension that you want to convert into the inspection dimension. Select the dimensions and press the ENTER key; the **Inspection Dimension** dialog box will be displayed again. All the selected dimensions will display the same value of the label and inspection dimension. Next, specify the shape of the boundary around the inspection dimension from the **Shape** area. Figure 8-50 displays the shapes available in the **Inspection Dimension** dialog box. Next, enter the values for the label and inspection rate in their respective edit boxes in the **Label/Inspection rate** area and choose the **OK** button. You can also control the display of label and inspection rate in the inspection dimension by clearing or selecting the **Label** and **Inspection rate** check boxes respectively. To remove the inspection dimension of the selected dimension, choose the **Remove Inspection** button from the **Inspection Dimension** dialog box and then choose the **OK** button.

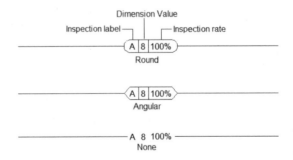

Figure 8-50 *Shapes and components of inspection dimension*

WORKING WITH TRUE ASSOCIATIVE DIMENSIONS

The true associative dimensions are the dimensions that are automatically modified when the objects to which they are associated are modified. By default, all dimensions in AutoCAD are true associative dimensions. If the dimension attached to the object is true associative, then it will be modified automatically when the object is modified. In this case, you do not have to select the definition points of the dimensions. Any dimension in AutoCAD can be converted into a disassociated dimension and then back to the true associative dimension. This is discussed next.

Removing the Dimension Associativity

Command: DIMDISASSOCIATE

The **DIMDISASSOCIATE** command is used to remove the associativity of the dimensions from the object to which they are associated. When you invoke this command, you will be prompted to select the dimensions to be disassociated. The true association of the selected dimensions is automatically removed once you exit this command. The number of dimensions disassociated is displayed in the Command prompt.

Converting a Dimension into a True Associative Dimension

| **Ribbon:** Annotate > Dimensions > Reassociate | **Command:** DIMREASSOCIATE |
| **Menu bar:** Dimension > Reassociate Dimension | |

The **Reassociate** tool is used to create a true associative dimension by associating the selected dimension to the specified object. When you invoke this command, you will be prompted to select the objects. These objects are the dimensions to be associated. Select the dimensions to be associated and press ENTER; a cross is displayed and you are prompted to select the feature location. This cross implies that the dimension is not associated. You can define a new association point for the dimensions by selecting the objects or by using the object snaps. If you select a dimension that has already been associated to an object, the cross will be displayed inside a box. The prompt sequence that is displayed varies depending upon the type of dimension selected. In case of linear, aligned, radius, and diameter dimensions, you can directly select the object to associate the dimension. If the arcs or circles are assigned angular dimensions using three points, then also you can select these arcs or circles directly for associating the dimensions. For rest of the dimension types, you can use the object snaps to specify the point to associate the dimensions.

> **Note**
> *If you have edited a dimension by using the **Mtext** or **Text** option, then on using the **DIMREASSOCIATE** command, the overridden value will be replaced by the original value.*

EXERCISE 5

Draw and dimension the object, as shown in Figure 8-51.

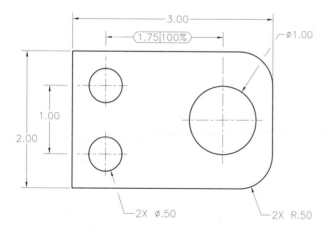

Figure 8-51 *Drawing for Exercise 5*

DRAWING LEADERS

Command: QLEADER

The leader line is used to attach annotations to an object or when the user wants to show a dimension without using another dimensioning command. Sometimes, leaders of the dimensions of circles or arcs are so complicated that you need to construct a leader of your own. The leaders can be created using the **QLEADER** command. The leaders drawn by using this command create the arrow and the leader lines as a single object. The text is created as a separate object. This command can create multiline annotations and offer several options such as copying existing annotations and so on. You can customize the leader and annotation by selecting the **Settings** option at the **Specify first leader point, or [Settings] <Settings>** prompt. The prompt sequence that will follow when you use this command is given next.

Specify first leader point, or [Settings] <Settings>: *Specify the start point of the leader.*
Specify next point: *Specify the endpoint of the leader.*
Specify next point: *Specify the next point.*
Specify text width <current>: *Enter the text width of multiline text.*
Enter first line of annotation text <Mtext>: *Press ENTER; AutoCAD displays the* **Text Editor.**
Enter text in the dialog box and then choose **Close Text Editor** *to exit from the* **Ribbon.**

If you press ENTER at the **Specify first leader point, or [Settings] <Settings>** prompt, then the **Leader Settings** dialog box is displayed. The **Leader Settings** dialog box gives you a number of options for the leader line and the text attached to it. It has the following tabs:

Annotation Tab

This tab provides you with various options to control annotation features, see Figure 8-52.

Annotation Type Area

The options in **Annotation Type** area and their usage are discussed next.

MText. When this radio button is selected, AutoCAD uses the **Text Editor** to create an annotation. Therefore, on selecting this radio button, the options in the **MText options** area become available.

Copy an Object. This option allows you to copy an existing annotation object (like multiline text, single line text, tolerance, or block) and attach it at the end of the leader. For example, if you have a text string in the drawing that you want to place at the end of the leader, you can use the **Copy an Object** option to place it at the end of the leader.

Tolerance. Select the **Tolerance** radio button and choose the **OK** button to exit from the **Leader Settings** dialog box. Then, specify the next point to complete the leader line; AutoCAD will display the **Geometric Tolerance** dialog box. In this dialog box, specify the tolerance and choose **OK** to exit it. AutoCAD will place the specified geometric tolerance with the feature control frame at the end of the leader (Figure 8-53).

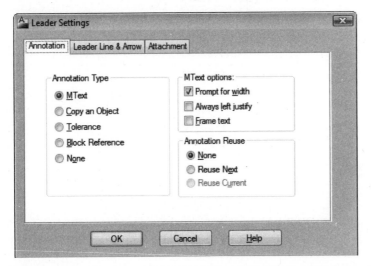

Figure 8-52 The **Annotation** *tab of the* **Leader Settings** *dialog box*

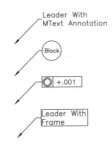

Figure 8-53 Leaders with different annotation types

Block Reference. The **Block Reference** radio button allows you to insert a predefined block at the end of the leader. When you select this option, AutoCAD will prompt you to enter the block name and insertion point.

None. This option creates a leader without placing any annotation at the end of the leader.

MText options Area

The options under this area will be available only if the **MText** radio button is selected from the **Annotation Type** area. This area provides you with the following options.

Prompt for width. Selecting this check box allows you to specify the width of the multiline text annotation.

Always left justify. Select this check box to left justify the multiline text annotation in all situations. Selecting this check box makes the Prompt for the width option unavailable.

Frame text. Selecting this check box draws a box around the multiline text annotation.

Annotation Reuse Area

The options under this area allow you to reuse the annotation.

None. When selected, AutoCAD does not reuse the leader annotation.

Reuse Next. This radio button is used to reuse the annotation that you are going to create next for all subsequent leaders.

Reuse Current. This option allows you to reuse the current annotation for all subsequent leaders.

Leader Line & Arrow Tab

The options in the **Leader Line & Arrow** tab are related to the leader parameters, see Figure 8-54.

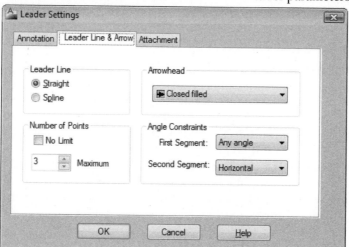

Figure 8-54 The Leader Line & Arrow tab of the Leader Settings dialog box

Leader Line Area

This area has the options for the leader line type such as straight or spline. The **Spline** option draws a spline through the specified leader points and the **Straight** option draws straight lines. Figure 8-55 shows the straight and spline leader lines.

Number of Points Area

The options provided under this area are used to specify the number of points in the leader.

No Limit. If this check box is selected, you can define as many number of points as you want in the leader line. AutoCAD will keep prompting you for the next point, until you press ENTER at this prompt.

Maximum. This spinner is used to specify the maximum number of points on the leader line. The default value in this spinner is 3; as a result, there will be only three points in the leader line. You can specify the number of points by using this spinner to control the shape of the leader. This has to be kept in mind that the start point of the leader is the first leader point. This spinner will be available only if the **No Limit** check box is clear. Figure 8-55 shows a leader line with five points.

Arrowhead Area

The drop-down list under this area allows you to define a leader arrowhead. The arrowhead is the same as the one for dimensioning. You can also use the user-defined arrows by selecting **User Arrow** from the drop-down list.

Angle Constraints Area

The options provided under this area are used to define the angle for the segments of the leader lines.

First Segment. This drop-down list is used to specify the angle at which the first leader line segment will be drawn. You can select the predefined values from this drop-down list.

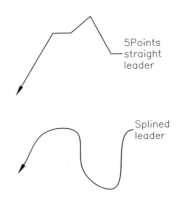

Figure 8-55 Straight and splined leaders

Second Segment. This drop-down list is used to specify the angle at which the second leader line segment will be drawn.

Attachment Tab

The **Attachment** tab (Figure 8-56) will be available only if you have selected **MText** from the **Annotation Type** area of the **Annotation** tab. The options in this tab are used to attach the multiline text to the leader. It has two columns: **Text on left side** and **Text on right side**. Both these columns have five radio buttons below them. Each radio button corresponds to the option of attaching the multiline text. If you draw a leader from the right to the left, AutoCAD uses the settings under **Text on left side**. Similarly, if you draw a leader from the left to the right, AutoCAD uses the settings as specified under **Text on right side**. This area also provides you with the **Underline bottom line** check box. If this check box is selected, then the last line of the multiline text will be underlined.

Note
*You can use the Command Line to create leaders with the help of the **LEADER** command. The options under this command are similar to those under the **QLEADER** command. The only difference is that the **LEADER** command uses the Command Line.*

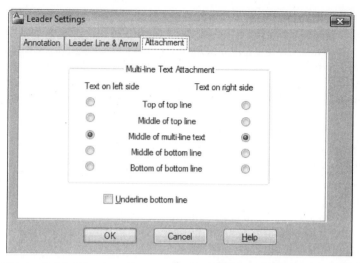

Figure 8-56 *The **Attachment** tab of the **Leader Settings** dialog box*

EXERCISE 6   *Qleader*

Make the drawing shown in Figure 8-57 and then use the **QLEADER** command to dimension the part accordingly. The distance between the dotted lines is 0.5 units.

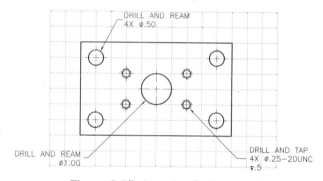

Figure 8-57 *Drawing for Exercise 6*

MULTILEADERS

Ribbon: Annotate > Leaders > Multileader	**Toolbar:** Multileader > Multileader
Command: MLEADER	

Multileaders are the enhanced leaders, wherein the leader and its content is part of the same object. A Multileader provides you enough flexibility so that you can create either the arrowhead or the tail end first. Alternatively, you can specify the content of the leader first and then draw the leader. Multileaders can have more than one leader so that a single comment can be pointed at more than one location. Besides this, multiple notes can be attached to a single leader so that more than one comment can be pointed at a single location. You can add or remove leaders from the previously created multileaders. You can also control and modify the appearance of the multileaders.

DRAWING MULTILEADERS

To draw a multileader, choose the **Multileader** tool from the **Leaders** panel. The prompt sequence that follows is given next.

Command: _**MLEADER**
Specify leader arrowhead location or [leader Landing first/Content first/Options]
<Options>: *Specify the location for the arrowhead from where the leader will start.*
Specify leader landing location: *Specify the leader landing location, see Figure 8-58.*

AutoCAD displays the **Text Editor***. Enter the text in the dialog box and click outside the editor to accept and exit the command. If you click outside the editor without entering any text and exit the command, no text will be attached to the multileader.*

The other options in the **MLEADER** command are discussed next.

leader Landing first

This option is used to locate the leader landing line first. Choose the **leader Landing first** option from the shortcut menu at the **Specify leader arrowhead location or [leader Landing first/ Content first/Options] <Options>** prompt to invoke this option. Alternatively, if the **Dynamic Input** button is turned on in the Status Bar, then you can choose this option from the dynamic input box associated with the cursor. On doing so, you will be prompted to specify a point where the leader landing should be placed. Specify the leader landing location; you will be prompted to specify the leader arrowhead location. The leader landing will automatically adjust itself to the leader

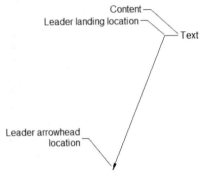

Figure 8-58 Image displaying the options to locate the leader start point

landing point, depending upon the direction in which you create the leader. For example, if you create the leader to the left of the leader landing location, the landing will automatically be created to the right of the leader landing location.

Content first

This option is used to specify the content (multiline text or block) of the leader first. Choose the **Content first** option from the shortcut menu or from the **Specify leader arrowhead location or [leader Landing first/Content first/Options] <Options>** prompt to invoke this option. Alternatively, if the **Dynamic Input** button is turned on in the Status Bar, you can choose this option from the dynamic input box associated with the cursor. After specifying the content to be attached with the leader, a horizontal leader landing will automatically be attached to the content and then you can specify the leader arrowhead location.

Note
If you have previously drawn a multileader by using any one of the above-mentioned options, then the succeeding multileaders will be created by using that option only, until otherwise modified.

Options

The suboptions within this option are used to control the leader type, content to be attached with the leader, size of the leader landing, and so on. These suboptions are discussed next.

Leader type

This option provides the suboptions for the leader line type such as straight, spline, or none. The **sPline** suboption draws a spline through the specified leader points and the **Straight** option draws the straight lines. Figure 8-55 shows the straight and spline leader lines. The **None** option enables you to draw the content of the leader without drawing the leader itself.

leader lAnding

This option provides you the suboptions to control the display of the horizontal landing line. The **No** suboption will create a leader with no horizontal landing and the **Yes** suboption will create a leader with the horizontal landing. If you choose the **Yes** suboption, then you will be prompted to specify the length of the horizontal landing. The new leader will be created with the specified landing length. Figure 8-59 shows the leaders created by using various suboptions of the **leader lAnding**.

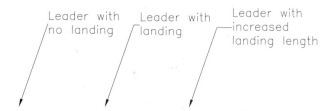

Figure 8-59 Leaders drawn by using various types of landing suboptions

Content type

This option is used to specify the type of content to be attached with the leader. The **Mtext** suboption uses the AutoCAD **Text Editor** to create the content. The **Block** suboption allows you to insert a predefined block at the end of the leader. When you select this option, AutoCAD will prompt you to enter the block name and the specified block will be attached to the new leader. The **None** suboption creates a leader without placing any content at the end of the leader.

Maxpoints

This option is used to specify the maximum number of points on the leader line. The default value is 2, which means there will be only two points on the leader line. You can specify the maximum number of points by entering the new value. This has to be kept in mind that the start point of the leader is the first leader point.

First angle

This option is used to specify the angle that the first leader line will measure from the horizontal at the start point. You can also specify a value which is the natural number multiple of 15-degree.

Second angle

This option is used to specify the angle that the second leader line will measure from the horizontal at the second point. You can only specify a value which is the natural number multiple of 15-degree.

eXit options

This option will return you back to the main options of the multileader.

Note
*All settings specified in the **Options** heading of the multileader will last only for that particular leader, and the next leader will be created with the default settings. You can also change the default settings of the multileader from the **Multileader Style Manager** dialog box that will be discussed in detail in Chapter 10.*

ADDING LEADERS TO EXISTING MULTILEADER

Ribbon: Annotate > Leaders > Add Leaders   **Toolbar:** Multileaders > Add Leaders
Command: MLEADEREDIT

The **Add Leaders** tool is used to add leaders to an existing multileader, so that you can point a single content to more than one location. To do so, choose the **Add Leaders** tool in the **Leaders** panel; you will be prompted to select the multileader to which you want to attach the new leader. After selecting the desired multileader, a new leader will branch out from the landing location of the existing multileader. Next, you will be prompted to specify the location of the arrowhead of the new leader. Next, you can attach any number of leaders to the selected multileader. To finish this command, press the ENTER key.

REMOVING LEADERS FROM EXISTING MULTILEADER

Ribbon: Annotate > Leaders > Remove Leader   **Toolbar:** Multileaders > Remove Leader
Command: MLEADEREDIT

The **Remove Leader** tool is used to remove leaders from an existing multileader. To remove a leader, choose the **Remove Leader** tool from the **Leaders** panel; you will be prompted to select the multileader from which you want to remove leaders. Select the multileader; you will be prompted to select the leaders to be removed from the selected multileader. Select the leaders and then press the ENTER key. You can also select all the leaders of the specified multileader. But, in such cases, all multileaders will get deleted from the drawing area.

ALIGNING MULTILEADERS

Ribbon: Annotate > Leaders > Align   **Toolbar:** Multileaders > Align Multileaders
Command: MLEADERALIGN

The **Align** tool is used to arrange the selected multileaders by aligning them about a line or making them parallel or by maintaining the spacing between the horizontal landing of the leaders. To do so, choose the **Align** tool from the **Leaders** panel; you will be prompted to select the multileaders to be aligned. After selecting the multileaders, press the ENTER key. Next, you will be prompted to select one of the multileaders with respect to which all other multileaders will get aligned. Also, you can choose the alignment type from the different options available at the Command prompt. The prompt sequence for the **Align** tool is given next.

Command: **MLEADERALIGN**
Select multileaders: *Select the multileaders to be aligned and* Enter.
Current mode: *Use current spacing.*

All further prompt sequences will vary with the alignment type used in aligning the leaders. The alignment type used previously will become the default alignment type for the current command.

Select multileader to align to or [Options]: *Enter **O** to choose the type of alignment.*
Enter an option [Distribute/make leader segments Parallel/specify Spacing/Use current spacing] <Use current spacing>: *Specify an option and* Enter.

The options in the above prompt sequence are discussed next.

Distribute

This option is used to accommodate all the selected multileaders at constant spacing between the two specified points. The prompt sequence that will be followed is given next.

Enter an option [Distribute/make leader segments Parallel/specify Spacing/Use current spacing] <Distribute>: **D** [Enter].
Specify first point or [Options]: *Specify the location of the first point.*
Specify second point: *Specify the location of the second point and its orientation with respect to the first point.*

The tail end of all the selected leaders landing will get aligned with equal spacing along the line joining the first and the second specified point. Figure 8-60 shows a drawing with its multileaders nonaligned. Figure 8-61 displays the same drawing with its multileaders aligned along an imaginary vertical line by using the **Distribute** option.

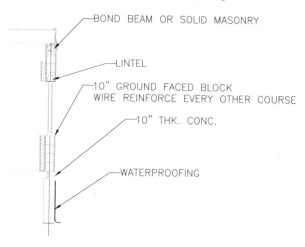

Figure 8-60 *Drawing with multileaders not aligned*

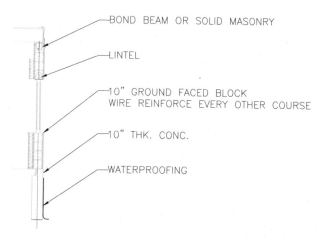

Figure 8-61 *Drawing with multileaders aligned using the* ***Distribute*** *option*

make leader segments Parallel

This option is used to make all the selected multileaders parallel to one of the leaders selected as reference. The prompt sequence that will be followed is given next.

Enter an option [Distribute/make leader segments Parallel/specify Spacing/Use current spacing] <current>: **P** Enter.
Select multileader to align to or [Options]: *Select the leader with which you want to make all other selected multileaders parallel.*

Figure 8-62 shows the drawing with its leader aligned parallel to the lowest leader of the drawing.

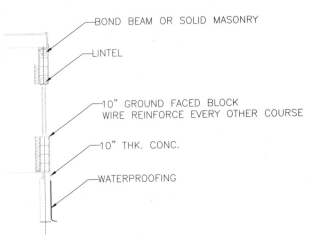

Figure 8-62 *Drawing with multileaders aligned using the **make leader segments Parallel** option*

specify Spacing

This option is used to align the selected multileaders in the specified direction, and also maintains a constant spacing between the landing lines. You need to specify one reference leader, with respect to which the specified interval and direction will be measured. The prompt sequence that will be followed is given next.

Enter an option [Distribute/make leader segments Parallel/specify Spacing/Use current spacing] <current>: **S** Enter.
Specify spacing <current>: *Enter the distance value to be maintained between the two consecutive leader landing lines.*
Select multileader to align to or [Options]: *Select one of the leaders with respect to which the specified distance will be measured.*
Specify direction: *Specify the desired direction in the drawing area for the alignment of multileaders.*

When you select a multileader to align with other multileaders, an imaginary line will be displayed in the drawing area. This line allows you to specify the direction of alignment. All the multileaders will get aligned to the selected multileader at the specified distance and direction. Figure 8-63 shows the drawing with its multileaders aligned vertical using the **specify Spacing** option.

Use current spacing

This option is used to align the selected multileaders in the specified direction and also to retain the distance between the landing lines intact. You need to specify one reference leader, with respect to which the distance will be measured in the specified direction. The prompt sequence that will be followed is given next.

Enter an option [Distribute/make leader segments Parallel/specify Spacing/Use current spacing] <specify Spacing>: **U** [Enter].

Select multileader to align to or [Options]: *Select one of the leaders with respect to which the direction to be specified is maintained.*

Specify direction: *Specify the desired direction in the drawing area for the alignment of the multileaders.*

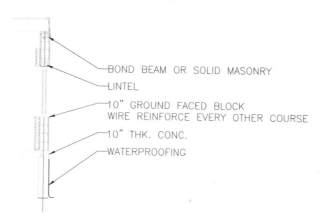

BOND BEAM OR SOLID MASONRY

LINTEL

10" GROUND FACED BLOCK
WIRE REINFORCE EVERY OTHER COURSE

10" THK. CONC.

WATERPROOFING

Figure 8-63 Drawing with multileaders aligned using the **specify Spacing** *option*

COLLECTING MULTIPLE NOTES TO SINGLE LANDING

Ribbon: Annotate > Leaders > Collect   **Toolbar:** Multileader > Collect Multileaders
Command: MLEADERCOLLECT

The **Collect** tool is used to collect the content of the selected multileaders and attach them to a single leader landing. Note that only the multileaders having predefined blocks as content can be collected by this command. To collect the multileaders, choose the **Collect** tool from the **Leaders** panel. Select the multileaders in the order in which you want them to be collected and press ENTER. The content of the last selected multileader will be retained to its landing position and the content of all other multileaders will move to the last multileader. Next, specify a point in the drawing area where the upper left-corner of the collected content of the multileaders will be placed. You can collect the content horizontally or vertically by placing the content one after another, or by specifying the space between which the contents will be placed, and the content exceeding the specified distance will be sent to the next row. The prompt sequence for the **Collect** tool is given next.

Command: **MLEADERCOLLECT**

Select multileaders: *Select the multileaders in the order in which you want to collect the content of these multileaders and press* [Enter].

Specify collected multileader location or [Vertical/Horizontal/Wrap] <current>: *Specify a point in the drawing area where the upper left-corner of the collected content will be placed or choose an option.*

If you choose **Horizontal** from the shortcut menu or from the **Specify collected multileader location or [Vertical/Horizontal/Wrap] <current>** prompt sequence, the contents of the selected multileaders will get collected horizontally. Figure 8-64 shows the multileaders without being collected, Figure 8-65(a) shows the same leaders after the horizontal alignment. If you choose **Vertical** from the shortcut menu or from the **Specify collected multileader location or [Vertical/**

Horizontal/Wrap] <current> prompt sequence, the contents of the selected multileaders will get collected vertically, see Figure 8-65(b).

If you choose the **Wrap** option from the shortcut menu displayed by right-clicking at the **Specify collected multileader location or [Vertical/Horizontal/Wrap] <current>** prompt sequence, you will be prompted to enter a value for the wrap width. The contents of the selected multileaders will be placed side by side and the contents that could not be accommodated into the specified wrap width will be sent to the next row. In Figure 8-65(c), three blocks get accommodated into the specified wrap width and the remaining fourth one is sent to the next row. Alternatively, you can also enter the number of blocks to be accommodated into one row and the remaining content will automatically be sent to the next row. Choose **Number** from the right-click shortcut menu displayed by right-clicking or from the **Specify wrap width or [Number]** prompt sequence. Next, enter the number of blocks to be accommodated into each row. In Figure 8-65(d), only two blocks are specified to be accommodated into each row, so the remaining two blocks have been sent to the second row.

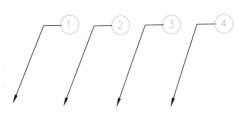 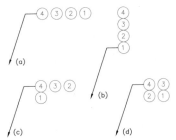

Figure 8-64 Multileaders prior to the collection   *Figure 8-65 Contents of the multileaders after being collected*

GEOMETRIC DIMENSIONING AND TOLERANCING

One of the most important parts of the design process is assigning the dimensions and tolerances to parts, since every part is manufactured from the dimensions given in the drawing. Therefore, every designer must understand and have a thorough knowledge of the standard practices used in the industry to make sure that the information given on the drawing is correct and can be understood by other people. Tolerancing is equally important, especially in the assembled parts. Tolerances and fits determine how the parts will fit. Incorrect tolerances could result in a product that is not usable. In addition to dimensioning and tolerancing, the function and the relationship that exists between the mating parts is important if the part is to perform the way it was designed. This aspect of the design process is addressed by geometric dimensioning and tolerancing, generally known as GDT.

Geometric dimensioning and tolerancing is a means to design and manufacture parts with respect to the actual function and relationship that exists between different features of the same part or the features of the mating parts. Therefore, a good design is not achieved by just giving dimensions and tolerances. The designer has to go beyond dimensioning and think of the intended function of the part and how the features of the part are going to affect its function. For example, Figure 8-66 shows a part with the required dimensions and tolerances. In this drawing, there is no mention of the relationship between the pin and the plate. Is the pin perpendicular to the plate? If it is, to what degree should it be perpendicular? Also, it does not mention on which surface the perpendicularity of the pin is to be measured. A design like this is open to individual interpretation based on intuition and experience. This is where geometric dimensioning and tolerancing play an important part in the product design process.

Figure 8-67 has been dimensioned using geometric dimensioning and tolerancing. The feature symbols define the datum (reference plane) and the permissible deviation in the perpendicularity of the pin with respect to the bottom surface. From a drawing like this, chances of making a mistake are minimized.

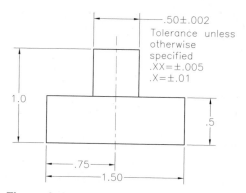

Figure 8-66 *Traditional dimensioning and tolerancing technique*

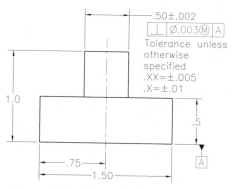

Figure 8-67 *Geometric dimensioning and tolerancing*

GEOMETRIC CHARACTERISTICS AND SYMBOLS

Before discussing the application of AutoCAD commands in geometric dimensioning and tolerancing, you need to understand the following feature symbols and tolerancing components. Figure 8-68 shows the geometric characteristics and symbols used in geometric dimensioning and tolerancing.

Kind of feature	Type of feature	Characteristics	
Related	Location	Position	⊕
		Concentricity or Coaxiality	◎
		Symmetry	⌖
	Orientation	Parallelism	//
		Perpendicularity	⊥
		Angularity	∠
Individual	Form	Cylindricity	⌭
		Flatness	⬦
		Circularity or Roundness	○
Individual or related	Profile	Straightness	—
		Surface Profile	⌓
		Line Profile	⌒
Related	Runout	Circular Runout	↗
		Total Runout	↗↗

Figure 8-68 *Characteristics and symbols used in Geometric Tolerancing*

Note
Symbols used in geometric dimensioning and tolerancing are the building blocks of geometric dimensioning and tolerancing.

ADDING GEOMETRIC TOLERANCE

Ribbon: Annotate > Dimensions > Tolerance	**Toolbar:** Dimension > Tolerance
Menu Bar: Dimension > Tolerance	**Command:** TOLERANCE

Geometric tolerance displays the deviations of profile, orientation, form, location, and runout of a feature. In AutoCAD, geometrical tolerancing is displayed by feature control frames. The frames contain all information about tolerances for a single dimension. To display feature control frames with various tolerancing parameters, you need to enter specifications in the **Geometric Tolerance** dialog box (Figure 8-69). You can invoke the **Geometric Tolerance** dialog box by choosing the **Tolerance** tool from the **Dimensions** panel.

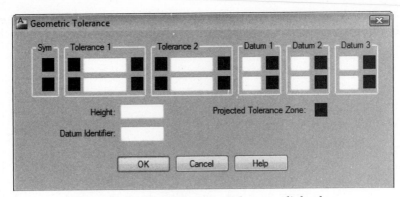

*Figure 8-69 The **Geometric Tolerance** dialog box*

Various components that constitute geometric tolerancing (GTOL) are shown in Figures 8-70 and 8-71.

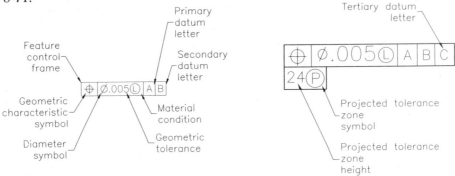

Figure 8-70 Components of GTOL

Figure 8-71 Components of GTOL

Feature Control Frame

The **feature control frame** is a rectangular box that contains the geometric characteristics symbols and tolerance definition. The box is automatically drawn to standard specifications; you do not need to specify its size. You can copy, move, erase, rotate, and scale the feature control frame. You can also snap to them using various Object snap modes. You can edit feature control frames using the **DDEDIT** command or you can also edit them using **GRIPS**. The system variable **DIMCLRD** controls the color of the feature control frame. The system variable **DIMGAP** controls the gap between the feature control frame and the text.

Geometric Characteristics Symbol

The geometric characteristics symbols indicate the characteristics such as straightness, flatness, perpendicularity, and so on of a feature. You can select the required symbol from the **Symbol** dialog box (Figure 8-72). This dialog box is displayed by selecting the box provided in the **Sym** area of the **Geometric Tolerance** dialog box. To select the required symbol, just pick the symbol using the left mouse button. The symbol will now be displayed in the box under the **Sym** area.

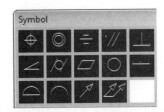

*Figure 8-72 The **Symbol** dialog box*

Tolerance Value and Tolerance Zone Descriptor

The tolerance value specifies the tolerance on the feature as indicated by the tolerance zone descriptor. For example, a value of .003 indicates that the feature must be within a 0.003 tolerance zone. Similarly, .003 indicates that this feature must be located at a true position within a 0.003 diameter. The tolerance value can be entered in the edit box provided under the **Tolerance 1** or the **Tolerance 2** area of the **Geometric Tolerance** dialog box. The tolerance zone descriptor can be invoked by selecting the box located to the left of the edit box. The system variable **DIMCLRT** controls the color of the tolerance text, variable **DIMTXT** controls the tolerance text size, and variable **DIMTXSTY** controls the style of the tolerance text. On using the **Projected Tolerance Zone**, the projected tolerance zone symbol, which is an encircled P will be inserted after the projected tolerance zone value.

Material Condition Modifier

The material condition modifier specifies the material condition when the tolerance value takes effect. For example, .003(M) indicates that this feature must be located at a true position within a 0.003 diameter at maximum material condition (MMC). The material condition modifier symbol can be selected from the **Material Condition** dialog box (Figure 8-73). This dialog box can be invoked by selecting the boxes located on the right side of the edit boxes under the **Tolerance 1**, **Tolerance 2**, **Datum 1**, **Datum 2**, and **Datum 3** areas of the **Geometric Tolerance** dialog box.

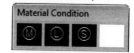

*Figure 8-73 The **Material Condition** dialog box*

Datum

The datum is the origin, surface, or feature from which the measurements are made. The datum is also used to establish the geometric characteristics of a feature. The datum feature symbol consists of a reference character enclosed in a feature control frame. You can create the datum feature symbol by entering characters (like -A-) in the **Datum Identifier** edit box in the **Geometric Tolerance** dialog box and then selecting a point where you want to establish this datum.

You can also combine datum references with geometric characteristics. AutoCAD automatically positions the datum references on the right end of the feature control frame.

COMPLEX FEATURE CONTROL FRAMES
Combining Geometric Characteristics

Sometimes, it is not possible to specify all geometric characteristics in one frame. For example, Figure 8-74 shows the drawing of a plate with a hole in the center.

In this part, it is determined that surface C must be perpendicular to surfaces A and B within 0.002 and 0.004, respectively. Therefore, we need two frames to specify the geometric characteristics of surface C. The first frame specifies the allowable deviation in perpendicularity of surface C with respect to surface A. The second frame specifies the allowable deviation in perpendicularity of surface C with respect to surface B. In addition to these two frames, we need a third frame that identifies datum surface C.

All the three feature control frames can be defined in one instance of the **TOLERANCE** command.

1. Choose the **Tolerance** tool to invoke the **Geometric Tolerance** dialog box. Select the box under the **Sym** area to display the **Symbol** dialog box. Select the **perpendicular** symbol. AutoCAD will display the selected symbol in the first row of the **Sym** area.

2. Enter **.002** in the edit box under the **Tolerance 1** area of the first row and enter **A** in the first row edit box under the **Datum 1** area.

3. Select the second row box under the **Sym** area to display the **Symbol** dialog box. Select the **perpendicular** symbol. AutoCAD will display the selected symbol in the second row box of the **Sym** area.

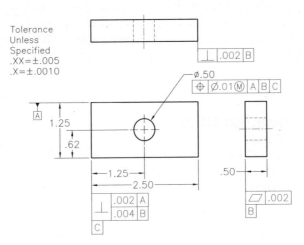

Figure 8-74 *Combining feature control frames*

4.   Enter **.004** in the second row edit box under the **Tolerance 1** area and enter **B** in the second row edit box under the **Datum 1** area.

5.   In the **Datum Identifier** edit box, enter **C** and then choose the **OK** button to exit the dialog box.

6.   In the graphics screen, select the position to place the frame.

7.   Similarly, create the remaining feature control frames.

Composite Position Tolerancing

Sometimes the accuracy required within a pattern is more important than the location of the pattern with respect to the datum surfaces. To specify such a condition, composite position tolerancing may be used. For example, Figure 8-75 shows four holes (pattern) of diameter 0.15. The design allows a maximum tolerance of 0.025 with respect to datum A, B, and C at the maximum material condition (holes are smallest). The designer wants to maintain a closer positional tolerance (0.010 at MMC) between the holes within the pattern. To specify this requirement, the designer must insert the second frame. This is generally known as composite position tolerancing. AutoCAD provides the facility to create two composite position tolerance frames by means of the **Geometric Tolerance** dialog box. The composite tolerance frames can be created as follows:

1.   Invoke the **Tolerance** tool to display the **Geometric Tolerance** dialog box. Select the box under the **Sym** area to display the **Symbol** dialog box. Select the **position** symbol. AutoCAD will display the selected symbol in the first row of the **Sym** area.

2.   In the first row of the **Geometric Tolerance** dialog box, enter the geometric characteristics and the datum references required for the first position tolerance frame.

3.   Next, select the box under the **Sym** area from the second row to display the **Symbol** dialog box. Select the position symbol; AutoCAD will display the selected symbol in the second row of the **Sym** area.

4.   In the second row of the **Geometric Tolerance** dialog box, enter the geometric characteristics and the datum references required for the second position tolerance frame.

5. When you have finished entering the values, choose the **OK** button in the **Geometric Tolerance** dialog box, and then select the point where you want to insert the frames. AutoCAD will create two frames and automatically align them with the common position symbol, as shown in Figure 8-75.

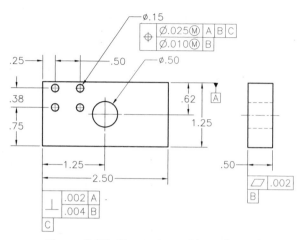

Figure 8-75 *Composite position tolerancing*

USING FEATURE CONTROL FRAMES WITH LEADERS

The **Leader Settings** dialog box that is invoked by using the **QLEADER** command has the **Tolerance** option, which allows you to create feature control frame and attach it to the end of the leader extension line, see Figure 8-76. The following is the prompt sequence for using the **QLEADER** command with the **Tolerance** option:

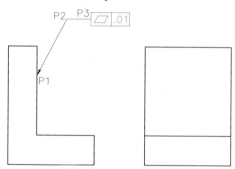

Figure 8-76 *Using the feature control frame with leader*

Specify first leader point, or [Settings] <Settings>: *Press ENTER to display the* **Leader Settings** *dialog box. Choose the* **Annotation** *tab. Now, select the* **Tolerance** *radio button from the* **Annotation Type** *area. Choose* **OK**.
Specify first leader point, or [Settings] <Settings>: *Specify the start point of the leader line.*
Specify next point: *Specify the second point.*
Specify next point: *Specify the third point to display the* **Geometric Tolerance** *dialog box.*

Now, specify the tolerance value and choose the **OK** button; the **Feature Control Frame** will be displayed with leader.

PROJECTED TOLERANCE ZONE

Figure 8-77 shows two parts joined with a bolt. The lower part is threaded, and the top part has a drilled hole. When these two parts are joined, the bolt that is threaded in the lower part will have the orientation error that exists in the threaded hole. In other words, the error in the threaded hole will extend beyond the part thickness, which might cause interference, and the parts may not assemble. To avoid this problem, projected tolerance is used. The projected tolerance establishes a tolerance zone that extends above the surface. In Figure 8-77, the position tolerance for the threaded hole is 0.010, which extends 0.1 above the surface (datum A). By using the projected tolerance, you can ensure that the bolt is within the tolerance zone up to the specified distance.

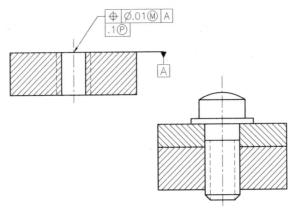

Figure 8-77 *Projected tolerance zone*

You can use the AutoCAD GDT feature to create feature control frames for the projected tolerance zone as discussed next.

1. Invoke the **QLEADER** command and then press ENTER at the **Specify first leader point, or [Settings] <Settings>** prompt to display the **Leader Settings** dialog box.

2. Choose the **Annotation** tab and then select the **Tolerance** radio button from the **Annotation Type** area. Choose **OK** to return to the command line. Specify the first, the second, and the third leader points, refer to Figure 8-77. On specifying the third point, the **Geometric Tolerance** dialog box will be displayed. Click once in the **Sym** area in the first row; the **Symbol** dialog box will be displayed. Choose the position symbol from the **Symbol** dialog box.

3. In the first row of the **Geometric Tolerance** dialog box, enter the geometric characteristics and the datum references required for the first position tolerance frame.

4. In the **Height** edit box, enter the height of the tolerance zone (.1 for the given drawing) and select the box on the right of **Projected Tolerance Zone**. The projected tolerance zone symbol will be displayed in the box.

5. Choose the **OK** button in the **Geometric Tolerance** dialog box. AutoCAD will create two frames and automatically align them, refer to Figure 8-77.

EXAMPLE 2   *Tolerance*

In this example, you will create a feature control frame to define the perpendicularity specification, see Figure 8-78.

1. Choose the **Tolerance** tool from the **Dimensions** panel of the **Annotate** tab to display the **Geometric Tolerance** dialog box. Choose the upper box from the **Sym** area to display the **Symbol** dialog box. Select the **perpendicularity** symbol. This symbol will be displayed in the **Sym** area.

2. Select the box on the left of the upper edit box under the **Tolerance 1** area; the diameter symbol will appear to denote a cylindrical tolerance zone.

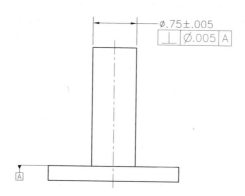

Figure 8-78 *Drawing for Example 2*

3. Enter **.005** in the upper edit box under the **Tolerance 1** area.

4. Enter **A** in the edit box under the **Datum 1** area. Choose the **OK** button to accept the changes made in the **Geometric Tolerance** dialog box.

5. The **Enter tolerance location** prompt is displayed in the Command line area and the **Feature Control Frame** is attached to the cursor at its middle left point. Select a point to insert the frame.

6. To attach the datum symbol, invoke the **Leader Settings** dialog box and select the **Tolerance** radio button as explained in step 1.

7. Choose the **Leader Line & Arrowhead** tab and select the **Datum triangle filled** option from the drop-down list in the **Arrowhead** area.

8. Set the number of points in the leader to **2** in the **Number of Points** spinner.

9. Choose **OK** to return to the command line. Specify the first and second leader points, refer to Figure 8-78. On specifying the second point, the **Geometric Tolerance** dialog box will be displayed.

10. Enter **A** in the **Datum Identifier** edit box. Choose **OK** to exit the **Geometric Tolerance** dialog box. On doing so, the datum symbol will be displayed, refer to Figure 8-78.

EXAMPLE 3   *Tolerance*

In this example, you will create a leader with combination feature control frames to control runout and cylindricity, see Figure 8-79.

1. Enter the **QLEADER** command and follow the prompt sequence given below.

Specify first leader point, or [Settings]
<Settings>: *Press ENTER to display the* ***Leader Settings*** *dialog box. Choose the* ***Annotation*** *tab. Select the* ***Tolerance*** *radio button from the* ***Annotation Type*** *area. Choose* ***OK***.
Specify first leader point, or [Settings]
<Settings>: *Specify the leader start point, as shown in Figure 8-79.*

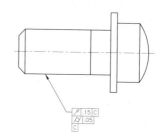

Figure 8-79 *Drawing for Example 3*

Specify next point: *Specify the second point of the leader.*
Specify next point: *Specify the third point of the leader line to display the **Tolerance** dialog box.*

2. Choose the runout symbol from the **Symbol** dialog box; the **runout** symbol will be displayed on the first row of the **Sym** area. Enter **.15** in the first row edit box under the **Tolerance 1** area.

3. Enter **C** in the edit box under the **Datum 1** area.

4. Select the edit box on the second row of the **Sym** area and select the **cylindricity** symbol; the **cylindricity** symbol will be displayed in the second row of the **Sym** area.

5. Enter **.05** in the second row edit box of the **Tolerance 1** area.

6. Enter **C** in the **Datum Identifier** edit box.

7. Choose the **OK** button to accept the changes in the **Geometric Tolerance** dialog box; the control frames will be automatically attached at the end of the leader.

CREATING ANNOTATIVE DIMENSIONS, TOLERANCES, LEADERS, AND MULTILEADERS

Annotative dimensions are drawn in the drawing by assigning annotative dimension style to the current drawing. To assign annotative dimension, select the **Dimension Style** (inclined arrow) from the **Dimensions** panel or from the **Dimension** toolbar. Note that the dimension style selected should have an annotative symbol next to it. You can also change the non-annotative dimensions to annotative by changing the dimension's **Annotative** property to **Yes** in the **Properties** palette.

For drawing the annotative geometric tolerances and leaders, you first need to add the simple tolerances to the drawing and then select all tolerances. Next, you need to modify the **Annotative** property under the **Misc** area of the **Properties** palette to **Yes**.

The annotative leaders are drawn by assigning annotative dimension style, whereas the annotative multileaders are drawn by assigning annotative multileader style. The leaders are created with two components: the leader and its content. Therefore, drawing annotative leaders does not ensure that the content attached to the leader will also be annotative. But, the multileaders are drawn as a single component, so the multileaders drawn with annotative style will have annotative content also.

Note
Creating and modifying dimension styles and multileader styles will be discussed in detail in Chapter 10.

Self-Evaluation Test

Answer the following questions and then compare them to those given at the end of this chapter:

1. You can specify dimension text or accept the measured value computed by AutoCAD. (T/F)

2. The **Adjust Space** tool is used to maintain equal spacing between linear dimensions or angular dimensions. (T/F)

3. The leaders drawn by using the **QLEADER** command create arrow, leader lines, and text as a single object. (T/F)

4. You cannot combine GTOL with leaders. (T/F)

5. The _____ tool is used to dimension only the *X* coordinate of a selected object.

6. The _____ symbols are the building blocks of geometric dimensioning and tolerances.

7. In the _____ dimensioning, the dimension text is aligned by default with the object being dimensioned.

8. The _____ dimensions are automatically updated when the object to which they are assigned is modified.

9. The _____ point is taken as the vertex point of the angular dimensions while dimensioning an arc or a circle.

10. Extra leaders can be added or removed from existing multileaders by using the _____ tool.

Review Questions

Answer the following questions:

1. Only inner angles (acute angles) can be dimensioned with angular dimensioning. (T/F)

2. In addition to the most recently drawn dimension (the default base dimension), you can use any other linear dimension as the base dimension. (T/F)

3. In the continued dimensions, the base point for the successive continued dimensions is the base dimension's first extension line. (T/F)

4. You cannot add dimension break to straight multileaders. (T/F)

5. Using the **Multileader** tool, you can specify the content of a multileader before drawing the leader itself. (T/F)

6. Which of the following tools can be used to convert dimensions into true associative dimensions?

   (a) **Reassociate**   (b) **Associate**
   (c) **Disassociate**   (d) None of the above

7. Which of the following tools can be used to dimension more than one object in a single effort?

   (a) **Linear**   (b) **Angluar**
   (c) **Quick Dimension**   (d) None of the above

8. Which of the following tools is used to dimension different points and features of a part with reference to a fixed point?

   (a) **Baseline**   (b) **Angular**
   (c) **Radius**   (d) **Aligned**

9. Which of the following tools can be used to add geometric dimensions and tolerance to the current drawing?

   (a) **GTOL**   (b) **Tolerance**
   (c) **Single Line**   (d) None of the above

10. Which of the following options of the **Align** tool can be used to maintain a constant spacing between the selected multileader landing lines?

   (a) **Distribute**   (b) **make leader segments Parallel**
   (c) **Specify Spacing**   (d) **Use current spacing**

11. Geometric dimensioning and tolerancing is generally known as _____.

12. Give three examples of geometric characteristics that indicate the characteristics of a feature.

   _____.

13. The three ways to return to the Command prompt from the **Dim:** prompt (dimensioning mode) are _____, _____, and _____.

14. The six fundamental dimensioning types which are provided by AutoCAD to dimension an object are_____.

15. Horizontal dimensions measure displacement along the _____.

EXERCISE 7

Draw the object shown in Figure 8-80 and then dimension it. Save the drawing as **DIMEXR7**.

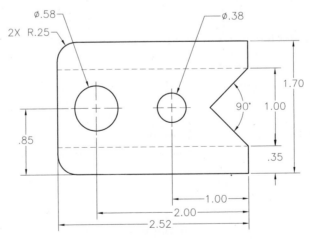

Figure 8-80 *Drawing for Exercise 7*

EXERCISE 8

Draw and dimension the object shown in Figure 8-81. Save the drawing as *DIMEXR8*.

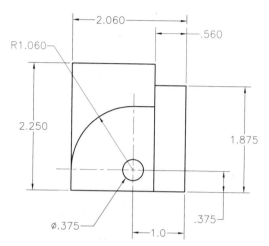

Figure 8-81 *Drawing for Exercise 8*

EXERCISE 9

Draw the object shown in Figure 8-82 and then dimension it. Save the drawing as *DIMEXR9*.

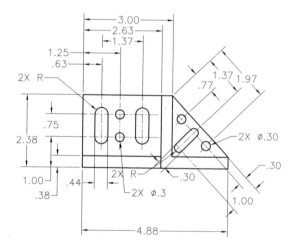

Figure 8-82 *Drawing for Exercise 9*

EXERCISE 10

Draw the object shown in Figure 8-83 and then dimension it. Save the drawing as *DIMEXR10*.

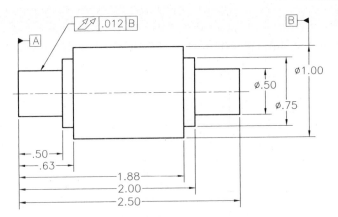

Figure 8-83 *Drawing for Exercise 10*

EXERCISE 11

Draw the object shown in Figure 8-84 and then dimension it. Save the drawing as *DIMEXR11*.

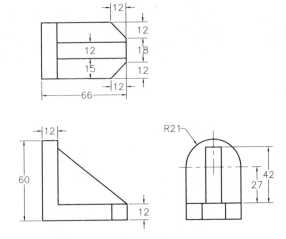

Figure 8-84 *Drawing for Exercise 11*

EXERCISE 12

Draw the object shown in Figure 8-85 and then dimension it. Save the drawing as *DIMEXR12*.

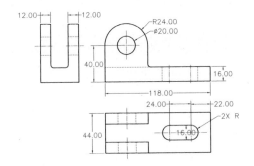

Figure 8-85 *Drawing for Exercise 12*

EXERCISE 13

Draw the object shown in Figure 8-86 and then dimension it. Save the drawing as *DIMEXR13*.

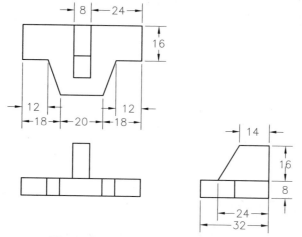

Figure 8-86 Drawing for Exercise 13

Problem-Solving Exercise 1

Draw the object shown in Figure 8-87 and dimension it, as shown in the drawing. Save the drawing as *DIMPSE1*.

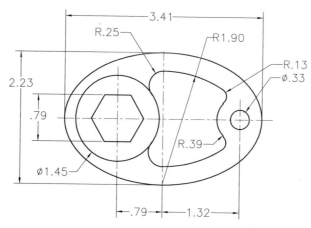

Figure 8-87 Drawing for Problem-Solving Exercise 1

Problem-Solving Exercise 2

Draw Figure 8-88 and then dimension it, as shown in the drawing. Save the drawing as *DIMPSE2*.

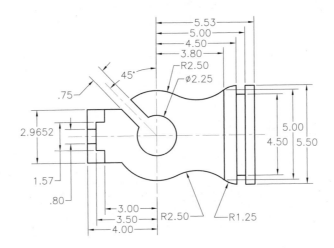

Figure 8-88 Drawing for Problem-Solving Exercise 2

Problem-Solving Exercise 3

Draw Figure 8-89 and then dimension it, as shown in the drawing. Save the drawing as *DIMPSE3*.

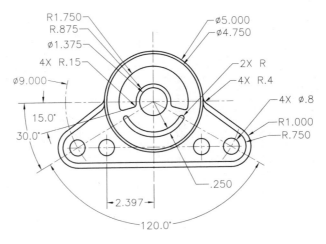

Figure 8-89 Drawing for Problem-Solving Exercise 3

Answers to Self-evaluation Test

1. T, **2.** T, **3.** F, **4.** F, **5. Ordinate**, **6.** geometric characteristics, **7.** aligned, **8.** true associative dimensions, **9.** center, **10. Remove Leader**

Chapter 9

Editing Dimensions

CHAPTER OBJECTIVES

In this chapter, you will learn to:
- *Edit dimensions.*
- *Stretch, extend, and trim dimensions.*
- *Use the Oblique and Text Angle command options to edit dimensions.*
- *Update dimensions using the Update command.*
- *Use the Properties palette to edit dimensions.*
- *Dimension in model space and paper space.*

KEY TERMS

- *Dimension editing tools*
- *Stretch*
- *Trim*

- *Extend*
- *Edit Dimension Text*
- *Update*

- *Dimension Properties Palette*
- *Multileader Properties Palette*

- *Model Space Dimension*
- *Paper Space Dimension*

EDITING DIMENSIONS USING EDITING TOOLS

For editing dimensions, AutoCAD has provided some special editing commands that work with dimensions. These editing commands can be used to define a new dimension text, return to the home text, create oblique dimensions, and rotate and update the dimension text. You can also use the Trim, Stretch, and Extend tools to edit the dimensions. In case the dimension assigned to the object is a true associative dimension, it will be automatically updated if the object is modified. However, if the dimension is not true associative dimension, you will have to include the dimension along with the object in the edit selection set. The properties of the dimensioned objects can also be changed using the **Properties** palette or the **Dimension Style Manager**.

Editing Dimensions by Stretching

You can edit a dimension by stretching it. However, to stretch a dimension, appropriate definition points must be included in the selection crossing or window. As the middle point of the dimension text is a definition point for all types of dimensions, you can easily stretch and move the dimension text to any location you want. When you stretch the dimension text, the gap in the dimension line gets filled automatically. While editing, the definition points of the dimension being edited must be included in the selection crossing box. The dimension is automatically calculated when you stretch the dimension.

Note

The dimension type remains the same after stretching. For example, the vertical dimension maintains itself as a vertical dimension and measures only the vertical distance even after the line it dimensions is modified and converted into an inclined line. The following example illustrates the stretching of object lines and dimensions.

EXAMPLE 1	*Edit by Stretching*

In this example, you will stretch the objects and dimensions shown in Figure 9-1 using grips. The new location of the lines and dimensions is at a distance of 0.5 unit in the positive *Y* axis direction. See Figure 9-2.

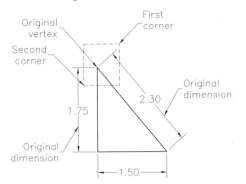

Figure 9-1 *Original location of lines and dimensions*

Figure 9-2 *New location of lines and dimensions*

1. Choose the **Stretch** tool from the **Modify** panel of the **Home** tab. The prompt sequence that will follow is given below:

    Select objects to stretch by crossing-window or crossing-polygon
    Select objects: Specify opposite corner: *Define a crossing window using the first and second corners, as shown in Figure 9-1.*
    Select objects: Enter
    Specify base point or [Displacement]<Displacement>: *Select original vertex using the osnaps as the base point.*

Specify second point of displacement or <use first point as displacement>: **@0.5<90**

2. The selected entities will be stretched to the new location. The dimension that was initially 1.75 will become 2.25 and the dimension that was initially 2.30 will become 2.70, see Figure 9-2. Press ESC to remove the grip points from the objects.

EXERCISE 1

The two dimensions given in Figure 9-3(a) are too close. Fix the drawing by stretching the dimension, as shown in Figure 9-3(b).

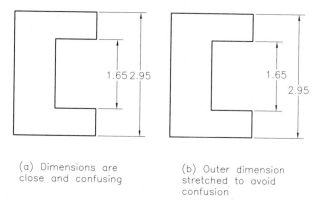

(a) Dimensions are close and confusing

(b) Outer dimension stretched to avoid confusion

Figure 9-3 *Drawing for Exercise 1, stretching dimensions*

1. Stretch the outer dimension to the right so that there is some distance between the two dimensions.
2. Stretch the dimension text of the outer dimension so that the dimension text is staggered (lower than the first dimension).

Editing Dimensions by Trimming and Extending

Trimming and extending operations can be carried out with all types of linear dimensions (horizontal, vertical, aligned and, rotated) and the ordinate dimension. Even if the dimensions are true associative, you can trim and extend them (see Figures 9-4 and 9-5). AutoCAD trims or

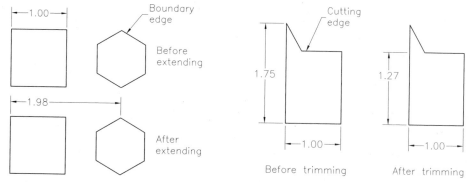

Figure 9-4 *Dimensions edited by extending*   ***Figure 9-5*** *Edgemode extended trimming*

extends a linear dimension between the extension line definition points and the object used as a boundary or trimming edge. To extend or trim an ordinate dimension, AutoCAD moves the feature location (location of the dimensioned coordinate) to the boundary edge. To retain the original ordinate value, the boundary edge to which the feature location point is moved should be orthogonal to the measured ordinate. In both cases, the imaginary line drawn between the two extension line definition points is trimmed or extended by AutoCAD, and the dimension is adjusted automatically.

EXERCISE 2

Use the **Edge > Extend** option of the **TRIM** command to trim the dimension given in Figure 9-6(a), so that it looks like Figure 9-6(b).

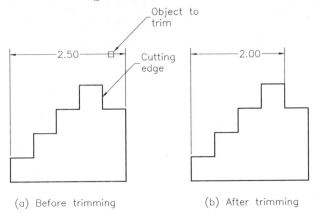

(a) Before trimming   (b) After trimming

Figure 9-6 *Drawing for Exercise 2*

1. Make the drawing and dimension it, as shown in Figure 9-6(a). Assume the missing dimensions.
2. Trim the dimensions by using the **Edge > Extend** option of the **TRIM** command.

Flipping Dimension Arrow

You can flip the arrowheads individually. To flip the arrow, select the dimension. Place the cursor on the grip corresponding to the arrowhead that you want to flip. When the color of the grip turns red, invoke the shortcut menu by right-clicking and choose the **Flip Arrow** option from the shortcut menu.

MODIFYING THE DIMENSIONS

Toolbar: Dimension > Dimension Edit   **Command:** DIMEDIT

The dimensions can be modified by choosing the **Dimension Edit** tool from the **Dimension** toolbar (Figure 9-7). Alternatively, you can use the **DIMEDIT** command to modify the dimensions. This command has four options: **New**, **Rotate**, **Home**, and **Oblique**. The prompt sequence that will follow when you choose this tool is given below:

Enter type of dimension editing (Home/New/Rotate/Oblique) <Home>: *Enter an option.*

Figure 9-7 *Invoking the **DIMEDIT** command from the **Dimension** toolbar*

New

The **New** option is used to replace the existing dimension with a new text string. When you invoke this option, the **Text Editor** will be displayed. By default, 0.0000 will be displayed. Using the **Text Editor**, enter the dimension or write the text string with which you want to replace the existing dimension. Once you have entered a new dimension in the editor and exit the **Text**

Editor, you will be prompted to select the dimension to be replaced. Select the dimension and press ENTER; it will be replaced with the new dimension.

Rotate

The **Rotate** option is used to position the dimension text at the specified angle. With this option, you can change the orientation (angle) of the dimension text of any number of associative dimensions. The angle can be specified by entering a value at the **Specify angle for dimension text** prompt or by specifying two points at the required angle. Once you have specified the angle, you will be prompted to select the dimension text to be rotated. Select the dimension and press ENTER; the text will rotate about its middle point, see Figure 9-8. You can also invoke this option by choosing the **Text Angle** tool from the **Dimensions** panel.

Home

The **Home** option restores the text of a dimension to its original (home/default) location if the position of the text has been changed by stretching or editing, see Figure 9-8.

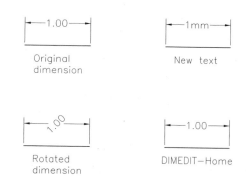

Oblique

In linear dimensions, extension lines are drawn perpendicular to the dimension line. The **Oblique** option bends the linear dimensions. It draws extension lines at an oblique angle (Figure 9-9). This option is particularly important to create isometric dimensions and can be used to resolve conflicting situations due to the overlapping of extension lines with

*Figure 9-8 Using the **DIMEDIT** command to edit dimensions*

other objects. Making an existing dimension oblique by specifying an angle oblique to it does not affect the generation of new linear dimensions. The oblique angle is maintained even after performing most editing operations. (See Chapter 13 for details about how to use this option) When you invoke this option, you will be prompted to select the dimension to be edited. After selecting it, you will be prompted to specify the obliquing angle. The extensions lines will be bent at the angle specified. You can also invoke this option by choosing the **Oblique** tool from the **Dimensions** panel.

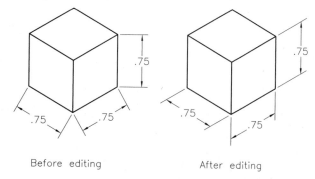

*Figure 9-9 Using the **Oblique** option to edit dimensions*

EDITING THE DIMENSION TEXT

Toolbar: Dimension > Dimension Text Edit **Command:** DIMTEDIT
Menu Bar: Dimension > Align Text

 The dimension text can be edited by using the **Dimension Text Edit** tool from the **Dimension** toolbar. This command is used to edit the placement and orientation of a single existing dimension. You can use this tool, in cases where dimension texts of two or more dimensions are too close together. In such cases, the **Dimension Text Edit** tool is invoked to move the dimension text to some other location so that there is no confusion. The prompt sequence that will follow when you choose the **Dimension Text Edit** tool is given next.

Select dimension: *Select the dimension to modify.*
Specify new location for dimension text or [Left/Right/Center/Home/Angle]:

Left

With this option, you can left-justify the dimension text along the dimension line. The vertical placement setting determines the position of the dimension text. In this setting, the horizontally aligned text is moved to the left and the vertically aligned text is moved down, see Figure 9-10. This option can be used only with the linear, diameter, or radial dimensions. You can also invoke this option by choosing **Left** tool from the **Dimensions** panel.

Right

With this option, you can right-justify the dimension text along the dimension line. Similar to the Left option, the vertical placement setting determines the position of the dimension text. The horizontally aligned text is moved to the right, and the vertically aligned text is moved up, see Figure 9-10. This option can be used only with linear diameter and radius dimensions. You can also invoke this option by choosing **Right** tool from the **Dimensions** panel.

Center

With this option, you can center-justify the dimension text for linear and aligned dimensions, see Figure 9-10. The vertical setting controls the vertical position of the dimension text. You can also invoke this option by choosing **Center** tool from the **Dimensions** panel.

Home

The **Home** option is used to restore (move) the dimension text of a dimension to its original (home/default) location, if the position of the text has changed, see Figure 9-10.

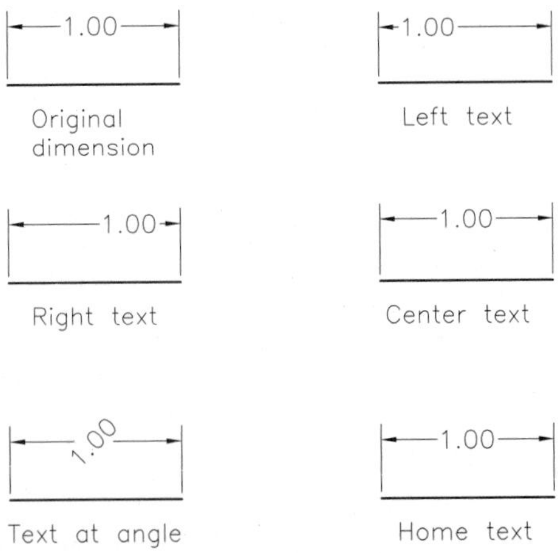

Angle

With the **Angle** option, you can position the dimension text at the angle you specify, see Figure 9-10. The angle can be specified by entering its value at the **Specify angle for dimension text** prompt or by specifying two points at the required angle. You will notice that the text rotates around its middle point. If the dimension text alignment is set to Orient Text Horizontally, the dimension text is aligned with the dimension line. If information about the

Figure 9-10 Using the **DIMTEDIT**
command to edit dimensions

dimension style is available on the selected dimension, AutoCAD uses it to redraw the dimension, or the prevailing dimension variable settings are used for the redrawing process. Entering 0-degree angle changes the text to its default orientation.

UPDATING DIMENSIONS

Ribbon: Annotate > Dimensions > Update	**Toolbar:** Dimension > Dimension Update
Menu Bar: Dimension > Update	**Command:** -DIMSTYLE

The **Update** tool is used to regenerates and updates the prevailing dimension entities (such as arrows heads and text height) using the current settings for the dimension variables, dimension style, text style, and units. On choosing this tool, you will be prompted to select the dimensions to be updated. You can select all the dimensions or specify those that should be updated.

EDITING DIMENSIONS WITH GRIPS

You can also edit dimensions by using the GRIP editing modes. GRIP editing is the easiest and quickest way to edit dimensions. You can perform the following operations with GRIPS.

1. Position the text anywhere along the dimension line. Note that you cannot move the text and position it above or below the dimension line.
2. Stretch a dimension to change the spacing between the dimension line and the object line.
3. Stretch the dimension along the length. When you stretch a dimension, the dimension text automatically changes.
4. Move, rotate, copy, mirror, or scale the dimensions.
5. Relocate the dimension origin.
6. Change properties such as color, layer, linetype, and linetype scale.
7. Load Web browser (if any *Universal Resource Locator* is associated with the object).

EDITING DIMENSIONS USING THE PROPERTIES PALETTE

Ribbon: View > Palettes > Properties	**Toolbar:** Standard > Properties
Menu Bar: Modify > Properties	**Command:** PROPERTIES

You can also modify a dimension or leader by using the **Properties** palette. The **Properties** palette is displayed when you choose the **Properties** button from the **Palettes** panel. Alternatively, you can invoke the **Properties** palette by choosing the **Properties** option from the **Palettes** cascade in the **Tools** menu. All the properties of the selected object are displayed in the **Properties** palette (see Figure 9-11). Select the dimension before invoking the **Properties** palette otherwise it would not give the description of the dimension.

Properties Palette (Dimension)

You can use the **Properties** palette (Figure 9-11) to change the properties, the dimension text style, or geometry, format, and annotation-related features of the selected dimension. The changes take place dynamically in the drawing. The **Properties** palette provides the following categories for the modification of dimensions.

General

In the general category, the various parameters displayed are **Color**, **Layer**, **Linetype**, **Linetype scale**, **Plot style**, **Lineweight**, **Hyperlink**, and **Associative** with their current values. To change the color of the selected object, select **Color** property and then select the required color from the drop-down list. Similarly, layer, plot style, linetype, and lineweight can be changed from the respective drop-down lists. The linetype scale can be changed manually in the corresponding cell.

Misc

This category displays the dimension style by name (for the **DIMSTYLE** system variable, use **SETVAR**) and also specifies whether the dimension text will be annotative or not. You can change the dimension style and the annotative property of the text from the respective drop-down lists of the selected dimension.

Lines & Arrows

The various parameters of the lines and arrows in the dimension such as arrowhead size, type, arrow lineweight, and so on can be changed in this category.

Text

The various parameters that control the text in the dimension object such as text color, text height, vertical position text offset, and so on can be changed in this category.

Fit

In the fit category, various parameters are **Dim line forced**, **Dim line inside**, **Dim scale overall**, **Fit**, **Text inside**, and **Text movement**. All parameters can be changed in the drop-down list, except **Dim scale overall** (which can be changed manually).

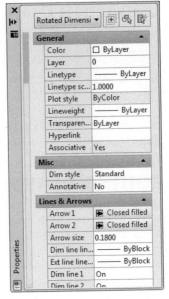

Figure 9-11 The **Properties** *palette for dimensions*

Primary Units

In the primary units category, the parameters displayed are **Decimal separator**, **Dim prefix**, **Dim suffix**, **Dim roundoff**, **Dim scale linear**, **Dim units**, **Suppress leading zeros**, **Suppress trailing zeros**, **Suppress zero feet**, **Suppress zero inches**, and **Precision**. Among these properties, **Dim units**, **Suppress leading zeros**, **Suppress trailing zeros**, **Suppress zero feet**, **Suppress zero inches**, and **Precision** properties can be changed with the help of corresponding drop-down lists. The other parameters can be changed manually.

Alternate Units

Alternate units are required when a drawing is to be read in two different units. For example, if an architectural drawing is to be read in both metric and feet-inches, you can turn the alternate units on. The primary units can be set to metric and the alternate units to architectural. As a result, the dimensions of the drawing will be displayed in metric units as well as in engineering. In the alternate unit category, there are various parameters for the alternate units. They can be changed only if the **Alt enabled** parameter is **on**. The parameters such as **Alt format**, **Alt precision**, **Alt suppress leading zeros**, **Alt suppress trailing zeros**, **Alt suppress zero feet**, and **Alt suppress zero inches** can be changed from the respective drop-down lists and others can be changed manually.

Tolerances

The parameters of this category can be changed only if the **Tolerances** display parameter has some mode of the tolerance selected. The parameters listed in this rollout depend on the mode of tolerance selected.

Properties Palette (Multileader)

The **Properties** palette for **Multileader** can be invoked by selecting a Leader and then choosing the **Properties** option from the **Palettes** panel. You can also invoke the **Properties** palette (Figure 9-12) from the shortcut menu by right-clicking in the drawing area and then choosing **Properties**. This palette can also be invoked by double-clicking in the leader to be edited. The various properties under the **Properties** palette (Multileader) are described next.

General

The parameters in the general category are the same as those discussed in the previous section (**Properties** palette for dimensions).

Misc

This category displays the **Overall scale, Multileader style, Annotative,** and **Annotative scale** of the **Multileader.** You can change the style, annotative property, and annotation scale of the Multileader by using the respective drop-down lists.

Leaders

This category controls the display of certain elements such as lines, arrowheads, landing distance, and so on. The **Multileader** is composed of these elements.

Text

This category displays the properties that are used to control the text appearance and location of the **Multileader** text. These properties can be changed by selecting the desired option from the respective drop-down lists. Now, you can modify the dimension text by double-clicking on the required dimension.

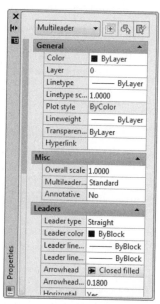

Figure 9-12 The Properties palette for modifying multileaders

---

EXAMPLE 2   *Modify Dimensions*

In this example, you will modify the dimensions given in Figure 9-13 so that they match the dimensions given in Figure 9-14. Assume the missing dimensions.

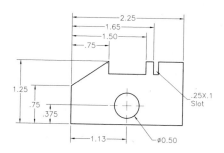

Figure 9-13 Drawing for Example 2

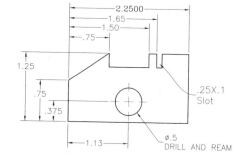

Figure 9-14 Drawing after editing the dimensions

1. Choose the **Text Style** (inclined arrow) tool from the **Text** panel of the **Annotate** tab and create a style with the name **ROMANC**. Select **romanc.shx** as the font for the style.

2. Select the dimension **2.25** and enter **PROP** in the command box to display the **Properties** palette.

3. In the **Text** category, select the **Text style** drop-down list and then select the **ROMANC** style from this drop-down list. The changes will take place dynamically.

4. Select **0.0000** from the **Precision** drop-down list in the **Primary Units** area.

5. Once all the required changes are made in the linear dimension, choose the **Select Object** button in the **Properties** palette; you will be prompted to select the object. Select the leader line and then press ENTER.

6. The **Properties** palette will display the leader options. Select **Spline** from the **Leader type** drop-down list in the **Leaders** category; the straight line will be dynamically converted into a spline with an arrow.

7. Close the **Properties** palette.

8. Choose the **Dimension Edit** tool from the **Dimension** toolbar. Enter **N** in the prompt sequence to display the **Text Editor**.

9. Enter **%%C.5 DRILL AND REAM** in the **Text Editor** and then click outside it to accept the changes. You will be prompted to select the object to be changed.

10. Select the diameter dimension and then press ENTER. The diameter dimension will be modified to the new value.

MODEL SPACE AND PAPER SPACE DIMENSIONING

Dimensioning objects can be drawn in the model space or paper space. If the drawings are in model space, associative dimensions should also be created in them. If the drawings are in the model space and the associative dimensions are in the paper space, the dimensions will not change when you perform such editing operations as stretching, trimming, and extending, or such display operations as zoom and pan in the model space viewport. The definition points of a dimension are located in the space where the drawing is drawn. You can select the **Scale dimensions to layout** radio button in the **Fit** tab (see Figure 9-15). This tab is available when you choose the **Modify, New,** or **Override** buttons from the **Dimension Style Manager** dialog box, depending on whether you want to modify the present style or you want to create a new style. Choose **OK** and then **Close** to exit the dialog boxes. AutoCAD calculates a scale factor that is compatible with the model space and the paper space viewports. Now, choose the **Update** tool from the **Dimension** toolbar and select the dimension objects for updating.

The drawing shown in Figure 9-16 uses paper space scaling. The main drawing and detail drawings are located in different floating viewports (paper space). The zoom scale factors for these viewports are different: 0.3XP, 1.0XP, and 0.5XP, respectively. When you use paper scaling,

AutoCAD automatically calculates the scale factor for dimensioning so that the dimensions are uniform in all the floating viewports (model space viewports).

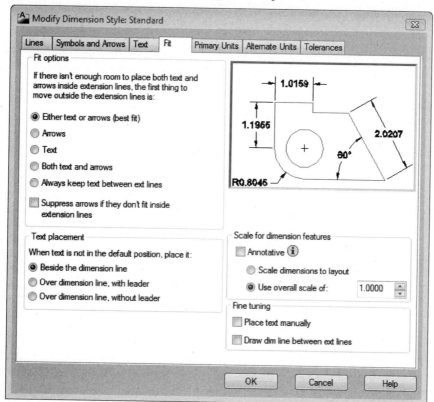

Figure 9-15 *Selecting paper space scaling in the **Modify Dimension Style** dialog box*

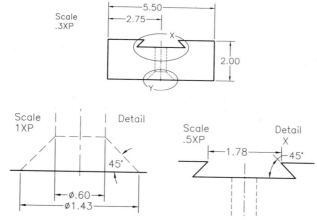

Figure 9-16 *Dimensioning in paper model space viewports using paper space scaling or setting **DIMSCALE** to 0*

Self-Evaluation Test

Answer the following questions and then compare them to those given at the end of this chapter:

1. In associative dimensioning, the items constituting a dimension (such as dimension lines, arrows, leaders, extension lines, and dimension text) are drawn as a single object. (T/F)

2. If the value of the variable **DIMDASSOC** is set to zero, the dimension lines, arrows, leaders, extension lines, and dimension text are drawn as independent objects. (T/F)

3. You cannot edit dimensions using grips. (T/F)

4. The true associative dimensions cannot be trimmed or extended. (T/F)

5. You can use the _____ command to break the dimensions into individual entities.

6. The _____ option of the **DIMTEDIT** command is used to justify the dimension text toward the left side.

7. The _____ option of the **DIMTEDIT** command is used to justify the dimension text to the center of the dimension.

8. The _____ option of the **DIMEDIT** command is used to create a new text string.

9. The _____ option of the **DIMEDIT** command is used to bend the extension lines through the specified angle.

10. The _____ tool from the **Dimension** panel is used to update the dimensions.

Review Questions

Answer the following questions:

1. The horizontal, vertical, aligned, and rotated dimensions cannot be edited using grips. (T/F)

2. Trimming and extending operations can be carried out with all types of linear (horizontal, vertical, aligned, and rotated) dimensions and with the ordinate dimension. (T/F)

3. To extend or trim an ordinate dimension, AutoCAD moves the feature location (location of the dimensioned coordinate) to the boundary edge. (T/F)

4. Once moved from the original location, the dimension text cannot be restored to its original position. (T/F)

5. With the _____ or _____ commands, you can edit the dimension text.

6. The _____ command is particularly important for creating isometric dimensions and is applicable in resolving conflicting situations due to overlapping of extension lines with other objects.

7. The _____ command is used to edit the placement and orientation of a single existing dimension.

8. The _____ command regenerates (updates) prevailing associative dimension objects (like arrows and text height) using the current settings for the dimension variables, dimension style, text style, and units.

9. Explain when to use the **Extend** tool and how it works with dimensions.

10. Explain the use and working of the **Properties** palette for editing dimensions.

EXERCISE 3   *Edit Dimensions*

1. Create the drawing shown in Figure 9-17. Assume the dimensions where necessary.
2. Dimension the drawing, as shown in Figure 9-17.
3. Edit the dimensions so that they match the dimensions shown in Figure 9-18.

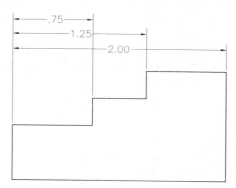
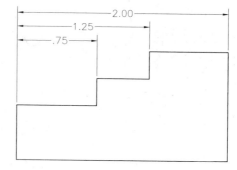

Figure 9-17 *Drawing for Exercise 3 before editing dimensions*

Figure 9-18 *Drawing for Exercise 3 after editing dimensions*

EXERCISE 4   *Edit Dimensions*

1. Draw the object shown in Figure 9-19(a). Assume the dimensions where necessary.
2. Dimension the drawing, as shown in Figure 9-19(a).
3. Edit the dimensions so that they match the dimensions shown in Figure 9-19(b).

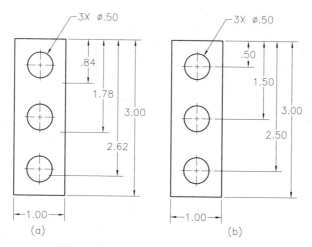

Figure 9-19 *Drawings for Exercise 4*

EXERCISE 5   *Edit Dimensions*

Create the drawing shown in Figure 9-20 and then dimension it. Assume the dimensions wherever necessary. After dimensioning the drawing, edit its dimensions so that they match the dimensions shown in Figure 9-20.

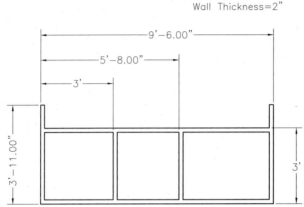

Figure 9-20 Drawing for Exercise 5

Problem-Solving Exercise 1

Create the drawing shown in Figure 9-21 and then dimension it. Edit the dimensions so that they are positioned as shown in the drawing. You can change the dimension text height and arrow size to 0.08 unit.

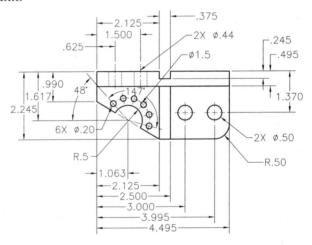

Figure 9-21 Drawing for Problem-Solving Exercise 1

Problem-Solving Exercise 2

Draw the front and side view of an object shown in Figure 9-22 and then dimension the two views. Edit the dimensions so that they are positioned as shown in the drawing. You may change the dimension text height and arrow size to 0.08 units.

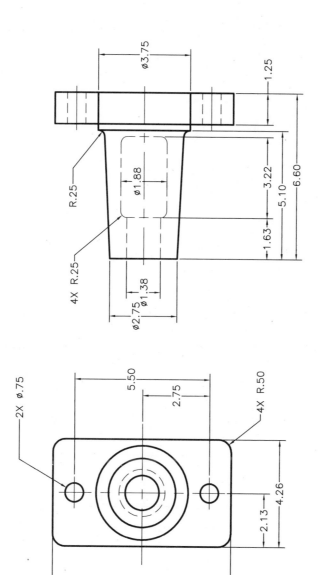

Figure 9-22 *Drawing for Problem-Solving Exercise 2*

Chapter 9

Problem-Solving Exercise 3

Create the drawing of the floor plan shown in Figure 9-23 and then apply dimensions to it. Edit the dimensions, if needed, so that they are positioned as shown in the drawing.

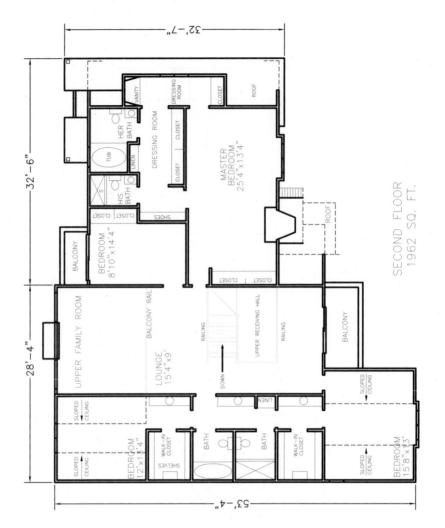

Figure 9-23 Drawing for Problem-Solving Exercise 3

Chapter 10

Dimension Styles, Multileader Styles, and System Variables

CHAPTER OBJECTIVES

In this chapter, you will learn:
- *To use styles and variables to control dimensions.*
- *To create dimension styles.*
- *To Set dimension variables using various tabs of the New, Modify, and Override Dimension Style.*
- *To use dimension style overrides.*
- *To compare and list dimension styles.*
- *To import externally referenced dimension styles.*
- *To create, restore, and modify multileader styles.*

KEY TERMS

- *DIMSTYLE*
- *DIMCEN*
- *Dimension Style Families*
- *Dimension Styles Overrides*
- *MLEADERSTYLE*

USING STYLES AND VARIABLES TO CONTROL DIMENSIONS

In AutoCAD, the appearance of dimensions on the drawing area and the manner in which they are saved in the drawing database are controlled by a set of dimension variables. The dimensioning commands use these variables as arguments. The variables that control the appearance of dimensions can be managed using dimension styles. You can use the **Dimension Style Manager** dialog box to control the dimension styles and dimension variables through a set of dialog boxes.

CREATING AND RESTORING DIMENSION STYLES

Ribbon: Annotate > Dimensions > Dimension, Dimension Style (The inclined arrow)
Toolbar: Dimension > Dimension Style or Styles > Dimension Style
Command: DIMSTYLE

The dimension styles control the appearance and positioning of dimensions and leaders in the drawing. If the default dimensioning style (Standard or Annotative) does not meet your requirements, you can select another existing dimensioning style or create a new one. The default names of the dimension style file are **Standard** and **Annotative**. Dimension styles can be created by using the **Dimension Style Manager** dialog box. Left-click on the inclined arrow in the **Dimensions** panel of the **Annotate** tab to invoke the **Dimension Style Manager** dialog box (Figure 10-1).

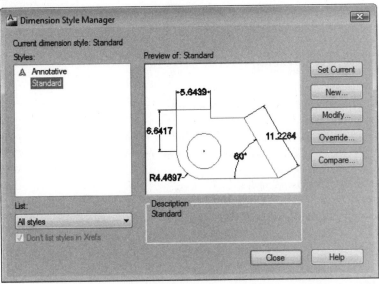

Figure 10-1 The **Dimension Style Manager** *dialog box*

From the **Dimension Style Manager** dialog box, choose the **New** button to display the **Create New Dimension Style** dialog box (Figure 10-2). Enter the dimension style name in the **New Style Name** text box and then select a style that you want to be basis of your style from the **Start With** drop-down list. Select the **Annotative** check box to make the new dimension style annotative. Choose the **i** (information) button to get the help and information about the annotative objects. The **Use for** drop-down list allows you to select the dimension type to which you want to apply the new dimension style. For example, if you wish to use the new style only for the diameter dimension, select **Diameter dimensions** from the **Use for** drop-down list. Choose the **Continue** button to display the **New Dimension Style** dialog box. The parameters of the **New Dimension Style** dialog box will be discussed later. After specifying the parameters of the new dimension style in the **New Dimension Style** dialog box, choose **OK**; the **Dimension Style Manager** dialog box will be displayed again. The options in this dialog box are discussed next.

In the **Dimension Style Manager** dialog box, the current dimension style name is shown in front of **Current dimension style** and is also shown highlighted in the **Styles** list box. A brief description of the current style (its differences from the default settings) is also displayed in the **Description** area. The **Dimension Style Manager** dialog box also has the **Preview of** window that displays the preview of the current dimension style.

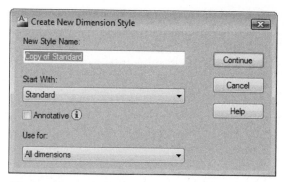

Figure 10-2 The **Create New Dimension Style** *dialog box*

A style can be made current (restored) by selecting the name of the dimension style you want to make current from the list of defined dimension styles and choosing the **Set Current** button. You can also make a style current by double-clicking on the style name in the **Styles** list box. The list of dimension styles displayed in the **Styles** list box is dependent on the option selected from the **List** drop down-list. If you select the **Styles in use** option, only the dimension styles in use will be listed in the **Style** list box. If you right-click on a style in the **Styles** list box, a shortcut menu is displayed that provides you with the options to **Set current**, **Rename**, or **Delete** a dimension style. Selecting the **Don't list styles in Xrefs** check box does not list the names of Xref styles in the **Styles** list box. Choosing the **Modify** button displays the **Modify Dimension Style** dialog box where you can modify an existing style. Choosing the **Override** button displays the **Override Current Style** dialog box where you can define overrides to an existing style (discussed later in this chapter). Both these dialog boxes along with the **New Dimension Style** dialog box have identical properties. Choosing the **Compare** button displays the **Compare Dimension Styles** dialog box (also discussed later in this chapter) that allows you to compare two existing dimension styles.

Note
The **Dimension Style** *drop-down list in the* **Dimensions** *panel under the* **Annotate** *tab also displays the dimension styles. Selecting a dimension style from this list also sets it current.*

NEW DIMENSION STYLE DIALOG BOX

The **New Dimension Style** dialog box can be used to specify the dimensioning attributes (variables) that affect the various properties of the dimensions. The various tabs provided under the **New Dimension Style** dialog box are discussed next.

Lines Tab

The options in the **Lines** tab (Figure 10-3) of the **New Dimension Style** dialog box are used to specify the dimensioning attributes (variables) that affect the format of the dimension lines. For example, the appearance and behavior of the dimension lines and extension lines can be changed with this tab. If the settings of the dimension variables have not been altered in the current editing session, the settings displayed in the dialog box are the default settings.

Dimension lines Area

This area provides you with the options of controlling the display of the dimension lines and leader lines. These options are discussed next.

Color. This drop-down list is used to set the colors for the dimension lines and arrowheads. Its dimension arrowheads have the same color as the dimension line because arrows constitute a part of the dimension line. The color you set here will also be assigned to the leader lines and arrows. The default color for the dimension lines and arrows is ByBlock. You can specify the color of the dimension line by selecting it from the **Color** drop-down list. You can also select **Select Color** from the **Color** drop-down list to display the **Select Color** dialog box where you can choose a specific color. The color number or the special color label is stored in the **DIMCLRD** variable, the default value is 0.

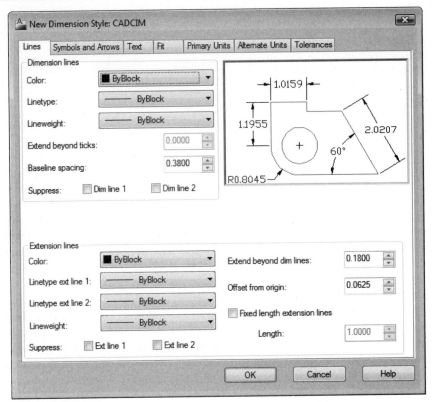

*Figure 10-3 The **Lines** tab of the **New Dimension Style** dialog box*

Linetype. This drop-down list is used to set the linetype for the dimension lines.

Lineweight. This drop-down list is used to specify the lineweight for the dimension line. You can select the required lineweight by selecting it from this drop-down list. This value is also stored in the **DIMLWD** variable. The default value is ByBlock. Remember that you cannot assign the lineweight to the arrowheads using this drop-down list.

Extend beyond ticks. The **Extend beyond ticks** spinner will be available only when you select the oblique, Architectural tick, or any such arrowhead type in the **First** and **Second** drop-down lists in the **Arrowheads** area from the **Symbols and Arrows** tab. This spinner is used to specify the distance by which the dimension line will extend beyond the extension line. The extension value, entered in the **Extend beyond ticks** edit box, gets stored in the **DIMDLE** variable. By default, this edit box is disabled because the oblique arrowhead type is not selected.

Baseline spacing. The **Baseline spacing** (baseline increment) spinner is used to control the spacing between successive dimension lines drawn using the baseline dimensioning, see Figure 10-4. You can specify the dimension line increment to your requirement by specifying the desired value using the **Baseline spacing** spinner. The default value displayed in the **Baseline spacing** spinner is 0.38 units. This spacing value is also stored in the **DIMDLI** variable.

Suppress. The **Suppress** check boxes control the display of the first and second dimension lines. By default, both dimension lines will be drawn. You can suppress one or both the dimension lines by selecting their corresponding check boxes. The values of these check boxes are stored in the **DIMSD1** and **DIMSD2** variables.

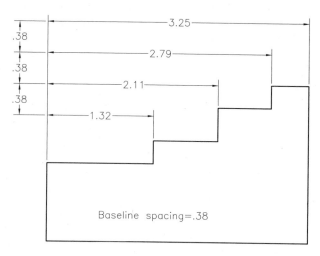

Figure 10-4 Baseline increment

Note
The first and second dimension lines are determined by how you select the extension line origins. If the first extension line origin is on the right, the first dimension line will also be on the right.

Extension lines Area

The options in this area are discussed next.

Color. This drop-down list is used to control the color of the extension lines. The default extension line color is ByBlock. You can assign a new color to the extension lines by selecting it from this drop-down list. The color number or the color label is saved in the **DIMCLRE** variable.

Linetype ext line 1. The options in this drop-down list are used to specify the linetype of extension line 1. By default, the line type for the extension line 1 is set to ByBlock. You can change the linetype by selecting a new value of linetype from the drop-down list.

Linetype ext line 2. The options in this drop-down list are same as for the Linetype ext line 1. The options are used to specify the line type of extension line 2.

Note
*You can specify different linetypes for the dimensions and the extensions lines. Use the **Linetype** drop-down list in the **Dimension Lines** area to specify the linetype for dimensions. Use the **Linetype ext line 1** and **Linetype ext line 2** drop-down lists to specify the linetype for the extension lines.*

Lineweight. This drop-down list is used to modify the lineweight of the extension lines. The default value is ByBlock. You can change the lineweight value by selecting a new value from this drop-down list. The value for lineweight is stored in the **DIMLWE** variable.

Extend beyond dim lines. It is the distance by which the extension lines extend past the dimension lines, see Figure 10-5. You can change the extension line offset using the **Extend beyond dim lines** spinner. This value is also stored in the **DIMEXE** variable. The default value for the extension distance is 0.1800 units.

Offset from origin. It is the distance by which the extension line is offset from the point you specify as the origin of the extension line, see Figure 10-6. You may need to override this setting for specific dimensions while dimensioning curves and angled lines. You can specify an offset distance of your choice using this spinner. AutoCAD stores this value in the **DIMEXO** variable. The default value for this distance is 0.0625.

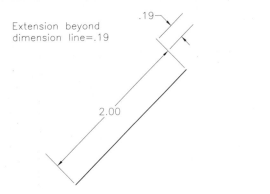

Figure 10-5 Extension beyond dimension lines

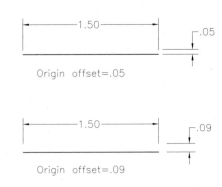

Figure 10-6 The offset from origin

Fixed length extension lines. With the selection of the **Fixed length extension lines** check box, you can specify a fixed length (starting from the dimension line to the origin) for the extension lines in the **Length** edit box. By default, the check box is cleared and the length spinner is not available. Once the check box is selected, you can specify the length of the extension line either by entering a numerical value in the **Length** edit box or using a spinner. Figure 10-7 displays a drawing with full length extension lines and Figure 10-8 displays the same drawing with fixed length extension lines.

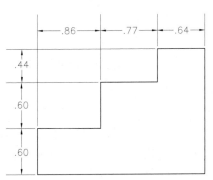

Figure 10-7 Drawings with full length extension lines

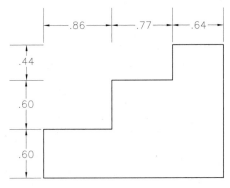

Figure 10-8 Drawing with fixed length extension lines

Suppress. The **Suppress** check boxes are used to control the display of the extension lines. By default, both extension lines will be drawn. You can suppress one or both of them by selecting the corresponding check boxes (Figure 10-9). The values of these check boxes are stored in the **DIMSE1** and **DIMSE2** variables.

 Note
The first and second extension lines are determined by how you select the extension line origins. If the first extension line origin is on the right, the first extension line will also be on the right.

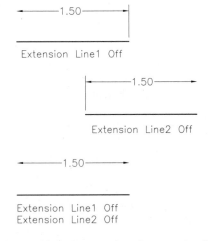

Figure 10-9 Suppressing the extension lines

Symbols and Arrows Tab

The options in the **Symbols and Arrows** tab (Figure 10-10) of the **New Dimension Style** dialog box are used to specify the variables and attributes that affect the format of the symbols and arrows. You can change the appearance of symbols and arrows.

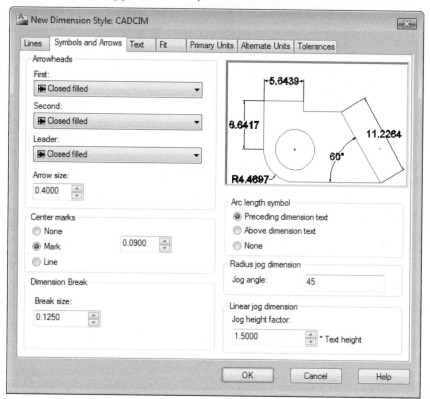

*Figure 10-10 The **Symbols and Arrows** tab of the **New Dimension Style** dialog box*

Arrowheads Area

The options in this area are used to specify the arrowheads and they are discussed next.

First/Second. When you create a dimension, AutoCAD draws the terminator symbols at the two ends of the dimension line. These terminator symbols, generally referred to as **arrowheads**, represent the beginning and end of a dimension. AutoCAD has provided nineteen standard termination symbols you can apply at each end of the dimension line. In addition to these, you can create your own arrows or terminator symbols. By default, the same arrowhead type is applied at both ends of the dimension line. If you select the first arrowhead, it is automatically applied to the second end by default. However, if you want to specify a different arrowhead at the second dimension line endpoint, you must select the desired arrowhead type from the **Second** drop-down list. The first endpoint of the dimension line is the intersection point of the first extension line and the dimension line. The first extension line is determined by the first extension line origin. However, in angular dimensioning, the second endpoint is located in a counterclockwise direction from the first point, regardless of how the points were selected when creating the angular dimension. The specified arrowhead types are selected from the **First** and **Second** drop-down lists. The first arrowhead type is saved in the **DIMBLK1** system variable and the second arrowhead type is saved in the **DIMBLK2** system variable.

AutoCAD provides you with an option of specifying a user-defined arrowhead. To define a user-defined arrow, you must create one as a block. (See Chapter 16 for information regarding blocks.) Now, from the **First** or the **Second** drop-down list, select **User Arrow**; the **Select Custom Arrow Block** dialog box will be displayed, see Figure 10-11.

Figure 10-11 *The Select Custom Arrow Block dialog box*

All the blocks in the current drawing will be available in the **Select from Drawing Blocks** drop-down list. You can select the desired block and it will become the current arrowhead.

Creating an Arrowhead Block

1. To create a block for an arrowhead, you will use a 1 X 1 box, refer to Figure 10-12. AutoCAD automatically scales the X and Y scale factors of the block to the arrowhead size multiplied by the overall scale. You can specify the arrow size in the **Arrow size** spinner. The **DIMASZ** variable controls the length of the arrowhead. For example, if **DIMASZ** is set to 0.25, the length of the arrow will be 0.25 units. Also, if the length of the arrow is not 1 unit, it will leave a gap between the dimension line and the arrowhead block.

2. The arrowhead must be drawn as it would appear on the right side of the dimension line. Choose the **Create** tool from **Block** panel in the **Home** tab to convert it into a block.

3. The insertion point of the arrowhead block must be the point that will coincide with the extension line, see Figure 10-12.

Leader. The **Leader** drop-down list displays the arrowhead types for the Leader arrow. Here, also, you can either select the standard arrowheads from the drop-down list or select **User Arrow** that allows you to define and use a user-defined arrowhead type.

Arrow size. This spinner is used to define the size of the arrowhead, see Figure 10-13. The default value is 0.18 units, which is stored in the **DIMASZ** system variable.

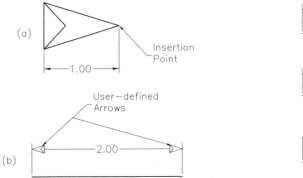

Figure 10-12 *Creating user-defined arrows*

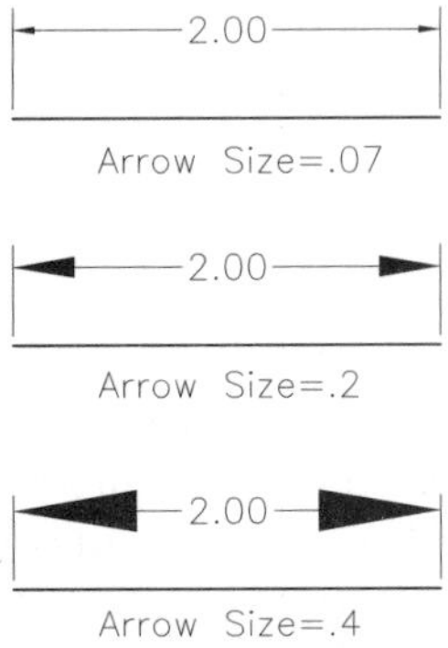

Figure 10-13 *Defining arrow sizes*

Center marks Area

This area deals with the options that control the appearance of the center marks and centerlines in the radius and diameter dimensioning. However, keep in mind that the center marks or the centerlines will be drawn only when dimensions are placed outside the circle.

None. If you select the **None** radio button, no mark or line will be drawn at the center of the circle.

Mark. If you select the **Mark** radio button, a mark will be drawn at the center of the circle.

Line. If you select the **Line** radio button, the centerlines will be drawn at the center of the circles.

The spinner in the **Center marks** area is used to set the size of the center marks or centerlines. This value is stored in the **DIMCEN** variable. The default value of this variable is 0.09.

Note

*If you use the **DIMCEN** command, a positive value will create a center mark, whereas a negative value will create a centerline. If the value is 0, AutoCAD does not create center marks or centerlines.*

Dimension Break Area

The **Break size** spinner in this area is used to control the default break length in the dimension while applying the break dimensions. The default value of **Break size** is 0.125. Figure 10-14 shows a drawing with dimensioning breaks.

Arc length symbol Area

The radio buttons in this area are used to specify the position of arc when applying the arc length dimension.

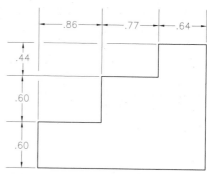

Figure 10-14 *Drawing with dimension breaks*

Preceding dimension text. If you select the **Preceding dimension text** radio button, the arc symbol will appear before the dimension text while applying the arc length dimension.

Above dimension text. With the **Above dimension text** radio button selected, the arc symbol in the arc length dimension will appear above the dimension text.

None. Select the **None** radio button, if you do not want the arc symbol to appear with the arc length dimension.

Radius jog dimension Area

The **Jog angle** edit box in this area is used to specify the angle of jog that appears while applying the jogged radius dimension. By default, its value is 45-degree.

Linear jog dimension Area

The option in this area is used to control the height of the jog (vertical height from the vertex of one jog angle to another) that appears while applying the jogged linear dimension. The jog height is always maintained with respect to the text height of the dimension. The **Jog height factor** spinner is used to change the factor that will be multiplied with the text height to calculate the linear jog height.

Note

*Unlike specifying a negative value for the **DIMCEN** variable, you cannot enter a negative value in the **Size** spinner. Selecting **Line** from the **Type** drop-down list automatically treats the value in the **Size** spinner as the size for the centerlines and sets **DIMCEN** to the negative of the value shown.*

EXERCISE 1

Draw and dimension the drawing, as shown in Figure 10-15. Assume the missing dimensions.

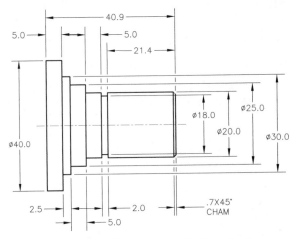

Figure 10-15 *Drawing for Exercise 1*

CONTROLLING THE DIMENSION TEXT FORMAT

Text Tab

You can control the dimension text format through the **Text** tab of the **New Dimension Style** dialog box (Figure 10-16). In the **Text** tab, you can control the parameters such as the placement, appearance, horizontal and vertical alignment of the dimension text, and so on. For example, you can force AutoCAD to align the dimension text along the dimension line. You can also force the dimension text to be displayed at the top of the dimension line. You can save the settings in a

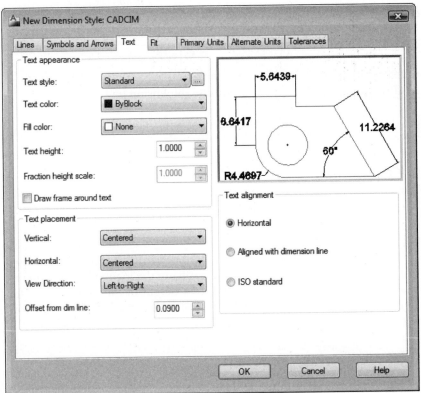

Figure 10-16 *The **Text** tab of the **New Dimension Style** dialog box*

dimension style file for future use. The **New Dimension Style** dialog box has a **Preview** window that updates dynamically to display the text placement as the settings are changed. Individual items of the **Text** tab and the related dimension variables are described next.

Text appearance Area

The options in this area are discussed next.

Text style. The **Text style** drop-down list displays the names of the predefined text styles. From this list, you can select the style name that you want to use for dimensioning. You must define the text style before you can use it in dimensioning (see "**Creating Text Styles**" in Chapter 7). Choosing the [...] button displays the **Text Style** dialog box that allows you to create a new or modify an existing text style. The value of this setting is stored in the **DIMTXSTY** system variable. The change in the dimension text style does not affect the text style you are using to draw the other text in the drawing.

Text color. This drop-down list is used to modify the color of the dimension text. The default color is **ByBlock**. If you choose the **Select Color** option from the **Text color** drop-down list, the **Select Color** dialog box is displayed, where you can choose a specific color. This color or color number is stored in the **DIMCLRT** variable.

Fill color. This drop-down list is used to set the fill color of the dimension text. A box of the selected color will be placed around the dimension text.

Text height. This spinner is used to modify the height of the dimension text, see Figure 10-17. You can change the dimension text height only when the current text style does not have a fixed height. In other words, the text height specified in the **STYLE** command should be zero. This is because a predefined text height (specified in the **STYLE** command) overrides any other setting for the dimension text height. This value is stored in the **DIMTXT** variable. The default text height is 0.1800 units.

Fraction height scale. This spinner is used to set the scale of the fractional units in relation to the dimension text height. This spinner will be available only when you select a format for the primary units, in which you can define the values in fractions, such as architectural or fractional. This value is stored in the **DIMTFAC** variable.

Draw frame around text. Select this check box to draw a frame around the dimension text, see Figure 10-18. This value is stored as a negative value in the **DIMGAP** system variable.

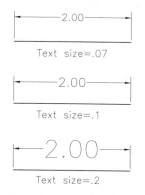

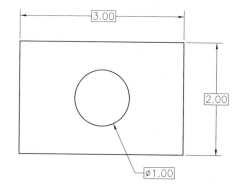

Figure 10-17 *Changing the dimension height*   ***Figure 10-18*** *Dimension text inside the frame*

Text placement Area

The options in this area are discussed next.

Vertical. The **Vertical** drop-down list displays the options that control the vertical placement of the dimension text. The current setting is highlighted. Controlling the vertical placement of the dimension text is possible only when the dimension text is drawn in its normal (default) location. This setting is stored in the **DIMTAD** system variable. The options in this drop-down list are discussed next.

Centered. If this option is selected, the dimension text gets positioned on the dimension line in such a way that the dimension line is split to allow for the placement of the text, see Figure 10-19. If the **1st** or **2nd Extension Line** option is selected in the **Horizontal** drop-down list, this centered setting will position the text on the extension line, not on the dimension line.

Above. If this option is selected, the dimension text is placed above the dimension line, except when the dimension line is not horizontal and the dimension text inside the extension lines is horizontal. The distance of the dimension text from the dimension line is controlled by the **DIMGAP** variable. This results in an unbroken solid dimension line under the dimension text, see Figure 10-19.

Outside. This option places the dimension text on the side of the dimension line.

JIS. This option lets you place the dimension text to conform to the **JIS** (Japanese Industrial Standards) representation.

Below. This option places the dimension text below the dimension line.

Note
*The horizontal and vertical placement options selected are reflected in the dimensions shown in the **Preview** window.*

Horizontal. This drop-down list is used to control the horizontal placement of the dimension text. You can select the required horizontal placement from this list. However, remember that these options will be useful only when the **Place text manually** check box in the **Fine Tuning** area of the **Fit** tab is cleared. The options in this drop-down list are discussed next.

Centered. This option is used to the dimension text between the extension lines. This is the default option.

At Ext Line 1. This option is used to place the text near the first extension line along the dimension line, see Figure 10-20.

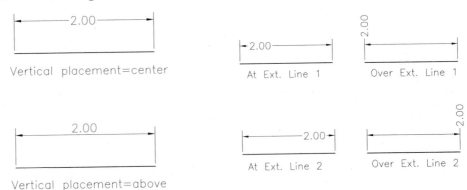

Figure 10-19 *Vertical text placement*   **Figure 10-20** *Horizontal text placement*

At Ext Line 2. This option is selected to place the text near the second extension line along the dimension line, see Figure 10-20.

Over Ext Line 1. This option is selected to place the text over the first extension line and also along the first extension line, see Figure 10-20.

Over Ext Line 2. This option is selected to place the text over the second extension line and also along the second extension line, see Figure 10-20.

Offset from dim line. This spinner is used to specify the distance between the dimension line and the dimension text (Figure 10-21). You can set the text gap you need using this spinner. The text gap value is also used as the measure of minimum length for the segments of the dimension line and in basic tolerance. The default value specified in this box is 0.09 units. The value of this setting is stored in the **DIMGAP** system variable.

Text alignment Area

The options in this area are discussed next.

Horizontal. This is the default option and if selected, the dimension text is drawn horizontally with respect to the current UCS (user coordinate system). The alignment of the dimension line does not affect the text alignment. Selecting this radio button turns both the **DIMTIH** and **DIMTOH** system variables **on**. The text is drawn horizontally even if the dimension line is at an angle.

Aligned with dimension line. If this radio button is selected then the text aligns with the dimension line (Figure 10-22) and both the system variables **DIMTIH** and **DIMTOH** are turned off.

ISO standard. If you select the **ISO Standard** radio button, the dimension text is aligned with the dimension line, only when the dimension text is inside the extension lines. Selecting this option turns the system variable **DIMTOH** on, that is, the dimension text outside the extension line is horizontal, regardless of the angle of the dimension line.

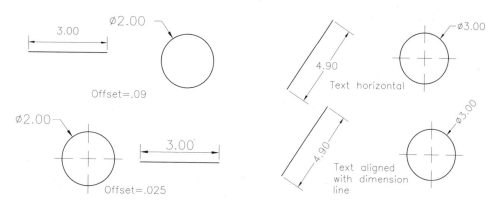

Figure 10-21 *Offset from the dimension line*   **Figure 10-22** *Specifying the text alignment*

EXERCISE 2

Draw Figure 10-23 and then set the values in the **Lines**, **Symbols and Arrows**, and **Text** tabs of the **New Dimension Style** dialog box to dimension the drawing, as shown in the figure. (Baseline spacing = 0.25, Extension beyond dimension lines = 0.10, Offset from origin = 0.05, Arrow size = 0.09, Text height = 0.08.)

Chapter 10

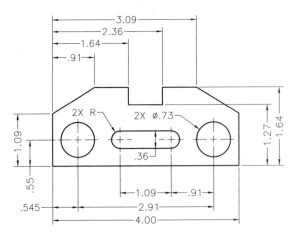

Figure 10-23 *Drawing for Exercise 2*

FITTING DIMENSION TEXT AND ARROWHEADS
Fit Tab
The options in this area are discussed next.

The **Fit** tab provides you with the options that are used to control the placement of dimension lines, arrowheads, leader lines, text, and the overall dimension scale (Figure 10-24).

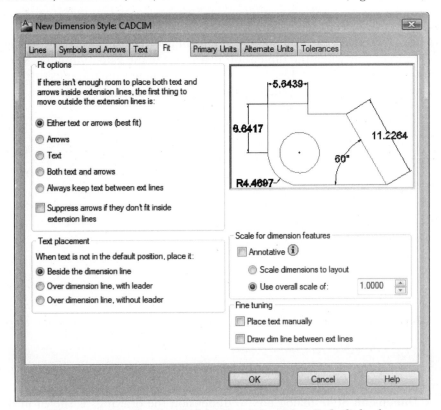

Figure 10-24 *The **Fit** tab of the **New Dimension Style** dialog box*

Fit options Area
These options are used to set the priorities for moving the text and arrowheads outside the extension lines, if the space between the extension lines is not enough to fit both of them.

Either text or arrows (best fit). This is the default option. In this option, AutoCAD places the dimension where it fits best between the extension lines.

Arrows. When you select this option, AutoCAD places the text and arrowheads inside the extension lines if there is enough space to fit both. If the space is not available, the arrows are moved outside the extension lines. If there is not enough space for text, both text and arrowheads are placed outside the extension lines.

Text. When you select this option, AutoCAD places the text and arrowheads inside the extension lines, if there is enough space to fit both. If there is enough space to fit the arrows, the arrows will be placed inside and the dimension text moves outside the extension lines. However, if there is not enough space for either the text or the arrowheads, both are placed outside the extension lines.

Both text and arrows. If you select this option, AutoCAD will place the arrows and dimension text between the extension lines, if there is enough space available to fit both. Otherwise, both the text and arrowheads are placed outside the extension lines.

Always keep text between ext lines. This option always keeps the text between the extension lines even in cases where AutoCAD would not do so. Selecting this radio button does not affect the radius and diameter dimensions. The value is stored in the **DIMTIX** variable and the default value is off.

Suppress arrows if they don't fit inside extension lines. If you select this check box, the arrowheads are suppressed if the space between the extension lines is not enough to adjust them. The value is stored in the **DIMSOXD** variable and the default value is off.

Text placement Area

This area provides you with the options to position the dimension text when it is moved from the default position. The value is stored in the **DIMTMOVE** variable. The options in this area are as follows:

Beside the dimension line. This option places the dimension text beside the dimension line.

Over dimension line, with leader. Selecting this option places the dimension text away from the dimension line and a leader line is created, which connects the text to the dimension line. But, if the dimension line is too close to the text, a leader is not drawn. The Horizontal placement decides whether the text is placed to the right or left of the leader.

Over dimension line, without leader. In this option, AutoCAD does not create a leader line, if there is insufficient space to fit the dimension text between the extension lines. The dimension text can be moved freely, independent of the dimension line.

Scale for dimension features Area

The options under this area are used to set the value for the overall dimension scale or scaling to the paper space.

Annotative. This check box is used to specify that the selected dimension style is annotative. With this option, you can also convert an existing non-annotative dimension style to annotative and vice-versa. If you select the **Annotative** check box from the **Create New Dimension Style** dialog box, then the **Annotative** check box will also be selected by default.

Use overall scale of. The current general scaling factor that pertains to all of the size-related dimension variables, such as text size, center mark size, and arrowhead size, is displayed in the **Use overall scale of** spinner. You can alter the scaling factor to your requirement by entering the scaling factor of your choice in this spinner. Altering the contents of this box alters the value of the **DIMSCALE** variable, since the current scaling factor is stored in it. The overall scale (**DIMSCALE**) is not applied to the measured lengths, coordinates, angles, or tolerance. The default value for this variable is 1.0. In this condition, the dimensioning variables assume their preset values and the drawing is plotted at full scale. The scale factor is the reciprocal of the drawing size and so the drawing is to be plotted at the half size. The overall scale factor (**DIMSCALE**) will be the reciprocal of ½, which is 2.

Note

*If you are in the middle of the dimensioning process and you change the **DIMSCALE** value and save the changed setting in a dimension style file, the dimensions with that style will be updated.*

Tip

*When you increase the limits of the drawing, you need to increase the overall scale of the drawing using the **Use overall scale of** spinner before dimensioning. This will save the time required in changing the individual scale factors of all the dimension parameters.*

Scale dimensions to layout. If you select the **Scale dimensions to layout** radio button, the scale factor between the current model space viewport and the floating viewport is computed automatically. Also, by selecting this radio button, you disable the **Use overall scale of** spinner (it is disabled in the dialog box) and the overall scale factor is set to 0. When the overall scale factor is assigned a value of 0, AutoCAD calculates an acceptable default value based on the scaling between the current model space viewport and the paper space. If you are in the paper space (**TILEMODE=0**), or are not using the **Scale dimensions to layout** feature, AutoCAD sets the overall scale factor to 1; otherwise, AutoCAD calculates a scale factor that makes it possible to plot text sizes, arrow sizes, and other scaled distances at the values, in which they have been previously set. (For further details regarding model space and layouts, refer to Chapter 15.)

Fine tuning Area

The **Fine tuning** area provides additional options governing placement of the dimension text. The options are as follows.

Place text manually. When you dimension, AutoCAD places the dimension text in the middle of the dimension line (if there is enough space). If you select the **Place text manually** check box, you can position the dimension text anywhere along the dimension line. You will also notice that when you select this check box, the **Horizontal Justification** is ignored. This setting is saved in the **DIMUPT** system variable. The default value of this variable is **off**. Selecting this check box enables you to position the dimension text anywhere along the dimension line.

Draw dim line between ext lines. This check box is selected when you want the dimension line to appear between the extension lines, even if the text and dimension lines are placed outside the extension lines. When you select this option in the radius and diameter dimensions (when default text placement is horizontal), the dimension line and arrows are drawn inside the circle or arc, and the text and leader are drawn outside. If you select the **Draw dim line between ext lines** check box, the **DIMTOFL** variable is set to on by AutoCAD. Its default setting is off.

FORMATTING PRIMARY DIMENSION UNITS
Primary Units Tab

You can use the **Primary Units** tab of the **New Dimension Style** dialog box to control the dimension text format and precision values (Figure 10-25). You can use the options under this tab to control Units, Dimension Precision, and Zero Suppression for dimension measurements. AutoCAD lets you attach a user-defined prefix or suffix to the dimension text. For example, you can define the diameter symbol as a prefix by entering %%C in the **Prefix** edit box; AutoCAD will automatically attach the diameter symbol in front of the dimension text. Similarly, you can define a unit type, such as **mm**, as a suffix; AutoCAD will then attach **mm** at the end of every dimension text. This tab also enables you to define zero suppression, precision, and dimension text format.

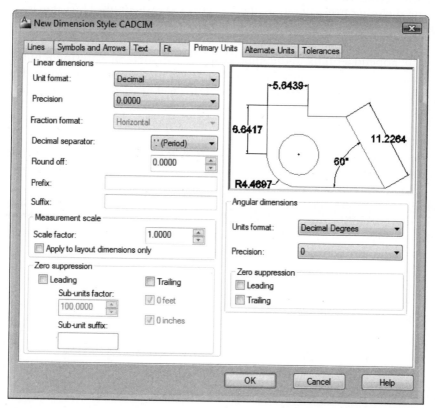

*Figure 10-25 The **Primary Units** tab of the **New Dimension Style** dialog box*

Linear dimensions Area

The options in this area are discussed next.

Unit format. This drop-down list provides you with the options of specifying the units for the primary dimensions. The formats include **Decimal**, **Scientific**, **Architectural**, **Engineering**, **Fraction**, and **Windows Desktop**. Remember that by selecting a dimension unit format, the drawing units (which you might have selected by using the **UNITS** command) are not affected. The unit setting for linear dimensions is stored in the **DIMLUNIT** system variable.

Precision. This drop-down list is used to control the number of decimal places for the primary units. The setting for precision (number of decimal places) is saved in the **DIMDEC** variable.

Fraction format. This drop-down list is used to set the fraction format. The options are Diagonal, Horizontal, and not stacked. This drop-down list will be available only when you select **Architectural** or **Fractional** from the **Unit format** drop-down list. The value is stored in the **DIMFRAC** variable.

Decimal separator. This drop-down list is used to select an option that will be used as the decimal separator. For example, Period [.], Comma [,] or Space [ ]. If you have selected Windows desktop units in the **Unit Format** drop-down list, AutoCAD uses the Decimal symbol settings. The value is stored in the **DIMDSEP** variable.

Round off. The **Round off** spinner is used to set the value for rounding off the dimension values. The number of decimal places of the round off value should always be less than or equal to the value in the **Precision** edit box. For example, if the **Round off** spinner is set to 0.05, all dimensions will be rounded off to the nearest 0.05 unit. Therefore, the value 1.06 will round off to 1.05 and the value 1.09 will round off to 1.10, see Figure 10-26. The value is stored in the **DIMRND** variable and the default value in the **Round off** edit box is 0.

Prefix. You can append a prefix to the dimension measurement by entering it in this edit box. The dimension text is converted into **Prefix<dimension measurement>** format. For example, if you enter the text "Abs" in the **Prefix** edit box, "Abs" will be placed in front of the dimension text (Figure 10-27). The prefix string is saved in the **DIMPOST** system variable.

Note

*Once you specify a prefix, default prefixes such as **R** in radius dimensioning and ∅ in diameter dimensioning are cancelled.*

Suffix. Just like appending a prefix, you can append a suffix to the dimension measurement by entering the desired suffix in this edit box. For example, if you enter the text **mm** in the **Suffix** edit box, the dimension text will have <dimension measurement>mm format, see Figure 10-25. AutoCAD stores the suffix string in the **DIMPOST** variable.

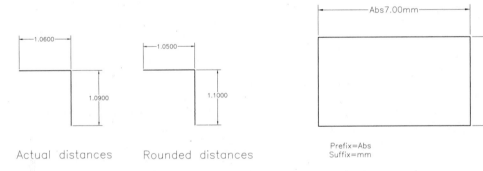

Figure 10-26 *Rounding off the dimension measurements*

Figure 10-27 *Adding prefix and suffix to the dimensions*

Tip

*The **DIMPOST** variable is used to append both prefix and suffix to the dimension text. This variable takes a string value as its argument. For example, if you want to have a suffix for centimeters, set **DIMPOST** to cm. To establish a prefix to a dimension text, type the prefix text string and then "<>".*

Measurement scale Area

The options in this area are discussed next.

Scale factor. You can specify a global scale factor for the linear dimension measurements by setting the desired scale factor in the **Scale factor** spinner. All the linear distances measured by dimensions, which include radii, diameters, and coordinates, are multiplied by the existing value in this spinner. For example, if the value of the **Scale factor** spinner is set to 2, two unit segments will be dimensioned as 4 units (2 X 2). However, the angular dimensions are not affected. In this manner, the value of the linear scaling factor affects the contents of the default (original) dimension text (Figure 10-28). The default value for linear scaling is 1. With the default value, the dimension text generated is the actual measurement of the object being dimensioned. The linear scaling value is saved in the **DIMLFAC** variable.

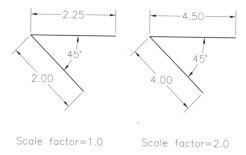

Scale factor=1.0   Scale factor=2.0

Figure 10-28 *Identical figures dimensioned using different scale factors*

Note
The linear scaling value is not exercised on rounding a value, or on plus or minus tolerance value. Therefore, changing the linear scaling factor will not affect the tolerance values.

Apply to layout dimensions only. When you select the **Apply to layout dimensions only** check box, the scale factor value is applied only to the dimensions in the layout. The value is stored as a negative value in the **DIMLFAC** variable. If you change the **DIMLFAC** variable from the **Dim:** prompt, AutoCAD displays the viewport option to calculate the **DIMLFAC** variable. First, set the **TILEMODE** to 0 (paper space), and then invoke the **MVIEW** command to get the **Viewport** option.

Zero suppression Area

The options in this area are used to suppress the leading or trailing zeros in the dimensioning. This area provides you with four check boxes. These check boxes can be selected to suppress the leading or trailing zeros or zeros in the feet and inches. The **0 feet** and the **0 inches** check boxes will be available only when you select **Engineering** or **Architectural** from the **Unit format** drop-down list. When the Architectural units are being used, the **Leading** and **Trailing** check boxes are disabled. For example, if you select the **0 feet** check box, the dimension text 0'-8 ¾" becomes 8 ¾". By default, the 0 feet and 0 inches value is suppressed. If you want to suppress the inches part of a feet-and-inches dimension when the distance in the feet portion is an integer value and the inches portion is zero, select the **0 inches** check box. For example, if you select the **0 inches** check box, the dimension text 3'-0" becomes 3'. Similarly, if you select the **Leading** check box, the dimension that was initially 0.53 will become .53. If you select the **Trailing** check box, the dimension that was initially 2.0 will become 2.

Angular dimensions Area

This area provides you with the options to control the units format, precision, and zero suppression for the Angular units.

Units format. The **Units format** drop-down list displays a list of unit formats for the angular dimensions. The default value, in which the angular dimensions are displayed, is **Decimal Degrees**. The value governing the unit setting for the angular dimensions is stored in the **DIMAUNIT** variable.

Precision. You can select the number of decimal places for the angular dimensions from this drop-down list. This value is stored in the **DIMADEC** variable.

Zero suppression Area. Similar to the linear dimensions, you can suppress the **leading**, **trailing**, neither, or both zeros in the angular dimensions by selecting the respective check boxes in this area. The value is stored in the **DIMAZIN** variable.

FORMATTING ALTERNATE DIMENSION UNITS
Alternate Units Tab

By default, the options in the **Alternate Units** tab of the **New Dimension Style** dialog box is disabled and the value of the **DIMALT** variable is turned off. If you want to perform alternate units dimensioning, select the **Display alternate units** check box. By doing so, AutoCAD activates various options in this area (Figure 10-29). This tab sets the format, precision, angles, placement, scale, and so on for the alternate units in use. In this tab, you can specify the values that will be applied to the alternate dimensions.

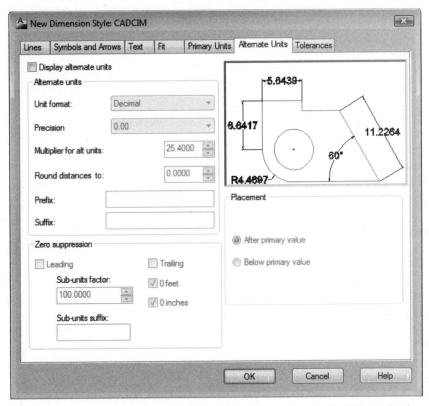

*Figure 10-29 The **Alternate Units** tab of the **New Dimension Style** dialog box*

Alternate units Area

The options in this area are identical to those under the **Linear dimensions** area of the **Primary Units** tab. This area provides you with the options to set the format for all dimension types.

Unit format. You can select a unit format to apply to the alternate dimensions from this drop-down list. The options under this drop-down list include Scientific, Decimal, Engineering, Architectural stacked, Fractional stacked, Architectural, Fractional, and Windows Desktop. The value is stored in the **DIMALTU** variable. The relative size of fractions is governed by the **DIMTFAC** variable.

Precision. You can select the number of decimal places for the alternate units from the **Precision** drop-down list. The value is stored in the **DIMALTD** variable.

Multiplier for alt units. To generate a value in the alternate system of measurement, you need a factor with which all the linear dimensions will be multiplied. The value for this factor can be set using the **Multiplier for alt units** spinner. The default value of 25.4 is for dimensioning in inches with the alternate units in millimeters. This scaling value (contents of the **Multiplier for alt units** spinner) is stored in the **DIMALTF** variable.

Round distances to. This spinner is used to set a value to which you want all your measurements (made in alternate units) to be rounded off. This value is stored in the **DIMALTRND** system variable. For example, if you set the value of the **Round distances to** spinner to 0.25, all the alternate dimensions get rounded off to the nearest .25 unit.

Prefix/Suffix. The **Prefix** and **Suffix** edit boxes are similar to the edit boxes in the **Linear dimensions** area of the **Primary Units** tab. You can enter the text or symbols that you want to precede or follow the alternate dimension text. The value is stored in the **DIMAPOST** variable. You can also use control codes and special characters to display special symbols.

Zero suppression Area
This area allows you to suppress the leading or trailing zeros in decimal unit dimensions by selecting either, both, or none of the **Trailing** and **Leading** check boxes. Similarly, selecting the **0 feet** check box suppresses the zeros in the feet area of the dimension, when the dimension value is less than a foot. Selecting the **0 inches** check box suppresses the zeros in the inches area of the dimension. For example, 1'-0" becomes 1'. The **DIMALTZ** variable controls the suppression of zeros for alternate unit dimension values. The values that are between 0 and 3 affect the feet-and-inch dimensions only.

Placement Area
This area provides the options that control the positioning of the Alternate units. The value is stored in the **DIMAPOST** variable.

After primary value. Selecting the **After primary value** radio button places the alternate units dimension text after the primary units. This is the default option, see Figure 10-30.

Below primary value. Selecting the **Below primary value** radio button places the alternate units dimension text below the primary units, see Figure 10-30.

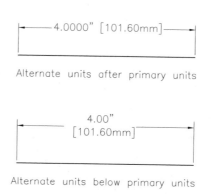

Figure 10-30 Placement of alternate units

FORMATTING THE TOLERANCES
Tolerances Tab
The **Tolerances** tab (Figure 10-31) allows you to set the parameters for options that control the format and display of the tolerance dimension text. These include the alternate unit tolerance dimension text.

Tolerance format Area
The **Tolerance format** area of the **Tolerances** tab (Figure 10-31) is used to specify the tolerance method, tolerance value, position of tolerance text, and precision and height of the tolerance text. For example, if you do not want a dimension to deviate more than plus 0.01 and minus 0.02, you can specify this by selecting **Deviation** from the **Method** drop-down list and then specifying the plus and minus deviation in the **Upper Value** and the **Lower Value** edit boxes. When you

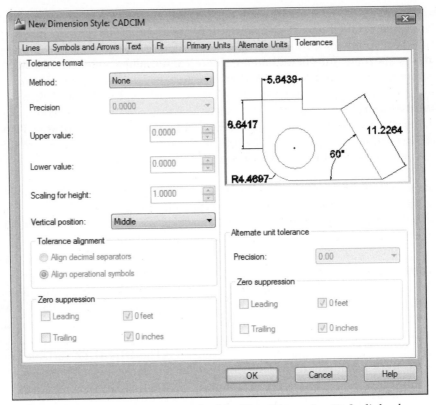

Figure 10-31 *The **Tolerances** tab of the **New Dimension Style** dialog box*

dimension, AutoCAD will automatically append the tolerance to it. The **DIMTP** variable sets the maximum (or upper) tolerance limit for the dimension text and **DIMTM** variable sets the minimum (or lower) tolerance limit for the dimension text. Different settings and their effects on relevant dimension variables are explained in the following sections.

Method. The **Method** drop-down list lets you select the tolerance method. The tolerance methods supported by AutoCAD are **Symmetrical**, **Deviation**, **Limits**, and **Basic**. These tolerance methods are described next.

None. Selecting the **None** option sets the **DIMTOL** variable to 0 and does not add tolerance values to the dimension text, that is, the **Tolerances** tab is disabled.

Symmetrical. This option is used to specify the symmetrical tolerances. When you select this option, the **Lower Value** spinner is disabled and the value specified in the **Upper Value** spinner is applied to both plus and minus tolerances. For example, if the value specified in the **Upper Value** spinner is 0.05, the tolerance appended to the dimension text is ±0.05, see Figure 10-32. The value of **DIMTOL** is set to 1 and the value of **DIMLIM** is set to 0.

Deviation. If you select the **Deviation** tolerance method, the values in the **Upper Value** and **Lower Value** spinners will be displayed as plus and minus dimension tolerances. If you enter values for the plus and minus tolerances, AutoCAD appends a plus sign (+) to the positive values of the tolerance and a negative sign (–) to the negative values of the tolerance. For example, if the upper value of the tolerance is 0.005 and the lower value of the tolerance is 0.002, the resulting dimension text generated will have a positive tolerance of 0.005 and a negative tolerance of 0.002 (Figure 10-32). Even if one of the tolerance values is 0, a sign is appended to it. On specifying the deviation tolerance, AutoCAD sets the **DIMTOL** variable value to 1 and the **DIMLIM** variable value to 0. The values in the **Upper Value** and **Lower Value** edit boxes are saved in the **DIMTP** and **DIMTM** system variables, respectively.

Limits. If you select the **Limits** tolerance method from the **Method** drop-down list, AutoCAD adds the upper value (contents of the **Upper Value** spinner) to the dimension text (actual measurement) and subtracts the lower value (contents of the **Lower Value** spinner) from the dimension text. The resulting values are displayed as the dimension text, see Figure 10-32. Selecting the **Limits** tolerance method results in setting the **DIMLIM** variable value to 1 and the **DIMTOL** variable value to 0. The numeral values in the **Upper Value** and **Lower Value** edit boxes are saved in the **DIMTP** and **DIMTM** system variables, respectively.

Basic. A basic dimension text is a dimension text with a box drawn around it (Figure 10-32). The basic dimension is also called a reference dimension. Reference dimensions are used primarily in geometric dimensioning and tolerances. The basic dimension can be realized by selecting the basic tolerance method. The distance provided around the dimension text (distance between dimension text and the rectangular box) is stored as a negative value in the **DIMGAP** variable. The negative value signifies the basic dimension. The default setting is off, resulting in the generation of dimensions without the box around the dimension text.

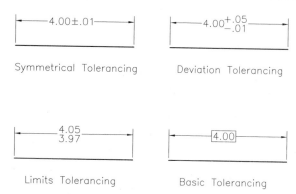

Figure 10-32 Specifying tolerance using various tolerancing methods

Precision. The **Precision** drop-down list is used to select the number of decimal places for the tolerance dimension text. The value is stored in **DIMTDEC** variable.

Upper value/Lower value. In the **Upper value** spinner, the positive upper or maximum value is specified. If the method of tolerances is symmetrical, the same value is used as the lower value also. The value is stored in the **DIMTP** variable. In the **Lower** spinner, the lower or minimum value is specified. The value is stored in the **DIMTM** variable.

Scaling for height. The **Scaling for height** spinner is used to specify the height of the dimension tolerance text relative to the dimension text height. The default value is 1, which means the height of the tolerance text is the same as the dimension text height. If you want the tolerance text to be 75 percent of the dimension height text, enter 0.75 in the **Scaling for height** edit box. The ratio of the tolerance height to the dimension text height is calculated by AutoCAD and then stored in the **DIMTFAC** variable. **DIMTFAC = Tolerance Height/Text Height**.

Vertical position. This drop-down list allows you to specify the location of the tolerance text for deviation and symmetrical methods only. The three alignments that are possible are with the **Bottom**, **Middle**, or **Top** of the main dimension text. The settings are saved in the **DIMTOLJ** system variable (Bottom=0, Middle=1, and Top=2).

Tolerance alignment Area

The options in this area are used to control the alignment of the tolerance value text when they are placed in stacked condition. These options get highlighted only when you select the **Deviation** or **Limit** option from the **Method** drop-down list. Select the **Align decimal separators** radio button to align the tolerance text vertically along the decimal point, see Figure 10-33(a). Select the **Align operational symbols** radio button to align the tolerance text vertically along the plus sign (+) and negative sign(-), see Figure 10-33(b).

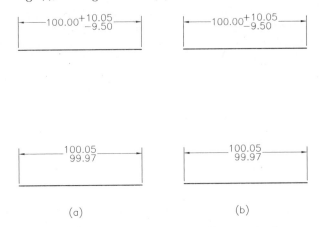

(a)   *(b)*

Figure 10-33 *Various tolerance alignment options*

Zero suppression Area

This area controls the zero suppression in the tolerance text depending on which one of the check boxes is selected. Selecting the **Leading** check box suppresses the leading zeros in all the decimal tolerance text. For example, 0.2000 becomes .2000. Selecting the **Trailing** check box suppresses the trailing zeros in all the decimal tolerance text. For example, 0.5000 becomes 0.5. Similarly, selecting both the boxes suppresses both the trailing and leading zeros and selecting none, suppresses none. If you select **0 feet** check box, the zeros in the feet portion of the tolerance dimension text are suppressed if the dimension value is less than one foot. Similarly, selecting the **0 inches** check box suppresses the zeros in the inches portion of the dimension text. The value is stored in the **DIMTZIN** variable.

Alternate unit tolerance Area

The options in this area define the precision and zero suppression settings for the Alternate unit tolerance values. The options under this area will be available only when you display the alternate units along with the primary units.

Precision. This drop-down list is used to set the number of decimal places to be displayed in the tolerance text of the alternate dimensions. This value is stored in the **DIMALTTD** variable.

Zero suppression Area

Selecting the respective check boxes controls the suppression of the **Leading** and **Trailing** zeros in decimal values and the suppression of zeros in the Feet and Inches portions for dimensions in the feet and inches format. The value is stored in the **DIMALTTZ** variable.

EXERCISE 3   *Dimension Style*

Draw Figure 10-34 and then set the values in various tabs of the **New Dimension Style** dialog box to dimension it, as shown. (Baseline spacing = 0.25, Extension beyond dim lines = 0.10, Offset from origin = 0.05, Arrowhead size =0.07, Text height = 0.08.) Assume the missing dimensions.

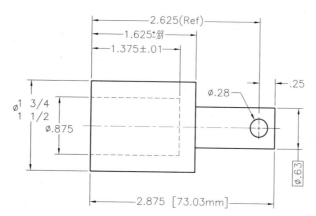

Figure 10-34 *Drawing for Exercise 3*

DIMENSION STYLE FAMILIES

The dimension style feature of AutoCAD lets the user define a dimension style with values that are common to all dimensions. For example, the arrow size, dimension text height, and color of the dimension line are generally the same in all types of dimensioning such as linear, radial, diameter, and angular. These dimensioning types belong to the same family because they have some characteristics in common. In AutoCAD, this is called a **dimension style family,** and the values assigned to the family are called **dimension style family values.**

After you have defined the dimension style family values, you can specify variations on it for other types of dimensions such as radial and diameter. For example, if you want to limit the number of decimal places to two in radial dimensioning, you can specify that value for radial dimensioning. The other values will stay the same as the family values to which this dimension type belongs. When you use the radial dimension, AutoCAD automatically uses the style that was defined for radial dimensioning; otherwise, it creates a radial dimension with the values as defined for the family. After you have created a dimension style family, any changes in the parent style are applied to family members, if the particular property is the same. Special suffix codes are appended to the dimension style family names that correspond to different dimension types. For example, if the dimension style family name is MYSTYLE and you define a diameter type of dimension, AutoCAD will append $4 at the end of the dimension style family name. The name of the diameter type of dimension will be MYSTYLE$4. The following are the suffix codes for different types of dimensioning.

Suffix Code	Dimension Type	Suffix Code	Dimension Type
0	Linear	2	Angular
3	Radius	4	Diameter
6	Ordinate	7	Leader

EXAMPLE 1   *Dimension Style Family*

The following example illustrates the concepts of dimension style families, refer to Figure 10-35.

1. Specify the values for the dimension style family.
2. Specify the values for the linear dimensions.
3. Specify the values for the diameter and radius dimensions.
4. After creating the dimension style, use it to dimension the given drawing.

Chapter 10

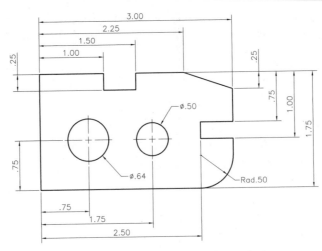

Figure 10-35 Drawing for Example 1

1. Start a new file in the **Drafting & Annotation** workspace and draw an object, as shown in Figure 10-35.

2. Left-click on the inclined arrow in the **Dimensions** panel of the **Annotate** tab; the **Dimension Style Manager** dialog box is displayed. The **Standard** and **Annotative** options are displayed in the **Styles** list box. Select the **Annotative** style from the **Styles** list box.

3. Choose the **New** button to display the **Create New Dimension Style** dialog box. In this dialog box, enter **MyStyle** in the **New Style Name** edit box. Select **Annotative** from the **Start With** drop-down list. Also select **All dimensions** from the **Use for** drop-down list. Now, choose the **Continue** button to display the **New Dimension Style: MyStyle** dialog box. In this dialog box, first choose the **Lines** tab and then enter the following values in their respective options. Next, choose the **Symbols and Arrows** tab and enter the following values.

Lines tab	**Symbols and Arrows** tab
Baseline Spacing: 0.15	Arrow size: 0.09
Offset from origin: 0.03	Center marks: Select the **Line** radio button
Extend beyond dim lines: 0.07	Enter **0.05** in the spinner in the **Center marks** area

4. Choose the **Text** tab and change the following values:

Text Height: 0.09	Offset from dimension line: 0.03

5. Choose the **Fit** tab and make sure the **Annotative** check box is selected.

6. After entering the values, choose the **OK** button to return to the **Dimension Style Manager** dialog box. This dimension style contains the values that are common to all dimension types.

7. Now, choose the **New** button again in the **Dimension Style Manager** dialog box to display the **Create New Dimension Style** dialog box. AutoCAD displays **Copy of MyStyle** in the **New Style Name** edit box. Select **MyStyle** from the **Start With** drop-down list if it is not already selected. From the **Use for** drop-down list, select **Linear dimensions**. Choose the **Continue** button to display the **New Dimension Style: MyStyle: Linear** dialog box and set the following values in the **Text** tab:

a. Select the **Aligned with dimension line** radio button in the **Text alignment** area.
b. In the **Text placement** area, select **Above** from the **Vertical** drop-down list.

8. In the **Primary Units** tab, set the precision to two decimal places. Select the **Leading** check box in the **Zero suppression** area. Next, choose the **OK** button to return to the **Dimension Style Manager** dialog box.

9. Choose the **New** button again to display the **Create New Dimension Style** dialog box. Select **MyStyle** from the **Start With** drop-down list and the **Diameter dimensions** type from the **Use for** drop-down list. Next, choose the **Continue** button; the **New dimension Style: MyStyle: Diameter** dialog box is displayed.

10. Choose the **Text** tab and select the **Centered** option from the **Vertical** drop-down list in the **Text placement** area. Then select the **Horizontal** radio button from the **Text alignment** area.

11. Choose the **Primary Units** tab and then set the **Precision** to two decimal places. Select the **Leading** check box in the **Zero suppression** area. Next, choose the **OK** button to return to the **Dimension Style Manager** dialog box.

12. In this dialog box, choose the **New** button to display the **Create New Dimension Style** dialog box. Select **MyStyle** from the **Start With** drop-down list and **Radius dimensions** from the **Use for** drop-down list. Choose the **Continue** button to display the **New Dimension Style: MyStyle: Radial** dialog box.

13. Choose the **Primary Units** tab and then set the precision to two decimal places. Select the **Leading** check box in the **Zero suppression** area. Next, enter **Rad** in the **Prefix** edit box.

14. In the **Fit** tab, select the **Text** radio button in the **Fit options** area. Next, choose the **OK** button to return to the **Dimension Style Manager** dialog box.

15. Select **MyStyle** from the **Styles** list box and choose the **Set current** button. Choose the **Close** button to exit the dialog box.

16. Choose the **Linear** button from the **Dimensions** panel; the **Select Annotation Scale** dialog box is displayed. Select the **1:1** option from the drop-down list and choose the **OK** button.

17. Use the linear and baseline dimensions to draw the linear dimensions refer to Figure 10-35. While entering linear dimensions, you will notice that AutoCAD automatically uses the values that were defined for the linear type of dimensioning.

18. Use the diameter dimensioning to dimension the circles, refer to Figure 10-35. Again, notice that the dimensions are drawn based on the values specified for the diameter type of dimensioning.

19. Now, use the radius dimensioning to dimension the fillet, refer to Figure 10-35.

USING DIMENSION STYLE OVERRIDES*

Most of the dimension characteristics are common in a production drawing. The values that are common to different dimensioning types can be defined in the dimension style family. However, at times, you might have different dimensions. For example, you may need two types of linear dimensioning: one with tolerance and one without. One way to draw these dimensions is to create two dimensioning styles. You can also use the dimension variable overrides to override the existing values. For example, you can define a dimension style (**MyStyle**) that draws dimensions without tolerance. Now, to draw a dimension with tolerance or update an existing dimension, you can override the previously defined value through the **Dimension Style Manager** dialog box or by setting the variable values at the Command prompt. Now, you can remove style override of any selected dimension by choosing the **Remove Style Overrides** option from the shortcut menu displayed on right clicking. The following example illustrates how to use the dimension style overrides.

EXAMPLE 2   *Dimension Style Override*

In this example, you will update the overall dimension (3.00) so that the tolerance is displayed with the dimension. You will also add linear dimensions, as shown in Figure 10-36.

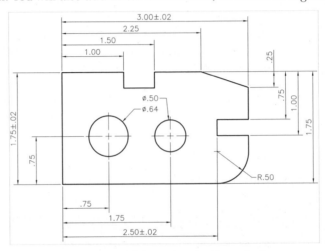

Figure 10-36 Drawing for Example 2

This problem can be solved by using dimension style overrides as well as by using the **Properties** palette. However, here only the dimension style overrides method is discussed.

1. Invoke the **Dimension Style Manager** dialog box. Select **MyStyle** from the **Styles** list box and choose the **Override** button to display the **Override Current Style: MyStyle** dialog box. The options in this dialog box are same to the **New Dimension Style** dialog box discussed earlier in this chapter.

2. Choose the **Tolerances** tab and select **Symmetrical** from the **Method** drop-down list.

3. Set the value of the **Precision** spinner to two decimal places. Set the value of the **Upper value** spinner to **0.02** and select the **Leading** check box in the **Zero suppression** area. Next, choose the **OK** button to exit the dialog box (this does not save the style). On doing so, you will notice that **<style overrides>** is displayed under **MyStyle** in the **Styles** list box, indicating that the style overrides the **MyStyle** dimension style.

4. The **<style overrides>** is displayed until you save it under a new name or under the style it is displayed in, or until you delete it. Select **<style overrides>** and right-click to display the shortcut menu. Choose the **Save to current style** option from the shortcut menu to save the overrides to the current style. Choosing the **Rename** option allows you to rename the style override and save it as a new style.

5. Choose the **Update** tool from the **Dimension** panel in the **Annotate** tab and select the dimension that measures **3.00**. The dimension now displays the symmetrical tolerance.

6. Draw the remaining two linear dimensions. They will automatically appear with the tolerances, see Figure 10-36.

Tip
*You can also use the **DIMOVERRIDE** command to apply the change to the existing dimensions. Apply the changes to the **DIMTOL**, **DIMTP**, and **DIMTM** variables.*

COMPARING AND LISTING DIMENSION STYLES

Choosing the **Compare** button in the **Dimension Style Manager** dialog box displays the **Compare Dimension Styles** dialog box where you can compare the settings of two dimensions styles or list all the settings of one of them (Figure 10-37).

The **Compare** and the **With** drop-down lists display the dimension styles in the current drawing. Selecting the dimension styles from the respective lists compare the two styles. In the **With** drop-down list, if you select **<none>** or the same style as selected from the **Compare** drop-down list, all the properties of the selected style are displayed. The comparison results are displayed under four headings: **Description** of the Dimension Style property, the **System variable** controlling a particular setting, and the **values of the variable for both the dimension styles** which differ in the two styles in comparison. The number of differences between the selected dimension styles are displayed below the **With** drop-down list. The button provided in this dialog box prints the comparison results to the Windows clipboard from where they can be pasted to other Windows applications.

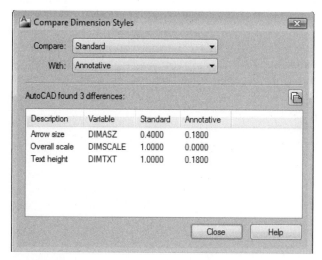

Figure 10-37 *The **Compare Dimension Styles** dialog box*

USING EXTERNALLY REFERENCED DIMENSION STYLES

The externally referenced dimensions cannot be used directly in the current drawing. When you Xref a drawing, the drawing name is appended to the style name and the two are separated by the vertical bar (|) symbol. It uses the same syntax as the other externally dependent symbols. For example, if the drawing (FLOOR) has a dimension style called DIM1 and you Xref this drawing in the current drawing, AutoCAD will rename the dimension style to FLOOR|DIM1, as shown in Figure 10-38. You cannot make this dimension style current, nor can you modify or override it. However, you can use it as a template to create a new style. To accomplish this, invoke the **Dimension Style Manager** dialog box. If the **Don't list styles in Xrefs** check box is selected, the styles in the Xref are not displayed. Clear this check box to display the Xref dimension styles and choose the **New** button. In the **New Style Name** edit box of the **Create**

New Dimension Style dialog box, enter the name of the dimension style. AutoCAD will create a new dimension style with the same values as those of the externally referenced dimension style (FLOOR|DIM1).

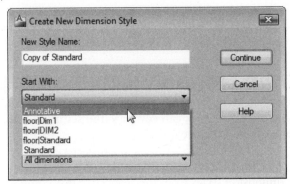

Figure 10-38 The ***Create New Dimension Style*** *dialog box*

CREATING AND RESTORING MULTILEADER STYLES

Ribbon: Annotate > Leaders > Multileader Style Manager (*Inclined arrow*)
Toolbar: Multileader > Multileader Style or Styles > Multileader Style
Command: MLEADERSTYLE

The multileader styles control the appearance and positioning of multileaders in the drawing. If the default multileader styles (**Standard** and **Annotative**) do not meet your requirement, you can select any other existing multileader style as per your requirement. The default multileader style file names are **Standard** and **Annotative**. Left-click on the inclined arrow in the **Multileaders** panel of the **Annotate** tab; the **Multileader Style Manager** dialog box will be displayed, as shown in Figure 10-39.

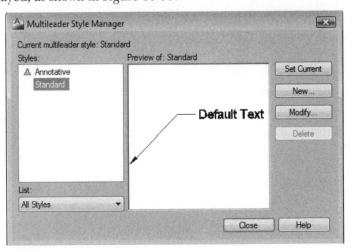

Figure 10-39 The ***Multileader Style Manager*** *dialog box*

From the **Multileader Style Manager** dialog box, choose the **New** button to display the **Create New Multileader Style** dialog box, see Figure 10-39. Enter the multileader style name in the **New style name** text box and then select a style from the **Start with** drop-down list on which you want to base your current style. Select the **Annotative** check box to specify that the new dimension style should be annotative. Click on the **i** (information) button to get the help and information about the annotative objects. Choose the **Continue** button to display the **Modify Multileader Style** dialog box in which you can define the new style. After defining the new style, choose the **OK** button. The parameters of the **New Dimension Style** dialog box are discussed in the next section.

In the **Multileader Style Manager** dialog box, the current multileader style name is shown in front of the **Current multileader style** option and is also highlighted in the **Styles** list box. The **Multileader Style Manager** dialog box also has a **Preview of** window that displays a preview of the current multileader style. A style can be made current (restored) by selecting the name of the multileader style that you want to make current from the list of defined multileader styles and choosing the **Set Current** button. You can also make a style current by double-clicking on the style name in the **Styles** list box. The **Multileader Style Control** drop-down list in the **Multileaders** panel also displays the multileader styles. Select the required multileader style from this list to set it as current. The list of multileader styles displayed in the **Styles** list box depends on the option selected from the **List** drop down-list. If you select the **Styles in use** option, only the multileader styles in use will be listed in the **Styles** list box. If you right-click on a style in the **Styles** list box, a shortcut menu will be displayed to provide you with the options such as **Set current**, **Modify**, **Rename**, and **Delete**. Choose the **Modify** button to display the **Modify Multileader Style** dialog box in which you can modify an existing style. Choose the **Delete** button to delete the selected multileader style that has not been used in the drawing.

MODIFY MULTILEADER STYLE DIALOG BOX

The **Modify Multileader Style** dialog box can be used to specify the multileader attributes (variables) that affect various properties of the multileader. The various tabs provided under the **Modify Multileader Style** dialog box are discussed next.

Leader Format

The options in the **Leader Format** tab (Figure 10-40) of the **Modify Multileader Style** dialog box are used to specify the multileader attributes that affect the format of the multileader lines. For example, the appearance and behavior of the multileader lines and the arrow head can be changed with this tab.

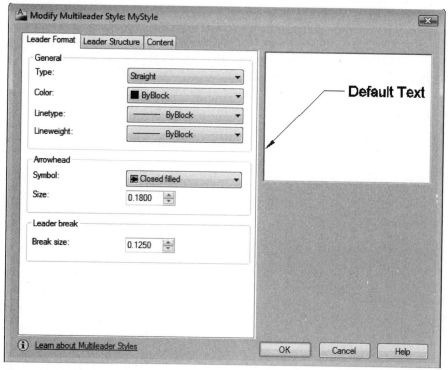

Figure 10-40 *The **Leader Format** tab of the **Modify Multileader Style** dialog box*

Chapter 10

General Area

This area provides you the options for controlling the display of the multileader lines. These options are discussed next.

Type. This option is used to specify the type of lines to be used for creating the multileaders. You can select the **Straight**, **Spline**, or **None** option from the **Type** drop-down list. Select the **None** option to create the multileader with no leader lines. You will only be able to draw the content of that multileader. The default line type for the multileaders is **Straight**.

Color. This drop-down list is used to set the colors for the leader lines and arrowheads. The multileader arrowheads have the same color as the multileader lines because the arrows are a part of the multileader lines. The default color for the multileader lines and arrows is **ByBlock**. Select the **Select Color** option from the **Color** drop-down list to display the **Select Color** dialog box from where you can choose a specific color.

Linetype. This drop-down list is used to set the linetype for the multileader lines.

Lineweight. This drop-down list is used to specify the lineweight for the multileader lines. You can select the required lineweight from this drop-down list. The default value is **ByBlock**. Note that you cannot assign the lineweight to the arrowheads using this drop-down list.

Arrowhead Area

This area provides you the options for controlling the shape and size of the arrowhead. The options in this area are discussed next.

Symbol. When you create a multileader, AutoCAD draws the terminator symbol at the starting point of the multileader line. This terminator symbol is generally referred to as the arrowhead, and it represents the beginning of the multileader. AutoCAD provides you nineteen standard termination symbols that can be selected from the **Symbol** drop-down list of the **Arrowhead** area. In addition to this, you can create your own arrows or terminator symbols.

Size. This spinner is used to specify the size of the arrowhead.

Leader break Area

The **Break size** spinner in this area is used to specify the break length in the multileader while applying the dimension breaks. The default value of **Break size** is 0.125.

Leader Structure Tab

The options in the **Leader Structure** tab (Figure 10-41) of the **Modify Multileader Style** dialog box are used to specify the dimensioning attributes that affect the structure of the multileader lines. The attributes that can be controlled by using the **Leader Structure** tab are the number of lines to be drawn before adding the content, adding landing before the content, length of the landing line, multiline to be annotative or not, and so on.

Constraints Area

This area provides you the options to control the number of multileader points and the direction of the multileader.

Maximum leader points. This check box is used to specify the maximum number of points in the multileader line. The default value is **2**, which means that there will be only two points in the leader line. Select the check box to specify the maximum number of points. You can change the value of the maximum leader points in the spinner in front of this option. Note that the start point of the multileader is the first multileader point and it should also be included in the counting.

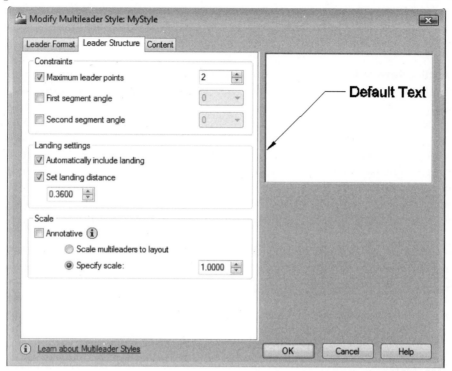

*Figure 10-41 The **Leader Structure** tab of the **Modify Multileader Style** dialog box*

First segment angle. This check box is used to specify the angle of the first multileader line from the horizontal. Select this check box to specify a value in the drop-down list displayed in front of it.

Second segment angle. This check box is used to specify the angle of the second multileader line from the horizontal. Select this check box to specify a value in the drop-down list displayed in front of it.

Note

*The angle values for the **First segment angle** and the **Second segment angle** can be in multiples of the specified angle in the respective spinners.*

Landing settings Area

This area provides you the options to control the inclusion of landing line and its length.

Automatically include landing. Select this check box to attach a landing line to the multileader. By default, this check box is selected and is used to attach the landing line to the multileaders.

Set landing distance. By default, this check box is selected and used to specify the length of the landing line to be attached with the multileader. You can specify the value in the spinner that is below this check box. This spinner gets highlighted only when you select the **Set landing distance** check box. The default value of the landing length is set to 0.36.

Scale Area

The options under this area are used to set the value for the overall multileader scale or the scale of the multileader in the paper space.

Annotative. This check box is used to set the selected multileader style to be annotative. With this option, you can also convert an existing non-annotative multileader style to annotative and vice-versa. If you select the **Annotative** check box from the **Create New Multileader Style** dialog box, then the **Annotative** check box in the **Modify Multileader Style** dialog box will also be selected by default.

Scale multileaders to layout. If you select the **Scale multileaders to layout** radio button, the scale factor between the current model space viewport and the floating viewport (paper space) will be computed automatically. Also, you can disable the **Specify scale** spinner (it is disabled in the dialog box) and the overall scale factor is set to 0 by selecting this radio button. When the overall scale factor is assigned a value of 0, AutoCAD calculates an acceptable default value based on the scaling between the current model space viewport and the paper space. AutoCAD sets the overall scale factor to 1 if you are in the paper space or not using the **Scale multileaders to layout** feature. Otherwise, AutoCAD calculates a scale factor that makes it possible to plot the multileaders at the values in which they have been previously set. For further details regarding the model space and layouts, refer to Chapter 12.

Specify scale. All the current multileaders are scaled with a value specified in the **Specify scale** spinner. You can alter the scaling factor as per your requirement by entering the scaling factor of your choice in this spinner. The default value for this variable is **1.0**. With this value, AutoCAD assumes its preset value and the drawing is plotted in a full scale. The scale factor is the reciprocal of the drawing size. So, for plotting the drawing at the one-fourth size, the overall scale factor will be 4.

Content Tab

The options in the **Content** tab (Figure 10-42) of the **Modify Multileader Style** dialog box are used to specify the multileader attributes that affect the content and the format of the text or block to be attached with the multileader.

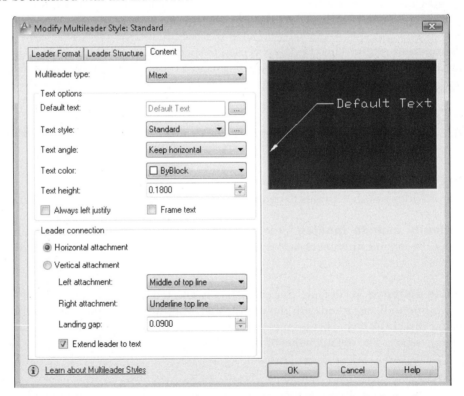

*Figure 10-42 The **Content** tab of the **Modify Multileader Style** dialog box*

You can attach the multiline text or block to the multileader. You can also attach nothing as a content to the multileader. Select the desired multileader type from the **Multileader type** drop-down list; the other options in the **Content** tab of the **Modify Multileader Style** dialog box will vary with the change in the multileader type. Figure 10-42 displays the options in the **Modify Multileader Style** dialog box for the **Mtext** multileader type and these options are discussed next.

Text options Area
The options in this area are discussed next.

Default text. This option is used to specify the default text to be attached with the multileader. Choose the **[...]** button; the **In-Place Text Editor** will be displayed. Enter the text that you want to display with the multileader by default and then, click outside the text editor window; the preview window in the **Content** tab will display the multileader with the default text attached to it. While creating a new multileader, AutoCAD prompts you to specify whether you want to retain the default text specified or overwrite it with the new text.

Text style. This drop-down list is used to select the text style to be used for writing the content of the multileader. Only the predefined and default text styles are displayed in the drop-down list.

Text angle. This drop-down list is used to control the rotation of the multiline text with respect to the landing line.

Text color. This drop-down list is used to select the color to be used for writing the content of the multileader.

Text height. This spinner is used to specify the height of the multiline text to be attached with the multileader.

Always left justify. Select this check box to left justify the text attached to the multileader. On clearing the **Always left justify** check box, a multileader with the multileader style shown in Figure 10-43 (a) is displayed. Whereas, on selecting this check box, a multileader with the multileader style as shown in Figure 10-43 (b) is displayed.

Frame text. Select this check box, it will enable you to draw a rectangular frame around the multiline text to be attached with the multileader, see Figure 10-44.

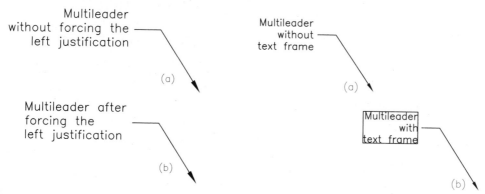

Figure 10-43 *Multileaders displaying the effect of the **Always left justify** option*

Figure 10-44 *Multileaders displaying the effect of the **Frame text** option*

Chapter 10

Leader connection Area

The options provided in this area are used for controlling the attachment of the multiline text to the multileader.

Horizontal attachment. Select this radio button to place the leader to the right or left of the text. If you select this radio button, you need to select an appropriate option from the **Left Attachment** and **Right attachment** drop-down lists to attach the multiline text to the landing line.

Vertical attachment. Select this radio button to place the leader at the top or bottom of the text. If you select this radio button, you need to select an appropriate option from the **Top attachment** and **Bottom attachment** drop-down lists to attach the multiline text with the landing line.

If you select **Block** from the **Multileader type** drop-down list, the options will change accordingly. Figure 10-45 displays the options in the **Modify Multileader Style** dialog box for the **Block** option of the **Multileader type** drop-down list and these options are discussed next.

Landing gap. This spinner is used to specify the gap between the landing line and the multiline text at the point of attachment. It is availabe in both the cases; Horizontal attachment or Vertical attachment. The default value in the spinner is 0.09.

Extend leader to text. Select this check box to extend the leader line to text. If you deselect it then the leader can be extended upto the frame only.

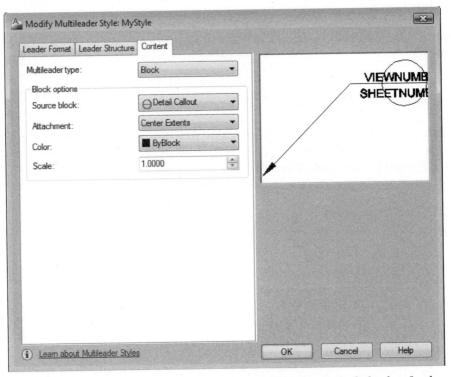

*Figure 10-45 The **Content** tab of the **Modify Multileader Style** dialog box for the **Block** option of the **Multileader type** drop-down list*

Block options Area

Source block. This drop-down list is used to select the block to be attached with the multileader as content. You can attach the standard block listed in the drop-down list or attach the user defined blocks to the multileader by selecting **User Block** from the drop-down list. Six standard shapes of block are available in this drop-down list with their previews on the side of block name. While creating the multileader, if you select the **Detail Callout** option from the **Source block** drop-down list, you will be prompted to specify the **View number** and **Sheet number** to be displayed within the block in stacked form. For the other remaining standard blocks, you will be prompted to specify the **Tag number** only to be displayed within the block. To attach the user-defined block to the multileader, select **User Block** from the **Source block** drop-down list; the **Select Custom Content Block** dialog box will be displayed with the list of all the blocks defined in the current drawing under the **Select from Drawing Blocks** drop-down list. Select the desired block and choose the **OK** button; the selected block will be added to the **Source block** drop-down list for the current drawing. Also, it will be attached to the multileader as content.

Attachment. This drop-down list is used to specify the point of attachment of the block to the leader. Select **Center Extents** from the drop-down list to attach the leader landing line in the middle of the outer boundary of the block. Select **Insertion point** to attach the leader landing line to the insertion point specified while defining the block.

Color. This drop down list is used to specify the color of the block to be attached with the leader landing line.

Self-Evaluation Test

Answer the following questions and then compare them to those given at the end of this chapter:

1. You can invoke the **Dimension Style Manager** dialog box using both the **Annotate** tab and the **Home** tab. (T/F)

2. The size of the arrow block is determined by the value specified in the **Arrow size** edit box. (T/F)

3. A block can be set as default content to be attached with the multileader. (T/F)

4. The size of the tolerance text with respect to dimensions can be defined. (T/F)

5. The **DIMTVP** variable is used to control the _____ position of the dimension text.

6. When you select the **Arrows** option from the **Fit** tab of the **Dimension Style** dialog box, AutoCAD places the text and arrowheads _____.

7. A basic dimension text is the dimension text with a _____ drawn around it.

8. The **Suppress** check boxes in the **Dimension Lines** area are used to control the display of _____ and _____.

9. You can specify the tolerance using _____ methods.

10. The _____ button in the **Dimension Style Manager** dialog box is used to override the current dimension style.

Review Questions

Answer the following questions:

1. You cannot replace the default arrowheads at the end of dimension lines. (T/F)

2. When the **DIMTVP** variable has a negative value, the dimension text is placed below the dimension line. (T/F)

3. The length of a multileader landing line cannot be changed. (T/F)

4. The named dimension style associated with the dimension being updated by overriding is not updated. (T/F)

5. Which of the following buttons can be used to make a dimension style active for dimensioning?

    (a) **Set Current**　　　　(b) **New**
    (c) **Override**　　　　　 (d) **Modify**

6. Which of the following tabs in the **Dimension Style Manager** dialog box is used to add the suffix **mm** to dimensions?

    (a) **Fit**　　　　　　　　(b) **Text**
    (c) **Primary Units**　　　 (d) **Alternate Units**

7. Which tab of the **Dimension Style Manager** dialog box will be used, if you want to place the dimension text manually every time you create a dimension?

    (a) **Fit**　　　　　　　　(b) **Text**
    (c) **Primary Units**　　　 (d) **Alternate Units**

8. The size of the _____ is determined by the value stored in the **Arrow size** edit box.

9. When **DIMSCALE** is assigned the value _____, AutoCAD calculates an acceptable default value based on the scaling between the current model space viewport and the paper space.

10. If you use the **DIMCEN** command, a positive value will create a center mark, whereas a negative value will create a _____.

11. If you select the _____ check box, you can position the dimension text anywhere along the dimension line.

12. You can append a prefix to the dimension measurement by entering the desired prefix in the **Prefix** edit box of the _____ dialog box.

13. If you select the **Limits** tolerance method from the **Method** drop-down list, AutoCAD _____ the upper value to the dimension and _____ the lower value from the dimension text.

14. You can also use the _____ command to override a dimension value.

15. What is the dimension style family and how does it help in dimensioning?

EXERCISES 4 Through 9

Create the drawings shown in Figures 10-46 through 10-51. You must create dimension style and multileader style files and specify the values for the different dimension types such as linear, radial, diameter, and ordinate. Assume missing dimensions.

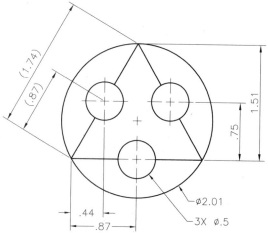

Figure 10-46 *Drawing for Exercise 4*

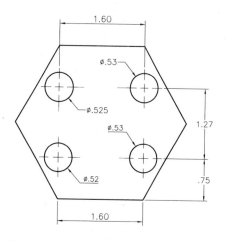

Figure 10-47 *Drawing for Exercise 5*

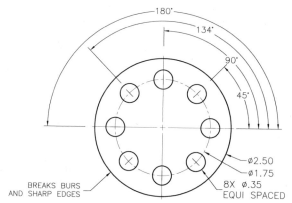

Figure 10-48 *Drawing for Exercise 6*

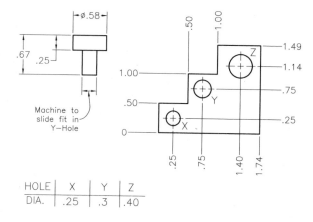

HOLE	X	Y	Z
DIA.	.25	.3	.40

Figure 10-49 *Drawing for Exercise 7*

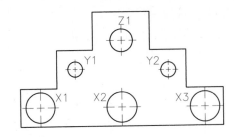

HOLE	X1	X2	X3	Y1	Y2	Z1
DIM.	R.2	R.2	R.2	R.1	R.1	R.15
QTY.	1	1	1	1	1	1
X	.25	1.375	2.50	.75	2.0	1.375
Y	.25	.25	.25	.75	.75	1.125
Z	THRU	THRU	THRU	1.0	1.0	THRU

Figure 10-50 Drawing for Exercise 8

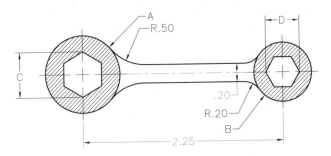

SPANNER NO.	A	B	C	D
S1	.85	.65	.50	.38
S2	1.00	.75	.59	.44
S3	1.15	.88	.67	.52
S4	1.25	.95	.74	.56

Figure 10-51 Drawing for Exercise 9

EXERCISE 10

Draw the sketch shown in Figure 10-52. You must create the dimension style and multileader style. Specify different dimensioning parameters. Also, suppress the leading and trailing zeros in the dimension style. Assume missing dimensions.

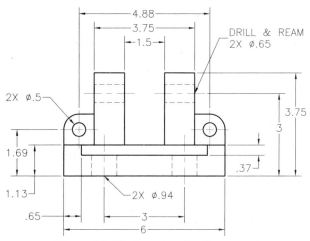

Figure 10-52 *Drawing for Exercise 10*

EXERCISE 11

Draw the sketch shown in Figure 10-53. You must create the dimension style and specify different dimensioning parameters in the dimension style. Assume missing dimensions.

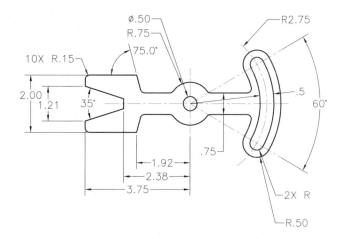

Figure 10-53 *Drawing for Exercise 11*

EXERCISE 12   *Dimension Style*

Draw the sketch shown in Figure 10-54. You must create the dimension style and specify different dimensioning parameters in the dimension style. Assume missing dimensions.

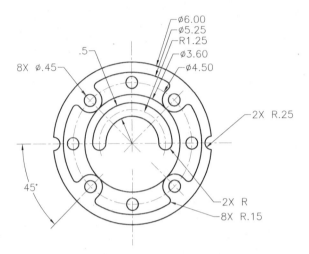

Figure 10-54 Drawing for Exercise 12

Problem-Solving Exercise 1   *Dimension Style*

Create the drawing shown in Figure 10-55. You must create the dimension style and specify different dimensioning parameters in the dimension style. Also, suppress the leading and trailing zeros in the dimension style. Assume missing dimensions.

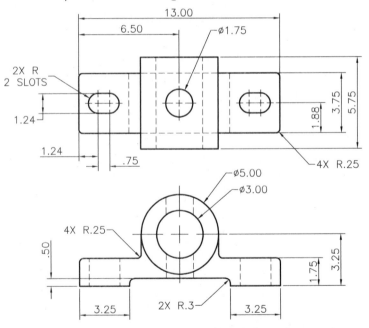

Figure 10-55 Drawing for Problem-Solving Exercise 1

Problem-Solving Exercise 2   *Dimension Style*

Draw the shaft shown in Figure 10-56. You must create the dimension style and specify the dimensioning parameters based on the given drawing. Assume missing dimensions.

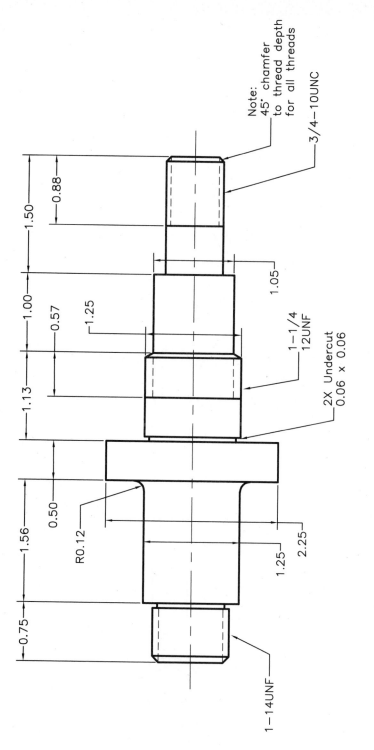

Figure 10-56 *Drawing for Problem-Solving Exercise 2*

Problem-Solving Exercise 3   *Dimension Style*

Draw the connecting rod shown in Figure 10-57. You must create the dimension style and specify the dimensioning parameters based on the given drawing. Assume missing dimensions.

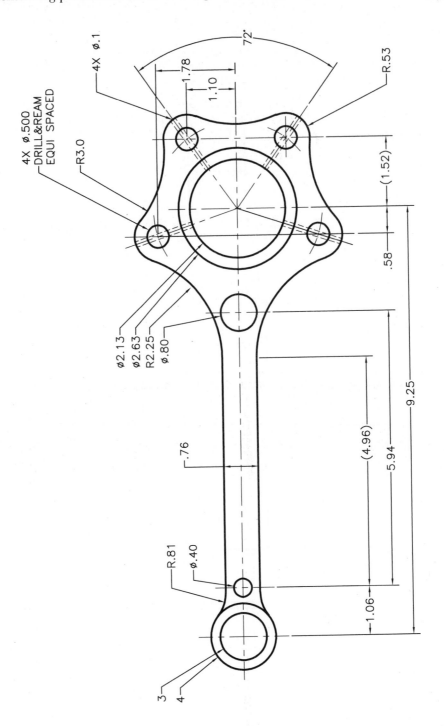

Figure 10-57 *Drawing for Problem-Solving Exercise 3*

Problem-Solving Exercise 4

Create the drawing shown in Figure 10-58. Assume missing dimensions.

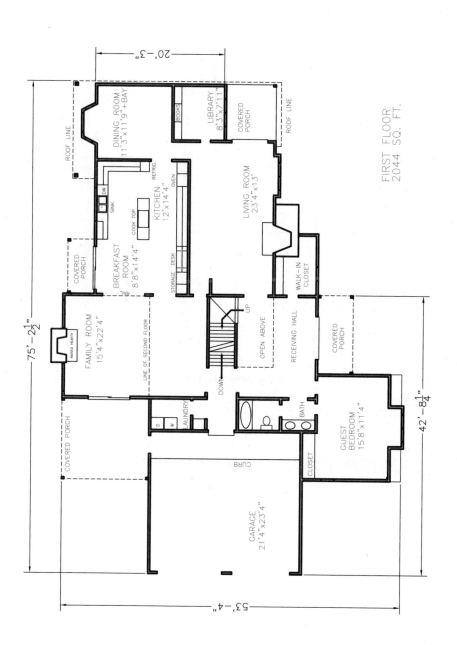

Figure 10-58 *Drawing for Problem-Solving Exercise 4*

Problem-Solving Exercise 5

Create the drawing shown in Figure 10-59. Assume missing dimensions.

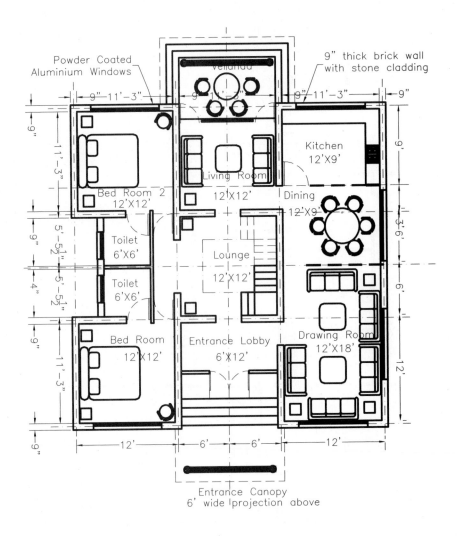

Figure 10-59 Drawing for Problem-Solving Exercise 5

Answers to Self-evaluation Test

1. T, **2.** T, **3.** T, **4.** T, **5.** vertical, **6.** inside, **7.** frame, **8.** first, second dimension lines, **9.** four, **10.** Override

Chapter *11*

Adding Constraints to Sketches

CHAPTER OBJECTIVES

In this chapter, you will learn to:

- *Add geometric constraints to sketches.*
- *Control the display of constraints.*
- *Apply constraints automatically.*
- *Apply dimensional constraints to sketches.*
- *Convert a dimensional constraint into an annotational constraint.*
- *Create equations.*

KEY TERMS

- *Horizontal*
- *Vertical*
- *Coincident*
- *Fix*
- *Collinear*
- *Parallel*

- *Concentric*
- *Tangent*
- *Symmetric*
- *Equal*
- *Smooth*
- *Infer Constraints*

- *Dimensional Constraints*
- *Annotational Constraints*
- *Dynamic Constraint Mode*

- *Annotational Constraint Mode*
- *Fully-Defined Sketch*
- *Under-Defined Sketch*
- *Over-Defined Sketch*
- *Parameters Manager*

INTRODUCTION

In AutoCAD, a wide set of tools are available that helps you create parametric drawings. These tools are available in the **Parametric** tab of the **Ribbon** as well as in the **Parametric** toolbar. The parametric nature of a drawing implies that the shape and size of a geometry can be defined using standard properties or parameters. The main function of parametric property is to drive a new size or shape to the selected geometry without considering its original dimensions. Also, you can modify the shape and size of any entity of a sketch at any stage of the design process. This property makes the designing process very easy.

To make a drawing parametric, you need to apply constraints. When you apply constraints to sketches, it restricts their degrees of freedom, thereby making them stable. The stability ensures that the size, shape, and location of sketches do not change unexpectedly with respect to surroundings. You can apply two types of constraints to sketches, geometric and dimensional. Geometric constraints are applied between two entities in a sketch and they control the shape of a drawing, whereas dimensional constraints are applied to a single entity and they control the size of the object in a drawing. Remember that dimensions applied and dimensional constraints applied are different. You will learn about the difference in this chapter.

If you need to draw a sketch of parts such as bolt, nut, valves, and so on in which the size of one entity is driven by one basic dimension, then in AutoCAD, you can draw these parts quickly and easily using the **Parameters Manager**.

ADDING GEOMETRIC CONSTRAINTS

AutoCAD provides you with twelve types of geometric constraints. Usually, these constraints are applied between two entities in a sketch and control the shape of a drawing. However, there are constraints, such as Fix, that are applied to a single entity. Whenever you apply a geometric constraint between two entities, the entity, which is selected second is affected. For example, if you apply the parallel constraint between two lines, the line selected second will orient toward the first line. It is recommended to first apply geometric constraints so that the shape of a sketched entity does not change on applying dimensional constraints. If the **Infer Constraints** button is chosen in the Status Bar, some of the geometric constraints are applied automatically while sketching or editing. For example, if you draw a horizontal line then the **Horizontal** constraint will be applied automatically and the corresponding constraint bar will be displayed above the line. In the following section, different geometric constraints and the procedure to apply them manually are discussed.

Applying the Horizontal Constraint

Ribbon: Parametric > Geometric > Horizontal
Menu Bar: Parametric > Geometric Constraints > Horizontal

The **Horizontal** constraint forces a selected line, polyline, or two points to become horizontal. To apply this constraint using the Command window, enter **GEOMCONSTRAINT** at the Command prompt. The prompt sequence to apply this constraint is given next.

Command: **GEOMCONSTRAINT**
Enter constraint type
[Horizontal/Vertical/Perpendicular/PArallel/Tangent/SMooth/Coincident/CONcentric/COLlinear/Symmetric/Equal/Fix] <Horizontal>: Enter
Select an object or [2Points] <2Points>: *Select a line or polyline.*

Remember that if you apply this constraint to a line on which no other constraint is applied, then the line will orient such that the endpoint nearer to the location you clicked on will become the fixed point and the other endpoint will become horizontal, thereby making the line horizontal.

If you need to make two points horizontal, press ENTER when the **Select an object or [2Points] <2Points>** prompt is displayed. The following prompt sequence will be displayed:

Select an object or [2Points] <2Points>:
Select first point: *Select the first point*
Select second point: *Select the second point*

You can also apply this constraint by choosing the **Horizontal** tool from the **Geometric** panel of the **Parametric** tab.

Applying the Vertical Constraint

> **Ribbon:** Parametric > Geometric > Vertical
> **Menu Bar:** Parametric > Geometric Constraints > Vertical

The **Vertical** constraint forces a selected line, polyline, or two points to become vertical. To apply this constraint using the Command window, enter **GEOMCONSTRAINT** at the Command prompt. The prompt sequence to apply this constraint is given next.

Command: **GEOMCONSTRAINT**
Enter constraint type
[Horizontal/Vertical/Perpendicular/PArallel/Tangent/SMooth/Coincident/CONcentric
/COLlinear/Symmetric/Equal/Fix] <Horizontal>: *Enter V and* ⏎
Select an object or [2Points] <2Points>: *Select a line or polyline*

Remember that if you apply this constraint to a line on which no other constraint is applied, then the line will orient such that the endpoint nearer to the location you clicked on will become the fixed point and the other endpoint will become vertical, thereby making the line vertical.

If you need to make two points vertical, press ENTER when the **Select an object or [2Points] <2Points>** prompt is displayed. The following prompt sequence will be displayed:

Select an object or [2Points] <2Points>: ⏎
Select first point: *Select the first point*
Select second point: *Select the second point*

You can also apply this constraint by choosing the the **Parametric** tab.

Note
On applying a constraint to an entity, the corresponding symbol is displayed near the entity, see Figure 11-1. If you move the cursor over the symbol, the symbol will be highlighted and a cross mark (X) will be displayed. Click on the cross mark to hide the symbol. However, if you place the cursor on the entity on which a constraint has already been applied, a glyph will be displayed, stating that a constraint is applied.

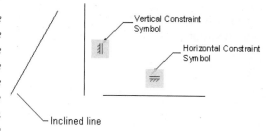

Figure 11-1 *The Vertical and Horizontal Constraint symbols*

Note
*You can apply a constraint by invoking a particular command using the tools in the **Parametric** tab of the **Ribbon** or by entering **GEOMCONSTRAINT** at the Command prompt and choosing the required option from the Command prompt. In this chapter, you will learn how to apply a constraint using the tools from the **Parametric** tab of the **Ribbon**. (The options in the Command prompt will be discussed, wherever necessary).*

Applying the Coincident Constraint

Ribbon: Parametric > Geometric > Coincident
Menu Bar: Parametric > Geometric Constraints > Coincident

The coincident constraint forces a selected point or keypoint of an entity to be coincident with the point or keypoint of another entity. The entity selected can be a line, polyline, circle, an arc or ellipse. Similarly, the selected point can be a point, or keypoints such as an endpoints or midpoint of a line, an arc or a polyline, or the centerpoint of an arc, ellipse, or a circle. To apply this constraint, choose the **Coincident** tool from the **Geometric** panel; the cursor will be replaced by a selection box along with the coincident symbol. Also, you will be prompted to select an object. Move the cursor over an entity; the keypoint that is close to the cursor will be highlighted in red. Next, click to select it; you will be prompted to select the second object to be coincident with the first object. Select the keypoint of the second entity; the keypoint of the second entity will move and coincide with that of the first entity.

Even if you have drawn the complete sketch by snapping keypoints, the coincident constraint will not be applied automatically. To apply the coincident constraints at all intersections at the same time, invoke the **Coincident** tool and enter **A** at the **Select first point or [Object/Autoconstrain] <Object>:** prompt; you will be prompted to select objects. Select the objects individually or by using the selection window and press ENTER; the coincident constraint will be applied to all selected entities automatically. The Command prompt sequence to apply the **Autoconstrain** is given next.

Command: **GEOMCONSTRAINT**
Enter constraint type [Horizontal/Vertical/Perpendicular/PArallel/Tangent/SMooth/Coincident/CONcentric/COLlinear/Symmetric/Equal/Fix] <Coincident>: Enter
Select first point or [Object/Autoconstrain] <Object>: *Enter **A** to apply **Autoconstrain**,* Enter
Select objects: *Select objects individually or by using the selection window:*
Select objects: *Select more objects or* Enter

Note
*The options to change the default settings for the **Autoconstrain** option will be discussed later in this chapter.*

Applying the Fix Constraint

Ribbon: Parametric > Geometric > Fix
Menu Bar: Parametric > Geometric Constraints > Fix

The **Fix** constraint forces the selected entity to become fixed at a position. If you apply this constraint to a line or to an arc, its location will be fixed with respect to the selected keypoint. But, you can change the size of the entity by dragging its endpoints. To apply the **Fix** constraint, choose the **Fix** tool from the **Geometric** panel; you will be prompted to select a point or an object. Next, move the cursor over the entity; the point closer to the cursor will be highlighted. Click to fix the highlighted keypoint. Now, you cannot move the fixed keypoint. However, you can drag and relocate other keypoints.

To fix all keypoints of an object, invoke the **Fix** constraint and press ENTER when the **Select point or [Object] <Object>** Command prompt is displayed; you will be prompted to select an object. Select the object; all its keypoints will be fixed. Now, you can drag the keypoints to resize or rotate the object, but you cannot relocate it.

Applying the Perpendicular Constraint

Ribbon: Parametric > Geometric > Perpendicular
Menu Bar: Parametric > Geometric Constraints > Perpendicular

The **Perpendicular** constraint forces the selected lines to become perpendicular to each other. To apply this constraint, choose the **Perpendicular** tool from the **Geometric** panel; you will be prompted to select an object. Select a line or a polyline; you will be prompted to select the second object. Select another line or a polyline. The object selected second will become perpendicular to the object selected first. Figure 11-2 shows two lines before and after applying the perpendicular constraint.

Applying the Parallel Constraint

Ribbon: Parametric > Geometric > Parallel
Menu Bar: Parametric > Geometric Constraints > Parallel

The **Parallel** constraint forces the selected lines to become parallel to each other. To apply this constraint, choose the **Parallel** tool from the **Geometric** panel; you will be prompted to select the first object. Select a line or a polyline; you will be prompted to select the second object. Select another line or a polyline; the object selected second will become parallel to the object selected first. Figure 11-3 shows two lines before and after applying the parallel constraint.

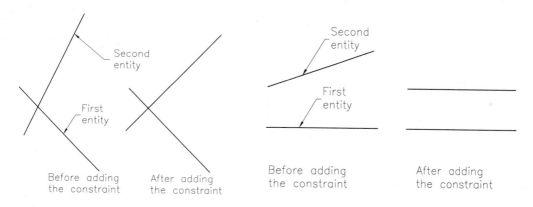

Figure 11-2 *Entities before and after applying the perpendicular constraint*

Figure 11-3 *Entities before and after applying the parallel constraint*

Applying the Collinear Constraint

Ribbon: Parametric > Geometric > Collinear
Menu Bar: Parametric > Geometric Constraints > Collinear

The **Collinear** constraint forces the selected lines to lie on the same infinite line. To apply this constraint, choose the **Collinear** tool from the **Geometric** panel; you will be prompted to select the first object. Select a line; you will be prompted to select the second object. Select another line; the line selected later will become collinear with the first line.

To apply the collinear constraint to multiple lines, choose the **Collinear** tool; the **Select first object or [Multiple]** prompt will be displayed at the Command prompt. Enter **M** and press ENTER; you will be prompted to select the first object. Select a line; you will be prompted to select the second object. Select the lines that need to be made collinear with the line selected first and then press ENTER; the selected lines will become collinear to the line selected first.

Applying the Concentric Constraint

Ribbon: Parametric > Geometric > Concentric
Menu Bar: Parametric > Geometric Constraints > Concentric

The concentric constraint forces the selected arc or circle to share the center point of another arc, circle, or ellipse. To apply this constraint, choose the **Concentric** tool from the **Geometric** panel; you will be prompted to select the first object. Select an arc, ellipse, or a circle; you will be prompted to select the second object. Select another arc, ellipse, or circle; the object selected second will become concentric to the object selected first.

Applying the Tangent Constraint

Ribbon: Parametric > Geometric > Tangent
Menu Bar: Parametric > Geometric Constraints > Tangent

The tangent constraint forces the selected arc, circle, spline, or ellipse to become tangent to another arc, circle, spline, ellipse, or line. To apply this constraint, choose the **Tangent** tool from the **Geometric** panel; you will be prompted to select an object. Select an arc, circle, spline, line, or an ellipse; you will be prompted to select the second object. Select another arc, circle, spline, ellipse, or line; the object selected second will become tangent to the object selected first. Figure 11-4 shows a line and a circle before and after applying the **Tangent** constraint. Figure 11-5 shows two arcs before and after applying the **Tangent** constraint.

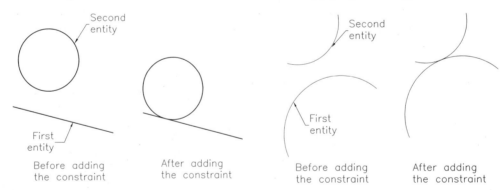

Figure 11-4 A line and a circle before and after applying the tangent constraint

Figure 11-5 Two arcs before and after applying the tangent constraint

Applying the Symmetric Constraint

Ribbon: Parametric > Geometric > Symmetric
Menu Bar: Parametric > Geometric Constraints > Symmetric

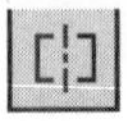

The symmetric constraint forces two selected lines, arcs, circles, or ellipses to remain equidistant from the centerline. To apply this constraint, choose the **Symmetric** tool from the **Geometric** panel; you will be prompted to select the first object or two points. Select an arc, an ellipse, a line, or a circle; you will be prompted to select the second object. Select another arc, ellipse, circle, or line; you will be prompted to select the symmetry line. Select a line; the selected objects will become symmetric with respect to the selected line. You can also make two points symmetric. To do so, press ENTER at the **Select first object or (2 points)** Command prompt. Next, select two points and the symmetric line; the selected points will become symmetric to the selected line.

Applying the Equal Constraint

Ribbon: Parametric > Geometric > Equal
Menu Bar: Parametric > Geometric Constraints > Equal

The equal constraint forces the selected lines or polylines to have equal length, or the selected circles, arcs, or an arc and a circle to have equal radii. To apply this constraint, choose the **Equal** tool from the **Geometric** panel; you will be prompted to select an object. Select the first object; you will be prompted to select the second object. Select another object; the size of the second object will become equal to that of the object selected first.

If you need to apply the equal constraint to multiple objects, choose the **Equal** tool from the **Geometric** panel; the **Select first object or [Multiple]** prompt will be displayed at the Command prompt. Enter **M** and press ENTER; you will be prompted to select the first object. Select an object; you will be prompted to select the second object. Select the other objects that need to be equal and press ENTER; the size of the objects selected later will become equal to that of the object selected first.

Applying the Smooth Constraint

Ribbon: Parametric > Geometric > Smooth
Menu Bar: Parametric > Geometric Constraints > Smooth

The smooth constraint is used to apply a curvature continuity between a spline and an entity connected together. The entities that can be connected to a spline include a line, spline, polyline, or an arc. To apply this constraint, choose the **Smooth** tool from the **Geometric** panel; you will be prompted to select the first object. Select a spline as the first object; you will be prompted to select the second object. Select another object; the smooth constraint will be applied between the selected objects.

Tip
*In AutoCAD, you can apply the constraints while sketching or editing. To do so, choose the **Infer Constraints** button from the Status Bar before sketching or editing.*

CONTROLLING THE DISPLAY OF CONSTRAINTS

To control the display of the constraints, click on the inclined arrow in the **Geometric** panel of the **Parametric** tab; the **Constraint Settings** dialog box will be displayed, as shown in Figure 11-6. In this dialog box, the **Geometric** tab will be chosen and the check boxes of all constraints will be selected. Also, the **Show constraint bars after applying constraints to selected objects** check box will be selected, by default. Therefore, on applying a constraint, a constraint bar will be displayed above the sketch and the symbol of that particular constraint will be displayed in the constraint bar, by default. If you do not need any particular constraint symbol to be displayed then in the **Constraint bar display settings** area, clear the check box of the constraints that you do not want to be displayed. If you do not need the constraint bar to be displayed on applying the constraints, then clear the **Show constraint bars after applying constraints to selected objects** check box. However, if the **Show constraint bars when objects are selected** check box is selected, then the constraint bars will be displayed when you select an object. You can control the transparency of the constraint bars by setting the value in the **Constraint bar Transparency** area of the **Constraint Settings** dialog box. In AutoCAD, you can apply the constraints while sketching or editing. To do so, select the **Infer geometric constraints** check box. You can also choose the **Infer Constraints** button from the Status Bar to apply constraints while sketching or editing. After specifying the parameters, choose **OK**.

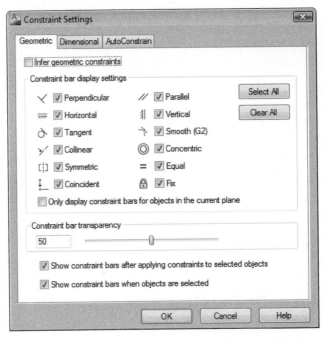

*Figure 11-6 The **Constraint Settings** dialog box with the **Geometric** tab chosen*

On moving the cursor over a constraint bar, the constraint symbol and entities on which the constraint was applied, will be highlighted. If a sketch has multiple constraints and all are displayed, it will be confusing. Therefore, you need to hide or display the constraint symbols. To hide a particular constraint, move the cursor over that constraint; the constraint bar will be highlighted and a cross-mark will be displayed on it. Click on the cross-mark to hide that constraint. To hide all constraints, choose the **Hide All** button in the **Geometric** panel. To display all hidden constraints, choose the **Show All** button from the **Geometric** panel. If you need to display constraints of a particular entity, choose the **Show/Hide** button from the **Geometric** panel; you will be prompted to select an entity. Select the entity from the drawing area and press ENTER; a menu will be displayed and you will be prompted to select an option. Select the **Show** option; the constraints of that particular entity will be displayed. Remember that hiding a constraint does not delete that constraint. To delete a constraint, move the cursor over the constraint bar and press the DELETE key. If you need to delete multiple applied constraints, choose the **Delete Constraints** button in the **Manage** panel of the **Parametric** tab of the **Ribbon**; you will be prompted to select entities. Select the entities individually or by using the selection box, and then press ENTER; the constraints applied on the selected entities will be deleted.

Tip
If you move the cursor over a constraint bar and right-click, a shortcut menu will be displayed. Choose the corresponding options to delete or hide a constraint. Also, you can choose the **Constraint Bar Settings** *option to display the* **Constraint Settings** *dialog box.*

EXAMPLE 1

Draw the sketch shown in Figure 11-7 and apply appropriate constraints to it. Do not dimension the sketch.

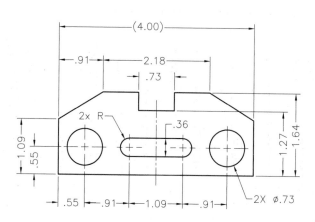

Figure 11-7 *Sketch for Example 1*

In this example, you will select the **Infer geometric constraints** check box and draw the sketch. Therefore, some of the constraints will be applied while sketching. Then, you will apply the constraints that are not applied automatically while sketching. First, you will create the outer loop of the sketch using the **Line** tool.

1. Start a new file with the *Acad.dwt* template in the **Drafting & Annotation** workspace.

2. Ensure that the **Infer Constraints** button is chosen in the Status Bar. By using the **Line** tool, draw the outer profile similar to that shown in Figure 11-7. Make sure that the endpoints are snapped so that they form a closed sketch, as shown in Figure 11-8. Note that in your drawing the constraints applied by default may be different from that shown in Figure 11-8, as it depends upon the start point and the procedure used to draw the lines. Also, if the ortho mode is on while drawing the horizontal and vertical lines, the corresponding constraints will be applied where ever possible.

   Next, you need to apply the constraints that are not applied by default. First, you will apply the **Equal** constraint.

3. Choose the **Equal** tool from the **Geometric** panel in the **Parametric** tab; the cursor changes to a selection box with the **Equal** constraint's symbol attached to it. Also, you are prompted to select the first object.

4. Select the vertical line on the left as the first object; you are prompted to select the second object.

5. Select the vertical line on the right as the second object; the equal constraint is applied and the constraint bar is displayed.

6. Similarly, apply the equal constraint between the two inclined lines, two small vertical lines, and two small horizontal lines. The sketch after applying all constraints is shown in Figure 11-9.

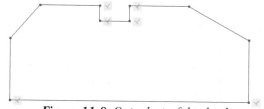

Figure 11-8 *Outer loop of the sketch*

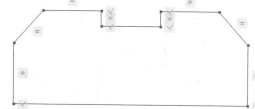

Figure 11-9 *Sketch after applying all constraints*

Chapter 11

Next, you need to apply the symmetric constraint between entities on either side of the centerline.

7. Choose the **Infer Constraints** button from the Status Bar to turn off the infer constraints

8. Draw a vertical centerline that passes through the midpoint of the horizontal line at the bottom, as shown in Figure 11-10.

9. Choose the **Symmetric** tool from the **Geometric** panel; the cursor changes to a selection box with symmetric symbol attached to it. Also, you are prompted to select the first object.

10. Select the vertical line on the left as the first object; you are prompted to select the second object.

11. Select the vertical line on the right as the second object; you are prompted to select the symmetry line.

12. Select the centerline as the symmetry line; the vertical lines become symmetric to the centerline and the symmetric symbol is displayed, as shown in Figure 11-11. Note that in this figure, other constraints are hidden for clarity.

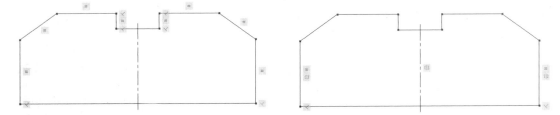

Figure 11-10 Sketch after drawing the centerline   *Figure 11-11 Sketch symmetric constraint symbol*

13. Similarly, apply the symmetric constraints between the two inclined lines and then between the two small horizontal lines. The three symmetric constraints applied are shown in Figure 11-12.

14. Choose the **Hide All** button from the **Parametric** tab of the **Geometric** panel to hide all constraint bars. Figure 11-13 shows the sketch after hiding all constraint bars.

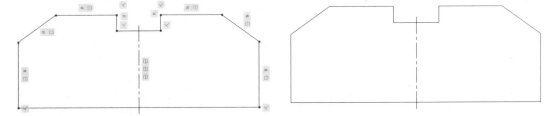

Figure 11-12 Sketch with symmetric constraints   *Figure 11-13 Sketch after hiding all constraints*

Next, you need to create the inner loop using the **Arc**, **Circle**, and **Line** tools at an arbitrary location. But, the profile should be similar to that given in Figure 11-7. Then, you need to apply the constraints.

15. Draw two circles of any radius inside the outer loop.

16. Choose the **Infer Constraints** button from the Status Bar to turn on the infer constraints

17. Draw a slot using the **Center, Start, End** and **Line** tools. The sketch after drawing the circles and the slot is shown in Figure 11-14. You can also draw the profile of the slot using the **Polyline** tool. However, the sketch in this tutorial has been drawn using the **Center, Start, End** and **Line** tools.

18. Apply the equal constraint first between two circles and then between two arcs.

19. Apply the symmetric constraint between the circles as well as between the arcs.

20. Choose the **Tangent** tool from the **Geometric** panel; the cursor changes to a selection box with the tangent symbol attached to it. Also, you are prompted to select the first object.

21. Select an arc; you are prompted to select the second object.

22. Select a line; the tangent constraint will be applied at one end of the line and the tangent symbol will be displayed.

23. Similarly, apply the tangent constraint at all other places in the slot, as shown in Figure 11-15. You need to choose the **Show/Hide** button to display all constraints.

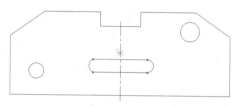

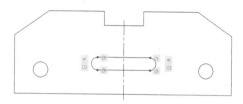

Figure 11-14 *Sketch after applying the coincident constraint*

Figure 11-15 *Sketch after applying all tangent constraints*

Next, you need to make the centerpoint of the circles and the slot to be in a straight line.

24. Choose the **Horizontal** tool from the **Geometric** panel; the prompt sequence that follows is given next.

Select an object or [2Points] <2Points>: Enter
Select first point: *Select the circle on the left side* Enter
Select second point: *cen* Enter
of: *Select the center of the arc on the left side* Enter

The sketch after applying the horizontal constraint is shown in Figure 11-16.

25. Choose the **Show/Hide** button from the **Parametric** tab and hide the display of all constraint bars and save the drawing. The sketch after applying the horizontal constraint is shown in Figure 11-17.

Chapter 11

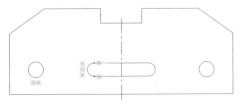

Figure 11-16 *Sketch after applying the horizontal constraint between a circle and an arc*

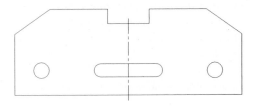

Figure 11-17 *Sketch after applying and hiding all constraints*

APPLYING CONSTRAINTS AUTOMATICALLY

In the previous section, twelve types of constraints are discussed. Of these, eight constraints can be applied automatically, provided you have a closed sketch. To apply constraints automatically, choose the **AutoConstrain** button in the **Geometric** panel of the **Parametric** tab of the **Ribbon**; you will be prompted to select an object. Select the entities individually or by using the selection box, and then press ENTER; all possible constraints will be applied automatically.

You can also specify the settings of the constraints that need to be applied automatically. To do so, choose the **AutoConstrain** button; the **Select objects or [Settings]** prompt will be displayed at the Command prompt. Enter **S** at the Command prompt and press the ENTER key; the **Constraint Settings** dialog box will be displayed, as shown in Figure 11-18.

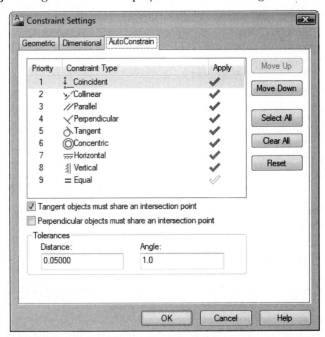

Figure 11-18 *The **Constraint Settings** dialog box with the **AutoConstrain** tab chosen*

In this dialog box, the **AutoConstrain** tab is chosen by default and the eight constraints that can be applied automatically are displayed. You can rearrange these constraints according to your priorities by choosing the **Move Up** or **Move Down** button. If you do not need a particular constraint to be applied automatically to the sketch, click on the green tick mark corresponding to the constraint; the tick mark will turn white indicating that the corresponding constraints will not be applied automatically. If the **Tangent objects must share an intersection point** check box is selected, the tangent constraint will be applied, only if the two entities intersect at a point. If you clear this check box, the tangent constraint will be applied, only if the distance between the endpoints satisfies the value specified in the **Tolerances** area. Similarly, if the **Perpendicular objects must share an intersection point** check box is selected, the perpendicular constraint will be applied, only if the two entities intersect at a point. If you clear this check box, the perpendicular constraint will be applied, only if the distance between the endpoints satisfies the tolerances value specified in the **Tolerances** area.

EXERCISE 1

Draw the sketch shown in Figure 11-19 and apply the appropriate constraints to it. Do not dimension the sketch.

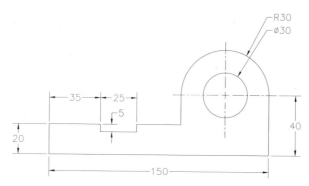

Figure 11-19 *Sketch for Exercise 1*

APPLYING DIMENSIONAL CONSTRAINTS

The older versions of AutoCAD were mainly used during the documentation stage of a design. However, the introduction of dimensional constraint has made AutoCAD a complete design package. In AutoCAD, you can apply dimensional constraint to an entity and change the size of the entity with ease. Therefore, now you can use AutoCAD at the initial design stage itself. Remember that the dimensions, which you apply using the tools in the **Dimensions** panel of the **Annotate** tab of the **Ribbon**, do not change the size of the entity. However, because of their associative nature, if you change the size of an entity by dragging grips, the dimension will get changed.

You can apply the horizontal, vertical, aligned, angular, radius, and diameter dimensional constraint to a sketch. The procedure to apply a dimensional constraint to an entity is similar to that of applying dimension to an entity. For example, to apply the horizontal dimensional constraint, invoke the **Linear** tool from **Parametric > Dimensional > Dimensional Constraints** drop-down; you will be prompted to select the first constraint point. Move the cursor near one of the keypoints; it will be highlighted. Click to select the keypoint; you will be prompted to select the second constraint point. Move the cursor near the other keypoint and click, when it gets highlighted; the dimension will be attached to the cursor and you will be prompted to place the dimension. On placing the dimension, the dimension value will be displayed in a textbox. Press ENTER to accept the value. If you need to change the value, enter a new value in the dimension box and press ENTER. A lock symbol will be displayed next to the dimension value indicating that it is a dimensional constraint. In this way, you can apply other dimensional constraints as well.

Instead of selecting keypoints to apply the dimensional constraint, you can also select the entity to be dimensioned. To do so, invoke the **Linear** tool from **Parametric > Dimensional > Dimensional Constraints** drop-down and press ENTER when the **Specify first constraint point** or **[Object]** **<Object>** prompt is displayed; you will be prompted to select an object. Select the object; the dimension will be attached to the cursor. Place it at the required position.

Tip
You can also enter the arithmetic functions directly in the text box that is displayed on placing a dimensional constraint. For example, if you have an arithmetic function such as (22 *5.5)-30+5, which is equal to 96, you do not need to calculate this function. Just enter the function in the text box and press ENTER; AutoCAD will automatically solve the mathematical expression. However, only the mathematical expression will be displayed in the text box and not the result.

CONVERTING A DIMENSIONAL CONSTRAINT INTO AN ANNOTATIONAL CONSTRAINT

As discussed earlier, the dimensional constraint applied and the dimensions applied are different. The dimensional constraints are applied by using the tools available in the **Dimensional** panel of the **Parametric** tab. Whereas, dimensions are applied using the tools in the **Dimensions** panel of the **Annotate** tab. The dimensions that are applied using the tools from the **Dimensions** panel are called annotational constraints as they are used only for annotational purpose and cannot control the size of an entity.

If you have applied a dimension (annotational constraint) to an entity, you can convert it into a dimensional constraint. To do so, choose the **Convert** tool from the **Parametric** tab; you will be prompted to select a dimension. Select the dimension and press ENTER; the dimension (annotational constraint) will be converted into the dimensional constraint.

You can also convert a dimensional constraint into an annotational constraint. To do so, select a dimensional constraint and invoke the **Properties** palette. Select **Annotational** from the **Constraint** field in the **Constraint** rollout; the dimensional constraint will change to the annotational constraint and the options in the **Properties** rollout will change accordingly.

A dimensional constraint will not be plotted when you plot the drawing using the **Plot** tool. Therefore, you need to convert a dimensional constraint into an annotational constraint, as discussed above. You can also place the dimensional constraint in the annotational constraint mode. To do so, expand the **Dimensional** panel in the **Parametric** tab. You will notice that the **Dynamic Constraint Mode** is chosen by default. Choose the **Annotational Constraint Mode**. Now, if you place the dimensional constraint, you can plot them.

You can also make a dimensional constraint as a reference constraint. A reference constraint is displayed in brackets and is driven by a dimensional or annotational constraint. To make a dimensional constraint as reference, invoke the **Properties** palette of a dimensional constraint and select **Yes** in the **Reference** field of the **Constraint** rollout. If you have changed a dimensional constraint to an annotational constraint, you can make this annotational constraint as a reference constraint using the **Properties** palette.

CONCEPT OF A FULLY-DEFINED SKETCH

It is important for you to understand the concept of fully-defined sketches. If you have created a sketch in the design stage, you need to add the required geometrical and dimensional constraints to the sketch to constrain its degrees of freedom with respect to the surrounding. After applying the required constraints, the sketch may exist in any of the following three states:

1. Under-defined
2. Fully-defined
3. Over-defined

Under-defined

An under-defined sketch is the one in which some of the geometrical constraints or dimensional constraints are not defined. As a result, the degrees of freedom of the sketch are not fully constrained, and therefore, the entities may move or change their size unexpectedly. So, you need to add some geometrical constraints or dimensional constraints to it to make it fully-defined.

Fully-defined

A fully-defined sketch is the one in which all entities and their positions are fully-defined by geometrical constraints, dimensional constraints, or both. In a fully-defined sketch, all degrees of freedom of a sketch are constrained, and therefore, the sketched entities cannot move or change their size and location unexpectedly.

Over-defined

An over-defined sketch is the one in which some of the geometrical constraints, dimensional constraints, or both are conflicting. Also, if the geometrical constraints or the dimensional constraints in a sketch exceed the required number, the sketch will be over-defined. If you try to apply more geometrical constraints or dimensional constraints to a fully-defined entity, a message box will be displayed with the message that applying this constraint will over define the entity, as shown in Figures 11-20 and 11-21.

Figure 11-20 The **Geometric Constraints** *message box displayed on applying more than the required geometric constraints*

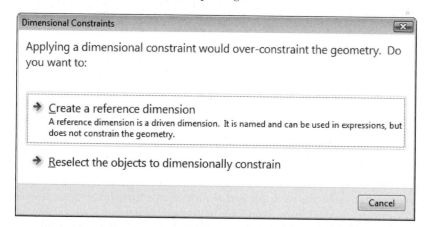

Figure 11-21 The **Dimensional Constraints** *message box displayed on applying more than the required dimensional constraints*

Note

*If you apply a dimensional constraint to a fully-defined entity, the **Dimensional Constraints** message box will be displayed, refer to Figure 11-21. Choose the **Create a reference dimension** option from the message box to convert that particular dimension into a reference dimension.*

CONTROLLING THE DISPLAY OF THE DIMENSIONAL CONSTRAINT

On applying a dimensional constraint to an entity, a value will be displayed along with a name. If you have also applied annotational constraints to the sketch, then the sketch will be cluttered with dimensions. Therefore, you need to hide these constraints. To do so, choose the **Show All** button from the **Dimensional** panel of the **Parametric** tab. Remember that this is a toggle button, therefore, you can choose this button again to display all dimensional constraints. To hide the annotational constraint, turn off the corresponding layer from the **Layer Properties Manager**.

When you apply a dimensional constraint to an entity, a default name and a lock symbol will be displayed along with the value. If you need to display only the value, click on the inclined arrow on the **Dimensional** panel of the **Parametric** tab; the **Constraint Settings** dialog box will be displayed with the **Dimensional** tab chosen, as shown in Figure 11-22. Specify the format of the dimensional constraint to be displayed in the **Dimension name format** drop-down list available in the **Dimensional constraint format** area of this dialog box.

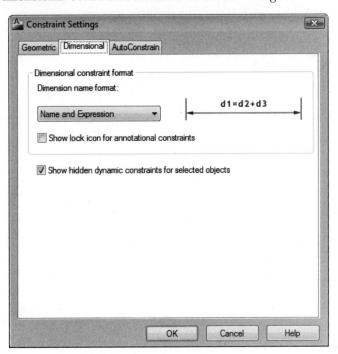

Figure 11-22 The **Constraint Settings** *dialog box with the **Dimensional** tab chosen*

If you convert a dimensional constraint into an annotational constraint, a lock symbol will be displayed. If you do not need the lock symbol to be displayed for an annotational constraint, clear the **Show lock icon for annotational constraints** check box from the **Constraint Settings** dialog box. In case, you want to view the dynamic constraints of an entity whose dimensional constraints are hidden by choosing the **Show Dynamic Constraints** button, select the **Show hidden dynamic constraints for selected objects** check box in the **Constraint Settings** dialog box. Now, if you select an entity whose dynamic constraints are hidden, the dynamic constraints will be displayed.

EXAMPLE 2

In this example, you will draw the sketch shown in Figure 11-23, and then apply the geometric and dimensional constraints to it.

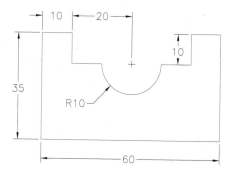

Figure 11-23 *Sketch for Example 2*

First, you will draw the profile and apply the geometrical constraints.

1. Start a new file with the *Acad.dwt* template in the **Drafting & Annotation** workspace and draw the outer profile with dimensions close to that given in Figure 11-23. Make sure that the endpoints are snapped so that they form a closed sketch, as shown in Figure 11-24.

2. Choose the **AutoConstrain** button from the **Geometric** panel of the **Parametric** tab, enter **S** at the Command prompt, and press ENTER; the **Constraint Settings** dialog box is displayed.

3. Make sure that all constraints are selected. As there is no need to apply the concentric or tangent constraints in this sketch, select the **Concentric** and **Tangent** constraints individually and move them down using the **Move Down** button.

4. Choose the **OK** button; you are prompted to select objects. Select all entities of the sketch by dragging a selection box. Then, press ENTER; all possible constraints are applied to the sketch, as shown in Figure 11-25.

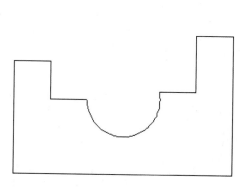

Figure 11-24 *Closed sketch drawn*

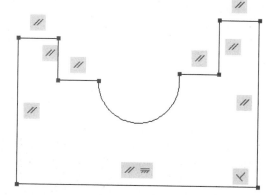

Figure 11-25 *Constraints applied automatically*

Note

*The constraints applied using the **AutoConstrain** tool may vary depending upon the profile. In your case, if the constraints are different from that shown in Figure 11-25, apply the missing constraints. Also, choose the **Show All** button from the **Geometric** panel, if the constraints are not visible by default.*

Chapter 11

5. Now, you need to apply the equal constraint. To do so, choose the **Equal** tool from the **Geometric** panel of the **Parametric** tab and apply the equal constraint between the lines that are supposed to have equal dimensions. For dimensions, refer to Figure 11-23. The equal constraints need to be applied to the two pairs of horizontal and two pairs of vertical lines in the sketch. The sketch after applying the equal constraint is shown in Figure 11-26.

Note

If you do not want to apply the equal constraint to vertical lines, you need to apply collinear constraint to the two pairs of horizontal lines. But, after applying the equal constraint, if you apply the collinear constraint, a message box will be displayed stating that this constraint cannot be applied and conflicts the existing constraints or is over-constraining the geometry.

Next, you need to make the centerpoint of the circle and the horizontal line to be in a line.

6. Choose the **Horizontal** tool from the **Geometric** panel of the **Parametric** tab; the prompt sequence that will follow is given next.

Select an object or [2Points] <2Points>: [Enter]
Select first point: *Move the cursor over the horizontal line and click when the endpoint near the arc is highlighted* [Enter]
Select second point: *cen* [Enter]   *and select the arc on the left side.*

The sketch after applying the horizontal constraint is shown in Figure 11-27.

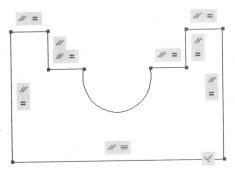

Figure 11-26 *Sketch after applying the equal constraint*

Figure 11-27 *Sketch after applying the horizontal constraint*

7. Choose the **Hide All** button from the **Geometric** panel in the **Parametric** tab; all constraint bars disappear.

Now, you need to apply dimensional constraints. Before that, expand the **Dimensional Constraint** panel and ensure that the **Dynamic Constraint Mode** option is chosen.

8. Choose the **Linear** tool from **Parametric** > **Dimensional** > **Dimensional Constraint** drop-down and follow the prompt sequence given next.

Specify first constraint point or [Object] <Object>: [Enter]
Select object: *Select the horizontal line at the bottom.*
Specify dimension line location: *Move the cursor down and place the horizontal dimension at the required location. Type* **60** *in the text box and press ENTER.*

9. Press the ENTER key again or choose the **Linear** tool from the **Dimensional** panel of the **Parametric** tab and follow the prompt sequence given next.

   Specify first constraint point or [Object] <Object>: [Enter]
   Select object: *Select the vertical line on the left*
   Specify dimension line location: *Move the cursor on the left and place the vertical dimension at the required location. Enter 35 in the textbox.*

10. Similarly, apply the dimension to the horizontal and vertical lines (refer to Figure 11-28) and change the value to **10**.

11. Choose the **Radius** tool from the **Dimensional** panel of the **Parametric** tab; you are prompted to select the arc or the circle.

12. Select the arc and enter **10** as radius.

Now, you need to apply the dimension between the center of the arc and the vertical line on the left, as shown in Figure 11-23.

13. Invoke the **Linear** tool from the **Parametric > Dimensional > Dimensional Constraints** drop-down and follow the prompt sequence given next.

   Specify first constraint point or [Object] <Object>: *Select the keypoint at the top of the smaller vertical line on the left.*
   Specify second constraint point: *cen* [Enter] *and select the arc.*
   Specify dimension line location: *Place the dimension at the required location and enter 20; the* **Dimensional Constraints** *message box is displayed.*

14. Choose the **Create a reference dimension** option from the message box. The value **20** is displayed in parenthesis stating that it is a reference dimension. The sketch after applying all dimensional constraints is shown in Figure 11-28.

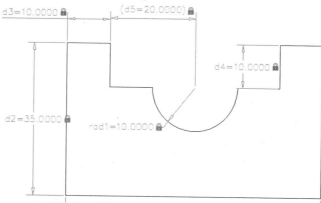

Figure 11-28 Sketch with all dimensional constraints

WORKING WITH EQUATIONS

Equations are mathematical relations between dimensions. Consider the sketch of a bolt. All dimensions of a bolt are driven by the base diameter. In AutoCAD, if you need to draw the sketch of an object whose dimensions are driven by a base dimension, you can draw it easily by using equations in which all dimensions are equated with base dimensions. If you want to relate one or two dimensions with base dimension, you can do so by applying a dimensional constraint.

But if most of the dimensions need to be represented as equations, then you need to use the **Parameters Manager**. Both these methods are discussed next.

Adding Equations while Applying Dimensional Constraints

When you apply the dimensional constraint using the tools from the **Dimensional** panel, the dimensions are displayed along with a default name. For the next dimensional constraint to be placed, you can represent it as an equation with the name of the first constraint. For example, to create a rectangle of length *l* units and width *w* units, which is half the length, you need to follow the steps given below.

1. Draw a rectangle and apply the required geometric constraints to it.
2. Choose the **Linear** tool from the **Dimensional** panel and select one of the lines representing the length and place the dimension; a text box will be displayed.
3. Enter **10** in the text box and press ENTER; a new dimension will be displayed with the default name d1=10.0000.
4. Choose the **Linear** tool from the **Dimensional** panel and select one of the lines representing the width and place the dimension; a text box will be displayed.
5. Enter **0.5*d1** in the text box and press ENTER; the rectangle will resize and a new dimension will be displayed with the default name fx: d2=0.5*d1.
6. Now, if you change the dimension of the length, the width will also change automatically.

Note that, if you change the default name, d1, the name in the expression d2 will also change.

Adding Equations Using the Parameters Manager

Adding equations to dimensions while applying constraints may look simpler, but in actual practice, when there are more dimensions, it is quite difficult. This problem can be solved by applying equations to dimensions using the **Parameters Manager**. After applying dimensional constraints, choose the **Parameters Manager** button from the **Manage** panel of the **Parametric** tab; the **Parameters Manager** palette will be displayed with all dimensions and expressions, as shown in Figure 11-29. The names of the dimensional constraints applied will be listed under the **Name** column in the **Dimensional Constraint Parameters** node. Also, the expression of the corresponding constraints will be listed in the **Expression** column.

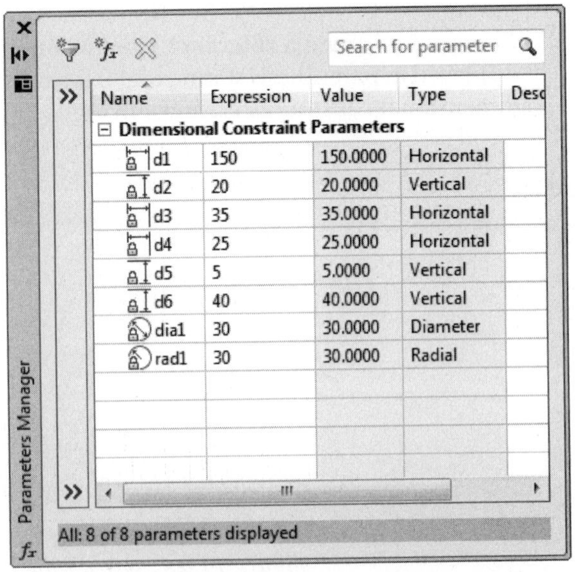

Now, you can change the existing expressions or add new expressions. To do so, double-click in the corresponding **Expression** field

Name	Expression	Value	Type	Desc
Dimensional Constraint Parameters				
d1	150	150.0000	Horizontal	
d2	20	20.0000	Vertical	
d3	35	35.0000	Horizontal	
d4	25	25.0000	Horizontal	
d5	5	5.0000	Vertical	
d6	40	40.0000	Vertical	
dia1	30	30.0000	Diameter	
rad1	30	30.0000	Radial	

All: 8 of 8 parameters displayed

*Figure 11-29 The **Parameters Manager** palette*

and add a new expression. If you want to add a list of new user-defined expressions, choose the **Creates a new user parameter** button from the **Parameters Manager** palette; a new field will be added under the **User Parameters** node. You can add user-defined expression in the field and specify the constraints as expression in any one of the fields under the **Dimensional Constraint Parameters** node.

EXAMPLE 3

In this example, you will open the sketch created in Exercise 1 of this chapter. You had applied geometrical constraints to this sketch. Now, you will apply dimensional constraints and then equate all dimensions, except the dimension 20 mm, with respect to the base length, 150 mm. To the dimension 20 mm, you will apply user-defined expressions as d150 = 20, d75 = 10, d300 = 40, and d100 = 13.3.

1. Choose the **Open** button from the **Quick Access Toolbar** and open the drawing created in Exercise 1 of this chapter.

2. Make sure that you have applied all required geometrical constraints, see Figure 11-30.

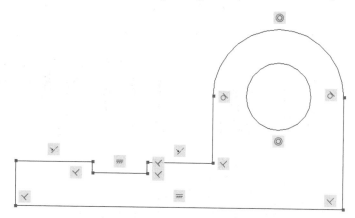

Figure 11-30 Sketch with all constraints applied

Note
If you have not applied all constraints to the sketch, it may get deformed while applying dimensional constraints.

3. Invoke the **Linear** tool from **Parametric > Dimensional > Dimensional Constraints** drop-down and follow the prompt sequence given next.

    Specify first constraint point or [Object] <Object>: Enter
    Select object: *Select the horizontal line at the bottom.*
    Specify dimension line location: *Move the cursor down and place the horizontal dimension at the required location. Type **150** in the text box and press ENTER.*

4. Press ENTER again or choose the **Linear** tool from the **Dimensional** panel and follow the prompt sequence given next.

    Specify first constraint point or [Object] <Object>: Enter
    Select object: *Select the vertical line on the left.*
    Specify dimension line location: *Move the cursor to left side and place the vertical dimension at the required location. Enter **20** in the text box and press ENTER.*

5. Similarly, apply dimensions to other horizontal and vertical lines.

6. Choose the **Radius** tool from the **Dimensional** panel of the **Parametric** tab; you are prompted to select an arc or a circle.

7. Select the arc and enter **30** as the value of radius.

8. Choose the **Diameter** tool from the **Dimensional** panel of the **Parametric** tab; you are prompted to select the circle.

9. Select the circle and enter **30** as diameter.

10. Choose the **Parameters Manager** button from the **Manage** panel; the **Parameters Manager** palette will be displayed with the names of dimensional constraints and their values, as shown in Figure 11-31.

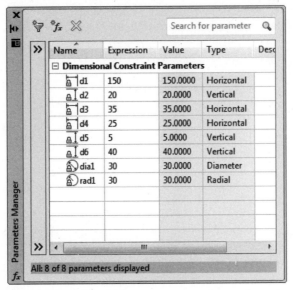

Figure 11-31 *The **Parameters Manager** palette after applying all dimensions*

11. Double-click on **40** in the **Expression** column of the **Parameters Manager** palette and enter **d1/3.75**.

12. Resize the palette. Double-click in the **Description** column corresponding to the value **40** and enter **Hole Location**.

13. Similarly, add the expression and description to all dimensions, except the vertical dimension 20, as given below.

Value to be selected from the Expression column	Expression to be added	Description to be added
35	d1/30*7	Slot Location
25	d1/6	Slot Width
5	d1/30	Slot Depth
30 (Diameter value)	d1/5	Hole Diameter
30 (Radius value)	d1/5	Radius

Note
*If the **Type** and **Description** columns are not added by default, right-click on the title of a column and choose the appropriate options.*

The **Parameters Manager** palette after you have changed all expressions is shown in Figure 11-32.

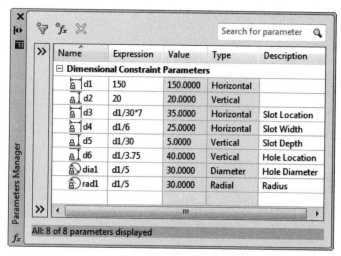

*Figure 11-32 The **Parameters Manager** palette after adding all expressions*

For the value 20, you need to create a user-defined expression as follows:

14. Choose the **Creates a new user parameter** button in the **Parameters Manager** palette; the **User Parameters** node is added with the default name, user1 selected.

15. Change the name user1 to **d150**, the corresponding expression to **20**, and the description to **If d1 = 150**.

16. Similarly, add three more user parameters and their corresponding expressions and descriptions as given below.

User parameter to be added	Expression to be added	Description to be added
d75	10	If d1 = 75
d300	40	If d1 = 300
d100	13.3	If d1 = 100

Note
*You can also write a suitable expression for the dimension 20 in the **Dimensional Constraints** node. For the purpose of example, the expression for the dimension 20 is given as a user variable.*

17. If the dimensional constraints are displayed, choose the **Show/Hide Dynamic Constraints** button from the **Dimensional** panel of the **Ribbon** to hide the dimensional constraints.

18. Now, change the dimensional constraints **d1** and **d2**, as specified below in the **Parameters Manager** palette. Observe the change in the sketch in the drawing area and the dimension value in the **Value** column of the **Parameters Manager** palette.

$$d1 = 75, \qquad d2 = d75$$
$$d1 = 300, \qquad d2 = d300$$
$$d1 = 100, \qquad d2 = d100$$
$$d1 = 150, \qquad d2 = 20$$

Self-Evaluation Test

Answer the following questions and then compare them to those given at the end of this chapter:

1. AutoCAD provides you _____ types of geometric constraints.

2. The _____ constraints are applied between two entities in a sketch and they control the shape of the drawing.

3. Whenever you apply a geometric constraint between two entities, the entity selected first will be affected. (T/F)

4. You cannot apply the coincident constraint to all keypoints at a time. (T/F)

5. It is always recommended to first apply dimensional constraints so that the shape of the sketched entity does not change on applying geometric constraints. (T/F)

6. You can add equations while applying dimensional constraints. (T/F)

7. Which of the following tools should be chosen in the **Dimensional** panel to apply both horizontal and linear dimensional constraints?

    (a) **Aligned**   (b) **Linear**
    (c) **Lateral**   (d) None of these

8. In which of the following states will a sketch exist, if only few constraints are defined?

    (a) Under-defined   (b) Fully-defined
    (c) Over-defined   (d) All of these

9. The _____ constraint forces the selected lines to lie on the same infinite line.

10. Invoke the _____ palette to add equations to dimensional constraints.

Review Questions

Answer the following questions:

1. The _____ makes you to change or modify the shape and size of an entity of a sketch at any stage of the design process.

2. The _____ constraint is used to apply a curvature continuity between a spline and an entity connected together.

3. A _____ constraint is displayed in brackets and is driven by a dimensional or annotational constraint.

4. The dimensions that you apply using the tools in the _____ panel do not change the size of an entity.

5. The symmetric constraint forces two selected lines, arcs, circles, or splines to remain equidistant from the centerline. (T/F)

6. Using the fix constraint, you can only fix the points. (T/F)

7. You can also enter arithmetic symbols directly in the text box that is displayed on placing a dimensional constraint. (T/F)

8. If you hide the display of a constraint, the constraint applied gets deleted. (T/F)

9. How many geometric constraints can be applied automatically?

    (a) Six   (b) Ten
    (c) Eight   (d) Twelve

10. Which of the following constraints forces two lines or polylines to have equal length and the selected circles, arcs, or an arc and circle to have equal radii?

    (a) Concentric   (b) Equal
    (c) Collinear   (d) Same

EXERCISE 2

Draw the sketches shown in Figures 11-33 and 11-34. Then, add geometric and dimensional constraints to them. **

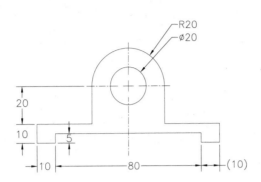

Figure 11-33 *Sketch for Exercise 2*

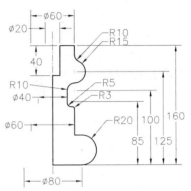

Figure 11-34 *Sketch for Exercise 2*

Answers to Self-Evaluation Test

1. twelve, **2.** Geometric, **3.** F, **4.** F, **5.** F, **6.** T, **7.** **Linear**, **8.** Under-defined, **9.** collinear, **10.** **Parameters Manager**

Chapter 12

Technical Drawing with AutoCAD

CHAPTER OBJECTIVES

In this chapter, you will learn:
- *The concepts of multiview drawings.*
- *About X, Y, and Z axes, XY, YZ and XZ planes, and parallel planes.*
- *To draw orthographic projections and position views.*
- *To dimension a drawing.*
- *About basic dimensioning rules.*
- *To draw sectional views using different types of sections.*
- *To hatch sectioned surfaces.*
- *To use and draw auxiliary views.*
- *To create assembly and detail drawings.*

KEY TERMS

- *Orthographic Projections*
- *Dimensioning Components*
- *Dimensioning Rules*

- *Full Section*
- *Half Section*
- *Broken Section*
- *Revolved Section*

- *Offset Section*
- *Aligned Section*
- *Hatch Lines*
- *Auxiliary Views*

- *Detail Drawing*
- *Assembly Drawing*
- *Bill of Materials*

MULTIVIEW DRAWINGS

When designers design a product, they visualize its shape in their minds. To represent that shape on paper or to communicate the idea to people, they must draw a picture of the product or its orthographic views. Pictorial drawings, such as isometric drawings, convey the shape of the object, but it is difficult to show all of its features and dimensions in an isometric drawing. Therefore, in industry, multiview drawings are the accepted standards for representing products. Multiview drawings are also known as orthographic projection drawings. To draw different views of an object, it is very important to visualize the shape of the product. The same is true when you look at different views of an object to determine its shape. To facilitate visualizing the shapes, you must picture the object in 3D space with reference to the *X*, *Y*, and *Z* axes. These reference axes can then be used to project the image into different planes. This process of visualizing objects with reference to different axes is, to some extent, natural in human beings. You might have noticed that sometimes, when looking at objects that are at an angle, people tilt their heads. This is a natural reaction, an effort to position the object with respect to an imaginary reference frame (*X*, *Y*, and *Z* axes).

UNDERSTANDING X, Y, AND Z AXES

To understand the *X*, *Y*, and *Z* axes, imagine a flat sheet of paper on the table. The horizontal edge represents the positive *X* axis, and the other edge along the width of the sheet, represents the positive *Y* axis. The point where these two axes intersect is the origin. Now, if you draw a line perpendicular to the sheet passing through the origin, the line defines the positive *Z* axis (Figure 12-1). If you project the *X*, *Y*, and *Z* axes in the opposite direction beyond the origin, you will get the negative *X*, *Y*, and *Z* axes (Figure 12-2).

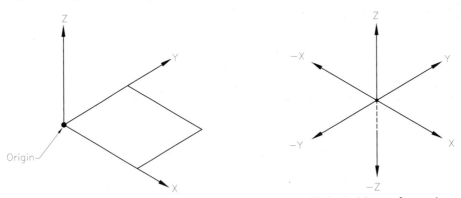

Figure 12-1 *X, Y, and Z axes*   *Figure 12-2* *Positive and negative axes*

The space between the *X* and *Y* axes is called the *XY* plane. Similarly, the space between the *Y* and *Z* axes is called the *YZ* plane, and the space between the *X* and *Z* axes is called the *XZ* plane (Figure 12-3). A plane parallel to these planes is called a parallel plane (Figure 12-4).

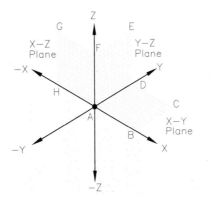   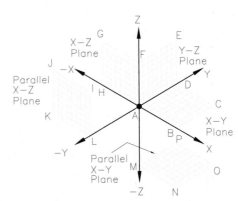

Figure 12-3 *XY, YZ, and XZ planes*   *Figure 12-4* *Parallel planes*

ORTHOGRAPHIC PROJECTIONS

The first step in drawing an orthographic projection is to position the object along the imaginary X, Y, and Z axes. For example, if you want to draw orthographic projections of the step block shown in Figure 12-5, position the block so that the far left corner coincides with the origin, and then align it with the X, Y, and Z axes.

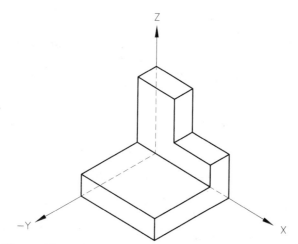

Figure 12-5 *Aligning the object with the X, Y, and Z axes*

Now you can look at the object from different directions. Looking at the object along the negative Y axis and toward the origin is called the front view. Similarly, looking at it from the positive X direction is called the right side view. To get the top view, you can look at the object from the positive Z axis. See Figure 12-6.

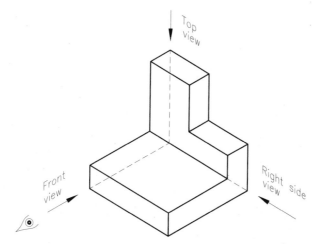

Figure 12-6 *Viewing directions for Front, Side, and Top views*

To draw the front, side, and top views, project the points onto the parallel planes. For example, to draw the front view of the step block, imagine a plane parallel to the XZ plane located at a certain distance in front of the object. Now, project the points from the object onto the parallel plane (Figure 12-7), and join them to complete the front view.

Repeat the same process for the side and top views. To represent these views on paper, position them, as shown in Figure 12-8.

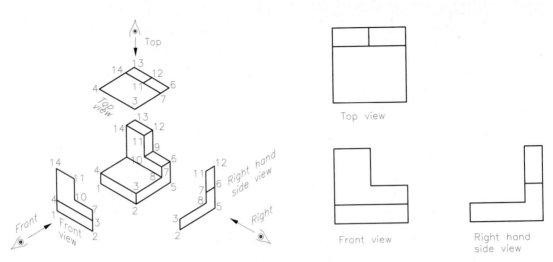

Figure 12-7 *Projecting points onto parallel planes*

Figure 12-8 *Representing views on paper*

Another way of visualizing different views is to imagine the object enclosed in a glass box (Figure 12-9).

Now, look at the object along the negative *Y* axis and draw the front view on the front glass panel. Repeat the process by looking along the positive *X* and *Z* axes, and draw the views on the right side and the top panel of the box (Figure 12-10).

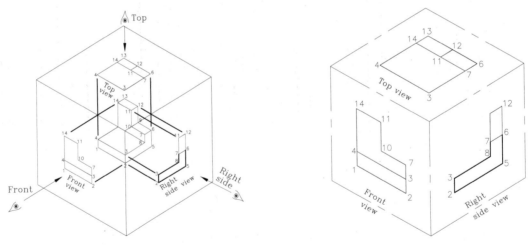

Figure 12-9 *Objects inside a glass box*

Figure 12-10 *Front, top, and side views*

To represent the front, side, and top views on paper, open the side and the top panels of the glass box (Figure 12-11). The front panel is assumed to be stationary.

After opening the panels through 90-degree, the orthographic views will appear, as shown in Figure 12-12.

POSITIONING ORTHOGRAPHIC VIEWS

Orthographic views must be positioned, as shown in Figure 12-13. The right side view must be positioned directly on the right side of the front view. Similarly, the top view must be directly above the front view. If the object requires additional views, they must be positioned, as shown in Figure 12-14.

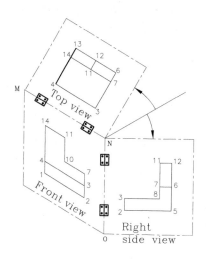

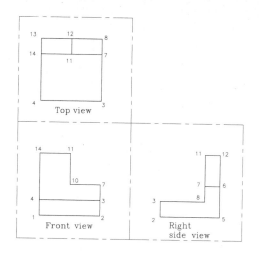

Figure 12-11 *Opening the side and the top panels*

Figure12-12 *Views after opening the box*

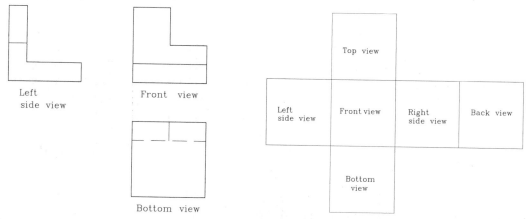

Figure 12-13 *Positioning orthographic views*

Figure 12-14 *Standard placement of orthographic views*

The different views of the step block are shown in Figure 12-15.

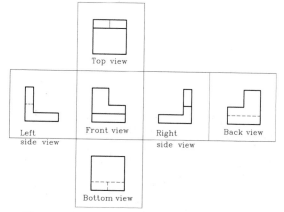

Figure 12-15 *Different views of the step block*

In this example, you will draw the required orthographic views of the object shown in Figure 12-16. Assume the dimensions of the drawing.

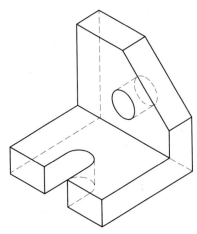

***Figure 12-16** Step block with hole and slot*

Drawing the orthographic views of an object involves the following steps:

1. Look at the object, and determine the number of views required to show all of its features. For example, the object in Figure 12-16 will require three views only (front, side, and top).

2. Based on the shape of the object, select the side that you want to show as the front view. In this example, the front view, i.e. the view along -y axis, is the one that shows the maximum number of features or that gives a better idea about the shape of the object. Sometimes, the front view is determined by how the part will be positioned in an assembly.

3. Picture the object in your mind, and align it along the imaginary *X*, *Y*, and *Z* axes (Figure 12-17).

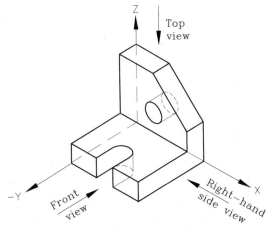

***Figure 12-17** Align the object with the imaginary X, Y, and Z axes*

4. Look at the object along the negative *Y* axis and project the image on the imaginary *XZ* parallel plane (Figure 12-18).

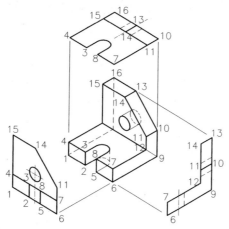

Figure 12-18 *Project the image onto the parallel planes*

5. Use the projection on imaginary XZ parallel plane to draw the front view of the object. If there are any hidden features, they must be drawn with hidden lines. The holes and slots must be shown with center lines.

6. To draw the right side view, look at the object along the positive *X* axis and project the image onto the imaginary *YZ* parallel plane.

7. Use the projection on imaginary YZ parallel plane to draw the right side view of the object. If there are any hidden features, they must be drawn with hidden lines. The holes and slots, when shown in the side view, must have one center line.

8. Similarly, draw the top view to complete the drawing. Figure 12-19 shows different views of the given object.

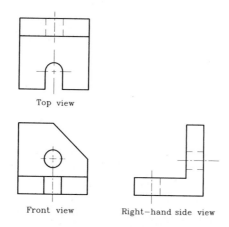

Figure 12-19 *Front, side, and top views*

EXERCISES 1 Through 4   *Orthographic Views*

Draw the required orthographic views of the objects shown in Figure 12-20 through Figure 12-23. The distance between the dotted lines is 0.5 units.

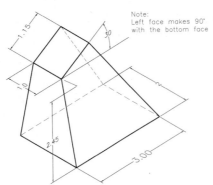

Note:
Left face makes 90°
with the bottom face

Figure 12-20 *Drawing for Exercise 1 (the object is shown as a surface wireframe model)*

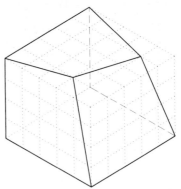

Figure 12-21 *Drawing for Exercise 2*

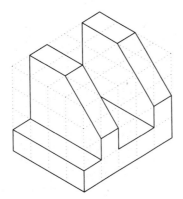

Figure 12-22 *Drawing for Exercise 3*

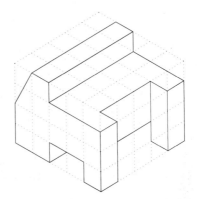

Figure 12-23 *Drawing for Exercise 4*

DIMENSIONING

Dimensioning is one of the most important elements in a drawing. When you dimension, you not only give the size of a part, you give a series of instructions to a machinist, an engineer, or an architect. The way the part is positioned in a machine, the sequence of machining operations, and the location of different features of the part depend on how you dimension it. For example, the number of decimal places in a dimension (2.000) determines the type of machine, which will be used to do that machining operation. The machining cost of such an operation is significantly higher than a dimension that has only one digit after the decimal (2.0). If you are using a computer numerical control (CNC) machine, locating a feature may not be a problem, but the number of pieces you can machine without changing the tool depends on the tolerance assigned to a dimension. A closer tolerance (+.0001, -.0005) will definitely increase the tooling cost and ultimately the cost of the product. Similarly, if a part is to be forged or cast, the radius of the edges and the tolerance you provide to these dimensions determine the cost of the product, the number of defective parts, and the number of parts you get from the die.

While dimensioning, you must consider the manufacturing process involved in making a part and the relationships that exist among different parts in an assembly. If you are not familiar with any operation, get help. You must not assume things when dimensioning or making a piece part drawing. The success of a product, to a large extent, depends on the way you dimension a part. Therefore, never underestimate the importance of dimensioning in a drawing.

Dimensioning Components

A dimension consists of the following components (Figure 12-24):

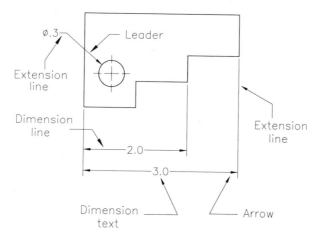

Figure 12-24 Dimensioning components

Extension line
Arrows or tick marks
Dimension line
Leader lines
Dimension text

The extension lines are drawn to extend the points that are dimensioned. The length of the extension lines is determined by the number of dimensions and the placement of the dimension lines. These lines are generally drawn perpendicular to the surface. The dimension lines are drawn between the extension lines, at a specified distance from the object lines. The dimension text is a numerical value that represents the distance between the points. The dimension text can also consist of a variable (A, B, X, Y, Z12,...), in which case the value assigned to it is defined in a separate table. The dimension text can be centered around or on the top of the dimension line. Arrows or tick marks are drawn at the end of the dimension line to indicate the start and end of the dimension. Leader lines are used when dimensioning a circle, arc, or any nonlinear element of a drawing. They are also used to attach a note to a feature or to give the part numbers in an assembly drawing.

Basic Dimensioning Rules

1. You should make the dimensions in a separate layer/layers. This makes it easy to edit or control the display of dimensions (freeze, thaw, lock, and unlock). Also, the dimension layer/layers should be assigned a unique color so that at the time of plotting, you can assign the desired pen to plot the dimensions. This helps to control the line width and the contrast of dimensions at the time of plotting.

2. The distance of the first dimension line should be at least 0.375 units (10 units for metric drawing) from the object line. In CAD drawing, this distance may be 0.75 to 1.0 units (19 to 25 units for metric drawings). Once you decide on the spacing, it should be maintained throughout the drawing.

3. The distance between the first dimension line and the second dimension line must be at least 0.25 units. In CAD drawings, this distance may be 0.25 to 0.5 units (6 to 12 units for metric drawings). If there are more dimension lines (parallel dimensions), the distances between them must be the same (0.25 to 0.5 units). Once you decide on the spacing (0.25 to

0.5), the same spacing should be maintained throughout the drawing. An appropriate snap setting is useful for maintaining this spacing. If you are using baseline dimensioning, you can use AutoCAD's **DIMDLI** variable to set the spacing. You must present the dimensions so that they are not crowded, especially when there is not much space (Figure 12-25).

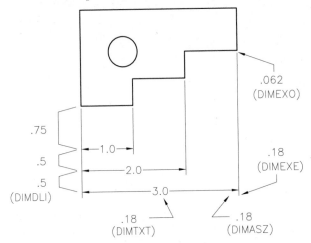

Figure 12-25 Arrow size, text height, and spacing between dimension lines

4. For parallel dimension lines, the dimension text can be staggered if there is not enough room between the dimension lines to place the dimension text. You can use the AutoCAD Object Grips feature or the **DIMTEDIT** command to stagger the dimension text (Figure 12-26).

 Note
You can change the grid, snap or UCS origin, and snap increment to make it easier to place the dimensions. You can also add the following lines to the AutoCAD menu file or toolbar buttons.

SNAP;R;\0;SNAP;0.25;GRID;0.25
SNAP;R;0,0;0;SNAP;0.25;GRID;0.5

The first line sets the snap to 0.25 units and allows the user to define the new origin of snap and grid display. The second line sets the grid to 0.5 and snap to 0.25 units. It also sets the origin for grid and snap to (0,0).

5. All dimensions should be given outside the view. However, the dimensions can be shown inside the view if they can be easily understood there and cause no confusion with the other dimensions or details (Figure 12-27).

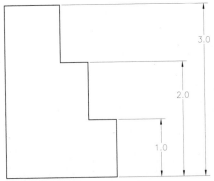

Figure 12-26 Staggered dimensions

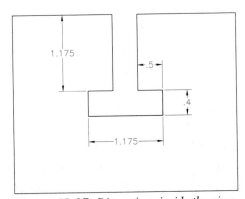

Figure 12-27 Dimensions inside the view

6. Dimension lines should not cross the extension lines (Figure 12-28). You can accomplish this by giving the smallest dimension first and then the next largest dimension (Figure 12-29).

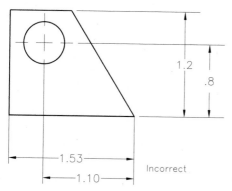

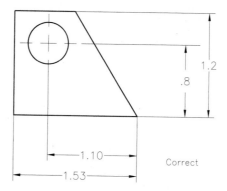

Figure 12-28 *Dimension lines crossing each other*   **Figure 12-29** *Dimensions placed correctly*

7. If you decide to have the dimension text aligned with the dimension line, then the entire dimension text in the drawing must be aligned (Figure 12-30). Similarly, if you decide to have the dimension text horizontal or above the dimension line, then to maintain uniformity in the drawing, the entire dimension text must be horizontal (Figure 12-31) or above the dimension line (Figure 12-32).

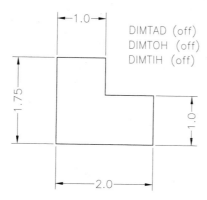

Figure 12-30 *Dimension text aligned with the dimension line*

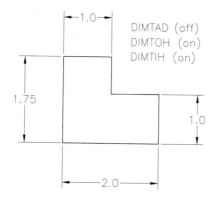

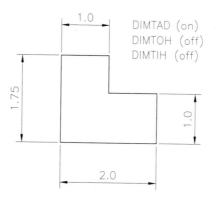

Figure 12-31 *Dimension text horizontal*   **Figure 12-32** *Dimension text above the dimension line*

8. If you have a series of continuous dimensions, they should be placed in a continuous line (Figure 12-33). Sometimes you may not be able to give the dimensions in a continuous line even after adjusting the dimension variables. In that case, give the dimensions that are parallel (Figure 12-34).

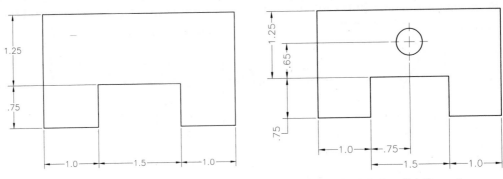

Figure 12-33 *Continuous Dimensions*   **Figure 12-34** *Parallel dimensions*

9. You should not dimension the hidden lines. The dimension should be given where the feature is visible (Figure 12-35). However, in some complicated drawings, you might be justified to dimension a detail with a hidden line.

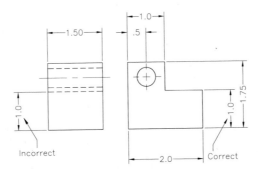

Figure 12-35 *Hidden lines not dimensioned*

10. The dimensions must be given where the feature that you are dimensioning is obvious and shows the contour of the feature (Figure 12-36).

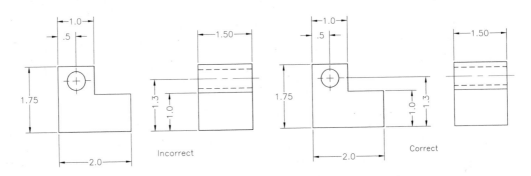

Figure 12-36 *Dimensions should be given where they are obvious*

11. The dimensions must not be repeated; this makes it difficult to update a dimension (Figure 12-37).

12. The dimensions must be given depending on how the part will be machined, and also on the relationship that exists between the different features of the part (Figure 12-38).

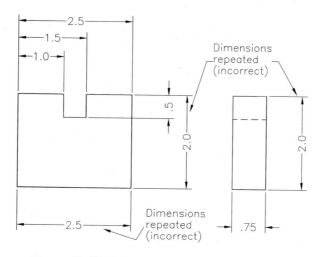

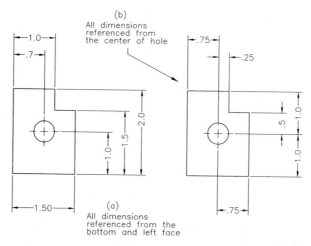

Figure 12-37 *Dimension must not be repeated*

Figure 12-38 *When dimensioning, consider the machining processes involved*

13. When a dimension is not required but you want to give it for reference, it must be a reference dimension. The reference dimension must be enclosed with parentheses (Figure 12-39).

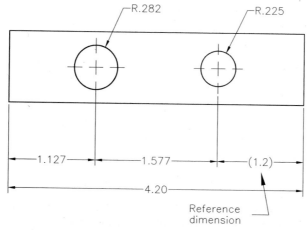

Figure 12-39 *Reference dimension*

14. If you give continuous (incremental) dimensions for dimensioning various features of a part, the overall dimension must be omitted or given as a reference dimension (Figure 12-40). Similarly, if you give the overall dimension, one of the continuous (incremental) dimensions must be omitted or given as a reference dimension. Otherwise, there will be a conflict in tolerances. For example, the total positive tolerance on the three incremental dimensions shown in Figure 12-40 is 0.06. Therefore, the maximum size based on the incremental dimensions is $(1 + 0.02) + (1 + 0.02) + (1 + 0.02) = 3.06$. Also, the positive tolerance on the overall 3.0 dimension is 0.02. Based on this dimension, the overall size, of the part must not exceed 3.02. This causes a conflict in tolerances: with incremental dimensions, the total tolerance is 0.06, whereas with the overall dimension, the total tolerance is only 0.02.

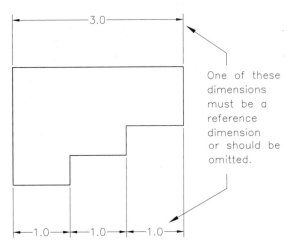

Figure 12-40 *Referencing or omitting a dimension*

15. If the dimension of a feature appears in a section view, you must not hatch the dimension text (Figure 12-41). You can accomplish it by selecting the dimension object when defining the hatch boundary. You can also accomplish this by drawing a rectangle around the dimension text and then hatching the area after excluding the rectangle from the hatch boundary. (You can also use the **Explode** tool to explode the dimension and then exclude the dimension text from hatching. This is not recommended because by exploding a dimension, the associativity of the selected dimension is lost.)

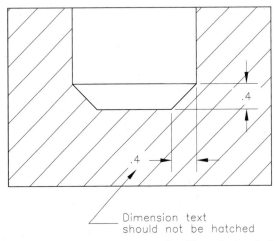

Figure 12-41 *Dimension text should not be hatched*

16. When dimensioning a circle, the diameter should be preceded by the diameter symbol (Figure 12-42). AutoCAD automatically puts the diameter symbol in front of the diameter value. However, if you override the default diameter value, you can use %%c followed by the value of the diameter (%%c1.25) to put the diameter symbol in front of the diameter dimension.

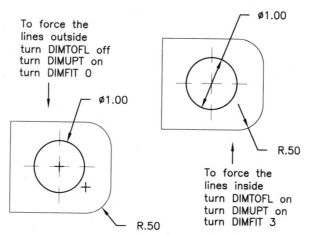

Figure 12-42 *Diameter should be preceded by the diameter symbol*

17. The circle must be dimensioned as a diameter, never as a radius. The dimension of an arc must be preceded by the abbreviation R (R1.25), and the center of the arc should be indicated by drawing a small cross. You can use the AutoCAD **DIMCEN** variable to control the size of the cross. If the value of this variable is 0, AutoCAD does not draw the cross in the center when you dimension an arc or a circle. You can also use the **DIMCENTER** command or the **CENTER** option of the **DIM** command to draw a cross at the center of the arc or circle.

18. When dimensioning an arc or a circle, the dimension line (leader) must be radial. Also, you should place the dimension text horizontally (Figure 12-43).

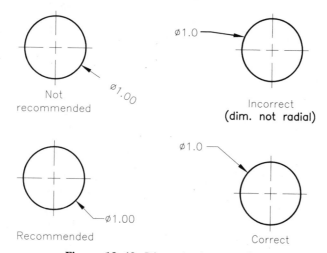

Figure 12-43 *Dimensioning a circle*

19. A chamfer can be dimensioned by specifying the chamfer angle and the distance of the chamfer from the edge. It can also be dimensioned by specifying the distances, as shown in Figure 12-44.

Chapter 12

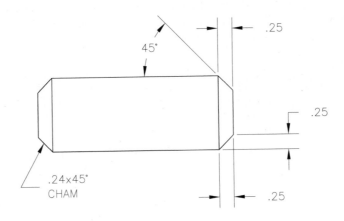

Figure 12-44 Different ways of specifying a chamfer

20. A dimension that is not in scale should be indicated by drawing a straight line under the dimension text (Figure 12-45). This line can be drawn by using the **DDEDIT** command. If you invoke this command, AutoCAD will prompt you to select an annotative object. Select the object; the **Text Editor** will be displayed and a new **Text Editor** tab will be added to the **Ribbon**. Select the text (< >) , and then choose the **U** button in the **Formatting** panel of the **Text Editor** tab to underline the text.

21. A bolt circle should be dimensioned by specifying the diameter of the bolt circle, the diameter of the holes, and the number of holes in the bolt circle (Figure 12-46).

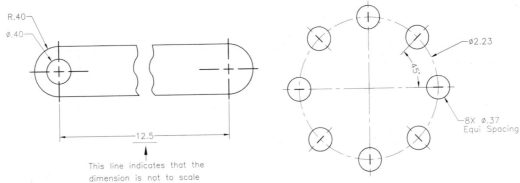

Figure 12-45 Specifying dimensions that are not in scale

Figure 12-46 Dimensioning a bolt circle

EXERCISES 5 through 10   Orthographic Views

Draw the required orthographic views of the following objects, and then give the dimensions (refer to Figures 12-47 through 12-52). The distance between the grid lines is 0.5 units.

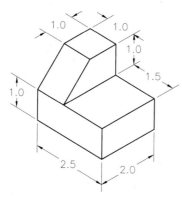

Figure 12-47 *Drawing for Exercise 5*

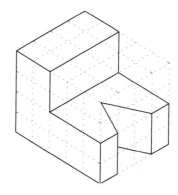

Figure 12-48 *Drawing for Exercise 6*

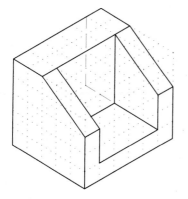

Figure 12-49 *Drawing for Exercise 7*

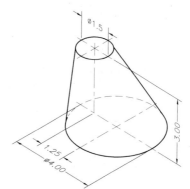

Figure 12-50 *Drawing for Exercise 8*

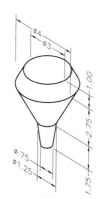

Figure 12-51 *Drawing for Exercise 9*

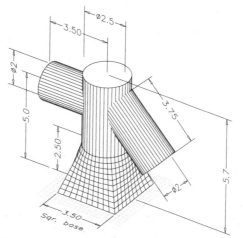

Figure 12-52 *Drawing for Exercise 10 (assume the missing dimensions)*

SECTIONAL VIEWS

In the principal orthographic views, the hidden features are generally shown by hidden lines. In some objects, the hidden lines may not be sufficient to represent the actual shape of the hidden feature. In such situations, sectional views can be used to show the features of the object that are not visible from outside. The location of the section and the direction of sight depend on the shape of the object and the features that need to be shown. Several ways to cut a section in the object are discussed next.

Chapter 12

Full Section

Figure 12-53 shows an object that has a drilled hole, counterbore, and taper. In the orthographic views, these features will be shown by hidden lines (Figure 12-54).

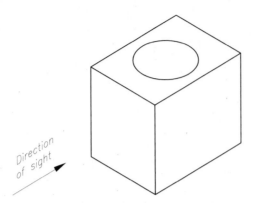

Figure 12-53 Rectangular object with hole

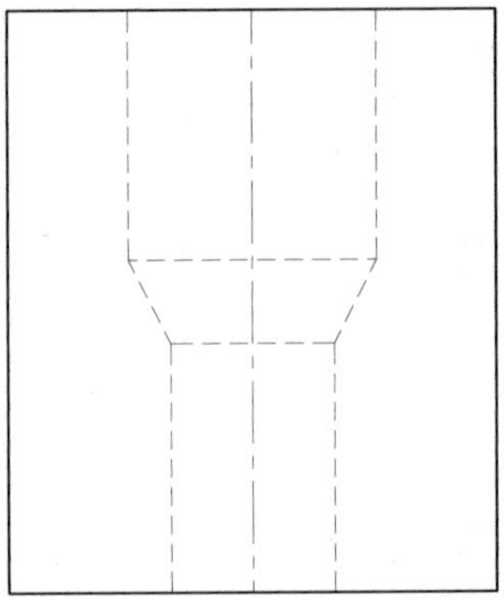

Figure 12-54 Front view without section

To better represent the hidden features, the object must be cut so that the hidden features are visible. In the full section, the object is cut along its entire length. To get a better idea of a full section, imagine that the object is cut into two halves along the centerline, as shown in Figure 12-55. Now remove the left half and look at the right half in the direction that is perpendicular to the sectioned surface. The view you get after cutting the section is called a full section view (Figure 12-56).

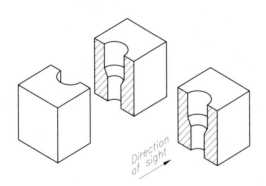

Figure 12-55 One-half of the object removed

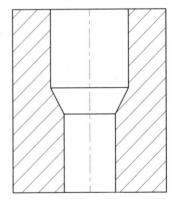

Figure 12-56 Full section view

In this section view, the features that would be hidden in a normal orthographic view are visible. Also, the part of the object where the material is actually cut is indicated by section lines. If the material is not cut, the section lines are not drawn. For example, if there is a hole, no material is cut when the part is sectioned, and so the section lines must not be drawn through that area of the section view.

Half Section

If the object is symmetrical, it is not necessary to draw a full section view. For example, in Figure 12-57 the object is symmetrical with respect to the centerline of the hole, so a full section is not required. Also, in some objects it may help to understand and visualize the shape of the hidden details better to draw the view in half section. In half section, one-quarter of the object is cut, as shown in Figure 12-58. To draw a view in half section, imagine one-quarter of the object removed, and then look in the direction that is perpendicular to the sectioned surface.

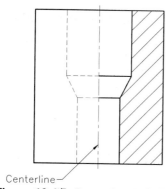

Figure 12-57 *Front view in half section*

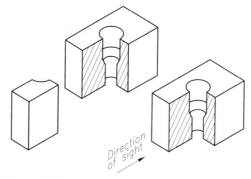

Figure 12-58 *One-quarter of the object removed*

You can also show the front view with a solid line in the middle, as shown in Figure 12-59. Sometimes the hidden lines, representing the remaining part of the hidden feature, are not drawn, as in Figure 12-60.

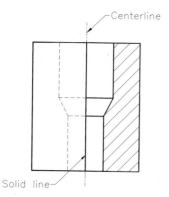

Figure 12-59 *Front view in half section*

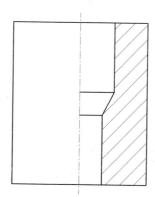

Figure 12-60 *Front view in half section with the hidden features not drawn*

Broken Section

In the broken section, only a small portion of the object is cut to expose the features that need to be drawn in the section. The broken section is designated by drawing a thick zigzag line in the section view (Figure 12-61).

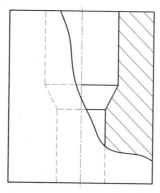

Figure 12-61 *Front view with broken section*

Revolved Section

The revolved section is used to show the true shape of the object at the point where the section is cut. The revolved section is used when it is not possible to show the features clearly in any principal view. For example, for the object in Figure 12-62, it is not possible to show the actual shape of the middle section in the front, side, or top view. Therefore, a revolved section is required to show the shape of the middle section.

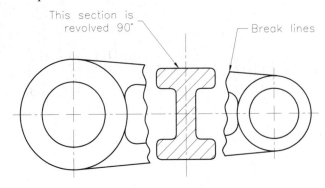

Figure 12-62 *Front view with the revolved section*

The revolved section involves cutting an imaginary section through the object and then looking at the sectioned surface in a direction that is perpendicular to it. To represent the shape, the view is revolved through 90-degree and drawn in the plane of the paper, as shown Figure 12-62. Depending on the shape of the object, and for clarity, it is recommended to provide a break in the object so that its lines do not interfere with the revolved section.

Removed Section

The removed section is similar to the revolved section, except that it is shown outside the object. The removed section is recommended when there is not enough space in the view to show it or if the scale of the section is different from the parent object. The removed section can be shown by drawing a line through the object at the point where the revolved section is desired and then drawing the shape of the section, as shown in Figure 12-63.

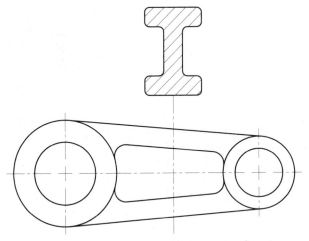

Figure 12-63 *Front view with the removed section*

The other way of showing a removed section is to draw a cutting plane line through the object where you want to cut the section. The arrows should point in the direction, in which you are looking at the sectioned surface. The section can then be drawn at a convenient place in the drawing. The removed section must be labeled, as shown in Figure 12-64. If the scale has been changed, it must be mentioned with the view description.

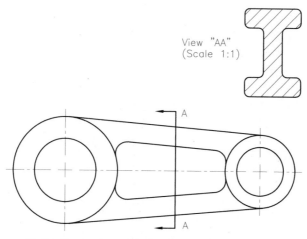

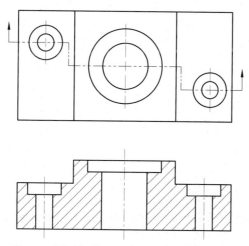

Figure 12-64 *Front view with the removed section drawn at a convenient place*

Offset Section

The offset section is used when the features of the object that you want to section are not in one plane. The offset section is designated by drawing a cutting plane line that is offset through the center of the features that need to be shown in the section (Figure 12-65). The arrows indicate the direction in which the section is viewed.

Figure 12-65 *Front view with offset section*

Aligned Section

In some objects, cutting a straight section might cause confusion in visualizing the shape of the section. Therefore, the aligned section is used to represent the shape along the cutting plane (Figure 12-66). Such sections are widely used in circular objects that have spokes, ribs, or holes.

Cutting Plane Lines

Cutting plane lines are thicker than object lines (Figure 12-67). You can use the **Polyline** tool to draw the polylines of the desired width, generally 0.005 to 0.01. However, for drawings that need to be plotted, you should assign a unique color to the cutting plane lines and then assign that color to the slot of the plotter that carries a pen of the required tip width. (For details, see Chapter 16, Plotting Drawings.)

Chapter 12

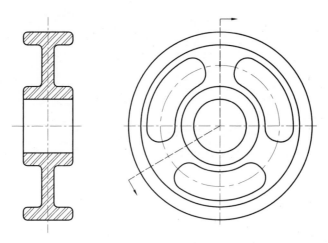

Figure 12-66 Side view in section (aligned section)

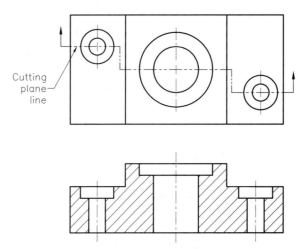

Cutting
plane
line

Figure 12-67 Cutting plane line

In the industry, generally three types of lines are used to show the cutting plane for sectioning. The first line consists of a series of dashes 0.25 units long. The second type consists of a series of long dashes separated by two short dashes (Figure 12-68). The length of the long dash can vary from 0.75 to 1.5 units, and the short dashes are about 0.12 units long. The space between the dashes should be about 0.03 units. Third, sometimes the cutting plane lines might clutter the drawing or cause confusion with other lines in the drawing. To avoid this problem, you can show the cutting plane by drawing a short line at the end of the section (Figure 12-68 and Figure 12-69). The line should be about 0.5 units long.

Note

*In AutoCAD, you can define a new linetype that you can use to draw the cutting plane lines. Refer to the Chapter 35 (Creating Linetypes and Hatch Patterns), for more information on defining linetypes. Add the following lines to the **ACLT.LIN** file, and then load the linetypes before assigning it to an object or a layer.*

*CPLANE1,___ ___ ___ ___
A,0.25,-0.03
*CPLANE2,___ ___ ___ ___
A,1.0,-0.03,0.12,-0.03,0.12,-0.03

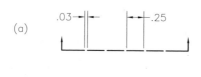

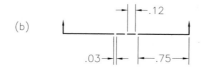

Figure 12-68 Cutting plane lines

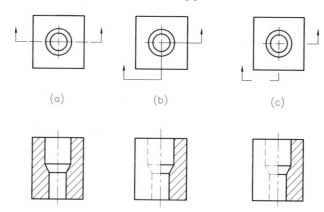

Figure 12-69 Application of cutting plane lines

Spacing for Hatch Lines

The spacing between the hatch (section) lines is determined by the space that is being hatched (Figure 12-70). If the hatch area is small, the spacing between the hatch lines should be smaller compared to a large hatch area.

In AutoCAD, you can control the spacing between the hatch lines by specifying the scale factor at the time of hatching. If the scale factor is 1, the spacing between the hatch lines is the same as defined in the hatch pattern file for that particular hatch. For example, in the following hatch pattern definition, the distance between the lines is 0.125.

> ***ANSI31, ANSI Iron, Brick, Stone masonry**
> **45, 0, 0, 0, .125**

When the hatch scale factor is 1, the line spacing will be 0.125; if the scale factor is 2, the spacing between the lines will be 0.125 x 2 = 0.25.

Direction of Hatch Lines

The angle for the hatch lines should be 45-degree. However, if there are two or more hatch areas next to one another representing different parts, the hatch angle must be changed so that the hatched areas look different (Figure 12-71).

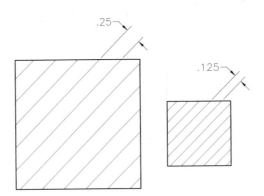

***Figure 12-70** Hatch line spacing*

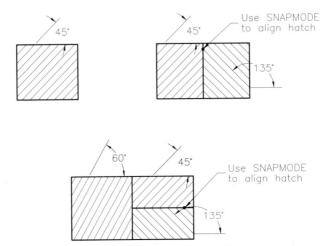

***Figure 12-71** Hatch angle for adjacent parts*

Also, if the hatch lines fall parallel to any edge of the hatch area, the hatch angle should be changed so that the lines are not parallel or perpendicular to any object line (Figure 12-72).

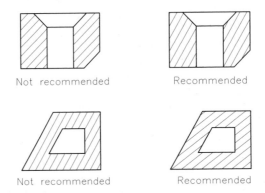

***Figure 12-72** Hatch angle*

Points to Remember

1. Some parts, such as bolts, nuts, shafts, ball bearings, fasteners, ribs, spokes, keys, and other similar items, if sectioned, should not be shown in the section. In case of ribs, if the cutting plane passes through the center plane of the feature, they should not be section-lined. If the cutting plane passes crosswise through ribs they must be section-lined.

2. Hidden details should not be shown in the section view unless the hidden lines represent an important detail or help the viewer to understand the shape of the object.

3. The section lines (hatch lines) must be thinner than the object lines. You can accomplish this by assigning a unique color to the hatch lines and then assigning the color to that slot on the plotter that carries a pen with a thinner tip.

4. The section lines must be drawn on a separate layer for display and editing purposes.

EXERCISES 11 and 12   *Section Views*

In the following drawings (Figures 12-73 and 12-74), the views have been drawn without a section. Draw these views in the section as indicated by the cutting plane lines in each object.

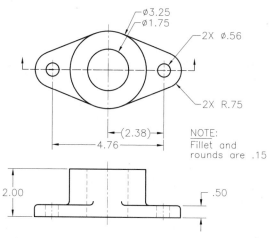

Figure 12-73 *Drawing for Exercise 11: draw the front view in section*

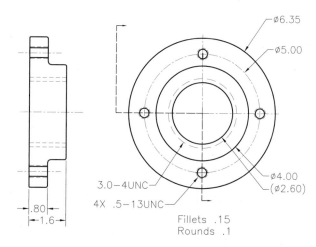

Figure 12-74 *Drawing for Exercise 12: draw the left side view in section*

EXERCISES 13 and 14   *Section Views*

Draw the orthographic views of the objects shown in Figures 12-75 and 12-76. The required view should be the front section view of the object such that the cutting plane passes through the holes. Also, draw the cutting plane lines in the top view.

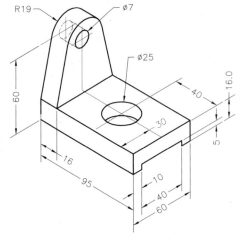

Figure 12-75 *Drawing for Exercise 13: draw the front view in full section*

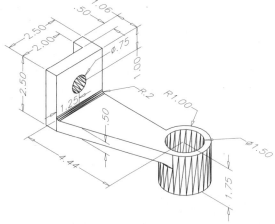

Figure 12-76 *Drawing for Exercise 14: draw the front view with offset section*

EXERCISE 15   *Section View*

Draw the required orthographic views of the objects in Figures 12-77 and 12-78 with the front view in the section. Also, draw the cutting plane lines in the top view to show the cutting plane. The material thickness is 0.25 units. (The object has been drawn as a surfaced 3D wiremesh model.)

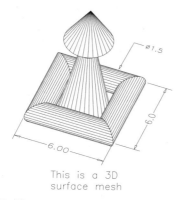

This is a 3D
surface mesh

Figure 12-77 *Draw the front view in half section*

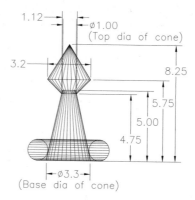

Figure 12-78 *Draw the front view in half section*

AUXILIARY VIEWS

As discussed earlier, most objects generally require three principal views (front view, side view, and top view) to show all features of the object. Round objects may require just two views. Some objects have inclined surfaces. It may not be possible to show the actual shape of the inclined surface in one of the principal views. To get the true view of the inclined surface, you must look at the surface in a direction that is perpendicular to the inclined surface. Then you can project the points onto the imaginary auxiliary plane that is parallel to the inclined surface. The view you get after projecting the points is called the auxiliary view, as shown in Figures 12-79 and 12-80.

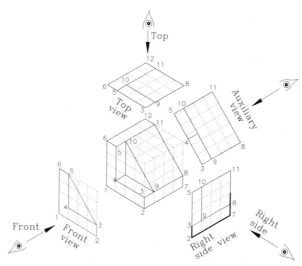

Figure 12-79 *Project points onto the auxiliary plane*

The auxiliary view in Figure 12-80 shows all the features of the object as seen from the auxiliary view direction. For example, the bottom left edge is shown as a hidden line. Similarly, the lower right and upper left edges are shown as continuous lines. Although these lines are technically correct, the purpose of the auxiliary view is to show the features of the inclined surface. Therefore, in the auxiliary plane, you should draw only those features that are on the inclined face, as shown in Figure 12-81. Other details that will help you to understand the shape of the object may also be included in the auxiliary view. The remaining lines should be ignored because they tend to cause confusion in visualizing the shape of the object.

Chapter 12

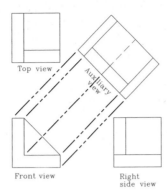

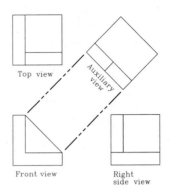

Figure 12-80 *Auxiliary, front, side, and top views*

Figure 12-81 *Auxiliary, front, side, and top views*

How to Draw Auxiliary Views

The following example illustrates how to use AutoCAD to generate an auxiliary view.

Draw the required views of the hollow triangular block with a hole in the inclined face, as shown in Figure 12-82. (The block has been drawn as a solid model.)

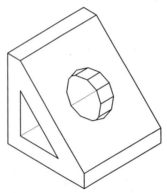

Figure 12-82 *Hollow triangular block*

The following steps are involved in drawing different views of this object.

1. Draw the required orthographic views: the front view, side view, and the top view as shown in Figure 12-83(a). The circles on the inclined surface appear like ellipses in the front and top views. These ellipses may not be shown in the orthographic views because they tend to clutter them [Figure 12-83(b)].

Figure 12-83(a) *Front, side, and top views*

Figure 12-83(b) *The ellipses are not visible in the orthographic views*

2. Determine the angle of the inclined surface. In this example, the angle is 45-degree. Use the **SNAP** command to rotate the snap by 45-degree (Figure 12-84) or rotate the UCS by 45-degree around the Z axis.

    Command: **SNAP** ⏎
    Specify snap spacing or [ON/OFF/Aspect/Style/Type] <0.5000>: **R** ⏎
    Specify base point <0.0000,0.0000>: *Select P1.*
    Specify rotation angle <0>: **45** ⏎

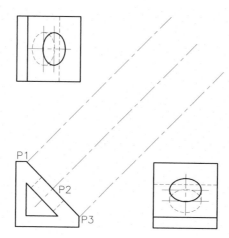

Figure 12-84 Grid lines at 45-degree

Using the rotate option, the snap will be rotated by 45-degree and the grid lines will also be displayed at 45-degree (if the **GRID** is on). Also, one of the grid lines will pass through point P1 because it was defined as the base point.

3. Turn **ORTHO** on, and project points P1, P2, and P3 from the front view onto the auxiliary plane. Now you can complete the auxiliary view and give the dimensions. The projection lines can be erased after the auxiliary view is drawn (Figure 12-85).

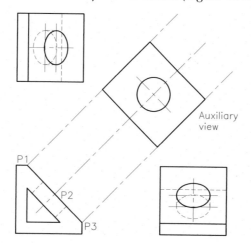

Figure 12-85 Auxiliary, front, side, and top views

EXERCISE 16   *Orthographic Views*

Draw the delete orthographic and auxiliary views of the object in Figure 12-86. The object is drawn as a surfaced 3D wiremesh model.

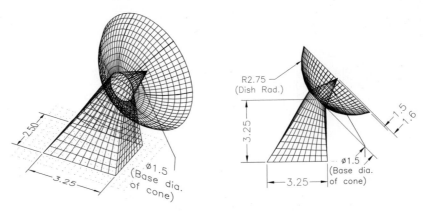

Figure 12-86 Drawing for Exercise 16

EXERCISES 17 and 18   *Orthographic Views*

Draw the orthographic and auxiliary views of the objects shown in Figure 12-87 and Figure 12-88. The objects are drawn as 3D solid models and are shown at different angles (viewpoints are different).

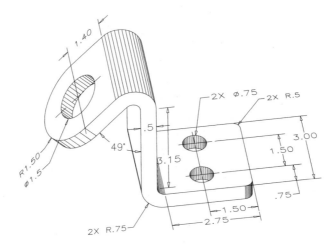

Figure 12-87 Drawing for Exercise 17

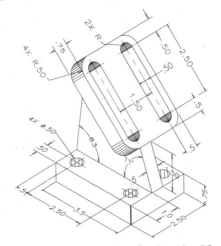

Figure 12-88 Drawing for Exercise 18

DETAIL DRAWING, ASSEMBLY DRAWING, AND BILL OF MATERIALS

Detail Drawing

A detail drawing is a drawing of an individual component that is a part of the assembled product. Detail drawings are also called **piece part drawings**. Each detail drawing must be drawn and dimensioned to completely describe the size and shape of the part. It should also contain information that might be needed in manufacturing the part. The finished surfaces should be indicated by using symbols or notes and all the necessary operations on the drawing. The material of which the part is made and the number of parts that are required for the production of the assembled product must be given in the title block. Detail drawings should also contain part numbers. This information is used in the bill of materials and the assembly drawing. The part numbers make it easier to locate the drawing of a part. You should make a detail drawing of each part, regardless of the part's size, on a separate drawing. When required, these detail drawings can be inserted in the assembly drawing by using the **XREF** command.

Assembly Drawing

The assembly drawing is used to show the parts and their relative positions in an assembled product or a machine unit. The assembly drawing should be drawn so that all parts can be shown in one drawing. This is generally called the main view. The main view may be drawn in full section so that the assembly drawing shows nearly all the individual parts and their locations. Additional views should be drawn only when some of the parts cannot be seen in the main view. The hidden lines, as far as possible, should be omitted from the assembly drawing because they clutter it and might cause confusion. However, a hidden line may be drawn if it helps to understand the product. Only assembly dimensions should be shown in the assembly drawing. Each part should be identified on the assembly drawing by the number used in the detail drawing and in the bill of materials. The part numbers should be given as shown in Figure 12-89. It consists of a text string for the detail number, a circle (balloon), a leader line, and an arrow or dot. The text should be made at least 0.2 inches (5 mm) high and enclosed in a 0.4 inch (10 mm) circle (balloon). The center of the circle must be located not less than 0.8 inches (20 mm) from the nearest line on the drawing. Also, the leader line should be radial with respect to the circle (balloon). The assembly drawing may also contain an exploded isometric or isometric view of the assembled unit.

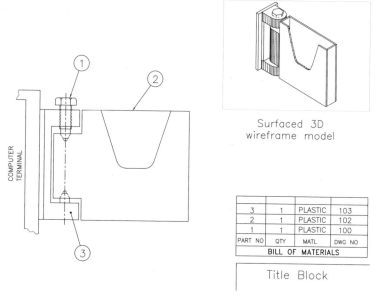

Surfaced 3D
wireframe model

3	1	PLASTIC	103
2	1	PLASTIC	102
1	1	PLASTIC	100
PART NO	QTY	MATL	DWG NO
BILL OF MATERIALS			

Title Block

Figure 12-89 *Assembly drawing with title block, bill of materials, and surfaced 3D wireframe model*

Bill of Materials

A bill of materials is a list of parts placed on an assembly drawing just above the title block (Figure 12-89). The bill of materials contains the part number, part description, material, quantity required, and drawing numbers of the detail drawings (Figure 12-90). If the bill of materials is placed above the title block, the parts should be listed in ascending order so that the first part is at the bottom of the table (Figure 12-91). The bill of materials may also be placed at the top of the drawing. In that case, the parts must be listed in descending order with the first part at the top of the table. This structure allows room for any additional items that may be added to the list.

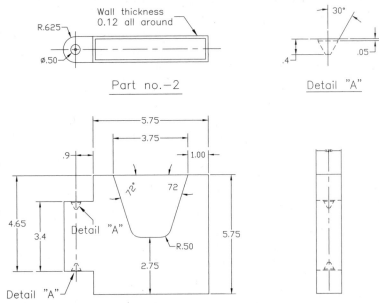

Figure 12-90 Detail drawing (piece part drawing) of Part Number 2

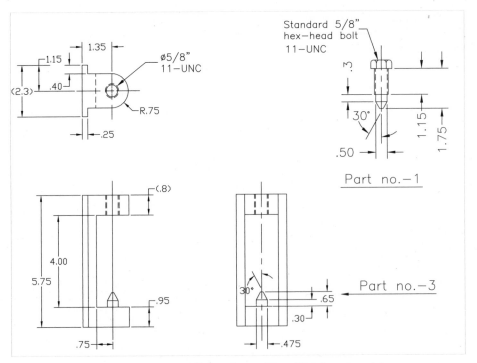

Figure 12-91 Detail drawings of Part Numbers 1 and 3

Self-Evaluation Test

Answer the following questions and then compare them to those given at the end of this chapter:

1. A line perpendicular to the *X* and *Y* axes defines the _____ axis.

2. The front view shows the maximum number of features or gives a better idea about the shape of the object. (T/F)

3. The number of decimal places in a dimension (for example, 2.000) determines the type of machine that will be used to do that machining operation. (T/F)

4. The dimension layer/layers should be assigned a unique color so that at the time of plotting you can assign a desired pen to plot the dimensions. (T/F)

5. All dimensions should be placed inside the view. (T/F)

6. The dimensions can be shown inside the view if they can be easily understood there and cause no confusion with the other object lines. (T/F)

7. The circle must be dimensioned as a _____, never as a radius.

8. If you give continuous (incremental) dimensioning for dimensioning various features of a part, the overall dimension must be omitted or given as a(n) _____ dimension.

9. If the object is symmetrical, it is not necessary to draw a full section view. (T/F)

10. The removed section is similar to the _____ section, except that it is shown outside the object.

11. In AutoCAD, you can control the spacing between hatch lines by specifying the _____ at the time of hatching.

12. When dimensioning, you must consider the manufacturing process involved in making a part and the relationship that exists between different parts in an assembly. (T/F)

13. The distance between the first dimension line and the second dimension line must be _____ to _____ units.

14. The reference dimension must be enclosed in _____.

15. A dimension must be given where the feature is visible. However, in some complicated drawings, you might be justified to dimension a detail with a hidden line. (T/F)

Review Questions

Answer the following questions:

1. Multiview drawings are also known as _____ drawings.

2. The space between the *X* and *Y* axes is called the *XY* plane. (T/F)

3. A plane that is parallel to the *XY* plane is called a parallel plane. (T/F)

4. If you look at the object along the negative *Y* axis and toward the origin, you will get the side view. (T/F)

5. The top view must be directly below the front view. (T/F)

6. Before drawing orthographic views, you must look at the object and determine the number of views required to show all features of the object. (T/F)

7. By dimensioning, you not only give the size of a part, but also give a series of instructions to a machinist, an engineer, or an architect. (T/F)

8. What are the components of a dimension?

9. Why should you make the dimensions in a separate layer/layers?

10. The distance of the first dimension line should be _____ to _____ units from the object line.

11. You can change grid, snap or UCS origin, and snap increments to make it easier to place the dimensions. (T/F)

12. For parallel dimension lines, the dimension text can be _____ if there is not enough room between the dimension lines to place the dimension text.

13. You should not dimension with hidden lines. (T/F)

14. A dimension that is not to scale should be indicated by drawing a straight line under the dimension text. (T/F)

15. Dimensions must be given where the feature that you are dimensioning is obvious and shows the contour of the feature. (T/F)

16. When dimensioning a circle, the diameter should be preceded by the _____.

17. When dimensioning an arc or a circle, the dimension line (leader) must be _____.

18. In radial dimensioning, you should place the dimension text vertically. (T/F)

19. If the dimension of a feature appears in a section view, you must hatch the dimension text. (T/F)

20. Dimensions must not be repeated; this makes it difficult to update dimensions, and they might get confusing. (T/F)

21. A bolt circle should be dimensioned by specifying the _____ of the bolt circle, the _____ of the holes, and the _____ of the holes in the bolt circle.

22. In the _____ section, the object is cut along the entire length of the object.

23. The part of the object where the material is actually cut is indicated by drawing _____ lines.

24. The _____ section is used to show the true shape of the object at the point where the section is cut.

25. The _____ section is used when the features of the object that you want to section are not in one plane.

26. Cutting plane lines are thinner than object lines. You can use the Polyline tool to draw the polylines of the desired width. (T/F)

27. The spacing between the hatch (section) lines is determined by the space being hatched. (T/F)

28. The angle for the hatch lines should be 45-degree. However, if there are two or more hatch areas next to one another, representing different parts, the hatch angle must be changed so that the hatched areas look different. (T/F)

29. Some parts, such as bolts, nuts, shafts, ball bearings, fasteners, ribs, spokes, keys, and other similar items that do not show any important feature, if sectioned, must be shown in section. (T/F)

30. Section lines (hatch lines) must be thicker than object lines. You can accomplish this by assigning a unique color to the hatch lines and then assigning the color to that slot on the plotter carrying a pen with a thicker tip. (T/F)

31. The section lines must be drawn on a separate layer for _____ and _____ purposes.

32. In a broken section, only a _____ of the object is cut to _____ the features that need to be drawn in section.

33. The assembly drawing is used to show the _____ and their _____ in an assembled product or a machine unit.

34. Why shouldn't hidden lines be shown in an assembly drawing? _____
_____.

EXERCISES 19 through 24

Draw the required orthographic views of the following objects (the isometric view of each object is given in Figures 12-92 through 12-97). The dimensions can be determined by counting the number of grid lines. The distance between the isometric grid lines is assumed to be 0.5 units. Also dimension the drawings.

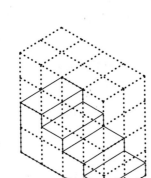

Figure 12-92 *Drawing for Exercise 19*

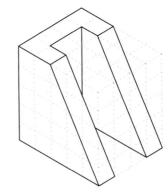

Figure 12-93 *Drawing for Exercise 20*

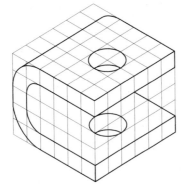

Figure 12-94 *Drawing for Exercise 21*

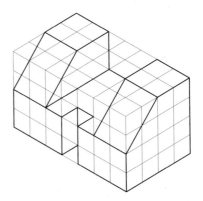

Figure 12-95 *Drawing for Exercise 22*

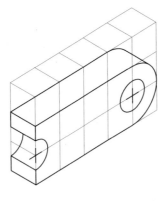

Figure 22-96 *Drawing for Exercise 23*

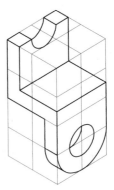

Figure 12-97 *Drawing for Exercise 24*

Answers to Self-Evaluation Test

1. Z, **2.** T, **3.** T, **4.** T, **5.** F, **6.** T, **7.** diameter, **8.** reference, **9.** T, **10.** revolved, **11.** scale factor, **12.** T, **13.** 0.25 to 0.5, **14.** parentheses, **15.** T

Chapter **13**

Isometric Drawings

CHAPTER OBJECTIVES

In this chapter, you will learn:
- *To set up layers and assign colors and line type to layers.*
- *About isometric drawings, isometric axes, and isometric planes.*
- *To set isometric grid and snap.*
- *To draw isometric circles in different isoplanes.*
- *To dimension isometric objects.*
- *To write text with isometric styles.*

KEY TERMS

- *Isometric Planes*
- *Isometric Axes*
- *Isometric Snap*
- *Isometric Grid*
- *Iso Circles*
- *Isometric Text*

ISOMETRIC DRAWINGS

Isometric drawings are generally used to help visualize the shape of an object. For example, if you are given the orthographic views of an object, as shown in Figure 13-1, it takes time to put information together to visualize the shape. However, if an isometric drawing is given, as shown in Figure 13-2, it is much easier to conceive the shape of the object. Thus, isometric drawings are widely used in industry to help in understanding products and their features.

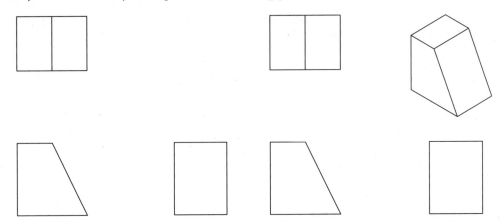

Figure 13-1 *Orthographic views of an object*   ***Figure 13-2*** *Orthographic views with an isometric drawing*

An isometric drawing should not be confused with a three-dimensional (3D) drawing. An isometric drawing is just a two-dimensional (2D) representation of a 3D drawing on a 2D plane. A 3D drawing is the 3D model of an object on the X, Y, and Z axes. In other words, an isometric drawing is a 3D drawing on a 2D plane, whereas a 3D drawing is a true 3D model of the object. The model can be rotated and viewed from any direction. A 3D model can be a wireframe model, surface model, or solid model.

ISOMETRIC PROJECTIONS

The word "isometric" means equal measurement. The angles between the three principal axes of an isometric drawing are 120° each (Figure 13-3). An isometric view is obtained by rotating the object 45° around the imaginary vertical axis, and then tilting the object forward through a 35°16' angle. If you project the points and edges on the frontal plane, the projected length of the edges will be approximately 81 percent (isometric length/actual length = 9/11), which is shorter than the actual length of the edges. However, isometric drawings are always drawn to a full scale because their purpose is to help the user visualize the shape of the object. Isometric drawings are not meant to describe the actual size of the object. The actual dimensions, tolerances, and feature symbols must be shown in the orthographic views. Also, you should avoid showing any hidden lines in the isometric drawings, unless they show an important feature of the object or help in understanding its shape.

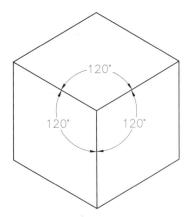

Figure 13-3 *Principal axes of an isometric drawing*

ISOMETRIC AXES AND PLANES

Isometric drawings have three axes: right horizontal axis (P0,P1), vertical axis (P0,P2), and left horizontal axis (P0,P3). The two horizontal axes are inclined at 30° to the horizontal or X axis (X1,X2). The vertical axis is at 90°, as shown in Figure 13-4.

When you draw an isometric drawing, the horizontal object lines are drawn along or parallel to the horizontal axis. Similarly, the vertical lines are drawn along or parallel to the vertical axis. For example, to make an isometric drawing of a rectangular block, the vertical edges of the block are drawn parallel to the vertical axis. The horizontal edges on the right side of the block are drawn parallel to the right horizontal axis (P0,P1), and the horizontal edges on the left side of the block are drawn parallel to the left horizontal axis (P0,P3). It is important to remember that the angles do not appear true in isometric drawings. Therefore, the edges or surfaces that are at an angle are drawn by locating their endpoints. The lines that are parallel to the isometric axes are called isometric lines. The lines that are not parallel to the isometric axes are called non isometric lines.

Similarly, the planes can be isometric planes or non isometric planes.

Isometric drawings have three principal planes, isoplane right, isoplane top, and isoplane left, as shown in Figure 13-5. The isoplane right (P0,P4,P10,P6) is defined by the vertical axis and the right horizontal axis. The isoplane top (P6,P10,P9,P7) is defined by the right and left horizontal axes. Similarly, the isoplane left (P0,P6,P7,P8) is defined by the vertical axis and the left horizontal axis.

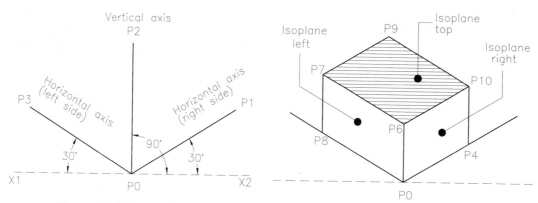

Figure 13-4 *Isometric axes*   **Figure 13-5** *Isometric planes*

SETTING THE ISOMETRIC GRID AND SNAP

You can use the **SNAP** command to set the isometric grid and snap. The isometric grid lines are displayed at 30° to the horizontal axis. Also, the distance between the grid lines is determined by the vertical spacing, which can be specified by using the **GRID** or **SNAP** command. The grid lines coincide with three isometric axes, which make it easier to create isometric drawings. The following command sequence and Figure 13-6 illustrate the use of the **SNAP** command to set the isometric grid and snap of 0.5 unit:

Command: **SNAP**
Specify snap spacing or [ON/OFF/Aspect/Style/Type] <0.5000>: **S**
Enter snap grid style [Standard/Isometric] <S>: **I**
Specify vertical spacing <0.5000>: *Enter a new snap distance.*

Note

*When you use the **SNAP** command to set the isometric grid, the grid lines may not be displayed. To display the grid lines, turn the grid on by choosing the **Grid Display** button from the Status Bar or press F7.*

You cannot set the aspect ratio for the isometric grid. Therefore, the spacing between the isometric grid lines will be the same.

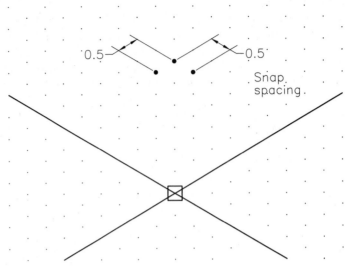

Figure 13-6 Setting the isometric grid and snap in dotted grid

You can also set the isometric grid and snap by using the **Drafting Settings** dialog box shown in Figure 13-7. This dialog box can be invoked by entering **DSETTINGS** at the Command prompt.

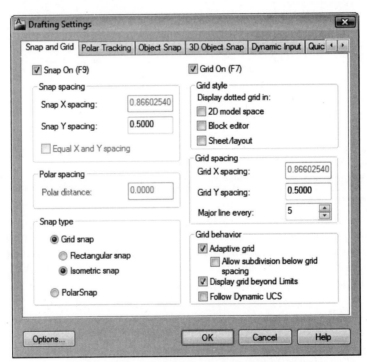

*Figure 13-7 The **Drafting Settings** dialog box*

You can also invoke this dialog box by right-clicking on **Snap Mode**, **Grid Display**, **Polar tracking,** or **Object Snap** button available in the Status Bar and choosing **Settings** from the shortcut menu.

The isometric snap and grid can be turned on/off by choosing the **Grid On (F7)** check box located in the **Snap and Grid** tab of the **Drafting Settings** dialog box. The **Snap and Grid** tab also contains the radio buttons to set the snap type and style. To display the grid on the screen, make sure the grid is turned on.

When you set the isometric grid, the display of crosshairs also changes. The crosshairs are displayed at an isometric angle and their orientation depends on current isoplane. You can toggle among isoplane right, isoplane left, and isoplane top by pressing the CTRL and E keys (CTRL+E) simultaneously or using the function key, F5. You can also toggle among different isoplanes by using the **Drafting Settings** dialog box or by entering the **ISOPLANE** command at the Command prompt:

> Command line: **ISOPLANE**
> Enter isometric plane setting [Left/Top/Right] <Top>: **T**
> Current Isoplane: **Top**

The Ortho mode is often useful when drawing in the Isometric mode. In the Isometric mode, Ortho aligns with the axes of the current isoplane.

EXAMPLE 1   Isometric Drawing

In this example, you will create the isometric drawing shown in Figure 13-8.

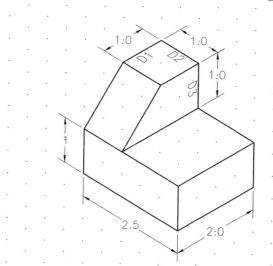

Figure 13-8 *Isometric drawing for Example 1*

1. Use the **SNAP** command to set the isometric grid and snap. The snap value is 0.5 unit.

   Command: **SNAP**
   Specify snap spacing or [ON/OFF/Aspect/Style/Type] <0.5000>: **S**
   Enter snap grid style [Standard/Isometric] <S>: **I**
   Specify vertical spacing <0.5000>: **0.5**   (or press ENTER.)

2. Change the isoplane to the isoplane left by pressing the F5 key. Choose the **Line** tool and draw lines between the points P1, P2, P3, P4, and P1, as shown in Figure 13-9.

   Tip
   *You can change the limits and increase the size of the crosshairs using the **Crosshair size** slider bar in the **Display** tab of the **Options** dialog box.*

3. Change the isoplane to the isoplane right by pressing the F5 key. Invoke the **Line** tool and draw the lines, as shown in Figure 13-10.

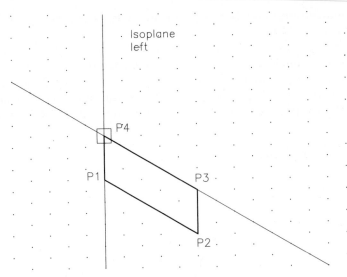

Figure 13-9 *Drawing the bottom left face*

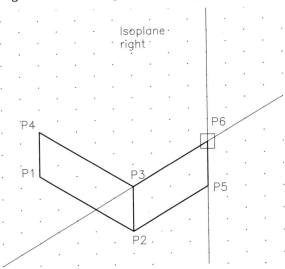

Figure 13-10 *Drawing the bottom right face*

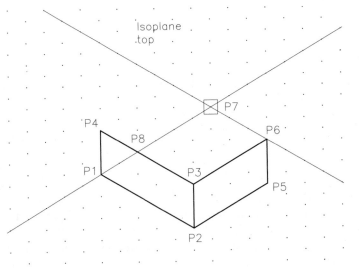

Figure 13-11 *Drawing the top face*

4. Change the isoplane to the isoplane top by pressing the F5 key. Invoke the **Line** tool and draw the lines, as shown in Figure 13-11.

5. Similarly, draw the remaining lines, as shown in Figure 13-12.

6. The front left end of the object is tapered at an angle. In isometric drawings, oblique surfaces (surfaces at an angle to the isometric axis) cannot be drawn like other lines. You must first check if Endpoint Object Snap is on, and then locate the endpoints of the lines that define the oblique surface. Next, draw the lines between those points. To complete the drawing shown in Figure 13-8, draw a line from P10 to P8 and from P11 to P4, see Figure 13-13.

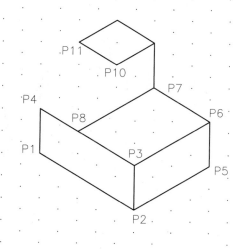

Figure 13-12 *Drawing the remaining lines*

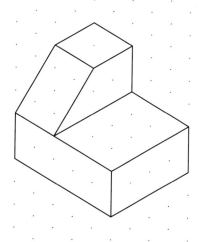

Figure 13-13 *Isometric drawing with the tapered face*

DRAWING ISOMETRIC CIRCLES

The isometric circles are drawn by using the **ELLIPSE** command and then selecting the **Isocircle** option. You must have the isometric snap on for the **ELLIPSE** command to display the **Isocircle** option. If the isometric snap is not on, you cannot draw an isometric circle. Before entering the radius or diameter of the isometric circle, you must make sure that you are in the required isoplane.

For example, to draw a circle in the right isoplane, you must toggle through the isoplanes until the required isoplane (right isoplane) is displayed. You can also set the required isoplane current before entering the **ELLIPSE** command. The crosshairs and the shape of the isometric circle will automatically change as you toggle through different isoplanes. As you enter the radius or diameter of the circle, AutoCAD draws the isometric circle in the selected plane, see Figure 13-14.

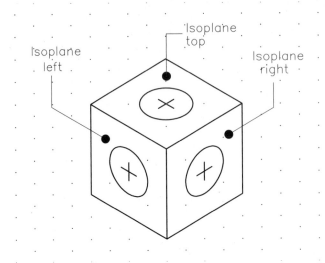

Figure 13-14 Drawing isometric circles

The prompt sequence to draw an isometric circle is as follow:

Command: **ELLIPSE**
Specify axis endpoint of ellipse or [Arc/Center/Isocircle]: **I**
Specify center of isocircle: *Select a point.*
Specify radius of isocircle or [Diameter]: *Enter circle radius.*

Creating Fillets in Isometric Drawings

To create fillets in isometric drawings, you first need to create an isometric circle and then trim its unwanted portion. Remember that there is no other method to directly create an isometric fillet.

DIMENSIONING ISOMETRIC OBJECTS

Isometric dimensioning involves two steps: (1) dimensioning the drawing using the standard dimensioning commands; (2) editing the dimensions to change them to oblique dimensions.

EXAMPLE 2	Dimensioning

The following example illustrates the process involved in dimensioning an isometric drawing. In this example, you will dimension the isometric drawing created in Example 1.

1. Dimension the drawing of Example 1, as shown in Figure 13-15. You can use the aligned or linear dimensions to dimension the drawing. Remember that when you select the points, you must use the **Intersection** or **Endpoint** object snap to snap the endpoints of the object you are dimensioning. AutoCAD automatically leaves a gap between the object line and the extension line, as specified by the **DIMGAP** variable.

2. The next step is to edit the dimensions. You can choose in the **Oblique** tool the **Dimensions** panel in the **Annotate** tab. After selecting the dimension that you want to edit, you are prompted to enter the obliquing angle. The obliquing angle is determined by the angle that the extension line of the isometric dimension makes with the positive *X* axis. The following prompt sequence is displayed when you invoke this option from the **Ribbon**:

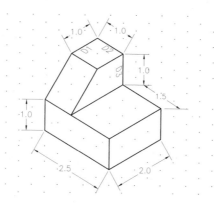

Figure 13-15 *The dimensioned isometric drawing before using the* **Oblique** *tool*

Select object: *Select the dimension (D1).*
Select object: *Press ENTER.*
Enter obliquing angle (Press ENTER for none): **150**

For example, the extension line of the dimension labeled D1 makes a 150° angle with the positive *X* axis [Figure 13-16(a)], therefore, the oblique angle is 150°. Similarly, the extension lines of the dimension labeled D2 make a 30° angle with the positive *X* axis [Figure 13-16(b)], therefore, the oblique angle is 30°. After you edit all dimensions, the drawing should appear as shown in Figure 13-17.

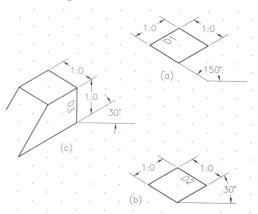

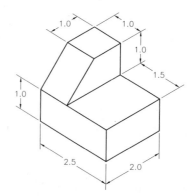

Figure 13-16 *Determining the oblique angle*   **Figure 13-17** *Object with isometric dimensions*

ISOMETRIC TEXT

You cannot use regular text when placing the text in an isometric drawing because the text in an isometric drawing is obliqued at a positive or negative 30°. Therefore, you must create two text styles with oblique angles of 30-degree and negative 30°. You can use the **-STYLE** command or the **Text Style** dialog box to create a new text style. Figure 13-18 shows the **Text Style** dialog box with a new text style, **ISOTEXT1** and its corresponding values. (For more details, refer to "Creating Text Styles" in Chapter 7.)

Similarly, you can create another text style, **ISOTEXT2**, with a negative 30° oblique angle. When you place the text in an isometric drawing, you must also specify the rotation angle for the text. The text style and the text rotation angle depend on the placement of the text in the isometric drawing, as shown in Figure 13-19.

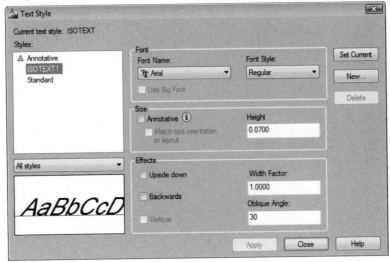

Figure 13-18 The **Text Style** *dialog box*

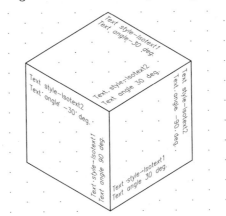

Figure 13-19 Text style and rotation angle for isometric text

Self-Evaluation Test

Answer the following questions and then compare them to those given at the end of this chapter:

1. The word isometric means "_____". The three angles between the three principal axes of an isometric drawing are _____ degree each.

2. The ratio of isometric length to the actual length in an isometric drawing is approximately _____.

3. The angle between the right isometric horizontal axis and the X axis is _____ degrees.

4. Isometric drawings have three principal planes: isoplane right, isoplane top, and _____.

5. To toggle among isoplane right, isoplane left, and isoplane top, you can use _____ key combination or _____ function key.

6. You can only use the aligned dimension option to dimension an isometric drawing. (T/F)

7. The isometric snap must be turned on to display the **Isocircle** option with the **ELLIPSE** command. (T/F)

8. When placing text in an isometric drawing, you need to specify the rotation angles. (T/F)

9. The lines that are not parallel to the isometric axes are called _____.

10. You should avoid showing any hidden lines in isometric drawings. (T/F)

Review Questions

Answer the following questions:

1. Isometric drawings are generally used to help in _____ the shape of an object.

2. An isometric view is obtained by rotating an object _____ degree around the imaginary vertical axis and then tilting the object forward through a _____ angle.

3. If you project the points and edges onto the frontal plane, the projected length of the edges will be approximately _____ percent shorter than their actual length.

4. You should avoid showing hidden lines in isometric drawings unless they help in understanding its shape. (T/F)

5. Isometric drawings have three axes: right horizontal axis, vertical axis, and _____.

6. Angles do not appear true in isometric drawings. (T/F)

7. The lines parallel to the isometric axis are called _____.

8. You can use the _____ command to set the isometric grid and snap.

9. Isometric grid lines are displayed at _____ degrees to the horizontal axis.

10. It is possible to set the aspect ratio for the isometric grid. (T/F)

11. You can also set the isometric grid and snap by using the **Drafting Settings** dialog box, which can be invoked by entering _____ at the Command prompt.

12. Isometric circles are drawn by using the **ELLIPSE** command and then selecting the _____ option.

13. You can draw an isometric circle without turning the isometric snap on. (T/F)

14. Only the aligned dimensions can be edited to change them to oblique dimensions. (T/F)

15. To place text in an isometric drawing, you must create two text styles with oblique angles of _____ degrees and negative _____ degrees.

EXERCISES 1 through 6

Draw the following isometric drawings (Figures 13-20 through 13-23). The dimensions can be determined by counting the number of grid lines. The distance between the isometric grid lines is assumed to be 0.5 unit. Dimension the odd-numbered drawings.

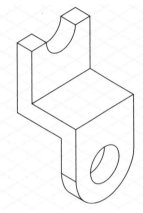

Figure 13-20 Drawing for Exercise 1

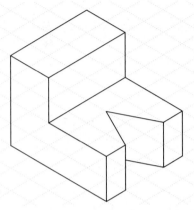

Figure 13-21 Drawing for Exercise 2

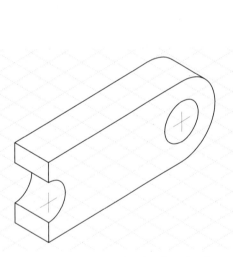

Figure 13-22 Drawing for Exercise 3

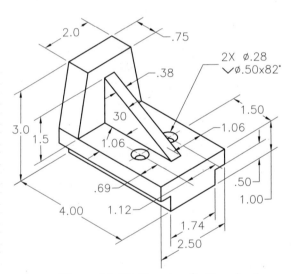

Figure 13-23 Drawing for Exercise 4

Answers to Self-Evaluation Test

1. equal measurement, 120, **2**. 9/11, **3**. 30, **4**. isoplane left, **5**. CTRL+E, or F5, **6**. F, **7**. T, **8**. T, **9**. non isometric lines, **10**. T

Chapter 14

Hatching Drawings

CHAPTER OBJECTIVES

In this chapter, you will learn:

- *To hatch an area by using the Hatch tool.*
- *To use boundary hatch with Predefined, User defined, and Custom hatch patterns as options.*
- *To specify pattern properties.*
- *To preview and apply hatching.*
- *To create annotative hatching.*
- *To edit associative hatch and hatch boundary.*
- *To hatch inserted blocks.*
- *To align hatch lines in adjacent hatch areas.*

KEY TERMS

- *Hatching*
- *Pattern*
- *Gradient*
- *Boundaries*
- *Islands*
- *Edit Hatch*

HATCHING

In many drawings, such as sections of solids, the sectioned area needs to be filled with some pattern. Different filling patterns make it possible to distinguish between different parts or components of an object. Also, the material of which an object is made can be indicated by the filling pattern. You can also use these filling patterns in graphics for rendering architectural elevations of buildings, or indicating the different levels in terrain and contour maps. Filling objects with a pattern is known as hatching (Figure 14-1). This hatching process can be accomplished by using the **Hatch** tool in the **Draw** panel of **Home** tab or the **Tool Palettes**.

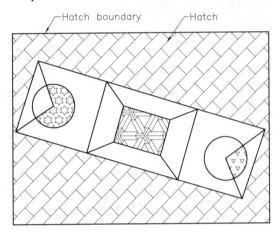

Figure 14-1 Illustration of hatching

Before using the **Hatch** tool, you need to understand some terms related to hatching. The following subsection describes some of the terms.

Hatch Patterns

AutoCAD supports a variety of hatch patterns (Figure 14-2). Every hatch pattern consists of one or more hatch lines or a solid fill. The lines are placed at specified angles and spacing. You can change the angle and the spacing between the hatch lines. These lines may be broken into dots and dashes, or may be continuous, as required. The hatch pattern is trimmed or repeated, as required, to fill exactly the specified area. The lines comprising the hatch are drawn in the current drawing plane. The basic mechanism behind hatching is that the line objects of the pattern you have specified are generated and incorporated in the desired area in the drawing. Although a hatch can contain many lines, AutoCAD normally groups them together into an internally generated object and treats them as such for all practical purposes. For example, if

Figure 14-2 Some hatch patterns

you want to perform an editing operation, such as erasing the hatch, all you need to do is select any point on the hatch and press ENTER; the entire pattern gets deleted. If you want to break a pattern into individual lines to edit them, you can use the **EXPLODE** command.

Hatch Boundary

Hatching can be used on parts of a drawing enclosed by a boundary. This boundary may be lines, circles, arcs, polylines, 3D faces, or other objects, and at least part of each bounding object must be displayed within the active viewport.

HATCHING DRAWINGS USING THE HATCH TOOL

Ribbon: Draw > Hatch   **Toolbar:** Draw > Hatch
Command: HATCH or H

The **Hatch** tool is used to hatch a region enclosed within a boundary (closed area) by selecting a point inside the boundary or by selecting the objects to be hatched. This tool automatically designates a boundary and ignores other objects (whole or partial) that may not be a part of this boundary. When you choose the **Hatch** tool, the **Hatch Creation** tab will be displayed, as shown in Figure 14-3. Also, you will be prompted to pick internal point. Select the type of hatch pattern in the **Pattern** panel, set the properties of the hatch pattern in the **Properties** panel, and move the cursor inside a closed profile; the preview of the hatch pattern will be displayed. Now, pick the internal point; the hatch will be applied. Alternatively, when you are prompted to pick an internal point, select the type of hatch pattern in the **Pattern** panel, set the properties of the hatch pattern in the **Properties** panel, and then pick an internal point; the hatch will be applied.

*Figure 14-3 The **Hatch Creation** tab*

EXAMPLE 1   *Hatch*

In this example, you will hatch a circle using the default hatch settings. Later, in the chapter you will learn how to change the settings to get the desired hatch pattern.

1. Choose the **Hatch** tool from the **Draw** panel; the **Hatch Creation** tab is displayed and you are prompted to pick an internal point. Move the cursor over the circle; the preview of the hatch pattern is displayed.

2. Select a point inside the circle (P1) (Figure 14-4) and press ENTER; the hatch is applied, as shown in Figure 14-5.

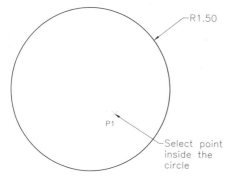

Figure 14-4 Specifying a point to hatch the circle

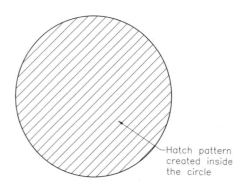

Figure 14-5 Drawing after hatching

Tip

*The **FILLMODE** system variable is set to On by default (value of the variable is 1) and hence the hatch patterns are displayed. In case of large hatching areas, you can set the **FILLMODE** to Off (value of the variable is 0), so that the hatch pattern is not displayed and the regeneration time is saved.*

You can also hatch a selected face of a solid model. However, to hatch a face of the solid model, you need to align the UCS at that face. You will learn more about this in later chapters.

PANELS IN THE HATCH CREATION TAB

Before or after specifying the pick points, you can change the parameters of a hatch pattern using various options in the panels of the **Hatch Creation** tab. These panels and their options are discussed next.

Boundaries Panel

The options in the **Boundaries** panel are used to define the hatch boundary. This is done by selecting a point inside a closed area to be hatched or by selecting the objects. You can also remove islands, create hatch boundary, and define the boundary set by using the options in this panel, as discussed next.

Pick Points

This option is chosen by default and used to define a boundary from the objects that form a closed area. After invoking the **Hatch** tool, move the cursor over a closed region; the preview of the hatch will be displayed. Click inside the closed object; a boundary will be defined around the selected point and the hatch will be applied, as shown in Figure 14-6. The following prompts appear when you click inside the closed object.

Pick internal point or [Select objects/remove Boundaries]: *Select a point inside the object to hatch.*
Selecting everything visible...
Analyzing the selected data...
Analyzing internal islands...
Pick internal point or [Select objects/remove Boundaries]: *Select another internal point or press ENTER to end selection.*

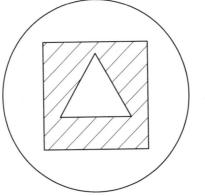

1. Invoke the Hatch command.
2. Select a point inside the square (between square and triangle)

Figure 14-6 Defining multiple hatch boundaries by selecting a point

Tip

*Suppose, you have selected a point within an area for hatching and later you realize that you have selected the wrong area, in such a case, you can enter **U** or **Undo** at the Command prompt to undo the last selection. You can also undo a hatch pattern that you have already applied to an area by entering **Undo** at the Command prompt.*

Boundary Definition Error. Sometimes, while selecting the boundary to be hatched, AutoCAD displays an error. This error may occur due to various reasons. AutoCAD displays different types of Boundary Definition Error message boxes, depending on the kind of error occurred while selecting the boundary. For example, if you pick a point inside any boundary that is not closed, a message box will be displayed, as shown in Figure 14-7, informing that the hatch boundary is not valid. Choose the **Close** button and then create a closed area as boundary. You can also specify the gap tolerance value in the **Gap Tolerance** edit box in the **Options** panel so that hatch will be created if the gap is within the permissible limit. If you select the same boundary twice, the message that the boundary duplicates an existing boundary will be displayed.

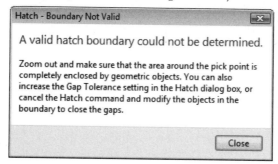

*Figure 14-7 The **Boundary Not Valid** message box*

Select

This option lets you select objects that form the boundary for hatching. It is useful when you have to hatch an object and disregard objects that lie inside it or intersect with it. When you select this option, AutoCAD will prompt you to select objects. You can select the objects individually or use the other object selection methods. In Figure 14-8, the **Select** option is used to select the triangle. It uses the triangle as the hatch boundary, and everything inside it is hatched. The text inside the triangle, in this case, also gets hatched. To avoid hatching of such internal objects, select them at the next **Select objects** prompt. In Figure 14-9, the text is selected to exclude it from hatching. When using the **Pick points** option, the text automatically gets selected to be excluded from the hatching.

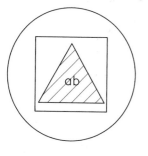

1. Choose the Select button.
2. Select the triangle for hatching and press ENTER.

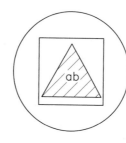

1. Choose the Select button.
2. Select the triangle for hatching
3. Select the text.
4. Press ENTER

*Figure 14-8 Using the **Select Objects** button to specify the hatching boundary*

*Figure 14-9 Using the **Select Objects** button to exclude the text from hatching*

Tip

If you have an area to be hatched that has many objects intersecting with it, it is easier to select the entire object to be hatched rather than choosing the internal points within each of the smaller regions created by the intersection.

You can also turn off layers that contain text or lines that make it difficult for you to select hatch boundaries.

Remove

This option is used to remove boundaries and islands from the hatching area. Boundaries inside another boundary are known as islands. If you choose the **Pick Points** option to hatch an area, the inside boundaries, which are known as islands, will not be hatched by default. But, if you want to hatch the islands, you can choose the **Select** option and then the **Remove** option in the **Boundaries** panel. For example, if you have a rectangle with circles in it, as shown in Figure 14-10, you can use the **Pick points** option to select a point inside the rectangle. On choosing this option, AutoCAD will select both the circles and the rectangle. To remove the circles (islands), you can use the **Remove** option. When you select this option, AutoCAD will prompt you to select the islands to be removed. Select the islands to be removed and press ENTER to return to the dialog box. Choose **OK** to apply the hatch. Similarly, you can remove boundaries.

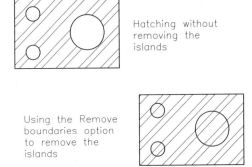

Hatching without removing the islands

Using the Remove boundaries option to remove the islands

*Figure 14-10 Using the **Remove Boundary** option to remove islands from the hatch area*

Tip

*It may be a good idea to use the **Select Objects** option to select the object containing islands, if you want to remove the islands from the hatching area.*

Recreate

This option is available while editing a hatch and is used to recreate boundaries around the existing hatch pattern. On choosing this option, you will be prompted to specify whether you want to recreate boundaries as a region or as a polyline. You can also associate the hatch to the new boundary.

Retain Boundary Objects drop-down list

When you select an internal point in a region to be hatched, a boundary is created around it. By default, this boundary will be removed as soon as the hatch pattern is applied. The **Retain Boundary Objects** drop-down list (see Figure 14-11) provides options to specify whether the boundary is to be retained as object or not. You can also specify the type of object it can be saved as.

Select the **Don't Retain Boundaries** option, if you do not need the boundary to be saved. In case. if you need to retain the boundary, you can retain it as a polyline or region. If you select the **Retain Boundaries - Polyline** option the boundary created around the hatch area will be a polyline. Similarly, when you select **Retain Boundaries - Region**, the boundary of the hatch area will be a region. Regions are two-dimensional areas that can be created from closed shapes or loops.

Display Boundary Objects drop-down list

If you have chosen any one of the **Retain Boundaries** option then you can choose this option to display the resulting boundary, while editing. However, the boundary will merge with the profile of the drawing object. You can use the **Move** command to view the resulting hatch boundary.

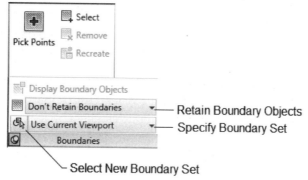

*Figure 14-11 Options in the expanded **Boundaries** panel*

Defining Boundary Set Area

When you invoke the **Hatch** tool and you have not formed a boundary set, there is only one option, **Use Current Viewport**, will be available in the **Specify Boundary Set** drop-down list (see Figure 14-11). The benefit of creating a selection set is that when you select a point or select the objects to define the hatch boundary, AutoCAD will search only for the objects that are in the selection set.

The default boundary set is **Use Current Viewport**, which comprises everything that is visible in the current viewport. Hatching is made faster by specifying a boundary set because, in this case, AutoCAD does not have to examine everything on screen. This option allows you to define a boundary area so that only a specific portion of the drawing is considered for hatching. You use this option to create a new boundary set.

When you choose the **Select New Boundary Set** button, you are prompted to select the objects to be included in the new boundary set. While constructing the boundary set, AutoCAD uses only those objects that you select and that are hatchable. If a boundary set already exists, it is replaced by the new one. If you do not select any hatchable objects, no new boundary set is created. AutoCAD retains the current set, if there is any. Once you have selected the objects to form the boundary set, press ENTER; you will notice that **Use Boundary Set** gets added to the drop-down list.

Remember that by confining the search to the objects in the selection set, the hatching process is faster. If you select an object that is not a part of the selection set, AutoCAD ignores it. When a boundary set is formed, it becomes the default for hatching until you exit the **Hatch** tool or select the **Use Current Viewport** option from the **Specify Boundary Set** drop-down list.

Tip
To improve the hatching speed in large drawings, you should zoom into the area to be hatched, so that defining the boundary to be hatched is easier. Also, since AutoCAD does not have to search the entire drawing to find hatch boundaries, the hatch process gets faster.

Pattern Panel

The **Pattern** panel displays all predefined patterns available in AutoCAD. However, the list depends upon the option selected in the **Hatch Type** drop-down list in the **Properties** panel. By default, the **Pattern** option is selected in this drop-down list and the predefined patterns are listed in the **Pattern** panel. A predefined pattern consists of ANSI, ISO, and other pattern types. The hatch pattern selected in this panel will be applied to the object. The selected pattern will be stored in the **HPNAME** system variable. ANSI31 is the default pattern in the **HPNAME** system variable. If you need to hatch a solid by using a solid color then choose the **Solid** option and specify the color in the **Hatch Color** option in the **Properties** panel. Similarly, if you need to create a user-defined pattern, choose the **User** option from the **Pattern** panel and set the properties.

Properties Panel

The options in this panel (see Figure 14-12) are used to set the properties of the pattern selected in the **Pattern** panel. The different options are discussed next.

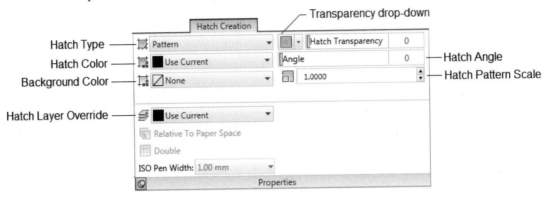

*Figure 14-12 Options in the **Properties** panel*

Hatch Type

The **Hatch Type** drop-down list displays the types of patterns that can be used for hatching drawing objects. The four types of hatch patterns available are **Solid, Gradient, Pattern**, and **User defined**. The predefined type of patterns come with AutoCAD and are stored in the *acad.pat* and *acadiso.pat* files. The **Pattern** type of hatch pattern is the default type. If you select the **Gradient** option from this drop-down list, the corresponding options will be listed in the **Hatch Creation** tab. These options are discussed later.

Hatch Color

Specify the color of the hatch in this drop-down list. On selecting the **Use Current** option, the color set in the **Properties** panel of the **Home** tab will be applied to the hatch.

Background Color

If you need to apply a background color to a hatch, then set a color in the **Background Color** drop-down list.

Hatch Transparency

If you need a hatch to be displayed in a transparent mode, set the transparency value in this edit box. You can also set the value by using the slider or by double-clicking in the edit box. By default it uses the transparency value set in the **Properties** panel of the **Home** tab. If you need to change the value, then select the required option in the **Transparency** drop-down. The other options available are **By Layer Transparency** and **By Block Transparency**.

Hatch Angle

The **Hatch Angle** slider is used to set the angle by which you can rotate the hatch pattern with respect to the *X* axis of the current UCS. The angle value is stored in the **HPANG** system variable. The angle of hatch lines of a particular hatch pattern is governed by the values specified in the hatch definition. For example, in the **ANSI31** hatch pattern definition, the specified angle of hatch lines is 45-degree. If you select an angle of 0, the angle of hatch lines will be 45-degree. If you enter an angle of 45-degree, the angle of the hatch lines will be 90-degree.

Hatch Pattern Scale

The **Hatch Pattern Scale** drop-down list is used to set the scale factors by which you can expand or contract the selected hatch pattern. You can enter the scale factor of your choice in the edit box by double-clicking in it. The scale value is stored in the **HPSCALE** system variable. The value 1 does not mean that the distance between the hatch lines is 1 unit. The distance between the hatch lines and other parameters of a hatch pattern is governed by the values specified in the hatch definition. For example, in the ANSI31 hatch pattern definition, the specified distance between the hatch lines is 0.125. If you select a scale factor of 1, the distance between the lines will be 0.125. If you enter a scale factor of 0.5, the distance between the hatch lines will be 0.5 X 0.125 = 0.0625.

Double

This option is available only for user-defined patterns. When you choose this option, AutoCAD doubles the original pattern by drawing a second set of lines at right angle to the original lines in the hatch pattern. For example, if you have a parallel set of lines as a user-defined pattern and if you select the **Double** option, the resulting pattern will have two sets of lines intersecting at 90-degree. You can notice the effect of selecting the **Double** option in Figure 14-13. If the **Double** option is selected, the **HPDOUBLE** system variable is set to 1.

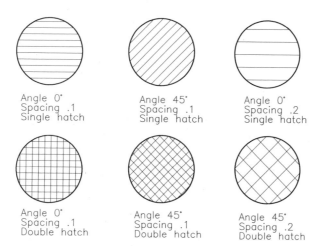

Figure 14-13 Specifying angle and spacing for user-defined hatch patterns

Relative to Paper Space

This option is available only in a layout. If this option is selected, then AutoCAD will automatically scale the hatched pattern relative to the paper space units. This option can be used to display the hatch pattern at a scale that is appropriate for your layout.

ISO Pen Width

The **ISO Pen Width** drop-down list is available only for ISO hatch patterns. You can select the desired pen width value from the **ISO Pen Width** drop-down list. The value selected specifies the ISO-related pattern scaling.

Origin Panel

Hatch pattern alignment is an important feature of hatching, as on many occasions, you need to hatch adjacent areas with similar or sometimes identical hatch patterns while keeping the adjacent hatch patterns properly aligned. Proper alignment of hatch patterns is taken care of automatically by generating all lines of every hatch pattern from the same reference point. The reference point is normally at the origin point (0,0). Figure 14-14 shows two adjacent hatch areas. The area on the right is hatched using the pattern ANSI32 at an angle of 0-degree and the area on the left is hatched using the same pattern at an angle of 90-degree. When you hatch these areas, the hatch lines may not be aligned, as shown in Figure 14-14(a). The options in the **Origin** panel allow you to specify the origin of hatch so that they get properly aligned, as shown in Figure 14-14(b). The options in this panel are discussed next.

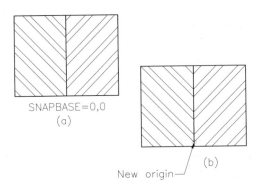

Figure 14-14 *Aligning hatch patterns using the* **SNAPBASE** *variable*

By default, the **Use Current Origin** button is chosen. This implies that the origin of the hatch pattern to be created is the origin of the current drawing. On choosing **Set Origin**, you need to specify the origin point for the hatch pattern in the drawing area.

You can also choose the other buttons to set the origin. For example, if you choose the **Bottom left** button from this panel, the origin of the hatch pattern will be at the bottom left corner of the boundary. Choose the **Store as Default Origin** button to store the origin just selected as the default origin for all the hatch patterns to be created now onwards.

Note
The reference point for hatching can also be changed using the **SNAPBASE** *system variable.*

Options Panel
The options in this area allow you to specify the draw order and some commonly used properties of the hatch pattern.

Associative
This button is chosen by default; therefore, when you modify the boundary of a hatch object, the hatch patterns will be automatically updated to fill up the new area. But, if the hatch boundary is a region, you cannot edit the shape of the hatch boundary. One of the major advantages with the associative hatch feature is that you can edit the hatch pattern or edit the geometry that is hatched without having to modify the associated pattern or boundary separately. After editing, AutoCAD automatically regenerates the hatch and the hatch geometry to reflect the changes. To edit the hatch boundary, select it and edit it by using grips.

Annotative
Choose this button to create an annotative hatch. An annotative hatch is defined relative to the paper space size. The scale of the annotative hatch objects changes in the viewport or layout according to the annotation scale assigned to the hatch objects and the annotation scale specified for that particular layout or viewport.

Gap tolerance

The **Gap tolerance** edit box is used to set the value up to which the open area will be considered closed when selected for hatching, using the **Pick Points** method. The default value of the gap tolerance is 0. As a result, an open area will not be selected for hatching. You can set the value of the gap tolerance using the **Gap Tolerance** edit box. If the gap in the open area is less than the value specified in this edit box, the area will be considered closed and will be selected for hatching.

Create Separate Hatches

If you hatch multiple closed areas that are not nested together, then on selecting this option, a separate hatch will be created for each closed area. As a result, you can edit the hatches separately. If this option is not selected, the hatch created in all the selected closed areas will be treated as single entity and can be edited together.

Island Detection Style

The drop-down list below the **Create Separate Hatches** option is used to select the style of the island detection during hatching. There are four styles available in this drop-down list. The effect of using a particular style is displayed in the form of an illustration in the image tile placed before the name and is also shown in Figure 14-15. To set a particular style, select the corresponding option. The four styles available in this drop-down list are discussed next.

Note
The selection of an island detection style carries meaning only if the objects to be hatched are nested (that is, one or more selected boundaries are within another boundary).

Normal Island Detection. This style is selected by default. This style hatches inward starting at the outermost boundary. If it encounters an internal boundary, it turns off the hatching. An internal boundary causes the hatching to turn off until another boundary is encountered. In this manner, alternate areas of the selected object are hatched, starting with the outermost area. Thus, areas separated from the outside of the hatched area by an odd number of boundaries are hatched, while those separated by an even number of boundaries are not.

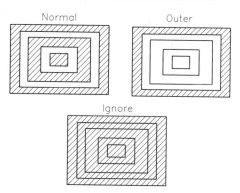

***Figure 14-15** Using hatching styles*

Outer Island Detection. This particular option also lets you hatch inward from an outermost boundary, but the hatching is turned off if an internal boundary is encountered. Unlike the previous case, it does not turn on the hatching again. The hatching process, in this case, starts from both ends of each hatch line; only the outermost level of the structure is hatched, hence, the name **Outer Island**.

Ignore Island Detection. In this option, all areas bounded by the outermost boundary are hatched. The option ignores any hatch boundaries that are within the outer boundary. All islands are completely ignored and everything within the selected boundary is hatched.

No Island Detection. In this option, even if the object is nested, the individual closed boundary is hatched separately.

Tip
*It is also possible to set the pattern and the island detection style at the same time by using the **HPNAME** system variable. AutoCAD stores the Normal style code by adding **N** to the pattern name. Similarly, **O** is added for the Outer style, and **I** for the Ignore style to the value of the **HPNAME** system variable. For example, if you want to apply the BOX pattern using the outer style of island detection, you should enter the **HPNAME** value to the **BOX, O**. Now, when you apply the hatch pattern, the BOX pattern is applied using the outer style of hatching.*

Match Properties

The **Match Properties** drop-down list is used to hatch the specified boundaries using the properties of an existing hatch. On selecting the down arrow, two options will be displayed. These two options are used to set the hatch origin with respect to an existing hatch. By default, the **Use current origin** option is selected; as a result, the current origin will be taken as the hatch origin. If you select the **Use source hatch origin** option, the hatch pattern will accept the origin of the source hatch as the origin of the current hatch pattern.

On selecting any one of the options, the following command sequence appears. Note that you must have applied hatch to an object before invoking this option.

> Pick internal point or [Select objects/seTtings]: _MA
> Select hatch object: *Select the object from which you have to inherit the hatch pattern*
> Pick internal point or [Select objects/seTtings]: *Pick a point inside the object to hatch.*
> Selecting everything...
> Selecting everything visible...
> Analyzing the selected data...
> Pick internal point or [Select objects/seTtings]: [Enter] *(The selected pattern now becomes the current hatch pattern. If you then want to adjust the hatch properties, such as the angle or scale of the pattern, you can do so.)*

Specifying the Draw Order

The drop-down list below the Island detection is used to assign a draw order to the hatch. If you want to send the hatch behind all the entities, select the **Send to Back** option. Similarly, if you want to place the hatch in front of all the entities, select the **Bring to Front** option. If you want to place the hatch behind the hatch boundary, select **Send Behind Boundary**. Similarly, if you want to place the hatch in front of the boundary, select **Bring in Front of Boundary**. You can also select the **Do Not Assign** option, if you do not want to assign the draw order to the hatch.

Hatch Settings

You can also set the parameters discussed above using the **Hatch and Gradient** dialog box, as shown in Figure 14-16. To invoke this dialog box, choose the inclined arrow (**Hatch Settings**) in the **Options** panel of the **Hatch Creation** tab or after invoking the **HATCH** command type **T** and press ENTER when you are prompted to pick internal point. The default appearance of the **Hatch and Gradient** dialog box is slightly different from the one shown in Figure 14-16. This is because options like **Islands**, **Boundary retention** are not visible by default. To make these options visible choose the **More Options** button available at the lower right corner of the **Hatch and Gradient** dialog box.

Close

Choose the **Close** button to close the **Hatch Creation** tab. You can also click once in the drawing area after completing the hatching process to exit the **Hatch Creation** tab.

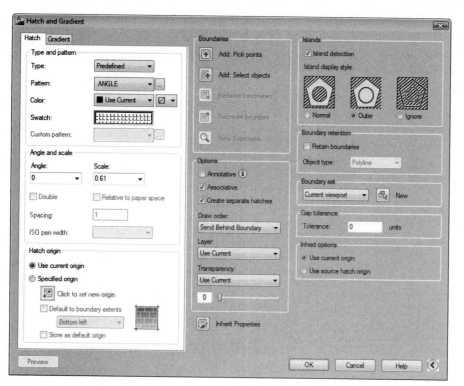

*Figure 14-16 The **Hatch and Gradient** dialog box*

EXERCISE 1   *Hatch Scale & Hatch Angle*

In this exercise, you will hatch the given drawing using the hatch pattern named **STEEL**. Set the scale and the angle to match the drawing shown in Figure 14-17.

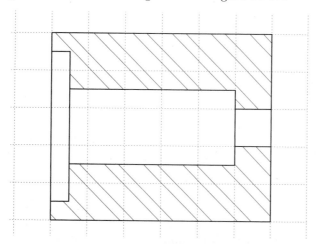

Figure 14-17 Drawing for Exercise 1

EXERCISE 2   *Hatch Pattern*

In this exercise, you will hatch the front section view of the drawing in Figure 14-18 using the hatch pattern for brass. Two views, top, and front are shown. In the top view, the cutting plane indicates how the section is cut and the front view shows the full section view of the object. The section lines must be drawn only where the material is actually cut.

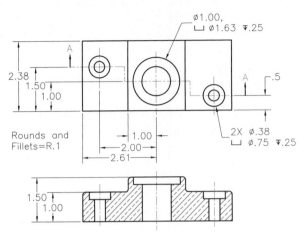

Figure 14-18 Drawing for Exercise 2

EXERCISE 3   *Align Hatch*

In this exercise, you will hatch the given drawing using the hatch pattern ANSI31. Align the hatch lines as shown in the drawing (Figure 14-19).

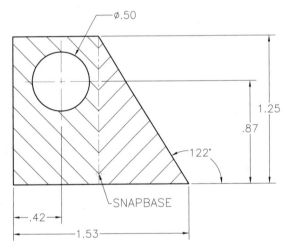

Figure 14-19 Drawing for Exercise 3

Setting the Parameters for Gradient Pattern

During hatching, you can select the **Gradient** option in the **Hatch Type** drop-down list to fill the boundary in a set pattern of colors. On selecting the **Gradient** option in the **Hatch Type** drop-down list, nine fixed patterns will be listed in the **Pattern** panel and their corresponding option will be displayed in the **Hatch Creation** tab, as shown in Figure 14-20. You can select the required gradient or specify the gradient pattern by entering its value in the **GFNAME** system variable. The default value of this variable is 1. As a result, the first gradient pattern is selected.

Figure 14-20 The **Hatch Creation** tab with the options for the **Gradient** pattern

You can set the colors for the Gradient in the **Gradient Color 1** drop-down list available in the **Properties** panel. By default, you can set one color to the gradient. You can set the angle and the tint of the color by using the corresponding sliders. If you need to set two colors then choose the **Gradient Colors** button available on the left of the **Gradient Colors 2** drop-down list. On doing so, the **Gradient Colors 2** drop-down list will be enabled. Now, you can set the other color in this drop-down list. In this case, the **Tint** slider will not be available.

The **Gradient Angle** slider displays the angles that you can rotate the gradient fill with respect to the *X* axis of the current UCS. You can select any angle from this slider or you can also double-click and enter an angle of your choice in the edit box. The angle value is stored in the **GFANG** system variable.

The symmetricity of the gradient is set by choosing the **Centered** button in the **Origin** panel. If this button is not chosen, then the gradient fill is moved up and to the left.

Note
The other options are same as discussed earlier.

Tip
*To snap the hatch objects, set the **OSOPTIONS** system variable to **1**. You can also invoke the **Options** dialog box and choose the **Drafting** tab. Then, clear the **Ignore hatch objects** check box from the **Object Snap Options** area.*

CREATING ANNOTATIVE HATCH

You can create annotative hatch having similar annotative properties like text and dimensions. The annotative hatch are defined relative to the viewport scaling, you only have to specify the hatch scale according to its display on the sheet. The display size of the hatch in the model space will be controlled by the current annotation scale multiplied by the paper space height. You can also control the annotative properties of the hatch pattern as discussed in Chapter 7.

To create an annotative hatch, choose the **Hatch** tool from the **Draw** panel; the **Hatch Creation** tab will be displayed. Choose the **Select** option from the **Boundaries** panel, select the objects that you want to hatch, and press the ENTER key; the **Hatch and Gradient** dialog box will be displayed again. In the **Options** area of this dialog box, select the **Annotative** check box and choose the **OK** button. You can also convert the existing non-annotative hatch into the annotative hatch. To do so, select the non-annotative hatch in the drawing and select **Yes** from the **Annotative** drop-down list in the **Pattern** list of the **Properties** palette.

Note
*While creating the hatch the annotative hatch pattern will only be displayed in the drawing area for the annotative scale that is set current. To display that hatch pattern at any other annotative scale, you have to add that scale to the selected hatch using the **OBJECTSCALE** command.*

HATCHING THE DRAWING USING THE TOOL PALETTES

Ribbon: View > Palette > Tool Palettes   **Command:** TOOLPALETTES
Toolbar: Standard Annotation > Tool Palettes Window

You can use the **Tool Palettes** window shown in Figure 14-21 to insert predefined hatch patterns and blocks in the drawings. By default, AutoCAD displays **Tool Palettes** as a window on the right of the drawing area. A number of tabs such as **Command Tool Samples**, **Hatches and Fills**, **Civil**, **Structural**, **Electrical**, and so on are available in this window. In this chapter, you will learn to insert **Imperial Hatches**, **ISO Hatches** and **Gradient Samples** in the **Hatches and Fills** tab of the **Tool Palettes** window. The **Imperial Hatches** list provides the options to insert the hatch patterns that are created using the Imperial units. The **ISO Hatches** list provides the options to insert the hatch patterns that are created using the Metric units. You will notice that the hatch patterns provided in both these lists are similar. The basic difference between the two lists is their scale factor. The **Gradient Samples** list provides option to add the color gradients to objects as hatch.

Tip
The **Tool Palettes** *window can be turned on and off by pressing the CTRL+3 keys.*

AutoCAD provides two methods to insert the predefined hatch patterns from the **Tool Palettes** window: **Drag and Drop** method and **Select and Place** method. Both these methods are discussed next.

Drag and Drop Method

To insert the predefined hatch pattern from the **Tool Palettes** using this method, move the cursor over the desired predefined pattern in the **Tool Palettes**. You will notice that as you move the cursor over the hatch pattern, the hatch icon gets converted into a 3D icon. Also, a tooltip is displayed that shows the name of the hatch pattern. Press and hold the left mouse button and drag the cursor within the area to be hatched. Release the left mouse button, and you will notice that the selected predefined hatch pattern is added to the drawing.

Select and Place Method

You can also add the predefined hatch patterns to the drawings using the select and place method. To add the hatch pattern, move the cursor over the desired pattern in the **Tool Palettes**; the pattern icon will be changed to a 3D icon. Next, click the left mouse button; the selected hatch pattern will be attached to the cursor and you will be prompted to specify the insertion point of the hatch pattern. Now, move the cursor within the area to be hatched and click the left mouse button; the selected hatch pattern will be inserted at the specified location.

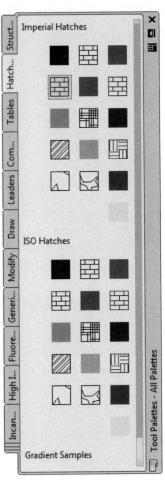

*Figure 14-21 The **Hatch** tab of the **Tool Palettes***

Modifying the Properties of the Predefined Patterns available in Tool Palettes

To modify the properties of the predefined hatch patterns, move the cursor over the hatch pattern in the **Tool Palettes** and right-click on it to display the shortcut menu. Using the options available in this shortcut menu, you can cut or copy the desired hatch pattern available in one tab of **Tool Palettes** and paste it on the other tab. You can also delete and rename a selected hatch pattern using the **Delete** and **Rename** options, respectively. To update and change the image displayed on the selected hatch pattern, choose the **Update tool image** or **Specify image** option. To modify the properties of the selected hatch pattern, choose **Properties** from the shortcut menu, as shown in Figure 14-22; the **Tool Properties** dialog box will be displayed, as shown in Figure 14-23.

In the **Tool Properties** dialog box, the name of the selected hatch pattern is displayed in the **Name** edit box. You can also rename the hatch pattern by entering a new name in the **Name** edit box. The **Image** area available on the left of the **Name** edit box displays the image of the selected hatch pattern. You can change the displayed image by choosing the **Specify image** option

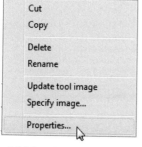

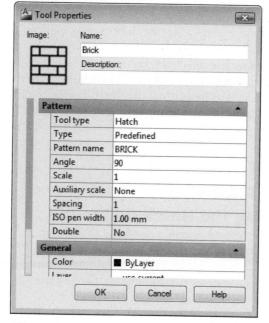

*Figure 14-22 Choosing **Properties** from the shortcut menu*

*Figure 14-23 The **Tool Properties** dialog box*

from the shortcut menu that will be displayed when you right-click on the image. If you enter a description of the hatch pattern in the **Description** text box, it is stored with the hatch definition in the **Tool Palettes**. Now, when you move the cursor over the hatch pattern in **Tool Palettes** and pause for a second, the description of the hatch pattern appears along with its name in the tooltip. The **Tool Properties** dialog box displays the properties of the selected hatch pattern under the following categories.

Pattern

In this category, you can change the pattern type, pattern name, angle, and scale of the selected pattern. You can modify the spacing of the **User defined** patterns in the **Spacing** text box. Also, the ISO pen width of the ISO patterns can be redefined in the **ISO pen width** text box. The **Double** drop-down list is available only for the **User-defined** hatch patterns. You can select Yes or No from the **Double** drop-down list to determine the hatch pattern to be doubled at right angles to the original pattern or not. When you choose the [...] button in the **Type** property field, AutoCAD displays the **Hatch Pattern Type** dialog box. You can select the type of hatch pattern from the **Pattern Type** drop-down list. If the **Predefined** pattern type is selected, you can specify the pattern name by either selecting it from the **Pattern** drop-down list or from the **Hatch Pattern Palette** dialog box displayed on choosing the **Pattern** button. Similarly, if you select the **Custom** pattern type, you can enter the name of the custom pattern in the **Custom Pattern** edit box. If you select the **User defined** pattern type, both the **Pattern** drop-down list and the **Custom Pattern** edit box are not available.

General

In this category, you can specify the general properties of the hatch pattern such as color, layer, linetype, plot style, transparency, and line weight for the selected hatch pattern. The properties of a particular field can be modified from the drop-down list available on selecting that field. Choose **OK** to apply the changes and close the dialog box.

HATCHING AROUND TEXT, DIMENSIONS, AND ATTRIBUTES

When you select a point within a boundary to be hatched and if it contains text, dimensions, and attributes then, by default, the hatch lines do not pass through the text, dimensions, and attributes present in the object being hatched by default. AutoCAD places an imaginary box around these objects that does not allow the hatch lines to pass through it. Remember that if you are using the select objects option to select objects to hatch, you must select the text/attribute/shape along with the object in which it is placed when defining the hatch boundary. If multiple line text is to be written, the **Multiline Text** tool is used. You can also select both the boundary and the text when using the window selection method. Figure 14-24 shows you how hatching takes place around multiline text, attributes, and dimensions.

EDITING HATCH PATTERNS

Using the Hatch Editor Tab

On selecting a hatch pattern, the **Hatch Editor** tab will be displayed in the **Ribbon**, as shown in Figure 14-25. The panels and their options in this tab are similar to that of the **Hatch** tab. Change the parameters of the hatch pattern in the corresponding panels; the hatch pattern will be modified instantaneously. Press ESC to exit the **Hatch Editor** tab.

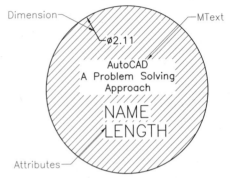

Figure 14-24 Hatching around dimension, multiline text, and attributes

*Figure 14-25 The **Hatch Editor** tab with the options for the **Gradient** pattern*

Using the HATCHEDIT Command

Ribbon: Home > Modify > Edit Hatch	**Tool Palette:** Modify > Edit Hatch
Toolbar: Modify II > Edit Hatch	**Command:** HATCHEDIT

The **Edit Hatch** tool is used to edit a hatch pattern. When you invoke this tool and select the hatch for editing, the **Hatch Edit** dialog box will be displayed, as shown in Figure 14-26.

The **Hatch Edit** dialog box is used to change or modify a hatch pattern. This dialog box is the same as the **Hatch and Gradient** dialog box, except that only the options that control the hatch pattern are available. The available options work in the same way as they do in the **Hatch and Gradient** dialog box.

The **Hatch Edit** dialog box has two tabs, just like the **Hatch** and **Gradient** dialog box. They are **Hatch** and **Gradient**. In the **Hatch** tab, you can redefine the type of hatch pattern by selecting another type from the **Type** drop-down list. If you are using the **Predefined** pattern, you can select a new hatch pattern name from the **Pattern** drop-down list. You can also change the scale or angle by entering new values in the **Scale** or **Angle** edit box. When using the **User defined** pattern, you can redefine the spacing between the lines in the hatch pattern by entering a new

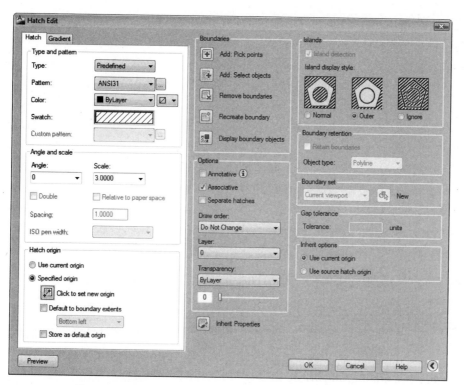

*Figure 14-26 The **Hatch** tab of the **Hatch Edit** dialog box*

value in the **Spacing** edit box. If you are using the **Custom** pattern type, you can select another pattern from the **Custom pattern** drop-down list. You can also redefine the island detection style by selecting either of the **Normal**, **Outer**, or **Ignore** styles from the **Islands** area of the dialog box. You can also convert a non-annotative hatch into an annotative hatch and vice-versa. If you want to copy the properties from an existing hatch pattern, choose the **Inherit Properties** button, and then select the hatch whose properties you want to be inherited. You can also change the draw order of the hatch pattern using the options available in the **Draw order** area. If the **Associative** radio button in the **Options** area of the dialog box is selected, the pattern is associative. This implies that whenever you modify the hatch boundary, the hatch pattern is automatically updated. The appearance of the gradient fill can be specified using the **Gradient** tab of the dialog box. You can specify the gradient fill comprising different shades and tints of a single color or double colors by selecting the **One color** and **Two color** radio buttons, respectively. You can also specify the color of the gradient fill or the shade and tint of a single color using the color swatch and the **Shade and Tint** Slider. If the **Centered** button is selected, AutoCAD will apply a symmetrical gradient fill. Also, AutoCAD provides you with an option to select the desired display pattern of the gradient fill by selecting any one of the nine fixed patterns for the gradient fill.

Note
If a hatch pattern is associative, the hatch boundary can be edited using grips and editing commands and the associated pattern is modified accordingly. This is discussed later.

In Figure 14-27, the object is hatched using the **ANSI31** hatch pattern. With the **HATCHEDIT** command, you can edit the hatch using the **Hatch Edit** dialog box, refer to Figure 14-28. You can also edit an existing hatch through the command line by entering **-HATCHEDIT** at the Command prompt.

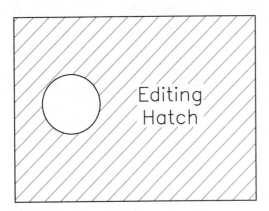

Figure 14-27 ANSI31 hatch pattern

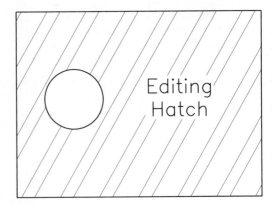

Figure 14-28 The modified hatch pattern using the Edit Hatch tool

Using the PROPERTIES Command

Quick Access Toolbar: Properties (*Customize to Add*)   **Command:** PROPERTIES
Ribbon: View > Palettes > Properties

You can also use the **PROPERTIES** command to edit a hatch pattern. When you select a hatch pattern for editing, and invoke the **PROPERTIES** command, AutoCAD displays the **Properties** palette for the hatch, see Figure 14-29. You can also invoke the **Properties** palette by selecting the hatch pattern and right-clicking to display a shortcut menu. Choose **Properties** from the shortcut menu; the **Properties** palette is displayed. This palette displays the properties of the selected pattern under the following categories.

General

In this category, you can change the general pattern properties like color, layer, linetype, linetype scale, lineweight, and so on. When you click on the field corresponding to a particular property, a drop-down list is displayed from where you can select an option or value. Whenever a drop-down list does not get displayed, you can enter a value in the field.

Pattern

In this category, you can change the pattern type, pattern name, annotative property, annotation scale, angle, scale, spacing, associativity, and island detection style properties of the pattern. In the case of ISO patterns, you can redefine the ISO pen width of the pattern modify the spacing of the **User defined** patterns. You can also determine whether you want to have a **User defined** pattern as double or not. When you choose the [**...**] button in the **Type** property field, AutoCAD displays the **Hatch Pattern Type** dialog box, see Figure 14-30. Here, you can select the type of hatch pattern from the **Pattern Type** drop-down list. If you have selected a **Predefined**

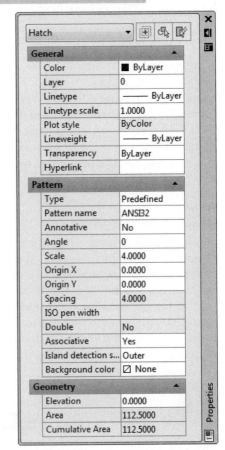

Figure 14-29 The Properties palette (Hatch)

pattern type, you can select a pattern name from the **Pattern** drop-down list or choose the **Pattern** button to display the **Hatch Pattern Palette** dialog box. You can then select a pattern here in one of the tabs and then choose **OK** to exit the dialog box. Now, choose **OK** in the **Hatch Pattern Type** dialog box to return to the **Properties** palette. You will notice that the pattern name you have selected is displayed in the **Pattern** name field. Similarly, if you select the **Custom** pattern type, you can enter the name of the custom pattern in the **Custom Pattern** edit box. If you select a **User defined** pattern type, both the **Pattern Type** drop-down list and the **Custom Pattern** edit box will not be available. When you

Figure 14-30 The Hatch Pattern Type dialog box

click in the **Pattern name** property field, the [...] button is displayed. When you choose the [...] button, the **Hatch Pattern Palette** dialog box is displayed with the ANSI31 pattern selected. Select the **Yes** or **No** option from the **Annotative** drop-down list to convert the existing hatch into annotative or non-annotative respectively. When you click on the **Annotation scale** property field, the [...] button will be displayed. On choosing the [...] button, the **Annotation Object Scale** dialog box will be invoked. You can add other annotation scales to the selected hatch with the help of this dialog box. The values of the angle, scale, and spacing properties can be entered in the corresponding fields. A value for the ISO pen width can be selected from the **ISO pen width** drop-down list. For the user-defined hatch patterns, you can select **Yes** or **No** from the **Double** drop-down list to specify if you want the hatch pattern to double at right angle to the original pattern or not. You can also select **Yes** or **No** from the **Associative** drop-down list to determine if a pattern is associative or not. Similarly, you can select an island detection style from the **Island detection style** drop-down list.

Tip
*It is convenient to choose the [...] button in the **Pattern Type** edit box to display the **Hatch Pattern Type** dialog box and enter both the type and name of the pattern at the same time.*

Geometry

In this category, the elevation of the hatch pattern can be changed. A new value of elevation for the hatch pattern can be entered in the **Elevation** field. This category also displays the area and the cumulative area of the hatch.

EDITING THE HATCH BOUNDARY
Using Grips

One of the ways you can edit the hatch boundary is by using grips. You can select the hatch pattern or hatch boundaries. If you select the hatch pattern, the hatch highlights and a grip is displayed at the centroid of the hatch. A centroid for a region that is coplanar with the *XY* plane is the center of that particular area. However, if you select an object that defines the hatch boundary, the object grips are displayed at the defining points of the selected object. Once you change the boundary definition, and if the hatch pattern is associative, AutoCAD will re-evaluate the hatch boundary and then hatch the new area. When you edit the hatch boundary, make sure that there are no open spaces in it. AutoCAD will not create a hatch if the outer boundary is not closed. Figure 14-31 shows the result of moving the circle and text, and shortening the bottom edge of the hatch boundary of the object shown in Figure 14-28. The objects were edited by using grips and since the pattern is associative, when modifications are made to the boundary object, the pattern automatically fills up the new area. If you select the hatch pattern, AutoCAD will display the hatch object grip at the centroid of the hatch. You can select the grip to make it hot and then edit the hatch object. You can stretch, move, scale, mirror, or rotate the hatch pattern. Once an editing operation takes place on the hatch pattern then, the associativity is lost and AutoCAD displays the message: "**Hatch boundary associativity removed**" in the prompt window.

Chapter 14

While creating the hatch, the **Don't Retain Boundaries** option is selected in the **Retain Boundary Objects** drop-down list of the **Hatch Creation** panel. Therefore, all objects selected or found by the selection or point selection processes during hatch creation are associated with the hatch as boundary objects. And, editing any of them may affect the hatch. If, however, the **Retain Boundaries - Polyline** or the **Retain Boundaries - Region** check boxes was selected when the hatch was created, only the retained polyline boundary created by **HATCH** is associated with the hatch as its boundary object. Editing it will affect the hatch, and editing the objects selected or found by selection or point

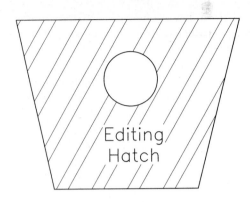

Figure 14-31 The result of using grips to edit the hatch boundary of the object shown in Figure 14-28

selection processes during the hatch created will have no effect on the hatch. You can, of course, select and simultaneously edit the objects selected or found by the selection or point selection processes, as well as the retained polyline boundary created by the **Hatch** tool.

Trimming the Hatch Patterns

You can trim the hatch patterns by using a cutting edge. For example, Figure 14-32 shows a drawing before trimming the hatch. In this drawing, the outer loop was selected as the object to be hatched. This is the reason the space between the two vertical lines on the right is also hatched. Figure 14-33 shows the same drawing after trimming the hatch using the vertical lines as the cutting edge. You will notice that even after trimming some of the portions of the hatch, it is a single entity.

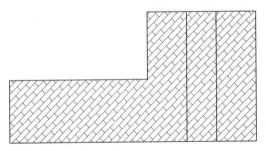

Figure 14-32 Before trimming the hatch

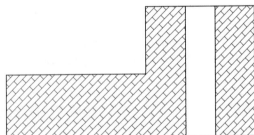

Figure 14-33 After trimming the hatch by using the vertical lines as the cutting edge

EXAMPLE 2   *Hatch*

In this example, you will hatch the drawing, as shown in the Figure 14-34, using various hatch patterns.

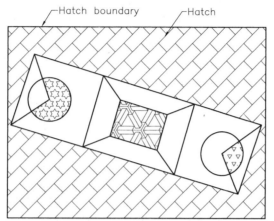

Figure 14-34 *Example 2 for Hatching*

1. Make a drawing as shown in Figure 14-34.

2. Choose the **Hatch** tool from the **Draw** panel in the **Home** tab; the **Hatch Creation** tab is displayed.

3. Select the **AR-BRSTD** pattern from the **Pattern** panel and click inside the outer rectangle's boundary to select the internal pick point.

4. Next, you need to change the angle of the hatch pattern to 45°. To do so, drag the **Angle** slider in the **Properties** panel. Also, adjust the **Hatch Pattern Scale** using the **Hatch Pattern Scale** spinner.

5. Select the **Outer Island Detection** from the **Island Detection** drop-down list in the **Options** panel if it is not selected by default.

6. Exit the **Hatch Creation** tab by pressing ENTER.

7. Again choose the **Hatch** tool and select the **STARS** pattern from the **Pattern** panel; you are prompted to pick an internal point in the object. Select a point in the right section of the left circle, refer to Figure 14-34; preview of the pattern is displayed. You have to specify the values of the **Hatch Angle** and **Hatch Pattern Scale** options. After making the required changes, exit the **Hatch Creation** tab.

8. Choose the **Hatch** tool and select the **ESCHER** pattern from the **Pattern** panel; you are prompted to pick an internal point in the object. Select a point in the innermost rectangle in the middle, refer to Figure 14-34; a preview will be displayed. Adjust the scale of the hatch pattern, if required and then exit.

9. Invoke the **Hatch** tool once again and select **TRIANG** from the **Pattern** panel; you are prompted to pick an internal point in the object. Select a point in the right section of the right circle, refer to Figure 14-34; preview of the pattern is displayed. Set the desired values for **Hatch Angle** and **Hatch Pattern Scale** and then exit.

Chapter 14

EXAMPLE 3   *Hatch*

In this example, you will hatch the drawing as shown in Figure 14-35. But, the hatching should be in two entities.

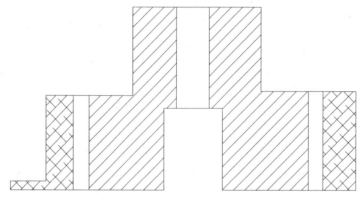

Figure 14-35 *Example 3 for hatching*

To make this kind of hatching, you have to follow the steps given below:

1. Make a drawing as shown in Figure 14-35.

2. Choose the **Hatch** tool from the **Draw** panel in the **Home** tab; the **Hatch Creation** tab is displayed. Make sure that the entities having different hatch patterns should be drawn as different entities.

3. Select **ANSI38** from the **Pattern** panel; you will be prompted to pick an internal point. Select any internal point of the two outermost entities. Set the scale of the pattern, if required.

4. Press SPACEBAR to exit the command.

5. Choose the **Hatch** tool and select the **ANSI31** pattern from the **Pattern** panel; you will be prompted to pick an internal point in the object.

6. Select the internal points of the rest two entities; a preview is displayed. Set the scale of pattern, if required. After making required adjustments, exit the command by choosing the **Close Hatch Creation** button in the **Close** panel.

Using AutoCAD Editing Tools

When you use the editing tools such as **Move**, **Rotate**, **Scale**, and **Stretch**, associativity is maintained, provided all objects that define the boundary are selected for editing. If any of the boundary-defining objects is missing, the associativity will be lost and AutoCAD will display the message **Hatch boundary associativity removed**. When you rotate or scale an associative hatch, the new rotation angle and the scale factor are saved with the hatch object data. This data is then used to update the hatch. When using the **Array**, **Copy**, and **Mirror** tools, you can array, copy, or mirror just the hatch boundary, without the hatch pattern. Similarly, you can erase just the hatch boundary, using the **Erase** tool and associativity is removed. If you explode an associative hatch pattern, the associativity between the hatch pattern and the defining boundary is removed. Also, when the hatch object is exploded, each line in the hatch pattern becomes a separate object.

When the original boundary objects are being edited, the associated hatch gets updated, but new boundary objects do not have any effect. For example, when you have a square island within a circular boundary and then you erase the square, the hatch pattern is updated to fill up the entire circle. But, once the island is removed, another island cannot be added, since it was not calculated as a part of the original set of boundary objects.

HATCHING BLOCKS AND XREF DRAWINGS

When you apply hatching to inserted blocks and xref drawings, their internal structure is treated as if the block or xref drawing were composed of independent objects. This means that if you have inserted a block that consists of a circle within a rectangle and you want the internal circle to be hatched, you need to invoke the **Hatch** tool and then specify a point within the circle to generate the hatch, refer to Figure 14-36. However, if you choose the **Select** option from the **Boundaries** panel of the **Hatch Creation** tab; you will be prompted to select an object. Select an object; the entire block will be selected and a hatch will be created, as shown in Figure 14-36.

When you xref a drawing, you can hatch any part of the drawing that is visible. Also, if you hatch an xref drawing and then use the **XCLIP** command to clip it, the hatch pattern is not clipped, although the hatch boundary associativity is removed. Similarly, when you detach the xref drawing, the hatch pattern and its boundaries are not detached, although the hatch boundary associativity is removed.

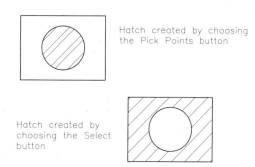

Figure 14-36 Hatching inserted blocks

CREATING A BOUNDARY USING CLOSED LOOPS

Ribbon: Home > Draw > Boundary	**Command:** BOUNDARY

The **Boundary** tool is used to create a polyline or region around a selected point within a closed area, in a manner similar to the one used for defining a hatch boundary. When this tool is invoked, AutoCAD displays the **Boundary Creation** dialog box, as shown in Figure 14-37.

The options in the **Boundary Creation** dialog box are similar to those in the **Boundaries** tab of the **Hatch Creation** tab discussed earlier. Although, only the options related to boundary selections are available, the **Pick Points** button in the **Boundary Creation** dialog box is used to create a boundary by selecting a point inside the object, whose boundary you want to create.

When you choose the **Pick Points** button, the dialog box disappears temporarily and you are prompted to select the internal point. Select a point that lies within the boundary of the object. Once you select an internal point, a polyline or a region will be formed around the boundary (Figure 14-38). To end this process, press ENTER or right-click at the **Pick internal point** prompt. Whether the boundary created is a polyline or a region is determined by the option you have selected from the **Object type** drop-down list in the **Boundary retention** area. **Polyline** is the default option. You can edit the boundary that has been created with the editing commands. The boundary is selected by using the **Last** object selection option.

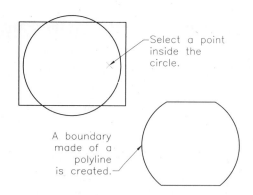

*Figure 14-37 The **Boundary Creation** dialog box*

*Figure 14-38 Using the **BOUNDARY** command to create a polyline boundary*

Similar to the **Hatch** tool, you can also define a boundary set here by choosing the **New** button in the **Boundary set** area of the dialog box. The dialog box will temporarily close to allow you to select objects to be used to create a boundary. The default boundary set is **current viewport**, which means everything visible in the current viewport consists of the boundary set. As discussed earlier, the boundary set option is especially useful when you are working with large and complex drawings, where examining everything on the screen becomes a time-consuming process.

The **Boundary Creation** dialog box also provides you the option of determining the island detection method. The **Island detection** check box is available below the **Pick Points** button and is selected by default. As a result, the island detection method is Flood **t**ype and islands are included as boundary objects. If you clear this check box, the Ray casting method of island detection is enabled, where rays shall be cast from the selected internal point in all directions (by default, using the dialog box). The nearest closed object it encounters is used for boundary creation. If the selected point does not lie within the encountered boundary, a **Boundary Definition Error** dialog box is displayed and you have to select another internal point.

Note
*The **HPBOUND** system variable controls the type of boundary object created using the **Boundary** tool. If the value is 1, the object created is a polyline and if the value is 0, the object created is a region.*

OTHER FEATURES OF HATCHING

1. In AutoCAD, the hatch patterns are separate objects. The objects store the information about the hatch pattern boundary with reference to the geometry that defines the pattern for each hatch object.

2. When you save a drawing, the hatch patterns are stored with the drawing. Therefore, the hatch patterns can be updated even if the hatch pattern file that contains the definitions of hatch patterns is not available.

3. If the system variable **FILLMODE** is 0 (Off), the hatch patterns are not displayed. To see the effect of a changed **FILLMODE** value, you must use the **REGEN** command to regenerate the drawing after the value has been changed.

4. You can edit the boundary of a hatch pattern even when the hatch pattern is not visible (**FILLMODE**=0).

5. The hatch patterns created in the earlier releases are automatically converted into AutoCAD 2010 hatch objects when the hatch pattern is edited.

6. When you save an AutoCAD 2012 drawing in Release 12 format (DXF), the hatch objects are automatically converted to Release 12 hatch blocks.

7. You can use the **CONVERT** command to change the pre-Release hatch patterns to AutoCAD 2012 objects. You can also use this command to change the pre-Release 14 polylines to AutoCAD 2012 optimized format to save memory and disk space.

EXERCISE 4   *Boundary*

See Figure 14-39 and then create the drawing shown in Figure 14-40 using the **Boundary** tool to create a hatch boundary. Copy the boundary from the drawing shown in Figure 14-40, and then hatch.

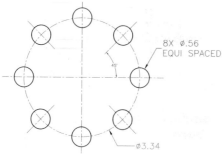

Figure 14-39 *Drawing for Exercise 4*

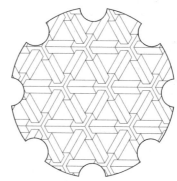

Figure 14-40 *Final drawing for Exercise 4*

Self-Evaluation Test

Answer the following questions and then compare them to those given at the end of this chapter:

1. Different filling patterns make it possible to distinguish between different parts or components of an object. (T/F)

2. If AutoCAD does not locate the entered pattern in the *acad.pat* file, it searches for it in a file with the same name as the pattern. (T/F)

3. When a boundary set is formed, it does not become the default boundary set for hatching. (T/F)

4. You can edit the hatch boundary and the associated hatch pattern by using grips. (T/F)

5. If you need to hatch a drawing that has a gap, then specify the gap value in the _____ edit box in the **Hatch Creation** tab.

6. One of the ways to specify a hatch pattern from the group of stored hatch patterns is by selecting one from the _____ drop-down list.

7. The value that you enter in the _____ edit box lets you rotate a hatch pattern with respect to the *X* axis of the current UCS.

8. The _____ option in the **Hatch Creation** tab lets you select the objects that form a boundary for hatching.

9. While hatching, you can select the _____ option in the **Hatch Type** drop-down list to fill the boundary in a set pattern of colors.

10. The **ISO Pen Width** drop-down list is available in the **Hatch Creation** tab only for the _____ patterns.

Review Questions

Answer the following questions:

1. The **Hatch** tool does not allow you to hatch a region enclosed within a boundary (closed area) by selecting a point inside the boundary. (T/F)

2. A Boundary Definition Error message box is displayed if AutoCAD finds that the selected point is not inside the boundary or that the boundary is not closed. (T/F)

3. The **Tolerance** edit box is used to set the value up to which the open area will be considered closed when selected for hatching using the **Pick Points** method. (T/F)

4. When you use editing tools such as **Move**, **Scale**, **Stretch**, and **Rotate**, associativity is lost, even if all the boundary objects are selected. (T/F)

5. The hatching procedure in AutoCAD does not work on inserted blocks. (T/F)

6. Patterns drawn using the **Hatch** tool are associative. (T/F)

7. Which of the following system variables has to be set to 1, if the **Double** hatch box is selected?

    (a) **HPSPACE**   (b) **HPDOUBLE**
    (c) **HPANG**   (d) **HPSCALE**

8. For which of the following hatch patterns will the **ISO pen width** drop-down list be available?

    (a) **ANSI**   (b) **Predefined**
    (c) **Custom**   (d) **ISO**

9. Which of the following options should be chosen if you want to have the same hatching pattern, style, and properties as that of the existing hatch?

    (a) **Match Properties**   (b) **Select**
    (c) **Pick points**   (d) **Inherit Properties**

10. Which of the following variables can be used to align the hatches in adjacent hatch areas?

    (a) **SNAPBASE**   (b) **FILLMODE**
    (c) **SNAPANG**   (d) **HPANG**

11. In which of the following system variables will the specified hatch spacing value be stored?

    (a) **HPDOUBLE**   (b) **HPANG**
    (c) **HPSCALE**   (d) **HPSPACE**

12. One of the advantages of using the _____ tool is that you don't have to select each object comprising the boundary of the area you want to hatch, as in the case of the _____ tool.

13. To select a custom hatch pattern, first select **Custom** from the **Hatch Type** drop-down list and then select the name of a previously stored hatch pattern from the _____ drop-down list.

14. There are three hatching styles available in AutoCAD. They are solid, _____, and _____.

15. If you select the _____ style, all areas bounded by the outermost boundary are hatched, ignoring any hatch boundaries that lie within the outer boundary.

EXERCISE 5   *Hatch Pattern*

Hatch the drawings shown in Figures 14-41 and 14-42 using the hatch pattern to match.

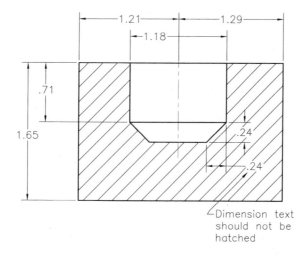

Figure 14-41 *Drawing for Exercise 5*

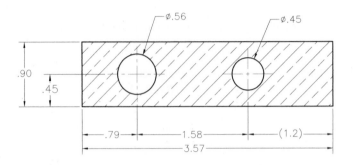

Figure 14-42 Drawing for Exercise 5

EXERCISE 6   *Hatch Align*

Hatch the drawing, as shown in Figure 14-43, using the hatch pattern **ANSI31**. Use the **SNAPBASE** variable to align the hatch lines, as shown in the drawing.

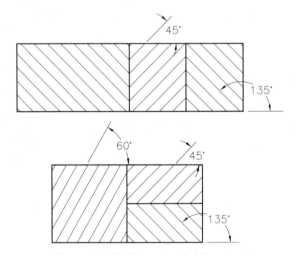

Figure 14-43 Drawing for Exercise 6

EXERCISE 7   *Hatch*

Figure 14-44 shows the top and front views of an object. It also shows the cutting plane line. Based on the cutting plane line, hatch the front views in section. Use the hatch pattern of your choice.

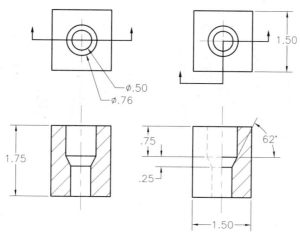

Figure 14-44 *Drawing for Exercise 7*

EXERCISE 8   *Hatch*

Figure 14-45 shows the top and front views of an object. Hatch the front view in full section. Use the hatch pattern of your choice.

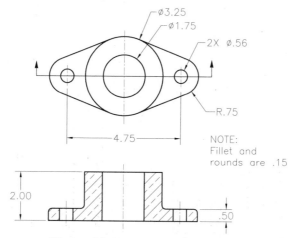

Figure 14-45 *Drawing for Exercise 8*

EXERCISE 9   *Hatch*

Figure 14-46 shows the front and side views of an object. Hatch the side view in half section. Use the hatch pattern of your choice.

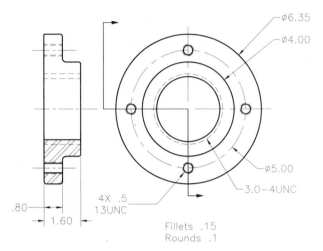

Figure 14-46 *Drawing for Exercise 9*

EXERCISE 10   *Hatch*

Figure 14-47 shows the front view with the broken section and top views of an object. Hatch the front view as shown. Use the hatch pattern of your choice.

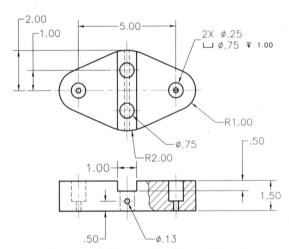

Figure 14-47 *Drawing for Exercise 10*

EXERCISE 11   *Hatch*

Figure 14-48 shows the front, top, and side views of an object. Hatch the side view in section using the hatch pattern of your choice.

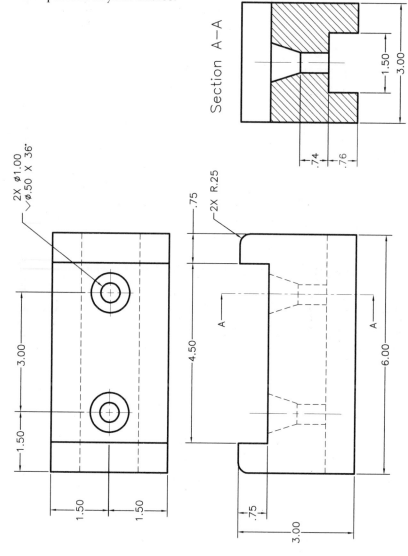

Figure 14-48 *Drawing for Exercise 11*

Problem-Solving Exercise 1

Figure 14-49 shows an object with front and side views with aligned section. Hatch the side view using the hatch pattern of your choice.

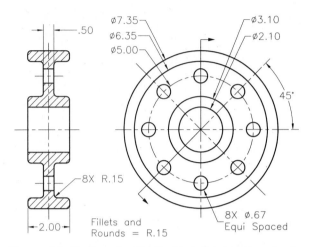

Figure 14-49 Drawing for Problem-Solving Exercise 1

Problem-Solving Exercise 2

Figure 14-50 shows the front view, side view, and the detail "A" of an object. Hatch the side view and draw the detail drawing as shown.

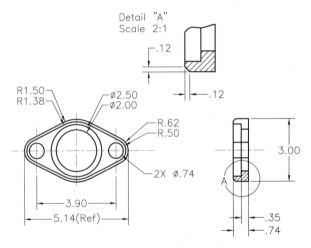

Figure 14-50 Drawing for Problem-Solving Exercise 2

Problem-Solving Exercise 3

Create the drawing shown in Figure 14-51. Assume the missing dimensions.

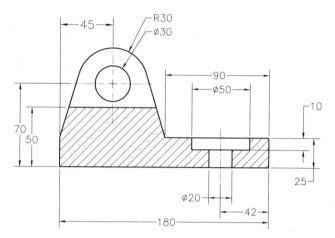

Figure 14-51 *Drawing for Problem-Solving Exercise 3*

Problem-Solving Exercise 4

Create the drawing shown in Figure 14-52. Assume the missing dimensions.

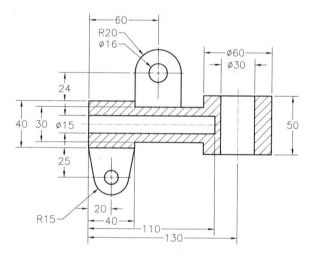

Figure 14-52 *Drawing for Problem-Solving Exercise 4*

Problem-Solving Exercise 5

Draw a detail drawing whose top, side, and section views are given in Figure 14-53. Then, hatch the section view. **

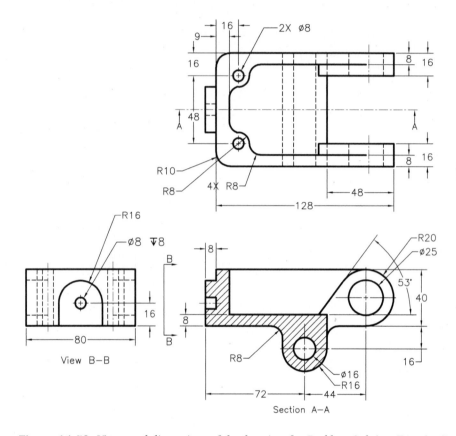

Figure 14-53 *Views and dimensions of the drawing for Problem-Solving Exercise 5*

Answers to Self-evaluation Test

1. T, **2.** T, **3.** F, **4.** T, **5. Tolerance**, **6. Pattern**, **7. Angle**, **8. Select objects**, **9. Gradient**, **10.** ISO hatch

Chapter 15

Model Space Viewports, Paper Space Viewports, and Layouts

CHAPTER OBJECTIVES

In this chapter, you will learn:
- *The concepts of model space and paper space.*
- *To create tiled viewports in the model space.*
- *To create floating viewports.*
- *The use of MSPACE and PSPACE commands.*
- *To control the visibility of viewport layers.*
- *To set the linetype scaling in paper space.*
- *To control the display of annotative objects in viewports.*

KEY TERMS

- *Viewports*
- *Model Space*
- *Paper Space*
- *Tiled Viewport*
- *Temporary Model Space*
- *Rectangular Viewport*
- *Polygonal Viewport*
- *Maximize Viewport*
- *MVIEW*
- *VPLAYER*
- *PAGESETUP*
- *MVSETUP*
- *SPACETRANS*

MODEL SPACE AND PAPER SPACE/LAYOUTS

For ease in designing, AutoCAD provides two different types of environments, model space, and paper space. The paper space is also called layout. The model space is basically used for designing or drafting work. This is the default environment that is active when you start AutoCAD. Almost the entire design is created in the model space. The paper space is used for plotting drawings or generating drawing views for the solid models. By default, the paper space provides you with two layouts. A layout can be considered a sheet of paper on which you can place the design created in the model space and then print it. You can also assign different plotting parameters to these layouts for plotting. Almost all the commands of the model space also work in the layouts. Note that you cannot select the drawing objects created in model space when they are displayed in the viewports in layouts. However, you can snap on to the different points of the drawing objects such as the endpoints, midpoints, center points, and so on, using the **OSNAP** options.

You can shift from one environment to the other by choosing the **Model** tab or the **Layout1**/**Layout2** tabs at the bottom of the drawing area. If these tabs are not available by default then, the **Model** and **Layout1** buttons will be available on the Status Bar. Right-click on any of these buttons and choose **Display Layout and Model Tabs** option from the shortcut menu; the **Model**, **Layout 1**, and **Layout 2** tabs will be displayed at the bottom of the drawing area. You can also invoke the model/layout tabs using the **Quick View Layouts** and **Quick View Drawings** buttons available at the bottom right of the Status Bar. The **Quick View Layouts** button enables a quick display of all the layouts and the model of the current drawing in a series of thumbnails. You can select a thumbnail to view the corresponding layout. Each thumbnail contains two buttons, **Plot** and **Publish**. Using these buttons, you can plot or publish the selected layout. On choosing the **Quick View Layouts** button, a control panel will be displayed along with a series of thumbnails of the existing model and layout tabs. Choose the **New Layout** button in this control panel to create a new layout. On doing so, a new thumbnail is attached to the end of series of thumbnails. The **Publish** button of the control panel enables you to publish all the layouts. You can pin the control panel for frequent use and close when not required. The shortcut menu displayed on right-clicking any thumbnail is the same as that of the **Model** and **Layout** tabs.

The function of the **Quick View Drawings** button is similar to the **Quick View Layouts** button. On choosing this button, all the opened drawings are displayed with their model space and layouts. These opened drawings are displayed in larger thumbnails whereas their layouts are displayed in smaller thumbnails. However, they are enlarged when you move the cursor over them. It allows you to quickly switch between the opened drawings and their layouts. The opened drawing thumbnails contain two buttons; **Save** and **Close**. These buttons enable you to save and close the opened drawings without actually opening them. The control panel displayed on choosing the **Quick View Drawings** button allows you to create a new drawing or open an existing one. In this case also, you can pin the control panel for frequent use, and close it, if no longer required.

You can also shift from one environment to the other using the **TILEMODE** system variable. The default value of this variable is **1**. If the value of this system variable is set to **0**, you will be shifted to the layouts and if its value is set to **1**, you will be shifted to model space. The viewports created in the model space are called tiled viewports and viewports in layouts are called floating viewports.

MODEL SPACE VIEWPORTS (TILED VIEWPORTS)

Menu Bar: View > Viewports > New Viewports
Toolbar: Viewports > Display Viewports Dialog or Layouts > Display Viewports Dialog
Command: VPORTS   **Ribbon:** View > Viewports > Viewport Configurations List

A viewport in the model space is defined as a rectangular area of the drawing window in which you can create the design. When you start AutoCAD, only one viewport is displayed in the model space. You can create multiple non-overlapping viewports in the model space to display different views of the same object, see Figure 15-1. Each of these viewports will act as individual drawing area. This is generally used while creating solid models. You can view the same solid model from different positions by creating the tiled viewports and defining the distinct coordinate system configuration for each viewport. You can also use the **Pan** or **Zoom** tool to display different portions or different levels of the detail of the drawing in each viewport. The tiled viewports can be created using the **New** tool available in the **Viewports** panel.

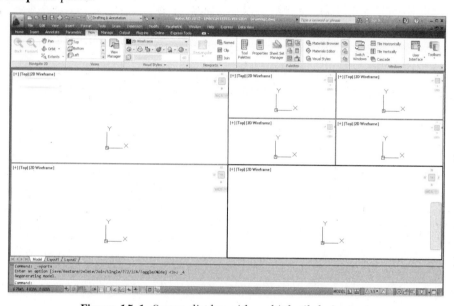

Figure 15-1 Screen display with multiple tiled viewports

Creating Tiled Viewports

As mentioned earlier, the display screen in the model space can be divided into multiple non-overlapping tiled viewports. All these viewports are created only in the rectangular shape. This number depends on the equipment and the operating system on which AutoCAD is running. Each tiled viewport contains a view of the drawing. The tiled viewports touch each other at the edges without overlapping. While using tiled viewports, you are not allowed to edit, rearrange, or turn individual viewports on or off. These viewports are created using the **New** tool when the system variable **TILEMODE** is set to 1 or the **Model** tab is active. When you choose the **Named** or **New** button from the **Viewports** flyout in the **View** menu, the **Viewports** dialog box is displayed. You can use this dialog box to create new viewport configurations and save them. The options under both the tabs of the **Viewports** dialog box are discussed next.

New Viewports Tab

The **New Viewports** tab of the **Viewports** dialog box (Figure 15-2) provides the options related to standard viewport configurations. You can also save a user-defined configuration using this tab. The name for the new viewport configuration can be specified in the **New name** edit box. If you do not enter a name in this edit box, the viewport configuration you create is not saved and, therefore, cannot be used later. A list of standard viewport configurations is listed in the **Standard viewports** list box. This list also contains the ***Active Model Configuration***, which is the current viewport configuration. From the **Standard viewports** list, you can select and apply any one of the listed standard viewport configurations. A preview image of the selected configuration is displayed in the **Preview** window. The **Apply to** drop-down list has the **Display** and **Current viewport** options. Selecting the **Display** option applies the selected viewport configuration to the entire display and selecting the **Current viewport** option applies the selected viewport configuration to only the current viewport. With this option, you can create more viewports inside the existing viewports. The changes will be applied to the current viewport and the new viewports will be created inside the current viewport.

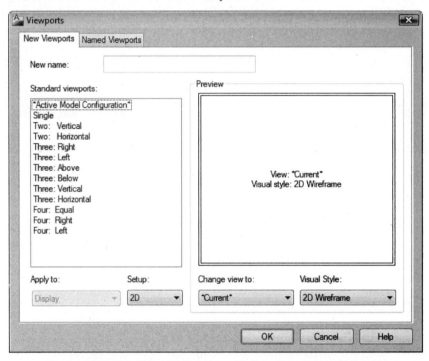

*Figure 15-2 The **New Viewports** tab of the **Viewports** dialog box*

From the **Setup** drop-down list, you can select **2D** or **3D**. When you select the **2D** option, it creates the new viewport configuration with the current view of the drawing in all the viewports initially. Using the **3D** option applies the standard orthogonal and isometric views to the viewports. For example, if a configuration has four viewports, they are assigned the **Top**, **Front**, **Right**, and **South East Isometric** views, respectively. You can also modify these standard orthogonal and isometric views by selecting from the **Change view to** drop-down list and replacing the existing view in the selected viewport. For example, you can select the viewport that is assigned the **Top** view and then choose **Bottom** view from the **Change view to** drop-down list to replace it. The preview image in the **Preview** window reflects the changes you make. If you use the **2D** option, you can select a named viewport configuration to replace the selected one. From the **Visual Style** drop-down list, you can select the visualization mode of your model. For example, you can select a viewport and then choose **3D Hidden** from the **Visual Style** drop-down list. Now, the model will be displayed with the hidden edges invisible in this viewport. Choose **OK** to exit the dialog box and apply the created or selected configuration to the current display in the drawing. When you save a new viewport configuration, it saves the information about the number and position of viewports, the viewing direction and zoom factor, and the grid, snap, coordinate system, and UCS icon settings.

Named Viewports Tab

The **Named Viewports** tab of the **Viewports** dialog box (Figure 15-3) displays the name of the current viewport next to **Current name**. The names of all the saved viewport configurations in a drawing are displayed in the **Named viewports** list box. You can select any one of the named viewport configurations and apply it to the current display. A preview image of the selected configuration is displayed in the **Preview** window. Choose **OK** to exit the dialog box and apply the selected viewport configuration to the current display. In the **Named viewports** list box, you can select a name and right-click to display a shortcut menu. Choosing **Delete** deletes the selected viewport configuration and choosing **Rename** allows you to rename the selected viewport configuration.

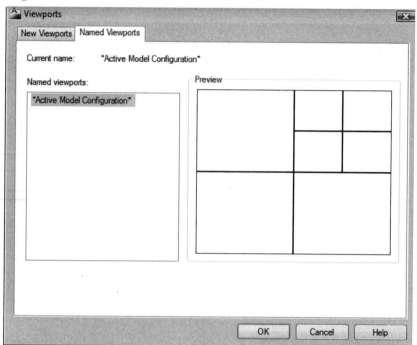

*Figure 15-3 The **Named Viewports** tab of the **Viewports** dialog box*

MAKING A VIEWPORT CURRENT

The viewport you are currently working in is called the current viewport. You can display several model space viewports on the screen, but you can work in only one of them at a time. You can switch from one viewport to another even when you are in the middle of a command. For example, you can specify the start point of the line in one viewport and the endpoint of the line in the other viewport. The current viewport is indicated by a border that is heavy compared to the borders of the other viewports. Also, the graphics cursor appears as a drawing cursor (screen crosshairs) only when it is within the current viewport. Outside the current viewport this cursor appears as an arrow cursor. You can enter points and select objects only from the current viewport. To make a viewport current, you can select it with the pointing device. Another method of making a viewport current is by assigning its identification number to the **CVPORT** system variable. The identification numbers of the named viewport configurations are not listed in the display.

JOINING TWO ADJACENT VIEWPORTS

AutoCAD provides you with an option of joining two adjacent viewports. However, remember that the viewports you wish to join should result in a rectangular-shaped viewport only. As mentioned earlier, the viewports in the model space can only be in rectangular shape. Therefore, you will not be able to join two viewports, in case they do not result in a rectangular shape. The viewports can be joined by using the **Join Viewports** tool available in the **Viewports** panel. On invoking this tool, you will be prompted to select the dominant viewport. A dominant viewport is one whose display will be retained after joining. After selecting the dominant viewport, you will be prompted to select the viewport to be joined. Figure 15-4 shows the viewport configuration before joining and Figure 15-5 shows the viewport configuration after joining.

Figure 15-4 Selecting the dominant viewport and the viewport to be joined

Figure 15-5 Viewports after joining

Note
*You can also use the Command line to create, save, restore, delete, or join the viewport configurations. This is done using the **-VPORTS** command.*

PAPER SPACE VIEWPORTS (FLOATING VIEWPORTS)*

As mentioned earlier, the viewports created in the layouts are called floating viewports. This is because unlike in model space, the viewports in the layouts can be overlapping and of any shape.

In layouts, there is no restriction of the shape of the viewports. You can even convert a closed object into a viewport in the layouts. Figure 15-6 shows a layout with floating viewports.

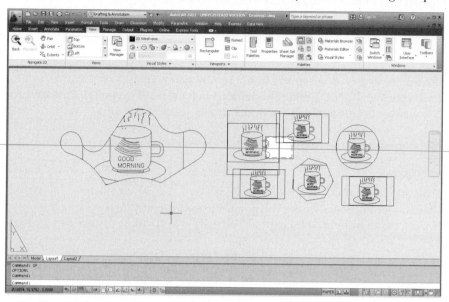

Figure 15-6 *Screen display with multiple floating viewports*

The method of creating floating viewports is discussed next.

Creating Floating Viewports

Ribbon: View > Viewports > Named   **Menu Bar:** View > Viewports > New Viewports
Toolbar: Viewports > Display Viewports Dialog or Layouts > Display Viewports Dialog
Command: VPORTS

This tool is used to create the floating viewports in layouts. However, when you invoke this tool in the layouts, the dialog box displayed is slightly modified. Instead of the **Apply to** drop-down list in the **New Viewports** tab, the **Viewport Spacing** spinner is displayed, see Figure 15-7. This spinner is used to set the spacing between the adjacent viewports. The rest of the options in both the **New Viewports** and the **Named Viewports** tabs of the **Viewports** dialog box are the same as those discussed under the tiled viewports. When you select a viewport configuration and choose **OK**, you will be prompted to specify the first and the second corner of a box that will act as a reference for placing the viewports. You will also be given an option of **Fit**. This option fits the configuration of viewports such that they fit exactly in the current display.

> **Note**
> *You can also use the **+VPORTS** command to display the **Viewports** dialog box. When you invoke this command, you will be prompted to specify the **Tab Index**. Enter **0** to display the **New Viewports** tab and enter **1** to display the **Named Viewports** tab.*

Creating Rectangular Viewports

Ribbon: View > Viewports > Create Viewport drop-down > Create Rectangular
Menu Bar: View > Viewports > New Viewports
Toolbar: Viewports > Polygonal Viewports   **Command:** -VPORTS

To create a rectangular viewport, choose the **Create Rectangular** tool from **View > Viewports > Create Viewports** drop-down (See Figure 15-8) in the **Ribbon**. The prompt sequence that will follow is given next.

Specify corner of viewport or
[ON/OFF/Fit/Shadeplot/Lock/Object/Polygonal/Restore/LAyer/2/3/4] <Fit>: *Specify the start point of the viewport.*
Specify opposite corner: *Specify the end point of the viewport.*
Regenerating model.

You can also create 2, 3 or 4 viewports in one go by entering 2, 3 or 4 at the prompt **Specify corner of viewport or [ON/OFF/Fit/Shadeplot/Lock/Object/Polygonal/Restore/LAyer/2/3/4] <Fit>** . The viewports automatically fit in the drawn rectangular area.

The prompt sequence for the option 2 Viewports is given next.

Command: **-VPORTS**
Specify corner of viewport or
[ON/OFF/Fit/Shadeplot/Lock/Object/Polygonal/Restore/LAyer/2/3/4] <Fit>: 2
Enter viewport arrangement [Horizontal/Vertical] <Vertical>: *Specify the orientation of the viewport.*
Specify first corner or [Fit] <Fit>: *Specify the start point of the viewport.*
Specify opposite corner: *Specify the end point of the viewport.*

The prompt sequence for creating 3 viewports is given next.

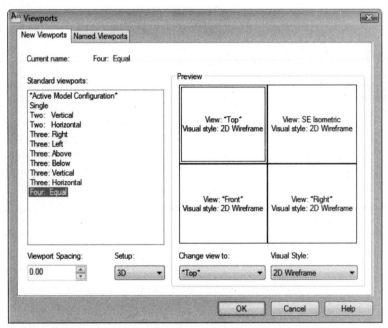

Figure 15-7 *The **New Viewports** tab of the **Viewports** dialog box displayed in layouts*

Command: **-VPORTS**
Specify corner of viewport or
[ON/OFF/Fit/Shadeplot/Lock/Object/Polygonal/Restore/LAyer/2/3/4] <Fit>: 3
Enter viewport arrangement
[Horizontal/Vertical/Above/Below/Left/Right] <Right>: *Specify the orientation of the viewport using Horizontal or Vertical option. You can also specify the position of the largest viewport in the three using Above, Below, Left or Right option.*
Specify first corner or [Fit] <Fit>: *Specify the start point of the viewport.*
Specify opposite corner: *Specify the end point of the viewport.*

The prompt sequence for creating 4 viewports is given next.

Command: **-VPORTS**
Specify corner of viewport or
[ON/OFF/Fit/Shadeplot/Lock/Object/Polygonal/Restore/LAyer/2/3/4] <Fit>: 4
Specify first corner or [Fit] <Fit>: *Specify the start point of the viewport.*
Specify opposite corner:*Specify the end point of the viewport.*

Creating Polygonal Viewports

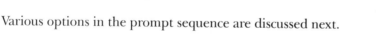

Ribbon: View > Viewports > Create Viewport drop-down > Create Polygonal
Menu Bar: View > Viewports > New Viewports
Toolbar: Viewports > Polygonal Viewports   **Command: -VPORTS**

As mentioned earlier, you can create floating viewports of any closed shape. The viewports thus created can even be self-intersecting in shape. To create a polygonal viewport, choose the **Create Polygonal** tool from **View > Viewports > Create Viewports** drop-down (See Figure 15-8). The prompt sequence that will follow when you choose this tool is given next.

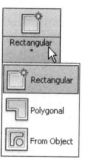

Command: **-VPORTS**
Specify corner of viewport or
[ON/OFF/Fit/Shadeplot/Lock/Object/Polygonal/Restore/
LAyer/2/3/4] <Fit>: Polygonal
Specify start point: *Specify the start point of the viewport.*
Specify next point or [Arc/Length/Undo]: *Specify the next point or select an option.*
Specify next point or [Arc/Close/Length/Undo]: *Specify the next point or select an option.*

*Figure 15-8 Tools in the **Create Viewports** drop-down*

Various options in the prompt sequence are discussed next.

Arc
This option is used to switch to the arc mode for creating the viewports. When you invoke this option, the options for creating the arcs will be displayed. You can switch back to the line mode by choosing the **Line** option.

Length
This option is used to specify the length of the next line of the viewport. The line will be drawn in the direction of the last drawn line segment. In case the last drawn segment was an arc, the line will be drawn tangent to it.

Undo
This option is used to undo the last drawn segment of the polygonal viewport.

Close
This option is used to close the polygon and create the viewport. The last entity that will be used to close the polygon will depend upon whether you were in the arc mode or in the line mode. If you were in the line mode, the last entity will be a line. If you were in the arc mode, the last entity will be an arc.

Figure 15-9 shows the polygonal viewport created using the combination of lines and arcs.

Converting an Existing Closed Object into a Viewport

Figure 15-9 *A polygonal viewport*

Ribbon: View > Viewports > Create Viewport drop-down > Create From Object
Toolbar: Viewports > Convert Object to Viewport
Command: VPORTS

This option allows you to convert an existing closed object into a viewport. However, remember that the object selected should be a single entity. The objects that can be converted into a viewport include polygons drawn using the **Polygon** tool, rectangles drawn using the **Rectangle** tool, polylines (last segment closed using the **Close** option), circles, ellipses, closed splines, or regions. To convert any of these objects into a viewport, choose the **Create from Object** tool from the **Viewports** panel. You will be prompted to select the object that need to be converted into a viewport. Figure 15-10 shows a viewport created using a polygon of nine sides and Figure 15-11 shows a viewport created using a closed spline.

Figure 15-10 *A viewport created using a polygon of nine sides*

Figure 15-11 *A viewport created using a closed spline*

Note
*When you shift to the layouts, the **Page Setup** dialog box is displayed for printing and a rectangular viewport is created that fits the drawing area. If you want, you can retain or delete this viewport using the **Erase** tool.*

TEMPORARY MODEL SPACE

Sometimes, when you create a floating viewport in the layout, the drawing is not displayed completely inside it, see Figures 15-10 and 15-11. In such cases, you need to zoom or pan the drawings to fit them in the viewport. But when you invoke any of the **Zoom** or the **Pan** tools in the layouts, the display of the entire layout is modified instead of the display inside of the viewport. Now, to change the display of the viewports, you will have to switch to the temporary model space. The temporary model space is defined as a state when the model space is activated in the layouts. The temporary model space is exactly similar to the actual model space and you can make any kind of modifications in the drawing from temporary model space. Therefore, the main reason for invoking the temporary model space is that you can modify the display of the drawing. The temporary model space can be invoked by choosing the **Paper** button from the status bar. You can also switch to the temporary model space by double-clicking inside the viewports. You will see that the model space UCS icon is automatically displayed when you switch to the temporary model space. Also, the extents of the viewport become the extents of the drawing. You can use the **Zoom** and **Pan** tools to fit the model inside the viewport. The temporary model space can also be invoked using the **MSPACE** command.

Once you have modified the display of the drawing in the temporary model space, you have to switch back to the paper space. This is done by choosing the **Model** button from the status bar. You can also switch back to the paper space by double-clicking anywhere in the layout outside the viewport, or by using the **PSPACE** command.

EXAMPLE 1   *Create Viewports*

In this example, you will draw the object shown in Figure 15-12 and then create a floating viewport of the shape shown in Figure 15-13 to display the object in the layout. The dimensions of the viewport are in the paper space. Do not dimension the object.

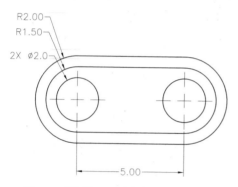

Figure 15-12 *Model for Example 1*   **Figure 15-13** *Shape of the floating viewport*

1. Start a new drawing and then draw the object shown in Figure 15-12.

2. Choose the **Layout1** tab to switch to the layout; a rectangular viewport will be displayed in this layout.

3. Choose the **Erase** tool from the **Modify** panel in the **Home** tab; you will be prompted to select the object. Type **L** in this prompt to delete the last object, that is, the viewport, in this case.

4. Choose the **Polygon** tool from the **Draw** panel in the **Home** tab and create the required hexagon.

5. Choose the **Convert Object to Viewport** tool from **View > Viewports > Create Viewports** drop-down; you will be prompted to select an object. Select the hexagon; it will be converted into a viewport. You will see that the complete object is not displayed inside the viewport. Therefore, you need to modify its display.

6. Double-click inside the viewport to switch to the temporary model space. The border of the viewport will become thick, indicating that you have switched to the temporary model space.

7. Now, using the **Zoom** and the **Pan** tools, fit the drawing inside the viewport.

8. Choose the **Model** button from the status bar to switch back to the paper space. The drawing will be displayed fully inside the viewport, see Figure 15-14.

EDITING VIEWPORTS

You can perform various editing operations on the viewports. For example, you can control the visibility of the objects in the viewports, lock their display, clip the existing viewports using an object, and so on. All these editing operations are discussed next.

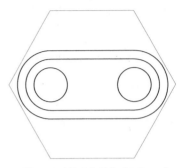

Figure 15-14 *Displaying the drawing inside the polygonal viewport*

Controlling the Display of Objects in Viewports

The display of the objects in the viewports can be turned on or off. If the display is turned off, the objects will not be displayed in the viewport. However, the object in the model space is not affected by this editing operation. To control the display of the objects, select the viewport entity in the layout and right-click to display the shortcut menu. In this menu, choose **Display Viewport Objects > No**. If there is only one viewport, you will be prompted to confirm whether you really want to turn off all active viewports. Enter **Y** to turn off the display. However, if there are more than one viewports, the visibility of the selected viewports will be automatically turned off when you choose **Display Viewport Objects > No** from the shortcut menu. Similarly, you can again turn on the display of the objects by choosing **Display Viewport Objects > Yes** from the shortcut menu. This shortcut menu is displayed on selecting the viewport and right-clicking. This editing operation can also be done using the **OFF** option of the **MVIEW** or the **-VPORTS** command.

Locking the Display of Objects in Viewports

To avoid accidental modification in the display of objects in the viewports, you can lock their display. If the display of a viewport is locked, the tools such as **Zoom** and **Pan** do not work in it. Also, you cannot modify the view in the locked viewport. For example, if the display of a viewport is locked, you cannot zoom or pan the display or change the view in that viewport even if you switch to the temporary model space. To lock the display of the viewports, select it and right-click on it to display the shortcut menu. In this menu, choose **Display Locked > Yes**. Now, the display of this viewport will not be modified. However, you can draw objects or delete objects in this viewport by switching to the temporary model space. Similarly, you can unlock the display of the objects in the viewports by choosing **Display Locked > No** from the shortcut menu. You can also lock or unlock the display of the viewports using the **Lock** option of the **MVIEW** command or the **-VPORTS** command.

Controlling the Display of Hidden Lines in Viewports

While working with three-dimensional solid or surface models, there are a number of occasions where you have to plot the solid models such that the hidden lines are not displayed. Plotting solid models in the model space (Tilemode=1) can be easily done by selecting the **Hidden** option from the **Shade plot** drop-down list. This drop-down list is available in the **Shaded viewport options** area in **Plot** dialog box. If this area is not available by default, you need to choose the **More Options** button in the dialog box. This option is not available in layouts. In this case, you will have to control the display of the hidden lines in the viewports. To control the display of the hidden lines, select the viewport and right-click to display the shortcut menu. In this menu, choose **Shade plot > Hidden**. Although the hidden lines will be displayed in the viewports, now they will not be displayed in the printouts. The display of the hidden lines can also be controlled using the **Shadeplot** option of the **MVIEW** command or the **-VPORTS** command.

Note
*You can also use the other options in the **Shade plot** drop-down list. These options are explained in detail in Chapter 16 (Plotting Drawings).*

Tip
Apart from the previously mentioned editing operations, you can also move, copy, rotate, stretch, scale, or trim the viewports using the respective commands. You can also use the grips to edit the viewports.

Clipping Existing Viewports

Ribbon: View > Viewports > Clip	**Command:** VCLIP
Toolbar: Viewports > Clip existing Viewport	**Menu Bar:** Modify > Clip > Viewport

You can modify the shape of the existing viewport by clipping it using an object or by defining the clipping boundary. The viewports can be clipped using the **VPCLIP** command. This command can also be invoked by choosing the **Clip** tool from the **Viewports** panel in the **View** tab. The prompt sequence that will follow when you invoke this tool is given next.

Select viewport to clip: *Select the viewport to be clipped.*
Select clipping object or [Polygonal/Delete] <Polygonal>: *Select an object for clipping the viewport or specify an option.*

Select clipping object Option
This option is used to clip the viewport using a selected closed loop. The objects that can be used for clipping the viewports include circles, ellipses, closed polylines, closed splines, and regions. As soon as you select the clipping object, the original viewport will be deleted and the selected object will be converted into a viewport. The portion of the display that was common to both the original viewport and the object selected will be displayed. You can, however, change the display of the viewport using the **Zoom** tool or the **Pan** tool. Figure 15-15 shows a viewport and an object that will be used to clip the viewport and Figure 15-16 shows the new viewport created after clipping.

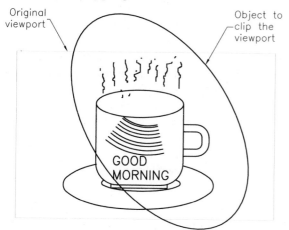

Figure 15-15 Selecting the object for clipping the viewport

Figure 15-16 New viewport created after clipping

Polygonal
This option is used to create a polygonal boundary for clipping the viewports. When you invoke this option, the options for creating a polygonal boundary will be displayed. You can draw a polygonal boundary for clipping the viewport using these options.

Delete

This option is used to delete the new clipping boundary created using an object or using the **Polygonal** option. The original viewport is restored when you invoke this option, which will be available only if the viewport has been clipped at least once. If the viewport is clipped more than once, you can restore only the last viewport clipping boundary.

Maximizing Viewports

Status Bar: Maximize Viewport

While working with floating viewports, you may need to invoke the temporary model space to modify the drawing. One of the options is that you double-click inside the viewport to invoke the temporary model space and make the changes in the drawing. But in this case, the shape and size of the floating viewport will control the area of the temporary model space. If the viewport is polygonal and small in size, you may have to zoom and pan the drawing a number of times. To avoid this, AutoCAD allows you to maximize a viewport on the screen. This provides you all the space in the drawing area to make the changes in the drawing.

To maximize a floating viewport, choose the **Maximize Viewport** button from the Status bar. The viewport is automatically maximized in the drawing area and the **Maximize Viewport** button is replaced by the **Minimize Viewport** button. If there are more than one floating viewports, two arrows will be displayed on either side of the **Minimize Viewport** button. These arrows can be used to switch to the display in the other floating viewports. After making the changes in the drawing, choose the **Minimize Viewport** button to restore the original display of the layout. When you do so, the view and the magnification in all the viewports is the same as that before maximizing them. Also, the visibility of the layers remains the same as that before maximizing the viewport.

CONTROLLING THE PROPERTIES OF VIEWPORT LAYERS

Command: VPLAYER

You can control the properties of layers inside a floating viewport using the **VPLAYER** command or by using the **Layer Properties Manager**. The **VPLAYER** command is used to control the visibility of layers in individual floating viewports. For example, you can use the **VPLAYER** command to freeze a layer in the selected viewport. The contents of this layer will not be displayed in the selected viewport, although in the other viewports, they will be displayed. This command can be used from either temporary model space or paper space. The only restriction is that **TILEMODE** is set to 0 (Off); that is, you can use this command only in the **Layout** tab.

Command: **VPLAYER** [Enter]
Enter an option [?/Color/Ltype/LWeight/TRansparency/Freeze/Thaw/Reset/Newfrz/Vpvisdflt]:

? Option

You can use this option to obtain a listing of the frozen layers in the selected viewport. When you enter ?, you will be prompted to select the viewport. Select the view port, AutoCAD text window will be displayed showing all the layers that are frozen in the current layer. On invoking this command in temporary model space, AutoCAD will temporarily shift you to the paper space to let you select the viewport.

Color Option

The **Color** option is used to assign a color to a layer (or layers) in the viewports. You can either use True colors or you can specify a Color book, if you have previously installed any. The changes can be applied on the current, selected, or all the viewports.

Ltype Option

You can use this option to specify any linetype to a layer(or layers) in the current, selected, or all the viewports.

LWeight Option

The **LWeight** option is used to specify a line width to a particular layer or layers. You can specify the width between 0.0mm to 2.11 mm. Then you are prompted to specify the name(s) of the layer(s) to assign the line width. In this case also, you can apply the changes to the current, selected, or all the viewports.

Transparency Option

The **Transparency** option is used to set the transparency level of a layer (or layers) in one or more viewports. When you select this option, you will be prompted to specify the transparency level. Specify the transparency level; you will be prompted to specify the layer for which you need to set the transparency level. If you want to specify more than one layer, the names of the layers must be separated by commas. If you are working on a temporary model space, you can also select an object whose layer you want to set the transparency level. On specifying the name of the layer(s), AutoCAD will prompt you to select the viewport(s) to change the transparency. Select one or all viewports; the transparency level will be applied to the specified layers in the viewport after you exit this command.

Freeze Option

The **Freeze** option is used to freeze a layer (or layers) in one or more viewports. When you select this option, you will be prompted to specify the name(s) of the layer(s) you want to freeze. If you want to specify more than one layer, the layer names must be separated by commas. You can also select an object whose layer you want to freeze. Once you have specified the name of the layer(s), AutoCAD prompts you to select the viewport(s), to freeze the specified layer(s). You can select one or all the viewports. The layers will be frozen after you exit this command.

Thaw Option

With this option, you can thaw the layers that have been frozen in the viewports. Layers that have been frozen, thawed, turned on, or turned off globally are not affected by **VPLAYER Thaw**. For example, if a layer has been frozen, the objects on it are not regenerated on any viewport, even if **VPLAYER Thaw** is used to thaw that layer in any viewport. If you want to thaw more than one layer, they must be separated by commas. You can thaw the specified layers in the current, selected, or all the viewports.

Reset Option

With the **Reset** option, you can set the visibility of the layer(s) in the specified viewports to their current default setting. The visibility defaults of a layer can be set by using the **Vpvisdflt** option of the **VPLAYER** command. When you invoke the **Reset** option, you will be prompted to specify the names of the layers to be reset. You can reset the layers in the current viewport, in selected viewports, or in all the viewports.

Chapter 15

Newfrz (New Freeze) Option

With this option, you can create new layers that are frozen in all the viewports. This option is used mainly where you need a layer that is visible only in one viewport. This can be accomplished by creating the layer with the **Newfrz** option and then thawing it in the viewport where you want to make the layer visible. On invoking this option, you will be prompted to specify the name(s) of the new layer(s) that will be frozen in all the viewports. To specify more than one layer, separate the layer names with commas. After you specify the name(s) of the layer(s), AutoCAD creates frozen layers in all viewports. Also, the default visibility setting of the new layer(s) is set to Frozen; therefore, if you create any new viewports, the layers created with **VPLAYER Newfrz** are also frozen in them.

Vpvisdflt (Viewport Visibility Default) Option

With this option, you can set a default for the visibility of the layer(s) in the subsequent new viewports. When a new viewport is created, the frozen/thawed status of any layer depends on the **Vpvisdflt** setting for that particular layer. When you invoke this option, you will be prompted to specify the names of the layer(s) whose visibility is to be changed. After specifying the name(s), you will be prompted to specify whether the layers should be frozen or thawed in the new viewports.

CONTROLLING THE LAYERS IN VIEWPORTS USING THE LAYER PROPERTIES MANAGER DIALOG BOX

You can use the **Layer Properties Manager** dialog box to control the layer display properties in viewports. When you invoke the **Layer Properties Manager** dialog box in the layout, some additional properties are added to it, see Figure 15-17. These properties are used to override the global property of the layer in a particular viewport, while retaining the global layer properties in other floating viewports and the model space. To set a property of a viewport double-click inside it to activate the temporary model space and then set the properties. The properties are discussed next.

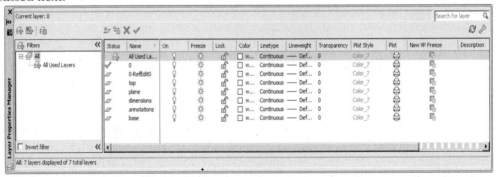

*Figure 15-17 Controlling the visibility of layers in the viewports using the **Layer Properties Manager** dialog box*

VP Freeze

When the **TILEMODE** option is turned off, you can freeze or thaw the selected layers in the current floating viewport by selecting the **VP Freeze** option. You can freeze a layer in the current floating viewport if it is thawed in the model space but you cannot thaw a layer in the current viewport if it is frozen in the model space. The frozen layers still remain visible in other viewports.

VP Color

The swatch under the **VP Color** column is used to change the display color of the objects in the selected layer in the current floating viewport. In all other floating viewports and model space, the color of the objects remains unaffected.

VP Linetype

The field under the **VP Linetype** column is used to change the linetype defined for the selected layer in the active floating viewport. In all other floating viewports and model space, the linetype remains unaffected.

VP Lineweight

The field under the **VP Lineweight** column is used to override the lineweight defined for the selected layer in the active floating viewport. In all other floating viewports and the model space, the lineweight remains unaffected.

VP Transparency

The field under the **VP Transparency** column is used to override the transparency defined for the layer selected in the active floating viewport. In all other floating viewports and the model space, the transparency remains unaffected.

VP Plot Style

The field under the **VP Plot Style** column is used to override the plot style settings assigned to the selected layer in the active floating viewport. The plot style for all other floating viewports and model space remains unaffected. You cannot override the plot styles for the color dependent plot styles. The override for the plot style also does not affect the plot if the visual style is set to **Realistic** or **Conceptual**.

Note
*By default, all the overrides that you have defined for the current active viewport will get highlighted in light blue color in the **Layer Properties Manager** dialog box. Also, the icon of the selected layer under the **Status** column will change to inform you that some of the properties of the selected layer have been overridden by the new one.*

Removing Viewport Overrides

To remove the viewport overrides defined for the active viewports, open the **Layer Properties Manager** dialog box in the active viewport and right-click on the layer property from which you want to remove the viewport override; a shortcut menu will be displayed. Next, choose the **Remove Viewport Overrides for** option from the shortcut menu to display the cascading menu. You can make a choice to remove the override for a single property / all properties of the selected layer / all layers in the current viewport / all viewports, see Figure 15-18. Choose the required option from the cascading menu to remove the viewport overrides. You can also remove the layer property overrides for all layers of the selected viewports from the paper space. To do so, make the paper space current; select the viewports boundaries of the viewports for which you want to remove the override and choose the **Remove Viewport Overrides for All Layers** option from the shortcut menu.

*Figure 15-18 Suboptions in the **Remove Viewport Override for** option of the **Layer** shortcut menu*

Note
*For more information about the **Layer Properties Manager** dialog box, see Chapter 4.*

PAPER SPACE LINETYPE SCALING (PSLTSCALE SYSTEM VARIABLE)

By default, the linetype scaling is controlled by the **LTSCALE** system variable. Therefore, the display size of the dashes depends on the **LTSCALE** factor, on the drawing limits, or on the drawing units. If you have different viewports with different zoom (XP) factors, the size of the dashes will be different for these viewports. Figure 15-19 shows three viewports with different sizes and different zoom (XP) factors. You will notice that the dash length is different in each of these three viewports.

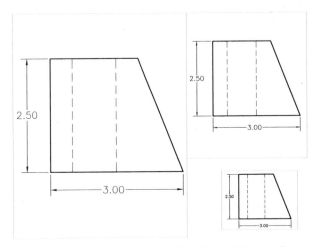

Figure 15-19 *Varying sizes of the dashed lines with* **PSLTSCALE** = 0

Generally, it is desirable to have identical line spacing in all viewports. This can be achieved with the **PSLTSCALE** system variable. By default, **PSLTSCALE** is set to 0. In this case, the size of the dashes depends on the **LTSCALE** system variable and on the zoom (XP) factor of the viewport where the objects have been drawn. If you set **PSLTSCALE** to 1 and **TILEMODE** to 0, the size of the dashes for objects in the model space are scaled to match the **LTSCALE** of objects in the paper space viewport, regardless of their zoom scale. In other words, if **PSLTSCALE** is set to 1, even if the viewports are zoomed to different sizes, the length of the dashes will be identical in all viewports. Figure 15-20 shows three viewports with different sizes. Notice that the dash length is identical in all the three viewports.

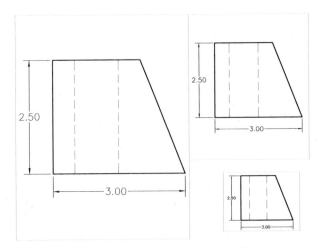

Figure 15-20 *Varying sizes of the dashed lines with* **PSLTSCALE** = 1

Tip
*You can control the scale factor for displaying the objects in the viewports using the **Viewport Scale Control** drop-down list. This drop-down list is available in the **Viewports** toolbar.*

INSERTING LAYOUTS

Status Bar: Quick View Layouts > New Layout	**Command:** LAYOUT
Menu Bar: Insert > Layout > New Layout	**Toolbar:** Layouts > New Layout

 The **LAYOUT** command is used to create a new layout. It also allows you to rename, copy, save, and delete existing layouts. A drawing designed in the **Model** tab can be composed for plotting in the **Layout** tab. The prompt sequence is as follows.

Enter layout option [Copy/Delete/New/Template/Rename/SAveas/Set/?]<Set>:

The options in the prompt sequence are discussed next.

New Option
This option is used to create a new layout. On choosing this option, you will be prompted to specify the name of the new layout. A new tab with the new layout name will appear in the drawing. Alternatively, you can right-click on the **Model** or the **Layout** tab and choose **New layout** from the shortcut menu to add a new layout. You can also, choose the **Quick View Layouts** button in the Status Bar and then choose **New Layout** button available below the thumbnails to create a new layout. The new layout tab will be added at the end of the existing layout tabs with the default name **Layout N**, where **N** is a natural number starting from one and acquires an ascending value that has not been used in the layout names of the current drawing.

Copy Option
This option is used to copy a layout. When you invoke this option, you will be prompted to specify the layout that has to be copied. Upon specifying the layout, you will be prompted to specify the name of the new layout. If you do not enter a name, the name of the copied layout is assumed with an incremental number in the brackets next to it. For example, Layout 1 is copied as Layout1 (2). The name of the new layout appears as a new tab next to the copied layout tab. Alternatively, you can right-click on the **Model** or **Layout** tab or on the **Quick View Layouts** tile and then choose **Move or Copy** from the shortcut menu to move or copy the

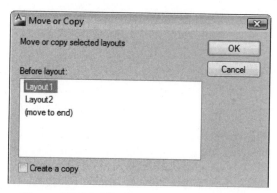

*Figure 15-21 The **Move or Copy** dialog box*

selected layout; the **Move or Copy** dialog box will be displayed, see Figure 15-21. Select the layout from the **Before layout** area to move it above the previous layout and then choose the **OK** button. For example, if you want to move the Layout 2 before Layout 1, then select Layout 2 from the **Move or Copy** dialog box; the Layout 2 will be moved before Layout 1. To create a copy of the selected layout select the **Create a copy** check box.

Tip
You can also drag and drop layouts to move them. To create copies of the selected layouts, press the CTRL key as you drag and drop the layouts.

Delete Option

This option is used to delete an existing layout. On invoking this option, you will be prompted to specify the name of the layout to be deleted. The current layout is the default layout for deleting. Remember that the **Model** tab cannot be deleted. You can also right-click on the **Model** or the **Layout** tab or on the **Quick View Layouts** tiles and choose **Delete** from the shortcut menu; the AutoCAD alert window will be displayed. Choose the **OK** button to delete the selected layout.

Template Option

This option is used to create a new template based on the existing layout template in the *.dwg*, *.dwt*, or *.dxf* files. This option invokes the **Select Template From File** dialog box, see Figure 15-22. You can also invoke this dialog box by right-clicking on the **Model**, the **Layout** tab or on the **Quick View Layouts** tiles and choose the **From template** option from the shortcut menu.

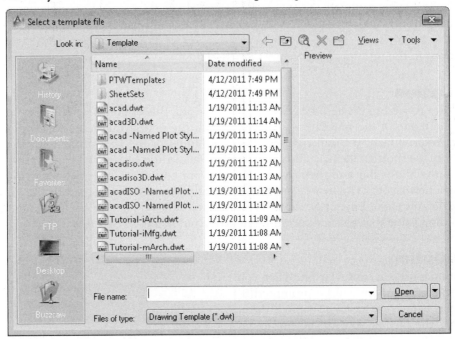

*Figure 15-22 The **Select Template From File** dialog box*

The layout and geometry from the specified template or drawing file is inserted into the current drawing. After the *dwt*, *dwg*, or *dxf* file is selected, the **Insert Layout(s)** dialog box is displayed, as shown in Figure 15-23.

Choose the **Layout from Template** button from the **Layouts** toolbar to create a layout using an existing template or a drawing file.

 Note
If you insert a template that has a title block, it will be inserted as a block and all the text in the title block will be inserted as attributes. You will learn more about blocks and attributes in later chapters.

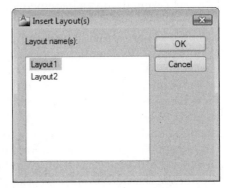

*Figure 15-23 The **Insert Layout(s)** dialog box*

Rename Option

This option allows you to rename a layout. On choosing this option, you will be prompted to specify the name of the layout to be renamed. On specifying the name, you will be prompted to specify the new name of the layout. The layout names have to be unique and can contain up to 255 characters, out of which only 32 are displayed in the tab. The characters in the name are not case-sensitive. You can also right-click on the **Model** or the **Layout** tab or on the **Quick View Layouts** tiles and choose **Rename** from the shortcut menu to rename the selected layout.

Tip
*You can also double-click on the **Layout** tab to rename it.*

SAveas Option

This option is used to save a layout in the drawing template file. On choosing this option, you will be prompted to specify the layout that has to be saved. If you specify the name of the layout to be saved, the **Create Drawing File** dialog box will be displayed. In this dialog box, you can enter the name of the template in the **File name** edit box. The layout templates can be saved in the *.dwg*, *.dwt*, or the *.dxf* format.

Set Option

This option is used to set a layout as the current layout. When you invoke this option, you will be prompted to specify the name of the layout that has to be made current.

? Option

This option is used to list all the layouts available in the current drawing. The list is displayed in the Command line. You can open the AutoCAD Text Window to view the list by pressing the F2 key.

IMPORTING LAYOUTS TO SHEET SETS

To add a layout to the current sheet set, right-click on the **Model** or the **Layout** tab and then choose the **Import Layout as Sheet** option from the shortcut menu; the **Import Layout as Sheet** dialog box will be invoked. This dialog box displays all the layouts available in the selected drawing. Select the check box on the left of the drawing file name under the **Drawing Name** column and choose the **Import Checked** button to import the selected layouts into the current sheet set.

Note
A layout can be imported to only one sheet set. If a layout already belongs to a sheet set, you have to create a copy of the drawing to import its layout to another sheet set.

Tip
You can also drag and drop the layout from a drawing to add it to the sheet set. But, make sure that you save the drawing file before you drag and drop the layout.

INSERTING A LAYOUT USING THE WIZARD

Command: LAYOUTWIZARD

This command displays the **Layout Wizard** that guides you step-by-step through the process of creating a new layout.

DEFINING PAGE SETTINGS

Ribbon: Output > Plot > Page Setup Manager   **Command:** PAGESETUP
Toolbar: Layouts > Page Setup Manager

This tool is used to specify the layout and plot device settings for each new layout. You can also right-click on the **Model** or the current **Layout** tab and choose **Page Setup Manager** from the shortcut menu to invoke this command. When you invoke this command, the **Page Setup Manager** dialog box is displayed, which will be discussed in Chapter 14.

EXAMPLE 2   *Plot*

In this example, you will create a drawing in the model space and then use the paper space to plot the drawing. The drawing to be plotted is shown in Figure 15-24.

1. Increase the limits to 75, 75 and then draw the sketch shown in Figure 15-24.

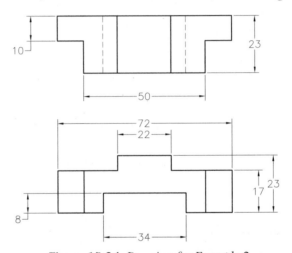

Figure 15-24 Drawing for Example 2

2. Choose the **Layout1** tab; AutoCAD displays **Layout1** with the default viewport. Delete this viewport. Right-click on the **Layout1** tab and then choose **Page Setup Manager** from the shortcut menu to display the **Page Setup Manager** dialog box. **Layout1** is automatically selected in the **Current page setup** list box.

3. Choose the **Modify** button to display the **Page Setup - Layout1** dialog box. Select the printer or plotter from the **Name** drop-down list in the **Printer/plotter** area. In this example, **HP Lasejet4000** is used. From the drop-down list in the **Paper size** area, select the paper size that is supported by your plotting device. In this example, the paper size is **A4 (210 x 297mm)**. Choose the **OK** button to accept the settings and exit the dialog box. Close the **Page Setup Manager** dialog box.

4. Choose the **New** tool from the **Viewports** flyout in the **View** menu; the **Viewports** dialog box is displayed. Select the **Single** option from the **Standard viewports** list box and choose **OK**. The prompt sequence is as follows:

    Tab index <0>: 0
    Specify first corner or [Fit] <Fit>: *Specify the first corner of the viewport near the bottom left corner of the drawing window.*
    Specify opposite corner: **@210,297**
    Regenerating model.

5. Use the **ZOOM** command in the paper space to zoom to the extents of the viewport.

6. Double-click in the viewport to switch to the temporary model space and use the **ZOOM** command to zoom the drawing to 2XP. In this example, it is assumed that the scale factor is 2:1; therefore, the zoom factor is 2XP.

7. Create the dimension style with the text height of 1.5 and the arrowhead height of 1.25. Define all the other parameters based on the text and arrowhead heights and then select the **Annotative** check box and the **Scale dimensions to layout** radio button from the **Scale for dimension features** area of the **Fit** tab in the **New Dimension Style** dialog box.

8. Using the new dimension style, dimension the drawing. Make sure that you do not change the scale factor. You can use the **Pan** tool to adjust the display.

9. Double-click in the paper space to switch back to the paper space. Choose the **Plot** tool from the **Quick Access** toolbar to display the **Plot** dialog box.

10. Choose the **Window** option from the **What to plot** drop-down list in the **Plot area**; the dialog box will be closed temporarily and you will be prompted to specify the first and second corners of the window. Define a window close to the boundary of the viewport such that the viewport is not included in it.

11. As soon as you define both the corners of the window, the **Plot** dialog box will be redisplayed on the screen. Select **1:1** from the **Scale** drop-down list of the **Plot scale** area.

12. Select the **Center the plot** check box from the **Plot offset (origin set to printable area)** area.

13. Choose the **Preview** button to display the plot preview. You can make any adjustments, if required, by redefining the window.

14. After you are satisfied with the preview, right-click and choose **Plot** from the shortcut menu. The drawing will be printed with the scale of 2:1. This means that two plotted units will be equal to one actual unit. Save this drawing with the name *Example2.dwg*.

WORKING WITH THE MVSETUP COMMAND

The **MVSETUP** command is a very versatile command and can be used both on the **Model** tab and on the **Layout** tab. This means that this command can be used when the **Tilemode** is set to **1** or when it is set to **0**. The uses of this command in both the drawing environments are discussed next.

Using the MVSETUP Command on the Model Tab

On the **Model** tab, this command is used to set the units, scale factor for the drawing, and the size of the paper. On invoking this command, you will be prompted to specify whether or not you want to enable the paper space. Enter **NO** at this prompt to use this command on the **Model** tab. The prompt sequence that will follow when you enter **NO** is given next.

Enter units type [Scientific/Decimal/Engineering/Architectural/Metric]: *Specify the unit. In this case, **Architectural** is selected.*

Architectural Scales
====================
(480) 1/40"=1'
(240) 1/20"=1'

(192) 1/16"=1'
(96) 1/8"=1'
(48) 1/4"=1'
(24) 1/2"=1'
(16) 3/4"=1'
(12) 1"=1'
(4) 3"=1'
(2) 6"=1'
(1) FULL

Enter the scale factor: *Specify the scale factor for displaying the drawing.*
Enter the paper width: *Specify the width of the paper.*
Enter the paper height: *Specify the height of the paper.*

A box of the specified width and height will be drawn and the drawing will be displayed inside the box. The display of the drawing will depend upon the scale factor you have specified.

Using the MVSETUP Command on the Layout Tab

The way this command works on the **Layout** tab is entirely different from that on the **Model** tab. On the **Layout** tab it is used to insert a title block, create an array of viewports, align the objects in the viewports, and so on. The prompt sequence that will follow when you invoke this command on the **Layout** tab is given next.

[Align/Create/Scale viewports/Options/Title block/Undo]: *Select an option.*

Align Option

This option is used to align the objects in one of the viewports with that of another viewport. You can specify horizontal, vertical, angled, or rotated alignment. After you have selected the option you will be prompted to select the base point. This base point is the point which will be used as the reference point for moving the objects in the specified viewport. Once you have specified the base point, you will be prompted to specify the point in the viewport that will be panned. This point will be aligned with the base point and thus the objects in this viewport will be moved.

Create Option

This option is used to create the viewports in the layouts. On invoking this option, you will be prompted to specify whether you want to create a viewport, delete the objects in it, or undo the changes of this option. The default option is **Create**. If this option is selected, the AutoCAD Text Window is displayed. This window provides you with four options. These options are discussed next.

0:	None	1:	Single
2:	Std. Engineering	3:	Array of Viewports

If you enter **0**, the viewports will not be created and the previous prompt will be displayed again. If you enter **1**, you will be prompted to specify two points for the bounding box of the viewport. A new viewport will be created inside the specified bounding box. If you enter **2**, you will be prompted to specify two corners of the bounding box. Inside the specified bounding box, four standard engineering viewports will be created. The first viewport will display the model from the top view, the second will display the model from the SE isometric view, the third will display the front view of the model (*ZX* plane of the current UCS), and the fourth will display the right-side view of the model (*YZ* plane of the current UCS). This option is generally used while working with 3D models. You will also be prompted to specify the distances between the viewports in the X direction and in the Y direction.

Note

The details about the UCS and different views will be discussed in Chapters 24 (The User Coordinate System) and 25 (Getting Started with 3D), respectively.

If you enter **3**, you will be allowed to create an array of viewports along the *X* axis and the *Y* axis. You will be prompted to define a bounding box, in which the viewports will be created. After you have defined the bounding box, you will be prompted to specify the number of viewports in the X and Y directions. Then you will be prompted to specify the distances between the viewports in the X and Y directions. Figure 15-25 shows an array of viewports in the layout with the distance between the viewports in the *X* axis and *Y* axis directions set to zero.

Figure 15-25 Array of viewports

Scale viewports Option

This option is used to modify the scale factors of the viewports. This scale factor is defined by the ratio of the paper space units to model space units. You can select one or more than one viewports to modify the scale factors. If you select more than one viewport, you will be prompted to specify whether you want to define a uniform scale factor for the viewports or define individual scale factors. You will be first prompted to specify the number of the paper space units in the viewport and then the model space units in it. The display in the viewports will be automatically modified once you have defined both the values.

Options Option

This option is used to set the parameters for inserting the title block. The parameters that can be set include layers, limits, units, and external reference options. Using the **Layers** option, you can predefine the layer into which the title block will be inserted. Using the **Limits** option, you can specify whether or not the limits should be reset to the extents of the title block after it is inserted. Using the **Units** option, you can set the units in which the size of the title block will be translated. Using the **Xref** option, you can set whether the title block, after inserting, will become the entity of the current drawing or remain as an Xref object.

Title block Option

This option is used to insert the title block of the desired size in the current layout, delete the entities from the current layout, or reset the origin of the current layout. The default option is that for inserting a title block. If you select this option, the **AutoCAD Text Window** will be invoked, displaying the different sizes and formats of the title blocks that can be inserted. The various options that will be displayed are given next.

0:	None	1:	ISO A4 Size(mm)
2:	ISO A3 Size(mm)	3:	ISO A2 Size(mm)
4:	ISO A1 Size(mm)	5:	ISO A0 Size(mm)
6:	ANSI-V Size(in)	7:	ANSI-A Size(in)
8:	ANSI-B Size(in)	9:	ANSI-C Size(in)
10:	ANSI-D Size(in)	11:	ANSI-E Size(in)
12:	Arch/Engineering (24 x 36in)		
13:	Generic D size Sheet (24 x 36in)		

You can enter the number corresponding to a title block to insert that title block in the current layout. The properties of this layout will depend on the parameters set using the **Options** option. You can also add or delete the title blocks in this default list. Figure 15-26 shows a layout with an A4 size title block and a viewport created in it.

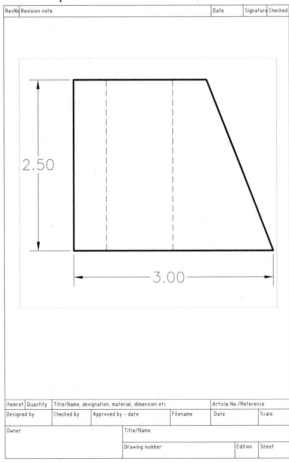

Figure 15-26 *Layout with the title block*

CONVERTING THE DISTANCE BETWEEN MODEL SPACE AND PAPER SPACE

Toolbar: Text > Convert distance between spaces   **Command:** SPACETRANS

 While working with drawings in layouts, you may need to find a distance value that is equivalent to a specific distance in the model space. For example, you may need to write text whose height should be equal to a similar text written in the model space. To convert these distances between the model space and layouts, AutoCAD provides the **SPACETRANS** command. Note that this command does not work in the model space. This command works only in layouts or in the temporary model space invoked from the layouts, but there should be at least one viewport in the drawing. When you invoke this command in the paper space, AutoCAD prompts you to select the viewport. After selecting the viewport, AutoCAD prompts you to specify the model space distance. Enter the original distance value that was measured in the model space; AutoCAD displays the paper space equivalent to the specified distance.

Similarly, when you invoke this command in the temporary model space, AutoCAD prompts you to specify the paper space distance. Enter the distance value measured in the paper space. AutoCAD displays the model space equivalent to the specified distance.

CONTROLLING THE DISPLAY OF ANNOTATIVE OBJECTS IN VIEWPORTS

Some new buttons have been added to the status bar to control the annotation scale in viewports and model space separately. When you activate a floating viewport, a new **Viewport Scale** button is added to the status bar. The **Viewport Scale** button lists the same set of scales as the **Annotation Scale**. You can set an annotation display scale either from the **Viewport Scale** button or the **Annotation Scale** button, and the other scales will be updated accordingly. The viewport will zoom to an appropriate scale so that the annotation objects can be displayed at the specified scale.

Also, when you activate a floating viewport, the **Lock/Unlock Viewport** button is added to the status bar. With the help of this button, you can toggle between the lock and unlock states of the viewport. The **Viewport Scale** and **Annotation Scale** buttons are not accessible when the viewport is locked. If the viewport is unlocked and you zoom the drawing instead of specifying the viewport scale, the annotation scale will not be changed and the current scale representation will remain intact and visible. But, the viewport scale will be changed to display the actual scale of the viewport.

The example given next will explain various concepts related to creating and controlling the display of annotations in viewports.

Note
When the paper space is active, the annotation scale is always 1:1 and it cannot be modified.

EXAMPLE 3   *Annotative Text*

In this example, you will draw the object, as shown in Figure 15-27, and create the floating viewports, as shown in Figure 15-28. The left viewport has a scale of 3/8"=1'-0" and the right viewport has a scale of 1-1/2"=1'-0". All the annotations that are created in the model space should be displayed with a text height of 0.08" on the sheet even if the same annotation appears in multiple viewports.

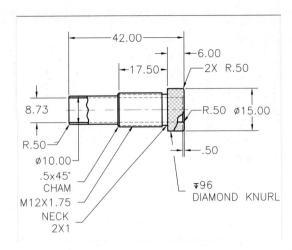

Figure 15-27 Model for Example 3

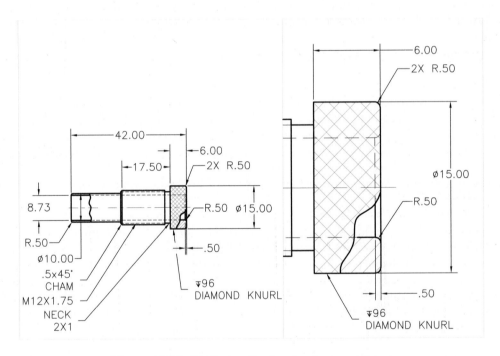

Figure 15-28 Displaying the drawing in two viewports

1. Start a new file in the **Drafting & Annotation** workspace and draw the object as shown in Figure 15-27.

2. Set the **Annotation Scale** to 3/8"=1'-0" from the Status Bar.

3. Create an annotative text style with the **Paper Text Height** equal to **0.08**.

4. Create an annotative dimension style with the values given next and draw the dimensions, refer to Figure 15-27.

   Arrow size = **0.07**
   Text height = **0.08**

5. Create an annotative multileader style with the values given next and draw the multileaders, refer to Figure 15-27.

    Landing distance = **0.075**
    Arrowhead size = **0.07**
    Text height = **0.08**

6. Change the annotation scale to 1-1/2"=1'-0", select all the dimensions from the right of the object, and then choose the **Add Current Scale** tool from the **Annotation Scaling** panel of the **Annotate** tab; the annotation scale of 1-1/2"=1'-0" is added to the selected dimensions. You will notice that the selected dimensions and multileaders appear smaller than the other annotations because they reflect the current annotation scale of 1-1/2"=1'-0".

7. Turn the **Annotation Visibility** button off. Now you can see the annotations that support a scale of 1-1/2"=1'-0". This helps you to find out the dimensions to which the scale of 1-1/2"=1'-0" has been assigned.

8. Select the annotations from the right of the object and adjust their locations with the help of grips, such that the placement of dimensions remains similar to the one shown in Figure 15-27. Notice that when you select the annotation object, its different scale representations are displayed with faded dashed lines.

9. From the Status Bar, set the annotation scale to 3/8"=1'-0" and then change it back to 1-1/2"=1'-0". Now you will notice that the same annotation objects not only change the size, but also change the location.

10. Choose the **Layout1** tab to switch to the layouts.

11. A rectangular viewport is automatically created in this layout. Choose the **Erase** tool from the **Modify** panel of the **Home** tab; you will be prompted to select the object. Enter **L** in this prompt to delete the last object, which in this case is the viewport.

12. Draw two rectangles of dimensions **3'X3.5'** and **2.25'X3.5'** side-by-side, refer to Figure 15-28.

13. Choose the **Convert Object to Viewport** tool from **View > Viewports > Create Viewports** drop-down; you will be prompted to select the object. Select one of the rectangles; it will be converted into a viewport. Similarly, convert the second rectangle into a viewport.

14. Activate the viewport on the left and set its annotation scale to 3/8"=1'-0" from the Status Bar.

15. Similarly, activate the viewport on the right and set its annotation scale to 1-1/2"=1'-0". Pan the view so that it looks similar to Figure 15-28.

Self-Evaluation Test

Answer the following questions and then compare them to those given at the end of this chapter:

1. Viewports in the model space can be of any shape. (T/F)

2. Viewports in the model space can overlap each other. (T/F)

3. You can join two different tiled viewports. (T/F)

4. You cannot insert any additional layout in the current drawing. (T/F)

5. The _____ tool is used to create the tiled viewports.

6. The viewports in layouts are called _____ viewports.

7. The two working environments provided by AutoCAD are _____ and _____.

8. The _____ command is used to insert a title block in the current layout.

9. When you join two adjacent viewports, the resultant viewport is _____ in shape.

10. The default viewport that is created in a new layout is _____ in shape.

Review Questions

Answer the following questions:

1. Only the viewports that are created in layouts can be polygonal in shape. (T/F)

2. You cannot lock the display of a floating viewport. (T/F)

3. An existing closed loop can be converted into a viewport in the model space. (T/F)

4. You can create an array of viewports using the **MVSETUP** command in the model space. (T/F)

5. Which of the following commands can be used to control the display of objects in viewports?

   (a) **MVIEW**　　　　　　　(b) **DVIEW**
   (c) **LAYOUT**　　　　　　 (d) **MSPACE**

6. Which of the following commands can be used to switch to the temporary model space?

   (a) **MVIEW**　　　　　　　(b) **DVIEW**
   (c) **LAYOUT**　　　　　　 (d) **MSPACE**

7. Which of the following options of the **MVIEW** command in the paper space can be used to hide the hidden lines of solid models in printing?

   (a) **Hide**　　　　　　　　(b) **Shadeplot**
   (c) **Create**　　　　　　　(d) **None**

8. Which of the following commands can be used to clip an existing floating viewport?

   (a) **MVIEW**　　　　　　　(b) **DVIEW**
   (c) **VPCLIP**　　　　　　 (d) **MSPACE**

9. Which of the following options of the **VPLAYER** command is used to create a layer that will be frozen in all viewports?

    (a) **Freeze**   (b) **Thaw**
    (c) **Newfrz**   (d) **Reset**

10. The _____ option of the **MVSETUP** command can be used to set the parameters related to the inserting of the title blocks.

11. You can work only in the _____ tiled viewport.

12. The _____ command can be used to set similar linetype scale for all the viewports.

13. The _____ command is used to switch back to the paper space from the temporary model space.

14. The _____ dialog box is used to save a viewport configuration in the model space.

15. Layers that have been frozen, thawed, switched on, or switched off globally are not affected by the _____ command.

EXERCISE 1   *Tiled Viewport*

In this exercise, you will perform the following operations:

a. In the model space, make the drawing of the shaft shown in Figure 15-29.
b. Create three tiled viewports in the model space and then display the drawing in all the three tiled viewports.
c. Create a new layout with the name **Title Block** and insert a title block of ANSI A size in this layout.
d. Create two viewports, one for the drawing and one for the detail "A". See Figure 15-29 for the approximate size and location. The dimensions in detail "A" viewport must not be shown in the other viewport. Also, adjust the LTSCALE factor for hidden and center lines.
e. Plot the drawing.

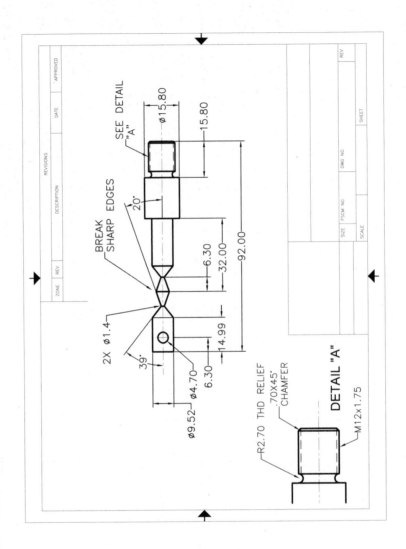

Figure 15-29 Drawing for Exercise 1

Answers to Self-evaluation Test

1. F, **2. F**, **3.** T, **4.** F, **5. VPORTS**, **6.** floating, **7.** model space and paper space/layouts, **8. MVSETUP**, **9.** rectangular, **10.** rectangular

Chapter 16

Plotting Drawings

CHAPTER OBJECTIVES

In this chapter, you will learn:
- *To set plotter specifications and plot drawings.*
- *To configure plotters and edit their configuration files.*
- *To create, use, and modify plot styles and plot style tables.*
- *To plot sheets in a sheet set.*

KEY TERMS

- *Plot*
- *Page Setup*
- *Plot Style*
- *Plotter Manager*
- *Named Plot Style*
- *Styles Manager*
- *Color Dependent Plot Style*
- *Plot Style Table Editor*

PLOTTING DRAWINGS IN AutoCAD

After you have completed a drawing, you can store it on the computer storage device such as the hard drive or diskette. However, to get its hard copy, you should plot the drawing on a sheet of paper using a plotter or printer. A hard copy is a handy reference for professionals working on site. With pen plotters, you can obtain a high-resolution drawing. Basic plotting has already been discussed in Chapter 2, Getting Started with AutoCAD. You can plot drawings in the **Model** tab or any of the layout tabs. A drawing has a **Model** and two layout tabs (**Layout1**, **Layout2**) by default. Each of these tabs has its own settings and can be used to create different plots. You can also create new layout tabs using the **LAYOUT** command. This is discussed in Chapter 15.

PLOTTING DRAWINGS USING THE PLOT DIALOG BOX

Quick Access Toolbar: Plot	**Toolbar:** Standard > Plot	**Command:** PLOT
Application Menu: Print > Plot	**Ribbon:** Output > Plot > Plot	

The **Plot** tool is used to plot a drawing. When you invoke this tool, the **Plot** dialog box is displayed. This dialog box can also be invoked by right-clicking on the **Model** tab or any of the **Layout** tabs and choosing the **Plot** option from the shortcut menu. Figure 16-1 shows the expanded **Plot** dialog box.

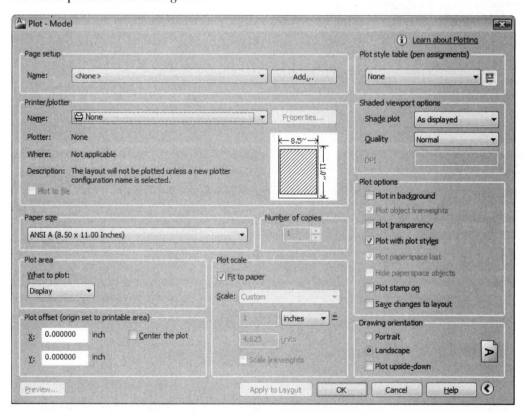

*Figure 16-1 The expanded **Plot** dialog box*

In this dialog box, some values were set when AutoCAD was first configured. You can examine these values and if they conform to your requirements, you can start plotting directly. Otherwise, you can alter these values to define plot specifications by using the options in the **Plot** dialog box. The available plot options are discussed next.

Page setup Area

The **Name** drop-down list in this area displays all the saved and named page setups. A page setup contains the settings required to plot a drawing on a sheet of paper to create a layout. It consists of all the settings related to the plotting of a drawing such as the scale, pen settings, and so on, and also includes the plot devices being used. These settings can be saved as a named page setup, which can be later selected from this drop-down list and used for plotting a drawing. If you select **Previous plot** from the drop-down list, the settings used for the last drawing plotted are applied to the current drawing. You can also import the existing page setup from the other files by

selecting the **Import** option from the drop-down list. The page setup can be imported from any drawing file(*.dwg*), from a template file(*.dwt*), or from a drawing interchange format file(*.dxf*). You can choose the base for the page setup on a named page setup, or you can add a new named page setup by choosing the **Add** button, which is located next to the drop-down list. When you choose this button, AutoCAD displays the **Add Page Setup** dialog box, as shown in Figure 16-2.

*Figure 16-2 The **Add Page Setup** dialog box*

Enter the name of the new page setup in this dialog box and choose **OK**. All the settings that you configure in the current **Plot** dialog box will be saved under this page setup.

Tip
*To create a new page setup based on the existing one, select the existing page setup from the **Name** drop-down list and make modifications in it. Next, choose the **Add** button; the modified page setup will be saved.*

Printer/plotter Area

This area displays all information about the configured printers and plotters currently selected from the **Name** drop-down list. It displays the plotter driver and the printer port being used. It also displays the physical location and some description text about the selected plotter or printer. All the currently configured plotters are displayed in the **Name** drop-down list.

Note
*To add plotters and printers to the **Name** drop-down list, choose the **Plotter Manager** tool from the **Plot** panel in the **Output** tab; a window will be displayed. Double-click on the **Add-A-Plotter Wizard** icon in this window to display the **Add Plotter** wizard. You can use this wizard to add a plotter to the list of configured plotters. Once the plotter is added, the plotter configuration file (PC3) for the plotter will be created. This file consists of all settings needed by the specific plotter to plot the drawing. (The **Plotters** window will be discussed later in this chapter in the section Adding Plotters).*

Properties

To check information about a configured printer or plotter, choose the **Properties** button. When you choose this button, the **Plotter Configuration Editor** dialog box will be displayed. This dialog box lists all the details of the selected plotter under three tabs: **General**, **Ports**, and **Device and Document Settings**. The **Plotter Configuration Editor** will be discussed later in the "Editing Plotter Configuration" section of this chapter.

Plot to file

If you select this check box, AutoCAD plots the output to a file rather than to the plotter.

Depending on the plotter selected, the file can be plotted in the *.dwf*, *.plt*, *.jpg*, or *.png* format. On selecting this check box and choosing **OK** from the **Plot** dialog box, the **Browse for Plot File** dialog box will be displayed. Specify the file name and its location in this dialog box.

Partial Preview Window

The window displayed below the **Properties** button is called the **Partial Preview** window. The preview in this window dynamically changes as you modify the parameters in the **Plot** dialog box. The outer rectangle in this window is the paper you selected. It also shows the size of the paper. The inner hatched rectangle is the section of the paper that is used by the image. If the image extends beyond the paper, a red border is displayed around the paper.

Paper size Area

The drop-down list in this area displays all standard paper sizes for the selected plotting device. You can select any size from the list to make it current. If **None** has been selected currently from the **Name** drop-down list in the **Printer/plotter** area, AutoCAD will display the list of all standard paper sizes.

Number of copies Area

You can use the spinner available in this area to specify the number of copies that you want to plot. If multiple layouts and copies are selected and some of the layouts are set for plotting to a file or AutoSpool, they will produce a single plot. AutoSpool allows you to send a file for plotting while you are working on another program.

Plot area Area

Using the **What to plot** drop-down list provided in this area, you can specify the portion of the drawing to be plotted. You can also control the way the plotting will be carried out. The options in the **What to plot** drop-down list are described next.

Display

If you select this option, the portion of the drawing that is currently being displayed on the screen is plotted.

Extents

This option resembles the **Extents** option of the **ZOOM** command and prints the drawing to the extents of the objects. If you add more objects to the drawing, they are also included in the plot and the extents of the drawing are recalculated. If you reduce the drawing extents by erasing, moving, or scaling the objects, the extents of the drawing are again recalculated. You can use the **Extents** option of the **ZOOM** command to determine which objects shall be plotted. If you use the **Extents** option when the perspective view is on and the position of the camera is not outside the drawing extents, the following message is displayed: **The annotation scale is not equal to the plot scale.**

Limits

This option is available only if you plot from the **Model** tab. Selecting this option plots the complete area defined within the drawing limits.

Note
*To be able to clearly view the differences between the three previously listed plotting options, it may be a good idea to make sure that the default scale options have been selected. Alternatively, select the **Fit to paper** check box from the **Plot scale** area of the dialog box if you are in the **Model** tab, and select **1:1** if you are working in any one of the layout tabs.*

Window

On selecting this option, you need to specify the section of the drawing to be plotted by defining a window. To define a window, select the **Window** option from the **What to plot** drop-down list. The **Plot** dialog box will be temporarily closed and you will be prompted to specify two points on the screen that define a window; the area within this window will be plotted. Once you have defined the window, the **Plot** dialog box is redisplayed on the screen. You will notice that the **Window** button is now displayed on the right of the **What to plot** drop-down list. To reselect the area to be plotted, choose the **Window** button. If the **Model** tab is chosen, you will notice that the previously selected area is displayed in white and the remaining area is displayed in gray. After selecting the area to be plotted, you can choose the **OK** button in the dialog box to plot the drawing.

Note

*Sometimes, while using the **Window** option, the area you have selected may appear clipped off. This may happen because the objects are too close to the window you have defined on the screen. You need to redefine the window in this situation. Such errors can be avoided by using the **Preview** option discussed later.*

View

Selecting the **View** option enables you to plot a view that was created with the **VIEW** command. The view must be defined in the current drawing. If no view has been created, the **View** option is not displayed. When you select this option, a drop-down list is displayed in this area. You can select a view for plotting from this drop-down list and then choose **OK** in the **Plot** dialog box. While using the **View** option, the specifications of the plot will depend on the specifications of the named view.

Layout

This option is available only when you are plotting from the layout. This option prints the entire content of the drawing that lies inside the printable area of the paper selected from the drop-down list in the **Paper size** area.

Plot offset (origin set to printable area) Area

This area allows you to specify an offset of the plotting area from the lower left corner of the paper. The lower left corner of the specified plot area is positioned at the lower left margin of the paper by default. If you select the **Center the plot** check box, AutoCAD automatically centers the plot on the paper by calculating the X and Y offset values. You can specify an offset from the origin by entering positive or negative values in the **X** and **Y** edit boxes. For example, if you want the drawing to be plotted 4 units to the right and 4 units above the origin point, enter **4** in both the **X** and **Y** edit boxes. Depending on the units you have specified in the **Paper size** area of the dialog box, the offset values are either in inches, in millimeters, or in pixels.

Plot scale Area

This area controls the drawing scale of the plot area. Apart from **Custom** option, the **Scale** drop-down list has thirty-three architectural and decimal scales by default. The default scale setting is **1:1** when you are plotting a layout. However, if you are plotting in a **Model** tab, the **Fit to paper** check box is selected. The **Fit to paper** option allows you to automatically fit the entire drawing on the paper. It is useful when you have to print a large drawing using a printer that uses a smaller size paper.

Whenever you select a standard scale from the drop-down list, the scale is displayed in the edit boxes as a ratio of the plotted units to the drawing units. You can also change the scale factor manually in these edit boxes. When you do so, the **Scale** edit box displays **Custom**. For example, for an architectural drawing, which is to be plotted at the scale 1/4"=1'-0", you can enter either 1/4"=1'-0" or 1=48 in the edit boxes.

Note

*The **PSLTSCALE** system variable controls the paper space linetype scaling and has a default value of 1. This implies that irrespective of the zoom scale of the viewports, the linetype scale of the objects in the viewports remains the same. If you want the linetype scale of the objects in different viewports with different magnification factors to appear different, you should set the value of the **PSLTSCALE** variable to 0. This has been discussed in detail in Chapter 18 (Model Space Viewports, Paper Space Viewports, and Layouts).*

The **Scale lineweights** check box is available only if you are plotting in a layout. This option is not available in the **Model** tab. If you select the **Scale lineweights** check box, you can scale lineweights in proportion to the plot scale. Lineweights generally specify the linewidth of the printable objects and are plotted with the original lineweight size, regardless of the plot scale.

Note

*You can save the custom scales that you use during plotting or in layouts in the **Edit Drawing Scales** dialog box. To invoke the **Edit Drawing Scales** dialog box, choose the **Scale List** tool from the **Annotation Scaling** panel in the **Annotate** tab. The **Scale List** area lists the default scales in AutoCAD. Choose the **Add** button to invoke the **Add Scale** dialog box. Enter a name for the custom scale in the **Name appearing in the scale list** edit box. Next, enter the scale in the **Scale Properties** area and choose the **OK** button; the name specified for the custom scale will be listed in the **Scale List** area. In this way, you can keep a track of the custom scales used.*

Plot style table (pen assignments) Area

This area will be available when you expand the **Plot** dialog box. This area allows you to view and select a plot style table, edit the current plot style table, or create a new plot style table. A plot style table is a collection of plot styles. A plot style is a group of pen settings that are assigned to an object or layer and that determine the color, linetype, thickness, line ending, line joining and the fill style of drawing objects when they are plotted. It is a named file that allows you to control the pen settings for a plotted drawing.

You can select the required plot style from the drop-down list in this area. Whenever you select a plot style, AutoCAD displays the **Question** message box prompting you to specify whether the selected plot style should be assigned to all layouts or not. If you choose **Yes**, the selected plot style will be used to plot from all layouts. You can also select **None** from the drop-down list to plot a drawing without using any plot styles. You can assign different plot style tables to a drawing and plot the same drawing differently each time. The use of plot styles will be discussed later in this chapter.

You can also select **New** from the drop-down list to create a new plot style. When you select this option, a wizard will be started that will guide you through the process of creating a new plot style.

Note

You will learn more about creating plot styles later in this chapter.

Edit

You can edit a plot style table you have selected from the **Plot style table** drop-down list by choosing the **Edit** button. This button is not available if you have selected **None** from the drop-down list. When you choose the **Edit** button, AutoCAD displays the **Plot Style Table Editor**, where you can edit the selected plot style table. This dialog box has three tabs: **General**, **Table View**, and **Form View**. The **Plot Style Table Editor** will be discussed later in the "Using Plot Styles" section of this chapter.

Shaded viewport options Area

The options in this area are used to print a shaded or a rendered image and are discussed next.

Shade plot

This drop-down list is used to select a technique that will be used for plotting the drawings. This drop-down list will be available only while plotting from the model space. If you select **As displayed** from this drop-down list, the drawing will be plotted as it is displayed on the screen. If the drawing is hidden, shaded, or rendered, it will be printed as it is. The hidden geometry consists of objects that lie behind the facing geometry and displays the object as it would be seen in reality. If you select the **Legacy wireframe** option, the model will be printed in wireframe mode displaying all hidden geometries, even if it is shaded in the drawing. Selecting the **Legacy hidden** option will plot the drawing with the hidden lines of the object suppressed. On selecting the **Hidden**, **Wireframe**, **Conceptual**, **Realistic**, **Shaded**, **Shaded with edges**, **Shades of gray**, **Sketchy**, and **X-Ray** options, the drawing will be plotted in the corresponding visual style, even if the model is displayed in some other visual styles. Similarly, selecting the **Rendered** option will plot the rendered image of the drawing. The other options in the drop-down list are used to specify the quality for plotting the shaded and rendered model. But if you are plotting a drawing from the paper space, the display settings of the viewport will be used for plotting it. To set the display settings of a viewport, select the viewport and right-click; a shortcut menu will be displayed. Choose the **Shade plot** option from the shortcut menu and then select the desired shading option.

Quality

This drop-down list is used to select printing quality in terms of dots per inch (dpi) for the printed drawing. The **Draft** option prints the drawing with 0 dpi, which results in the wireframe printout. The **Preview** option prints the drawing at 50 dpi, the **Normal** option prints the drawing at 100 dpi, the **Presentation** option prints the drawing at 200 dpi, the **Maximum** option prints the drawing at the selected plotting device's maximum dpi. You can also specify a custom dpi by selecting the **Custom** option from this drop-down list. The custom value of dpi can be specified in the **DPI** drop-down list, which is enabled below the **Quality** drop-down list on selecting the **Custom** option.

> **Note**
> *Selecting the **Draft** option from the **Quality** drop-down list prints the drawing in wireframe mode even if you select **Rendered** from the **Shade plot** drop-down list. Also, to plot the drawings in layouts with the hidden lines suppressed, you need to use the **Shade plot** option discussed in Chapter 18 (Model Space Viewports, Paper Space Viewports, and Layouts). You will learn more about wireframe, hidden, shaded, and rendered models in later chapters.*

Plot options Area

This area displays six options that can be selected as per the plot requirements. They are described next.

Plot in background

Select this check box to perform the printing and plotting operation in the background. While plotting in the background, you can return immediately to your drawing without waiting for the printing or plotting operation to finish. By default, the background plotting is turned off for plotting and it is turned on for publishing. The **BACKGROUNDPLOT** system variable is used to control the default settings for background plotting.

Plot object lineweights

This check box is not available if the **Plot with plot styles** check box is selected. To activate this option, clear the **Plot with plot styles** check box. Select this check box, if you need to plot the drawing with specified lineweights. To plot the drawing without specified lineweights, clear this check box.

Plot Transparency

When you select this check box, AutoCAD will plot using the transparency applied to the objects.

Plot with plot styles

When you select the **Plot with plot styles** check box, AutoCAD plots using the plot styles applied to the objects in the drawing and defined in the plot style table. The different property characteristics associated with the different style definitions are stored in the plot style tables and can be easily attached to the geometry. This setting replaces the pen mapping used in earlier versions of AutoCAD.

Plot paperspace last

This check box is not available when you are in the **Model** tab because no paper space objects are present in the **Model** tab. This option is available when you are working in the layout tab. By selecting the **Plot paperspace last** check box, you can plot the model space geometry before paper space objects. Usually the paper space geometry is plotted before the model space geometry. This option is also useful when there are multiple tabs selected for plotting and you want to plot the model space geometry before the layout tabs.

Hide paperspace objects

This check box is also available in the **Layout** tab and is used to specify whether or not the objects drawn in layouts will be hidden while plotting. If this check box is selected, the objects created in layouts will be hidden.

Plot stamp on

This check box is selected to turn the plot stamp on. The plot stamp is a user-defined information that will be displayed on the sheet while plotting. You can set the plot stamp information when you select this check box. When you select this check box, the **Plot Stamp Settings** button is displayed on the right of this check box. You can choose this button to display the **Plot Stamp** dialog box to set the parameters for the plot stamp.

Save changes to layout

This check box is selected to save the changes made using the **Plot** dialog box and apply them to the layout selected to be plotted.

Drawing orientation Area

This area provides options that help you specify the orientation of the drawing on the paper for the plotters that support landscape or portrait orientation. You can change the drawing orientation by selecting the **Portrait** or **Landscape** radio button, with or without selecting the **Plot upside-down** check box. The paper icon displayed on the right side of this area indicates the media orientation of the selected paper and the letter icon (A) on it indicates the orientation of the drawing on the page. The **Landscape** radio button is selected by default for AutoCAD drawings and orients the length of the paper along the X axis. If we assume this orientation to be at a rotation angle of 0-degree, then while selecting the **Portrait** radio button, the plot is oriented with the width along the X axis, which is equivalent to the plot being rotated through a rotation angle of 90-degree. Similarly, if you select both the **Landscape** radio button and the **Plot upside-down** check box at the same time, the plot gets rotated through a rotation angle of 180-degree and if you select both the **Portrait** radio button and the **Plot upside-down** check box at the same time, the plot gets rotated through a rotation angle of 270-degree. The AutoCAD screen conforms to the landscape orientation by default.

Preview

When you choose the **Preview** button, AutoCAD displays the drawing on the screen just as it would be plotted on the paper. Once the regeneration is performed, the dialog boxes on the screen disappear, and an outline of the paper size is shown. In the plot preview (Figure 16-3), the cursor is replaced by the **Zoom Realtime** icon. This icon can be used to zoom in and out interactively by pressing and moving the left mouse button. You can right-click to display a shortcut menu and then choose **Exit** to exit the preview or press the ENTER or ESC key to return to the dialog box. You can also choose **Plot** to plot the drawing right away or choose the other zooming options available.

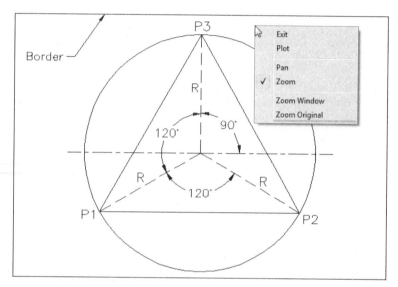

Figure 16-3 *Full plot preview with the shortcut menu*

Tip
*You can also choose **Plot Preview** from the **Quick Access Toolbar** to preview the plot, and at the same time, bypass the **Plot** dialog box.*

After finishing all the settings and other parameters, if you choose the **OK** button in the **Plot** dialog box, AutoCAD starts plotting the drawing in the file or plotters as specified. Also, the **Plot Job Progress** dialog box, where you can view the actual progress of plotting will be displayed.

ADDING PLOTTERS

Application Menu: Print > Manage Plotters	**Command:** PLOTTERMANAGER
Ribbon: Output > Plot > Plotter Manager	

In AutoCAD, the plotters are added using the **Plotter Manager** tool. This tool is discussed next.

The Plotter Manager Tool

When you invoke the **Plotter Manager** tool, AutoCAD will display the **Plotters** window, see Figure 16-4.

The **Plotters** window is basically a Windows Explorer window. It displays all the configured plotters and the **Add-A-Plotter Wizard** icon. You can right-click on any one of the icons belonging to the plotters that have already been configured to display a shortcut menu. You can choose **Delete** from the shortcut menu to remove a plotter from the list of available plotters in the

Chapter 16

Name drop-down list in the **Plot or Page Setup** dialog box. You can also choose **Rename** from the shortcut menu to rename the plotter configuration file or choose **Properties** to view the properties of the configured device.

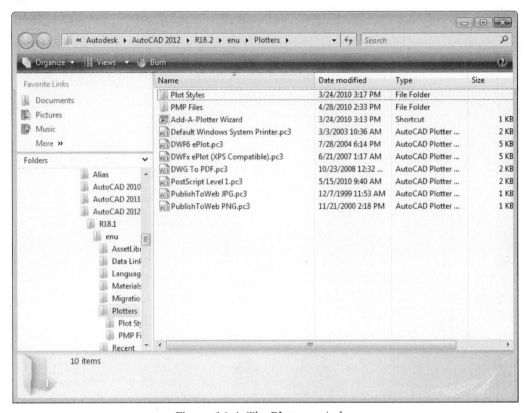

*Figure 16-4 The **Plotters** window*

Add-A-Plotter Wizard

If you double-click on the **Add-A-Plotter Wizard** icon in the **Plotters** window, AutoCAD guides you to configure a nonsystem plotter for plotting your drawing files. AutoCAD stores all the information of a configured plotter in configured plot (PC3) files. The PC3 files are stored in the *C:\Users\<owner>\AppData\Roaming\Autodesk\AutoCAD 2012\R18.2\enu\Plotters* folder by default. In AutoCAD 2012, you can add your own folder to store information of a configured plotter. To do so, add path of the folder in the **Support Files Search Path** given in the **Files** tab of the **Options** dialog box. The steps for configuring a new plotter using the **Add-A-Plotter Wizard** are as follows:

1. Open the **Plotters** window by choosing the **Plotter Manager** tool from the **Plot** panel and double-click on the **Add-A-Plotter Wizard** icon.

2. In the **Add Plotter - Introduction Page**, read the introduction carefully, and then choose the **Next** button to advance to the **Add Plotter - Begin** page.

3. On the **Add Plotter - Begin** page, the **My Computer** radio button is selected by default. Choose the **Next** button; the **Add Plotter - Plotter Model** page is displayed.

4. On this page, select a manufacturer and model of your nonsystem plotter from the **Manufacturers** and **Models** list boxes, respectively. Now, choose the **Next** button. The **Add Plotter - Import Pcp or Pc2** page is displayed.

If your plotter is not present in the list of available plotters, and you have a driver disk for your plotter, choose the **Have Disk** button to locate the *.hif* file from the driver disk, and install the driver supplied with your plotter.

5. In the **Add Plotter - Import Pcp or Pc2** page, if you want to import configuring information from a PCP or a PC2 file created with a previous version of AutoCAD, you can choose the **Import File** button and select the file. Otherwise, simply choose the **Next** button to advance to the next page.

6. On the **Add Plotter - Ports** page, select the port from the list to be used while plotting, and choose **Next**.

7. On the **Add Plotter - Plotter Name** page, you can specify the name of the currently configured plotter or the default name will be entered automatically. Choose **Next**.

8. When you reach the **Add Plotter - Finish** page, you can choose the **Finish** button to exit the **Add-A-Plotter Wizard**.

   You can also choose the **Edit Plotter Configuration** button to display the **Plotter Configuration Editor** dialog box where you can edit the current plotter's configuration. Also, in this page, you can choose the **Calibrate Plotter** button to display the **Calibrate Plotter** wizard. This wizard allows you to calibrate your plotter by setting up a test measurement. After test plotting, it compares the plot measurements with the actual measurements and computes a correction factor.

Once you have chosen **Finish** to exit the wizard, a PC3 file for the newly configured plotter will be displayed in the **Plotters** window. This PC3 file contains all the settings needed by the plotter to plot. Also, the newly configured plotter name is added to the **Name** drop-down list in the **Plotter configuration** area of the **Plot Device** tab in the **Plot** dialog box. You can now use the plotter for plotting.

EDITING THE PLOTTER CONFIGURATION

You can modify the properties of the selected plot device by using the **Plotter Configuration Editor** dialog box. This dialog box can be invoked in several ways. As discussed earlier, when using the **Plot** or **Page Setup** dialog box, you can choose the **Properties** button in the **Printer/plotter** area to display the **Plotter Configuration Editor** dialog box, see Figure 16-5.

You can modify the default settings of a plotter while configuring it by choosing the **Edit Plotter Configuration** button on the **Add Plotter - Finish** page of the **Add Plotter** wizard. You can also select the PC3 file for editing in the **Plotters** window by using Windows Explorer (by default, PC3 files are stored in the *\Documents and Settings\<owner>\Application Data\Autodesk\AutoCAD 2012\R18.2\enu\Plotters* folder) and double-click on the file or right-click on the file and choose **Open** from the shortcut menu. The three tabs in the **Plotter Configuration Editor** dialog box are discussed next.

General Tab

This tab contains basic information about the configured plotter or the PC3 file. You can make changes only in the **Description** area. The rest of the information in the tab is read only. This tab contains information on the configured plotter file name, plotter driver type, HDI driver file version number, name of the system printer (if any), and the location and name of the PMP file (if any calibration file is attached to the PC3 file).

Chapter 16

Ports Tab

This tab contains information about the communication between the plotting device and your computer. You can choose between a serial (local), parallel (local), or network port. The default settings for parallel and serial ports are **LPT1** and **COM1**, respectively. You can also change the port name, if your device is connected to a different port. You can also select the **Plot to File** radio button, if you want to save the plot as a file. You can select **AutoSpool**, if you want plotting to occur automatically, while you continue to work on another application.

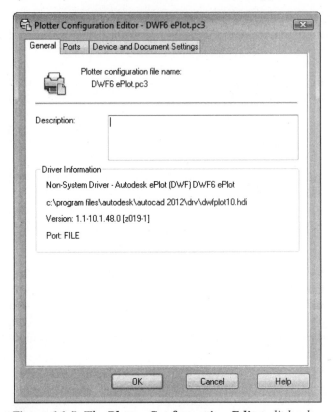

Figure 16-5 *The **Plotter Configuration Editor** dialog box*

Device and Document Settings Tab

This tab contains the plotting options specific to the selected plotter and are displayed as a tree view in the window. For example, if you configure a nonsystem plotter, you have the option to modify the pen characteristics. You can select any plotter properties from the tree view displayed in the window to change the values as required. Whenever you select an icon from the tree view in the window, the corresponding information is displayed in an area below. For example, if you select **PMP File Name <None>** in the tree view of the window, a **PMP file** area is displayed below. This area contains the current settings and the options to modify it. The information that is displayed within brackets (<>) can be modified. By default, **Custom properties** is displayed as highlighted in the window because it contains the properties that are modified commonly. The **Access Custom Dialog** area is displayed below the window. On choosing the **Custom Properties** button in this area, a dialog box specific to the selected plotter will be displayed. This dialog box has several properties of the selected plotter grouped and displayed under various areas and can be modified here.

Once you have made the desired changes, choose **OK** to exit the dialog box and then choose **Save As** in the **Plotter Configuration Editor** dialog box to save the changes you just made to the PC3 file. You can also import old plot configuration files (PCP or PC2) from the previous releases of AutoCAD using the **Import** button and save some of the information from these settings as a new PC3 file. When you choose the **Import** button, the **Plotting Components** dialog box is displayed. This dialog box displays what to use while importing AutoCAD 14 information into an AutoCAD 2012 drawing. For example, it tells you that to import PCP or PC2 file setting into a drawing in AutoCAD 2012, you should use the **Import PCP or PC2 Plot Settings** wizard. Choose **OK** to exit this dialog box; the **Import** dialog box is displayed. Here, you can select the file to be imported and then choose the **Import** button.

Tip
It is better to create a new PC3 file for a plotter and keep the original file as it is so that you encounter no error while using the specific printer later. The PC3 files determine the proper function of a plotter and any modifications may lead to errors.

IMPORTING PCP/PC2 CONFIGURATION FILES

If you want to import a PCP or PC2 configuration file or plot settings created by previous releases of AutoCAD into the **Model** tab or the current layout for the drawing, you can use the **PCINWIZARD** command to display the **Import PCP or PC2 Plot Settings** Wizard. All information from a PCP or PC2 file regarding plot area, rotation, plot offset, plot optimization, plot to file, paper size, plot scale, and pen mapping can be imported. Read the **Introduction** page of the wizard that is displayed carefully and then choose the **Next** button. The **Browse File name** page is displayed. Here, you can either enter the name of the PCP or PC2 file directly in the **PC2 or PCP file name** edit box or choose the **Browse** button to display the **Import** dialog box, where you can select the file to be imported. After you specify the file for importing, choose **Import** to return to the wizard. Choose the **Next** button to display the **Finish** page. After importing the files, you can modify the rest of the plot settings for the current layout.

SETTING PLOT PARAMETERS

Before starting with the drawing, you can set various plotting parameters in the **Model** tab or in the Layouts. The plot parameters that can be set include the plotter to be used, for example, plot style table, the paper size, units, and so on. All these parameters can be set using the **Page Setup Manager** tool as discussed next.

Working with Page Setups

Toolbar: Layouts > Page Setup Manager	**Application Menu:** Print > Page Setup
Ribbon: Output > Plot > Page Setup Manager	**Command:** PAGESETUP

As discussed earlier, a page setup contains the settings required to plot a drawing. Each layout, as well as the **Model** tab, can have a unique page setup attached to it. You can use the **Page Setup Manager** tool to create named page setups that can be used later. A page setup consists of specifications for the layout page, plotting device, paper size, and settings for the layouts to be plotted. This tool can also be invoked from the shortcut menu by right-clicking on the current **Model** or **Layout** tab and choosing **Page Setup Manager**. Remember that the **Page Setup Manager** option will be available in the shortcut menu only for the current **Model** or **Layout** tab.

When you invoke the **Page Setup Manager** tool, AutoCAD displays the **Page Setup Manager** dialog box. The names of the tabs displayed in the **Current page setup** list box of the **Page setups** area depend on the tab from which you invoke this dialog box. For example, if you invoke this dialog box from the **Model** tab, it displays only **Model** in this list box. However, if you invoke this dialog box from the **Layout** tab, it displays the list of all the layouts that are invoked at least once. Figure 16-6 shows the **Page Setup Manager** dialog box invoked from the **Layout** tab. In this case, only the **Layout1** was activated at least once.You can use this dialog box to create a new page setup, modify the existing page setup, or import a page setup from an existing file.

Creating a New Page Setup

To create a new page setup, choose the **New** button from the **Page Setup Manager** dialog box. The **New Page Setup** dialog box will be displayed, as shown in Figure 16-7. Enter the name of the new page setup in the **New page setup name** text box. The existing page setups with which

*Figure 16-6 The **Page Setup Manager** dialog box invoked from the **Layout1** tab*

*Figure 16-7 The **New Page Setup** dialog box*

you can start working are shown in the **Start with** area. You can select any of the page setups listed in this area and choose **OK** to proceed.

When you choose **OK**, the **Page Setup** dialog box will be displayed. This dialog box is similar to the **Plot** dialog box, see Figure 16-8.

Modifying the Page Setup

To modify the page setup, select the page setup from the **Current page setup** list box and choose the **Modify** button. The **Page Setup** dialog box will be displayed. Modify the parameters in this dialog box and exit.

Note
*If you select the **Display when creating a new layout** check box, the **Page Setup Manager** will be displayed whenever you invoke a layout for the first time.*

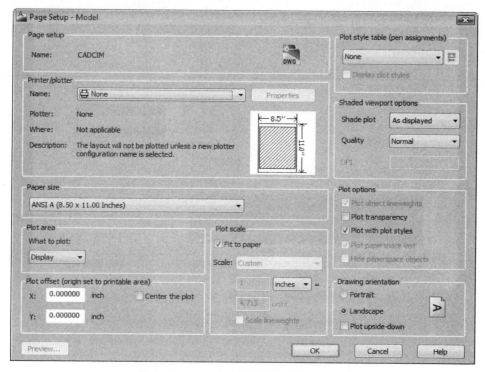

*Figure 16-8 The **Page Setup** dialog box*

Importing a Page Setup

Command: PSETUPIN

AutoCAD allows you to import a user-defined page setup from an existing drawing and use it in the current drawing or base the current page setup for the drawing on it. This option is available on choosing the **Import** button from the **Page Setup Manager** dialog box. It is also possible to bypass this dialog box and directly import a page setup from an existing drawing into a new drawing layout by using the **PSETUPIN** command. This command facilitates importing a saved and named page setup from a drawing into a new drawing. The settings of the named page setup can be applied to layouts in the new drawing. When you choose the **Import** button from the **Page Setup Manager** dialog box or invoke the **PSETUPIN** command, the **Select Page Setup From File** dialog box is displayed, as shown in Figure 16-9. You can use this dialog box to locate a *.dwg*, *.dwt*, or *.dwf* file whose page setups have to be imported. After you select the file, AutoCAD displays the **Import Page Setups** dialog box, as shown in Figure 16-10. You can also enter **-PSETUPIN** at the Command prompt to display prompts at the command line.

Note
*If a page setup with the same name already exists in the current file, the **AutoCAD Alert** box is displayed and you will be informed that "A page setup with the same name already exists in the current file, do you want to redefine it?" If you choose **Yes** in this message box, the current page setup will be redefined.*

USING PLOT STYLES

The plot styles can change the complete look of a plotted drawing. You can use this feature to

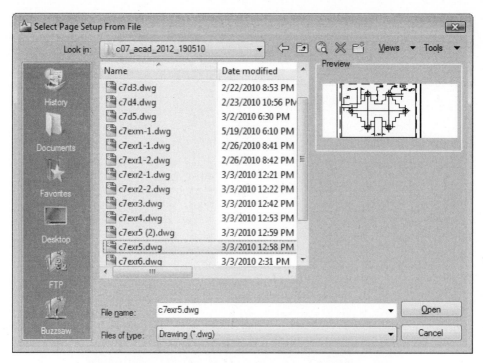

*Figure 16-9 The **Select Page Setup From File** dialog box*

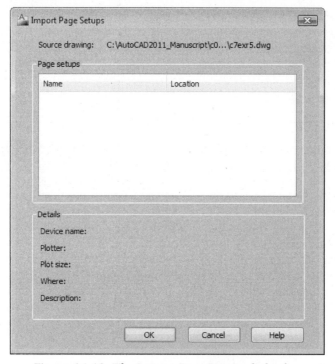

*Figure 16-10 The **Import Page Setups** dialog box*

override a drawing object's color, linetype, and lineweight. For example, if an object is drawn on a layer that is assigned the red color and no plot style is assigned to it, the object will be plotted as red. However, if you have assigned a plot style to the object with the color blue, the object will be plotted as blue, irrespective of the layer color it was drawn on. Similarly, you can change the Linetype, Lineweight, end, join, and fill styles of the drawing, and also change the output effects such as dithering, grayscales, pen assignments, and screening. Basically, you can use **Plot Styles** effectively to plot the same drawing in various ways.

Every object and layer in the drawing has a plot style property. The plot style characteristics are defined in the plot style tables attached to the **Model** tab, layouts, and viewports within the layouts. You can attach and detach different plot style tables to get different looks for your plots. Generally, there are two plot style modes. They are **Color-dependent** and **Named**. The **Color-dependent** plot styles are based on object color and there are **255** color-dependent plot styles. It is possible to assign each color in the plot style a value for the different plotting properties and these settings are then saved in a color-dependent plot style table file that has a *.ctb* extension. Similarly, **Named** plot styles are independent of the object color and you can assign any plot style to any object regardless of that object's color. These settings are saved in a named plot style table file that has *.stb* extension. Every drawing in AutoCAD is in either of the plot style modes.

Adding a Plot Style

Application Menu: Print > Manage Plot Styles   **Command:** STYLESMANAGER

All plot styles are saved in the *C:\Users <owner>\AppData\Roaming\Autodesk\AutoCAD 2012\R18.2\ enu\Plot Styles* folder. On choosing **Print > Manage Plot Styles** from the **Application Menu**, the **Plot Styles** window will be displayed, see Figure 16-11. This window displays icons for all available plot styles, in addition to the **Add-A-Plot Style Table Wizard** icon. You can double-click on any of the plot style icons to display the **Plot Style Table Editor** dialog box and edit the selected plot style. When you double-click on the **Add-A-Plot Style Table Wizard** icon, the **Add Plot Style Table** wizard is displayed and you can use it to create a new plot style.

*Figure 16-11 The **Plot Styles** window*

Add-A-Plot Style Table Wizard

If you want to add a new plot style table to your drawing, double-click on the **Add-A-Plot Style Table Wizard** in the **Plot Styles** window to display the **Add Plot Style Table** wizard. The following steps are required for creating a new plot style table using the wizard.

1. Read the introduction page carefully and choose the **Next** button.

2. In the **Begin** page, select the **Start from scratch** radio button and choose **Next**. Selecting this option creates a new plot style table. Therefore, the **Browse File** page is not available.

In addition to the **Start from scratch** option, this page has three more options: **Use an existing plot style table**, **Use My R14 Plotter Configuration (CFG)**, and **Use a PCP or PC2 file**. When you use the **Use an existing plot style table** option, an existing plot style table is used as a base for the new plot style table you are creating. In such a situation, the **Table Type** page of the wizard is not available and is not displayed because the table type will be based on the existing plot style table you are using to create a new one. With the **Use My R14 Plotter Configuration (CFG)** option, the pen assignments from the Release 14 *acad.cfg* file are used as a base for the new table you are creating. If you are using the **Use a PCP or PC2** option, the pen assignments saved earlier in the Release 14 PCP or PC2 file are used to create the new plot style.

3. The **Pick Plot Style Table** page allows you to select the **Color-Dependent Plot Style Table** or the **Named Plot Style Table** according to your requirement. Select the **Color-Dependent Plot Style Table** radio button and then choose the **Next** button.

4. Since you have selected the **Start from scratch** radio button in the **Begin** page, the **Browse File** page is not available and the **File name** page is displayed. However, if you had selected any of the other three options on the **Begin** page, the **Browse File** page would have been displayed. You can select an existing file from the drop-down list available in this page or choose the **Browse** button to display the **Select File** dialog box. You can then browse and select a file from a specific folder and choose **Select** to return to the wizard. You can also enter the name of the existing plot style table on which you want to base the new plot style table, directly in the edit box. After you have specified the file name, choose **Next** to display the **File name** page of the wizard. In the **File name** page, enter a file name for the new plot style table and choose **Next**. The **Finish** page is displayed.

Note

*If you are using the pen assignments from Release 14 acad.cfg file to define the new plot style table, you also have to specify the printer or plotter to use from the drop-down list available in the **Browse File** page of the wizard.*

5. The **Finish** page gives you the option of choosing the **Plot Style Table Editor** button to display the **Plot Style Table Editor** and then edit the plot style table you have created. If you select the **Use this plot style table for new and pre-AutoCAD 2010 drawings** check box in this page of the wizard, the plot style table that you have created will become the default plot style table for all the drawings you create. This check box is available only if the plot style mode you have selected in the wizard is the same as the default plot style mode specified by you in the **Default plot style behavior for new drawings** area of the **Plot Style Table Settings** dialog box. This dialog box can be invoked by choosing the **Plot Style Table Settings** button from the **Plot and Publish** tab of the **Options** dialog box. Choose **Finish** in the **Finish** page of the wizard to exit the wizard. A new plot style table gets added to the **Plot Styles** window and can be used for plotting.

Note

*You can also choose **Add Plot Style Table/ Add Color-Dependent Plot Style Table** from the **Tools > Wizards** menu to display wizards that are similar to the **Add Plot Style Table** wizard However, in the **Begin** page, the **Use an existing plot style table** option is not available. Also the **Table Type** page is not available and the **Finish** page has an additional option to use the new plot style table for the current drawing.*

Plot Style Table Editor

When you double-click on any of the plot style table icons in the **Plot Styles** window, the **Plot Style Table Editor** is displayed, where you can edit the particular plot style table. You can also choose the **Plot Style Table Editor** button in the **Finish** page to display the **Plot Style Table Editor**. Alternatively, expand the **Plot** dialog box and choose the button adjacent to the drop-down list in the **Plot style table (pen assignments)** area to display the **Plot Style Table Editor** dialog box.

The **Plot Style Table Editor** has three tabs: **General**, **Table View**, and **Form View**. You can edit all properties of an existing plot style table using these tabs. The description of the tabs is as follows.

General Tab

This tab provides information about the file name, location of the file, version, and scale factor. All information except the description is read only. You can enter a description about the plot style table in the **Description** text box here. If you select the **Apply global scale factor to non-ISO linetypes** check box, all the non-ISO linetypes in a drawing are scaled by the scale factor specified in the **Scale factor** edit box below the check box. If this check box is cleared, the **Scale factor** edit box is not available.

Table View Tab

This tab displays all plot styles along with their properties in a tabular form and they can be edited individually in the table, see Figure 16-12.

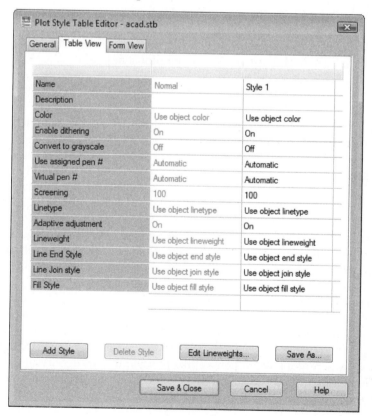

Figure 16-12 *The **Plot Style Table Editor** with the **Table View** tab chosen*

In case of a named plot style table, you can edit existing styles or add new styles by choosing the **Add Style** button. On doing so, a new column with the default style name **Style1** will be added in the table. You can change the style name if you want. To edit various properties in the table, select the particular value to be modified; the corresponding drop-down list will be displayed. You can select a value from this drop-down list. This manner of editing is similar to the one you use when editing the properties of objects using the **Properties** palette. The **Normal** plot style table is not available, and therefore, it cannot be edited. This plot style is assigned to layers by default.

You can select a particular plot style by clicking on the gray bar above the column and the entire column gets highlighted. You can select several plot styles by pressing the SHIFT key and selecting more plot styles. All the plot styles that are selected are highlighted. If you choose the **Delete Style** button now, the selected plot styles are removed from the table.

On choosing the **Edit Lineweights** button, the **Edit Lineweights** dialog box will be displayed. You can select the units for specifying the lineweights in the **Units for Listing** area of the dialog box, and can use either **Millimeters (mm)** or **Inches (in)** for specifying units. You can also edit the value of a lineweight by selecting it in the **Lineweights** list box and choosing the **Edit Lineweight** button. After you have edited the lineweights, choose the **Sort Lineweights** button to rearrange the lineweight values in the list box. Choose **OK** to exit the **Edit Lineweights** dialog box and return to **Plot Style Table Editor**.

With a color-dependent plot style table, the **Table View** tab displays all 255 plot styles, one for each color, and the properties can be edited in the table in the same way as discussed for named plot styles. The only difference is that you cannot add a new plot style or delete an existing one, and therefore, the **Delete Style** and **Add Style** buttons are not available. The properties that can be defined in a plot style are discussed next.

Name. This field displays the name of the color in the case of the color-dependent plot styles and the name of the style in the case of the named plot styles.

Description. Here you can enter the description about the plot style.

Color. The color you assign to the plot style overrides the color of the object in the drawing. The default value is Use object color.

Enable dithering. Dithering is described as a mixing of various colored dots to produce a new color. You can enable or disable this property by selecting from the drop-down list that is available in this field. Dithering is enabled by default and is independent of the color selected.

Convert to grayscale. If you are using a plotter that supports grayscaling, selecting **On** from the drop-down list in this field applies a grayscale to the objects color. By default, Off is selected and the object's color is used.

Use assigned pen #. This property is applied only to pen plotters. The pens range from 1 to 32. The default value is Automatic, which implies that the pen used will be based on the plotter configuration.

Virtual pen #. Nonpen plotters can behave like pen plotters using virtual pens. The value of this property lies between 1 and 255. The default value is Automatic. This implies that AutoCAD will assign a virtual pen automatically from the AutoCAD Color Index (ACI).

Screening. This property of a plot style indicates the amount of ink used while plotting. The value ranges between 0 and 100. The default value is 100, which creates the plot in its full intensity. Similarly, a value of 0 produces white color. This may be useful when plotting on a colored background.

Linetype. Like the property of color, the linetype assigned to a plotstyle overrides the object linetype on plotting. The default value is Use object linetype.

Adaptive adjustment. This property is applied by default and implies that the linetype scale of a linetype is applied such that on plotting, the linetype pattern will be completed.

Lineweight. The lineweight value assigned here overrides the value of the object lineweight on plotting. The default value is Use object lineweight.

Line End Style. This determines the manner in which a plotted line ends. The effect of this property is more noticeable when the thickness of the line is substantial. The line can end in a **Butt**, **Square**, **Round**, or **Diamond** shape. The default value is Use object end style.

Line Join style. You can select the manner in which two lines join in a plotted drawing. The available options are **Miter**, **Bevel**, **Round**, and **Diamond**. The default value is Use object join style.

Fill Style. The fill style assigned to a plot style overrides the objects fill style, when plotted. The available options are **Solid**, **Checkerboard**, **Crosshatch**, **Diamonds**, **Horizontal Bars**, **Slant Left**, **Slant Right**, **Square Dots**, and **Vertical Bars**.

Form View Tab

This tab displays all properties in one form, see Figure 16-13. Also, it displays all available plot styles in the **Plot styles** list box. You can select any style from the list box and then edit its properties in the **Properties** area.

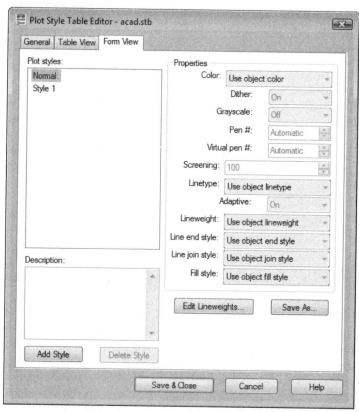

Figure 16-13 The **Plot Style Table Editor** *with the* **Form View** *tab chosen*

If you want to add a new plot style to the named plot style table, choose the **Add Style** button when the **Form View** tab is chosen; the **Add Plot Style** dialog box with the default plot style name **Style 1** will be displayed. You can change the name of a plot style in the **Plot Style** edit box, see Figure 16-14. Now, when you choose the **OK** button, the new style will be added in the **Plot Styles** list box and you can select it for editing. If you want to delete a style, select it and then choose the **Delete Style** button. With color-dependent plot styles, the **Form View** tab also does not provide options that allow you to add or delete plot styles.

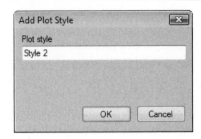

Figure 16-14   The Add Plot Style dialog box

After editing, choose the **Save & Close** button to save and return to the **Plot Styles** window. You can also choose the **Save As** button to display the **Save As** dialog box and save a plot style table with another name. To remove the changes you made to a plot style table, choose the **Cancel** button.

Applying Plot Styles

The **Model**, or any of the layout tabs, can be assigned a plot style table. The plot style table can be either named or color-dependent, as discussed earlier. These plot style modes for a new drawing can be determined from the **Default plot style behavior for new drawings** area of the **Plot Style Table Settings** dialog box. This dialog box can be invoked by choosing the **Plot Style Table Settings** button from the **Plot and Publish** tab of the **Options** dialog box. The **Use color dependent plot styles** option is selected by default and the drawings are assigned a color-dependent plot style. If you select the **Use named plot styles** radio button, the new drawings (not the current drawing) will be assigned a named plot style. The **PSTYLEPOLICY** system variable also controls the default plot style modes of the new drawings. A value of 0 implies a named plot style mode and a value of 1 implies a color-dependent plot style mode.

You can select a plot style table that you want to use as a default for drawings from the **Current plot style table settings** area of the **Plot Style Table Settings** dialog box. If the **None** option is selected, the drawing will be plotted with the object properties as displayed on the screen. When you select the **Use named plot styles** radio button from the **Default plot style behavior for new drawings** area of the **Plot Style Table Settings** dialog box, the **Default plot style for layer 0** and **Default plot style for objects** drop-down lists will be available. You can select the default plot styles that you want to assign to Layer 0 and to the objects in a drawing from these drop-down lists, respectively. If you choose the **Add or Edit Plot Style Tables** button from the **Plot Style Table Settings** dialog box, the **Plot Styles** window will be displayed. Here, you can double-click on any plot style table icon available and edit it using the **Plot Style Table Editor** that is displayed.

To change the plot style table for a current layout, you have to invoke the **Page Setup** or **Plot** dialog box and then select a plot style table from the **Plot styles** drop-down list in the **Plot style table (pen assignments)** area of the dialog box. A color-dependent plot style table can be selected and applied to a tab only if the default plot style mode has already been set to color dependent. Similarly, to apply a named plot style table to a tab, select the **Use named plot styles** radio button from the **Plot Style Table Settings** dialog box.

A color-dependent plot style cannot be applied to objects or layers and therefore the **Plot Style Control** drop-down list is not available in the **Properties** toolbar when a drawing has a color-dependent plot style mode. The plot styles also appear grayed out in the **Layer Properties Manager** dialog box and cannot be selected and changed. However, named plot styles can be applied to objects and layers. A plot style applied to an object overrides the plot style applied to the layer on which the object is drawn. To apply a plot style to a layer, invoke the **Layer Properties Manager** dialog box where all the layers in the selected tab are displayed. Select a layer to which you want to apply a plot style and select the default plot style (Normal) currently applied to the layer; the **Select Plot Style** dialog box will be displayed, see Figure 16-15. The **Plot styles** list box in this dialog box displays all the plot styles present in the plot style table attached to the current tab. You can select another plot style table to be attached to the current tab from the **Active plot style table** drop-down list. You will notice that all the plot styles in the selected plot style table are displayed in the list box now. You can also choose the **Editor** button to display the **Plot Style Table Editor** to edit plot style tables, as discussed earlier. The **Select Plot Style** dialog box also displays the name of the original plot style assigned to the object adjacent to **Original**. Also, the new plot style to be assigned to the selected object is displayed next to **New**.

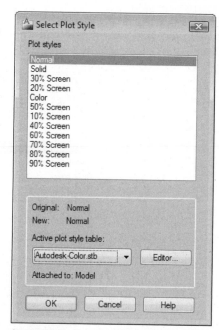

*Figure 16-15 The **Select Plot Style** dialog box*

You can apply a named plot style to an object using the **Plot Style Control** drop-down list in the **Properties** toolbar or the **Properties** palette. The process is the same as that applied for layers, colors, linetypes, and lineweights. This named plot style is applied to an object irrespective of the tab on which it is drawn. If the plot style assigned to an object is present in the plot style table of the tab in which it is present, the object is plotted with the specified plot style. However, if the plot style assigned to the object is not present in the plot style table assigned to the tab on which it is drawn, the object will be plotted with the properties that are displayed on the screen. The default plot style assigned to an object is **Normal** and the default plot style assigned to a layer is **ByLayer**.

Setting the Current Plot Style

You can use the **PLOTSTYLE** command to set the current plot style of new objects or of the selected object. You can invoke this command by choosing **Print > Edit Plot Style Tables** from the **Application Menu**, and if no object has been selected in the drawing, AutoCAD displays the **Current Plot Style** dialog box, see Figure 16-16. However, if any object selection is there in the drawing, then AutoCAD displays the **Select Plot Style** dialog box (Figure 16-15), which has been discussed earlier. You can select a plot style from the list box and choose **OK** to assign it to the selected objects in the drawing. All the plot styles present in the current plot style table that are assigned to the current tab are displayed in the list box in the **Current Plot Style** dialog box. You can select any one of these plot styles and choose **OK**. Now, if you create new objects, they will have the plot style that you had set current in the **Current Plot Style** dialog box. The parameters of this dialog box are described next.

Current plot style

The name of the current plot style is displayed adjacent to this label.

Plot style list box

This list box lists all the available plot styles that can be assigned to an object, including the default plot style, **Normal**.

Active plot style table

This drop-down list displays the names of all the available plot style tables. The current plot style table attached to the current layout or viewport is displayed in the edit box.

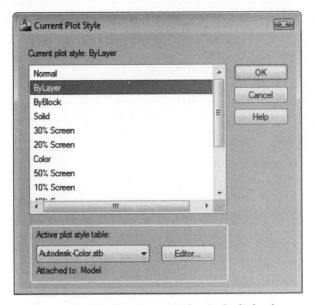

Figure 16-16 The **Current Plot Style** *dialog box*

Editor

If you choose the **Editor** button, adjacent to the **Active plot style table** drop-down list, AutoCAD displays the **Plot Style Table Editor** to edit the selected plot style table.

Attached to

The tab to which the selected plot style table is attached, such as **Model** or any one of the layout tabs, is displayed next to this label.

Note
*As soon as the drawing is plotted, the **Plot/Publish Details Report Available** button is displayed in the Status Bar tray. Choose this button to see the plot and publish details. This option has already been discussed in Chapter 1.*

EXAMPLE 1   *Plot Style*

Create the drawing shown in Figure 16-17. Next, create a named plot style with the name *My First Table.stb* with a plot style having the following parameters:

1. Screening   :   65%
2. Object line weight :   0.6000 mm
3. Line Type   :   ISO Dash
4. Object Color   :   Blue (5)

Additionally, plot the drawing using the **Date and Time** and **Paper size** stamps.

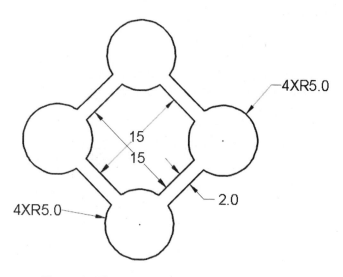

Figure 16-17 Drawing for Example 1

1. Create the drawing, as shown in Figure 16-17. You can create the drawing by making an octagon and then converting its alternate edge into an arc with the given radius.

2. Choose the **Print > Manage Plot Styles** from the **Application Menu**; a window will be displayed with all plot styles available. Double-click on the **Add-A-Plot Style Table Wizard** shortcut icon; the **Add Plot Style Table** wizard will be displayed. Choose the **Next** button; the **Begin** page will be displayed.

3. Select the **Start from scratch** radio button and choose the **Next** button; the **Pick Plot Style Table** page will be displayed. Now, select the **Named Plot Style Table** radio button and then choose the **Next** button. A window will be displayed prompting for the file name. Enter *My First Table* in the **File name** text box and then choose the **Next** button.

4. In the **Add Plot Style Table - Finish** window, select **Plot Style Table Editor** button; a window named **Plot Style Table Editor - My First Table.stb** is displayed with the **Table View** tab chosen by default.

5. Choose the **Add Style** button in the bottom left of the window; a new column named **Style 1** will be added in the **Add Style** area. Now, enter the following values in the database:

Row	Value
• Screening	65
• Lineweight	0.6000 mm
• Linetype	ISO Dash
• Color	Blue

Next, choose the **Save & Close** button and then choose the **Finish** button in the displayed window.

6. Now, set the **PSTYLEPOLICY** to 0 to make **Named Plot Style Table** as default. You have to restart the AutoCAD to select the *My First Table* plot style. After restarting, type the command **PLOT** at the Command prompt; a plot window will be displayed. Now, select the plot style **My First Table.stb** in the **Plot style table** drop-down list; the **Question** window will be displayed. Choose the **Yes** button.

7. Select the layout to be printed/plotted.

8. Select the **Plot stamp on** check box; the **Plot Stamp Settings** button will be displayed on the right of the check box. Choose this button; the **Plot Stamp** dialog box will be displayed. In this dialog box, select the **Date and Time** and **Paper size** check boxes; clear all other check boxes. Choose the **OK** button to exit the dialog box, and then choose the **Preview** button from the **Plot-Model** dialog box to preview the plot. If the plot seems to be fine, right-click to invoke the shortcut menu. Choose the **Plot** option from the menu to take print.

EXERCISE 1   *Plot Style*

Create the drawing shown in Figure 16-18. Create a named plot style table *My Named Table. stb* with three plot styles: Style 1, Style 2, and Style 3, in addition to the Normal plot style. The Normal plot style is used for plotting the object lines. These three styles have the following specifications:

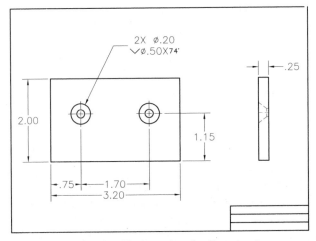

Figure 16-18 Drawing for Exercise 1

Style 1. This style has a value of Screening = 50. The dimensions, dimension lines, and the text in the drawing must be plotted with this style.

Style 2. This style has a value of Lineweight = 0.800. The border and title block must be plotted with this style.

Style 3. This style has a linetype of Medium Dash. The centerlines must be plotted with this plot style.

PLOTTING SHEETS IN A SHEET SET

Using the **Sheet Set Manager**, you can easily plot all the sheets available in a sheet set. However, before plotting the sheets in a sheet set, you need to make sure that you have selected the required printer in the page setup of all the sheets in the sheet set. This is because the printer set in the page setup of the sheet will be automatically selected to plot the sheet.

To print the sheets after setting the page setup, right-click on the name of the sheet set in the **Sheet Set Manager** and choose **Publish > Publish to Plotter** from the shortcut menu. If the value of the **BACKGROUNDPLOT** system variable is set to **2**, which is the default value, all the sheets will be automatically plotted in the background and you can continue working on the drawings.

Self-Evaluation Test

Answer the following questions and then compare them to those given at the end of this chapter:

1. All settings of a plotter are saved in the *.PC3* file. (T/F)

2. Different objects in the same drawing can be plotted in different colors with different linetypes and line widths. (T/F)

3. You can partially or fully preview a drawing before plotting. (T/F)

4. The **PSLTSCALE** system variable controls the paper space linetype scaling and has a default value of 1. (T/F)

5. The size of a plot can be specified by selecting any paper size from the _____ drop-down list in the **Plot** dialog box.

6. If you want to store a plot in a file and not have it printed directly on a plotter, select the _____ check box in the **Printer/plotter** area.

7. The scale for a plot can be specified in the _____ edit box in the **Plot** dialog box.

8. If you select the _____ option from the **What to plot** drop-down list, the portion of the drawing that is in the current display is plotted.

9. Before you plot sheets in the sheet set, it is important that you set the _____ for the individual sheets in their page setups.

10. You can set the quality of a plot using the _____ drop-down list in the **Shaded viewport options** area.

Review Questions

Answer the following questions:

1. The **Page Setup Manager** dialog box is displayed when you invoke the **Page Setup Manager** tool. (T/F)

2. By selecting the **View** option from the **What to plot** drop-down list in the **Plot** area, you can plot a view that was created with the **VIEW** command in the current drawing. (T/F)

3. If you do not want the hidden lines of a 3D object created in the **Model** tab, you can select the **Hidden** option from the **Shade plot** drop-down list in the **Shaded viewport options** area. (T/F)

4. The orientation of a drawing can be changed using the **Plot** dialog box. (T/F)

5. Which of the following check boxes is available in the **Plot options** area of the **Plot** dialog box, while plotting in a **Layout** tab?

    (a) **Plot paperspace last**   (b) **Hide objects**
    (c) **Plot with plot styles**   (d) None of these

Chapter 16

6. On invoking which of the following tools, the **Plotters** window will be displayed?

    (a) **Plot**   (b) **Manage Plotters**
    (c) **Plot Style**   (d) None of these

7. With which command is it possible to bypass the **Plot/Page Setup** dialog box and directly import a page setup from an existing drawing into a new drawing layout?

    (a) **PSETUPIN**   (b) **PLOTTERMANAGER**
    (c) **PLOTSTYLE**   (d) None of these

8. Which of the following commands is used to create a new plot style?

    (a) **Style**   (b) **Manage Plot Styles**
    (c) **Plot Style**   (d) None of these

9. Which of the following commands can be used to import a PCP file or PC2 files?

    (a) **PCINWIZARD**   (b) **PCIN**
    (c) **PLOTSTYLE**   (d) None of these

10. You can view a plot on the specified paper size before actually plotting it by choosing the _____ button from the **Plot** dialog box.

EXERCISE 2

Create the drawing shown in Figure 16-19 and plot it according to the following specifications. Also, create and use a plot style table with the specified plot styles.

1. The drawing is to be plotted on 10 X 8 inch paper.
2. The object lines must be plotted with a plot style Style 1. Style 1 must have a lineweight = 0.800 mm.
3. The dimension lines must be plotted with plot style Style 2. Style 2 must have a value of screening = 50.
4. The centerlines must be plotted with plot style Style 3. Style 3 must have a linetype of Medium Dash and screening = 50.
5. The border and title block must be plotted with plot style Style 4. The value of the lineweight should be =0.25 mm.

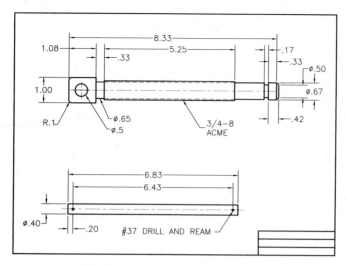

Figure 16-19 *Drawing for Exercise 2*

EXERCISE 3

Create the drawing shown in Figure 16-20 and then plot it according to your specifications.

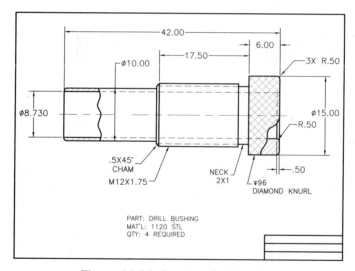

Figure 16-20 *Drawing for Exercise 3*

Problem-Solving Exercise 1

Make the drawing shown in Figure 16-21 and plot it according to your specifications.

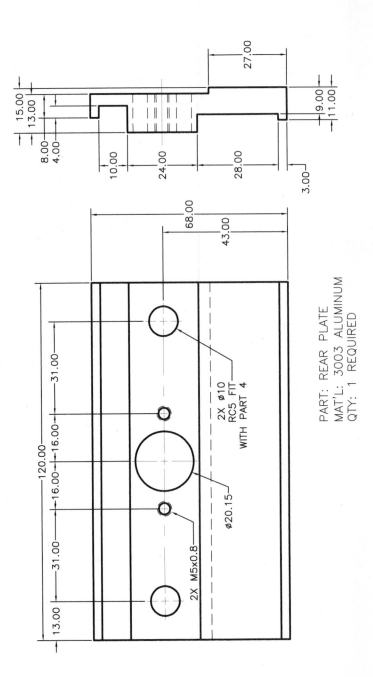

Figure 16-21 *Drawing for Problem-Solving Exercise 1*

Answers to Self-Evaluation Test

1. T, **2.** T, **3.** F, **4.** T, **5. Paper size**, **6. Plot to file**, **7. Custom**, **8. Display**, **9.** Paper size **10. Quality**

Chapter 17

Template Drawings

CHAPTER OBJECTIVES

After completing this chapter, you will learn:
* *To create template drawings.*
* *To load template drawings using dialog boxes and the Command line.*
* *To do an initial drawing setup.*
* *To customize drawings with layers and dimensioning specifications.*
* *To customize drawings with layouts, viewports, and paper space.*

KEY TERMS

* *Template*
* *Plot Size*
* *Layout*

* *Template Files (*.dwt)*
* *LTSCALE*
* *Viewport*

* *Drawing Scale*
* *DIMSCALE*
* *STARTUP*

* *TEXTSIZE*

CREATING TEMPLATE DRAWINGS

One way to customize AutoCAD is to create template drawings that contain initial drawing setup information and if desired, visible objects and text. When the user starts a new drawing, the settings associated with the template drawing are automatically loaded. If you start a new drawing from the scratch, AutoCAD loads default setup values. For example, the default limits are (0.0,0.0), (12.0,9.0) and the default layer is 0 with white color and a continuous linetype. Generally, these default parameters need to be reset before generating a drawing on the computer using AutoCAD. A considerable amount of time is required to set up the layers, colors, linetypes, lineweights, limits, snaps, units, text height, dimensioning variables, and other parameters. Sometimes, border lines and a title block may also be needed.

In production drawings, most of the drawing setup values remain the same. For example, the company title block, border, layers, linetypes, dimension variables, text height, LTSCALE, and other drawing setup values do not change. You will save considerable time if you save these values and reload them when starting a new drawing. You can do this by creating template drawings that contain the initial drawing setup information configured according to the company specifications. They can also contain a border, title block, tolerance table, block definitions, floating viewports in the paper space, and perhaps some notes and instructions that are common to all drawings.

STANDARD TEMPLATE DRAWINGS

AutoCAD comes with standard template drawings like *Acad.dwt*, *Acadiso.dwt*, *Acad3D.dwt*, *Acadiso3D.dwt*, *Acad-named plot styles.dwt*, *Acadiso-named plot styles.dwt*, and so on. The iso template drawings are based on the drawing standards developed by ISO (International Organization for Standardization). When you start a new drawing with **STARTUP** system variable set to 1, the **Create New Drawing** dialog box will be displayed. To load the template drawing, choose the **Use a Template** button; the list of standard template drawings is displayed. From this list, you can select any template drawing according to your requirements. If you want to start a drawing with the default settings, select the **Start from Scratch** button in the **Create New Drawing** dialog box. The following are some of the system variables, with the default values that are assigned to the new drawing.

System variable Name	Default Value
CHAMFERA	0.0000
CHAMFERB	0.0000
COLOR	Bylayer
DIMALT	OFF
DIMALTD	2
DIMALTF	25.4000
DIMPOST	None
DIMASO	ON
DIMASZ	0.1800
FILLETRAD	0.0000
GRID	0.5000
GRIDMODE	0
ISOPLANE	Top
LIMMIN	0.0000,0.0000
LIMMAX	12.0000,9.0000
LTSCALE	1.0000
MIRRTEXT	0 (Text not mirrored like other objects)
TILEMODE	1 (OFF)

EXAMPLE 1	*Advance Setup Wizard*

Create a drawing template using the **Advanced Setup** wizard of the **Use a Wizard** button with the following specifications and save it with the name *proto1.dwt*.

Units	Engineering with precision 0'-0.00"
Angle	Decimal degrees with precision 0
Angle Direction	Counterclockwise
Area	144'x96'

Step 1: Setting the STARTUP system variable

Set the value of the **STARTUP** system variable to 1. Choose the **New** tool from the **Quick Access** toolbar to display the **Create New Drawing** dialog box. Choose the **Use a Wizard** button and choose the **Advanced Setup** option, as shown in Figure 17-1. Choose **OK**. The **Units** page of the **Advanced Setup** dialog box is displayed, as shown in Figure 17-2.

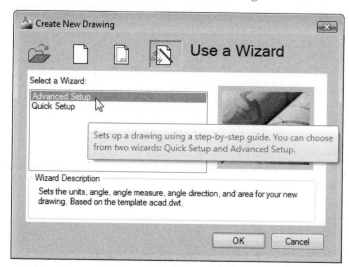

*Figure 17-1 The **Advanced Setup** wizard in the **Create New Drawing** dialog box*

Step 2: Setting the units of the drawing file

Select the **Engineering** radio button. Select **0'-0.00"** precision from the **Precision** drop-down list, refer to Figure 17-2, and then choose the **Next** button; the **Angle** page of the **Advanced Setup** dialog box is displayed.

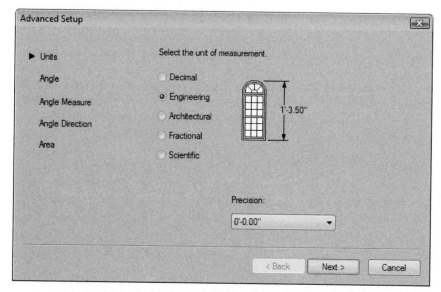

*Figure 17-2 The **Units** page of the **Advanced Setup** dialog box*

Chapter 17

Step 3: Setting the angle measurement system

In the **Angle** page, select the **Decimal Degrees** radio button and select **0** from the **Precision** drop-down list, as shown in Figure 17-3. Choose the **Next** button; the **Angle Measure** page of the **Advanced Setup** dialog box is displayed.

Step 4: Setting the horizontal axis for angle measurement

In the **Angle Measure** page, select the **East** radio button. Choose the **Next** button to display the **Angle Direction** page.

Step 5: Setting the angle measurement direction and drawing area

Select the **Counter-Clockwise** radio button and then choose the **Next** button. The **Area** page is displayed. Specify the area as 144' and 96' by entering the value of the width and length as **144'** and **96'** in the **Width** and **Length** edit boxes and then choose the **Finish** button. Use the **All** option of the **ZOOM** command to display new limits on the screen.

Step 6: Saving the drawing as template file

Now, save the drawing as *proto1.dwt* using the **Save** tool from the **Quick Access** toolbar. You need to select **AutoCAD Drawing Template (*dwt)** from the **Files of type** drop-down list and enter **proto1** in the **File name** edit box in the **Save Drawing As** dialog box. Next, choose the **Save** button; the **Template Options** dialog box will be displayed on the screen, as shown in Figure 17-4. Enter the description about the template in the **Description** edit box and choose the **OK** button. Now, the drawing will be saved as *proto1.dwt* on the default drive.

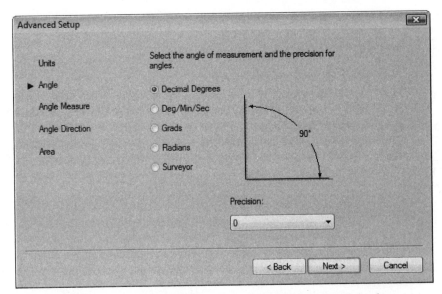

*Figure 17-3 The **Angle** page of the **Advanced Setup** dialog box*

Note
*To customize only the units and area, you can use the **Quick Setup** option in the **Create New Drawing** dialog box.*

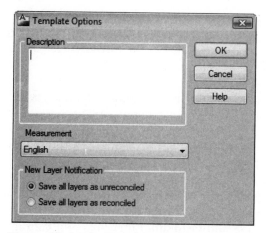

Figure 17-4 *The **Template Options** dialog box*

EXAMPLE 2   *Start from Scratch Option*

Create a drawing template using the following specifications. The template should be saved with the name *proto2.dwt*.

Limits	18.0,12.0
Snap	0.25
Grid	0.50
Text height	0.125
Units	3 digits to the right of decimal point
	Decimal degrees
	2 digits to the right of decimal point
	0 angle along positive X axis (east)
	Angle positive if measured counterclockwise

Step 1: Starting a new drawing

Start AutoCAD and choose the **Start from Scratch** button from the **Create New Drawing** dialog box. From the **Default Settings** area, select the **Imperial (feet and inches)** radio button, as shown in Figure 17-5. Choose **OK** to open a new file.

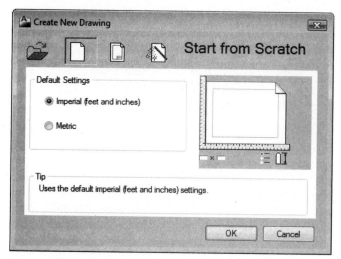

Figure 17-5 *The **Start from Scratch** button of the **Create New Drawing** dialog box*

Step 2: Setting limits, snap, grid, and text size

The **LIMITS** command can be invoked by entering **LIMITS** at the Command prompt.

> Command: **LIMITS**
> Reset Model space limits:
> Specify lower left corner or [ON/OFF] <0.0000,0.0000>: Enter
> Specify upper right corner <12.0000,9.0000>: **18,12**

After setting the limits, the next step is to expand the drawing display area. Use the **ZOOM** command with the **All** option to display new limits on the screen.

Now, right-click on the **Snap Mode** or **Grid Display** button in the Status Bar to display a shortcut menu. Choose the **Settings** option from the shortcut menu to display the **Drafting Settings** dialog box. Choose the **Snap and Grid** tab. Enter **0.25** and **0.25** in the **Snap X spacing** and **Snap Y spacing** edit boxes in **Snap spacing** area, respectively. Enter **0.5** and **0.5** in the **Grid X spacing** and **Grid Y spacing** edit boxes, respectively. Then, choose **OK**.

Note
*You can also use **SNAP** and **GRID** commands to set these values.*

The size of the text can be changed by entering **TEXTSIZE** at the Command prompt.

> Command: **TEXTSIZE**
> Enter new value for TEXTSIZE <0.2000>: **0.125**

Step 3: Setting units

Choose the **Units** tool from **Application Menu > Drawing Utilities** or enter **UNITS** at the Command prompt to invoke the **Drawing Units** dialog box, as shown in Figure 17-6. In the **Length** area, select **0.000** from the **Precision** drop-down list. In the **Angle** area, select **Decimal Degrees** from the **Type** drop-down list and **0.00** from the **Precision** drop-down list. Also make sure the **Clockwise** check box in the **Angle** area is not selected.

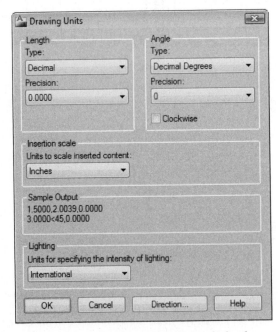

*Figure 17-6 The **Drawing Units** dialog box*

Choose the **Direction** button from the **Drawing Units** dialog box to display the **Direction Control** dialog box (Figure 17-7) and then select the **East** radio button. Exit both the dialog boxes.

*Figure 17-7 The **Direction Control** dialog box*

Step 4: Saving the drawing as template file

Now, save the drawing as *proto2.dwt* using the **Save** tool from the **Quick Access** toolbar. You need to select **AutoCAD Drawing Template (*dwt)** from the **Files of type** drop-down list and enter **proto2** in the **File name** edit box in the **Save Drawing As** dialog box. Next, choose the **Save** button; the **Template Options** dialog box will be displayed on the screen. Enter the description about the template in the **Description** edit box and choose the **OK** button. The drawing will be saved as *proto2.dwt* on the default drive. You can also save this drawing to some other location by specifying other location from the **Save in** drop-down list of the **Save Drawing As** dialog box.

LOADING A TEMPLATE DRAWING

You can use the template drawing to start a new drawing file. To use the preset values of the template drawing, start AutoCAD or choose the **New** tool from the **Quick Access** toolbar. The dialog box that appears will depend on whether you have set the **STARTUP** system variable to **1** or **0**. If you have set this value as **1**, the **Create New Drawing** dialog box will appear. Choose the **Use a Template** option. All templates that are saved in the default **Template** directory will be shown in the **Select a Template** list box, see Figure 17-8. If you have saved the template in any other location, choose the **Browse** button. The **Select a template file** dialog box will be displayed. You can use this dialog box to browse the directory in which the template file is saved.

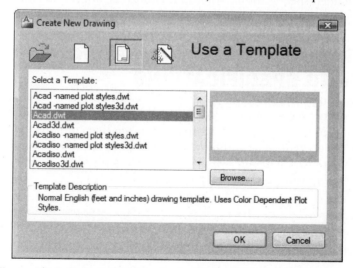

*Figure 17-8 Default templates in the **Create New Drawing** dialog box*

If you have set the **STARTUP** system variable to 0, the **Select template** dialog box appears when you choose the **New** tool. This dialog box also displays the default **Template** folder and all template files saved in it, see Figure 17-9. You can use this dialog box to select the template file that you want to open.

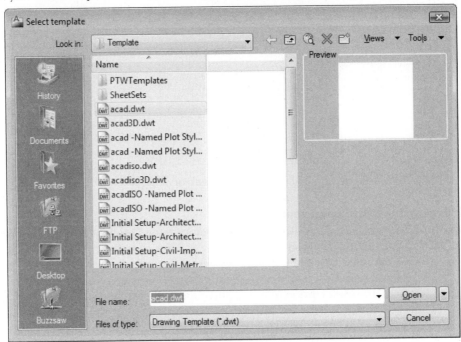

Figure 17-9 *The **Select template** dialog box that appears while starting a new drawing file, when the value of **STARTUP** variable is set to **0***

Using any of the previously mentioned dialog boxes, select the *proto1.dwt* template drawing. AutoCAD will start a new drawing that will have the same setup as that of the template drawing *proto1.dwt*.

You can have several template drawings, each with a different setup. For example, **PROTOB** for a 18" X 12" drawing, **PROTOC** for a 24" X 18" drawing. Each template drawing can be created according to user-defined specifications. You can then load any of these template drawings, as discussed previously.

CUSTOMIZING DRAWINGS WITH LAYERS AND DIMENSIONING SPECIFICATIONS

Most production drawings need multiple layers for different groups of objects. In addition to layers, it is a good practice to assign different colors to different layers to control the line width at the time of plotting. You can generate a template drawing that contains the desired number of layers with linetypes and colors according to your company specifications. You can then use this template drawing to make a new drawing. The next example illustrates the procedure used for customizing a drawing with layers, linetypes, and colors.

EXAMPLE 3   *Template with Title Block & Layers*

Create a template drawing *proto3.dwt* that has a border and the company's title block, as shown in Figure 17-10.

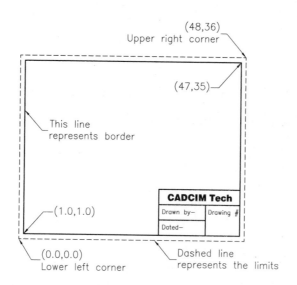

Figure 17-10 The template drawing for Example 3

This template drawing will have the following initial drawing setup:

Limits	48.0,36.0
Text height	0.25
Border line lineweight	0.012"
Ltscale	4.0

DIMENSIONS
Overall dimension scale factor 4.0
Dimension text above the extension line
Dimension text aligned with dimension line

LAYERS

Layer Name	Line Type	Color
0	Continuous	White
OBJ	Continuous	Red
CEN	Center	Yellow
HID	Hidden	Blue
DIM	Continuous	Green
BOR	Continuous	Magenta

Step 1: Setting limits, text size, polyline width, polyline, and linetype scaling

Start a new drawing with default parameters by selecting the **Start from Scratch** option in the **Create New Drawings** dialog box. In the new drawing file, use the AutoCAD commands to set up the values as given for this example. Also, draw a border and a title block, as shown in Figure 17-10. In this figure, the hidden lines indicate drawing limits. The border lines are 1.0 units inside the drawing limits. For border lines, increase the lineweight to a value of 0.012".

Use the following procedure to produce the prototype drawing for Example 3.

1. Invoke the **LIMITS** command by entering **LIMITS** at the Command prompt. The prompt sequence is given next.

Command: **LIMITS**
Reset Model space limits:
Specify lower left corner or [ON/OFF] <0.0000,0.0000>: Enter
Specify upper right corner <12.0000,9.0000>: **48,36**

2. Increase the drawing display area by choosing the **All** option of the **ZOOM** command.

3. Enter **TEXTSIZE** at the Command prompt to change the text size.

Command: **TEXTSIZE**
Enter new value for TEXTSIZE <0.2000>: **0.25**

4. Next, draw the border using the **RECTANGLE** tool. The prompt sequence to draw the rectangle is:

Command: *Choose the* **RECTANGLE** *tool from the* **Draw** *panel*
Specify first corner point or [Chamfer/Elevation/Fillet/Thickness/Width]: **1.0,1.0**
Specify other corner point or [Area/Dimensions/Rotation]: **47.0,35.0**

5. Now, select the rectangle and change its lineweight to **0.012"**. Make sure that the **Show/ Hide Lineweight** button is chosen in the Status Bar.

6. Enter **LTSCALE** at the Command prompt to change the linetype scale.

Command: **LTSCALE**
Enter new linetype scale factor<Current>: **4.0**

Step 2: Setting dimensioning parameters

You can use the **Dimension Style Manager** dialog box to set the dimension variables. Click the inclined arrow displayed on the **Dimensions** panel title bar in the **Annotate** tab; the **Dimension Style Manager** dialog box will be displayed, as shown in Figure 17-11.

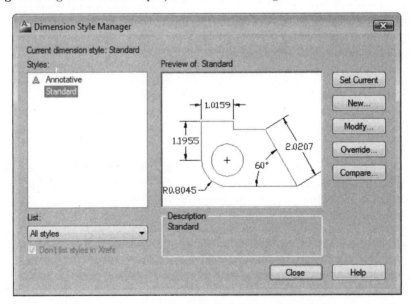

Figure 17-11 The **Dimension Style Manager** *dialog box*

You can also invoke this dialog box by entering **DIMSTYLE** at the Command prompt. Choose the **New** button from the **Dimension Style Manager** dialog box; the **Create New Dimension Style** dialog box will be displayed. Specify the new style name as **MYDIM1** in the **New Style Name** edit box, as shown in the Figure 17-12 and then choose the **Continue** button. The **New Dimension Style:MYDIM1** dialog box is displayed.

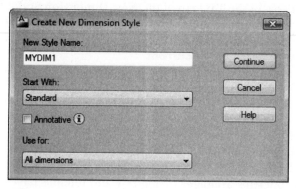

Figure 17-12 *The **Create New Dimension Style** dialog box*

Specifying Overall dimension scale factor

To specify a dimension scale factor, choose the **Fit** tab of the **New Dimension Style: MYDIM1** dialog box. Set the value in the **Use overall scale of** as 4 in the **Scale for dimension features** area, as shown in Figure 17-13.

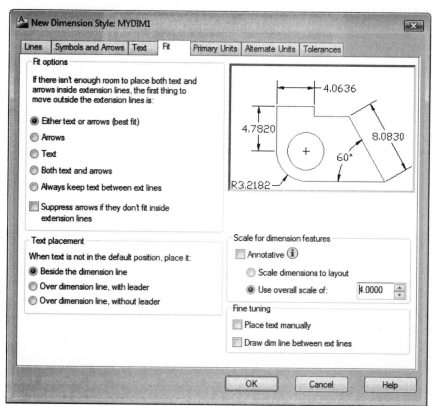

Figure 17-13 *The **Fit** tab of the **New Dimension Style: MYDIM1** dialog box*

Placing the dimension text over the dimension line

In the **Fit** tab of the **New Dimension Style: MYDIM1** dialog box. Select the **Over dimension line, with leader** radio button from the **Text placement** area.

Dimensioning text aligned with the dimension line

In the **Text alignment** area of the **Text** tab of the **New Dimension Style: MYDIM1** dialog box, select the **Aligned with dimension line** radio button and then choose **OK** button.

Setting the new dimension style to current

A new dimension style with the name **MYDIM1** is shown in the **Styles** area of the **Dimension Style Manager** dialog box. Select this dimension style and then choose the **Set Current** button to make it the current dimension style. Choose the **Close** button to exit this dialog box.

Step 3: Setting layers

Choose the **Layer Properties** tool from the **Layers** panel or choose the **Layer** tool from the **Format** menu bar to invoke the **Layer Properties Manager** dialog box. Choose the **New Layer** button in the **Layer Properties Manager** dialog box and rename **Layer1** as **OBJ**. Choose the color swatch of the OBJ layer to display the **Select Color** dialog box. Select the **Red** color and choose **OK**; the red color is assigned to the OBJ layer. Again, choose the **New Layer** button in the **Layer Properties Manager** dialog box and rename the **Layer1** as **CEN**. Choose the linetype swatch to display the **Select Linetype** dialog box.

If different linetypes are not already loaded, choose the **Load** button to display the **Load or Reload Linetypes** dialog box. Select the **CENTER** linetype from the **Available Linetypes** area and choose **OK**; the **Select Linetype** dialog box will reappear. Select the **CENTER** linetype from the **Loaded linetypes** area and choose **OK**. Choose the color swatch to display the **Select Color** dialog box. Select the **Yellow** color and choose **OK**; the color yellow and linetype center will be assigned to the layer CEN.

Similarly, different linetypes and different colors can be set for different layers mentioned in the statement of this example, as shown in Figure 17-14.

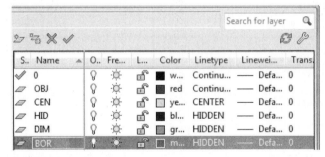

*Figure 17-14 Partial view of the **Layer Properties Manager** dialog box*

Step 4: Adding title block

Next, add the title block and the text, as shown in Figure 17-10. After completing the drawing, save it as *proto3.dwt*. You have created a template drawing (proto3) that contains all the information given in Example 3.

CUSTOMIZING A DRAWING WITH LAYOUT

The Layout (paper space) provides a convenient way to plot multiple views of a three-dimensional (3D) drawing or multiple views of a regular two-dimensional (2D) drawing. It takes quite some time to set up the viewports in the model space with different vpoints and scale factors. You can create prototype drawings that contain predefined viewport settings, with vpoint and the other desired information. If you create a new drawing or insert a drawing, the views are automatically generated. The following example illustrates the procedure for generating a prototype drawing with paper space and model space viewports:

EXAMPLE 4   *Template with Layouts*

Create a drawing template, as shown in Figure 17-15, with four views in Layout3 (Paper space) that display front, top, side, and 3D views of the object. The plot size is 10.5 by 8 inches. The plot scale is 0.5 or 1/2" = 1". The paper space viewports should have the following vpoint settings:

Viewports	Vpoint	View
Top right	1,-1,1	3D view
Top left	0,0,1	Top view
Lower right	1,0,0	Right side view
Lower left	0,-1,0	Front view

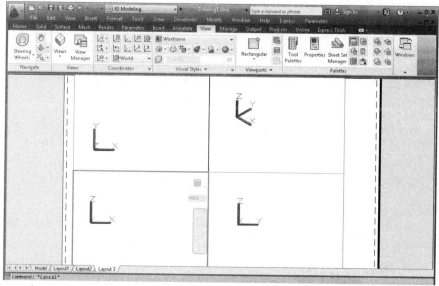

Figure 17-15 Paper space with four viewports

Start AutoCAD and create a new drawing. Use the following commands and options to set various parameters:

Step 1: Creating a new layout

First, you need to create a new layout using the **LAYOUT** command. A new layout is automatically created with the default name. Alternatively, you can also use the **LAYOUT** command to create a new layout. The Command prompt that is displayed is given next.

> **Note**
> *You can also choose **New Layout** tool from the **Quick View Layouts** panel available on choosing the **Quick View Layouts** button on the Status Bar.*

Command: **LAYOUT**
Enter layout option [Copy/Delete/New/Template/Rename/SAveas/Set/?] <set>: **N**
Enter new Layout name <Layout3>: Enter

Step 2: Specifying the page setup for the new layout

Next, choose the new layout (**Layout3**) tab. The new layout (**Layout3**) will be displayed with the default viewport. Delete this existing viewport. To invoke the **Page Setup Manager** dialog box, right-click on the **Quick View Layouts** button in the **Status Bar**; a shortcut menu will be displayed. Choose **Page Setup Manager** from the shortcut menu; the **Page Setup Manager** dialog box will be displayed. Choose the **Modify** button to modify the default page setup. In the **Page Setup - Layout3** dialog box, select the required printer from the **Printer/plotter** area. In this example, HP LaserJet 4000 has been used.

Now, from the **Paper size** area, select the paper size that is supported by the selected plotting device. In this example, the paper size is A4. Choose the **OK** button to accept the settings and return to the **Page Setup Manager** dialog box. Choose the **Close** button to close the **Page Setup Manager** dialog box.

Step 3: Creating new layer for the viewports object

Next, you need to set up a layer with the name VIEW for viewports object and assign it green color. Invoke the **Layer Properties Manager** dialog box. Choose the **New Layer** button and rename the Layer1 as VIEW. Choose the color swatch of the VIEW layer to display the **Select Color** dialog box. Select the color **Green** and choose the **OK** button. This color will be assigned to VIEW layer. Also, make the VIEW layer current and then exit the **Layer Properties Manager** dialog box.

Step 4: Creating viewports

To create four viewports, use the **MVIEW** command. You can choose the **4Viewports** tool from **View > Viewports** in the menu bar or enter **MVIEW** command at the Command prompt. The following is the prompt sequence when you invoke the **MVIEW** command.

    Command: **MVIEW**
    Specify corner of viewport or
    [ON/OFF/Fit/Shadeplot/Lock/Object/Polygonal/Restore/LAyer/2/3/4] <Fit>: **4**
    Specify first corner or [Fit] <Fit>: **0.25,0.25**
    Specify opposite corner: **10.25,7.75**

Step 5: Setting the required viewpoint in all layers

Choose the **PAPER** button in the Status Bar to activate the model space or enter **MSPACE** at the Command prompt.

    Command: **MSPACE** (or **MS**)

Make the lower left viewport active by clicking on it. Next, you need to change the viewpoints of different paper space viewports using the **VPOINT** command. To invoke this command, choose the **Viewpoint** tool from **View > 3D Views** in the menu bar or enter **VPOINT** at the Command prompt. The viewpoint values for different viewports are given in the statement of Example 4. To set the view point for the lower left viewport the Command prompt sequence is given next.

    Command: *Choose the **Viewpoint** tool from **View > 3D Views** in the menu bar*
    Current view direction: VIEWDIR= 0.0000,0.0000,1.0000
    Specify a view point or [Rotate] <display compass and tripod>: **0,-1,0**

Similarly, use the **VPOINT** command to set the viewpoint of the other viewports.

    Note
    You can also use the ViewCube or the SteeringWheel to orient the view in any viewport.

Make the top left viewport active by selecting a point in the viewport and then use the **ZOOM** command to specify the paper space scale factor to 0.5. The **ZOOM** command can be invoked by choosing **View > Zoom > Scale** from the **Application Menu** or by entering **ZOOM** at the Command prompt. The Command prompt sequence is given next.

Command: **ZOOM**
Specify corner of window, enter a scale factor (nX or nXP), or
[All/Center/Dynamic/Extents/Previous/Scale/Window/Object] <real time>: **0.5XP**

Now, make the next viewport active and specify the zoom scale factor. Do the same for the remaining viewports.

Step 6: Creating border line

Use the **MODEL** button in the Status Bar to change to paper space and then set a new layer PBORDER with yellow color. Make the PBORDER layer current, draw a border, and if needed, a title block using the **PLINE** command. You can also change to paper space by entering **PSPACE** at the Command prompt.

The **PLINE** command can be invoked by choosing the **Polyline** tool from the **Draw** menu or by choosing the **Polyline** tool from the **Draw** panel. The **PLINE** command can also be invoked by entering **PLINE** at the Command prompt. While specifying the coordinate values for the **PLINE** command, make sure that the **Dynamic Input** button is turned off in the Status Bar.

Command: *Choose the **Polyline** tool from the **Draw** panel*
Specify start point: **0,0**
Current line-width is 0.0000
Specify next point or [Arc/Halfwidth/Length/Undo/Width]: **0,8.0**
Specify next point or [Arc/Close/Halfwidth/Length/Undo/Width]: **10.5,8.0**
Specify next point or [Arc/Close/Halfwidth/Length/Undo/Width]: **10.5,0**
Specify next point or [Arc/Close/Halfwidth/Length/Undo/Width]: **C**

Step 7: Saving and testing the template file

The last step is to choose the **Model** tab (or change the **TILEMODE** to 1), if not already active and save the prototype drawing as template. To test the layout that you just created, make the 3D drawing, as shown in Figure 17-19 or make any 3D object. Switch to the **Layout 3** tab; you will find four different views of the object (Figure 17-16). If the object views do not appear in the viewports, use the **PAN** commands to position the views in the viewports. You can freeze the VIEW layer so that the viewports do not appear on the drawing. You can plot this drawing from the Layout3 with a plot scale factor of 1:1 and the size of the plot will be exactly as specified.

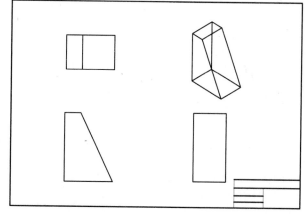

Figure 17-16 *Four views of a 3D object in paper space*

CUSTOMIZING DRAWINGS WITH VIEWPORTS

In certain applications, you may need multiple model space viewport configurations to display different views of an object. This involves setting up the desired viewports and then changing the viewpoint for different viewports. You can create a prototype drawing that contains a required number of viewports and the viewpoint information. If you insert a 3D object in one of the viewports of the prototype drawing, you will automatically get different views of the object without setting viewports or viewpoints. The following example illustrates the procedure for creating a prototype drawing with a standard number (four) of viewports and viewpoints.

EXAMPLE 5   *Template with Viewports*

Create a prototype drawing with four viewports, as shown in Figure 17-17. The viewports should have the following viewpoints:

Viewports	Vpoint	View
Top right	1,-1,1	3D view
Top left	0,0,1	Top view
Lower right	1,0,0	Right side view
Lower left	0,-1,0	Front view

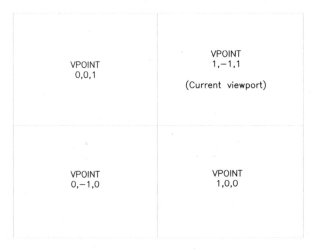

Figure 17-17 Viewports with different viewpoints

Step 1
Start AutoCAD and create a new drawing from scratch.

Step 2
Setting viewports in Model tab
Viewports and corresponding viewpoints can be set with the **VPORTS** command. You can also choose the **New Viewports** tool from **View > Viewports** in the menu bar or choose the **New** tool from **View > Viewports** panel to display the **Viewports** dialog box, as shown in Figure 17-18. Select **Four: Equal** from the **Standard viewports** area. In the **Preview** area, four equal viewports are displayed. Select **3D** from the **Setup** drop-down list. The four viewports with the different viewpoints will be displayed in the **Preview** area as Top, Front, Right and SE Isometric. **Top** represents the viewpoints as (0,0,1), **Front** represents the viewpoints as (0,-1,0), **Right** represents the viewpoints as (1,0,0) and **SE Isometric**, represents the viewpoints as (1,-1,1) respectively. Choose the **OK** button. Save the drawing as *proto5.dwt*.

Viewports and viewpoints can also be set by entering **-VPORTS** and **VPOINT** at the Command prompt, respectively.

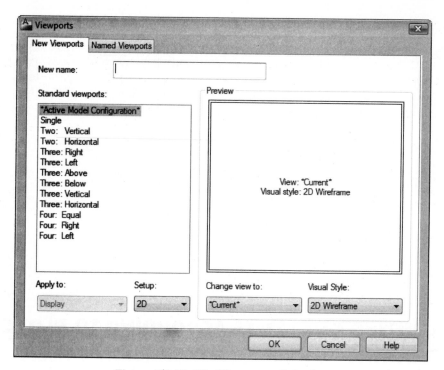

Figure 17-18 *The* **Viewports** *dialog box*

Step 3

Start a new drawing and draw the 3D tapered block, as shown in Figure 17-19.

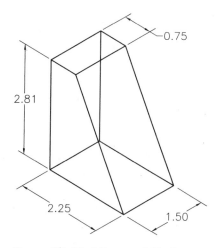

Figure 17-19 *3D tapered block*

Step 4

Again, start a new drawing, TEST, using the prototype drawing *proto5.dwt*. Make the top right viewport current and then insert or create a drawing, refer to Figure 17-19. Four different views will be automatically displayed on the screen, as shown in Figure 17-20.

Note

The method for creating 3D models is discussed in Chapter 25 (Getting Started with 3D).

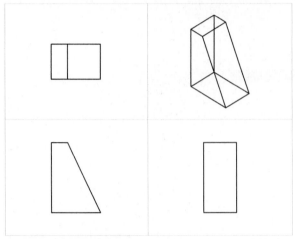

Figure 17-20 *Different views of a 3D tapered block*

CUSTOMIZING DRAWINGS ACCORDING TO PLOT SIZE AND DRAWING SCALE

For controlling the plot area, it is recommended to use layouts. You can make the drawing of any size, use the layout to specify the sheet size, and then draw the border and title block. However, you can also plot a drawing in the model space and set up the system variables so that the plotted drawing is according to your specifications. You can generate a template drawing according to plot size and scale. For example, if the scale is 1/16" = 1' and the drawing is to be plotted on a 36" by 24" area, you can calculate drawing parameters like limits, **DIMSCALE**, and **LTSCALE** and save them in a template drawing. This will save considerable time in the initial drawing setup and provide uniformity in the drawings. The next example explains the procedure involved in customizing a drawing according to a certain plot size and scale.

Note
You can also use the paper space to specify the paper size and scale.

EXAMPLE 6 *Template with Plot Size and Drawing Scale*

Create a drawing template (PROTO6) with the following specifications:

Plotted sheet size	36" by 24" (Figure 17-21)
Scale	1/8" = 1.0'
Snap	3'
Grid	6'
Text height	1/4" on plotted drawing
Linetype scale	Calculate
Dimscale factor	Calculate
Units	Architectural
	Precision, 16-denominator of the smallest fraction
	Angle in degrees/minutes/seconds
	Precision, 0d00'
	Direction control, base angle, east
	Angle positive, if measured counterclockwise
Border	Border should be 1" inside the edges of the plotted drawing sheet, using PLINE 1/32" wide when plotted (Figure 17-21)

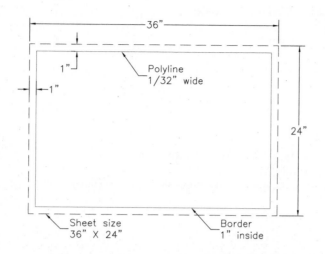

Figure 17-21 *Border of the template drawing*

Step 1: Calculating limits, text height, linetype scale, dimension scale, and polyline width

In this example, you need to calculate some values before you set the parameters. The limits of the drawing depend on the plotted size of the drawing and its scale. Similarly, **LTSCALE** and **DIMSCALE** depend on the plot scale of the drawing. The following calculations explain the procedure for finding the values of limits, ltscale, dimscale, and text height.

Limits

> *Given:*
> Sheet size   36" x 24"
> Scale   1/8" = 1'
>   or 1" = 8'

> *Calculate:*
> X Limit
> Y Limit
> Since sheet size is 36" x 24" and scale is 1/8"=1'
> Therefore,   X Limit = 36 x 8 = 288'
>   Y Limit = 24 x 8 = 192'

Text height

> *Given:*
> Text height when plotted = 1/4"
> Scale   1/8" = 1'

> *Calculate:*
> Text height
> Since scale is 1/8" = 1'
>   or 1/8" = 12"
>   or   1" = 96"
> Therefore, scale factor = 96
>   Text height = 1/4" x 96
>   = 24" = 2'

Linetype scale and dimension scale

Known:
Since scale is 1/8" = 1'
or 1/8" = 12"
or 1" = 96"

Calculate:
LTSCALE and DIMSCALE
Since scale factor = 96
Therefore, LTSCALE = Scale factor = 96
Similarly, DIMSCALE = 96
(All dimension variables, like DIMTXT and DIMASZ, will be multiplied by 96.)

Polyline Width

Given:
Scale is 1/8" = 1'

Calculate:
PLINE width
Since scale is 1/8" = 1'
or 1" = 8'
or 1" = 96"

Therefore,
PLINE width = 1/32 x 96
= 3"

After calculating the parameters, use the following AutoCAD commands to set up the drawing and save the drawing as *proto6.dwt*.

Step 2: Setting units

Start a new drawing and enter **UNITS** at the Command prompt to display the **Drawing Units** dialog box. Select **Architectural** from the **Type** drop-down list in the **Length** area. Select **0'-01/16"** from the **Precision** drop-down list. Make sure the **Clockwise** check box in the **Angle** area is not selected. Select **Deg/Min/Sec** from the **Type** drop-down list and select **0d00'** from the **Precision** drop-down list in the **Angle** area. Now, choose the **Direction** button to display the **Directional Control** dialog box. Select the **East** radio button, if it is not selected, in the **Base Angle** area and then choose the **OK** button twice to close both the dialog boxes.

Step 3: Setting limits, snap and grid, textsize, linetype scale, dimension scale, dimension style and pline

To set limits, enter **LIMITS** at the Command prompt.

Command: **LIMITS**
Specify lower left corner or [ON/OFF] <0'-0",0'-0">:**0,0**
Specify upper right corner <1'-0",0'-9">: **288',192'**

Invoke the **All** option of the **ZOOM** command to increase the drawing display area.

Right-click on the **Snap Mode** or **Grid Display** button in the Status Bar to invoke the shortcut menu. In the shortcut menu, choose **Settings** to display the **Drafting Settings** dialog box. In this dialog box, enter **3'** and **3'** in the **Snap X spacing** and **Snap Y spacing** edit boxes, respectively. Similarly, enter **6'** and **6'** in the **Grid X spacing** and **Grid Y spacing** edit boxes, respectively. Then, choose **OK**.

You can also set these values by entering **SNAP** and **GRID** at the Command prompt.

The size of the text can be changed by entering **TEXTSIZE** at the Command prompt.

> Command: **TEXTSIZE**
> Enter new value for TEXTSIZE <current>: **2'**

To set the **LTSCALE**, choose the **Other** option from **Home > Properties > Linetype** drop-down list, or choose the **Linetype** option from the **Format** menu bar, or enter **LINETYPE** at the Command prompt; the **Linetype Manager** dialog box will be invoked. Choose the **Show details** button; the **Linetype Manager** dialog box will be expanded. Now, specify the **Global scale factor** as **96** in the **Global scale factor** edit box. Choose the **OK** button to accept the changes and exit the **Linetype Manager** dialog box.

You can also change the scale of the linetype by entering **LTSCALE** at the Command prompt.

Choose the inclined arrow displayed on the **Dimensions** panel title bar from the **Annotate** tab to invoke the **Dimension Style Manager** dialog box, as shown in Figure 17-11. Choose the **New** button from the **Dimension Style Manager** dialog box to invoke the **Create New Dimension Style** dialog box. Specify the new style name as **MYDIM2** in the **New Style Name** edit box and then choose the **Continue** button. The **New Dimension Style: MYDIM2** dialog box will be displayed. Choose the **Fit** tab and set the value in the **Use overall scale of** spinner to **96** in the **Scale for dimension features** area. Now, choose the **OK** button to again display the **Dimension Style Manager** dialog box. Choose **Close** button to exit the dialog box.

You can invoke the **PLINE** command by choosing the **Polyline** tool from the **Draw** panel or entering **PLINE** at the Command prompt. While specifying the coordinate values for the **PLINE** command, make sure that the **Dynamic Input** button is turned off in the Status Bar.

> Command: *Choose the **Polyline** tool from the **Draw** panel*
> Specify start point: **8',8'**
> Current line-width is 0'-0"
> Specify next point or [Arc/Halfwidth/Length/Undo/Width]: **W**
> Specify starting width<0'-00">: **3**
> Specify ending width<0'-3">: Enter
> Specify next point or [Arc/Halfwidth/Length/Undo/Width]: **280,0**
> Specify next point or [Arc/Close/Halfwidth/Length/Undo/Width]: **0,184**
> Specify next point or [Arc/Close/Halfwidth/Length/Undo/Width]: **-280,0**
> Specify next point or [Arc/Close/Halfwidth/Length/Undo/Width]: **C**

Now, save the drawing as *proto6.dwt*.

Self-Evaluation Test

Answer the following questions and then compare them to those given at the end of this chapter:

1. The template drawings are saved in the _____ format.

2. To use a template file, select the _____ option in the **Create New Drawing** dialog box.

3. To start a drawing with the default setup, select the _____ option in the **Create New Drawing** dialog box.

4. If the plot size is 36" x 24", and the scale is 1/2" = 1', then X Limit = _____ and Y Limit = _____ .

5. You can use AutoCAD's _____ command to set up a viewport in the paper space.

6. You can use AutoCAD's _____ command to change to model space.

7. The values that can be assigned to **TILEMODE** are _____ and _____ .

Review Questions

Answer the following questions:

1. The default value of **DIMSCALE** is _____ .

2. The default value of **DIMTXT** is _____ .

3. The default value of **SNAP** is _____ .

4. Architectural units can be selected by using the _____ command or the _____ command.

5. Name three standard template drawings that come with AutoCAD software _____ , _____ , and _____ .

6. If the plot size is 24" x 18", and the scale is 1:20, the X Limit = _____ and Y Limit = _____ .

7. If the plot size is 200 x 150 and limits are (0.00,0.00) and (600.00,450.00), the **LTSCALE** factor = _____ .

8. _____ provides a convenient way to plot multiple views of a 3D drawing or multiple views of a regular 2D drawing.

9. You can use the _____ command to change to paper space.

10. In the model space, if you want to reduce the display size by half, the scale factor you enter in the **ZOOM > Scale** option is _____ .

EXERCISE 1   *Template with Limits*

Create a drawing template (*protoe1.dwt*) with the following specifications:

Units	Architectural with precision 0'-0 1/16
Angle	Decimal Degrees with precision 0
Base angle	East
Angle direction	Counterclockwise
Limits	48' x 36'

EXERCISE 2   *Template with Limits, Text Height, & Units*

Create a drawing template (*protoe2.dwt*) with the following specifications:

Limits	36.0,24.0
Snap	0.5
Grid	1.0
Text height	0.25
Units	Decimal
	Precision 0.00
	Decimal degrees
	Precision 0
	Base angle, East
	Angle positive if measured counterclockwise

EXERCISE 3   *Template with Layers, LTSCALE, & DIMSCALE*

Create a drawing template (*protoe3.dwt*) with the following specifications:

Limits	48.0,36.0
Text height	0.25
PLINE width	0.03
LTSCALE	4.0
DIMSCALE	4.0
Plot size	10.5 x 8

LAYERS

Layer Name	Line Type	Color
0	Continuous	White
OBJECT	Continuous	Green
CENTER	Center	Magenta
HIDDEN	Hidden	Blue
DIM	Continuous	Red
BORDER	Continuous	Cyan

EXERCISE 4   *Relative Rectangular & Absolute Coordinates*

Create a prototype drawing (*protoe4.dwt*) with the following specifications:

Limits	36.0,24,0
Border	35.0,23.0
Grid	1.0
Snap	0.5
Text height	0.15
Units	Decimal (up to 2 places)
LTSCALE	1
Current layer	Object

LAYERS

Layer Name	Linetype	Color
0	Continuous	White
Object	Continuous	Red
Hidden	Hidden	Yellow
Center	Center	Green
Dim	Continuous	Blue
Border	Continuous	Magenta
Notes	Continuous	White

This prototype drawing should have a border line and a title block as shown in Figure 17-22.

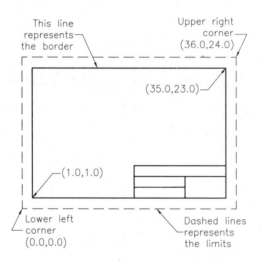

Figure 17-22 Prototype drawing

EXERCISE 5   *Template with Plot Sheet Size & Title Block*

Create a template drawing shown in Figure 17-23 with the following specifications and save it with the name *protoe5.dwt*:

Plotted sheet size	36" x 24" (Figure 17-23)
Scale	1/2" = 1.0'
Text height	1/4" on plotted drawing
LTSCALE	24
DIMSCALE	24
Units	Architectural
	32-denominator of smallest fraction to display
	Angle in degrees/minutes/seconds
	Precision 0d00"00"
	Angle positive if measured counterclockwise
Border	Border is 1-1/2" inside the edges of the plotted drawing sheet, using the PLINE 1/32" wide when plotted.

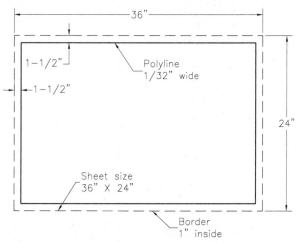

Figure 17-23 *Drawing for Exercise 5*

EXERCISE 6   *Template with Title Block & Dimension Style*

Create a prototype drawing, as shown in Figure 17-24 with the following specifications (the name of the drawing is *protoe6.dwt*):

Plotted sheet size	24" x 18" (Figure 17-24)
Scale	1/2"=1.0'
Border	The border is 1" inside the edges of the plotted drawing sheet, using the PLINE 0.05" wide when plotted (Figure 17-24)

Dimension text over the dimension line
Dimensions aligned with the dimension line
Calculate overall dimension scale factor
Enable the display of alternate units
Dimensions to be associative.

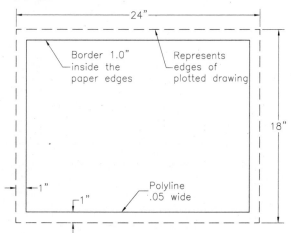

Figure 17-24 *Prototype drawing*

Chapter 17

EXERCISE 7   *Creating & using Template for Generating Drawing*

Create a template with the necessary specifications to draw the Articulated Rod shown in Figure 17-25, with the following specifications.

Limits	36.0,24,0
Border	35.0,23.0
Grid	1.0
Snap	0.5
Text height	0.15
Units	Decimal (up to 2 places)
LTSCALE	1
Current layer	Object

LAYERS

Layer Name	Linetype	Color
0	Continuous	White
Object	Continuous	Red
Hidden	Hidden	Yellow
Center	Center	Green
Dim	Continuous	Blue
Border	Continuous	Magenta
Notes	Continuous	White

Also, use this template and the layers created in it to draw the views of the Articulated Rod.

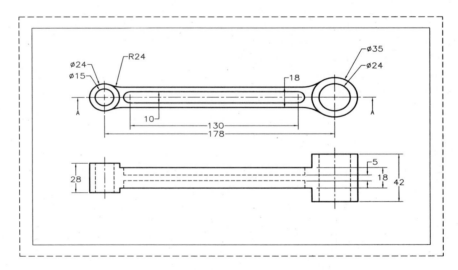

Figure 17-25 Views and dimensions of the Articulated Rod

Answers to Self-Evaluation Test

1. *.dwt*, **2. Use a Template, 3. Start from Scratch, 4. 72",48", 5. MVIEW, 6. MSPACE, 7. 0, 1.**

AutoCAD PART II

Author's Website

For Faculty: Please contact the author at **Stickoo@purduecal.edu** or **tickoo525@gmail.com** to access the website that contains the following:

1. PowerPoint presentations, program listings, and drawings used in this textbook.
2. Syllabus, chapter objectives and hints, and questions with answers for every chapter.

For Students: To download drawings, exercises, tutorials, and programs, please visit the author's website: *http://www.cadcim.com*. Alternatively, you can visit: *www.cengage.com/cad/autodeskpress*, choose the *Online Companions* option, and then click on the ***AutoCAD 2012: A Problem Solving Approach*** link to download student resources.

Chapter *18*

Working with Blocks

CHAPTER OBJECTIVES

In this chapter, you will learn:
- *To create and insert blocks.*
- *To add parameters and assign action to the blocks to make them dynamic blocks.*
- *To create drawing files by using the Write Block dialog box.*
- *To use the DesignCenter to locate, preview, copy, or insert blocks and existing drawings.*
- *To use the Tool Palettes to insert blocks.*
- *To edit blocks.*
- *To split a block.*
- *To rename blocks and delete unused blocks.*

KEY TERMS

- *Blocks*
- *Annotative Blocks*
- *Dynamic Blocks*
- *Parameters*
- *Actions*
- *Nesting of Blocks*
- *WBLOCK*
- *XPLODE*
- *Constraints to Blocks*

THE CONCEPT OF BLOCKS

The ability to store parts of a drawing, or the entire drawing, such that they need not be redrawn when required in the same drawing or another drawing is a great benefit to the user. These parts of a drawing, entire drawings, or symbols (also known as **blocks**) can be placed (inserted) in a drawing at the location of your choice, with the desired orientation, and scale factor. A block is given a name (block name) and is referenced (inserted) by its name. All objects within a block are treated as a single object. You can move, erase, or list the block as a single object, that is, you can select the entire block simply by selecting anywhere on it. As for the edit and inquiry commands, the internal structure of a block is immaterial, since a block is treated as a primitive object, like a polygon. If a block definition is changed, all references to the block in the drawing are updated to incorporate the changes.

A block is created by using the **Create Block** tool in the **Block Definition** panel. You can also save objects in a drawing, or an entire drawing as a drawing file using the **WBLOCK** command. The main difference between the two is that a wblock can be inserted in any other drawing, but a block can be inserted only in the drawing file, in which it was created.

One of the important feature of AutoCAD is the annotative blocks that can be used as symbol to annotate your drawing. The annotative blocks are defined in terms of paper space height. The annotative blocks are displayed in floating viewports and their respective size in model space is calculated by multiplying the current annotation scale set for those spaces and the paper space height of the block.

Another feature of AutoCAD is that instead of inserting a symbol as a block (which results in the addition of the content of the referenced drawing to the drawing in which it is inserted), you can reference the other drawings (Xref) in the current file. This means that the contents of the referenced drawing are not added to the current drawing file, although they become part of that drawing on the screen. This is explained in detail in Chapter 20.

Advantages of Using Blocks

Blocks offer the following advantages:

1. Drawings often have some repetitive features. Instead of drawing the same feature again and again, you can create its block and insert it wherever required. This helps you to reduce drawing time and better organize your work.
2. Blocks can be drawn and stored for future use. You can thus create a custom library of objects required for different applications. For example, if your drawings are concerned with gears, you could create their blocks and then integrate them with custom menus. In this manner, you could create an application environment of your own.
3. The size of a drawing file increases as you add objects to it. AutoCAD keeps track of information about the size and position of each object in the drawing, for example, the points, scale factors, annotation scale, radii, and so on. If you combine several objects into a single object by forming a block by using the **Create Block** tool in the **Block Definition** panel, there will be a single scale factor, annotation scale, rotation angle, position, and so on, for all objects in the block, thereby saving storage space. Each object, repeated in the multiple block insertions, needs to be defined only once in the block definition. Ten insertions of a gear, made of forty-three lines and forty-one arcs, require only ninety-four objects (10+43+41), while ten individually drawn gears require 840 objects [10 X (43+41)].
4. If the specification for an object changes, the drawing needs to be modified. This is a very tedious task, if you need to detect each location where the change is to be made and edit it individually. But, if this object has been defined as a block, you can redefine it. Also, where ever that object appears, it is revised automatically. This has been dealt in Chapter 19, Defining Block Attributes.

5. Attributes (textual information) can be included in blocks. Different attribute values can be changed in each insertion of a block.

6. You can create symbols and then store them as blocks by using the **Create Block** tool. Later on, with the **Insert** tool, you can insert the blocks in the drawing, in which they were defined. There is no limit to the number of times you can insert a block in a drawing.

7. You can store symbols as drawing files using the **WBLOCK** command and later insert them into any other drawing, using the **Insert** tool.

8. The X, Y, and Z scale factors and rotation angles of blocks may vary from one insertion to another (Figure 18-1).

9. You can insert multiple copies of a block about a specified path, using the **Divide** or **Measure** tool from the **Point** drop-down in the **Draw** panel.

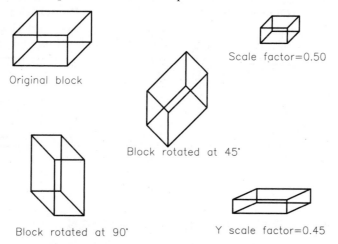

Figure 18-1 Blocks with different specifications

FORMATION OF BLOCKS

Drawing Objects for Blocks

The first step for creating blocks is to draw the object(s) to be converted into a block. You can consider any symbol, shape, or view that may be used more than once for conversion into a block (Figure 18-2). Even a drawing, that is to be used more than once, can be inserted as a block.

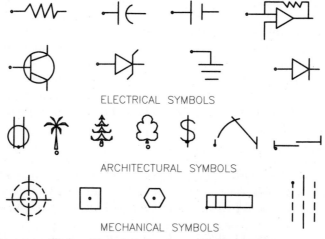

Figure 18-2 Some common drafting symbols

Tip
Examine the sketch of the drawing you are about to begin and carefully look for any shapes, symbols, and components that are to be used more than once. They can then be drawn once and stored as blocks.

Checking Layers

You should be particularly careful about the layers on which the objects to be converted into a block are drawn. The objects must be drawn on layer 0, if you want the block to inherit the linetype and color of the layer on which it is inserted. If you want the block to retain the linetype and color of the layer on which it was drawn, draw the objects defining the block on that layer. For example, if you want the block to have the linetype and color of the layer OBJ, draw the objects comprising the block on layer OBJ. Now, even if you insert the block on any other layer, the linetype and color of the block will be that of the layer OBJ.

If the objects in a block are drawn with the color, linetype, and lineweight as **ByBlock**, this block when inserted, assumes the properties of the current layer. To set the color to ByBlock, select **ByBlock** from the **Object Color** drop-down list of the **Properties** palette or from the **Properties** panel in the **Home** tab. To set the linetype to ByBlock, select **ByBlock** from the **Selects a linetype** drop-down list in the **Properties** panel of the **Home** tab. Similarly, you can specify the lineweight as **ByBlock**.

Note

*As discussed earlier, before creating a block, you should check the layer on which the objects will be drawn because a layer determines the color and linetype of the block. To change the property of a layer associated with an object, invoke the **Properties** palette. To do so, choose the **Properties** button from the **Palettes** panel of the **View** tab, or from the **Standard** toolbar, or press CTRL+1, or from the shortcut menu. Next, change the properties as desired. You can also change the properties by entering the **Properties** option at the **CHANGE** command or the **CHPROP** command.*

CONVERTING ENTITIES INTO A BLOCK

Ribbon: Insert > Block Definition > Create Block	
Toolbar: Draw > Make Block	**Command:** BLOCK

You can convert the entities in the drawing window into a block by using the **BLOCK** command or by choosing the **Create Block** tool from the **Block Definition** panel in the **Insert** tab or the **Home** tab. Alternatively, you can do so by choosing the **Make Block** tool from the **Draw** toolbar. When you invoke this tool, the **Block Definition** dialog box will be displayed, as shown in Figure 18-3. You can use the **Block Definition** dialog box to save any object or objects as a block.

In the **Name** edit box of the **Block Definition** dialog box, enter the name of the block you want to create. All the block names present in the current drawing are displayed in the **Name** drop-down list in an alphabetical and numerical order. This way you can verify whether a block you have defined has been saved. By default, the block name can have up to 255 characters. Also, it can contain letters, digits, blank spaces as well as any special characters including the $ (dollar sign), - (hyphen), and _ (underscore), provided they are not being used for any other purpose by AutoCAD or Windows.

Note

*The block name is controlled by the **EXTNAMES** system variable and its default value is 1. If the value is set to 0, the block name will be only thirty-one characters long and will not include spaces or any other special characters, apart from a $ (dollar sign), - (hyphen), or _ (underscore).*

If a block already exists, with the block name that you have specified in the **Name** edit box, and you choose **OK** in the **Block Definition** dialog box, the **Block - Redefine Block** message box will be displayed informing you that the block with this name is already defined. It will also prompt you to specify if you want to redefine the block. In this dialog box, you can either redefine the existing block by choosing the **Redefine** button or you can exit it by choosing the **No** button. You can then use another name for the block in the dialog box.

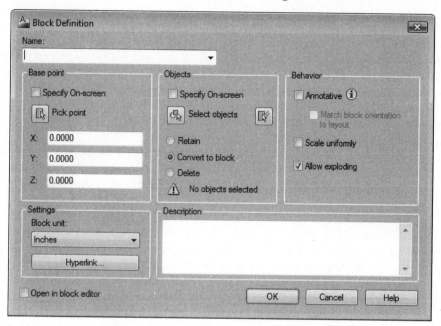

*Figure 18-3 The **Block Definition** dialog box*

After you have specified a block name, you are required to specify the insertion base point, which will be used as a reference point to insert the block. Usually, either the center or the lower left corner of the block is defined as the insertion base point. Later on, when you insert the block, you will notice that it appears at the insertion point, and you can insert it with reference to this point. The point you specify as the insertion base point is taken as the origin point of the block's coordinate system. You can specify the insertion point by choosing the **Pick point** button in the **Base point** area of the dialog box. The dialog box is temporarily removed, and you can select a base point on the screen. Alternatively, you can also enter the coordinates in the **X**, **Y**, and **Z** edit boxes. Once the insertion base point is specified, the dialog box reappears and the *X*, *Y*, and *Z* coordinates of the insertion base point are displayed in the **X**, **Y**, and **Z** edit boxes respectively. If no insertion base point is selected, AutoCAD assumes the insertion point coordinates to be 0,0,0, which are the default coordinates.

Tip
In drawings where the insertion point of a block is important, for example, in the case of inserting the block of a door symbol in architectural plans, it may be a good idea to first specify the requisite scale factors and rotation angles and then the insertion point. This way, you are able to preview the block symbol as it is to be inserted.

After specifying the insertion base point, you are required to select the objects that will constitute the block. Until the objects are selected, AutoCAD displays a warning: **No objects selected**, at the bottom of the **Objects** area. Choose the **Select objects** button; the dialog box is temporarily removed from the screen. You can select the objects on the screen using any selection method. After completing the selection process, right-click or press the ENTER key to return to the dialog box. The number of objects selected is displayed at the bottom of the **Objects** area of the dialog box.

The **Objects** area of the **Block Definition** dialog box also has a **QuickSelect** button. On choosing this button, the **Quick Select** dialog box is displayed, which allows you to define a selection set based on the properties of objects. **Quick Select** is used in cases where the drawings are very large and complex. The **Quick Select** dialog box has been discussed earlier in Chapter 6, Editing Sketched Objects-II.

In the **Objects** area, if you select the **Retain** radio button, the selected objects that form the block, are retained on the screen as individual objects. If you select the **Convert to block** radio button, the selected objects are converted into a block and will not be displayed on the screen after the block has been defined. Rather, the created block will be displayed in place of the original objects. If you select the **Delete** radio button, the selected objects will be deleted from the drawing, after the block has been defined.

Tip
*When you select the **Delete** radio button from the **Objects** area while defining a block, the selected objects are removed from the drawing. To restore them, you can use the **OOPS** command. On entering **U** at the Command prompt or choosing the **Undo** button from the **Standard** toolbar, the block definition will be restored from the drawing.*

If you select the **Annotative** check box in the **Behavior** area, the block created will become annotative. The block will now acquire annotative properties like text, dimension, hatch, and so on. On selecting the **Annotative** check box, the **Match block orientation to layout** check box will get highlighted. If you select the **Match block orientation to layout** check box, the orientation of the block in paper space viewport will always remain aligned to the orientation of the layout, while inserting the block. If you select the **Scale uniformly** check box, the block can only be scaled uniformly in all directions, while inserting. In such cases, if you create an annotative block, you can scale it only uniformly, while inserting. The **Scale uniformly** check box will be selected by default and remains unavailable for modification when you select the **Annotative** check box. If you select the **Allow exploding** check box, the block can be exploded into separate entities using the **Explode** tool, whenever required.

In the **Settings** Area, the **Block unit** drop-down list displays the units that will be used when inserting the current block. For example, if the block is created with inches as the units and you select **Feet** from the **Block unit** drop-down list, the block will be scaled to feet. You can also invoke the **Insert Hyperlink** dialog box by choosing the **Hyperlink** button in the **Settings** Area. This dialog box allows you to link any specific files, websites, or named views with the current drawing. You can also link the drawing to a default email address or you can select one from the recently used email addresses.

The **Open in block editor** check box is available at the bottom of **Block Definition** dialog box. If this check box is selected, the **Block Editor** will be opened as soon as you close the **Block Definition** dialog box. Also, the entities of the block will be displayed in the authoring area of the block editor. You will learn more about the **Block Editor** in the next section.

After setting all parameters in the **Block Definition** dialog box, choose the **OK** button to complete defining the block.

Tip
It is a good idea to create a unit block, especially for drawings where the same block is to be inserted with various scale factors. A unit block can be defined as a block that is created within a unit square area (1 unit x 1 unit). This way, every time you specify a scale factor, the block simply gets enlarged or reduced by the said amount directly, since the multiplication factor is 1.

Note
*You can also create blocks by entering **-BLOCK** at the Command prompt.*

EXERCISE 1 *Block*

Draw a circle of 1unit radius and then draw multiple circles inside it, as shown in Figure 18-4. Next, convert them into a block and name the entity as CIRCLE. Refer to the following figure for details.

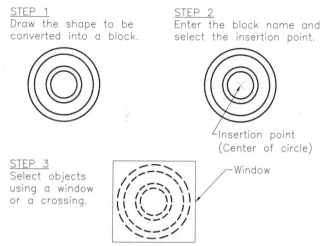

STEP 1
Draw the shape to be converted into a block.

STEP 2
Enter the block name and select the insertion point.

Insertion point
(Center of circle)

STEP 3
Select objects using a window or a crossing.

Window

*Figure 18-4 Using the **BLOCK** command*

INSERTING BLOCKS

Ribbon: Insert > Block > Insert
Toolbar: Insert > Insert Block or Draw > Insert Block   **Command:** INSERT

The blocks created in the current drawing are inserted using the **Insert** tool. An inserted block is called a block reference. You should determine the layer and location to insert the block, and also the angle by which you want the block to be rotated prior to its insertion. If the layer on which you want to insert the block is not the current layer, select the appropriate option from the **Layer** drop-down list in the **Layers** panel of the **Home** tab or choose the **Set Current** button in the **Layer Properties Manager** palette to make it current. When you invoke the **Insert** tool, the **Insert** dialog box is displayed, as shown in Figure 18-5. You can specify different parameters of the block or external file to be inserted in the **Insert** dialog box.

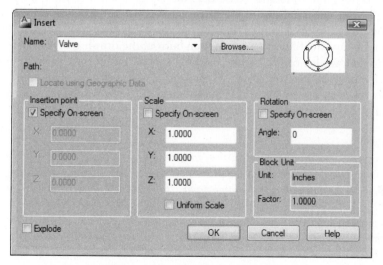

*Figure 18-5 The **Insert** dialog box*

Name

This drop-down list is used to specify the name of the block to be inserted. Select the name from this drop-down list. You can also enter a name for it. All blocks, created in the current drawing, are available in the **Name** drop-down list.

Note
*The last block name inserted becomes the default name for the next insertion and is displayed in the **Name** drop-down list.*

The **Browse** button is used to insert external files. When you choose this button, the **Select Drawing File** dialog box is displayed, as shown in Figure 18-6.

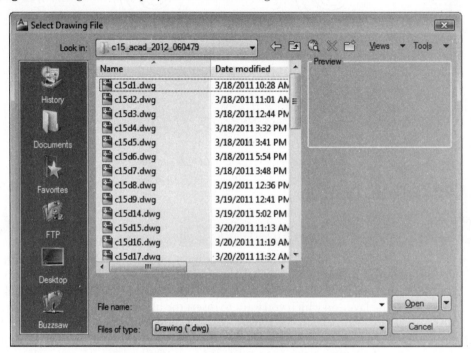

*Figure 18-6 The **Select Drawing File** dialog box*

This dialog box is similar to the standard **Select File** dialog box, which has been discussed earlier in Chapter 1. You can select a drawing file from the files listed in the current directory. You can also change the directory by selecting the desired directory from the **Look in** drop-down list. Once you select the drawing file, choose the **Open** button; the drawing file name is displayed next to the **Name** drop-down list of the **Insert** dialog box. The **Path** option displays the path of the external file selected to be inserted. Now, if you want to change the block name, just change the name in the **Name** drop-down list. In this manner, the drawing can be inserted with a different name. Changing the original drawing does not affect the inserted drawing.

Note
*If the name you have specified in the **Name** edit box does not exist as a block in the current drawing, AutoCAD will search the drives and directories on the path (specified in the **Options** dialog box) for a drawing of the same name. If the block is found, it will be inserted.*

*Also, suppose you have inserted a block in a drawing and then you want to insert a drawing with the same name as the block, AutoCAD will display a message saying that XX is already defined as a block and asks what do you want to do. If you choose **Redefine**, the block in the drawing will get replaced by the drawing with the same name, that is, the block gets redefined.*

Tip

*You can create a block in the current drawing from an existing drawing file. This saves the time, by avoiding redrawing the object as a block. Locate and select an existing drawing file using the **Browse** button. Next, choose the **Open** button after selecting the existing drawing. Next, choose the **OK** button; the **Insert** dialog box will be closed, prompting you to specify the insertion point. Now, instead of specifying the insertion point, press the ESC key; the selected file will be converted into a block, but will not be inserted into the drawing.*

Insertion point Area

When a block is inserted, its coordinate system is aligned parallel to the current UCS. In the **Insertion point** area, you can specify the X, Y, and Z coordinate locations of the block insertion point in the **X**, **Y**, and **Z** edit boxes, respectively. If you select the **Specify On-screen** check box, the **X**, **Y**, and **Z** edit boxes will not be available. You can specify the insertion point on the screen. By default, the **Specify On-screen** check box is selected, and hence, you can specify the insertion point on the screen.

Scale Area

In this area, you can specify the X, Y, and Z scale factors of the block to be inserted in the **X**, **Y**, and **Z** edit boxes, respectively. By selecting the **Specify On-screen** check box, you can specify the scale of the block at which it has to be inserted on the screen. The **Specify On-screen** check box is cleared by default and the block is inserted with the scale factors of 1 along the three axes. Also, if the **Uniform Scale** check box is selected, the X scale factor value is assumed for the Y and Z scale factors also. This means that if this check box is selected, you need to specify only the X scale factor. All dimensions in the block are multiplied by the same scale factors that you specify. These scale factors allow you to stretch or compress a block along the *X* and *Y* axes, respectively. You can also insert 3D objects into a drawing by specifying the third scale factor (since 3D objects have three dimensions), the Z scale factor. Figure 18-7 shows a block with different X and Y scale factors.

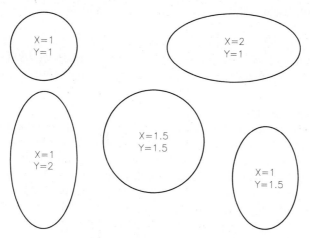

Figure 18-7 *Block inserted with different scale factors*

Tip

By specifying negative scale factors, you can insert a mirror image of a block along a particular axis. A negative scale factor for both the X and Y axes is the same as rotating the block reference through 180-degree, since it mirrors the block reference in the opposite quadrant of the coordinate system. The effect of the negative scale factor on the block (DOOR) can be marked by a change in position of the insertion point marked with an (X), as shown in Figure 18-8.

Also, specifying a scale factor of less than 1 inserts the block reference smaller than the original size. A scale factor greater than 1 inserts the block reference larger than its original size.

X = Insertion point

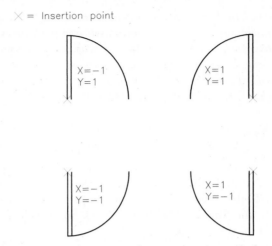

Figure 18-8 Block inserted using negative scale factors

Rotation Area

You can enter the angle of rotation for the block to be inserted in the **Angle** edit box. The insertion point is taken as the location about which the rotation takes place. Selecting the **Specify On-screen** check box enables you to specify the angle of rotation on the screen. This check box is cleared by default and the block is inserted at an angle of zero-degree.

Block Unit Area

This area displays the information related to the block unit specified in the **Block Definition** dialog box and the scale factor used for the unit to insert the block. For example, if you have defined a block with the block unit as feet and you insert the same block in a drawing whose unit is inches, the **Block Unit** area will display **Unit** as feet and **Factor** as 12. The information displayed in this area is read-only.

 Note
*You can control the units for the insertion of the block through the **INSUNITS** system variable. While doing so, the **Factor** information displayed in the **Block Unit** area will change accordingly.*

Explode Check Box

By selecting this check box, the block is inserted as a collection of individual objects. The function of the **Explode** check box is identical to that of the **Explode** tool. Once a block is exploded, the X, Y, and Z scale factors become identical. Therefore, you are provided access to only one scale factor edit box (**X** edit box), the **Y** and **Z** edit boxes are not available, and the **Uniform Scale** check box also gets selected. The X scale factor is assigned to the Y and Z scale factors too. You must enter a positive scale factor value.

 Note
*If you select the **Explode** check box before insertion, the block reference will be exploded and the objects will be inserted with their original properties such as layer, color, and linetype.*

Once you have entered the relevant information in the dialog box, choose the **OK** button. The **Insert** dialog box is removed from the screen. If you have selected the **Specify On-Screen** check boxes, you can specify the insertion point, scale, and angle of rotation with a pointing device. Whenever the insertion point is to be specified on the screen (by default), you can specify the scale factors and rotation angle values. These values will override the values specified in the dialog box using the command line. Before specifying the insertion point on the screen, you can right-click to display the shortcut menu, which has all the options available through the command line. You can choose the options for insertion from the dynamic preview. On being prompted to specify the insertion point, press the down arrow key to see the options in the dynamic preview. However, if you have already specified an insertion point in the dialog box and have selected the **Specify On-screen** check boxes in the **Scale** or **Rotation** areas of the **Insert** dialog box, you are allowed to specify only the scale factors or the rotation angle at the Command line. The following Command prompt will be displayed when all three **Specify On-screen** check boxes are selected:

Specify insertion point or [Basepoint/Scale/X/Y/Z/Rotate]: *You can specify an insertion point on the screen or select an option.*
Enter X scale factor, specify opposite corner, or [Corner/XYZ] <1>: *Specify scale factor in X- axis or select an option or ENTER to accept the default scale factor of 1.*
Enter Y scale factor <use X scale factor>: *Specify scale factor in Y- axis or press ENTER to use the X scale factor.*
Specify rotation angle <0>: *Specify the angle of rotation of the inserted block.*

Tip
Using the Command line, you can override scale factors and rotation angles already specified in the dialog box.

The Command line options are discussed next.

Basepoint

If you enter **B** at the **Specify insertion point** prompt, or select the option from the dynamic preview, you are again prompted to specify the base point. Specify the base point, by clicking the cursor at the desired location. This point will now act as the insertion point. All prompts are also displayed at the cursor input so there is no need to always take a look on the command prompt. The prompt sequence for this option is given next.

Specify insertion point or
[Basepoint/Scale/X/Y/Z/Rotate]: **B** `Enter`
Specify base point: *Specify base point*
Specify insertion point or [Basepoint/Scale/X/Y/Z/Rotate]: *Specify insertion point*

Scale

On entering **S** at the **Specify insertion point** prompt, you are asked to enter the scale factor. After entering it, the block assumes the specified scale factor, and AutoCAD lets you drag the block until you locate the insertion point on the screen. The *X*, *Y*, and *Z* axes are uniformly scaled by the specified scale factor. The prompt sequence for this option is given next.

Specify insertion point or [Basepoint/Scale/X/Y/Z/Rotate]: **S**
Specify scale factor for XYZ axes<1>: *Enter a value to preset general scale factor.*
Specify insertion point or [Basepoint/Scale/X/Y/Z/Rotate]: *Specify insertion point*

X, Y, Z

With X, Y, or Z as the response to the **Specify insertion point** prompt, you can specify the X, Y, or Z scale factors before specifying the insertion point. These options are only available when the **Uniform Scale** check box is cleared from the **Insert** dialog box. The prompt sequence when you use the **X** option is given next.

Specify insertion point or [Basepoint/Scale/X/Y/Z/Rotate]: **X**
Specify X scale factor<1>: *Enter a value for the X scale factor.*

Specify insertion point or [Basepoint/Scale/X/Y/Z/Rotate]: *Select the insertion point.*
Specify rotation angle <0>: *Specify angle of rotation.*

Rotate

On entering **R** at the **Specify insertion point** prompt, you are asked to specify the rotation angle. You can enter the angle of rotation or specify it by specifying two points on the screen. Dragging is resumed only when you specify the angle. Also, the block assumes the specified rotation angle. The prompt sequence for this option is given next.

Specify insertion point or [Basepoint/Scale/X/Y/Z/Rotate]: **R**
Specify rotation angle<0>: *Enter a value for the rotation angle.*
Specify insertion point or [Basepoint/Scale/X/Y/Z/Rotate]: *Select the insertion point.*

The XYZ and Corner Options

When you specify the insertion point and scale factors on the screen, the following prompt appears after you have selected an insertion point.

Enter X scale factor, specify opposite corner or [Corner/XYZ]<1>: *Enter X scale factor, specify opposite corner or select an option.*

Here, the **XYZ** option can be used to enter a 3D block reference and the successive prompts allow you to specify the X, Y, and Z scale factors individually.

With the **Corner** option, you can specify both X and Y scale factors at the same time. When you invoke this option, and **DRAGMODE** is turned off, and you are prompted to specify the other corner (the first corner of the box is the insertion point). You can also enter a coordinate value, instead of moving the cursor. The length and breadth of the box are taken as the X and Y scale factors for the block. For example, if the X and Y dimensions (length and width) of the box are the same as that of the block, the block will be drawn without any change. The points selected should be above and to the right of the insertion block. Mirror images will be produced if points selected are below or to the left of the insertion point. The prompt sequence is given next.

Enter X scale factor, specify opposite corner or [Corner/XYZ]<1>: **C**
Specify opposite corner: *Select a point as the corner.*

The **Corner** option also allows you to use a dynamic scaling technique when **DRAGMODE** is turned to **Auto (default)**. You can also move the cursor at the **Enter X scale factor** prompt to change the block size dynamically and, when the size meets your requirements, select a point.

Note
*1. You should avoid the use of the **Corner** option, if you want to have the same X and Y scale factors, because it is difficult to select a point whose X distance equals its Y distance, unless the **SNAP** mode is on. If you use the **Corner** option, it is better to specify the X and Y scale factors explicitly or select the corner by entering coordinates.*
*2. If the **Uniform Scale** check box is selected in the **Insert** dialog box, then the **X/Y/Z** options will not be available at the Command prompt.*
*3. You can also insert blocks using the command window either by entering the **-INSERT** command or by simply writing **-I** .*

EXERCISE 2   *Basepoint Insertion*

Create a block with the name SQUARE at the insertion base point 1,2. Insert this block in the drawing. The X scale factor is 2 units, the Y scale factor is 2 units, and the angle of rotation is 35-degree. It is assumed that the block SQUARE is already defined in the current drawing.

EXERCISE 3   *Scaled Insertion*

a. Insert the block CIRCLE created in Exercise 1. Use different X and Y scale factors to get different shapes after inserting this block.

b. Insert the block CIRCLE created in Exercise 1. Use the **Corner** option to specify the scale factor.

EXERCISE 4   *Scaled Insertion*

a. Construct a triangle and form a block of it. Name the block as TRIANGLE. Now, set the Y scale factor of the inserted block as 2.

b. Insert the block TRIANGLE with a rotation angle of 45-degree. After defining the insertion point, enter the X and Y scale factor of 2.

CREATING AND INSERTING ANNOTATIVE BLOCKS

You can create an annotative block by selecting the **Annotative** check box from the **Behavior** area of the **Block Definition** dialog box. The annotative block acquires the annotative properties like text, dimension, hatch, and so on. To convert an existing non-annotative block into an annotative one, choose the **Create Block** tool from the **Block Definition** panel of the **Insert** tab. Select the block to be converted to annotative from the **Name** drop-down list of the **Block Definition** dialog box. Select the **Annotative** check box from the **Behavior** area and then choose the **OK** button. AutoCAD will display a message box stating that the selected block is already defined as a block. If you want to redefine the block, choose the **Redefine** button; the specified block will be converted to annotative.

The annotative blocks are indicated by an annotative symbol attached to the block in the preview displayed in the **Insert** dialog box. While inserting the annotative block in the AutoCAD session for the first time, the **Select Annotation Scale** dialog box will be displayed on the screen, see Figure 18-9. Select the annotation scale from the drop-down list and then choose the **OK** button; the block will be inserted at the specified annotation scale. The Annotative blocks are inserted at a scale value decided by the multiplication of the current annotation scale and the block scale specified in the **Scale** area of the **Insert** dialog box.

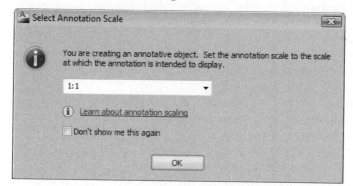

*Figure 18-9 The **Select Annotation Scale** dialog box*

Note

*While inserting the annotative blocks, the settings specified for the **INSUNITS** system variable are ignored. Also, the **Block Unit** area of the **Insert** dialog box will display no change in the factor value.*

EXAMPLE 1 — *Annotative Block*

In this example, you will draw the object shown in Figure 18-10, and convert it into an annotative block, named NOR Gate. Next, you will insert the NOR Gate block into the drawing at the annotation scales of 1:1, 1:2, and 1:8 and notice the changes in the size of the annotative blocks inserted in the drawing.

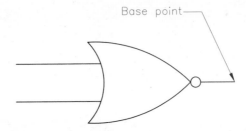

Base point

***Figure 18-10** Drawing of block for Example 1*

1. Start a new file in the **Drafting & Annotation** workspace and draw the object, as shown in Figure 18-10.

2. Invoke the **Block Definition** dialog box by choosing the **Create Block** tool from the **Block Definition** panel in the **Insert** tab.

3. Enter **NOR Gate** as the name of the block in the **Name** edit box. Next, choose the **Select objects** button from the **Objects** area; the **Block Definition** dialog box will disappear. Select the object drawn and press the ENTER key; the **Block Definition** dialog box will appear. Next, select the **Delete** radio button from the **Objects** area.

4. Choose the **Pick point** button from the **Base point** area; the **Block Definition** dialog box will disappear from the screen. Specify the base point, as shown in Figure 18-10.

5. Select the **Annotative** check box from the **Behavior** area. Next, choose the **OK** button from the **Block Definition** dialog box; the selected objects disappear from the screen and an annotative block is defined with the name **NOR Gate**.

Before proceeding further, ensure that the **Automatically Add Scales to annotative objects when the annotation scale changes** button is chosen in the Status Bar.

6. Now, you need to change the annotation scale of the drawing to 1:8. To do so, click on the down-arrow on the right of the **Annotation Scale** button in the Status Bar, and choose the scale 1:8 from the flyout displayed.

7. To insert the block into the drawing at the annotation scale of 1:8, choose the **Insert** tool from the **Block** panel in the **Insert** tab; the **Insert** dialog box is displayed.

8. Select the **NOR Gate** block from the **Name** drop-down list. Select the **Specify On-screen** check box from the **Insertion point** area and choose the **OK** button.

9. Specify the insertion point for the block by clicking on the screen; the block is inserted into the drawing at an annotation scale of 1:8.

10. Similarly, set the annotation scale to **1:2** in the Status Bar and insert the block. Then set the annotation scale to **1:1** and insert the block.

11. Notice the difference in the sizes of the inserted blocks, see Figure 18-11. This automated variation in the size occurs due to annotative blocks. You will notice that the **Annotation Visibility** button on the right of the **Annotation Scale** button is on.

12. Now, set the current annotation scale to **1:8**. Choose the **Add/Delete Scales** tool from the **Annotation Scaling** panel in the **Annotate** tab. Select the block displayed at the annotation scale of 1:8 and press ENTER; the **Annotation Object Scale** dialog box is displayed. Choose the **Add** button; the **Add Scales to Object** dialog box is displayed. Press and hold the CTRL key and select the scale of 1:2 and 1:1 from the **Scale List** area.

13. Next, choose **OK**; the selected annotation scales get associated with the selected block. In this way, you can associate different annotation scales to a single annotative block.

14. Select the block to which you had added the annotation scales; the preview of the block with all annotation scales associated to the block is displayed, as shown in Figure 18-12. The bigger block with the current annotation (1:8) scale is displayed with dark dotted lines and the smaller blocks with other annotation scales (1:1, 1:2) are displayed with the faded dashed lines.

15. To move the block with the current annotation scale (1:8) to the other locations, select the blue grip displayed at the insertion point and move the block to the desired location. This way you can place the block with different annotation scales to any desired location.

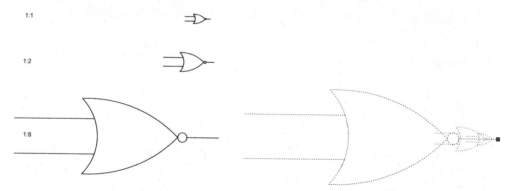

Figure 18-11 Annotative blocks inserted at different annotation scales

Figure 18-12 Different annotation scales associated to a single block

Block Editor

Ribbon: Insert > Block Definition > Block Editor or Home > Block > Edit
Toolbar: Standard > Block Editor   **Command:** BEDIT

Block Editor

The **Block Editor** is an important feature. This application is used to edit existing blocks or create new blocks. To invoke the **Block Editor**, choose the **Block Editor** tool from the **Block Definition** panel of the **Insert** tab or enter **BE** (shortcut for the **BEDIT** command) at the Command line. You can also double-click on the existing block to edit the block. On doing so, the **Edit Block Definition** dialog box will be displayed, as shown in Figure 18-13.

If you want to create a new block, enter its name in the **Block to create or edit** text box and choose **OK**; the **Block Editor** will be invoked where you can draw the entities in the new block. Similarly, if you want to edit a block, select it from the list box provided in this dialog box; its preview will be displayed in the **Preview** area. Also, its related description, if any, will be displayed in the **Description** area. Choose the **OK** button; the **Block Editor** is invoked.

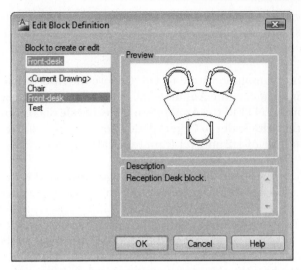

*Figure 18-13 The **Edit Block Definition** dialog box*

The default appearance of the **Block Editor** is shown in Figure 18-14. The drawing area of the **Block Editor** has a dull background and is known as the authoring area. You can edit existing entities or add new ones to the block in the authoring area. In addition to the authoring area,

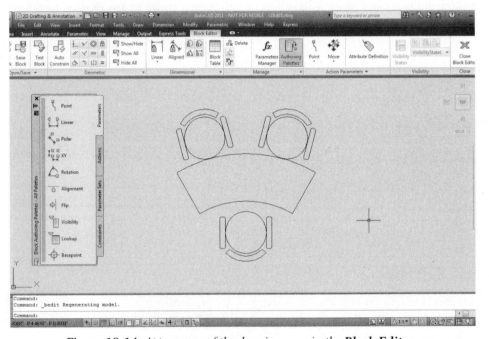

*Figure 18-14 Appearance of the drawing area in the **Block Editor***

the **Block Editor** tab and **Block Authoring Palettes** are provided in the **Block Editor** that contain tools to create dynamic blocks. This will be discussed in detail, later in this chapter. Now, you can edit a block using any tool, as you did in the drawing. When you are finished with the editing, choose the **Save Block** tool in the **Open/Save** panel of the **Block Editor** tab to save the changes. Then, choose the **Close Block Editor** button from the **Close** panel to return to the drawing again. Like this, you can edit the block at any time of your design process.

DYNAMIC BLOCKS

The concept of dynamic blocks is new in AutoCAD. It provides you the flexibility of modifying the geometry of the inserted blocks dynamically or using the **Properties** palette. You can create dynamic blocks by adding **Parameters** and **Actions** to existing blocks using the **Action Parameters** panel in the **Block Editor** tab (see Figure 18-15) or the **Block Authoring Palettes**.

*Figure 18-15 The **Block Editor** tab*

Block Editor Tab

The **Block Editor** tab, shown in Figure 18-15, provides the tools to create and modify dynamic blocks. Additionally, you can create visibility states for dynamic blocks using the **Visibility** panel of the **Block Editor** tab. All these tools are discussed next.

Edit Block

If there are multiple blocks to be edited, choose this tool to display the **Edit Block Definition** dialog box. Note that if you have modified the current block using the tools in the **Block Editor**, you will be prompted to specify whether you want to save the changes in the current block. On saving the changes, the **Edit Block Definition** dialog box will be displayed. You can select another block to edit or enter the name of the new block that you want to create. Next, proceed to the **Block Editor** to edit the block.

Save Block

This tool is chosen to save the changes that you have made to the block in the **Block Editor**.

Save Block As

This tool is used to save a block with a different name. Expand the **Open/Save** panel and choose this tool; the **Save Block As** dialog box will be displayed, as shown in Figure 18-16. Enter the name with which you want to save the current block in the **Block Name** edit box and then choose the **OK** button; the block will be saved with the specified name and it will become the current block in the **Block Editor**.

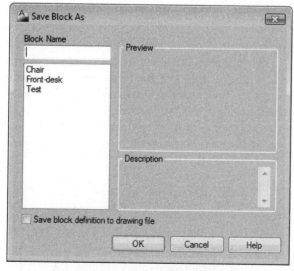

*Figure 18-16 The **Save Block As** dialog box*

Test Block

On choosing this button, the **Test Block Window** will be displayed. If you have made any changes to the block in the **Block Editor**, you can check them in this window without saving the block. After testing the block, choose the **Close Test Block Window** button to close the **Test Block Window**.

Authoring Palettes

The **Authoring Palettes** button is chosen from the **Manage** panel to toggle on/off the display of the **Block Authoring Palettes** in the authoring area. The **Block Authoring Palettes** contain the tools for creating dynamic blocks.

Parameters

Parameters are predefined rules that are applied to blocks to enable them to react dynamically based on the action or the set of actions assigned to them. To add parameters to the existing blocks, enter **BPARAMETER** at the Command prompt or select a tool from the **Parameter** drop-down and then choose one of the buttons from the flyout displayed, as shown in Figure 18-17. If you enter **BPARAMETER** at the Command prompt when the **Dynamic Input** button is chosen, all parameters will be listed in the dynamic preview. These parameters are discussed later in this chapter.

Actions

You can choose a tool from the **Actions** drop-down to assign actions to the parameters added to the dynamic blocks. Depending on the type of parameter selected, the possible actions are displayed in the dynamic preview. The types of actions in the **Block Editor** are discussed later in this chapter.

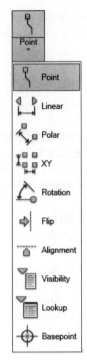

*Figure 18-17 Tools in the **Parameter** drop-down*

Attribute Definition

Choose the **Attribute Definition** tool from the **Action Parameters** panel of the **Block Editor** tab to invoke the **Attribute Definition** dialog box to assign attributes to the block. Attributes are discussed in detail in Chapter 19.

Update Fields

 Choose the **Update Fields** tool from the **Data** panel in the **Insert** tab to regenerate the dynamic block such that the text size of the parameters and actions are updated.

Close

When you choose this button, the **Block Editor** is closed and you return to the drawing environment. If this button is chosen before saving the changes in the current block, the **Block - Changes Not Saved** message box will be displayed, as shown in Figure 18-18 and you will be prompted to specify whether or not you want to save changes in the current block.

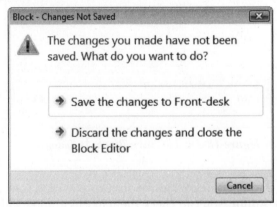

*Figure 18-18 The **Block - Changes Not Saved** message box*

Adding Parameters and Assigning Actions to Dynamic Blocks (Block Authoring Palettes)

The **Block Authoring Palettes**, shown in Figure 18-19, contains the tools to add parameters and actions to the dynamic blocks. You can also invoke these tools from the **Parameters** drop-down in the **Action Parameters** panel of the **Block Editor** tab, refer to Figure 18-17. On doing so, different types of parameters and actions will be displayed at the pointer input or the Command prompt. The command sequence will be the same in both the cases. The types of parameters and actions that can be added to dynamic blocks are discussed next.

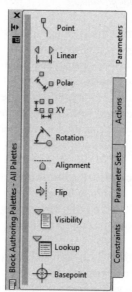

*Figure 18-19 The **Parameters** tab of the **Block Authoring Palettes***

Point Parameter

The **Point** parameter is used to assign the **Stretch** or **Move** action to a dynamic block. A grip point will be displayed at the location where you place the **Point** parameter. This grip point will be used as the base point when you assign an action to the **Point** parameter.

To assign a **Point** parameter, invoke the **Point** tool from the **Action Parameters** tab of the **Block Authoring Palettes**. The following prompt sequence will be displayed when you invoke this parameter from the **Action Parameters** tab:

Specify parameter location or [Name/Label/Chain/Description/Palette]: *Select a keypoint on the block to add the **Point** parameter.*
Specify label location: *Specify a point to place the label.*
Enter number of grips [0/1] <1>: *Enter 0 or 1* [Enter]

Tip
*You can double-click on the label to find out the actions that are assigned with that parameter. For example, if you double-click on the **Point** parameter, you can perform the move and stretch actions.*

The options available at the Command prompt are discussed next.

Name. The name of a parameter is displayed in the **Properties** palette when you select the **Point** parameter in the **Block Editor**. The default name for the point parameter is Point. However, you can change the name of the parameter using this option. The prompt sequence to modify the name of a parameter is given next.

Specify parameter location or [Name/Label/Chain/Description/Palette]: **N** [Enter]
Enter parameter name <Point>: *Enter new name for the parameter.*
Specify parameter location or [Name/Label/Chain/Description/Palette]: *Specify a location for the parameter.*

Label. Labels of a parameter are displayed in the authoring area, as shown in Figure 18-20, and in the **Properties** palette in the **Block Editor**. The default label of a point parameter is Position. Using the **Label** option, you can modify the label of the point parameter. The prompt sequence for modifying the label is as follows:

Specify parameter location or [Name/Label/Chain/Description/Palette]: **L** [Enter]
Enter position property label <Position>: *Enter the new label for the **Point** parameter*

Specify parameter location or [Name/Label/Chain/Description/Palette]: *Specify a location for the parameter*

Chain. This option is used when the current point parameter is assigned an action, which is also associated with other parameters. In this case, if the **Chain** option is set to **Yes** and the other parameters are edited, the point parameter is also edited with them. If this option is set to **No**, the point parameter is not edited, when the other parameters are edited.

Description. This option allows you to enter the description about a parameter. This description is displayed in the **Properties** palette in the **Block Editor**.

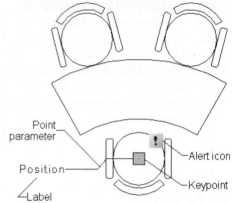

Figure 18-20 Labels of the Point Parameter

Palette. This option is used to specify whether or not the base X and base Y coordinate values of the keypoint where the parameter label is placed will be displayed in the **Properties** palette when you select the block outside the **Block Editor**.

Assigning the Move Action to the Point Parameter

To assign the move action to the **Point** parameter, double-click on the label and select the **Move** option from the dynamic preview or from the Command line. You can also choose the **Move** tool from the **Action Parameters** panel. Alternatively, you can choose the **Move** tool from the **Actions** tab of the **Block Authoring Palettes**. The prompt sequence that will follow when you invoke it from the **Block Authoring Palettes** is given next.

Select parameter: *Select the point parameter.*
Specify selection set for action
Select objects: *Select objects on which action will be applied.*
Select objects: Enter

You need to specify the distance multiplier and angle offset properties to the actions using the **Properties** palette. The **Move** action will not work until these options are set.

Distance Multiplier. This option is used during the dynamic modification of the blocks, using grips of the parameters. Using this option, you can specify the distance by which the selected entities of the dynamic block will move with respect to the one unit movement of the cursor, while dragging the cursor using the grips. To set the distance multiplier, select the action parameter and invoke the **Parameters** palette. Now, enter the required multiplier value in the **Distance multiplier** field in the **Overrides** rollout of the **Properties** palette. By default, the value of the distance multiplier factor is 1. As a result, the entities will move by the same distance as moved by the cursor. If you set this value to 0.5, the entities will move half the distance moved by the cursor, refer to Figure 18-21.

Note
*You can test the action parameters by invoking the **Test Block Window** and moving the block.*

Angle Offset. This option is used to specify the angular offset for the selected entities of the dynamic block. The angle offset works when the entities are dragged using the parameter grip. To set the angle offset, select the action parameter and invoke the **Parameters** palette. Now, enter the required value in the **Angle offset** field in the **Overrides** rollout of the **Properties** palette. By setting this value, you can ensure that the entities move at an angle from the trace line. For

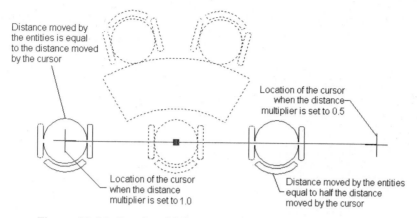

Figure 18-21 Results of different values of the distance multiplier

example, if you set the value of the angle offset to 30, the entities will be moved at an angle of 30-degree with respect to the trace line, which is displayed while dragging the entities using the parameter grips. Figure 18-22 shows the manipulation of entities in the drawing with the default angle offset of 0-degree and an angle offset of 30-degree.

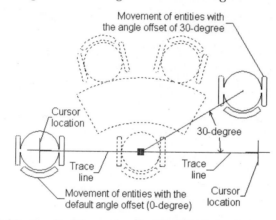

Figure 18-22 Result of the angle offset during the dynamic manipulation

Assigning the Stretch Action to the Point Parameter

You can invoke the **Stretch** action by double-clicking on the parameter or by choosing it from the **Block Authoring Palettes**. On invoking this action and selecting the parameter, you are prompted to specify the stretch frame. The entities lying inside the stretch frame will be moved during the dynamic manipulation. The entities that are crossed by the stretch frame will be stretched. The following prompt sequence will be displayed on invoking this action:

Select parameter: *Select the parameter to associate the action.*
Specify first corner of stretch frame or [CPolygon]: *Specify the first corner of stretch frame or enter CP to define a crossing polygon for specifying the stretch frame.*
Specify opposite corner: *Specify opposite corner of stretch frame.*
Specify objects to stretch
Select objects: *Select objects for stretching.*
Select objects: [Enter]

Figure 18-23(a) shows the stretch frame and the entities selected for the stretch action and Figure 18-23(b) shows the preview of the entities during dynamic manipulation.

Chapter 18

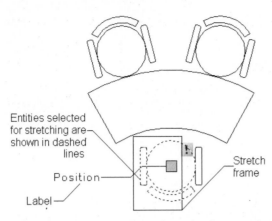

Entities selected for stretching are shown in dashed lines

Label

Position

Stretch frame

Figure 18-23(a) *Selecting entities for the* **Stretch** *action*

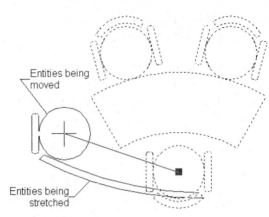

Entities being moved

Entities being stretched

Figure 18-23(b) *Dynamically stretching the entities of a block*

Linear Parameter

The linear parameter is displayed as the distance between two keypoints and is similar to a linear or an aligned dimension. This parameter can be associated with the Array, Move, Stretch, or Scale actions. Figure 18-24 shows a block with a linear parameter added to it. To add this parameter, choose the **Linear** tool from the **Block Authoring Palettes**; the following prompt sequence will be displayed:

> Specify start point or [Name/Label/Chain/ Description/Base/Palette/Value set]: *Specify start point for the parameter.*
> Specify endpoint: *Specify endpoint for the parameter.*
> Specify label location: *Specify a point where the label will be placed.*

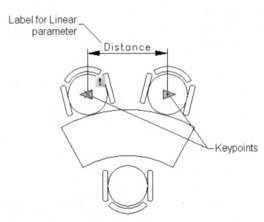

Label for Linear parameter

Distance

Keypoints

Figure 18-24 *The* **Linear** *parameter added to a block*

You can specify the base of the parameter and the distance within which the entities are to be moved at the Command prompt when you are prompted to specify the start point. Some of the options that will be available when you invoke the **Linear Parameter** tool have already been discussed in the **Point** parameter. The remaining options are discussed next.

Base. This option lets you to specify the base of the parameter. This base point will be stationary while editing the endpoint of the block. You can specify the option of keeping the start point or the midpoint stationary while using the **Base** option.

Value set. This option allows you to specify a set of values for parameters such that the manipulation will be in accordance with these specified values. The values for the parameter can be specified by providing a list or by specifying the increment and range. If there is a specified value set for a parameter, small vertical grey lines are displayed at those values in the **Block Editor**. These vertical grey lines are also displayed in the drawing when you select the parameter grip to manipulate the block.

Note
You can also set the **Base** *and* **Value Set** *options in the* **Misc** *and* **Value set** *rollouts of the* **Properties** *palette of a* **Linear Parameter**, *respectively.*

After adding the **Linear** parameter, you can perform the array, move, scale, and stretch actions. Assign the **Move** or **Stretch** action as discussed earlier.

Assigning the Scale Action to the Linear Parameter

Invoke the **Scale** action tool from the **Actions** tab of **Block Authoring Palettes**, or double-click on the linear parameter and select the **Scale** option. The prompt sequence that follows when you assign the **Scale** action is as follows:

Select parameter: *Select the linear parameter to associate the action.*
Specify selection set for action
Select objects: *Select the objects for scaling.*
Select objects: [Enter]

You can scale the blocks by using a property called **Base type**. To specify the **Base type** property, display the **Properties** palette for the **Scale** action. Next, specify the base type in the **Overrides** rollout. There are two types of base type, dependent and independent. The base type can be dependent on the original base point of the parameter to which the scale action is assigned. You can also make the base type independent so that the entities can be scaled based on the specified base point and the second point.

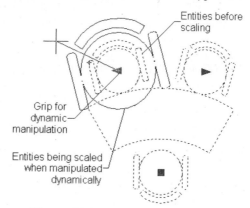

Figure 18-25 Entities being scaled by using dynamic manipulation

To test the action, choose the **Test Block** tool from the **Open/Save** panel and then select the block; the **Linear** parameter grips will be displayed. Click on a grip to select it; the **Specify point location or [Base Point/Undo/Exit]** prompt will be displayed. Specify the new base point, if needed. Moving the cursor in the second or third quadrant scales the entities by a factor more than 1, as shown in Figure 18-25. If you move the cursor in the first or fourth quadrant, the entities will be scaled by a factor less than 1.

Assigning the Array Action to the Linear Parameter

By adding array action to the entities in the block, you can create the array dynamically by using grips. Invoke the **Array** action tool from **Block Authoring Palettes** or double-click on the linear parameter and select the **Array** option. You will be prompted to select the objects to create the array. Next, you will be prompted to enter the distance between the columns. The number of instances in the array will depend on the distance by which you move the cursor after selecting the parameter grip. Figure 18-26 shows the preview of an array being created dynamically using the **Array** action.

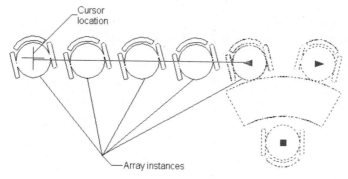

*Figure 18-26 Creating an array dynamically using the **Array** action*

Polar Parameter

The **Polar** parameter is defined by the distance between the two keypoints and an angle. The **Move**, **Scale**, **Stretch**, **Polar Stretch**, or **Array** actions can be associated with the **Polar** parameter. To add a polar parameter, choose **Point > Polar** from the **Action Parameters** panel of the **Block Editor**. The following prompt sequence will be displayed:

Specify base point or [Name/Label/Chain/Description/Palette/Value set]: *Specify start point for the parameter.*
Specify endpoint: *Specify endpoint for the Parameter.*
Specify label location: *Specify a point where Label will be placed.*

Note that the working of the **Value set** option is different in case of the **Polar** parameter. You can invoke the **Properties** palette for the **Polar** parameter and change the value sets in the **Value Set** parameters, or specify them when you are prompted to specify the base point. The prompt sequence for this option is given next.

Specify base point or [Name/Label/Chain/Description/Palette/Value set]: **V** `Enter`
Enter distance value set type [None/List/Increment] <None>: **L** `Enter`
Enter list of distance values (separated by commas): *Specify list of values for the Distance* `Enter`
Enter angle value set type [None/List/Increment] <None>: **L** `Enter`
Enter list of angle values (separated by commas): *Specify list of values for the* **Angle** *parameter.*
Specify base point or [Name/Label/Chain/Description/Palette/Value set]:

Similarly, you can use the **Increment** option to specify the increment and range for the distance and angle parameters.

Assigning the Polar Stretch Action to the Polar Parameter

By specifying the polar stretch action, you can stretch the entities and at the same time rotate them. As you move the cursor, the entities specified for rotation will rotate by the same angle that has been moved by the cursor. Remember that the polar stretch action can only be applied to the polar parameter. To add this action, choose the **Polar Stretch** action from the **Block Authoring Palettes** or double-click on the parameter and select the polar stretch option. The following is the prompt sequence:

Select parameter: *Select the parameter to associate the action.*
Specify parameter point to associate with action or enter [sTart point/Second point] <Second>: *Specify the point that will be associated with the action.*
Specify first corner of stretch frame or [CPolygon]: *Specify the first corner of stretch frame.*
Specify opposite corner: *Specify opposite corner of stretch frame.*
Specify objects to stretch
Select objects: *Select the object to stretch, refer to Figure 18-27.*
Select objects: `Enter`
Specify objects to rotate only
Select objects: *Select the object that needs to be rotated only, refer to Figure 18-27.*
Select objects: `Enter`

Set the **Distance multiplier** and **Angle offset** in the **Properties** palette, as discussed earlier. Next, invoke the **Test Block Window** and select the block; the polar parameter grips will be displayed.

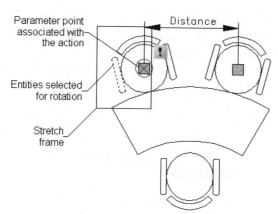

Figure 18-27 *Selections for adding the polar stretch action*

Click on the grip to select it; you will be prompted to specify the point location. Specify the new point; some of the entities will stretch, whereas the others will rotate, as shown in Figure 18-28. Exit the **Test Block Window** by choosing the **Close Test Block Window** button.

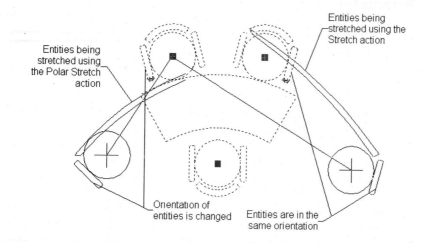

Figure 18-28 *Difference between the stretch action and the polar stretch action*

Tip
*While you assign different actions and parameters to the block entities in the **Block Editor**, there no need to exit the block editor and check the actions. Choose the **Test Block** tool in the **Open/ Save** panel in the **Block Editor**; a new test window will be displayed. Here you can check the changes applied to the block. To exit this new window, choose the **Close Test Block Window** button from the **Close** panel.*

XY Parameter

The XY parameter is used to define the X and Y distances between a specified base point and an endpoint in the dynamic block. You can assign the array, move, scale, and stretch actions to the XY parameter. To add this parameter, choose the **XY** parameters tool from the **Block Authoring Palettes**. The prompt sequence that will be displayed is given next.

Specify base point or [Name/Label/Chain/Description/Palette/Value set]: *Specify the base point from where the X and Y distance will be measured.*
Specify endpoint: *Specify endpoint up to which the distances will be measured.*

Figure 18-29 shows the XY parameter. Double-click on the parameter to display the dynamic preview. The dynamic preview lists the actions that can be assigned to this parameter.

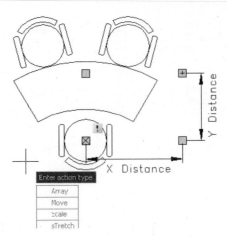

Figure 18-29 XY parameter assigned to a point and the actions that can be assigned to this parameter

Rotation Parameter

The rotation parameter is used to define an angular variation for the selected entities of the dynamic block. You can assign the rotate action to this parameter, which ensures that the selected entities of the dynamic block can be rotated around a predefined base point. To add this parameter, choose the **Rotation** parameter from the **Block Authoring Palettes**. The prompt sequence that will be displayed is as follows:

> Specify base point or [Name/Label/Chain/Description/Palette/Value set]: *Specify base point around which the selected entities of the dynamic block will be rotated.*
> Specify radius of parameter: *Specify the radius, which defines the imaginary circle along the circumference of which the entities will be rotated.*
> Specify default rotation angle or [Base angle] <0>: *Specify default rotation angle at which the grip point will be placed for rotating the entities dynamically.*
> Specify label location: *Specify a point where the label will be placed.*

When you specify the radius of the parameter, you will be prompted to specify the default rotation angle or base angle. The base angle defines the datum for measuring the rotation angle. By default, this value is 0-degree, which means that the angles will be measured from the positive X-axis. You can also modify the base angle to any other value, such that the angles are measured from the new base angle. For example, if you modify the base angle value to 30-degree, the default rotation angle will be measured from an axis defined at 30-degree from the X-axis, as shown in Figure 18-30.

Assigning the Rotate Action to the Rotation Parameter

The **Rotate** action can only be assigned to the **Rotation** parameter. This action allows you to rotate the selected components of the dynamic block using grips, as shown in Figure 18-31. To add the **Rotate** action, double-click on the **Rotation** parameter. The following prompt sequence will be displayed:

> Specify selection set for action
> Select objects: *Select the entities to be rotated.*
> Select objects: Enter

Next, invoke the **Test Block Window** and select the block; the **Rotation** parameter grips will be displayed on the block. Click on a grip to select it; you will be prompted to specify the rotation angle. Specify the angle; the entities will rotate as specified. Exit the **Test Block Window** by choosing the **Close Test Block Window** button.

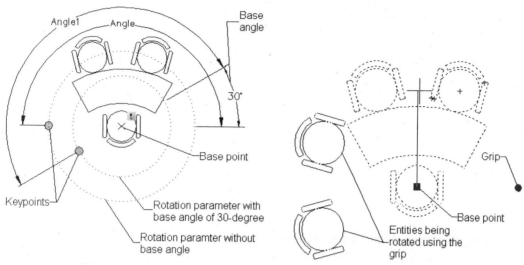

Figure 18-30 *The* **Rotation** *parameters with and without the base angle*

Figure 18-31 *The* **Dynamic** *manipulation of entities after adding the* **Rotate** *action*

Alignment Parameter

The **Alignment** parameter is used to align the entire dynamic block with the other entities in the drawing. The alignment is defined using the base point of alignment, at which the dynamic block grip is created. The block is rotated and moved while aligning. Remember that because the entire block is aligned using this parameter, you do not need to assign any action to this parameter. When you invoke this parameter, the following prompt sequence is displayed:

Specify base point of alignment or [Name]: *Specify base point to align the dynamic block.*
Alignment type = Perpendicular
Specify alignment direction or alignment type [Type] <Type>: *Specify the alignment direction by defining another point in the drawing window.*

Remember that the dynamic block will be aligned at the angle that is defined while defining the alignment direction in the **Specify alignment direction or alignment type [Type] <Type>** prompt.

Type. Enter **T** at the **Specify alignment direction or alignment type [Type] <Type>** prompt to specify the type of alignment. The types of alignment are discussed next.

Perpendicular. This alignment type ensures that the dynamic block is aligned perpendicular to the entities in the drawing.

Tangent. This alignment type ensures that the dynamic block is aligned tangent to the entities in the drawing.

After adding the alignment parameter, invoke the **Test Block Window** and select the block; the dynamic block grip is displayed at the base point of the alignment parameter. Select the grip and then move the dynamic block close to an existing entity in the drawing. The dynamic block aligns itself to the entity and the alignment angle depends on the alignment direction that you defined. Exit the **Test Block Window** by choosing the **Close Test Block Window** button.

Flip Parameter

The **Flip** parameter is used to add a flip action to the dynamic block. The flip action reverses

Chapter 18

the orientation of the dynamic block. Thus in simple terms, the flip parameter is used to create a mirrored image of the dynamic block and remove the original block. The mirror line is defined by the reflection line, which you specify while defining the flip parameter, see Figure 18-32. Remember that depending on the location of the reflection axis, the location of the flipped entities also changes. When you add the flip action, an arrow is displayed at the base point. You can click on this arrow to flip the orientation of the selected entities of the dynamic block. The following prompt sequence will be displayed when you invoke the **Flip** parameter tool from the **Block Authoring** palette:

> Specify base point of reflection line or [Name/Label/Description/Palette]: *Specify the base point from where the reflection line will start.*
> Specify endpoint of reflection line: *Specify the endpoint of the reflection line, thus defining its orientation.*
> Specify label location: *Specify location for label of the parameter.*

Assigning the Flip Action to the Flip Parameter
As mentioned earlier, you can assign the **Flip** action to the **Flip** parameter so as to reverse the orientation of the selected entities of the block. To assign the **Flip** action, double-click on the **Flip** parameter; you will be prompted to select the objects to be flipped. Select the objects and press ENTER; the flip action will be applied. Figure 18-33 shows the orientation and location of the selected entities before and after flipping.

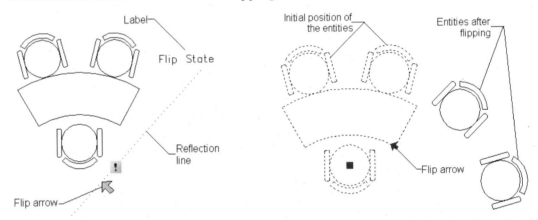

*Figure 18-32 Defining the **Flip** parameter*

Figure 18-33 Dynamically flipping the entities of the dynamic block

Visibility Parameter
The visibility parameter is used to create visibility states of a dynamic block. Using the visibility states, you can control the display of entities in the dynamic block and specify whether they would be visible or hidden. When you invoke this tool, you will be prompted to specify the location of the parameter. As soon as the visibility parameter is added, the following options will be available in the **Visibility** panel of the **Block Editor**.

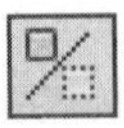

Visibility Mode. This button toggles the value of the **Visibility Mode** in between 0 and 1. The default value is set to 0 and so the entities that have been made invisible will not be shown in the authoring area. If the value of this mode is set to 1, the entities that you make invisible will also be displayed in the authoring area, but in gray color.

Make Invisible. This tool is used to make some of the entities invisible in the current visibility state. When you choose this tool, you are prompted to select the entities to be hidden.

Make Visible. This tool is chosen to make the entities visible that were made invisible in the current visibility state using the **Make Invisible** tool. When you choose this tool, all the entities that were hidden are displayed in the authoring area, even if the **Visibility Mode** is set to 0. Select the entities that you want to make visible; the selected entities will be made visible after you exit the **Make Visible** tool.

Visibility States. This tool is used to create an additional visibility state. On choosing this tool, the **Visibility States** dialog box will be displayed, as shown in Figure 18-34. By default, VisibilityState0 is available and listed in the **Visibility states** area. To create a new visibility state, choose the **New** button; the **New Visibility State** dialog box will be displayed, as shown in Figure 18-35. The options of this dialog box are discussed next.

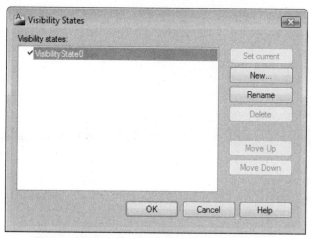

*Figure 18-34 The **Visibility States** dialog box*

*Figure 18-35 The **New Visibility State** dialog box*

Visibility state name. Specify the name of the new visibility state in this edit box.

Visibility options for new states. The options in this area are used to control the visibility of the new visibility state. By default, the **Leave visibility of existing objects unchanged in new state** radio button is selected, implying that the visibility of entities in the new state will be similar to the visibility of the objects in the current state. If you select the **Hide all existing objects in new state** radio button, all entities will be hidden in the new visibility state. If you select the **Show all existing objects in new state** radio button, all objects will be visible in the new visibility state.

The current visibility state is displayed in the drop-down list in the **Visibility** panel of the **Block Editor** tab. To make any other visibility state as the current visibility state, select it in the drop-down list. To rename an existing visibility state, select its name from the **Visibility states** list box available in the **Visibility states** dialog box and choose the **Rename** button.

After defining the visibility states, exit the **Block Editor**. Now, you can hide or show the entities as per the requirements of the drawing. When you select the dynamic block for which the visibility state has been defined, the lookup grip will be displayed, which resembles an arrow head. If you click on the lookup grip, a shortcut menu will be displayed, listing all available visibility states. In the list, the current visibility state is displayed with a check mark on its left. In Figure 18-36, **Visibility State0** is the current visibility state. Figure 18-37 shows the drawing in which the **Visibility State1** has been made as the current visibility state and therefore, the table is hidden in it.

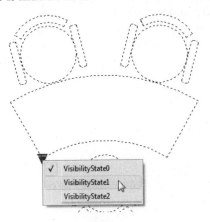

Figure 18-36 Shortcut menu displayed on clicking the lookup grip

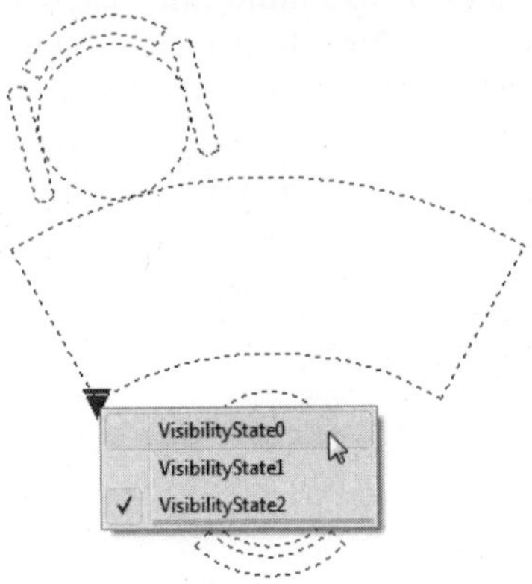

Figure 18-37 Changing the visibility state using the shortcut menu

Lookup Parameter

The **Lookup** parameter is added to assign a lookup action to the dynamic block. Using the lookup action, you can control the values of the parameters assigned to the dynamic block. To add this parameter, choose its tool from the **Block Authoring Palettes** and specify the location of the parameter.

Assigning the Lookup Action to the Lookup Parameter

To assign the lookup action to the lookup parameter, double-click on it; you will be prompted to specify the action location. On specifying it, the **Property Lookup Table** dialog box will be displayed, as shown in Figure 18-38.

To add a lookup property, choose the **Add Properties** tool; the **Add Parameter Properties** dialog box will be displayed, as shown in Figure 18-39. This dialog box displays the parameters assigned to the current dynamic block. Select single or multiple properties and choose **OK**; all properties will be added as rows in the **Input Properties** area of the **Properties Lookup Table** dialog box. Save the changes and close the **Block Editor**. Next, select the block; the lookup grips will be displayed. Select a lookup grip; a shortcut menu listing all property values will be displayed. Select the required values and perform the action.

Base Point Parameter

The **Base Point** parameter is used to create a base point for the block, which will be used as the insertion base point while inserting this block. Note that the blocks that are inserted are also modified, depending on the new base point that you specify using this parameter. To add this parameter, choose the **Base Point** tool from the **Block Authoring Palettes** and select the point that you want to use as the base point.

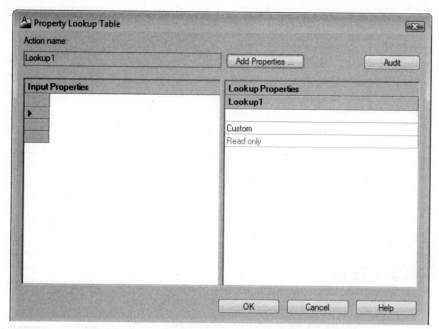

Figure 18-38 The *Property Lookup Table* dialog box

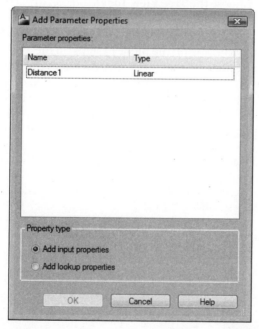

Figure 18-39 The *Add Parameter Properties* dialog box

ADDING PARAMETER AND ACTION SIMULTANEOUSLY USING PARAMETER SETS

In the previous section, you learned how to add a parameter and then assign an action to it. AutoCAD also allows you to add parameters and actions simultaneously using the **Parameters Sets** tab of the **Block Authoring Palettes**, see Figure 18-40. Various parameter-action combinations that can be defined for a dynamic block are available in this tab as individual sets. However, note that the entities that will be affected by the specified action are not automatically defined using the parameter sets. You need to double-click on the action to select the objects. For example, instead of first applying the **Point** parameter and then adding the move action to it, you can directly apply the **Point Move** set. After you specify the parameter and the action, double-click on **Move** in the authoring area and select the entities of the dynamic block that will be moved using this action.

Tip
*To delete any action applied to a parameter, right-click on the action bar displayed on the parameter. Next, choose the **Delete** option from the shortcut menu displayed. You can also create a new selection set or modify the existing selection set by choosing the **Action Selection Set** option. To rename an action, choose the **Rename Action** option from the shortcut menu.*

INSERTING BLOCKS USING THE DESIGNCENTER

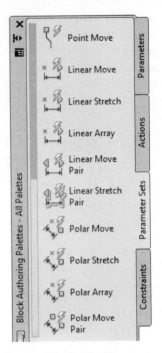

*Figure 18-40 The **Parameter Sets** tab of the **Block Authoring Palettes***

You can use the **DesignCenter** to locate, preview, copy, or insert blocks or existing drawings into the current drawing. To insert a block, choose the **DesignCenter** button from the **Palettes** panel in the **View** tab to display the **DesignCenter** window. By default, the tree pane is displayed. Choose the **Tree View Toggle** button to display the tree pane on the left side, if it is not already displayed. Expand **Computer** to display *C:/Program Files/AutoCAD 2012/ Sample/DesignCenter* folder by clicking on the plus (+) signs on the left of the respective folders. Click on the plus sign adjacent to the folder to display its contents. Select a drawing file you wish to use to insert blocks from and then click on the plus sign adjacent to the drawing again. All the icons depicting the components such as blocks, dimension styles, layers, linetypes, text styles, and so on, in the selected drawing are displayed. Select **Blocks** by clicking on it in the **tree pane**; the blocks in the drawing are displayed in the palette. Select the block you wish to insert and drag and drop it into the current drawing. Later, you can move it in the drawing to the desired location. You can also right-click on the block name to display the shortcut menu, as shown in Figure 18-41.

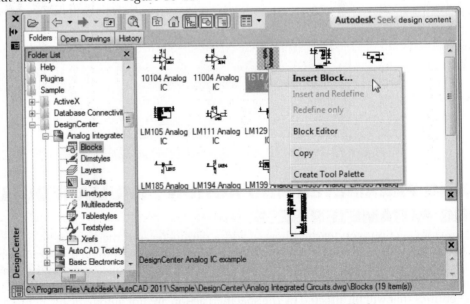

*Figure 18-41 Using the **DesignCenter** to insert blocks*

Choose **Insert Block** from the shortcut menu; the **Insert** dialog box is displayed. Here, you can specify the insertion point, scale, and rotation angle in the respective edit boxes. On selecting the **Specify On-screen** check box in the **Insert** dialog box, you will be allowed to specify these parameters on the screen. The selected block is inserted into the current drawing. You can also choose **Copy** from the shortcut menu; the selected block will be copied to the clipboard. Now, you can right-click in the drawing area to display a shortcut menu and choose **Paste**; the preview of the selected block will be attached to the cursor and as you move the cursor, the block also moves with it. You can now select a point to paste the block. The advantage of this option over drag and drop is that you can select an insertion point at the time of pasting. In the case of a drag and drop operation, you need to move the inserted block to a specific point after it has been dropped on the screen.

Note
*For a detailed explanation of the **DesignCenter**, see Chapter 6, Editing Sketched Objects-II.*

USING TOOL PALETTES TO INSERT BLOCKS

You can use the **Tool Palettes** (see Figure 18-42) to insert predefined blocks in the current drawing. The **Tool Palettes** window has many tabs. You can view the complete list of tabs by clicking on stacks of the tabs available at the end of the list of tabs. In this chapter, you will learn how to insert blocks using the tabs of the **Tool Palettes**.

Inserting Blocks in the Drawing

AutoCAD provides two methods to insert blocks from the **Tool Palettes**; **Drag and Drop** method and **Select and Place** method. Both these methods of inserting blocks using the **Tool Palettes** are discussed next.

Drag and Drop Method

To insert blocks from the **Tool Palettes** in the drawing, using this method, move the cursor over the desired predefined block in the **Tool Palettes**. You will notice that as you move the cursor over the block, the block icon gets converted into a 3D icon. Also, a tooltip is displayed that shows the name and description of the block. Press and hold the left mouse button and drag the cursor to the drawing area. Release the left mouse button, and you will notice that the selected block is inserted in the drawing. You may need to modify the drawing display area to view the block. Remember that when a block is inserted from the **Tool Palettes** using the drag and drop

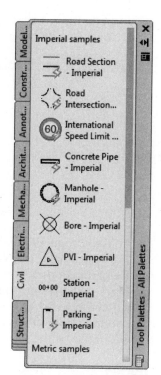

Figure 18-42 The Tool Palettes

method, you are not prompted to specify its rotation angle or scale. The blocks are automatically inserted with their default scale factor and rotation angle.

Select and Place Method

You can also insert the desired block in the drawings using the select and place method. To insert the block using this method, move the cursor over the desired block in the **Tool Palettes**; the block icon is changed to a 3D icon. Press the left mouse button; the selected block is attached to the cursor and the **Specify insertion point or [Basepoint/Scale/X/Y/Z/Rotate]** prompt is displayed. Modify any parameter using the prompt sequence and then move the cursor to the required location in the drawing area. Click on the left mouse button; the selected block is inserted at the specified location.

Modifying Properties of the Blocks in the Tool Palettes

To modify the properties of a block, move the cursor over it in the **Tool Palettes** and right-click to display the shortcut menu. Using the options in this shortcut menu, you can cut or copy the desired block available in one tab of **Tool Palettes** and paste it on the other tab. You can also delete and rename the selected block using the **Delete** and **Rename** options, respectively. To modify the properties of the block, choose **Properties** from the shortcut menu. The **Tool Properties** dialog box is displayed, as shown in Figure 18-43.

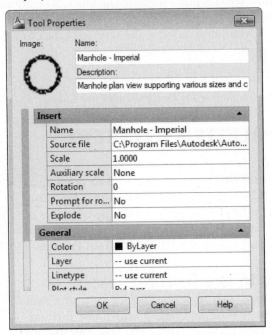

*Figure 18-43 The **Tool Properties** dialog box*

In the **Tool Properties** dialog box, the name of the selected block is displayed in the **Name** edit box. You can rename the block by entering a new name in the **Name** edit box. The **Image** area on the left of the **Name** edit box displays the image of the selected block. If you enter a description of the block in the **Description** text box, it is stored with the block definition in the **Tool Palettes**. Now, when you move the cursor over the block in **Tool Palettes** and pause for a second, the description along with its name appears in the tooltip. The **Tool Properties** dialog box displays the properties of the selected block under the following categories.

Insert

In this category, you can specify the insertion properties of the selected block such as its name, original location of the block file, scale, and rotation angle. The **Name** edit box specifies the name of the block. The **Source file** edit box displays the location of the file, in which the selected block is created. When you choose the [...] button in the **Source file** edit box, AutoCAD displays the location of the file in the **Select Linked Drawing** dialog box. The **Scale** edit box is used to specify the scale factor of the block. The block will be inserted in the drawing according to the scale factor specified in this edit box. You can enter the angle of rotation in the **Rotation** edit box. The **Explode** edit box is used to specify whether the block will be exploded while inserting or will be inserted as a single entity.

General

In this category, you can specify the general properties of the block such as the **Color**, **Layer**, **Linetype**, **Plot style**, and **Lineweight** for the selected block. The properties of a particular field can be modified from the drop-down list available on selecting that field.

Custom

In this category, you can specify the custom properties of the block, such as the dimensions and related parameters.

ADDING BLOCKS IN TOOL PALETTES

By default, the **Tool Palettes** window displays the predefined blocks in AutoCAD. You can also add the desired block and the drawing file to the **Tool Palettes** window. This is done using the **DesignCenter**. AutoCAD provides two methods for adding blocks from the **DesignCenter** to the **Tool Palettes**; **Drag and Drop** method and **Shortcut menu**. These two methods are discussed next.

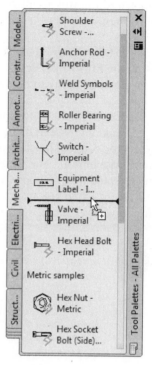

Drag and Drop Method

To add blocks from the **DesignCenter** in the **Tool Palettes**, move the cursor over the desired block in the **DesignCenter**. Press and hold the left mouse button on the block and drag the cursor to the **Tool Palettes** window. You will notice that a box with a + sign is attached to the cursor and a black line appears on the **Tool Palettes** window, as shown in Figure 18-44. If you move the cursor up and down in the **Tool Palettes**, the black line also moves between the two consecutive blocks. This line is used to define the position of the block to be inserted in the **Tool Palettes** window. Release the left mouse button and you will notice that the selected block is added to the location specified by the black line in the **Tool Palettes**.

Figure 18-44 Specifying the location for inserting a block in the Tool Palettes window

Shortcut Menu

You can also add the desired block from the **DesignCenter** to the **Tool Palettes** using the shortcut menu. To add the block, move the cursor over the desired block in the **DesignCenter** and right-click on it to display a shortcut menu. Choose **Create Tool Palette** from it. You will notice that a new tab with the name **New Palette** is added to the **Tool Palettes**. And, the block is added in the new tab of the **Tool Palettes**. Also, a text box appears that displays the current name of the tab. You can change its name by entering a new one in this text box.

You can also add a number of blocks in a drawing to the **Tool Palettes** using the following two methods:

Select the **Blocks** folder of any drawing in the **DesignCenter** and right-click in the **Tree View** area of the **DesignCenter**; a shortcut menu will be displayed. Choose **Create Tool Palette** from the shortcut menu; a new tab will be added to the **Tool Palettes** with the same name as that of the drawing file selected in the **DesignCenter**. This new tab contains all blocks which were in the folder that you selected from the **DesignCenter**.

You can also add all the blocks in a drawing by right-clicking on the drawing in the **tree view** of the **DesignCenter**; a shortcut menu is displayed. Choose **Create Tool Palette** from it; you will notice that a new tab is added to the **Tool Palettes**, which contains all the blocks that were available in the selected drawing. You will also notice that the new tab has the same name as that of the selected drawing.

MODIFYING EXISTING BLOCKS IN THE TOOL PALETTES

If you modify an existing block that was added to the **Tool Palettes** and then insert it using the **Tool Palettes** in the same or a new drawing, you will notice that the modified block is inserted and not the original block. However, if you insert the modified block from the **Tool Palettes** in the drawing in which the original block was already inserted, AutoCAD inserts the original block and not the modified one. This is because the file already has a block of the same name in its memory.

To insert the modified block, you first need to delete the original block from the current drawing, using the **Erase** tool. Next, you need to delete the block from the memory of the current drawing. The unused block can be deleted from the memory of the current drawing, using the **PURGE** command. To invoke this command, enter **PURGE** at the command window. The **Purge** dialog box is displayed. Choose the (+) sign located on the left of **Blocks** in the tree view available in the **Items not used in drawing** area. You will notice that a list of blocks in the drawing is shown. Select the original block to be deleted from the memory of the current drawing and then choose the **Purge** button. The **Confirm Purge** dialog box is displayed, which confirms the purging of the selected item. Choose **Yes** in it and then choose the **Close** button to exit the **Purge** dialog box. Next, when you insert the block using the **Tool Palettes**, the modified block is inserted in the drawing.

Note
*You will learn more about the **PURGE** command in Chapter 22, Grouping and Advanced Editing of Sketched Objects. Deleting unused blocks using the command line is, however, discussed later in this chapter.*

*If you create a block with the name that is defined in the **Tool Palettes**, the block in the **Tool Palettes** is redefined in the current drawing. However, when you open a new drawing and insert the block using the **Tool Palettes**, the original block will be inserted.*

LAYERS, COLORS, LINETYPES, AND LINEWEIGHTS FOR BLOCKS

A block possesses the properties of the layer on which it is drawn. The block may be composed of objects drawn on several different layers, with different colors, linetypes, and lineweights. All this information is preserved in the block. At the time of insertion, each object in the block is drawn on its original layer with the original linetype, lineweight, and color, irrespective of the current drawing layer, object color, object linetype, and object lineweights. You may want all instances of a block to have identical layers, linetype properties, lineweight, and color. This can be achieved by allocating all the properties explicitly to the objects forming the block. On the other hand, if you want the linetype and color of each instance of a block to be set according to the linetype and color of the layer on which it is inserted, draw all the objects forming the block on layer 0 and set the color, lineweight, and linetype to **By Layer**. Objects with a **By Layer** color, linetype, and lineweight can have their colors, linetypes, and lineweights changed after insertion by changing the layer settings. If you want the linetype, lineweight, and color of each instance of a block to be set according to the current explicit linetype, lineweight, and color at the time of insertion, set the color, lineweight, and linetype of its objects to **By Block**. You can use the **Properties** palette to change some of the characteristics associated with a block (such as layer).

Note
The block is inserted on the current layer, but the objects comprising the block are drawn on the layers on which they were drawn when the block was being defined.

For example, assume block B1 includes a square and a triangle that were originally drawn on layer X and layer Y, respectively. Let the color assigned to the layer X be red and to layer Y be green. Also, let the linetype assigned to layer X be continuous and for layer Y be hidden. Now, if we insert B1 on layer L1 with color yellow and linetype dot, block B1 will be on layer L1, but the square will be drawn on layer X with color red and linetype continuous. The triangle will be drawn on layer Y with the color green and the linetype hidden.

The **By Layer** option instructs AutoCAD to assign objects within the block the color and linetype of the layers on which they were created. There are three exceptions:

1. If objects are drawn on a special layer (layer 0), they are inserted on the current layer. These objects assume the characteristics of the current layer (the layer on which the block is inserted) at the time of insertion, and can be modified after insertion by changing that layer's settings.

2. Objects created with the special color **By Block** are generated with the color that is current at the time of insertion of the block. This color may be explicit or **By Layer**. You are thus allowed to construct blocks that assume the current object color.

3. Objects created with the special linetype **By Block** are generated with the linetype that is prevalent at the time the block is inserted. Blocks are thus constructed with the current object linetype, which may be **By Layer** or explicit.

Note

If a block is created on a layer that is frozen at the time of insertion, it is not shown on the screen.

Tip

*If you provide drawing files to others for their use, using only **BYLAYER** settings provide the greatest compatibility with varying office standards for layer/color/linetype/lineweight. This is because they can be changed more easily after insertion.*

NESTING OF BLOCKS

The concept of having one block within another block is known as the **nesting of blocks**. For example, you can insert several blocks by selecting them, and then, with the **Create Block** tool, create another block. Similarly, if you use the **Insert** tool to insert a drawing, containing several blocks, into the current drawing, it creates a block containing nested blocks in the current drawing. There is no limit to the degree of nesting. The only limitation in nesting of blocks is that blocks that reference themselves cannot be inserted. The nested blocks must have different block names. Nesting of blocks affects layers, colors, and linetypes. The general rule is given next.

If an inner block has objects on layer 0, or objects with linetype or color **By Block**, these objects may be said to behave like fluids. They "float up" through the nested block structure until they find an outer block with fixed color, layer, or linetype. These objects then assume the characteristics of the fixed layer. If a fixed layer is not found in the outer blocks, then the objects with color or linetype **By Block** are formed; that is, they assume the color white and the linetype CONTINUOUS.

EXAMPLE 2   *Nested Blocks*

To clarify the concept of nested blocks, let us do the following example.

1. Draw a rectangle on layer 0, and form a block with the named X using this block.

2. Change the current layer to OBJ, set its color to red, and linetype to hidden.
3. Draw a circle on OBJ layer.
4. Insert the block X in the OBJ layer.
5. Combine the circle with the block X (rectangle) to form a block Y.
6. Now, insert block Y in any layer (say, layer CEN) with the color green and linetype continuous.

You will notice that block Y is generated in red and its linetype is hidden. However, the block X, which is nested in block Y and created on layer 0, is not generated in the color (green) and linetype (continuous) of the layer CEN. This is because, the object (rectangle) on layer 0 floated up through the nested block structure and assumed the color and linetype of the first outer block (Y) with a fixed color (red), layer (OBJ), and linetype (hidden). If both the blocks (X and Y) were on layer 0, the objects in the block Y would assume the color and linetype of the layer on which the block was inserted.

EXAMPLE 3   *Nested Block*

1. Change the color of layer 0 to red.
2. Draw a circle with color **By Block** and then form its block, B1. It appears white because its color is set to **By Block** (Figure 18-45).
3. Set the color to **By Layer** and draw a square. The color of the square is red.
4. Insert block B1. Notice that the block B1 (circle) assumes red color.
5. Create another block B2 consisting of the Block B1 (circle) and square.
6. Create a layer L1 with green color and hidden linetype. Make it current. Insert block B2 in layer L1.
7. Explode block B2. Notice the change.
8. Explode block B1, circle. You will notice that the color of the circle changes to white because it was drawn with the color set to **By Block** .

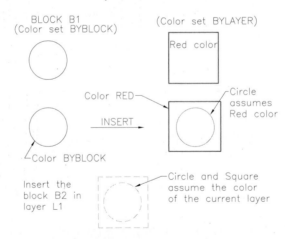

Figure 18-45 Blocks versus layers and colors

EXAMPLE 4   *Nested Block*

Part A
1. Draw a unit square on layer 0 and make it a block named B1.
2. Draw a circle of radius 0.5 and change it into a block named B2.
3. Insert block B1 into the drawing with an X scale factor of 3 and a Y scale factor of 4.
4. Now, insert the block B2 in the drawing and position it at the top of B1.
5. Make a block of the entire drawing and name it Plate.
6. Insert the block **Plate** in the current layer.
7. Create a new layer with different colors and linetypes and insert blocks B1, B2, and Plate.

Keep in mind the layers on which the individual blocks and the inserted block were made.

Part B
Try nesting the blocks drawn on different layers and with different linetypes.

Part C
Change the layers and colors of the different blocks you have drawn so far.

INSERTING MULTIPLE BLOCKS

Command: MINSERT

The **MINSERT** (multiple insert) command is used for the multiple insertion of a block. This command comprises features of the **Insert** and **Array** tools. **MINSERT** is similar to the **-INSERT** command because it is used to insert blocks at the command level. The difference between these two commands is that the **MINSERT** command inserts more than one copy of the block in a rectangular fashion, similar to the **Array** tool. Also, blocks inserted by using **MINSERT** cannot be exploded. Preceding the block name with an asterisk does not explode it on insertion. With the **MINSERT** command, only one block reference is created. But, in addition to the standard features of a block definition (insert point, X/Y scaling, rotation angle, and so on), this block has repeated row and column counts. In this manner, using this command saves time and disk space. The prompt sequence is very similar to that of the **-INSERT** and **-ARRAY** commands. Consider an example in which you have to array a block named BENCH, with the following specifications: number of rows = 4, number of columns = 3, unit cell distance between rows = 0.50, and unit cell distance between columns = 2.0. The prompt sequence for creating the arrangement of benches in an auditorium (Figure 18-46) is given next.

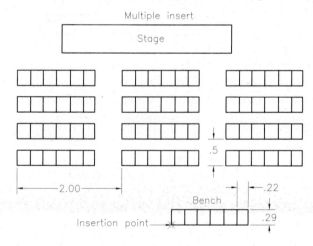

Figure 18-46 *Arrangement of benches in an auditorium*

Command: **MINSERT**
Enter block name or [?] <current>: **BENCH**
Units: Inches   Conversion: 1.0000
Specify insertion point or [Basepoint/Scale/Rotate]: *Press ENTER or specify a base point and press ENTER*
Specify scale factor <1>: *Press ENTER or specify a value and then press ENTER*
Specify rotation angle <0>: *Press ENTER or specify a value and then press ENTER*
Enter number of rows (---)<1>: **4**
Enter number of columns (|||) <1>: **3**
Enter distance between rows or specify unit cell (---): **0.50**
Specify distance between columns (|||): **2.0**

Note

*You cannot use an annotative block for the **MINSERT** command.*

All options of the **MINSERT** command are similar to the ones discussed earlier in the **-INSERT** command. The **?** option of the **Enter block name or [?]:** prompt lists the names of the blocks. When you specify the insertion point, AutoCAD prompts you to specify the scale factor. Once you specify the scale factor, you will be prompted to specify the rotation angle. If you need to insert the block at an angle, specify the angle and press ENTER. Next, you need to specify the number of rows, columns, distance between the rows, and distance between the columns in succession. The array of blocks will be created.

The prompts pertaining to the number of rows and columns require a positive nonzero integer. For the number of rows exceeding 1, you are prompted to specify the distance between the rows. Similarly, if the number of columns exceeds 1, you need to specify the distance between columns. The distance between rows and columns can be a positive or negative value. While specifying the distance between the rows and columns, you can also specify a unit cell, by selecting two points on the screen that form a box whose width and height represent the distance between columns and rows respectively. While inserting a block, if you specify the block rotation, each copy of the inserted block is rotated through that angle. If you specify the rotation angle in the **MINSERT** command, the whole **MINSERT** block (array) is rotated by the specified rotation angle. If both the block and the **MINSERT** block (array) are rotated through the same angle, the block does not seem rotated in the **MINSERT** pattern. This is the same as an array with the rotation angle equal to 0-degree. If you want to rotate individual blocks within the array, you should first generate a rotated block and then **MINSERT** it.

Note

*In the array of blocks generated with the use of the **MINSERT** command, you can alter the number of rows or columns or the spacing between them. This is done using the **Misc** rollout of the **Properties** palette.*

*When you right-click while specifying the insertion point at the **Specify insertion point** prompt, a shortcut menu is displayed where all the command line options are available.*

EXERCISE 5   MINSERT

1. Create a triangle with each side equal to 3 units.
2. Generate a block of the triangle.
3. Use the **MINSERT** command to insert the block to create a 3 by 3 array. The distance between the rows is 1 unit, and the distance between columns is 2 units.
4. Again, use the **MINSERT** command to insert the block to create a 3 by 3 array. The distance between the rows is 2 units and between the columns is 3 units. The array is rotated through 15-degree.

CREATING DRAWING FILES USING THE WRITE BLOCK DIALOG BOX

Command: WBLOCK

The blocks are symbols created by the **BLOCK** command and can be used only in the drawing, in which they were created. This is a shortcoming because you may need to use a particular block in different drawings. The **WBLOCK** command is used to export symbols by writing them to new drawing files that can then be inserted in any drawing. With the **WBLOCK** command, you can create a drawing file (*.dwg* extension) of the specified blocks, selected objects in the current drawing, or the entire drawing. All the used named objects (linetypes, layers, styles, and system variables) of the current drawing are inherited by the new drawing created with the **WBLOCK** command. This block can then be inserted in any drawing.

When you invoke the **WBLOCK** command, the **Write Block** dialog box is displayed, as shown in Figure 18-47. This dialog box converts the blocks into drawing files and also saves objects as drawing files. You can also save the entire current drawing as a new drawing file.

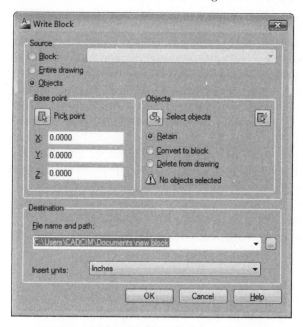

*Figure 18-47 The **Write Block** dialog box*

The **Write Block** dialog box has two main areas: **Source** and **Destination**. The **Source** area allows you to select objects and blocks, specify insertion base points and convert them into drawing files. In this area of the dialog box, different default settings are displayed, depending upon the selection you make. By default, the **Objects** radio button is selected. In the **Destination** area, the **File name and path** edit box displays *new block.dwg* as the new file name and its location. The **Block** drop-down list is also not available. Now, you can select objects in a drawing and save them as a wblock, and can enter a name and a path for the file. Sometimes the current drawing consists of blocks. To save a block as a wblock, you can select the **Block** radio button. When the **Block** radio button is selected, the **Block** drop-down list is available. The **Block** drop-down list displays all the block names in the current drawing and you can select a block name to convert it into a wblock. The **Base point** and **Objects** areas are not available, since the insertion points and objects have already been saved with the block definition. Also, you will notice that in the **Destination** area, by default, the **File name and path** edit box displays the name of the selected block. This means that you can keep the name of the wblock the same as the selected block or you can change it. Selecting the **Entire drawing** radio button, selects the current drawing as a block and saves it as a new file. When you use this option, the **Base point** and **Objects** areas are not available.

The **Base point** area allows you to specify the base point of a wblock, which is used as an insertion point. You can either enter values in the **X**, **Y**, and **Z** edit boxes or choose the **Pick point** button to select it on the screen. The default value is 0, 0, 0. The **Objects** area allows you to select objects to save as a file. You can use the **Select Objects** button to select objects or use the **QuickSelect** button to set parameters in the **Quick Select** dialog box to select objects in the current drawing. The number of objects selected is displayed at the bottom of the **Objects** area. If the **Retain** radio button is selected in the **Objects** area, the selected objects in the current drawing are kept as such, after they have been saved as a new file. If the **Convert to block** radio button is selected, the selected objects in the current drawing will be converted into a block with the same name as the new file, after being saved as a new file. Selecting the **Delete from drawing** radio button deletes the selected objects from the current drawing after they have been saved as a file.

Note

*Both the **Base point** and **Objects** areas are available in the **Write Block** dialog box only when the **Objects** radio button is selected in the **Source** area of the dialog box.*

The **Destination** area sets the file name, location, and units of the new file, in which the selected objects are saved. In the **File name and path** edit box, you can specify the file name and the path of the block or the selected objects. You can choose the [**...**] button to display the **Browse for Drawing File** dialog box, where you can specify the path where the new file will be saved. From the **Insert Units** drop-down list, you can select the units, the new file will use when inserted as a block. The settings for units are stored in the **INSUNITS** system variable and the default option Inches has a value of 1. On specifying the required information in the dialog box, choose **OK**. The objects or the block is saved as a new file in the path specified by you. A **WBLOCK Preview** window with the new file contents is displayed. This preview image is stored and displayed in the DesignCenter, when using it to insert drawings and blocks.

Note

Whenever a drawing is inserted into the current drawing, it acts as a single object. It cannot be edited unless exploded.

Tip

*Using the **Entire drawing** radio button in the **Write Block** dialog box to save the current drawing as a new drawing is a good way to reduce the drawing file size. This is because all unused blocks, layers, linetypes, text styles, dimension styles, multiline styles, shapes, and so on are removed from the drawing. The new drawing does not contain any information that is no longer needed. This **Entire drawing** option is faster than the **-PURGE** command (which also removes unused named objects from a drawing file) and can be used whenever you have completed a drawing and want to save it. The **-PURGE** command has been discussed later in this chapter.*

DEFINING THE INSERTION BASE POINT

Ribbon: Insert > Block Definition > Set Base Point or Home > Block > Set Base Point
Command: BASE

The **Set Base Point** tool lets you set the insertion base point for a drawing, just as you set the base insertion point using the **Create Block** tool. This base point is defined so that when you insert the drawing into some other drawings, the specified base point is placed on the insertion point. By default, the base point is at the origin (0,0,0). When a drawing is inserted on a current layer, it does not inherit the color, linetype, or thickness properties of the current layer. On invoking the **Set Base Point** tool, following prompt sequence is displayed.

Enter base point <0.0000,0.0000,0.0000>: *Specify a base point or press ENTER to accept the default value.*

Tip

*You can insert a drawing that you want to refer to for checking dimensions or certain features into the current drawing for reference. Later on, you can use the **UNDO** command to delete the inserted drawing. This way you can save stationery and time spent in printing.*

Note

The inserted drawings become a part of the current drawing and increase the file size. Also, when the current drawing is modified, the inserted drawing does not get updated. Hence, it is better to xref a drawing into the current drawing instead. External references are discussed in Chapter 20, Understanding External References.

EXERCISE 6	WBLOCK

1. Create a drawing file named CHAIR using the **WBLOCK** command. Make a listing of your *.dwg* files and make sure that *CHAIR.dwg* is listed. Quit the drawing editor.
2. Begin a new drawing and insert the drawing file into it. Save the drawing.

EDITING BLOCKS

Ribbon: Insert > Reference > Edit Reference	**Command:** REFEDIT
Toolbar: Refedit > Edit reference In place	

You can edit blocks by breaking them into parts and then making modifications and redefining them or by editing them in place. Both these methods have been discussed here.

Note

*The blocks inserted using the **MINSERT** command can be refedited. However, the non-uniformly scaled blocks cannot be refedited.*

Editing Blocks in Place

You can edit blocks in the current drawing by using the **Edit Reference** tool, referred to as the in-place reference editing. This feature of AutoCAD allows you to make minor changes to blocks, wblocks, or drawings that have been inserted in the current drawing without breaking them up into component parts or opening the original drawing and redefining them. This command saves valuable time of going back and forth between drawings and redefining blocks.

Note

*You cannot use the **Edit Reference** tool on blocks that have been inserted using the **MINSERT** command. This tool is discussed in reference to xrefs in Chapter 20, Understanding External References.*

You can invoke the **Edit Reference** tool from the **Reference** panel in the **Insert** tab, as shown in Figure 18-48, or by entering **REFEDIT** at the Command prompt. On doing so, the **Reference Edit** dialog box will be displayed. Alternatively, select the block, right-click on any block reference and then choose the **Edit Block In-Place** option from the shortcut menu displayed. When you invoke the **REFEDIT** command, the following prompt is displayed.

Select reference: *Select the block to edit.*

Once you have selected the block reference to be edited, the **Reference Edit** dialog box is displayed. The **Reference Edit** dialog box has two tabs: **Identify Reference** and **Settings**. The description of these tabs is as follows.

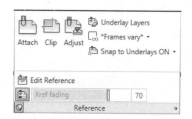

Figure 18-48 *The **Edit Reference** tool in the **Reference** panel of the **Insert** tab*

Identify Reference Tab

The **Identify Reference** tab, as shown in Figure 18-49, provides information for identifying the reference edit, selecting, and editing the references.

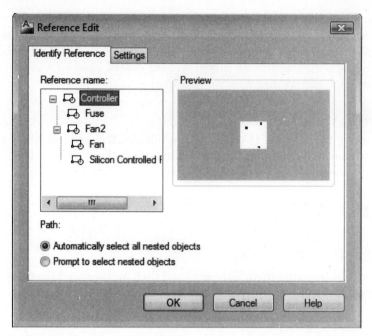

*Figure 18-49 The **Identify Reference** tab of the **Reference Edit** dialog box*

Reference name. The **Reference name** list box displays the name of the selected reference. A plus sign (+) next to the name implies that the selected block reference contains nested references. When you click on the plus sign, the nested reference is displayed in a tree view. If a block does not contain any nested blocks, this plus sign is not displayed next to the reference name in the list box. In the figure, the block Valve-2 contains a nested block Valve. An image of the selected block is displayed in the **Preview** box. If you want to see a preview of the nested block, select its name in the list box; the corresponding preview image is displayed in the **Preview** area.

Path. It displays the location of the selected reference. Note that if the selected reference is a block, no path is displayed.

The **Automatically select all nested objects** radio button is selected by default and allows you to automatically include all the selected objects in the reference editing session. The selected objects also include the nested objects.

The **Prompt to select nested objects** radio button, when selected, allows you to individually select the nested objects in the reference editing session. If this radio button is selected and you choose the **OK** button, the **Reference Edit** dialog box is closed. You are prompted to select the nested objects in the reference that you want to edit.

Settings Tab
The **Settings** tab, as shown in Figure 18-50, provides options for editing references. The **Create unique layer, style, and block names** check box is selected by default and allows you to control the layer and symbol names of the objects that are extracted from the selected reference. If you clear this check box, the names of the layers remain the same as in the reference drawing. Similarly, the **Display attribute definitions for editing** check box, when selected, allows you to edit attributes and attribute definitions associated with the block references. The modified attribute definitions are effective for future insertions, and no changes are made in the current

insertions. Attributes are explained in detail in the next chapter. The **Lock objects not in working set** check box, when selected, allows you to lock all the objects that are not in the working set.

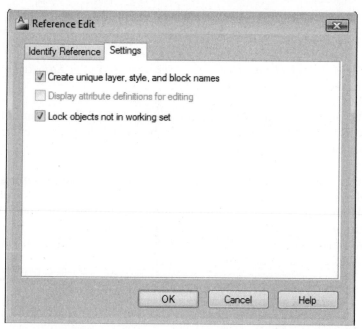

*Figure 18-50 The **Settings** tab of the **Reference Edit** dialog box*

After selecting the required block to edit, choose **OK** to exit the dialog box. If you have selected the **Prompt to select nested objects** radio button, the following prompt sequence will be displayed:

> Select nested objects: *Select objects within the block that you want to modify.*
> n entities added
> Select nested objects: Enter
> n items selected
> Use **REFCLOSE** or the **Refedit** toolbar to end reference editing session.

Once you have selected the objects to be edited and pressed ENTER, the total number of selected objects will be displayed. Also, you will notice that the **Edit Reference** panel has been added to the **Ribbon** and the objects, except the selected ones have faded out. You can control the fading intensity of the objects by using the options from the **Fade Control** area of the **Display** tab in the **Options** dialog box, see Figure 18-51. In this area, you can use the slider bar to increase or decrease the fading intensity or you can enter a value in the edit box. You can also control the fading intensity of other objects at the time of editing by using the **XFADECTL** variable. By default, the fading intensity value is set at 50.

The objects that have been selected for editing are referred to as the working set. You can use any drawing and editing command to modify this working set. Sometimes, you may need to add an external object from the current drawing to the block. To do so, choose the **Add to Working Set** button in the **Edit Reference** panel, see Figure 18-52. Next, select the objects to be added to the working set and press ENTER; these objects will then get deleted from the current drawing and added to the current working set. Similarly, to remove objects from a working set, choose the **Remove from Working set** button from the **Edit Reference** panel. You are prompted to select the objects from the working set that you want to remove. On completing the selection, press ENTER. The objects that have been removed from the working set appear faded and get

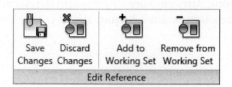

Figure 18-51 The **Fade control** *area of the* **Display** *tab in the* **Options** *dialog box*

Figure 18-52 The **Edit Reference** *panel*

added back to the current drawing. You can also use the **REFSET** command to add or remove objects from a working set. But this command can be used only when a reference has already been selected using the **REFEDIT** command.

After adding or removing objects from the existing set, choose the **Save Changes** tool in the **Edit Reference** panel to save the changes; the **AutoCAD** information box will be displayed informing that all reference edits will be saved. This information box also prompts you to specify whether you want to continue saving or cancel the command. Choose **OK** to save the modifications or choose **Cancel** to cancel the command. If you choose **OK**, the modifications are applied to the current block as well as to all future insertions.

Note
The Automatic Save feature of AutoCAD is disabled during reference editing.

To exit reference editing without saving any changes in the block reference, choose the **Discard Changes** button from the **Edit Reference** panel; the **AutoCAD** information box will be displayed, as shown in Figure 18-53, informing that all changes that have been made in the object will be discarded. To continue, choose **OK**, and to get back to reference editing, choose **Cancel**.

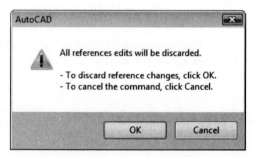

Figure 18-53 *AutoCAD information box displayed on choosing the* **Discard Changes** *or* **Save Changes** *button from the* **Edit Reference** *panel*

Note
You can also use the **-REFEDIT** *command to display prompts on the Command line.*

Exploding Blocks Using the XPLODE Command

Command: XPLODE

With the **XPLODE** command, you can explode a block or blocks into component objects and simultaneously control their properties such as layer, linetype, color, and lineweight. The scale factor of the object to be exploded should be equal. Note that if the scale factor of the objects to be exploded is not equal, you need to change the value of the **EXPLMODE** system variable to 1. Note that, if the **Allow exploding** check box in the **Behavior** area of the **Block Definition** dialog box was cleared while creating the block, you will not be able to explode the block. The command prompts for this command are as follows:

Command: **XPLODE**
Select objects to XPlode
Select objects: *Using any object selection method, select objects, and then press ENTER.*

On pressing ENTER, AutoCAD reports the total number of objects selected and also the number of objects that cannot be exploded. If you select multiple objects to explode, AutoCAD further prompts you to specify whether the changes in the properties of the component objects should be made individually or globally. The prompt is given next.

XPlode Individually/<Globally>: *Enter i, g, or press ENTER to accept the default option.*

If you enter **i** at the above prompt, AutoCAD will modify each object individually, one at a time. The next prompt is given below.

Enter an option [All/Color/LAyer/LType/LWeight/Inherit from parent block/Explode] <Explode>: *Select an option.*

The options available at the Command line are discussed next.

All

This option sets all the properties such as color, layer, linetype, and lineweight of the selected objects, after exploding them. AutoCAD prompts you to enter new color, linetype, lineweight, and layer name for the exploded component objects.

Color

This option sets the color of the exploded objects. The prompt is given next.

New color [Truecolor/COlorbook]<BYLAYER>: *Enter a color option or press ENTER.*

When you enter **BYLAYER**, the component objects take on the color of the exploded object's layer and when you enter **BYBLOCK**, they take on the color of the exploded object.

Layer

This option sets the layer of the exploded objects. The default option is inheriting the current layer. The command prompt is as follows.

Enter new layer name for exploded objects <current>: *Enter an existing layer name or press ENTER.*

LType

This option sets the linetype of the components of the exploded object. The command prompt is given next.

Enter new linetype name for exploded objects <ByLayer>: *Enter a linetype name or press ENTER to accept the default options.*

LWeight

This option sets the lineweight of the components of the exploded object. The command prompt is given next.

Enter new lineweight: *Enter a lineweight or press ENTER to accept the default option.*

Inherit from parent block

This option sets the properties of the component objects to that of the exploded parent object, provided the component objects are drawn on layer 0 and the color, lineweight, and linetype are **BYBLOCK**.

Explode

This option explodes the selected object exactly as in the **EXPLODE** command.

Selecting the **Globally** option, applies changes to all the selected objects at the same time. The options are similar to the ones discussed in the **Individually** option.

RENAMING BLOCKS

Command: RENAME

Blocks can be renamed using the **RENAME** command. To rename a block, enter **RENAME** at the Command prompt and press ENTER; the **Rename** dialog box will be displayed, see Figure 18-54. This dialog box allows you to modify the name of an existing block. In the **Rename** dialog box, the **Named Objects** list box displays the categories of object types that can be renamed, such as blocks, layers, dimension styles, linetypes, Multileader styles, material, table styles and text styles, UCSs, views, and viewports. You can rename all of these except layer 0 and continuous linetype. When you select **Blocks** from the **Named Objects** list, the **Items** list box displays all the block names in the current drawing. When you select a block name to rename from the **Items** list box, it is displayed in the **Old Name** edit box. Enter the new name to be assigned to the block in the **Rename To** edit box. Choosing the **Rename To** button applies the change in name to the old name. Choose **OK** to exit the dialog box. For example, to rename a block named Bracket to **Valve-3**, select **Bracket** from the **Items** list box; it is displayed in the **Old Name** edit box. Enter **Valve-3** in the **Rename To** edit box and choose the **Rename To** button; the **Valve-3** block appears in the **Items** list box. Now, choose **OK** to exit the dialog box.

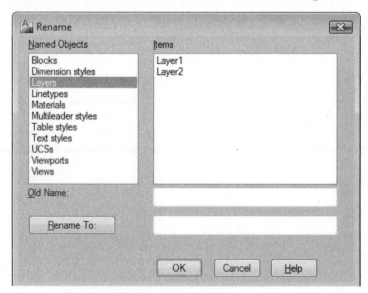

*Figure 18-54 The **Rename** dialog box*

Note

*The layer 0 and Continuous linetype cannot be renamed and therefore, they do not appear in the **Items** list box, when **Layers** and **Linetypes** are selected in the **Named Objects** list box.*

*To rename a block, enter **-RENAME or -REN** at the Command window.*

DELETING UNUSED BLOCKS

Sometimes, after completing a drawing, you may notice that the drawing contains several named objects, such as dimstyles, textstyles, layers, blocks, and so on that are not being used. Since these unused named objects unnecessarily occupy disk space, you may want to remove them. Unused blocks can be deleted with the **-PURGE** command. For example, to delete an unused block named Drawing2, the prompt sequence is given next.

Command: **-PURGE**
Enter type of unused objects to purge
[Blocks/Dimstyles/LAyers/LTypes/MAterial/MUltileaderstyles/Plotstyles/SHapes/textSTyles/
Mlinestyles/Tablestyles/Visualstyles/Regapps/Zero-length geometry/Empty Text Objects/
All]: **B**
Enter name(s) to purge <*>: **Drawing2**
Verify each name to be purged? [Yes/No] <Y>: Enter
Purge block "Drawing2"? <N>: **Y**

If there are no objects to be removed, AutoCAD displays a message "No unreferenced visual styles found.".

Note
*The unused blocks can also be deleted using the **PURGE** command discussed in Chapter 22 (Object Grouping and Editing Commands).*

*Preceding the name of the wblock with an (*) asterisk when entering the name of the wblock while using the **-WBLOCK** command has the same effect as while using the **-PURGE** command. Also, you can select the **Entire drawing** radio button in the **Write Block** dialog box when creating a block using the **WBLOCK** command to get the same effect. But, the **WBLOCK** command is faster and deletes the unused named objects automatically, while the **-PURGE** command allows you to select the type of named objects you want to delete, and it also gives you an option to verify the objects before the deletion occurs.*

APPLYING CONSTRAINTS TO BLOCKS

While creating assembly drawings, you may need some parts that have same geometry but different sizes. AutoCAD 2012 allows you to create the drawings of these parts by using geometric and dynamic constraints, if you could control these constraints in a block table. Note that you need to apply the geometric and dynamic constraints to blocks in the **Block Editor**.

To create the parts having same geometry but different sizes, draw the sketch of the parts and then convert them into a block. Next, invoke the **Block Editor** to edit the block. In the **Block Editor**, apply the necessary geometric and dimensional constraints and make the sketch a fully-defined sketch. Now, choose the **Block Table** tool from the **Dimensional** panel of the **Block Editor** tab; you will be prompted to specify the location of the parameter. Specify a suitable location; you will be prompted to specify the number of grips. If you enter **0**, no grip will be displayed in the drawing area. It is recommended to have atleast one grip. After specifying the number of grips, the **Block Properties Table** will be displayed, as shown in Figure 18-55.

In the **Block Properties Table**, choose the **Adds properties which appear as columns in the table** button to add all or some of the dimensional constraints added to the block. On choosing this button, the **Add Parameter Properties** dialog box will be displayed, as shown in Figure 18-56. Next, select the parameter properties to be added to the **Block Properties Table** by pressing and holding the CTRL key and then choose **OK**; the selected properties will be listed in the **Block Properties Table**. If you need to add any new parameter to the table, choose the **Creates a new user parameter and adds it to the table** button from the **Block Properties Table**; the **New Parameter** dialog box will be displayed, as shown in Figure 18-57. Specify the parameters and choose the **OK** button; the new parameter will be added to the **Block Properties Table**. You can also check the errors in the table. To do so, choose the **Audits the block property table for errors** button from the **Block Properties Table**. After specifying all parameters, choose the **OK** button from the **Block Properties Table**. Next, save the changes and exit the **Block Editor**. Now, if you select the block, the lookup grip will be displayed. Click on the grip; different values will be displayed, as shown in Figure 18-58. Next, choose the required value; the block will change accordingly. Note that if you have selected the **Block properties must match a row in the table** check box in the **Block Properties Table**, then the lookup grip will not be displayed and you cannot change the block sizes dynamically.

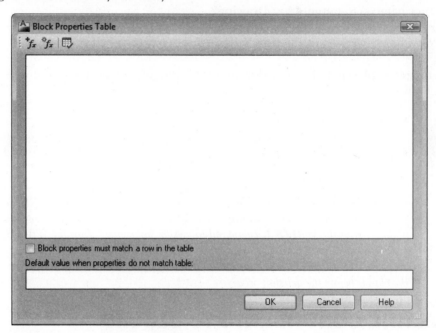

Figure 18-55 The **Block Properties Table**

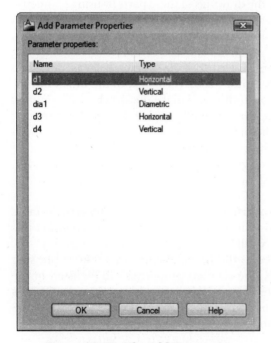

Figure 18-56 The **Add Parameter Properties** *dialog box*

Figure 18-57 The **New Parameter** *dialog box*

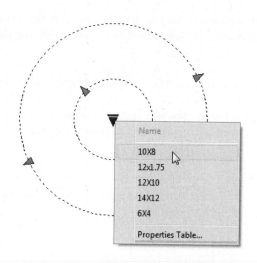

Figure 18-58 *Shortcut menu displayed on selecting the lookup grip*

EXAMPLE 5   *Constraints to Blocks*

In this example, you will create the sketch of a bolt, as shown in Figure 18-59. Then, you will convert it to a block and then apply constraints such that the bolts of diameter 6, 10, 12.5, 15, 18, 24, and 25 are created.

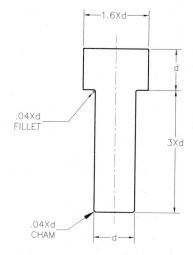

Figure 18-59 *Sketch for Example 5*

Note
The bolt shown in Figure 18-59 does not have any standard dimension or annotation. It is given for the purpose of example. However, you can follow the procedure given next to design a standard bolt.

1. Start a new file with the *acad.dwt* template in the **Drafting & Annotation** workspace. Draw the sketch assuming that the value of d=10 and satisfies other relations shown in Figure 18-59. Make sure that the **Infer Constraints** button is chosen in the Status bar and the endpoints are snapped so that it becomes a closed sketch, as shown in Figure 18-60.

2. Choose the **Create Block** tool from the **Block** panel in the **Home** tab and convert the sketch into a block. Name the block as **Bolt**.

3. Choose the **Block Editor** tool from the **Block** panel in the **Home** tab; the **Edit Block Definition** dialog box is displayed. Select **Bolt** and choose **OK**; the **Block Editor** environment is displayed.

4. Draw a vertical line passing through the midpoint of the bottom horizontal line.

5. Choose the **Construction Geometry** tool from the **Manage** panel of the **Block Editor** tab, select the vertical line drawn in the previous step, and press ENTER; you will be prompted to enter an option.

6. Enter **C** and press ENTER; the vertical line is converted to a construction line, as shown in Figure 18-61. Note that you can also create a centerline in the drawing mode but for the purpose of example, this method is given.

Figure 18-60 *The closed sketch of the bolt*   *Figure 18-61* *Block after converting the vertical line into a construction line*

7. Choose the **Equal** tool from the **Geometric** panel and apply the equal constraints between fillets, chamfers, vertical lines tangential to fillets, horizontal lines tangential to fillets, and vertical lines at the top.

8. Choose the **Symmetric** tool from the **Geometric** panel and apply the symmetric constraints between all entities on either side of the construction line.

9. Hide all geometric constraints by choosing the **Hide All** button from the **Geometric** panel of the **Block Editor**.

10. Apply the dimensional constraints by using the tools in the **Dimensional** panel of the **Block Editor**. Figure 18-62 shows the sketch with all dimensional constraints applied. Also, make sure that you snap the end points of the vertical line while applying constraints to the base diameter and to the length of the bolt.

You can observe the grips (arrow) for every dimensional constraint. You need to hide them.

11. Select all dimensional constraints and invoke the **Properties** palette. In the **Misc** rollout, change the **Number of Grips** to 0.

12. Choose the **Parameters Manager** button from the **Manage** panel; the **Parameters Manager** palette is displayed.

Next, you need to rename the constraints in the **Name** column of the **Properties Manager** palette, as shown in Figure 18-63.

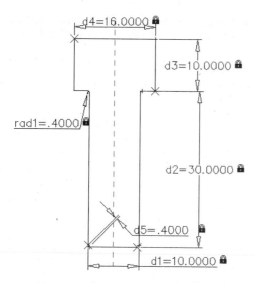

Figure 18-62 Sketch with all dimensional constraints

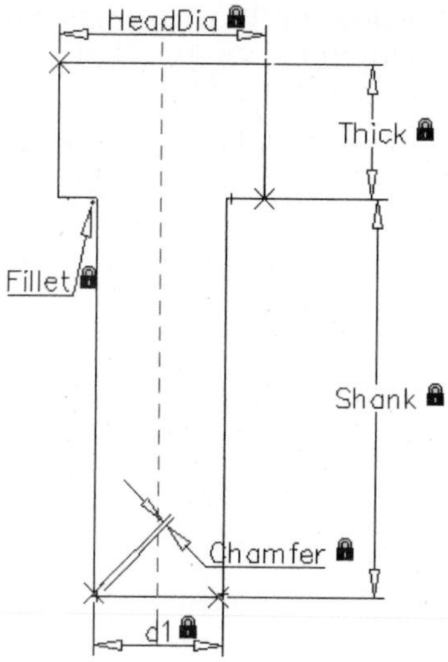

Figure 18-63 New name of the constraints

13. To rename a constraint, select it in the **Name** column and press the F2 key or double-click on the respective field; an edit box will be displayed. Change the name of the constraint.

14. Apply a new expression to each renamed constraint, as shown in Figure 18-63. The **Parameters Manager** palette after renaming the constraints and applying appropriate expressions is shown in Figure 18-64.

15. Choose the **Block Table** tool from the **Dimensional** panel of the **Block Editor**; you are prompted to specify the parameter location.

Search for parameter

Name	Expression	Value	Show
Dimensional Constraint Parameters			
Shank	3*d1	30.0000	Yes
d1	10.0000	10.0000	Yes
Thick	d1	10.0000	Yes
HeadDia	1.6*d1	16.0000	Yes
Chamfer	0.04*d1	0.4000	Yes
Fillet	0.04*d1	0.4000	Yes

All: 6 of 6 parameters displayed

Parameters Manager

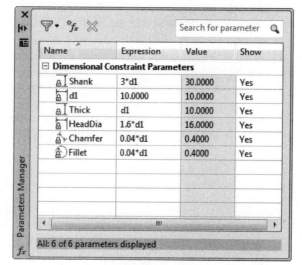

*Figure 18-64 The **Parameters Manager***

16. Select the midpoint of the horizontal line at the bottom; you are prompted to specify the number of grips.

17. Enter **1** and press ENTER; the **Block Properties Table** is displayed.

18. Choose the **Creates a new user parameter and adds it to the table** button; the **New Parameter** dialog box is displayed.

19. Enter **Size** in the **Name** edit box. Next, select the **String** option from the **Type** drop-down list and choose the **OK** button; a new column named **Size** is added to the **Block Properties Table**.

20. Choose the **Add properties which appear as columns in the table** button from the **Block Properties Table**; the **Add Parameter Properties** dialog box is displayed.

21. Select the **d1** parameter and choose the **OK** button; the selected parameter is added to the **Block Properties Table**.

Chapter 18

22. Clear the **Block properties must match a row in the table** check box, if not cleared already.

23. Enter the following values in the **Size** and **d1** columns:

Size	d1
M6	6
M10	10
M12.5	12.5
M15	15
M18	18
M24	24
M25	25

24. Choose the **Audits the block property table for errors** button; a message box is displayed stating that no errors have been found.

25. Close the message box and choose **OK** in the **Block Properties Table**; the **Size** parameter is added to the **Parameters Manager** palette under the **User Parameters** node.

26. Now, you need to test the block by invoking the **Test Block Window**. To do so, choose the **Test Block** tool from the **Open/Save** panel of the **Block Editor**; the **Test Block Window** is opened. You can view the name of the new window at the title bar.

27. Select the block; the block is displayed in dashed lines and two grips are also displayed, one at the insertion point of the block and other at the midpoint of the horizontal line at the bottom. The triangular grip at the midpoint is known as lookup grip.

28. Click on the lookup grip; a shortcut menu is displayed.

29. Choose different sizes and check whether the block resizes without any error.

30. After checking the block, choose the **Close Test Block Window** button from the **Close** panel; the **Block Editor** is displayed.

31. Choose the **Close Block Editor** button to close the block editor; a message box is displayed.

32. Choose the **Save changes and exit the Block Editor** option from the message box.

33. Select the block and click on the lookup grip; different possible sizes are displayed. Select the required size.

34. Save the drawing and exit.

Note
In this example, the block has been created without creating a fully-defined sketch. But it is recommended to create a fully-defined sketch while working with dynamic blocks.

Self-Evaluation Test

Answer the following questions and then compare them to those given at the end of this chapter:

1. Individual objects in a block cannot be erased using the **Block Editor**. (T/F)

2. A block can be mirrored by providing a scale factor of -1 for X. (T/F)

3. Blocks created by the **Create Block** tool in the **Block** panel can be used in any drawing. (T/F)

4. You cannot redefine any existing block. (T/F)

5. The _____ command lets you create a drawing file (.*dwg* extension) of a block defined in the current drawing.

6. The _____ command can be used to change the name of a block.

7. The _____ tool is used to place a previously created block in a drawing.

8. You can delete the unreferenced blocks using the _____ command.

9. You can use the _____ to locate, preview, copy, and insert blocks or existing drawings into the current drawing.

10. The _____ command is used for in-place reference editing.

Review Questions

Answer the following questions:

1. An entire drawing can be converted into a block. (T/F)

2. The objects in a block possess the properties of the layer on which they are drawn, such as color and linetype. (T/F)

3. If the objects forming a block were drawn on layer 0 with color and linetype **BYLAYER**, then at the time of the insertion, each object that makes up a block is drawn on the current layer with the current linetype and color. (T/F)

4. Objects created with the special color **BYBLOCK** are generated with the color that is current at the time the block was inserted. (T/F)

5. In the array generated with the **MINSERT** command, there is no way to change the number of rows or columns or the spacing between them, after insertion. The whole **MINSERT** pattern is considered as one object that cannot be exploded. (T/F)

6. Which one of the following actions cannot be assigned to the **Point** parameter?

 (a) **Move** (b) **Array**
 (c) **Stretch** (d) None of these

7. Which of the following commands should you use to get back the objects that consist of the block and have been removed from the drawing?

   (a) **OOPS** (b) **BLIPS**
   (c) **BLOCK** (d) **UNDO**

8. By what amount is a block rotated, if the value of both the X and Y scale factors is -1?

   (a) 90 (b) 180
   (c) 270 (d) 360

9. When you insert a drawing into a current drawing, how many of the blocks belonging to the inserted drawing are brought into the current drawing?

   (a) One (b) None
   (c) All (d) Two

10. The _____ command is used to create a rectangular array of a block?

11. The **Entire drawing** option of the **WBLOCK** command has the same effect as the **PURGE** command. (T/F)

12. The _____ tab of the **Block Authoring Palettes** allows you to insert a parameter set.

13. The automatic save feature is _____ during in-place reference editing.

14. You cannot use the **REFEDIT** command on blocks inserted using the _____ command.

15. The Layer 0 and the Continuous linetype _____ be renamed using the **RENAME** command.

EXERCISE 7   *Block*

Draw part (a) of Figure 18-65 and define it as a block named A. Then, using the **Insert** tool, insert the block in the plate, as shown.

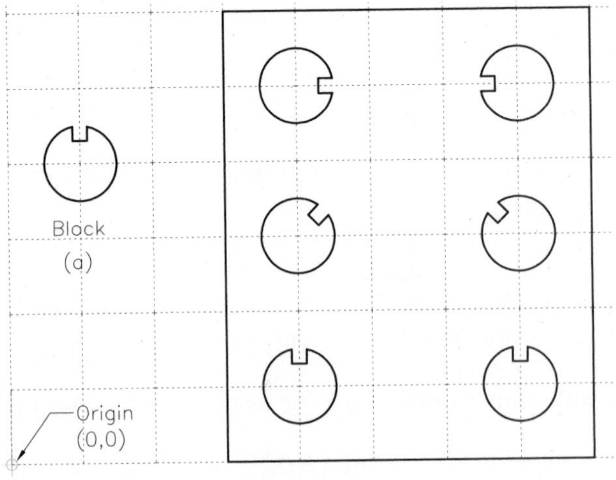

Figure 18-65 Drawing for Exercise 7

EXERCISE 8   *Block*

Draw the diagrams shown in Figure 18-66 using blocks.
a. Create a block for the valve, Figure 18-66(a).
b. Use a thick polyline for the flow lines.

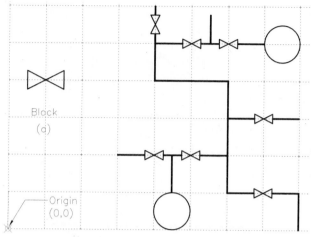

Figure 18-66 *Drawing for Exercise 8*

EXERCISE 9   *Block*

Draw part (a) of Figure 18-67 and define it as a block named B. Then, using the relevant insertion method, generate the pattern as shown. Note that the pattern is rotated at 30-degree.

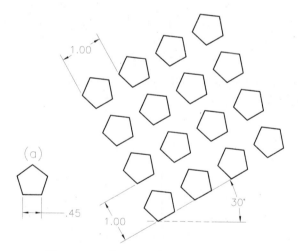

Figure 18-67 *Drawing for Exercise 9*

EXERCISE 10   *Block*

Draw Block A of Figure 18-68 and define it as a block named Chair. The dimensions of the chair can be referred from the Problem Solving Exercise 3 of Chapter 5. Then, using the block insert command, insert the chair around the table as shown in the same figure.

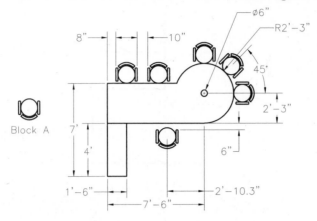

Figure 18-68 Drawing for Exercise 10

EXERCISE 11   *Constraints to Block*

Create the sketch of a bolt as shown in Figure 18-69. Then, convert it to a block and apply constraints such that you can create bolts of diameter 6, 10, 12.5, 15, 18, 24, and 25.

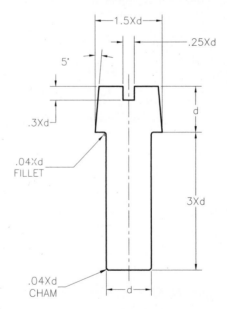

Figure 18-69 Drawing for Exercise 11

Answers to Self-Evaluation Test

1. F, **2.** T, **3.** F, **4.** F, **5.** WBLOCK, **6.** RENAME, **7.** Insert, **8.** PURGE, **9.** DesignCenter, **10.** REFEDIT

Chapter *19*

Defining Block Attributes

CHAPTER OBJECTIVES

In this chapter, you will learn:
- *About attributes and define them with a block.*
- *To create annotative block attributes.*
- *To edit attribute tag names.*
- *To insert blocks with attributes and assign values to attributes.*
- *To extract attribute values from inserted blocks.*
- *To control attribute visibility.*
- *To perform global and individual editing of attributes.*
- *To insert a text file in a drawing to create Bill Of Materials.*

KEY TERMS

- *Attribute Definition*
- *Block Attribute Manager*
- *Data Extraction*
- *ATTDISP*
- *EATTEDIT*
- *-ATTEDIT*
- *REFEDIT*

UNDERSTANDING ATTRIBUTES

AutoCAD has provided a facility that allows the user to attach information to blocks. This information can then be retrieved and processed by other programs for various purposes. For example, you can use this information to create a bill of material for a project, find the total number of computers in a building, or determine the location of each block in a drawing. Attributes can also be used to create blocks (such as title blocks) with prompted or preformatted text, to control text placement. The information associated with a block is known as attribute value or attribute. AutoCAD references the attributes with a block through tag names.

Before assigning attributes to a block, you must create an attribute definition by using the **Define Attributes** tool. The attribute definition describes the characteristics of the attribute. You can define several attribute definitions (tags) and include them in the block definition. Each time you insert the block, AutoCAD will prompt you to enter the value of the attribute. The attribute value automatically replaces the attribute tag name. The information (attribute values) assigned to a block can be extracted and written to a file by using AutoCAD's **EATTEXT** command. This file can then be inserted in the drawing as a table or processed by other programs to analyze the data. The attribute values can be edited by using the **EATTEDIT** command. The display of attributes can be controlled by using the **ATTDISP** command.

DEFINING ATTRIBUTES

Ribbon: Insert > Block Definition > Define Attributes	**Command:** ATTDEF

When you invoke the **Define Attributes** tool, the **Attribute Definition** dialog box is displayed, see Figure 19-1. The block attributes can be defined through this dialog box. When creating an attribute definition, you must define the mode, attributes, insertion point, and text information for each attribute. All this information can be entered in the dialog box. The following is the description of each area of the **Attribute Definition** dialog box:

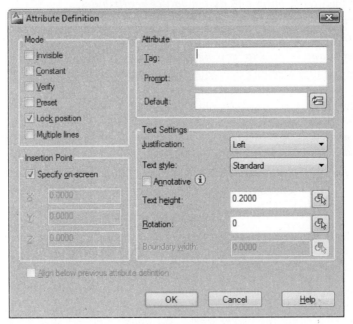

*Figure 19-1 The **Attribute Definition** dialog box*

Mode Area

The **Mode** area of the **Attribute Definition** dialog box has six check boxes: **Invisible**, **Constant**, **Verify**, **Preset**, **Lock position**, and **Multiple lines**. These determine the display and edit features of the block attributes. For example, if you select the **Invisible** check box, the attribute becomes invisible which means, it is not displayed on the screen. Similarly, if the **Constant** check box is selected, the attribute becomes constant. This means that its value is predefined and cannot be changed. These options are described next.

Invisible

This option lets you create an attribute that is not visible on the screen, by default. Select this check box if you want the attribute to be invisible.

Tip
*The **Invisible** mode is especially useful when you do not want the attribute values to be displayed on the screen to avoid cluttering the drawing. Also, if the attributes are invisible, it takes less time to regenerate the drawing.*

You can make the invisible attribute visible by using the **ATTDISP** command discussed later in this chapter, in the section "Controlling Attribute Visibility".

Constant

This option lets you create an attribute that has a fixed value and cannot be changed after block insertion. When you select this mode, the **Prompt** edit box and the **Verify** and **Preset** check boxes are disabled. Since the value is constant, there is no need to be prompted for new values. This check box is cleared by default and you can use different attribute values for the blocks.

Verify

This option allows you to verify the attribute value you have entered when inserting a block by asking you twice for the data. If the value is incorrect, you can correct it by entering the new value. If this check box is cleared, you are not prompted for verification of the attribute values.

Preset

This option allows you to create an attribute that is automatically set to the default value. The attribute values are not requested when you insert a block and the default values are used. But unlike a constant attribute, the preset attribute value can be edited later.

Lock position

This check box is selected by default and it is used to lock the position of the attributes with respect to the block object. In case of non-dynamic blocks, if the attributes are created with the **Lock position** check box cleared, then a separate grip will be displayed on that attribute. This grip can be used to modify the location of the attribute. When the dynamic blocks are created with the **Lock position** check box cleared, the separate grip does not get displayed on the attribute. Besides this, the attribute cannot be manipulated by the parameters and action.

Multiple lines

Select this check box to specify that the default value of the attribute may be a single-line or multiple line text.

Attribute Area

The **Attribute** area (Figure 19-2) of the **Attribute Definition** dialog box has three edit boxes: **Tag**, **Prompt**, and **Default**, where you can enter values. You can enter up to 256 characters in these edit boxes. If the first character to be entered in any one of these edit boxes is a space, you should start with a backslash (\\). But if the

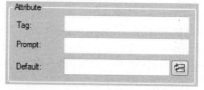

*Figure 19-2 The **Attribute** area of the **Attribute Definition** dialog box*

first character is a backslash (\), you should start the value to be entered with two backslashes (\\). The three edit boxes have been described next.

Tag

This is like a label that is used to identify an attribute. For example, the tag name COMPUTER can be used to identify an item. Here you can enter the tag names as uppercase, lowercase, or both, but all lowercase letters are automatically converted into uppercase when displayed. The tag name cannot be null. Also, it must not contain any blank spaces.

Tip
It is advisable to specify a tag name that reflects the contents of the item being tagged. For example, the tag name COMP or COMPUTER is an appropriate name for labeling computers.

Prompt

The text that you enter in the **Prompt** edit box is used as a prompt when you insert a block that contains the defined attribute. For example, if COMPUTER is the tag, you can enter WHAT IS THE MEMORY? or ENTER MEMORY: in the **Prompt** edit box. AutoCAD will then prompt you with this same statement when you insert the block with which the attribute is defined. If you have selected the **Constant** check box in the **Mode** area, the **Prompt** edit box is not available because no prompt is required if the attribute is constant. If you do not enter anything in the **Prompt** edit box, the entry made in the **Tag** edit box is used as the prompt.

Default

The entry in the **Default** edit box defines the default value of the specified attribute. If you do not enter a value, the attribute takes the value entered in the **Tag** edit box. The entry of a value is optional.

Insert Field

 This button is used to insert a field as the value of the attribute. When you choose this button, the **Field** dialog box is displayed that can be used to insert the required field.

Multiline Editor

 This button is displayed in place of the **Insert Field** button when the **Multiple lines** check box is selected from the **Mode** area. Choose this button to invoke the **Text Formatting** window on the screen. You can enter the text that you want to assign as the default value of the attribute in the text window.

Note
*The **Text Formatting** toolbar displayed here has only some commonly used options. The full **Text Formatting** toolbar can be displayed by changing the value of the **ATTIPE** system variable to 1.*

Insertion Point Area

The **Insertion Point** area of the **Attribute Definition** dialog box (Figure 19-3) lets you define the insertion point of the block attribute text. You can define the insertion point by entering the values in the **X**, **Y**, and **Z** edit boxes or by specifying it on the screen. To specify the insertion point on the screen, select the **Specify on-Screen** check box. Now, set the parameters in all the other fields and areas and then choose **OK**. You will be prompted to select the start point of the attribute.

Figure 19-3 The Insertion Point area

Just below the **Insertion Point** area of the dialog box is a check box labeled **Align below previous attribute definition**. This check box is not available when you use the **Attribute Definition** dialog box for the first time. After you have defined an attribute and when you press ENTER to display the **Attribute Definition** dialog box again, this check box is available. You can select this check box to place the subsequent attribute text just below the previously defined attribute automatically. When you select this check box, the **Insertion Point** area and the **Text Settings** areas of the dialog box are not available and AutoCAD assumes previously defined values for text such as text height, text style, text justification, and text rotation. The text is automatically placed on the following line.

Text Settings Area

The **Text Settings** area of the **Attribute Definition** dialog box (Figure 19-4) lets you control the justification, text style, annotation property, height, rotation, and boundary width of the attribute text. To set the text justification, select a justification type from the **Justification** drop-down list. The default option is **Left** for the single line text and **Top left** for the multiline text. Similarly, you can use the **Text style** drop-down list to select a text style. All the text styles defined in the current drawing are displayed in the **Text style** drop-down list. The default text style is **Standard**. To write annotative attributes, select the **Annotative** text style from the **Text style** drop-down list. Select the **Annotative** check box to assign the annotative property to the attribute text. The annotative attributes can be attached to both the annotative

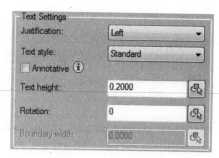

*Figure 19-4 The **Text Settings** area of the **Attribute Definition** dialog box*

or non-annotative blocks. Choose the information (**i**) button on the side of the **Annotative** check box to collect more information about the annotative object. You can specify the text height and text rotation in the **Text height** and **Rotation** edit boxes, respectively. You can also define the text height by choosing the **Text height** button. When you choose this button, AutoCAD temporarily exits the **Attribute Definition** dialog box and lets you enter the height value by selecting points on the screen or from the Command prompt. Once the height on the screen is defined, the **Attribute Definition** dialog box will reappear and the defined text height will be displayed in the edit box. Similarly, you can define the text rotation by choosing the **Rotation** button and then selecting points on the screen or by entering the rotation angle at the Command prompt. The **Boundary width** edit box is used to specify the maximum width for the multiline text. If the width of the text line is more than this value, the text will automatically go to the second line. This edit box will be available only when you select the **Multiple lines** check box from the **Mode** area.

Note
The text style must be defined before it can be used to specify the text.

*If you select a style that has the predefined height, AutoCAD automatically disables the **Text height** edit box.*

*If you have selected the **Align** option from the **Justification** drop-down list, the **Insertion Point** area and the **Text height** and **Rotation** edit boxes in the **Text Settings** area will be disabled.*

*If you have selected the **Fit** option from the **Justification** drop-down list, the **Insertion Point** area and the **Rotation** edit box are disabled. To specify the location of the attribute, choose **OK** from the dialog box. AutoCAD prompts you to specify the first and second endpoints of the text baseline. The text will fit between the two endpoints that you specify.*

After you complete the settings in the **Attribute Definition** dialog box and choose **OK**, the attribute tag text is inserted in the drawing at the specified insertion point. Now, you can use the **BLOCK** or **WBLOCK** commands to select all the objects and attributes to define a block.

Note

*You can use the -**ATTDEF** command to display the dynamic prompt in the drawing area. The options in the **Attribute Definition** dialog box are available through the dynamic prompt too. See Figure 19-5.*

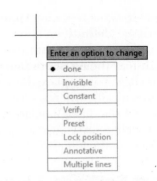

*Figure 19-5 Dynamic prompt for the -**ATTDEF** command*

EXAMPLE 1   *Attribute Definition*

In this example, you will define the following attributes for a computer and then create a block using the **BLOCK** command. The name of the block is COMP.

Mode	Tag name	Prompt	Default value
Constant	ITEM		Computer
Preset, Verify	MAKE	Enter make:	CAD-CIM
Verify	PROCESSOR	Enter processor type:	Unknown
Verify	HD	Enter Hard-Drive size:	40 GB
Invisible, Verify	RAM	Enter RAM:	256 MB

1. Draw the sketch of a computer, as shown in Figure 19-6. Assume the dimensions, or measure the dimensions of the computer you are using for AutoCAD.

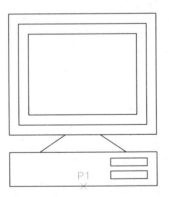

Figure 19-6 Drawing for Example 1

2. Choose the **Define Attributes** tool from the **Block Definition** panel in the **Insert** tab; the **Attribute Definition** dialog box is displayed.

3. Define the first attribute shown in the preceding table. Select the **Constant** check box in the **Mode** area because the mode of the first attribute is constant. In the **Tag** edit box, enter the tag name, **ITEM**. Similarly, enter **COMPUTER** in the **Default** edit box. Note that the **Prompt** edit box is not available because the mode is constant.

4. In the **Insertion Point** area, select the **Specify on-screen** check box to define the text insertion point.

5. In the **Text Settings** area, specify the justification, style, annotative property, height, and rotation of the text.

6. Choose the **OK** button once you have entered information in the **Attribute Definition** dialog box. Select a point below the insertion base point (P1) of the computer to place the text.

7. Press ENTER to invoke the **Attribute Definition** dialog box again. Enter the mode and attribute information for the second attribute as shown in the table at the beginning of Example 1. You need not define the insertion point and text options again. Select the **Align below previous attribute definition** check box that is located just below the **Insertion Point** area. You will notice that when you select this check box, the **Insertion Point** and **Text Settings** areas are not available. Now, choose the **OK** button. AutoCAD places the attribute text just below the previous attribute text.

8. Similarly, define the remaining attributes also (Figure 19-7).

ITEM

MAKE

PROCESSOR

HD

RAM

Figure 19-7 Attributes defined below the computer drawing

9. Now, use the **BLOCK** command to create a block. Make sure that the **Retain** radio button is selected in the **Objects** area. The name of the block is **COMP**, and the insertion point of the block is P1, midpoint of the base. When you select objects for the block, make sure you also select attributes.

10. Insert the block created. You will notice that the order of prompts is the same as the order of attributes selection.

EDITING ATTRIBUTE DEFINITION

Menu Bar: Modify > Object > Text > Edit   **Command:** DDEDIT

You can edit the text and attribute definitions before you define a block; using the **DDEDIT** command. After invoking this command, AutoCAD will prompt you to **Select an annotation object or [Undo]**. If you select an attribute created using the **Attribute Definition** dialog box, the **Edit Attribute Definition** dialog box is displayed (Figure 19-8).

You can also invoke the **Edit Attribute Definition** dialog box by double-clicking on the attribute definition. You can enter the new values in the respective edit boxes. Once you have entered the changed values, choose the **OK** button in the dialog box. After you exit the dialog box, AutoCAD will continue to prompt you to select another text or attribute object (Attribute tag).

If you have finished editing and do not want to select another attribute object to edit, press ENTER to exit the command.

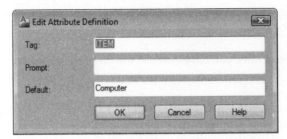

*Figure 19-8 The **Edit Attribute Definition** dialog box*

Using the Properties Palette

The **Properties** tool has been already discussed in Chapter 4, Working with Drawing Aids. It can also be used to edit defined attributes. Select the attribute to be modified and right-click to display a shortcut menu. Next, choose **Properties** from the shortcut menu; the **Properties** palette will be displayed, see Figure 19-9. You will notice that **Attribute Definition** is displayed in the text box located at the top of the palette. Also, you will find that all properties of the selected attribute are displayed under four headings. They are **General**, **3D Visualization**, **Text**, and **Misc**. You can change these values in their corresponding fields. For example, you can modify the color, layer, linetype, thickness, linetype scale, and so on, of the selected attribute under the **General** head. Similarly, you can modify the tag name, prompt, and value of the selected attribute in the **Tag**, **Prompt**, and **Value** fields under the **Text** heading. Under the **Text** head, you can also modify the text style, justification, annotation scale, text height, rotation angle, width factor, and make the angle values of the selected attribute oblique.

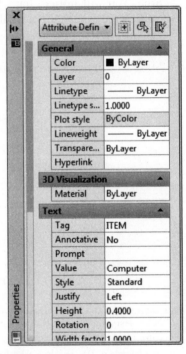

*Figure 19-9 The **Properties** palette to modify attributes*

You can also determine if you want the attribute text to appear upside down or backwards or not under the **Misc** heading. Here, you can also modify the attribute modes, which have been already defined. When you select a particular mode, for example **Invisible**, a drop-down list is available in the corresponding field. This list displays two options, **Yes** and **No**. If you select **Yes**, the attribute is made invisible and if you select **No**, the attribute is made visible. To select a particular mode, you should choose **Yes** from their corresponding drop-down lists.

Note
Remember that if you change the justification of the first attribute below which you have aligned the remaining attributes, the remaining attributes will not be updated automatically. You will have to change justifications of the remaining attributes individually to align them below the first attribute.

If you change the justification of the attributes from align to some other, the size of the text will also be changed.

*You can also use the **CHANGE** command to edit the attribute objects using the Command prompt.*

INSERTING BLOCKS WITH ATTRIBUTES

The value of the attributes can be specified during block insertion, either at the Command prompt or in the **Edit Attributes** dialog box, if the system variable **ATTDIA** is set to **1**. When you use the **Insert** tool or the **-INSERT** command to insert a block in a drawing (discussed earlier in Chapter 18, Working with Blocks), and after you have specified the insertion point, scale factors, and rotation angle, the **Edit Attributes** dialog box (Figure 19-10) will be displayed.

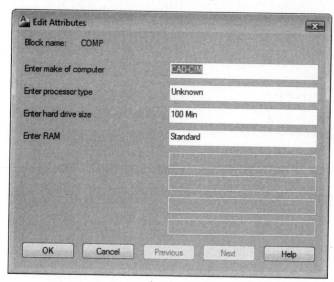

*Figure 19-10 The **Edit Attributes** dialog box*

The default value for **ATTDIA** is **0**, which disables the dialog box. The prompts and their default values, which you had specified with the attribute definition, are then displayed on the Command prompt after you have specified the insertion point, scale, and rotation angle for the block to be inserted.

In the **Edit Attributes** dialog box, the prompts that were entered at the time of attribute definition in the dialog box are displayed with their default values in the corresponding fields. If an attribute has been defined with the **Constant** mode, it is not displayed in the dialog box because a constant attribute value cannot be edited. You can enter the attribute values in the fields located next to the attribute prompt. If no new values are specified, the default values are displayed. Eight attribute values are displayed at a time in the dialog box. If there are more attributes, they can be accessed by using the **Next** or **Previous** buttons. The block name is displayed at the top of the dialog box. After entering the new attribute values, choose the **OK** button. AutoCAD will place these attribute values at the specified location.

Tip

*It is more convenient to use the **Edit Attributes** dialog box because you can view all the attribute values at a glance and can correct them before placing. Therefore, it is a good idea to set the **ATTDIA** value to 1, before you insert a block with attributes.*

*Attributes can also be defined from the Command prompt by setting the system variable **ATTDIA** to 0 (default value). Now, when you use the **Insert** tool, AutoCAD does not display the **Edit Attributes** dialog box. Instead, AutoCAD prompts you to enter the attribute values for various attributes that have been defined in the block at the pointer input.*

EXAMPLE 2   *ATTDIA*

In this example, you will insert the block (COMP) that was defined in Example 1. The following is the list of the attribute values for computers:

ITEM	MAKE	PROCESSOR	HD	RAM
Computer	Gateway	486-60	150 MB	16 MB
Computer	Zenith	486-30	100 MB	32 MB
Computer	IBM	386-30	80 MB	8 MB
Computer	Del	586-60	450 MB	64 MB
Computer	CAD-CIM	Pentium-90	100 Min	32 MB
Computer	CAD-CIM	Unknown	600 MB	Standard

1. Draw the floor plan shown in Figure 19-11 (assume the dimensions).

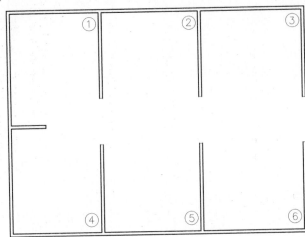

Figure 19-11 *Floor plan drawing for Example 2*

2. Set the system variable **ATTDIA** to 1. Use the **Insert** tool available in the **Blocks** panel in the **Insert** tab to insert the blocks. When you invoke the **Insert** tool, the **Insert** dialog box is displayed. Enter **COMP** in the **Name** edit box and choose **OK** to exit the dialog box. Select an insertion point on screen to insert the block. After you specify the insertion point, the **Edit Attributes** dialog box is displayed. In this dialog box, you can specify the attribute values, if you need to, in the different edit boxes.

3. Choose the **Insert** tool again to insert other blocks and define their attribute values, as shown in Figure 19-12.

4. Save the drawing for further use.

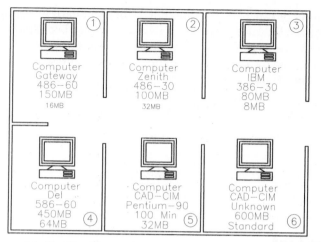

Figure 19-12 The floor plan after inserting blocks and defining their attributes

MANAGING ATTRIBUTES

Ribbon: Insert > Block Definition > Manage Attributes   **Command:** BATTMAN
Menu Bar: Modify > Object > Attribute > Block Attribute Manager

The **Manage Attributes** tool allows you to manage attribute definitions for blocks in the current drawing through the **Block Attribute Manager** dialog box. This dialog box is shown in Figure 19-13. By default, any attribute changes you make are applied to all existing block references in the current drawing. Immediate changes in the default value of the attribute in the existing drawing can only be seen upon regeneration, if the attribute that is edited is in the **Constant** mode. If the mode is other than **Constant**, the changes in the attribute can only be seen for new block insertions.

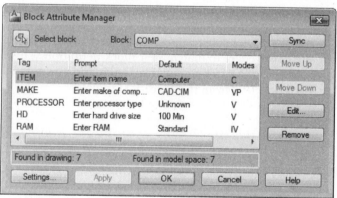

*Figure 19-13 The **Block Attribute Manager** dialog box*

When you invoke this tool, AutoCAD displays the number of blocks that exist in the current drawing below the list box in the dialog box. The block can be in the model space or in the layout. The names of the existing blocks in the current drawing are displayed in the **Block** drop-down list. Attributes of the selected block are displayed in the attribute list. By default, the **Tag**, **Prompt**, **Default**, **Modes**, and **Annotative** attribute properties are displayed in the attribute list. You can edit the attribute definitions in blocks, remove attributes from blocks, and change the order in which you were prompted for attribute values while inserting the block.

Select block

This button allows you to select a block from the drawing area whose attributes you want to edit. When you choose the **Select block** button, the dialog box closes temporarily until you select block from the drawing or cancel by pressing ESC. Once you have selected the block, the dialog box is displayed again. The name of the block you have selected is displayed in the **Block** drop-down list. If you modify the attributes of a block and then select a new block before saving the attribute changes, you will be prompted to save the changes before selecting another block through an alert message box (Figure 19-14).

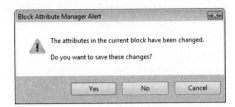

*Figure 19-14 The **Block Attribute Manager Alert** message box*

Block Drop-down List

This drop-down list in the **Block Attribute Manager** dialog box allows you to choose the block whose attributes you want to modify. It lists all the blocks in the current drawing that have attributes.

Sync

This button of the **Block Attribute Manager** dialog box allows you to update all instances of the selected block with the attribute properties currently defined. This does not affect any of the values assigned to attributes in each block when attributes were defined. On choosing this button, you can update attributes in all block references in the current drawing with the changes you made to the block definition. For example, you may have used the **Block Attribute Manager** dialog box to modify attribute properties of several block definitions in your drawing. But you do not want it to automatically update the existing block references when you made the changes. Now, when you are satisfied with the attribute changes you made, you can apply those changes to all blocks in the current drawing by using the **Sync** button.

 You can also use the **Synchronize** tool in the **Attributes** panel to update attribute properties in block references to match their block definition. This command can also be invoked by entering **ATTSYNC** at the Command prompt or by choosing the **Synchronize Attributes** tool from the **Modify II** toolbar.

 Note
*Changing the attribute properties of an existing block does not affect the values assigned to the block. For example, in a block containing an attribute whose **Tag** is Cost and default value is 200. The default value 200 remains unaffected, if the **Tag** is changed to UnitCost.*

Move Up

This button moves the selected attribute tag earlier in the prompt sequence. Now, when you insert the block, the attribute you have just moved up will appear earlier in the prompt sequence. The **Move Up** button is not available when a constant attribute is selected and if the attribute is at the topmost position in the attribute list.

Move Down

This button moves the selected attribute tag later in the prompt sequence. The **Move Down** button is not available when a constant attribute is selected and when the attribute is at the bottom position in the attribute list.

Tip
When you define a block, the order in which you select the attributes determines the order in which you are prompted for the attribute information when you insert the block. You can use the **Block Attribute Manager** *to change the order of prompts that request attribute values.*

Edit

This button of the **Block Attribute Manager** allows you to modify the attribute properties. To edit the attribute definition in blocks, select the attribute and then choose **Edit** from the dialog box; the **Edit Attribute** dialog box will be displayed. In the dialog box, **Active Block** displays the name of the block that you have selected and whose properties have to be edited.

Note
To edit an attribute, you can also double-click on the attribute or choose the **Single** *tool from* **Insert > Block > Edit Attributes** *drop-down. Then, select the attribute to display the* **Enhanced Attribute Editor** *dialog box.*

The options in the **Edit Attribute** dialog box are discussed next.

Attribute Tab

The options under the **Attribute** tab of the **Edit Attribute** dialog box are used to modify the mode and data of attributes (see Figure 19-15). The options under the **Mode** area can be selected for the block. The **Data** area of the dialog box has three edit boxes: **Tag**, **Prompt**, and **Default**, where you can enter the data to be changed.

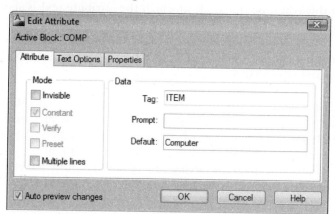

*Figure 19-15 The **Attribute** tab of the **Edit Attribute** dialog box*

Text Options Tab

The options under this tab (Figure 19-16) are used to set the properties that define the way an attribute's text is displayed in the drawing. The options in this tab are discussed next.

Text Style. This drop-down list specifies the text style for attribute text. Default values for this text style are assigned to the text properties displayed in this dialog box.

Justification. This drop-down list specifies how attribute text is justified.

Height. This edit box specifies the height of the attribute text. If the height is defined in the current text style, this edit box is not available.

Rotation. This edit box specifies the rotation angle of the attribute text.

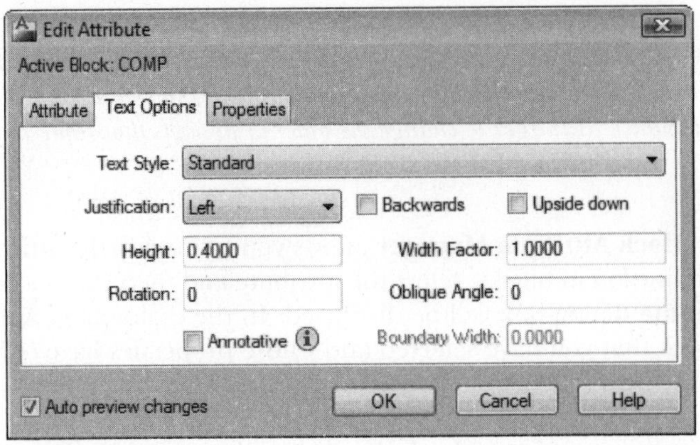

*Figure 19-16 The **Text Options** tab of the **Edit Attribute** dialog box*

Backwards. This check box specifies whether or not the text is displayed backwards. Select this box to display the text backwards.

Upside down. This check box specifies whether or not the text is displayed upside down. Select this box to display the text upside down.

Width Factor. This edit box sets the character spacing for attribute text. Entering a value less than 1.0 condenses the text and a value greater than 1.0 expands it.

Oblique Angle. In this edit box, the angle at which the attribute text is slanted away from its vertical axis is specified.

Annotative. This check box is used to convert the non-annotative attribute to the annotative attribute and vice-versa.

Boundary Width. This edit box sets the width of the boundary that will be displayed while writing the multiline text for the attributes' default value. The value specified here is the relative scale value. If you specify a value less than 1, it will decrease the boundary width. Whereas, if you specify a value more than 1, it will increase the boundary width. Accordingly, the paragraph width for the multiline text will increase or decrease. You can modify the value of the boundary width only when you select the **Multiple lines** check box from the **Mode** area of the **Attribute Definition** dialog box.

Note
*If you change the above-mentioned properties of the **Constant** attribute, you may need to regenerate the drawing to view the effects.*

Properties
The options under this tab of the **Edit Attribute** dialog box (Figure 19-17) are used to define the layer that the attribute is on and the color, lineweight, and linetype for the attribute's line. Also, if the drawing uses plot styles, you can assign a plot style to the attribute. The options of the **Properties** tab are discussed next.

Layer. This drop-down list specifies the layer that the attribute is on. The layer on which the attribute was defined is displayed here in this drop-down list. This cannot be changed using the dialog box.

Linetype. This drop-down list specifies the linetype of attribute text.

Color. This drop-down list specifies the attribute's text color.

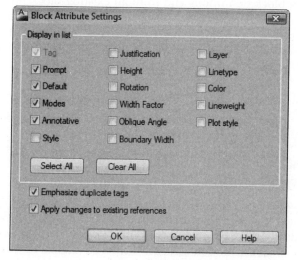

*Figure 19-17 The **Properties** tab of the **Edit Attribute** dialog box*

Lineweight. This drop-down list specifies the lineweight of attribute text. The changes you make to this option do not come into effect if the **LWDISPLAY** system variable is off.

Plot Style. This drop-down list specifies the plot style of the attribute. If the current drawing uses color-dependent plot styles, the **Plot style** list in this dialog box is not available.

Auto preview changes

This check box controls whether or not the drawing area is immediately updated to display any visible attribute changes you make. If the **Auto preview changes** check box is selected, changes are immediately visible. If this check box is cleared, changes are not immediately visible.

Tip
*The buttons such as **Move Up**, **Move Down**, **Edit**, **Remove**, and **Settings** can also be invoked by using the shortcut menu, which appears by right-clicking on the attributes in the attribute list in the **Block Attribute Manager** dialog box.*

Settings

Choosing this button of the **Block Attribute Manager** opens the **Block Attribute Settings** dialog box (Figure 19-18), where you can customize how attribute information is listed in the **Block Attribute Manager**.

*Figure 19-18 The **Block Attribute Settings** dialog box*

Display in list

You can specify the attribute properties you want to be displayed in the list by selecting the check boxes available in this area of the **Block Attribute Settings** dialog box. This area shows the check boxes of all the properties that are displayed in the attribute list. Only the check boxes of the properties selected are displayed in the attribute list. These properties can be viewed in the attribute list by scrolling it. The **Tag** check box is always selected by default.

This area of the dialog box has two buttons, **Select All** and **Clear All**, which select and clear all the check boxes, respectively.

Emphasize duplicate tags

This check box turns duplicate tag emphasis on and off. If this check box is selected, duplicate attribute tags are displayed in red type in the attribute list. For example, if a block has PRICE as a tag more than once, the attribute PRICE will be displayed in red in the attribute list. If this check box is cleared, duplicate tags are not emphasized in the attribute list.

Apply changes to existing references

This check box of the dialog box specifies whether or not to update all existing instances of the block whose attributes you are modifying. Therefore, if this check box is selected, all instances of the block with the new attribute definitions are updated. If this check box is cleared, only new insertions of the block with the new attribute definitions are displayed.

Use **Sync** in the **Block Attribute Manager** to apply changes immediately to existing block instances. This temporarily overrides the **Apply changes to existing references** option.

Tip
*If constant attributes or nested attributed blocks are affected by your changes, use the **REGEN** command to update the display of those blocks in the drawing area.*

Remove

This button removes the selected attribute from the block definition. If the **Apply changes to existing references** check box is selected in the **Block Attribute Settings** dialog box before you choose the **Remove** button, the attribute is removed from all instances of the block in the current drawing. If this check box is cleared before choosing the **Remove** button, the attribute is removed only from the attribute list and the blocks that will be inserted henceforth. You can remove attributes from block definitions and from all existing block references in the current drawing. Attributes removed from existing block references do not disappear in the drawing area until you regenerate the drawing using the **REGEN** command. You cannot remove all attributes from the block, at least one attribute should remain. However, if you want to remove all the attributes, you have to redefine the block.

Note that the **Remove** button is not available for blocks with only one attribute.

EXTRACTING ATTRIBUTES

> **Ribbon:** Insert > Linking & Extraction > Extract Data
> **Command:** EATTEXT or DATAEXTRACTION

The **Extract Data** tool allows you to extract the block attribute information and property information such as drawing summary from single or multiple drawings. The settings that were used to extract data are stored in the *data extraction (.dxe)* file format. When this tool is invoked, the **Data Extraction** wizard is displayed (Figure 19-19). This wizard includes the following pages:

Begin

The **Begin** page, as shown in Figure 19-19, allows you to save the settings that are meant to be used for extracting data to a new *.dxe* file or to modify an existing *.dxe* file. You can also use the saved *.dxe* or *.blk* file as the template of settings for extracting data.

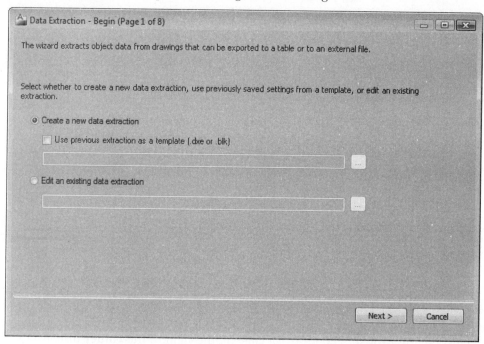

*Figure 19-19 The **Begin** page of the **Data Extraction** wizard*

This wizard page has two radio buttons: **Create a new data extraction** and **Edit an existing data extraction**. If you select the **Use previous extraction as a template [.dxe or .blk]** check box, you will be allowed to select an existing template file. To select a template file, choose the **Browse[...]** button available on the right of the template name edit box. A *.blk* template file is a text file that allows you to specify the attribute and property values that you want to extract and the information about the drawing that you want to retrieve. The file can be created by using any text editor such as Notepad or WordPad. While using a *.blk* template file, you will be prompted to save the data extraction to a *.dxe* file to extract the information. Select the **Edit an existing data extraction** check box to modify an existing *.dxe* file. Now, you will be allowed to select the existing data file. To select a template file, choose the **Browse[...]** button available on the right of the template name edit box.

Choose the **Next** button from the **Begin** page of the **Data Extraction** wizard; the **Save Data Extraction As** dialog box will be invoked. Next, specify the desired name and location in this dialog box to save the *.dxe* file and choose the **Save** button; the **Define Data Source** page will be displayed, as shown in Figure 19-20.

Define Data Source

The **Define Data Source** page is used to select an object either from a single drawing or from multiple drawings. The **Data source** area of this dialog box has two radio buttons: **Drawings/Sheet set** and **Select objects in the current drawing**.

Drawings/Sheet set

This radio button enables you to specify the drawings and folder from which you can extract the data. If this radio button is selected, the **Add Folder** and **Add Drawings** buttons will be enabled. The name of the drawings selected will be displayed with full path in the **Drawing files and folders** area.

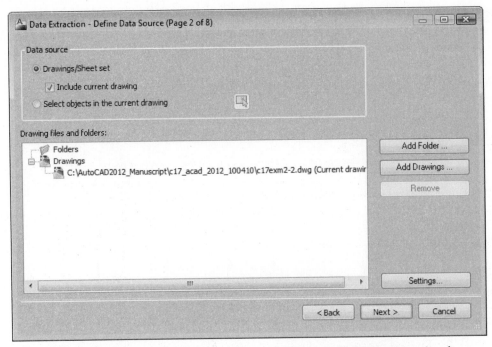

*Figure 19-20 The **Define Data Source** page of the **Data Extraction** wizard*

Choose the **Add Folder** button; the **Add Folder Options** dialog box will be invoked, see Figure 19-21. All drawings within the added folder will be considered for extracting data. To add a folder, choose the **Browse[...]** button available on the right of the **Folder** display box. Select the **Automatically include new drawings added in this folder to the data extraction** check box in the **Options** area to automatically consider the new drawing added to this folder for data extraction. Select the **Include subfolders** check box to include the drawings within the subfolders of the folder selected to be added. Select the **Utilize wild-card characters to select drawings** check box to include only the files matching the specified wild-card characters for data extraction. By default, *.dwg is used as the wild-card character and that is why, all the files with extension *.dwg*(all the drawing files) are considered for the data extraction. For example, if you want AutoCAD to consider all the drawing files having "computer" in their name, enter the wild-card character ***computer*.dwg**. Next, specify the desired settings and choose the **OK** button.

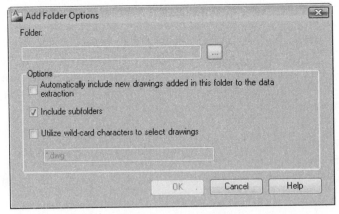

*Figure 19-21 The **Add Folder Options** dialog box*

Similarly, you can add a single drawing for the data extraction by choosing the **Add Drawings** button of the **Define Data Source** page of the **Data Extraction** wizard.

Select the **Include current drawing** check box in the **Define Data Source** page to add the current drawing for the data extraction including the folders and drawings.

Select objects in the current drawing

This radio button enables you to add the selected object from the current drawing for the data extraction. If you select this radio button, the **Select objects in the current drawing** button will be enabled. Choose this button to select the objects from the drawing.

Drawing files and folders

This area of the dialog box displays the tree view of all the drawing and folder names with their locations that are selected for the data extraction. All the drawing and folders are displayed under the **Folders** and **Drawings** heads respectively. Only the checked drawings under the folder are included for the data extraction. You can also remove the selected drawing or folder from the list of drawing to be used for data extraction. To remove the file or folder from the list of files and folders to be used for the data extraction, select them from the **Drawing files and folders** area and choose the **Remove** button.

Settings

Choose the **Settings** button from the **Define Data Source** page to invoke the **Data Extraction - Additional Settings** dialog box, as shown in Figure 19-22. This dialog box has two areas: **Extraction settings** and **Extract from**. By default, the **Extract objects from blocks** and **Extract objects from xrefs** check boxes are selected in the **Extraction settings** area. The first two check boxes allow you to specify whether you want to extract the block attribute information from the external reference drawings or the nested blocks, or both. If you select the **Include xrefs in block counts** check box, then the xrefs will automatically be included while counting the blocks. The two radio buttons available in the **Extract from** area allow you to control the settings for counting the blocks. If you select the **Objects in model space** radio button, only the blocks currently inserted into the model space will be counted. If you select the **All objects in drawing** radio button, all the blocks from the current drawing will be counted. Choose the **OK** button to exit the **Data Extraction - Additional Settings** dialog box. Choose the **Next** button from the **Define Data Source** page; the **Select Objects** page of the **Data Extraction** wizard will be displayed.

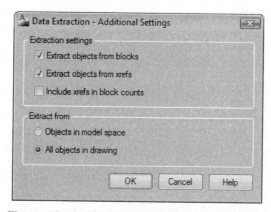

Figure 19-22 *The **Data Extraction - Additional Settings** dialog box*

Select Objects

The **Select Objects** page, as shown in Figure 19-23, allows you to filter out the objects to be included for data extraction. The objects that are checked in the **Objects** list box will only be considered for data extraction. The objects that are present in the current drawing are checked by default. Apart from the check box column, three more columns namely, **Object**, **Display Name**, and **Type** are displayed in the **Objects** list box. These columns display the information about the objects that are available to be selected and their type based on blocks and non-blocks.

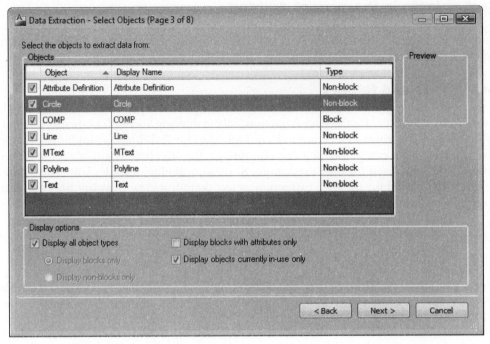

*Figure 19-23 The **Select Objects** page of the **Data Extraction** wizard*

The **Display options** area is used to control the objects to be displayed in the **Objects** list box. You can sort out the display from the blocks and non-blocks, blocks having attributes, and objects used in the current drawing. After specifying the objects to be displayed, choose the **Next** button from the **Select Objects** page to move to the **Select Properties** page.

Tip
*You can reverse the display order of the entities by clicking on the respective column head in the **Objects** list box. The direction of the arrow displayed at the column head of the selected column represents whether the data is arranged in the ascending or the descending order.*

Select Properties

This page, as shown in Figure 19-24, allows you to select the properties of the objects to be included for data extraction. Only the properties that are checked in the **Properties** list box are considered for data extraction. Apart from the check box column, three more columns namely, **Property**, **Display Name**, and **Category** are displayed in the **Properties** list box. These columns display the information about the available properties and their category.

The **Category filter** area is used to control the properties further to be displayed in the **Properties** list box. To do so, select the categories of the properties to be included in the **Properties** list box. For example, if you select the **Attribute** check box in the **Category filter** area, only the information about the attributes defined for the drawing will be extracted and displayed in the **Properties** list box. All the properties and categories displayed here are similar to the one displayed in the **Properties** palette.

You can also edit the names below the **Display Name** column. Select the row for which you want to change the display name and double-click on the **Display Name** column or right-click to display the shortcut menu. Next, choose the **Edit Display Name** option from the shortcut menu and enter a new name. After specifying the properties, choose the **Next** button from the **Select Properties** page to display the **Refine Data** page.

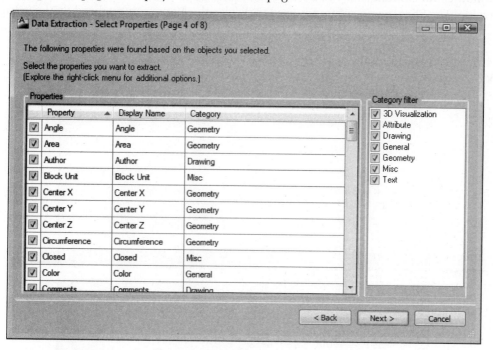

*Figure 19-24 The **Select Properties** page of the **Data Extraction** wizard*

Refine Data

This page, as shown in Figure 19-25, allows you to format the data extracted from a drawing. The properties selected from the **Select Properties** page of the **Data Extraction** wizard are displayed as the column head on the **Refine Data** page and the extracted data is displayed under these column heads of the table.

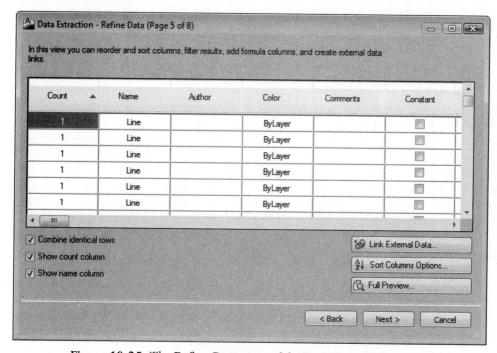

*Figure 19-25 The **Refine Data** page of the **Data Extraction** wizard*

The **Combine identical rows** check box is used to merge the rows having the same data. The sum of the merged rows having the same data will be displayed below the **Count** column head. The **Show count column** check box is used to display the **Count** column head on the extracted data table. The **Show name column** is used to display the **Name** column head on the extracted data table. All the check boxes explained above are selected by default.

You can also add columns from an excel sheet to the data extracted from a drawing. Choose the **Link External Data** button; the **Link External Data** dialog box will be displayed, see Figure 19-26. The options in the **Link External Data** dialog box are discussed next.

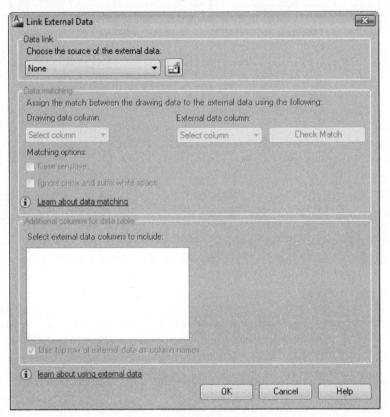

*Figure 19-26 The **Link External Data** dialog box*

Data link Area

This area is used to specify the excel sheet to be attached with the drawing. The **Choose the source of the external data** drop-down list displays all the external data links that are attached to the current drawing. Select the external data link from which you want to include the columns. To attach a new excel file to the drawing, choose the **Launch Data Link Manager** button; the **Data Link Manager** dialog box will be displayed on the screen. The other options in the **Data Link Manager** dialog box have been discussed earlier in Chapter 7 (Creating Text and Tables). The rest of the options in the **Link External Data** dialog box will be available as soon as you specify the external data to be linked with the drawing.

Data matching Area

The options in this area are used to establish a relationship between the data extracted from the drawing and the attached external excel sheet. There must be at least one cell with the matching data in the selected column of the drawing data and the external data column respectively. This matching data in the external and internal data acts as a key cell to specify the proper insertion of the external data into the extracted data.

The **Drawing data column** drop-down list displays all the property column heads of the data extracted by the drawing. Select the column head from this drop-down list that is to be used for matching. The **External data column** displays the column names from the external data. Select the column name from where you want to insert data into the extracted data. Select the **Case sensitive** check box to check the case of the alphabets while matching the data of the selected columns from the extracted and external data. Select the **Ignore prefix and suffix white space** check box to ignore the blank spaces in the front and back of the entries in the columns that are used for matching.

Specify preferences for matching the data and choose the **Check Match** button to compare the data of the specified columns from the extracted and external data. If the data matching is successful, the **Valid Key** information box will be displayed, see Figure 19-27. Otherwise, the **Check Key** information box will be displayed, suggesting the possible reasons of conflict between the two datums, see Figure 19-28. Unless there is a proper matching between the two datums, the merging of the external data into the extracted data will not take place.

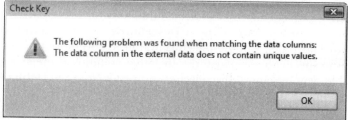

Figure 19-27 *The* **Valid Key** *information box*

Figure 19-28 *The* **Check Key** *information box*

Additional columns for data table Area

This area displays all columns of the external data. Select the check box on the left of each column entry to include that column into the data extraction. The **Use top row of external data as column names** check box will be used to specify the first row of the selected columns as the column head for the extracted data. Otherwise, the column heads are displayed as **Column 1**, **Column 2**, and so on. By default, this check box is selected.

Specify the preferences for the external data to be included in the current data extraction in the **Link External Data** dialog box and choose the **OK** button. The specified columns of the external data will be added to the extracted data and displayed in the **Refine Data** page of the **Data Extraction** wizard. The drawing data column used for matching and the external data column included in the extracted data will be represented with an external data link symbol. This symbol will be displayed on the side of the column head in the **Refine Data** page, as shown in Figure 19-29.

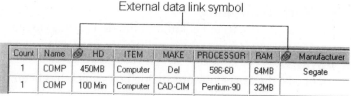

Figure 19-29 *The external data link symbol*

You can also change the display order of rows in the extracted data. To do so, choose the **Sort Columns Options** button from the **Refine Data** page; the **Sort Columns** dialog box will be displayed, see Figure 19-30. You can define the sort order for a single column or a group of columns at a time with this dialog box. The options in the **Sort Column** dialog box are discussed next.

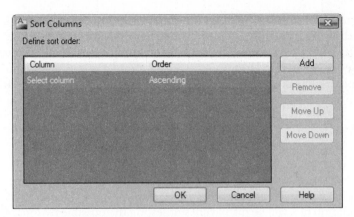

*Figure 19-30 The **Sort Columns** dialog box*

The **Define sort order** list box has two columns with the **Column** and **Order** column heads. Click on the drop-down arrow below the **Column** head to display all the column head entries for the extracted data. Select the column for which you want to change the display order of the rows. Choose the **Ascending** or **Descending** option from the **Order** drop-down list to specify the display order. Choose the **Add** button to add more rows in the **Define sort order** list box so as to specify the display order for other columns of the extracted data. The **Remove** button enables you to remove the selected sort order from the **Define sort order** list box. Choose the **Move Up** or **Move Down** button to move the selected row in the sort order, one level up or one level down respectively in the **Define sort order** list box.

Choose the **Full Preview** button; the preview will be displayed in the **Full Preview** window. A partial view of the **Full Preview** window is shown in Figure 19-31.

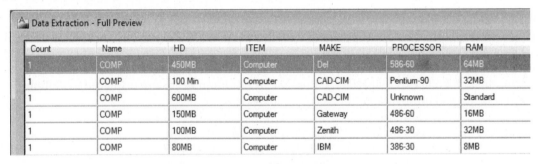

*Figure 19-31 The **Full Preview** window*

When you right-click on the table displayed on the **Refine Data** page, a shortcut menu will be displayed, see Figure 19-32. The options in the shortcut menu are discussed next.

Sort Descending/Sort Ascending

These options sort the data of the selected column in the descending/ ascending order.

Sort Columns Options

This option is used to invoke the **Sort Columns** dialog box, which has been discussed earlier.

Rename Column

This option is used to rename the column head of the selected column.

Hide Column

This option is used to hide the selected column from the table of the extracted data. The selected column is only removed from the view, it is not deleted.

Show Hidden Columns

This option is used to display the hidden columns. This option becomes available only when atleast one column is hidden from the view. You can choose to display all the hidden columns or only the selected one.

Set Column Data Format

This option is used to invoke the **Set Cell Format** dialog box. You can set the data format for the selected row through this dialog box.

Insert Formula Column

This option is used to insert a formula column in the extracted data. Choose the **Insert Formula Column** option to invoke the **Insert Formula Column** dialog box, see Figure 19-33. You can define and insert a formula column using this dialog box. The entries in the **Formula** column are governed by the mathematical formula defined by using one or more columns. The entry in a particular cell of the formula column can be calculated by using the formula and taking the value from the same row under the column that has been used to define the formula. Enter the column head name to be displayed on the formula column in the **Column name** edit box. The **Formula** area is used to define the

Figure 19-32 The shortcut menu for the Refine Data page

formula. To select a column to define the formula, double click on the column in the **Columns** list box, or drag and drop the selected column into the **Formula** area. To use an operator such as +,-,*,or / in the formula, choose the desired operator button from the **Formula** area. You can also make entry in the **Formula** area through the keyboard. Choose the **Validate** button to check the validity of the formula. If the defined formula is valid, AutoCAD will display the **Valid Equation** message box, as shown in Figure 19-34, and if the formula defined is invalid, the **Invalid Equation** message box will be displayed, as shown in Figure 19-35. The **Formula** column is always inserted to the right of the selected column and is indicated by a formula symbol on the side of the column head name, see Figure 19-36.

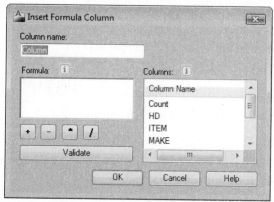

Figure 19-33 The Insert Formula Column dialog box

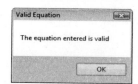

Figure 19-34 The Valid Equation message box

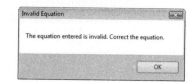

Figure 19-35 The Invalid Equation message window

Count	▲	Name	HD	fx	Quantity	ITEM	▼	MAKE
1		COMP	100 Min		2.0000	Computer		CAD-CIM
1		COMP	600MB		2.0000	Computer		CAD-CIM

*Figure 19-36 The **Formula** and **Filter** symbols*

Edit Formula Column

This option is used to edit the pre-defined formula. Choose this option to invoke the **Edit Formula Column** dialog box.

Remove Formula Column

This option is used to delete the formula column from the extracted data. The **Edit Formula Column** and the **Remove Formula Column** options are available in the shortcut menu only when you right-click on the formula column.

Combine Record Mode

This option is used to control the display of the numeric data into the selected columns by displaying them either as separate values or collapsing the identical property rows into one and displaying their total sum. If any formula column is added and you apply the **Combine Record Mode** option to the formula column, then in case of the **Separate Values** sub-option, the formula will be applied after combining the identical rows. But in case of **Sum Values** sub-option, the formula will be applied to the individual rows and then the identical rows will be combined together and the resultant value will be displayed as the sum of those individual columns. This option will be available only when the selected columns have numerical data and the **Combined identical rows** check box is selected.

Show Count Column/Show Name Column

These options are used to display the **Count** column and the **Name** column, respectively. These options are selected by default.

Insert Totals Footer

This option is used to add a footer row under the selected column. This row can display the sum, maximum or minimum value in the selected column, and the average of all values above the footer row in that column. This option is only available for the columns that have numerical data.

Remove Totals Footer

This option is used to remove the totals footer from the selected row.

Filter Options

This option is used to decide the filter criterion for the selected column. Choose this option to invoke the **Filter <column name>** dialog box, see Figure 19-37. Only the entries that fulfill the specified filter criterion will be displayed on the table. The column on which the filter is applied is represented by a Filter symbol on the side of the column name, refer to Figure 19-36.

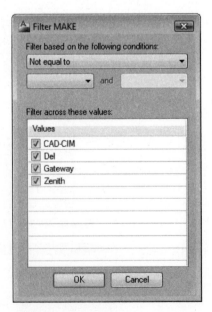

*Figure 19-37 The **Filter** <column name> dialog box*

Reset Filter/Reset All Filters

These options are used to remove the applied filters from the selected columns or from all the columns of the table at a time respectively.

Copy to Clipboard

This option is used to copy the selected table to the clipboard.

After formatting the data from the extracted files, choose the **Next** button to display the **Choose Output** page.

Choose Output

The **Choose Output** page, as shown in Figure 19-38, allows you to specify the format of the output file and the location to save it. Select the **Insert data extraction table into drawing** check box to insert the extracted data into the drawing as a table. You will be prompted to specify the insertion point of the table when the wizard is finished. Select the **Output data to external file (.xls .csv .mdb .txt)** check box to save the extracted data to a separate file at the end of the wizard. Select the **Browse [...]** button to specify the location to save this file. Choose the **Next** button from the **Choose Output** page to move to the **Table Style** page.

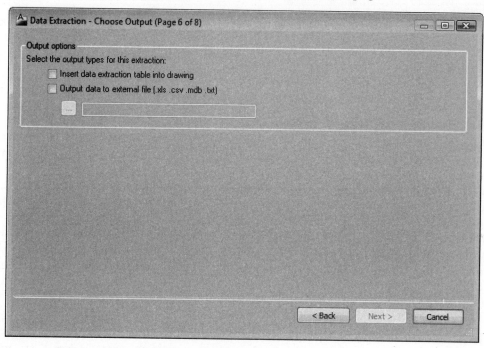

Figure 19-38 The ***Choose Output*** *page of the* ***Data Extraction*** *wizard*

Table Style

The **Table Style** page, as shown in Figure 19-39, will be available only if the **Insert data extraction table into drawing** check box is selected in the **Choose Output** page. The options in the **Table Style** page are discussed next.

Table style Area

You can select the table style from the **Select the table style to use for the inserted table** drop-down list. By default, only the **Standard** table style is available. To create a new table style, choose the **Table Style** button on the right of the **Select the table style to use for the inserted**

table drop-down list; the **Table Style** dialog box will be displayed. You can use this dialog box to create a new table style or modify an existing table style to be used for the extracted attributes.

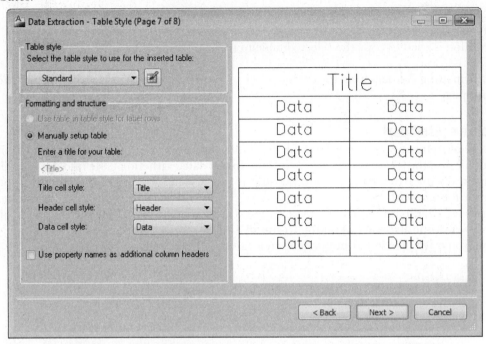

Figure 19-39 *The* ***Table Style*** *page of the* ***Data Extraction*** *wizard*

Formatting and structure Area

The **Use table in table style for label rows** radio button will be available only when the table style that you have specified contains a table template. Select this radio button to insert the extracted data into the pre-defined tables with the pre-defined header and footers. Select the **Manually setup table** radio button to enter the title, header, and data cell styles manually. Enter the title for the table in the **Enter a title for your table** text box. Select the title style defined in the specified table styles in the **Title cell style** drop-down list. Similarly, you can specify the header and data cell style in their respective drop-down lists. The **Use property names as additional column headers** check box is selected by default. This option enables you to include the column headers into the table and use the displayed name as the header row. The preview of the table style is displayed on the right of the **Table Style** page. After specifying the **Table Style** formatting and structure, choose the **Next** button to move to the **Finish** page.

Finish

The **Finish** page, as shown in Figure 19-40, completes the extraction of data. Choose the **Finish** button and specify the insertion point in the drawing area to insert the extracted data. Note that this is possible only if the **Insert data extraction table into drawing** check box is selected on the **Choose Output** page of the **Data Extraction** wizard.

The ATTEXT Command for Attribute Extraction

Command: ATTEXT

The **ATTEXT** command allows you to use the **Attribute Extraction** dialog box (Figure 19-41) for extracting attributes. The information about the **File Format**, **Template File**, and **Output File** must be entered in the dialog box to extract the defined attribute. Also, you must select the blocks whose attribute values you want to extract. If you do not specify a particular block, all the blocks in the drawing are used.

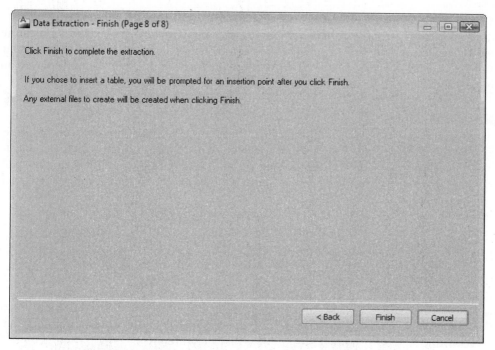

*Figure 19-40 The **Finish** page of the **Data Extraction** wizard*

File Format Area

This area of the dialog box lets you select the file format. You can select either of the three radio buttons available in this area. They are **Comma Delimited File (CDF)**, **Space Delimited File (SDF)**, and **DXF Format Extract File (DXX)**. The format selection is determined by the application that you plan to use to process data. Both CDF and SDF formats can be used with the database software. All of these formats are printable.

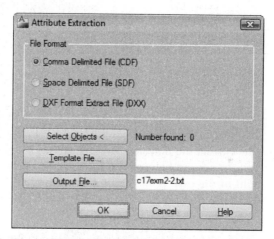

*Figure 19-41 The **Attribute Extraction** dialog box*

Comma Delimited File (CDF). When you select this radio button, the extracted attribute information is displayed in a CDF format. Here, each character field is enclosed in single quotes, and the records are separated by a delimiter (comma by default). A CDF file is a text file with the extension *.txt*. Of the three formats available, this is the most cumbersome.

Space Delimited File (SDF). In SDF format, the records are of fixed width as specified in the template file. The records are separated by spaces and the character fields are not enclosed in single quotes. The SDF file is a text file with the extension *.txt*. This file format is the most convenient and easy to use.

DXF Format Extract File (DXX). If you select this file format, you will notice that the **Template File** button and edit box in the **Attribute Extraction** dialog box are not available. This is because extraction in this file format does not require any template. The file created by this option contains only block references, attribute values, and end-of-sequence objects. This is the most complex of the three file formats available and is related to programming. The extension of these files is *.dxx*.

Select Objects

Select the blocks with attributes whose attribute information you want to extract. You can use any object selection method. Once you have selected the blocks that you want to use for attribute extraction, right-click or press ENTER. The **Attribute Extraction** dialog box is redisplayed and the number of objects you have selected is displayed adjacent to **Number found**.

Template File

When you choose the **Template File** button, the **Template File** dialog box is displayed, where you are allowed to select a template file that has been defined already. After you have selected a template file, choose the **Open** button to return to the **Attribute Extraction** dialog box. The name of the selected file is displayed in the **Template File** edit box. The template file is saved with the extension of the file as *.txt*. The following are the fields that you can specify in a template file (the comments given on the right are for explanation only; they must not be entered with the field description):

BL:LEVEL	Nwww000	(Block nesting level)
BL:NAME	Cwww000	(Block name)
BL:X	Nwwwddd	(X coordinate of block insertion point)
BL:Y	Nwwwddd	(Y coordinate of block insertion point)
BL:Z	Nwwwddd	(Z coordinate of block insertion point)
BL:NUMBER	Nwww000	(Block counter)
BL:HANDLE	Cwww000	(Block's handle)
BL:LAYER	Cwww000	(Block insertion layer name)
BL:ORIENT	Nwwwddd	(Block rotation angle)
BL:XSCALE	Nwwwddd	(X scale factor of block)
BL:YSCALE	Nwwwddd	(Y scale factor of block)
BL:ZSCALE	Nwwwddd	(Z scale factor of block)
BL:XEXTRUDE	Nwwwddd	(X component of block's extrusion direction)
BL:YEXTRUDE	Nwwwddd	(Y component of block's extrusion direction)
BL:ZEXTRUDE	Nwwwddd	(Z component of block's extrusion direction)
Attribute tag		(The tag name of the block attribute)

The extract file may contain several fields. For example, the first field might be the item name and the second field might be the price of the item. Each line in the template file specifies one field in the extract file. A line in a template file consists of the name of the field, the width of the field in characters, and its numerical precision (if applicable). For example, consider the case given next.

ITEM	N015<u>002</u>
<u>BL:NAME</u>	C015000

Where	**BL:NAME**	--Field name
	Blankspaces	---Blank spaces (must not include the tab character)
	C	------------------Designates a character field
	N	------------------Designates a numerical field
	015	---------------Width of field in characters
	002	---------------Numerical precision

BL:NAME or **ITEM**	Indicates the field names; can be of any length.
C	Designates a character field; that is, the field contains characters or it starts with characters. If the file contains numbers or starts with numbers, then C will be replaced by N. For example, **N015002**.

015 Designates a field that is fifteen characters long.

002 Designates the numerical precision. In this example, the numerical precision is 2, or two places following the decimal. The decimal point and the two digits following the decimal are **included in the field width**. In the next example, (000), the numerical precision, is not applicable because the field does not have any numerical value (the field contains letters only).

After creating a template file, when you choose the **Template File** button, the **Template File** dialog box is displayed, where you can browse and select a template file (Figure 19-42).

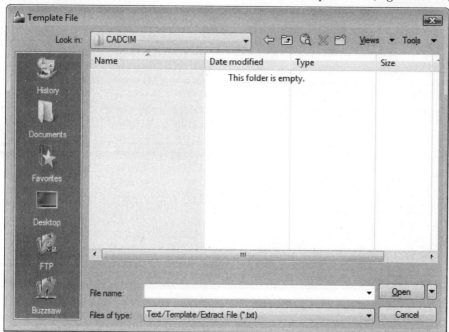

*Figure 19-42 The **Template File** dialog box to select a template file*

Note
You can put any number of spaces between the field name and the character C or N (ITEM N015002). However, you must not use the tab characters. Any alignment in the fields must be done by inserting spaces after the field name.

In the template file, a field name must not appear more than once. The template file name and the output file name must be different.

The template file must contain at least one field with an attribute tag name because the tag names determine which attribute values are to be extracted and from which blocks. If several blocks have different block names but the same attribute tag, AutoCAD will extract attribute values from all the selected blocks. For example, if there are two blocks in the drawing with the attribute tag PRICE, then when you extract the attribute values, AutoCAD will extract the value from both blocks (if both blocks were selected). To extract the value of an attribute, the tag name must match the field name specified in the template file. AutoCAD automatically converts the tag names and the field names to uppercase letters before making a comparison.

Output File
When you choose the **Output File** button, the **Output File** dialog box is displayed. You can select an existing file here, if you want the extracted or output file to be saved as an existing file. You can enter a name in the **File name** edit box in the **Output File** dialog box (Figure 19-43) and

then choose the **Save** button, if you want to save the output file as a new file. By default, the output file has the same name as the drawing name. For example, a drawing named *Drawing1. dwg* will have an output file by the name *Drawing1.txt* by default. Once a name for the output file is specified, it is displayed in the **Output File** edit box in the **Attribute Extraction** dialog box. You can also enter the file name in this edit box. As discussed earlier, AutoCAD appends *.txt* file extension for CDF or SDF files and *.dxx* file extension for DXF files.

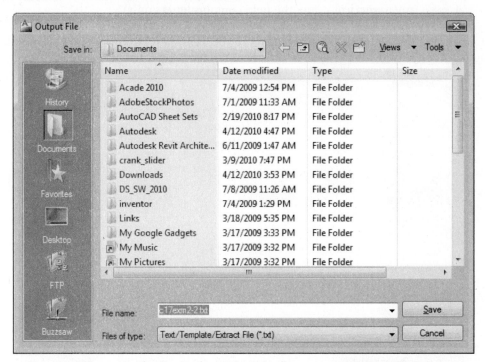

*Figure 19-43 The **Output File** dialog box to save an output file*

Note
*You can also use the **-ATTEXT** command to extract attributes using the Command prompt. Here, you are prompted to specify a file format for the extract information and you can also select specific blocks to extract their attribute information. You can specify a template file using the **Select Template File** dialog box and an extract file using the **Create extract file** dialog box.*

EXAMPLE 3

In this example, you will extract the attribute values defined in Example 2. Extract the values of NAME, COUNT, MAKE, PROCESSOR, HD, RAM, and ITEM. These attribute values must be saved in a *.txt* format named TEST and arranged, as shown in the table given next.

COMP	1	Del	586-60	450 MB	64 MB	Computer
COMP	1	CAD-CIM	Pentium-90	100 Min	32 MB	Computer
COMP	1	CAD-CIM	Unknown	600 MB	Standard	Computer
COMP	1	Gateway	486-60	150 MB	16 MB	Computer
COMP	1	Zenith	486-30	100 MB	32 MB	Computer
COMP	1	IBM	386-30	80 MB	8 MB	Computer

1. Open the drawing that you saved in Example 2.

2. Choose the **Extract Data** tool from the **Link & Extraction** panel of the **Insert** tab; the **Data Extraction** wizard is displayed. Select the **Create a new data extraction** radio button from the **Begin** page and choose the **Next** button.

3. Enter **TEST** in the **File name** edit box of the **Save Data Extraction As** dialog box and choose the **Save** button.

4. Select the **Drawings/Sheet set** radio button and make sure that the **Include current drawing** check box is selected on the **Define Data Source** page. Choose the **Next** button.

5. Clear all check boxes in the **Object** column except the **COMP** object and choose the **Next** button.

6. Clear all check boxes except the **Attribute** check box from the **Category filter** area of the **Select Properties** page of the **Data Extraction** wizard. Also, make sure that the check box on the left of the **HD**, **ITEM**, **MAKE**, **PROCESSOR**, and **RAM** properties under the **Property** column are selected. Choose the **Next** button.

7. Arrange the display order of the columns according to the table displayed above, by dragging and dropping the column heads to the desired position in the **Refine Data** page of the **Data Extraction** wizard. Make sure that the **Show count column** and **Show name column** check boxes are selected. Choose the **Full Preview** button and check the data displayed in the **Data Extraction - Full Preview** window and match it with the table displayed above. Close the window and choose the **Next** button from the **Refine Data** page of the **Data Extraction** wizard.

8. Select the **Output data to external file (.xls .csv .mdb .txt)** check box and choose the **Browse [...]** button. Specify the location to save the text file, select the **Text/Template/Extract File (*.txt)** option from the **Files of type** drop-down list, enter **TEST.txt** in the **File name** edit box, and choose the **Save** button. Choose the **Next** button from the **Choose Output** page.

9. Choose the **Finish** button from the **Finish** page of the **Data Extraction** wizard.

CONTROLLING ATTRIBUTE VISIBILITY

Ribbon: Block > Retain Attribute Display drop-down > Retain Attribute Display /Display All Attributes/Hide All Attributes   **Command:** ATTDISP

The **ATTDISP** command allows you to change the visibility of all attribute values. Normally, the attributes are visible unless they are defined invisible by using the **Invisible** mode. The invisible attributes are not displayed, but they are a part of the block definition. You can select any one of the options of the **ATTDISP** command to turn the display of the attributes completely on or off. If the attributes are created in the **Invisible** mode and you have selected the **Normal** option from the **ATTDISP** command, the attributes continue to be invisible. You can also select the options of the **ATTDISP** command from the **Block** panel in the **Insert** tab. The options that can be chosen are **Retain Attribute Display/Display All Attributes/Hide All Attributes** under **Retain Attribute Display** drop-down. The prompt sequence for the **ATTDISP** command is given next.

Command: **ATTDISP**
Enter attribute visibility setting [Normal/ON/OFF] <Current>: *Specify an option and press ENTER.*

When you choose **ON**, all attribute values will be displayed, including the attributes that are defined with the **Invisible** mode. You can also do it by choosing **Display All** tool from **Attributes** panel.

If you choose **OFF**, all attribute values will become invisible. You can also do it by choosing **Hide All** tool from the **Attributes** panel. Similarly, if you choose **Normal**, AutoCAD will display the attribute values the way they are defined, that is, the attributes that were defined invisible will stay invisible and the attributes that were defined visible will be visible. You can also do this by choosing the **Retain Display** tool from the **Attributes** panel.

In Example 2, the RAM attribute was defined with the **Invisible** mode. Therefore, the RAM values are not displayed with the block. If you want to make the RAM attribute values visible (Figure 19-44), choose the **Display All** tool from the **Attributes** panel.

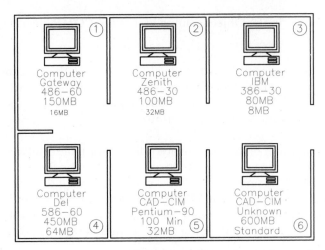

Figure 19-44 Using -ATTEDIT to change the attribute values

 Tip
*After you have defined the attribute values and saved them with the block definition, it may be a good idea to use the **OFF** option of the **ATTDISP** command. By doing this, the drawing is simplified and also the regeneration time is reduced.*

EDITING BLOCK ATTRIBUTES

Ribbon: Insert > Block > Edit Attribute drop-down > Single   **Command:** EATTEDIT
Toolbar: Modify II > Edit Attribute

 The block attributes can be edited using the **EATTEDIT** command.

Editing Attributes Using the Enhanced Attribute Editor
Choose the **Single** tool from the **Edit Attribute** drop-down in the **Attributes** panel to edit the block attribute values. When you invoke this tool, AutoCAD prompts you to select the block whose values you want to edit. After selecting the block, the **Enhanced Attribute Editor** dialog box is displayed (Figure 19-45). This dialog box allows you to change the **Attributes**, **Text Options**, and **Properties** of the block attribute. This dialog box can also be invoked by double-clicking on the block or the attribute in the current drawing. The **Enhanced Attribute Editor** dialog box displays all the attributes of the selected block and their properties. You can change the properties of the attribute you want. If an attribute has been defined with the **Constant** mode, it is not displayed in the dialog box because a constant attribute value cannot be edited. The dialog box displays the following tabs.

Attribute Tab

The options under this tab are used to change the value of the attribute. Also, here the **Mode** of the attribute is not displayed. The other attributes of the block such as **Tag**, **Prompt**, and **Value** are displayed, as shown in Figure 19-45.

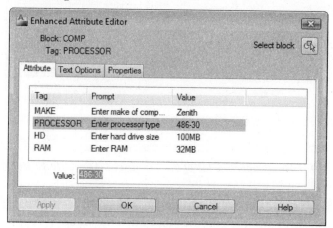

Figure 19-45 *The **Attribute** tab of the **Enhanced Attribute Editor** dialog box*

Text Options Tab

Under this tab, the options related to the text of attributes can be modified. This includes **Text Style**, **Justification**, **Height**, **Rotation**, **Backwards**, **Upside down**, **Width Factor**, **Oblique Angle**, **Annotative**, and **Boundary width**, as shown in Figure 19-46.

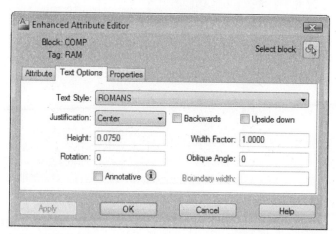

Figure 19-46 *The **Text Options** tab of the **Enhanced Attribute Editor** dialog box*

Properties Tab

This tab can be used to edit the properties of attributes such as **Layer**, **Linetype**, **Color**, **Lineweight**, and **Plot style**. Choose the **Properties** tab of **Enhanced Attribute Editor** to display the dialog box, as shown in Figure 19-47.

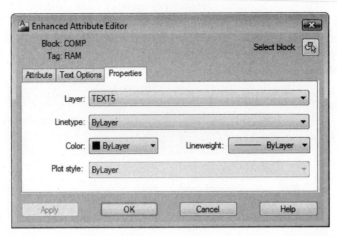

*Figure 19-47 The **Properties** tab of the **Enhanced Attribute Editor** dialog box*

Editing Attributes Using the Edit Attributes Dialog Box

The **ATTEDIT** command is used to edit the block attribute values through the **Edit Attributes** dialog box. Invoke this command; AutoCAD prompts you to select the block whose values are to be edited. After selecting the block, the **Edit Attributes** dialog box is displayed (Figure 19-48).

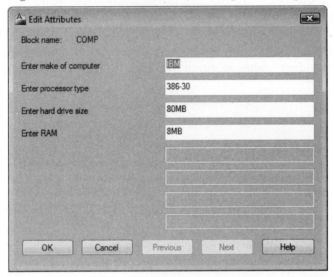

*Figure 19-48 The **Edit Attributes** dialog box*

The dialog box is similar to the **Enter Attributes** dialog box and shows the prompts and the attribute values of the selected block. If an attribute has been defined with the **Constant** mode, it is not displayed in the dialog box because a constant attribute value cannot be edited. To make any changes, select the existing value and enter a new value in the corresponding edit box. After you have made the modifications, choose the **OK** button. The attribute values are updated in the selected block.

If a selected block has no attributes, AutoCAD will display the alert message **That block has no editable attributes**. Similarly, if the selected object is not a block, AutoCAD again displays the alert message **That object is not a block**.

Note
*You cannot use the **ATTEDIT** command to perform the global editing of attribute values or to modify the position, height, or style of the attribute value. The global editing can be performed using the **-ATTEDIT** command, which is discussed in the next section.*

EXAMPLE 4   *EATTEDIT*

In this example, you will choose the **Single** tool from the **Edit Attribute** drop-down in the **Attributes** panel to change the attribute of the first computer (150 MB to 2.1 GB), which is located in Room-1.

1. Open the drawing created in Example 2. The drawing has six blocks with attributes. The name of the block is **COMP**, and it has five defined attributes, one of them is invisible. Zoom in the drawing so that the first computer is displayed on the screen (Figure 19-49).

2. Choose the **Single** tool from **Insert > Attributes > Edit Attribute** drop-down; you are prompted to select a block. Select the computer located in Room-1.

   AutoCAD will display the **Enhanced Attribute Editor** dialog box (refer to Figure 19-45) that shows the attribute prompts and the attribute values.

3. Edit the values, choose **Apply**, and then the **OK** button in the dialog box. When you exit the dialog box, the attribute values are updated.

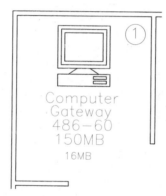

Figure 19-49 Zoomed view of the first computer

Tip
*You can also edit the attributes with the **FIND** command. This command can be invoked by entering **FIND** at the pointer input. When you invoke this command, the **Find and Replace** dialog box is displayed.*

*When you know that a set of attributes in a drawing may have to be changed in the future, you can make a group out of them by using the **GROUP** command. Then, you can choose the **Select objects** button in the **Find and Replace** dialog box, enter **G** at the **Select objects** prompt, and enter the name of the group; all the objects in the group will get selected.*

Global Editing of Attributes

Ribbon: Insert > Block > Edit Attribute drop-down > Multiple   **Command:** -ATTEDIT

Choose the **Multiple** tool from the **Edit Attribute** drop-down in the **Block** panel to edit the attribute values independently of the blocks that contain the attribute reference. For example, if there are two blocks, COMPUTER and TABLE, with the attribute value PRICE, you can globally edit this value (PRICE) independently of the block that references these values. You can also edit the attribute values one at a time. For example, you can edit the attribute value (PRICE) of the block TABLE without affecting the value of the other block, COMPUTER. On choosing the **Multiple** tool from the **Edit Attribute** drop-down, AutoCAD displays the following prompt:

Command: **-ATTEDIT**
Edit attributes one at a time? [Yes/No]<Y>: **N**
Performing global editing of attribute values

If you enter **N** at this prompt, it means that you want to do the global editing of the attributes. However, you can restrict the editing of attributes by block names, tag names, attribute values, and visibility of attributes on the screen.

Editing Visible Attributes Only

After you select global editing, AutoCAD will display the following prompt:

Edit only attributes visible on screen? [Yes/No] <Y>: **Y**

If you enter **Y** at this prompt, AutoCAD will edit only those attributes that are visible and displayed on the screen. The attributes might not have been defined with the **Invisible** mode, but if they are not displayed on the screen, they are not visible for editing. For example, if you zoom in, some of the attributes may not be displayed on the screen. Since the attributes are not displayed on the screen, they are invisible and cannot be selected for editing.

Editing All Attributes

If you enter **N** at the previously mentioned prompt, AutoCAD flips from graphics to text screen and displays the message stating: '**Drawing must be regenerated afterwards**'.

Now, AutoCAD will edit all attributes even if they are not visible or displayed on the screen. Also, changes that you make in the attribute values are not reflected immediately. Instead, the attribute values are updated and the drawing is regenerated after you are done with the command.

Editing Specific Blocks

Although you have selected global editing, you can confine the editing of attributes to specific blocks by entering the block name at the prompt. For example,

Enter block name specification <*>: **COMP**

When you enter the name of the block, AutoCAD will edit the attributes that have the given block (COMP) reference. You can also use the wild-card characters to specify the block names. If you want to edit attributes in all blocks that have attributes defined, press ENTER.

Editing Attributes with Specific Attribute Tag Names

Like blocks, you can confine attribute editing to those attribute values that have the specified tag name. For example, if you want to edit the attribute values that have the tag name MAKE, enter the tag name at the following AutoCAD prompt:

Enter attribute tag specification <*>: **MAKE**

When you specify the tag name, AutoCAD will not edit attributes that have a different tag name, even if the values being edited are the same. You can also use the wild-card characters to specify the tag names. If you want to edit attributes with any tag name, press ENTER.

Editing Attributes with a Specific Attribute Value

Like blocks and attribute tag names, you can confine attribute editing to a specified attribute value. For example, if you want to edit the attribute values that have the value 100 MB, enter the value at the following AutoCAD prompt:

Enter attribute value specification <*>: **100MB**

When you specify the attribute value, AutoCAD will not edit attributes that have a different value, even if the tag name and block specification are the same. You can also use the wild-card characters to specify the attribute value. If you want to edit attributes with any value, press ENTER.

Sometimes the value of an attribute is null, and these values are not visible. If you want to select the null values for editing, make sure you have not restricted the global editing to visible attributes. To edit the null attributes, enter \ at the following prompt:

Enter attribute value specification <*>: \

After you enter this information, AutoCAD will prompt you to select the attributes. You can select the attributes by selecting individual attributes or by using one of the object selection options (Window, Crossing, or individually).

Select Attributes: *Select the attribute values parallel to the current UCS only.*

After you select the attributes, AutoCAD will prompt you to enter the string you want to change and the new string. A string is a sequence of consecutive characters. It could also be a portion of the text. AutoCAD will retrieve the attribute information, edit it, and then update the attribute values.

Enter string to change: *Enter the value to be modified.*
Enter new string: *Enter the new value.*

The following is the complete command prompt sequence of the **-ATTEDIT** command. It is assumed that the editing is global and for visible attributes only.

Command: -**ATTEDIT**
Edit attributes one at a time? [Yes/No] <Y>: N
Performing global editing of attribute values.
Edit only attributes visible on screen? [Yes/No] <Y>: N
Drawing must be regenerated afterwards.
Enter block name specification <*>: `Enter`
Enter attribute tag specification <*>: `Enter`
Enter attribute value specification <*>: `Enter`
n attributes selected.
Enter string to change: *Enter the value to be modified.*
Enter new string: *Enter the new value.*

Note
AutoCAD regenerates the drawing at the end of the command automatically unless the system variable **REGENAUTO** *is off, which controls automatic regeneration of the drawing.*

If you select an attribute defined with the **Constant** *mode using the* **-ATTEDIT** *command, the prompt displays 0 found, since attributes with* **Constant** *mode are not editable.*

EXAMPLE 5   *-ATTEDIT*

In this example, you will use the drawing from Example 2 to edit the attribute values that are highlighted in the following table. The tag names are given at the top of the table (ITEM, MAKE, PROCESSOR, HD, RAM). The RAM values are invisible in the drawing.

	ITEM	MAKE	PROCESSOR	HD	RAM
COMP	Computer	Gateway	486-60	150 MB	**16 MB**
COMP	Computer	Zenith	486-30	100 MB	**32 MB**
COMP	Computer	IBM	386-30	80 MB	**8 MB**
COMP	Computer	Del	586-60	450 MB	**64 MB**
COMP	Computer	**CAD-CIM**	Pentium-90	100 Min	**32 MB**
COMP	Computer	**CAD-CIM**	**Unknown**	600 MB	Standard

Make the following changes in the highlighted attribute values shown in the above table.

1. Change Unknown to Pentium.
2. Change CAD-CIM to Compaq.
3. Change MB to Meg for all attribute values that have the tag name RAM. (No changes should be made to the values that have the tag name HD.)

The following are the steps to change the attribute value from **Unknown** to **Pentium**.

1. Choose the **Multiple** tool from **Insert > Block > Edit Attribute** drop-down and enter **N** at the next prompt and press ENTER as given below.

   Command: **-ATTEDIT**
   Edit attributes one at a time? [Yes/No] <Y>: **N**
   Performing global editing of attribute values.

2. You need to edit only those attributes that are visible on the screen, so press ENTER at the following prompt:

   Edit only attributes visible on screen? [Yes/No] <Y>: `Enter`

3. As shown in the table, the attributes belong to a single block, COMP. A drawing may have more blocks. To confine the attribute editing to the COMP block only, enter the name of the block (COMP) at the next prompt.

   Enter block name specification <*>: **COMP**

4. At the next two prompts, enter the attribute tag name and the attribute value specification. When you enter these two values, only those attributes that have the specified tag name and attribute value will be edited.

   Enter attribute tag specification<*>: **Processor**
   Enter attribute value specification<*>: **Unknown**

   On entering these values, AutoCAD will prompt you to select attributes. Use any object selection option to select all the blocks. AutoCAD will search for the attributes that satisfy the given criteria (attributes belong to the block COMP, the attributes have the tag name Processor, and the attribute value is Unknown). Once AutoCAD locates such attributes, they are highlighted.

5. At the next two prompts, enter the string that you want to change, and then enter the new string.

   Enter string to change: **Unknown**
   Enter new string: **Pentium**

6. The following is the Command prompt sequence to change the make of the computer from **CAD-CIM** to **Compaq**:

Command: **-ATTEDIT**
Edit attributes one at a time? [Yes/No] <Y>: **N**
Performing global editing of attribute values.
Edit only attributes visible on screen? [Yes/No] <Y>: [Enter]
Enter block name specification <*>: **COMP**
Enter attribute tag specification <*>: **MAKE**
Enter attribute value specification <*>: [Enter]
Select Attributes: *Use any selection method to select the attributes.*
n attributes selected.
Select Attributes: [Enter]
Enter string to change: **CAD-CIM**
Enter new string: **Compaq**

7. The following is the Command prompt sequence to change **MB** to **Meg**:

Command: **-ATTEDIT**
Edit attributes one at a time? [Yes/No] <Y>: **N**
Performing global editing of attribute values.
At the next prompt, you must enter **N** because the attributes you want to edit (tag name, RAM) are not visible on the screen.

Edit only attributes visible on screen? [Yes/No] <Y>: **N**
Drawing must be regenerated afterwards.
Enter block name specification <*>: **COMP**

At the next prompt, about the tag specification, you must specify the tag name because the text string MB also appears in the hard drive size (tag name, HD). If you do not enter the tag name, AutoCAD will change all MB attribute values to Meg.

Enter attribute tag specification <*>: **RAM**
Enter attribute value specification <*>:
n Attributes selected
Enter string to change: **MB**
Enter new string: **Meg**

8. Choose the **Display All Attributes** tool from the **Block** panel to display the invisible attributes.

Note
You can also use the -ATTEDIT command to edit the attribute values individually. Attribute value, position, height, angle, style, layer, and color can be changed using this command.

EXAMPLE 6   *EATTEDIT*

In this example, you will use the drawing of Example 5 to edit attributes individually (Figure 19-50).

Make the following changes in the attribute values:

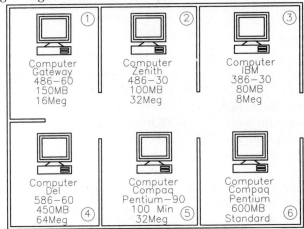

Figure 19-50 Using -ATTEDIT to change the attribute values individually

a. Change the attribute value 100 Min to 100 MB.

b. Change the height of all attributes with the tag name RAM to 0.075 units.

1. Load the drawing that you have saved in Example 5.

2. Double-click on the **100Min** attribute value in the drawing; the **Enhanced Attribute Editor** dialog box is displayed.

3. In the **Value** edit box, change **100 Min** to **100MB**. Choose **Apply** and then choose **OK**.

4. To change the height of attribute text, double-click on the value of **RAM** in the drawing; the **Enhanced Attribute Editor** dialog box is displayed.

5. In the dialog box, choose the **Text Options** tab and change the height to **0.075** in the **Height** edit box.

6. Choose **Apply** and then **OK**.

7. Repeat Steps 5 and 6 to change the height of the other attribute values.

Tip
When you are defining attributes and have certain attributes whose values are not known at that time, you should enter AAAA or something similar. Later, you can replace such text with the values that you have obtained, using the ATTEDIT or -ATTEDIT commands. This is easier than adding an attribute later after the block has already been defined.

REDEFINING A BLOCK WITH ATTRIBUTES

If a block with attributes has been inserted a number of times in a drawing and you wish to edit the block geometry, or add or delete attributes from them, it may take a lot of time. You can use the **ATTREDEF** command to redefine an existing block with attributes, and once redefined, all the copies of the block with attributes get updated automatically. But before using this command, you need to explode a copy of the block containing the attributes and make the changes or you may use the new geometry. If you do not explode a block and try to use this command, an error message will be displayed: **New block has no attributes**. The prompt sequence is as follows:

Command: **ATTREDEF**
Enter name of the block you wish to redefine: *Enter the name of the block to be redefined and press ENTER.*
Select objects for new Block...
Select objects: *Select the new block geometry with attributes and press ENTER.*
Specify insertion base point of the new Block: *Specify an insertion base point.*

After you select an insertion base point, the block is redefined and all the copies of the block in the drawing are updated. If you do not include an existing attribute when selecting objects for the new block, it will not be a part of the new block definition.

Note
*The **ATTREDEF** command removes all the formatting changes incorporated into the property by the **ATTEDIT** or **EATTEDIT** command.*

In-place Editing of Blocks with Attributes

Ribbon: Insert > Reference > Edit Reference   **Command:** REFEDIT
Toolbar: Refedit > Edit Reference In-Place

The **Edit Reference** tool has already been discussed in Chapter 18, Working with Blocks. This tool can also be used for in-place editing of blocks with attributes. After invoking the **Edit Reference** tool, select a block with attributes that you want to edit. The **Reference Edit** dialog box is displayed with the name of the block in the **Reference name** list box. To be able to edit attributes, you should select the **Display attribute definitions for editing** check box in the **Settings** tab of the dialog box and then choose **OK** to exit. When you select this check box, it displays the attributes and makes them available for editing. At the next prompt, **Select nested objects**, you can select any part of the block with attributes that need to be edited and press ENTER. If you want to edit some other objects that are part of the block, you can select them too. All the selected objects become a part of the working set. You will notice that when you select the block with attributes, the tags replace the assigned attribute values and objects in the drawing that were not selected appear faded.

You can now edit the block geometry or edit only the attributes. Enter the **DDEDIT** command at the Command prompt and select the attribute to be edited at the next prompt; the **Edit Attribute Definition** dialog box is displayed. Make changes in the values displayed in the **Tag**, **Prompt**, and **Default** edit boxes and choose **OK** to exit the dialog box. You can then select another attribute to edit and so on until you have made all the modifications in the attributes you intended and then press ENTER to exit the command. Now, choose the **Save Changes** tool in the **Edit Reference** panel. An AutoCAD message box is displayed prompting you whether you want to save the changes made or cancel the command and discard the changes. Choose **OK** to save the completed changes and exit the **REFEDIT** command.

Once the block has been redefined, the blocks that already exist in the drawing are not updated, unless a change is made to a default value of an attribute that has been defined with the **Constant** mode. The modifications made using the **REFEDIT** command can be seen in the blocks inserted later.

Note
*You can also use the **-REFEDIT** command to change block attributes. The prompts are displayed at the pointer and Command prompt. All options available through the **REFEDIT** command are available here also. The **DDEDIT** command is used to make modifications in the attribute tags. Then, you can either choose the **Save Changes** button in the **Edit Reference** panel of the **Ribbon** or enter the **REFCLOSE** command to save and exit the command.*

INSERTING TEXT FILES IN THE DRAWING

Ribbon: Annotate > Text > Text drop-down > Multiline Text   **Command:** MTEXT
Menu Bar: Draw > Text > Multiline Text

 After you have extracted attribute information to a file, you may want to insert this text into a drawing. You can insert this text file in a drawing by using the **Import Text** option that is displayed when you right-click in the text window of the **Text Editor**. The **Text Editor** is displayed on using the **Multiline Text** tool.

On invoking the **Multiline Text** tool, AutoCAD prompts you to enter the insertion point and other corner of the paragraph text box, within which the text file will be placed. After you specify these points, the **Text Editor** is displayed. To insert the text file *TEST.txt* (created in Example 3), right-click in the text window of the **Text Editor** to display the shortcut menu. In this shortcut menu, choose **Import Text**. AutoCAD displays the **Select File** dialog box.

In the **Select File** dialog box, you can select the text file **TEST** from the list box and then choose the **Open** button. The imported text is displayed in the text window of the **Text Editor** (Figure 19-51). Note that only the ASCII files are properly interpreted. Now, click on the screen, outside the text editor, to get the imported text in the selected area on the screen (Figure 19-52).

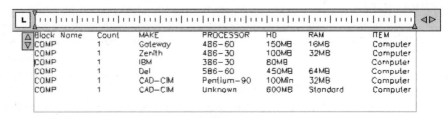

Block Name	Count	MAKE	PROCESSOR	HD	RAM	ITEM
COMP	1	Gateway	486-60	150MB	16MB	Computer
COMP	1	Zenith	486-30	100MB	32MB	Computer
COMP	1	IBM	386-30	80MB		Computer
COMP	1	Del	586-60	450MB	64MB	Computer
COMP	1	CAD-CIM	Pentium-90	100Min	32MB	Computer
COMP	1	CAD-CIM	Unknown	600MB	Standard	Computer

*Figure 19-51 The **Text Editor** displaying the imported text*

Block Name	Count	MAKE	PROCESSOR	HD	RAM	ITEM
COMP	1	Gateway	486-60	150MB	16MB	Computer
COMP	1	Zenith	486-30	100MB	32MB	Computer
COMP	1	IBM	386-30	80MB8MB		Computer
COMP	1	Del	586-60	450MB	64MB	Computer
COMP	1	CAD-CIM	Pentium-90	100Min	32MB	Computer
COMP	1	CAD-CIM	Unknown	600MB	Standard	Computer

Figure 19-52 The imported text file on the screen

You can also use the **Multiline Text** tool to change the text style, height, direction, width, rotation, line spacing, and attachment.

Self-Evaluation Test

Answer the following questions and then compare them to those given at the end of this chapter:

1. Like the **Constant** attribute, the **Preset** attribute cannot be edited. (T/F)

2. In case of tag names, any lowercase letter is automatically converted to an uppercase letter. (T/F)

3. You can choose the **Single** tool from the **Edit Attribute** drop-down in the **Block** panel to modify the justification, height, or style of an attribute value. (T/F)

4. If you select the **Display All** option in the **Attribute Display** drop-down list of the **Block** panel, then even the attributes defined with the **Invisible** mode are displayed. (T/F)

5. The entry in the **Value** edit box of the **Attribute Definition** dialog box defines the _____ of the specified attribute.

6. If you have selected the **Align** option from the **Justification** drop-down list in the **Text Options** tab of the **Enhanced Attribute Editor** dialog box, the **Height** and **Rotation** edit boxes will be _____.

7. You can use the _____ and _____ commands or the **Properties** palette to edit text or attribute definitions.

8. If the default value of the **ATTDIA** variable is _____, it disables the dialog box.

9. In the _____ file, the records are not separated by a comma and the character fields are not enclosed in single quotes.

10. The _____ command is used to redefine an existing block with attributes in a drawing, and once redefined, all copies of the block in the drawing are updated.

Review Questions

Answer the following questions:

1. If you do not enter anything in the **Prompt** edit box, the entry made in the **Tag** edit box will be displayed as prompt. (T/F)

2. You can also use the **Find and Replace** dialog box to modify attribute values. (T/F)

3. When using the **Edit Reference** tool for in-place editing of attributes, the existing blocks in a drawing do not get updated. Only the subsequent insertions display the modified block. (T/F)

4. You can use ? and * in the string value. When these characters are used in string value, AutoCAD does not interpret them as wild-card characters. (T/F)

5. The template file name and the output file name can be the same. (T/F)

6. Not selecting any of the check boxes in the **Mode** area of the **Attribute Definition** dialog box displays all the prompts at the Command prompt and the values will be visible on the screen. This is also referred to as which mode?

    (a) Formal   (b) Normal
    (c) Abnormal   (d) None of these

7. Which of the following system variables when set to 0 will suppress the display of prompts for the new values?

    (a) **ATTDIA**   (b) **ATTREQ**
    (c) **ATTMODE**   (d) **ATTDEF**

8. Selecting the **OFF** option of which command turns off the visibility of all attribute values?

   (a) **ATTDISP** (b) **ATTEXT**
   (c) **ATTEDIT** (d) **ATTDEF**

9. AutoCAD regenerates the drawing at the end of the **-ATTEDIT** command, unless which of the following is turned off?

   (a) **AUTOSNAP** (b) **REGENAUTO**
   (c) **ATTDIA** (d) **ATTMODE**

10. If you need a leading blank space in the string to be changed, the character string must start with which one of the following characters?

    (a) space ( ) (b) backlash (\\)
    (c) asterisk (*) (d) colon (:)

11. You can insert the text file in the drawing by choosing the _____ button in the **In-Place Text Editor** dialog box displayed when using the **Multi Line Text** tool.

12. The function of the **Preset** option is _____. _____

13. If you select the **Constant** check box in the **Mode** area of the **Attribute Definition** dialog box, the **Prompt** edit box will _____.

14. You should select the _____ check box in the **Attribute Definition** dialog box to automatically place the subsequent attribute text just below the previously defined attribute.

15. The attribute extract file is saved as a _____ file.

EXERCISE 1   *Attribute Definition*

In this exercise, you will define the following attributes for a resistor and then create a block using the **BLOCK** command. The name of the block is **RESIS**.

Mode	Tag name	Prompt	Default value
Verify	RNAME	Enter name	RX
Verify	RVALUE	Enter resistance	XX
Verify, Invisible	RPRICE	Enter price	00

1. Draw the resistor, as shown in Figure 19-53.

2. Enter **ATTDEF** at the AutoCAD Command prompt to invoke the **Attribute Definition** dialog box.

3. Define the attributes, as shown in the preceding table, and position the attribute text as shown in Figure 19-53.

4. Use the **BLOCK** command to create a block. The name of the block is **RESIS**, and the insertion point of the block is at the left end of the resistor. When you select the objects for the block, make sure you also select the attributes.

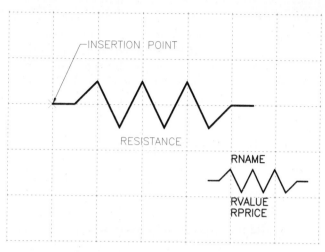

Figure 19-53 *Drawing of a resistor for Exercise 1*

EXERCISE 2   *ATTDIA*

In this exercise, you will use the **Insert** tool to insert the block that was defined in Exercise 1 (RESIS). The following is the list of the attribute values for the resistances in the electric circuit:

RNAME	RVALUE	RPRICE
R1	35	.32
R2	27	.25
R3	52	.40
R4	8	.21
RX	10	.21

1. Draw the electric circuit diagram, as shown in Figure 19-54 (assume the dimensions).

2. Set the system variable **ATTDIA** to **1**. Use the **Insert** tool to insert the blocks, and define the attribute values in the **Attribute Definition** dialog box.

3. Repeat the **Insert** tool to insert other blocks, and define their attribute values as given in the table. Save the drawing as *attexr2.dwg* (Figure 19-55).

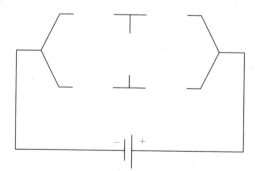

Figure 19-54 *Electric circuit diagram without resistors for Exercise 2*

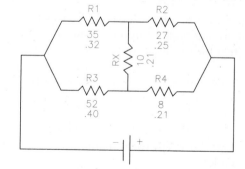

Figure 19-55 *Electric circuit diagram with resistors for Exercise 2*

EXERCISE 3 — Data Extraction

In this exercise, you will extract the attribute values that were defined in Exercise 2. Extract the values of RNAME, RVALUE, and RPRICE. These attribute values must be saved in a Tab Delimited File format named **RESISLST** and arranged, as shown in the following table.

RESIS	1	R1	35	.32
RESIS	1	R2	27	.25
RESIS	1	R3	52	.40
RESIS	1	R4	8	.21
RESIS	1	RX	10	.21

1. Load the drawing **ATTEXR2** that you saved in Exercise 2.

2. Use the **EATTEXT** command to invoke the **Data Extraction** wizard, and select the **Select Objects** radio button. Now, choose **Select Objects** button and select the objects in the current drawing, whose attributes have to be extracted. Now, choose **Next** in the wizard.

3. On the **Additional Settings** page, clear both the check boxes of the Xrefs and Nested blocks and choose **OK** and then **Next**.

4. On the **Use Template** page, select the **No Template** radio button and choose **Next**.

5. On the **Select Attribute** page, under the Block Name heading, select the check box named **RESIS**. In the attribute list, clear all the check boxes and select the **RNAME**, **RVALUE**, and **RPRICE** check boxes. Choose **Next**.

6. The **View output** page shows the result of your query. On this page, you can choose the view in which you want the data to be displayed in the exported file. Select the **Next** button.

7. Ignore the **Save template** page and choose **Next**.

8. On the **Export** page, enter **RESISLST** as the name of the output file, and save it in **Tab Delimited File** format. Choose **Finish**. Use Notepad to list the output file, **RESISLST**. The output file will be similar to the file shown at the beginning of Exercise 3.

EXERCISE 4 — Edit Attribute

In this exercise, you will use the **EATTEDIT** command to change the attributes of the resistances that are highlighted in the following table. You will also extract the attribute values and insert the text file in the drawing.

1. Load the drawing **ATTEXR2** that was created in Exercise 2. The drawing has five resistances with attributes. The name of the block is RESIS, and it has three defined attributes, one of them is invisible.

2. Use the **EATTEDIT** command to edit the values that are highlighted in the following table.

RESIS	R1	**40**	.32
RESIS	R2	**29**	.25

RESIS	R3	52	**.45**
RESIS	R4	8	**.25**
RESIS	**R5**	10	.21

3. Extract the attribute values and write the values to a text file.

4. Use the **Multiline Text** tool to insert the text file in the drawing.

EXERCISE 5 — *Attribute Data Extraction*

Use the information given in Exercise 3 to extract the attribute values, and write the data to the output file. The data in the output file should be Comma Delimited CDF. Use the **ATTEXT** or **-ATTEXT** commands to extract the attribute values.

EXERCISE 6 — *Attribute Data Extraction*

In this exercise, you will create the blocks with the required attributes. Next, draw the circuit diagram shown in Figure 19-56 and then extract the attributes to create the Bill of Materials.

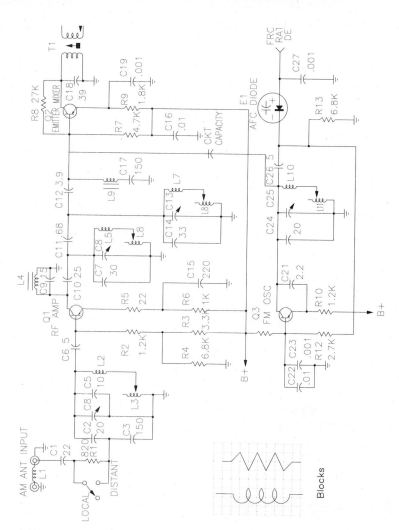

***Figure 19-56** Drawing of the circuit diagram for Exercise 6*

EXERCISE 7 — *Data Extraction*

In this exercise, you will create the blocks with the required attributes. Next, draw the circuit diagram shown in Figure 19-57 and then extract the attributes to create the Bill Of Materials.

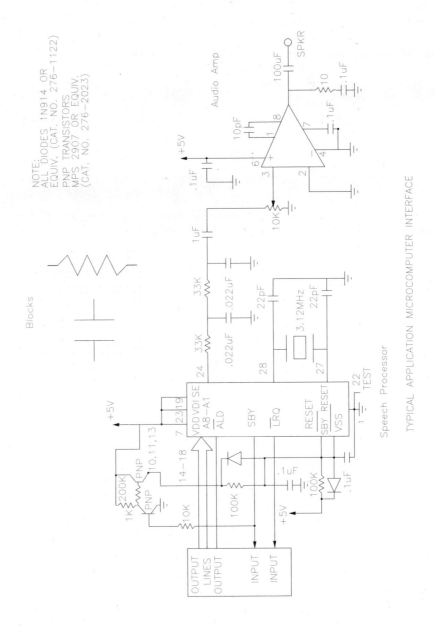

Figure 19-57 Drawing for Exercise 7

Answers to Self-Evaluation Test

1. F, **2**. T, **3**. T, **4**. T, **5**. default value, **6**. disabled, **7**. **EATTEDIT**, **CHANGE**, **8**. 0, **9**. Space Delimited, **10**. **ATTREDEF**

Chapter 20

Understanding External References

CHAPTER OBJECTIVES

In this chapter, you will learn:
- *About external references and their applications.*
- *About dependent symbols.*
- *To use the External References Palette.*
- *To use the Attach, Unload, Reload, Detach, and Bind options.*
- *To edit the path of an xref.*
- *The difference between the Overlay and Attachment options.*
- *The use of the Attach tool.*
- *To work with Underlays.*
- *To use the DesignCenter to attach a drawing as an xref.*
- *To use the Bind tool to add dependent symbols.*
- *To use the Clip tool to clip xref drawings.*
- *To use the REFEDIT command for in-place editing.*

KEY TERMS

- *XRef*
- *Attach*
- *Reload*
- *Detach*
- *Unload*
- *Bind*
- *Overlay*
- *Underlay*
- *Frame*
- *Clip*

EXTERNAL REFERENCES

The external reference feature allows you to reference an external drawing without making that drawing a permanent part of the existing drawing. For example, assume that we have an assembly drawing ASSEM1 that consists of two parts, SHAFT and BEARING. The SHAFT and BEARING are separate drawings created by two CAD operators or provided by two different vendors. We want to create an assembly drawing from these two parts. One way to create an assembly drawing is to insert these two drawings as blocks by using the **Insert** tool in the **Block** panel. Now assume that the design of BEARING has changed due to customer or product requirements. To update the assembly drawing, you have to make sure that you insert the BEARING drawing after the changes have been made. If you forget to update the assembly drawing, then the assembly drawing will not reflect the changes made in the piece part drawing. In a production environment, this could have serious consequences.

You can solve this problem by using the **external reference** facility, which lets you link the piece part drawings with the assembly drawing. If the xref drawings (piece part) get updated, the changes are automatically reflected in the assembly drawing. This way, the assembly drawing stays updated, no matter when the changes were made in the piece part drawings. There is no limit to the number of drawings that you can reference. You can also have **nested references**. For example, the piece part drawing BEARING could be referenced in the SHAFT drawing, then the SHAFT drawing could be referenced in the assembly drawing ASSEM1. When you open or plot the assembly drawing, AutoCAD automatically loads the referenced drawing SHAFT and the nested drawing BEARING. When using External References, several people working on the same project can reference the same drawing and all the changes made are displayed everywhere the particular drawing is being used.

If you use the **Insert** tool to insert the piece parts, the piece parts become a permanent part of the drawing, and therefore, the drawing file size increases. However, if you use the external reference feature to link the drawings, the piece part drawings are not saved with the assembly drawing. AutoCAD only saves the reference information with the assembly drawing; therefore, the size of the drawing is minimized. Like blocks, the xref drawings can be scaled, rotated, or positioned at any desired location, but they cannot be exploded. You can also use only a part of the Xref by making clipped boundary of Xrefs.

Tip
External referenced drawings are useful for creating parts or subassemblies and then putting them together in one drawing to create the main assembly. You can also use it for laying out the contents of a drawing with multiple views before plotting.

DEPENDENT SYMBOLS

When you use the **Insert** tool to insert a drawing, the information about the named objects is lost, if names are duplicated. However, if they are unique, they are imported. The named objects are entries such as blocks, text styles and layers. For example, if the assembly drawing has a layer Hidden of green color and HIDDEN linetype and the piece part Bearing has also a layer Hidden of blue color and HIDDEN2 linetype, the values set in the assembly drawing will override the values of the inserted drawing when the Bearing drawing is inserted in the assembly drawing (Figure 20-1). As a result, in the assembly drawing, the layer Hidden will retain green color and HIDDEN linetype, ignoring the layer settings of the inserted drawing. Only those layers that have the same names are affected. The remaining layers that have different layer names are added to the current drawing.

In the xref drawings, the information about the named objects is not lost because AutoCAD will create additional named objects such as specified layer settings, as shown in Figure 20-2.

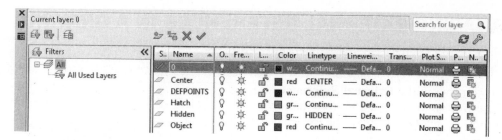

Figure 20-1 *Partial view of the layer settings of the current drawing in the* **Layer Properties Manager** *dialog box*

For xref drawings, these named objects become dependent symbols (features such as layers, linetypes, object color, text style, and so on).

S..	Name	O..	Fre..	L...	Color	Linetype	Linewei..	Trans...	Plot S...	P...	N...
	0	♀	☼	🔓	w...	Continu...	—— Defa...	0	Color_7	🖨	🖫
	Border	♀	☼	🔓	bl...	Continu...	—— Defa...	0	Color_5	🖨	🖫
	Border Text	♀	☼	🔓	m...	Continu...	—— Defa...	0	Color_6	🖨	🖫
	BOX	♀	☼	🔓	w...	Continu...	—— Defa...	0	Color_7	🖨	🖫
	c20d3|Bor...	♀	☼	🔓	bl...	Continu...	—— Defa...	0	Color_5	🖨	🖫
	c20d3|Bor...	♀	☼	🔓	m...	Continu...	—— Defa...	0	Color_6	🖨	🖫
	c20d3|BOX	♀	☼	🔓	w...	Continu...	—— Defa...	0	Color_7	🖨	🖫
	c20d3|CEN	♀	☼	🔓	w...	c16d3|C...	—— 0.15 ...	0	Color_7	🖨	🖫
	c20d3|DIM	♀	☼	🔓	w...	Continu...	—— 0.15 ...	0	Color_7	🖨	🖫
	c20d3|HA...	♀	☼	🔓	w...	Continu...	—— Defa...	0	Color_7	🖨	🖫
	c20d3|HID	♀	☼	🔓	w...	Continu...	—— 0.20 ...	0	Color_7	🖨	🖫
	c20d3|OBJ	♀	☼	🔓	w...	Continu...	▬▬ 0.80 ...	0	Color_7	🖨	🖫

Figure 20-2 *Xref creates additional layers*

The layer Hidden of the xref drawing (c20d3) is appended with the name of the xref drawing Bearing, and the two are separated by the vertical bar symbol (|). The names of these layers appear in light gray color in the **Layer Properties Manager** or **Layer** drop-down list of the **Layers** panel in the **Home** tab. These layers can neither be selected nor be made current. The layer name Border changes to c20d3|Border. Similarly, CEN is renamed as c20d3|CEN and OBJ is renamed as c20d3|OBJ (Figure 20-2). The information added to the current drawing is not permanent. It is added only when the xref drawing is loaded. If you detach the xref drawing, the dependent symbols are automatically erased from the current drawing.

When you xref a drawing, AutoCAD does not let you reference the symbols directly. For example, you cannot make the dependent layer, c20d3|Border, as current layer. Therefore, you cannot add any objects to that layer. However, you can change the color, linetype, lineweight, plotstyle, or visibility (on/off, freeze/thaw) of the layer in the current drawing. If the **Retain changes to Xref layers** check box in the **External References (Xrefs)** area of the **Open and Save** tab of the **Options** dialog box is cleared, which also implies that the system variable **VISRETAIN** is set to 0, the settings are retained only for the current drawing session. This means that when you save and exit the drawing, the changes are discarded and the layer settings return to their default status. If this check box is selected (default), which also implies that the **VISRETAIN** variable is set to 1, layer settings such as color, linetype, on/off, and freeze/thaw are retained, and they are saved with the drawing and used when you open the drawing the next time. Whenever you open or plot a drawing, AutoCAD reloads each Xref in the drawing and as a result, the latest updated version of the drawing is loaded automatically.

Note
You cannot make the xref-dependent layers current in a drawing. Only when the xref drawing is bound to the current drawing by using the **XBIND** *command, you can make the xref-dependent layers a permanent part of the current drawing and use them. The* **XBIND** *command will be discussed later in this chapter.*

MANAGING EXTERNAL REFERENCES IN A DRAWING

Ribbon: View > Palettes > External References Palette	
Menu Bar: Insert > External References	**Command:** XREF

When you choose the **External References Palette** button from the **Palettes** panel (Figure 20-3), AutoCAD displays the **External References** palette (Figure 20-4). The **External References** palette displays the status of each Xref in the current drawing and the relation between the various Xrefs. It allows you to attach a new xref, detach, unload, load an existing one, change an attachment to an overlay, or an overlay to an attachment. You can also open a reference drawing for editing from this palette. Additionally, it allows you to edit an xref's path and bind the xref definition to the drawing.

*Figure 20-3 The **Palettes** panel*

Apart from the methods mentioned in the command box, you can also invoke the **External References** palette by selecting an Xref in the current drawing and then right-clicking in the drawing area to display a shortcut menu. Next, choose **External References** from the shortcut menu; the **External References** palette is displayed.

The upper right corner of the palette has two buttons: **List View** and **Tree View**.

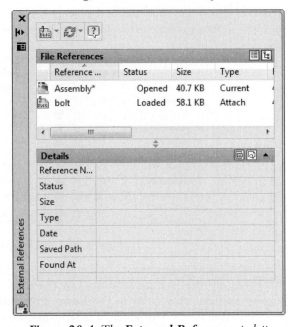

*Figure 20-4 The **External References** palette*

List View

Choosing the **List View** button displays the xrefs present in the drawing in alphabetical order. This is the default view. The list view displays information about xrefs in the current drawing under the following headings:

Reference Name

This column lists the name of all existing references in the current drawing.

Status

This column lists the current status of each xref in the drawing. It lists whether an xref is loaded, unloaded, unreferenced, not found, orphaned, unresolved, or marked to be reloaded. A loaded xref implies that the xref is attached to the current drawing. You can then unload it and then reload it using the options in the dialog box (this will be discussed later). An xref selected to be unloaded or reloaded displays **Unload** and **Reload**, respectively, under the **Status** column. If the xref has nested references that cannot be found, the status is **Unreferenced**, and if the parent of the nested reference gets unloaded, or cannot be found, its status is described as **Orphaned**. An unreferenced xref will not be displayed. If the xref is not found in the search paths defined, its status is **Not Found**. A missing xref or one that cannot be found is **Unresolved**.

Size

The file size of each xref is listed here.

Type
This column lists whether the xref is an attachment or overlay.

Date
This column lists the date on which the xref drawing was last saved.

Saved Path
This column lists the path of the xref, that is, the route taken to locate the particular referenced drawing.

Choose any of these headings, sorts and lists the Xrefs in the current drawing according to that particular title. For example, on choosing **Reference Name**, the xrefs are sorted and listed as per the name. The column widths can be increased or decreased as per requirements. When you place your cursor at the edge of a column title button, the cursor changes to a horizontal resizing cursor. Now, press the pick button of your mouse and drag the column edge to increase or decrease its width. After you increase the column widths, it is possible that the width of the columns extend beyond the list box width. In such a case, a horizontal scroll bar appears at the bottom of the list box. You can use the scroll bar to view the columns that extend beyond the width of the list box.

On choosing the **Tree View** button, displays the xrefs in the drawing in a hierarchical tree view in the drawing. It displays information on nested xrefs and their relationship with one another. Xrefs are indicated by an icon of a paper with a paper clip. This icon appears faded when the xref has been unloaded, and if there is a missing xref, a question mark appears. Similarly, an upward arrow indicates that the xref was reloaded and an arrow pointing downward indicates that the xref is unloaded. You can also choose **List View** and **Tree View** by pressing the F3 and F4 keys, respectively.

Attaching an Xref Drawing (Attach Option)
The **Attach** button is available at the upper left corner of the **External References** palette. If you choose the down arrow on this button, a flyout will be displayed. Select the **Attach DWG** option from the flyout to attach an xref drawing to the current drawing. This option can also be invoked by right-clicking on the **File References** area. The following examples illustrate the process of attaching an xref to the current drawing. In this example, it is assumed that there are two drawings, SHAFT and BEARING. SHAFT is the current drawing that is loaded on the screen (Figure 20-5) and the BEARING

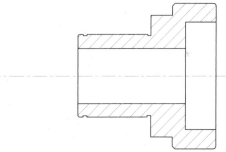

Figure 20-5 *Current drawing - SHAFT*

drawing is saved on the disk. If you want to xref the BEARING drawing in the SHAFT drawing, you need to follow the steps given below.

1. The first step is to make sure that the SHAFT drawing is on the screen (draw the shaft drawing with assumed dimensions).

Tip
No drawing needs to be on the screen. You could attach both drawings, BEARING and SHAFT to an existing drawing, even if it is a blank drawing.

2. Choose the **External References Palette** button from the **Palettes** panel to display the **External References** palette. In this palette, choose the **Attach DWG...** button; the **Select Reference File** dialog box will be displayed.

Select the drawing that you want to attach (BEARING) and then choose the **Open** button; the **Attach External Reference** dialog box will be displayed on the screen (Figure 20-6). In the **Attach External Reference** dialog box (Figure 20-6), the name of the file that you have selected to be attached to the current drawing as an xref is displayed in the **Name** edit box. You can also specify the name of the file to be attached from the **Name** drop-down list.

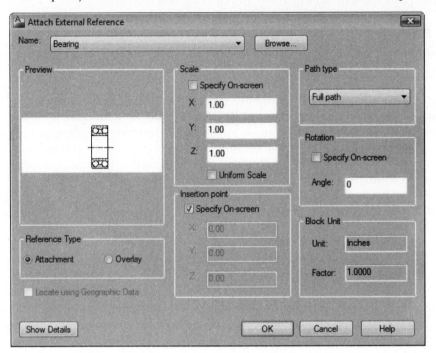

*Figure 20-6 The **Attach External Reference** dialog box*

Note
*AutoCAD also searches for the xref file in the paths defined in the **Project Files Search Path** folder in the **Files** tab of the **Options** dialog box. This folder does not have any path defined in it and displays **Empty** when the tree view is expanded. To define a search path, select **Empty** and choose the **Add** button in the dialog box. You can enter a project name here, if you want. Now, expand the tree view, select **Empty** again and choose the **Browse** button; the **Browse for Folder** window is displayed. Select the folder that is to be searched for the file and choose **OK**. Then, choose **Apply** and **OK** in the **Options** dialog box to exit it.*

In the **Reference Type** area, select the **Attachment** radio button, if it is not already selected (default option). The **Overlay** option is discussed later in this chapter. The **Path type** drop-down list is used to specify whether you want to attach a drawing with full path, relative path, or no path. If you select the **Full path** option, the precise location of the xreffed drawing is saved. If you select the **Relative path** option, the position of the xreffed drawing with reference to the host drawing is saved. If you select the **No path** option, AutoCAD will search for the xreffed drawing in only that folder in which the host drawing is saved. You can either specify the insertion point, scale factors, and rotation angle in the respective **X**, **Y**, **Z**, and **Angle** edit boxes or select the **Specify On-screen** check boxes to use the pointing device to specify them on the screen. By default, the scale factors in **X**, **Y**, and **Z** edit boxes is 1 and the rotation angle is 0. The **Block Unit** area provides information regarding the units of the inserted block. The **Unit** edit box displays the unit of the block. The **Factor** edit box displays the scale factor, depending on the unit of the block and that of the current drawing. If you choose the **Show Details** button, the path of the file is displayed adjacent to **Found in**. Also, the saved path of the file is displayed adjacent to **Saved path**. Choose the **OK** button from the **Attach External Reference** dialog box to accept the default values and exit the dialog box. Specify the insertion point; the drawing is attached to the current drawing, but it appears faded. After attaching the BEARING drawing as an xref, save the current drawing with the file name SHAFT (Figure 20-7). You can control the fading intensity of the attached objects by using the options from the **Fade Control** area of the **Display** tab in the **Options** dialog box. In this area, you can either use the slider bar to increase or decrease the fading intensity or you can enter a value in the edit box. You can also use the **XFADECTL** variable to control the fading intensity of other objects while editing them. By default, the fading intensity value is set to 50.

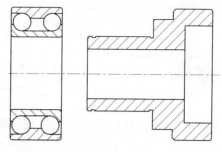

Figure 20-7 *Attaching the BEARING drawing as an xref*

Note
*You can also use the **-XREF** command to attach a drawing from the pointer input. Except the **Open** option, all the options available in the **External References** palette are available through the dynamic preview too.*

3. Open the drawing BEARING and make the changes shown in Figure 20-8 (draw polylines on the sides). Now, save the drawing with the file name BEARING.

4. Open the SHAFT drawing. In the **External References** palette, you will notice that the message **Needs Reloading** is displayed in the **Status** column of the BEARING drawing. Right-click on the BEARING and choose **Reload**; the modified BEARING drawing is automatically updated (Figure 20-9). This is the most useful feature of the **XREF** command. You could also have inserted the BEARING drawing as a block, but if you had updated the BEARING drawing, the drawing in which it was inserted would not have been updated automatically.

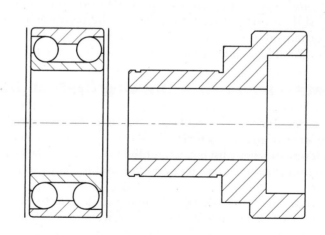

Figure 20-8 *The modified xref drawing BEARING*

Figure 20-9 *BEARING automatically updated after loading the SHAFT drawing*

When you attach an xref drawing, AutoCAD remembers the name of the attached drawing. If you xref the drawing again, AutoCAD displays a message as given next.

Xref "BEARING" has already been defined.
Using existing definition.

Specify insertion point or [Scale/X/Y/Z/Rotate/PScale/PX/PY/PZ/PRotate]: *Specify the location to place another copy of the xref.*

Note
*If the xref drawing that you want to attach is currently being edited, AutoCAD will attach the drawing that was last saved through the **SAVE**, **WBLOCK**, or **QUIT** command.*

Points to Remember about Xref

1. When you enter the name of the xref drawing, AutoCAD checks for block names and xref names. If a block exists with the same name as the name of the xref drawing in the current drawing, the **XREF** command is terminated and an error message is displayed.

2. When you xref a drawing, the objects that are in the model space are attached. Any objects that are in the paper space are not attached to the current drawing.

3. The layer 0, DEFPOINTS, and the linetype CONTINUOUS are treated differently. The current drawing layers 0, DEFPOINTS, and linetype CONTINUOUS will override the layers and linetypes of the xref drawing. For example, if the layer 0 of the current drawing is white and the layer 0 of the xref drawing is red, the white color will override the red.

4. The xref drawings can be nested. For example, if the BEARING drawing contains the reference INRACE and you xref the BEARING drawing to the current drawing, the INRACE drawing is automatically attached to the current drawing. If you detach the BEARING drawing, the INRACE drawing gets detached automatically.

5. You can rename an xref under the **File References** column name in the list box of the **External References** palette by highlighting the xref and then clicking on it again. You can now enter a new name. An AutoCAD warning is displayed: **Caution! "XXXX" is an externally referenced block. Renaming it will also rename its dependent symbols**.

6. When you xref a drawing, AutoCAD stores the name and path of the drawing by default. If the name of the xref drawing or the path where the drawing was originally stored has changed or you cannot find it in the path specified in the **Options** dialog box, AutoCAD cannot load the drawing, plot it, or use the **Reload** option of the **XREF** command.

Detaching an Xref Drawing (Detach Option)

The **Detach** option can be used to detach or remove the xref drawings. If there are any nested xref drawings defined with the xref drawings, they are also detached. Once a drawing is detached, it is erased from the screen. To detach an xref drawing, select the file name in the **External References** palette list box to highlight it and then right-click on it to select the **Detach** option. After selecting the **Detach** option, the xref is completely removed from the current drawing.

You can also use the **-XREF** command to detach the xref drawings. When AutoCAD prompts for an xref file name to be detached, you can enter the name of one xref drawing or the name of several drawings separated by commas. You can also enter * (asterisk), in which case, all referenced drawings, including the nested drawings, will be detached.

Updating an Xref Drawing (Reload Option)

When you load a drawing, AutoCAD automatically loads the referenced drawings. The **Reload** option of the **XREF** command lets you update the xref drawings and nested xref drawings at any time. You do not need to exit the drawing editor and then reload the drawing. To reload the xref drawings, invoke the **External References** palette, select the xreffed drawing in the list box, and then right-click on it to choose the **Reload** option from the shortcut menu. AutoCAD will scan for the referenced drawings and the nested xref drawings and load the most recently saved version of the drawing.

You can reload all the attached xref drawings at one time by selecting **Reload All References** from the drop-down list of the **Refresh** button on the upper left corner of the **External References** palette.

The **Reload** option is generally used when the xref drawings are currently being edited and you want to load the updated drawings. The xref drawings are updated based on what is saved on the disk. Therefore, before reloading an xref drawing, you should make sure that the xref drawings that are being edited have been saved. If AutoCAD encounters an error while loading the referenced drawings, the **XREF** command is terminated, and the entire reload operation is canceled.

You can also reload the xref drawings by using the **-XREF** command. When you enter the **-XREF** command, AutoCAD will prompt you to enter an option. Enter **R** for the **Reload** option. Next, you are prompted to enter the name of the xref drawing. You can enter the name of one xref drawing or the names of several drawings separated by commas. If you enter * (asterisk), AutoCAD will reload all xref and nested xref drawings.

Unloading an Xref Drawing (Unload Option)

The **Unload** option allows you to temporarily remove the definition of an xref drawing from the current drawing. However, AutoCAD retains the pointer to the xref drawings. When you unload the xref drawings, the drawings are not displayed on the screen. You can reload the xref drawings by using the **Reload** option.

Tip
It is recommended that you unload the referenced drawings if they are not being used. After unloading the xref drawings, the drawings load much faster and need lesser memory.

Adding an Xref Drawing (Bind Option)

The **Bind** option lets you convert the xref drawings to blocks in the current drawing. The bound drawings, including the nested xref drawings (that are no longer xrefs), become a permanent part of the current drawing. The bound drawing cannot be detached or reloaded. You can use this option when you want to send a copy of your drawing to a customer for review. Because the xref drawings are a part of the existing drawing, you do not need to include the xref drawings or the path information. You can also use this option to safeguard the master drawing from accidental editing of the piece parts. To bind the xref drawings, select the file names in the **External References** palette and then right-click and choose the **Bind** option from the shortcut menu; the **Bind Xrefs** dialog box (Figure 20-10) is displayed. AutoCAD provides two methods to bind the xref drawing in the **Bind Type** area of the dialog box. These methods are discussed next.

*Figure 20-10 The **Bind Xrefs** dialog box*

Bind

When you use the **Bind** radio button, AutoCAD binds the selected xref definition to the current drawing. All the xrefs are converted to blocks and the named objects are renamed. For example, if you xref the drawing BEARING with a layer named OBJECT, a new layer BEARING|OBJECT is created in the current drawing. When you bind this drawing, the xref dependent layer BEARING|OBJECT will become a locally defined layer BEARING0OBJECT (Figure 20-11).

If the BEARING0OBJECT layer already exits, AutoCAD will automatically increment the number, and the layer name becomes BEARING1OBJECT.

S..	Name	O..	Fre...	L...	Color	Linetype	Linewei...	Trans...	Plot S...
✓	0	♀	☼	⌂	■ w...	Continu...	—— Defa...	0	Normal
◹	Bearing0CEN	♀	☼	⌂	■ w...	Bearing|...	—— 0.15 ...	0	Normal
◹	Bearing0Center	♀	☼	⌂	■ red	Bearing|...	—— Defa...	0	Normal
◹	Bearing0DIM	♀	☼	⌂	■ w...	Continu...	—— 0.15 ...	0	Normal
◹	Bearing0Hatch	♀	☼	⌂	▨ gr...	Continu...	—— 0.20 ...	0	Normal
◹	Bearing0Hidden	♀	☼	⌂	■ bl...	Bearing|...	—— Defa...	0	Normal
◹	Bearing0OBJ	♀	☼	⌂	■ w...	Continu...	■■ 0.80 ...	0	Normal
◹	Bearing0Object	♀	☼	⌂	■ red	Continu...	—— Defa...	0	Normal
◹	CEN	♀	☼	⌂	■ w...	CENTER	—— 0.15 ...	0	Normal

*Figure 20-11 Partial view of the **Layer Properties Manager** dialog box*

Insert

When you use the **Insert** option, AutoCAD inserts the xref drawing. The xrefs get converted into blocks. For example, if you xref the drawing SHAFT with a layer named OBJECT, a new layer SHAFT|OBJECT is created in the current drawing. If you use the **Insert** option to bind the xref drawing, the layer name SHAFT|OBJECT is renamed as OBJECT. If the object layer already exists, then the values set in the current drawing override the values of the inserted drawing.

Note
It is possible to bind only an individual or several xref-dependent named objects instead of the entire drawing into the current drawing. This will be discussed later in this chapter.

*You can also use the **-XREF** command to bind the xref drawings. In such cases, prompts are displayed on the Command prompt.*

Editing an XREF's Path

By default, AutoCAD saves the path of the referenced drawing and displays it in the **Saved Path** column in the **External References** palette. As mentioned earlier, when AutoCAD loads the drawing containing a referenced file, and if it is not able to find the file at the location specified in the **Saved Path** column of the **External References** palette, it searches for the file in the current directory, and in the **Support File Search Path** locations specified in the **Files** tab of the **Options** dialog box. If a file with the same name is found here, it is loaded. Now, when you invoke the **External References** palette, you will notice that when you select an xref name in the list box to highlight it, the path displayed in the **Saved Path** column for the xref file is different from the one displayed in the **Found At** edit box. To update the path of the xref file, choose the **Save Path** button. The new path is saved and displayed in the **Saved Path** column.

If AutoCAD cannot locate the specified file even in the directories specified in the **Files** tab of the **Options** dialog box, it will display an error message saying that it cannot find the specified file. The path of the file is displayed as a marker text in the current drawing. Now, when you invoke the **External References** palette, the status of the drawing is shown as **Not Found**. To specify a new path for the xref file, select the xref file name in the list box of the dialog box and choose the **Browse** button. The **Select new path** dialog box is displayed where you can locate the drawing to be used as xref. Once you have found the file, choose the **Open** button to return to the **External References** palette. The new path is displayed in the **Saved Path** column and the **Found At** edit box. The specified xref file is reloaded and replaces the marker text in the drawing, when you choose the **OK** button in the **External References** palette. If you remember the new location of the xref file, you can also enter it in the **Found At** edit box. For example, if a drawing, which was originally in the C:\CAD\Proj1 subdirectory, has been moved to A:\Parts directory, the path must be edited so that AutoCAD is able to load the xref drawing.

THE OVERLAY OPTION

As discussed earlier, when you attach an xref to a drawing, the **Attach External References** dialog box is displayed. The **Reference Type** area of this dialog box has two radio buttons, **Attachment** and **Overlay**. You can select any of these radio buttons to xref a drawing. The **Attachment** radio button is selected by default. The advantage of selecting the **Overlay** radio button is that you can access the desired drawing instead of the drawing along with its xreffed attachments. For example, consider three people working on three different drawings that are a part of the same project. The first designer is working on the layout of walls of a room, the second designer is working on the furniture layout of the room, and the third on the electrical layout of that room. The names of the drawings are WALLS, FURNITURE, and ELECTRICAL, respectively. Assume that the designer working on the walls layout selects the **Attachment** radio button to xref the FURNITURE drawing so that he or she can check the furniture layout according to the wall structure. After insertion, the WALLS drawing will comprise the wall structure (current drawing) along with the furniture layout (xreffed drawing). Now, if the designer working on the electrical layout xrefs the WALLS drawing to check the location of electrical fittings with respect to the walls, he/she will get the drawing that has the furniture layout as well as the wall layout. This is because the FURNITURE drawing was xreffed in the WALLS drawing by selecting the **Attachment** radio button.

In the above example, the designer working on the ELECTRICAL drawing may not require the FURNITURE drawing. This is because at this stage, he/she is more interested in checking the electrical fittings with respect to the wall structure. So the furniture layout that is xreffed with the wall structure needs to be avoided. This can be done by selecting the **Overlay** radio button while X-referencing the FURNITURE drawing in the WALLS drawing. This means that the designer working on the wall structure needs to xref the furniture layout by selecting the **Overlay** radio button. Now, if the wall structure is xreffed in some other drawing, the furniture layout will not appear.

One of the problems of using the **Attachment** option is that you cannot have a circular reference. For example, assume you are designing the plant layout of a manufacturing unit. One person is working on the floor plan (see Figure 20-12), and the second person is working on the furniture layout in the offices (Figure 20-13). The names of the drawings are FLOORPLN and OFFICES, respectively. The person working on the office layout uses the **Attachment** option to insert the FLOORPLN drawing so that he or she has the latest floor plan drawing. The person who is working on the floor plan wants to reference the OFFICES drawing.

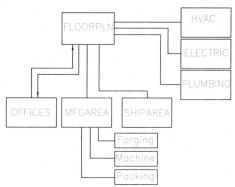

Figure 20-12 *Hierarchy of FLOORPLAN drawing*

Now, if the **Attachment** option is used to reference the drawings, AutoCAD displays an error message. This is because by attaching the OFFICES drawing, a circular reference is created. The AutoCAD message displayed is "**Circular references detected. Continue?**" (Figure 20-14). If you choose the **No** button, the **XREF** command is canceled and no drawing is referenced. But if you choose the **Yes** button, the following message is displayed **Breaking circular reference from "offices" to "current drawing"** and the particular file you wanted to reference is referenced.

However, to overcome this problem of circular reference, you can use the **Overlay** option to overlay the OFFICES drawing. This is a very useful option because the **Overlay** option lets different operators avoid circular reference and share the drawing data without affecting the

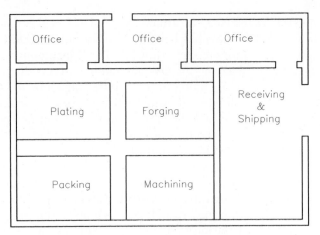

Figure 20-13 *Hierarchy of OFFICES drawing*

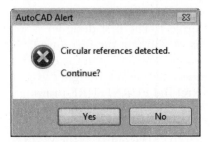

Figure 20-14 *The **AutoCAD Alert** message box*

drawing. Overlaying allows you to view a referenced drawing without having to attach it to the current drawing. This option can be invoked by selecting the **Overlay** radio button in the **Attach External References** dialog box, which is displayed after you have selected a drawing to reference. Also, when a drawing that has a nested overlay is overlaid, the nested overlay is not visible in the current drawing. This is another difference between attaching an xref and overlaying an xref to a drawing. This feature is especially useful when you want to reference a drawing that another user who is referencing your drawing does not need. Although the attachment will reference the nested reference too, the overlay option ignores nested references.

EXAMPLE 1   *Attachment and Overlay*

In this example, you will use the **Attachment** and **Overlay** options to attach and reference the drawings. Two drawings, PLAN and PLANFORG are given. The PLAN drawing (Figure 20-15) consists of the floor plan layout, and the PLANFORG drawing (Figure 20-16) has the details of the forging section only. The CAD operator who is working on the PLANFORG drawing wants to xref the PLAN drawing for reference. Also, the CAD operator working on the PLAN drawing should be able to xref the PLANFORG drawing to complete the project. The following steps illustrate how to accomplish the defined task without creating a circular reference.

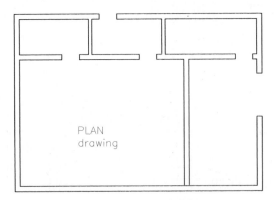

Figure 20-15 *The PLAN drawing*

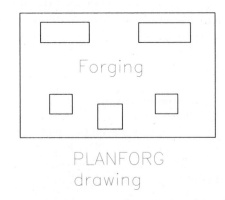

Figure 20-16 *The PLANFORG drawing*

How circular reference is caused?

1. Open the drawing PLANFORG, and then choose the **External References** tool from the **Palettes** panel in the **View** tab. Next, choose the **Attach DWG** button from the **External References** palette; the **Select Reference File** dialog box is displayed. Select the PLAN drawing from the list box of the **Select Reference File** dialog box and choose the **Open** button; the **Attach External Reference** dialog box is displayed. In this dialog box, the name of the PLAN drawing is displayed in the **Name** edit box, and the **Attachment** radio button is selected by default in the **Reference Type** area. Choose **OK** to exit the dialog box and specify an insertion point on the screen. Now, the drawing consists of the PLANFORG and PLAN. Save the drawing.

2. Open the drawing file PLAN. Next, choose the **External References Palette** button and attach the PLANFORG drawing to the PLAN drawing using the same steps as described in Step 1. AutoCAD will display the message that the circular reference has been detected and will ask you if you want to continue. If you choose **Yes** in the AutoCAD message box, the circular reference is broken and you are allowed to reference the specific drawing.

    The possible solution for the operator working on the PLANFORG drawing is to detach the PLAN drawing. This way the PLANFORG drawing does not contain any reference to the PLAN drawing and would not cause any circular reference. The other solution is to use the **Overlay** option, as follows :

How to prevent circular reference?

3. Open the PLANFORG drawing (Figure 20-17) and select the **Overlay** radio button in the **Attach External References** dialog box, which is displayed after you have selected the PLAN drawing to reference. The PLAN drawing is overlaid on the PLANFORG drawing (Figure 20-18).

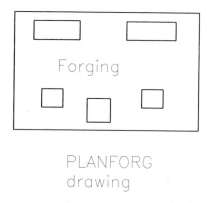

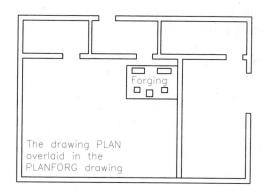

Figure 20-17 The PLANFORG drawing

Figure 20-18 The PLANFORG drawing after overlaying the PLAN drawing

4. Open the drawing file PLAN (Figure 20-19), and select the **Attachment** radio button in the **Attach External Reference** dialog box, which is displayed when you select the PLANFORG drawing in the **Select Reference File** dialog box to attach it as an xref to the PLAN drawing. You will notice that only the PLANFORG drawing is attached (Figure 20-20). The drawing that was overlaid in the PLANFORG drawing (PLAN) does not appear in the current drawing. This way, the CAD operator working on the PLANFORG drawing can overlay the PLAN drawing, and the CAD operator working on the PLAN drawing can attach the PLANFORG drawing, without causing a circular reference.

Chapter 20

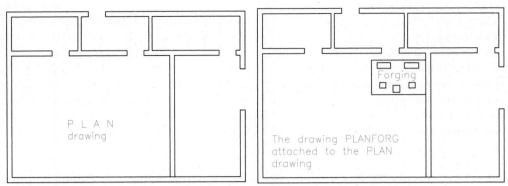

Figure 20-19 *The PLAN drawing*

Figure 20-20 *The PLAN drawing after attaching the PLANFORG drawing*

ATTACHING FILES TO A DRAWING

Ribbon: Insert > Reference > Attach	**Command:** ATTACH

You can use the **Attach** tool in the **Reference** panel (Figure 20-21) to attach a DWG, DGN, DWF, PDF, or image file without invoking the **External References** palette. Using this tool, you can attach a drawing file easily, since most of the xref operations involve simply attaching a drawing file. When you invoke this tool, AutoCAD displays the **Select Reference File** dialog box. To attach a *.dwg, .dgn, .dwf, .pdf,* or image file, specify it in the **Files of type** drop-down list in the **Select Reference** dialog box; the corresponding files will be listed in the dialog box. Select the drawing file to be attached and choose the **Open** button; the **Attach External Reference** dialog box is displayed. Select the **Attachment** radio button under the **Reference Type** area. You can specify the insertion point, scale, and rotation angle on screen or in the respective edit boxes.

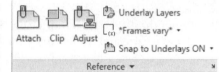

Figure 20-21 *The **Attach** tool in the **Reference** panel*

Tip

*When you attach or reference a drawing that has a drawing order created by using the **DRAWORDER** command, the drawing order is not maintained in the xref. To maintain the drawing order, first open the xref drawing and specify the drawing order in it. Now, use the **WBLOCK** command to convert it into a drawing file and the **XATTACH** command to attach the newly created drawing file to the current drawing. In this way, the drawing order will be maintained.*

Note

*To select a .dgn, .dwf, .pdf, or image file, specify the file type in the **Files of type** drop-down list in the **Select Reference File** dialog box; the files will be attached as an underlay. Underlays are discussed in the next section.*

*AutoCAD maintains a log file (.xlg) for xref drawings if the **XREFCTL** system variable is set to 1. This file lists information about the date and time of loading and other xref operations to be completed. This .xlg file is saved in the current drawing with the same name as the current drawing and is updated each time the drawing is loaded or any xref operations are carried out.*

WORKING WITH UNDERLAYS

You can attach a DWF, DGN, or PDF file as an underlay to the current drawing file. The underlay files are not a part of original drawing files. Therefore, if you add a file as an underlay, it does not increase the file size of the current drawing. The procedure to add a file as an underlay is similar to attaching a drawing file using the **Attach** tool. After you select the file to attach, the **Attach <XXXX> Underlay** dialog box will be displayed, where <XXXX> is the file type. Figure 20-22 shows the **Attach PDF Underlay** dialog box that is displayed on selecting a pdf file from the **Select Reference File** dialog box.

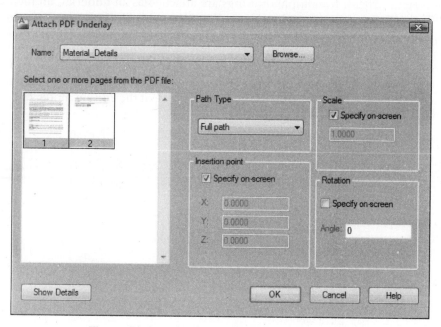

*Figure 20-22 The **Attach PDF Underlay** dialog box*

If the selected *.pdf* file has multiple pages, all pages of the pdf file are listed in the **Select one or more pages from the PDF file** area. Select the pages to be attached from this area. If you have selected multiple pages as well as the **Specify on-screen** check box from the **Insertion point** area, then you are prompted to specify different insertion points for different pages.

The files attached as an underlay behave like blocks. The general modify commands like move, copy, rotate, mirror, and so on can be applied on them. However, you cannot bind a file that is attached as an underlay or modify the attached file in the current drawing file.

Editing an Underlay

As discussed earlier, you cannot edit a file that is attached as an underlay. However, you can control the appearance of the underlay by adjusting the contrast, fade, and monochrome display. To do so, select the attached object; a contextual tab will be displayed. Figure 20-23 shows the **DGN Underlay** tab that is displayed on selecting a DGN file.

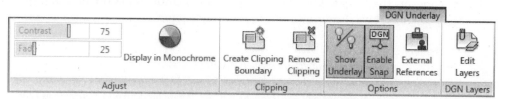

*Figure 20-23 The **DGN Underlay** tab*

You can use the options in this contextual tab to adjust the fade, contrast, and monochrome display of the underlay. You can create a new clipping boundary similar to that of an attached drawing. By default, the **Enable Snap** button is chosen in the **Options** panel. Therefore, you can snap entities in the underlaid object. If you deselect this button, you cannot snap entities in the underlaid objects. You can use the **Show Underlay** button from the **Options** panel to display/hide the underlaid objects. Both **Enable Snap** and **Show Underlay** are toggle buttons. If the attached underlay has layers, you can hide/display the selected layers. To do so, choose the **Edit Layers** button from the contextual tab; the **Underlay Layers** dialog box will be displayed, as shown in Figure 20-24. If multiple drawings are attached as an underlay, all file names will be listed in the **Reference Name** drop-down list. Select the file from the **Reference Name** drop-down list for which you need to hide the layers; the corresponding layers will be listed in the list box. Click on the bulb icon of the layer to be hidden. Alternatively, select the layer from the list box and right-click; a shortcut menu will be displayed. Choose the **Layer(s) Off** option to hide the selected layer. To hide multiple layers, press the CTRL key and select the layers to be hidden. Next, choose **Layer(s) Off**. Then, press ESC to exit the editing mode; the contextual tab will disappear.

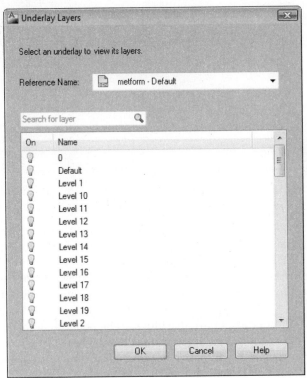

*Figure 20-24 The **Underlay Layers** dialog box*

If you have attached a file as an underlay, you can invoke the **Underlay Layers** dialog box without invoking the contextual tab by choosing the **Underlay Layers** tool from the **Reference** panel in the **Insert** tab. Similarly, you can switch on/off the OSNAP settings for the drawing that are underlaid by choosing the **Snap to Underlays ON / Snap to Underlays OFF** button respectively from the **Snap to Underlays** drop-down in the **Reference** panel.

Note
*If there are more underlays of different file types and you have set different osnap setting for each file type, then the **Underlay Osnaps Vary** button will be displayed in the **Snap to Underlays** drop-down in the **Reference** panel. However, if you choose the **Snap to Underlays ON / Snap to Underlays OFF** button, the corresponding osnap setting will be applied to all underlays.*

OPENING AN XREFFED OBJECT IN A SEPARATE WINDOW

If you are in the host drawing and you want to open a selected xreffed object in a separate window without using the **Select File** dialog box, you can use the **XOPEN** command. When you invoke this command, you are prompted to select the xref. Select the xreffed object that you want to open in a separate window. When you select the desired xref, AutoCAD opens the DWG file of the xreffed object in a separate window. You can now make the desired changes in the DWG file of the xreffed object. Save the changes and then close the drawing. When you open the host drawing, you will notice that the **Communication Center** displays a message that the external reference file has changed and the xreffed drawing needs to be reloaded. Reload the xreffed drawing using the **External References** palette and you will notice that the host drawing is updated.

USING THE DesignCenter TO ATTACH A DRAWING AS AN XREF

The **DesignCenter** can also be used to attach an xref to a drawing. To do so, choose the **DesignCenter** button from the **Palettes** panel in the **View** tab; the **DesignCenter** will be displayed. In the **DesignCenter**, choose the **Tree View Toggle** button, if it is not chosen already, to display the tree pane. Expand the **Tree view** and double-click on the folder whose contents you want to view. The contents of the selected folder are displayed in the palette. From the thumbnails of drawings displayed on the right-side of the palette, right-click on the drawing to be attached as an xref; a shortcut menu is displayed, as shown in Figure 20-25. Choose **Attach as Xref** from the shortcut menu; the **Attach External Reference** dialog box will be displayed. Alternatively, you can also use the right or left mouse button to drag and drop the drawing into the current drawing; a shortcut menu will be displayed again. Choose **Attach as Xref**; the **Attach External Reference** dialog box will be displayed.

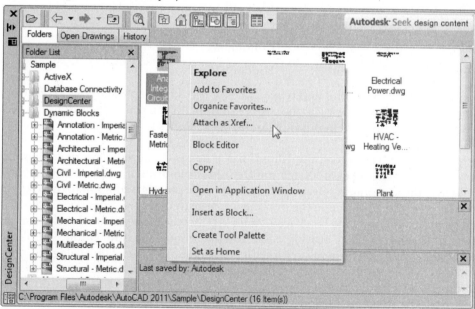

Figure 20-25 The **DesignCenter**

In this dialog box, the **Name** edit box displays the name of the selected file to be inserted as an xref. Select the **Attachment** radio button in the **Reference type** area, if not already selected. Specify the **Insertion point**, **Scale**, and **Rotation** in the respective edit boxes or select the **Specify On-screen** check boxes to specify this information on the screen. Choose **OK** to exit the dialog box; the selected object is attached as an xref to the current drawing.

ADDING XREF DEPENDENT NAMED OBJECTS

Menu Bar: Modify > Object > External Reference > Bind
Toolbar: Reference > Xbind   **Command:** XBIND

You can use the **Xbind** tool (Figure 20-26) to add the selected named objects such as blocks, dimension styles, layers, line types, and text styles of the xref drawing to the current drawing. The following example describes the use of this command.

*Figure 20-26 The **Xbind** tool in the **Reference** toolbar*

1. Load the drawing Bearing that was created earlier when the **Attach** option of the **XREF** command was discussed. Make sure the drawing has the following layer setup. Otherwise, create the following layers using the **Layer Properties Manager** dialog box.

Layer Name	Color	Linetype
0	White	Continuous
Object	Red	Continuous
Hidden	Blue	Hidden2
Center	White	Center2
Hatch	Green	Continuous

2. Draw a circle and create it as a block with the name SIDE. Save the drawing as BEARING.

3. Start a new drawing with the following layer setup:

Layer Name	Color	Linetype
0	White	Continuous
Object	Red	Continuous
Hidden	Green	Hidden

4. Use the **XAttach** tool in the **Reference** panel or the **Attach** option of the **XREF** command to attach the Bearing drawing to the current drawing. When you xref the drawing, the layers will be added to the current drawing, as discussed earlier in this chapter.

5. Now, invoke the **Xbind** tool; the **Xbind** dialog box is displayed. This dialog box has two areas with list boxes. They are **Xrefs** and **Definitions to Bind**. If you want to bind the blocks defined in the xref drawing Bearing, first click on the plus sign adjacent to the xref Bearing. Icons for the named objects in the drawing are displayed in a tree view (Figure 20-27). Click on the plus sign next to the Block icon. AutoCAD lists the blocks defined in the xref drawing (Bearing). Select the block Bearing|SIDE and then choose the **Add** button. The block name will be added to the **Definitions to Bind** list box. Choose **OK** to exit the dialog box. AutoCAD will bind the block with the current drawing and a message is displayed at the Command prompt: **1 Block(s) bound**. The name of the block will change to Bearing0SIDE. You can invoke the **Block Definition** dialog box by choosing the **Create Block** tool in the **Block Definition** panel and check the **Name** drop-down list to see if the block with the name Bearing0SIDE has been added to the drawing. If you want to insert the block, you must enter the new block name (Bearing0SIDE). You can also rename the block to another name that is easier to use.

If the block contains a reference to another xref drawing, AutoCAD binds that xref drawing and all its dependent symbols to the current drawing also. Once you **XBIND** the dependent symbols, AutoCAD does not delete them, not even when the xref drawing is detached. For example, the block Bearing0SIDE will not be deleted even when you detach the xref drawing or end the drawing session.

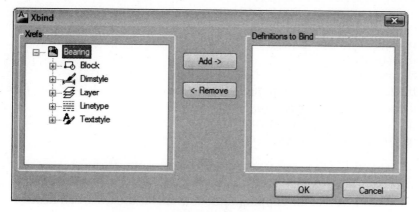

*Figure 20-27 The **Xbind** dialog box*

You can also use the **-XBIND** command to bind the selected dependent symbols of the xref drawing using the Command prompt.

6. Similarly, you can bind the dependent symbols, Bearing|STANDARD (textstyle), Bearing|Hidden, and Bearing|Object layers of the xref drawing. Click on the plus signs adjacent to the respective icons to display the contents and then select the layer or textstyle you want to bind and choose the **Add** button. The selected named objects are displayed in the **Definitions to Bind** list box. If you have selected a named object that you do not want to bind, select it in the **Definitions to Bind** list box and choose the **Remove** button. Once you have finished selecting the named objects that you want to bind to the current drawing, choose **OK**. A message indicating the number of named objects that are bound to the current drawing is displayed at the Command prompt.

    Once bound, the layer names will change to Bearing0Hidden and Bearing0Object. If the layer name Bearing0Hidden was already there, the layer will be named Bearing1Hidden. These two layers become a permanent part of the current drawing. Even if the xref drawing is detached or the current drawing session is closed, the layers are not discarded.

CLIPPING EXTERNAL REFERENCES

Ribbon: Insert > Reference > Clip   **Command:** CLIP

The **Clip** tool (Figure 20-28) is used to trim an xref after it has been attached to a drawing to display only a portion of the drawing (Figure 20-29). After you have attached an xref to a drawing, you can trim it to display only a portion of the drawing by using the **Clip** tool. However, clipping xref does not, in any way, modify the referenced drawing. On invoking the **Clip** tool, you are prompted to select an object to clip. Select a DWG, DGN, IMAGE, or PDF; the respective clipping options will be displayed. You can also invoke the **Clip** tool

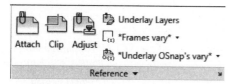

*Figure 20-28 The **Clip** tool in the*
***Reference** panel*

*Figure 20-29 Using the **CLIP** command to clip*

by selecting an xref and then right-clicking in the drawing area to display a shortcut menu. Next, choose the corresponding clip command from it. For example, to clip an attached *.dwg* file, select the attached *.dwg* file, right-click, and choose the **CLIP Xref** option; the following prompt sequence will be displayed.

Command: **_XCLIP** *1 found*
Select objects: *Select the xref that needs to be clipped.*
Select objects: *Press ENTER*
Enter clipping option
[ON/OFF/Clipdepth/Delete/generate Polyline/New boundary] <New>: *Press ENTER to specify a new clip boundary or enter an option.*
[Select polyline/Polygonal/Rectangular/Invert clip] <Rectangular>: *Press ENTER to select the* **Rectangular** *option, and then specify two corners of the rectangular boundary.*

You can also right-click in the drawing area at the **Enter clipping option** prompt to display a shortcut menu. This shortcut menu displays all the options available at the Command prompt that can be selected here. All the clipping options available are discussed next.

New boundary

The clipping boundary can be specified by using the **Rectangular**, **Polygonal**, or **Select polyline** option. The **Rectangular** option generates a rectangular boundary and the **Polygonal** option allows you to specify a boundary of any shape. You can draw a boundary using the **Polyline** tool or the **Polygon** tool and then, use the **Select polyline** option to select this polyline as the clipping boundary. The **Invert clip** option inverts the direction of the clipping. For example, if the objects outside the clipping boundary are hidden, by default, the **Invert clip** option will hide the objects inside the clipping boundary and display the objects outside the boundary. After selecting the **Invert clip** option, you need to redefine the clipping boundary using any one of the options discussed above. The **Invert clip** option can only be used with the 2D Wireframe visual style. If a boundary already exists, AutoCAD will ask you if the old boundary should be deleted. You can enter **YES** if you want to delete the old boundary and define a new one. If you enter **NO**, the old boundary is retained and the **XCLIP** command ends.

ON/OFF

The **ON/OFF** option allows you to specify whether to display the clipped portion or not. When the clipping boundary is off, you can see the complete xref drawing and when it is on, the drawing that is within the clipping polygon is displayed.

Delete

The **Delete** option completely deletes the predefined clipping boundary and the entire xref gets displayed. The **Erase** tool cannot be used to delete the clipping boundary.

generate Polyline

If you have a clipped boundary, select this option to create a polyline coinciding with the boundary of the clipped xref drawing. You can edit the polyline boundary without affecting the clipped drawing. For example, if you stretch the boundary, it does not affect the xref drawing. The edited boundary can be used later to specify a new clipping boundary.

Note
Polyline generated using the **generate Polyline** *option of the* **XCLIP** *command should not be confused with the clipping boundary. Although both coincide with each other, you can select and edit only the polyline. When you select the clipping boundary, the xref object is also selected.*

Clipdepth

This option allows you to define the front and back clipping planes for 3D xref objects or blocks. The objects between the front and back planes will be displayed (Figures 20-30 and 20-31). You will be able to understand this option better after you have dealt with 3D drawings later in the book. In the figures that follow, the clipping boundary is defined aligned to the front face of the object, and therefore, the clipping planes will get defined parallel to it. The xref must contain a clip boundary before specifying a clip depth. If you use the **Clipdepth** option, it prompts you to specify a front clip point and a back clip point. Specifying these points, creates a clipping frame passing through them and parallel to the clipping boundary. If the front clipping plane is specified behind the back clipping plane, AutoCAD displays an error message and the clipdepth is also not applied. The **Distance** option for specifying the front or back clip points creates a clipping plane at a specified distance from the clipping boundary parallel to it. The **Remove** option removes both the front and back clipping planes and the entire object is visible.

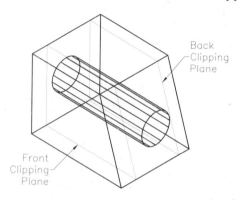

Figure 20-30 *3D xref object before clipping*

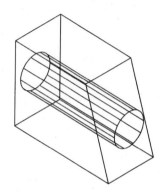

Figure 20-31 *3D xref object after clipping*

Similarly, you can clip the other underlaid xrefs. The options available for clipping the underlaid Xrefs are **ON**, **OFF**, **Delete**, and **New Boundary**. These options are the same as discussed above.

DISPLAYING CLIPPING FRAME

You can use the **FRAME** system variable to turn on or off the display of clipping boundary of all clipped Xrefs. When the value of **FRAME** system variable is 0 (default), the clipping boundary will not be displayed. When it is 1, the clipping boundary will be displayed and plotted. When it is 2, the clipping boundary will be displayed, but not plotted. To control the display of frames of individual attached Xrefs, you need to invoke the individual system variables like **XCLIPFRAME**, **DGNFRAME**, **PDFFRAME**, and **IMAGEFRAME**. Alternatively, choose the appropriate option from the **Frames** drop-down in the **Reference** panel of the **Insert** tab (Figure 20-32). If there are multiple underlays of different file types and you have set different frame settings for each file type, then the **Frames Vary** button will be displayed in the **Reference** panel of the **Insert** tab.

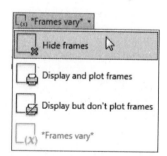

Figure 20-32 *Options in the Frame drop-down list*

DEMAND LOADING

The demand loading feature loads only that part of the referenced drawing that is required in the existing drawing. For example, demand loading provides a mechanism by which objects on frozen layers are not loaded. Also, only the clipped portion of the referenced drawing can be loaded. This makes the xref operation more efficient since less disk space is used, especially when the drawing is reopened. Demand loading is enabled by default. You can modify its settings in the **External References (Xrefs)** area of the **Open and Save** tab of the **Options** dialog box (Figure 20-33). You can select any of the three settings available in the **Demand load Xrefs** drop-down list in this dialog box. They are **Disabled**, **Enabled**, and **Enabled with copy**. These options correspond to a value of **0**, **1**, and **2** of the **XLOADCTL** system variable, respectively, and are discussed next.

Setting	Value of XLOADCTL	Features
Disabled	0	1. Turns off demand loading. 2. Loads entire xref drawing file. 3. The file is available on the server and other users can edit the xref drawing.
Enabled	1	1. Turns on demand loading. 2. The referenced file is kept open. 3. Makes the referenced file read-only for other users.
Enabled with copy	2	1. Turns on demand loading with the copy option. 2. A copy of a referenced drawing is opened. 3. Other users can access and edit the original referenced drawing file.

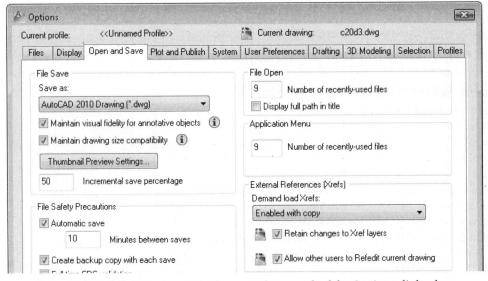

*Figure 20-33 Partial view of the **Open and Save** tab of the **Options** dialog box*

You can also set the value of **XLOADCTL** at the Command prompt. When you are using the **Enabled with copy** option of demand loading, the temporary copies of the xref are saved in the AutoCAD temporary files directory (defined in the **Temporary External Reference File Location** folder in the **Files** tab of the **Options** dialog box) or in a user-specified directory. The **XLOADPATH** system variable creates a temporary path to store demand loaded Xrefs.

Spatial and Layer Indexes

As mentioned earlier, the demand loading improves performance when the drawing contains referenced files. To make it work effectively and to take full advantage of demand loading, you must store a drawing with Layer and Spatial indexes. The layer index maintains a list of objects in different layers and the spatial index contains lists of objects based on their location in 3D space.

Layer and spatial indexes are created using the **Save Drawing As** dialog box. Choose the **Tools** button available on the upper right corner of the dialog box to display a shortcut menu. Choose the **Options** option from the shortcut menu to display the **Saveas Options** dialog box. Choose the **DWG Options** tab if it is not already chosen and select the type of index you want to save the file with from the **Index type** drop-down list. **None** is the default option, that is, no indexes are created. The other options available are: **Layer**, **Spatial**, and **Layer & Spatial**. Once you have selected the type of index to create, choose the **OK** button to exit the dialog box and choose the **Save** button in the **Save Drawing As** dialog box to save the drawing with the indexes. The **INDEXCTL** variable also controls the creation of layer index and spatial index and its value can be set using the Command prompt. The following are the settings of the **INDEXCTL** system variable, and they correspond to the **Index type** options available in the **DWG Options** tab of the **Save Options** dialog box.

Setting	Features	Index Type option
0	No index created	None
1	Layer index created	Layer
2	Spatial index created	Spatial
3	Layer and spatial index created	Layer and Spatial

EDITING REFERENCES IN-PLACE

Ribbon: Insert > Reference > Edit Reference   **Command:** REFEDIT
Toolbar: Refedit > Edit Reference In-Place

Often you have to make minor changes in the xref drawing if you want to save yourself from the trouble of going back and forth between drawings. In this situation, you can use in-place reference editing to select the xref, modify it, and then save it after modifications. This feature of AutoCAD has already been discussed in detail with reference to blocks in Chapter 18, Working with Blocks. The following steps explain the referencing editing process.

1. To edit in-place, select an external referenced object; the **External Reference** tab will be added to the **Ribbon**. Choose the **Edit Reference In-Place** tool from the **Edit** panel; the **Reference Edit** dialog box will be displayed. After selecting the external referenced object, you can also invoke the shortcut menu and choose the **Edit Xref In-place** option from it to invoke the **Reference Edit** dialog box.

2. In the **Reference name** list box of the **Reference Edit** dialog box, the selected reference and its nested references are listed. Select the specific xref you wish to edit from the list.

3. In the **Settings** tab of the dialog box, select both the **Create unique layer, style, and block names** and the **Lock objects not in working set** check boxes. The **Display attribute definitions for editing** check box is disabled by default. In the **Identify Reference** tab of the dialog box, select the **Prompt to select nested objects** radio button. Choose **OK** to return to the current drawing. At the **Select nested objects:** prompt, select the objects you want to modify and press ENTER. The **Edit Reference** panel is added to the **Ribbon** (Figure 20-34),

and the objects that have not been selected for modifications become faded. The fading is controlled by the **XFADECTL** system variable or by using the **Xref display** slider bar in the **Display** tab of the **Options** dialog box. You can add or remove objects to the working set by choosing the respective tools in the **Edit Reference** panel in the **External Reference** tab or by using the **REFSET** command. Once you have defined a working set, all the standard AutoCAD commands can be used to modify them.

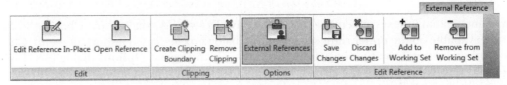

*Figure 20-34 The **Edit Reference** panel added to the **Ribbon***

4. When you choose the **Save Changes** tool in the **Edit Reference** panel, AutoCAD displays a message: **All reference edits will be saved**. Choose **OK** to return to the drawing area. The modifications made are saved in the drawing as well as in the current drawing as a reference. On choosing the **Discard Changes** button in the **Edit Reference** panel, the changes made will not be saved and the current drawing is redisplayed.

Note
*In the **External References (Xrefs)** area of the **Open and Save** tab of the **Options** dialog box, if you select the **Allow other users to Refedit current drawing** check box, the current drawing can be edited in-place by other users even when it is open and is being referenced by another file. This option is selected by default and can also be controlled by the **XEDIT** system variable. The default value of **XEDIT** is 1 and can be changed using the Command prompt.*

Self-Evaluation Test

Answer the following questions and then compare them to those given at the end of this chapter:

1. If an assembly drawing has been created by inserting a drawing, the drawing will be updated automatically, if a change is made in the drawing that was inserted. (T/F)

2. The external reference facility helps you keep a drawing updated no matter when the changes were made in the piece part drawings. (T/F)

3. Objects can be added to a dependent layer. (T/F)

4. While referencing a drawing, if you use the **Attachment** option, the drawing will reference the nested references too, while the **Overlay** option ignores the nested references. (T/F).

5. The _____ are entries such as blocks, layers, and text styles.

6. If you use the **Insert** tool to insert a drawing, the information about the named objects is lost if the names are _____, and if the names are _____, the drawing will be imported.

7. The _____ button of the **External References** palette is used to attach an xref drawing to the current drawing.

8. The _____ feature loads only that part of the referenced drawing that is required in the existing drawing.

9. The _____ option can be used to overcome the problem of circular reference.

10. If the **Retain changes to Xref layers** check box is _____ in the **External References (Xrefs)** area of the **Open and Save** tab of the **Options** dialog box, the layer settings such as color, linetype, on/off, and freeze/thaw are retained. The settings are saved with the drawing and are used when you xref the drawing the next time.

Review Questions

Answer the following questions:

1. If the xref drawings get updated, changes are not automatically reflected in the assembly drawing when you open an assembly drawing. (T/F)

2. There is a limit to the number of drawings you can reference. (T/F)

3. It is not possible to have nested references. (T/F)

4. Like blocks, the xref drawings can be scaled, rotated, or positioned at any desired location. (T/F)

5. You can change the color, linetype, or visibility (on/off, freeze/thaw) of a dependent layer. (T/F)

6. Which of the following features lets you reference an external drawing without making this drawing a permanent part of the existing drawing?

   (a) demand loading (b) external reference
   (c) external clipping (d) insert drawing

7. If an xref has nested references that cannot be found, which of the following will be displayed under the status heading of the **List View** button in the **External References** palette?

   (a) **Orphaned** (b) **Not found**
   (c) **Unreferenced** (d) **Unresolved**

8. Which of the following commands can be used from the Command prompt to attach a drawing?

   (a) **-XBIND** (b) **-XREF**
   (c) **XCLIP** (d) None of these

9. Which of the following system variables, when set to 1, will allow AutoCAD to maintain a log file (*.xlg*) for xref drawings?

   (a) **XLOADCTL** (b) **XLOADPATH**
   (c) **XREFCTL** (d) **INDEXCTL**

10. Which of the following system variables, when set to 0, will not allow the clipping boundary to be displayed?

    (a) **FRAME**   (b) **XLOADCTL**
    (c) **INDEXCTL**   (d) **XREFCTL**

11. In the _____ drawings, the information regarding dependent symbols is not lost.

12. What is the function of the **FRAME** system variable? Explain.

13. AutoCAD maintains a log file for xref drawings, if the _____ variable is set to 1.

14. It is possible to edit xrefs in-place using the _____ command.

15. You can use the _____ command to add the selected dependent symbols from the xref drawing to the current drawing.

EXERCISE 1	*XBIND*

In this exercise, you will start a new drawing and xref the drawings as Part-1 and Part-2, refer to Figure 20-35 and Figure 20-36. You will also edit one of the piece parts to correct the size and use the **XBIND** command to bind some of the dependent symbols to the current drawing. The parameters of layers for Part-1, Part-2 and ASSEM1 are as follows:

For Part-1, set up the following layers:

Layer Name	Color	Linetype
0	White	Continuous
Object	Red	Continuous
Hidden	Blue	Hidden2
Center	White	Center2
Dim-Part1	Green	Continuous

For Part-2, set up the following layers:

Layer Name	Color	Linetype
0	White	Continuous
Object	Red	Continuous
Hidden	Blue	Hidden
Center	White	Center
Dim-Part2	Green	Continuous
Hatch	Magenta	Continuous

For ASSEM1, set up the following layers:

Layer Name	Color	Linetype
0	White	Continuous
Object	Blue	Continuous
Hidden	Yellow	Hidden

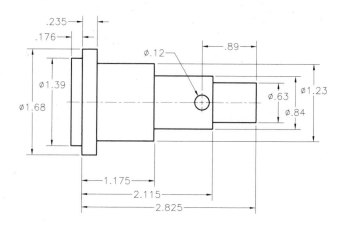

Figure 20-35 *Drawing of Part-1*

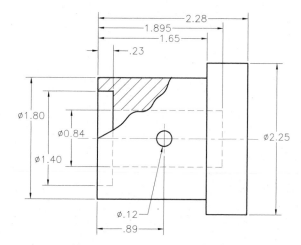

Figure 20-36 *Drawing of Part-2*

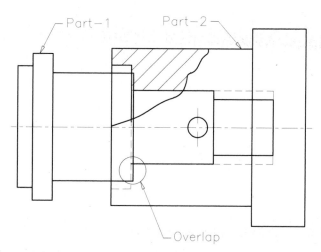

Figure 20-37 *Assembly drawing after attaching Part-1 and Part-2*

Hint:

1. *Xref the two drawings as Part-1 and Part-2 such that the centers of the two drilled holes coincide. Notice the overlap, refer to Figure 20-37. Save the assembly drawing as ASSEM1.*

2. *Use the XBIND command to bind the Object and Hidden layers that belong to the drawing Part-1. Check again to see if you can make one of these layers current.*

3. *Use the Detach option to detach the xref drawing Part-1. Study the layer again, and notice that the layers that were edited with the XBIND command have not been erased. Other layers belonging to Part-1 are erased.*

4. *Use the Bind option of the XREF command to bind the xref drawing Part-2 with the assembly drawing ASSEM1. Open the xref drawing Part-1 and add a border or make any other changes in the drawing. Now, open the assembly drawing ASSEM1 and check to see if the drawing is updated.*

Answers to Self-Evaluation Test

1. F, **2.** T, **3.** F, **4.** T, **5.** named objects, **6.** duplicated, unique, **7. Attach**, **8.** demand loading, **9. Overlay**, **10.** selected

Chapter 21

Working with Advanced Drawing Options

CHAPTER OBJECTIVES

In this chapter, you will learn:
- *To define multiline style and specify the properties of multilines using the MLSTYLE command.*
- *To draw various types of multilines using the MLINE command.*
- *To edit multilines using the MLEDIT command.*
- *To draw NURBS splines using the Spline tool.*
- *To edit NURBS splines using the Edit Spline tool.*

KEY TERMS

- *Multiline*
- *MLEDIT*
- *Revision Cloud*
- *Wipeout*
- *Spline*
- *Fit Points*
- *CV*
- *Knots*
- *Kink*
- *Weight*

UNDERSTANDING THE USE OF MULTILINES

The **Multiline** feature allows you to draw composite lines that consist of multiple parallel lines. Multilines consist of parallel lines, which are called elements. You can draw these lines with the **MLINE** command. Before drawing multilines, you need to set the multiline styles. This can be accomplished by using the **MLSTYLE** command. Also, editing of the multilines is made possible by the **MLEDIT** command.

DEFINING THE MULTILINE STYLE

Menu Bar: Format > Multiline Style	**Command:** MLSTYLE

You can create the style of multilines by using the **MLSTYLE** command. You can specify the number of elements in a multiline and the properties of each element. The style also controls the start and end caps, the start and end lines, and the color of multilines and fill. When you invoke this command, AutoCAD displays the **Multiline Style** dialog box (Figure 21-1). In this dialog box, you can set the spacing between parallel lines, specify linetype pattern, set colors, solid fill, and capping arrangements. By default, the multiline style (STANDARD) has two lines that are offset at 0.5 and -0.5.

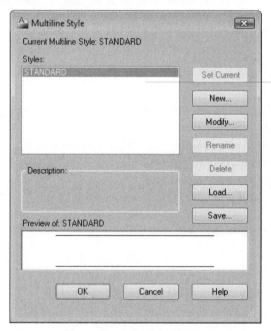

*Figure 21-1 The **Multiline Style** dialog box*

Note
You cannot modify the style of a multiline that is used in the current drawing. You first need to delete the multiline from all the places where it is used and then you can modify its style.

Styles List Box

The **Styles** list box displays the names of the styles. By default, the **STANDARD** style is available in this list box. If several styles have been defined, all of them will be listed in this list box.

Set Current

If several styles have been defined, select a style from the **Styles** list box and choose the **Set Current** button to make the style as the current style. The current multiline style is saved in the **CMLSTYLE** system variable.

New

The **New** button is chosen to create a new multiline style. On choosing this button, the **Create New Multiline Style** dialog box will be displayed, as shown in Figure 21-2. The **New Style Name** edit box is used to specify the name of the new multiline style you want to define. The **Start With** drop-down list is available only if some styles have already been defined. You can select an existing style as a reference to create the new style. The new style will start with the properties associated with the style selected from the **Start**

Figure 21-2 The **Create New Multiline Style** *dialog box*

With drop-down list. As soon as you specify the name of the new style, the **Continue** button is highlighted. To create a new style, enter the name of the new style in the **New Style Name** edit box and then choose the **Continue** button; the **New Multiline Style** dialog box will be displayed, as shown in Figure 21-3.

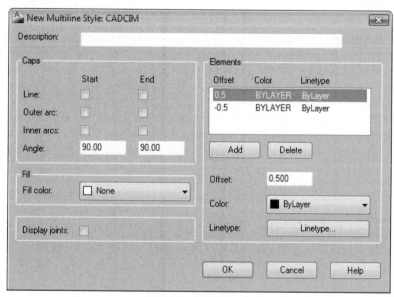

Figure 21-3 The **New Multiline Style** *dialog box*

The title bar of the **New Multiline Style** dialog box displays the name of the new style. By default, some properties in the dialog box are defined at some preset values. These values correspond to the selection made in the **Start With** drop-down list.

Description

This edit box used to specify the description of the new multiline style. The description, including spaces, can be up to 255 characters long. This is an optional edit box and you can leave it blank if you want.

Caps Area

The options under this area are used to set the type and location of the start and end caps of the multilines. These options are discussed next.

Line. Selecting the **Start** and **End** check boxes of this option caps the multiline with a line at the start point and the endpoint (Figure 21-4).

Outer arc. Selecting the **Start** and **End** check boxes of this option draws a semicircular arc between the endpoints of the outermost lines (Figure 21-4).

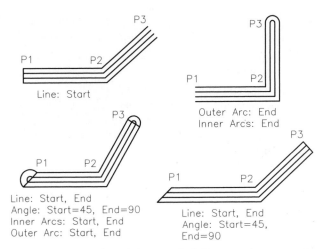

Figure 21-4 *Drawing multilines with different end cap specifications*

Inner arcs. This option is used to draw a semicircular arc between the two innermost lines at the start point and endpoint of a multiline. The arc is drawn between the even-numbered inner lines (Figure 21-4). For example, if there are two inner lines, an arc will be drawn at their ends. On the other hand, if there are three inner lines, the two outer lines are capped with the arc and the middle line is not capped.

Angle. By default, the end caps of a multiline are drawn tangentially to the multiline elements, that is, the angle the cap makes with the multiline elements is 90-degree. You can control the cap angle at the start and end of a multiline by entering angles in the corresponding **Angle** edit boxes (Figure 21-4). This angle can be between 10-degree and 170-degree.

Fill Area

The options under this area are used to fill the multiline with a specified color. By default, **None** is selected from the **Fill color** drop-down list. As a result, there is no fill effect in the multiline. To specify a fill color, select it from this drop-down list. Selecting the **Select Color** option displays the **Select Color** dialog box, using which you can select additional colors.

Display joints

If you select the **Display joints** check box, AutoCAD will draw a line across all elements of the multiline at the vertex points, where the multiline changes direction. The joints of the multiline are also called miter joints. If you draw only one multiline segment, no line is drawn at the miter joint.

Elements Area

The options in this area modify the element properties of the multiline style. The element properties include the number of lines, color, and linetype of the lines comprising the multiline. The following options are available for setting the properties of the individual lines (elements) that constitute the multiline.

Offset, Color, and Linetype

This list box displays the offset, color, and linetype of each line that constitutes the current multiline style. The lines are always listed in the descending order, based on the offset distance. For example, a line with a 0.5 offset will be listed first and a line with a 0.25 offset will be listed next. You can select an element to be modified from the list box; the selected element is highlighted.

Add

This button allows you to add new lines to the current line style. The maximum number of lines that you can add is 16. When you choose the **Add** button, AutoCAD inserts a line with the offset distance of 0.0, **Color** = BYLAYER, and **Linetype** = Bylayer. The new element gets added between two existing elements. After the line is added, you can change its offset distance, color, or linetype. The offset distance can be entered in the **Offset** edit box. Similarly, you can select the options in the **Color** drop-down list to specify a color for the selected element of the multiline. Also, you can choose the **Linetype** button to display the **Select Linetype** dialog box. You can use this dialog box and its options to load and select the linetype you want to apply to the selected element of a multiline.

Delete

This button allows you to delete a selected element from the multiline style. The highlighted element is removed from the **Elements** list box.

Offset

This edit box allows you to change the offset distance of the selected line in the **Elements** list box. The offset distance is defined with respect to the origin (0,0). The offset distance can be a positive or a negative value, which enables you to center the lines.

Color

This drop-down list allows you to assign a color to the selected line. When you choose the **Select Color** option from the drop-down list, AutoCAD displays the **Select Color** dialog box. You can select a color from the dialog box or enter a color number or name in the **Color** edit box.

Linetype

This button allows you to assign a linetype to the selected line. When you choose this button, AutoCAD displays the **Select Linetype** dialog box. After selecting a linetype in this dialog box, choose **OK** to exit. You can also choose the **Load** button in the **Select Linetype** dialog box to display the **Load or Reload Linetypes** dialog box, where you can select a linetype you want to use. The selected linetype is displayed in the **Loaded Linetype** list box of the **Select Linetype** dialog box. You can now select the specific linetype and apply it to a selected element of a multiline.

After you have added the elements to a multiline and set their properties, choose **OK** button to exit the **New Multiline Style** dialog box. An image of the current multiline style is displayed in the **Preview of** area in the **Multiline Style** dialog box. The new style is now listed in the **Style** list box.

Load

Choose this button from the **Multiline Style** dialog box to load a multiline style from an external multiline library file (*acad.mln*). On choosing this button, the **Load Multiline Styles** dialog box will be displayed, as shown in Figure 21-5. This dialog box displays all multiline styles available in the current *mln* directory. If the style is saved in another directory or file, choose the **File** button to display the **Load Multiline Style From File** dialog box and then select the multiline file that you want to load. Once the file is loaded, all styles saved under it are displayed in the list box. You can select the style and then choose **OK** to make it current.

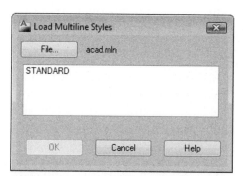

*Figure 21-5 The **Load Multiline Styles** dialog box*

Save

This button lets you save a multiline style to an external file (.*mln*). When you choose this button, AutoCAD displays the **Save Multiline Style** dialog box. This dialog box lists the names of the available predefined multiline style files (.*mln*). From the file listing, select or enter the name of the file where you want to save the current multiline style and choose the **Save** button. By default, the multiline styles get saved in the *acad.mln* file at the location *C:\Users\<user name>\AppData\Roaming\Autodesk\AutoCAD 2012\R18.2\enu\Support*.

Rename

This button allows you to rename the current multiline style. To do so, select the multiline style to be renamed and choose the **Rename** button; the text will be displayed in an edit box. Rename the text and click once on the list box. Note you cannot rename the **STANDARD** multiline style.

Preview of Area

The **Multiline Style** dialog box also displays the current multiline configuration in the **Preview of** area. The panel will display the color, linetype, caps and relative spacing of the lines. As you select another multiline style, the image in the preview area also gets updated accordingly. The name of the multiline style whose preview is being shown is displayed above the **Preview of** area.

DRAWING MULTILINES

Menu Bar: Draw > Multilines	**Command:** MLINE or ML

The **MLINE** command is used to draw multilines. The following is the prompt sequence for the **MLINE** command:

Current settings: Justification = Top, Scale = 1.00, Style = STANDARD
Specify start point or [Justification/Scale/STyle]: *Select a start point or enter an option.*
Specify next point: *Select the second point.*
Specify next point or [Undo]: *Select the next point or enter U for undo.*
Specify next point or [Close/Undo]: *Select the next point, enter U, or enter C for close. You can also right-click to display the shortcut menu and choose **Enter**, **Cancel**, **Recent Input**, **Dynamic Input Close**, **Undo**, **Snap overrides**, **Pan**, **Zoom**, **SteeringWheels**, or **Quick calc**.*

When you invoke the **MLINE** command, it always displays the current status of the multiline justification, scale, and style name. Remember that all line segments created in a single **MLINE** command are a single entity. The options provided under this command are discussed next.

Justification Option

The justification determines how a multiline is drawn between the specified points. Three justifications are available for the **MLINE** command. They are **Top**, **Zero**, and **Bottom**.

Top

This justification produces a multiline in which the top line coincides with the points you selected on the screen. Since the line offsets in a multiline are arranged in descending order, the line with the largest positive offset will coincide with the selected points. This is the default justification option, see Figure 21-6.

Zero

This option will produce a multiline in which the zero offset position of the multiline coincides with the selected points. Multilines will be centered if the positive and negative offsets are equal, see Figure 21-6.

Bottom

This option will produce a multiline in which the bottom line (the line with the least offset distance) coincides with the selected point when the line is drawn from the left to right, see Figure 21-6.

Scale Option

The **Scale** option is used to change the scale of a multiline. For example, if the scale factor is 0.5, the distance between the lines (offset distance) will be reduced to half, as shown in Figure 21-7. Therefore, the width of the multiline will be half of what was defined in the multiline style. A negative scale factor will reverse the order of the offset lines. The multilines will now be drawn such that the line with the maximum negative offset distance is at the top and the line with the maximum positive offset distance is at the bottom. For example, if you enter a scale factor of -0.5, the order in which the lines are drawn will be reversed, and the offset distances will be reduced by half. Here it is assumed that the lines are drawn from left to right. If the lines are drawn from right to left, the offsets are reversed. Also, if the scale factor is 0, AutoCAD forces the multiline into a single line. However, the line will still possess the properties of a multiline. This scale factor does not affect the linetype scale factor (**LTSCALE**).

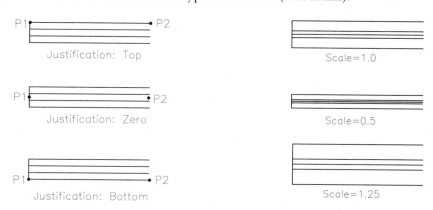

Figure 21-6 *Justifications in multilines*   **Figure 21-7** *Different multiline scales*

STyle Option

The **STyle** option allows you to change the current multiline style. The style must be defined before using the **STyle** option to change the style.

Note

All options explained above are available at the dynamic prompt. To access the options through dynamic prompt, press the down arrow key from the keyboard whenever dynamic prompting is available at the cursor. Now, select the option using the cursor.

EDITING MULTILINES BY USING GRIPS

Multilines can be easily edited using grips. When you select a multiline, the grips appear at the endpoints, based on the justification used when drawing multilines. For example, if the multilines are top-justified, the grips will be displayed at the endpoint of the first (top) line segment. Similarly, for zero and bottom-justified multilines, the grips are displayed on the centerline and bottom line, respectively.

Note

*Multilines do not support certain editing commands such as **BREAK**, **CHAMFER**, and **FILLET**. However, commands such as **COPY**, **MOVE**, **ROTATE**, **MIRROR**, **STRETCH**, **SCALE**, and **EXPLODE** as well as most object snap modes can be used with multilines.*

Chapter 21

Tip
*You must use the **MLEDIT** command to edit multilines. The **MLEDIT** command has several options that make it easier to edit these lines.*

EDITING MULTILINES BY USING THE DIALOG BOX

Menu Bar: Modify > Object > Multiline   **Command:** MLEDIT

When you invoke the **MLEDIT** command, AutoCAD displays the **Multilines Edit Tools** dialog box, see Figure 21-8. This dialog box contains five basic editing tools. To edit a multiline, first select the editing operation you want to perform by clicking on the image tile. Once you have selected the editing option, AutoCAD will prompt you to select the first and second multilines. After editing one set of multilines, you will be prompted again to select the first multiline or undo the last editing. Press ENTER to exit this command after you have finished the editing.

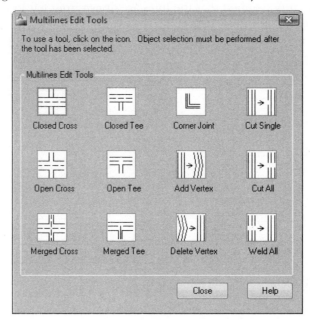

*Figure 21-8 The **Multilines Edit Tools** dialog box*

The following is the list of options for editing multilines:

Cross Intersection
Closed Cross (CC), Open Cross (OC), Merged Cross (MC)

Tee Intersection
Closed Tee (CT), Open Tee (OT), Merged Tee (MT)

Corner Joint (CJ)

Adding and Deleting Vertices
Add Vertex (AV), Delete Vertex (DV)

Cutting and Welding Multilines
Cut Single (CS), Cut All (CA), Weld All (WA)

Cross Intersection (CC/OC/MC)

With the **MLEDIT** command options, you can create three types of cross intersections: closed, open, and merged. You must be careful about the order in which you select multilines for editing. The order in which multilines are selected determines the edited shape of a multiline (Figure 21-9). The cross intersection can belong to a self-intersecting multiline or two different multilines.

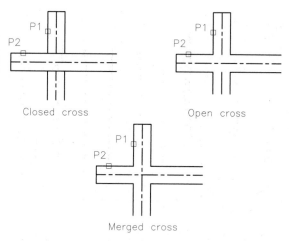

Closed cross   Open cross

Merged cross

Figure 21-9 *Three types of cross intersections*

When you invoke the **MLEDIT** command, the **Multilines Edit Tools** dialog box is displayed. Choose the **Closed Cross** image tile; the dialog box disappears from the screen and the following prompts appear in the Command prompt:

Select first mline: *Select the first multiline.*
Select second mline: *Select the second multiline.*
Select first mline or [Undo]: *Select another multiline, press ENTER to end the command, or enter* **U** *to undo the last operation.*

If you enter **U**, AutoCAD will undo the last editing operation and prompt you to select the first multiline. However, if you do not enter **U** and select another multiline, AutoCAD will prompt you to select the second multiline.

Tee Intersection (CT/OT/MT)

With the **MLEDIT** command options, you can create three types of tee-shaped intersections: closed, open, and merged. As with the cross intersection, you must be careful how you select the objects because the order in which you select them determines the edited shape of a multiline (Figure 21-10). The prompt sequence for the tee intersection is the same as that for the cross intersection.

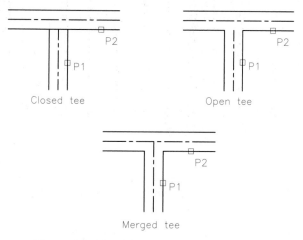

Closed tee   Open tee

Merged tee

Figure 21-10 *Three types of tee intersections*

When you invoke the **MLEDIT** command, the **Multilines Edit Tools** dialog box is displayed. Choose the **Closed Tee** image tile; the dialog box disappears from the screen and the following prompts appear in the Command prompt.

Select first mline: *Select the first multiline.*
Select second mline: *Select the intersecting multiline.*
Select first mline or [Undo]: *Select another multiline, enter **U** to undo the last operation, or press ENTER to end the command.*

Corner Joint (CJ)

The **Corner Joint** option is used to create a corner joint between the two selected multilines. The multilines must be two separate objects (multilines) or self-intersecting multilines (Figure 21-11). When you specify two multilines, AutoCAD trims or extends the first multiline to intersect with the second one. The prompt sequence that follows when you invoke this option is given next.

Select first mline: *Select the multiline to trim or extend.*
Select second mline: *Select the intersecting multiline.*
Select first mline or [Undo]: *Select another multiline, enter **U** to undo the last operation, or press ENTER to end the command.*

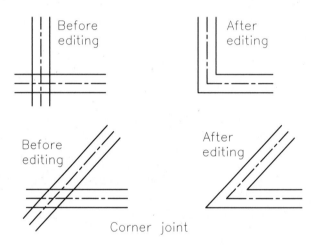

Figure 21-11 Editing corner joints

Adding and Deleting Vertices (AV/DV)

You can use the **MLEDIT** command to add or delete the vertices of a multiline (Figure 21-12). When you select a multiline for adding a vertex, AutoCAD inserts a vertex point at the point where you clicked on the object while selecting it. Later, you can move the vertex by using grips. Similarly, you can also use the **MLEDIT** command to delete vertices by selecting the object whose vertex point you want to delete. AutoCAD removes the vertex that is nearest to the point where you click to select the multiline segment.

When you invoke the **MLEDIT** command, the **Multilines Edit Tools** dialog box is displayed. Choose the **Add Vertex** image tile; the dialog box is removed from the screen and the following prompts appear in the Command prompt.

Select mline: *Select the multiline for adding vertex.*
Select mline or [Undo]: *Select another multiline for modifications; enter **U** to undo the last operation, or press ENTER to exit the command.*

A same prompt sequence appears at the Command prompt when you use the **Delete Vertex** option.

Cutting and Welding Multilines (CS/CA/WA)

You can also use the **MLEDIT** command to cut or weld multilines. When you cut a multiline, it does not create two separate multilines. They continue to be a part of the same object (multiline). You can cut away an area from any of the elements of a multiline or from the entire multiline. Also, the points selected for cutting the multiline do not have to be on the same element of the multiline (Figure 21-13). You can select the two points that define the cut on different elements of the multiline. When cutting a single element of the multiline, if you select the two cut points on different elements, the element you selected first gets cut by the distance defined by the two points.

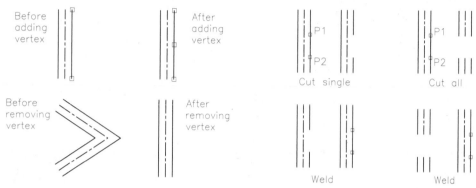

Figure 21-12 Adding and deleting vertices　　　*Figure 21-13 Cutting and welding multilines*

In the **Multilines Edit Tools** dialog box, choose the **Cut Single** image tile; the dialog box is removed from the screen, and the following prompts appear in the Command prompt.

Select mline: Select the multiline.
(The point where you select the multiline specifies the first cut point.)
Select second point: *Select the second cut point.*
Select mline or [Undo]: *Select another multiline to cut, enter **U** to undo the last operation, or press ENTER to exit the command.*

The **Weld** option reconnects the multilines that have been cut by using the **Cut Single** or **Cut All** option of the **MLEDIT** command (Figure 21-13). You cannot weld or join separate multilines. In the **Multilines Edit Tools** dialog box, choose the **Weld All** image tile; the dialog box is removed from the screen, and the following prompts appear in the Command prompt.

Select mline: *Select the multiline.*
Select second point: *Select a point on the other side of the cut.*
Select mline or [Undo]: *Select another multiline to modify, enter U to undo the last operation, or press ENTER to exit the command.*

Note
*The following system variables store values for various aspects of multilines. The **CMLJUST** variable stores the justification of the current multiline (0-Top, 1-Middle, 2-Bottom). The **CMLSCALE** variable stores the scale of the current multiline (Default scale = 1.0000) and the **CMLSTYLE** variable stores the name of the current multiline style (Default style = STANDARD).*

Tip
*You can also edit multilines by entering the **-MLEDIT** command at the pointer input. All options are displayed at the dynamic prompt, see Figure 21-14.*

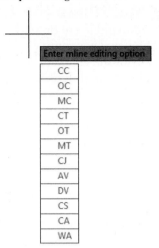

Figure 21-14 *Dynamic prompt for the **-MLEDIT** command*

EXAMPLE 1   **MLINE**

In this example, you will create a multiline style that represents a wood-frame wall system. The wall system consists of ½" wallboard, 4 X ½" wood stud, and ½" wallboard.

1. Enter **MLSTYLE** at the Command prompt to display the **Multiline Style** dialog box. Also, the current style, **STANDARD** is displayed in the **Styles** list box.

2. Choose the **New** button to invoke the **Create New Multiline Style** dialog box. Enter **MYSTYLE** in the **New Style Name** edit box and choose **Continue** to display the **New Multiline Style: MYSTYLE** dialog box.

3. Enter **Wood-frame Wall System** in the **Description** edit box.

4. Select the **0.5** line definition in the **Elements** list box. In the **Offset** edit box, replace **0.500** with **1.00**. This redefines the first line which will now be 1.00" above the centerline of the wall.

5. Select the **-0.5** line definition in the **Elements** list box. In the **Offset** edit box, replace **-0.500** with **-1.00**. This redefines the second line which will now be placed 1.00" below the centerline of the wall.

6. Choose the **Add** button; a new line is added with offset value of 0 and is highlighted in the **Elements** list box.

7. In the **Offset** edit box, replace **0.000** with **1.50**.

8. Select **Yellow** from the **Color** drop-down list.

9. Repeat steps 6 through 8, but this time use the value **-1.50** in place of **1.50** in step 7 to add another line to the current multiline style. The color of this line should be red.

10. Choose the **OK** button and return to the **Multiline Style** dialog box. The new multiline style is displayed in the **Preview of** area.

11. Choose the **Set Current** button to make the **MYSTYLE** as current style.

12. Choose the **OK** button from the **Multiline Style** dialog box to return to the drawing area. To test the new multiline style, use the **MLINE** command and draw a series of lines (Figure 21-15).

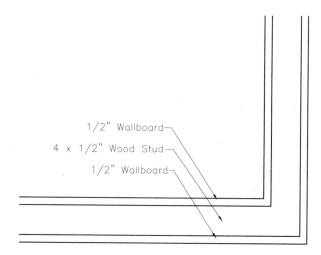

1/2" Wallboard

4 x 1/2" Wood Stud

1/2" Wallboard

Figure 21-15 Multiline style created for Example 1

CREATING REVISION CLOUDS

Ribbon: Home > Draw > Revision Cloud or Annotate > MarkUp > Revision Cloud	
Tool Palettes: Draw > Revision Cloud	**Toolbar:** Draw > Revision Cloud
Menu Bar: Draw > Revision Cloud	**Command:** REVCLOUD

The **Revision Cloud** tool in the **Draw** panel is used to create a cloud-shaped polyline, as shown in Figure 21-16. This tool can be used to highlight the details of a drawing. The prompt sequence that will follow when you choose this tool from the **Draw** panel is as follows:

Minimum arc length: 0.5000   Maximum arc length: 0.5000   Style: Normal
Specify start point or [Arc length/Object/Style] <Object>: *Specify the start point of the revision cloud.*
Guide crosshairs along cloud path...

As you move the cursor, different arcs of the cloud with varied radii are drawn. When the start point and endpoint meet, the revision cloud is completed and you get a message.

Revision cloud finished.

You can define the length of the arcs to be drawn using the **Arc length** option. Also, all the arcs drawn are of constant length. You can convert a closed loop into a revision cloud using the **Object** option. Note that the selected closed loop should be a single entity such as an ellipse, a circle, a rectangle, a polyline, and so on. The **Style** option is used to define the arc style for the revision cloud. The default style is **Normal** that creates a revision cloud similar to the one shown in Figure 21-16. You can change the style to **Calligraphy** to create a revision cloud similar to the one shown in Figure 21-17.

Note

*The **REVCLOUD** stores the last used arc length in the system registry. This value is multiplied by the **DIMSCALE** value to provide consistency when the program is used with drawings that have different scale factors.*

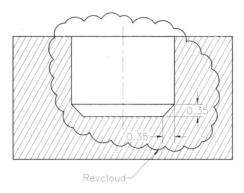

Figure 21-16 *Creating revcloud*

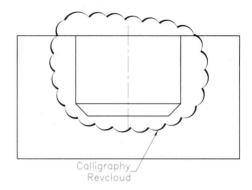

Figure 21-17 *Calligraphy revcloud*

CREATING WIPEOUTS

Ribbon: Home > Draw > Wipeout or Annotate > Markup > Wipeout
Command: WIPEOUT

The **Wipeout** tool is used to create a polygonal area to cover the existing objects with the current background color. The area defined by this command is governed by the wipeout frame. The frame can be turned on and off for editing and plotting the drawings, respectively. This tool can be used to add notes and details to a drawing. The prompt sequence that will follow when you choose the **Wipeout** tool from the **Draw** panel is as follows:

Specify first point or [Frames/Polyline] <Polyline>: *Specify the start point of the wipeout.*
Specify next point: *Specify the next point of the wipeout.*
Specify next point or [Undo]: *Specify the next point of the wipeout.*
Specify next point or [Close/Undo]: *Specify the next point of the wipeout.*

Figure 21-18 shows a drawing before creating a wipeout and Figure 21-19 shows a drawing after creating the wipeout.

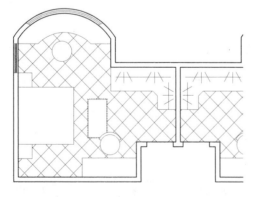

Figure 21-18 *Drawing before creating a wipeout*

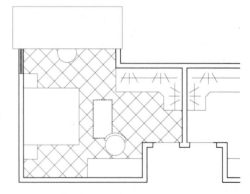

Figure 21-19 *Drawing after creating a wipeout*

If you do not want the frame of the wipeout to be displayed, enter **F** at the **Specify first point or [Frames/Polyline] <Polyline>** prompt and turn the frame off. The display of frames of all existing wipeouts will be turned off. Also, the display of frames of all new wipeouts will be turned off.

CREATING NURBS

Ribbon: Home > Draw > Spline Fit   **Menu Bar:** Draw > Spline > Spline Fit
Tool Palettes: Draw > Spline   **Toolbar:** Draw > Spline   **Command:** SPLINE

The NURBS is an acronym for **Non-Uniform Rational Bezier-Spline**. These splines are considered true splines. In AutoCAD, you can create NURBS using the **Spline** tool. The spline created with the **Spline** tool is different from the spline created using the **Polyline** tool. The nonuniform aspect of the spline enables the spline to have sharp corners because the spacing between the spline elements that constitute a spline can be irregular. Rational means that irregular geometry such as arcs, circles, and ellipses can be combined with free-form curves. The Bezier-spline (B-spline) is the core that enables accurate fitting of curves to input data with Bezier's curve-fitting interface. Not only are spline curves more accurate compared to smooth polyline curves, but they also use less disk space.

The prompt sequence to draw a spline with six fit points is given below.

Command: **_SPLINE**
Current settings: Method=Fit   Knots=Chord
Specify first point or [Method/Knots/Object]: *Select point P1*
Enter next point or [start Tangency/toLerance]: *Select point P2.*
Enter next point or [end Tangency/toLerance/Undo/Close]: *Select point P3.*
Enter next point or [end Tangency/toLerance/Undo/Close]: *Select point P4*
Enter next point or [end Tangency/toLerance/Undo/Close]: *Select point P5.*
Enter next point or [end Tangency/toLerance/Undo/Close]: *Select point P6 and press*
ENTER to end the process of point specification

Options for Creating Splines

The options under this command for creating splines are as follows:

Object

This option allows you to change a 2D or 3D spline fitted polyline into a NURBS. The original splined polyline is deleted if the system variable **DELOBJ** is set to 1, which is the default value of the variable. You can change a polyline into a splined polyline using the **Spline** option of the **Edit Polyline** tool.

Method

In AutoCAD, you can create a spline by specifying fit points or control vertices. By default, a spline is created by specifying the fit points. Fit points are the points through which the spline will pass. Control vertices determine the shape of the curve. To set the method to create a spline, enter **M** at the **Specify first point or [Method/Knots/Object]** prompt. Then, you need to specify the method at the **Enter spline creation method [Fit/CV]** prompt. Figure 21-20 shows the spline created by using the **Fit** method. Figure 21-21 shows the spline created by using the **CV** method.

Note
On selecting a spline, a grip will be displayed at the start point of the spline. Left-click on the grip; a shortcut menu will be displayed. Now, you can choose the appropriate option to display the fit points or the control vertices, as shown in Figure 21-22.

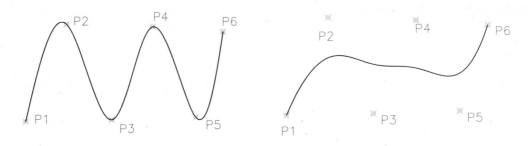

Figure 21-20 *Spline drawn by using the **Fit** method*

Figure 21-21 *Spline drawn by using the **CV** method*

Knots

If you have selected the **Fit** method, the **Knots** option will be displayed at the Command prompt, as given below.

> Enter spline creation method [Fit/CV] <CV>: **FIT**
> Current settings: Method=Fit   Knots=Uniform
> Specify first point or [Method/Knots/Object]:

You can specify three types of knots: Chord, Square root, and Uniform. Figure 21-23 shows the splines formed by using three different knot options.

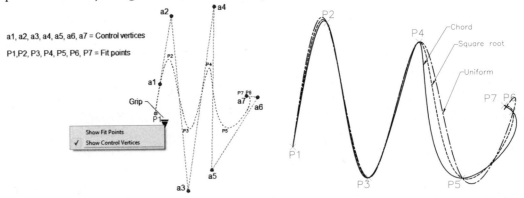

Figure 21-22 *Control vertices and fit points of a spline*

Figure 21-23 *Splines with different knot options*

Degree

If you have selected the **CV** method, then the **Specify first point or [Method/Degree/Object]** prompt will be displayed. Enter **D** at the Command prompt and specify the degree.

Close

This option allows you to create closed NURBS. When you use this option, AutoCAD will automatically join the endpoint of the spline with the start point, and you will be prompted to define the start tangent only.

Tolerance

This option is used to control the fit of the spline between specified points. By default, this value is zero; as a result, the spline passes through the points by which it is created. Using this option, you can specify the tolerance values that will govern the spline creation, see Figure 21-24. The splines will be offset from the specified point by a distance equal to the tolerance value. The smaller the value of the tolerance, the closer the spline will be to the specified points.

Start and End Tangency

This option is used to control the tangency of a spline at its start point and the endpoint. On entering **T** at the **Enter next point or [start Tangency/toLerance]** or **Enter next point or [end Tangency/toLerance/Undo/Close]** prompt, you will be prompted to specify the tangent. Specify a point; the tangency will be determined by the slope of the spline at the specified point. Remember that after specifying the start point of the spline, you need to specify the start tangency. If you specify the second point, the default value will be used for the tangency. Similarly, you need to specify the end tangency after specifying the last point of the spline. Figure 21-25 shows the spline with start and end tangencies. The virtual tangent line is for reference.

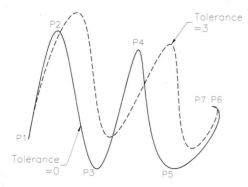

Figure 21-24 *Creating a spline with the* **Tolerance** *of 0 and 3*

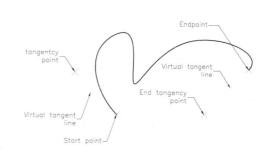

Figure 21-25 *Spline with start and end tangencies*

EDITING SPLINES

Ribbon: Home > Modify > Edit Spline	**Menu Bar:** Modify > Object > Spline
Tool Palettes: Modify > Edit spline	**Toolbar:** Modify II > Edit Spline
Command: SPLINEDIT	

The NURBS can be edited using the **Edit Spline** tool. Using this tool, you can fit data in a selected spline, close or open a spline, move vertex points, and refine or reverse a spline. Apart from the ways mentioned in the preceding command box, you can use the options available in **Spline** flyout from the shortcut menu that is displayed when you select a spline and right-click. The prompt sequence that will follow when you choose the **Edit Spline** tool in the **Modify** panel is given next.

> Select spline: *Select the spline to be edited if not selected already, using the above-mentioned shortcut menu.*
> Enter an option [Close/Join/Fit data/Edit vertex/convert to Polyline/Reverse/Undo/eXit] <eXit>: *Select any one of the options.*

Options for Editing Splines

The options under this command for editing the splines are described next.

Fit data*

When you draw a spline, the spline is fit to the specified points (data points). The **Fit data** option allows you to edit these points. You can add, delete, or move the data points. These data points or control points are also referred to as fit points. For example, if you want to redefine the start and end tangents of a spline, select the **Fit data** option, then select the **Tangents** option. The prompt sequence that will follow when you invoke this option is given next.

Select spline: *Select the spline to be edited.*
Enter an option [Close/Join/Fit data/Edit vertex/convert to Polyline/Reverse/Undo/eXit]
<eXit>: **F**
Enter a fit data option [Add/Close/Delete/Kink/Move/Purge/Tangents/
toLerance/eXit] <eXit>: **T**

The start and end tangent points can be selected or their coordinates can be entered. The options available within the **Fit data** option are as follows:

Add. You can use this option to add new fit points to the spline. When you invoke this option, you will be prompted to specify the existing fit point on the spline. After selecting the existing fit point, you will be prompted to specify the location of the new control points. The fit point you select and the next fit point appear as selected grips. You can now add a fit point between these two selected fit points, as shown in Figure 21-26. If you select the start point or endpoint of the spline, only those points are highlighted. When you select the start point of the spline, you are prompted to specify whether you want to add the new fit point before or after the start point of the spline. AutoCAD will continue prompting for the location of new control points until you press ENTER at the **Specify new point <exit>** prompt.

Close/Open. This option is used to close an open spline or open a closed spline, see Figure 21-27. If the spline is open, the **Close** option is available and if the spline is closed, the **Open** option is displayed.

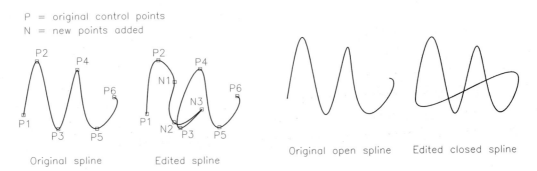

Figure 21-26 *The original spline and the edited spline after adding fit points*

Figure 21-27 *An open spline and a closed spline*

Delete. This option is used to delete a selected fit point from a spline. You can continue deleting fit points from a spline until only two fit points are left in the spline.

Kink. This option is used to add a knot and a fit point to the selected point. On selecting this option, you will be prompted to specify a point on the spline. Select a point; a kink will be added.

Move. You can move fit points by using this option. When you select this option, the start point of the spline is highlighted and the following prompt sequence appears:

Specify new location or [Next/Previous/Select point/eXit] <N>: *Select a new location for the start point using the mouse pick button or enter any one of these options.*

You can enter **N** if you want to select the next point, **P** if you want to select a previous point, or **S** if you want to select any other point. If you enter **X** at the preceding prompt, you can exit the command. Figure 21-28 shows the movement of data points in a spline. Remember that this option is used to move only the data points on the spline and not the control points on the Bezier control frame.

Purge. This option is used to remove fit point data from a spline. This reduces the file size, which is useful when a drawing, for example, a landscape contour map, contains large number of splines. Purging simplifies the spline definition and the drawing file size.

Tangents. This option is used to modify the tangents of the start and endpoints of a selected spline, see Figure 21-29. When you select this option, you will be prompted to specify the start and end tangents for the spline. You can specify the start and end tangents or use the systems default tangents.

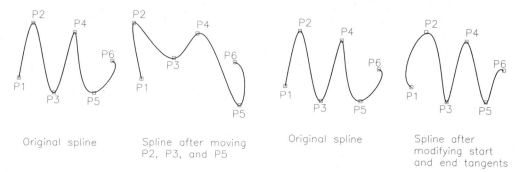

Figure 21-28 *The original spline and the edited spline after moving the data points*

Figure 21-29 *The original spline and the spline after modifying the start and end tangents*

toLerance. This option is used to modify the fit tolerance values of a selected spline. As discussed while creating the splines, different tolerance values produce different splines. A smaller tolerance value creates a spline that passes very closely through the definition points of a spline. When you invoke this option, you will be prompted to specify the tolerance value for the spline.

eXit. This option is used to exit the **Fit data** option of editing splines.

Close/Open

This option is used to close an open spline or open a closed spline. When you select the **Close** option, AutoCAD lets you open, edit the vertex, covert spline to poyline, or reverse the spline.

Edit Vertex

This option is used to refine a spline by adding more control points in it, elevating the order, or adding weight to vertex points. The prompt sequence that will follow when you invoke this option is given next.

> Enter an option [Open/Fit data/Edit vertex/convert to Polyline/Reverse/Undo/eXit] <eXit>: *E*
> Enter a vertex editing option [Add/Delete/Elevate order/Move/Weight/eXit] <eXit>:

The options in this prompt are discussed next.

Add. This option is used to add more control points on the spline. When you invoke this option, you will be prompted to specify the location of the new point on the spline. You can directly specify the location of the new point using the left mouse button. A new control point will be added at the specified location.

Delete. This option is used to delete a selected fit point from a spline.

Elevate order. This option is used to increase the order of a spline curve. An order of the curve can be defined as the highest power of the algebraic expression that defines the spline plus 1. For example, the order of a cubic spline will be 3 + 1 = 4. Using the **Elevate order** option, you can increase this order for a selected spline. This results in more control points on the curve and a greater possibility of controlling the spline. The value of the spline order varies from 4 to 26. You can only increase the order and not decrease it. For example, if a spline order is 18, you can elevate its order to any value greater than 18, but not less than 18.

Move. When you draw a spline, it is associated with the Bezier control frame. The **Move vertex** option allows you to move the vertices of the control frame. The **Move** option is similar to the **Move** option of the **Fit data** option of editing the splines. The prompt sequence that follows when you invoke this option is given next.

> Enter an option [Close/Join/Fit data/Edit vertex/convert to Polyline/Reverse/Undo/eXit] <eXit>: **E**
> Enter a vertex editing option [Add/Delete/Elevate order/add Kink/Move/Weight/eXit] <eXit>: **M**
> Specify new location or [Next/Previous/Select point/eXit] <N>:

Weight. You can also add weight to any of the control vertices of a spline by using this option. When weight is added to a particular vertex, the spline gets pulled more towards it. Similarly, a lower value of weight of a particular spline vertex will result in the spline getting pulled less towards that particular vertex. In other words, adding weight to a particular point will force the selected point to maintain its tangency with the point. The more weight added to the point, the more is the distance through which the spline will remain tangent to the point. By default, the spline gets pulled equally towards the vertices of the spline. The default value of weight provided to each control point is 1.0 and can have only positive values. Once you have added the weight to a point, you can proceed to the next point. You can also directly select the point to which the weight has to be added. The prompt sequence for using this option is given next.

> Enter an option [Close/Join/Fit data/Edit vertex/convert to Polyline/Reverse/Undo/eXit] <eXit>: **E**
> Enter a vertex editing option [Add/Delete/Elevate order/Move/Weight/eXit] <eXit>: **W**
> Spline is not rational. Will make it so.
> Enter new weight (current = 1.0000) or [Next/Previous/Select point/eXit] <N>: **S**
> Specify existing fit point on spline <exit>: *Select a fit point*
> Enter new weight (current = 1.0000) or [Next/Previous/Select point/eXit] <N>: **3**
> Enter new weight (current = 1.0000) or [Next/Previous/Select point/eXit] <N>: **Exit**

Join

If a spline, line, or an arc is joined to the spline that is being edited, then you can use this option to join those entities to the spline. On choosing this option you will be prompted to select an open curve. Select the curves to be joined and press ENTER; all curves will be joined to the spline, and knot and fit point will be added at the joining point. The prompt sequence for using this option is given next.

> Enter an option [Close/Join/Fit data/Edit vertex/convert to Polyline/Reverse/Undo/eXit] <eXit>: **J**
> Select any open curves to join to source: *Select an object. 1 found*
> Select any open curves to join to source: *Select an object. 1 found, 2 total*
> Select any open curves to join to source:
> 2 objects joined to source

Reverse

This option is used to reverse the spline direction. This implies that when you apply this option to a spline, its start point becomes its endpoint and vice versa.

convert to Polyline

This option is used to convert a spline into a polyline. On specifying this option, you will be prompted to specify a precision value. Specify a value between 0 and 99, and then press ENTER; the spline will be converted into a polyline. The precision value determines the accuracy of the resulting polyline.

Undo

This option will undo the previous editing operation applied to a spline within the current session of the **SPLINEDIT** command. You can continue to use this option till you reach the spline as it was when you started to edit it.

EDITING SPLINES USING 3D EDIT BAR*

Command: 3DEDITBAR

The **3DEDITBAR** command is used to edit a spline by using the Move gizmo. To edit a spline using this command, select the spline and right click; a shortcut menu will be displayed. Choose the **Spline>3D Edit Bar** option from the shortcut menu; you will be prompted to select a point on the curve. Select a point on the curve at the position where you want to modify the spline; the Move gizmo will be displayed with one extra arrow tangential to the curve at the selected point. This arrow can be used to change the tangent direction as well as the magnitude of tangency at the point. Using this arrow, you can dynamically edit the spline.

Self-Evaluation Test

Answer the following questions and then compare them to those given at the end of this chapter:

1. The name of the multiline style that you enter in the **New Style Name** edit box of the **Create New Multiline style** dialog box can have up to thirty-one characters including spaces. (T/F)

2. The **Zero** option is used to create a multiline so that the zero offset position of the multiline coincides with the selected points. (T/F)

3. Multilines cannot be edited using grips. (T/F)

4. When using the **Top** justification option to draw a multiline, the line with the largest positive offset coincides with the points selected on the screen. (T/F)

5. When a multiline is cut, two separate multilines are created. (T/F)

6. The maximum number of lines that can be added to a multiline style is _____.

7. When using the **Inner arcs** option for multilines, a capping arc is drawn between the _____ numbered inner lines.

8. The **Multilines Edit tools** dialog box consists of _____ basic editing tools.

9. The _____ option of the **Fit data** option of the **Edit Spline** tool is used to reduce the file size by removing the fit data of the selected splines.

10. The _____ system variable stores the justification of the current multiline.

Review Questions

Answer the following questions:

1. By default, the *.mln* files are saved at the location *Autodesk\AutoCAD 2012\R18.2\enu\Support*. (T/F)

2. Once the **Purge** option of the **Fit data** option of the **SPLINEDIT** command is used on a spline, it becomes difficult to edit the spline and the **Fit data** option of the **SPLINEDIT** command is no longer available for the particular spline. (T/F)

3. You can break or trim a multiline. (T/F)

4. You cannot change the lineweights of the lines that constitute a multiline. (T/F)

5. You cannot weld or join separate multilines. (T/F)

6. In which file are the multiline styles saved by default?

   (a) *acad.mln*   (b) *acad.pat*
   (c) *acad.mpr*   (d) *acad.plt*

7. Which of the following is the default value of the start and end angles for multilines?

   (a) 90   (b) 180
   (c) 45   (d) 0

8. Which of the following commands does not work on multilines?

   (a) **MIRROR**   (b) **FILLET**
   (c) **STRETCH**   (d) **EXPLODE**

9. Which of the following options can be used to reverse the direction of spline creation while editing a spline?

   (a) **Fit data**   (b) **Refine**
   (c) **Reverse**   (d) **Weight**

10. AutoCAD forces a multiline into a single line if the scale factor of the multiline is_____.

11. The current multiline style is saved in the _____ system variable and the default value is **STANDARD**.

12. Three justifications are available for the **MLINE** command. They are _____, _____, and _____.

13. Using the **MLEDIT** command options, you can create three types of cross intersections. They are _____, _____, and _____.

14. If value of **Fit tolerance** of a spline is _____, the spline passes exactly through the fit points of the spline.

15. You need to set the degree of a spline, if you are drawing the spline by using the _____ method.

EXERCISE 1

Create the drawing shown in Figure 21-30. Assume the missing dimensions.

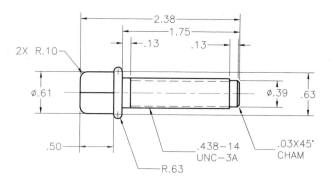

Figure 21-30 *Drawing for Exercise 1*

EXERCISE 2

Create the drawing shown in Figure 21-31. Use the **MULTILINE** command to draw the walls. The wall thickness is 12 inches. Assume the missing dimensions.

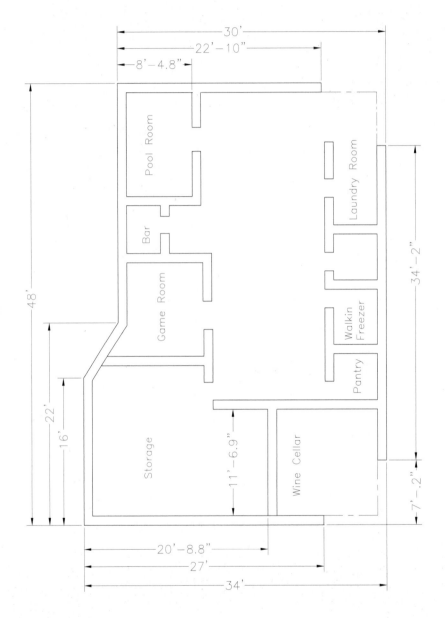

***Figure 21-31** Drawing for Exercise 2*

Problem-Solving Exercise 1

Create the drawing shown in Figure 21-32. Some of the reference dimensions are given in the drawing. Assume the missing dimensions so that the drawing looks similar to the one given below.

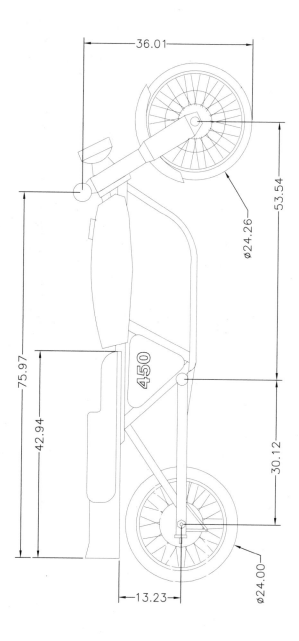

Figure 21-32 Drawing for Problem-Solving Exercise 1

Problem-Solving Exercise 2

Create the drawing shown in Figure 21-33. Some of the reference dimensions are given in the drawing. Assume the missing dimensions so that the drawing looks similar to the one given below.

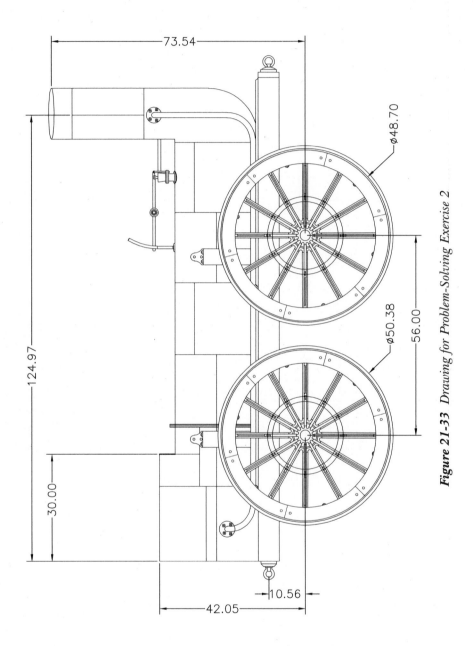

Figure 21-33 Drawing for Problem-Solving Exercise 2

Problem-Solving Exercise 3

Create the drawing shown in Figure 21-34. Some of the reference dimensions are given in the drawing. Assume the missing dimensions so that the drawing looks similar to the one given below.

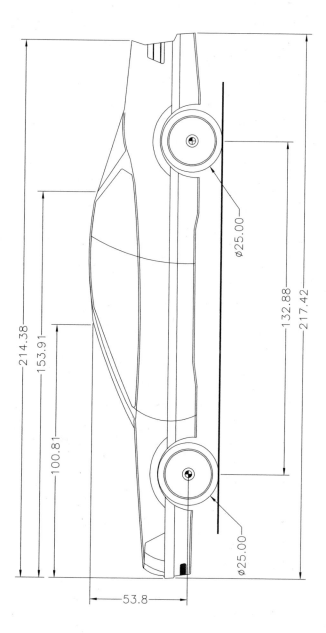

Figure 21-34 Drawing for Problem-Solving Exercise 3

Problem-Solving Exercise 4

Create the drawing shown in Figure 21-35. Some of the reference dimensions are given in the drawing. Assume the missing dimensions so that the drawing looks similar to the one given below.

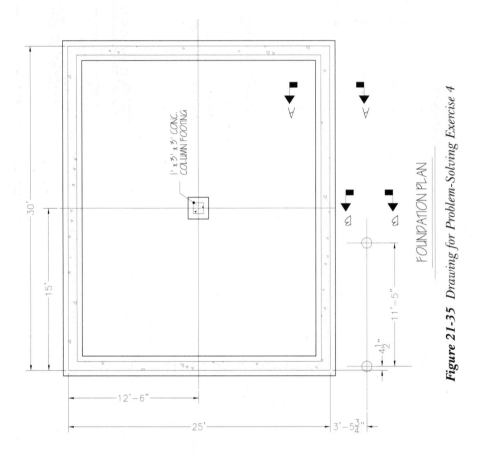

Figure 21-35 *Drawing for Problem-Solving Exercise 4*

Problem-Solving Exercise 5

Create the drawing shown in Figure 21-36. Some of the reference dimensions are given in the drawing. Assume the missing dimensions so that the drawing looks similar to the one given below.

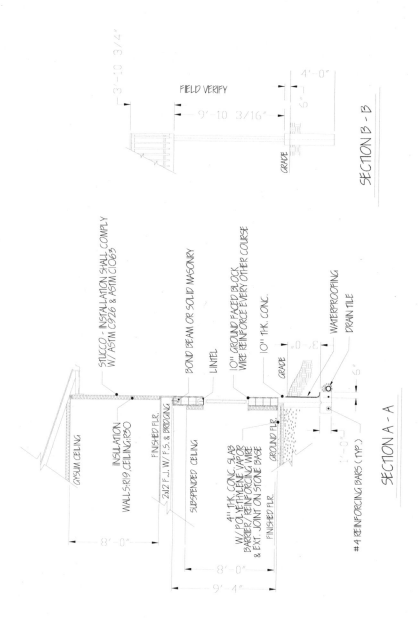

***Figure 21-36** Drawing for Problem-Solving Exercise 5*

Problem-Solving Exercise 6

Create the drawing shown in Figure 21-37. Some of the reference dimensions are given in the drawing. Assume the missing dimensions so that the drawing looks similar to the one given below.

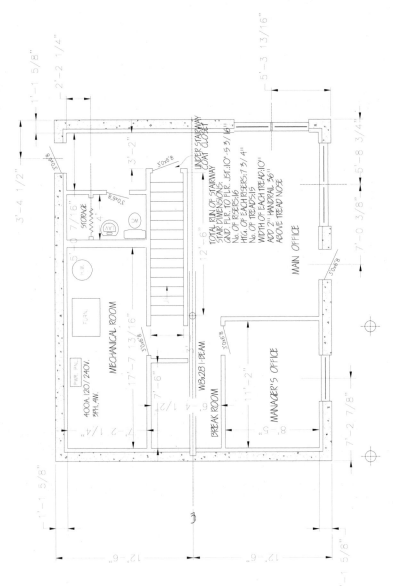

Figure 21-37 Drawing for Problem-Solving Exercise 6

Problem-Solving Exercise 7

Create the drawing shown in Figure 21-38. Some of the reference dimensions are given in the drawing. Assume the missing dimensions so that the drawing looks similar to the one given below.

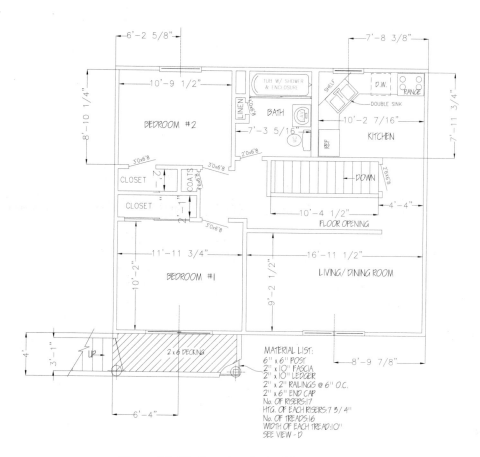

Figure 21-38 *Drawing for Problem-Solving Exercise 7*

Problem-Solving Exercise 8

Create the drawing shown in Figure 21-39. Some of the reference dimensions are given in the drawing. Assume the missing dimensions so that the drawing looks similar to the one given below.

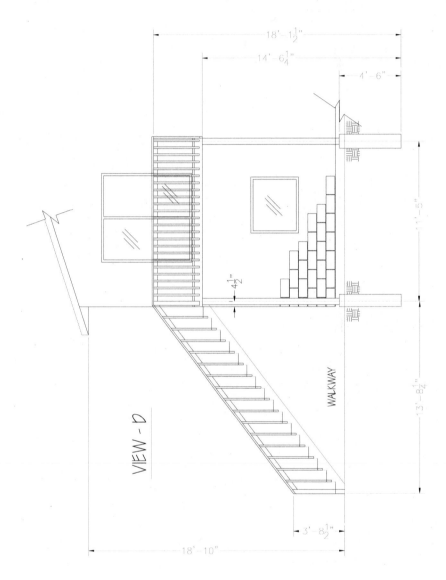

Figure 21-39 Drawing for Problem-Solving Exercise 8

Answers to Self-Evaluation Test

1. T, **2.** T, **3.** F, **4.** T, **5.** F, **6.** 16, **7.** even, **8.** five, **9. purge**, **10. CMLJUST**

Chapter 22

Grouping and Advanced Editing of Sketched Objects

CHAPTER OBJECTIVES

In this chapter, you will learn:
* *To group sketched objects.*
* *To select and cycle through defined groups.*
* *To change properties and location of sketched objects.*
* *To perform editing operations on polylines.*
* *To explode compound objects and undo previous commands.*
* *To rename named objects and remove unused named objects.*

KEY TERMS

* *Group*
* *Explode*
* *PEDIT*
* *SPLINETYPE*
* *UNDO*
* *REDO*
* *Rename*
* *Purge*

GROUPING SKETCHED OBJECTS USING THE OBJECT GROUPING DIALOG BOX

Ribbon: Home > Group > Group Manager   **Command:** CLASSICGROUP

To invoke the **Object Grouping** dialog box, you can choose the **Group Manager** button from the **Group** panel in the **Home** tab. You can use the **Object Grouping** dialog box (Figure 22-1) to group AutoCAD objects and assign a name to the group. Once you have created groups, you can select the objects by the group name. The individual characteristics of an object are not affected by forming groups. Groups enable you to select all the objects in a group together for editing. It makes the object selection process easier and faster. Objects can be members of several groups. Also, a group can contain several smaller group. This may be referred to as nested groups.

*Figure 22-1 The **Object Grouping** dialog box*

Although an object belongs to a group, you can still select an object as if it does not belong to any group. Groups can be selected by entering the group name or by selecting an object that belongs to the group. You can also highlight the objects in a group or sequentially highlight the groups to which an object belongs. The options in the **Object Grouping** dialog box are discussed next.

Group Name List Box

The **Group Name** list box in the **Object Grouping** dialog box displays the names of the existing groups. The list box also displays whether a group is selectable or not, under the **Selectable** column. A selectable group is the one in which all members of the group are selected on selecting a single member. The members that are in frozen or locked layers are not selected. You can also make a group non-selectable; this allows you to select the objects in the group individually.

Group Identification Area

The options under the **Group Identification** area are as follows:

Group Name

The **Group Name** edit box displays the name of the selected group. You can also use the **Group Name** edit box to enter the name of a new group. You can enter any name, but it is recommended that you use a name that reflects the type of objects in the group. For example, the group name

WALLS can include all lines that form walls. Group names can be up to thirty-one characters long and can include letters, numbers, and special characters ($, _, and -). Group names cannot have spaces.

Description

The **Description** edit box displays the description of the selected group. It can be used to enter the description of a group. The description text can be up to sixty-four characters long, including spaces.

Find Name

The **Find Name** button is used to find the group name or names associated with an object. When you select this button, the dialog box temporarily disappears from the screen and AutoCAD prompts you to select an object of the group. Once you select the object, the **Group Member List** dialog box (Figure 22-2) will be displayed, showing the group name(s) to which the selected object belongs.

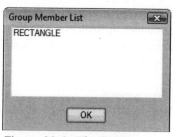

*Figure 22-2 The **Group Member List** dialog box*

Highlight

The **Highlight** button is used to highlight the objects in the selected group name. This button is available only when you select a group from the **Group Name** list box. When you choose the **Highlight** button, the objects that are members of the selected group are highlighted in the drawing and the **Object Grouping** message box (Figure 22-3) is displayed. Choose the **Continue** button or press ENTER to return to the **Object Grouping** dialog box.

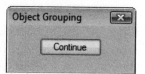

*Figure 22-3 The **Object Grouping** message box*

Include Unnamed

The **Include Unnamed** check box is used to display the names of the unnamed groups in the **Object Grouping** dialog box. The unnamed groups are created when you select the **Unnamed** check box in the **Create Group** area or when you copy the objects that belong to a group. AutoCAD automatically groups and assigns a name to the copied objects. The format of the name is An (for example, *A1, *A2, *A3). If you select the **Include Unnamed** check box, the unnamed group names (*A1, *A2, *A3, ...) will also be displayed in the **Group Name** list box of the dialog box. The unnamed groups can also be created by not assigning a name to the group (see **Unnamed** in the next section, "**Create Group** Area").

Create Group Area

The options under this area are discussed next.

New

The **New** button is used to create a new group. Enter the name of the group in the **Group Name** edit box and then choose this button. Once you choose the **New** button, the **Object Grouping** dialog box will be temporarily closed and you will be prompted to select the objects to be included in the group. After selecting the objects, right-click; the dialog box will be displayed again and a new group will be created from the selected objects. The new group name is now displayed in the **Group Name** list box of the dialog box. The group names in the **Group Name** list box are listed alphabetically.

Selectable

The **Selectable** check box allows you to define a group as selectable. A selectable group has a property that if you select any one object in the group, the entire group will be selected. If you clear this check box while defining a group, the new group created will not be selectable. Therefore, you need to select all members of the group individually as the entire group will not be selected upon selecting the individual entities of the group.

Even if a group is defined as selectable, and the **PICKSTYLE** system variable is set to 0, you will not be able to select the entire group by selecting one of its members. You will be able to individually select members of the group. You can also turn the group selectability on or off by selecting or clearing the **Object grouping** check box in the **Selection modes** area of the **Selection** tab of the **Options** dialog box or by using the SHIFT+CTRL+A toggle keys.

Unnamed

When you create a group, you can assign it a name or leave it unnamed. If you select the **Unnamed** check box, AutoCAD will automatically assign a group name to the selected objects and these unnamed groups will be listed in the **Group Name** list box. To view the name of the unnamed group, the **Include Unnamed** check box should be selected. The format of the name is A*n* (*A1, *A2, *A3 ...), where *n* is a number incremented with each new unnamed group.

Change Group Area

All the options under this area, other than the **Re-order** button, will be available only when you highlight a group name by selecting it in the **Group Name** list box of the dialog box. The options in this area are used to modify the properties of the existing group. These options are as follows.

Remove

The **Remove** button is used to remove objects from the selected group. Once you select the group name from the **Group Name** list box and then choose the **Remove** button; the dialog box will be temporarily closed and AutoCAD will display the following prompts at the Command prompt.

> Select objects to remove from group...
> Remove objects: *(Select the objects that you want to remove from the selected group.)*

Also, all objects belonging to the selected group will be highlighted. As you select the objects to be removed from the group, they no longer appear highlighted. Note that even if you remove all objects from the group, the group name will still exist, unless you use the **Explode** option to remove the group definition (discussed later in this section). If you remove objects and then add them later in the same drawing session, the objects retain their order in the group.

Add

The **Add** button is used to add objects to the selected group. When you select this option, AutoCAD will prompt you to select the objects you want to add to the selected group. The prompt sequence in this case will be similar to that of the **Remove** option. Similar to removing an object, all objects belonging to the selected group are highlighted and as soon as you select an object to be added to the group, it also gets highlighted.

Rename

The **Rename** button is used to rename the selected group. To rename a group, first select the group name from the **Group Name** list box, enter the new name in the **Group Name** edit box, and then choose the **Rename** button. AutoCAD will rename the specified group. You can use this option to rename unnamed groups that were discussed earlier.

Re-Order

The **Re-Order** button is used to change the order of the objects in the selected group. The objects are numbered in the order in which you select them when selecting objects for the group. Sometimes, when creating a tool path, you may want to change the order of these objects to get a continuous tool motion. You can do this using the **Re-Order** button. This option is also useful in some batch operations where one object needs to be on top of another object for display reasons. When you choose the **Re-Order** button, AutoCAD displays the **Order Group** dialog box (Figure 22-4) where you can change the order of the group members. The following example explains the use of the **Re-Order** button using the **Order Group** dialog box.

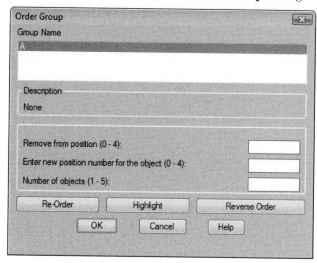

*Figure 22-4 The **Order Group** dialog box*

Description

The **Description** button is used to change the description of the selected group. To change the description of a group, first select the group name from the **Group Name** list box, enter the new description in the **Description** edit box, and then choose the **Description** button. AutoCAD will update the description of the specified group.

Explode

The **Explode** option is used to delete group definition of the selected group. The objects that were in the group become regular objects without a group reference. If you had made copies of the group that was exploded in a drawing, they shall remain as unnamed groups in the drawing. You can select the **Include Unnamed** check box in the **Group Identification** area of the **Object Grouping** dialog box to view the names of the unnamed groups in the **Group Name** list box. You can then select them in the list box and explode them or rename them as required.

Selectable

The **Selectable** button is used to change the selectable status of the selected group. To change the selectable status of the group, first select the group name from the **Group Name** list box, and then choose the **Selectable** button. This button acts like a toggle key between the options of selectable and non-selectable. If the selectable status for a group is displayed as "**Yes**", you can choose the **Selectable** button to change it to "**No**". Selectable implies that the entire group gets selected when one object in the group is selected. If the selectable status is "**No**", you cannot select the entire group by selecting one object in the group.

EXAMPLE 1   *Reorder Group*

In this example, you will draw a rectangle representing a toolpath. The rectangle should comprise of four individual lines. Group all lines by selecting them in the order shown in Figure 22-5(a). Then, highlight the order, and reorder the grouping, as shown in Figure 22-5(b).

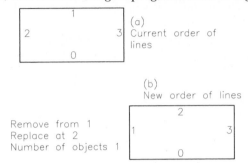

Figure 22-5 Changing the order of group objects

1. Draw four lines that are connected at the end points to form a rectangle.

2 Type **CLASSICGROUP** at the Command prompt and press ENTER; the **Object Grouping** dialog box is displayed.

3. Type **G1** in the **Group Name** edit box and choose the **New** button; the **Object Grouping** dialog box will disappear and you will be prompted to select the objects for grouping.

4. Select the lines in the order specified in the Figure 22-5(a) and right-click; the **Object Grouping** dialog box is displayed again.

5. Make the group (G1) current by selecting it in the **Group Name** list box and choose the **Re-Order** button in the dialog box; the **Order Group** dialog box is displayed.

6. Select the group G1 in the **Group Name** list box, if not already selected, and choose the **Highlight** button; the **Object Grouping** dialog box (Figure 22-6) is displayed. You can use the **Next** and **Previous** buttons to highlight the grouped objects in the order of selection.

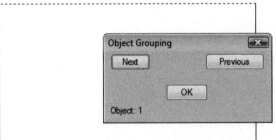

*Figure 22-6 The **Object Grouping** dialog box*

You will notice that the order of selection is not proper. As these lines represent a tool path, you need to reorder the objects, as shown in Figure 22-5(b).

7. To get a clockwise tool path, you must switch object numbers 1 and 2. You can do so by entering the necessary information in the **Order Group** dialog box. Enter **1** in the **Remove from position {0-3}** edit box and **2** in the **Enter new position number for the object {0-3}** edit box. Enter **1** in the **Number of objects (1-4)** edit box because there is only one object to be replaced.

8. After entering the information, choose the **Re-Order** button to define the new order of the objects. You can confirm the change by choosing the **Highlight** button again and cycling through the objects.

GROUPING SKETCHED OBJECTS USING THE GROUP BUTTON*

Ribbon: Home > Group > Group	**Command:** GROUP

Now, you can directly group objects in AutoCAD 2012 by using the **Group** button from the **Group** panel of the **Home** tab. Invoke the command by entering **GROUP** in the command box. The prompt sequence to create a group is given next:

Command: **GROUP**
Select objects or [Name/Description]: **N**
Enter a group name or [?]: *Enter the group name*
Select objects or [Name/Description]: *Select the objects to be grouped.*
Select objects or [Name/Description]: Enter
Group *"group name"* has been created.

You can also group the objects by using the shortcut menu. Select the objects to be grouped, and right-click in the drawing area; a shortcut menu will be displayed. Choose **Group>Group** from the shortcut menu; the selected objects will be grouped as *Unnamed group*.

Later on, you can ungroup them using **Ungroup** option from the same shortcut menu.

SELECTING GROUPS

You can select a group by name by entering **G** at the **Select objects** prompt. For example, if you have to move a particular group, choose the **Move** tool from the **Modify** panel in the **Home** tab and then enter the following prompt sequence:

Select objects: **G**
Enter group name: *Enter the group name*
n found
Select objects: Enter

Instead of entering **G** at the **Select Objects:** prompt, if you select any member of a selectable group, all the group members get selected. Make sure that the **PICKSTYLE** system variable is set to 1 or 3. Also, the group selection can be turned on or off by pressing SHIFT+CTRL+A. If the group selection is off and you want to erase a group from a drawing, choose the **Erase** tool from the **Modify** panel. The prompt sequence for erasing the group is given next.

Select objects: *Press SHIFT+CTRL+A*
<Groups on> *Select an object that belongs to a group. (If the group has been defined as selectable, all objects belonging to that group will be selected.)*

If the group has not been defined as selectable, you cannot select all objects in the group, even if you turn the group selection on by pressing the SHIFT+CTRL+A keys. This setting can also be changed in the **Selection** tab of the **Options** dialog box. You can select or clear the **Object grouping** check box in the **Selection modes** area of the dialog box to turn group selectability on or off.

Chapter 22

Tip
The combination of SHIFT+CTRL+A is used as a toggle to turn the group selection on or off.

CHANGING PROPERTIES OF AN OBJECT

AutoCAD provides you different options for changing the properties of an object. These options are discussed next.

Using the Properties Palette

Ribbon: View > Palettes > Properties	**Toolbar:** Standard > Properties
Quick Access Toolbar: Properties (*Customize to Add*)	**Command:** PROPERTIES or PR

The categories displayed in the **Properties** palette depend on the type of object selected. The **General** category displays the general properties of objects, such as color, layer, linetype, linetype scale, plot style, lineweight, hyperlink, and thickness. The procedure to change the general properties of objects using the **Properties** palette has already been discussed earlier in Chapter 4, Working with Drawing Aids. In this chapter, the **Geometry** category containing properties that control the geometry of an object will be discussed. Depending on the type of object selected, the **Geometry** category will contain a set of different properties. If you have selected many types of objects in a drawing, the edit box at the top of the **Properties** palette display **All**, **General**, and **3D Visualization** categories. If you select a type of object from the drop-down list, the corresponding categories of the object properties will be displayed in the palette. You can also invoke the **Properties** palette from the shortcut menu displayed on selecting the object.

Selecting a Line

If you have selected a line in the drawing, the **Properties** palette appears similar to what is shown in Figure 22-7, where you can change the properties of the selected object.

Start X/Start Y/Start Z. The **Start X/Start Y/Start Z** fields in the **Geometry** category display the start point coordinates of the selected line. You can enter new values in these fields or use the **Pick Point** button that appears in the field you click in. When you choose the **Pick Point** button, a rubber-band line gets attached to the cursor on its start point and the endpoint of the line is fixed. You can now move the cursor and specify a new start point for the line.

End X/End Y/End Z. Similar to the start point, you can also modify the location of the endpoint of a line by entering new coordinate values in the **End X**, **End Y**, **End Z** fields. You can also choose the **Pick Point** button in the respective fields and select a new location for the endpoint of the selected line. This **Pick Point** button appears in the fields when you click on them.

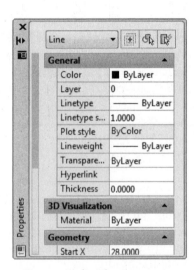

*Figure 22-7 The **Properties** palette*

Note
*The **Delta X**, **Delta Y**, **Delta Z**, **Length**, and **Angle** fields are not available for change but the values in these fields get updated as the start and endpoints of the line are changed.*

Selecting a Circle

When you select a circle, the **Geometry** category displays the categories as shown in Figure 22-8. You can modify the values in these fields. The effects of modifying the values are discussed next.

Center X/Center Y/Center Z. Here, under the **Center X/Center Y/Center Z** fields of the **Geometry** category, you can specify a new location of the center point of the circle by entering new coordinate values in the **Center X**, **Center Y**, and **Center Z** fields. When you click in any of these fields, the **Pick Point** button is displayed. You can choose this button to locate a new location of the center point of the circle. The circle will be drawn at the new location with the same radius as specified before. It is as if the circle has moved from the old location to a new one.

Geometry		
Center X	-7.0000	
Center Y	-10.0000	
Center Z	0.0000	
Radius	7.0000	
Diameter	14.0000	
Circumfer...	43.9823	
Area	153.9380	
Normal X	0.0000	
Normal Y	0.0000	
Normal Z	1.0000	

*Figure 22-8 The **Geometry** category of the **Properties** palette for a circle*

Radius/Diameter. You can enter a new value for the radius or diameter of the circle in the **Radius** or **Diameter** fields respectively. The circle's radius or diameter is modified as per the new values you have entered. As you modify the values of the radius or the diameter, the values in the **Circumference** and **Area** fields get modified accordingly.

Note
*You can also enter new values of the circumference or area in the **Circumference** and **Area** fields, respectively, if you want to create a circle with a given value of circumference or area. You will notice that the radius and diameter values are modified accordingly. The **Normal X**, **Normal Y**, and **Normal Z** fields are not available for modifications.*

Selecting the Multiline Text

If you select multiline text written in a drawing and invoke the **Properties** palette, apart from the **General** category, three more categories of properties are displayed: **Text**, **3D Visualization** and **Geometry**. The properties in the **Test** and **Geometry** categories (Figure 22-9) and how they can be modified are discussed next.

Contents. The current contents of the text are displayed in the **Contents** field under the **Text** category of the **Properties** palette. To modify the contents, click in the field and the [...] button is displayed at the right-corner of the field. You can then choose this button to display the **Text Editor** where you can make the necessary modifications using the options in the dialog box. Once you have made the changes, choose the **Close Text Editor** button from the **Text Editor** contextual tab to exit the text editing and return to the **Properties** palette. The changes you made are reflected in the value of the **Contents** field of the **Properties** palette and in the drawing.

Style. When you click in the **Style** field in the **Properties** palette, a drop-down list of the defined text styles is displayed. By default, only the Standard style is displayed. If you had defined more text styles in the drawing, they can be selected from the drop-down list and applied to the current text in the drawing.

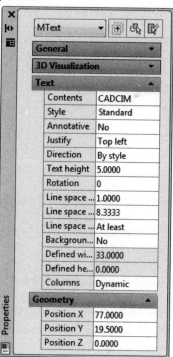

*Figure 22-9 The **Geometry** and **Text** categories of the **Properties** palette for text*

Chapter 22

Annotative. You can convert the non-annotative text to annotative and vice-versa by selecting the **Yes** and **No** options respectively from the **Annotative** drop-down list.

Justify. All the text justification options are available in the **Justify** drop-down list. This drop-down list is available when you click in the **Justify** field. By default, the Top left justification is applied to the text. You can select a text justification from the drop-down list to be applied to the selected text in the drawing.

Direction. This drop-down list displays the possible directions for the text. This option is generally available for the multiline text objects.

Text height. You can enter a new value for the height of the selected text in the drawing. The default text height is 0.2. But, you can make the text smaller or larger in the drawing by simply entering a new value in the **Text height** field of the **Properties** palette.

Rotation. This field displays the current angle of the selected text and is 0-degree by default. You can change the angle of rotation of the text by entering an angle by which you want to rotate the selected text in the **Rotation** field and then pressing ENTER. The effect is immediately visible in the drawing.

Line space factor. The effect of changing the value of this field is more evident when you have selected a paragraph or multiline text. This value controls the spacing between the lines in a paragraph or multiline text. A scale factor of 1 is the default value. You can increase or decrease the spacing between the lines of the selected paragraph text by entering a new scale factor of your choice and pressing ENTER. The result is immediately visible in the drawing.

Line space distance. This field is used to modify the distance between the spacing of each line of the multiline text. Click in this field and enter the new value in it.

Line space style. This field when selected displays two line spacing options in the corresponding drop-down list. They are **Exactly** and **At least**. **At least** is the default option.

Background mask. This field specifies whether or not any background mask is assigned to the multiline text.

Columns. Columns are of three types, None, Static, or Dynamic. To edit the column settings, choose the **[...]** button; the **Column Settings** dialog box will be displayed. You can change the column settings through this dialog box.

Defined width/Defined height. These fields display the defined width/height of the text window in the **In-Place Text Editor**. These fields will be available, if you select the **None** option from the **Columns** field. To change the values of these fields, click in the field and enter the new value in it.

Position X/Position Y/Position Z. You can change the location of the insertion point of the selected text by entering new *X/Y/Z* coordinates for the text in these fields. You can also use the **Pick Points** button that is displayed in these fields when you click in them. When you choose the **Pick Points** button, a rubber-band line is attached between the cursor and the selected text and you can move the cursor and specify a new location for the text on the screen. The values of the three fields get updated automatically.

Selecting the Annotative Block Reference

When you select a block in the drawing and invoke the **Properties** palette, apart from the **General** category, three more categories containing the properties of the block are available.

They are **Geometry**, **3D Visualization**, and **Misc**. The properties under these categories are discussed next.

Position X/Position Y/Position Z. Here, under the **Position X/Position Y/Position Z** fields of the **Geometry** category, you can enter new *X*, *Y*, and *Z* coordinates for the selected block. The block reference shall move to the new location as you specify the coordinates. You can also click in any one of the fields to display the **Pick Points** button and choose it to specify a new location for the block on the screen.

Scale X/Scale Y/Scale Z. You can specify new X, Y, and Z scale factors for the selected block in the **Scale X**, **Scale Y**, and **Scale Z** fields, respectively. The current scale factors for the block are displayed in the **Scale X**, **Scale Y**, and **Scale Z** fields and as you change them, the modifications are reflected in the drawing.

Name. This field under the **Misc** category displays the name of the selected block reference and is not available for modifications.

Rotation. This field specifies the current rotation angle of the selected block. You can specify a new angle of rotation for the selected block in this field. After specifying an angle, when you press ENTER, the selected block in the drawing is rotated through the specified angle.

Annotative. This field specifies whether the block is annotative or non-annotative. This information is read-only.

Annotative scale. This field displays the current annotation scale of the drawing. You can add more annotation scales to the drawing in the **Annotation Object Scale** dialog box that will be displayed when you choose the **[...]** button from the **Annotative scale** field.

Match orientation to layout. This field displays whether or not the orientation of the text used in the block is forced to align with the orientation of the floating viewports. This information is read-only.

Block Unit. This field specifies the unit of the block.

Unit factor. This field specifies the conversion factor between the block unit and the drawing unit.

Selecting the Annotative Attribute

When you select an attribute before converting it into a block, apart from the **General** category, the other four categories under which the properties of an attribute are displayed are **Text**, **General**, **3D Visualization**, and **Misc**. How to use the **Properties** palette to edit an attribute has already been discussed in Chapter 19, Defining Block Attributes.

Tag. You can modify the text of the tag by entering a new value in the **Tag** field under the **Text** category. After you have modified the tag text and pressed ENTER, the old attribute text in the drawing is replaced by the new text.

Annotative. This field specifies whether the attribute is annotative or non-annotative. Select the appropriate option from the **Annotative** drop-down list to change the annotative property of the attributes.

Annotative scale. This field displays the current annotation scale of the drawing. You can add more annotation scales to the drawing in the **Annotation Object Scale** dialog box that will be displayed when you choose the **[...]** button from the **Annotative scale** field.

Chapter 22

Prompt. You can also change the text of the prompt and enter a new prompt value in the **Prompt** field.

Value. This field displays the default value of the selected attribute. You can change this default value by entering a new value in the **Value** field.

Style. When you click in this field, the **Style** drop-down list is available. You can select a style for the attribute text from this drop-down list. Only the text styles that already have been defined are available in this drop-down list and can be selected.

Justify. You can select an attribute text justification from the **Justify** drop-down list.

Paper text height. This field displays the height of the annotative text that is to be maintained on the paper.

Model text height. This field displays the height of the text that is to be displayed in the model space. The **Model text height** value depends on the **Paper text height** value and the **Annotative scale**.

Height. You can change the height of the selected non-annotative attribute text by entering a new value in this field.

Rotation. You can rotate the selected attribute text by specifying an angle in the **Rotation** field.

Width factor. You can modify the width factor of the attribute text by entering a new value in the **Width factor** field of the **Properties** palette.

Obliquing. The attribute text can also be made slanting by entering an angle value in the **Obliquing** field of the **Properties** palette.

Direction. This field displays the flow direction of the attribute text. You can select the desired flow direction from the **Horizontal**, **Vertical**, or **By style** options (defined in the Textstyle) in the **Direction** drop-down list.

Boundary Width. This field displays the width of the text window in the **In-Place Text Editor**. The value displayed here is the relative value; specifying a value less than **1** will decrease the boundary width and a value more than **1** will increase the boundary width. The value in the **Boundary Width** will be available only if the **Multiple lines** check box is selected in the **Mode** area of the **Attribute Definition** dialog box. You can also select the **Yes** option from the **Multiple lines** drop-down list in the **Misc** area of the **Properties** Palette.

Text alignment X/Text alignment Y/Text alignment Z. The **Text alignment X/Text alignment Y/Text alignment Z** fields of the **Text** category display the *X*, *Y*, and *Z* coordinates of the alignment point of the selected attribute. Note that these fields will be available only if you modify the properties such as the text alignment of the attributes, while defining them. If these fields are available and you click on them, the **Pick Point** button appears in the field. You can use this button to specify the new location of the attribute.

Note

*If you modify the text alignment of the first attribute and define the remaining attributes below the previous, the remaining attributes will also have the **Text Alignment** fields active.*

Misc. As discussed earlier, under the **Misc** category, you can redefine the attribute modes that have been defined at the time of the attribute definition. You can also make the attribute text appear upside-down or backwards by selecting **Yes** from the **Upside down** drop-down list and the **Backward** drop-down list, respectively.

Note

*Choose the **QuickCalc** button from the respective fields to open the Quick Calculator. Using the Quick Calculator, you can enter the numerical values as mathematical equations. These equations may include scientific calculations. You can also perform units conversion using the Quick Calculator.*

EXERCISE 1 *Properties*

Draw a hexagon on layer OBJ in red color. Let the linetype be hidden. Now, use the **Properties** palette to change the layer of the hexagon to some other existing layer, the color to yellow, and the linetype to continuous. Also, in the **Properties** palette, under the **Geometry** category, specify a vertex in the **Vertex** field or select one using the arrow (Next or Previous) buttons and then relocate it. The values of the **Vertex X** and **Vertex Y** fields change as the coordinate values of the vertices change. Also, notice the change in the value in the **Area** field. You can also assign a start and end lineweight to each of the hexagon segments between the specified vertices. Use the **LIST** command to verify that the changes have taken place.

EXPLODING COMPOUND OBJECTS

Ribbon: Home > Modify > Explode	**Toolbar:** Modify > Explode
Menu Bar: Modify > Explode	**Command:** EXPLODE

The **Explode** tool is used to split compound objects such as blocks, polylines, regions, polyface meshes, polygon meshes, multilines, 3D solids, 3D meshes, bodies, or dimensions into the basic objects that make them up (Figure 22-10). For example, if you explode a polyline or a 3D polyline, the result will be ordinary lines or arcs (tangent specification and width are not considered). When a 3D polygon mesh is exploded, the result is 3D faces. Polyface meshes are turned into 3D faces, points, and lines. Upon exploding 3D solids, the planar surfaces of the 3D solid turn into regions, and nonplanar surfaces turn into bodies, multilines are changed to lines. Regions turn into lines, ellipses, splines, or arcs. On exploding, 2D polylines lose their width and tangent specifications and 3D polylines explode into lines. When a body is exploded, it changes into single-surface bodies, curves, or regions. When a leader is exploded, the components are lines, splines, solids, block inserts, text or multiline text, tolerance objects, and so on. Multiline text explodes into a single line text. This tool is especially useful when you have inserted an entire drawing and you need to alter a small detail. After you invoke the **Explode** tool, you are prompted to select the objects you want to explode. After selecting the objects, press ENTER or right-click to explode the selected objects and then end the command.

When a block or dimension is exploded, there is no visible change in the drawing. The drawing remains the same except that the color and linetype may have changed because of floating layers, colors, or linetypes. The exploded block is turned into a group of objects that can be modified separately. To check whether the explosion of the block has taken place, select any object that was a part of the block. If the block has been exploded, only that particular object will be highlighted. With the **EXPLODE** command, only one nesting level is exploded at a time. Hence, if there is a nested block or a polyline in a block and you explode it, the inner block or the polyline will not be exploded. Attribute values are deleted when a block is exploded, and the attribute definitions are redisplayed.

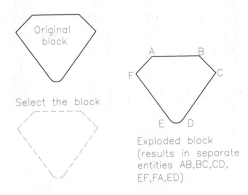

Figure 22-10 *Use of the* **EXPLODE** *command*

Note
Remember that the blocks inserted using the **MINSERT** *command and x-refs cannot be exploded. This command was discussed in relation to the blocks in Chapter 16.*

Tip
If you want to insert a block in the form of its separate components while using the **INSERT** *command, select the* **Explode** *check box in the* **Insert** *dialog box. Also, if you are using the* **-INSERT** *command, type * in front of the block name at the* **Enter block name or [?]:** *prompt, the block will be inserted in the drawing as separate component object, and not as an entire block.*

EDITING POLYLINES

A polyline can assume various characteristics such as width, linetype, joined polyline, and closed polyline. You can edit polylines, polygons, or rectangles to attain the desired characteristics using the **Edit Polyline** tool. In this section, we will be discussing how to edit simple 2D polylines. The following are the editing operations that can be performed on an existing polyline using the **Edit Polyline** tool. These operations are discussed in detail later in this chapter.

1. A polyline of varying widths can be converted to a polyline of uniform width.

2. An open polyline can be closed and a closed one can be opened.

3. You can remove bends and curved segments between two vertices to make a straight polyline.

4. A polyline can be split up into two and individual polylines or polyarcs connected to one another can be joined into a single polyline.

5. After invoking the **Edit Polyline** tool, if you select an entity that is not a polyline, you are prompted to specify whether you want the entity to be converted into a polyline. You can avoid this prompt by setting the value of the **PEDITACCEPT** variable to 1.

6. Multiple polylines can be edited.

7. The appearance of a polyline can be changed by moving and adding vertices.

8. Curves of arcs and B-spline curves can be fitted to all vertices in a polyline, with the specification of the tangent of each vertex being optional.

9. The linetype generation at the vertices of a polyline can be controlled.

10. Multiple polylines can be joined together to form a single polyline.

Editing Single Polyline

Ribbon: Home > Modify > Edit Polyline   **Toolbar:** Modify II > Edit Polyline
Menu Bar: Modify > Object > Polyline   **Command:** PEDIT

Apart from the methods displayed in the command box, you can also invoke the **Edit Polyline** tool by choosing **Polyline Edit** from the shortcut menu that is displayed when you select a polyline and right-click. You can use the **Edit Polyline** tool to edit any type of polyline. When you invoke this tool, the following prompt sequence is displayed:

Select polyline or [Multiple]:

If the selected entity is not a polyline, the following message will be displayed at the Command prompt.

Object selected is not a polyline.
Do you want to turn it into one? <Y>:

If you want to turn the object into a polyline, respond by entering Y and pressing ENTER or by simply pressing ENTER. To let the object be as it is, enter N. AutoCAD will then prompt you to select another polyline or object to edit. As mentioned earlier, you can avoid this prompt by setting the value of the **PEDITACCEPT** variable to 1. The subsequent prompts and editing options depend on the type of polyline that has been selected. AutoCAD provides you the option of either selecting a single polyline or multiple polylines. In this case, a single 2D polyline is selected, the next prompt displayed is given next.

Enter an option [Close/Join/Width/Edit vertex/Fit/Spline/Decurve/Ltype gen/Reverse/Undo]: *Enter an option or press ENTER to end command.*

Note
*Depending on the type of polyline selected, the **PEDIT** command options change. In this chapter, the **2D polyline** options have been discussed. The **3D polylines** and **3D Polygon Mesh** editing options will be discussed in Chapter 25, Getting Started with 3D.*

The options available in the **Single** polyline selection method are discussed next.

Close (C) Option
This option is used to close an open polyline. **Close** creates the segment that connects the last segment of the polyline to the first. You will get this option only if the polyline is not closed. Figure 22-11 illustrates this option.

Open (O) Option
If you close an open polyline, the **Close** option is replaced by the **Open** option. Enter **O** for open, the closing segment is removed, see Figure 22-11.

Join (J) Option

This option appends lines, polylines, or arcs whose endpoints meet a selected polyline at any of its endpoints and adds (joins) them to it (Figure 22-12). This option can be used only if a polyline is open. After this option has been selected, AutoCAD prompts you to select objects. Once you have chosen the objects to be joined to the original polyline, AutoCAD examines them to determine whether any of them has an endpoint in common with the current polyline, and joins such an object with the original polyline. The search is then repeated using new endpoints. They will not join if the endpoint of the object does not exactly meet the polyline. The line touching a polyline at its endpoint to form a T will not be joined. If two lines meet a polyline in a Y shape, only one of them will be selected, and this selection depends upon the sequence of selecting the lines. If you select the two lines, the first selected line will get joined to the polyline. To verify which lines have been added to the polyline, use the **LIST** command or select a part of the object. All the segments that are joined to the polyline will be highlighted.

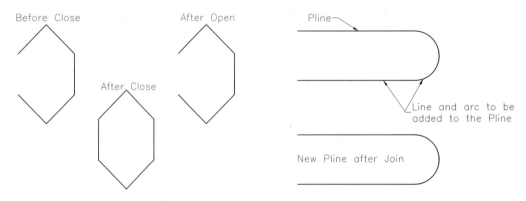

*Figure 22-11 The **Close** and **Open** options*   *Figure 22-12 Using the **Join** option*

Width (W) Option

The **W** option allows you to define a new, unvarying width for all segments of a polyline (Figure 22-13). It changes the width of a polyline with a constant or a varying width. The desired new width can be specified either by entering the width using the keyboard or by specifying the width as the distance between the two specified points. Once the width has been specified, the polyline assumes it. Here, you will change the width of the given polyline in figure from 0.02 to 0.05 (Figure 22-14). On invoking the **Edit Polyline** tool, the following prompt is displayed:

Select polyline or [Multiple]: *Select a polyline.*
Enter an option [Close/Join/Width/Edit vertex/Fit/Spline/Decurve/Ltype gen/Reverse/Undo]: **W**
Specify new width for all segments: **0.05**

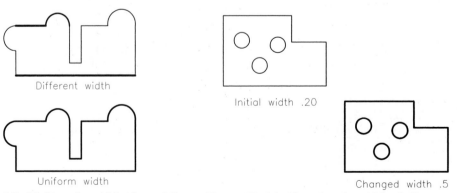

Figure 22-13 Making the width of a polyline uniform   *Figure 22-14 Changing the width of all segments*

Note

*Circles drawn using the **CIRCLE** command cannot be changed to polylines. However, for polycircles drawn by using the **Arc** option of the **Polyline** tool (by drawing two semicircular polyarcs) or by using the **Donut** tool, you can modify the thickness using the **Width** option of the **Polyline Edit** tool.*

Edit vertex (E) Option

The **Edit vertex** option lets you select a vertex of a polyline and perform different editing operations on the vertex and the segments following it. A polyline segment has two vertices. The first one is at the start point of the polyline segment; the other one is at the endpoint of the segment. When you invoke this option, an X marker appears on the screen at the first vertex of the selected polyline. If a tangent direction has been specified for this particular vertex, an arrow is generated in that direction. After this option has been selected, the next prompt appears with a list of options for this prompt. The prompt sequence, after you invoke the **Edit Polyline** tool, is given next.

> Select polyline or [Multiple]: *Select the polyline to be edited.*
> Enter an option [Close/Join/Width/Edit vertex/Fit/Spline/Decurve/Ltype gen/Reverse/Undo]: **E**
> Enter a vertex editing option
> [Next/Previous/Break/Insert/Move/Regen/Straighten/Tangent/Width/eXit]<N>: *Enter an editing option or press ENTER to accept default.*

All the options for the **Edit vertex** option are discussed next.

Next and Previous Options. These options move the X marker to the next or the previous vertex of a polyline. The default value in the **Edit vertex** option is one of these two options. The option that is selected as default is the one you chose last. In this manner, the **Next** and **Previous** options help you to move the X marker to any vertex of the polyline by selecting one of these two options, and then pressing ENTER to reach the desired vertex. These options cycle back and forth between the first and last vertices, but cannot move past the first or last vertex, even if the polyline is closed (Figure 22-15).

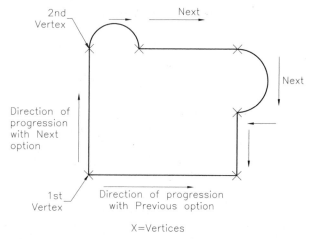

*Figure 22-15 The **Next** and **Previous** options*

The prompt sequence for using this option, after you invoke the **Edit Polyline** tool, is given next.

Select polyline or [Multiple]: *Select the polyline to be edited.*
Enter an option [Close/Join/Width/Edit vertex/Fit/Spline/Decurve/Ltype gen/Reverse/Undo]: **E**
Enter a vertex editing option
[Next/Previous/Break/Insert/Move/Regen/Straighten/Tangent/Width/eXit] <N>: *Enter N or P to move to the next or previous vertices respectively.*

Break Option. With the **Break** option, you can remove a portion of the polyline, as shown in Figure 22-16 or break it at a single point. You can specify the vertices within which the polyline would be removed. By specifying two different vertices, all the polyline segments and vertices between the specified vertices are erased. If one of the selected vertices is at the endpoint of the polyline, the **Break** option will erase all the segments between the first vertex and the endpoint of the polyline. The exception to this is that AutoCAD does not erase the entire polyline, if you specify the first vertex at the start point (first vertex) of the polyline and the second vertex at the endpoint (last vertex) of the polyline. If both vertices are at the endpoint of the polyline, or only one vertex is specified and its location is at the endpoint of the polyline, no change is made to the polyline. The last two selections of vertices are treated as invalid by AutoCAD, which acknowledges this by displaying the message ***Invalid***.

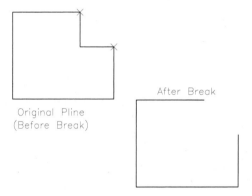

*Figure 22-16 Use of the **Break** option*

To use the **Break** option, first you need to move the marker to the first vertex where you want the split to start. The placement of the marker can be achieved with the help of the **Next** and **Previous** options. Once you have selected the first vertex to be used in the **Break** operation, invoke the **Break** option by entering **B** at the Command prompt. AutoCAD takes the vertex where the marker (X) is placed as the first point of the breakup. The next prompt asks you to specify the position of the next vertex for the breakup. You can enter **GO** at this prompt if you want to split the polyline at one vertex only. Otherwise, use the **Next** or **Previous** option to specify the position of the next vertex and then enter **GO**. On doing so, the polyline segments between the two selected vertices will be erased. The prompt sequence for using this option, after you have invoked the **Edit Polyline** tool, is given next.

Select polyline or [Multiple]: *Select the polyline to be edited.*
Enter an option [Close/Join/Width/Edit vertex/Fit/Spline/Decurve/Ltype gen/Reverse/Undo]: **E**
Enter a vertex editing option
[Next/Previous/Break/Insert/Move/Regen/Straighten/Tangent/Width/eXit]<N>: *Enter N or P to locate the first vertex for the **Break** option.*
Enter a vertex editing option

[Next/Previous/Break/Insert/Move/Regen/Straighten/Tangent/Width/eXit]<N>: **B**

Once you invoke the **Break** option, AutoCAD treats the vertex where the marker (X) is displayed as the first point for splitting the polyline. The next prompt issued is given next.

> Enter an option [Next/Previous/Go/eXit] <N>: *Enter **G** if you want to split the polyline at one vertex only or move the X marker using the **Next** or **Previous** option to specify the position of the next vertex for breakup.*

After you have specified the next position of the X marker using the **Next** and **Previous** options, entering **GO** deletes the polyline segment between the two markers specified. Now, exit the **Enter a vertex editing option** prompt using the **eXit** option.

Insert Option. The **Insert** option is used to define a new vertex and add it to the polyline (Figure 22-17). You can invoke this option by entering I for Insert. You should invoke this option only after you have moved the marker (X) to the vertex after which the new vertex is to be added. The new vertex is inserted immediately after the vertex with the X mark. After you invoke the **Edit Polyline** tool, the prompt sequence for using the **Insert** option will be as follows:

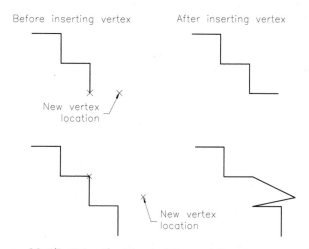

Figure 22-17 *Using the **Insert** option to define new vertex points*

> Select polyline or [Multiple]: *Select the polyline to be edited.*
> Enter an option [Close/Join/Width/Edit vertex/Fit/Spline/Decurve/Ltype gen/Reverse/Undo]: **E**
> Enter a vertex editing option
> [Next/Previous/Break/Insert/Move/Regen/Straighten/Tangent/Width/eXit]<N>: *Move the marker to the vertex after which the new vertex is to be inserted.*
> Enter a vertex editing option
> [Next/Previous/Break/Insert/Move/Regen/Straighten/Tangent/Width/eXit]<N>: **I**
> Specify location for new vertex: *Move the cursor and select to specify the location of the new vertex or enter the coordinates of the new location.*

Tip

*AutoCAD 2011 onwards, you can insert vertex by using the shortcut menu. To do so, select a polyline; the grips will be displayed. Move the cursor near one of the grips; a shortcut menu will be displayed. Choose the **Add Vertex** option and click on the drawing area to add new vertex. You can also stretch the selected arc or convert the line to arc or vice versa by selecting appropriate option from the shortcut menu.*

Move option. This option is used to move the X-marked vertex to a new position (Figure 22-18). Before invoking the **Move** option, you must move the X marker to the vertex you want to relocate by selecting the **Next** or **Previous** option. The prompt sequence for relocating a vertex after you invoke the **Edit Polyline** tool is given next.

> Select polyline or [Multiple]: *Select the polyline to be edited.*
> Enter an option [Close/Join/Width/Edit vertex/Fit/Spline/Decurve/Ltype gen/Reverse/Undo]: **E**
> Enter a vertex editing option
> [Next/Previous/Break/Insert/Move/Regen/Straighten/Tangent/Width/eXit]<N>: *Enter **N** or **P** to move the X marker to the vertex you want to relocate.*
> Enter a vertex editing option
> [Next/Previous/Break/Insert/Move/Regen/Straighten/Tangent/Width/eXit]<N>: **M**
> Specify new location for marked vertex: *Specify the new location for the selected vertex by using the pick button of your pointing device or by entering its coordinate values.*
> Enter a vertex editing option
> [Next/Previous/Break/Insert/Move/Regen/Straighten/Tangent/Width/eXit]<N>: *Enter an option or enter **X** to exit.*

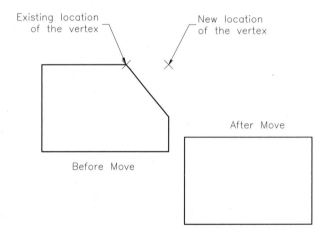

*Figure 22-18 The **Move** option*

Regen Option. The **Regen** option regenerates the polyline to display the effects of edits you have made, without having to exit the vertex mode editing. It is used most often with the **Width** option.

Straighten Option. The **Straighten** option is used to straighten polyline segments or arcs between specified vertices (Figure 22-19). It deletes the arcs, line segments, or vertices between the two specified vertices and substitutes them with one polyline segment.

The prompt sequence to use this option after you invoke the **Edit Polyline** tool is given next.

> Select polyline or [Multiple]: *Select the polyline to be edited.*
> Enter an option [Close/Join/Width/Edit vertex/Fit/Spline/Decurve/Ltype gen/Reverse/Undo]: **E**
> Enter a vertex editing option
> [Next/Previous/Break/Insert/Move/Regen/Straighten/Tangent/Width/eXit]<N>: *Move the marker to the desired vertex from where you want to start applying the **Straighten** option with the **Next** or **Previous** option.*
> Enter a vertex editing option
> [Next/Previous/Break/Insert/Move/Regen/Straighten/Tangent/Width/eXit]<N>: **S**

Enter an option [Next/Previous/Go/eXit] <N>: *Move the marker to the next desired vertex, until you reach the vertex you want to straighten.*
Enter an option [Next/Previous/Go/eXit] <N>: **G**

The polyline segments between two marker locations are replaced by a single straight line segment. If you specify a single vertex, the segment following the specified vertex is straightened, if it is curved.

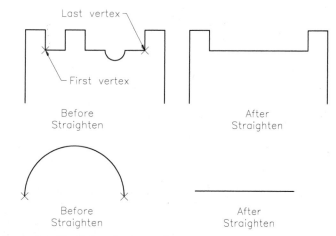

Figure 22-19 *Using the* **Straighten** *option to straighten polylines*

Tangent option. The **Tangent** option is used to associate a tangent direction to the current vertex (marked by X). The tangent direction is used in the curve fitting or the **Fit** option of the **Edit Polyline** tool. This option is discussed in detail in the subsequent section on curve fitting. The prompt issued on using the **Tangent** option is given next.

Specify direction of vertex tangent: *Specify a point or enter an angle.*

You can specify the direction of the vertex tangent by entering an angle at the previous prompt or by selecting a point to express the direction with respect to the current vertex. You can then move the marker to another vertex using the **Next** or **Previous** option and change its direction of tangent or enter X to exit the **Enter a vertex editing option** prompt.

Width option. The **Width** option lets you change the width at the start and end points of a polyline segment that follows the current vertex (Figure 22-20). By default, the ending width is equal to the starting width and hence, you can get a polyline segment of uniform width by pressing ENTER at the **Specify ending width for next segment <starting width>** prompt. You can also specify different starting and ending widths to get a varying-width polyline. The prompt sequence is given next.

Enter an option [Close/Join/Width/Edit vertex/Fit/Spline/Decurve/Ltype gen/Reverse/Undo]: **E**
Enter a vertex editing option
[Next/Previous/Break/Insert/Move/Regen/Straighten/Tangent/Width/eXit]<N>: *Move the marker to the starting vertex of the segment whose width is to be altered, using the* **Next** *or* **Previous** *option.*
Enter a vertex editing option
[Next/Previous/Break/Insert/Move/Regen/Straighten/Tangent/Width/eXit]<N>: **W**
Specify starting width for next segment <current>: *Enter the revised starting width.*
Specify ending width for next segment <starting width>: *Enter the revised ending width or press ENTER to accept the default option of keeping the ending width of the segment equal to the starting width.*

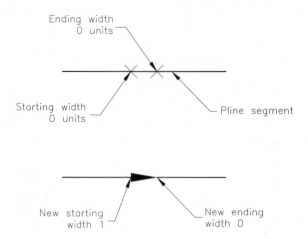

*Figure 22-20 Using the **Width** option to change the width of a polyline*

If no difference is noticed in the appearance of the polyline, you may need to use the **Regen** option. The segment with the revised widths is redrawn after invoking the **Regen** option or when you exit the vertex mode editing.

eXit Option. This option lets you exit the vertex mode editing and return to the **PEDIT** prompt.

Fit Option

The **Fit** or **Fit curve** option is used to generate a curve that passes through all the corners (vertices) of the polyline, using the tangent directions of the vertices (Figure 22-21).

The curve is composed of a series of arcs passing through the corners (vertices) of the polyline. This option is used when you draw a polyline with sharp corners and you need to convert it into a series of smooth curves. An example of this is a graph. In a graph, we need to show a curve by joining a series of plotted points. The process involved is called curve fitting; therefore, this option is known as **Fit**. The vertices of the polyline are also known as the control points. The closer these control points are, the smoother is the curve. Therefore, if the **Fit** option does not give optimum results, insert more vertices into the polyline or edit the tangent directions of vertices and then use the **Fit** option on the polyline. Before using this option, you may give each vertex a tangent direction.

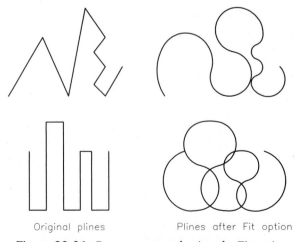

*Figure 22-21 Curves generated using the **Fit** option*

The curve is then constructed, according to the tangent directions you have specified. The following prompt sequence illustrates the **Fit** option after you invoke the **Edit Polyline** tool.

> Select polyline or [Multiple]: *Select the polyline to be edited.*
> Enter an option [Close/Join/Width/Edit vertex/Fit/Spline/Decurve/Ltype gen/Reverse/Undo]: **F**

If the tangent directions need to be edited, use the **Edit vertex** option of the **Edit Polyline** tool. Move the X marker to each of the vertices that need to be changed. Now, you can invoke the **Tangent** option, and either enter the tangent direction in degrees or select points. The chosen direction is expressed by an arrow placed at the vertex. The prompt sequence to use the **Tangent** option of the **Edit Polyline** tool is given next.

> Select polyline or [Multiple]: *Select the polyline to be edited.*
> Enter an option [Close/Join/Width/Edit vertex/Fit/Spline/Decurve/Ltype gen/Reverse/Undo]: **E**
> Enter a vertex editing option
> [Next/Previous/Break/Insert/Move/Regen/Straighten/Tangent/Width/eXit]<N>: **T**
> Specify direction of vertex tangent: *Specify a direction in + or - degrees, or select a point in the desired direction. Press ENTER.*

Once you specify the tangent direction, use the **eXit** option to return to the previous prompt and use its **Fit** option.

Tip
*AutoCAD 2011 onwards, after applying **Fit** option to a polyline, you can change the tangent direction by using the tool tip. To do so, select a polyline; the grips will be displayed. Move the cursor near one of the grips; a tooltip will be displayed. Choose the **Tangent Direction** option and specify the new direction in the drawing area.*

Spline Option

The **Spline** option (Figure 22-22) also smoothens the corners of a straight segment polyline, as does the **Fit** option, but the curve passes through only the first and the last control points (vertices), except in the case of a closed polyline. The spline curve is stretched toward the other control points (vertices) but does not pass through them, as in the case of the **Fit** option. The greater the number of control points, the greater the force with which the curve is stretched toward them. The prompt sequence on invoking the **Edit Polyline** tool is as follows:

> Select polyline or [Multiple]: *Select the polyline.*
> Enter an option [Close/Join/Width/Edit vertex/Fit/Spline/Decurve/Ltype gen/Reverse/Undo]: **S**

The generated curve is a B-spline curve. The frame is the original polyline without any curves in it. If the original polyline has arc segments, these segments are straightened when the spline's frame is formed. A frame that has width produces a spline curve that tapers smoothly from the width of the first vertex to that of the last. Any other width specification between the first width specification and the last is neglected. When a spline is formed from a polyline, the frame is displayed as a polyline with a zero width and a continuous linetype. Also, AutoCAD saves its frame so that it may be restored to its original form. Tangent specifications on control point vertices do not affect spline construction.

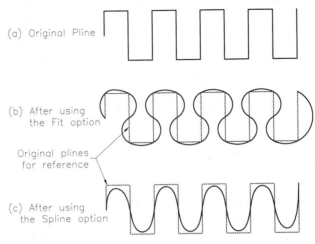

(a) Original Pline

(b) After using the Fit option

Original plines for reference

(c) After using the Spline option

Figure 22-22 *Smoothening the corners using the **Spline** option*

By default, the spline frames are not shown on the screen, but you may want them displayed for reference. In this case, the system variable **SPLFRAME** needs to be manipulated. The default value for this variable is zero. If you want to see the spline frame as well, set it to 1.

Now, whenever the **Spline** option is used on a polyline, the frame will also be displayed. Most editing tools such as **Move**, **Erase**, **Copy**, **Mirror**, **Rotate**, and **Scale** work similarly for both the Fit curves and Spline curves. They work on both the curve and its frame, whether the frame is visible or not. The **Extend** tool changes the frame by adding a vertex at the point the last segment intersects with the boundary. If you use any of the previously mentioned tools and then use the **Decurve** option to decurve the spline curve and later use the **Spline** option again, the same spline curve is generated. The **Break**, **Trim**, and **Explode** tools delete the frame. The **Divide**, **Measure**, **Fillet**, **Chamfer**, **Area**, and **Hatch** tools recognize only the spline curve and do not consider the frame. The **Stretch** tool first stretches the frame and then fits the spline curve to it.

When you use the **Join** option of the **PEDIT** command, the spline curve is decurved and the original spline information is lost. The **Next** and **Previous** options of the **Edit vertex** option of the **Edit Polyline** tool move the marker only to the points on the frame, whether visible or not. The **Break** option discards the spline curve. The **Insert**, **Move**, **Straighten**, and **Width** options refit the spline curve. Object Snaps consider the spline curve and not the frame; therefore, if you wish to snap to the frame control points, restore the original frame.

There are two types of spline curves: **Quadratic B-spline** and **Cubic B-spline**

These types of curves pass through the first and the last control points, which is the characteristic of the spline curve. Cubic curves are very smooth. A cubic curve passes through the first and last control points, and the curve is closer to the other control points. Quadratic curves are not as smooth as the cubic ones, but they are smoother than the curves produced by the **Fit curve** option. A quadratic curve passes through the first and last control points, and the rest of the curve is tangent to the polyline segments between the remaining control points (Figure 22-23).

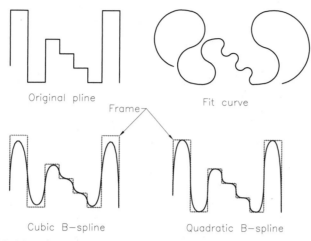

Figure 22-23 *Comparison of fit curve, quadratic B-spline, and cubic B-spline*

Generation of different types of spline curves. If you want to edit a polyline into a B-spline curve, you are first required to enter a relevant value in the **SPLINETYPE** system variable. A value of 5 produces the quadratic curve, whereas a value of 6 produces a cubic curve. The default value is 6, which implies that when we use the **Spline** option of the **Edit Polyline** tool, a cubic curve is produced by default. You can change the value of the **SPLINETYPE** variable using the Command prompt.

In case you are editing 3D polygon meshes, the smoothness of the meshes is controlled by the **SURFTYPE** variable. The value of the variable can be set on the Command prompt. You can select the type of polygon mesh you want when you are using the **Smooth Surface** option of the **Edit Polyline** tool from the **Set Spline Fit Variables** dialog box.

SPLINESEGS. The system variable **SPLINESEGS** governs the number of line segments used to construct the spline curves, and therefore you can use this variable to control the smoothness of the curve. The default value for this variable is 8. With this value, a reasonably smooth curve that does not need much regeneration time is generated. The greater the value of this variable, the smoother the curve, but greater the regeneration time, the more space occupied by the drawing file.

Figure 22-24 shows cubic curves with different values for the **SPLINESEGS** variable.

Decurve Option

The **Decurve** option straightens the curves generated after using the **Fit** or **Spline** option on a polyline. On using this option, the curves return to their original shape (Figure 22-25). The polyline segments are straightened using the **Decurve** option. The vertices inserted after using the **Fit** or **Spline** option are also removed. Information entered for the tangent reference is retained for use in future fit curve operations. You can also use this tool to straighten out any curve drawn with the help of the **Arc** option of the **Polyline** tool. Enter **D** at the **Enter an option [Close/Join/Width/Edit vertex/Fit/Spline/Decurve/Ltype gen/Reverse/Undo]** prompt to invoke the **Decurve** option of the **Edit Polyline** tool.

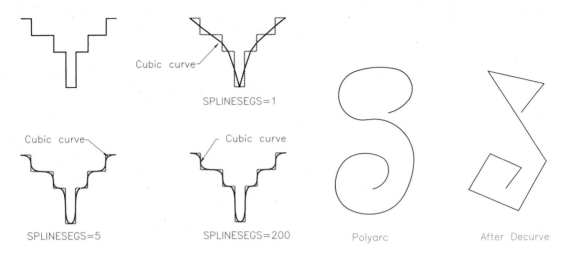

Figure 22-24 *Using the* **SPLINESEGS** *variable*   **Figure 22-25** *Using the* **Decurve** *option*

 Note
If you use editing tools such as the **Break** *or* **Trim** *on spline curves, the* **Decurve** *option cannot be used.*

Ltype gen Option

You can use this option to control the linetype pattern generation for linetypes other than Continuous with respect to the vertices of the polyline. This option has two modes: ON and OFF. If turned off, the break in the noncontinuous linetypes will be avoided at the vertices of the polyline and a continuous segment will be displayed at the vertices (Figure 22-26). If turned on, this option generates the linetype in a continuous pattern such that the gaps may be displayed at the vertices. This option is not applicable to polylines with tapered segments.

The command prompt displayed when you select a polyline and invoke the **Ltype gen** option of the **Edit Polyline** tool is given next.

Enter polyline linetype generation option [ON/OFF] <Off>: *Enter* **ON** *or press ENTER to accept the default value.*

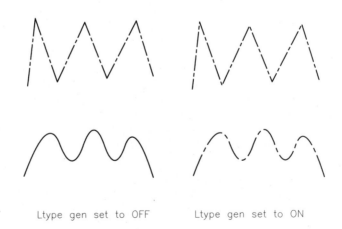

Ltype gen set to OFF   Ltype gen set to ON

Figure 22-26 *Using the* **Ltype gen** *option*

The linetype generation for new polylines can also be controlled with the help of the **PLINEGEN** system variable, which acts as a toggle function. The default value of this variable is off. Entering the value **ON** turns on the **Ltype gen** option. Alternatively, select the appropriate option in the **Linetype generation** field in the **Misc** area of the **Properties** palette with the polyline selected.

Reverse Option

The **Reverse** option is used to change the start point of the polyline. On entering **R** and pressing ENTER at the **Enter an option [Close/Join/Width/Edit vertex/Fit/Spline/Decurve/Ltype gen/Reverse/Undo]** Command prompt, the start point of the polyline will be changed. This can be confirmed by using the **Edit vertex** option of the **Edit Polyline** tool before and after using the **Reverse** option.

Undo (U) Option

The **Undo** option negates the effect of the most recent **PEDIT** operation. You can go back as far as you need to in the current **PEDIT** session by using the **Undo** option repeatedly until you get the desired screen. If you started editing by converting an object into a polyline, and you want to change the polyline back to the object from which it was created, the **Undo** option of the **Edit Polyline** tool will not work. In this case, you will have to exit to the Command prompt and use the **UNDO** command to undo the operation.

Editing Multiple Polylines

Selecting the **Multiple** option of the **Edit Polyline** tool allows you to select more than one polyline for editing. You can select the polylines using any of the objects selection techniques. After the objects to be edited are selected, press ENTER or right-click to proceed with the command. The prompt sequence that will follow is given next.

Enter an option [Close/Open/Join/Width/Fit/Spline/Decurve/Ltype gen/Reverse/Undo]:

Join

This option is used to join more than one polylines that may or may not be in contact with each other. Even the polylines that are not coincident can be joined using this option. The polylines are joined using a specified distance value called the **Fuzz distance**. After you select the **Join** option, the sequence of prompts is as follows.

Join Type = Extend
Enter fuzz distance or [Jointype] <current>: *Enter a distance or* **J** *for changing the* **Jointype***.*

The prompt sequence to follow when you enter **J** at the above prompt is given next.

Enter join type [Extend/Add/Both] <Extend>:

Extend. This option is used to extend or trim the selected polylines at the endpoints that are nearest to each other, see Figure 22-27. Keep in mind that the segments of the selected polylines that are nearest to each other should not be parallel. This means that the segments that are nearest to each other should intersect at some point when extended.

Add. This option is used to join the selected polylines by drawing a straight line between the nearest endpoints, see Figure 22-28. The fuzz distance should be greater than the actual distance between the two endpoints to be joined.

Both. This option joins the endpoints by extending or trimming if possible; otherwise, it adds a line segment between the two endpoints.

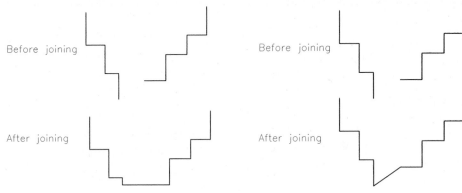

Figure 22-27 *Joining two polylines using the* **Extend** *option*

Figure 22-28 *Joining two polylines using the* **Add** *option*

EXERCISE 2 — Edit Polyline

a. Draw a line from point (0,0) to point (6,6). Convert the line into a polyline with a starting width 0.30 and a ending width 0.00.

b. Draw different polylines of varying width segments and then use the **Join** option to join all the segments into one polyline. After joining the segments, make their width uniform, with the **Width** option.

EXERCISE 3 — Edit Polyline

a. Draw a staircase-shaped polyline and use the different options of the **Edit Polyline** tool to generate a fit curve, a quadratic B-spline, and a cubic B-spline, and then convert the curves back to the original polyline.

b. Draw square-wave-shaped polylines, and use different options of the **Edit vertex** option to navigate around the polyline, split the polyline into two at the third vertex, insert more vertices at different locations in the original polyline, and convert the square-wave-shaped polyline into a straight-line polyline.

UNDOING PREVIOUS COMMANDS

Quick Access Toolbar: Undo   **Command:** UNDO

The **LINE** and **PLINE** commands have the **Undo** option, which can be used to undo (nullify) the changes made within these commands. The **UNDO** command is used to undo a previous command or to undo more than one command at a time. This command can be invoked by entering **UNDO** at the Command prompt. **Undo** is also available in the **Quick Access Toolbar**, the **Edit** menu, and the **Standard** toolbar. The **UNDO** button in the **Quick Access Toolbar** can only undo the previous command and only one command at a time. If you right-click in the drawing area; a shortcut menu will be displayed. Choose **Undo**. It is equivalent to **UNDO > 1** or to enter **U** at the Command prompt. The **U** command can undo only one operation at a time. You can use **U** as many times as you want until AutoCAD displays the message: **Nothing to undo**. When an operation cannot be undone, a message is displayed but no action takes place. External commands such as **PLOT** cannot be undone.

Undoing Previous Commands using the Quick Access Toolbar

Click the arrow on the **Undo** tool in the **Quick Access Toolbar**; the **Undo** list box will be displayed. All commands used recently will be listed in the **Undo** list box. Select the commands that you want to undo by placing the cursor in the list box and moving downward. The Commands selected are highlighted in blue color, as shown in Figure 22-29. The Status Bar below the **Undo** list box displays the number of commands selected. Click the left mouse button to confirm the selections.

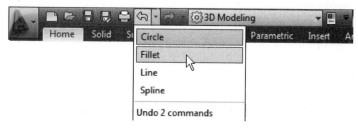

*Figure 22-29 The **Undo** list box*

Undoing Previous Commands using the Command prompt

To Undo previous commands using the Command Line, enter **UNDO** at the Command prompt (if you are working in dynamic mode, the command will be entered at pointer input). The prompt sequence for the **UNDO** command is given next.

Command: **UNDO**
Current settings: Auto = On, Control = All, Combine = Yes, Layer = Yes
Enter the number of operations to undo or [Auto/Control/BEgin/End/Mark/Back] <1>:
Enter a positive number, an option, or press ENTER to undo a single operation.

Various options of this command are discussed next.

Number (N) Option

This is the default option. This number represents the number of previous command sequences to be deleted. For example, if the number entered is 1, the previous command is undone; if the number entered is 4, the previous four commands are undone, and so on. This is identical to invoking the **U** command four times, except that only one regeneration takes place. AutoCAD lets you know which commands were undone by displaying a message after you press ENTER. The prompt sequence is given next.

Command: **UNDO**
Enter the number of operations to undo or [Auto/Control/BEgin/End/Mark/Back]<1>:3
PLINE LINE CIRCLE

Control (C) Option

This option lets you determine how many of the options are active in the **UNDO** command. You can disable the options you do not need. With this option, you can even disable the **UNDO** command. To access this option, type C. You will get the following prompt.

Enter an UNDO control option [ALL/None/One/Combine/Layer] <All>: *Enter an option or press ENTER to accept the default.*

Chapter 22

ALL

The **All** option activates all the features (options) of the **UNDO** command.

None (N)

This option turns off **UNDO** and the **U** command. If you have used the **BEgin** and **End** options or **Mark** and **Back** options to create **UNDO** information, all of that information is undone. The prompt sequence for invoking the **Control** option is as follows.

Command: **UNDO**
Enter the number of operations to undo [Auto/Control/BEgin/End/Mark/Back] <1>: **C**
Enter an UNDO control option [ALL/None/One/Combine/Layer] <All>: **N**

If you try to use the **U** command now, while the **UNDO** command has been disabled, AutoCAD gives you the following message.

Command: **U**
U command disabled. Use **UNDO** command to turn it on.

The prompt sequence for the **UNDO** command after issuing the **None** option is given next.

Command: **UNDO**
Enter an UNDO control option [All/None/One/Combine/Layer] <All>: *Enter O or press ENTER to accept the default.*

To enable the **UNDO** options again, you must enter the **All** or **One** (one mode) option.

One (O)

This option restrains the **UNDO** command to a single operation. All **UNDO** information saved earlier during editing is scrapped. The prompt sequence is as follows.

Enter the number of operations to undo or [Auto/Control/BEgin/End/Mark/Back] <1>: **C**
Enter an UNDO control option [All/None/One/Combine/Layer] <All>: **O**

If you then enter the **UNDO** command, you will get the prompt that is given next.

Command: **UNDO**
Enter an option [Control] <1>: *Press ENTER to undo the last operation or enter **C** to select another control option.*

In response to this last prompt, you can now either press ENTER to undo only the previous command, or go into the **Control** options by entering C. AutoCAD acknowledges undoing the previous command with messages such as given next.

Command: **UNDO**
Enter an option [Control] <1>: [Enter]
CIRCLE
Everything has been undone

Tip
*AutoCAD stores all information about all the **UNDO** operations performed in a drawing session. When you use the **None** control option of the **UNDO** command, this information is removed and valuable disk space is made available. In case of limited disk space, it may be a good idea to use the **One** suboption of the **UNDO Control** command. Using this option, you can use both the **U** and **UNDO** commands and at the same time to make the disk space available.*

Combine

This option allows you to combine all the consecutive **Zoom** and **Pan** operations as a single operation, while using the **UNDO** and **REDO** commands.

Layer

This option allows you to combine layer property actions as a single action.

Auto (A) Option

Enter A to invoke this option. The following prompt is displayed.

> Enter UNDO Auto mode [ON/OFF] <On>: *Select ON or OFF or press ENTER to accept the default setting.*

This option is on by default and every menu item is a single operation. When you use the **Begin** and **End** options to group a series of commands as a single operation, entering **U** at the Command prompt removes the objects individually and not as a group, if the **Auto** option is on. But if **Auto** is off, entering **U** at the Command prompt will remove the entire group of objects you had grouped using the **Begin** and **End** options in one single operation.

Also, if you have put a marker in between the start and end operations and if **Auto** is On, the group of commands till the marker, is undone altogether on using the **Back** option. This group of commands is considered to be a single group and is undone as a single command, although, entering **U** at the Command prompt removes the objects individually until the marker is encountered. But, if the **Auto** option is off, entering **U** at the Command prompt, or using the **Back** option of the **UNDO** command, undoes the entire group of operations irrespective of the marker being there.

BEgin (BE) and End (E) Options

A group of commands is treated as one command for the **U** and **UNDO** commands (provided **Auto** option is off) by embedding the commands between the **BEgin** and **End** options of the **UNDO** command. If you anticipate the removal of a group of successive commands later in a drawing, you can use this option, since all of the commands after the **BEgin** option and before the **End** option are treated as a single command by the **U** command (if the **Auto** option is Off). For example, the following sequence illustrates the possibility of removal of two commands.

> Command: **UNDO**
> Enter the number of operations to undo or [Auto/Control/BEgin/End/Mark/Back] <1>: **BE**
> Command: **CIRCLE**
> Specify center point for Circle or [3P/2P/Ttr (tan tan radius)]: *Specify the center.*
> Specify radius of circle or [Diameter] <current>: *Specify the radius of the circle.*
> Command: **PLINE**
> Specify start point: *Select the first point.*
> Specify next point or [Arc/Halfwidth/Length/Undo/Width]: *Select the next point.*
> Specify next point or [Arc/Close/Halfwidth/Length/Undo/Width]: *Press ENTER.*
> Command: **UNDO**
> Enter the number of operations to UNDO or [Auto/Control/BEgin/End/Mark/Back]: **E**
> Command: **U**

To start the next group once you have finished specifying the current group, use the **End** option to end this group. Another method is to enter the **BEgin** option to start the next group while the current group is active. This is equivalent to issuing the **End** option followed by the **BEgin** option. The group is complete only when the **End** option is invoked to match a **BEgin** option. If **U** or the **UNDO** command is issued after the **BEgin** option has been invoked and before the **End** option has been issued, only one command is undone at a time until it reaches the juncture where the **BEgin** option has been entered. If you want to undo the commands issued before the **BEgin** option was invoked, you must enter the **End** option so that the group is complete. This is demonstrated in the following example:

EXAMPLE 2   *UNDO*

Enter the following commands in the sequence given below, and notice the changes that take place on screen.

> **CIRCLE**
> **POLYGON**
> **UNDO BEgin (Make sure that the Auto is OFF)**
> **PLINE**
> **ELLIPSE**
> **U**
> **DONUT**
> **UNDO End**
> **TEXT**
> **U**
> **U**
> **U**
> **U**

The first **U** command will undo the **ELLIPSE** command. If you repeat the **U** command, the **PLINE** command will be undone. Any further invoking of the **U** command will not undo any previously drawn object (**POLYGON** and **CIRCLE**, in this case), because after the **PLINE** is undone, you have an **UNDO BEgin**. Only after you enter **UNDO End**, you can undo the **POLYGON** and the **CIRCLE**. In the example, the second **U** command will undo the **TEXT** command, the third **U** command will undo the **DONUT** and **PLINE** commands (these are enclosed in the group), the fourth **U** command will undo the **POLYGON** command, and the fifth **U** command will undo the **CIRCLE** command. Whenever the commands in a group are undone, the name of each command or operation is not displayed as it is undone. Only the name, **GROUP**, is displayed.

Tip
*You can use the **BEgin** and **End** options only when the **UNDO Control** is set to **All** and the **Auto** option is off.*

Mark (M) and Back Options

The **Mark** option installs a marker in the Undo file. The **Back** option lets you undo all the operations until the mark. In other words, the **Back** option returns the drawing to the point where the previous mark was inserted. For example, if you have completed a portion of your drawing and do not want anything up to this point to be deleted, you can insert a marker and then proceed. Then, even if you use the **UNDO Back** option, it will work only until the marker. You can insert multiple markers, and with the help of the **Back** option, you can return to the successive mark points. The following prompt sequence is used to introduce a mark point:

Command: **UNDO**
Enter the number of operations to undo or [Auto/Control/BEgin/End/Mark/Back]: **M**

The prompt sequence for using the **Back** option is given next.

Command: **UNDO**
Enter the number of operations to undo or [Auto/Control/BEgin/End/Mark/Back]: **B**

Once all the marks have been exhausted with the successive **Back** options, any further invoking of the **Back** option displays the message: **This will undo everything. OK? <Y>**. If you enter Y (Yes) at this prompt, all operations carried out since you entered the current drawing session will be undone. If you enter N (No) at this prompt, the **Back** option will be disregarded.

You cannot undo certain commands and system variables, for example, **DIST**, **LIST**, **DELAY**, **NEW**, **OPEN**, **QUIT**, **AREA**, **HELP**, **PLOT**, **QSAVE** and **SAVE**, among many more. Actually, these commands have no effect that can be undone. Commands that change operating modes (**GRID**, **UNITS**, **SNAP**, **ORTHO**) can be undone, though the effect may not be apparent at first. This is the reason why AutoCAD displays the command names as they are undone.

REVERSING THE UNDO OPERATION

Quick Access Toolbar: Redo   **Toolbar:** Standard > Redo   **Command:** REDO

If you right-click in the drawing area, a shortcut menu is displayed. Choose **Redo** to invoke the **REDO** command. The **REDO** command brings back the process you removed previously using the **U** and **UNDO** commands. This command undoes the last **UNDO** command performed, provided it is entered immediately after the **UNDO** command. On using this command, the objects previously undone reappear on the screen. Click on the **Redo** button in the **Quick Access Toolbar**; the **Redo** list box (Figure 22-30) is displayed. It lists the commands that have been recently undone. Select the commands that you want to **Redo** and click the left mouse button to confirm the selection.

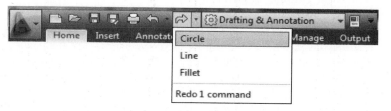

*Figure 22-30 The **Redo** list box*

Note
*Consecutive **Zoom** and **Pan** operations can be undone or redone by using the **Undo** or **Redo** command only once so that the drawing returns to the initial zoom state. To disable this feature, clear the **Combine zoom and pan commands** check box in the **Undo/Redo** area of the **User Preferences** tab of the **Options** dialog box.*

RENAMING NAMED OBJECTS

Menu Bar: Format > Rename   **Command:** RENAME

You can edit the names of the named objects such as blocks, dimension styles, layers, linetypes, styles, UCS, views, and viewports using the **Rename** dialog box. You can select the named object from the list in the **Named Objects** area of the dialog box. The corresponding names of all the objects of the specified type that can be renamed are displayed in the **Items** area. For example, if you want to rename the layer named HID to HIDDEN, the process will be as follows:

1. Enter **RENAME** at the Command prompt and then press ENTER; the **Rename** dialog box will be displayed, as shown in Figure 22-31.

2. Select **Layers** in the **Named Objects** list box. All the layer names in the current drawing that can be renamed are displayed in the **Items** list box.

3. Select **HID** in the **Items** list box to highlight it. HID will be displayed in the **Old Name** edit box.

4. Enter **HIDDEN** in the **Rename To** edit box, and choose the **OK** button.

Now, the layer named **HID** will be renamed to **HIDDEN**. You can view the change in the **Layer** drop-down list in the **Layers** panel of the **Home** tab or invoke the **Layer Properties Manager** dialog box to notice this change. You can rename blocks, dimension styles, linetypes, styles, UCS, views, and viewports in the same way.

Note

*The **RENAME** command cannot be used to rename drawing files created using the **WBLOCK** command. You can change the name of a drawing file by using the **Rename** option of your Operating System.*

*The Layer **0** and the **Continuous** linetype cannot be renamed and, therefore, do not appear in the **Items** list box of the **Rename** dialog box.*

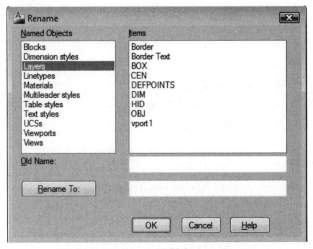

*Figure 22-31 The **Rename** dialog box*

REMOVING UNUSED NAMED OBJECTS

Application Menu: File > Drawing Utilities > Purge	**Command:** PURGE

This is another editing operation used for deletion and it was discussed earlier in relation to blocks. You can delete unused named objects such as blocks, layers, dimension styles, linetypes, text styles, shapes, and so on with the help of the **PURGE** command. When you create a new drawing or open an existing one, AutoCAD records the named objects in that drawing and notes other drawings that reference the named objects. Usually only a few of the named objects in the drawing (such as layers, linetypes, and blocks) are used. For example, when you create a new drawing, the prototype drawing settings may contain various text styles, blocks, and layers which you do not want to use. Also, you may want to delete particular unused named objects such as unused blocks in an existing drawing. Deleting inactive named objects is important and useful because doing so reduces the space occupied by the drawing. With the **PURGE** command, you can select the named objects you want to delete. You can use this command at any time in the drawing session. When you invoke the **PURGE** command, the **Purge** dialog box is displayed (Figure 22-32).

View items you can purge

When this radio button is selected, AutoCAD displays the tree view of all the named objects that are in the current drawing and those can be purged. When this radio button is selected, the **Items not used in drawing** area, and the two check boxes, **Confirm each item to be purged** and **Purge nested items**, are displayed in the dialog box. These are discussed next.

Items not used in drawing

This area lists all the named objects that can be purged. You can list the items of any object type by choosing the plus (+) sign or by double-clicking on the named object in the tree view. Select them to be purged. You can select more than one item by holding down the SHIFT or the CTRL key and selecting the items.

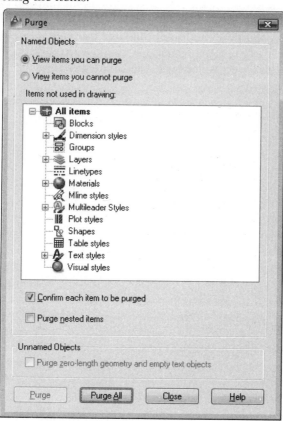

Figure 22-32 *The **Purge** dialog box with items that can be purged*

Confirm each item to be purged

This check box is selected to confirm before purging the selected item. The **Confirm Purge** dialog box is displayed when items are selected to be purged and after the **Purge** or **Purge All** button is chosen. You can confirm or skip the items to be purged in this dialog box.

Purge nested items

This check box, when selected, removes all the nested items not in use. This removes the nested items only when you choose the **Purge All** button or select blocks.

Purge zero - length geometry and empty text objects

While drawing a line, arc, or polyline if the same point is selected as the start and the end points by mistake, then a zero length geometry will be created. Also, if there is only empty space created by mistake while creating a text, then the **Purge zero-length geometry and empty text objects** check box will be available. You can purge the zero length geometry or the empty spaces by selecting this check box.

View items you cannot purge

This radio button is used to display the tree view of the items that you cannot purge. These are the items that are used in the current drawing or are the default items that cannot be removed (Figure 22-33). Choosing this radio button displays the tree view in the **Items currently used in drawing** area of the dialog box. When you select an object in this tree view, information about why you cannot purge this object is displayed below the tree view.

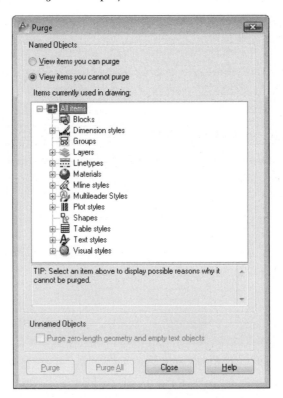

Figure 22-33 The **Purge** dialog box with items that cannot be removed

Note

*The **Entire Drawing** option of the **Write Block** dialog box, which is invoked on using the **WBLOCK** command or the -**WBLOCK** option, has the same effect as the **PURGE All** command.*

The only difference is that in this case the unused named objects are removed automatically without any verification, though this method is much faster as compared to the previous one.

*Standard objects created by AutoCAD (such as layer 0, STANDARD text style, Dimension style, and linetype CONTINUOUS) cannot be removed by the **PURGE** command, even if these objects are not used.*

SETTING SELECTION MODES USING THE OPTIONS DIALOG BOX *

Application Menu: Options **Command:** OPTIONS

When you select a number of objects, the selected objects form a selection set. Selection of the objects is controlled in the **Options** dialog box (Figure 22-34) that is invoked by choosing the **Options** button from the **Application Menu**. Six selection modes are provided in the **Selection** tab of this dialog box. You can select any one of these modes or a combination of various modes.

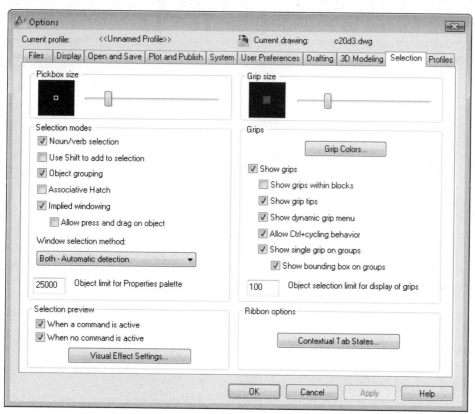

Figure 22-34 *The **Selection** tab of the **Options** dialog box*

Noun/verb selection

By selecting the **Noun/verb selection** check box, you can select the objects (noun) first and then specify the operation (verb) (command) to be performed on the selection set. This mode is active by default. For example, if you want to move some objects when the mode **Noun/verb selection** is enabled, first select the objects to be moved, and then invoke the **MOVE** command. The objects selected are highlighted automatically when the **MOVE** command is invoked, and AutoCAD does not issue any **Select objects** prompt. The following are some of the commands that can be used on the selected objects when the **Noun/verb selection** mode is active.

ARRAY	BLOCK	CHANGE	CHPROP	COPY
DVIEW	EXPLODE	ERASE	LIST	MIRROR
MOVE	Properties	ROTATE	SCALE	WBLOCK

The following are some of the commands that are not affected by the **Noun/verb selection** mode. You are required to specify the objects (noun) on which an operation (command/verb) is to be performed, after specifying the command (verb).

BREAK	CHAMFER	DIVIDE	EDGESURF	EXTEND
FILLET	MEASURE	OFFSET	REVSURF	RULESURF
TABSURF	TRIM			

When the **Noun/verb selection** mode is active, the **PICKFIRST** system variable is set to 1 (On). In other words, you can also activate the **Noun/verb selection** mode by setting the **PICKFIRST** system variable to 1 (On).

Use Shift to add to selection Option

The next option in the **Selection modes** area of the **Selection** tab of the **Options** dialog box is **Use Shift to add to selection**. Selecting this option establishes the additive selection mode, which is the normal method of most Windows programs. In this mode, you have to hold down the SHIFT key when you want to add objects to the selection set. For example, suppose X, Y, and Z are three objects on the screen and you have selected the **Use Shift to add to selection** check box in the **Selection modes** area of the **Options** dialog box. Select object X. It is highlighted and put in the selection set. After selecting X, and while selecting object Y, if you do not hold down the SHIFT key, only object Y is highlighted and it replaces object X in the selection set. On the other hand, if you hold down the SHIFT key while selecting Y, it is added to the selection set (which contains X), and the resulting selection set contains both X and Y. Also, both X and Y are highlighted. To summarize the concept, objects are added to the selection set only when the SHIFT key is held down while objects are selected. Objects can be discarded from the selection set by reselecting these objects while the SHIFT key is held down. If you want to clear an entire selection set quickly, draw a blank selection window anywhere in a blank drawing area. You can also right-click to display a shortcut menu and choose **Deselect All**. All selected objects in the selection set are discarded from it.

When the **Use Shift to add to selection** mode is active, the **PICKADD** system variable is set to 0 (Off). In other words, you can activate the **Use Shift to add to selection** mode by setting the **PICKADD** system variable to off.

Implied windowing

By selecting this option, you can automatically create a Window or Crossing selection when the **Select objects** prompt is issued. The selection window or crossing window, in this case, is created in the following manner: At the **Select objects** prompt, select a point in the empty space on the screen. This becomes the first corner point of the selection window. After this, AutoCAD asks you to specify the other corner point of the selection window. If the first corner point is to the right of the second corner point, a Crossing selection is defined; if the first corner point is to the left of the second corner point, a Window selection is defined. The **Implied windowing** check box is selected by default.

If this option is not active, or if you need to select the first corner in a crowded area where selecting would select an object, you need to specify Window or Crossing at the **Select objects** prompt, depending on your requirement.

When the **Implied windowing** mode is active, the **PICKAUTO** system variable is set to 1 (On).

In other words, you can activate the **Implied windowing** mode by setting the value of the **PICKAUTO** system variable to 1 (On).

Press and drag

This selection mode is used to govern the way you can define a selection window or a crossing window. When you select this option, you can create the window by pressing the pick button to select one corner of the window and continuing to hold down the pick button and dragging the cursor to define the other diagonal point of the window. When you have the window you want, release the pick button. If the **Press and drag** mode is not active, you have to select twice to specify the two diagonal corners of the window to be defined. This mode is not active by default. This implies that to define a selection window or a crossing window, you have to select twice to define their two opposite corner points.

When the **Press and drag** mode is active, the **PICKDRAG** system variable is set to 1 (On). In other words, you can activate the **Press and drag** mode by setting **PICKDRAG** to On.

Object grouping

This turns the automatic group selection on and off. When this option is on and you select a member of a group, the whole group is selected. You can also activate this option by setting the value of the **PICKSTYLE** system variable to 1. (Groups were discussed earlier in this chapter.)

Tip
*You can clear the **Object grouping** check box in the **Selection Modes** area of the **Selection** tab of the **Options** dialog box to be able to select the objects of a group individually for editing without having to explode the group.*

Associative Hatch

If the **Associative Hatch** check box is selected in the **Selection modes** area of the **Selection** tab of the **Options** dialog box, the boundary object is also selected when an associative hatch is selected. You can also select this option by setting the value of the **PICKSTYLE** system variable to 2 or 3. It is recommended that you select the **Associative Hatch** check box for most drawings.

Pickbox Size

The **Pickbox** slider bar controls the size of the pickbox. The size ranges from 0 to 20. The default size is 3. You can also use the **PICKBOX** system variable.

Window selection method

This drop-down list can be used to control the selection method. During the window selection, you can control the window behaviour in the following two ways:

Click and Click. Using this method, you can define the selection window by clicking in the screen to specify the start point and endpoint of the window.

Press and drag. Using this method, you can define the selection window by clicking on any position in the drawing screen and dragging the cursor according to requirement.

You can also use both options by selecting **Both - Automatic detection** from the **Window selection method** drop-down.

Self-Evaluation Test

Answer the following questions and then compare them to those given at the end of this chapter:

1. Even if a group is defined as selectable and the **PICKSTYLE** system variable is set to 0, you cannot select the entire group by selecting one of its members. (T/F)

2. The **Properties** palette is used to change the properties associated with an object. (T/F)

3. The **Properties** palette can be used to modify the geometry of objects apart from their general properties. (T/F)

4. The **RENAME** command can be used to change the name of a drawing file created by using the **WBLOCK** command. (T/F)

5. A group of commands is treated as one command for the **U** and **UNDO** commands by embedding them between the _____ and _____ options of the **UNDO** command.

6. If the **Press and Drag** mode is not active, you have to select _____ to specify the two diagonal points of a selection window.

7. The _____ option of the **Edit Polyline** tool's main prompt can be used to change the starting and ending widths of a polyline separately to a desired value.

8. The _____ option of the **Edit Polyline** tool is used to select more than one polyline.

9. You can move past the first and last vertices in a closed polyline by using either the _____ option or the _____ option.

10. While using the _____ tool, if you select a line or an arc, AutoCAD provides you with the option of converting them into a polyline first.

Review Questions

Answer the following questions:

1. When you use the **GROUP** command to form object groups, AutoCAD lets you sequentially highlight the groups of which the selected object is a member. (T/F)

2. The standard objects created by AutoCAD (such as layer 0, STANDARD text style, and linetype CONTINUOUS) cannot be removed by using the **PURGE** command. (T/F)

3. If you explode an object, it remains identical to its original geometry; only the color and linetype may change because of floating layers, colors, or linetypes. (T/F)

4. The **PURGE** command has the same effect as that of the **WBLOCK Entire drawing** or the **-WBLOCK** method. The only difference is that with the **PURGE** command, deletion takes place automatically. (T/F)

5. If you have made a copy of a group without naming the copy, its name is displayed in which of the following notations in the **Group Name** list box?

   (a) $A1   (b) @A1
   (c) %A1   (d) *A1

6. Which of the following options is used delete a selected group from a drawing?

   (a) **Remove**   (b) **Reorder**
   (c) **Explode**   (d) **Rename**

7. Which of the following system variables controls the smoothness of a curve?

   (a) **SPLINETYPE**   (b) **SPLINESEGS**
   (c) **PLINEGEN**   (d) **PICKFIRST**

8. Which of the following options of the **Edit Polyline** tool can straighten a curve drawn with the help of the **Polyline** tool?

   (a) **Join**   (b) **Close**
   (c) **Decurve**   (d) None of these

9. Which of the following properties cannot be changed using the **CHANGE** command?

   (a) **Lineweight**   (b) **Color**
   (c) **Plotstyle**   (d) **Thickness**

10. The _____ command can be used to change the name of a block.

11. If the _____ option of the **UNDO** command is off, any group of commands grouped together using the **Begin** and **End** options of the **UNDO** command are undone together.

12. Circles drawn using the **CIRCLE** command can be changed to _____.

13. The _____ option in the **GROUP** command lets you change the order of the objects in the selected group.

14. You can select a group by entering _____ at the **Select objects**: prompt.

15. The _____ option of the **UNDO** command disables the **UNDO** and **U** commands entirely.

EXERCISE 4   *Properties and Edit Polyline*

Draw part (a) in Figure 22-35 and then using the **Properties** and relevant **Edit Polyline** tool options, convert it into parts (b), (c), and (d). The linetype used in part (d) is HIDDEN.

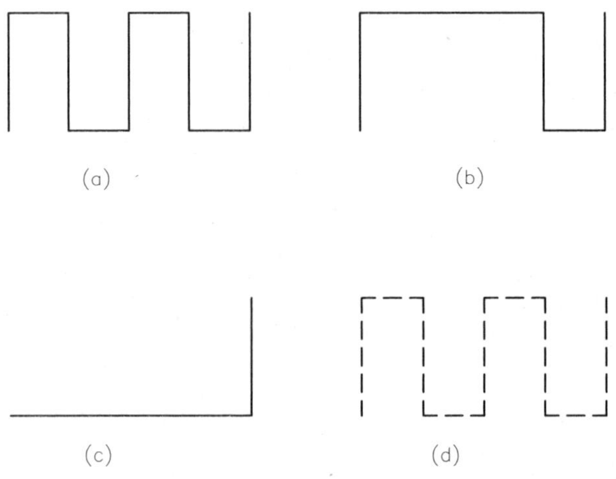

(a)   (b)

(c)   (d)

Figure 22-35 Drawing for Exercise 4

EXERCISE 5 — *Edit Polyline*

Draw the sketch shown in Figure 22-36 using the **Line** tool. Change the object to a polyline with a width of 0.05.

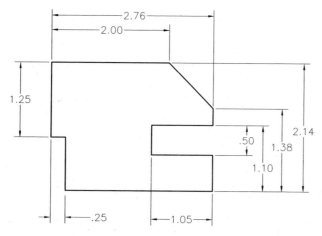

Figure 22-36 Drawing for Exercise 5

EXERCISE 6 — *Edit Polyline*

Draw part (a) of Figure 22-37 and then using the relevant **Edit Polyline** tool options, convert it into drawing (b).

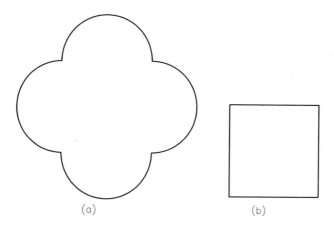

Figure 22-37 Drawing for Exercise 6

EXERCISE 7   *Edit Polyline*

Draw part (a) of Figure 22-38 and then using the relevant **Edit Polyline** tool options, convert it into drawings (b), (c), and (d). Identify the types of curves.

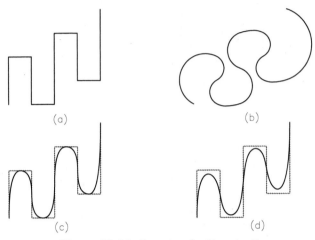

Figure 22-38 *Drawing for Exercise 7*

EXERCISE 8   *Edit Polyline*

Draw part (a) of Figure 22-39 and then using the relevant **Edit Polyline** tool options, convert it into drawing (b).

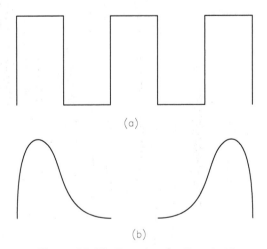

Figure 22-39 *Drawing for Exercise 8*

EXERCISE 9   *Edit Polyline*

Draw part (a) of Figure 22-40 and then using the relevant **Edit Polyline** tool options, convert it into part (b) of the drawing.

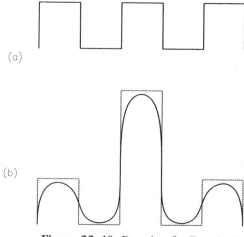

Figure 22-40 Drawing for Exercise 9

EXERCISE 10   *Edit Polyline*

Draw the part (a) of Figure 22-41 and change it into the figure shown in part (b) by using the appropriate commands. You can also use grips to obtain this illustration. Assume the missing dimensions.

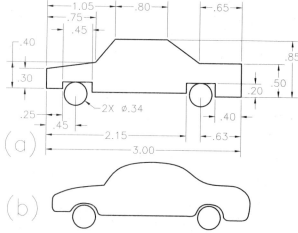

Figure 22-41 Drawing for Exercise 10

EXERCISE 11   *Edit Polyline*

Create the drawing shown in Figure 22-42. Use the splined polylines to draw the break lines. Dimension the drawing as shown.

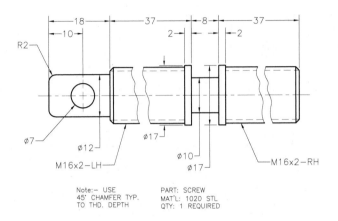

Figure 22-42 *Drawing for Exercise 11*

Chapter 23

Working with Data Exchange & Object Linking and Embedding

CHAPTER OBJECTIVES

In this chapter, you will learn:
- *To import and export .dxf files using the SAVEAS and OPEN commands.*
- *To convert scanned drawings into the drawing editor using the DXB file format.*
- *To attach the raster images to the current drawing.*
- *To edit raster images.*
- *About the embedding and linking functions of the OLE feature of Windows.*
- *About the DWG Convert and Content Explorer*

KEY TERMS

- *DXF Files*
- *DXB Files*
- *ACIS Files*

- *DWG Convert*
- *3D Studio Files*
- *DGN Files*

- *BMP Files*
- *Attach Image*
- *EPS Files*

- *OLE Objects*
- *Draw Order*
- *Content Explorer*

UNDERSTANDING THE CONCEPT OF DATA EXCHANGE IN AutoCAD

Different companies have developed different software for applications such as CAD, desktop publishing, and rendering. This non-standardization of software has led to the development of various data exchange formats that enable the transfer (translation) of data from one data processing software to the other. This chapter will cover various data exchange formats provided in AutoCAD. AutoCAD uses the *.dwg* format to store drawing files. This format is not recognized by most other CAD software such as Intergraph, CADKEY, and Microstation. To eliminate this problem so that the files created in AutoCAD can be transferred to other CAD software for further use, AutoCAD provides various data exchange formats such as DXF (data interchange file) and DXB (binary drawing interchange).

CREATING DATA INTERCHANGE (DXF) FILES

The DXF file format generates a text file in ASCII code from the original drawing. This allows any computer system to manipulate (read/write) data in a DXF file. Usually, the DXF format is used for CAD packages based on microcomputers. For example, packages like SmartCAM use DXF files. Some desktop publishing packages, such as PageMaker and Ventura Publisher, also use DXF files.

Creating a Data Interchange File

The **SAVE** or **SAVEAS** commands are used to create an ASCII file with a *.dxf* extension from an AutoCAD drawing file. Once you invoke any of these commands, the **Save Drawing As** dialog box is displayed. You can also use the **-WBLOCK** command. The **Create Drawing File** dialog box is displayed, where you can enter the name of the file in the **File name** edit box and select DXF [*dxf] from the **Files of type** drop-down list. By default, the DXF file to be created assumes the name of the drawing file from which it will be created. However, you can specify any file name for the DXF file by typing it in the **File name** edit box. Select the extension as DXF [*dxf] from the **Files of type** drop-down list.

Next, choose the **Save As** button from the **File** menu; the **Save Drawing As** dialog box will be displayed. In the **Save Drawing As** dialog box, choose the **Tools** button to display the flyout. In the flyout, choose **Options** to display the **Saveas Options** dialog box, see Figure 23-1. In this dialog box, choose the **DXF Options** tab and enter the degree of accuracy for the numeric values. The default value for the degree of accuracy is sixteen decimal places. You can enter a value between 0 and 16 decimal places.

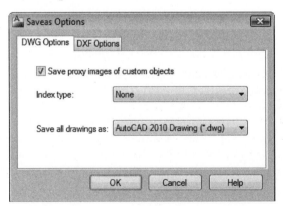

*Figure 23-1 The **Saveas Options** dialog box*

In the **DXF Options** tab, you can also select the **Select objects** check box, which allows you to specify the objects to be included in the DXF file. In this case, the definitions of the named objects such as block definitions, text styles, and so on are not exported. Selecting the **Save thumbnail preview image** check box saves a preview image with the file that can be previewed in the **Preview** window of the **Save Drawing As** dialog box.

Select the **ASCII** radio button. Choose **OK** to return to the **Save Drawing As** dialog box. Choose the **Save** button here. Now an ASCII file with a *.dxf* extension has been created, and this file can be accessed by other CAD systems. This file contains data on the objects specified. By default, DXF files are created in the ASCII format. However, you can also create binary format files by selecting the **BINARY** radio button in the **Saveas Options** dialog box. Binary DXF files are more efficient and occupy only 75 percent of the ASCII DXF file. You can access a binary format file more quickly than an ASCII format file.

Information in a DXF File

The DXF file contains data of the objects which are specified using the **Select objects** check box in the **Saveas Options** dialog box. You can change the data in this file to your requirement. To examine the data in this file, load the ASCII file in the word processing software. A DXF file is composed of the following parts.

Header

In this part of the drawing database, all variables in the drawing and their values are displayed.

Classes

This section deals with the internal database.

Tables

All named objects such as linetypes, layers, blocks, text styles, dimension styles, and views are listed in this part.

Blocks

The objects that define blocks and their respective values are displayed in this part.

Entities

All entities in the drawing, such as lines, circles, and so on, are listed in this part.

Objects

Objects in the drawing are listed in this part.

Converting DXF Files into Drawing Files

You can import a DXF file into a new AutoCAD drawing file with the **Open** tool. After you invoke the **Open** tool, the **Select File** dialog box is displayed. From the **Files of type** drop-down list, select **DXF [*.dxf]**. In the **File name** edit box, enter the name of the file you want to import into AutoCAD or select the file from the list. Choose the **Open** button. Once this is done, the specified DXF file is converted into a standard DWG file, regeneration is carried out, and the file is inserted into the new drawing. Now, you can perform different operations on this file just as with the other drawing files.

Tip
*A data interchange file (DXF) can also be inserted into the current drawing using the **Insert** tool.*

Importing CAD Files*

Ribbon: Insert > Import > Import
Menu Bar: File > Import   **Command:** IMPORT

You can import any surface, solid, 2D and 3D wire geometry from the supported file formats into AutoCAD. In AutoCAD 2012, you can import IGES, CATIA, Rhino, Pro/Engineer and many other file formats. To import a CAD geometry into AutoCAD, enter **IMPORT** in the Command box; the **Import File** dialog box will be displayed. From the **Files of type** drop-down list, select the required format. Next, browse to the required file and then choose the **Open** button from the dialog box; the **Import Processing Background Job** message box will be displayed stating that the import job is processing in the background. Next, close the message box. The details of the progress can be seen on the bottom-right corner of the screen. After the processing is complete, a link will be displayed as pop-up. Click on the link to import the CAD geometry into AutoCAD; the inserted part geometry will be placed as a block reference.

OTHER DATA EXCHANGE FORMATS

The other formats that can be used to exchange data from one data processing format to the other are discussed next.

DXB File Format

Menu Bar: Insert > Drawing Exchange Binary   **Command:** DXBIN

AutoCAD offers another file format for data interchange, DXB. This format is much more compressed than the binary DXF format and is used when you want to translate a large amount of data from one CAD system to another. For example, when you scan a drawing, a DXB file is created. This file has a huge amount of data in it. To import a DXB file, enter the **DXBIN** command at the Command prompt; the **Select DXB File** dialog box will be displayed. In the **File name** edit box, enter the name of the file (in DXB format) you want to import into AutoCAD. Choose the **Open** button. Once this is done, the specified DXB file will be converted into a standard DWG file and is inserted into the current drawing.

Note
Before importing a DXB file, you must create a new drawing file. No editing or drawing setup (limits, units, and so on) can be performed in this file. This is because if you import a DXB file into an old drawing, the settings (definitions) of layers, blocks, and so on, of the file being imported are overruled by the settings of the file into which you are importing the DXB file.

Creating and Using an ACIS File

Application Menu: Export > Other Formats   **Command:** EXPORT, ACISOUT

Trimmed nurbs surfaces, regions, and solids can be exported to an ACIS file with an ASCII format. To do this, choose **Export > Other Formats** from the **Application Menu**; the **Export Data** dialog box will be displayed. Enter a file name in the **File name** edit box. From the **Files of type** drop-down list, select **ACIS [*.sat]** and then choose the **Save** button; you will be prompted to select the solids, regions, or ACIS bodies that you want to add to the file. Select the objects you wish to save as an ACIS file and press ENTER. AutoCAD appends the file extension *.sat* to the file name. You can also use the **ACISOUT** command to perform this function. It prompts you to select objects before the **Create ACIS File** dialog box is displayed on the screen. Keep in mind that only solids, regions, or ACIS body can be selected to add to the *.sat* file.

To import an ACIS file, choose the **Import** tool from **Import** panel in the **Insert** tab; the **Import File** dialog box will be displayed. In the **Files of type** drop-down list, select the *.sat* file format. Next, select a file to be imported and choose the **Open** button; the file will be inserted. Alternatively, enter **ACISIN** at the Command prompt and import the file using the **Select ACIS File** dialog box.

Importing 3D Studio Files

Menu Bar: Insert > 3D Studio   **Command:** 3DSIN

You can import 3D geometry, views, lights, and objects with surface characteristics created in 3D Studio and saved in *.3ds* format. The **3DSIN** command displays the **3D Studio File Import** dialog box. Select the file you wish to import and choose the **Open** button. You can also invoke this command by choosing **3D Studio** from the **Insert** menu.

Creating and Using a Windows Metafile

Application Menu: Export > Other Formats   **Command:** EXPORT, WMFOUT

The Windows Metafile format (WMF) file contains screen vector and raster graphics format. In the **Export Data** dialog box or in the **Create WMF File** dialog box, enter the file name. Select **Metafile [*.wmf]** from the **Files of type** drop-down list. The extension *.wmf* is appended to the file name. Next, save the settings and then select the objects you want to save in this file format.

The **WMFIN** command displays the **Import WMF** dialog box. Window metafiles are imported as blocks in AutoCAD. Select the *.wmf* file you want to import and choose the **Open** button. Specify an insertion point, rotation angle, and scale factor. Specify the scaling by entering a scale factor using the **corner** option to specify an imaginary box whose dimensions correspond to the scale factor, or by entering **xyz** to specify 3D scale factors. You can also import a *.wmf* file by using the **Import File** dialog box that will be displayed on choosing the **Import** tool from **Import** panel in the **Insert** tab.

Creating and Using a V8 DGN File

Application Menu: Export > DGN   **Command:** EXPORT, DGNEXPORT

The **V8 DGN** (*.DGN*) file format is used by the Microstation software. A **.dwg* file created in AutoCAD can be exported to Microstation using this format. To do so, choose **Export > DGN** from the **Application Menu**; the **Export DGN File** dialog box will be displayed. Enter a file name in the **File name** edit box. From the **Files of type** drop-down list, select **V8 DGN (*.dgn)** and then choose the **Save** button; the **Export DGN Settings** dialog box will be displayed. In this dialog box, you can specify how the externally referenced files in the current drawing are taken care of when converting the current drawing into the DGN file. You can also specify the seed files (standard template files containing the settings for the units to be used for conversion, working units, resolution of the drawing, and so on) to be used for conversion in this dialog box.

To import a DGN file, choose **Open > DGN** from the **Application Menu**; the **Import DGN File** dialog box will be displayed. Alternatively, you can use the **DGNIMPORT** command to import a DGN file. From the **Files of type** drop-down list, select the **MicroStation DGN [*.dgn]** option. Next, select or enter the name of the design file and choose the **Open** button; the **Import DGN Settings** dialog box will be invoked. You can exchange the data of the drawing files and design files using the **Export/Import DGN Settings** dialog box. For example, you can translate DGN elements to DWG objects, Levels to Layers, Line Styles to Linetypes, Cells to Blocks, Sheet Model to DWG Layouts, etc. You can save and reuse, modify, rename or delete the mapping translations using the **DGNMAPPING** command.

Chapter 23

Creating a BMP File

Application Menu: Export > Other Formats	**Command:** EXPORT, BMPOUT

This is used to create bitmap images of the objects in your drawing. In the **Export Data** dialog box, enter the name of the file, select **Bitmap [*.bmp]** from the **Files of type** drop-down list and then choose **Save**. Select the objects you wish to save as bitmap and press ENTER. Entering **BMPOUT** displays the **Create Raster File** dialog box. Enter the file name and choose **Save**. Select the objects to be saved as bitmap. The file extension *.bmp* is appended to the file name.

RASTER IMAGES

A raster image consists of small square-shaped dots known as pixels. In a colored image, the color is determined by the color of pixels. The raster images can be moved, copied, or clipped, and used as a cutting edge with the **Trim** tool. They can also be modified by using grips. You can also control the image contrast, transparency, and quality of the image. AutoCAD stores images in a special temporary image swap file whose default location is the Windows **Temp** directory. You can change the location of this file by modifying it under **Temporary Drawing File Location** in the **Files** tab of the **Options** dialog box.

The images can be 8-bit gray, 8-bit color, 24-bit color, or bitonal. When the image transparency is set to On, the image file formats with transparent pixels is recognized by AutoCAD and transparency is allowed. The transparent images can be in color or grayscale. AutoCAD supports the formats given next.

Image Type	File Extension	Description
BMP	.bmp, .dib, .rle	Windows and OS/2 Bitmap Format
CALS-I	.gp4, .mil, .rst	Mil-R-Raster I
FLIC	.flc, .fli	Flic Autodesk Animator Animation
GEOSPOT	.bil	GeoSPOT (BIL files must be accompanied with HDR and PAL files with connection data in the same directory.)
IG4	.ig4	Image Systems group 4
IGS	.igs	Image Systems Grayscale
JFIF or JPEG	.jpg, .jpeg	Joint Photographics Expert group
PCX	.pcx	Picture PC Paintbrush Picture
PICT	.pct	Picture Macintosh Picture
PNG	.png	Portable Network Graphic
RLC	.rlc	Run-length Compressed
TARGA	.tga	True Vision Raster based Data format
TIFF/LZW	.tif, .tiff	Tagged image file format

When you store images as Tiled Images, that is, in the Tagged Image File Format (TIFF), you can edit or modify any portion of the image; only the modified portion is regenerated, thus saving time. Tiled images load much faster as compared to the untiled images.

Attaching Raster Images

Ribbon: Insert > Reference > Attach   **Tool Palettes:** Draw > Attach Image
Toolbar: Reference > Attach Image or Insert > Attach Image   **Command:** IMAGEATTACH

Attaching Raster images does not make them part of a drawing. To attach an image, choose the **Attach** tool from the **Reference** panel; the **Select Reference File** dialog box will be displayed, as shown in Figure 23-2. You can also invoke this dialog box by using the **IMAGEATTACH** command. Select the image to be attached; the preview of the selected image will be displayed in the **Preview** area of the dialog box. Next, choose the **Open** button; the **Attach Image** dialog box will be displayed, as shown in Figure 23-3. The name of the selected image is displayed in the **Name** drop-down list. You can select another file by using the **Browse** button. The **Name** drop-down list displays the names of all the images in the current drawing. Select the **Specify on-screen** check boxes to specify the **Insertion point**, the **Scale**, and **Rotation** Angle on the screen. Alternatively, you can clear these check boxes and enter values in the respective edit boxes. Choosing the **Show Details** button expands the **Image** dialog box and provides the image information such as the Horizontal and Vertical resolution values, current AutoCAD unit, and Image size in pixels and units. Choose **OK** to return to the drawing screen.

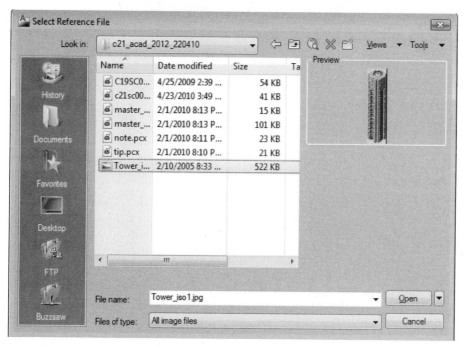

*Figure 23-2 The **Select Reference File** dialog box*

You can also attach and scale an image from the Internet. In the **Select Reference File** dialog box, choose the **Search the Web** button. Once you have located the image file you wish to attach to the current drawing, enter the URL address in the **Look in** drop-down list and the image file name in the **Name or URL** edit box. Choose the **Open** button twice. Specify the insertion point, scale, and rotation angle in the **Attach Image** dialog box. You can also right-click on the image you wish to attach to the current drawing and choose **Properties** from the shortcut menu to display the **Properties** window. From here, you can select, cut, and paste the URL address in the **File name** edit box of the **Select Reference File** dialog box.

Chapter 23

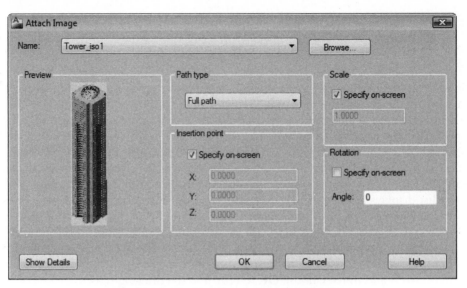

*Figure 23-3 The **Attach Image** dialog box*

Managing Raster Images

Ribbon: View > Palette > External References Palette	**Command:** IMAGE
Toolbar: Reference > External References	**Menu Bar:** Insert > External References

Choose the **External References Palette** button from the **Palettes** panel; the **External References** palette will be displayed (Figure 23-4). If the image has not been inserted earlier, right-click in the **File References** area and choose **Attach Image** from the shortcut menu. You can view the image information either as a list or as a tree view by choosing the respective buttons located at the right corner of the **File References** head of the **External References** palette. The **List View** displays the names, loading status, size, date last modified on, and search path of all images in the drawing. The **Tree View** displays images in hierarchy, which shows its nesting levels within blocks and Xrefs. The **Tree View** does not display the status, size, or any other information about an image file. You can rename an image file in this dialog box.

Preview

To display the **Preview** area, choose the **Preview** button located at the right corner of the **Details** head. In this area, you can see the preview of the image. This area displays the preview of the image file attached to the current drawing and the file details of the attached image file. To get the file details, choose the **Details** button located adjacent to the **Preview** button. The information about the image file such as file name, saved path, and so on will be listed, as shown in Figure 23-5. Once you have referenced the current drawing to an image file, you can

*Figure 23-4 The **External References** palette*

replace it with another image file by using the **Found At** edit box. To do so, choose the **Browse [...]** button in the corner of the edit box to display the **Select Image File** dialog box. Browse and select the image file and then choose the **Open** button. Now, the new path will be displayed in the **Found At** edit box. Also, the location of the image file will automatically be updated in the **Saved Path** edit box.

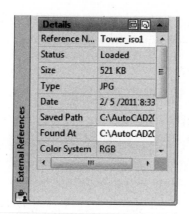

Figure 23-5 *The* **Details** *area of* **External References** *palette*

Tip
You can also use the Command prompt for managing the image files using the **-IMAGE** *command. Except the* **Open** *option, all the other options of the* **External References** *palette will be available in the Command prompt.*

When you select the name of the attached image and right-click on it, the shortcut menu will be displayed. The options in this shortcut menu are discussed next.

Open
This option enables you to open the attached image in its default viewing software from its current location in the AutoCAD file.

Attach
This option allows you to attach an image file with the current drawing. You can also attach an image file by choosing the **Attach Image** option from the flyout on the upper left corner of the **External References** palette; the **Select Reference File** dialog box will be displayed. Select the file that you want to attach to the drawing and the selected image is displayed in the **Preview** box. Choose the **Open** button; the **Image** dialog box is displayed. The name of the file and its path and extension are displayed in the **Attach Image** dialog box. You can use the **Browse** button to select another file. Choose the **OK** button in the **Attach Image** dialog box and specify a point to insert the image. Next, AutoCAD will prompt you to enter the scale factor. You can enter the scale factor or specify a point on the screen.

Detach
This option detaches the selected and all associated raster images from the current drawing. Once the image is detached, no information about the image is retained in the drawing.

Reload
Reload simply reloads an image. The changes made to the image since the last insert will be loaded on the screen.

Unload
The **Unload** option unloads an image. An image is unloaded when it is not needed in the current drawing session. When you unload an image, AutoCAD retains information about the location and size of the image in the database of the current file. Unloading a file does not unlink the file from the drawing and the image frame is displayed. If you reload the image, the image will appear at the same point and in the same size as the image was before unloading. Unloading

Chapter 23

the raster images enhances AutoCAD performance. Also, the unloaded images are not plotted. If multiple images are to be loaded and the memory is insufficient, AutoCAD automatically unloads them.

EDITING RASTER IMAGE FILES
Clipping Raster Images

Ribbon: Insert > Reference > Clip	**Toolbar:** Reference > Clip Image
Menu Bar: Modify > Clip > Image	**Command:** IMAGECLIP

This command is used to clip the boundaries of images and to provide a desired shape to them. To invoke this command, select the image to be clipped; the **Image** tab will be added to the **Ribbon**. Choose the **Create Clipping Boundary** tool; you will be prompted to specify the boundary. Specify a rectangular or polygonal boundary; the image will be clipped at the specified boundary. Also, an arrow will be displayed pointing inside the clipping boundary. Click on the arrow to flip the clipping direction.

You can also invoke this command by selecting an image to be clipped, right-clicking, and choosing **Image > Clip** from the shortcut menu.

On invoking this command by entering **IMAGECLIP** at the Command prompt and pressing ENTER, AutoCAD will prompt you to select the image to be clipped. Select the raster image by selecting the boundary edge of the image. The image boundary must be visible to select it. The clipping boundary must be specified in the same plane as the image or a plane that is parallel to it. The following prompt sequence is displayed when you choose the **Clip** tool from the **Reference** panel.

Select image to clip: *Select the image.*
Enter image clipping option [ON/OFF/Delete/New boundary] <New>:

New boundary
This option is used to define a boundary for clipping the image. When you invoke this option, you will be prompted to specify whether or not you want to delete the old boundary, if a boundary already exists. If you enter **No** at the **Delete old boundary? [No/Yes] <Yes>** prompt, this command will end. You can enter **Yes** to specify the new clipping boundary. The boundary can be defined in a rectangular shape or a polygonal shape.

ON
This option is used to turn the image clipping on. The image will be clipped using the last boundary.

OFF
This option is used to turn the image clipping off. If you have a clipped image, you can redisplay the entire image.

Delete
This option is used to delete the existing clipping boundary.

One such clipping is shown in Figures 23-6 and 23-7. Figure 23-6 shows the actual image and Figure 23-7 shows the clipped image where the clipping boundary has become the midpoint of all the four sides of the image.

Figure 23-6 Image before clipping

Figure 23-7 Image after clipping

Tip
*You can improve the performance by turning on or off the image highlighting that is visible when you select an image. This can be done by selecting or clearing the **Highlight raster image frame only** check box in the **Display** tab of the **Options** dialog box. You can also use the IMAGEHLT system variable. By default, **IMAGEHLT** is set to 0, that is, only the raster image frame will be highlighted.*

Adjusting Raster Images

Ribbon: Insert > Reference > Adjust   **Menu bar:** Modify > Object > Image > Adjust
Toolbar: Reference > Adjust Image   **Command:** IMAGEADJUST/ADJUST

The **IMAGEADJUST** command allows you to adjust the brightness, contrast, and fade of a raster image. On selecting an attached image, the **Image** tab will be added to the **Ribbon**. Adjust the brightness, contrast, and fade using the corresponding option in the **Adjust** panel of the **Image** tab.

Alternatively, choose **Adjust Image** from the **Reference** toolbar to invoke the **Image Adjust** dialog box (Figure 23-8). This dialog box is used to adjust an image. As you adjust the brightness, contrast, or fade, the changes are dynamically made in the image. If you choose the **Reset** button, the values will be returned to the default values (Brightness= 50, Contrast= 50, and Fade= 0). You can also choose **Properties** in the shortcut menu to display the **Properties** palette. In this palette, you can enter new values of brightness, contrast, and fade. If you choose the **Adjust** tool from the **Reference** panel, you need set the values in the Command prompt.

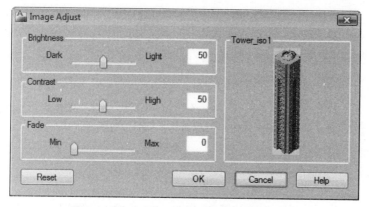

*Figure 23-8 The **Image Adjust** dialog box*

Chapter 23

Modifying the Image Quality

Menu Bar: Modify > Object > Image > Quality	**Command:** IMAGEQUALITY
Toolbar: Reference > Image Quality	

 The **IMAGEQUALITY** command allows you to control the quality of the image that affects the display performance. A high-quality image takes a longer time to display. When you change the quality, the display changes immediately without causing a **REGEN**. The images are always plotted using a high-quality display. Although draft quality images appear grainier, they are displayed more quickly.

Modifying the Transparency of an Image

Toolbar: Reference > Image Transparency	**Command:** TRANSPARENCY
Menu Bar: Modify > Object > Image > Transparency	

When you attach an image with transparent background to a drawing, the transparent background turns opaque. The **TRANSPARENCY** command is used to control the transparency of an image created in the transparent background. To control the transparency of the attached image, select it; the **Image** tab will be added to the **Ribbon**. Choose the **Background Transparency** button from the **Image** tab to turn on the transparency of the image; the background of the image will become transparent. Figure 23-9 shows two images one over the other with the transparency of the top image turned off. Figure 23-10 shows the figure with the transparency of the top image turned on. You can also right-click on an image and choose **Properties** from the shortcut menu displayed. In the **Properties** palette, select an option from the **Background Transparency** drop-down list displayed under the **Misc** list.

 Note
You have to choose images in such a way that they can give the effect of transparency. Dull and opaque images will not give the effect of transparency.

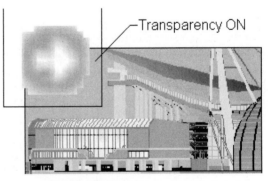

Figure 23-9 *Image before turning the transparency on*

Figure 23-10 *Image after turning the transparency on*

Controlling the Display of Image Frames

Menu Bar: Modify > Object > Image > Frame	**Command:** IMAGEFRAME
Toolbar: Reference > Image Frame	

The **IMAGEFRAME** command is used to turn the image boundary on or off. If the image boundary is off, the image cannot be selected with the pointing device and therefore cannot be accidentally moved or modified.

CHANGING THE DISPLAY ORDER

Command: DRAWORDER **Toolbar:** Draw Order
Ribbon: Home > Modify > Draw Order drop-down > Bring to Front

The tools in the **Draw Order** drop-down (Figure 23-11) of the **Modify** panel are used to change the display order of the selected images and other objects.

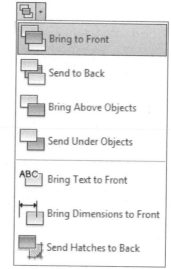

If you choose the **Bring to Front** tool, the selected object will be moved to the front of all entities. If you choose the **Send to Back** tool, the selected object will be moved behind all entities. If you choose the **Bring Above Objects** tool, the selected object will be moved above the referenced objects. If you choose the **Send Under Objects** tool, the selected object will be moved under the referenced objects.

The **Bring Text to Front** tool can be used to bring all text to the front of all entities. If you choose **Bring Dimensions to Front**, all dimensions will come to front of all the entities. The **Send Hatches to Back** tool is used to send all the hatches to the back of all entities.

Figure 23-11 *Tools list in the **Draw Order** drop-down*

Other Editing Operations

You can also perform other editing operations such as copy, move, and stretch to edit a raster image. Remember that you can also use the image as the trimming edge for trimming objects. However, you cannot trim an image. You can insert the raster image several times or make multiple copies of it. Each copy can have a different clipping boundary. You can also edit the image using grips. You can use the **Properties** window to change the image layer, boundary linetype, and color, and perform other editing commands such as changing the scale, rotation, width, and height by entering new values in the respective edit boxes. You can also display an image or turn off the display by selecting options from the **Show Image** drop-down list in the **Misc** field of the **Properties** window. If you select **Yes**, the image is displayed and if you select **No**, the display of the image is turned off. You can turn off the display of the image when you do not need it, thus improving the performance.

Scaling Raster Images

The scale of the inserted image is determined by the actual size of the image and the unit of measurement (inches, feet, and so on). For example, if the image is 1" X 1.26" and you insert this image with a scale factor of 1, its size on the screen will be 1 AutoCAD unit by 1.26 AutoCAD units. If the scale factor is 5, the image will be five times larger. The image that you want to insert must contain the resolution information (DPI). If the image does not contain this information, AutoCAD treats the width of the image as one unit.

DWG CONVERT*

Application Button: Save As > DWG Convert **Command:** DWGCONVERT
Menu Bar: File > DWG Convert

The **DWG Convert** tool is a new tool introduced in AutoCAD 2012. This tool is used to convert a single or batch of the AutoCAD drawing files into any previous AutoCAD drawing format. You can do the conversion by following the procedure given next.

Invoke the **DWG Convert** dialog box by choosing the **DWG Convert** tool from the **File** menu; the **DWG Convert** dialog box will be displayed, as shown in Figure 23-12. Choose the **Add Files** button from the bottom of the **Files Tree** area; the **Select File** dialog box will be displayed. Next, browse to the required location, select the required file/files, and then choose the **Open** button; the selected files will be displayed in tree form in the **Files Tree** area of the **DWG Convert** dialog box. Also, a checkbox will be displayed on the left of each drawing file. By default, all check boxes are selected. You can clear the check box corresponding to the files that are not to be converted. You can also view the selected files along with their details such as their path and location, version, size, date, and so on in the form of a table by choosing the **Files Table** tab. If you want to save the current file list, then choose the **Save list** button; the **Save Conversion List** dialog box will be displayed. Save the file list in the desired folder.

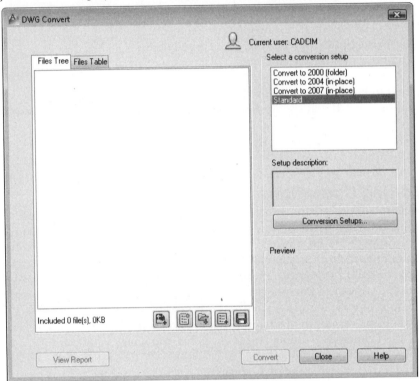

*Figure 23-12 The **DWG Convert** dialog box*

Next, select the version of output file from the **Select a conversion setup** area; the description of the conversion setup will be displayed in the **Setup description** area. You can also specify the conversion settings, if required, by choosing the **Conversion Setups** button. This is discussed in the next head. Now, choose the **Convert** button at the lower right of the dialog box, the **Conversion Package Creation is in Progress** window will be displayed and in a few moments the files will be converted. Exit the dialog box by choosing the **Close** button.

You can also open any previously saved file list by using the **Open** list button. If you choose this button; the **Open Conversion List** dialog box will be displayed. Browse to the required file and choose the **Open** button; all files in the conversion list will be imported in the **Files Tree** area of the **DWG Convert** dialog box. Later on, you can replace this list with a new list using the **Append List** button.

Conversion Setup Options

The **Conversion Setup** options are used to control the output of the files to be converted. To modify the conversion settings, choose the **Conversion Setups** button from the **DWG Convert** dialog box; the **Conversion Setups** dialog box will be displayed, as shown in Figure 23-13. The options in this dialog box are discussed next.

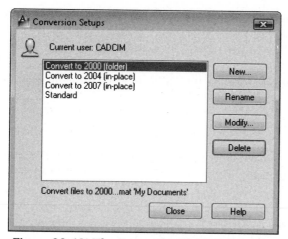

Figure 23-13 *The Conversion Setups dialog box*

New

This button is be used to create a new conversion scheme. On choosing this button, the **New Conversion Setup** window will be displayed, as shown in Figure 23-14. You can enter the name of new conversion setup in the **New Conversion setup name** edit box. Next, select an option from the **Based on** drop-down list. Finally, choose the **Continue** button; the **Modify Conversion Setup** dialog box will be displayed. This dialog box is discussed under the **Modify** button topic.

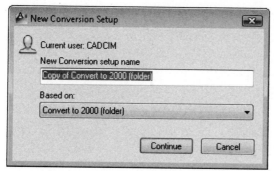

Figure 23-14 *The New Conversion Setup dialog box*

Rename

This button is used to rename the selected conversion setup.

Modify

This button is used to modify the existing conversion settings according to the desired output. On choosing the **Modify** button; the **Modify Conversion Setup** dialog box will be displayed, as shown in Figure 23-15. Using this dialog box, you can modify conversion settings. The options in this dialog box are discussed next.

Conversion type and location Area

The options in this area are discussed next.

In-place (overwrite files). This option is used if you want the converted files to be in the same directory, in which the source files are present. If you want to overwrite the existing files, this option also allows the software to overwrite the previous files with the converted new files.

Folder (set of files). This option is used if you want the converted files to be saved in the specified directory. On selecting this option, the Conversion file folder drop-down list gets activated. Using this drop-down list, you can browse to the desired folder where you want to save the converted files.

Chapter 23

Self-extracting executable(*.exe). This option is used to bind all the converted files into a *.exe extractable file. Using this executable file, all converted files can be extracted into the required folder.

Zip(*.zip). This option is used to bind all converted drawing files into a zip file, which can be uncompressed to the specific location later.

Path options Area

The **Path options** area will be enabled when you select the **Folder (set of files)** option from the **Conversion package type** drop-down list. In the **Path options** area, there are three radio buttons that are explained next.

Use organized folder structure. On selecting this radio button, you can store the converted files in the specified folder in same organized way as they are present in the source folder (in **Source root folder** drop-down).

Place all files in one folder. On selecting this radio button, you can save all converted files in one folder which is specified as output folder.

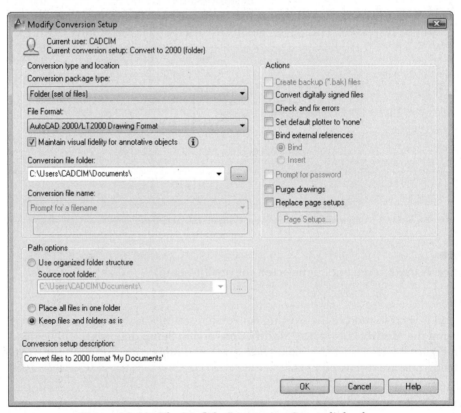

*Figure 23-15 The **Modify Conversion Setup** dialog box*

Keep files and folder as is. On selecting this radio button, the files are saved in the way they were originally present. In other word, the hierarchy remains the same as of original file location.

Actions Area

In the **Actions** area, there are 8 check boxes. Description of these check boxes is given next.
Create backup(*.bak) files. This check box will be enabled only when the **In-place (overwrite files)** conversion package type is selected. If you select this check box then the backup file will also be created with the converted file.

Convert digitally signed files. On selecting this check box, the digitally signed files will be included in the conversion list.

Check and fix errors. If you select this check box, the errors will automatically be fixed during the conversion.

Set default plotter to 'none'. If you select this check box, the plotter for all converted files will be set to **None**.

Bind external references. You can bind the external referenced files with the converted drawing files using this check box. This option has two radio buttons, **Bind** and **Insert**. **Bind** is used to bind the external referenced file as external object. **Insert** is used to bind the external reference as internal object of the drawing.

Prompt for password. This check box is enabled only if you select **Self-extracting executable(*.exe) or Zip(*.zip)** from the **Conversion package type** drop-down list. If this check box is selected then a password will be required to extract the drawing file from the executable or zip file. After choosing the **Convert** button from the **DWG Convert** dialog box; a dialog box will be displayed to save the file created by conversion. As you choose the **Save** button from this dialog box; a dialog box named **Archive - Set Password** will be displayed. Specify a password in the edit box to make the converted files password-protected.

Purge drawings. Select this check box if you want to purge the drawings.

Replace page setups. Select this check box if you want to replace the page setup of the converted drawing files with a pre-stored template. To get a pre-stored template, choose the **Page Setups** button after selecting the check box; the **Page Setups** dialog box will be displayed. Choose the **Browse** button, and then select a pre-stored template.

Conversion setup description Area

This area is located at the bottom of the dialog box. You can enter the description of the conversion setup in the **Conversion setup description** edit box in this area.

Delete

This button is used to delete any of the conversion setup in the list.

WORKING WITH POSTSCRIPT FILES

Menu Bar: File > Export	**Application Menu:** Export > Other Format
Command: EXPORT, PSOUT	

PostScript is a page description language developed by Adobe Systems. It is mostly used in DTP (desktop publishing) applications. AutoCAD allows you to work with PostScript files. You can create and export PostScript files from AutoCAD, so that these files can be used for DTP applications. PostScript images have a higher resolution than raster images. The extension for these files is *.eps* (Encapsulated PostScript).

Creating PostScript Files

As just mentioned, any AutoCAD drawing file can be converted into a PostScript file. This can be accomplished by using the **PSOUT** command. When you invoke this command, the **Create PostScript File** dialog box is displayed, as shown in Figure 23-16. You can also use the **Export** tool to display the **Export Data** dialog box. The options are similar in both the dialog boxes.

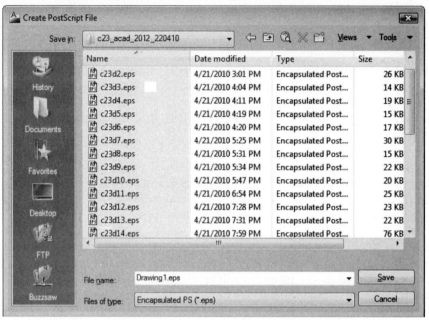

*Figure 23-16 The **Create PostScript File** dialog box*

In the **File name** edit box, enter the name of the PostScript (EPS) file you want to create. Select **Encapsulated PS (*.eps)** from the **Files of type** drop-down list. Next, choose the **Save** button to accept the default setting and create the PostScript file. You can also choose **Options** from the **Tools** flyout of the **Create PostScript File** dialog box to change the settings through the **PostScript Out Options** dialog box (Figure 23-17) and then save the file. The **PostScript Out Options** dialog box provides the following options.

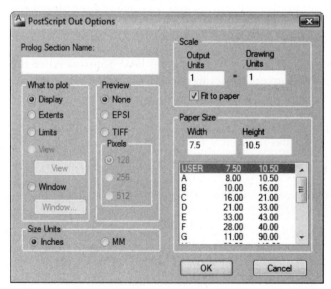

*Figure 23-17 The **PostScript Out Options** dialog box*

Prolog Section Name

In this edit box, you can assign a name for a prolog section to be read from the *acad.psf* file.

What to plot Area

The **What to plot** area of the dialog box has the following options.

Display. If you specify this option when you are in the model space, the display in the current viewport is saved in the specified EPS file. Similarly, if you are in the layouts, the current view is saved in the specified EPS file.

Extents. If you use this option, the PostScript file will contain the section of the AutoCAD drawing that currently holds objects. In this way, this option is similar to the **ZOOM Extents** option. If you add objects to the drawing, they are also included in the PostScript file to be created because the extents of the drawing are also altered. If you reduce the drawing extents by erasing, moving, or scaling objects, then you must use the **ZOOM Extents** or **ZOOM All** option. Only then does the **Extents** option of the **PSOUT** command understand the extents of the drawing to be exported. If you are in the model space, the PostScript file is created in relation to the model space extents; if you are in the paper space, the PostScript file is created in relation to the paper space extents.

Limits. With this option, you can export the whole area specified by the drawing limits. If the current view is not the plan view [viewpoint (0,0,1)], the **Limits** option exports the area just as the **Extents** option would.

View. Any view created with the **VIEW** command can be exported with this option. This option will be available only when you have created a view using the **VIEW** command. Choose the **View** tool to display the **View Name** dialog box from where you can select the view.

Window. In this option, you need to specify the area to be exported with the help of a window. When this radio button is selected, the **Window** button is also available. Choose the **Window** button to display the **Window Selection** dialog box where you can select the **Pick** button and then specify the two corners of the window on the screen. You can also enter the coordinates of the two corners in the **Window Selection** dialog box.

Preview Area

The **Preview** area of the dialog box has two types of formats to preview images: **EPSI** and **TIFF**. If you want a preview image with no format, select the **None** radio button. If you select **TIFF** or **EPSI**, you are required to enter the pixel resolution of the screen preview in the **Pixels** area. You can select a preview image size of 128 X 128, 256 X 256, or 512 X 512.

Size Units Area

In this area, you can set the paper size units to **Inches** or **MM** by selecting their corresponding radio buttons.

Scale Area

In this area, you can set an explicit scale by specifying how many drawing units are to be output per unit. You can select the **Fit to paper** check box so that the view to be exported is made as large as possible for the specified paper size.

Paper Size Area

You can select a size from the list or enter a new size in the **Width** and **Height** edit boxes to specify a paper size for the exported PostScript image.

OBJECT LINKING AND EMBEDDING (OLE)

With Windows, it is possible to work with different Windows-based applications by transferring information between them. You can edit and modify the information in the original Windows application, and then update this information in other applications. This is made possible by creating links between different applications and then updating those links, which in turn update or modify the information in the corresponding applications. This linking is a function of the OLE feature of Microsoft Windows. The OLE feature can also join together separate pieces of information from different applications into a single document. AutoCAD and other Windows-based applications such as Microsoft Word, Notepad, and Windows WordPad support the Windows OLE feature.

For the OLE feature, you should have a source document where the actual object is created in the form of a drawing or a document. This document is created in an application called a **server** application. AutoCAD for Windows and Paintbrush can be used as server applications. Now this source document is to be linked to (or embedded in) the **compound** (destination) document, which is created in a different application, known as the **container** application. AutoCAD for Windows, Microsoft Word, and Windows WordPad can be used as container applications.

Clipboard

The transfer of a drawing from one Windows application to another is performed by copying the drawing or the document from the server application to the Clipboard. The drawing or document is then pasted in the container application from the Clipboard. Hence, a Clipboard is used as a medium for storing the documents while transferring them from one Windows application to another. The drawing or the document on the Clipboard stays there until you copy a new drawing, which overwrites the previous one, or until you exit Windows. You can save the information present on the Clipboard with the *.clp* extension.

Object Embedding

You can use the embedding function of the OLE feature when you want to ensure that there is no effect on the source document even if the destination document has been changed through the server application. Once a document is embedded, it has no connection with the source. Although editing is always done in the server application, the source document remains unchanged. Embedding can be accomplished by means of the following steps. In this example, AutoCAD for Windows is the server application and MS Word is the container application.

1. Create a drawing in AutoCAD (the server application).

2. Open MS Word (container application).

3. It is preferable to arrange both the container and server windows so that both are visible, as shown in Figure 23-18.

4. In AutoCAD, copy the drawing using the CTRL+C keys; the **Copy Clip** tool is invoked. This tool can be used to embed drawings. You can also invoke this tool from the **Edit** menu (Choose **Copy**), by entering **COPYCLIP** at the Command prompt. The next prompt, **Select objects**, allows you to select the entities you want to transfer. You can either select the full drawing by entering **ALL** or select some of the entities by selecting them. You can use any of

the object selection methods for selecting objects. With this command, the selected objects are automatically copied to the Windows Clipboard.

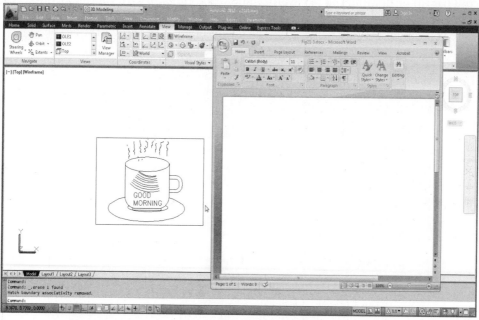

Figure 23-18 *AutoCAD graphics screen with the MS Word window*

5. After the objects are copied to the Clipboard, make the MS Word window active. To get the drawing from the Clipboard to the MS Word application (client), you need to paste it into the Word application. Choose the **Paste Special** tool from **Home > Clipboard > Paste** drop-down (Figure 23-19); the **Paste Special** dialog box is displayed, as shown in Figure 23-20. In this dialog box, select the **Paste** radio button (default) for embedding, and then choose **OK**. The drawing is now embedded in the MS Word window.

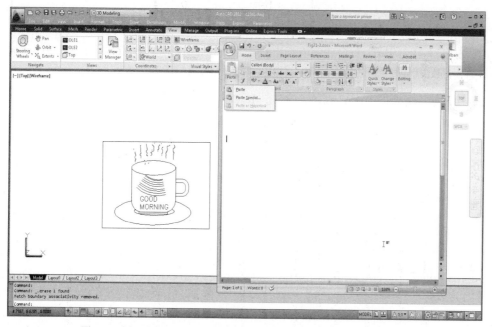

Figure 23-19 *Pasting a drawing into the MS Word application*

6. Your drawing is now displayed in the MS Word window, but it may not be displayed at the proper position. You can get the drawing displayed in the current viewport by moving the scroll bar up or down in the MS Word window. You can also save your embedded drawing by choosing the **Save** button in MS Word. On doing so, the **Save As** dialog box is displayed where you can enter a file name. Save the file and then exit AutoCAD.

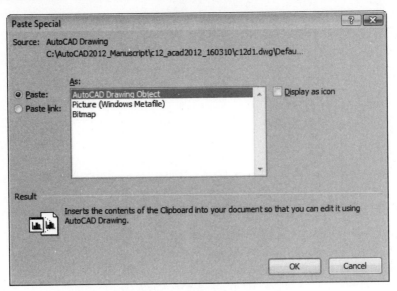

Figure 23-20 The ***Paste Special*** *dialog box*

7. Now, you need to edit the embedded drawing. Editing is performed in the server application, which in this case is AutoCAD. You can get the embedded drawing into the server application (AutoCAD) directly from the container application (MS Word) by double-clicking on the drawing in MS Word. The other method of editing an embedded drawing is by right-clicking on the it and then choosing **AutoCAD Drawing Object > Edit** from the shortcut menu.

8. Now you are in AutoCAD, with your embedded drawing displayed on the screen, but as a temporary file with a file name such as [Drawing in Document]. Here, you can edit the drawing by changing the color and linetype or by adding and deleting the text, entities, and so on. In this example, the cup and plate have been hatched.

9. After you have finished modifying your drawing, choose **File > Update Microsoft Word** from the menu bar in the server (AutoCAD), see Figure 23-21. To display the menu bar, click on the down arrow in the **Quick Access Toolbar** and select **Show Menu Bar**. AutoCAD automatically updates the drawing in MS Word (container application). Now, you can exit this temporary file in AutoCAD.

Note
Do not zoom or pan the drawing in the temporary file. If you do so, this will be included in the updating and the new file will display only that portion of the drawing that lies inside the original area.

10. This completes the embedding function. Now, so you can exit the container application. While exiting it, a message box asking whether or not to save changes in MS Word is displayed.

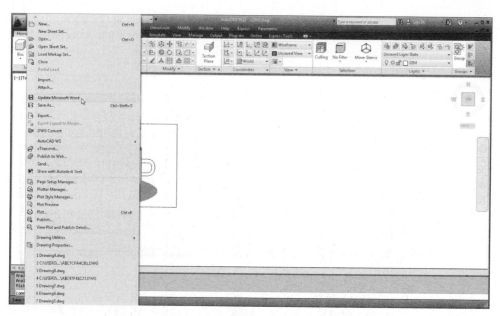

Figure 23-21 *Updating the drawing in AutoCAD*

Linking Objects

The linking function of OLE is similar to the embedding function. The only difference is that here a link is created between the source document and the destination document. If you edit the source, you can simply update the link, which automatically updates the client. This allows you to place the same document in a number of applications, and if you make a change in the source document, the clients will also change by simply updating the corresponding links. Consider AutoCAD for Windows to be the server application and MS Word to be the container application. Linking can be performed by means discussed next.

1. Open a drawing in the server application (AutoCAD). If you have created a new drawing, then you must save it before you can link it with the container application. Then, open MS Word (container application).

2. It is preferable to arrange both the container and the server windows so that both are visible.

3. In AutoCAD, choose **Copy Link** in the **Edit** menu of the menu bar to invoke the **COPYLINK** command. This command can be used for linking the drawing. This command copies all objects that are displayed in the current viewport directly to the Clipboard. Here, you cannot select objects for linking. If you want only a portion of the drawing to be linked, you can zoom into that view so that it is displayed in the current viewport prior to invoking the **COPYLINK** command. This command also creates a new view of the drawing having a name OLE1.

4. Make the MS Word window active. To get the drawing from the Clipboard to the MS Word (container) application, choose the **Paste Special** tool from **Home > Clipboard > Paste** drop-down; the **Paste Special** dialog box will be displayed. In this dialog box, select the **Paste link** radio button for linking. Choose **OK**. Note that the **Paste link** radio button in the **Paste Special** dialog box will not be available if you have not saved the drawing.

5. The drawing is now displayed in the MS Word window and is linked to the original drawing. You can also save your linked drawing by choosing **Save** from the **Quick Access Toolbar** in MS Word. It displays the **Save As** dialog box where you can enter a file name.

Chapter 23

6. You can now edit your original drawing. Editing can be performed in the server application, which, in this case, is AutoCAD for Windows. You can edit the drawing by changing the color and linetype or by adding and deleting text, entities, and so on. Next, save your drawing in AutoCAD by using the **Save** tool. You can now exit AutoCAD.

7. You will notice that the drawing is automatically updated, and the changes made in the source drawing are present in the destination drawing also. If the drawing is not updated automatically, right-click on the drawing and then choose the **Update Link** option from the shortcut menu displayed. The automatic updating is dependent on the selection of the **Automatic** radio button (default) in the **Links** dialog box (Figure 23-22). The **Links** dialog box can be invoked by choosing **Linked AutoCAD Drawing Object > Links** from the shortcut menu displayed on right clicking on the drawing in the Word application.

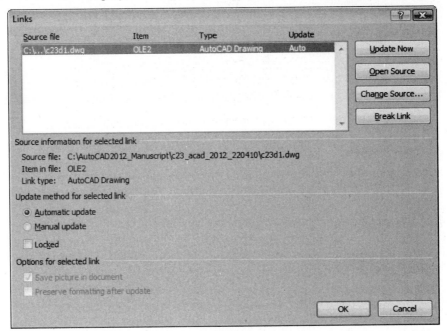

*Figure 23-22 The **Links** dialog box*

8. Exit the container application after saving the updated file.

Note
The dialog box may be different in your case as it depends on the version of MS Word being used currently.

Linking Information to AutoCAD
Similarly, you can also embed and link information from a server application into an AutoCAD drawing. You can also drag selected OLE objects from another application into AutoCAD, provided this application supports Microsoft ActiveX and the application is running and visible on the screen. Dragging and dropping is like cutting and pasting. If you press the CTRL key while you drag the object, it is copied to AutoCAD. Dragging and dropping an OLE object into AutoCAD embeds it into AutoCAD.

Linking Objects to AutoCAD
Start any server application such as MS Word and open a document in it. Select the information you wish to use in AutoCAD with your pointing device and choose the **Copy** tool to copy this data to the Clipboard. Open the AutoCAD drawing to which you want to link this data. Choose

the **Paste Special** tool from **Home > Clipboard > Paste** drop-down or use the **PASTESPEC** command; the **Paste Special** dialog box will be displayed (Figure 23-23). In the **As** list box, select the data format you wish to use. For example, for an MS Word document, select **Microsoft Office Word Document**. Picture format uses a Metafile format. Select the **Paste Link** radio button to paste the contents of the Clipboard into the current drawing. If you select the **Paste** radio button, the data will be embedded and not linked. Choose **OK** to exit the dialog box. Specify the insertion point; the data will be displayed in the drawing.

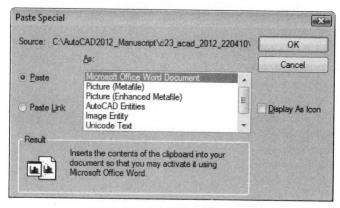

Figure 23-23 The **Paste Special** *dialog box*

You can also invoke the **INSERTOBJ** command by entering **INSERTOBJ** at the Command prompt, or by choosing **OLE Object** from the **Data** panel of the **Insert** tab, or by choosing the **OLE Object** tool in the **Insert** toolbar. This command links an entire file to a drawing from within AutoCAD. On invoking this command, the **Insert Object** dialog box is displayed, as shown in Figure 23-24.

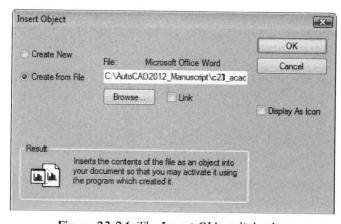

Figure 23-24 The **Insert Object** *dialog box*

Select the **Create from File** radio button. Also, select the **Link** check box. Choosing the **Browse** button displays the **Browse** dialog box. Select a file you want to link from the list box or enter a name in the **File name** edit box and choose the **Open** button. The path of the file is displayed in the **File** edit box of the **Insert Object** dialog box. If you select the **Display As Icon** check box, an icon is displayed in the dialog box as well as in the drawing. Choose **OK** to exit the dialog box; the selected file is linked to the AutoCAD drawing.

Whenever the server document changes, AutoCAD updates links automatically by default. But, you can use the **OLELINKS** command to display the **Links** dialog box (Figure 23-25) where you can change these settings. This dialog box can also be displayed by choosing **OLE Links** from the **Edit** menu. In the **Links** dialog box, select the link you want to update and then choose

Chapter 23

the **Update Now** button. Then choose the **Close** button. If the server file location changes or if it is renamed, you can choose the **Change Source** button in the **Links** dialog box to display the **Change Source** dialog box. In this dialog box, locate the server file name and location and choose the **Open** button. You can also choose the **Break Link** button in the **Links** dialog box to disconnect the inserted information from the server application. This is done when the linking between the two is no longer required.

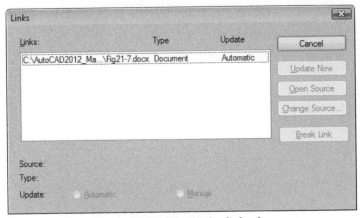

*Figure 23-25 The **Links** dialog box*

Embedding Objects into AutoCAD

Open the server application and select the data you want to embed into the AutoCAD drawing. Copy this data to the Clipboard by choosing the **Copy** tool in the toolbar or **Copy** from the **Edit** menu. Open the AutoCAD drawing and choose **Paste** from the AutoCAD **Edit** menu. You can use the **PASTECLIP** command also. Specify the insertion point; the selected information is embedded into the AutoCAD drawing.

You can also create and embed an object into an AutoCAD drawing starting from AutoCAD itself by using the **INSERTOBJ** command. Alternatively, choose the **OLE Object** tool from the **Data** panel of the **Insert** tab to invoke the **Insert Object** dialog box (Figure 23-26). In this dialog box, select the **Create New** radio button and select the application you wish to use from the **Object Type** list box. Choose **OK**. The selected application opens. Now, you can create the information you wish to insert in the AutoCAD drawing and save it before closing the application. You can edit information embedded in the AutoCAD drawing by double-clicking on the inserted OLE object; the server application will open. The modifications made in the server application get automatically reflected in the AutoCAD drawing.

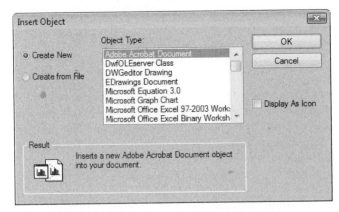

*Figure 23-26 The **Insert Object** dialog box*

Working with OLE Objects

Select an OLE object and right-click to display a shortcut menu. Next, choose **Properties** from the shortcut menu; the **Properties** window will be displayed. Specify a new height value and a new width value in the **Height** and **Width** edit boxes in the **Geometry** area. Enter a value in percentage of the current values in the **Scale height %** and **Scale width %** edit boxes. If you select the **Yes** option in the **Lock Aspect** field, whenever you change the height or the width under **Scale** or **Size**, the respective width or height changes automatically to maintain the aspect ratio. If you want to change only the height or the width, select **No** for this field.

The pointing device can be used to modify and scale an OLE object in the AutoCAD drawing. Selecting the object displays the object frame. The move cursor allows you to select and drag an object to a new location. The middle handle allows you to select a frame and stretch it. It does not scale objects proportionally. The corner handles scale an object proportionally.

Select an OLE object and right-click to display a shortcut menu. Choosing **Clipboard > Cut** removes an object from a drawing and pastes it on the Clipboard. The **Copy** option is used to place a copy of the selected object on the Clipboard and the **Erase** option is used to remove an object from a drawing and does not place it on the Clipboard. Choosing **OLE** displays the **Open**, **Reset**, **Text Size**, and **Convert** options. Choosing **Convert** displays the **Convert** dialog box where you can convert objects from one type to another and choosing **Open** opens an object in the Server application where you can edit and update it in the current drawing. Choosing the **Reset** option restores the selected OLE objects to their original size, that is, the size they had when inserted. Choose the **Text Size** option to resize the text. You can set the draw order by choosing the corresponding option from **Draw Order** in the shortcut menu.

If you want to change the layer of an OLE object, select the object, and then select the required layer from the **Layer** drop-down list in the **Layers** panel.

The **OLEHIDE** system variable controls the display of OLE objects in AutoCAD. The default value is 0, which makes all OLE objects visible. The different values and their effects are as follows:

0   All OLE objects are visible
1   OLE objects are visible in paper space only
2   OLE objects are visible in model space only
3   No OLE objects are visible

The **OLEHIDE** system variable affects both screen display and printing.

CONTENT EXPLORER*

Ribbon Bar: Plug ins > Content > Explore	**Command:** CONTENTEXPLORER

Content Explorer is the latest feature added in AutoCAD 2012. It is an extension of **Design Center**. When you search any content using this tool, the tool searches for the content in local files as well as in Autodesk Seek. Autodesk Seek is a community available online for Autodesk files. To invoke **Content Explorer**, enter **CONTENTEXPLORER** in the command prompt; a window named **Content Explorer** is displayed, as shown in Figure 23-27. In this window, you can add search locations by choosing the **Add Watched Folder** button. You can browse different categories of design files on Autodesk Seek by selecting **Autodesk Seek** from the **Home** drop-down list located on the top left of the window. To find the desired DWG file, double-click on the required category; all available products from different manufacturers will be displayed in the window. Now, double-click on any of the products; the details of the product will be displayed. Choose the **Available Files** button on the right of the thumbnail and then double-click on any of the file; the file will get attached to the cursor as a block and a process indicator will be shown with the thumbnail of the selected file. As the opening process completes, the file will open in a new window.

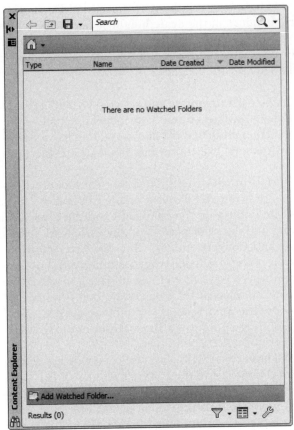

Figure 23-27 *The* **Content Explorer** *window*

Self-Evaluation Test

Answer the following questions and then compare them to those given at the end of this chapter:

1. The DXF file format generates a text file in ASCII code. (T/F)

2. You can directly open a DXF file in AutoCAD by using the **OPEN** command. (T/F)

3. An image attached to AutoCAD cannot be detached. (T/F)

4. A file in the container application is automatically modified only when a link is maintained between the server application and the container application. (T/F)

5. You can import a DXB file into an AutoCAD drawing file by using the _____ command.

6. The _____ system variable controls the display of OLE objects in AutoCAD.

7. If you import a DXB file into an old drawing, the settings of layers, blocks, and so on of the file being imported are _____ by the settings of the file into which you are importing the DXB file.

8. The _____ is used as a medium of storing documents while transferring them from one Windows application to the other.

9. You can get an embedded drawing into the server application directly from the container application by _____ on the drawing.

10. The **COPYLINK** command copies a drawing in the _____ to the Clipboard.

Review Questions

Answer the following questions:

1. In AutoCAD, you can clip the boundary of an image and change it to the desired shape. (T/F)

2. The scanned files can be imported into the current session of AutoCAD using the **DXBIN** command. (T/F)

3. You can export an AutoCAD drawing into 3D Studio MAX. (T/F)

4. Raster images can be attached using the **IMAGEATTACH** command or the **IMAGE** command. (T/F)

5. Which of the following commands can be used to open a DXF format file?

    (a) **OPEN**   (b) **NEW**
    (c) **START**   (d) None of these

6. Which of the following commands is used to detach an attached image file?

    (a) **IMAGE**   (b) **IMAGEATTACH**
    (c) **IMAGECLIP**   (d) **IMAGEADJUST**

7. Which of the following commands is used to modify the brightness of an image?

    (a) **IMAGE**   (b) **IMAGEATTACH**
    (c) **IMAGECLIP**   (d) **IMAGEADJUST**

8. The _____ command is used to modify frames by retaining only the desired portion of images.

9. The _____ command is used to create an ASCII format file with the *.dxf* extension from AutoCAD drawing files.

10. Binary DXF files are _____ efficient and occupy only 75 percent of the ASCII DXF file.

11. File access for files in binary format is _____ than for the same file in ASCII format.

12. If the **OLEHIDE** system variable value is _____, the OLE objects are visible in the paper space only.

13. When you select an image file for attaching it, the _____ dialog box is displayed where you can define the insertion point, scale factor, and rotation angle for the image.

14. The clipping boundary for the raster images can be of _____ or the _____ shape.

EXERCISE 1

In this exercise, you will create a cup and a plate, refer to Figure 23-28 for dimensions. Assume the missing dimensions. Below the cup and plate, enter the following text in MS Word: **These objects are drawn in AutoCAD 2012**. Then using OLE, paste the text into the current drawing.

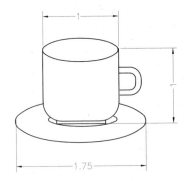

Figure 23-28 Dimensions for Exercise 1

Chapter 24

User Coordinate System

CHAPTER OBJECTIVES

In this chapter, you will learn:
- *About the concept of World Coordinate System (WCS).*
- *About the concept of User Coordinate System (UCS).*
- *To control the display of UCS icon.*
- *To change the current UCS icon type.*
- *To use the UCS command.*
- *To dynamically move and align the UCS*
- *About different options of changing UCS using the UCS command.*
- *To change UCS using the Dynamic UCS button.*
- *To manage UCS through a dialog box using the UCSMAN command.*
- *About different system variables related to the UCS and the UCS icon.*

KEY TERMS

- *UCSICON*
- *UCS*
- *UCSMAN*

THE WORLD COORDINATE SYSTEM (WCS)

As discussed earlier, when you start a new AutoCAD drawing, by default, the world coordinate system (WCS) is established. The objects you have drawn until now use the WCS. In the WCS, the *X*, *Y*, and *Z* coordinates of any point are measured with respect to the fixed origin (0,0,0). By default, this origin is located at the lower left corner of the screen. This coordinate system is fixed and cannot be moved. The WCS is generally used in 2D drawings, wireframe models, and surface models. However, it is not possible to create a solid model keeping the origin and the orientation of the *X*, *Y*, and *Z* axes at the same place. The reason for this is that in case you want to create a feature on the top face of an existing model, you can do it easily by shifting the working plane using the **Elevation** option of the **ELEV** command, see Figure 24-1. But, using the **ELEV** command, it is not possible to create a feature on the faces other than the top and bottom faces of an existing model.

This problem can be solved using a concept called the user coordinate system (UCS). Using the UCS, you can relocate and reorient the origin and *X*, *Y*, and *Z* axes and establish your own coordinate system, depending on your requirement. The UCS is mostly used in 3D drawings, where you may need to specify points that vary from each other along the *X*, *Y*, and *Z* axes. It is also useful for relocating the origin or rotating the *X* and *Y* axes in 2D work, such as ordinate dimensioning, drawing auxiliary views, or controlling the hatch alignment. The UCS and its icon can be modified using the **UCSICON** and **UCS** commands. After reorienting the UCS, you can create a feature on any face of an existing model; see Figure 24-2.

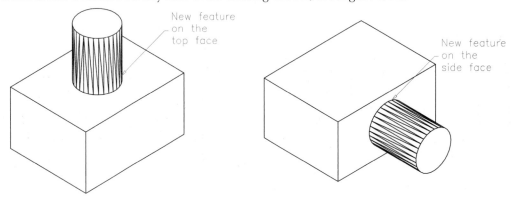

Figure 24-1 *Creating a new feature on the top face*

Figure 24-2 *Creating a new feature on the side face*

CONTROLLING THE VISIBILITY OF THE UCS ICON*

Ribbon: View > Coordinates > Show UCS Icon / Show UCS Icon at Origin/Hide UCS Icon
Menu Bar: View > Display > UCS Icon   **Command:** UCSICON

This command is used to control the visibility and location of the UCS icon, which is a geometric representation of the directions of the current *X*, *Y*, and *Z* axes. AutoCAD displays different UCS icons in model space and paper space, as shown in Figures 24-3 and 24-4. By default, the UCS icon is displayed near the bottom left corner of the drawing area. You can change the location and visibility of this icon using the **UCSICON** command. The prompt sequence for the **UCSICON** command is given next.

> Enter an option [ON/OFF/All/Noorigin/ORigin/Selectable/Properties] <ON>: *You can specify any option or press ENTER to accept the default option.*

ON

This option is used to display the UCS icon on the screen. You can also display the UCS icon by choosing the **Show UCS Icon** button from the **Coordinates** panel in the **View** tab.

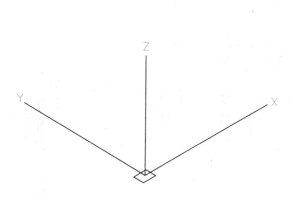
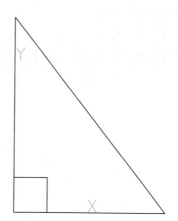

Figure 24-3 *The UCS icon in the Model space*   **Figure 24-4** *The UCS icon in the Paper space*

OFF

This option is used to make the UCS icon invisible from the screen. When you choose this option, the UCS icon will no longer be displayed on the screen. You can again turn on the display using the **On** option of the **UCSICON** command. Alternatively, choose the **Hide UCS Icon** button from the **Coordinates** panel in the **View** tab to make the UCS icon invisible.

All

This option is used to apply changes to the UCS icon in all the active viewports. If this option is not used, the changes will be applied only to the current viewport.

Noorigin

This option is used to display the UCS icon at the lower left corner of the viewport, irrespective of the actual location of the origin of the current UCS. Alternatively, choose the **Show UCS Icon at Origin** button from the **Coordinates** panel in the **View** tab.

ORigin

This option is used to place the UCS icon at the origin of the current UCS.

Selectable

This option allows you to control the selection of UCS. By default, UCS is selectable.

Properties

When you invoke this command, the **UCS Icon** dialog box will be displayed, as shown in Figure 24-5. You can also display this dialog box by choosing the **UCS Icon, Properties** tool from the **Coordinates** panel in the **View** tab. The options in this dialog box are discussed next.

UCS icon style Area

2D. If this radio button is selected, the 2D UCS icon will be displayed instead of the 3D UCS icon, see Figure 24-6.

3D. If this radio button is selected, the 3D UCS icon will be displayed. This is the default option.

Line width. This drop-down list provides the width values that can be assigned to the 3D UCS icon. The default value for the line width is 1. This drop-down list will not be available if the **2D** radio button is selected.

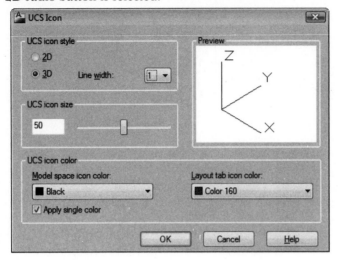

Figure 24-5 The **UCS Icon** dialog box

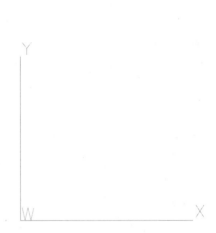

Figure 24-6 2D UCS icon at the World position

UCS icon color Area

The **UCS icon color** area in this dialog box has two drop-down lists, **Model space icon color** and **Layout tab icon color**. The options of these drop-down lists are used to change the color of the UCS icon. By default, the color in the Model space is black and in the Layout is blue. You can assign any color to the UCS icon. By default, there are seven colors in these drop-down lists. However, you can also select a color from the **Select Color** dialog box which will be displayed after you select **Select Color** from the **Model space icon color** drop-down list or the **Layout tab icon color** drop-down list.

DEFINING THE NEW UCS*

Ribbon: View > Coordinates > UCS	Toolbar: UCS
Menu Bar: Tools > New UCS	Command: UCS

The **UCS** command is used to set a new coordinate system by shifting the working plane (*XY* plane) to the desired location. For certain views of the drawing, it is better to have the origin of measurements at some other point on or relative to your drawing objects. This makes locating the features and dimensioning the objects easier. The change in the UCS can be viewed by the change in the position and orientation of the UCS icon, which is placed by default at the lower left corner of the drawing window. The origin and orientation of a coordinate system can be redefined by using the **UCS** command. Alternatively, choose the **UCS** tool from the **Coordinates** panel. The prompt sequence is as follows:

Command: **UCS**
Current ucs name: *WORLD*
Specify origin of UCS or [Face/NAmed/OBject/Previous/View/World/X/Y/Z/ZAxis]
<World>: *Select an option*.

If the **UCSFOLLOW** system variable is set to 0, any change in the UCS will not affect the drawing view.

If you choose the default option of the **UCS** command, you can establish a new coordinate system by specifying a new origin point, a point on the positive side of the new *X* axis, and a point on the positive side of the new *Y* axis. The direction of the *Z* axis is determined by applying the right-hand rule, about which you will learn in the next chapter. This option changes the orientation of the UCS to any angled surface. The prompt sequence that will follow is given next.

Specify origin of UCS or [Face/NAmed/OBject/Previous/View/World/X/Y/Z/ZAxis] <World>: *Specify the origin point of the new UCS.*
Specify point on X-axis or <accept>: *Specify a point on the positive portion of the X axis.*
Specify point on the XY plane or <accept>: *Specify a point on the positive portion of the Y axis to define the orientation of the UCS completely.*

In AutoCAD 2012, you can directly manipulate UCS as per your requirement. This implies that, you can easily move the origin of UCS, align the UCS with objects, or rotate it around X, Y, and Z axis without invoking the UCS command. You can also move the UCS and place it at the desired location without using any command. To do so, select the UCS; the grips are displayed on it. Place the cursor on the rectangular grip displayed at the intersection of three axes of UCS; a shortcut menu will be displayed, as shown in Figure 24-7.

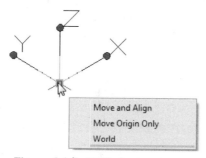

Figure 24-7 UCS shortcut menu

Choose the **Move and Align** option from the shortcut menu; the UCS will be displayed attached to the cursor. You can move the UCS to any point but you can align it only to the face of a 3D object, surface, or mesh. To align the UCS, move it to any point on the face of 3D object, surface, or mesh; the face gets highlighted and the UCS automatically aligns to the orientation of that face, surface, or mesh, as shown in Figure 24-8. In case of a curved surface, the UCS will get aligned in such a way that the Z-axis becomes normal to the surface at the specified point. After aligning the UCS to the selected face, you can change the direction of X axis, Y axis or Z axis dynamically. You can also use the shortcut menu displayed on placing the cursor on the X, Y, or Z grip.

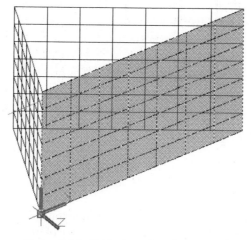

Figure 24-8 UCS aligned to the highlighted face

If you choose **World** from the UCS shortcut menu, then the UCS will move and align to the World Coordinate System. The world coordinate system is discussed next.

Note

*Depending upon the type of UCS required, various commands can be invoked by choosing the corresponding tools from the **Coordinates** panel in the **View** tab.*

A null response to the point on the *X* or *Y* axis prompt will lead to a coordinate system, in which the *X* or *Y* axis of the new UCS is parallel to that of the previous UCS.

W (World) Option

With this option, you can set the current UCS back to the WCS, which is the default coordinate system. When the UCS is placed at the world position, a small rectangle is displayed at the point where all the three axes meet in the UCS icon, see Figure 24-9 (a). If the UCS is moved from its default position, this rectangle is no longer displayed, indicating that the UCS is not at the world position, as shown in Figure 24-9(b).

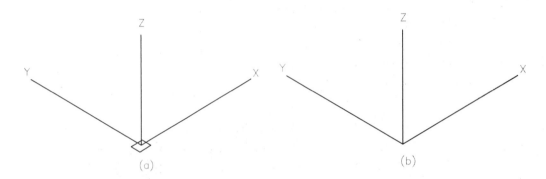

Figure 24-9 UCS at the world position and UCS not at the world position

Tip

In case you have selected the 2D UCS instead of the 3D UCS icon, the W will not be displayed, if the UCS is not at the world position.

F (Face) Option

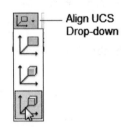

This option is used to align a new UCS with the selected face of a solid object. You can also invoke this option by choosing the **Face** tool from the **Align UCS** drop-down in the **Coordinates** panel, as shown in Figure 24-10. The prompt sequence that will follow when you choose this tool is given next.

Select face of solid object: *Select the face to align the UCS.*
Enter an option [Next/Xflip/Yflip] <accept>: *Select an option or accept the selected face to align.*

*Figure 24-10 The **Face** tool in the **Align UCS** drop-down*

The **Next** option is used to locate the new UCS on the adjacent face or the back face of the selected edge. **Xflip** rotates the new UCS by 180-degree about the *X* axis and **Yflip** rotates it about the *Y* axis. Pressing ENTER at the **Enter an option [Next/Xflip/Yflip] <accept>** accepts the location of the new UCS as specified.

OB (OBject) Option

With the **OB (OBject)** option of the **UCS** command, you can establish a new coordinate system by selecting a object in a drawing. However, some of the objects such as 3D polyline,

3D mesh, viewport object, or xline cannot be used for defining a UCS. The positive *Z* axis of the new UCS is in the same direction as the positive *Z* axis of the object selected. If the *X* and *Z* axes are given, the new *Y* axis is determined by the right-hand rule. You can also invoke this invoke this option by choosing the **Object** tool from the **Align UCS** drop-down in the **Coordinates** panel. The prompt sequence that will follow when you choose this tool is given next.

Select object to align UCS: *Select the object to align the UCS.*

In Figure 24-11, the UCS is relocated using the **OBject** option and is aligned to the circle. The origin and the *X* axis of the new UCS are determined by the following rules.

Arc.

When you select an arc, its center becomes the origin for the new UCS. The *X* axis passes through the endpoint of the arc that is closest to the point selected on the object.

Circle/Cylinder/Ellipse.

The center of the circle becomes the origin of the new UCS, and the *X* axis passes through the point selected on the object (Figure 24-12).

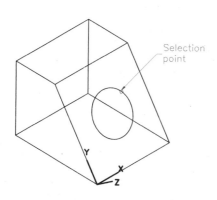

Figure 24-11 *Relocating the UCS using a circle*   **Figure 24-12** *UCS at a new location*

Line/Mline/Ray/Leader.

The new UCS origin is the endpoint of the line nearest to the point selected on the line. The *X* axis is defined so that the line lies on the *XY* plane of the new UCS. Therefore, in the new UCS, the *Y* coordinate of the second endpoint of the line is 0.

Note
The linear edges of solid models or regions are considered as individual lines when selected to align a UCS.

Spline.

The origin of the new UCS is the endpoint of the spline that is nearest to the point selected on the spline. An imaginary line will be drawn between the two endpoints of the spline and the *X* axis will be aligned along this imaginary line.

Trace.

The origin of the new UCS is the "**start point**" of the trace. The new *X* axis lies along the direction of the selected trace.

Dimension.

The middle point of the dimension text becomes the new origin. The *X* axis direction is identical to the direction of the *X* axis of the UCS that existed when the dimension was drawn.

Point.

The position of the point is the origin of the new UCS. The directions of the *X*, *Y*, and *Z* axes will be same as those of the previous UCS.

Solid.

The origin of the new UCS is the first point of the solid. The *X* axis of the new UCS lies along the line between the first and second points of the solid.

2D Polyline.

The start point of the polyline or polyarc is treated as the new UCS origin. The *X* axis extends from the start point to the next vertex.

3D Face.

The first point of the 3D face determines the new UCS origin. The *X* axis is determined by using the first two points, and the positive side of the *Y* axis is determined by the first and fourth points for a flat rectangular face. For a flat planar face, the X axis will always be parallel to the edge through which the UCS is moved on the face. The *Z* axis is determined by applying the right-hand rule.

Shape/Text/Insert/Attribute/Attribute Definition.

The insertion point of the object becomes the new UCS origin. The new *X* axis is defined by the rotation of the object around its positive *Z* axis. Therefore, the object you select will have a rotation angle of zero in the new UCS.

Tip
Except for 3D faces, the XY plane of the new UCS will be parallel to the XY plane existing when the object was drawn; however, X and Y axes may be rotated.

V (View) Option

The **V** (**View**) option of the **UCS** command is used to define a new UCS whose *XY* plane is parallel to the current viewing plane. The current viewing plane, in this case is the screen of the monitor. Therefore, a new UCS is defined that is parallel to the screen of the monitor. The origin of the UCS defined in this option remains unaltered. This option is used mostly to view a drawing from an oblique viewing direction or to write text for the objects on the screen. You can also invoke this option by choosing the **View** tool from the **Align UCS** drop-down in the **Coordinates** panel. As soon as you choose this tool, a new UCS is defined parallel to the screen of the monitor.

X/Y/Z Options

With these options, you can rotate the current UCS around a desired axis. You can specify the angle by entering the angle value at the required prompt or by selecting two points on the screen with the help of a pointing device. You can specify a positive or a negative angle. The new angle is taken relative to the *X* axis of the existing UCS. The **UCSAXISANG** system variable stores the default angle by which the UCS is rotated around the specified axis, by using the **X**/ **Y**/ **Z** options of the **New** option of the **UCS** command. The right-hand thumb rule is used to determine the positive direction

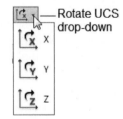

Figure 24-13 Tools in the Rotate UCS drop-down

of rotation of the selected axis. You can also invoke the corresponding option by choosing the **X/Y/Z** button from the **Rotate UCS** drop-down in the **Coordinates** panel, as shown in Figure 24-13. However, in AutoCAD 2012, you can rotate the UCS dynamically using the grips and shortcut menu.

X Option

In Figure 24-14, the UCS is rotated using the **X** option by specifying an angle about the X axis. The first model shows the UCS setting before the UCS was relocated and the second model shows the relocated UCS. The prompt sequence that will follow when you choose the **X** tool from the **Rotate UCS** drop-down in the **Coordinates** panel is given next.

Specify rotation angle about X axis <90>: *Specify the rotation angle.*

Alternatively, you can dynamically rotate the UCS about the X axis by following the steps given below:

1. Click on the UCS in the drawing area to select it.
2. Move the cursor to the grip of the Y or Z axis, as shown in Figure 24-15.
3. Choose the **Rotate Around X Axis** option from the shortcut menu and then rotate the UCS dynamically. You can also enter the desired rotation angle in the edit box attached with the cursor.

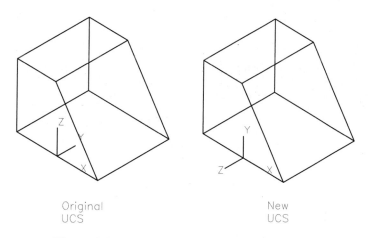

Figure 24-14 *Rotating the UCS about the X axis*

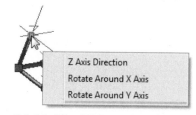

Figure 24-15 *The UCS rotation shortcut menu*

Y Option

In Figure 24-16, the UCS is rotated using the **Y** option by specifying an angle about the Y axis. The first model shows the UCS setting before the UCS was relocated and the second model shows the relocated UCS. The prompt sequence that will follow when you choose the **Y** tool from the **Rotate UCS** drop-down is given next.

> Specify rotation angle about Y axis <90>: *Specify the angle.*

Alternatively, you can dynamically rotate the UCS about the Y axis by following the steps given below:

1. Click on the UCS in the drawing area to select it.
2. Move the cursor to the grip of the X or Z axis.
3. Choose the **Rotate Around Y Axis** option from the shortcut menu and rotate the UCS dynamically or specify the desired angle.

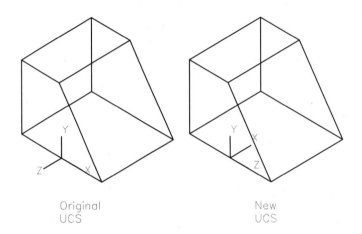

Original
UCS

New
UCS

Figure 24-16 Rotating the UCS about the Y axis

Z Option

In Figure 24-17, the UCS is rotated using the **Z** option by specifying an angle about the Z axis. The first model shows the UCS setting before the UCS was rotated and the second model shows the rotated UCS. The prompt sequence that will follow when you choose the **Z** tool from the **Rotate UCS** drop-down is given next.

> Specify rotation angle about Z axis <90>: *Specify the angle.*

Alternatively, you can dynamically rotate the UCS about the Z axis by following the steps given below:

1. Click on the UCS in the drawing area to select it.
2. Move the cursor to the grip of the X or Y axis.
3. Choose the **Rotate Around Z Axis** option from the shortcut menu and rotate the UCS dynamically or specify the desired rotation angle.

Note
*You can also rotate the UCS about any axis by using the options in the shortcut menu displayed on right-clicking on the UCS. To do so, move the cursor on the **Rotate Axis** sub-menu in the shortcut menu and then choose the required option.*

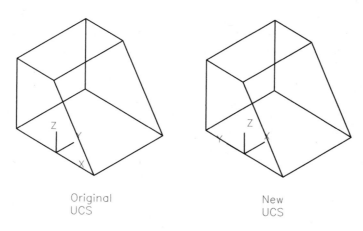

Figure 24-17 Rotating the UCS about the Z axis

ZAxis Option

This option is used to change the coordinate system by selecting the origin point of the *XY* plane and a point on the positive *Z* axis. After you specify a point on the *Z* axis, AutoCAD determines the *X* and *Y* axes of the new coordinate system accordingly. The prompt sequence that will follow when you choose the **Z Axis Vector** tool from the **Coordinates** panel is given next.

> Specify new origin point or [Object] <0,0,0>: *Specify the origin point, as shown in Figure 24-18.*
> Specify point on positive portion of Z-axis <default>: **@ 0,-1,0**

Now, the front face of the model will become the new work plane (Figure 24-19) and all new objects will be oriented accordingly. If you give a null response to the **Specify point on positive portion of Z axis <current>** prompt, the *Z* axis of the new coordinate system will be parallel to (in the same direction as) the *Z* axis of the previous coordinate system. Null responses to the origin point and the point on the positive *Z* axis establish a new coordinate system, in which the direction of the *Z* axis is identical to that of the previous coordinate system; however, the *X* and *Y* axes may be rotated around the *Z* axis. The positive *Z* axis direction is also known as the extrusion direction.

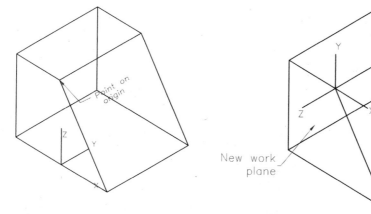

Figure 24-18 Specifying a point on the origin

*Figure 24-19 Relocating the UCS using the **ZA** option*

When the **Object** option is selected at the **Specify new origin point or [Object] <0,0,0>** prompt, then you will be prompted to select an object. Select an open object and the UCS will be placed at the end point nearest to the point of the object selection, with its Z-axis along the tangent direction at that end point and pointing away from the object.

Previous Option

The **Previous** option is used to restore the current UCS settings to the previous UCS settings. AutoCAD saves the last ten UCS settings. You can go back to the previous ten UCS settings in the current space using the **Previous** option. If the **TILEMODE** is off, the last ten coordinate systems in paper space and in model space are saved. You can also invoke this option by choosing the **UCS, Previous** tool from the **Coordinates** panel. When you choose this tool, the previous UCS settings are automatically restored.

NAmed Option

This option is used to name and save the current UCS settings, restore a previously saved UCS setting, view a previously saved UCS list, and delete the saved UCS from the list. The prompt sequence for this option is given next.

Command: **UCS**
Current ucs name: *WORLD*
Specify origin of UCS or [Face/NAmed/OBject/Previous/View/World/X/Y/Z/ZAxis] <World>: **NA**
Enter an option [Restore/Save/Delete/?]: *Enter an option.*

Depending on your requirement, you can select any one of the following options:

Restore Option

The **Restore** option of the **UCS** command is used to restore a previously saved named UCS. Once it is restored, it becomes the current UCS. However, the viewing direction of the saved UCS is not restored. You can also restore a named UCS by selecting it from the **UCS** dialog box that appears when you invoke the **UCS, UCS Named** tool from the **Coordinates** panel. As this option does not have a button to restore the Named UCS, you need to choose this option by using the **UCS** command at the Command prompt. The prompt sequence that will follow when you invoke the **UCS** command is given next.

Enter an option [Restore/Save/Delete/?]: *R*
Enter name of UCS to restore or [?]: *Specify the name of UCS or ENTER to list all the UCS*

You can specify the name of the UCS to be restored or list the UCSs that can be restored by pressing ENTER at the previous prompt. The prompt sequence that will be followed is given next.

Enter UCS name(s) to list <*>: *Specify the name of the UCS to list or give a null response to list all available UCSs.*

If you give a null response at the previously mentioned prompt, then the AutoCAD text window will be opened listing all the available UCSs.

Save Option

With this option, you can name and save the current UCS settings. The following points should be kept in mind while naming the UCS:

1. The name can be up to 255 characters long.

2.   The name can contain letters, digits, blank spaces and the special characters $ (dollar), - (hyphen), and _ (underscore).

As this option does not have a tool, you can invoke this option using the Command prompt. The prompt sequence that will follow when you invoke the **UCS** command is given next.

> Current ucs name: *WORLD*
> Specify origin of UCS or [Face/NAmed/OBject/Previous/View/World/X/Y/Z/ZAxis]: **NA**
> Enter an option [Restore/Save/Delete/?]: **S**
> Enter name to save current UCS or [?]: *Specify a name to the UCS or press ENTER to list.*

Enter a valid name for the UCS at this prompt. AutoCAD saves it as a UCS. You can also list the previously saved UCSs by entering **?** at this prompt. The next prompt sequence is:

> Enter UCS name(s) to list <*>:

Enter the name of the UCS to list or give a null response to list all the available UCSs.

Delete Option

The **Delete** option is used to delete the selected UCS from the list of saved coordinate systems. This option also does not have a button. As this option does not have a button, you can invoke this option using the Command prompt. The prompt sequence that will follow when you invoke the **UCS** command is given next.

> Current ucs name: *WORLD*
> Specify origin of UCS or [Face/NAmed/OBject/Previous/View/World/X/Y/Z/ZAxis]: **NA**
> Enter an option [Restore/Save/Delete/?]: **D**
> Enter UCS name(s) to delete <none>: *Specify the name of the UCS to delete.*

The UCS name you enter at this prompt is deleted. You can delete more than one UCS by entering the UCS names separated with commas or by using wild cards.

? Option

By invoking this option, you can list the name of the specified UCS. This option gives you the name, origin, and X, Y, and Z axes of all the coordinate systems relative to the existing UCS. If the current UCS has no name, it is listed as WORLD or UNNAMED. The choice between these two names depends on whether the current UCS is the same as the WCS. The prompt sequence that will be displayed when you invoke this option using the Command prompt is given next.

> Current ucs name: *WORLD*
> Specify origin of UCS or [Face/NAmed/OBject/Previous/View/World/X/Y/Z/ZAxis]: **NA**
> Enter an option [Restore/Save/Delete/?]: **?**
> Enter UCS name(s) to list <*>:

Apply Option

The **Apply** option applies the current UCS settings to a specified viewport or to all the active viewports in a drawing session. If the **UCSVP** system variable is set to 1, each viewport saves its UCS settings. The **Apply** option is not displayed at the Command prompt as an option for the **UCS** command. Therefore, you have to choose the **Apply** button from the **UCS** toolbar. The prompt sequence that will follow when you choose this button is given next.

> Pick viewport to apply current UCS or [All]<current>: *Pick inside the viewport or enter **A** to apply the current UCS settings to all viewports.*

Origin Option

 This option is used to define a new UCS by changing the origin of the current UCS, without changing the directions of the X, Y, Z axes, see Figures 24-20 and 24-21. The new point defined becomes the origin point (0,0,0) for all the coordinate entries from this point on, until the origin is changed again. You can specify the coordinates of the new origin or click on the screen using the pointing device. Alternatively, choose the **Origin** tool from the **Coordinates** panel to specify the coordinates of the new origin. The prompt sequence that will follow when you choose this tool is given next.

Specify new origin point <0,0,0>: *Specify the origin point, as shown in Figure 24-20.*

Tip
If you do not provide the Z coordinate for the origin, the current Z coordinate value will be assigned to the UCS.

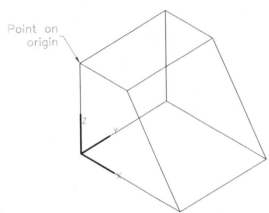

Figure 24-20 *Defining a new origin for the UCS*   **Figure 24-21** *Relocating the UCS to the new origin*

3 (3-Point) Option

The **3-Point** option is the default option of the **UCS** command, which changes the orientation of the UCS to any angled surface. With this option, you can establish a new coordinate system by specifying a new origin point, a point on the positive side of the new X axis, and a point on the positive side of the new Y axis (Figure 24-22). The direction of the Z axis is determined by applying the right-hand rule about which you will learn in the next chapter. The prompt sequence that will follow when you choose the **3-Point** tool from the **Coordinates** panel is given next.

Specify new origin point <0,0,0>: *Specify the origin point of the new UCS.*
Specify point on X-axis <or accept>: *Specify a point on the positive portion of the X axis.*
Specify point on XY plane <or accept>: *Specify a point on the positive portion of the Y axis.*

A null response to the **Specify new origin point <0,0,0>** prompt will lead to a coordinate system in which the origin of the new UCS is identical to that of the previous UCS. Similarly, null responses to the point on the X or Y axis prompt will lead to a coordinate system, in which the X or Y axis of the new UCS is parallel to that of the previous UCS. In Figure 24-23, the UCS has been relocated by specifying three points (the origin point, a point on the positive portion of the X axis, and a point on the positive portion of the Y axis).

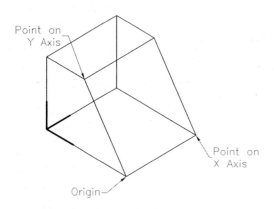

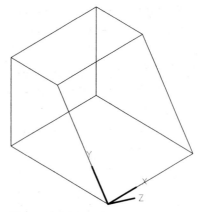

Figure 24-22 *Relocating the UCS using the 3-Point option*

Figure 24-23 *UCS at a new position*

EXAMPLE 1   3-Point

In this example, you will draw a tapered rectangular block. After drawing it, you will align the UCS to the inclined face of the block by using the dynamic alignment. Next, you will draw a circle on the inclined face. The dimensions for the block and the circle are shown in Figure 24-24.

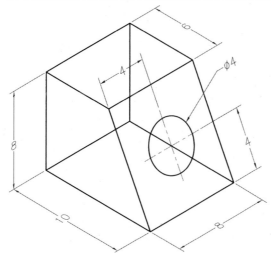

Figure 24-24 *Model for Example 1*

The steps to do this example are given next.

- Draw the edges of the bottom face of the tapered block.
- Specify a new origin.
- Draw the edges of the top face of the tapered block.
- Draw the remaining edges of the block.
- Align the UCS to the inclined face.
- Draw the circle on the inclined face.

These steps are discussed next.

1. Start a new drawing and choose the **Line** tool. The prompt sequence is given next.

    Specify first point: **1,1**
    Specify next point or [Undo]: **@10,0**

Specify next point or [Undo]: **@0,8**
Specify next point or [Close/Undo]: **@-10,0**
Specify next point or [Close/Undo]: **C**

2. As the top face of the model is at a distance of 8 units from the bottom face in the positive Z direction, you need to define a UCS at a distance of 8 units to create the top face. To do so, select the UCS and then move the cursor to the rectangular grip displayed at the intersection of the three axes; a shortcut menu will be displayed. Choose the **Move Origin Only** option; AutoCAD will prompt you to specify a new origin point. The prompt sequence is as follows:

Specify the origin point : **0,0,8**

3. Choose the **Line** tool from the **Draw** panel. The prompt sequence is as follows:

Specify first point: **1,1**
Specify next point or [Undo]: **@6,0**
Specify next point or [Undo]: **@0,8**
Specify next point or [Close/Undo]: **@-6,0**
Specify next point or [Close/Undo]: **C**

4. When you open a new drawing, you view the model from the top view by default. Therefore, when viewing from the top, the three edges of the top side overlap with the corresponding bottom edges. As a result, you will not be able to see all edges. However, you can change the viewpoint to clearly view the model in 3D. Choose **SE Isometric** from the **View Controls** list in the **In-Canvas Viewport Controls** to orient the view to the SE isometric view. Now, you can see the 3D view of the objects, see Figure 24-25.

5. Join the remaining edges of the model using the **Line** tool. The model after joining all edges should look similar to the one shown in Figure 24-26.

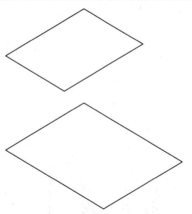 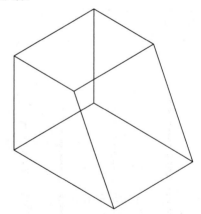

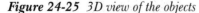

Figure 24-25 *3D view of the objects*   *Figure 24-26* *Model after joining all the edges*

Now, dynamic alignment of UCS can be done only with a surface, a mesh, and a solid face but the inclined portion is a wire frame. So, you need to convert the wire frame into a surface by using the **REGION** command.

6. Invoke the **REGION** command; AutoCAD will prompt you to select objects for region. Next, select the four edges bounding the inclined face and press ENTER; a region will be generated.

7. Next, select the UCS and place the cursor on the rectangular grip; a shortcut menu will be displayed. Choose the **Move and Align** option from the shortcut menu; you will be prompted to select a face or origin point.

8. Select the inclined face by moving through the edge to which you want the X axis to be parallel or you can align the axes by dynamically rotating them. In this case, the X axis is parallel to P1P2 line; *refer to Figure 24-27.*

9. Choose the **Center, Radius** tool from the **Circle** drop-down in the **Draw** panel and draw the circle on the inclined face.

The final model should look similar to the one shown in Figure 24-28.

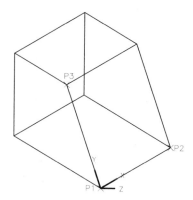
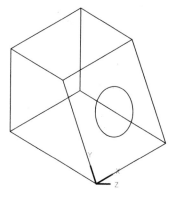

Figure 24-27 *Model after aligning the UCS on the inclined face*   **Figure 24-28** *Final model for Example 1*

MANAGING THE UCS THROUGH THE DIALOG BOX

Ribbon: View > Coordinates > UCS, Named UCS   **Command:** UCSMAN
Menu Bar: Tools > Named UCS

In AutoCAD, you can save and restore the UCS by using the **UCS, Named UCS** tool from the **Coordinates** panel. Choosing this tool displays the **UCS** dialog box (Figure 24-29). This dialog box is used to restore the saved and orthographic UCSs, specify the UCS icon settings, and rename UCSs. This dialog box has three tabs: **Named UCSs**, **Orthographic UCSs**, and **Settings**.

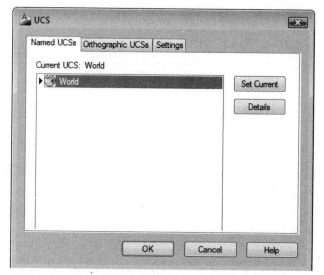

Figure 24-29 *The **Named UCSs** tab of the **UCS** dialog box*

Named UCSs Tab

The list of all coordinate systems defined (saved) on your system is displayed in the list box of the **View** tab. The first entry in this list is **World**, which means WCS. The next entry will be **Previous**, if you have defined any other coordinate systems in the current editing session. Selecting the **Previous** entry and then choosing the **OK** button repeatedly allow you to go backward through the coordinate systems defined in the current editing session. If you have created a new UCS as the current coordinate system and not named it, then **Unnamed** will be the first entry in the list. Double-click on the name and rename it. Else, the unnamed UCS will disappear on selecting other UCS. If there are a number of viewports and unnamed settings, only the current viewport UCS name will be displayed in the list. The current coordinate system is indicated by a small pointer icon to the left of the coordinate system's name. The name of the current UCS is also displayed next to **Current UCS** on top of the **UCS** dialog box. To make other coordinate system current, select its name in the **UCS Names** list, choose the **Set Current** button, and choose **OK**. Alternatively, select the required UCS from the **Named UCS Combo Control** drop-down list available in the coordinate panel of the **View** tab. To delete a coordinate system created previously, select its name, and then right-click to display a shortcut menu. Choose the **Delete** button to delete the selected UCS name. To rename a coordinate system, select its name and then right-click to display a shortcut menu and choose **Rename**. Now, enter the new name. You can also double-click on the name you need to modify, and then change the name. The other options in the shortcut menu are **Set Current** and **Details**.

Tip
*All the changes and updating of the UCS information in the drawing are carried out only after you choose the **OK** button.*

If you want to check the current coordinate system's origin and X, Y, and Z axis values, select a UCS from the list and then choose the **Details** button; the **UCS Details** dialog box (Figure 24-30) containing that information is displayed. Alternatively, right click on the name of a specific UCS in the list box and then choose the **Details** option from the shortcut menu.

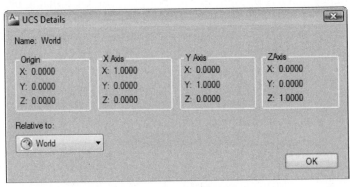

*Figure 24-30 The **UCS Details** dialog box*

Setting the UCS to Preset Orthographic UCSs Using the Orthographic UCSs Tab

Different orthographic views are listed in the **Orthographic UCSs** tab, as shown in Figure 24-31. You can select any one of the six orthographic views of the WCS or the user defined coordinate system using the **UCS** dialog box. To define an orthographic coordinate system relative to a UCS, select it from the **Relative to** drop-down list, select the orthographic view from the list of preset orthographic views, and then choose the **Set Current** button; the UCS icon is placed at the corresponding view. You can also right-click on the specific orthographic UCS name in the list and choose **Set Current** from the shortcut menu. The name of the current UCS is displayed above the list of orthographic view. If the current settings have not been saved and named, the **Current UCS** name displayed is **Unnamed**.

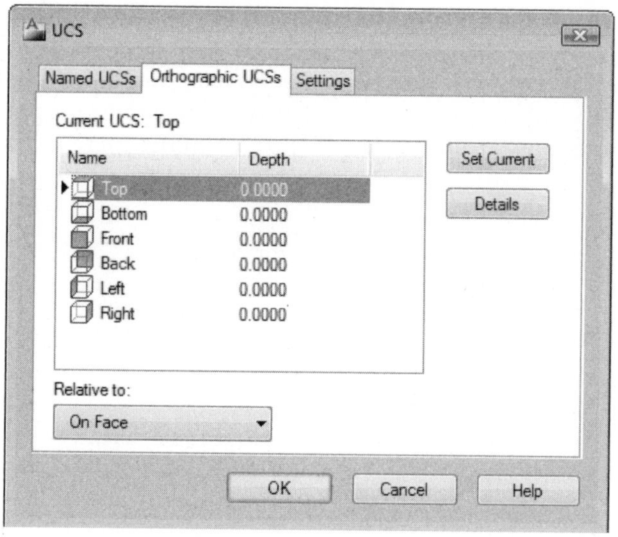

Figure 24-31 *The **Orthographic UCSs** tab of the **UCS** dialog box*

The **Depth** field in the list box of this dialog box displays the distance between the *XY* plane of the selected orthographic UCS setting and a parallel plane passing through the origin of the UCS base setting. The **UCSBASE** system variable stores the name of the UCS, which is considered as the base setting; that is, it defines the origin and orientation. You can enter or modify the values of depth by double-clicking on the depth value of the selected UCS in the list box. On doing so, the **Orthographic UCS depth** dialog box (Figure 24-32) will be displayed, where you can enter new depth values. You can also right-click on a specific orthographic UCS in the list box to display a shortcut menu. Choose **Depth** to display the **Orthographic UCS depth** dialog box. Enter a value in the edit box. If you need to pick a position from the drawing area, choose the **Select new origin** button available next to the edit box; the dialog box will disappear, allowing you to specify a new origin or depth on the screen.

Figure 24-32 *The **Orthographic UCS depth** dialog box*

Choosing the **Details** button displays the **UCS Details** dialog box with the origin and the *X, Y, Z* coordinate values of the selected UCS. You can also choose **Details** from the shortcut menu that is displayed on right-clicking on a UCS in the list box. The shortcut menu also has a **Reset** option. Choosing **Reset** from the shortcut menu restores the selected UCS to its default settings. Also, the depth value, if assigned, becomes zero.

Settings Tab

The **Settings** tab of the **UCS** dialog box (Figure 24-33) is used to display and modify the UCS and UCS icon settings of a specified viewport. If you choose the inclined arrow on the **Coordinates** panel, the **UCS** dialog box will be displayed with the **Settings** tab chosen. The options in this dialog box are discussed next.

UCS Icon settings Area

Selecting the **On** check box displays the UCS icon in the current viewport. This process is similar to using the **UCSICON** command and enables you to set the display of the UCS icon to on or

Chapter 24

off. If you select the **Display at UCS origin point** check box, the UCS icon is displayed at the origin point of the coordinate system in use in the current viewport. If the origin point is not visible in the current viewport, or if the check box is cleared, the UCS icon is displayed in the lower left corner of the viewport. Selecting the **Apply to all active viewports** check box applies the current UCS icon settings to all the active viewports in the current drawing.

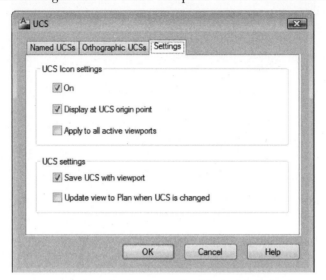

*Figure 24-33 The **Settings** tab of the **UCS** dialog box*

UCS settings Area

This area of the **Settings** tab in the **UCS** dialog box specifies the UCS settings for the current viewport. Selecting the **Save UCS with viewport** check box saves the UCS with the viewport. This setting is stored in the **UCSVP** system variable. If you clear this check box, the **UCSVP** variable will be set to 0 and the UCS of the viewport will reflect the UCS of the current viewport. When you select the **Update view to Plan when UCS is changed** check box, the plan view is restored after the UCS in the viewport is changed. When the selected UCS is restored, the plan view is also restored. The value is stored in the **UCSFOLLOW** system variable.

SYSTEM VARIABLES

The coordinate value of the origin of the current UCS is held in the **UCSORG** system variable. The X and Y axis directions of the current UCS are held in the **UCSXDIR** and **UCSYDIR** system variables, respectively. The name of the current UCS is held in the **UCSNAME** variable. All these variables are read-only. If the current UCS is identical to the WCS, the **WORLDUCS** system variable is set to 1; otherwise, it holds the value 0. The current UCS icon setting can be examined and manipulated with the help of the **UCSICON** system variable. This variable holds the UCS icon setting of the current viewport. If more than one viewport is active, each one can have a different value for the **UCSICON** variable. If you are in paper space, the **UCSICON** variable will contain the setting for the UCS icon of the paper space. The **UCSFOLLOW** system variable controls the automatic display of a plan view when you switch from one UCS to another. If **UCSFOLLOW** is set to 1 and you switch from one UCS to another, a plan view is automatically displayed. The **UCSAXISANG** variable stores the default angle value for the X, Y, or Z axis around which the UCS is rotated using the **X, Y, Z** options of the **New** option of the **UCS** command. The **UCSBASE** variable stores the name of the UCS that acts as the base; that is, it defines the origin and orientation of the orthographic UCS setting. The **UCSVP** variable decides whether the UCS settings are stored with the viewport or not.

Self-Evaluation Test

Answer the following questions and then compare them to those given at the end of this chapter:

1. The World Coordinate System can be moved from its position. (T/F)

2. The **View** tool in the **Align UCS** drop-down of the **Coordinates** panel is used to define a new UCS whose *XY* plane is parallel to the current viewing plane. (T/F)

3. An ellipse cannot be used as an object for defining a new UCS. (T/F)

4. If you do not specify the *Z* coordinate of the new point while moving the UCS, the previous *Z* coordinate will be taken for the new value. (T/F)

5. By default, the _____ is established as the coordinate system in the AutoCAD environment.

6. The _____ coordinate system can be moved to and rotated about any desired location.

7. Once a saved UCS is restored, it becomes the _____ UCS.

8. The _____ command controls the display of the UCS icon.

9. You can change the 3D UCS icon to a 2D UCS icon using the _____ option of the **UCSICON** command.

10. The _____ point of the 3D face defines the origin of the new UCS.

11. In dynamic alignment of UCS for a flat planar face, the _____ will always be parallel to the edge through which the UCS is moved on the face.

REVIEW QUESTIONS

Answer the following questions:

1. Selection of the UCS can be controlled by using the **Selectable** option of the **UCSICON** command. (T/F)

2. The line width of the UCS icon can be changed. (T/F)

3. The size of the UCS icon can vary between 5 and 100. (T/F)

4. The name of the current UCS is stored in the **UCSNAME** variable. (T/F)

5. Which of the following options is used to restore the previous UCS settings?

    (a) **ZAxis**   (b) **Restore**
    (c) **Previous**   (d) **Save**

6. Which of the following options is used to list the names of all the saved UCSs?

    (a) **ZAxis**   (b) **Restore**
    (c) **?**   (d) **Save**

7. Which of the following system variables is used to control the automatic display of the plan view when you switch to the new UCS?

    (a) **UCSORG**   (b) **UCSFOLLOW**
    (c) **UCS**   (d) **UCSBASE**

8. Which of the following system variables is used to store the coordinates of the origin of the current UCS?

    (a) **UCSICON**   (b) **UCSORG**
    (c) **UCSVP**   (d) **UCSFOLLOW**

9. Which of the following commands is used to manage the UCS using a dialog box?

    (a) **UCS**   (b) **UCSMAN**
    (c) **UCSICON**   (d) **UCSBASE**

10. The _____ variable is used to control the *X* axis direction of the current UCS.

11. The _____ direction is known as the extrusion direction.

12. The lineweight of the UCS icon can be changed using the _____ option of the **UCSICON** command.

13. The _____ option of the **UCS** command is used to change the origin of the current UCS.

14. The _____ variable is used to control the *Y* axis direction of the current UCS.

15. The _____ variable is used to store the default angle value by which the UCS will be rotated.

EXERCISES 1 through 4   *UCS*

Draw the objects shown in Figures 24-34 through 24-37. Assume the missing dimensions. Use the **UCS** command to align the UCS icon and then draw the objects.

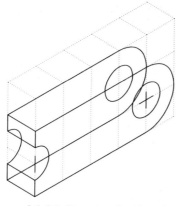

Figure 24-34 *Drawing for Exercise 1*

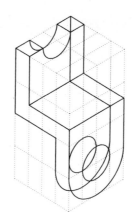

Figure 24-35 *Drawing for Exercise 2*

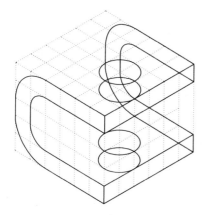

Figure 24-36 *Drawing for Exercise 3*

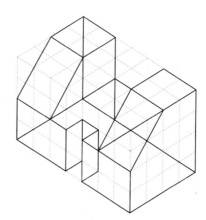

Figure 24-37 *Drawing for Exercise 4*

EXERCISE 5   UCS

Create the drawing shown in Figure 24-38. Assume its missing dimensions. Use the **UCS** command to align the UCS icon and then draw the objects. Note that in this drawing, the left side of the transition should make a 90-degree angle with the bottom.

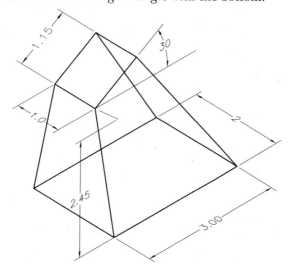

Figure 24-38 *Model for Exercise 5*

EXERCISE 6   UCS

Create the drawing shown in Figure 24-39. Assume its missing dimensions. Use the **UCS** command to position the UCS icon and then draw the objects. Note that in this drawing, the center of the top polygon is offset at 0.75 units from the center of the bottom polygon.

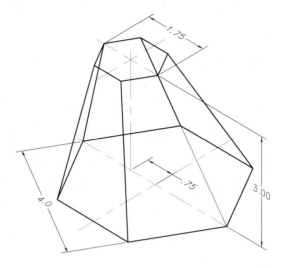

Figure 24-39 *Model for Exercise 6*

EXERCISES 7 and 8   UCS

In these exercises, you will create the drawing shown in Figures 24-40 through 24-41. Assume the missing dimensions.

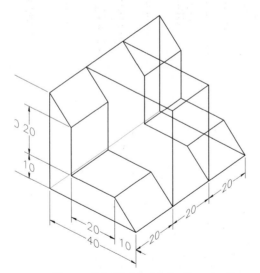

Figure 24-40 *Model for Exercise 7*

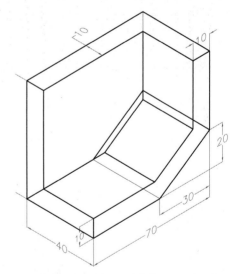

Figure 24-41 *Model for Exercise 8*

Answers to Self-Evaluation Test

1. F, **2.** T, **3.** F, **4.** T, **5.** World Coordinate System, **6.** user, **7.** current, **8.** **UCSICON**, **9.** **Properties**, **10.** first, 11. X axis

Chapter 25

Getting Started with 3D

CHAPTER OBJECTIVES

In this chapter, you will learn:
- *To start 3D workspace in AutoCAD for creating models.*
- *About the use of three-dimensional (3D) drawing.*
- *About various types of 3D models.*
- *About conventions in AutoCAD.*
- *To use the DDVPOINT command to view objects in 3D space.*
- *To use the VPOINT command to view objects in 3D space.*
- *To use the ViewCube to view objects easily in 3D space.*
- *About various types of 3D coordinate systems.*
- *To set thickness and elevation for objects.*
- *To dynamically view 3D objects using the 3DORBIT command.*
- *To use the DVIEW command for viewing a model.*
- *To use the SteeringWheel for viewing a model.*

KEY TERMS

- *Wireframe*
- *Surface*
- *Solids*
- *ViewCube*
- *DDVPOINT*

- *3D Coordinate System*
- *ELEV*
- *Hide*
- *3D Polyline*

- *3D Face*
- *PFACE*
- *3DMESH*
- *Edit Polyline*
- *SteeringWheel*

- *Orbit*
- *3DCLIP*
- *Nudge*

STARTING THREE DIMENSIONAL (3D) MODELING IN AutoCAD

In AutoCAD, you can start the 3D Modeling in a separate workspace. All tools required to create the 3D design are displayed in this workspace, by default. To start a new file in the 3D workspace, invoke the **Select template** dialog box and select the *Acad3d.dwt* template. This template file supports the 3D environment. The other template files that support the 3D workspace are *Acadiso3D.dwt*, *Acad-named plot styles3d.dwt*, and *Acadiso-named plot styles3d.dwt*. Next, select the **3D Modeling** option from the **Workspace** drop-down list available at the top left corner of the title bar. Figure 25-1 shows the **3D Modeling** workspace of AutoCAD.

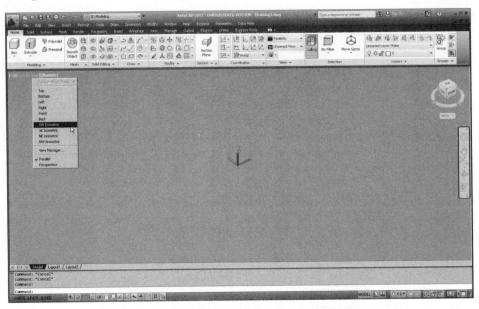

*Figure 25-1 The **3D Modeling** workspace of AutoCAD*

USE OF THREE-DIMENSIONAL DRAWING

The real world, and literally whatever you see around in it, is three-dimensional (3D). Each and every design idea you think of is 3D. Before the development of 3D software packages, it was not possible to materialize these ideas in the design industry due to the lack of the third dimension. This is because the drawings were made on the drawing sheets, which are two dimensional objects with only *X* and *Y* coordinates. Therefore, it was not possible to draw the 3D objects. As prototyping is a very long and costly affair, therefore, the designers had to suppress their ideas and convert the 3D designs into 2D drawings.

The use of computers and the CADD (Computer Aided Design & Drafting) Technology has brought a significant improvement in materializing the design ideas and creating the engineering drawings. You can create a proper 3D object in the computer using the CADD Technology. This technology allows you to create models with the third coordinate called the *Z* coordinate, in addition to the *X* and *Y* coordinates. Apart from materializing the design ideas, the 3D models have the following advantages:

1. **To generate the drawing views**. Once you have created a 3D model, its 2D drawing views can be automatically generated.

2. **To provide realistic effects**. You can provide realistic effects to the 3D models by assigning a material and providing light effects to them. For example, an architectural drawing can be

made more realistic and presentable by assigning the material to the walls and interiors and adding lights to it. You can also create its bitmap image and use it in presentations.

3. **To create the assemblies and check them for interference**. You can assemble various 3D models and create the assemblies. Once the components are assembled, you can check them for interference to reduce the errors and material loss during manufacturing. You can also generate the 2D drawing views of the assemblies.

4. **To create an animation of the assemblies**. You can animate the assemblies and view the animation to provide the clear display of the mating parts.

5. **To apply Boolean operations**. You can apply Boolean operations to the 3D models.

6. **To calculate the mass properties**. You can calculate the mass properties of the 3D models.

7. **Cut Sections**. You can cut sections through the solid models to view the shape at that cross-section.

8. **NC Programs**. You can generate an NC program with the help of 3D models by using a CAM software.

TYPES OF 3D MODELS

In AutoCAD, depending on their characteristics, the 3D models are divided into the following three categories.

Wireframe Models

These models are created using simple AutoCAD entities such as lines, polylines, rectangles, polygons, or some other entities such as 3D faces, and so on. To understand the wireframe models, consider a 3D model made up of match sticks or wires. These models consist of only the edges and hence you can see through them. You cannot apply the Boolean operations on these models and cannot calculate their mass properties. Wireframe models are generally used in frame building of vehicles. Figure 25-2 shows a wireframe model.

Surface Models

These models are made up of one or more surfaces. They have a negligible wall thickness and are hollow inside. To understand these models, consider a wireframe model with a cloth wrapped around it. You cannot see through it. These models are used in the plastic molding industry, shoe manufacturing, utensils manufacturing, and so on.

You can directly create a surface or a mesh. Alternatively, you can create wireframe model and then convert it into a mesh or surface model. Remember that you cannot perform the Boolean operations in surface models. Figure 25-3 shows a surface model created using a single surface or a mesh and Figure 25-4 shows a surface model created using a combination of surfaces.

Solid Models

These are the solid filled models having mass properties. To understand a solid model, consider a model made up of metal or wood. You can perform the Boolean operations on these models, cut a hole through them, or even cut them into slices. Figure 25-5 shows a solid model.

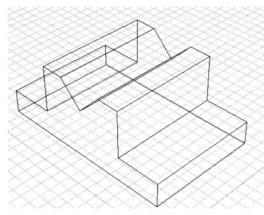

Figure 25-2 *A wireframe model*

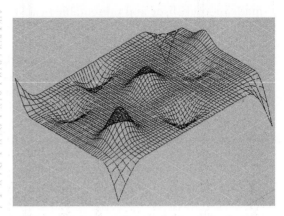

Figure 25-3 *A surface model created using a single surface*

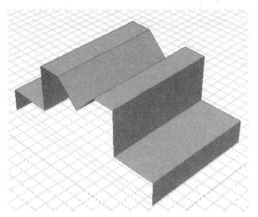

Figure 25-4 *A semi-finished surface model created using more than one surface*

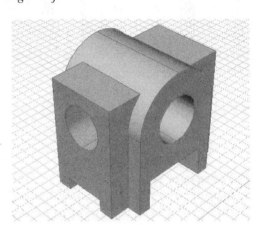

Figure 25-5 *A solid model*

CONVENTIONS IN AutoCAD

It is important for you to know the conventions in AutoCAD before you proceed with 3D. AutoCAD follows these three conventions:

1. By default, any drawing you create in AutoCAD will be in the world XY plane.

2. The right-hand thumb rule is followed in AutoCAD to identify the *X*, *Y*, and *Z* axis directions. This rule states that if you keep the thumb, index finger, and middle finger of the right hand mutually perpendicular to each other, as shown in Figure 25-6(a), then the thumb of the right hand will represent the direction of the positive *X* axis, the index finger will represent the direction of the positive *Y* axis, and the middle finger of the right hand will represent the direction of the positive *Z* axis, see Figure 25-6(b).

3. The right-hand thumb rule is followed in AutoCAD to determine the direction of rotation or revolution in the 3D space. The right-hand thumb rule states that if the thumb of the right hand points in the direction of the axis of rotation, then the direction of the curled fingers will define the positive direction of rotation, see Figure 25-6(c).

This rule will be used during the rotation of 3D models or the UCS and also, during the creation of revolved surfaces and revolved solids.

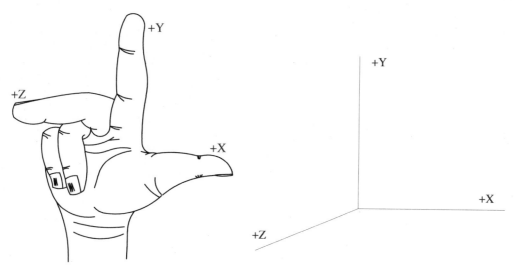

Figure 25-6(a) *The orientation of fingers as per the right-hand thumb rule*

Figure 25-6(b) *Orientation of the X, Y, and Z axes*

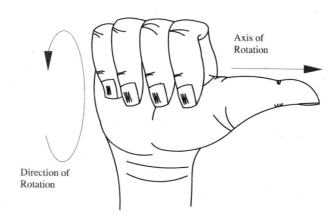

Figure 25-6(c) *Right-hand thumb rule showing the axis and direction of rotation*

CHANGING THE VIEWPOINT TO VIEW 3D MODELS

Until now, you have been drawing only the 2D entities in the *XY* plane and in the Plan view. But in the Plan view, it is very difficult to find out whether the object displayed is a 2D entity or a 3D model, see Figure 25-7. In this view, it appears as the cuboid displayed is a rectangle. The reason for this is that, by default, you view the objects in the Plan view from the direction of the positive *Z* axis. You can avoid this confusion by viewing the object from a direction that also displays the *Z* axis. In order to do that, you need to change the viewpoint so that the object is displayed in the space with all three axes, as shown in Figure 25-8. In this view, you can also see the *Z* axis of the model along with the *X* and *Y* axes. It will now be clear that the original object is not a rectangle but a 3D model. The viewpoint can be changed by using the **View Cube** available in the drawing area, by using the options in the **View** panel, **DDVPOINT** command, or the **VPOINT** command.

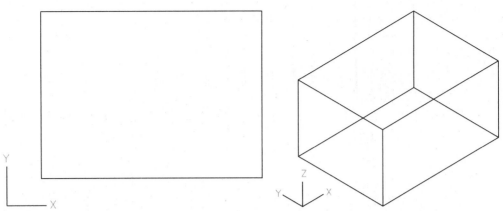

Figure 25-7 *Cuboid looking like a rectangle in the Plan view*

Figure 25-8 *Viewing the 3D model after changing the viewpoint*

Tip
Remember that changing the viewpoint does not affect the dimensions or the location of the model. When you change the viewpoint, the model is not moved from its original location. Changing the viewpoint only changes the direction from which you view the model.

Changing the Viewpoint Using the ViewCube

You can change the view of the models easily and quickly using the ViewCube. The ViewCube is displayed when the **3D Modeling** workspace is enabled and it allows you to switch between the standard and isometric views, roll the current view, or return to the **Home** view of a model. ViewCube consists of Cube, Home, Compass, and WCS menu. When you move the cursor over the ViewCube, it becomes active and the area at the pointer tip gets highlighted. The highlighted area can be a face, a corner, or an edge of the ViewCube. These highlighted areas are called hotspots. There are 6 faces, 12 edges, and 8 corners on a cube, see Figure 25-9. So, you can get 26 views by using the ViewCube. You can select the required hotspot to restore a view. You can also go back to the Home view by clicking on the Home icon of the ViewCube. You can set any existing view as the Home view by choosing the **Set Current View as Home** option from the shortcut menu that is displayed on right-clicking on the ViewCube. By default, a model is displayed in the perspective view. To display a model in parallel projection, right-click on the ViewCube and choose **Parallel** from the shortcut menu.

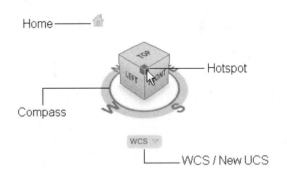

Figure 25-9 *The ViewCube*

The compass on the ViewCube indicates the geographic location of a model. The **N**, **E**, **S**, and **W** alphabets on the compass indicate the North, East, South, and West directions of a 3D model. You can pick and drag the compass ring to rotate the current view in the same plane. You can choose between the **UCS** and **WCS** from the WCS menu and even create a new UCS from the menu. When you right-click on the ViewCube, a shortcut menu will be displayed. Choose the **ViewCube Settings** option from it; the **ViewCube Settings** dialog box will be displayed. (See Figure 25-10). This dialog box allows you to adjust the display of the compass ring and UCS menu, size, appearance, and location of the ViewCube.

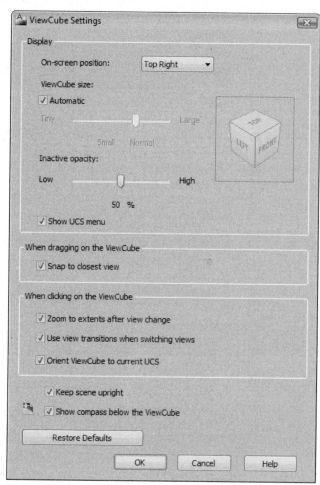

*Figure 25-10 The **ViewCube Settings** dialog box*

Note

*In AutoCAD 2012, you can change the current viewport by using the **In-canvas Viewport controls**, displayed as **[-][Custom View][Realistic]**. It is available at the top left corner of the drawing area. When you choose the **[Custom View]** option; a flyout is displayed, refer to Figure 25-1. You can select the desired view from the flyout.*

Changing the Viewpoint Using the Ribbon or the Toolbar

You can also set the viewpoint either by using the **Views** panel in the **View** tab of the **Ribbon** or by using the **View** toolbar.

The **View** drop-down list (Figure 25-11) in the **View** panel consists of six preset orthographic views: **Top, Bottom, Left, Right, Front**, and **Back**; and four preset isometric views: **SW Isometric, SE Isometric, NE Isometric**, and **NW Isometric**. This drop-down list also allows you to retrieve the previously set views or any previously created named views. All the above-mentioned controls are also available in the **View Manager** dialog box that is invoked on selecting the **View Manager** option from the **View** drop-down list.

Choose the **New** button from the **View Manager** dialog box to create a new view; the **New View / Shot Properties** dialog box will be displayed, as shown in Figure 25-12.

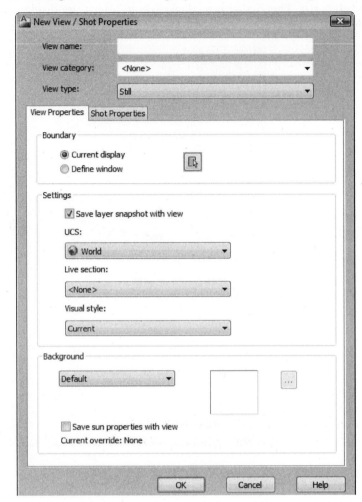

Figure 25-11 *The View drop-down list*

Figure 25-12 *The New View / Shot Properties dialog box*

Enter a name in the **View name** edit box to save the view. You can save different types of views into different categories using the options in this dialog box. The different types of views available in the **View type** drop-down list are **Still**, **Cinematic,** and **Recorded Walk**. But only **Still** is used for creating views and the other two are used for creating shots. The **View Properties** tab of this dialog box has three main areas: **Boundary**, **Settings** and **Background**. The **Boundary** area allows you to save the current display or you can select an area of display by choosing the **Define window** radio button. In the **Settings** area, you can control the position of UCS to save the view, select the section of the view, and control the display of the model. The drop-down list in the **Background** area is used to select the type of background to save the view. The preview of the selected background is displayed in the preview area next to the drop-down list. You need to select the **Save sun properties with view** check box to save the view with the sun properties.

Tip
*When you select any of the preset orthographic views from the **View** drop-down list, the UCS is also aligned to that view.*

Changing the Viewpoint Using the Viewpoint Presets Dialog Box

Menu Bar: View > 3D Views > Viewpoint Presets   **Command:** DDVPOINT

The **DDVPOINT** command is used to invoke the **Viewpoint Presets** dialog box for setting the viewpoint to view 3D models. The viewpoint is set with respect to the angle from the X axis and the angle from the XY plane. Figure 25-13 shows different view direction parameters.

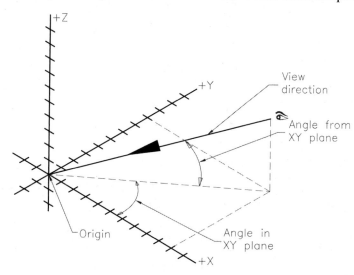

Figure 25-13 View direction parameters

Viewpoint Presets Dialog Box Options
The options in the **Viewpoint Presets** dialog box (Figure 25-14) are discussed next.

Absolute to WCS. This radio button is selected to set the viewpoint with respect to the world coordinate system (WCS).

Relative to UCS. This radio button is selected to set the viewpoint with respect to the current user coordinate system (UCS).

Note
The WCS and the UCS have already been discussed in Chapter 24.

From: X Axis. This edit box is used to specify the angle in the *XY* plane from the *X* axis. You can directly enter the required angle in this edit box or select the desired angle from the image tile (Figure 25-15). There are two arms in this image tile, and by default, they are placed one over the other. The gray arm displays the current viewing angle and the black one displays the new viewing angle. On selecting a new angle from the image tile, it is automatically displayed in the **X Axis** edit box. This value can vary between 0 and 359.9.

From: XY Plane. This edit box is used to specify the angle from the *XY* plane. You can directly enter the desired angle in it or select the angle from the image tile, as shown in Figure 25-16. This value can vary from -90 to +90.

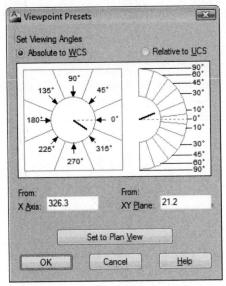

Figure 25-14 The **Viewpoint Presets** dialog box

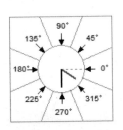

Figure 25-15 The image tile for selecting the angle in the XY plane

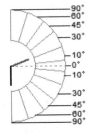

Figure 25-16 The image tile for selecting the angle from the XY plane

Set to Plan View. This button is chosen to set the viewpoint to the Plan view of the WCS or the UCS. If the **Absolute to WCS** radio button is selected, the viewpoint will be set to the Plan view of the WCS. If the **Relative to UCS** radio button is selected, the viewpoint will be set to the Plan view of the current UCS. Figures 25-17 and 25-18 show the 3D models from different viewing angles.

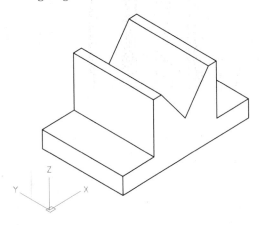

Figure 25-17 Viewing the 3D model with the angle in the XY plane as 225 and the angle from the XY plane as 30

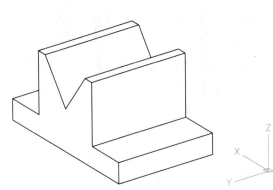

Figure 25-18 Viewing the 3D model with the angle in the XY plane as 145 and the angle from the XY plane as 25

Tip
If you set a negative angle from the XY plane, the Z axis will be displayed as a dotted line, indicating a negative Z direction, see Figures 25-19 and 25-20.

In case of confusion in identifying the direction of X, Y, and Z axes, use the right-hand thumb rule.

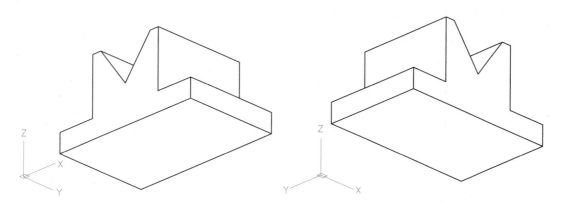

Figure 25-19 *Viewing the 3D model with the angle in the XY plane as 315 and the angle from the XY plane as -25*

Figure 25-20 *Viewing the 3D model with the angle in the XY plane as 225 and the angle from the XY plane as -25*

Changing the Viewpoint Using the Command

Menu Bar: View > 3D Views > Viewpoint   **Command:** VPOINT

The **VPOINT** command is also used to set the viewpoint for viewing the 3D models. Using this command, the user can specify a point in the 3D space that will act as the viewpoint. The prompt sequence that will follow when you invoke this command is given next.

Current view direction: VIEWDIR=0.0000,0.0000,1.0000
Specify a view point or [Rotate] <display compass and tripod>:

Specifying a Viewpoint

This is the default option and is used to set a viewpoint by specifying its location (of viewer) using the *X*, *Y*, and *Z* coordinates of that particular point. AutoCAD follows a convention of the sides of the 3D model for specifying the viewpoint. The convention states that if the UCS is at the World position (default position), then

1. The side at the positive *X* axis direction will be taken as the right side of the model.
2. The side at the negative *X* axis direction will be taken as the left side of the model.
3. The side at the negative *Y* axis direction will be taken as the front side of the model.
4. The side at the positive *Y* axis direction will be taken as the back side of the model.
5. The side at the positive *Z* axis direction will be taken as the top side of the model.
6. The side at the negative *Z* axis direction will be taken as the bottom side of the model.

Some standard viewpoint coordinates and the view they display are given next.

Value	View	Value	View	Value	View
1,0,0	Right side	-1,0,0	Left side	0,1,0	Back
0,-1,0	Front	0,0,1	Top view	0,0,-1	Bottom view
1,1,1	Right, Back, Top	-1,1,1	Left, Back, Top	1,-1,1	Right, Front, Top
-1,-1,1	Left, Front, Top	1,1,-1	Right, Back, Bottom	-1,1,-1	Left, Back, Bottom
1,-1,-1	Right, Front, Bottom	-1,-1,-1	Left, Front, Bottom		

You can also enter the values in decimal, but the resultant views will not be the standard views. Figures 25-21 through 25-24 show the 3D model from different viewpoints.

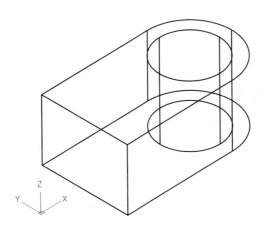

Figure 25-21 *Viewing the model from the viewpoint -1,-1,1*

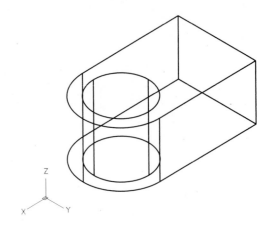

Figure 25-22 *Viewing the model from the viewpoint 1,1,1*

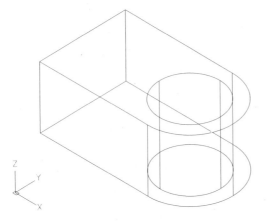

Figure 25-23 *Viewing the model from the viewpoint 1,-1,1*

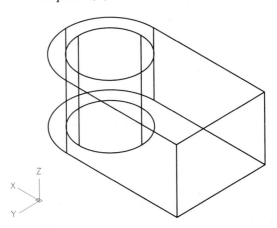

Figure 25-24 *Viewing the model from the viewpoint -1,1,1*

Compass and Tripod

When you press ENTER at the **Specify a viewpoint or [Rotate] <display compass and tripod>** prompt, a compass and an axis tripod are displayed, as shown in Figure 25-25. You can directly set the viewpoint using this compass. The crosshairs appear on the compass. You can select any point on this compass to specify the viewpoint. The compass consists of two circles, a smaller circle and a bigger circle. Both these circles are divided into four quadrants: first, second, third, and fourth. The resultant view will depend upon the quadrant and the circle, on which you select the point. In the first quadrant, both the *X* and *Y* axes are positive; in the second quadrant, the *X* axis is negative and the *Y* axis is positive; in the third quadrant, both the *X* and *Y* axes are negative; and in the fourth quadrant, the *X* axis is positive and the *Y* axis is negative. Now, if you select a point inside the inner circle, it will be in the positive *Z* axis direction. If you select the point outside the inner circle and inside the outer circle, it will be in the negative *Z* axis direction. Therefore, if the previously mentioned statements are added, the following is concluded.

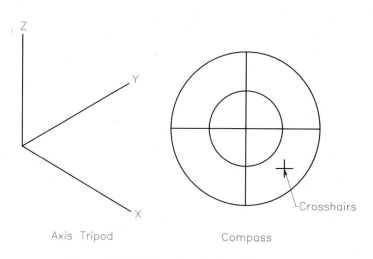

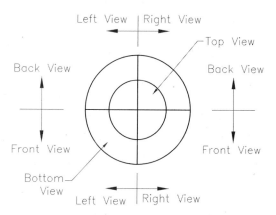

Figure 25-25 *The axis tripod and the compass*

1. If you select a point inside the smaller circle in the first quadrant, the resultant view will be the Right, Back, or Top view. If you select a point outside the smaller circle and inside the bigger circle in this quadrant, the resultant view will be the Right, Back, or Bottom view (Figure 25-26).

2. If you select a point inside the smaller circle in the second quadrant, the resultant view will be the Left, Back, or Top view. If you select a point outside the smaller circle and inside the bigger circle in this quadrant, the resultant view will be the Left, Back, or Bottom view (Figure 25-26).

Figure 25-26 *The directions of the compass*

3. If you select a point inside the smaller circle in the third quadrant, the resultant view will be the Left, Front, or Top view. If you select a point outside the smaller circle and inside the bigger circle in this quadrant, the resultant view will be the Left, Front, or Bottom view (Figure 25-26).

4. If you select a point inside the smaller circle in the fourth quadrant, the resultant view will be the Right, Front, or Top view. If you select a point outside the smaller circle and inside the bigger circle in this quadrant, the resultant view will be the Right, Front, or Bottom view (Figure 25-26).

Rotate

This option is similar to setting the viewpoint in the **Viewpoint Presets** dialog box, which is displayed by invoking the **DDVPOINT** command. When you select this option, you will be prompted to specify the angle in the *XY* plane and the angle from the *XY* plane. The angle in the *XY* plane will be taken from the *X* axis.

IN-CANVAS VIEWPORT CONTROL*

In-canvas Viewport control is a new feature introduced in AutoCAD 2012. It enables you to control the viewport as well as visual style. By default, the **In-canvas Viewport control** is displayed as [-][Top][2D Wireframe] in the top left corner of the drawing area. The tools available in this viewport control are discussed next.

[-]

When you choose the [-] button in the **In-canvas Viewport control**, a flyout will be displayed, as shown in Figure 25-27. The options in this flyout are discussed next.

Restore Viewport

When you choose this option, the drawing screen is displayed with four equal partitions. In each of these partitions a [+] button is displayed. When you click on the [+] button; a flyout will be displayed with the **Maximize Viewport** option. Using this option, you can maximize the partition window.

Figure 25-27 The [-] flyout

Viewport Configuration List

When you choose this option, a flyout will be displayed, as shown in the Figure 25-28. In this flyout, 12 options are available for viewport settings. Using these options, you can set how many partitions need to be created and their position in the drawing area.

ViewCube, SteeringWheels, and Navigation Bar

ViewCube is used to minimize your time in changing the view of the model while the SteeringWheels and Navigation Bar are used to minimize the navigation time. Using the ViewCube from the Viewport Controls flyout; you can toggle the display of the ViewCube. As you choose the **SteeringWheels** option; SteeringWheels will get attached to the cursor. SteeringWheels will be discussed later in this chapter. Using the **Navigation Bar** option; you can toggle the display of the Navigation Bar in the right of the drawing screen. Navigation Bar is discussed in chapter 1.

Figure 25-28 The Viewport Configuration List flyout

View Controls

View Controls flyout contains various options to change your current view or to create a new view. When you choose [**Top**] from the **In-canvas Viewport control**; a flyout will be displayed, as shown in Figure 25-29. The options available in this flyout are discussed next.

- **Custom Model Views.** This flyout consists of custom views created by using the **View Manager**.

- **Views.** There are 10 view options available in this flyout to set 10 different views. When you change the view, the UCS gets reoriented according to the view selected.

- **View Manager.** Using this option, you can create a new view or edit the view settings of an existing view. **View Manager** is already discussed in chapter 6.

Figure 25-29 The View Controls flyout

- **Parallel and Perspective.** These options are used to control the projection. If you choose the **Parallel** option, the drawing will be displayed parallel to the screen. If you choose the **Perspective** option, the drawing will be displayed as a perspective view.

Visual Style Controls

The **Visual Style Controls** flyout are used to control the visual style of the model. On choosing **[2D Wireframe]** from the **In-canvas Viewport control**; a flyout will be displayed, as shown in Figure 25-30. In this flyout, the following options are available:

- **Custom Visual Styles**. When you click on this option a flyout is displayed. This flyout consists of the visual styles you created.

- **Visual Styles**. There are 10 options that enable you to select the required visual style.

- **Visual Style Manager**. This option is used to create a new visual style as well as to change the settings of the current visual style.

Figure 25-30 The Visual Style Controls flyout

3D COORDINATE SYSTEMS

Similar to 2D coordinate systems, there are two types of coordinate systems. They are discussed next.

Absolute Coordinate System

This type of coordinate system is similar to the 2D absolute coordinate system in which the coordinates of the point are calculated from the origin (0,0). The only difference is that here, along with the *X* and *Y* coordinates, the *Z* coordinate is also included. For example, to draw a line in 3D space starting from the origin to a point say 10,6,6, the procedure to be followed is given next.

> Command: *Choose the **Line** tool from the **Draw** panel.*
> Specify first point: **0,0,0**
> Specify next point or [Undo]: **10,6,6**
> Specify next point or [Undo]: [Enter]

Figure 25-31 shows the model drawn using the absolute coordinate system.

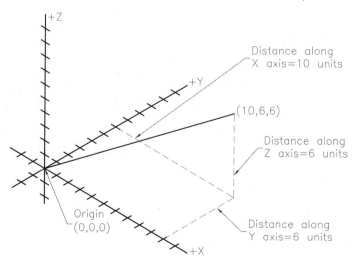

Figure 25-31 Drawing a line from origin to 10,6,6

Relative Coordinate System

The following are the three types of relative coordinate systems in 3D:

Relative Rectangular Coordinate System

This coordinate system is similar to the relative rectangular coordinate system of 2D, except that in 3D you also have to enter the *Z* coordinate along with the *X* and *Y* coordinates. The syntax of the relative rectangular system for the 3D is **@X coordinate, Y coordinate, Z coordinate**.

Relative Cylindrical Coordinate System

In this coordinate system, you can locate the point by specifying its distance from the reference point, the angle in the *XY* plane, and its distance from the *XY* plane. The syntax of the relative cylindrical coordinate system is **@Distance from the reference point in the XY plane < Angle in the XY plane from the X axis, Distance from the XY plane along the Z axis**. Figure 25-32 shows the components of the relative cylindrical coordinate system.

Relative Spherical Coordinate System

In this coordinate system, you can locate the point by specifying its distance from the reference point, the angle in the *XY* plane, and the angle from the *XY* plane. The syntax of the relative spherical coordinate system is **@Length of the line from the reference point < Angle in the XY plane from the X axis < Angle from the XY plane**. Figure 25-33 shows the components of the relative spherical coordinate system.

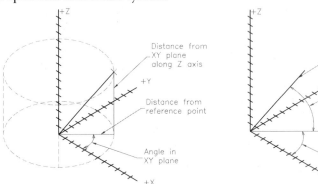

Figure 25-32 Various components of the relative cylindrical coordinate system

Figure 25-33 Various components of the relative spherical coordinate system

Tip

The major difference between the relative cylindrical and relative spherical coordinate system is that in the relative cylindrical coordinate system, the specified distance is the distance from the reference point in the XY plane. On the other hand, in the relative spherical coordinate system, the specified distance is the total length of the line in the 3D space.

EXAMPLE 1   *Relative Coordinate System*

In this example, you will draw the 3D wireframe model shown in Figure 25-34. Its dimensions are given in Figures 25-35 and 25-36.

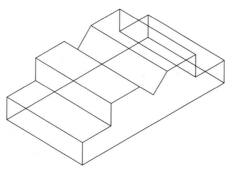

Figure 25-34 *Wireframe model for Example 1*

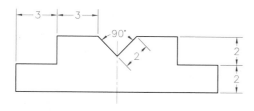

Figure 25-35 *Front view of the model*

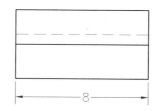

Figure 25-36 *Side view of the model*

1. Start a new file in the *Acad3d.dwt* template and select the **SW Isometric** option from the **View** list in the **View** tab. Alternatively, select the Southwest top corner hotspot on the ViewCube (refer to the compass of the ViewCube to get the Southwest view of the model).

2. As the start point of the model is not given, you can start the model from any point. Choose the **Line** tool from the **Draw** panel and follow the prompt sequence given next.

   Specify first point: **4,2,0**
   Specify next point or [Undo]: **@0,0,2**
   Specify next point or [Undo]: **@3,0,0**
   Specify next point or [Close/Undo]: **@0,0,2**
   Specify next point or [Close/Undo]: **@3,0,0**
   Specify next point or [Close/Undo]: **@2<0<315**
   Specify next point or [Close/Undo]: **@2<0<45**
   Specify next point or [Close/Undo]: **@3,0,0**
   Specify next point or [Close/Undo]: **@0,0,-2**
   Specify next point or [Close/Undo]: **@3,0,0**
   Specify next point or [Close/Undo]: **@0,0,-2**
   Specify next point or [Close/Undo]: **C**

3. Choose the **Line** tool again and follow the prompt sequence given next.

   Specify first point: **4,2,0**
   Specify next point or [Undo]: **@0,8,0**
   Specify next point or [Undo]: **@0,0,2**
   Specify next point or [Close/Undo]: **@3,0,0**
   Specify next point or [Close/Undo]: **@0,0,2**
   Specify next point or [Close/Undo]: **@3,0,0**
   Specify next point or [Close/Undo]: **@2<0<315**
   Specify next point or [Close/Undo]: **@2<0<45**

Specify next point or [Close/Undo]: **@3,0,0**
Specify next point or [Close/Undo]: **@0,0,-2**
Specify next point or [Close/Undo]: **@3,0,0**
Specify next point or [Close/Undo]: **@0,0,-2**
Specify next point or [Close/Undo]: **@0,-8,0**
Specify next point or [Close/Undo]: Enter

4. Complete the model by joining the remaining edges using the **Line** tool.

5. The final 3D model should look similar to the one shown in Figure 25-37.

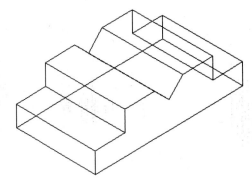

Figure 25-37 Final 3D wireframe model for Example 1

Direct Distance Entry Method

In AutoCAD 2012, you can directly create models in 3D space. This method is similar to that in 2D. The easiest way to draw a model by using the **Line** tool in 3D space of AutoCAD is by using the Direct Distance Entry method. Before drawing a model by using this method, ensure that the **Dynamic Input** button is chosen in the Status Bar. Next, choose the **Line** tool; you will be prompted to specify the start point. Enter the coordinate values in the text box and press ENTER; you will be prompted to specify the next point. Move the cursor along the direction in which you need to draw the line and enter the absolute length of the line and its angle in the corresponding text boxes, with respect to the previous point. Note that you can use the TAB key to toggle between the text boxes.

To draw a line along the Z axis, move the cursor along the Z axis; you will notice that the tool tip displays **Ortho: (current length) < +Z**, if the ortho mode is on. Type the absolute distance and the angle and press ENTER. If the ortho mode is not chosen, position the cursor at the desired angle, type the length at the Command prompt, and then press ENTER. If necessary change the UCS as discussed earlier.

Note
*If you choose the Home option of the ViewCube in the **Drafting & Annotation** workspace, then you can also draw the profiles in 3D space by using the 3D coordinates.*

EXAMPLE 2	Direct Distance Entry

In this example, you will draw the 3D wireframe model shown in Figure 25-38.

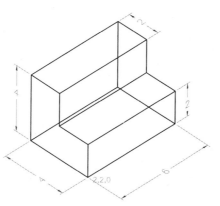

Figure 25-38 3D wireframe model for Example 2

1. Start a new file with the *Acad3d.dwt* template in the **3D Modeling** workspace and make sure the **Ortho Mode** and **Dynamic Input** buttons are chosen. Then, change the drawing view by using the ViewCube. The view needed is **SW Isometric**. The Southwest view of the ViewCube has three hotspots: bottom, edge, and top. You need to select the top hotspot to get the required view. Refer to Figure 25-39 for the SW Isometeric hotspot. However, you can check different hotspots and view the model in different angles by using the ViewCube.

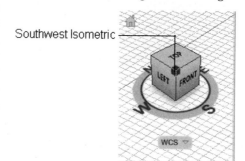

Figure 25-39 *ViewCube showing the SW Isometric*

2. Choose the **Line** tool from the **Draw** panel and specify the start point at 2,2,0.

3. Move the cursor along the positive X axis. Then, type **6** and press ENTER; a line of length 6 units is drawn along the X axis.

4. Move the cursor along the positive Y axis, type **4**, and press ENTER; a line of length 4 units is drawn along the Y axis.

5. Draw the other two lines to create the closed profile at the bottom of the model.

6. From the start point of the model, move the cursor along the positive Z axis. You will notice that the tooltip displays **Ortho: (current length) < +Z**, as shown in Figure 25-40.

7. Type **2** at the Command prompt and press ENTER; a line of length 2 units is drawn.

8. Move the cursor along the positive X axis, type **6**, and press ENTER; a line of length 6 units is drawn.

9. Similarly, draw other lines to complete the model. The final model should look similar to the one shown in Figure 25-41.

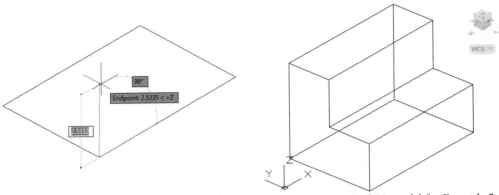

Figure 25-40 *Tooltip displayed on moving the cursor along the positive Z axis*

Figure 25-41 *Final 3D model for Example 2*

EXERCISE 1

In this exercise, you will draw the 3D wireframe model shown in Figure 25-42. Its dimensions are given in the figure itself. You can start the model from any point.

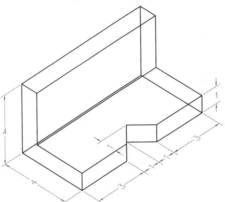

Figure 25-42 *Wireframe model for Exercise 1*

TRIM, EXTEND, AND FILLET TOOLS IN 3D

As discussed in earlier chapters, you can use the options of these tools in 3D space. You can also use the **PROJMODE** and **EDGEMODE** variables. The **PROJMODE** variable sets the projection mode for trimming and extending. A value of 0 implies **True 3D mode**, that is, no projection. In this case, the objects must intersect in 3D space to be trimmed or extended. It is similar to using the **None** option of the **Project** option of the **Extend** and **Trim** tools. A value of 1 projects onto the *XY* plane of the current UCS and is similar to using the **UCS** option of the **Project** option of the **Trim** or **Extend** tools. A value of 2 projects onto the current view plane and is like using the **View** option of the **Project** option of the **Trim** or **Extend** tools. The value of the **EDGEMODE** system variable controls the cutting or trimming boundaries. A value of 0 considers the selected edge with no extension. You can also use the **No extend** option of the **Edge** option of the **Trim** or **Extend** tools. A value of 1 considers an imaginary extension of the selected edge. This is similar to using the **Extend** option of the **Edge** option of the **Trim** or **Extend** tools.

You can fillet coplanar objects whose extrusion directions are not parallel to the *Z* axis of the current UCS, by using the **Fillet** tool. The fillet arc exists on the same plane and has the same extrusion direction as the coplanar objects. If the coplanar objects to be filleted exist in opposite directions, the fillet arc will be on the same plane but will be inclined toward the positive Z axis.

SETTING THICKNESS AND ELEVATION FOR NEW OBJECTS

You can create the objects with a preset elevation or thickness using the **ELEV** command. This command is discussed next.

The ELEV Command

> **Command:** ELEV

This is a transparent command and is used to set elevation and thickness for new objects. The following prompt sequence is displayed when you invoke this command:

> Specify new default elevation <0.0000>: *Enter the new elevation value.*
> Specify new default thickness <0.0000>: *Enter the new thickness value.*

Elevation

This option is used to specify the elevation value for new objects. Setting the elevation is nothing but moving the working plane from its default position. By default, the working plane is on the world *XY* plane. You can move this working plane using the **Elevation** option of the **ELEV** command. However, remember that the working plane can be moved only along the *Z* axis (Figure 25-43). The default elevation value is 0.0 and you can set any positive or negative value. All objects that will be drawn hereafter will be with the specified elevation. The **Elevation** option sets the elevation only for the new objects, and the existing objects are not modified using this option.

Thickness

This option is used to specify the thickness values for new objects. It can be considered as another method of creating surface models. Specifying the thickness creates extruded surface models. The thickness is always taken along the *Z* axis direction (Figure 25-44). The 3D faces will be automatically applied on the vertical faces of the objects drawn with a thickness. The **Thickness** option sets the thickness only for the new objects and the existing objects are not modified.

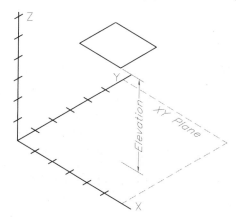

Figure 25-43 *Object drawn with elevation*

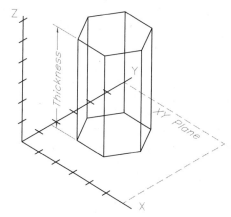

Figure 25-44 *Object drawn with thickness*

Note
*By default, the model will be displayed in the **Realistic** visual style. To change it to wireframe, select the **2D Wireframe** option from the **Visual Styles Controls** flyout in the **In-canvas Viewport Controls**. You can suppress the hidden edges in the model by entering **HIDE** at the Command prompt. The **HIDE** command is discussed in the next section.*

Tip
The elevation value will be reset to 0.0 when you change the coordinate system to WCS.

*The rectangles drawn using the **Rectangle** tool do not consider the thickness value set by using the **ELEV** command. To draw a rectangle with thickness, use the **Thickness** option of the **Rectangle** tool.*

Figure 25-45 shows a point, line, polygon, circle, and ellipse drawn with a thickness of 5 units.

To write text with thickness, first write a single line text and then change its thickness using any of the commands that modify the properties (Figure 25-46).

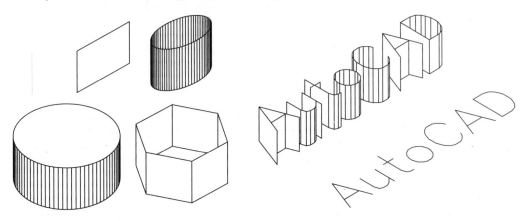

Figure 25-45 *A point, line, polygon, circle, and an ellipse drawn with a thickness of 5 units*   **Figure 25-46** *Text with and without thickness*

SUPPRESSING THE HIDDEN EDGES

Menu Bar: View > Hide	**Toolbar:** Render > Hide	**Command:** HIDE

Whenever you create a surface or a solid model, the edges that lie at the back (also called the hidden edges) are also visible. As a result, the model appears like a wireframe model. You need to manually suppress the hidden lines in the 3D model using the **HIDE** command. When you invoke this command, all objects on the screen are regenerated. Also, the 3D models are displayed with the hidden edges suppressed. If the value of the **DISPSILH** system variable is set to **1**, the model is displayed only with the silhouette edges. The internal edges that have facets will not be displayed. The hidden lines are again included in the drawing when the next regeneration takes place. Figure 25-47 shows the surface models displaying the hidden edges. Figure 25-48 shows surface models without the hidden edges.

Tip
*The **HIDE** command considers Circles, Solids, Traces, Polylines with width, Regions, and 3D Faces as opaque surfaces that will hide objects.*

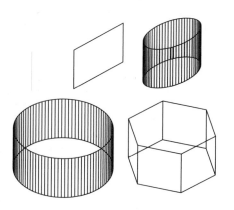
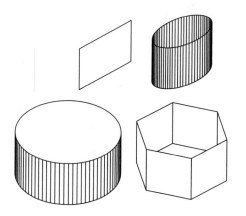

Figure 25-47 *3D models displaying all edges*   **Figure 25-48** *Models without the hidden edges*

CREATING 3D POLYLINES

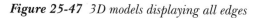

Ribbon: Home > Draw > 3d Polylines   **Menu Bar:** Draw > 3D Polyline
Command: 3DPOLY

 The **3dPolylines** tool is used to draw straight polylines in a 2D plane or a 3D space. This tool is similar to the **Polyline** tool except that you can draw polylines in a plane other than the *XY* plane also. However, this tool does not provide the **Width** option or the **Arc** option; and therefore you cannot create a 3D polyline with a width or an arc. The **Close** and the **Undo** options of this tool are similar to those of the **Polyline** tool.

> **Tip**
> *You can fit a spline curve about the 3D polyline using the **Edit Polyline** tool. Use the **Spline curve** option to fit the spline curve and the **Decurve** option to remove it.*

CONVERTING WIREFRAME MODELS INTO SURFACE MODELS

You can convert a wireframe model into a surface model using the **3DFACE** command and the **PFACE** command. These commands are discussed next.

Creating 3D Faces

Menu Bar: Draw > Modeling > Meshes > 3D Face   **Command:** 3DFACE

This command is used to create 3D faces in space. You can create three-sided or four-sided faces using this command. You can specify the same or a different *Z* coordinate value for each point of the face. On invoking this command, you will be prompted to specify the first, second, third, and fourth points of the 3D face. Once you have specified the first four points, it will again prompt you to specify the third and fourth points. In this case, it will take the previous third and fourth points as the first and second points, respectively. This process will continue until you press ENTER at the **Specify third point or [Invisible] <exit>** prompt. Keep in mind that the points must be specified in the natural clockwise or the counterclockwise direction to create a proper 3D face. Figure 25-49 shows a simple 3D wireframe model.

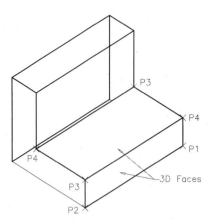

Figure 25-49 *3D faces applied to the wireframe model*

However, in case of complex wireframe models (Figure 25-50), it is not easy to apply 3D faces. This is because when you apply a 3D face, it generates edges about all points that you specify. These edges will be displayed on the wireframe model, as shown in Figure 25-50. You can avoid these unwanted visible edges by applying the 3D faces with invisible edges using the **Invisible** option. It is very important to note here that the edge that will be created after you enter **I** at the **Specify next point or [Invisible] <exit>** prompt will not be the invisible edge. The edge that will be created after this edge will be the invisible edge. Therefore, to make the P3, P4 edge invisible, specify P1 as the first point and P2 as the second point. Before specifying P3 as the third point, enter **I** at the **Specify third point or [Invisible] <exit>** prompt. Similarly, follow the same procedure for creating the other invisible edges (Figure 25-51).

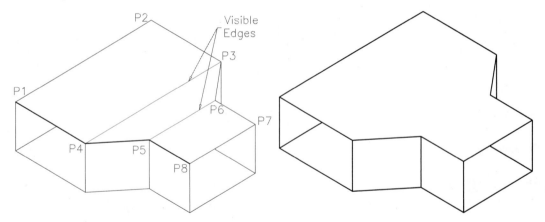

Figure 25-50 *Wireframe model with the edges*   ***Figure 25-51*** *Wireframe model without the edges*

Creating Polyface Meshes

Command: PFACE

The **PFACE** command is similar to the **3DFACE** command. This command also allows you to create a mesh of any desired surface shape by specifying the coordinates of the vertices and assigning the vertices to the faces in the mesh. The difference between them is that on using the **3DFACE** command, you do not select the vertices that join another face twice. With the **PFACE** command, you need to select all the vertices of a face, even if they are coincident with the vertices of another face. In this way, you can avoid generating unrelated 3D faces that have coincident vertices. Also, with this command, there is no restriction on the number of faces and vertices the mesh can have.

Command: **PFACE**
Specify location for vertex 1: *Specify the location of the first vertex.*
Specify location for vertex 2 or <define faces>: *Specify the location of the second vertex.*
Specify location for vertex 3 or <define faces>: *Specify the location of the third vertex.*
Specify location for vertex 4 or <define faces>: *Specify the location of the fourth vertex.*
Specify location for vertex 5 or <define faces>: *Specify the location of the fifth vertex.*
Specify location for vertex 6 or <define faces>: *Specify the location of the sixth vertex.*
Specify location for vertex 7 or <define faces>: *Specify the location of the seventh vertex.*
Specify location for vertex 8 or <define faces>: *Specify the location of the eighth vertex.*
Specify location for vertex 9 or <define faces>: Enter

After defining the locations of all the vertices, press ENTER and assign vertices to the first face.

Face 1, vertex 1:
Enter a vertex number or [Color/Layer]: **1**
Face 1, vertex 2:
Enter a vertex number or [Color/Layer] <next face>: **2**
Face 1, vertex 3:
Enter a vertex number or [Color/Layer] <next face>: **3**
Face 1, vertex 4:
Enter a vertex number or [Color/Layer] <next face>: **4**

Once you have assigned the vertices to the first face, give a null response at the next prompt.

Face 1, vertex 5:
Enter a vertex number or [Color/Layer] <next face>:

You can also change the color and layer of the first face with the **PFACE** command. For example, to change the color of the first face to red, following is the prompt sequence.

Face 1, vertex 5:
Enter a vertex number or [Color/Layer] <next face>: **C**
New Color [Truecolor/COlorbook] <BYLAYER>: **RED**

Now, the first face will be red.

After assigning the vertices, give a null response at the next prompt. AutoCAD provides you with the following prompt sequence, in which you need to assign vertices to the second face.

Face 2, vertex 1:
Enter a vertex number or [Color/Layer]: **1**
Face 2, vertex 2:
Enter a vertex number or [Color/Layer]: **4**
Face 2, vertex 3:
Enter a vertex number or [Color/Layer]: **5**
Face 2, vertex 4:
Enter a vertex number or [Color/Layer]: **4**

Once you have assigned the vertices to the second face, give a null response at the next prompt.

Face 2, vertex 5:
Enter a vertex number or [Color/Layer]:

Face 3, vertex 1:
Enter a vertex number or [Color/Layer]:

Similarly, assign vertices to all faces (Figure 25-52). To view the faces, select the **Realistic** option from the **Visual Styles** drop-down list in the **Visual Styles** panel of the **View** tab. To make an edge invisible, specify a negative number for its first vertex. The display of the invisible edges of 3D solid surfaces is governed by the **SPLFRAME** system variable. If **SPLFRAME** is set to 0, invisible edges are not displayed. If this variable is set to a number other than 0, all invisible edges are displayed after regeneration using the **REGENALL** command.

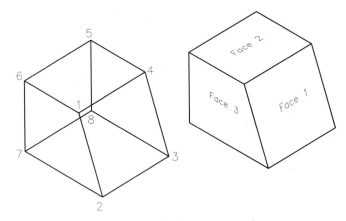

Figure 25-52 Faces of block and assigned vertices

Controlling the Visibility of the 3D Face Edges

Command: EDGE

This command is used to control the visibility of the edges created by using the **3DFACE** command. You can hide the edges of the 3D faces or display them using this command (Figures 25-53 and 25-54). On invoking this command, you will be prompted to specify the 3D face edge to toggle its visibility or display. To hide an edge, select it. To display the edges, enter **D** at the **Specify edge of 3dface to toggle visibility or [Display]** prompt. You can display all invisible 3D face edges or the selected edges using this option.

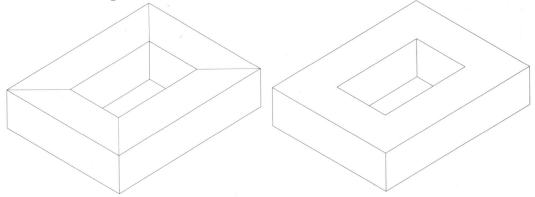

Figure 25-53 Model with the visible 3D face edges   *Figure 25-54 Model after hiding the edges*

EXERCISE 2   3DFace

In this exercise, you will apply the 3D faces to the 3D wireframe model created in Example 2.

CREATING PLANAR SURFACES

Ribbon: Surface > Create > Planar   **Command:** PLANESURF
Toolbar: Modeling > Planar Surface

The **Planar** tool is used to generate 2D surfaces on the working plane by specifying two diagonally opposite points. These two points specify the rectangular shape area to be covered by this planar surface, see Figure 25-55. The prompt sequence displayed on choosing the **Planar** tool from the **Create** panel is discussed next.

Specify first corner or [Object] <Object>: *Specify the first corner point of the planar surface.*
Specify other corner: *Specify the diagonally opposite point of the planar surface.*

You can also convert an existing object to a surface by entering **O** at the **Specify first corner or [Object] <Object>** prompt. While selecting the object, you can directly select a closed boundary or a number of individual objects that result in a closed boundary, see Figure 25-56.

The number of lines displayed in the surface is controlled by the **SURFU** and **SURFV** system variables. The **SURFU** variable controls the number of lines to be displayed in the M direction and the **SURFV** variable controls the number of lines to be displayed in the N direction.

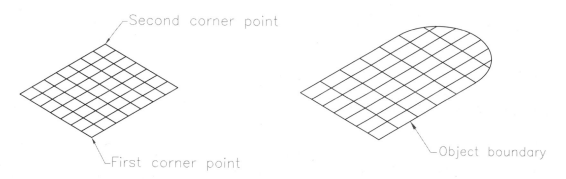

Figure 25-55 *Planar surface created by specifying corner points*

Figure 25-56 *Planar surface created by selecting an object*

THE 3DMESH COMMAND

Command: 3DMESH

This command is used to create a free-form polygon mesh by using a matrix of M X N size. When you invoke this command, you will be prompted to specify the size of the mesh in the M direction and the N direction, where M X N is the total number of vertices in the mesh. The values of M and N can be 2 to 256. Once you have specified the size, you need to specify the coordinates of each and every vertex individually. For example, if the size of the mesh in the M and N directions is 12 X 12, then you will have to specify the coordinates of all the 144 vertices. You can specify any 2D or 3D coordinates for the vertices. Figure 25-55 shows a 4 X 4 freeform mesh. This command is mostly used for programming. For a general drawing, use the **Mesh** option of the **3D** command. The 3D polygon mesh is always open in the M and N directions. You can edit this mesh or the mesh created using the **3D** command or the **Edit Polyline** tool.

EDITING THE SURFACE MESH

The polyface or polygon meshes can be edited using the **Edit Polyline** tool as discussed next.

The Edit Polyline Tool

Ribbon: Home > Modify > Edit Polyline	**Menu Bar:** Modify > Object > Polyline
Toolbar: Modify II > Edit Polyline	**Command:** PEDIT

To edit a polyface or polygon mesh, invoke the **Edit Polyline** tool from the **Modify** panel. You can also select the **Polyline Edit** option from the shortcut menu displayed on selecting the surface mesh and right-clicking on it, to edit a polyface or polygon face. The prompt sequence that will be followed on choosing the **Edit Polyline** tool and selecting the mesh is given next.

> Select polyline or [Multiple]: *Select the surface mesh.*
> Enter an option [Edit vertex/Smooth surface/Desmooth/Mclose/Nclose/Undo]: *Select an option.*

Edit Vertex

This option is used for the individual editing of vertices of the mesh in the M direction or the N direction. On invoking this option, a cross will appear on the first point of the mesh and the following sub-options will be provided. In AutoCAD 2012, the surfaces will be displayed in the **Realistic** visual style by default. Change the visual style to Wireframe by selecting the **Wireframe** option from the **Visual Styles** drop-down list in the **View** panel of the **Home** tab, so that you can view the vertices clearly.

Next. This sub-option is used to move the cross to the next vertex.

Previous. This sub-option is used to move the cross to the previous vertex.

Left. This sub-option is used to move the cross to the previous vertex in the N direction of the mesh.

Right. This sub-option is used to move the cross to the next vertex in the N direction of the mesh.

Up. This sub-option is used to move the cross to the next vertex in the M direction of the mesh.

Down. This sub-option is used to move the cross to the previous vertex in the M direction of the mesh.

Move. This sub-option is used to redefine the location of the current vertex. You can define a new location by using any of the coordinate systems or can directly select the location on the screen, see Figures 25-57 and 25-58.

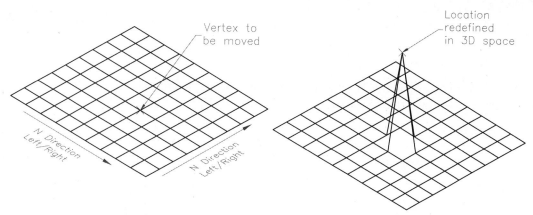

Figure 25-57 *Vertex before moving*   **Figure 25-58** *Vertex after moving*

REgen. This sub-option is used to regenerate the 3D mesh.

eXit. This sub-option is used to exit the **Edit Vertex** option.

Smooth surface

This option is used to smoothen the surface of the 3D mesh by fitting a smooth surface, as shown in Figures 25-59 and 25-60. The smoothness of the surface will depend upon the **SURFU** and **SURFV** system variables. The value of these system variables can vary between 2 and 200.

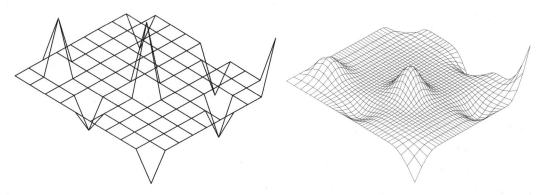

Figure 25-59 *3D mesh before fitting the smooth surface*   **Figure 25-60** *3D mesh after fitting the smooth surface*

The type of the smooth surface to be fitted is controlled by the **SURFTYPE** system variable. If you set the value of the **SURFTYPE** system variable to **5**, it will fit a Quadratic B-spline surface. If you set the value to **6**, it will fit a Cubic B-spline surface. If you set the value to **7**, it will fit a Bezier surface.

Desmooth

This option is used to remove the smooth surface that was fitted on the 3D mesh using the **Smooth surface** option and restore to the original mesh.

Mclose

This option is used to close the 3D mesh in the M direction, as shown in Figure 25-61 and Figure 25-62.

Nclose

This option is used to close the 3D mesh in the N direction, as shown in Figures 25-61 and 25-62.

Mopen

This option will be available if the 3D mesh is closed in the M direction. It is used to open the 3D mesh in the M direction (Figures 25-61 and 25-62).

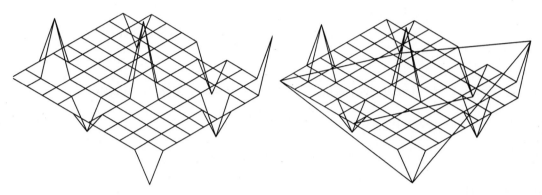

Figure 25-61 *3D mesh open in the M and N directions*

Figure 25-62 *3D mesh closed in the M and N directions*

Nopen

This option will be available if the 3D mesh is closed in the N direction. It is used to open the 3D mesh in the N direction (Figures 25-61 and 25-62).

Undo

This option is used to undo all the operations to the start of the **Edit Polyline** tool.

eXit

This option is used to exit the **Edit Polyline** tool.

DYNAMIC VIEWING OF 3D OBJECTS

Navigator Toolbar: Full Navigation Wheel	**Command:** NAVSWHEEL
Ribbon: View > Navigate > SteeringWheels	

The options available in the **Full Navigation Wheel** make the entire process of the solid modeling very interesting. You can shade the solid models and dynamically rotate them on the screen to view them from different angles. You can also define clipping planes and then rotate the solid models such that whenever the models pass through the cutting planes, they are sectioned just to be viewed. The original model can be restored as soon as you exit this commands. The tools and its options that are used to perform all these functions are discussed next.

Using the SteeringWheels

The SteeringWheels enables you to easily view and navigate through your 3D models. It is an ideal tool for navigating through your models. It includes not only the common navigating tools such as **Zoom, Pan,** and **Orbit** but also an option to move the user specified center of the model to the center of the display window. Using this tool, you can also rewind and get back your previous views. It is a time saving tool, since it combines many common navigation tools into a single tool.

The SteeringWheel is divided into different sections known as Wedges. Each wedge on a wheel represents a command. You can activate a particular navigation option by clicking on the respective wedges and then modify the view of the model by dragging the cursor, refer to Figure 25-63. The navigation tools available in the SteeringWheels are discussed next.

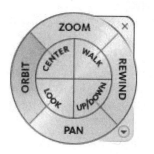

Figure 25-63 The SteeringWheel

ZOOM

This wedge is used in the same way as you use camera's zoom lens. It makes the object appear closer or far away, without changing the position of the camera. To invoke the **ZOOM** tool, move the cursor to the **ZOOM** wedge and drag. A magnifier is attached to the cursor that allows you to zoom in or out with reference to a base point. However, you can change the base point or pivot point to zoom in or zoom out of the model by choosing the **CENTER** wedge. You can even enable the incremental zoom of 25 percent on a single click from the **Zoom** area of the **SteeringWheels Settings** dialog box. To invoke this dialog box, right-click on the SteeringWheels; a shortcut menu will be displayed. Choose the **SteeringWheels Settings** option from the shortcut menu; the **SteeringWheels Settings** dialog box will be displayed. You can even choose the **Fit to Window** option from the shortcut menu of the SteeringWheels to view the entire drawing in the viewport.

REWIND

This wedge enables you to quickly restore the previous views. To invoke this tool, move the cursor to the **REWIND** wedge and click; a series of frames from the previous view orientation are displayed. You can now quickly pick the required view and control the thumbnail previews generated on changing various views using the **Rewind thumbnail** area of the **SteeringWheels Settings** dialog box.

PAN

This wedge allows you to drag the view to a new location in the drawing window. To invoke this tool, move the cursor to the **PAN** wedge and drag. A plus arrow is attached to the cursor, enabling you to drag the view in any direction. You can choose the **Fit to Window** option from the shortcut menu of the SteeringWheels to view the entire drawing in the viewport.

ORBIT

This wedge allows you to visually maneuver around the 3D objects to obtain different views. When you move your cursor to the **ORBIT** wedge and drag, a circular arrow is attached to the cursor. Now, you can rotate the model with respect to a base point. Similar to the **ZOOM** option, in this case also, you can specify the required center point of the view and rotate the model using the **CENTER** wedge option. You can also set the viewpoint upside down using the options in the **SteeringWheels Settings** dialog box. Moreover, you can choose the **Fit to Window** option to view the full drawing view.

CENTER

This wedge helps you specify the pivot or center point of the view to orbit or zoom a model. When you move your cursor to this wedge and drag, a pivot ball is attached to the cursor. You can now adjust your center or target point by placing the cursor at the required point. However, you can restore the center of the viewport by choosing the **Restore Original Center** option from the shortcut menu of the SteeringWheels. You can also restore the home view of the model by choosing the **Go Home** option from the shortcut menu.

LOOK

This tool works similar to turning the camera on a tripod without changing the distance between the camera and the target. This means the source remains the same, but the target changes. To invoke this tool, move the cursor to the **LOOK** wedge; an arc is attached to the cursor which enables you to change your view direction. You can also invert the vertical direction of the **Look** tool by selecting the **Invert vertical axis for Look tool** check box from the SteeringWheels **Settings** dialog box. You can choose the **Fit to Window** option to view the entire drawing in the viewport.

WALK

This tool moves the camera around the model as if you are walking away or towards the focus direction of the camera. In the walk 3D navigation, the camera moves parallel to the XY plane and gives the effect as if you are walking. To invoke this tool, move the cursor to the **WALK** wedge. An arrow is attached to the cursor showing the direction of the camera. You can drag the cursor to move around the model and also control the walk speed using the options in the **Walk Tool** area of the **SteeringWheels Settings** dialog box. Using this dialog box, you can also constraint the walk angle to the ground plane.

UP/DOWN

This option slides the view along the Y-axis of the screen (which is like being in an elevator). When you move your cursor to the **UP/DOWN** wedge, a vertical slider is attached to the cursor. You can slide from up to down in the slider to view the model from up to down. Using the options in the shortcut menu of the SteeringWheels, you can go back to the original view, restore the original center, or fit the entire drawing in the viewport.

Modes of SteeringWheels

There are four working modes of SteeringWheels, see Figure 25-64. These working modes are discussed next.

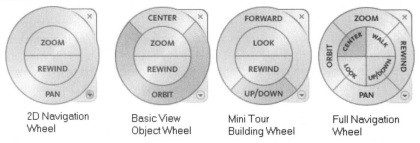

Figure 25-64 Different modes of the SteeringWheels

2D Navigation Wheel

You can invoke the SteeringWheel that has only the 2D navigation tools. To do so, choose the **2D** tool from **View > Navigate > SteeringWheels** drop-down; the 2D SteeringWheel with the **PAN**, **ZOOM**, and **REWIND** tools will be displayed.

Basic View Object Wheel

This mode contains only the viewing tools such as **CENTER**, **ZOOM**, **REWIND**, and **ORBIT**. You can invoke this mode by choosing **Basic Wheels > View Object Wheel** from the shortcut menu of the SteeringWheels or by choosing **Basic View Object Wheel** from the flyout that is displayed on clicking the downarrow in the **SteeringWheels** group in the **Navigation bar**.

Mini Tour Building Wheel

This mode is ideal for navigating inside a building. It contains the **FORWARD, REWIND, LOOK,** and **UP/DOWN** tools. You can invoke this mode by choosing **Basic Wheels > Tour Building**

Wheel from the shortcut menu of the SteeringWheels or by choosing **Mini Tour Building Wheel** from the flyout that is displayed on clicking the downarrow in the **SteeringWheels** group in the **Navigation bar**.

Full Navigation Wheel

This mode contains both the view and the navigating tools. It enables you to select a variety of tools from a single wheel. You can invoke this mode by choosing the **Full Navigation Wheel** option from the shortcut menu of the SteeringWheels or from the **Navigation bar**.

The SteeringWheels are of two types: Big Wheels and Mini Wheels. Both the wheels have the same functioning modes. The Big Wheels mode is displayed by default, whereas the Mini Wheels mode is invoked by choosing the **Mini View Object Wheel**, **Mini Tour Building Wheel**, or **Mini Full Navigation Wheel** option from the shortcut menu of the SteeringWheels or by choosing option from the **SteeringWheels** drop-down in the **Navigation bar** or by choosing option from the **SteeringWheels** drop-down in the **Navigate panel** of the **View** tab. You can customize the size, appearance, and behavior of the SteeringWheel using the **SteeringWheels Settings** dialog box, refer to Figure 25-65. You can control the size and opacity of the Big Wheels and Mini Wheels with the help of a slider available next to the wheels. Also, using this dialog box you can control the tooltips and messages that are displayed when you move the cursor over any wedge.

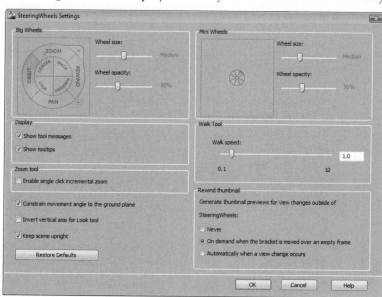

Figure 25-65 The **SteeringWheels Settings** *dialog box*

Dynamically Rotating the View of a Model

| **Ribbon:** View > Navigate > Orbit | **Navigation Bar:** Constrained Orbit |
| **Toolbar:** 3D Navigation > Constrained Orbit | **Command:** 3DORBIT |

The **Orbit** tool allows you to visually maneuver around 3D objects to obtain different views. This is one of the most important tools for the advanced 3D viewing options. All other advanced 3D viewing options can be invoked inside this tool or using the **3D Navigation** toolbar. This tool activates a 3D Orbit view in the current viewport. You can click and drag your pointing device to view the 3D object from different angles. In 3D orbit viewing, the target is considered stationary and the camera location is considered to be moving around it. The cursor looks like a sphere encircled by two arc shaped arrows. This is known as **Orbit mode** cursor, and clicking and dragging the pointing device allows you to rotate the view freely. You can move the **Orbit mode** cursor horizontally, vertically, and diagonally. If you drag the pointing device horizontally, the camera will move parallel to the XY plane of the WCS. If you move your pointing device vertically, then the camera will move along the Z axis. This tool is a transparent command and can be invoked inside any other command. You can select individual objects or the entire drawing to view before you invoke the **Orbit** tool.

Tip
Press and hold the SHIFT key and the middle mouse button to temporarily enter the constrained orbit mode.

You can invoke the other advanced viewing options using the shortcut menu that is displayed on right-clicking in the drawing window when you are inside the **Orbit** tool. The shortcut menu with the required options is shown in Figure 25-66. The options of the shortcut menu are discussed next.

Exit

This option is used to exit the **Orbit** tool. You can also exit this tool by pressing ESC.

Current Mode

This option displays the currently active navigation mode in which you obtained this shortcut menu.

Other Navigation Modes

This option enables you to choose the other navigation modes of the **3D Navigation** toolbar. In this way, you can toggle between the navigation modes without exiting the tool. The suboptions provided by the **Other Navigation Modes** are as follows:

Free Orbit. This suboption can be invoked by entering the **3DFORBIT** command or by choosing the **Free Orbit** tool from **View > Navigate > Orbit** drop-down in the Ribbon (Figure 25-67); 3D orbit view will be activated in the current viewport. When you invoke the **3DFORBIT** command, an arcball appears. This arcball is a circle with four smaller circles placed such that they divide the bigger circle into quadrants. The UCS icon is replaced by a shaded 3D UCS icon. In the 3D orbit viewing, the target is considered stationary and the camera location is considered to be moving around it. The target is the center of the arcball. The **3DFORBIT** command is a transparent command and can be invoked inside any other command.

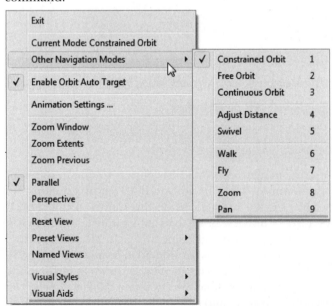

*Figure 25-66 The shortcut menu displayed when the **3DORBIT** command is active*

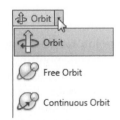

*Figure 25-67 The tools in the **Orbit** drop-down*

The cursor icon changes as you move the cursor over the 3D Orbit view. The different icons indicate the different directions in which the view is being rotated. They are as follows.

Orbit mode. When you move the cursor within the arcball, the cursor looks like a sphere encircled by two lines. This icon is the **Orbit mode** cursor, and clicking and dragging the pointing device allows you to rotate the view freely as if the cursor was grabbing a sphere surrounding the objects and moving it around the target point. You can move the **Orbit mode** cursor horizontally, vertically, and diagonally.

Roll mode. When you move the cursor outside the arcball, it changes to look like a sphere encircled by a circular arrow. When you click and drag, the view is rotated around an axis that extends through the center of the arcball, perpendicular to the screen.

Orbit Left-Right. When you move the cursor to the small circles placed on the left and right of the arcball, it changes into a sphere surrounded by a horizontal ellipse. When you click and drag, the view is rotated around the Y axis of the screen.

Orbit Up-down. When you move the cursor to the circles placed in the top or bottom of the arcball, it changes into a sphere surrounded by a vertical ellipse. Clicking and dragging the cursor rotates the view around the horizontal axis or the X axis of the screen, and passes it through the middle of the arcball.

Figure 25-68 shows a model being rotated using the **3DFORBIT** command.

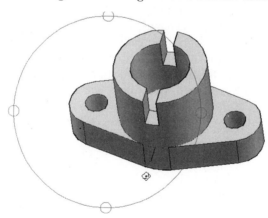

Figure 25-68 *3D orbit view and arcball with circular graphics*

Tip
*Press and hold the SHIFT+CTRL keys and the middle button of the mouse to temporarily enter the **Free Orbit** navigation mode.*

*If you are in the **3D Orbit** navigation mode, press and hold the SHIFT key to temporarily enter the **Free Orbit** navigation mode.*

Continuous Orbit. This suboption can be invoked by entering the **3DCORBIT** command. It allows you to set the objects you select in a 3D view into continuous motion in a free-form orbit. You can also invoke this command by choosing the **Continuous Orbit** tool from **View > Navigate > Orbit** drop-down. While using this suboption, the cursor icon changes to the continuous orbit cursor. Clicking in the drawing area and dragging the pointing device in any direction starts to move the object(s) in the direction you have dragged it. When you release the pointing device button, that is, stop clicking and dragging, the object continues to move in its orbit in the direction you specified. The speed at which the cursor is moved determines the speed at which the objects spin. You can click and drag again to specify another direction for rotation in the continuous orbit. While using this suboption, you can right-click in the drawing area to display the shortcut menu and choose the other suboptions without exiting the command. Choosing **Pan, Zoom, Orbit, Adjust Clipping Plane** ends the continuous orbit.

Adjust Distance. This suboption can be invoked by entering the **3DDISTANCE** command or by choosing **Other Navigation Modes > Adjust Distance** from the shortcut menu. Alternatively, choose **View > Camera > Adjust Distance > 3D Adjust Distance** from the menu bar. This suboption is similar to taking the camera closer to or farther away from the target object. Therefore, it makes the object appear closer or farther away. You can press the left button and drag the mouse to adjust the distance between the target and the camera. This option is used to reduce the distance between the camera and the object when the object rotates by a large angle in the **3DORBIT** command.

Swivel Camera. This suboption can be invoked by entering the **3DSWIVEL** command or by choosing **View > Camera > Swivel** from the menu bar. It works similar to turning the camera on a tripod without changing the distance between the camera and the target. You can press the left button and drag the mouse to swivel the camera.

Walk. This suboption can be invoked by entering the **3DWALK** command or by choosing the **Walk** tool from **Render > Animations > Walk** drop-down after customizing. It is similar to the **Walk** option in the SteeringWheel. This suboption is used to move the camera around a model with the help of the keyboard, as if you are walking away or toward the focus direction of the camera. In walk 3D navigation, the camera moves parallel to the XY plane and this seems as if you are walking. Choose the **Walk** button; the **POSITION LOCATOR** window, showing the top view of the model with the location of the camera indicated by a red spot, will be displayed. The **POSITION LOCATOR** window shows a dynamic preview of the location and movement of the camera with respect to the object, see Figure 25-69. The following keys are used to control the location of the camera with respect to the object.

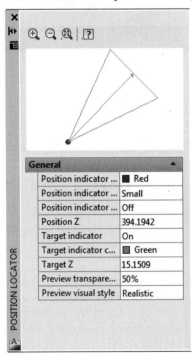

*Figure 25-69 The **POSITION LOCATOR** window*

Motion Type	Key
Move the camera in the forward direction	UP ARROW or W key
Move the camera in the backward direction	DOWN ARROW or S key
Move the camera in the left direction	LEFT ARROW or A key
Move the camera in the right direction	RIGHT ARROW or D key
Change the focus direction of the camera	Press and hold the left mouse button in the drawing area and drag the mouse in the desired direction
Change the orientation of the XY plane	Press and hold the middle mouse button in the drawing area and drag the mouse in the desired direction
Change the distance of the camera	Move the cursor to the green from the object triangular area in the **POSITION LOCATOR** window and pan to the desired location
Change the coverage area of the camera	Move the cursor onto the camera in the **POSITION LOCATOR** window and pan the camera

 Fly. This suboption can be invoked by entering the **3DFLY** command or by choosing the **Fly** tool from **Render > Animations > Walk** drop-down after customizing. This suboption is used to move the camera through a model with the help of the keyboard as if you are flying away or toward the focus direction (Figure 25-69) of the camera. In the fly 3D navigation, the camera can even move at an inclination with the XY plane and this seems as if you are flying around the object. The **POSITION LOCATOR** window will be displayed on the screen in the **3DFLY** command also. The position of the camera with respect to the object can be controlled by using the same procedure as discussed in the **3DWALK** command.

Note

*To control the settings related to walk and fly, invoke the command and right-click in the drawing area. Choose the **Walk and Fly Settings** option; the **Walk and Fly Settings** dialog box will be displayed. You can specify the settings in this dialog box.*

*The **Walk** and **Fly** 3D navigation modes can be activated only in the perspective projection. Activating the **Walk** or **Fly** 3D navigation mode in the parallel projection will display the **Warning** window, as shown in Figure 25-70.*

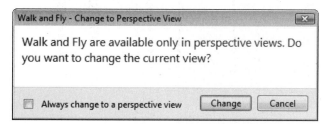

***Figure 25-70** The **Warning** window for activating the **Walk** and **Fly** 3D navigation modes in the parallel projection*

 Tip

You can toggle between the 3D navigation modes from the shortcut menu or by entering the number displayed in front of each option in the shortcut menu.

Zoom. This suboption can be invoked by entering the **3DZOOM** command. This command can also be invoked by choosing **View > Zoom** from the menu bar. It is similar to the **ZOOM** option in the SteeringWheel and functions like the camera's zoom lens. You can invoke the other options of the Zoom by entering **3DZOOM** at the Command prompt. Alternatively, the options can also be displayed by right-clicking in the drawing area when you are inside this command.

Pan. The **Pan** suboption is invoked by entering the **3DPAN** command. This command can also be invoked by choosing **View > Pan** from the menu bar. It is similar to the **PAN** option in the SteeringWheel. This command also starts the 3D interactive viewing. You can right-click in the drawing area, when you are inside this command, to display a shortcut menu that displays all the 3D Orbit options.

Enable Orbit Auto Target

This option, if chosen, ensures that the target point is focused on the object while rotating the view. If this option is cleared, the target point is set at the center of the viewport, in which the **3DORBIT** command is invoked.

Animation Settings

This option, if chosen, displays the **Animation Settings** dialog box, see Figure 25-71. Using this dialog box, you can change the settings for visual style, resolution, number of frames to be captured in one second, and the format in which you want to save the file while creating the animation of 3D navigation. Creating animations will be discussed in the next chapter.

Zoom Window/Extents/Previous

These options are used to zoom the solid model using a window, zoom the objects to the extents, or retrieve the previous zoomed views of the model.

Parallel

This option is chosen to display the selected model using the parallel projection method. It is the default method when a new AutoCAD file is started with 2D templates. In this method, no parallel lines in the model converge at any point.

Perspective

This option is chosen to use the one point perspective method to display a model. It is the default method when a new AutoCAD file is started with 3D templates. In this method, all parallel lines in the model converge at a single point, thus providing a realistic view of the model when viewed with the naked eye, see Figure 25-72.

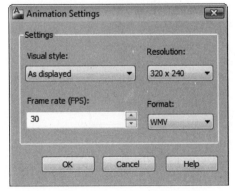

*Figure 25-71 The **Animation Settings** dialog box*

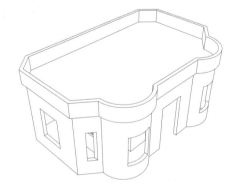

Figure 25-72 Viewing the model using the perspective projection

Reset View

This option, when chosen automatically, restores the view that was current when you started rotating the solid model in the 3D orbit. Note that you will not exit the **3DORBIT** command if you choose this option.

Preset Views

This option is used to make current any of the six standard orthogonal views or four isometric views.

Named Views

If you have created some views using the **VIEW** command, then the existing named views will be displayed as the suboptions under the **Named Views** option in the **3D Orbit** shortcut menu. Use this option to activate the saved view.

Visual Styles

This option is invoked by the **SHADEMODE** command and the suboptions provided under this option are used to specify the mode of shading of the model while it is being rotated in the 3D orbit. The suboptions used to shade the solid model are discussed next.

Conceptual. This suboption is used to display different faces of the model with a combination of two different colors. The color on the faces of a model keeps changing as you change the view of the model. The edges between the faces are smooth and at the same time, the tangent edges are hidden. The **Conceptual** visual style gives a better visualization of different faces on the model, see Figure 25-73.

Hidden. This suboption is used to display a solid model after hiding the hidden lines. You will not be able to see through the solid model.

Realistic. This suboption is used to apply the color on all faces of a solid model. The color of the shade will depend on the shade of the solid model. The edges between the faces are smooth and the tangent edges are made visible (Figure 25-74). In this visual style, the material that you attach to the object is also displayed.

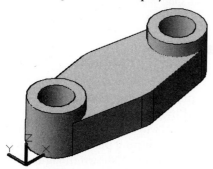

Figure 25-73 *Shading using the*
Conceptual *shade*

Figure 25-74 *Shading using the*
Realistic *shade*

Shaded. This suboption applies smooth shading to the faces of the model. In this case, the edges of the visible faces will not be visible.

Shaded with Edges. This suboption applies smooth shading to the faces of the model and the edges of the visible faces will be visible.

Shades of Grey. This suboption applies a single shade (grey) to all faces of the model.

Sketchy. This suboption is selected to display the model as if it is hand sketched.

Wireframe. This suboption is used to display the solid model as a wireframe model. You can see through the solid model while the model is being rotated in the 3D orbit view.

X Ray. This suboption is used to display the model shading and mild transparency. In this case the hidden lines will be visible.

Visual Aids

The **3DORBIT** command provides you with three visual aids for the ease of visualizing the solid model in the 3D orbit view. These visual aids are **Compass**, **Grid**, and **UCS Icon**. A check mark is displayed in front of the options you have selected. You can select none, one, two, or all three of the visual aids. Choosing **Compass** displays a sphere drawn within the arcball with three lines. These lines indicate the *X*, *Y*, and *Z* axis directions. Choosing the **Grid** displays a grid at the current *XY* plane. You can specify the height by using the **ELEVATION** system variable. The **GRID** system variable controls the display options of the **Grid** before using the **3DORBIT** command. The size of the grid will depend upon the limits of the drawing. Choosing the **UCS Icon** displays the UCS icon. If this option is not selected, then AutoCAD turns off the display of the UCS icon. While using any of the 3D Orbit options, interactive 3D viewing is on and a shaded 3D view UCS icon is displayed. The *X* axis is red, the *Y* axis is green, and the *Z* axis is blue or cyan. Figure 25-75 shows a realistic shaded solid model displaying the compass, grid, and UCS icon.

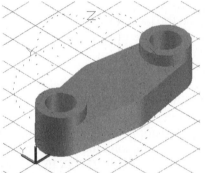

Figure 25-75 *Realistic shaded solid model showing the compass, grid, and UCS icon*

Clipping the View of a Model Dynamically

Command: 3DCLIP

The **3DCLIP** command is used to adjust the clipping planes for sectioning the selected solid model. Note that the actual solid model is not modified by this command. This command is used only for the purpose of viewing. The original solid model will be restored when you exit this command. With this option, you can adjust the location of the clipping planes. When you choose this option, the **Adjust Clipping Planes** window will be displayed, as shown in Figure 25-76. In this window, the object appears rotated at 90-degree to the top of the current view in the window. This facilitates the display of cutting planes. Setting the location of the front and back clipping planes in this window is reflected in the result in the current view. The various options of the **3DCLIP** command are displayed in the **Adjust Clipping Planes** toolbar in the window. There are seven buttons provided in this window. The functions of these buttons are discussed next.

 Adjust Front Clipping. This button is chosen to locate the front clipping plane, which is defined by the line located in the lower end of the **Adjust Clipping Planes** window. You can see the result in the 3D Orbit view if the front clipping is on.

 Adjust Back Clipping. This button is chosen to locate the back clipping plane, which is defined by the line located in the upper end of the **Adjust Clipping Planes** window. You can see the result in the 3D Orbit view if the back clipping is on.

 Create Slice. Choosing this button causes both the front and back clipping planes to move together. The slice is created between the two clipping planes and is displayed in the current 3D view. You can first adjust the front and the back clipping planes individually by choosing the respective options. Next, by choosing the **Create Slice** button, activate both the clipping planes simultaneously and display the result in the current view.

 Pan. Displays the pan cursor that you can use to pan the object in **Adjust Clipping Planes** window. Hold down the left mouse button and drag it in any direction. The pan cursor stays active until you click another button.

 Zoom. Displays a magnifying-glass cursor that you can use to enlarge or reduce the clipping plane. To enlarge the image in the **Adjust Clipping Planes** window.

 Front Clipping On/Off. Turns on or off the sectioning of a solid model using the front clipping plane, see Figure 25-77. You can toggle between the front clipping on and off by choosing the **Front Clip On/Off** button in the **3D Navigation** toolbar.

Back Clipping On/Off. Turns on or off the sectioning of the solid model using the back clipping plane. You can toggle between back clipping on and off by choosing the **Back Clip On/Off** button from the **3D Navigation** toolbar.

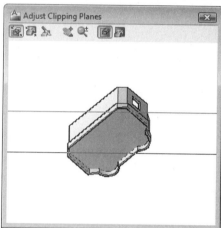

*Figure 25-76 The **Adjust Clipping Planes** window*

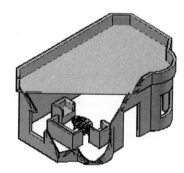

Figure 25-77 The interior of a house displayed by setting the front clipping on and then hiding the hidden lines

NUDGE FUNCTIONALITY*

This is one of the features added in AutoCAD 2012. Nudge functionality gives you the freedom to move any kind of object by 2 pixels in any direction orthogonal to screen. When the snap mode is ON, the object moves to the distance specified in the **Snap spacing** edit box in the **Drafting Setting** dialog box. To nudge any object, select the object and press CTRL + arrow keycorresponding to the direction in which you want to move the object.

Self-Evaluation Test

Answer the following questions and then compare them to those given at the end of this chapter:

1. The **DDVPOINT** command is used to change the viewpoint to view a solid model. (T/F)

2. The wireframe models can be converted into surface models as well as solid models. (T/F)

3. In AutoCAD, the right-hand thumb rule is followed to identify the direction of the X, Y, and Z axes. (T/F)

4. In AutoCAD, the right-hand thumb rule is followed to find the direction of rotation in the 3D space. (T/F)

5. The _____ and _____ commands can be used to change the viewpoint for viewing models in the 3D space.

6. Changing the viewpoint moves a 3D model from its default position. (T/F)

7. Using the **DDVPOINT** command, you can set the viewpoint with respect to _____ and _____.

8. Various types of 3D coordinate systems are _____.

9. If you set the clipping options in the **3DCLIP** command, the same options are also set in the **DVIEW** command. (T/F)

10. The _____ command is used to draw standard 3D surface primitives.

Review Questions

Answer the following questions:

1. A 3D model can be rotated continuously using the **3D Orbit** tool. (T/F)

2. The **ELEV** command is a transparent command. (T/F)

3. You can directly write a text with thickness. (T/F)

4. In which of the following views is a model displayed by default in 3D Modeling workspace?

    (a) **Perspective View**   (b) **Parallel View**
    (c) **Top View**   (d) None of these

5. Which of the following tools is used to create a 3D polyline?

    (a) **Polyline**   (b) **3D Poly**
    (c) **3D Polyline**   (d) **Polyline 3D**

6. Which of the following commands is used to set the elevation and thickness for new objects?

    (a) **ELEVATION**   (b) **THICKNESS**
    (c) **ELEV**   (d) **THICK**

7. Which of the following commands is used to suppress the hidden edges in a 3D model?

    (a) **SUPPRESS**   (b) **HIDE**
    (c) **3DHIDE**   (d) **EDGE**

8. In which of the following views are you by default when you open a new drawing?

    (a) SE Isometric View   (b) SW Isometric View
    (c) Plan View   (d) Bottom View

9   For the **Sphere** and **Torus** meshes, if the **MESHTYPE** system variable is set to 0, the resulting surface will be a _____ mesh.

10. The highlighted area in a ViewCube that is used to change a view is called _____.

EXERCISES 3 AND 4

In these exercises, you will create the models shown in Figures 25-78 and 25-79. Assume the missing dimensions.

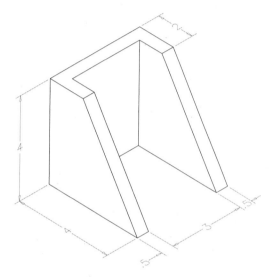

Figure 25-78 *Model for Exercise 3*

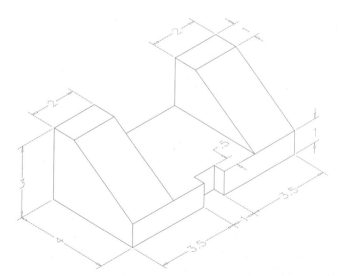

Figure 25-79 *Model for Exercise 4*

EXERCISE 5

In this exercise, you will create the model shown in Figure 25-80. The dimensions for the model are shown in Figure 25-81. Hidden lines are suppressed for clarity[**]

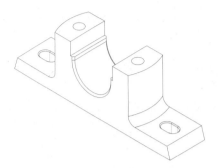

Figure 25-76 *Solid model of the Plummer Block Casting*

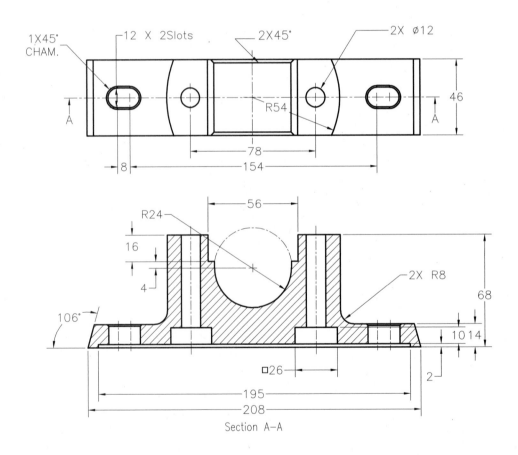

Figure 25-77 *Dimensions of the Plummer Block Casting*

Answers to Self-Evaluation Test

1. T, **2.** F, **3.** F, **4.** F, **5. DDVPOINT, VPOINT**, **6.** F, **7.** the angle in the *XY* plane from the *X* axis, angle from the *XY* plane, **8.** absolute, relative rectangular, relative cylindrical, and relative spherical, **9.** T, **10. 3D**

Chapter 26

Creating Solid Models

CHAPTER OBJECTIVES

In this chapter, you will learn:
* *About solid modeling.*
* *To create standard solid primitives and polysolids.*
* *About regions and create them.*
* *About the use of Boolean operations.*
* *To modify the visual styles of solids.*
* *To define new UCS using the DUCS, ViewCube, and Ribbon.*
* *To use the Extrude and Revolve tools to create solid models.*
* *To use the Sweep and Loft tools to create complex solid models.*
* *To use the Presspull tool.*

KEY TERMS

• *Solid Primitives*	• *Pyramid*	• *Intersect*	• *Sweep*
• *Box*	• *Polysolid*	• *INTERFERE*	• *Loft*
• *Cone*	• *Helix*	• *3D Snap*	• *Solid*
• *Cylinder*	• *Visual Styles*	• *Dynamic UCS*	• *Surface*
• *Sphere*	• *Region*	• *Extrude*	• *Presspull*
• *Torus*	• *Union*	• *Mode*	
• *Wedge*	• *Intersect*	• *Revolve*	

WHAT IS SOLID MODELING?

Solid modeling is the process of building objects that have all attributes of an actual solid object. For example, if you draw a wireframe or a surface model of a bushing, it is sufficient to define the shape and size of the object. However, in engineering, the shape and size alone are not enough to describe an object. For engineering analysis, you need more information such as volume, mass, moment of inertia, and material properties (density, Young's modulus, Poisson's ratio, thermal conductivity, and so on). When you know these physical attributes of an object, it can be subjected to various tests to make sure that it performs as required by the product specifications. It eliminates the need for building expensive prototypes and makes the product development cycle shorter. Solid models also make it easy to visualize the objects because you always think of and see the objects as solids. With computers getting faster and software getting more sophisticated and affordable, solid modeling has become the core of the manufacturing process. AutoCAD solid modeling is based on the ACIS solid modeler, which is a part of the core technology.

CREATING PREDEFINED SOLID PRIMITIVES

The Solid primitives form the basic building blocks for a complex solid. ACIS has seven predefined solid primitives that can be used to construct a solid model such as Box, Wedge, Cone, Cylinder, Pyramid, Sphere, and Torus. The number of lines in a solid primitive is controlled by the value assigned to the **ISOLINES** variable. These lines are called tessellation lines. The number of lines determines the number of computations needed to generate a solid. If the value is high, it will take significantly more time to generate a solid on the screen. Therefore, the value assigned to the **ISOLINES** variable should be realistic. When you enter commands for creating solid primitives, AutoCAD Solids will prompt you to enter information about the part geometry. The height of the primitive is always along the positive *Z* axis and perpendicular to the construction plane. Similar to surface meshes, solids are also displayed as wireframe models unless you hide, render, or shade them. The **FACETRES** system variable controls the smoothness in the shaded and rendered objects and its value can go up to 10. The tools to create solid primitives are grouped in the **Solid Primitives** drop-down in the **Modeling** panel, as shown in Figure 26-1.

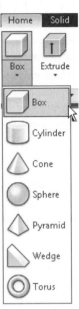

Figure 26-1 Tools in the Solid Primitives drop-down

Creating a Solid Box

Ribbon: Home > Modeling > Solid Primitives drop-down > Box	**Command:** BOX
Toolbar: Modeling > Box	**Menu Bar:** Draw > Modeling > Box

 The **BOX** command is used to create a solid rectangular box or a cube. Start a new file with the *Acad3d.dwt* template. In 3D drawing templates, you can dynamically preview the operations that you perform. The methods to create a solid box are discussed next.

Creating a Box Dynamically

To create a box dynamically, choose the **Box** tool from the **Modeling** panel. Specify the first corner point of the box and then specify the diagonally opposite corner point for defining the base of the box, as shown in Figure 26-2. Now, specify the height of the box by moving the cursor away from the base. After getting the desired height, click in the drawing window to create the box or specify the height of the box in the dynamic input box, see Figure 26-2.

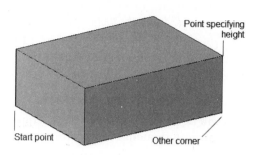

Figure 26-2 Box created dynamically

Two Corner Option

This is the default option. You can use this option to create a solid box by defining the first corner of the box and then its other corner (Figure 26-3). Note that the length of the box will always be taken along the *X* axis, the width along the *Y* axis, and the height along the *Z* axis. Therefore, in this case, when you specify the other corner, the value along the *X* axis will be taken as the length of the box, the value along the *Y* axis will be taken as the width of the box, and the value along Z axis will be taken as height of the box. Given below is the prompt sequence to draw a box of length 5 units and height 4 units.

> Specify first corner or [Center] <0,0,0>: *Specify start point.*
> Specify other corner or [Cube/Length]: *Drag the cursor in any one of the quadrants and type* **5**
> *(Planar face of length 5 units is displayed).*
> Specify height or [2Point]: *4*

Center-Length Option

The center of the box is the point where the center of gravity of the box lies. This option is used to create a box by specifying the center of the box followed by the length, width, and height of the box (Figure 26-4). The following prompt sequence is displayed when you choose the **Box** button:

> Specify first corner or [Center] <0,0,0>: **C**
> Specify center <0,0,0>: **4,4**
> Specify corner or [Cube/Length]: **L**
> Specify length: **8**
> Specify width: **6**
> Specify height or [2point]: **3**

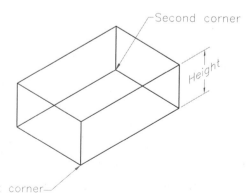

Figure 26-3 Creating a box using the Two Corner option

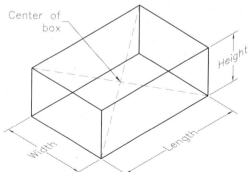

Figure 26-4 Creating a box using the Center-Length option

Tip
*The **Corner-Length** option is similar to the **Center-Length** option, except that in the **Corner-Length** option, you will define the first corner of the box and then the length, width, and height of the box. The **2point** option is used to specify the height of the box by specifying 2 points using the pointing device.*

Corner-Cube Option

This option is used to create a cube starting from a specified corner, as shown in Figure 26-5. You are creating a cube, and so you will be prompted to enter the length of the cube, that will also act as the width and height. The prompt sequence for creating the cube using the Corner-Cube option is given next.

Specify first corner or [Center] <0,0,0>: **2,2**
Specify other corner or [Cube/Length]: **C**
Specify length: **5**

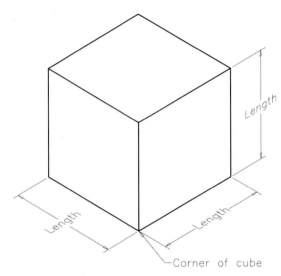

Figure 26-5 Creating a cube using the Corner-Cube option

Tip
*The **Center-Cube** option is similar to the **Corner-Cube** option, except that in the **Center-Cube** option, you will have to define the center and the length of the cube.*

Creating a Solid Cone

Ribbon: Home > Modeling > Solid Primitives drop-down > Cone	Command: CONE
Toolbar: Modeling > Cone	Menu Bar: Draw > Modeling > Cone

The **CONE** command creates a solid cone with an elliptical or circular base. This command provides you with the option of defining the cone height or the location of the cone apex. Defining the location of the apex will also define the height of the cone and the orientation of the cone base from the *XY* plane.

To create a cone, start a new file with the *Acad3d.dwt* template file and choose the **Cone** tool from the **Modeling** panel and specify the center point and the radius of the base, as shown in Figure 26-6. Next, specify the height of cone by moving the cursor away from the base. After getting the desired height, click in the drawing window to create the cone. You can also enter all parameters defining the cone in the dynamic input box. Figure 26-6 shows a cone created dynamically.

Different methods to create a cone are discussed next.

Creating a Cone with Circular Base

The options for defining the circular base are Center Radius, Center Diameter, 3 Point, 2 Point, and Tangent Tangent Radius. The prompt sequence that will be displayed when you choose the **Cone** tool is given next.

> Specify center point for base or [3P/2P/Ttr/Elliptical]: *Specify the center of the base.*

The default option to define a cone with circular base is Center Radius. In this option, after specifying the center point, enter the radius value and then the height of the cone. If you need to enter the diameter value, then enter **D** at the **Specify center point for base or [3P/2P/Ttr/Elliptical]** prompt and specify the diameter value followed by the height. The prompt sequence for using this option is given below.

> Specify center point for base or [3P/2P/Ttr/Elliptical]: *Specify the center of the base.*
> Specify base radius or [Diameter] <default>: *Specify the radius or enter* **D** *to specify the diameter of the cone.*
> Specify height or [2Point/Axis endpoint/Top radius]<default>: *Specify the height of the cone or enter an option or press the ENTER key to accept the default value.*

If you want to create a cone by specifying two ends of the circular base, then type **2P** at the **Specify base radius or [Diameter]** prompt and then specify the two ends of the diameter. Similarly, if the base of the cone is tangent to two circles, type **Ttr** at the **Specify base radius or [Diameter]** prompt, select the circles in succession, and then specify the radius.

Creating a Cone with Elliptical Base

To create a cone with elliptical base, select the **Elliptical** option at the **Specify center point for base of cone or [3P/2P/Ttr/Elliptical]** prompt. On selecting this option, the **Specify endpoint of first axis or [Center]** prompt will be displayed. Now, you can create the elliptical base by specifying the endpoints of axis, the location of the minor axis, and then the height. If you know the centerpoint of the ellipse, enter **C** at the **Specify endpoint of first axis or [Center]** prompt and specify the centerpoint, endpoint of the other two axes in succession, and finally the height; a cone with elliptical base will be created, as shown in Figure 26-7.

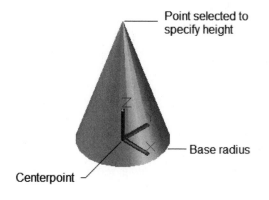

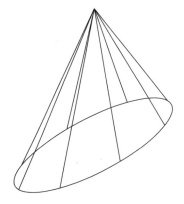

Figure 26-6 Cone created dynamically   *Figure 26-7* An elliptical cone

Creating a Cone with Apex

To create a cone with apex, first specify the base of the cone. Then, specify the height of the cone by choosing an option at the **Specify height or [2Point/Axis endpoint/Top radius]** prompt. By default, you will be prompted to specify the height of the cone. Enter the height of the cone or specify a location in the drawing area; the cone will be created.

To specify the height using the **2Point** option, enter **2P** at this prompt and specify two points in the drawing area. The distance between these two specified points will be considered as the height of the cone. The prompt sequence that will follow is given next.

> Specify height or [2Point/Axis endpoint/Top radius] <current>: **2P**
> Specify first point: *Specify the first point.*
> Specify second point: *Specify the second point on the screen.*

To specify the height using the **Axis endpoint** option, select the **Axis endpoint** option at the **Specify height or [2Point/Axis endpoint/Top radius]** prompt. On selecting this option, you can specify the endpoint of the axis, whose start point is the center of the base. Apart from fixing the height of the cone, this option is also used to specify the orientation of the cone in 3D space. This means you can also create an inclined cone, see Figure 26-8. The prompt sequence is as follows:

> Specify axis endpoint: *Specify the location of the apex or axis endpoint of the right circular cone in 3D space.*

Creating a Frustum of a Cone

To create a frustum of a cone, select the **Top radius** option at the **Specify height or [2Point/ Axis endpoint/Top radius]** prompt; you will be prompted to specify the top radius. Specify the radius of the top of the frustum as the top radius; you will be prompted to specify the height. Specify the height as discussed earlier; a frustum of a cone will be created, see Figure 26-9. The prompt to create a frustum of a cone is given next.

> Specify top radius <default>: *Specify the radius for the top of the cone frustum.*
> Specify height or [2Point/Axis endpoint] <default>: *Specify the height of the frustum or choose an option.*

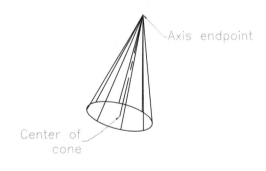

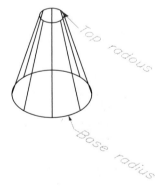

*Figure 26-8 Creating a cone using the **Axis endpoint** option*

*Figure 26-9 Creating a cone using the **Top radius** option*

Tip
In AutoCAD 2012, when you select a primitive, grips are displayed on it. You can modify the dimensions of the primitive by dragging these grips.

Creating a Solid Cylinder

Ribbon: Home > Modeling > Solid Primitives drop-down > Cylinder   **Command:** CYLINDER
Menu Bar: Draw > Modeling > Cylinder   **Toolbar:** Modeling > Cylinder

To create a cylinder, start a new file with the *Acad3d.dwt* template file and choose the **Cylinder** tool from the **Modeling** panel. Next, specify the center point of the base and then specify the radius of the base, as shown in Figure 26-10. Now, specify the height of the cylinder by moving the cursor away from the base. After getting the desired height, click in the drawing window to create the cylinder. You can also enter all parameters in the dynamic input box to define the cylinder.

Similar to the **Cone** tool, this tool provides you with two options for creating the cylinder: **circular cylinder** and **elliptical cylinder** (Figures 26-10 and 26-11). This tool also allows you to define the height of a cylinder dynamically or specify the height by using the **2Point** or **Axis endpoint** options. On selecting the **Axis endpoint** option, you can specify the endpoint of the axis, whose start point is the center of the base. Apart from fixing the height of the cylinder, this option is also used to specify the orientation of the cylinder in 3D space. This means you can also create an inclined cylinder, as shown in Figures 26-12 and 26-13.

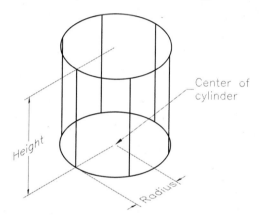

Figure 26-10 *Creating a circular cylinder*

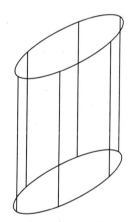

Figure 26-11 *Creating an elliptical cylinder*

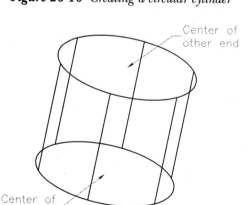

Figure 26-12 *Creating an inclined cylinder with circular base*

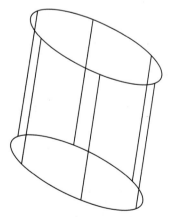

Figure 26-13 *Creating an inclined cylinder with elliptical base*

Creating a Solid Sphere

Ribbon: Home > Modeling > Solid Primitives drop-down > Sphere	**Command:** SPHERE
Menu Bar: Draw > Modeling > Sphere	**Toolbar:** Modeling > Sphere

To create a sphere, choose the **Sphere** tool from the **Modeling** panel; you will be prompted to specify the center of the sphere. After specifying the center, you can create the sphere by defining its radius or diameter, as shown in Figure 26-14. Instead of specifying the center of the sphere, you can also specify its circumference by choosing any one of the **3P/ 2P/ Ttr** options as discussed earlier.

Figure 26-14 Solid sphere created

Creating a Solid Torus

Ribbon: Home > Modeling > Solid Primitives drop-down > Torus	**Command:** TORUS
Menu Bar: Draw > Modeling > Torus	**Toolbar:** Modeling > Torus

You can use the **Torus** tool to create a torus (a tyre-tube like shape), as shown in Figure 26-15. A torus is centered on the construction plane. The top half of the torus is above the construction plane and the other half is below it. To create a torus, choose the **Torus** tool from the **Modeling** panel; you will be prompted to enter the diameter or the radius of the torus and the diameter or the radius of tube (Figure 26-16). The radius of torus is the distance from the center of the torus to the center-line of the tube. This radius can have a positive or a negative value. If the value is negative, the torus will have a rugby-ball like shape (Figure 26-17). A torus can be self-intersecting. If both the radii of the tube and the torus are positive and the radius of the tube is greater than the radius of the torus, the resulting solid looks like an apple (Figure 26-18). Instead of specifying the radius or diameter of the torus, you can specify the points through which the circumference of the torus will pass by choosing any one of the **3P**, **2P**, and **Ttr** options. These options are discussed earlier.

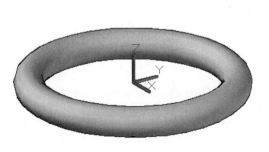

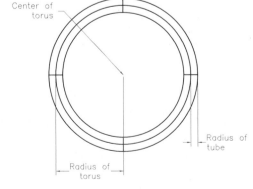

Figure 26-15 Torus created dynamically   *Figure 26-16 Parameters associated with a torus*

Figure 26-17 *Torus with a negative radius value*   **Figure 26-18** *Torus with radius of tube more than the radius of torus*

Creating a Solid Wedge

Ribbon: Home > Modeling > Solid Primitives drop-down > Wedge	**Command:** WEDGE
Menu Bar: Draw > Modeling > Wedge	**Toolbar:** Modeling > Wedge

 You can create a solid wedge by using the **Wedge** tool. To create a wedge by using this tool, you need to specify the start point, the diagonally opposite point, and the height of the wedge, as shown in Figure 26-19. The other options in this tool are similar to that of the **Box** tool. Also, the other options to create a box and a wedge are similar.

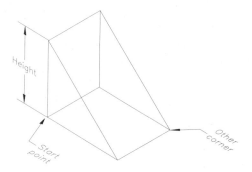

Figure 26-19 *Parameters of a wedge*

Creating a Pyramid

Ribbon: Home > Modeling > Solid Primitives drop-down > Pyramid	**Command:** PYRAMID
Menu Bar: Draw > Modeling > Pyramid	**Toolbar:** Modeling > Pyramid

The **Pyramid** tool is used to create a solid pyramid where all faces, other than the base, are triangular and converge at a point called apex (Figure 26-20). The base of a pyramid can be any polygon, but is typically a square. To create a pyramid, choose the **Pyramid** tool from the **Modeling** panel and follow the prompt sequence given next.

Command: *pyramid 4 sides circumscribed*
Specify center point of base or [Edge/Sides]: *Specify the center point.*
Specify base radius or [Inscribed] <Current Value>: *Specify the base radius.*
Specify height or [2Point/Axis endpoint/Top radius] <Current Value>: *Specify the height to create the pyramid,* refer to Figure 26-20.

Different methods for creating a solid pyramid are discussed below.

On invoking the **Pyramid** tool, you need to specify the center point of the base polygon or the length of the edges, or the number of sides of the polygon. If you select the **Edge** option in the Command prompt, you will be prompted to pick two points from the drawing area that determine the length and orientation of the edge of the base polygon. If you select the **Sides** option, you will be prompted to specify the number of sides of the base polygon. Note that in AutoCAD, the number of sides of a pyramid can vary from 3 to 32.

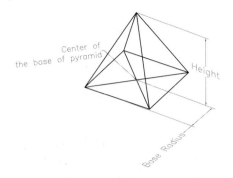

Figure 26-20 *Creating a pyramid dynamically*

Next, you need to specify the base radius of the circle in which the base polygon is created. The base polygon is either circumscribed around a circle or inscribed within a circle. By default, the base of the pyramid is circumscribed. After specifying the base radius, you need to specify the height of the pyramid or select an option.

The options to specify the height are same as for cone or cylinder.

You can also create the frustum of a pyramid by selecting the **Top radius** option at the **Specify height or [2Point/Axis endpoint/Top radius]** Command prompt. On selecting this option, you are prompted to specify the radius of the top of the pyramid frustum. Then, you need to specify the height of the frustum either dynamically or by selecting the **2Point** or the **Axis endpoint** option.

Creating a Polysolid

Ribbon: Home > Modeling > Polysolid	**Toolbar:** Modeling > Polysolid
Menu Bar: Draw > Modeling > Polysolid	**Command:** POLYSOLID or PSOLID

The **Polysolid** tool is similar to the **Polyline** tool with the only difference that this tool is used to create a solid with a rectangular cross-section of a specified width and height. This command can also be used to convert existing lines, 2D polylines, arcs, and circles to a polysolid feature. To create a polysolid, choose the **Polysolid** tool from the **Modeling** panel and follow the prompt sequence given next.

Command: _Polysolid
Height = current, Width = current, Justification = current
Specify start point or [Object/Height/Width/Justify]<Object>: *Specify the start point for the profile of the polysolid. Else, press the ENTER key to select an object to convert to a polysolid or enter an option.*

The options available at the Command prompt are discussed next.

Next Point of Polysolid
When you specify the start point for the profile of the solid, this option is displayed. This option is used to specify the next point of the current polysolid segment. If additional polysolid segments are added to the first polysolid, AutoCAD automatically makes the endpoint of the previous polysolid, the start point of the next polysolid segment. The prompt sequence is given next.

Command: _Polysolid
Height = current, Width = current, Justification = current
Specify start point or [Object/Height/Width/Justify] <Object>: *Specify the start point.*

Specify next point or [Arc/Undo]: *Specify the endpoint of the first polysolid segment.*
Specify next point or [Arc/Close/Undo]: *Specify the endpoint of the second polysolid segment or press the ENTER key to exit the command.*

Arc. This option is used to switch from drawing linear polysolid segments to drawing curved polysolid segments. The prompt sequence that will follow is given next.

Specify next point or [Arc/Close/Undo]: **A** Enter
Specify end point of the arc or [Close/Direction/Line/Second point/Undo]: *Specify the endpoint of the arc or choose an option to create the arc.*

The **Close** option closes the polysolid by creating an arc shaped trajectory from the most recent endpoint to the initial start point and exits the **POLYSOLID** command. Usually, the arc drawn with the **POLYSOLID** command is tangent to the previous polysolid segment. This means that the starting direction of the arc is the ending direction of the previous segment. The **Direction** option allows you to specify the tangent direction of your choice for the arc segment to be drawn. You can specify the direction by specifying a point. The **Line** option switches the command to the **Line** mode. The **Second point** option selects the second point of the arc in the **3P** arc option. The **Undo** option reverses the changes made in the previously drawn polysolid.

Close. This option is available when at least two segments of the polysolid are drawn. It closes the polysolid segment by creating a linear polysolid from the recent endpoint to the initial point.

Undo. This option is used to erase the most recently drawn polysolid segment. You can use this option repeatedly until you reach the start point of the first polysolid segment.

Object

This option is used to convert an existing 2D object into a polysolid. The 2D objects that can be converted into a polysolid include lines, 2D polylines, arcs, and circles. Figure 26-21 shows a polysolid object created by converting a 2D polyline. Note that the polysolid is displayed in the **Realistic** visual style by default, but for clarity, it has been displayed in the **Wireframe** visual style. The prompt sequence for using this option is given next.

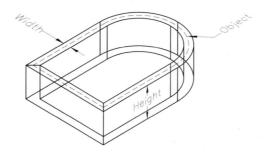

Figure 26-21 *Object converted to a polysolid by using the **Object** option*

Command: **POLYSOLID**
Height = current, Width = current, Justification = current
Specify start point or [Object/Height/Width/Justify]<Object>: Enter
Select object: *Select an object to convert it into a polysolid.*

Height

This option is used to specify the height of the rectangular cross-section of the polysolid. The value that you specify will be set as the default value for this command. The prompt sequence for this option is given next.

> Command: _Polysolid
> Height = current, Width = current, Justification = current
> Specify start point [Object/Height/Width/Justify]<Object>: **H** [Enter]
> Specify height <current>: *Specify the value of the height or press the ENTER key to accept the current value.*
> Specify start point [Object/Height/Width/Justify]<Object>: *Specify the start point of the polysolid or choose an option.*

Width

This option is used to specify the width of the rectangular cross-section of a polysolid. The prompt sequence for this option is given next.

> Command: _Polysolid
> Height = current, Width = current, Justification = current
> Specify start point [Object/Height/Width/Justify]<Object>: **W** [Enter]
> Specify width <current>: *Specify the width or press the ENTER key to accept the current value.*
> Specify start point [Object/Height/Width/Justify]<Object>: *Specify the start point of the polysolid or choose an option.*

Justify

The justification option determines the position of the created polysolid with respect to the starting point. The three justifications that are available for a **POLYSOLID** command are **Left**, **Center**, and **Right**.

Left. This justification produces a polysolid with its left extent fixed at the start point and the width added to the right of the start point, when you view the polysolid from the direction of its creation, see Figure 26-22.

Center. This justification produces a polysolid with its center fixed at the start point and the width added equally to both the sides of the start point when you view the polysolid from the direction of its creation, see Figure 26-22.

Right. This justification produces a polysolid with its right extent fixed at the start point and the width added to the left of the start point, when you view the polysolid from the direction of its creation, see Figure 26-22.

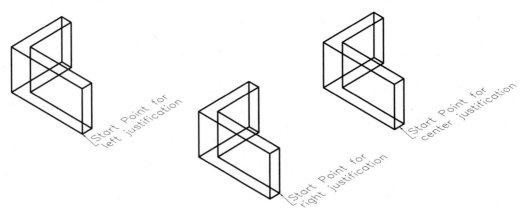

***Figure 26-22** Start point for the **Left**, **Right**, and **Center** justifications of a polysolid*

Creating a Helix

Ribbon: Home > Draw > Helix	**Toolbar:** Modeling > Helix
Menu Bar: Draw > Helix	**Command:** HELIX

The **Helix** tool is used to create a 3D helical curve. To generate a helical curve, you need to first specify the center for the base of the helix. Next, you need to specify the base radius, radius at the top of the helix, and height of the helix. A helix of converging or diverging shape can also be generated by specifying different values for the base and top radius. You can also control the number of turns in a helix and specify whether the twist will be in the clockwise direction or counter clockwise direction. Figure 26-23 shows a helix with the parameters to be defined.

The prompt sequence displayed on choosing the **Helix** tool is given next.

Command: _Helix
Number of turns = current   Twist = CCW
Specify center point of base: *Specify a point for the center of the helix base.*
Specify base radius or [Diameter] <1.0000>: *Specify the value for the base radius, enter **D** to specify the diameter, or press the ENTER key to select the default value for the radius.*
Specify top radius or [Diameter] <current>: *Specify the value for the top radius of the helix, enter **D** to specify the diameter, or press the ENTER key to select the current value of the base radius.*
Specify helix height or [Axis endpoint/Turns/turn Height/tWist] <default>: *Specify the height of the helix, press the ENTER key to select the default value or choose an option.*

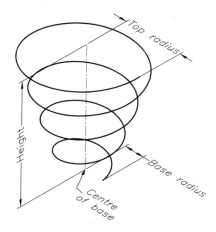

Figure 26-23 Helix with associated parameters

The options available at the command prompt are discussed next.

Axis endpoint

To specify the height using the **Axis endpoint** option, choose the **Axis endpoint** option at the **Specify helix height or [Axis endpoint/Turns/turn Height/tWist]** prompt. Using this option, you can specify the endpoint of the axis, whose start point is assumed to be at the center of the base of the helix. Apart from specifying the height of the helix, this option is also used to specify the orientation of the helix in 3D space or in other words, you can also create an inclined helix, see Figure 26-24.

Turns

To specify the number of turns in the helix, choose the **Turns** option at the **Specify helix height or [Axis endpoint/Turns/turn Height/tWist]** prompt. Using this option, you can change the number of turns in the helical curve. The default number of turns for the **HELIX** command is three.

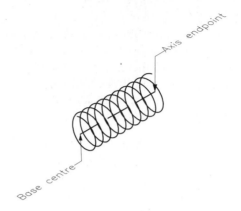

*Figure 26-24 Helix created using the **Axis endpoint** option*

Note
In a helical curve, the number of turns cannot be more than 500.

turn Height

To specify the pitch of the helix, choose the **turn Height** option at the **Specify helix height or [Axis endpoint/Turns/turn Height/tWist]** prompt. Using this option, instead of specifying the whole height of the helix, you can specify the height of one complete single turn of the helix. The helical curve will be generated with the total height equal to the turn height multiplied by the number of turns.

tWist

To specify the starting direction of the helical curve, choose the **tWist** option from the shortcut menu at the **Specify helix height or [Axis endpoint/Turns/turn Height/tWist]** prompt. By using this option, you can specify whether the curves will be generated in the clockwise or the counterclockwise direction. Figure 26-25 shows a helix with counterclockwise turns and Figure 26-26 shows a helix with clockwise turns. The default value for the starting direction of the helical curve is counterclockwise.

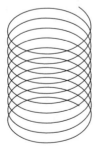

*Figure 26-25 Helix created using the **CCW** option*

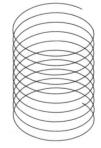

*Figure 26-26 Helix created using the **CW** option*

Tip
*You can use a generated helix as a trajectory for the **SWEEP** command, which will be discussed later in this chapter.*

MODIFYING THE VISUAL STYLES OF SOLIDS*

Ribbon: Home > View > Visual Styles > Visual Styles Manager
Command: VISUALSTYLES   **Menu Bar:** Tools > Palettes > Visual Styles

AutoCAD provides you with various predefined modes of visual styles. These visual styles are grouped together in the **Visual Styles Manager**, see Figure 26-27. To invoke the **Visual Styles Manager**, select the **Visual Styles Manager** option from the **Visual Styles** drop-down list in the **View** panel. Alternatively, click on the inclined arrow at the bottom right of the **Visual Styles** panel in the **View** tab. There are ten visual styles available in the **Available Visual Styles in Drawing** rollout of the **Visual Styles Manager**. To apply a visual style, you need to double-click on it. Alternatively, select the desired visual style from the **Visual Styles** drop-down list in the **View** panel; the model will be update automatically to the new visual style.

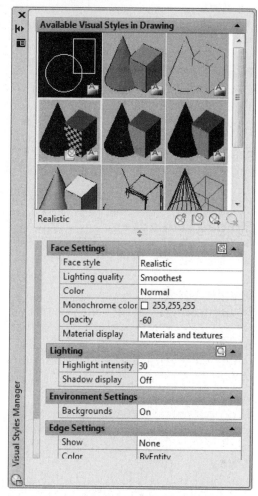

The properties of the selected visual style will be displayed below the **Available Visual Styles in Drawing** rollout. You can modify these properties depending upon your requirement. You can also create a user-defined visual style based on the current visual style applied to the model. To do so, choose the **Create New Visual Style** button from the tool strip at the bottom of the **Available Visual Styles in Drawing** rollout; the **Create New Visual Style** dialog box will be displayed. Enter a new name and description for the newly created visual style. Next, choose **OK**; a new visual style will be created. Set the properties of the new visual style by modifying the default properties. Alternatively, set the properties of a visual style and then select **Save as a New Visual Style** from the **Visual Styles** drop-down list on the **View** tab; you will

*Figure 26-27 The **Visual Styles Manager** palette*

be prompted to specify a name to save the current visual style. Enter a name and press ENTER; the new visual style will be saved.

Tip
*In AutoCAD 2012, you can also use the **In-canvas Viewport controls** displayed on the top left corner of the drawing area for changing the visual style of a solid.*

Available Visual Styles in Drawing

The options available in this rollout are discussed next.

2D/3D Wireframe

On applying this visual style, all hidden lines will be displayed along with the visible lines in the model. Sometimes, it becomes difficult to recognize the visible lines and hidden lines, if you set this visual style for complex models.

Conceptual

In this visual style, the model will be displayed as shaded and the edges of the visible faces of the model will also be displayed.

Hidden

On applying this visual style, the hidden lines in the model will not be displayed, and only the edges of the faces that are visible in the current viewport will be displayed.

Realistic

This visual style is the same as **Conceptual**, but with a more realistic appearance. Moreover, if materials are applied to a model, this visual style will display the model along with the materials applied.

Shaded

In this visual style, smooth shading is applied to the faces of a model. In this case, the edges of the visible faces are not visible.

Shaded with edges

In this visual style, smooth shading is applied to the faces of a model and the edges of the visible faces are visible.

Shades of Gray

In this visual style, a single shade (grey) is applied to all faces of a model.

Sketchy

In this visual style, a model appears as if it is hand sketched.

Wireframe

This visual style is used to display a solid model as a wireframe model. You can see through the solid model while the model is being rotated in the 3D orbit view.

X-Ray

This visual style is used to display the model with shading and mild transparency. In this case, the hidden lines will be visible.

2D Wireframe options

The options in this area are discussed next.

Contour Lines

The lines that are displayed on the curved surface of a solid are known as contour lines. The default value for the number of contours to be displayed on each curved surface is 4 and it is controlled by system variable **ISOLINES**. Figure 26-28 shows a cylinder with 4 contour lines and Figure 26-29 shows a cylinder with 8 contour lines.

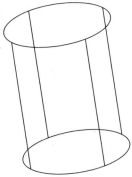

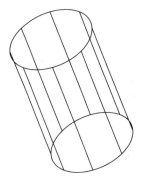

Figure 26-28 *Cylinder with 4 contour lines*   **Figure 26-29** *Cylinder with 8 contour lines*

Draw true silhouettes

This option is used to specify whether or not to display the true silhouette edges of a model when it is hidden by using the **HIDE** command. If you select **Yes** from this drop-down list, the silhouette edges will be displayed when you hide the model, see Figure 26-30. However, when you select **No** from this drop-down list, the silhouette edges will not be displayed when you hide the model, see Figure 26-31.

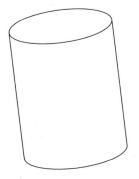

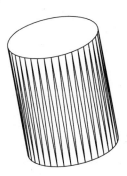

Figure 26-30 *Model with the true silhouette edges on*   **Figure 26-31** *Model with the true silhouette edges off*

2D Hide - Occluded Lines

The options in this area are used to set the linetype and color of the hidden lines (also called occluded lines).

Linetype

This drop-down list is used to set the linetype for the obscured lines. You can select the desired linetype from this drop-down list. The linetype of the hidden lines can also be modified using the **OBSCUREDLTYPE** system variable. The default value of this variable is 0. As a result, the hidden lines are suppressed when you invoke the **HIDE** command. You can set any value between 0 and 11 for this system variable. The following table gives the details of the different values of this system variable and the corresponding linetypes that will be assigned to the hidden lines:

Value	Linetype	Sample
0	Off	None
1	Solid	————————————

2	Dashed	‒ ‒ ‒ ‒ ‒ ‒ ‒ ‒ ‒ ‒ ‒
3	Dotted	··
4	Short Dash	‒ ‒ ‒ ‒ ‒ ‒
5	Medium Dash	‒ ‒ ‒ ‒ ‒ ‒
6	Long Dash	‒‒ ‒‒ ‒‒ ‒‒ ‒‒
7	Double Short Dash	‒ ‒ ‒
8	Double Medium Dash	‒‒ ‒‒ ‒‒
9	Double Long Dash	‒‒ ‒‒ ‒‒ ‒‒
10	Medium Long Dash	‒‒ ‒‒ ‒‒ ‒‒ ‒‒
11	Sparse Dot	▪ ▪ ▪ ▪ ▪ ▪

Figure 26-32 shows a model with a hidden linetype changed to dashed lines and Figure 26-33 shows the same model with hidden lines changed to dotted lines.

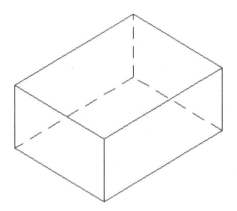

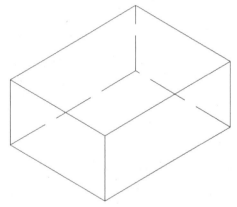

Figure 26-32 *Model with hidden lines changed to dashed lines*

Figure 26-33 *Model with hidden lines changed to double long dashed lines*

Color

The **Color** drop-down list is used to define the color for the obscured lines. If you define a separate color, the hidden lines will be displayed with that color, when you invoke the **HIDE** command. You can select the required color from this drop-down list. This can also be done using the **OBSCUREDCOLOR** system variable. Using this variable, you can define the color to be assigned to the hidden lines. The default value is 257. This value corresponds to the **ByEntity** color. You can enter the number of any color at the sequence that will follow when you enter this system variable. For example, if you set the value of the **OBSCUREDLTYPE** variable to **2**, and that of the **OBSCUREDCOLOR** to **1**, the hidden lines will appear in red dashed lines, when you invoke the **HIDE** command.

> **Note**
> *The linetype and the color for the hidden lines defined using the previously mentioned variables are valid only when you invoke the **HIDE** command. They do not work if the model is regenerated.*

2D - Intersection Edges

The options in this area are used to display a 3D curve at the intersection of two surfaces. Note that the 3D curve is displayed only after invoking the **HIDE** command. These options are discussed next.

Show

If you select **Yes** from the **Visible** drop-down list, a 3D curve will define the intersecting portion of the 3D surfaces or solid models.

Color

The **Color** drop-down list is used to specify the color of the 3D curve that is displayed at the intersection of the 3D surfaces or solid models.

2D Hide - Miscellaneous

This head in the **Visual Styles Manager** is used to set the percentage for halo gap.

Halo gap %

The **Halo gap %** area is used to specify the distance in terms of percentage of unit length by which the edges that are hidden by a surface or solid model will be shortened, see Figures 26-34 and 26-35.

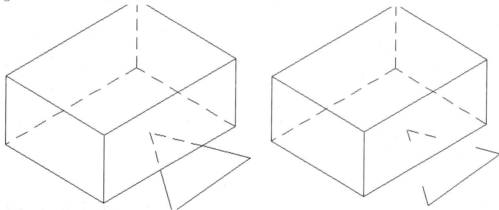

Figure 26-34   *Hiding lines with halo gap % as **0***

Figure 26-35   *Hiding lines with halo gap % as **40***

Note

The obscured linetype in Figures 26-34 and 26-35 is changed to dashed.

Similarly, you can specify the settings for faces, edges, and environment required for other visual styles. These settings are discussed next.

Controlling the Settings of Edges

You can control display settings of the edges by using the **Edge Settings** rollout. Figure 26-36 shows different types of edge display. The **Show** field in this rollout has three options: **None**, **Isolines**, and **Facet Edges**. On selecting the **Isolines** option, the contour lines on the surface of the solid will be displayed. Also, the **Always on Top** field will be available in the **Show** field. On selecting the **Yes** option in this field, all isolines in the model will be displayed. If you do not want the edges to be displayed, choose **None** in the **Show** field. On selecting the **Facet Edges** option in the **Show** field, the edges between the planar surfaces will be displayed. You can also set other options in the sub rollout. Note that some of the sub rollouts will not be available for the **Shaded** and **Realistic** visual styles.

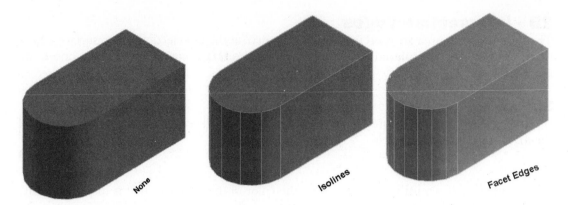

Figure 26-36 Different types of edge display

The **Occluded Edges** sub rollout is used to control the display of hidden edges. Select the **Yes** option in the **Show** filed to display the hidden edges. You can also set the color and the line type of the hidden edges in the **Color** and **Linetype** fields, respectively.

The options in the **Intersection Edges** sub rollout are used to control the display of the resulting curve when two surfaces intersect. Select the **Yes** option in the **Show** filed to display the curve at the intersection. You can also set the color and the line type of the resulting curve in the **Color** and **Linetype** fields, respectively.

The **Silhouette Edges** sub rollout is used to control the display of the silhouette edges. Select the **Yes** option in the **Show** filed to display the silhouette edges. You can set the width of the silhouette edges in the **Width** field. This value ranges from 1 to 25.

The **Edge Modifiers** sub rollout is used to assign special effects such as **Line Extensions** and **Jitter** to edges. Both the effects exhibit a hand-drawn effect with the only difference that the sketches drawn in **Line Extensions** extend beyond the model, whereas in **Jitter**, the edges look as if they were drawn using pencil and has multiple lines. Note that you need to choose the **Line Extension edges** and **Jitter edges** buttons from the title bar of the subrollout to enable these options. You can set the value for the extension line from the vertices by using the **Line Extensions** spinner. Similarly, you can control the effect of jitter by using the options in the **Jitter** drop-down list. Figure 26-37 shows the effect of line extensions and jitter edges. The **Crease angle** field will be displayed only if you choose the **Facets Edges** option in the **Edge Settings** rollout. Note that the Facets edges will be displayed only if the angle between the facets is smaller than the specified crease angle.

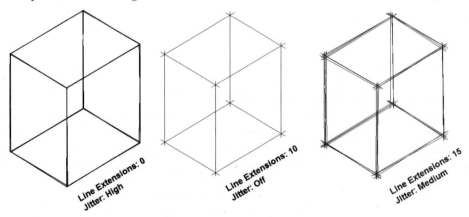

*Figure 26-37 The **Line Extensions** and **Jitter** effects of the edges*

Controlling the Face Display

AutoCAD enables you to control the display of faces by providing options to apply color effects and shading to solids or surfaces. The shading of models can be controlled by using the **Face style** field. The options in this field are **Realistic**, **Gooch**, and **None**, refer to Figure 26-38. By default, the **Realistic** option is chosen and therefore gives a real world effect to the model. The **Gooch** effect enhances the display by using light colors, instead of dark colors. If the **None** option is chosen, then no style will be applied to the model and only the edges will be displayed.

You can also control the appearance of faces in solids or surfaces using the **Lighting quality** field. The options available in this field are **Faceted**, **Smooth**, and **Smoothest**. On selecting the **Faceted** option, the color is applied to each face, thereby making the object to appear flat. On selecting the **Smooth** option, the color applied is the gradient between the two vertices of the face. This option is selected by default. On selecting the **Smoothest** option, the color is calculated for per pixel lighting.

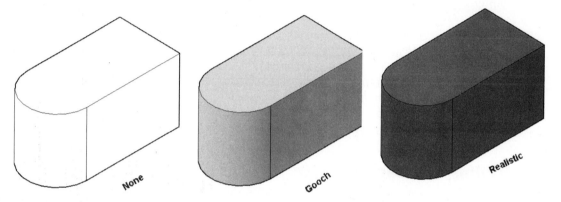

Figure 26-38 Different face styles

You can apply different color settings to faces by selecting an option from the **Color** drop-down list in the **Face Settings** rollout. The options in this drop-down list are **Normal**, **Monochrome**, **Tint**, and **Desaturate**. No face color will be applied if the **Normal** option is selected. If you select the **Monochrome** option, all faces will be shaded with a single color. If you want to use the same color to shade all faces by changing the hue and saturation values of the color, then select the **Tint** option. The hue and saturation values can be changed by using the **Tint** drop-down list. Select the **Desaturate** option to soften the color by reducing the saturation by 30%.

You can also control the opacity of solid models by choosing the **Opacity** button in the title bar of the **Facet Settings** rollout. On doing so, the **Opacity** field will be enabled. Specify the desired transparency level. Figure 26-39 shows an object with different opacity values.

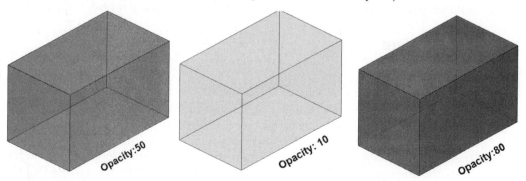

Figure 26-39 An object with different opacity value

The shininess of an object is set by using the options in the **Lighting** rollout. To set the shininess, choose the **Highlight intensity** button on the title bar of this rollout and set the required value in the **Highlight intensity** spinner. Note that you cannot set this value when a material is applied on it.

Shadows provide a real effect at rendering. Shadows are generated when light is applied. Lights can be user-defined lights or the sunlight. The options in this drop-down list are **Off**, **Ground Shadow**, and **Mapped Object shadows**. The Ground Shadows are the shadows that an object casts on the ground. The Mapped Object shadows are the shadows cast by one object on other objects. Shadows appear darker when they overlap. Use of shadows can slow down the system performance. So, you need to turn off the shadows while working, and turn them on when you need shadows. You can set the display of the shadow by using the **Properties** palette. The options under the **Shadow** area of the **Properties** palette are **Casts and Receives Shadows**, **Casts shadows**, **Receives shadows** and **Ignore shadows**. The shadows will be explained in detail in later chapters.

Note
*The **Mapped Object shadows** cannot be displayed when the **Enhanced 3D Performance** option is off. To turn off this option, enter **3dconfig** at the Command prompt to invoke the **Adaptive Degradation and Performance Tuning** dialog box. Then, choose the **Manual Tune** button; the **Manual Performance Tuning** dialog box will be displayed. Select the **Enable hardware acceleration** check box and click on the **ON** option in the **Value** field of the **Enhanced 3D Performance** and select **OFF** to turn off the effect.*

Controlling the Backgrounds

You can control the display of background by using the option in the **Environment Settings** rollout. You can turn on or off the background by using the **Backgrounds** drop-down list in this rollout. The background can be a single solid color, a gradient fill, an image, or the sun and the sky. This is explained in detail in Chapter 31 (Rendering and Animating Designs).

Note
*In AutoCAD, the default visual style applied to a model is **Realistic**. However, in this textbook, the **Wireframe** visual style is applied to models for the clarity of printing.*

CREATING COMPLEX SOLID MODELS

Until now you have learned how to create simple solid models using the standard solid primitives. However, the real-time designs are not just the simple solid primitives, but complex solid models. These complex solid models can be created by modifying the standard solid primitives with the Boolean operations or directly by creating complex solid models by extruding or revolving the regions. All these options of creating complex solid models are discussed next.

Creating Regions

Ribbon: Home > Draw > Region	**Toolbar:** Draw > Region
Menu Bar: Draw > Region	**Command:** REGION

The **Region** tool is used to create regions from the selected loops or closed entities. Regions are the 2D entities with properties of 3D solids. You can apply the Boolean operation on the regions and you can also calculate their mass properties. Bear in mind that the 2D entity you want to convert into a region should be a closed loop. Once you have created regions, the original object is deleted automatically. However, if the value of the **DELOBJ** system variable is set to **0**, the original object is retained. The valid selection set for creating the regions are closed polylines, lines, arcs, splines, circles, or ellipses. The current color, layer, linetype, and lineweight will be applied to the regions.

CREATING COMPLEX SOLID MODELS BY APPLYING BOOLEAN OPERATIONS

You can create complex solid models by applying the Boolean operations on the standard solid primitives. The Boolean operations that can be performed are union, subtract, intersect, and interfere. The commands used to apply these Boolean operations are discussed next.

Combining Solid Models

Ribbon: Home > Solid Editing > Solid, Union	**Toolbar:** Modeling > Union
Menu Bar: Modify > Solid Editing > Union	**Command:** UNION

The **Solid, Union** tool is used to apply the Union Boolean operations on the selected set of solids or regions. You can create a composite solid or region by combining them using this command. You can combine any number of solids or regions. When you invoke this tool, you will be asked to select the solids or regions to be added. Figure 26-40 shows two solid models before union and Figure 26-41 shows the composite solid created after union.

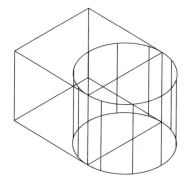

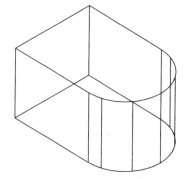

Figure 26-40 Solid models before union   *Figure 26-41 Composite solid created after union*

Subtracting One Solid From the Other

Ribbon: Home > Solid Editing > Solid, Subtract	**Toolbar:** Modeling > Subtract
Menu Bar: Modify > Solid Editing > Subtract	**Command:** SUBTRACT

This tool is used to create a composite solid by removing the material common to the selected set of solids or regions. On invoking this tool, you will be prompted to select the set of solids or regions to subtract from. Once you have selected it, you will be prompted to select the solids or regions to subtract. The material common to the first selection set and the second selection set is removed from the first selection set. The resultant object will be a single composite solid, see Figures 26-42 and 26-43.

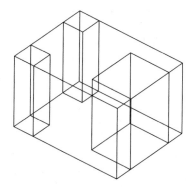

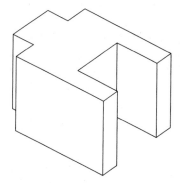

Figure 26-42 Solid models before subtracting   *Figure 26-43 Composite solid created after subtracting*

Intersecting Solid Models

Ribbon: Home > Solid Editing > Solid, Intersect   **Command:** INTERSECT
Menu Bar: Modify > Solid Editing > Intersect   **Toolbar:** Modeling > intersect

The **Solid, Intersect** tool is used to create a composite solid or region by retaining the material common to the selected set of solids or regions. When you invoke this tool, you will be asked to select the solids or regions to intersect. The material common to all the selected solids or regions will be retained to create a new composite solid (Figures 26-44 and 26-45).

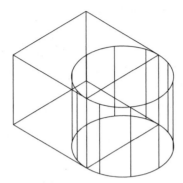

Figure 26-44 Solid models before intersecting

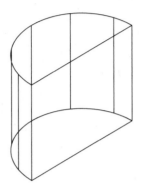

Figure 26-45 Solid model created after intersecting

Checking Interference in Solids

Ribbon: Home > Solid Editing > Interfere   **Command:** INTERFERE
Menu Bar: Modify > 3d Operations > Interference Checking

The **Interfere** tool is used to create a composite solid model by retaining the material common to the selected sets of solids. This tool is generally used for analyzing interference between the mating parts of an assembly. On invoking this tool, you will be prompted to select the first set of objects. Once you have selected them, you will be prompted to select the second set of objects. Select the second set of objects and press ENTER; the **Interference Checking** dialog box will be displayed. Also, the interference will be highlighted in the drawing area. The **Interfering objects** area in this dialog box displays the number of objects selected in both the first and second sets and the number of interferences found between them. Choose the **Previous** or the **Next** button in the **Highlight** area to view the previous or next interferences. If the **Zoom to pair** check box is selected in this area, the interference is zoomed while cycling through the interference objects. You can also use the navigation tools in the **Interference Checking** dialog box to view the interfering area clearly. Clear the **Delete interference objects created on close** check box, if you need to retain the interference objects after closing the dialog box. Close this dialog box by choosing the **Close** button.

If you clear the **Delete interference objects created on Close** check box and then close the **Interference Checking** dialog box, you can move the interfering objects by using the **Move** tool. Figure 26-46 shows two mating components of an assembly with interference between them and Figure 26-47 shows the interference solid created and moved out.

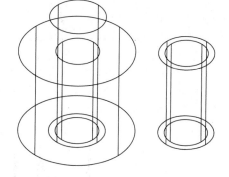

Figure 26-46 *Two mating components with interference*

Figure 26-47 *Interference solid created using the* ***INTERFERE*** *command*

You can also specify the visual style and color for the interfering objects. To do so, invoke the **Interfere** tool, type **S** at the **Select first set of Objects or [Nested selection/Settings]** prompt and press ENTER; the **Interference Settings** dialog box will be displayed. In this dialog box, specify the required parameters.

EXAMPLE 1

In this example, you will create the solid model shown in Figure 26-48.

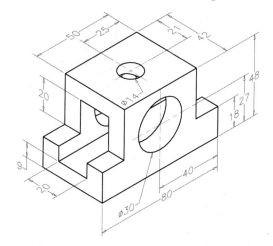

Figure 26-48 *Solid model for Example 1*

1. Increase the drawing limits to 100,100. Zoom to the limits of the drawing.

2. Enter **UCSICON** at the Command prompt. The following prompt sequence is displayed:

    Enter an option [ON/OFF/All/Noorigin/ORigin/Properties] <ON>: **N**

3. Right-click on the ViewCube and choose the **Parallel** option.

4. Invoke the ortho mode, if it is not turned on.

5. Select **Wireframe** in **Home > View > Visual Styles** drop-down to set the visual style as Wireframe.

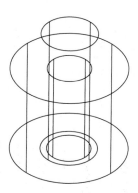

6. Choose the **Box** tool from **Home > Modeling > Solid Primitives** drop-down and follow the prompt sequence given next.

   Specify first corner or [Center] <0,0,0>: **10,10**
   Specify other corner or [Cube/Length]: **L**
   Specify length: *(Move the cursor along Y axis)* **80**
   Specify width: **42**
   Specify height: **48**

7. Again, invoke the **Box** tool and follow the prompt sequence given below.

   Specify first corner or [Center] <0,0,0>: *Specify a point on the screen.*
   Specify other corner or [Cube/Length]: **L**
   Specify length: *(Move the cursor along X axis)* **42**
   Specify width: **15**
   Specify height: **30**

8. Right-click on the **3D Object Snap** button in the Status Bar and select the **Midpoint on edge** option. Clear the other options in this shortcut menu.

9. Turn **3D Object Snap** on by choosing it, if it is not already on.

10. Move the new box inside the old box by snapping the midpoint on edge of both the boxes.

11. By using the **Copy** tool, and the **Vertex** option of the **3D Object Snap**, copy the new box to the other side of the box, refer to Figure 26-49.

12. Choose the **Solid, Subtract** tool from the **Solid Editing** panel and follow the prompt sequence given next.

    Select solids and regions to subtract from.
    Select objects: *Select the bigger box.*
    Select objects: Enter
    Select solids and regions to subtract.
    Select objects: *Select one of the smaller boxes.*
    Select objects: *Select other smaller boxes.*
    Select objects: Enter

13. Change the visual style to Hidden from the **In-canvas Visual Style controls**. The model after subtracting the boxes should look similar to the one shown in Figure 26-50.

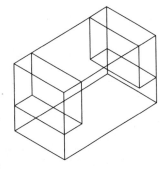

Figure 26-49 *Boxes moved inside the bigger box using midpoints*

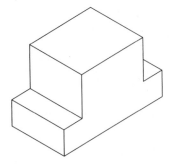

Figure 26-50 *Model after subtracting the smaller boxes*

14. Again, change the visual style to **Wireframe**. Then, choose the **Box** tool and follow the prompt sequence given next.

    Specify first corner or [Center] <0,0,0>: *Specify a point on the screen.*
    Specify other corner or [Cube/Length]: **L**
    Specify length: *(Move the cursor along the X axis)* **20**
    Specify width: **80**
    Specify height: **29**

15. Invoke the **Move** tool and select the box created in the previous step.

16. Specify the midpoint of the lower edge as the base point; the **Specify second point or <use first point as displacement>** prompt is displayed.

17. Enter **From** and press ENTER; you are prompted to specify the base point.

18. Specify the midpoint of the lower edge of the solid created by using the **Subtract** tool.

19. Move the cursor vertically upward. Enter **9** at the Command prompt and press ENTER; the box is moved, as shown in Figure 26-51.

20. Choose the **Solid, Subtract** tool from the **Solid Editing** panel and follow the prompt sequence.

    Select solids and regions to subtract from.
    Select objects: *Select the model.*
    Select objects: Enter
    Select solids and regions to subtract.
    Select objects: *Select the box.*
    Select objects: Enter

   The model after changing the visual style to **Hidden** should look similar to the one shown in Figure 26-52.

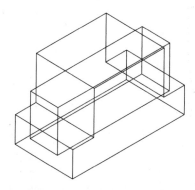

Figure 26-51 *Model before subtraction*

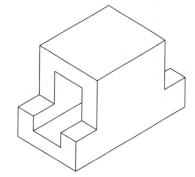

Figure 26-52 *Model after subtraction*

21. Right-click on the **3D Object Snap** button in the Status Bar and choose the **Center of face** option. Clear the other options in this shortcut menu.

22. Choose the **Cylinder** tool from **Home > Modeling > Solid Primitives** drop-down and follow the command sequence given next to create the cylinder, as shown in Figure 26-53.

Specify center point of base or [3P / 2P / Ttr/ Elliptical]: *Move the cursor on top face of the model and snap the center of the top face.*
Specify base radius or [Diameter]: **7**
Specify height or [2Point / Axis endpoint]: *Move the cursor vertically downward and enter* **10**

23. Subtract this cylinder from the existing model, refer to Figure 26-54.

24. Create a new UCS on the front face of the model.

25. Choose **Home > Modeling > Cylinder** from the **Ribbon**. The following command sequence will be displayed:

Specify center point of base or [3P / 2P / Ttr/ Elliptical]: *Use the **From** option to specify the center point of the cylinder at a distance of* **27** *units from the midpoint of the longer edge of the base.*
Specify base radius or [Diameter]: **15**
Specify height or [2Point / Axis endpoint]: **-42**

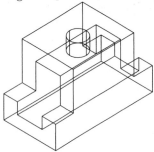
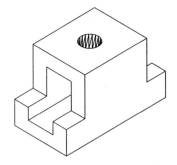

Figure 26-53 *Model before subtraction*   ***Figure 26-54*** *Model after subtraction*

26. Subtract this cylinder from the model and change the visual style to **Shades of Gray**. The final model should look similar to the one shown in Figure 26-55.

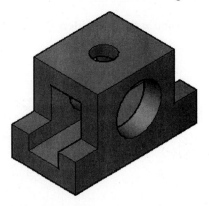

Figure 26-55 *Final solid model for Example 1*

EXERCISE 1

In this exercise, you will create the solid model shown in Figure 26-56. Save this drawing with the name *\Ch-26\Exercise1.dwg*.

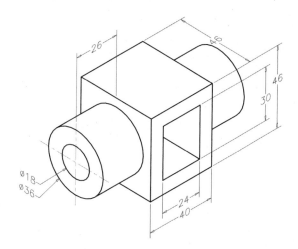

Figure 26-56 Solid model for Exercise 1

DYNAMIC UCS

This option is used to temporarily align the XY plane of the current UCS with the selected face of an existing solid. Choose the **Allow / Disallow Dynamic UCS** toggle button from the Status Bar to turn it on. Now, invoke a command and move the cursor near the face on which you want to create the new sketch or solid; the face will be highlighted in dashed lines. Click the left mouse button on the desired face; the XY plane of the UCS will automatically be aligned with the selected face. Now, you can create the sketch or solid at the selected face. After the completion of the sketch or feature, the UCS will again move automatically to its original position.

For example, if you want to create a cylinder on the slant face of the model shown in Figure 26-57, choose the **Dynamic UCS** button from the Status Bar to turn it on. Then, choose the **Cylinder** tool from the **Modeling** panel. Next, move the cursor near the slant face of the model; the slant face of the model will be highlighted. Next, left-click on that face and create the cylinder, as shown in Figure 26-58.

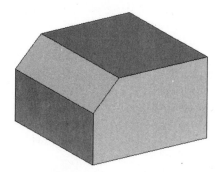

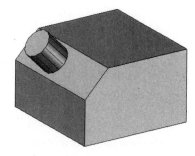

Figure 26-57 Model with UCS at the origin

Figure 26-58 UCS dynamically moved on the slant face of the model

If you want a better visualization of how the X, Y, and Z axes of the UCS will be oriented, right-click on the **DUCS** button and choose the **Display crosshair labels** option from the shortcut menu. This will enable you to see the X, Y and Z labels attached with the axes and they will keep on changing as you move the cursor to different faces.

Note
*The option of **DUCS** works only when any command is active.*

Tip
*To toggle between the on and off states of the **DUCS** button, use the F6 function key. To temporarily deactivate the **DUCS** button, press and hold the SHIFT+Z keys.*

DEFINING THE NEW UCS USING THE ViewCube AND THE RIBBON

In the earlier chapters, you learned how to create new UCS using the **UCS** command. You can also define a new UCS is parallel to the current viewing plane by using the ViewCube and the **Ribbon**. To define a new UCS whose XY plane is parallel to the current viewing plane, click on a hotspot in the ViewCube; the view corresponding to the hotspot will become normal to screen. Next, choose **WCS > New UCS** from the **WCS** flyout in the ViewCube; you will be prompted to specify the origin. Enter **V** at the Command prompt and press ENTER; a new UCS will be created at the current viewing plane. Now, if you draw a sketch, it will be in the new UCS. To save the new UCS, choose the **UCS, Named UCS** tool from the **Coordinates** panel in the **View** tab; the **UCS** dialog box will be displayed with the **Unnamed** option. Rename it and choose **OK**; the new name of the UCS will be saved and also displayed in the **WCS** flyout of the ViewCube.

To define a new UCS using the **Ribbon**, choose any one of the predefined orthographic views from the **Views** drop-down list in the **Views** panel of the **View** tab; the model will orient to that view. Also, you will notice that the ViewCube is also oriented according to the selected view, but **Top** is displayed as the view and **Unnamed** is displayed at the **WCS** flyout. Now, if you draw a sketch, it will be in the new UCS. However, if there are different faces parallel to the new UCS, you can draw sketches on those faces.

CREATING EXTRUDED SOLIDS

Ribbon: Home > Modeling > Solid Creation drop-down > Extrude Or Solid > Solid > Extrude	
Command: EXTRUDE	**Menu Bar:** Draw > Modeling > Extrude
Toolbar: Modeling > Extrude	

Sometimes, the shape of a solid model is such that it cannot be created by just applying the Boolean operations on the standard solid primitives. In such cases, you can use the tools from the **Solid Creation** drop-down, as shown in Figure 26-59. These tools are also available in the **Solid** panel of the **Solid** tab. Using these tools, you can create solid models of any complex shape.

The **Extrude** tool is used to create a complex solid/surface model by extruding a 2D entity or a region along the Z axis or any specified direction or about a specified path. On invoking this tool, following prompt sequence will be displayed.

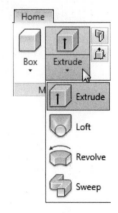

*Figure 26-59 Tools in the **Solid Creation** drop-down*

Command: _extrude
Current wire frame density: ISOLINES=4, Closed profiles creation mode = Solid

Select objects to extrude or [MOde]: _MO Closed profiles creation mode
[SOlid/SUrface] <Solid>: _SO
Select objects to extrude or [MOde]:

In AutoCAD, you can specify mode to extrude an object. There are two modes available: **Solid** and **Surface**. If you need to change the mode, enter **MO** at the **Select objects to extrude or [MOde]** prompt and specify the mode. Note that if the original entity is a closed loop or a region, you can extrude it as a solid/surface model. If the 2D entity is an open loop, then you can extrude it as a surface only. After specifying the mode, select the objects to be extruded and press ENTER; the **Specify height of extrusion or [Direction/Path/Taper angle/Expression]** prompt will be displayed. Specify the depth of extrusion or select other options to specify the depth. Different methods to extrude an object by using the options at the Command prompt are discussed next.

Extruding along the Normal

This is the default option and is used to create a model by extruding a 2D entity or a region along the normal. Figure 26-60 shows the region to be converted into an extruded solid and Figure 26-61 shows the solid created on extruding the region. The prompt sequence that will be displayed when you choose the **Extrude** tool is given next.

Select objects to extrude: *Select the region.*
Select objects to extrude: [Enter]
Specify height of extrusion or [Direction /Path /Taper angle]: *Specify the height.*

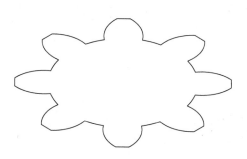

Figure 26-60 *Base object for extruding*

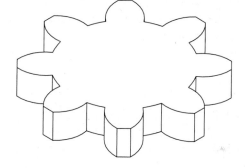

Figure 26-61 *Solid created on extruding*

Extruding with a Taper Angle

You can specify the taper angle for an extruded solid/surface by selecting the **Taper angle** option at the **Specify height of extrusion or [Direction/Path/Taper angle]** prompt. The positive value of the taper angle will taper in from the base object and the negative value will taper out of the base object (Figure 26-62).

Extruding along a Direction

This option is used to create a solid by extruding a 2D entity or a region in any desired direction by specifying two points. You can create an inclined extruded object by selecting a start point and an endpoint. The distance between the start point and endpoint acts as the height of the extrusion. Figure 26-63 shows an object extruding along a direction.

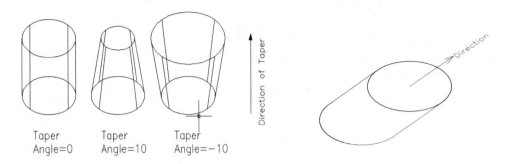

Taper
Angle=0

Taper
Angle=10

Taper
Angle=−10

Figure 26-62 *Results of various taper angles*

Figure 26-63 *Object extruding along a direction*

Extruding along a Path

This option is used to extrude a 2D entity or a region about a specified path. Bear in mind that the path of extrusion should be normal to the plane of the base object. If the path consists of more than one entity, all of them should be first joined using the **PEDIT** command so that the path remains a single entity. This option is generally used for creating complex pipelines and also by architects and interior designers for creating beadings. The path used for extrusion can be a closed entity or an open entity and the valid entities that can be used as path are lines, circles, ellipses, polygons, arcs, polylines, or splines. You cannot specify the taper angle when you use a path. Figure 26-64 shows a base object and a path about which the base object has to be extruded and Figure 26-65 shows the solid created upon extruding the base entity about the specified path.

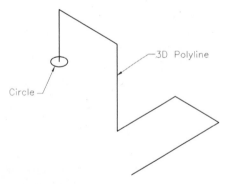

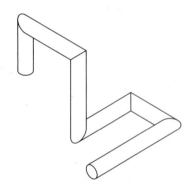

Figure 26-64 *The base object and the path for extrusion*

Figure 26-65 *Solid model created on extruding the base entity about the specified path*

Extruding using Expressions

This option is used to specify the extrusion depth in terms of formula or equations, as you have created parametric drawings.

CREATING REVOLVED SOLIDS

Ribbon: Home > Modeling > Solid Creation drop-down > Revolve
        Or Solid > Solid > Revolve
Menu Bar: Draw > Modeling > Revolve
Command: REVOLVE   **Toolbar:** Modify > Revolve

The **Revolve** tool is used to create a complex model by revolving 2D entities or regions about a specified revolution axis. The entire 2D entity should be on one side of the revolution axis. Self-intersecting and crossed entities cannot be revolved using this tool. Remember that the direction of revolution is determined using the right-hand thumb rule. The axis of rotation is defined by specifying two points, using the *X* or *Y* axis of the current UCS, or using an existing object. Similar to the **Extrude** tool, you can create a solid/surface model by specifying the mode after invoking the **Revolve** tool. On specifying the mode and selecting the object, the **Specify axis start point or define axis by [Object/X/Y/Z]** prompt will be displayed. Specify the axis of revolution. After setting the revolution axis option, you will be prompted to specify the angle of revolution. The default value is 360°. You can enter the required value at this prompt.

The options that are used to specify the axis are discussed next.

Start Point for the Axis of Revolution

This option is used to define the axis of revolution using two points: the start point and the endpoint of the axis of revolution. The positive direction of the axis will be from the start point to the endpoint and the direction of revolution will be defined using the right-hand thumb rule. Before revolving, make sure the complete 2D entity is on one side of the axis of revolution.

Object

This option is used to create a revolved model by revolving a selected 2D entity or a region about a specified object. The valid entities that can be used as an object for defining the axis of revolution can be a line or a single segment of a polyline. If the polyline selected as the object consists of more than one entity, then AutoCAD draws an imaginary line from the start point of the first segment to the endpoint of the last segment. This imaginary line is then taken as the object for revolution.

X

This option uses the positive direction of the *X* axis of the current UCS for revolving the selected entity. If the selected entity is not completely on one side of the *X* axis, it will give you an error message that it cannot revolve the object.

Y

This option uses the positive direction of the *Y* axis of the current UCS as the axis of revolution for creating the revolved solid.

Z

This option uses the positive direction of the *Z* axis of the current UCS as the axis of revolution for creating the revolved solid.

Figure 26-66 shows the entity to be revolved and the revolution axis along with the revolved solid. Figure 26-67 shows the same solid revolved through an angle of 270°.

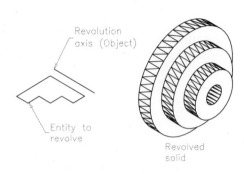

Figure 26-66 *Creating a revolved solid*

Figure 26-67 *Solid revolved to an angle*

CREATING SWEPT SOLIDS

Ribbon: Home > Modeling > Solid Creation drop-down > Sweep
Or Solid > Solid > Sweep/Loft Drop-down > Sweep
Command: SWEEP **Menu Bar:** Draw > Modeling > Sweep
Toolbar: Modeling > Sweep

The **Sweep** tool is used to sweep an open or a closed profile along a path. This tool is also available in the **Solid** panel of the **Solid** tab. To create this model by using this tool, you need an object and a path. The object is a cross-section for the sweep feature and the path is the course taken by the object while creating the swept solid. Figure 26-68 shows the object to be swept and the path to be followed and Figure 26-69 shows the resulting sweep feature.

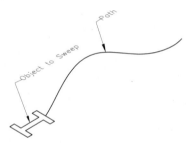

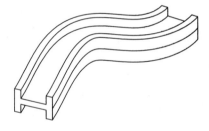

Figure 26-68 *Object to be swept and path*

Figure 26-69 *Resulting sweep feature*

The path specified for the profile and the object for the cross-section of the swept model can be open or closed. If the object for the cross-section is open, or closed but not forming a single region, the generated feature will result in a surface sweep. If the object for the section is closed and it forms a single region, the generated feature will be a solid. Figure 26-70 shows the object and the closed path and Figure 26-71 shows the resulting sweep feature.

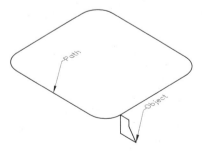

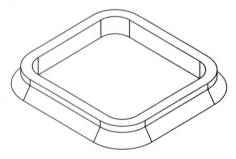

Figure 26-70 *Object to be swept with a closed path*

Figure 26-71 *Resulting sweep feature*

Note

*The path selected for the **Sweep** tool should not form a self-intersecting loop.*

You can select more than one section at a time to be swept along a path, but all these sections should be drawn on the same plane. To create a sweep draw the cross section profile, path, and select the Sweep tool. Then, follow the prompt sequence given next.

> Command: _sweep
> Current wire frame density: ISOLINES=4, Closed profiles creation mode = Solid
> Select objects to sweep or [MOde]: *Select the object to be swept.*
> Select objects to sweep or [MOde]: *Select more objects to be swept or press ENTER.*
> Select sweep path or [Alignment/Base point/Scale/Twist]: *Select the path to be followed by the object or enter an option.*

Depending on your requirement, you can invoke the following options at the **Select sweep path or [Alignment/Base point/Scale/Twist]** prompt:

Alignment

This option is used to specify whether the object will be oriented normal to the curve at the start point. By default, the sweep feature is created with a section normal to the path at the start point. To avoid this, choose **No** from the shortcut menu at the **Align sweep object perpendicular to path before sweep [Yes/No]** prompt. Figure 26-72 shows the object and the path for the sweep command. Figure 26-73 shows the resulting sweep feature with the aligned option **Yes** and Figure 26-74 shows the resulting sweep feature with the aligned option **No**.

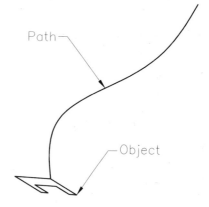

Base Point

This option is used to specify a point on the object that will be attached to the path while sweeping it. You can specify the location of the cross-section while sweeping

Figure 26-72 Object and path for sweep command

it when prompted to specify the base point. Figure 26-75 shows the path and the cross-section with base points P1 and P2. Figure 26-76 shows the resulting sweep feature with P1 as the base point and Figure 26-77 shows the resulting sweep feature with P2 as the base point.

*Figure 26-73 Swept solid with the aligned option **Yes***

*Figure 26-74 Swept solid with the aligned option **No***

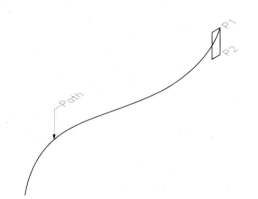

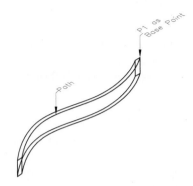

Figure 26-75 *Path with object and P1 and P2 as base points*

Figure 26-76 *Sweep with P1 as base point*

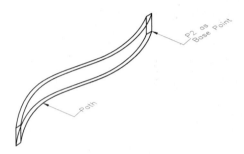

Figure 26-77 *Sweep with P2 as the base point*

Scale

This option is used to scale an object uniformly from the start of the path to the end of the path while sweeping it. Note that the size of the start section will remain the same and then end section will be scaled by the specified value. Also, the transition from the beginning to the end is smooth. Figure 26-78 shows a pentagon swept along a path with a scale factor of 0.25.

Twist

You can rotate an object about a path uniformly from the start to the end with a specified angle. The value of the twist should be less than 360°. Figure 26-79 shows a sweep feature with 300-degree of twist.

 Note

*A helix can also be used as the path for the **Sweep** tool. Figure 26-80 shows the object and a helical path generated using the **Helix** tool and Figure 26-81 shows the resulting swept solid.*

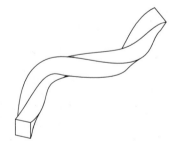

Figure 26-78 *Sweep with the **Scale** option*

Figure 26-79 *Sweep with the **Twist** option*

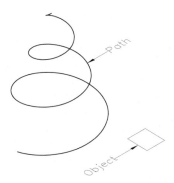

Figure 26-80 *Helical path and object to be swept*

Figure 26-81 *Sweep feature with the helical path*

CREATING LOFTED SOLIDS

Ribbon: Home > Modeling > Solid Creation drop-down > Loft
Or Solid > Solid > Sweep/Loft Drop-down > Sweep
Command: LOFT
Menu Bar: Draw > Modeling > Loft   **Toolbar:** Modeling > Loft

The **Loft** tool is used to create a feature by blending two or more similar or dissimilar cross-sections together to get a free form shape. These similar or dissimilar cross-sections may or may not be parallel to each other. To create a loft feature, choose the **Loft** tool from the **Solid** panel (Figure 26-82); the following prompt sequence will be displayed:

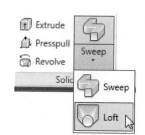

Figure 26-82 *Selecting the **Loft** tool from the **Sweep/Loft** drop-down in the **Solid** panel*

```
Command: _loft
Current wire frame density:  ISOLINES=4, Closed profiles creation mode = Solid
Select cross sections in lofting order or [POint/Join multiple edges/MOde]: _MO
Closed profiles creation mode [SOlid/SUrface] <Solid>: _SO
Select cross sections in lofting order or [POint/Join multiple edges/MOde]
```

Next, select at least two sections in a sequence in which you want to blend them. The prompt sequence that will be followed after choosing the **Loft** button is given next.

```
Select cross sections in lofting order or [POint/Join multiple edges/MOde]: 1 found
Select cross sections in lofting order or [POint/Join multiple edges/MOde]: 1 found, 2
total
Select cross sections in lofting order or [POint/Join multiple edges/MOde]:
 2 cross sections selected
Enter an option [Guides/Path/Cross sections only/Settings] <Cross sections only>: Enter
```

Note

*As you are prompted to select cross-sections while creating a loft, you can select open or closed sections. Note that if you select an open section first, all subsequent sections should be open. The model thus created will be a surface. Similarly, if you select a closed section, all subsequent sections should be closed. The model created using these sections can be solid/surface. By default, a loft model is solid. Therefore, if cross sections selected are closed profiles and you need a surface model, then first you need to change the mode by entering **MO** at the **Select cross sections in lofting order or [POint/Join multiple edges/MOde]** prompt.*

After specifying the mode, you can select the cross sections, an imaginary point as the start point of the loft, or multiple edges on a solid model. These options are discussed next.

POint

In AutoCAD, you can start a loft such that it begins from an imaginary point and passes through different sections. To do so, enter **PO** at the **Select cross sections in lofting order or [POint/Join multiple edges/MOde]** prompt and specify the start point of the loft, as shown in Figure 26-83. Next, select the cross sections in succession and press ENTER after selecting all cross sections; the preview of the loft feature will be displayed. Also, a Look up grip near the last cross section and the **Enter an option [Guides/Path/Cross sections only/Settings/COntinuity/Bulge magnitude] <Cross sections only>** prompt will be displayed. Select the **Cross sections only** option and press ENTER; the loft will be created, as shown in Figure 26-84.

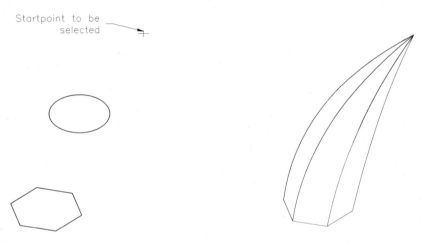

Figure 26-83 *Sections and the imaginary start point to be selected*

Figure 26-84 *Resulting loft feature by using the PO option*

Join multiple edges

In AutoCAD, you can create a loft feature between the edges of a solid model and a profile. To do so, enter **J** at the **Select cross sections in lofting order or [POint/Join multiple edges/ MOde]** prompt; the cursor will have the symbol of solid and you will be prompted to select edges that are into be joined to a single cross section. Select the edges in succession (See Figure 26-85)and press ENTER. Then, select the cross sections and press ENTER; the preview of the loft will be displayed. One look up grip at the start of the loft surface and other look up grip near the base of the model will be displayed. The Look up grip at the start of the loft will have the surface continuity symbol (See Figure 26-86) and the other Look up grip will have the Loft symbol. Click on the Look up grip that has the surface continuity symbol and select the required option from the flyout displayed. The options in the other look up grip will be discussed later.

Given below is the prompt sequence to select the edges on the top face of the cuboid and a circle as the cross section.

> Select cross sections in lofting order or [POint/Join multiple edges/MOde]: **J** `Enter`
> Select edges that are to be joined into a single cross section: (*Select the first edge*)1 found
> Select edges that are to be joined into a single cross section: (*Select the second edge*) 1 found, 2 total
> Select edges that are to be joined into a single cross section: (*Select the third edge*) 1 found, 3 total
> Select edges that are to be joined into a single cross section: (*Select the fourth edge*) 1 found, 4 total

Select edges that are to be joined into a single cross section: [Enter]
Select cross sections in lofting order or [POint/Join multiple edges/MOde]: 1 found
Select cross sections in lofting order or [POint/Join multiple edges/MOde]: [Enter]
2 cross sections selected [Enter]
Enter an option [Guides/Path/Cross sections only/Settings/COntinuity/Bulge magnitude] <Cross sections only>: [Enter] *(Preview of the loft will be displayed with grips (See Figure 26-86). Press ENTER; the loft surface will be created, (Figure 26-87).*

After selecting all cross sections and on pressing ENTER, the **Enter an option [Guides/Path/ Cross sections only/Settings/COntinuity/Bulge magnitude] <Cross sections only>** prompt will be displayed. The **COntinuity** and **Bulge magnitude** options will be displayed in the Command prompt only if you have selected the edges of a model, as one of the cross sections. Depending on requirement, you can invoke any one of those options. These options are discussed next.

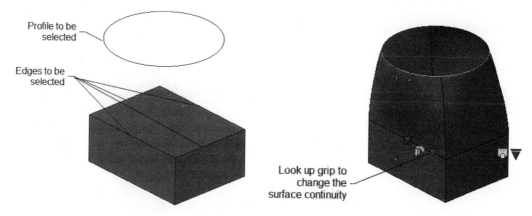

Figure 26-85 *Edges and the profile to be selected*   *Figure 26-86* *Look up grips displayed*

Figure 26-87 *Loft surface created*

Guide

AutoCAD enables you to select guide curves between the sections of the loft feature. These guide curves define the shape of the material addition between the selected sections for the loft feature. Guide curves also specify the points of different sections to be joined while adding material between them. This reduces the possibility of an unnecessary twist in the resulting feature. You can select as many guide curves as you require. But these guide curves should pass through or should touch each section that you selected for loft feature. Also the guides should start from the first section and terminate at the last section. Figure 26-88 shows three sections for the loft feature. Figure 26-89 shows the resulting loft feature. Figure 26-90 shows the sections to be lofted and the guide curves and Figure 26-91 shows the resulting model.

Figure 26-88 *Three section curves for the loft feature*

Figure 26-89 *Resulting loft feature*

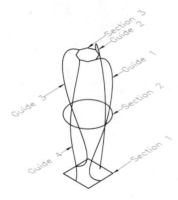

Figure 26-90 *Sections and guide curves for the loft feature*

Figure 26-91 *The loft feature created by using the guide curve*

Path

The path option is used to create a loft feature by blending more than one section along the direction specified by the path curve. Note that you can select only one path curve. The selected path curve may or may not touch all sections, but it should pass through all sketching planes of the sections. Figure 26-92 shows the sections and the path along which the blending of material will propagate and Figure 26-93 shows the resulting loft feature.

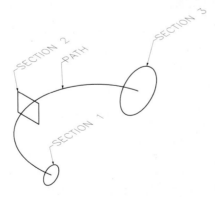

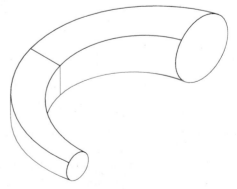

Figure 26-92 *Sections and path to be used for the **Loft** option*

Figure 26-93 *Resulting loft feature after using the **Path** option*

Cross sections only

This is the default option displayed at the **Enter an option [Guides/Path/Cross sections only/ Settings/COntinuity/Bulge magnitude] <Cross sections only>** prompt. If you need to create a loft without a guide or path, press ENTER; the loft feature will be created by using the cross sections only.

Settings

When you select **Settings**, the **Loft Settings** dialog box will be displayed, as shown in Figure 26-94. This dialog box is used to control the shape of the material addition between sections without specifying any guide curve or path. You can also select these options from the Look up grip displayed along with the preview. Various options in the **Loft Settings** dialog box are discussed next.

Ruled

If you select the **Ruled** radio button, a straight blend will be created by connecting different sections with straight lines. The loft feature created by this option will have sharp edges at the sections. Figure 26-95 shows the loft feature when the **Ruled** radio button is selected.

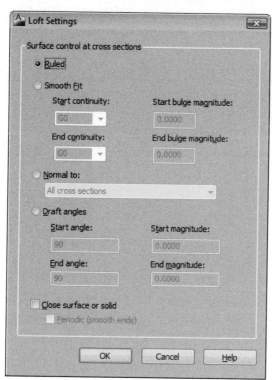

Figure 26-94 *The* **Loft Settings** *dialog box*

Smooth Fit

If you select the **Smooth Fit** radio button, a smooth loft will be created between sections. The loft feature created by this option will have smooth edges at the intermediate section.

Figure 26-96 shows the loft feature when the **Smooth Fit** radio button is selected.

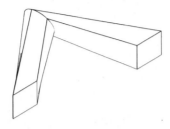

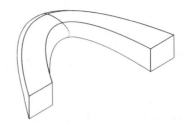

Figure 26-95 *Loft solid with the **Ruled** radio button selected*

Figure 26-96 *Loft solid with the **Smooth Fit** radio button selected*

Normal to

This option controls the shape of the loft between the sections. When you select this radio button, the drop-down list will be enabled. The following options are available in this drop-down list:

Start cross section. In this option, the normal of the lofted feature is normal to the start section. The loft feature starts normal to the start section and follows a smooth polyline as it approaches the next section. Figure 26-97 shows the section selected to create the loft and Figure 26-98 shows the loft created by specifying the **Start cross section** option.

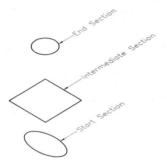

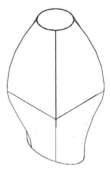

Figure 26-97 *Start and end sections to create loft*

Figure 26-98 *Loft solid created by blending material normal to the start section*

End cross section. In this option, the normal to the blending material is normal to the end section. At the second last section, the blending material starts along a spline and while reaching up to end section, it becomes normal to it, as shown in Figure 26-99.

Start and End cross sections. In this option, the normal to the blending material is normal to both the start and end section.

All cross sections. In this option, the normal to the blending material always remains normal to all cross-sections, as shown in Figure 26-100.

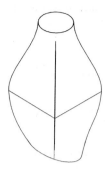

Figure 26-99 *Loft solid created by blending material normal to the end section*

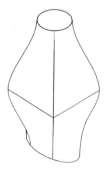

Figure 26-100 *Loft solid created by blending material normal to all sections*

Draft angles

In this option, you can define the angle at the start section and the end section of the loft. Remember that you cannot define an angle for the intermediate sketches. You can also specify the distance up to which the blending material will follow this draft angle before bending toward the intermediate sections. The various options for specifying the draft angle are discussed next.

Start angle. This edit box is used to specify the draft angle at the start section of the loft. Figure 26-101 shows the two sections to be blended and Figure 26-102 shows the loft with the start and end angles as 0-degree. Figures 26-103 and 26-104 show the loft feature with the start angle as 90-degree and 180-degree, respectively.

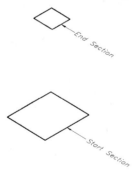

Figure 26-101 *Start and end sections of the loft feature*

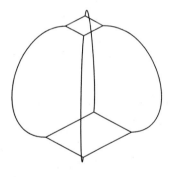

Figure 26-102 *Loft feature with 0-degree start and end angles*

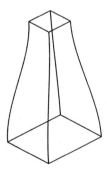

Figure 26-103 *Loft feature with 90-degree start and end angles*

Figure 26-104 *Loft feature with 180-degree start angle and 90-degree end angle*

Start magnitude. The value specified in this region signifies the distance up to which the blending material will follow the start angle before bending toward the next section. Figure 26-105 shows a loft feature with 180-degree as the start and end angles, and 15 as the start magnitude. Figure 26-106 shows a loft feature with 180-degree as the start and end angles and 2 as the start magnitude. Notice the difference in the extent up to which the blended material follows the start angle in the two figures.

End angles. This edit box is used to specify the draft angle at the end section of the loft. Figure 26-101 shows two sections to be blended and Figure 26-102 shows the loft with the start and end angles as 0-degree. Figure 26-103 and Figure 26-104 show the loft feature with the end angle as 90-degree.

End magnitude. The value specified in this region signifies the distance up to which the blending material will follow the end angle before approaching the last section. Figure 26-105 shows a loft feature with 180-degree as the start and end angles, and 2 as the end magnitude. Figure 26-106 shows a loft feature with 180-degree as the start and end angles, and 10 as the end magnitude. Notice the difference in the extent up to which the blended material follows the end angle in the two figures.

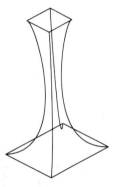

Figure 26-105 Loft feature with 180-degree as the start and end angles, 15 as the start magnitude, and 2 as the start magnitude

Figure 26-106 Loft feature with 180-degree as the start and end angles, 2 as the start magnitude, and 10 as the end magnitude

Close surface or solid

The **Close surface or solid** check box is selected to close a loft feature by joining the end section with the start section. Figure 26-107 shows a loft feature created with this check box cleared and Figure 26-108 shows the same sections lofted by this check box selected.

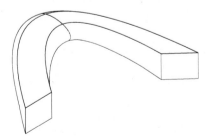

*Figure 26-107 Loft solid with the **Close surface or solid** check box cleared*

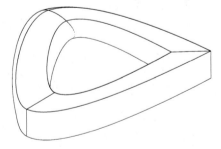

*Figure 26-108 Loft solid with the **Close surface or solid** check box selected*

Note

*To visualize the effect of the **Close surface or solid** option, the loft feature should contain at least three sections. Otherwise, the extra blending material will be added at the same location where the loft feature would have been created without selecting this check box, thereby resulting in no change.*

Periodic (smooth ends). If this check box is selected, the seam of the closed surface will not buckle, if you alter the shape of the surface.

Tip

You can also use a point as a section for the loft feature, but it should always be the start section or the end section of the loft feature.

Continuity

Continuity defines the smoothness of the loft surface between two cross sections, if one of the cross sections is an edge of a model. Continuity can be set to **G0** (position continuity), **G1** (tangency continuity), or **G2** (curvature continuity). **G0** maintains the position of the selected edges, **G1** maintains tangency of the loft surface where it meets edges, and **G2** maintains the same curvature. By default, the continuity is set to G0. To edit the continuity setting of a loft, enter **CON** at the **Enter an option [Guides/Path/Cross sections only/Settings/COntinuity/Bulge magnitude] <Cross sections only>** prompt and press ENTER; you will be prompted to specify the continuity. Enter the required option and press ENTER.

You can also set the surface continuity by using the Continuity grip.

Bulge magnitude

The bulge magnitude determines the roundness or the amount of bulge between two cross sections, if one of the cross sections is an edge of a model. By default, the bulge magnitude is set to 0.5. If you want to edit the bulge magnitude, enter **B** at the **Enter an option [Guides/Path/Cross sections only/Settings/COntinuity/Bulge magnitude] <Cross sections only>** prompt and press ENTER; you will be prompted to enter the loft bulge magnitude at the start. Enter a value and press ENTER. If there is only one solid edge as a cross section, then the bulge will be created. If there are two solid edges as cross sections (one at the start and the other at the end), then you will be prompted to enter the bulge magnitude at the start and end. Type a value and press ENTER; the bulge will be created. If you need to change the bulge magnitude while creating, then click the Bulk magnitude Look up grip; a dot with a leader will be displayed. Drag the dot dynamically to change the bulk magnitude.

You can modify the bulk magnitude value after the loft is created. To do so, select the loft; three grips will be displayed on it. Right-click on the loft feature and choose the **Properties** option from the shortcut menu displayed; the **Properties** palette will be displayed with the settings of the loft feature. Using this palette, you can modify the bulge magnitude and other parameters.

Figure 26-109 shows the edges on the top faces of two solids as well as the region which need to be selected for creating a loft feature. Figure 26-110 shows the preview of the loft feature with grips. Figure 26-111 shows the loft feature with bulge magnitude added at the start and the end.

Note

*You can also set the surface continuity and the bulge continuity in the **Loft Settings** dialog box.*

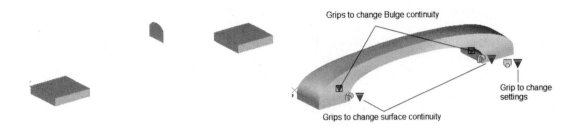

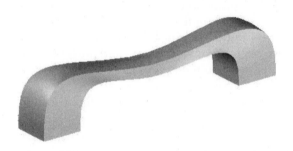

Figure 26-109 *Two solids and a region to create a loft*

Figure 26-110 *Preview of the loft and the grips displayed*

Figure 26-111 *Loft surface created*

CREATING PRESSPULL SOLIDS

Ribbon: Home > Modeling > Presspull	**Toolbar:** Modeling > Presspull
Command: PRESSPULL	

The **Presspull** tool is used to remove or add material of any desired shape in a model. As the name presspull indicates, if you press a closed boundary inside an existing model, it will remove the material of the shape of the closed boundary. However, if you pull a closed boundary outside an existing model, it will add material to the model. To perform this operation, choose the **Presspull** tool from the **Modeling** panel. Next, click inside a closed boundary that you want to press or pull and then move the mouse in the desired direction. You can specify the desired height by entering a value or by clicking the mouse. Figure 26-112 shows the solid model with the closed area to be presspulled. Figure 26-113 shows the model created by pressing the closed region and Figure 26-114 shows the model created by pulling the closed area.

If the sketch that you select to press or pull has nested loops, as shown in Figure 26-115, the inner closed loops will be deleted from the outer loop, as shown in Figure 26-116. The closed area you select for the presspull should consist of lines, polylines, 3D face, edges, regions, and faces of a 3D solid. But all these entities should form a coplanar closed area.

Tip
*To perform the presspull operation without choosing the **Presspull** button, press and hold the CTRL+ALT keys and then click inside the closed area to be pressed or pulled.*

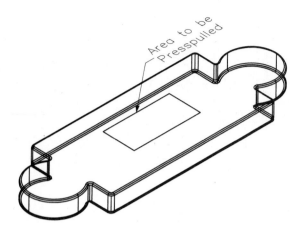

Figure 26-112 *Model with closed area to be presspulled*

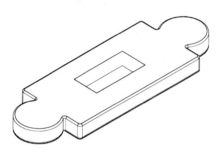

Figure 26-113 *Model with the closed area pressed*

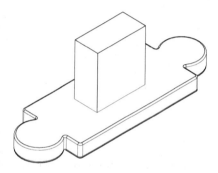

Figure 26-114 *Model with closed area pulled*

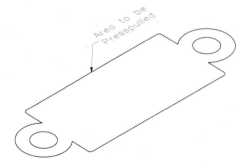

Figure 26-115 *Area to be presspulled with closed boundary inside it*

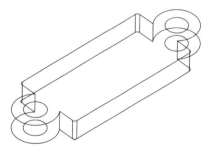

Figure 26-116 *The model after the presspull operation is performed*

EXAMPLE 2

In this example, you will create the solid model shown in Figure 26-117.

1. Start a new file using the *Acad3d.dwt* template.

2. Turn off the grid display, increase the limits to 120,120 and then zoom to the limits of the drawing.

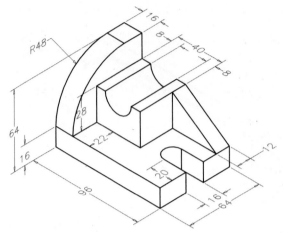

Figure 26-117 Solid model for Example 2

3. Set the view to SE Isometric by using the ViewCube. Then, right-click on the ViewCube and choose the **Parallel** option.

4. Switch on the ortho mode. Using lines and arc, create the base of the model with the dimensions shown in Figure 26-118.

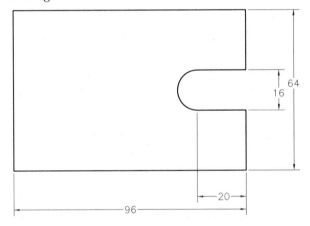

Figure 26-118 Base for the solid model

5. Choose the **Region** tool from the **Draw** panel in the **Home** tab and follow the prompt sequence given next to convert the sketch drawn into a region.

   Select objects: *Select the complete base.*
   Select objects: Enter
   1 loop extracted.
   1 Region created.

 Note
*You can also use the **PLINE** command to create this sketch. In that case, you do not need to convert it into a region.*

6. Choose the **Extrude** tool from **Home > Modeling > Solid Creation** drop-down and follow the prompt sequence given next to extrude the region.

Command: _extrude
Current wire frame density: ISOLINES=4, Closed profiles creation mode = Solid
Select objects to extrude or [MOde]: _MO Closed profiles creation mode [SOlid/SUrface]
<Solid>: _SO
Select objects to extrude or [MOde]: *Select the region* [Enter]
1 found
Select objects to extrude or [MOde]: [Enter]
Specify height of extrusion or [Direction/Path/Taper angle/Expression]: **16**

7. Change the visual style to Wireframe by choosing the **Wireframe** option from the **Visual Styles** drop-down list in the **In-canvas Viewport control**.

8. Choose the **Wedge** tool from **Home > Modeling > Solid Primitives** drop-down and follow the prompt sequence given next to create the next feature.

   Specify first corner or [Center]: *Pick a point on the screen.*
   Specify other corner or [Cube/Length]: **L**
   Specify length: **40**
   Specify width: **12**
   Specify height or [2Point]: **28**

9. Move this wedge and align it with the base. Use the **Vertex** option of the **3D Object Snap** to snap the vertex. The model after applying the **Hidden** visual style is shown in Figure 26-119.

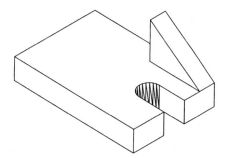

Figure 26-119 Wedge aligned with the base

10. Orient the view to **Front** by clicking on **Front** in the ViewCube and choose **WCS > New UCS** from the drop-down list below it; you are prompted to specify the origin.

11. Enter **V** at the **Specify origin of UCS or [Face/NAmed/OBject/Previous/View/World/X/Y/Z/ ZAxis] <World>** prompt and press ENTER.

12. Draw the profile shown in Figure 26-120 at any place in the drawing area and then convert it into a region.

13. Choose the **Extrude** tool from **Home > Modeling > Solid Creation** drop-down and extrude the region to a depth of 42 units.

14. Move this extrude feature and align it with the wedge feature. Use the **Vertex** option of the **3D Object Snap** to snap the vertex of the wedge feature.

   The isometric view of the model after aligning the extrude feature is shown in Figure 26-121.

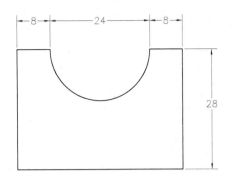

Figure 26-120 *The region to be extruded*

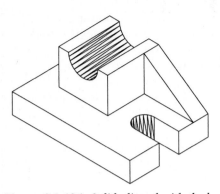

Figure 26-121 *Solid aligned with the base*

15. Orient the view to Right by clicking **Right** in the ViewCube and choose **WCS > New UCS** from the drop-down list below it; you are prompted to specify the origin.

16. Enter **V** at the **Specify origin of UCS or [Face/NAmed/OBject/Previous/View/World/X/Y/Z/ZAxis] <World>** prompt and press ENTER.

17. Draw the profile shown in Figure 26-122 at any place in the drawing area and then convert it into a region.

18. Extrude the region to a depth of 16 units and then move it so that it is properly aligned with the base, see Figure 26-123.

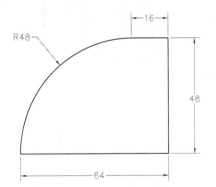

Figure 26-122 *The region before extrusion*

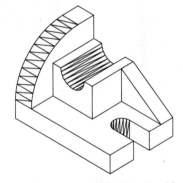

Figure 26-123 *Next solid aligned with the base*

19. Choose the **Solid Union** tool from the **Solid Editing** panel and follow the prompt sequence given next:

    Select objects: *Select all objects.*
    Select objects: Enter

20. Change the visual style by selecting the **Shades of Gray** option from the **Visual Styles** drop-down list in the **In-canvas Viewport Controls**. The final solid model should look similar to the one shown in Figure 26-124.

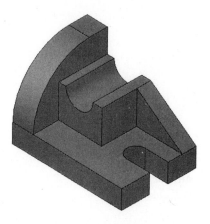

Figure 26-124 Solid model for Example 2

EXAMPLE 3

In this example, you will create the solid model shown in Figure 26-125.

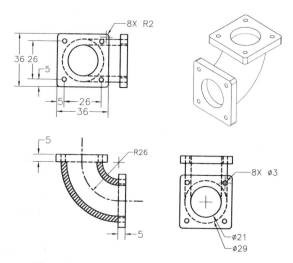

Figure 26-125 Solid model for Example 3

1. Start a new file in the *Acad3d.dwt* template.

2. Turn off the grid display, increase the limits to 75,75, and then zoom to the limits of the drawing.

3. Orient the view to Left by choosing **Left** in the ViewCube and choose **WCS > New UCS** from the drop-down list below it; you are prompted to specify the origin.

4. Enter **V** at the **Specify origin of UCS or [Face/NAmed/OBject/Previous/View/World/X/Y/Z/ZAxis] <World>** prompt and press ENTER.

5. Right-click on the ViewCube and choose the **Parallel** option.

6. Draw the cross section and path of the sweep feature, as shown in Figure 26-126.

7. Choose the **Region** tool from the **Draw** panel in the **Home** tab and follow the prompt sequence given next to convert the sketch into a region.

Select objects: *Select both the circles.*
Select objects: [Enter]
2 loop extracted.
2 Region created.

8. Choose the **Subtract** tool from the **Solid Editing** panel in the **Home** tab and follow the prompt sequence:

   Select objects: *Select the region of diameter 29.*
   Select objects: [Enter]
   Select solids and regions to subtract.
   Select objects: *Select the region of diameter 21.*
   Select objects: [Enter]

9. Change the visual style to Wireframe by choosing the **Wireframe** option from the **Visual Styles** drop-down list in the **View** panel.

10. Choose the **Sweep** tool from **Home > Modeling > Solid Creation** drop-down and follow the prompt sequence given next to sweep the region along the path.

    Current wire frame density: ISOLINES=4, Closed profiles creation mode = Solid
    Select objects to sweep or [MOde]: _MO Closed profiles creation mode [SOlid/SUrface] <Solid>: _SO
    Select objects to sweep or [MOde]: *Select the region created in Step 8*
    1 found
    Select objects to sweep or [MOde]: [Enter]
    Select sweep path or [Alignment/Base point/Scale/Twist]:*Select the arc created in Step 6.*

11. Set the view to **SE Isometric** by choosing the appropriate hotspot in the ViewCube. Also, select the **Realistic** option in the **Visual Styles** drop-down list in the **View** panel. The resulting model should look similar to the one shown in Figure 26-127.

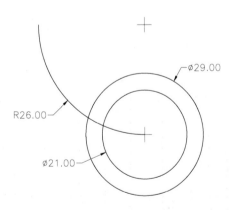

Figure 26-126 *Sketch for base pipe feature*

Figure 26-127 *Resulting sweep feature*

12. Choose the **Top** option from the **Views** panel in the **View** tab and draw the sketch with the dimensions shown in Figure 26-128.

13. Change the view to **SE Isometric**. The model after drawing the sketch and turning off the dimension is shown in Figure 26-129.

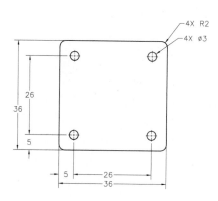

Figure 26-128 *Sketch for the area to Press/Pull*

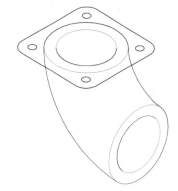

Figure 26-129 *Location of the sketch*

14. Choose the **Presspull** tool from the **Modeling** panel and then click inside the rectangular area at the location shown as Area A in Figure 26-130. Next, move the mouse in the upward direction, enter the value **5** at the Command prompt, and press ENTER.

15. Again, choose the **Presspull** tool and pull the area shown as Area B in Figure 26-131 by the same value as specified in Step 8. The resulting model should look similar to the one shown in Figure 26-131.

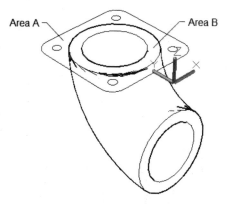

Figure 26-130 *Two areas to presspull*

Figure 26-131 *Resulting model after Step 10*

16. Similarly, create the flange for the other face of the pipe. The resulting model should be similar to the one shown in Figure 26-132.

17. Choose the **Solid Union** tool from the **Solid Editing** panel in the **Home** tab and follow the prompt sequence given next.

Select objects: *Select all objects.*
Select objects: Enter

The final solid model should look similar to the one shown in Figure 26-133.

Figure 26-132 *Resulting model after Step 16*   **Figure 26-133** *Final model after Union*

Self-Evaluation Test

Answer the following questions and then compare them to those given at the end of this chapter:

1. The **Cone** tool can be used to create a solid cone with a circular and an elliptical base. (T/F)

2. Open entities can be converted into regions. (T/F)

3. An ellipse can be converted into a polysolid. (T/F)

4. You can twist a cross-section while sweeping it. (T/F)

5. The _____ tool can be used to add as well as remove material from a model.

6. In AutoCAD, you can create a pyramid of maximum 42 sides. (T/F)

7. The _____ tool is used to generate a helical curve.

8. The number of paths for a loft feature cannot be more than _____ .

9. At least _____ cross-sections are required to create a loft feature.

10. You can extrude the selected region about a path using the _____ option of the **Extrude** tool.

Review Questions

Answer the following questions:

1. You can revolve an open entity. (T/F)

2. You cannot apply Boolean operations on regions. (T/F)

3. The entity to be revolved should lie completely on one side of the revolution axis. (T/F)

4. You cannot apply the **Presspull** tool on an open object. (T/F)

5. There cannot be more than one guide for the loft feature. (T/F)

6. A curve generated by the **Helix** tool can be used as a path for the swept solids. (T/F)

7. Which of the following values for the taper angle is used to taper the extruded model from the base?

   (a) Positive   (b) Negative
   (c) Zero   (d) None of these

8. Which of the following tools is used to create a cube?

   (a) **Box**   (b) **Cuboid**
   (c) **Polygon**   (d) **Cylinder**

9. Which of the following commands is used to check the interference between the selected solid models?

   (a) **INTERFERE**   (b) **INTERSECT**
   (c) **INTERFERENCE**   (d) **CHECK**

10. The **Presspull** tool is used for which of the following functions?

    (a) Add material   (b) Remove material
    (c) Both add and remove material (d) Neither add nor remove material

11. The _____ direction is known as the extrusion direction.

12. The _____ command is used to create an apple-like structure.

13. The _____ command is used to create a revolved solid.

14. Guides for the **Loft** tool should always _____ the cross-section.

15. The _____ option of the **Revolve** tool is used to select a 2D entity as the revolution axis.

EXERCISE 2

In this exercise, you will create the solid model shown in Figure 26-134. Assume the missing dimensions.

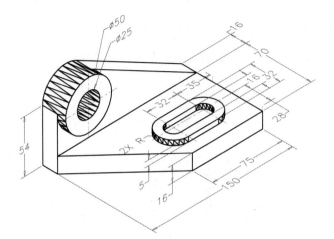

Figure 26-134 *Model for Exercise 2*

EXERCISE 3

In this exercise, you will create the solid model shown in Figure 26-135. The dimensions for the model are given in the same figure. Assume the missing dimensions.

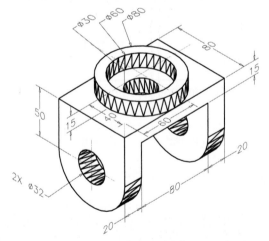

Figure 26-135 *Solid model for Exercise 3*

EXERCISE 4

In this exercise, you will create the solid model shown in Figure 26-136. The dimensions for the model are shown in Figures 26-137 and 26-138.

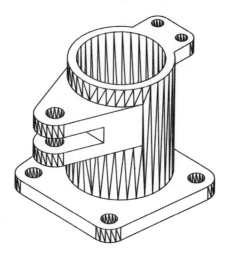

Figure 26-136 *Model for Exercise 4*

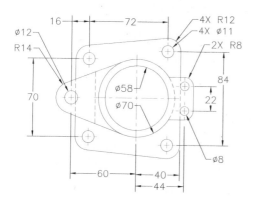

Figure 26-137 *Top view of the model*

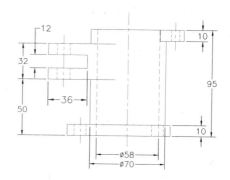

Figure 26-138 *Front view of the model*

Problem-Solving Exercise 1

In this exercise, you will create the solid model with dimensions shown in Figure 26-139 and Figure 26-140.

Chapter 26

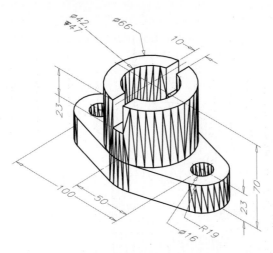

Figure 26-139 *Solid model for Problem-Solving Exercise 1*

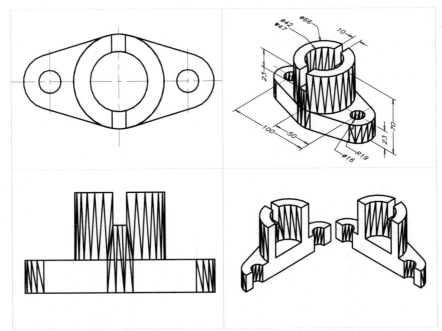

Figure 26-140 *Viewports for Problem-Solving Exercise 1*

Answers to Self-Evaluation Test

1. T, **2**. F, **3**. F, **4**. T, **5**. **PRESSPULL**, **6**. **F**, **7**. **HELIX**, **8**. one, **9**. two, **10**. **Path**

Chapter 27

Modifying 3D Objects

CHAPTER OBJECTIVES

In this chapter, you will learn:
- *To create fillets and chamfers in the solid models.*
- *To rotate and mirror solid models in 3D space.*
- *To create an array in 3D space.*
- *To align the solid models.*
- *To extract the edges of a solid model.*
- *To slice the solid models and create cross-sections.*
- *To convert objects to surfaces and solids.*
- *To convert surfaces to solids.*

KEY TERMS

- *Fillet Edges*
- *Chamfer Edges*
- *Rotate 3D*
- *Rotate Gizmo*

- *3D Mirror*
- *3D Move*
- *Move Gizmo*
- *3DARRAY*

- *Align*
- *3D Align*
- *Extract Edges*
- *Convert to Surface*

- *Convert to Solid*
- *Thicken*
- *Slice*
- *SECTION*

FILLETING SOLID MODELS

Ribbon: Solid > Solid Editing > Fillet/Chamfer Edge drop-down > Fillet Edge
Toolbar: Modify > Fillet Edges   **Command:** FILLETEDGE

Like the **Fillet** tool in the **Modify** panel is used to fillet the sharp corners in a sketch, the **Fillet Edge** tool in the **Fillet/Chamfer** drop-down (see Figure 27-1) is used to round the edges of a model. On invoking this tool, following prompt sequence will be displayed:

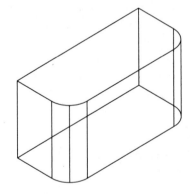

Command: _FILLETEDGE
Radius = 1.0000
Select an edge or [Chain/Loop/Radius]:

*Figure 27-1 Tools in the **Fillet/Chamfer** drop-down*

Also, you are prompted to select an edge. Select an edge; the preview of the fillet with default radius will be displayed and you will be prompted to select another edge. If you need to fillet multiple edges, select the edges in succession (see Figure 27-2). After selecting the edge or edges, press ENTER to end the selection. On doing so, you will be prompted to press ENTER again to accept the default radius or enter a new radius. Also, a grip will be displayed on the fillet. Enter **R** for a new radius; specify radius value and press ENTER, or drag the grip to edit the fillet radius and press ENTER; the fillet will be created, as shown in Figure 27-3.

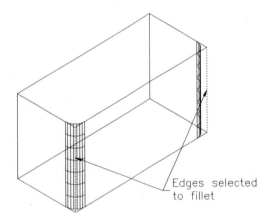

Figure 27-2 The edges to be filleted   *Figure 27-3 Model after filleting the edges*

Other options in the above given Command prompt are discussed next.

Chain

On choosing this option, you can select all tangential edges on a face of the solid model in a single attempt, as shown in Figure 27-4. Figure 27-5 shows the resulting fillet.

Loop

With this option, you can select any of the loops attached to the selected edge. On choosing this option,; preview of the selected loop will be displayed and a contextual tab will open. In this tab, you can accept the loop displayed or select next to check the next possible loop.

Radius

This option is used to redefine the fillet radius. You can also define the radius using an expression.

Note
Although the **Fillet** *tool in the* **Modify** *panel is used to fillet the sketched entities, you can also fillet the edges of the solid model by using the* **Fillet Edge** *tool. To do so, invoke this tool and select an edge. Then, specify the fillet radius; the* **Select an edge or [Chain/Loop/Radius]** *prompt will be displayed. Now, you can fillet the edges, as discussed earlier. However, in this case, the edit grip will not be displayed.*

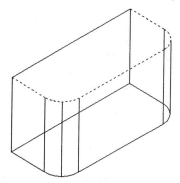

Figure 27-4 *Edges that will be selected using the* **Chain** *option*

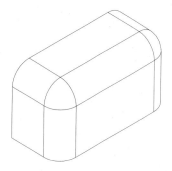

Figure 27-5 *The resulting fillet*

CHAMFERING SOLID MODELS

Ribbon: Solid > Solid Editing > Fillet/Chamfer Edge drop-down > Chamfer Edge
Toolbar: Modify > Chamfer Edges   **Command:** CHAMFEREDGE

 The **Chamfer Edge** tool is used to bevel the edges of solid models. The working of this tool is similar to that of the **Fillet Edge** tool. To create a chamfer, invoke the **Chamfer Edge** tool from the **Solid Editing** panel (Figure 27-6) and then select the edges of the same face in succession; the preview of the chamfer will be displayed. Press ENTER to complete the selection; you will be prompted to specify a value for the new chamfer distance or press ENTER to accept the default value. Also, two grips will be displayed at the edge selected to chamfer. You can enter the chamfer distances or drag the grips on each surface to change the chamfer distance. Then, press ENTER to create the chamfer.

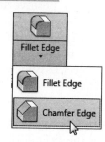

Figure 27-6 *Choosing the* **Chamfer Edge** *tool from the* **Fillet/Chamfer** *drop-down*

Figures 27-7 and 27-8 show the solid model before and after chamfering, respectively.

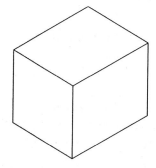

Figure 27-7 *Solid model before chamfering*

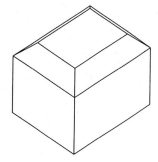

Figure 27-8 *Solid model after chamfering*

Loop

This option is used to select all edges that make a closed loop on the selected face of the solid model. To select edges using the loop option, enter **L** at the **Select an edge or [Loop/Distance]** prompt and select any one of the edges that forms the loop; all edges will be selected.

Distance

This option is used to define the chamfer distances. When the preview is displayed, you can also drag the grips displayed on the edge to be chamfered and change the chamfer distance.

Note
*You can also chamfer the edges by using the **Chamfer** tool in the **Modify** panel. On selecting an edge after invoking this tool, you will be prompted to specify the surface to be selected. Select the surface and press ENTER to accept the surface selected or select the next surface and press ENTER; you will be prompted to specify the distance. Enter a suitable value; the chamfer will be created. In this case, the edit grips will not be displayed.*

ROTATING SOLID MODELS IN 3D SPACE

Command: ROTATE3D

The **ROTATE3D** command is used to rotate the selected solid model in the 3D space about a specified axis. Once again the right-hand thumb rule will be used to determine the direction of rotation of the solid model in 3D space. The prompt sequence that will follow when you choose this command is given next.

Current positive angle: ANGDIR=counterclockwise ANGBASE=0
Select objects: *Select the solid model.*
Select objects: `Enter`
Specify first point on axis or define axis by
[Object/Last/View/Xaxis/Yaxis/Zaxis/2Points]: *Specify a point on the axis of rotation or select an option.*

2points Option

This is the default option for rotating solid models. This option allows you to rotate the solid model about an axis specified by two points. The direction of the axis will be from the first point to the second point. Using this direction of the axis, you can calculate the direction of rotation of the solid model by applying the right-hand thumb rule. The prompt sequence that will follow when you invoke this option is given next.

[Object/Last/View/Xaxis/Yaxis/Zaxis/2points]:2P `Enter`
Specify first point on axis: *Specify the first point of the rotation axis. See Figure 27-9.*
Specify second point on axis: *Specify the second point of the rotation axis. See Figure 27-9.*
Specify rotation angle or [Reference]: *Specify the angle of rotation.*

Object Option

This option is used to rotate the solid model in the 3D space using a 2D entity. The 2D entities that can be used are lines, circles, arcs, or 2D polyline segments. If the selected entity is a line or a straight polyline segment, then it will be directly taken as the rotation axis. However, if the selected entity is an arc or a circle, then an imaginary axis will be drawn starting from the center and normal to the plane, in which the arc or circle is drawn. The object will then be rotated about this imaginary axis. The prompt sequence that will follow when you invoke this option is given next.

[Object/Last/View/Xaxis/Yaxis/Zaxis/2points]: **O** `Enter`

Select a line, circle, arc, or 2D-polyline segment: *Select the 2D entity, as shown in Figure 27-10.*
Specify rotation angle or [Reference]: *Specify the angle of rotation.*

Last Option

This option uses the same axis that was last selected to rotate the solid model.

View Option

This option is used to rotate the solid model about the viewing plane. In this case, the viewing

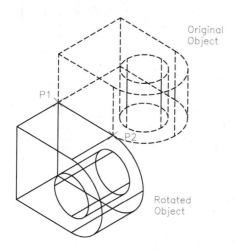

Figure 27-9 *Rotating the solid model using the **2points** option*

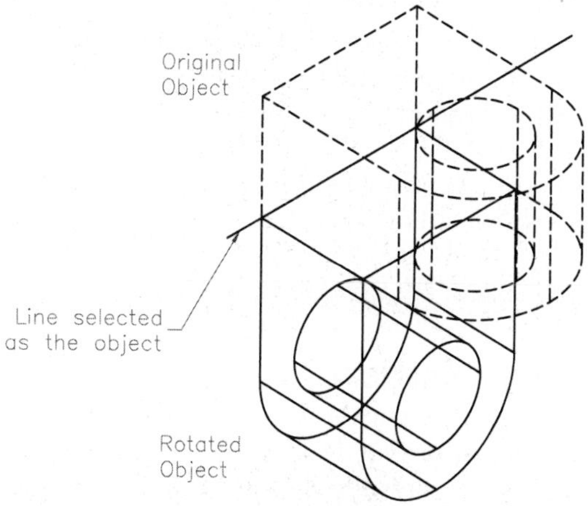

Figure 27-10 *Rotating the solid model using the **Object** option*

plane is the screen of the computer. This option draws an imaginary axis starting from the specified point and continues normal to the viewing plane. The model is then rotated about this axis. The prompt sequence that will follow when you invoke this option is given next.

Specify first point on axis or define axis by
[Object/Last/View/Xaxis/Yaxis/Zaxis/2points]: **V**
Specify a point on the view direction axis <0,0,0>: *Specify the point on the view plane, as shown in Figure 27-11.*
Specify rotation angle or [Reference]: *Specify the angle of rotation.*

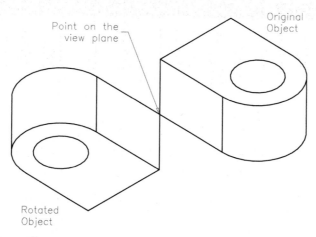

Figure 27-11 *Rotating the solid model using the **View** option through an angle of 180°*

Xaxis Option

This option is used to rotate the solid model about the positive *X* axis of the current UCS. On invoking this option, you will be prompted to select a point on the *X* axis. The prompt sequence that will follow when you invoke this command is given next.

> Specify first point on axis or define axis by
> [Object/Last/View/Xaxis/Yaxis/Zaxis/2points]: **X**
> Specify a point on the X axis <0,0,0>: *Specify the point on the X axis.*
> Specify rotation angle or [Reference]: *Specify the angle of rotation.*

Yaxis Option

This option is used to rotate the solid model about the positive *Y* axis of the current UCS. When you invoke this option, you will be prompted to select a point on the *Y* axis. The prompt sequence that will follow when you invoke this command is given next.

> Specify first point on axis or define axis by
> [Object/Last/View/Xaxis/Yaxis/Zaxis/2points]: **Y**
> Specify a point on the Y axis <0,0,0>: *Specify the point on the Y axis.*
> Specify rotation angle or [Reference]: *Specify the angle of rotation.*

Zaxis Option

This option is used to rotate the solid model about the positive *Z* axis of the current UCS. When you invoke this option, you will be prompted to select a point on the *Z* axis. The prompt sequence that will follow when you invoke this command is given next.

> Specify first point on axis or define axis by
> [Object/Last/View/Xaxis/Yaxis/Zaxis/2points]: **Z**
> Specify a point on the Z axis <0,0,0>: *Specify the point on the Z axis.*
> Specify rotation angle or [Reference]: *Specify the angle of rotation.*

ROTATING SOLID MODELS ABOUT AN AXIS

Ribbon: Home > Modify > 3D Rotate	**Toolbar:** Modeling > 3D Rotate
Menu Bar: Modify > 3D Operations > 3D Rotate	**Command:** 3DROTATE

This tool is used to dynamically rotate an object or subobjects such as edges and faces with respect to a specified base point and about a selected axis. To perform this operation, choose the **3D Rotate** tool from the **Modify** panel and select the object or subobject to be rotated. Then press ENTER; the **Rotate Gizmo** will be displayed (Figure 27-12) at the center of the object. Specify a new base point, if you need to rotate the object with respect to a new point, else give a null response. You need to select the axis about which you want to rotate the object. To do so, move the cursor close to the required axis handle; the axis gets highlighted and an axis passing through the center of the gizmo will be displayed, as shown in Figure 27-13. Click once and drag the cursor or enter a value at the Command prompt.

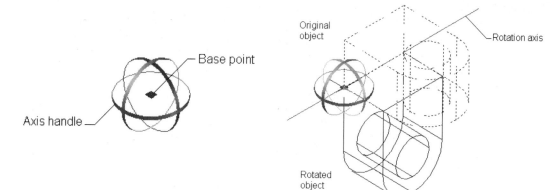

Figure 27-12 *The* **Rotate Gizmo**

Figure 27-13 *Model rotated by 90° about the X axis*

> **Tip**
> *If you select a model in* **3D Modeling** *workspace with 3D visual style, the* **Move Gizmo** *will be displayed on it. This is because, the default gizmo displayed in the* **Subobject** *panel is the* **Move Gizmo**. *If you select the* **Rotate Gizmo** *option from the* **Gizmo** *drop-down and select a model, the* **Rotate Gizmo** *will be displayed. Using this gizmo, you can directly rotate the model in 3D space.*

MIRRORING SOLID MODELS IN 3D SPACE

Ribbon: Home > Modify > 3D Mirror   **Command:** MIRROR3D
Menu Bar: Modify > 3D Operations > 3D Mirror

 The **3D Mirror** tool is used to mirror the solid models about a specified plane in the space. To mirror a solid model, choose the **3D Mirror** tool from the **Modify** panel and follow the prompt sequence given next.

Select objects: *Select the solid model to be mirrored.*
Select objects: Enter
Specify first point of mirror plane (3 points) or
[Object/Last/Zaxis/View/XY/YZ/ZX/3points] <3points>: *Specify the first point on the mirror plane or select an option.*

The options in this prompt are discussed next.

3points

This is the default option for mirroring the solid models. As discussed earlier, a line can be defined by specifying two points through which the line will pass. Similarly, a plane can be defined by specifying three points through which it passes. This option allows you to specify the three points through which the mirroring plane passes. The prompt sequence that follows when you invoke this command is given next.

Specify first point of mirror plane (3 points) or
[Object/Last/Zaxis/View/XY/YZ/ZX/3points] <3points>: Enter
Specify first point on mirror plane: *Specify the first point on the plane. See Figure 27-14.*
Specify second point on mirror plane: *Specify the second point on the plane. See Figure 27-14.*
Specify third point on mirror plane: *Specify the third point on the plane. See Figure 27-14.*
Delete source objects? [Yes/No] <N>: Enter

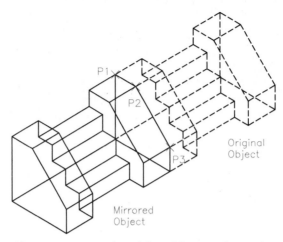

*Figure 27-14 Mirroring the solid model using the **3points** option*

Object

This option is used to mirror the solid model using a 2D entity, refer to Figure 27-15. The 2D entities that can be used to mirror the solids are circles, arcs, and 2D polyline segments. The prompt sequence that follows when you invoke this option is given next.

Specify first point of mirror plane (3 points) or
[Object/Last/Zaxis/View/XY/YZ/ZX/3points] <3points>: **O**
Select a circle, arc, or 2D-polyline segment: *Select the 2D entity, as shown in Figure 27-15.*
Delete source objects? [Yes/No] <N>: Enter

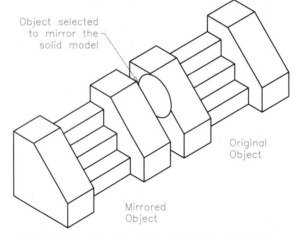

*Figure 27-15 Mirroring the solid model using the **Object** option*

Zaxis

This option allows you to define a mirroring plane using two points. The first point is the point on the mirroring plane and the second point is a point on the positive direction of the Z axis of that plane. The prompt sequence that follows when you invoke this option is given next.

Specify first point of mirror plane (3 points) or
[Object/Last/Zaxis/View/XY/YZ/ZX/3points] <3points>: **Z**
Specify point on mirror plane: *Specify the point on the plane, as shown in Figure 27-16.*

Specify point on Z-axis (normal) of mirror plane: *Specify the point on the Z direction of the plane, as shown in Figure 27-16.*
Delete source objects? [Yes/No] <N>: Enter

View Option

This is one of the interesting options provided for mirroring solid models. This option is used to mirror the selected solid model about the viewing plane. The viewing plane, in this case, is the screen of the monitor (Figure 27-17). The prompt sequence that will follow when you invoke this command is given next.

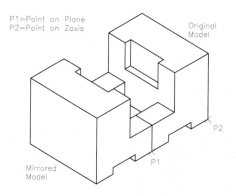

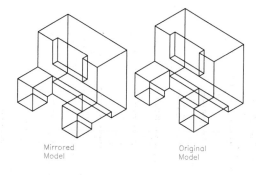

Figure 27-16 *Mirroring the solid model using the* **Zaxis** *option*

Figure 27-17 *A solid model mirrored using the* **View** *option and then moved*

Specify first point of mirror plane (3 points) or
[Object/Last/Zaxis/View/XY/YZ/ZX/3points] <3points>: **V**
Specify point on view plane <0,0,0>: *Specify the point on the view plane.*
Delete source objects? [Yes/No] <N>: Enter

Tip
As the model is mirrored about the view plane, the new model will be placed over the original model when you view it from the current viewpoint. The best option to view the mirrored model is to hide the hidden lines using the **HIDE** *command. You can also move the new model away from the last model using the* **Move** *tool or change the viewpoint for viewing both the models simultaneously.*

XY/YZ/ZX

These options are used to mirror the solid model about the *XY*, *YZ*, or *ZX* plane of the current UCS.

Tip
All these planes will be considered with reference to the current orientation of the UCS and not with its world position. This means that when you select the **XY** *option to mirror the solid model, then the XY plane of the current UCS will be considered to mirror the model and not the XY plane of the world UCS.*

EXAMPLE 1

In this example, you will create the solid model shown in Figure 27-18. Assume the missing dimensions.

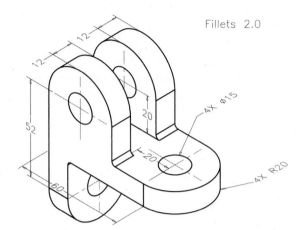

Fillets 2.0

Figure 27-18 *Solid model for Example 1*

1. Start a new 3D template drawing file in **3D Modeling** workspace. Set the UCS to the top view and make it current. Increase the limit to 100,100.

2. Create the base of the model and then change the view to **SE Isometric** view using the ViewCube, see Figure 27-19.

3. Set the UCS in the **Front** view.

4. Draw the sketch for the next object using the **Rectangle** tool. Convert one side of the rectangle to an arc, by using the shortcut menu that is displayed when you hover the cursor on the middle grip.. Then, create the object by using the **Region** and **Extrude** tools. The resulting object is shown in Figure 27-20.

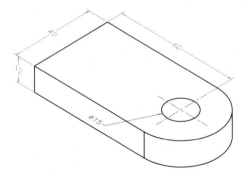

Figure 27-19 *Base of the model*

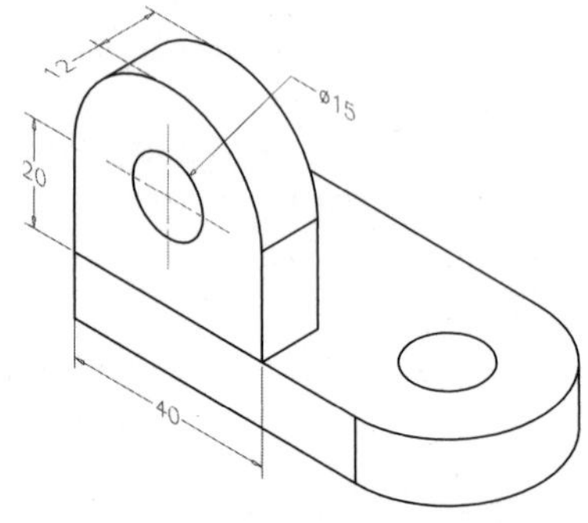

Figure 27-20 *Model after creating the next object and moving it*

5. Invoke the **3D Mirror** tool from the **Modify** panel and follow the prompt sequence given next.

    Select objects: *Select the last object.*
    Select objects: [Enter]
    Specify first point of mirror plane (3 points) or
    [Object/Last/Zaxis/View/XY/YZ/ZX/3points] <3points>: **Z**
    Specify point on mirror plane: *Select P1, as shown in Figure 27-21.*

Specify point on Z-axis (normal) of mirror plane: *Select P2, as shown in Figure 27-21.*
Delete source objects? [Yes/No] <N>: ⏎

6. The object after mirroring and hiding the hidden lines should look similar to that shown in Figure 27-22.

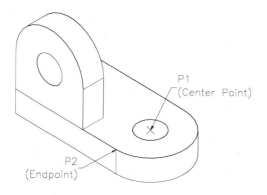

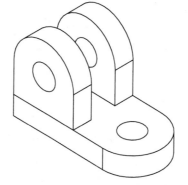

Figure 27-21 Selecting the points to be mirrored   *Figure 27-22 Model after mirroring*

7. Set the UCS in the **Left** view and create the feature at the bottom, as shown in Figure 27-23.

8. Choose the **Union** tool from the **Solid Editing** panel and then unite all objects, see Figure 27-24.

9. Choose the **Fillet Edge** tool from **Solid > Solid Editing > Fillet/Chamfer** drop-down and follow the prompt sequence given below.

    Command: _FILLETEDGE
    Radius = 1.0000
    Select an edge or [Chain/Loop/Radius]: *Select first edge, as shown in Figure 27-25.*

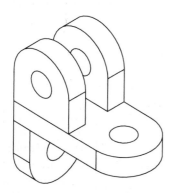

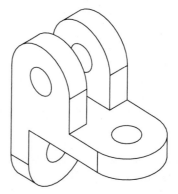

Figure 27-23 Model after creating the next object at the bottom   *Figure 27-24 Model after creating the union*

    Select an edge or [Chain/Loop/Radius]: *Select the second edge, as shown in Figure 27-25.*
    Select an edge or [Chain/Loop/Radius]: *Select the third edge, as shown in Figure 27-25.*
    Select an edge or [Chain/Loop/Radius]: ⏎
    3 edge(s) selected for fillet
    Press Enter to accept the fillet or [Radius]: **R**
    Specify Radius or [Expression]<1.0000>: **2**
    Select an edge or [Chain/Loop/Radius]: ⏎

10. The final model for Example 3 should look similar to the one shown in Figure 27-26.

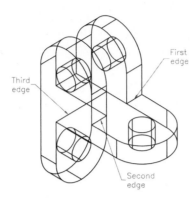

Figure 27-25 *Selecting the edges for filleting*

Figure 27-26 *The final model for Example 3*

MOVING MODELS IN 3D SPACE

Ribbon: Home > Modify > 3D Move	**Toolbar:** Modeling > 3D Move
Menu Bar: Modify > 3D Operations > 3D Move	**Command:** 3DMOVE

Sometimes, you need to move objects such as edges, or faces of a solid model or the entire solid in 3D space. In such cases, you can use the **3D MOVE** tool to translate them to the desired location. You can move the selected object or subobject at an orientation, parallel to a plane or along an axial direction. To perform this operation, choose the **3D Move** tool from the **Modify** panel and select the object or subobject to be moved; the **Move Gizmo** will be displayed on the selected object, see Figure 27-27. To move the object in 3D space with respect to a base point other than

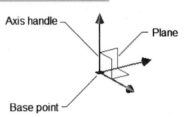

Figure 27-27 *The Move Gizmo*

the default one, click on the drawing area; the new base point will be specified. Then, specify the displacement or drag the cursor and place the component.

To move an object along a particular direction, move the cursor near the axis handle corresponding to that direction; a line of the same color as the selected axis will be displayed. Click when the axis is highlighted and drag the cursor to move the object along that direction.

To move a face or an edge of a solid model in a particular plane, invoke the **3D Move** tool, press the CTRL key, and then select a face or an edge; the move gizmo will be displayed on the selected object. Next, move the cursor near the desired plane in the **Move Gizmo**; the plane, including the corresponding axis handle, will turn golden in color. Press and hold the left mouse button when the plane is highlighted and then drag; the selected object will move only parallel to the selected plane.

Tip
*By default, the **Move Gizmo** option is selected in the **Default Gizmo** drop-down available in the **Subobject** panel. Therefore, on selecting an object, the **Move Gizmo** will be displayed on the object and you can translate the selected object. Also, by default, the **No Filter** option is selected in the **Filter** drop-down. As a result, you can select a face, an edge, or a vertex of a model. However, if any other option is selected in the **Filter** drop-down, then only the corresponding entity can be selected.*

CREATING ARRAYS IN 3D SPACE

Menu Bar: Modify > 3D Operations > 3D Array   **Command:** 3DARRAY

As mentioned earlier, the arrays are defined as the method of creating multiple copies of the selected object in a rectangular or a polar fashion. The 3D arrays can also be created similar to the 2D arrays. The only difference is that in a 3D array, another factor called the *Z* axis is also taken into consideration. There are two types of 3D arrays. Both these types are discussed next.

3D Rectangular Array

This is the method of arranging the solid model along the edges of a box. In this type of array, you need to specify three parameters. They are the rows (along the Y axis), the columns (along the *X* axis), and the levels (along the *Z* axis). You will also have to specify the distances between the rows, columns, and levels. The 3D rectangular array can be easily understood by taking an example shown in Figure 27-28. This figure shows two floors of a building. Initially, only one chair is placed on the ground floor. Now, if you create a 2D rectangular array, the chairs will be arranged only on the ground floor. However, when you create the 3D rectangular array, then the chairs will be arranged on the first floor (along the *Z* axis) as well as the ground floor, see Figure 27-29. In this example, the number of rows is three, columns is four, and levels is two.

The prompt sequence that will follow when you choose this tool from the **Modify** panel is given next.

> Select objects: *Select the object to array.*
> Select objects: `Enter`
> Enter the type of array [Rectangular/Polar] <R>: `Enter`
> Enter the number of rows (---) <1>: *Specify the number of rows along the X axis.*
> Enter the number of columns (|||) <1>: *Specify the number of columns along the Y axis.*
> Enter the number of levels (...) <1>: *Specify the number of levels along the Z axis.*
> Specify the distance between rows (---): *Specify the distance between the rows.*

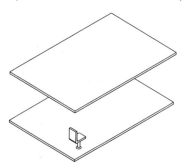

Figure 27-28 *The model before creating the rectangular array*

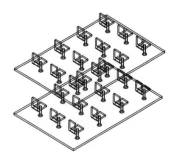

Figure 27-29 *The model after creating the array in 3D space*

> Specify the distance between columns (|||): *Specify the distance between the columns.*
> Specify the distance between levels (...): *Specify the distance between the levels.*

3D Polar Array

The 3D polar arrays are similar to the 2D polar arrays. The only difference between them is that in 3D, you will have to specify an axis about which the solid models will be arranged. For example, consider the solid models shown in Figure 27-30. This figure shows a circular plate that has four holes. Now, to place the bolts in this plate, you can use the 3D polar array, as shown in

Figure 27-31. The axis for an array is defined using the centers at the top and the bottom faces of the circular plate. The prompt sequence that will follow is given next.

    Select objects: *Select the object to array.*
    Select objects: [Enter]
    Enter the type of array [Rectangular/Polar] <R>: **P**
    Enter the number of items in the array: *Specify the number of items.*
    Specify the angle to fill (+=ccw, -=cw) <360>: [Enter]
    Rotate arrayed objects? [Yes/No] <Y>: [Enter]
    Specify center point of array: *Specify the first point on the axis.*
    Specify second point on axis of rotation: *Specify the second point on the axis.*

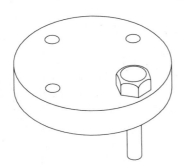 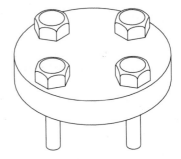

Figure 27-30 *Model before creating an array*   ***Figure 27-31*** *Model after creating an array*

ALIGNING SOLID MODELS

Ribbon: Home > Modify > Align	**Command:** ALIGN
Menu Bar: Modify > 3D Operations > Align	

The **Align** tool is a very versatile and highly effective tool. It is extensively used in solid modeling. As the name suggests, this tool is used to align the selected solid model with another solid model. In addition to this, it can also be used to translate, rotate, and scale the selected solid model. This command uses pairs of source and destination points to align the solid model. The source point is a point with which you want to align the object. The destination point is a point on the destination object at which you want to place the source object. You can specify one, two, or three pairs of points to align the objects. However, the working of this command will be different in all the three cases. All these three cases are discussed next.

Aligning the Objects Using One Pair of Points

When you align the object using just one pair of points, the working of this command will be similar to the **Move** tool. This means that the source object will be moved as it is from its original location and will be placed on the destination object. Here, the source point will work as the first point of displacement and the destination point will work as the second point of displacement. Also, a reference line will be drawn between the source and the destination points. This line will disappear once you exit the command. The prompt sequence is as follows:

    Select objects: *Select the object to align.*
    Select objects: [Enter]
    Specify first source point: *Select S1, as shown in Figure 27-32.*
    Specify first destination point: *Select D1, as shown in Figure 27-32.*
    Specify second source point: [Enter]

Figure 27-33 shows the models after aligning.

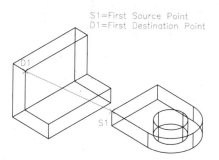

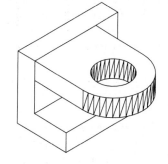

Figure 27-32 *Models before aligning*   **Figure 27-33** *Models after aligning*

Aligning the Objects Using Two Pairs of Points

The second case is to align the objects using two pairs of source and destination points. This method of aligning the objects forces the selected object to translate and rotate once. You are also provided with an option of scaling the source object to align with the destination object. The prompt sequence that follows is given next.

Select objects: *Select the object to align.*
Select objects: Enter
Specify first source point: *Select S1, as shown in Figure 27-34.*
Specify first destination point: *Select D1, as shown in Figure 27-34.*
Specify second source point: *Select S2, as shown in Figure 27-34.*
Specify second destination point: *Select D2, as shown in Figure 27-34.*
Specify third source point or <continue>: Enter
Scale objects based on alignment points? [Yes/No] <N>: **Y** *(You can also enter **N** at this prompt, if you do not want to scale the object).*

Figure 27-35 shows the objects after aligning and scaling.

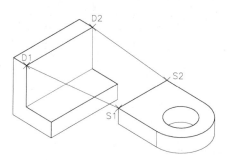

Figure 27-34 *Objects before aligning*   **Figure 27-35** *Objects after aligning and scaling*

Aligning the Objects Using Three Pairs of Points

The third case is to align the objects using three pairs of points. This option forces the selected object to rotate twice and then translate, see Figures 27-36 and 27-37. The first pair of source and destination points is used to specify the base point of alignment, the second pair of source and destination points is used to specify the first rotation angle, and the third pair of source and destination points is used to specify the second rotation angle. In this case, you will not be allowed to scale the object. The prompt sequence that follows is given next.

Select objects: *Select the object to align.*
Select objects: Enter
Specify first source point: *Select S1, as shown in Figure 27-36.*
Specify first destination point: *Select D1, as shown in Figure 27-36.*

Specify second source point: *Select S2, as shown in Figure 27-36.*
Specify second destination point: *Select D2, as shown in Figure 27-36.*
Specify third source point or <continue>: *Select S3, as shown in Figure 27-36.*
Specify third destination point: *Select D3, as shown in Figure 27-36.*

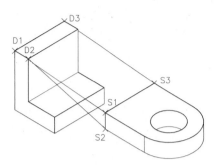

Figure 27-36 *Objects before aligning*

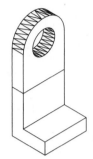

Figure 27-37 *Objects after aligning and rotating*

ALIGNING SOLIDS BY DEFINING AN ALIGNMENT PLANE

Ribbon: Home > Modify > 3D Align	**Toolbar:** Modeling > 3D Align
Menu Bar: Modify > 3D Operations > 3D Align	**Command:** 3DALIGN

The **3D Align** tool is used to align the selected solids or the copy of those solids with another solid object. To do so, define a plane for the source and destination objects that are to be aligned to each other. In this command, you are prompted to specify the first, second, and third points continuously on the source to define a plane for the source. Then, you will be prompted to specify the first, second, and third points on the destination object to define a plane for the destination. As a result, these planes will be automatically aligned to each other. While specifying the points on the destination object, you get a dynamic preview for the placement of the source object. The following is the significance of each point that you specify on the source and destination objects.

First point on source. This is the base point on the source object that later coincides with the first point of the destination object.

Second point on source. This point defines the direction of the X axis on the new XY plane that will be created on the source object.

Third point on source. This point is specified on the new XY plane to define the direction of the Y axis. This completely defines the new XY plane of the source. After specifying this point, the source object is rotated to make this new XY plane parallel to the XY plane of the current UCS.

First point on destination. This is the base point of the destination. The base point of the source will coincide with this point.

Second point on destination. This point defines the direction of the X axis on the new XY plane of the destination object. The X axis of the source will get aligned to the X axis of the destination object.

Third point on destination. This point is specified on the new XY plane to define the direction of the Y axis. This completely defines the new XY plane of the destination.

The prompt sequence that will be followed when you choose the **3D Align** tool from the **Modeling** toolbar is given next.

> Select objects: *Select the objects to align.*
> Select objects: [Enter]
> Specify source plane and orientation ...
> Specify base point or [Copy]: *Select S1, as shown in Figure 27-38, or enter **C** to align the copy of the selected object and leave the original one as it is.*
> Specify second point or [Continue] <C>: *Select S2, as shown in Figure 27-38.*
> Specify third point or [Continue] <C>: *Select S3, as shown in Figure 27-38.*
> Specify destination plane and orientation ...
> Specify first destination point: *Select D1, as shown in Figure 27-38.*
> Specify second destination point or [eXit] <X>: *Select D2, as shown in Figure 27-38.*
> Specify third destination point or [eXit] <X>: *Select D3, as shown in Figure 27-38.*

Figure 27-39 shows the model after aligning.

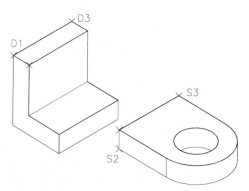

Figure 27-38 *Objects before aligning*

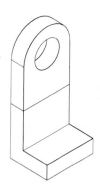

Figure 27-39 *Objects after aligning*

If you press the ENTER key at the **Specify second point or [Continue] <C>** prompt, then AutoCAD assumes the X and Y axes of the new plane to be parallel and in same direction of the XY plane of the current UCS. Also, you will be prompted to select the first destination point.

If you press the ENTER key at the **Specify second destination point or [eXit] <X>** prompt, then AutoCAD assumes the X and Y axes of the new plane for the destination to be parallel and in the same direction as the XY plane of the current UCS. Also, the XY plane of the source object will get aligned to this plane.

EXTRACTING EDGES OF A SOLID MODEL

> **Ribbon:** Home > Solid Editing > Edge Editing drop-down > Extract Edges   **Command:** XEDGES
> **Menu Bar:** Modify > 3D Operations > Extract Edges

This tool is used to automatically generate the wireframe from a previously created solid model. The wireframe will be generated at the same location and with the same orientation where the parent model was located. The prompt sequence that will follow when you choose the **Extract Edges** tool from the **Solid Editing** panel (see Figure 27-40) is given next.

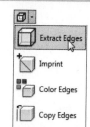

Figure 27-40 *The* **Extract Edges** *tool in the* **Edge Editing** *drop-down*

> Command: _xedges
> Select objects: *Select object from which you want to extract the edges.*
> Select objects: *Select more objects or press* [Enter]

Figure 27-41 shows a model from which the edges are to be extracted and the wireframe extracted from that model after moving it to another location.

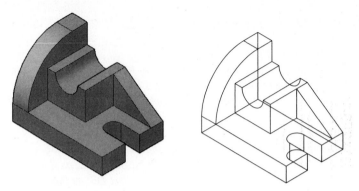

Figure 27-41 *Model and the resulting edges after extraction*

 Note
*The **XEDGES** command can only be applied to regions, surfaces, and solids.*

 Tip
To extract edges from a particular edge or face of the model, press and hold the CTRL key during the selection of the objects.

CONVERTING OBJECTS TO SURFACES

Ribbon: Home > Solid Editing > Convert to Surface	**Command:** CONVTOSURFACE
Menu Bar: Modify > 3D Operations > Convert to Surface	

The **Convert to Surface** tool is used to convert objects into surfaces. The objects that can be converted to surfaces include arcs and lines with nonzero thickness, open zero width polyline having a nonzero thickness, 2D solids, donuts, boundary, regions, and 3D faces created on a plane.

To convert objects to surfaces, choose the **Convert to Surface** tool from the **Solid Editing** panel and then select the objects to be converted into a surface. Figure 27-42 shows a boundary to be converted into a surface and Figure 27-43 shows the resulting surface.

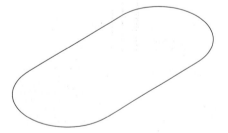

Figure 27-42 *The boundary to be converted into a surface*

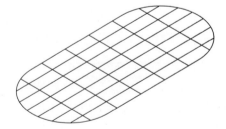

Figure 27-43 *The resulting surface*

 Tip
*You can convert 3D solids with curved faces to surfaces by using the **EXPLODE** command. On invoking this command, the curved faces of the solid will be converted into surfaces and the planar faces will be converted into regions.*

CONVERTING OBJECTS TO SOLIDS

Ribbon: Home > Solid Editing > Convert to Solid   **Command:** CONVTOSOLID
Menu Bar: Modify > 3D Operation > Convert to Solid

This tool is used to convert objects with some thickness into solid models. The solid generated by using this option will be similar to an extruded solid having height equal to the thickness of the object. The objects that can be converted into solids are open polylines with nonzero uniform width and thickness, closed polylines with thickness, and circles with thickness.

To convert objects into solids, choose the **Convert to Solid** tool from the **Solid Editing** panel and then select the objects to be converted into solids. Figure 27-44 shows a closed zero width polyline having a nonzero thickness and Figure 27-45 shows the resulting solid.

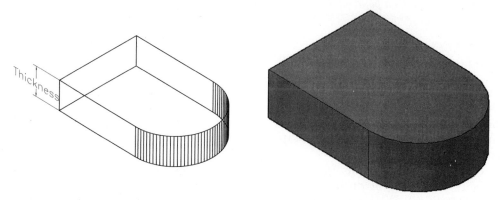

Figure 27-44   *Polyline with thickness*   *Figure 27-45*   *Resulting solid*

CONVERTING SURFACES TO SOLIDS

Ribbon: Home > Solid Editing > Thicken   **Command:** THICKEN
Menu Bar: Modify > 3D Operation > Thicken

This tool is used to convert any surface into a solid by adding material of a specified thickness to the surface. The material addition takes place in the direction normal to the surface. To convert surfaces into solids, choose the **Thicken** tool from the **Solid Editing** panel. Next, select the surface to thicken and then specify the thickness of the material to be added to the surface. Figure 27-46 shows the surface to be thickened and Figure 27-47 shows the resulting solid after thickening.

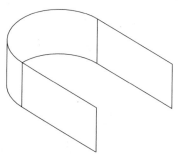 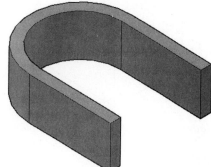

Figure 27-46   *Surface to be thickened*   *Figure 27-47*   *Resulting solid after thickening*

Note
You cannot thicken 3D primitive surfaces generated by using the 3D command.

SLICING SOLID MODELS

Ribbon: Home > Solid Editing > Slice	**Command:** SLICE
Menu Bar: Modify > 3D Operation > Slice	

 As the name suggests, the **Slice** tool is used to slice the selected solid with the help of a specified plane. You will be given an option to select the portion of the sliced solid that has to be retained. You can also retain both the portions of the sliced solids.

The prompt sequence that follows is given next.

> Select objects to slice: *Select the object to slice.*
> Select objects to slice: Enter
> Specify start point of slicing plane or [planar Object/Surface/Zaxis/View/XY/YZ/ZX/3points] <3points>: *Specify a point on the slicing plane or select an option.*

3points

This option is used to slice a solid by using a plane defined by three points, see Figures 27-48 and 27-49. The prompt sequence that follows is given next.

> Select objects to slice: *Select the object to be sliced.*
> Select objects to slice: Enter
> Specify start point of slicing plane or [planar Object/Surface/Zaxis/View/XY/YZ/ZX/3points] <3points>: Enter
> Specify first point on plane: *Specify the point P1 on the slicing plane, as shown in Figure 27-48.*
> Specify second point on plane: *Specify the point P2 on the slicing plane, as shown in Figure 27-48.*
> Specify third point on plane: *Specify the point P3 on the slicing plane, as shown in Figure 27-48.*
> Specify a point on desired side of the plane or [keep Both sides]: *Select the portion of the solid to retain or enter B to retain both the portions of the sliced solid.*

The model after slicing is shown in Figure 27-49.

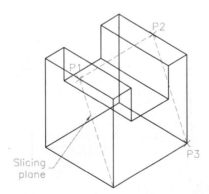

Figure 27-48 *Defining the slicing plane*

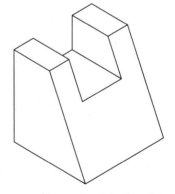

Figure 27-49 *Model after slicing*

Object

This option is used to slice the solid model using an object. The objects that can be used to slice the solid model include arcs, circles, ellipses, 2D polylines, and splines. The prompt sequence that will follow is given next.

> Select objects to slice: *Select the object to slice.*

Select objects to slice: [Enter]
Specify start point of slicing plane or [planar Object/ Surface/Zaxis/View/XY/YZ/ ZX/3points] <3points>: **O**
Select a circle, ellipse, arc, 2D-spline, 2D-polyline to define the slicing plane: *Select the object to slice the solid.*
Specify a point on desired side or [keep Both sides]: *Specify the portion of the solid to retain.*

Surface

This option is used to slice the solid model using a surface. Note that the ruled, tabulated, revolved, or edge surfaces cannot be used for slicing the solids. Figure 27-50 shows the object to be sliced and the slicing surface, and Figure 27-51 shows the model after slicing. The prompt sequence that will follow is given next.

Select objects to slice: *Select the object to slice.*
Select objects to slice: [Enter]
Specify start point of slicing plane or [planar Object/Surface/Zaxis/View/XY/YZ/ ZX/3points] <3points>: **S**
Select a surface: *Select the surface to slice the solid.*
Select solid to keep or [keep Both sides] <Both>: *Specify the portion of the solid to retain.*

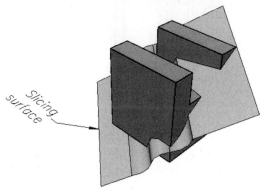

Figure 27-50 *Model with slicing surface*

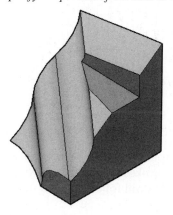

Figure 27-51 *Model after slicing*

Zaxis

This option is used to slice the solid using a plane defined by two points. The first point is the point on the section plane and the second point is the point in the direction of the Z axis of the plane. The prompt sequence that follows is given next.

Select objects to slice: *Select the object to slice.*
Select objects to slice: [Enter]
Specify start point of slicing plane or [planar Object/Surface/Zaxis/View/XY/YZ/ ZX/3points] <3points>: **Z**
Specify a point on the section plane: *Specify the point on the section plane.*
Specify a point on the Z-axis (normal) of the plane: *Specify the point in the direction of the Z axis of the section plane.*
Specify a point on desired side or [keep Both sides]: *Select the portion to retain.*

View

This option is used to slice the selected solid about the viewing plane, see Figures 27-52 and 27-53. The viewing plane in this case will be the screen of the monitor. On selecting this option, you will be prompted to specify a point on the solid model through which the viewing plane will pass. The prompt sequence that will follow is given next.

Select objects to slice: *Select the object to slice.*
Select objects to slice: Enter
Specify start point of slicing plane or [planar Object/Surface/Zaxis/View/XY/YZ/ZX/3points] <3points>: **V**
Specify a point on the current view plane <0,0,0>: *Specify the point P1, as shown in Figure 27-52.*
Specify a point on the desired side or [keep Both sides] <Both>: *Specify the portion of the solid to retain.*

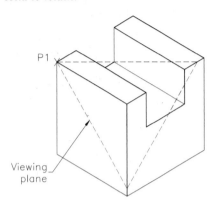

Figure 27-52 *Defining the slicing plane*

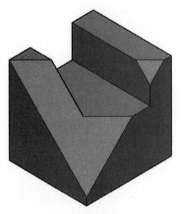

Figure 27-53 *Model after slicing*

XY, YZ, ZX

These options are used to slice the selected solid about the *XY, YZ,* or the *ZX* plane respectively. When you invoke this option, you will be prompted to select the point on the plane. The prompt sequence that will follow is given next.

Select objects to slice: *Select the object to slice.*
Select objects to slice: Enter
Specify start point of slicing plane or [planar Object/ Surface/Zaxis/View/XY/YZ/ZX/3points] <3points>: *Select the XY, YZ, or the ZX plane.*
Specify a point on the XY-plane <0,0,0>: *Specify the point on the slicing plane.*
Specify a point on desired side or [keep Both sides] <Both>: *Specify the portion to retain.*

CREATING THE CROSS-SECTIONS OF SOLIDS

Command: SECTION

The **SECTION** command works similar to the **Slice** tool. The only difference between these two is that this command does not chop the solid. Instead, it creates a cross-section along the selected section plane. The cross-section thus created is a region. You need to move the region created to view the cross section. If needed, you can hatch the region. Note that the regions are created in the current layer and not in the layer in which the sectioned solid is stored. The prompt sequence that follows when you enter the **SECTION** command is given next.

Select objects: *Select the object to section.*
Select objects: Enter
Specify first point on section plane by [Object/Zaxis/View/XY/YZ/ZX/3points] <3points>:

3points

This option is used to define the section plane using three points.

Object

This option is used to specify the section plane using a planar object. The objects that can be used to create the sections are arcs, circles, ellipses, splines, or 2D polylines.

View

This option uses the current viewing plane to define the section plane. You will be prompted to specify the point on the current view plane. It will then automatically create a cross-section parallel to the current viewing plane and passing through the specified point.

XY, YZ, ZX

These options are used to define the section planes that are parallel to the *XY*, *YZ*, or the *ZX* planes, respectively. On invoking this option, you will be prompted to specify the point through which the selected plane will pass, see Figures 27-54 and 27-55.

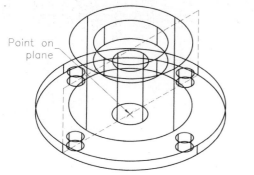

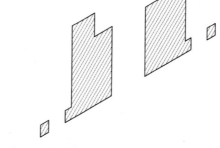

Figure 27-54 *Creating the cross-section along the YZ plane*

Figure 27-55 *Cross-section created, isolated, and hatched for clarity*

Tip

The number of regions created as cross-sections will be equal to the number of solids selected to create the cross-section.

*You can also hatch the cross-section created using the **SECTION** command. Select the entire region as the object for hatching. You may have to define a new UCS based on the section to hatch it.*

EXAMPLE 2

In this example, you will draw the solid model shown in Figure 27-56. The dimensions for the model are shown in Figures 27-57 and 27-58. After creating it, slice it as shown in Figure 27-59.

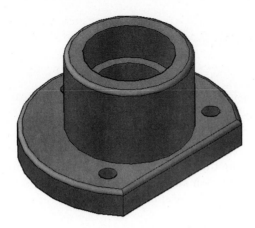

Figure 27-56 *Model for Example 2*

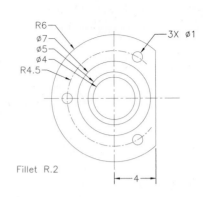

Figure 27-57 *Top view of the model*

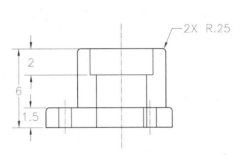

Figure 27-58 *Front view of the model*

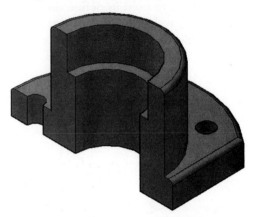

Figure 27-59 *The sliced solid*

1. Start a new 3D template drawing file and then draw the sketch for the base of the model in the **Top** view, as shown in Figure 27-60. Choose the **Region** tool from the **Draw** panel in the **Home** tab and convert the sketch to a region.

2. Choose the **Extrude** tool from **Home > Modeling > Solid Creation** drop-down and extrude the region to a distance of 1.5.

3. Set the view to **SW Isometric** by using the ViewCube. The model should look similar to the one shown in Figure 27-61.

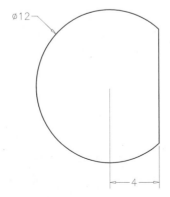

Figure 27-60 *Sketch for the base of the model*

Figure 27-61 *Base of the model*

4. Set the view to **Front** and set the UCS aligned to this view. Draw the sketch for the next feature anywhere in the drawing area and convert it into a region, see Figure 27-62.

5. Choose the **Revolve** tool from **Home > Modeling > Solid Creation** drop-down and then revolve the sketch to an angle of 360°.

6. Right-click on the **3D Object Snap** button in the Status Bar and select the **Center of face** option. Also, clear the other snap options.

7. Ensure that the **3D Object Snap** is on and **Object Snap** is off. Move the revolved feature by snapping the center point of the lower face and place it at the center of the upper face of the base of the model.

8. Choose the **Union** tool from the **Solid Editing** panel in the **Home** tab and then unite both the objects, see Figure 27-63.

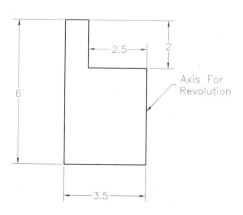

Figure 27-62 *Sketch for the next feature*   **Figure 27-63** *Model after union*

9. Create a new cylinder of diameter 1 unit of same height as that of the base.

10. Draw a circle of radius 4.5 units on the top face of the base.

11. Move the cylinder by snapping the center of the top face and place it on the quadrant point along the X-axis of the circle.

12. Choose the **3D Array** tool from the **Modify** panel. The prompt sequence that follows is given next:

   Initializing... 3DARRAY loaded.
   Select objects: *Select the cylinder.*
   Select objects: [Enter]
   Enter the type of array [Rectangular/Polar] <R>: **P**
   Enter the number of items in the array: **3**
   Specify the angle to fill (+=ccw, -=cw) <360>: [Enter]
   Rotate arrayed objects? [Yes/No] <Y>: [Enter]
   Specify center point of array: *Select the center of the base of the model.*
   Specify second point on axis of rotation: *Select the center of the top face of the model.*

13. Subtract all three cylinders from the model by choosing the **Solid, Subtract** tool from the **Solid Editing** panel.

14. Create a cylinder of diameter 4 units and height 6 units and subtract it from the model to create the central hole.

15. Choose the **Fillet Edge** tool from **Solid > Solid Editing > Fillet/Chamfer** drop-down; you are prompted to select the edges to be filleted.

16. Select the edges to fillet and press ENTER; you are prompted to enter the radius. Specify the fillet radius as 0.25 units. The model should now look similar to the one shown in Figure 27-64.

17. Relocate the UCS at the WCS.

18. Choose the **Slice** tool from the **Solid Editing** panel to cut the solid model to half. The prompt sequence is as follows:

    Select objects: *Select the model.*
    Select objects: Enter
    Specify first point on slicing plane by [Object/Zaxis/View/XY/YZ/ZX/3points] <3points>: **ZX**
    Specify a point on the ZX-plane <0,0,0>: *Select the center of the top face of the model.*
    Specify a point on desired side of the plane or [keep Both sides]: *Specify a point on the back side of the model to retain it.*

19. The final model for Example 2 should look similar to the one shown in Figure 27-65.

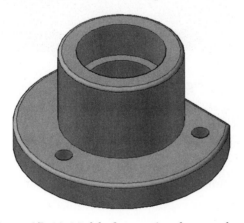

Figure 27-64 *Model after creating the central hole fillet*

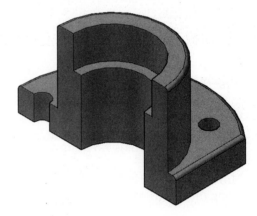

Figure 27-65 *Final model after slicing*

Self-Evaluation Test

Answer the following questions and then compare them to those given at the end of this chapter:

1. A model can be dynamically moved using the **3D MOVE** command. (T/F)

2. The edges of a solid model can be filleted using the **3D Fillet** tool. (T/F)

3. You can select an ellipse as an object to rotate the solid model using the **Rotate 3D** tool. (T/F)

4. The **Convert to Surface** tool can be used to convert an open zero-width polyline having a nonzero thickness to a surface. (T/F)

5. The **SECTION** command is used to create a cross-section and also to slice a model. (T/F)

6. The _____ rule is used to determine the direction of rotation of the solid model in 3D space.

7. The _____ tool is used to extract edges from a solid model.

8. The _____ tool is used to move, rotate, and scale the solid model in a single attempt.

9. The **SECTION** command is used to create a _____ along the plane of a section.

10. The _____ tool is used to align solids by defining an alignment plane.

Review Questions

Answer the following questions:

1. A surface can be used to slice a model using the **Slice** tool. (T/F)

2. The **Convert to Solid** tool can convert objects with a zero thickness into solid. (T/F)

3. You can also extract an individual edge of a solid model using the **Extract Edges** tool. (T/F)

4. You can select all tangential edges of the selected solid model for filleting using the **Chain** option of the **Fillet Edges** tool. (T/F)

5. Regions can be converted to surfaces by using the **Convert to Surface** tool. (T/F)

6. Which of the following options of the **3D Rotate** tool is used to select a 2D entity as an object for rotating the solid models?

   (a) **2D**　　　　　　　　(b) **Last**
   (c) **Object**　　　　　　 (d) **Entity**

7. Which of the following commands is used to convert a surface to a solid?

   (a) **CONVTOSOLID**　　　(b) **CONVQUILT**
   (c) **CONVTOSURFACE**　　(d) **THICKEN**

8. Which of the following objects can be converted into a solid using the **Convert to Solid** tool?

   (a) Rectangle   (b) Circle
   (c) Ellipse   (d) Polygon

9. Which of the following options of the **3D Mirror** command is used to mirror the object about the view plane?

   (a) **Object**   (b) **Last**
   (c) **View**   (d) **Zaxis**

10. What is the least number of points that you have to define on the source object to align it using the **3D Align** tool?

    (a) One   (b) Two
    (c) Three   (d) Four

11. The _____ command gives a dynamic preview while aligning the solid model.

12. The _____ option of the **3D Rotate** tool is used to select the same axis that was last selected to rotate the solid model.

13. In the **3D Array** command, the levels are arranged along the _____ axis.

14. The object having some _____ can be converted into a solid using the **Convert to Solid** tool.

15. The _____ option is used to slice a model by using a sketched entity.

EXERCISE 1

In this exercise, you will create the solid model shown in Figure 27-66. The dimensions of the model are shown in Figures 27-67 and 27-68. After creating it, slice it to get the model shown in Figure 27-69.

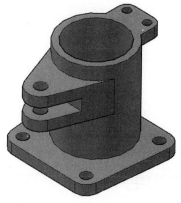

Figure 27-66 *Model for Exercise 1*

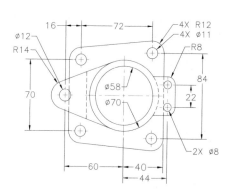

Figure 27-67 *Top view of the model*

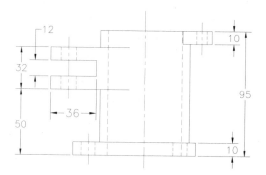

Figure 27-68 *Front view of the model*

Figure 27-69 *Model after slicing*

EXERCISE 2

In this exercise, you will create the solid model shown in Figure 27-70. The dimensions of the model are given in the drawing shown in the same figure. The fillet radius is 5mm.

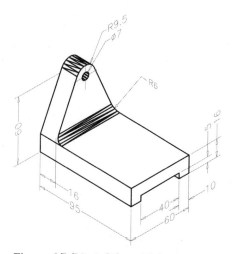

Figure 27-70 *Solid model for Exercise 2*

EXERCISE 3

In this exercise, you will create the solid model shown in Figure 27-71. The dimensions for the model are given in the same figure. The fillet radius is 0.13 units.

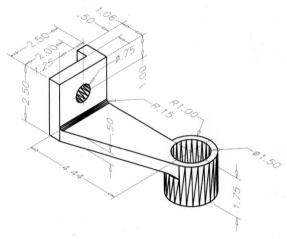

Figure 27-71 *Solid model for Exercise 3*

EXERCISE 4

In this exercise, you will create the solid model shown in Figure 27-72. The dimensions for the model are given in the same figure. Assume the missing dimensions.

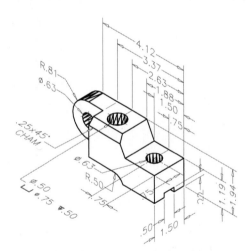

Figure 27-72 *Solid model for Exercise 4*

Problem-Solving Exercise 1

For the Gear Puller shown in Figure 27-73, draw the following parts:

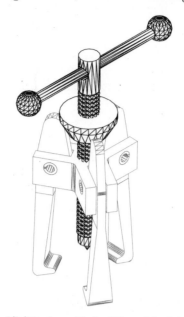

Figure 27-73 *Assembled solid model of the Gear Puller*

Piece part drawings are shown in Figures 27-74 through 27-78, assembly drawing with the Bill of Materials is shown in Figure 27-79, and the exploded view of the solid model is shown in Figure 27-80.

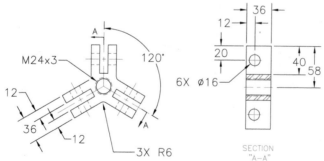

P1: Yoke

Figure 27-74 *Front and side views of the Yoke*

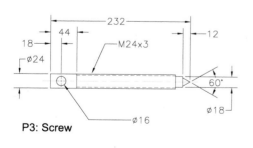

P3: Screw

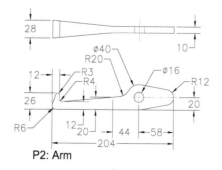

P2: Arm

Figure 27-75 *Front view of the Screw*   **Figure 27-76** *Front and top views of the Arm*

Chapter 27

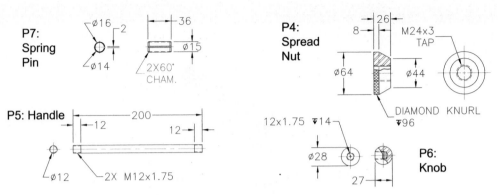

Figure 27-77 *Views of the Spring Pin and Handle*

Figure 27-78 *Views of the Spread Nut and Knob*

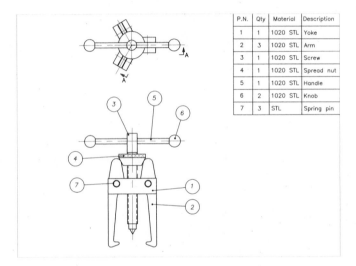

P.N.	Qty	Material	Description
1	1	1020 STL	Yoke
2	3	1020 STL	Arm
3	1	1020 STL	Screw
4	1	1020 STL	Spread nut
5	1	1020 STL	Handle
6	2	1020 STL	Knob
7	3	STL	Spring pin

Figure 27-79 *Assembled view and BOM of the Gear Puller*

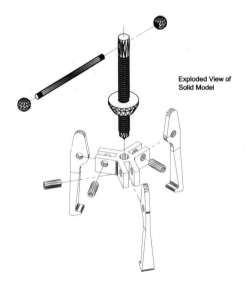

Exploded View of
Solid Model

Figure 27-80 *Exploded solid model of the Gear Puller*

Problem-Solving Exercise 2

Create the model shown in Figure 27-81. The dimensions of the model are given in Figure 27-82. **

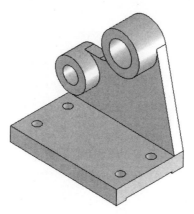

Figure 27-81 *Solid model for Problem-Solving Exercise 2*

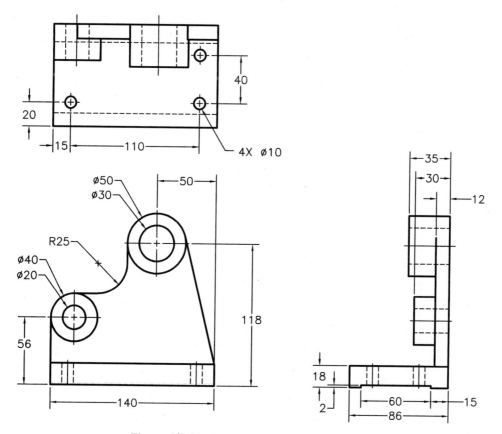

Figure 27-82 *Dimensions for the solid model*

Problem-Solving Exercise 3

Create the model shown in Figure 27-83. The dimensions of the model are given in Figure 27-84. **

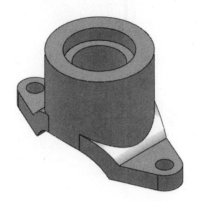

Figure 27-83 *Solid model for Problem-Solving Exercise 3*

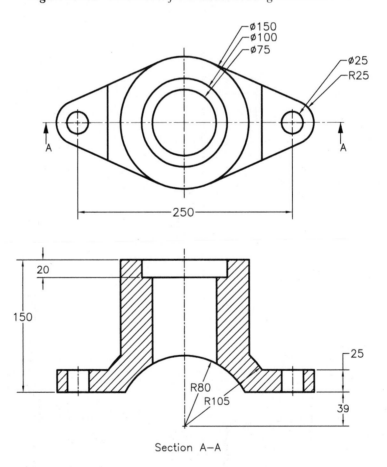

Figure 27-84 *Dimensions for the solid model*

Problem-Solving Exercise 4

Create the model shown in Figure 27-85. The sectioned view and dimensions of the model are given in Figure 27-86. **

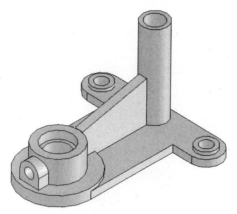

Figure 27-85 *Solid model for Problem-Solving Exercise 4*

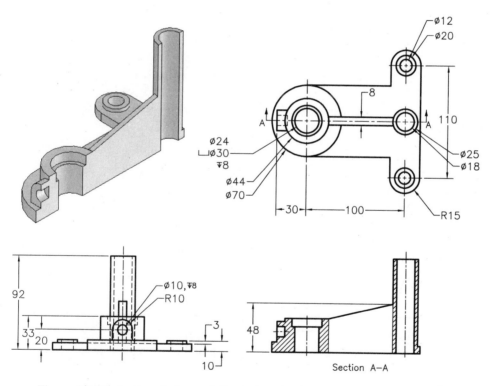

Figure 27-86 *Sectioned view and dimensions for the solid model*

Problem-Solving Exercise 5

Create the model shown in Figure 27-87. The sectioned view and dimensions of the model are given in Figure 27-88. Note that the hidden lines have been removed for clarity. **

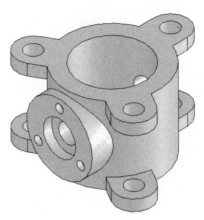

Figure 27-87 Solid model for Problem-Solving Exercise 5

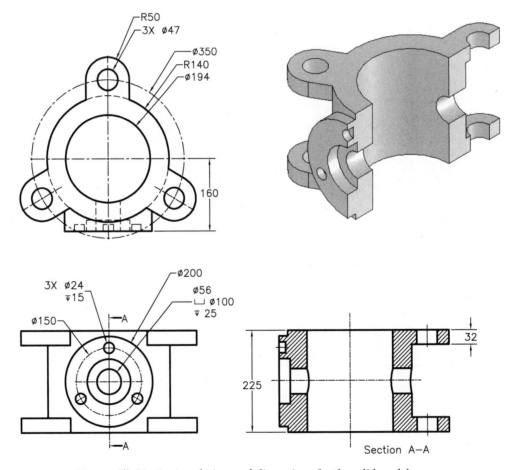

Figure 27-88 Sectioned view and dimensions for the solid model

Problem-Solving Exercise 6

Create the model shown in Figure 27-89. The dimensions of the model are given in Figure 27-90. **

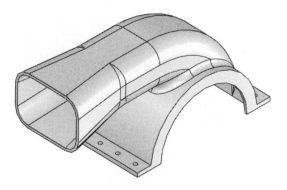

Figure 27-89 *Solid model for Problem-Solving Exercise 6*

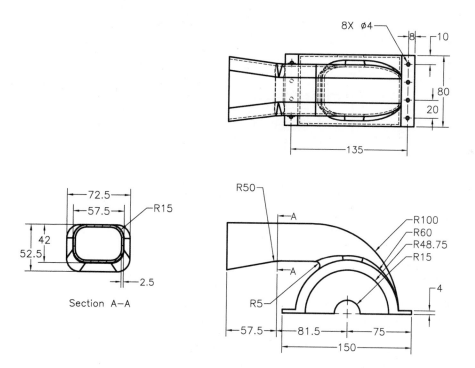

Figure 27-90 *Dimensions for the solid model*

Problem-Solving Exercise 7

Create the model shown in Figure 27-91. The dimensions of the model are given in Figure 27-92. **

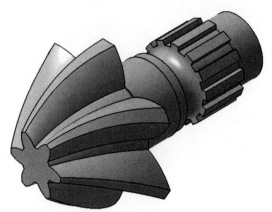

Figure 27-91 Solid model for Problem-Solving Exercise 7

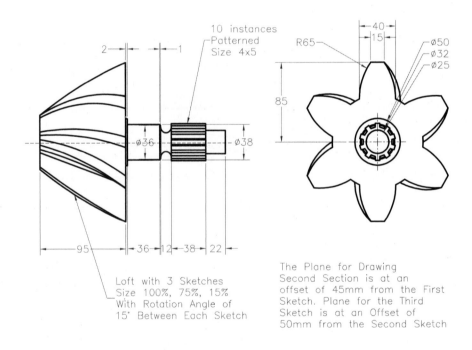

Figure 27-92 Dimensions for the solid model

Answers to Self-Evaluation Test

1. T, **2**. F, **3**. F, **4**. T, **5**. F, **6**. right-hand thumb, **7**. **Extract Edges**, **8**. **Align**, **9**. cross-section, **10**. **3D Align**

Chapter 28

Editing 3D Objects

CHAPTER OBJECTIVES

In this chapter, you will learn:
- *To edit solid models using the SOLIDEDIT command.*
- *To generate sections of a model.*
- *To modify a composite solid using the Solid History tool.*
- *To generate drawing views of a solid model using the SOLVIEW command.*
- *To generate profiles and section in drawing views using the SOLDRAW command.*
- *To create the profile images of a solid model using the SOLPROF command.*
- *To calculate the mass properties of the solid models using the MASSPROP command.*
- *About the use of Action Recorder.*
- *To use the ShowMotion tool for presentation.*
- *To edit models in Autodesk Inventor Fusion.*
- *To generate drawing views.*

KEY TERMS

- *Extrude Faces*
- *Taper Faces*
- *Move Faces*
- *Copy Faces*
- *Offset Faces*
- *Delete Faces*

- *Rotate Faces*
- *Color Faces*
- *Extract Edges*
- *Color Edges*
- *Imprint*
- *Copy Edges*

- *Offset Edge*
- *Section Plane*
- *Live Section*
- *Solid History*
- *SOLVIEW*
- *SOLPROF*

- *Drawing Views*
- *Mass Properties*
- *Record*
- *ShowMotion*
- *Edit in Fusion*

EDITING SOLID MODELS

Ribbon: Home > Solid Editing	**Menu Bar:** Modify > Solid Editing
Toolbar: Solid Editing	**Command:** SOLIDEDIT

One of the major enhancements in the recent releases of AutoCAD is editing the solid models. This has made solid modeling in AutoCAD very user-friendly. In AutoCAD, you can edit the solid models using the tools available in the **Solid Editing** panel. These tools can be used to edit the selected faces or selected edges, or the entire body of a solid model. Various editing processes and the tools used to perform them are discussed next.

Editing Faces of a Solid Model

You can edit the faces of the solid models by using the tools in the **Face Editing** drop-down of the **Solid Editing** panel, refer to Figure 28-1. The tools available in this drop-down are **Extrude Faces**, **Taper Faces**, **Move Faces**, **Copy Faces**, **Offset Faces**, **Delete Faces**, **Rotate Faces**, and **Color Faces**. These editing tools are discussed next.

Extrude Faces

The **Extrude Faces** tool is used to extrude the selected faces of a solid model to a specific height or along a selected path (Figure 28-2). To extrude faces, choose the **Extrude Faces** tool from the **Face Editing** drop-down of the **Solid Editing** panel in the **Home** tab. The prompt sequence that will follow when you invoke this tool is given next.

Command: **SOLIDEDIT**
Solids editing automatic checking: SOLIDCHECK=1
Enter a solids editing option [Face/Edge/Body/Undo/ eXit] <eXit>: _face
Enter a face editing option [Extrude/Move/Rotate/Offset/Taper/Delete/Copy/coLor/ mAterial/Undo/eXit] <eXit>: _rotate
Select faces or [Undo/Remove]: *Select faces to extrude or enter an option.*
Select faces or [Undo/Remove/ALL]: *Select another face, an option, or* Enter.

*Figure 28-1 The **Face Editing** drop-down*

The **Undo** option cancels the selection of the most recently selected face. The **Remove** option allows you to remove a previously selected face from the selection set for extrusion. The **ALL** option selects all the faces of the specified solid. After you have selected a face for extrusion, the next prompt is given below:

Specify height of extrusion or [Path]: *Enter a height value or enter **P** to select a path.*

The **Path** option allows you to select a path for extrusion based on a specified line or curve. At the **Select extrusion path** prompt, select the line, circle, arc, ellipse, elliptical arc, polyline, or spline to be specified as the extrusion path. Note that this path should not lie on the same plane as the selected face and should not have areas of high curvature.

A positive value for the height of extrusion extrudes the selected face outward, whereas a negative value extrudes the selected face inward. On specifying the height of extrusion, the next prompt is given next.

Specify angle of taper for extrusion <0>: *Specify a value between -90 and +90-degree or press ENTER to accept the default value.*

A positive angle value tapers the selected face inward and a negative value tapers the selected face outward (Figure 28-2).

Move Faces

The **Move Faces** tool is used to move the selected faces from one location to the other without changing the orientation of the solid (Figure 28-3). For example, you can move holes from one location to the other without actually modifying the solid model. To move faces, choose the **Move Faces** tool from the **Face Editing** drop-down of the **Solid Editing** panel. The prompt sequence that will follow when you invoke this tool is given next.

Command: **SOLIDEDIT**
Solids editing automatic checking: SOLIDCHECK=1
Enter a solids editing option [Face/Edge/Body/Undo/eXit] <eXit>: _face
Enter a face editing option
[Extrude/Move/Rotate/Offset/Taper/Delete/Copy/coLor/mAterial/Undo/eXit] <eXit>: _move
Select faces or [Undo/Remove]: *Select one or more faces or enter an option.*
Select faces or [Undo/Remove/ALL]: *Select one or more faces, enter an option, or* Enter.
Specify a base point or displacement: *Specify a base point.*
Specify a second point of displacement: *Specify a point or* Enter.

The face moves to the specified location.

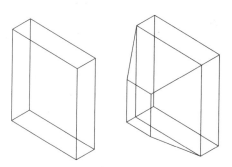

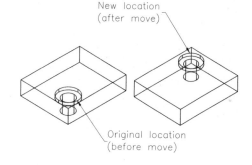

New location
(after move)

Original location
(before move)

*Figure 28-2 Using the **Extrude Faces** option to extrude a solid face with a positive taper angle*

*Figure 28-3 Using the **Move Faces** option to move the hole from the original position*

Tip
If the feature you want to move consists of more than one face, then you must select all faces to move the entire feature.

Offset Faces

The **Offset Faces** tool is used to offset the selected faces of a solid model uniformly through a specified distance on the model. Offsetting takes place in the direction of the positive side of the face. For example, you can offset holes to a larger or smaller size in a 3D solid through a specified distance. In Figure 28-4, the hole on the solid has been offset to a larger size. To offset faces, choose the **Offset Faces** tool from the **Face Editing** drop-down of the **Solid Editing** panel. The prompt sequence that will follow when you choose this tool is given next.

Chapter 28

Command: **SOLIDEDIT**
Solids editing automatic checking: SOLIDCHECK=1
Enter a solids editing option [Face/Edge/Body/Undo/eXit] <eXit>: _face
Enter a face editing option
[Extrude/Move/Rotate/Offset/Taper/Delete/Copy/coLor/mAterial/Undo/eXit] <eXit>:
_offset
Select faces or [Undo/Remove]: *Select one or more faces or enter an option.*
Select faces or [Undo/Remove/ALL]: *Select one or more faces, enter an option, or* Enter.
Specify the offset distance: *Specify the offset distance.*

Rotate Faces

The **Rotate Faces** tool is used to rotate the selected faces of a solid model through a specified angle. For example, you can rotate cuts or slots around a base point through an absolute or relative angle, refer to Figure 28-5. The direction in which the rotation takes place is determined by the right-hand thumb rule. To rotate faces, choose the **Rotate Faces** tool from the **Face Editing** drop-down of the **Solid Editing** panel. The prompt sequence that will follow when you choose this tool is given next.

Command: **SOLIDEDIT**
Solids editing automatic checking: SOLIDCHECK=1
Enter a solids editing option [Face/Edge/Body/Undo/eXit] <eXit>: _face
Enter a face editing option
[Extrude/Move/Rotate/Offset/Taper/Delete/Copy/coLor/mAterial/Undo/eXit] <eXit>:
_rotate
Select faces or [Undo/Remove]: *Select face(s) or enter an option.*
Select faces or [Undo/Remove/ALL]: *Select face(s), or enter an option, or* Enter.
Specify an axis point or [Axis by object/View/Xaxis/Yaxis/Zaxis] <2points>: *Select a point, enter an option, or* Enter.

If you press ENTER to use the **2points** option, you will be prompted to select the two points that will define the axis of rotation. The prompt sequence is given next.

Specify the first point on the rotation axis: *Specify the first point.*
Specify the second point on the rotation axis: *Specify the second point.*
Specify a rotation angle or [Reference]: *Specify the angle of rotation or enter **R** to use the reference option.*

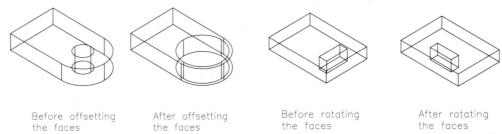

Before offsetting the faces   After offsetting the faces   Before rotating the faces   After rotating the faces

Figure 28-4 *Using the **Offset Faces** option to offset a solid face with a positive taper angle*

Figure 28-5 *Using the **Rotate Faces** option to rotate the faces from the original position*

The **Axis by object** option is used to select an object to define the axis of rotation. The axis of rotation is aligned with the selected object; for example, if you select a circle, the axis of rotation will be aligned along the axis that passes through the center of the circle and will be normal to the plane of the circle. The selected objects can also be a line, ellipse, arc, 2D, 3D, or LW polyline or even a spline. The prompt sequence is given next.

Select a curve to be used for the axis: *Select the object to be used to define the axis of rotation.*
Specify a rotation angle or [Reference]: *Specify rotation angle or use the reference angle option.*

The **Xaxis**, **Yaxis**, **Zaxis** options align the axis of rotation with the axis that passes through the specified point. The Command prompt is given next:

Specify the origin of rotation <0,0,0>: *Select a point through which the axis of rotation will pass.*
Specify a rotation angle or [Reference]: *Specify rotation angle or use the reference angle option.*

Taper Faces

The **Taper Faces** tool is used to taper selected face(s) through a specified angle (Figure 28-6). Tapering takes place depending on the selection sequence of the base point and the second point along the selected face. It also depends on whether the taper angle is positive or negative. A positive value tapers the face inward and a negative value tapers the face outward. You can enter any value between -90-degree and +90-degree. The default value 0-degree extrudes the face along the axis that is perpendicular to the selected face. If you have selected more than one face, they will be tapered through the same angle. To create taper faces, choose the **Taper Faces** tool from the **Face Editing** drop-down of the **Solid Editing** panel. The prompt sequence that will follow when you choose this tool is given next.

Command: **SOLIDEDIT**
Solids editing automatic checking: SOLIDCHECK=1
Enter a solids editing option [Face/Edge/Body/Undo/eXit] <eXit>: _face
Enter a face editing option
[Extrude/Move/Rotate/Offset/Taper/Delete/Copy/coLor/mAterial/Undo/eXit] <eXit>: _taper
Select faces or [Undo/Remove/ALL]: *Select one or more faces, enter an option, or* Enter.
Specify the base point: *Specify a base point.*
Specify another point along the axis of tapering: *Specify a point.*
Specify the taper angle: *Specify an angle between –90 and +90-degree.*

Delete Faces

The **Delete Faces** tool is used to remove the selected face(s) from the specific 3D solid. It also removes chamfers and fillets (Figure 28-7). To remove faces, choose the **Delete Faces** tool from the **Face Editing** drop-down of the **Solid Editing** panel. The prompt sequence that will follow when you choose this tool is given next.

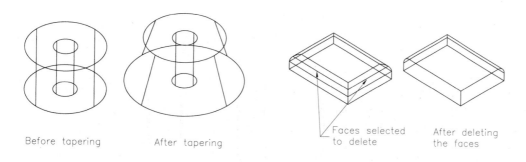

Before tapering   After tapering   Faces selected to delete   After deleting the faces

Figure 28-6 *Tapering the faces using the **Taper Faces** option*   **Figure 28-7** *Deleting the faces using the **Delete Faces** option*

Command: **SOLIDEDIT**
Solids editing automatic checking: SOLIDCHECK=1
Enter a solids editing option [Face/Edge/Body/Undo/eXit] <eXit>: _face
Enter a face editing option
[Extrude/Move/Rotate/Offset/Taper/Delete/Copy/coLor/mAterial/Undo/eXit] <eXit>:
_delete
Select faces or [Undo/Remove]: *Select the face to remove.*
Select faces or [Undo/Remove/ALL]: *Select one or more faces, enter an option, or* [Enter].
Solid validation started.
Solid validation completed.

Copy Faces

The **Copy Faces** tool is used to copy a face as a body or a region (Figure 28-8). To copy
faces, choose the **Copy Faces** tool from the **Face Editing** drop-down of the **Solid Editing**
panel. The prompt sequence that will follow when you choose this tool is given next.

Command: **SOLIDEDIT**
Solids editing automatic checking: SOLIDCHECK=1
Enter a solids editing option [Face/Edge/Body/Undo/eXit] <eXit>: _face
Enter a face editing option
[Extrude/Move/Rotate/Offset/Taper/Delete/Copy/coLor/mAterial/Undo/eXit] <eXit>:
_copy
Select faces or [Undo/Remove]: *Select the faces to copy.*
Select faces or [Undo/Remove/ALL]: *Select one or more faces, enter an option, or* [Enter].
Specify a base point or displacement: *Specify a base point.*
Specify a second point of displacement: *Specify the second point.*

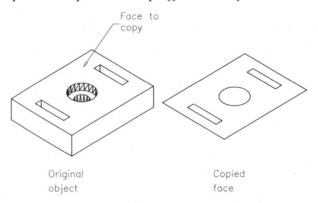

*Figure 28-8 Copying the face using the **Copy Faces** option*

Color Faces

The **Color Faces** tool is used to change the color of a selected face. To do so, choose the
Color Faces tool from the **Face Editing** drop-down of the **Solid Editing** panel. The
prompt sequence that will follow when you choose this tool is given next.

Select faces or [Undo/Remove/ALL]: *Select one or more faces, enter an option, or* [Enter].

When you press ENTER at the previous prompt, the **Select Color** dialog box will be displayed.
You can assign the color to the selected face using this dialog box.

The above-mentioned face editing tools can also be assessed using the **Face** option of the

SOIDEDIT command. The other options available through this command are **mAterial**, **Undo** and **eXit**. These options are discussed next.

mAterial

This option is used to apply material to a selected face and is invoked using the **SOLIDEDIT** command. The prompt sequence that will follow is given next.

> Command: **SOLIDEDIT**
> Solids editing automatic checking: SOLIDCHECK=1
> Enter a solids editing option [Face/Edge/Body/Undo/eXit] <eXit>: *Enter* **F**
> Enter a face editing option
> [Extrude/Move/Rotate/Offset/Taper/Delete/Copy/coLor/mAterial/Undo/eXit] <eXit>:
> *Enter* **A**
> Select faces or [Undo/Remove]: *Select one or more faces to apply the material.*
> Select faces or [Undo/Remove/ALL]: *Select one or more faces, enter an option, or* Enter
> Enter new Material name <ByLayer>: *Enter the name of the material or* Enter

Undo

This option is used to cancel the changes made to the faces of a solid model. This option will be available only when you invoke the **SOLIDEDIT** command by using the Command prompt.

eXit

This option is used to exit the face editing operations. This option will be available only when you invoke the **SOLIDEDIT** command using the Command prompt.

> **Tip**
> *You can select a face, edge, or solid and perform the editing operation. To select an entity with ease, set the appropriate option from the **Filter** drop-down in the **Subobject** panel.*

Editing Edges of a Solid Model

To modify the properties of the edges of a solid, you can use the tools available in the **Edge Editing** drop-down of the **Solid Editing** panel, refer to Figure 28-9. The edge editing tools available in this drop-down are: **Extract Edges**, **Imprint**, **Color Edges,** and **Copy Edges**. These editing tools are discussed next.

Extract Edges

The **Extract Edges** tool is used to extract all the edges of a solid, a surface, or a mesh. The edges are copied as lines, polylines, arcs, circles, ellipses, or splines depending upon the model selected. To extract the edges, choose the **Extract Edges** tool from the **Solid Editing** panel. The prompt sequence that will follow when you choose this tool is given next.

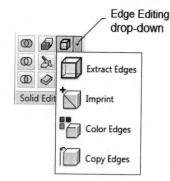

Figure 28-9 *The Edge Editing drop-down*

> Command: _xedges
> Select objects: *Select the solid from which edges are to be extracted*
> Select objects: 1 found
> Select objects: Enter

All the edges of the solid will be extracted and you can be individually selected.

Imprint

 The **Imprint** tool is used to imprint an object on a 3D solid object. Remember that the object to be imprinted should intersect one or more faces of a solid object. The objects that can be imprinted are arcs, circles, lines, 2D and 3D polylines, ellipses, splines, regions, bodies, and 3D solids. To invoke this tool, choose the **Imprint** tool from the **Solid Editing** panel. The prompt sequence that will follow when you choose this tool is given next.

> Command: **IMPRINT**
> Select a 3D solid or surface: *Select an object.*
> Select an object to imprint: *Select an object to be imprinted.*
> Delete the source object [Yes/No] <N>: *Select an option or* Enter*.*
> Select an object to imprint: *Select a new object or* Enter*.*

The above-mentioned body editing tools can also be accessed using the **Body** option of the **SOIDEDIT** command. Other body editing options available for this command are **Undo** and **eXit**. The **Undo** option cancels the modifications made during the **Body** option of the **SOLIDEDIT** command, and **eXit** option is used to exit the **Body** option of this command.

Copy Edges

 The **Copy Edges** tool is used to copy the individual edges of a solid model. The edges are copied as lines, arcs, circles, ellipses, or splines. To invoke this tool, choose the **Copy Edges** tool from the **Solid Editing** panel. The prompt sequence that will follow when you choose this tool is given next.

> Command: **SOLIDEDIT**
> Solids editing automatic checking: SOLIDCHECK=1
> Enter a solids editing option [Face/Edge/Body/Undo/eXit] <eXit>: _edge
> Enter an edge editing option [Copy/coLor/Undo/eXit] <eXit>: _copy
> Select edges or [Undo/Remove]: *Select the edges to be copied.*
> Select edges or [Undo/Remove]: Enter
> Specify a base point or displacement: *Specify the first point of displacement.*
> Specify a second point of displacement: *Specify the second point of displacement.*

Color Edges

The **Color Edges** tool is used to change the color of a selected edge of a 3D solid. To invoke this tool, choose the **Copy Edges** tool from the **Solid Editing** panel. The prompt sequence that will follow when you choose this tool is given next.

> Command: **SOLIDEDIT**
> Solids editing automatic checking: SOLIDCHECK=1
> Enter a solids editing option [Face/Edge/Body/Undo/eXit] <eXit>: _edge
> Enter an edge editing option [Copy/coLor/Undo/eXit] <eXit>: _color
> Select edges or [Undo/Remove]: *Select one or more edges or enter an option.*
> Select edges or [Undo/Remove]: Enter

After you select edges and press ENTER, the **Select Color** dialog box will be displayed. You can assign the color to the selected edges using this dialog box. You can also assign color by entering their names or numbers. Individual edges can be assigned individual colors.

The above-mentioned edge editing tools can also be assessed using the **Edge** option of the **SOIDEDIT** command. The other edge editing options available for this command are **Undo** and **eXit**. The **Undo** option cancels the modifications made during the **Edge** editing of the SOLIDEDIT command, and **eXit** option is used to exit the **Edge** editing of this command.

Offset Edge

The **Offset Edge** tool is used to offset the boundary edges of a planar solid face or a surface. You can offset the edges either dynamically or by setting the offset distance. Note that all edges should lie on the same plane. Depending upon the position of cursor, the selected edges will be offset inwards or outwards. During the operation, you can choose the **Corner** option to make the corners of the offset edges sharp or rounded. To offset the boundary edges, choose the **Offset Edge** tool from the **Solid Editing** panel of the **Solid** tab. The prompt sequence for the **Offset Edge** command is given next.

Command: _offsetedge
Corner = Sharp
Select face: *Select the face whose edges are to be offset.*
Specify through point or [Distance/Corner]: *Specify the point or [distance].*
Select face: [Enter]

If you enter **Corner** at the **Specify through point or [Distance/Corner]** prompt, then the resulting offset profile will have filleted corners.

If you enter **Distance** at the **Specify through point or [Distance/Corner]** prompt, then you can specify the distance at which the offset edges will be created.

Editing Entire Body of a Solid Model

To edit the entire body of a solid model, you can use the tools available in the **Body Editing** drop-down of the **Solid Editing** panel, refer to Figure 28-10. The tools available in this drop-down are: **Separate**, **Clean**, **Shell**, and **Check**. Besides these tools, the **Imprint** tool available in the **Edge Editing** drop-down of the **Solid Editing** panel can also be used for editing a solid body, refer to Figure 28-9. The body editing tools are discussed next.

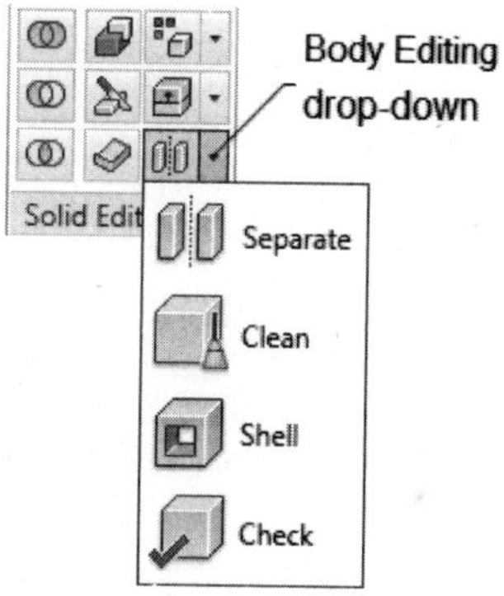

Figure 28-10 The Body Editing drop-down

Separate

The **Separate** tool is used to separate 3D solids with disjointed volumes into separate 3D solids. Solids with disjointed volumes are created by the union of two solids that are not in contact with each other. Note that the composite solids created by performing the Boolean operations and share the same volume cannot be separated using this option. To invoke this tool, choose the **Separate** tool from the **Solid Editing** panel. The prompt sequence that will follow when you choose this tool is given next.

Command: **SOLIDEDIT**
Solids editing automatic checking: SOLIDCHECK=1
Enter a solids editing option [Face/Edge/Body/Undo/eXit] <eXit>: _body
Enter a body editing option
[Imprint/seParate solids/Shell/cLean/Check/Undo/eXit] <eXit>: _separate
Select a 3D solid: *Select a solid object with multiple lumps.*

Chapter 28

Shell

The **Shell** tool is one of the most extensively used tools for body editing. This tool is used to create a shell of a specified 3D solid model. Shelling is defined as a process of scooping out material from a solid model in such a manner that the walls with some thickness are left. One 3D object can have only one shell. You can exclude certain faces before the scooping out material. This means the selected face will not be included during the shelling, see Figure 28-11. To invoke this tool, choose the **Shell** tool from the **Solid Editing** panel. The prompt sequence that will follow when you choose this tool is given next.

Command: **SOLIDEDIT**
Solids editing automatic checking: SOLIDCHECK=1
Enter a solids editing option [Face/Edge/Body/Undo/eXit] <eXit>: _body
Enter a body editing option
[Imprint/seParate solids/Shell/cLean/Check/Undo/eXit] <eXit>: _shell
Select a 3D solid: *Select an object.*
Remove faces or [Undo/Add]: *Select one or more faces or enter an option.*
Remove faces or [Undo/Add/ALL]: *Select a face, enter an option, or* Enter
Enter the shell offset distance: *Specify the wall thickness.*

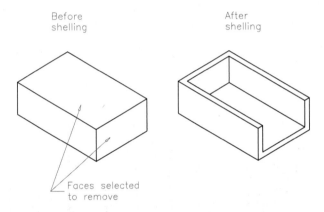

Figure 28-11 Creating a shell in the solid model

Tip
If you have specified a positive distance, the shell will be created on the inside perimeter of a solid, and if it is negative, the shell will be created on the outside perimeter of the 3D solid.

Clean

The **Clean** tool is used to remove the shared edges or vertices sharing the same surface or curve definition on either side of the edge or vertex. It removes all redundant edges and vertices, imprinted as well as unused geometry. It also merges faces that share the same surface. To invoke this tool choose the **Clean** tool from the **Body Editing** drop-down of the **Solid Editing** panel. The prompt sequence that will follow when you choose this tool is given next.

Command: **SOLIDEDIT**
Solids editing automatic checking: SOLIDCHECK=1
Enter a solids editing option [Face/Edge/Body/Undo/eXit] <eXit>: _body
Enter a body editing option
[Imprint/seParate solids/Shell/cLean/Check/Undo/eXit] <eXit>: _clean
Select a 3D solid: *Select the solid model to clean.*

Check

The **Check** tool can be used to check if a solid object created is a valid 3D solid object. This option is independent of the settings of the **SOLIDCHECK** variable. The **SOLIDCHECK** variable turns on or off the solid validation for the current drawing session. The solid validation is turned on when the value of the **SOLIDCHECK** variable is set to 1, which is the default value. If a solid object is a valid 3D solid object, you can use solid editing options to modify it without AutoCAD displaying ACIS error messages. Solid editing can take place only on valid 3D solids. To invoke this tool, choose the **Check** tool from the **Solid Editing** panel. The prompt sequence that will follow when you choose this tool is given next.

Command: **SOLIDEDIT**
Solids editing automatic checking: SOLIDCHECK=1
Enter a solids editing option [Face/Edge/Body/Undo/eXit] <eXit>: _body
Enter a body editing option
[Imprint/sePerate solids/Shell/cLean/Check/Undo/eXit] <eXit>: _check
Select a 3D solid: *Select the solid you want to check.*
This object is a valid ShapeManager solid.

EXAMPLE 1 *Solid Editing*

In this example, you will create the solid model shown in Figure 28-12. After creating the model, modify it, as shown in Figure 28-13, using the tools in the **Solid Editing** panel.

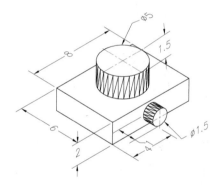

Figure 28-12 *Model for Example 1*

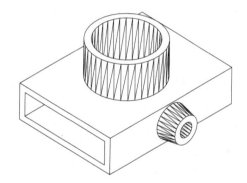

Figure 28-13 *Model after modification*

1. Create a box and two cylinders at appropriate position, refer to Figure 28-12.

2. Combine all objects to make them a solid model.

3. Choose the **Shell** tool from **Home > Solid Editing > Body Editing** drop-down and follow the prompt sequence given below:

Command: **SOLIDEDIT**
Solids editing automatic checking: SOLIDCHECK=1
Enter a solids editing option [Face/Edge/Body/Undo/eXit] <eXit>: _body
Enter a body editing option
[Imprint/sePerate solids/Shell/cLean/Check/Undo/eXit] <eXit>: _shell
Select a 3D solid: *Select the solid.*
Remove faces or [Undo/Add/ALL]: *Click on the top face of the bigger cylinder.*
Remove faces or [Undo/Add/ALL]: *Click on the front face of the smaller cylinder.*
Remove faces or [Undo/Add/ALL]: *Click on the left face of the box.*

Remove faces or [Undo/Add/ALL]: [Enter]
Enter the shell offset distance: **0.25** [Enter]
Solid validation started
Solid validation completed
Enter a body editing option
[Imprint/seParate solids/Shell/cLean/Check/Undo/eXit] <eXit>: [Enter]
Enter a solids editing option [Face/Edge/Body/Undo/eXit] <eXit>: *Press ESC.*

4. Choose the **Taper Faces** tool from **Home > Solid Editing > Face Editing** drop-down and follow the prompt sequence given below:
Solids editing automatic checking: SOLIDCHECK=1
Enter a solids editing option [Face/Edge/Body/Undo/eXit] <eXit>: _face
Enter a face editing option
[Extrude/Move/Rotate/Offset/Taper/Delete/Copy/coLor/Undo/eXit] <eXit>: _taper
Select faces or [Undo/Remove]: *Click on the cylindrical faces of the smaller cylinder*
Remove faces or [Undo/Add/ALL]: [Enter]
Specify the base point: *Select the center of the front face of the smaller cylinder*
Specify another point along the axis of tapering: *Select the center at the back face of the smaller cylinder*
Specify the taper angle: **-10** [Enter]
Solid validation started.
Solid validation completed.
Enter a face editing option
[Extrude/Move/Rotate/Offset/Taper/Delete/Copy/coLor/Undo/eXit] <eXit>: [Enter]
Enter a solids editing option [Face/Edge/Body/Undo/eXit] <eXit>: *Press ESC*

The final model should look similar to the one shown in Figure 28-13.

GENERATING A SECTION BY DEFINING A SECTION PLANE

Ribbon: Home > Section > Section Plane   **Command:** SECTIONPLANE
Menu Bar: Draw > Modeling > Section Plane

The **Section Plane** tool is used to create a sectioned object by defining a sectioning plane. By default, the section plane cuts all the 3D objects, surfaces, and regions lying in the line of this plane even if the section plane does not actually pass through them. Various methods of defining the section plane are discussed next.

Defining a Section Plane by Selecting a Face

In this method, the section plane will be aligned to the plane of the selected face of the 3D solid. Choose the **Section Plane** tool from the **Section** panel and select the desired face on the solid object; a section plane aligned with the selected face will be displayed, as shown in Figure 28-14. The prompt sequence is given next.

Command: **SECTIONPLANE**
Select face or any point to locate section line or [Draw section/Orthographic]: *Select a face on the 3D solid where you want to align the section plane*

Now, you can move the section plane along normal direction to place it at the intended location. To do so, select a section line visible on that section plane, the **Move Gizmo** will be displayed at the center of the section line. Use the **Move Gizmo** to move the plane in any desired direction. Even if the display of gizmo is turned off, you can move the section plane along normal direction by using the triangular arrow grip displayed to the midpoint of the section line, see Figure 28-15. After moving the section plane at the desired location, click once to place the plane; the preview of the remaining section of the model will be displayed. Figure 28-16 shows the remaining section of the solid model after moving the section plane shown in Figure 28-14, to the center of the model.

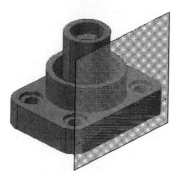

Figure 28-14 *Model with a section plane*

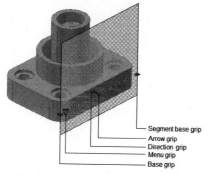

Segment base grip
Arrow grip
Direction grip
Menu grip
Base grip

Figure 28-15 *Various grips visible on the section line*

You can also flip the direction of the sectioned portion of the model. To do this, click on the section line and then click on the direction grip, see Figure 28-17. To rotate the section plane, right-click on the section line and then choose the **Rotate** option from the shortcut menu. Next, select the base point about which you want to rotate the section plane and specify the rotation angle. Figure 28-18 shows the remaining section of the model after rotating the section plane by -90-degree about the midpoint of the section line.

Figure 28-16 *Remaining section of the model after moving the section plane*

Figure 28-17 *Resulting model after flipping the direction of the section plane*

Instead of displaying only the section plane, you can show the section boundary with the section plane. The section boundary displays a two dimensional rectangular sketch. Only the object lying within this boundary will be displayed as the remaining section. To create a section boundary, click on the menu grip and choose the **Section Boundary** option. You can also show the section volume with the section boundary and the section plane. The section volume displays a three dimensional box and only the object lying within this boundary will be displayed as the remaining section. To create a section volume, choose the **Section Volume** option from the menu grip. Figure 28-19 shows a model with a section plane, section boundary, and section volume. The size of the section boundary and the section plane can also be edited with the help of grips.

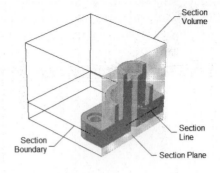

Figure 28-18 *Resulting model after rotating the section plane by -90°*

Figure 28-19 *Model after choosing the **Section Volume** option from the menu grip*

Generating a section by a section plane does not affect the original model. The original model is hidden and only the remaining section of the model is displayed. If you delete the section line from the drawing, the model will be redisplayed without any section.

 AutoCAD also allows you to display both the sectioned and unsectioned portions of a 3D solid. To do so, choose the **Live Section** button from the **Section** panel to activate or deactivate the live sectioning of the solid model. Alternatively, select the section line and right-click to display the shortcut menu. Clear the check mark displayed on the left of the **Activate live sectioning** option by choosing it from the shortcut menu.

Even if the live sectioning is on, you can display the sectioned portion of the solid model. To do so, select the section line, invoke the shortcut menu, and choose the **Show cut-away geometry** option from it; a translucent and red colored sectioned model will be displayed.

You can add jogs to the section plane. A jog provides a 90-degree bend to the section plane for a certain distance and after that the section plane again continues in the original direction. To create a jog, choose the **Add Jog** tool from the **Section** panel; you will be prompted to select the section object. Select the section line from the drawing area. Next, you will be prompted to specify a point on the section line where you want to add the jog. Specify a location on the section line to create the jog. Alternatively, select the section line and choose **Add jog to section** from the shortcut menu to add a jog in the section line. After adding the jog, you can edit its length and location with the help of grips. You can add multiple jogs to the section plane.

Note
The procedure to modify the location and appearance of a section plane, displaying the sectioned portion of the model, and adding jogs to the section plane by grips remains the same in all the other options of defining the section plane that are discussed next.

Defining a Section Plane by Specifying Two Points

In this method, a section plane is generated such that it passes through the two points that specified in the drawing area. If you select a face on the 3D object, it generates a section plane at the selected face, but if you specify the first point at any location in the drawing area or on the 3D object, this point acts as the base point about which the section plane can rotate. Specifying the second point will fix the through point of the section plane. A section plane will be generated passing through these two points, see Figure 28-20. The prompt sequence for this option is given next.

Command: **SECTIONPLANE**
Select face or any point to locate section line or [Draw section/Orthographic]: *Specify a point on the screen or on the 3D object.*
Specify through point: *Specify the other point through which the section line will pass.*

Once you generate the section plane, you will not get a preview of the sectioned object. This is because the **Activate live sectioning** option is not active by default when you generate a section plane by specifying two points. To activate the live sectioning, choose the **Live Section** button from the **Section** panel. Figure 28-20 displays a model sectioned by specifying two points with the live sectioning turned on. You can also activate the live sectioning by entering **LIVESECTION** at the Command prompt. The prompt sequence for this command is given next.

Command: **LIVESECTION**
Select section object: *Select the section line on the section plane.*

Defining a Section Plane by Drawing the Section Line

In this method, you can draw a section line and generate a section plane that passes along it. You can section a 3D object in many planes by picking points through which the section plane will pass. To do so, choose the **Section Plane** tool from **Section** panel. Next, right-click in the drawing area and choose the **Draw section** option from the shortcut menu. Now, specify the point from which the section line will start and then specify the points through which it will pass. After specifying all the points, press the ENTER key and then specify the side of the object to retain after sectioning. By default, the section plane in this case will be displayed as the section boundary with the live sectioning deactivated. Figure 28-21 shows the model sectioned by drawing a section line with the live sectioning activated.

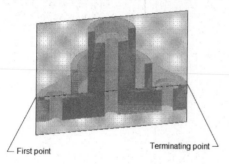

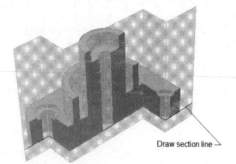

Figure 28-20 Model sectioned by specifying two points

Figure 28-21 Model sectioned by drawing the section line

Note
A section line should be drawn only on one sketching plane. Otherwise, you will not be able to visualize the sectioned model after activating the live sectioning. Also, the drawn section line should not be self-intersecting.

Defining a Section Plane by Selecting a Preset Orthographic Plane

In this option, a section plane will be generated parallel to preset orthographic planes such as front, top, right so on, and passing through an imaginary center of a box. This box is calculated by enveloping all the solids, surfaces, and regions present in the drawing. Choose the **Section Plane** tool from the **Section** panel. Next, right-click in the drawing area and choose the **Orthographic** option from the shortcut menu; you will be prompted to select the predefined orthographic plane with which the generated section plane will be aligned. On selecting this, you get a sectioned solid whose material is chopped from the side you selected as the preset orthographic plane. In this option, you will instantaneously get the sectioned object because live sectioning is activated automatically.

Generating 2D and 3D Sections

Generate Section   As mentioned earlier, sectioning a model by generating a section plane does not affect the original model. The original model is hidden and only the remaining section of the sectioned model is displayed. On deleting the section plane, the sectioned object will get deleted. If you want to retain the generated section as a 2D or 3D section with the original model, you can use the **Generate Section** tool.

To generate a 2D or a 3D section, select a section line and choose the **Generate Section** tool from the **Section** panel; the **Generate Section/Elevation** dialog box will be displayed. Alternatively, choose the **Generate 2D/3D section** from the shortcut menu to display the **Generate Section/Elevation** dialog box. Choose the **Show Detail** button to expand this dialog box, as shown in Figure 28-22. The options in this dialog box are discussed next.

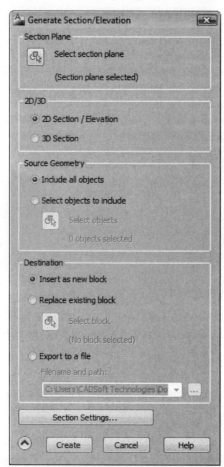

Figure 28-22 *The expanded **Generate Section/Elevation** dialog box*

Section Plane Area

If there are multiple section planes in the model, choose the **Select section plane** button from this area and select the section plane for which you need to create 2D/3D sections.

2D/3D Area

This area allows you to specify the type of section you want to generate. The **2D Section / Elevation** option is used to generate a two dimensional sketch of the sectioned object on the XY plane of the current UCS. The **3D section** option generates a 3D sketch of the sectioned object on a plane parallel to the section plane. Figure 28-23 shows the model with its 2D section and Figure 28-24 shows the model with its 3D section.

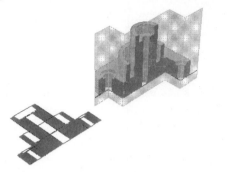

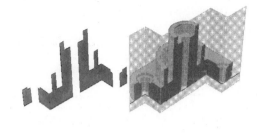

Figure 28-23 *Model with 2D section*   **Figure 28-24** *Model with 3D section*

Source Geometry Area

This area allows you to specify the objects of the drawing that you want to include while creating the 2D or 3D section. The **Include all objects** radio button, if selected, includes all the 3D objects, surfaces, and regions in the drawing. It also includes the external reference objects and blocks. The **Select objects to include** option allows you to select the objects to be included in the section generation. If you select the **Select objects to include** radio button and then choose the **Select objects** button, the **Generate Section/Elevation** dialog box will disappear allowing to select the solids. After selecting the solids, surfaces, or regions, press the ENTER key; the **Generate Section/Elevation** dialog box will reappear, displaying the number of selected objects below the **Select objects** button.

Destination Area

This area provides the option to specify how you want to use the generated section. The **Insert as new block** option inserts the generated section in the same drawing file as a block. The **Replace existing block** option inserts the generated section by replacing an existing block. Choose the **Select block** button to select the block to be replaced by the newly generated section block. The **Export to a file** option allows you to create a new drawing file with the generated section. To save it as a new file, choose the **Browse [...]** button to specify the location where you want to save the new drawing file.

Section Settings

This button is used to invoke the **Section Settings** dialog box, see Figure 28-25. You can also invoke this dialog box by clicking on the inclined arrow in the title bar of the **Section** panel. If you select the **2D Section/Elevation** option in the 2D/3D area, the **Section Settings** dialog box will be displayed with the **2D section / elevation block creation settings** radio button selected by default. Similarly, if you select the **3D Section/Elevation** option in the 2D/3D area, the **Section Settings** dialog box will be displayed with the **3D section block creation settings** option selected by default. The **Section Settings** dialog box allows you to modify the properties of the sections generated. These properties depend on the radio button selected.

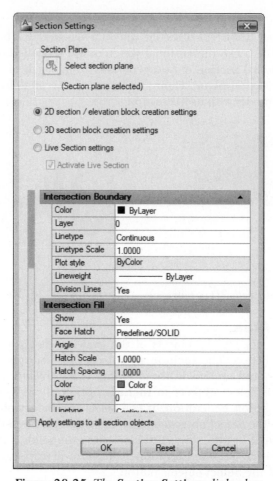

Figure 28-25 *The **Section Settings** dialog box*

Intersection Boundary. This option modifies the display of the lines that are created to outline the boundary of the intersected solid with the section plane.

Intersection Fill. This option modifies the display of the filled portion representing the presence of material at the intersection plane. The **Show** drop-down list allows you to select whether you want to display the section filled. Selecting the **Select hatch pattern type** option from the **Face Hatch** drop-down list, displays the **Hatch Pattern Type** dialog box, see Figure 28-26. Choose the **Pattern** button from this dialog box to select the predefined standard hatch patterns to be displayed as the intersection fill.

Figure 28-26 *The **Hatch Pattern Type** dialog box*

Background Lines. This option modifies the display of the hidden edges of the model. In case of a 2D section, you can choose to display the background lines.

Cut-away Geometry. This option modifies the display of the cut-away solid part in the generated section, if you select **Yes** in the **Show** drop-down list of the **Cut-away Geometry** category.

Curve Tangency Lines. This option modifies the display of curved lines that are tangent to the section plane. This option is only available when you generate a 2D section.

Create

Choosing this button from the **Generate Section/Elevation** dialog box finishes the section generation process. If you insert the generated section as a new block, you will be prompted to specify the insertion point, scale factor, and the rotation angle for the new generated section to be inserted as a block. If you replace an existing block, the generated section will be inserted with the same values of insertion point, scale factor, and rotation angle as for the block selected to be replaced. If you choose to export the generated section to a new file, a new file will be created and the section will be created at the same location with the same scale and orientation as the model in the previous file.

Live Section settings

This option is used to control the dynamic display of a sectioned solid in the drawing. Select the section line and choose **Live Section settings** from the shortcut menu; the **Section Settings** dialog box will be displayed with the **Live Section settings** option selected by default. The options available to control the display of the live sectioning of the model are similar to the options discussed in the **Section Settings** dialog box. The **Face Transparency** and **Edge Transparency** options in the **Cut-away Geometry** area are only available for the live section settings. These two options control the transparency of the edge and the face of the cut-away solid for a better view of the section and the remaining portion of the model. The value of transparency signifies the amount of light that passes through the cut-away part.

SOLID HISTORY*

In AutoCAD 2012, **Solid History** tool has been enhanced and is available in the **Primitive** panel of the **Solid** tab. By default, this button is not chosen. As a result, you can directly edit faces, edges and vertices of a solid. If you turn choose the **Solid History** tool, the history of individual solid object is recorded. So, if a composite solid model is created by performing a Boolean operation, the features used to create solid will be saved with the composite solid. Therefore, you can modify the shape of the composite solid by modifying the original features.

Consider a case in which a composite solid is created by subtracting a wedge from the cylinder, as shown in Figures 28-27 and 28-28. To modify this composite solid, select the **Solid History** option from **Solid > Selection > Subobject** drop-down; the solid history symbol will be attached to the cursor. Move the cursor near the composite solid and select the wedge when its preview is displayed (see Figure 28-29). You can also use the **Selection list** box to select the wedge. When the grips of the wedge are displayed, drag the grips and modify the shape of the wedge; the shape of the composite solid gets changed automatically, as shown in Figure 28-30.

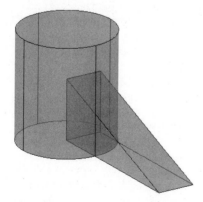

Figure 28-27 *A cylinder and a wedge as two separate solid bodies*

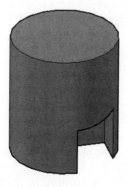

Figure 28-28 *Resulting composite solid model after subtracting the wedge from the cylinder*

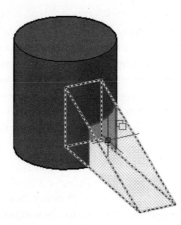

Figure 28-29 *History of the wedge being selected*   **Figure 28-30** *Composite solid model after editing*

GENERATING DRAWING VIEWS OF A SOLID MODEL

As mentioned earlier, one of the major advantages of working with a solid models is that the drawing views of a solid model are automatically generated once you have the solid model with you. All you have to do is to specify the type of drawing view required. This is done using the **Solid View**, **Solid Draw**, and **Solid Profile** tools. All these tools are discussed next.

Solid View

Ribbon: Home > Modeling > Solid View	**Command:** SOLVIEW
Menu Bar: Draw > Modeling > Setup > View	

The **Solid View** tool is an externally defined ARX application (*solids.arx*). This tool is used to create the documentation of the solid models in the form of the drawing views. This command guides you through the process of creating orthographic or auxiliary views. The **Solid View** tool creates floating viewports using orthographic projections to layout sectional view and multiview drawings of solids, while you are in the layout tab.

For the ease of documentation, this tool automatically creates layers for visible lines, hidden lines, and section hatching for each view. This tool also creates layers for dimensioning that are visible in individual viewports. You can use these layers to draw dimensions in individual viewports. The following is the naming convention for layers.

Layer name	Object type
View name-VIS	Visible lines
View name-HID	Hidden lines
View name-DIM	Dimensions
View name-HAT	Hatch patterns (for sections)

The view name is the user-defined layer name given to the view when created.

Note
*The information stored on these layers is deleted and updated when you use the **Solid Draw** tool. Therefore, do not place any permanent drawing information on these layers.*

To invoke the **Solid View** tool, choose the **Solid View** tool from the **Modeling** panel. The prompt sequence that will be displayed is given next.

> Command: **SOLVIEW**
> Enter an option [Ucs/Ortho/Auxiliary/Section]: *Select an option to generate the drawing view.*

These four options are discussed next.

Ucs Option

This option is used to create a profile view relative to a UCS. This is the first view that has to be generated. The reason for this is that the other options require existing viewports. The UCS can either be the current UCS or any other named UCS. The view generated using this option will be parallel to the *XY* plane of the current UCS. After selecting this option the next prompt sequence is:

> Enter an option [Named/World/?/Current] <Current>: *Enter an option or* Enter *to accept the current UCS.*

The **Named** option uses a named UCS to create the profile view. The prompt sequence is as follows:

> Enter name of UCS to restore: *Enter the name of the UCS you want to use.*
> Enter view scale <1.0>: *Enter a positive scale factor.*

This view scale value you enter is equivalent to zooming into your viewport by a factor relative to the paper space. The default ratio is 1:1. The next prompt sequence is as follows:

> Specify view center: *Specify the location of the view.*
> Specify view center <specify viewport>: Enter

This point is based on the Model space extents. The next prompt sequence is as follows.

> Specify the first corner of viewport: *Specify a point.*
> Specify opposite corner of viewport: *Specify a point.*
> Enter view name: *Enter a name for the view.*

The **World** option uses the *XY* plane of the WCS to create a profile view. The prompt sequences are the same as the **Named** option.

The **?** option lists all the named UCSs. The prompt sequence is given next.

> Enter UCS names to list<*>: *Enter a name or press ENTER to list all the named UCSs.*

The **Current** option uses the *XY* plane of the current UCS to generate the profile view. This is also the default option.

Ortho Option

An orthographic view is generated by projecting the lines on to a plane normal to an existing view. The **Ortho** option generates an orthographic view from the existing view. Keep in mind that the orthographic views created in AutoCAD are folded views and not the unfolded views. AutoCAD prompts you to specify the side of the viewport from where you want to project the view. The resultant view will depend on the side of the viewport selected. The prompt sequence is given next.

Specify side of viewport to project: *Select the edge of a viewport.*
Specify view center: *Specify the location of the orthographic view.*
Specify view center <specify viewport>: [Enter]
Specify first corner of viewport: *Select the first point.*
Specify opposite corner of viewport: *Select the second point.*
Enter view name: *Specify the view name.*

Auxiliary Option

This option is used to create an auxiliary view from an existing view. An auxiliary view is the one that is generated by projecting the lines from an existing view on to a plane that is inclined at a certain angle. AutoCAD prompts you to define the inclined plane used for the auxiliary projection. The prompt sequence is as follows.

Specify first point of inclined plane: *Specify the first point on the inclined plane.*
Specify second point of inclined plane: *Specify the second point on the inclined plane.*
Specify side to view from: *Specify a point.*
Specify view center: *Specify the location of the view.*
Specify view center <specify viewport>: [Enter]
Specify first corner of viewport: *Specify the first corner of the viewport.*
Specify opposite corner of viewport: *Specify the second corner of the viewport.*
Enter view name: *Enter the view name.*

Section Option

This option is used to create the section views of the solid model. The type of the section view depends on the alignment of the section plane and the side from which you view the model. Initially, the crosshatching is not displayed in the section view. To display the crosshatching, invoke the **SOLDRAW** command. AutoCAD prompts you to define the cutting plane, and the prompt sequence is as follows.

Specify first point of cutting plane: *Specify the first point on the cutting plane.*
Specify second point of cutting plane: *Specify the second point on the cutting plane.*

Once you have defined the cutting plane, you can specify the side to view from and the view scale, relative to paper space. The prompt sequence is given next.

Specify side to view from: *Specify the point in the direction from which the section is viewed.*
Enter view scale <current>: *Enter a positive number.*
Specify view center: *Specify the location for the view.*
Specify view center <specify viewport>: [Enter]
Specify first corner of viewport: *Specify the first corner of the viewport.*
Specify opposite corner of viewport: *Specify the second corner of the viewport.*
Enter view name: *Enter the view name.*

Solid Drawing

| **Ribbon:** Home > Modeling > Solid Drawing | **Command:** SOLDRAW |
| **Menu Bar:** Draw > Modeling > Setup > Drawing | |

The **Solid Drawing** tool generates profiles and sections in the viewports that were created with the **Solid View** tool. The basic function of this command is to clean up the viewports created by the **Solid View** tool. It creates the visible and hidden lines that represent the silhouettes and edges of solids in a viewport, and then projects them in a plane that is perpendicular to the viewing direction. When you use the **Solid Drawing** tool with the sectional view, a temporary copy of the solid is created. This temporary copy is then chopped at the section plane defined by you. Then the **Solid Drawing** tool creates the visible half as the section and discards the copy of the solid model. The crosshatching is automatically created using the current hatch pattern (HPNAME), hatch angle (HPANG), and hatch scale (HPSCALE). In this case, the layer **View name- HID** is frozen. If there are any existing profiles and sections in the selected viewports, they are deleted and replaced with new ones. The prompt sequence is as follows.

Select viewports to draw...
Select objects: *Select viewports to be drawn.*

Solid Profile

Ribbon: Home > Modeling > Solid Profile   **Command:** SOLPROF
Menu Bar: Draw > Modeling > Setup > Profile

The **Solid Profile** tool creates profile images of 3D solids by displaying only the edges and silhouettes of the curved surfaces of the solids in the current view. Note that before you invoke this tool, you must be in the temporary model space of the layout. This means that the value of the **TILEMODE** variable should be set to 0, but you should switch to the model space using the **MSPACE** command. The prompt sequence that follows when you invoke the **Solid Profile** tool is given next.

Select objects: *Select the object in the viewport.*
Select objects: Enter
Display hidden profile lines on separate layer? [Yes/No] <Y>: *Specify an option.*
Project profile lines onto a plane? [Yes/No] <Y>: *Specify an option.*
Delete tangential edges? [Yes/No] <Y>: *Specify an option.*
One solid selected.

DRAWING VIEWS*

In AutoCAD 2012, drafting has been made easier then before by introducing a new panel called **Drawing Views**. This panel can be accessed from the **Annotate** tab in the Ribbon. Tools available in this panel can be used only in the drafting space (Layouts). With the help of these tools, you can generate a base view or a projected view of the solid or surface present in the model space. You can also generate views of a model created in Autodesk Inventor. The tools available in the **Drawing Views** panel are discussed next.

Base View

Ribbon: Annotate > Drawing Views > Base View   **Command:** VIEWBASE

The **Base View** tool is used to create a base view of all entities present in the model space or the model present in the Autodesk Inventor file. On choosing the **Base View** tool, all solids or surfaces present in the model will be imported in the layout. Note that only the visible entities can be imported in this process. T. If the model space is empty and you invoke this tool, then the **Select File** window will be displayed and you will be prompted to select an Autodesk Inventor file.

Figure 28-31 shows a solid model. To create the base view of this model, you have to follow the procedure given next.

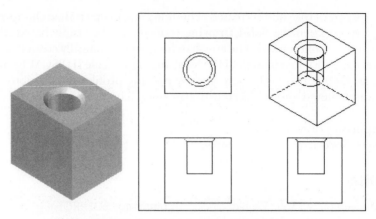

Figure 28-31 Model and its drawing views

Choose the Layout workspace and then delete the default viewport. Next, choose the **Base View** tool from the **Drawing Views** panel in the **Annotate** tab. On doing so, a new tab, **Drawing View Creation** tab will be displayed in the **Ribbon**, as shown in Figure 28-32. Select the desired orientation of the model for the drawing view from the **Orientation** list box. You can also choose the visual style of the base view from the **View Style** drop-down in the **Appearance** panel. You can specify the absolute scale of the object using the **Scale** drop-down list. You can control the visibility of the objects by selecting the check boxes provided in the **Object Visibility** drop-down list. The check boxes available in the **Object Visibility** drop-down list are discussed next.

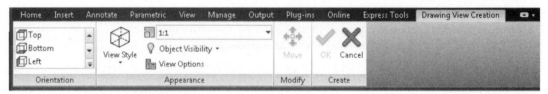

*Figure 28-32 The **Drawing View Creation** tab*

Interference edges. This check box is used to toggle the visibility of both hidden and visible edges which are not displayed due to an intersection condition.

Tangent edges. This check box is used to toggle the visibility of the edges formed by tangential intersection of surfaces.

Tangent edges foreshortened. This check box is used to shorten the length of tangential edges to differentiate them from the visible edges.

Bend extents. This check box is used to toggle the display of the sheet metal extent lines. If there is a bend in the sheet metal part, then the **Bend extents** tool will be used to display the extent line of the sheet metal.

Thread features. This check box is used to toggle the display of thread lines on screws or tapped holes.

Presentation trails. This check box is used to display the lines that indicate the directions along which the components will move when assembled.

You can also specify the justification of a view by using the **View Options** button in the **Appearance** panel. When you specify the scale of the view, the drawing view stretches or shrinks. If you select **Fixed** from the **View Justification** drop-down list, the alignment will remain unchanged. But if you select **Centered**, then the drawing view will be adjusted to the center position.

You can move the view to the required position by choosing the **Move** button in the **Modify** panel of the **Drawing View Creation** tab. This tab will become active only after placing the base view.

After specifying all the properties, click on the screen at the desired location; the base view will be created. Choose the **OK** button to place the base view; the **Drawing View Creation** tab will disappear and you will be prompted to specify the location of the projected view. The orientation of the projected view depend upon its location. Specify the location of projected view and exit by pressing ENTER.

Projected View

Ribbon: Annotate > Drawing Views > Projected View	Command: VIEWPROJ

 The **Projected View** tool is used to create a projected view from the base view present in the layout workspace. By default, there are eight projected, four orthogonal, and four isometric views. You can create a projected view by using the procedure discussed next.

Choose the **Projected View** tool from the **Drawing Views** panel; you will be prompted to select the parent view. Select the existing view in the layout; the projected view will be attached to the cursor and you will be prompted to specify the location of the projected view. Specify the location; you will be prompted to specify the location of the projected view again. You can specify the location or you can exit by pressing ENTER.

Edit View

Ribbon: Annotate > Drawing Views > Edit View	Command: VIEWEDIT

 The **Edit View** tool is used to change the settings of a view such as representation, visibility, view style scale, and so on. On invoking this tool, the **Drawing View Editor** tab will be added in the **Ribbon**. You can also move the view using this tool, but you cannot reorient this view by using the tool.

Update View

Ribbon: Annotate > Drawing Views > Update View	Command: VIEWUPDATE

 The **Update View** tool is used to update the layout. If you change the viewport, the drawing view will not be updated automatically. Therefore, you need to invoke the **Update View** tool to update the views. If you make changes to the model from which the views are derived, then the views in the layout will be displayed with a red mark on each corner. This red mark indicates that the model has been modified or removed. Also, a popup window is displayed at the bottom notifying that the model has been changed. You can update the views by choosing the **Update View** tool from the **Drawing Views** panel in the **Annotate** tab. However, note that the **Update View** tool updates only the selected view. If you want to update all the views in the layout then you have to select the **Update all Views** tool from the **Update view** drop-down in the **Drawing Views** panel.

Drafting Standard

Ribbon: Annotate > Drawing Views > Drafting Standard	Command: VIEWSTD

The **Drafting Standard** tool is used to set the default settings for drafting. These settings will be applied to the newly generated views. To invoke this tool, click on the inclined arrow located on the right of the **Drawing View** panel; the **Drafting Standard** dialog box will be displayed, as shown in Figure 28-33. The three areas in this dialog box, **Projection type**, **Thread style**, and **Shading/Preview**, are discussed next.

Projection type. In this area, there are two buttons: **First angle** and **Third angle**. These buttons represent the first angle projection and the third angle projection, respectively. By default, **Third angle** is chosen.

Thread style. In this area, there are two radio buttons: **Partial circular thread edge** and **Full circular thread edge**. You can select the desired radio button to represent the threads.

Shading/Preview. In this area, there are two drop-down lists: **Shaded view quality** and **Preview type**. You can set the dot per inch quality of the shaded view in the drawing view by selecting the desired option from the **Shaded view quality** drop-down list. From the **Preview type** drop-down list, you can either select the **Shaded** preview or the **Bounding box** preview. If you select the **Shaded**, the model is displayed shaded. If you select the **Bounding box,** only the bounding box will be displayed in the preview, not the model.

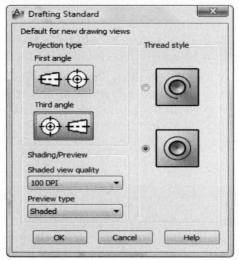

Figure 28-33 *The Drafting Standard window*

CREATING FLATSHOT

Ribbon: Home > Section > Flatshot OR Solid > Section > Flatshot
Command: FLATSHOT

The **Flatshot** tool creates a 2D sketch of a model by projecting the model in the current view on the XY plane of the current UCS. The created sketch is inserted in the drawing as a block. A flatshot can be created in any preset orthographic view or created view in the parallel projection. If you create a flatshot in the perspective view, a 3D wireframe of the entire model will be created by projecting the model in the current view on the XY plane of the current UCS. Choose the **Flatshot** tool from the **Section** panel; the **Flatshot** dialog box will be displayed, see Figure 28-34. The options in this dialog box are discussed next.

Destination Area
The options in this area are similar to those discussed in the **Generating 2D and 3D Sections** heading.

Foreground lines Area
The options in this area control the color and line type of the edges in the front at the time of the flatshot creation.

Obscured lines Area
The options in this area control the display, color, and line type of the hidden edges at the time of the flatshot creation.

Figure 28-34 *The Flatshot dialog box*

Create
Choosing this button finishes the flatshot creation process.

Include tangential edges
Select this check box to include the tangential edges of the curved surface into the flatshot.

Note
The 3D objects that are sectioned by the section plane will be flatshot as if they are not sectioned.

Tip
All 3D objects in the current drawing are included in the flatshot creation. If you do not want to include some objects in the flatshot creation, transfer them to a separate layer and freeze the layer before creating the flatshot.

EXAMPLE 2   *Solid View & Solid Drawing*

In this example, you will use the **Solid View** and **Solid Drawing** tools to generate the top view and the sectioned front view of the solid model shown in Figure 28-35. Assume the missing dimensions of the model.

1. Create the solid model by assuming the dimensions.

2. Switch to **Layout1**, the default viewport is displayed.

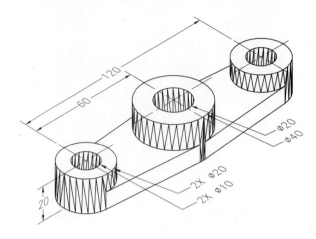

Figure 28-35 Solid model to generate drawing views

3. Delete the default vport. To do so, invoke the **ERASE** command, select the Viewport, and press ENTER.

4. Choose the **Solid View** tool from the **Modeling** panel and follow the prompt sequence given below:

   Enter an option [Ucs/Ortho/Auxiliary/Section]: **U**
   Enter an option [Named/World/?/Current] <Current>: **W**
   Enter view scale <1.0000>: [Enter]
   Specify view center: *Specify the view location close to the top left corner of the drawing window.*
   Specify view center <specify viewport>: [Enter]
   Specify first corner of viewport: *Specify the point of the viewport, P1, see Figure 28-36.*
   Specify opposite corner of viewport: *Specify the second point, P2, see Figure 28-36.*
   Enter view name: **Top**
   Enter an option [Ucs/Ortho/Auxiliary/Section]: [Enter]

5. Double-click in the viewport to switch to the temporary model space. Zoom the drawing to the extents using the **Extents** option of the **ZOOM** command. Now, switch back to the paper space by choosing the **MODEL** button displayed in the Status Bar. The layout after generating the top view should look similar to the one shown in Figure 28-37.

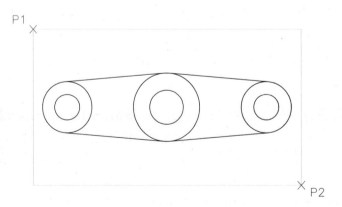

Figure 28-36 Generating the top view

6. Again, choose the **Solid View** tool from the **Modeling** panel and follow the prompt sequence given below:

Enter an option [Ucs/Ortho/Auxiliary/Section]: **S**
Specify first point of cutting plane: *Select the left quadrant of the left cylindrical portion in the top view.*
Specify second point of cutting plane: *Select the right quadrant of the right cylindrical portion in the top view.*
Specify side to view from: *Select a point below the top view in the viewport.*
Enter view scale <current>: Enter
Specify view center: *Specify the location of the view below the top view.*
Specify view center <specify viewport>: Enter
Specify first corner of viewport: *Select first corner of the viewport.*
Specify opposite corner of viewport: *Select the second corner of the viewport.*
Enter view name: **FRONT**
Enter an option [Ucs/Ortho/Auxiliary/Section]: Enter

7. Choose the **Solid Drawing** tool from the **Modeling** panel and follow the prompt sequence given next.

Select viewports to draw..
Select objects: *Select the first viewport.*
Select objects: *Select the second viewport.*
Select objects: Enter

8. The drawing views should look similar to those shown in Figure 28-37.

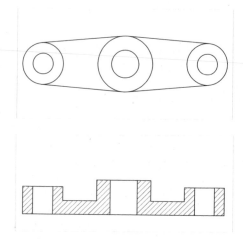

Figure 28-37 *Layout after generating the drawing views*

Tip
*You can change the scale and the type of hatch pattern using the **Properties** palette. You can also set different linetypes and colors to different lines in the views by selecting their respective layers using the **LAYER** command.*

Chapter 28

CALCULATING THE MASS PROPERTIES OF SOLID MODELS

Menu Bar: Tools > Inquiry > Region/Mass Properties   **Command:** MASSPROP
Toolbar: Inquiry > Region/Mass Properties

The mass properties of the solids or the regions can be automatically calculated using the **MASSPROP** command. Various mass properties are displayed in the AutoCAD Text Window. You can also write these properties to a file. This file will be in the *.mpr* format. The properties thus calculated can be used in analyzing the solid models. Note that the mass properties are calculated by taking the density of the material of the solid model as one unit.

It is very important for you to first understand the various properties that will be calculated using this command. The properties are:

Mass. This property provides the measure of mass of a solid. This property will not be available for the regions.

Area. This property provides the measure of area of the region. This property will be available only for the regions.

Volume. This property tells you the measure of the space occupied by the solid. Since AutoCAD assigns a density of 1 to solids, the mass and volume of a solid are equal. This property will be available only for the solids.

Perimeter. This property, providing the measure of the perimeter of the region, will be available only for the regions.

Bounding Box. If the solid were to be enclosed in a 3D box, the coordinates of the diagonally opposite corners would be provided by this property. Similarly, for regions, the *X* and the *Y* coordinates of the bounding box are displayed.

Centroid. This property provides the coordinates of the center of mass for the selected solid. The density of a solid is assumed to be unvarying.

Moments of Inertia. This property provides the mass moments of inertia of a solid about the three axes. The equation used to calculate this value is given next.

$$\text{mass\_moments\_of\_inertia} = \text{object\_mass} * (\text{radius of axis})^2$$

The radius of axis is nothing but the radius of gyration. The values obtained are used to calculate the force required to rotate an object about the three axes.

Products of Inertia. The value obtained with this property helps you to determine the force resulting in the motion of the object. The equation used to calculate this value is:

$$\text{product \_of\_inertia YX,XZ} = \text{mass} * \text{dist centroid\_to\_YZ} * \text{dist centroid\_to\_XZ}$$

Radii of Gyration. The equation used to calculate this value is given next.

$$\text{gyration\_radii} = (\text{moments\_of\_inertia/body\_mass})^{1/2}$$

Principal Moments and X-Y-Z Directions about Centroid. This property provides you with the highest, lowest, and middle value for the moment of inertia about an axis passing through the centroid of the object.

The prompt sequence that follows when you invoke the **MASSPROP** command is given next.

Select objects: *Select the solid model.*
Select objects: [Enter]

---------------- SOLIDS ----------------

Mass:	67278.5917
Volume:	67278.5917
Bounding box:	X: 1.4012 -- 151.4012
	Y: 118.6406 -- 158.6406
	Z: 0.0000 -- 25.0000
Centroid:	X: 76.4012
	Y: 138.6406
	Z: 9.2026
Moments of inertia:	X: 1308864825.5181
	Y: 507244127.3062
	Z: 1799135898.4724
Products of inertia:	XY: 712635884.7490
	YZ: 85837337.1138
	ZX: 47302745.3493
Radii of gyration:	X: 139.4790
	Y: 86.8301
	Z: 163.5285

Principal moments and X-Y-Z directions about centroid:
Press ENTER to continue: [Enter]

I: 9991363.2928 along [1.0000 0.0000 0.0000]
J: 108831209.8607 along [0.0000 1.0000 0.0000]
K: 113244795.9294 along [0.0000 0.0000 1.0000]

Write analysis to a file? [Yes/No] <N>: *Specify an option.*

If you enter **Y** at the last prompt, the **Create Mass and Area Properties File** dialog box will be displayed. This dialog box is used to specify the name of the *.mpr* file. The mass properties will then be written in this file.

RECORDING THE DRAWING STEPS BY USING THE ACTION RECORDER

Ribbon: Manage > Action Recorder > Record   **Command:** ACTRECORD
Menu Bar: Tools > Action Recorder > Record

The **Record** tool helps you to record the drawing steps quickly and easily. These steps are called actions. These actions are saved to a named macro for repeated use. To record an action, choose the **Record** tool from the **Action Recorder** panel of the **Manage** tab. On invoking this tool, a red circle appears near the cursor, indicating that the actions are recorded. Now, if you perform any action, it will be recorded. Also, the **Action Tree** will be expanded and the actions you perform will update simultaneously, as shown in Figure 28-38. If you need any message to be displayed during playback, choose the **Insert Message** tool from the **Action Recorder** panel; the **Insert User Message** dialog box will be displayed. Enter the message and choose the **OK** button; the message will be added in the **Action Tree**. This message will be displayed during playback. Similarly, if you want the user to input any values while playback, you can choose the **Pause for User Input** tool from the **Action Recorder** panel and continue recording the action.

After recording the actions, choose the **Stop** button in the **Action Recorder** panel to stop the recording; the **Action Macro** dialog box will be displayed, as shown in Figure 28-39. Specify a name for the action macro in the **Action Macro Command Name** edit box. If you want to give any description of the action, enter it in the **Description** edit box. The check boxes in the **Restore pre-playback view** area are used to invoke the playback message box when you play the action macro. If you select the **Check for inconsistencies when playback begins** check box and if there is inconsistency in the workspace, template, or any other, while you playback the action macro, then a warning box will be displayed. Remember that if you invoke any dialog boxes such as **Hatch and Gradient** or **OSnap** and change any settings during recording, then those settings will not be recorded. As a result, during the playback, the corresponding dialog box will be displayed again and you need to set the parameters again in it.

Figure 28-38 *The* **Action Recorder** *panel recording the actions performed*

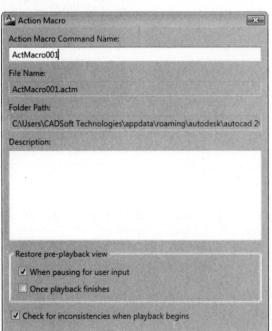

Figure 28-39 *The* **Action Macro** *dialog box*

The **Folder Path** display box in the **Action Macro** dialog box displays the default path where the action macro is saved. The **ACTRECPATH** command is used to change the path where the action macro is stored, by default. On entering this command, the following prompt sequence will be displayed:

Command: **ACTRECPATH**
Enter new value for ACTRECPATH, or . for none <"C:\Documents and Settings\User\Application Data\Autodesk\A...">: *Enter the new location within the quotes (" ")*

To playback the recorded action macro, select the action macro from the **Available Action Macro** drop-down list in the **Action Recorder** panel and choose the **Play** button; the action macro will start playing. You can use an existing action macro in a new drawing. To do so, open a new drawing file in the same template as used for the action macro. Next, choose the **Play** button from the **Action Recorder** panel.

USING SHOWMOTION FOR PRESENTATION

Ribbon: View > Windows > User Interface drop-down > ShowMotion
Navigation Bar: ShowMotion   **Command:** NAVSMOTION

The **ShowMotion** tool is used to create the animated views of the current drawing. This can be used to quickly step through different named views in the drawing, enabling you to play the animated views for presentations or to review the drawings. This tool can be invoked by selecting the **ShowMotion** check box from **User Interface** drop-down of the **Windows** panel.

On invoking this tool, a control panel for **ShowMotion** with five buttons will be displayed above the **Status Bar**, refer to Figure 28-40. The **Play all** button is used to play all the saved animations, which include shots and view categories, in a sequence. While playing the animations, the **Play all** button automatically gets converted into the **Pause** button, enabling you to pause the playback and vice-versa. The **Stop** button stops

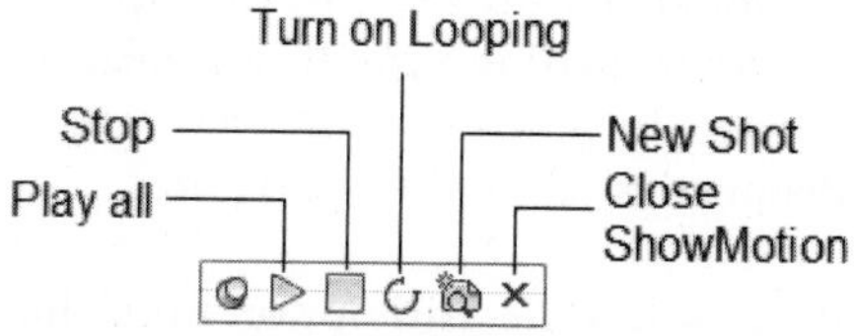

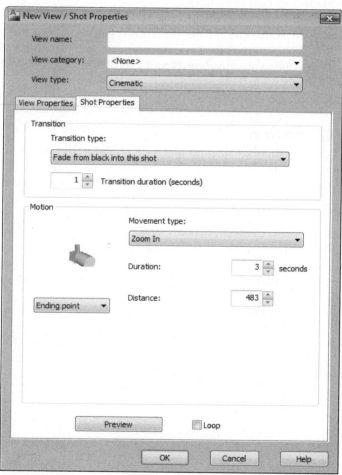

Figure 28-40 The ShowMotion control panel

playing the animation. The **Turn on Looping** button of the **ShowMotion** allows the animation to continue playing until you choose the **Stop** button or press the ESC key. A new animated view or a shot is created by choosing the **New Shot** button. On choosing this button, the **New View / Shot Properties** dialog box will be displayed, refer to Figure 28-41. The options in this dialog box are discussed next.

Figure 28-41 The New View/Shot Properties dialog box

View name. You can specify the name for the shot in the **View name** edit box. These names will be displayed below the respective shot thumbnails.

View category. You can enter the name of the category in this edit box. You can also group a sequence of shots of the same view categories. The different view categories are displayed as a series of thumbnails at the bottom of the drawing area.

View type. You can take three different types of shots using this drop-down list. These are **Still, Cinematic** and **Recorded Walk**. You can give transition effects between two shots, and the length or duration of the effects can be specified in the **Transition** area. Select an option in the **Transition type** drop-down list to specify the transition effects between two shots. You can see the preview of any view type in the preview box available in the **Motion** area of the **Shot Properties** tab of the dialog box.

If you select the **Still** option from the **View type** drop-down list, the position of camera will be fixed and you can set the duration of the shot in the **Motion** area of the dialog box.

If you select the **Cinematic** option from the **View type** drop-down list, you can move the camera position and set the duration of the recording. Select the type of movement in the **Movement type** drop-down list and specify the duration and distance in the respective spinners. On selecting the **Crane up** or **Crane down** option in the **Movement type** drop-down list, the **Shift left** check box becomes available. Consequently, you can specify the amount by which the camera shifts to left or right during the movement. On selecting the **Always look at camera pivot point** check box, the view is always focused at the camera.

If you select the **Recorded Walk** option from the **View type** drop-down list, you can create an animated view. To do so, choose the **Start Recording** button available in the **Motion** area of the **Shot Properties** tab. The dialog box disappears and a navigation tool is attached to the cursor. Next, press and hold the left mouse button and move the cursor. Once you release the left mouse button, the dialog box reappears. Choose the **OK** button to exit the dialog box.

Playing the Animation

Each **View Category** and its shots are displayed as thumbnails at the bottom of the drawing area. Each **View Category** thumbnail contains its corresponding name at its bottom. The series of shots under each category are displayed as smaller thumbnails above the **View category** thumbnails, as shown in Figure 28-42. However, the thumbnails are enlarged when the cursor moves over them. Each thumbnail contains two buttons, **Play** and **Go**. Choosing the **Play** button on the shot thumbnail plays that particular animated shot whereas choosing the **Play** button on the **View Category** thumbnail plays the sequence of shots under that particular category. On choosing the **Go** button available on the right side of the thumbnails, you can move the camera to the camera position of the first shot. To play all shots, choose the **Play all** button from the control panel.

You can rename, delete, recapture, or update a shot using the shortcut menu displayed on right-clicking on a thumbnail. You can also re-order the sequence of shots by moving them to the left or right using the options in the shortcut menu, refer to Figure 28-42. To create a new shot, choose the **New View / Shot** option from the shortcut menu. To modify the shots, choose the **Properties** option from the shortcut menu; the **View / Shot Properties** dialog box is displayed. Now, you can change the required properties in this dialog box.

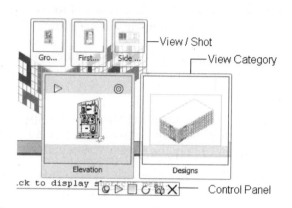

Figure 28-42 *The* ***ShowMotion*** *control panel and the* ***View Category*** *thumbnails*

Note

On invoking the ***NEWSHOT*** *command in the* ***Layout*** *tab, the* ***Recorded Walk*** *option will not be available in the* ***View type*** *drop-down list. Only the* ***Cut to Shot*** *option will be available in the* ***Transition type*** *drop-down list of the* ***New View /Shot Properties*** *dialog box.*

EDITING IN FUSION*

Ribbon: Plug-ins > Inventor Fusion > Edit in Fusion	
Menu Bar: Modify > Solid Editing/Surface Editing > Edit in Fusion	**Command:** EDITINFUSION

The **Edit in Fusion** tool is used to edit an object(solid / surface) created in AutoCAD. This tool is available in the **Inventor Fusion** panel of the **Plug-ins** tab. Note that this tool is available in the **Inventor Fusion** panel only if you have installed Fusion during the installation of AutoCAD 2012. The **Edit in Fusion** tool is also available in the shortcut menu that is displayed when you select any solid or surface from the drawing area and then right click on it. This tool will be available in the right click shortcut menu only if you have invoked this tool from the **Ribbon** atleast once. The **Edit in Fusion** tool will be added to the active tab of the **Ribbon** when you select any solid or surface. As soon as you invoke this tool, you will be prompted to select objects (if not selected). Select the desired object; the **Inventor Fusion** software will be launched and the **Autodesk Inventor Fusion** window will be displayed. In this window, you can edit the solid or surface easily by using both parametric as well as non-parametric modeling methodology. If you follow parametric methodology, you can edit the object by using the editing operations. In this methodology, you need to specify dimensions to edit the object. If you follow non-parametric methodology, you can edit the shape and size of the object dynamically. Some of the important tools in **Autodesk Inventor Fusion** that enable modifications with ease are discussed next.

Sculpt. This tool is available in the **Solid** panel. It is used to add, remove, or create volume using bounded volume formed by the intersection of solids, surfaces, or workplanes.

Edit Edge. This tool is available in the **Form Edit** panel. Using this tool, you can freely change the shape of an edge using pick points that will be available on selecting the edge. You can also add more pick points to the edge according to your requirement.

Assign Symmetry. This tool is available in the **Form Edit** panel. Using this tool, you can assign a condition to the edges so that they remain symmetric with respect to the selected plane, irrespective of the changes being applied to them.

Plane. There are 8 tools available in the **Plane** drop-down of the **Construction** panel to create planes. Using these tools, you can perform solid or surface editing operations easily.

Axis. Axis is used as a directional vector for tools such as extrude, revolve, sweep, or so on. There are 5 tools available in the **Axis** drop-down that can be used to create axes. This drop-down is available in the **Construction** panel.

Point. Points are used to specify location for features like holes, workplanes, as well as for creating a trajectory for sweep feature, loft feature, and so on. There are 4 tools available in the **Point** drop-down. This drop-down is available in the **Construction** panel.

Look at. The **Look at** tool is available in the **Navigate** panel of the **View** tab. Using this tool, you can position the selected planar face parallel to the screen making the selected edge horizontal to the screen.

The user interface of Inventor Fusion is shown in Figure 28-43.

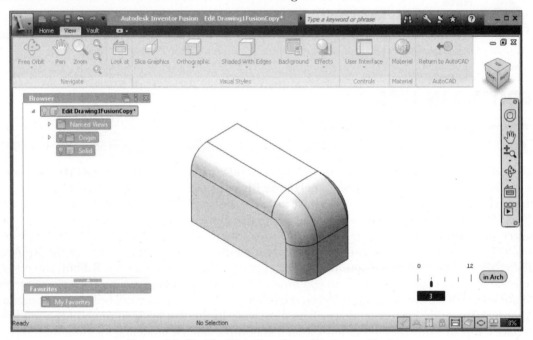

*Figure 28-43 The User Interface of **Inventor Fusion***

After performing desired modifications in Inventor Fusion, choose the **Return to AutoCAD** tool from the **AutoCAD** panel to return to AutoCAD. The Autodesk Inventor Fusion window closes and modified model is displayed in AutoCAD.

Note
You cannot work in Autodesk Inventor Fusion and AutoCAD simultaneously.

Self-Evaluation Test

Answer the following questions and then compare them to those given at the end of this chapter:

1. Solid models can be edited using the **EDITSOLID** command. (T/F)

2. You can create a shell in a solid model using the **Shell** tool. (T/F)

3. A section plane cuts all 3D objects, surfaces, and regions only if the section plane actually passes through them. (T/F)

4. The **Record** tool in the **Action Recorder** panel is used to create the animated views of the current drawing. (T/F)

5. The history of a composite solid created by performing the Boolean operation is saved along with the composite solid. (T/F)

6. The _____ tool is used to generate the drawing views of the solid model.

7. The mass properties of a selected solid model can be written into a file of _____ format.

8. The _____ tool creates a 2D sketch of a model by projecting the model in the current view on the XY plane of the current UCS.

9. Choose the _____ tool to separate 3D solids from the 3D solids with disjointed volumes.

10. The impression on a solid models can be removed using the _____ tool.

Review Questions

Answer the following questions:

1. You can edit a wireframe models by using the **SOLIDEDIT** command. (T/F)

2. You can move a selected face of a solid model from one location to the other. (T/F)

3. You can assign different colors to different faces of a solid model. (T/F)

4. After adding a jog, you can edit its length and location with the help of grips. (T/F)

5. The sketch created using the **Flatshot** tool is inserted into a drawing as a block. (T/F)

6. Which of the following tools is used to generate the profiles and sections in the viewports created using the **Solid View** tool?

    (a) **Solid Draw**   (b) **Solid Profile**
    (c) **Solid Edit**   (d) None of these

7. Which of the following methods is used to create a section plane?

    (a) by Specifying Two Points   (b) by Selecting a Preset Orthographic Plane
    (c) by Drawing the Section Line   (d) All of the above

8. Which option of the **SOLIDEDIT** command is used to edit the faces of a solid model?

    (a) **Body**   (b) **Face**
    (c) **Solid**   (d) None of these

9. Which of the following options of the **Solid View** tool is used to generate the drawing view relative the current UCS?

    (a) **New**   (b) **Section**
    (c) **UCS**   (d) **Ortho**

10. Which of the following commands is used to calculate mass properties of the solid model?

    (a) **PROPERTIES**   (b) **MASSPROP**
    (c) **CALCULATE**   (d) **MASSPROPERTIES**

EXERCISE 1

In this exercise, you will create the solid model shown in Figure 28-44. The dimensions for the solid model are shown in Figures 28-45 and 28-46. After creating the model, generate its top and sectioned front view. Also, set the front clipping, and then shade the model by using the Realistic shading and then dynamically rotate it using the SteeringWheel.

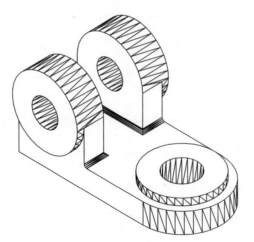

Figure 28-44 *Solid model for Exercise 1*

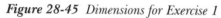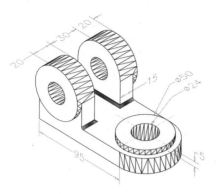

Figure 28-45 *Dimensions for Exercise 1*

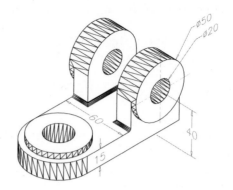

Figure 28-46 *Dimensions for Exercise 1*

EXERCISE 2

In this exercise, you will create the solid model shown in Figure 28-47. After creating the model, shade and rotate it in the 3D space using the **3DORBIT** command. Assume the missing dimensions.

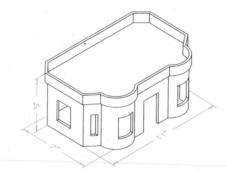

Figure 28-47 *Solid model for Exercise 2*

Problem-Solving Exercise 1

Draw the Piece part drawings shown in Figure 28-48. Then assemble them and create the Bill of Materials, as shown in Figure 28-49. The solid model of the assembled Tool Organizer is shown in Figure 28-50.

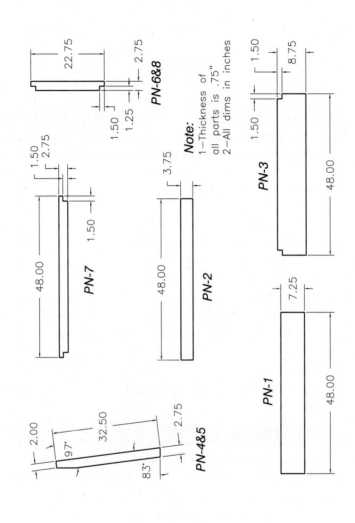

Note:
1–Thickness of all parts is .75"
2–All dims in inches

Figure 28-48 *Piece part drawing of the Tool Organizer*

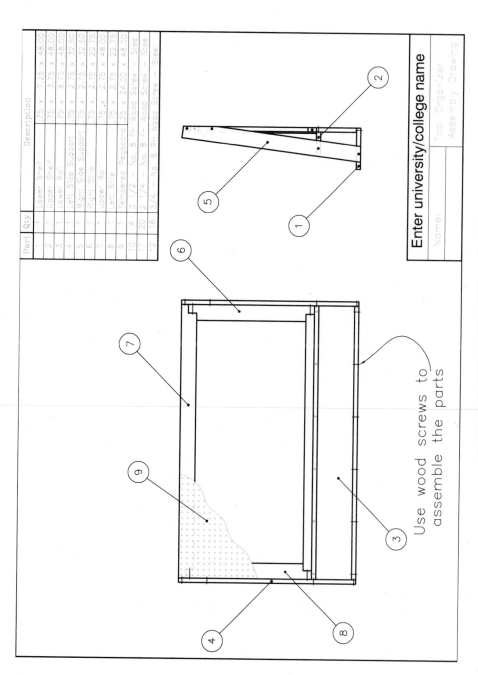

Figure 28-49b *Assembly drawing and BOM of the Tool Organizer*

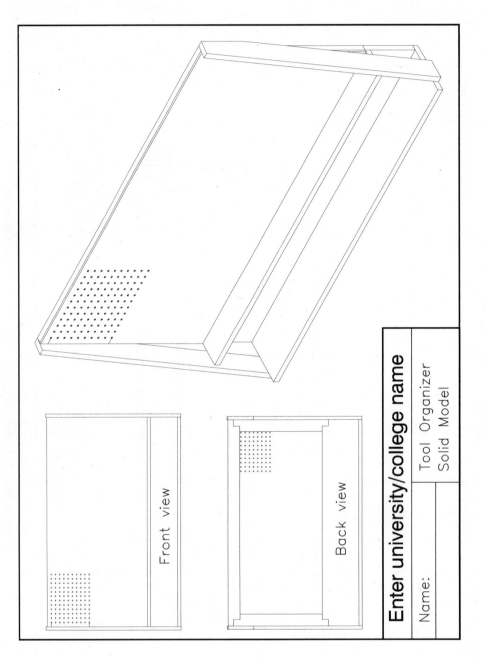

Front view

Back view

Enter university/college name

Tool Organizer
Solid Model

Name:

***Figure 28-50** Solid model of the Tool Organizer*

Problem-Solving Exercise 2

Draw the Piece part drawings shown in Figure 28-50. Then assemble them and create the Bill of Materials, as shown in Figure 28-51. The solid model of the assembled Work Bench is shown in Figure 28-52.

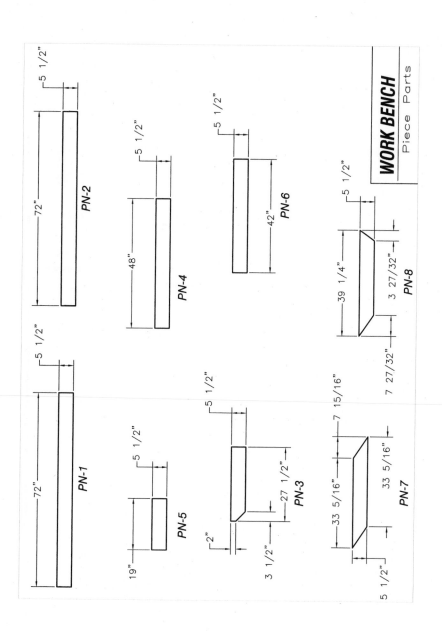

Figure 28-51 Piece part drawings of the Work Bench

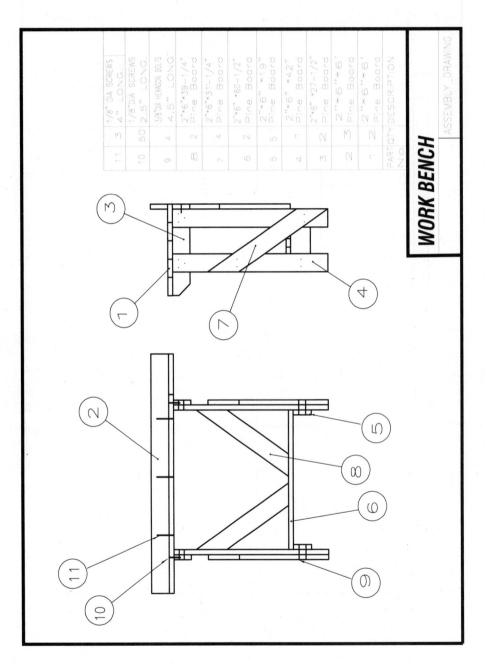

PART NO.	QTY	DESCRIPTION
11	3	1/8" DIA SCREWS 4" LONG.
10	50	1/8"DIA SCREWS 2.5" LONG.
9	4	3/8"DIA HEXAGON BOLTS 4.5" LONG.
8	2	2"×6"×39-1/4" Pine Board
7	4	2"×6"×31-1/4" Pine Board
6	2	2"×6"×60-1/2" Pine Board
5	5	2"×6"×19" Pine Board
4	1	2"×6"×42" Pine Board
3	2	2"×6"×27-1/2" Pine Board
2	3	2"×6"×6' Pine Board
1	2	2"×6"×6' Pine Board

WORK BENCH

ASSEMBLY DRAWING

Figure 28-52 Assembly drawing and BOM of the Work Bench

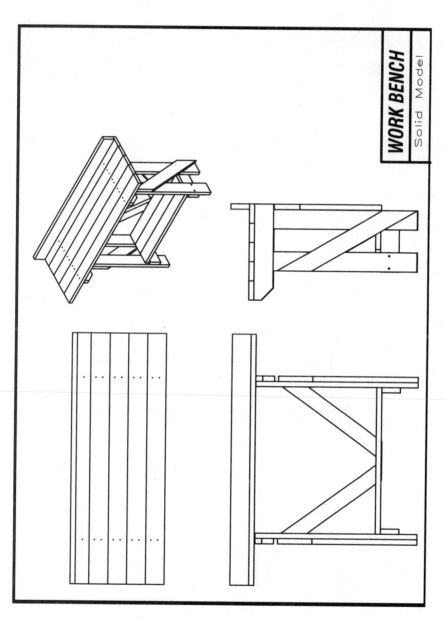

WORK BENCH

Solid Model

Figure 28-53 *Solid model of the Work Bench*

Problem-Solving Exercise 3

Create the model shown in Figure 28-54. Also, create the section view of the model as shown in Figure 28-55. The dimensions of the model are given in Figure 28-56. **

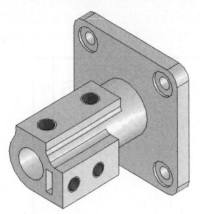

Figure 28-54 Solid model

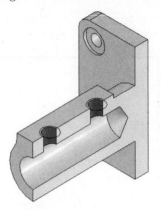

Figure 28-55 Section view of the model

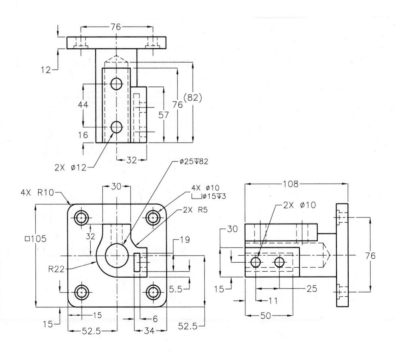

Figure 28-56 Dimensions of the model for Problem-Solving Exercise 3

Problem-Solving Exercise 4

Create various components of the Pipe Vice assembly and then assemble them by moving, rotating, and aligning. The Pipe Vice assembly is shown in Figure 28-57. For your reference, the exploded view of the assembly is shown in Figure 28-58. The dimensions of the individual components are shown in Figures 28-59 and 28-60.

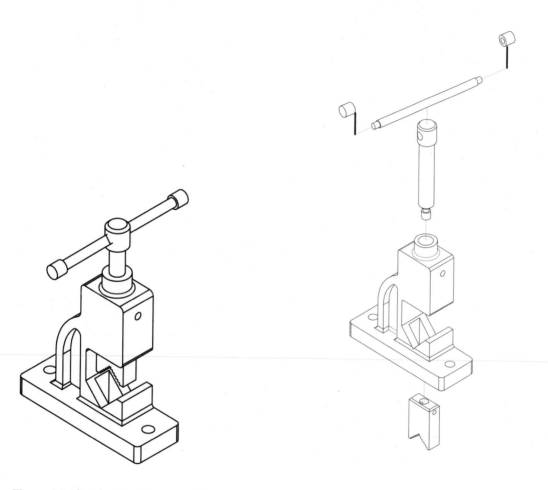

Figure 28-57 *The Pipe Vice assembly*   **Figure 28-58** *Exploded view of the Pipe Vice assembly*

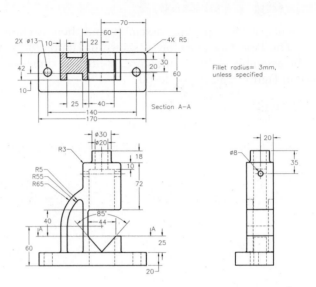

Figure 28-59 *Dimensions of the Base*

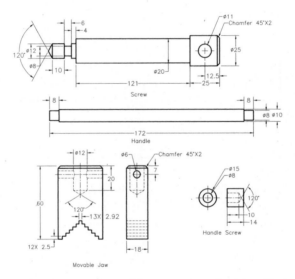

Figure 28-60 *Dimensions of the Screw, Handle, Movable Jaw, and Handle Stop*

Chapter 29

Surface Modeling

CHAPTER OBJECTIVES

In this chapter, you will learn to:
- *Create wireframe elements.*
- *Create extruded, revolved, and loft surfaces.*
- *Create planar and network surfaces.*
- *Create blend, patch, and offset surfaces.*
- *Fillet, trim, untrim, extend, and sculpt surfaces.*
- *Edit, add, remove, and rebuild the CVs of a NURBS surface.*
- *Project geometries onto a surface.*
- *Perform the zebra, curvature, and draft angle analyses on the surfaces.*

KEY TERMS

- *SPLINE*
- *Spline Freehand*
- *PLANESURF*
- *SURFNETWORK*
- *Surface Associativity*
- *NURBS Creation*
- *Continuity*
- *Bulge magnitude*

- *SURFPATCH*
- *Constraining Geometry*
- *Solid*
- *Connect*
- *Surffillet*
- *Extend*
- *Projection Direction*

- *Untrim*
- *Sculpt*
- *Convert to NURBS*
- *CV Edit Bar*
- *Insert Knots*
- *Rebuild*
- *AutoTRIM*
- *Project to UCS*

- *Project to View*
- *Project to 2 Points*
- *Zebra Analysis*
- *Curvature Analysis*
- *Draft Analysis*

SURFACE MODELING

Surface models are three-dimensional (3D) models of zero thickness. Surface modeling is a technique that is used for creating planar or non-planar geometries of zero thickness.

Most of the real world models are solid models. However, some models are complex; therefore, first you need to create surfaces to get the desired complex shapes. After creating the required shape of a model using surfaces, you can convert it into a solid model. The techniques of creating a surface using the surface modeling tools are explained in this chapter. If you are familiar with the concept of solid modeling, it will be easier for you to learn surface modeling.

Surface modeling is used to create procedural as well as NURBS surfaces. These surfaces are discussed later in this chapter. In AutoCAD, **Surface** tab contains all tools for creating and editing surfaces. In this chapter, you will learn to use all these tools.

CREATING WIREFRAME ELEMENTS

Wireframe elements help in creating surfaces. These elements can be used as profiles, guide curves, paths, and so on. The tools for constructing these wireframe elements are available in the **Curves** panel of the **Surface** tab. These tools include **Spline CV**, **Spline Fit**, **3D/2D Polyline**, **Line**, and so on. The **Spline** tools are grouped together in the **Spline** drop-down. These tools have the options compatible for creating NURBS surface. The tools in the **Spline** drop-down are discussed next.

Spline CV

Ribbon: Surface > Curves > Spline drop-down > Spline CV	**Command:** SPLINE

The **Spline CV** tool is used to draw a 3D spline by defining specific points that do not lie on that curve. These points act as control vertices for the curve, see Figure 29-1. You can modify the position of any control vertex. When you do so, only that part of the curve related to the corresponding CV will be affected. You can also add, remove, or refine control points. These processes are discussed later in this chapter. To create a spline CV, choose the **Spline CV** tool from the **Spline** drop-down, see Figure 29-2; the **Specify first point or [Method/Degree/Object]:** prompt will be displayed. Now, you can specify the location of CV in a sequence for creation of curve. You can also specify the method of spline creation, set the degree of curve, or select a polyline object to convert it into a CV spline. These methods have already been discussed in Chapter 21.

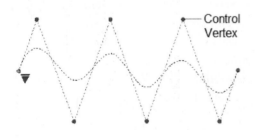

Figure 29-1 Spline CV curve

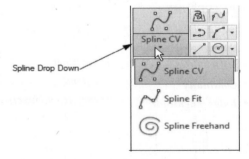

*Figure 29-2 The **Spline** drop-down*

Spline Knot

Ribbon: Surface > Curves > Spline drop-down > Spline Fit	**Command:** SPLINE

The **Spline Fit** tool is used to draw a 3D spline by defining specific points that lie on that curve. These points are called knots or fit points. You can edit a curve directly by using these points. The method of drawing a spline knot is the same as that discussed in Chapter 21.

Spline Freehand

Ribbon: Surface > Curves > Spline drop-down > Spline Freehand   **Command:** SKETCH

The **Spline Freehand** tool is used to create a spline by dragging the mouse in the drawing area. After the desired shape of curve has been displayed on the screen, release the mouse and press ENTER; a spline will be created and the CV or fit points will be created automatically on it. You can switch between CV's and fit points as per your requirement.

Note
*After drawing a spline, if you select it, a menu grip will be displayed, see Figure 29-3. You can click on the grip and switch between control vertices and fit points. When you switch from **Show Fit Points** to **Show Control Vertices**, the curve shown in Figure 29-3 will be displayed.*

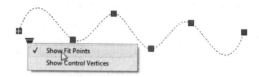

***Figure 29-3** Spline with fit points*

CREATING SURFACES BY USING PROFILES

The profile used to create surfaces can be open/close geometries, planar/non-planar geometries, or edges of an existing solid or surface. Open geometries always result in a surface, whereas closed geometries can result in a solid or a surface, depending on the tool selected in the **Solid** or **Surface** tab. You can also create a surface when the **Solid** tab is active by changing the mode of creation. The profile surfaces include extruded surface, revolved surface, lofted surface, and swept surface. The functions of the tools in the **Surface** tab are similar to that of the **Solid Modeling** tools discussed in Chapter 26.

Creating an Extruded Surface

Ribbon: Surface > Create > Extrude   **Command:** EXTRUDE

The **Extrude** tool is used to create solids as well as surfaces. Extruded surfaces are similar to extruded solids and are created by extending a profile in 3D space. As discussed earlier, if the profile drawn is an open curve, the resultant extruded model will be a surface only. But if the profile is a closed curve, the resultant model will be a surface, if this tool is selected from the **Surface** tab. To create an extruded surface, choose the **Extrude** tool from the **Create** panel. The command sequence that will follow when you choose this tool is given next.

Command: _EXTRUDE
Current wire frame density: ISOLINES=4, Closed profiles creation mode = Solid
Select objects to extrude or [MOde]: _mo Closed profiles creation mode
[SOlid/SUrface] <Solid>: _su
Select objects to extrude or [MOde]: mo
Closed profiles creation mode [SOlid/SUrface] <Solid>: **SU**
Select objects to extrude or [MOde]: *Select an object to extrude*
Select objects to extrude or [MOde]: *Select other object to extrude or* [Enter]
Specify height of extrusion or [Direction/Path/Taper angle/Expression] <Current>:

Specify the height of extrusion. You can also specify the direction, path, or taper angle of extrusion. These options have already been discussed in Chapter 26.

Note
*If you want to create a surface model by using the **Extrude** tool in the **Solid** tab, enter **MO** at the **Select objects to extrude or [MODE]** prompt; two options, **Solid** and **Surface**, will be displayed. Choose **Surface** to create the surface model. Next, select the required entities and press ENTER; you will be prompted to define the height of extrusion.*

Figure 29-4 shows the open profile to be extruded and Figure 29-5 shows the resulting extruded surface. Figure 29-6 shows the closed profile to be extruded and Figure 29-7 shows the resulting extruded surface.

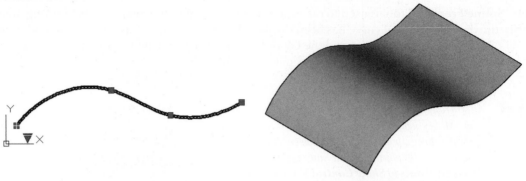

Figure 29-4 *Open profile to be extruded*   *Figure 29-5* *The resulting extruded surface*

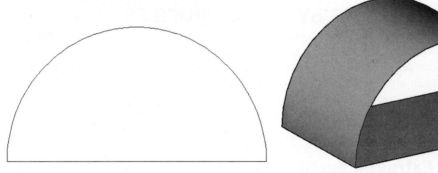

Figure 29-6 *Closed profile to be extruded*   *Figure 29-7* *The resulting extruded surface*

Select the extruded surface created; the grips will be displayed on it. You can drag the grip that is along the direction of extrusion to increase the extrusion depth. And, you can drag the grip that is normal to the direction of extrusion to provide a draft to the extruded surface. Figure 29-8 shows how to change distance by dragging one of the grips and Figure 29-9 shows how to change the angle by dragging another grip. Apart from these two grips, other grips will be displayed at the base and they are used to change the shape of the surface.

Figure 29-8 *Changing the distance*   *Figure 29-9* *Changing the angle*

Creating a Revolved Surface

| **Ribbon:** Surface > Create > Revolve | **Command:** REVOLVE |

You can create revolved surfaces by revolving a profile curve around an axis. To create a revolved surface, first draw the profile and then choose the **Revolve** tool from the **Create** panel. Next, select the objects to be revolved and then press ENTER; you will be prompted to specify the axis of revolution. Specify the axis of revolution; the **Specify angle of revolution or [STart angle/ Reverse/Expression] <360>:** prompt will be displayed. Now, using the options available in this prompt, you can specify the angle of revolution, specify the start angle of the revolution, and reverse the direction of revolution.

Figure 29-10 shows a profile and an axis of revolution to create a revolved surface. Figure 29-11 shows the surface created by revolving the profile through an angle of 180 degrees.

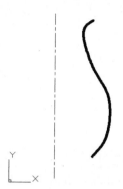

Figure 29-10 *The profile and the axis of revolution*

Figure 29-11 *Surface revolved through an angle of 180 degrees*

Creating a Loft Surface

Ribbon: Surface > Create > Loft   **Command:** LOFT

You can blend more than one cross-section together to form a loft surface. These cross-sections can be open profiles, closed profiles, or edge subobjects. Also, the cross-sections may or may not be parallel to each other. The procedure to create a loft surface is the same as that of a loft solid, which has already been discussed in Chapter 26.

Creating a Sweep Surface

Ribbon: Surface > Create > Sweep   **Command:** SWEEP

The procedure to create sweep surfaces is similar to that of sweep solids, which has been discussed in Chapter 26.

Creating a Planar Surface

Ribbon: Surface > Create > Planar   **Command:** PLANESURF
Menu Bar: Draw > Modeling > Surfaces > Planar   **Toolbar:** Modeling

The **Planar** tool is used to generate 2D surfaces by specifying two diagonally opposite points. These two points specify the rectangular area to be covered by this planar surface, see Figure 29-12. The prompt sequence for this command is discussed next.

Command: _**PLANESURF**
Specify first corner or [Object] <Object>: *Specify the first corner point of the planar surface.*
Specify other corner: *Specify the diagonally opposite point of the planar surface.*

You can also convert an existing object into a surface by entering **O** at the **Specify first corner or [Object] <Object>** prompt. While selecting the object, you can directly select a region or a number of individual objects that result in a closed boundary, see Figure 29-13. The number of lines displayed on the surface is controlled by the **SURFU** and **SURFV** system variables.

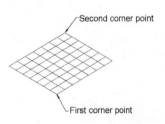

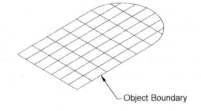

Figure 29-12 *Planar surface created by specifying corner points*

Figure 29-13 *Planar surface created by selecting an object*

Creating a Network Surface

Ribbon: Surface > Create > Network	**Command:** SURFNETWORK

The **Network** tool is used to create a surface between a number of boundary elements. These boundary elements can be 2D curves or existing edges of surfaces, solids, or regions. However, while creating a surface, there should not be any gap between the consecutive elements. Before creating a Network surface, you need to draw the boundary elements. After drawing them, choose the **Network** tool from the **Create** panel; you will be prompted to select objects in the first direction. Select objects in the U or V direction and then press ENTER; you will be prompted to select objects in the second direction. Select objects in the other direction. Figure 29-14 shows the elements to be selected. Here, Curve 1 and Curve 2 are selected in the first direction and Curve 3 and Curve 4 are selected in the second direction. Figure 29-15 shows the resulting network surface.

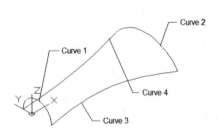

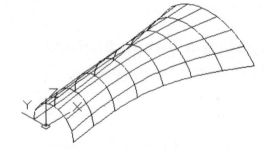

Figure 29-14 *Curves to be selected*

Figure 29-15 *Network surface created*

CREATING SURFACES FROM OTHER SURFACES

The tools discussed till now are used to create simple surfaces. But, most of the time, the product and industrial designers come up with some unique designs to make the products more attractive and presentable. In order to create such designs, you need some advanced tools. AutoCAD has some advanced tools like **Blend**, **Patch**, and **Offset**, to create new surfaces by using the subobjects of other surfaces. These tools are discussed next.

Before you learn the usage of these tools, you need to learn about the procedural and NURBS surfaces in AutoCAD. The procedural surfaces are the surfaces that maintain relationship with other surfaces, even after getting edited. These are associative and get modified as a group. Therefore, the procedural surfaces are mostly used, when associativity is required to be maintained. To create a procedural surface, choose the **Surface**

Associativity button in the **Create** panel and then invoke any surface creation tool; a procedural surface will be created. By default, this button is active and you need to choose it again to turn off the associativity.

A NURBS surface is created when the shape of the surface needs to be modified. These surfaces are mostly used in ship building industries. These surfaces are parametric in nature and have multiple knots. These knots can be adjusted to modify the shape of the surface. To create a NURBS surface, choose the **NURBS Creation** button in the **Create** panel and then create a surface; the surface thus created will be a NURBS surface. You need to choose this button again to turn off the NURBS creation. Note that you can keep both the buttons, **Surface Associativity** and **NURBS Creation**, on or off at the same time.

In the next section, you will learn about the tools that are used to create surfaces from other surfaces.

Creating a Blend Surface

Ribbon: Surface > Create > Blend   **Command:** SURFBLEND

The **Blend** tool is used to create a new surface between two existing surfaces, solids, or regions. To create a blend surface, invoke the **Blend** tool from the **Create** panel; you will be prompted to select the first surface edge to blend. Select the surface edge/s and press ENTER; you will be prompted to select the second surface edge/s to blend. Select it and press ENTER again; the preview of the blend surface will be displayed with the default settings. Figure 29-16 shows the entities selected to blend and Figure 29-17 shows the preview of the blend surface. The default settings include the surface continuity and magnitude of bulge at both ends. Next, press ENTER if the preview seems to be fine, else edit the continuity setting or the bulge magnitude. After adjusting the settings according to your requirement, press ENTER to create the blend surface. The method to edit the continuity setting and the bulge magnitude is discussed next.

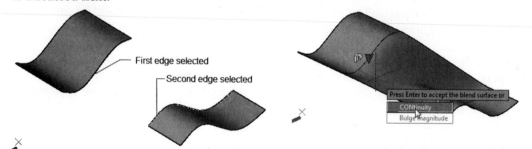

Figure 29-16   *The edges selected*   *Figure 29-17*   *Preview of the resulting blend surface*

Continuity

Continuity defines the smoothness of a blend surface at a point where it meets the other surface. Continuity can be set to **G0** (position), **G1** (Tangency), or **G2** (continuity). **G0** maintains the position of the selected edges, **G1** maintains the tangency of the blend surface where it meets the edges at both ends, and **G2** maintains the same curvature.

On selecting the **CONtinuity** option from the preview displayed, the continuity options for the first edge will be displayed. Choose the required option; you will be prompted to set the continuity of the second edge. Choose the required option; the preview of the blend surface will be updated. Also, you will be prompted to press ENTER to accept or edit the settings. Edit or accept the settings to create the blend surface.

You can also edit the continuity of a blend surface after it is created. To do so, select the blend surface, grips will be displayed on both the edges of the surface. Now, if you choose a grip displayed on the surface, a flyout with the continuity options will be displayed, see Figure 29-18. By default, **G1** is selected in this flyout. You can select any other option from the flyout to set continuity as per your requirement. Similarly, you can set the continuity of the other edge by clicking on it.

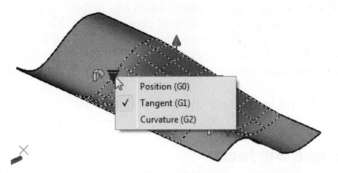

Figure 29-18 Continuity options displayed on clicking on a grip

Bulge Magnitude

Bulge magnitude defines the roundness or the amount of bulge between the blend surface and the existing surface. In other words, it is the amount of curvature where two surfaces meet. This magnitude can only be applied to the surfaces that have the G1 or G2 continuity.

By default, the bulge magnitude is set to 0.5. If you want to edit the bulge magnitude, choose the respective option from the preview displayed or enter **B** at the **Press Enter to accept the blend surface or [CONtinuity/Bulge magnitude]:** Command prompt; you will be prompted to define the bulge magnitude for the first edge. Define the value of bulge for the first edge and then the value of bulge at the second edge. The bulge magnitude can vary between 0 and 1. On entering 0 value, the surface will be flattened and on entering 1, the surface will be curved the most. The bulge magnitude has already been discussed in detail in Chapter 26.

You can modify a blend surface after it is created. To do so, double-click on the blend surface; the **Properties** palette with the settings of the blend surface will be displayed. Using this palette, you can modify the continuity setting and the bulge magnitude at each edge, the appearance of the surface, and so on.

Creating a Patch Surface

Ribbon: Surface > Create > Patch	**Command:** SURFPATCH

Using the **Patch** tool, you can close the open edges of an existing surface. To create a patch surface, invoke the **Patch** tool from the **Create** panel; you will be prompted to select the edges to create a patch. Select the edges one by one and then press ENTER; the preview of the patch with the default settings will be displayed. Also, you will be prompted to either accept the preview, or change the settings for continuity between the patch surface and the edges selected, define the bulge magnitude, or define the geometry to be constrained. Figure 29-19 shows the model with open edge and Figure 29-20 shows the patch created on the open edge.

By default, the continuity is set to G0 and the bulge magnitude is set to 0.5. Continuity and bulge magnitude have already been discussed in the previous topic, and the method to specify the geometry to be constrained is discussed next.

Figure 29-19 *Model before creating the patch*

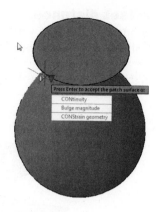

Figure 29-20 *Model after creating the patch*

Constraining Geometry

You can use some additional curves as guide curves to create the patch. Guide curves are used to guide the shape of the patch as per the geometry of the guide curve. Note that you need to draw the guide curve first and then invoke the **Patch** tool. Choose this option when the **Press Enter to accept the patch surface or [CONtinuity/Bulge magnitude/CONStrain geometry]:** prompt is displayed. Next, select the guide curve and press ENTER; a patch will be created. If you select the patch surface after it is created, a grip will be displayed. On clicking the grip, the options for continuity will be displayed. You can choose an option to specify the required continuity. Figure 29-21 shows the guide curve created for constraining geometry and Figure 29-22 shows the patch created using the guide curve.

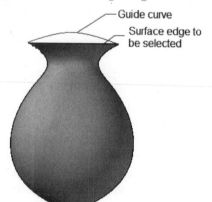

Figure 29-21 *Model before creating the patch*

Figure 29-22 *Model after creating the patch*

You can modify a patch surface by using the **Properties** palette. Invoke the **Properties** palette; the patch settings will be displayed in the **Geometry** area. You can modify the continuity option, the bulge magnitude, the number of isolines in the U and V directions, and so on using this palette.

Creating an Offset Surface

Ribbon: Surface > Create > Offset   **Command:** SURFOFFSET

The **Offset** tool is used to create a surface at an offset distance from an existing surface. To do so, choose the **Offset** tool from the **Create** panel; you will be prompted to select an existing surface. Select a surface and press ENTER; the direction arrows indicating the direction of the offset and the following prompt **Specify offset distance or [Flip direction/Both sides/Solid/Connect] <Current>:** will be displayed. The options displayed in this prompt are discussed next.

Specify offset distance

If the direction of offset is as you want, then specify the required offset distance and press ENTER; an offset surface will be created. Figure 29-23 shows the original surface and the offset surface.

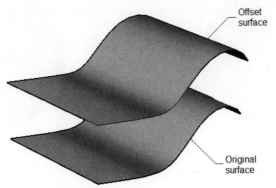

Figure 29-23 *The original and offset surfaces*

Flip direction

If you want to change the direction of offset surface creation, choose the **Flip direction** option in the prompt; the direction will be changed. Next, you can specify the offset distance.

Both sides

To create offset surfaces in both sides of the selected surface, choose the **Both sides** option. On doing so, you will notice that the direction arrows are displayed on both sides of the selected surface. Specify the offset distance.

Solid

The **Solid** option is used to create a solid by offsetting an existing surface. After selecting the surface, choose the **Solid** option; you will be prompted to specify the offset distance. Specify the value of the offset distance and press ENTER; the resulting entity will be a solid.

Connect

The **Connect** option is used when you select multiple connected surfaces to offset and you need the offset surfaces to be connected to each other. Select the existing connected surfaces, choose the **Connect** option, and specify the offset distance; the offset surfaces will be connected to each other.

If you select an offset surface, a grip will be displayed on it. You can click on the grip and drag it to adjust the offset distance as per your need. An edit box is also displayed while dragging the grip. You can enter the required offset distance in this edit box and press ENTER. On doing so, the offset surface will move to the specified distance.

You can control the display of the offset surface using the **Properties** palette. Invoke the **Properties** palette to set the isolines of the offset surface along U and V, the offset distance value, etc.

EDITING SURFACES

After creating the surfaces, you may need to modify them. AutoCAD provides you with the surface operation tools to modify surfaces. The surface operation tools are used regularly while creating surface models. These tools are used to enhance the surfaces and help in creating complex geometries. They save a lot of modeling time as well.

Various surface editing operations are discussed next.

Creating Fillets

Ribbon: Surface > Edit > Fillet   **Command:** SURFFILLET

The **Fillet** tool is used to create a rounded surface between two existing surfaces. To create a fillet surface, invoke the **Fillet** tool from the **Edit** panel; you will be prompted to select the surfaces between which you want the fillet to be created. Select the surfaces one by one and press ENTER; a fillet will be created with the default settings. Also, you will be prompted to press ENTER to accept the fillet created, or set the new fillet radius, or set the trim surface option. Press ENTER to accept the fillet, or choose the **Radius** option to enter the new radius for the fillet. You can also specify whether or not to trim the surfaces after creating the fillet by choosing the **Trim Surface** option.

If you select the fillet surface after creating it, a grip will be displayed. Click on the grip and drag it to modify the fillet radius dynamically. Figure 29-24 shows the model before creating the fillet and Figure 29-25 shows the model after creating the fillet.

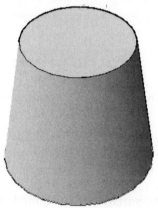

Figure 29-24 Model before creating the fillet

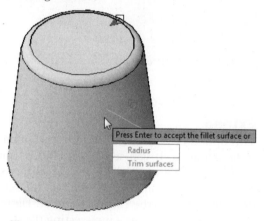

Figure 29-25 Model after creating the fillet

Trimming Surfaces

Ribbon: Surface > Edit > Trim   **Command:** SURFTRIM
Menu Bar: Modify > Surface Editing > Trim   **Toolbar:** Surface Editing

The **Trim** tool is used to trim a portion of a surface or a region by using an intersecting object. The lines, arcs, ellipses, polylines, splines, curves, 3D polylines, 3D splines, helix, regions, and surfaces can be used as intersecting objects. You can invoke the **Trim** tool from the **Edit** panel. But before invoking this tool, you need to have the object to be trimmed and the object to be used as the cutting element. On invoking this tool, you will be prompted to select the objects to be trimmed. Select the object/s to be trimmed and press ENTER; you will be prompted to select the cutting element to be used for trimming. Select the cutting element/s; you will be prompted to click on the portion to be removed. Click on the portion/s of the object to be trimmed; the selected object will be trimmed with the default settings. You can also specify the extension type and the projection direction. Figure 29-26 shows the surface before trimming and Figure 29-27 shows the surface after trimming.

Select a trimmed surface and then right-click on it; a shortcut menu will be displayed. Choose the **Properties** option from this shortcut menu; the Properties palette will be displayed. Expand the **Trims** rollout in this window; various parameters in this rollout will be displayed. These parameters are discussed next.

• **Trimmed surface**. It displays if the surface is trimmed or not.

- **Trimming edges**. It displays the number of trimming edges.

- **Edge**. It displays the edge number for which the associativity is being checked.

- **Associative Trim**. It indicates if the curren edge is Associative or not.

The options displayed in the Command prompt on invoking the **Trim** tool are discussed next.

Figure 29-26 The untrimmed surface   *Figure 29-27 The trimmed surface*

Extend

This option is used when the cutting element is a surface and it does not intersect the surface to be trimmed. If you set the **Extend** option to **Yes** at the Command prompt displayed on invoking the **Trim** tool, the selected surface gets trimmed at the imaginary intersection of the cutting element. But if it is set to **No**, the selected surface does not get trimmed as the surface does not intersect the cutting element. Figure 29-28 shows the surface to be trimmed and the cutting surface that is not intersecting the surface to be trimmed. Figure 29-29 shows the surface trimmed with the **Extend** option set to **Yes**.

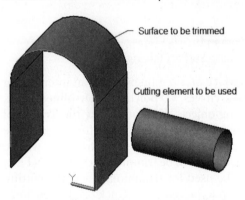
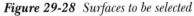
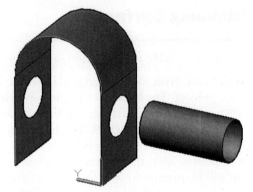

Figure 29-28 Surfaces to be selected   *Figure 29-29 Surface after trimming*

Projection direction

This option controls the direction of projection of the cutting element on the element to be trimmed. By default, the **Automatic** option is set. As a result, the direction of projection is set automatically depending on the view at which the trimming takes place. The other options in the projection direction are **View**, **UCS**, and **None**. If you choose the **View** option, the cutting geometry gets projected on the element to be trimmed along the current view direction. If you choose the **UCS** option, the geometry is projected in the +/- Z direction of the current UCS. But if you choose the **None** option, the geometry will not be projected and the trimming will take place only when the geometry will lie on the surface.

Note

If you select the objects after performing the trimming operation, grips will be displayed on both the trimmed object as well as the trimming object. Click on the grip and drag; the trimmed object will move and the trim location will be updated automatically. Similarly, you can drag the grip of the cutting element to resize it. Figure 29-30 shows the model after moving the cutting element to its bottom and Figure 29-31 shows the model after resizing the cutting element.

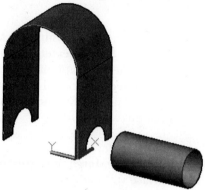

Figure 29-30 *Model after moving the cutting element*

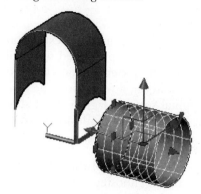

Figure 29-31 *Model after resizing the cutting element*

Untrimming Surfaces

Ribbon: Surface > Edit > Untrim	**Command:** SURFUNTRIM
Menu Bar: Modify > Surface Editing > Untrim	**Toolbar:** Surface Editing

The **Untrim** tool is used to untrim the trimmed edges or surfaces. To untrim a trimmed object invoke the **Untrim** tool from the **Edit** panel; you will be prompted to select the trimmed edges or the trimmed surface. If you select the trimmed edges, only the portion removed at the selected edges will be regained. However, if you choose the **SURface** option from the Command prompt, then the portions removed from the selected surface will be regained. Figure 29-32 shows the edge selected from the surface to be untrimmed and Figure 29-33 shows the resultant untrimmed surface.

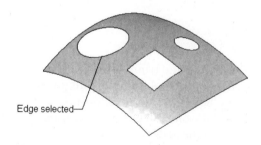

Edge selected

Figure 29-32 *Edge selected to be untrimmed*

Figure 29-33 *Resultant untrimmed surface*

Extending Surfaces

Ribbon: Surface > Edit > Extend	**Command:** SURFEXTEND
Menu Bar: Modify > Surface Editing > Extend	**Toolbar:** Surface Creation

You can extend or lengthen the edge/s of one or more surfaces to a certain distance. To extend a surface, invoke the **Extend** tool from the **Edit** panel; you will be prompted to select the edges of the surfaces to extend. Select the edges to extend and press ENTER; you will be prompted to enter a distance value or specify the mode of extension. The options displayed in the prompt on invoking the **Extend** tool are discussed next.

Specify extension distance

Enter a value to specify the extent of extension and press ENTER; the surface will extend.

Expression

You can enter a formula or an equation at the Command prompt to control the geometry of extension.

Modes

You can specify the mode of extension by using the **Modes** option. On choosing the **Modes** option, you will be prompted to specify either the **Extend** or the **Stretch** extension type. If you choose the **Extend** option, the extended surface will resemble the shape of the original surface. However, if you choose the **Stretch** option, then the extended surface will not resemble the original surface. Figure 29-34 shows the surface extended using the **Extend** option and Figure 29-35 shows the surface extended using the **Stretch** option.

Creation type

After you choose any of the modes and press ENTER, two options, **Merge** and **Append**, are displayed. On choosing the **Merge** option, the extended surface will merge with the existing surface, and on choosing the **Append** option, a new surface will be added to the existing surface, thus resulting in two surfaces.

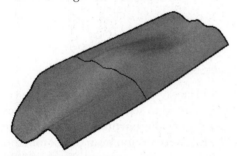

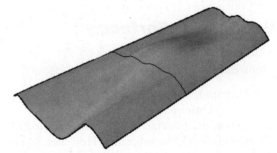

Figure 29-34 Surface extended using the **Extend** option

Figure 29-35 Surface Extended using the **Stretch** option

Sculpting Surfaces

Ribbon: Surface > Edit > Sculpt	**Command:** SURFSCULPT
Toolbar: Surface Editing	

The **Sculpt** tool is used to create a 3D solid by combining and trimming the surfaces that form a closed volume. To sculpt surfaces, invoke the **Sculpt** tool from the **Edit** panel; you will be prompted to select the continuous surfaces. Select the surfaces forming a sealed area and press ENTER; the surfaces will get trimmed and a 3D solid will be formed. Figure 29-36 shows the surfaces enclosing a volume and Figure 29-37 shows the 3D solid created after sculpting the surfaces.

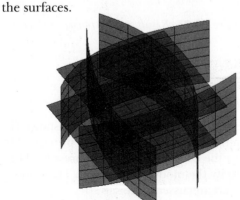

Figure 29-36 Surfaces enclosing a volume

Figure 29-37 3D solid created after sculpting

Extracting Intersections

Ribbon: Surface > Edit > Extract Intersections **Command:** INTERFERE
Toolbar: SurfaceEditing

Extract Intersections is a new tool added in AutoCAD 2012. This tool is used to create a solid, surface, or spline by extracting the intersections of two solids, a solid and a surface, or two surfaces, respectively. To extract intersection of two surfaces, invoke the **Extract Intersections** tool from the **Surface Edit** panel of the **Surface** tab in the Ribbon. You will be prompted to select two objects. Select two intersecting surfaces and press ENTER; you will be asked if you want to create an intersecting object or not. Select **Yes** or type **Y** and press ENTER; the spline will be created. Figure 29-38 shows two intersecting surfaces and Figure 29-39 shows the spline generated.

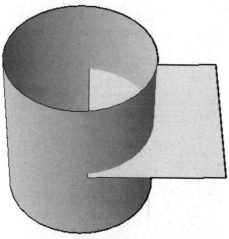

Figure 29-38 *Two intersecting surfaces*

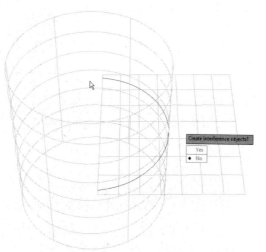

Figure 29-39 *Interference spline*

EXAMPLE 1   *Surface Creation*

In this example, you will create the model shown in Figure 29-40. Its views and dimensions are shown in Figure 29-41. The dimensions are given for your reference only. You can use your own dimensions.

Figure 29-40 *The isometric view of the model*

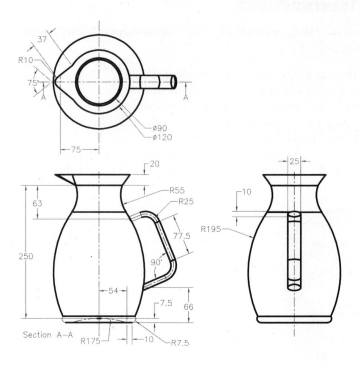

Figure 29-41 *Orthographic views and dimensions of the model*

1. Start a new drawing with **acad3D.dwt** template and draw the sketch of a circle with its center at the origin and diameter 90. This sketch is the first cross-section for the loft.

2. Draw another sketch for the second cross-section, as shown in Figure 29-42. Then, convert the sketch to polyline by using the **PEDIT** command.

3. Move the second sketch to a distance of 20 units along the Z axis.

4. Choose the **Loft** tool from the **Create** panel in the **Surface** tab; you are prompted to select the cross-sections that you want to loft. Select the two sketches one by one and press ENTER; the preview of the loft surface is displayed.

5. Press ENTER to create the loft surface, as shown in Figure 29-43.

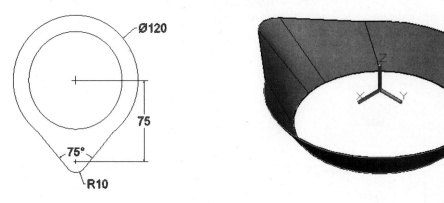

Figure 29-42 *The sketch for the second cross-section*   **Figure 29-43** *The resulting loft surface*

Next, you need to create the revolved surface.

6. Change the current viewport to **Right** from the **In-canvas View Controls**.

7. Draw the sketch for the revolved feature, as shown in Figure 29-44, using the lines and arcs in the right view. Refer to Figure 29-41 for the missing dimensions. However, the profile may not be of exact dimensions.

8. Convert the profile into a polyline by using the **JOIN** command.

9. Choose the **Revolve** tool from the **Create** panel in the **Surface** tab; you are prompted to select the profile to be revolved.

10. Select the sketch and press ENTER; you are prompted to select the axis of revolution. You can select the axis of co-ordinate system in this case or draw a vertical line at the center for axis of revolution. Select the axis of revolution and press ENTER; the revolved surface is created, as shown in Figure 29-45. Next, you need to create the handle of the jug by sweeping a profile along a guide curve.

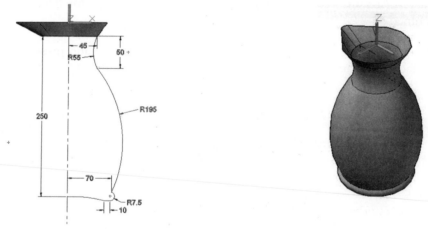

Figure 29-44 *The sketch for creating the revolved surface*

Figure 29-45 *The model after creating the revolved surface*

11. Draw the path for the sweep surface in the right view by invoking the **Polyline** tool, as shown in Figure 29-46. Note that the profile is rotated by 20 degrees.

12. Draw the profile of an ellipse, as shown in Figure 29-47, for creating the sweep surface. This profile can be drawn anywhere in the drawing area.

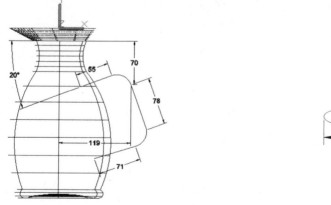

Figure 29-46 *The path for the sweep surface*

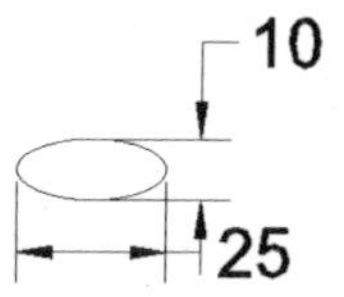

Figure 29-47 *The profile for the sweep surface*

13. Choose the **Sweep** tool from the **Create** panel in the **Surface** tab; you are prompted to select the profile to be swept.

14. Select the ellipse and press ENTER; you are prompted to select the sweep path.

15. Select the sweep path created in Step 11 and press ENTER; the sweep surface is created.

16. Create a new layer for profiles, move profiles to the new layer, and then freeze the layer so that the profiles are not displayed on the model.

    Next, the handle of the jug needs to be trimmed so that it does not penetrate the model.

17. Choose the **Trim** tool from the **Edit** panel in the **Surface** tab; you are prompted to select the surfaces to trim.

18. Select the sweep surface of the handle and press ENTER; you are prompted to select the cutting elements.

19. Select the revolved surface of the jug and press ENTER; you are prompted to select the area to be trimmed.

20. Rotate the model, select the inner portion of the sweep surface, and press ENTER; the sweep surface penetrating the revolved surface is trimmed and the final model of the jug is created.

EXERCISE 1   *Surface Creation*

In this exercise, you will create the model shown in Figure 29-48. Its drawing views and dimensions are shown in Figure 29-49.

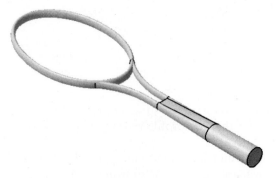

Figure 29-48   *The isometric view of the model*

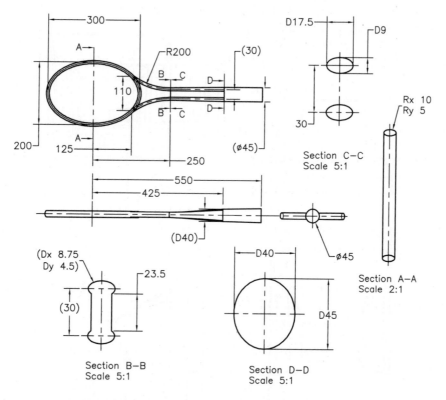

Figure 29-49 *The views and dimensions of the model*

EXAMPLE 2   *Surface Editing*

In this example, you will create the model of the back cover of a toy monitor, as shown in Figure 29-50. You will create this model using the editing tools. After creating the surface model, you need to convert it into a solid body. The orthographic views and dimensions of the model are shown in Figure 29-51.

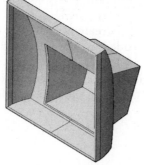

Figure 29-50 *Back cover of the toy monitor*

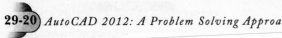

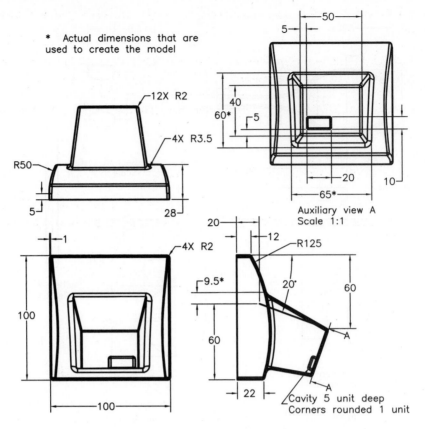

Figure 29-51 Orthographic views and dimensions of the model for Example 2

1. Start a new drawing file with **acad3D.dwt** template and change the current viewport to the **Right** by using **In-canvas Viewport Controls**.

2. Draw the sketch of the monitor, as shown in Figure 29-52, using the **Line** and **Arc** tools. Next, convert it into a polyline using the **JOIN** command.

3. Choose the **Extrude** tool from the **Create** panel of the **Surface** tab and extrude the sketch to a distance of 50 units; an extruded surface is created.

4. Change the view to **SE Isometric**. Choose the **Extend** tool from the **Edit** panel of the **Surface** tab and extend the surface to 50 units in the opposite direction; the base surface is created, as shown in Figure 29-53.

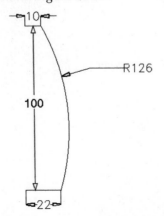

Figure 29-52 Sketch of the base surface

Figure 29-53 Resulting base surface

Next, you will create the loft surface. Therefore, you need to create the path and cross-sections for the loft.

5. Draw the sketch to create the path of the loft in the Right view, as shown in Figure 29-54. Move it to the middle of the base surface.

6. Tilt the UCS using the **ZA** option of UCS command such that the cross-sections are aligned to the path created.

Note
You can also tilt the UCS dynamically as discussed earlier in Chapter 24.

7. Draw two rectangular cross-sections, refer to Figure 29-55 for the dimensions. Next, snap the midpoint of the bigger rectangle to the left endpoint of the path, and midpoint of the smaller rectangle to the right endpoint of the path, refer to Figure 29-54.

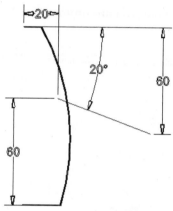

Figure 29-54 Path of the loft

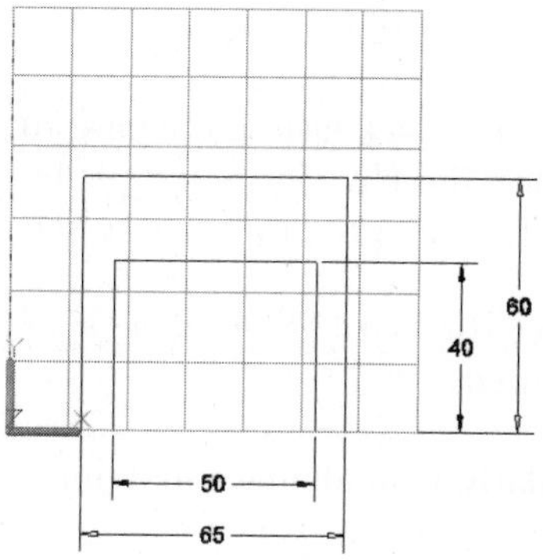

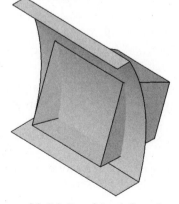

Figure 29-55 Cross-sections for the loft

8. Choose the **Loft** tool from the **Create** panel in the **Surface** tab; you are prompted to select the cross-sections.

9. Select the cross-sections one by one and press ENTER; the preview of the loft is displayed and you are prompted to press ENTER to accept or select a guide curve or a path.

10. Choose the **Path** option and select the line segment; the loft is created, as shown in Figure 29-56. Next, press ENTER.

11. Draw the sketch on both sides on the top face of the base surface, as shown in Figure 29-57.

12. Choose the **Extrude** tool and extrude the sketch to coincide it with the bottom part of the first surface, see Figure 29-58. Use the 3D object snap to snap the extrusion height.

Next, you need to trim the unwanted portions of the surfaces.

Figure 29-56 Resulting loft surface

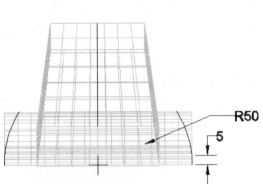

Figure 29-57 *Sketch to be drawn*

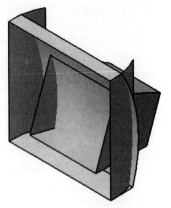

Figure 29-58 *Resulting extruded surface*

13. Choose the **Trim** tool from the **Edit** panel of the **Surface** tab; you are prompted to select the surfaces to be trimmed.

14. Select the back face of the base surface and press ENTER; you are prompted to select the surfaces used to trim. Select the left and right extruded surfaces, see Figure 29-59, and press ENTER; you are prompted to select the portions to trim.

15. Select the part of the base surface that is projecting outward; the base surface gets trimmed.

16. Similarly, trim all unwanted portions of the surfaces, as shown in Figure 29-60.

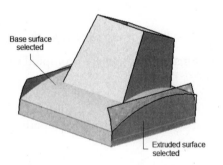

Figure 29-59 *Surfaces selected for trimming*

Figure 29-60 *The model after trimming the surfaces*

17. Create a new layer for the profiles, move all profiles to the new layer, and then freeze the layer so that the profiles are not displayed on the model.

    Next, you need to close the back face of the loft.

18. Invoke the **Patch** tool from the **Create** panel of the **Surface** tab; you are prompted to select the edges of the surfaces to patch.

19. Choose the **CHain** option and select all the four edges of the back face of the lofted surface and press ENTER; the preview of the patch surface is displayed. Press ENTER to accept it.

    Next, you need to create the frame of the monitor.

20. Enter the **SURFOFFSET** command at the Command prompt. Select the extruded and extended surface and press ENTER; the direction arrows are displayed. If the direction arrows are pointing outward, flip them. When the arrows point inward, set the offset value to 2.

21. Similarly, offset the other two sides inward to a distance of 2 units, refer to Figure 29-61.

22. Trim the four surfaces to form corners, see Figure 29-61.

    Next, you need to create a panel by closing the offset faces and the original faces.

23. Invoke the **Blend** tool from the **Create** panel in the **Surface** tab; you are prompted to select the first surface edge to blend.

24. Select the top edge of the extruded and extended surface, and then press ENTER; you are prompted to select the second surface edge to blend.

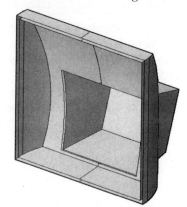

Figure 29-61 *Model after offsetting and trimming surfaces*

25. Select the offset surface below the top edges and press ENTER; the preview of the blend surface is created.

26. Press ENTER to accept the settings of the blend surface. Similarly, blend all surfaces. The final model of the toy monitor after blending all surfaces is shown in Figure 29-62.

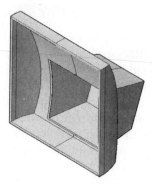

Figure 29-62 *Back cover of the toy monitor*

Editing the NURBS Surfaces

NURBS is an acronym for 'Non-Uniform Rational B-Splines' and encompasses all characteristics of bezier curves or splines. However, in AutoCAD, splines are optimized to create NURBS. You can adjust the control points, fit points, and degree of a spline to control its shape, thereby controlling the NURBS surface. NURBS surface consists of control vertices and fit points, which can be used to modify the overall shape of a surface. While creating any surface, if you choose the **NURBS Creation** button; the NURBS surface will be created.

Convert to
NURBS

You can also convert an existing surface into a NURBS surface. To do so, choose the **Convert to NURBS** tool from the **Control Vertices** panel; you will be prompted to select the object to be converted into a NURBS surface. Select the surface and press ENTER; the selected surface will be converted into a NURBS surface. Figure 29-63 shows a NURBS surface with its control vertices. If the CVs are not displayed on the NURBS surface, you need to invoke the **Show CV** tool. This tool is discussed next.

Displaying CVs

Ribbon: Surface > Control Vertices > Show CV/Hide CV   **Command:** CVSHOW/CVHIDE
Menu Bar: Modify > Surface Editing > NURBS Surface Editing > Show CV/Hide CV

Show CV

Hide CV

You can edit the shape of a NURBS surface by dragging its control vertices or fit points. By default, control vertices are not displayed on the NURBS surface. To view the control vertices on the surface, choose the **Show CV** tool from the **Control Vertices** panel; the CVs will be displayed on the surface. You can drag these control vertices to new locations to edit the surface, see Figure 29-64. After editing the surface, if you want to hide the CVs from the NURBS surface, choose the **Hide CV** tool.

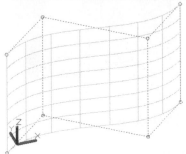

Figure 29-63 NURBS surface

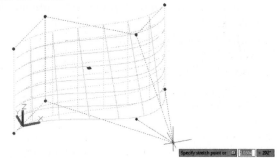

Figure 29-64 Editing NURBS surface

Editing CVs

Ribbon: Surface > Control Vertices > CV Edit Bar   **Command:** 3DEDITBAR

CV Edit Bar

While converting an existing surface into a NURBS surface, the CVs are formed at the corners of the NURBS surface. However, you can specify a point on the NURBS surface to create a CV at that particular point. To do so, choose the **CV Edit Bar** tool from the **Control Vertices** panel; you will be prompted to select surfaces. Select the surfaces; the cursor will turn red. Now, you can specify the direction of creation of the CV on the surface. After specifying a point, a move gizmo along with two handles will be displayed on the specified point, see Figure 29-65. Using these handles, you can edit the shape of the NURBS surface. The round handle is used to specify the magnitude of drag and the square handle is used to move the selected point to a new place.

You can set the tangency conditions and position of a control vertex using the triangular grip displayed on the gizmo, see Figure 29-65. Right-click on the triangular grip; a shortcut menu will be displayed, as shown in Figure 29-66. The options in the shortcut menu are discussed next.

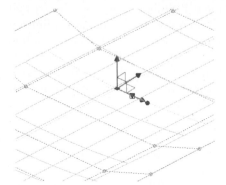

Figure 29-65 Editing CVs

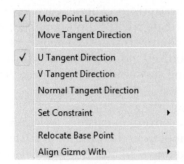

Figure 29-66 Shortcut menu displayed at the triangular grip

By default, the **Move Point Location** option is selected in this shortcut menu. As a result, the NURBS surface will be modified with respect to the point but the tangency will not be modified. If you select the **Move Tangent Direction** option, the tangency of the curve will be modified, but not its shape. If you select the **U Tangency Direction** or **V Tangency Direction** option, the tangency will be modified only in the U-direction or V-direction, respectively. The **Normal Tangent Direction** option is used to edit the tangency of the curve normal to the current UCS. You can also constrain the change in tangency or point location to a particular axis using the **Set Constraint** option. You can also relocate the base point or realign the gizmo using the **Relocate Base Point** and **Align Gizmo With** options, respectively in this shortcut menu.

Note
You cannot edit surfaces that have less than 3 degrees. Therefore, you need to rebuild such surfaces. These surfaces are discussed later.

Adding CVs

Ribbon: Surface > Control Vertices > Add	**Command:** CVADD
Menu Bar: Modify > Surface Editing > NURBS Surface Editing > Add CV	**Toolbar:** Surface Editing

When you create a NURBS surface, some CVs are displayed on it by default. However, you can add more CVs to it if you need. To add CVs, choose the **Add** tool from the **Control Vertices** panel; you will be prompted to select a NURBS surface where you want to add CVs. Select the surface; the preview of the CVs will be attached to the cursor, see Figure 29-67. By default, the CVs get added in the U-direction. If you want to add CVs in the V-direction, choose the **Direction** option from the Command prompt; the CVs will be added in the V-direction.

If you want to create knots on the surface, choose the **insert Knots** option from the Command prompt; knots or fit points will be created on the surface instead of CVs.

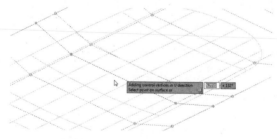

Figure 29-67 *Preview of CVs*

Rebuilding a NURBS Surface

Ribbon: Surface > Control Vertices > Rebuild	**Command:** CVREBUILD
Menu Bar: Modify > Surface Editing > NURBS Surface Editing > Rebuild	**Toolbar:** Surface Editing

You can reconstruct a NURBS surface by defining the number of CVs created on it and the degree of the CV curve. For example, the surfaces created by line, arc, or circle are of less than 3 degrees. You may need to change the degree of these surfaces. You can do so by invoking the **Rebuild** tool from the **Control Vertices** panel. On doing so, you will be prompted to select the surface to be rebuilt. Select the required surface; the **Rebuild Surface** dialog box will be displayed, as shown in Figure 29-68.

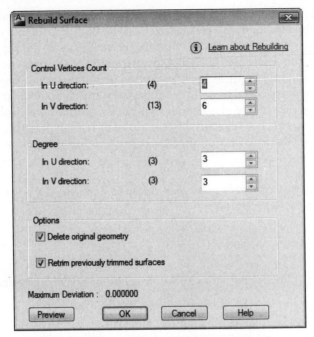

*Figure 29-68 The **Rebuild Surface** dialog box*

The areas in the dialog box are discussed next.

Control Vertices Count

In this area, you can specify the number of CVs to be created on the surface in the U and V directions.

Degree

In this area, you can specify the degree of a CV curve along the U and V directions.

Options

In this area, you can specify whether to delete the original geometry of a CV curve or re-trim a surface that has already been trimmed.

Preview Button

You can preview the CVs created on a surface. To do so, first set the required values in the **Rebuild Surface** dialog box and then choose the **Preview** button; the preview of the CVs created will be displayed. If the preview of the CVs created is satisfactory, press ENTER. If you want to change the values, press the ESC key.

Removing CVs

Ribbon: Surface > Control Vertices > Remove	**Command:** CVREMOVE
Menu Bar: Modify > Surface Editing > NURBS Surface Editing > Remove CV	**Toolbar:** Surface Editing

You can remove the unwanted CVs from a NURBS surface. To do so, choose the **Remove** tool from the **Control Vertices** panel; you will be prompted to select a NURBS surface to remove the CVs. Select a NURBS surface and press ENTER; you will be prompted to select a point on the surface to remove vertices in the U-direction. Select a point on the surface near the CV that you want to remove. If you want to change the direction of removal of CVs, use the **Direction** option; the CVs will be set to be removed from the V-direction. Note that the minimum number of CVs in each direction is 2. At least one CV must be available on the surface; else, an error message will be displayed.

Projecting Geometries

The Projection tools are used to project one or more elements on a surface. These tools are available in the **Project Geometry** panel of the **Surface** tab. You can project curves, lines, points, 2D polyline, 3D polyline, or helix onto a solid or a surface. When you project a curve, a duplicate of that curve is created. You can move, modify, change, delete this curve without affecting the original curve. You can also trim the surface onto which the curve has been projected. This can be done by using the **Auto Trim** button. When this button is active, the surface will be trimmed automatically on projecting the curve onto the surface. The projection of curves can be normal to the surface along a specified direction, normal to the current view, or along a specified path. These projection tools are discussed next.

Project to UCS

Using this tool, you can project a curve on a surface along the +z or -z axis of the current UCS.

Project to View

This tool enables you to project a curve on a surface based on the current view.

Project to 2 Points

Using this tool, you can project a curve along a path defined by two points.

Figure 29-69 shows the surface and the curve to be projected. Figure 29-70 shows the curve projected using the **Project to UCS** tool with the **Auto Trim** button chosen and Figure 29-71 shows the curve projected using the **Project to View** tool with the **Auto Trim** button chosen. Figure 29-72 shows the curve projected on the surface by specifying the 2 endpoints of the line segment as the direction of projection.

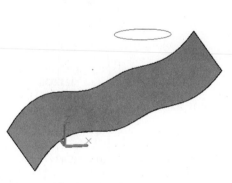

Figure 29-69 *Surface and curve to be projected*

Figure 29-70 *Curve projected using the* **Project to UCS** *tool*

Figure 29-71 *Curve projected using the* **Project to View** *tool*

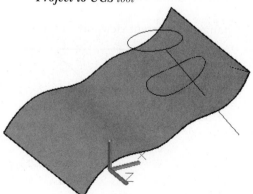

Figure 29-72 *Curve projected using the* **Project to 2 Points** *tool*

Chapter 29

EXAMPLE 3   *Editing Tools*

In this example, you will create the model of eyeglasses, as shown in Figure 29-73.

Figure 29-73 Final model of the eyeglasses.

1.   Start a new drawing file in **Acad3d.dwt** template.

2.   Change the current viewport to **Right** from the **In-canvas Viewport Controls** with **Parallel** projection.

3.   Draw the spline using the **Spline CV** tool, as shown in Figure 29-74.

4.   You need to array this spline along Z-axis. To do so, invoke the **COPY** command and create two more instances of spline by using the prompt sequence given next.

Command: **COPY**
Select objects: *Select the spline.*
Select objects: [Enter]
Current settings: Copy mode = Multiple
Specify base point or [Displacement/mOde]
<Displacement>:*Select any point on the spline.*
Specify second point or [Array] <use first point as displacement>: **ARRAY** [Enter]
Enter number of items to array: **3** [Enter]
Specify second point or [Fit]: **@0,0,60** [Enter]
Specify second point or [Array/Exit/Undo] <Exit>:
[Enter]

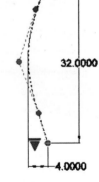

Figure 29-74 Cross-section for frame

Next, you need to create guide curves.

5.   Change the current viewport to **Front** and draw guide curves, as shown in Figure 29-75.

Figure 29-75 Guide curves for frame

6. Invoke the **Network** tool from the **Create** panel of the **Surface** tab; you are prompted to select the curves in first direction.

7. Select the three splines created earlier and press ENTER; you are prompted to select the curves in the second direction.

8. Select the two joining guide curves drawn previously and press ENTER; the network surface is created. Change the current viewport to **SE Isometric** and then change the visual style, the surface will be displayed as shown in Figure 29-76.

Figure 29-76 Network surface created

Next, you need to draw the shape of the frame and then trim the unwanted portion.

9. Align the UCS on the network surface and draw the sketch using the **Line** and **Arc** tools in the **Front** viewport, as shown in Figure 29-77. Refer to Figure 29-78 for dimensions.

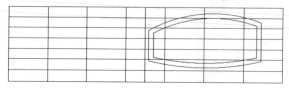   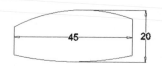

Figure 29-77 Sketch to be drawn   *Figure 29-78 Dimensions for sketch*

Note
The sketch created should be joined using the JOIN command.

10. Invoke the **Trim** tool from the **Edit** panel in the **Surface** tab; you are prompted to select the objects to trim.

11. Select the network surface and press ENTER; you are prompted to select the cutting curves.

12. Select the two joined curves and press ENTER; you are prompted to select the area to be trimmed.

13. Select the innermost and outermost areas of the curves and press ENTER; the surface is created and displayed, as shown in Figure 29-79.

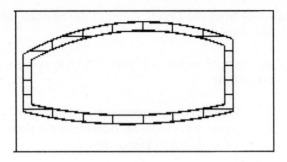

Figure 29-79 *Partial view of glass frame*

14. Create a new layer, move all profiles to the new layer, and then freeze the layer so that the profiles are not displayed on the model.

15. Change the current viewport to **SE Isometric**.

16. Invoke the **Extrude** tool, select the surface, and then extrude it by 2 units.

Next, you need to create a surface for glasses.

17. Invoke the **Patch** tool from the **Create** panel of the **Surface** tab; you are prompted to select the surface edges. Select the innermost edge of the frame and press ENTER twice; the surface will be created, as shown in Figure 29-80.

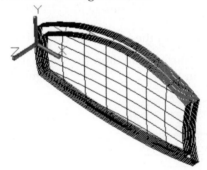

Figure 29-80 *The patch surface created*

Next, you need to create the side arm.

18. Change the current viewport to **Front** using the **In-canvas Viewport Controls** and then create a spline by using the cv method, refer to Figure 29-81.

19. Invoke the **Extrude** tool and then extrude this spline to 100 units along negative Z-axis.

20. Change the current viewport to **Right** and align the UCS to the extruded surface created in Step 19.

21 Create a profile for side arm and join it using the **JOIN** command, as shown in Figure 29-82.

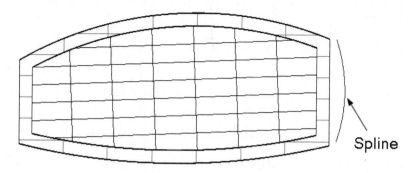

Figure 29-81 *Spline created for arm*

Figure 29-82 *Profile for arm*

22. Invoke the **Trim** tool from the **Edit** panel in the **Surface** tab; you are prompted to select the objects to trim.

23. Select the extruded surface and press ENTER; you are prompted to select the curves for trimming the object.

24. Select the profile created for the arm and press ENTER; you are prompted to select the area to be trimmed.

25. Select the outer area of the profile and press ENTER; the surface is created.

26. Move the profiles to the frozen layer.

27. Select the newly created surface and extrude it to 1 unit along negative Z-axis.

28. Change the current viewport to **Front** and align the UCS to the front face of the frame.

29. Create two circles of radius 0.45 each and a spline, as shown in the Figure 29-83.

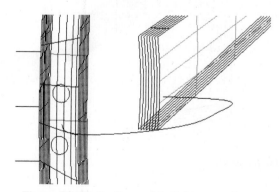

Figure 29-83 *Section and path for sweep feature*

30. Invoke the **Sweep** tool, select the two circles, and press ENTER; you are prompted to select the sweep path.

31. Select the spline curve; the sweep feature is created. Move the profiles to the frozen layer.

32. Change the current viewport to **Front** and mirror all the features using the **MIRROR** command about a line drawn at a distance of 7.5 units from the edge opposite to the edge joined with the arm.

    Next, you need to link the mirrored feature with the original feature.

33. Align the UCS to the side face of the frame, as shown in the Figure 29-84, and create a circle of radius 0.5 at **0,0** co-ordinate.

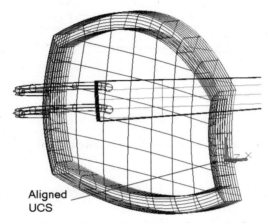

Aligned
UCS

Figure 29-84 Aligning the UCS and creating the circle

34. Set the UCS to WCS, change the viewport to **Front**, and create a profile for the joint, as shown in Figure 29-85.

35. Move the profile to the center of the circle created in Step 33.

36. Sweep the circle on the profile drawn for creating the joint.

The final model of the eyeglasses after applying the desired materials is shown in shaded view in Figure 29-86.

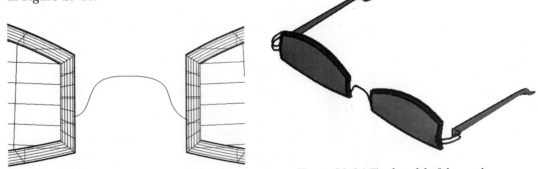

Figure 29-86 Final model of the eyeglasses

Figure 29-85 Profile for joint

Note
You will learn about applying materials in Chapter 31.

PERFORMING SURFACE ANALYSIS

AutoCAD allows you to analyze the surface continuity, curvature, and draft angles of surfaces using the surface analysis tools. You can also validate surfaces and curves before manufacturing using these tools. There are three types of surface analysis tools: **Zebra**, **Curvature**, and **Draft**. Note that these tools work only in 3D visual styles. These tools are discussed in detail in the next section.

Zebra

Ribbon: Surface > Analysis > Zebra	**Command:** ANALYSISZEBRA

The **Zebra** tool projects the zebra stripes on the selected model. This tool is used to visualize the curvature of surfaces and to determine if there are any surface discontinuities and inflections. To perform the **Zebra** analysis, invoke the **Zebra** tool from the **Analysis** panel of the **Surface** tab; you will be prompted to select the solid or surface on which you want to perform the analysis. Select the required surface and press ENTER; zebra strips will be displayed on the selected model. Figure 29-87 shows the zebra analysis performed on a binocular. The way these strips are aligned or curved on the model allows you to interpret the smoothness of the surfaces. This type of analysis is mostly used where one surfaces joins with other surface.

You can also change the settings of zebra stripes as per your requirements. To do so, choose the **Analysis Options** button from the **Analysis** panel; the **Analysis Options** dialog box will be displayed, as shown in Figure 29-88. By default, the **Zebra** tab is chosen in the dialog box.

Figure 29-87 *Zebra analysis performed*

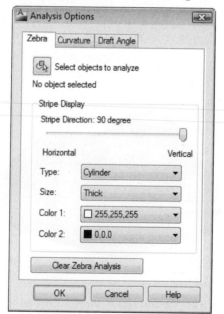

Figure 29-88 *The **Analysis Options** dialog box*

You can set the angle of stripes by using the slider available in the **Stripe Display** area of this dialog box. The value in the slider varies from 0 to 90. On setting 0, horizontal stripes are displayed on the model, whereas on setting 90, vertical stripes are displayed on the model. As you change the value on the slider bar, the zebra stripes will also change dynamically on screen. You can also set the type, color, thickness of zebra strips in the **Strip Display** area. Choose the **Clear Zebra Analysis** button in this dialog box to clear the zebra analysis. If you want to select objects for analysis, choose the **Select objects to analyze** button; the dialog box will disappear for a while, allowing you to select the objects from the screen.

Curvature

Ribbon: Surface > Analysis > Curvature   **Command:** ANALYSISCURVATURE

 The **Curvature** tool is used to display a color gradient or contours on the selected model. This color contour allows you to graphically visualize the radius of curvature of a model and evaluate the high and low areas of curvature. The contour also allows you to ensure that the model is within a certain range. The areas of high curvature or positive Gaussian value are displayed in red and the areas of low curvature or negative Gaussian value are displayed in blue. The mean curvature or a zero Gaussian value is displayed in red, which indicates a flat surface.

To perform the curvature analysis, choose the **Curvature** tool from the **Analysis** panel; you will be prompted to select the objects to perform the curvature analysis. Select the required objects and press ENTER; the color gradient will be displayed on the model. Figure 29-89 shows the curvature analysis performed on the binocular. You can change the settings for the curvature analysis by choosing the **Curvature** tab of the **Analysis Options** dialog box. Figure 29-90 shows the **Analysis Options** dialog box with the **Curvature** tab chosen. You can set the display style of curvature to be **Gaussian**, **Mean**, **Max radius**, or **Min radius** using the **Display Style** drop-down list in this dialog box. You can also set the minimum and maximum range in the corresponding edit boxes in the **Color mapping** area of the dialog box.

To clear the curvature analysis, choose the **Clear Curvature Analysis** button in the **Analysis Options** dialog box.

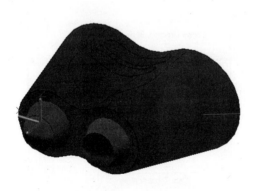

Figure 29-89 Curvature analysis performed

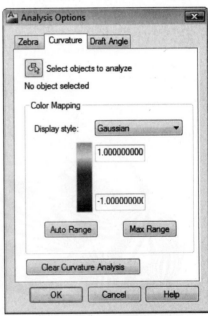

*Figure 29-90 The **Analysis Options** dialog box with the **Curvature** tab chosen*

Draft

Ribbon: Surface > Analysis > Draft   **Command:** ANALYSISDRAFT

The **Draft** tool is used to visualize the draft applied on the selected model. If you are designing a model that can be cast, then this tool is used to check whether a part can be removed from a mold or die. Then, it calculates whether there is sufficient draft between the part and the mold, based on the pull direction. To perform the draft analysis, choose the **Draft** tool from the **Analysis** panel; you will be prompted to select the object on which you want to perform the analysis. Select the object; the draft analysis will be performed on the object. Figure 29-91 shows the draft analysis performed on the binocular.

Surface angles are defined with respect to the draft plane. The colors displayed on the model after performing the draft analysis will depend on the angles defined. You can set these colors by using the **Draft Angle** tab of the **Analysis Options** dialog box, see Figure 29-92. Choose the **Clear Draft Angle Analysis** button to clear the draft angle analysis.

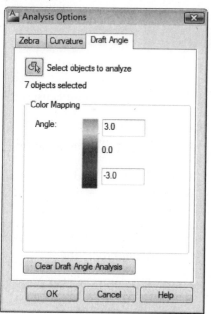

Figure 29-91 *Draft analysis performed*

Figure 29-92 *The **Analysis Options** dialog box with the **Draft Angle** tab chosen*

Self-Evaluation Test

Answer the following questions and then compare them to those given at the end of this chapter:

1. You can create CVs and knots in curves using the **Spline** tool. (T/F)

2. You can create solids as well as surfaces using the **Extrude** tool. (T/F)

3. You can create a solid by offsetting a surface. (T/F)

4. You can control the number of lines displayed in a Planar surface by using the _____ and _____ system variables.

5. To create procedural surfaces, the _____ button must be turned on.

6. Continuity can be set to _____, _____ or _____.

7. The _____ tool is used to fill the area formed between the open edges of a surface.

8. The _____ tool is used to create a 3D solid from a closed volume.

9. Which of the following tools can be used to edit a surface?

    (a) **Fillet**   (b) **Trim**
    (c) **Extend**   (d) All of these

10. The _____ button is used to trim a surface when a geometry is projected on that surface.

Review Questions

Answer the following questions:

1. The _____ tool is used to create a surface between the elements forming a closed boundary.

2. To create a NURBS surface, the _____ button must be turned on.

3. The _____ defines the smoothness of the blend surface where it meets the other surfaces.

4. The bulge magnitude varies between _____ and _____.

5. In the **Extend** tool, the _____ and _____ options are available.

6. Which of the following tools can be used to project geometries on a surface?

    (a) **Project to UCS**   (b) **Project to View**
    (c) **Project to 2 Points**   (d) All of these
    (e) None of these

7. You can convert an existing surface into a NURBS surface using the _____ tool.

8. The _____ dialog box can be used to reconstruct the shape of a NURBS surface.

9. You can add, remove, or rebuild the CVs of the surfaces. (T/F)

10. Which of the following analyses can be performed on a surface?

    (a) **Zebra**   (b) **Curvature**
    (c) **Draft**   (d) All of these

EXERCISE 2   *Web Camera*

In this exercise, you will create the model of a web camera, as shown in Figure 29-93. The reference dimensions for the model are given in Figure 29-94. Assume the missing dimensions.

Figure 29-93 *Model of a web camera*

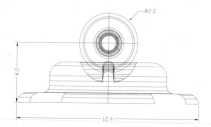

Figure 29-94 *Dimensions for the model*

EXERCISE 3 — *Surface Tools*

In this exercise, you will create the model of a ship, as shown in Figure 29-95. The other views of the ship are shown in Figure 29-96. Assume the dimensions.

Figure 29-95 *Model of a ship*

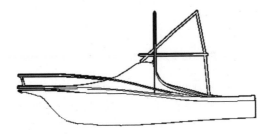

Figure 29-96 *Other views of the ship*

Answers to Self-Evaluation Test

1. T, **2.** T, **3.** F **4. SURFU, SURFV**, **5. Surface Associativity**, **6.** 0, 1, 2, **7. Patch**, **8. Sculpt**, **9.** (d), **10. Auto Trim**

Chapter 30

Mesh Modeling

CHAPTER OBJECTIVES

In this chapter, you will learn:
- *To create standard mesh primitives.*
- *To create mesh surfaces.*
- *To smoothen, refine, and crease mesh objects.*
- *To edit mesh faces.*
- *To convert mesh objects to solids and surfaces.*
- *To work with gizmos*

KEY TERMS

- *MESH*
- *REVSURF*
- *EDGESURF*
- *RULESURF*
- *TABSURF*

- *Smooth Object*
- *Refine Mesh*
- *MESHCREASE*
- *Split Face*
- *MESHEXTRUDE*

- *Merge Face tool*
- *MESHCAP tool*
- *MESHCOLLAPSE*
- *Spin Triangle Face*
- *CONVTOSOLID*

- *CONVTOSURFACE*
- *Move Gizmo*
- *Rotate Gizmo*
- *Scale Gizmo*

INTRODUCTION

A mesh model is similar to a solid model without mass and volume properties. The mesh modeling tools are available in the **3D Modeling** workspace. You can create mesh primitives as you create solid primitives and then create freeform designs from these mesh primitives. The mesh models are easily editable and can be smoothened to give a rounded effect. You can refine the meshes at required areas for precision. You can perform editing operations such as gizmo editing, extrusion on the mesh model, and so on to create freeform designs. Some of the important features of mesh modeling are that you can deform a mesh model using gizmos, add creases to sharpen the mesh model, split mesh faces to segment the meshes as required, and extrude mesh faces. You will learn all these techniques later in this chapter.

CREATING MESH PRIMITIVES

Mesh primitives are standard shapes that help you create complex models. These standard shapes include box, cone, cylinder, pyramid, sphere, wedge, and torus. You can create mesh primitives as you create solid models. By default, mesh primitives are not smooth. However, you can make these primitives smooth by adding smoothness to them. Note that as you smoothen the mesh, the model becomes complex and the system performance gets low. Therefore, you must perform the editing operations first and then smoothen the mesh to reduce the complexity of the model. The procedure for creating mesh primitives is given next.

Creating a Mesh Box

Ribbon: Mesh > Primitives > Mesh Primitives drop-down > Mesh Box	**Command:** Mesh

 You can create a 3D mesh box similar to a solid or surface box. To do so, invoke the **Mesh Box** tool from the **Primitives** panel and follow the command sequence given next.

Command: _MESH
Current smoothness level is set to: 0
Enter an option [Box/Cone/CYlinder/Pyramid/Sphere/Wedge/Torus/SEttings] <Box>: _BOX
Specify first corner or [Center]: *Specify the first corner*
Specify other corner or [Cube/Length]: *Specify the diagonally opposite corner*
Specify height or [2Point] <1>: *Specify the height of the mesh box*

After you specify the height of the mesh box, it will be created in the drawing area. Figure 30-1 shows the mesh box created.

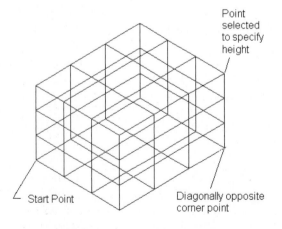

Figure 30-1 *The mesh box created*

You can create a mesh box in five ways using the above prompt. These ways are listed below:

1. Dynamically.
2. By first specifying the two corners of the rectangular base and then, the height of the box.
3. By first specifying the center, a corner of the box and then, the height of the box.
4. By specifying a corner and the length of the edge of cube.
5. By specifying the length, width, and height of the box.

The number of faces/divisions in a mesh box depends upon the values specified in the **Mesh Primitive Options** dialog box, see Figure 30-2. This dialog box can be invoked by clicking on the inclined arrow available in the **Primitives** panel. Specify the required values in the **Tessellation Divisions** rollout of this dialog box. Next, if you invoke the **MESH** command to create a new mesh, the model will be created based on the newly specified values. You can also change the appearance of the mesh by changing the level of smoothness of the model. Various areas of the **Mesh Primitive Options** dialog box are discussed next.

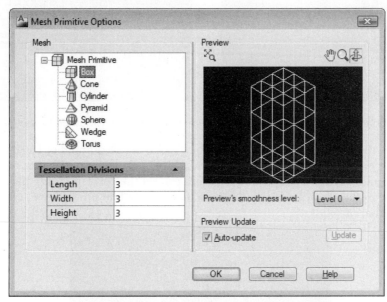

*Figure 30-2 The **Mesh Primitive Options** dialog box*

The Mesh Area

The **Mesh** area displays a list of mesh primitives and its corresponding parameters. You can select the required primitive from the **Mesh Primitive** list to control its appearance. The **Tessellation Divisions** rollout in this area displays the parameters of the selected primitive. You can edit the corresponding parameters to set the number of divisions on it. For example, if you choose **Cylinder** from the **Mesh Primitive** node in the **Mesh** area, the **Axis**, **Height,** and **Base** parameters will be available in the **Tessellation Divisions** rollout. Now, you can edit the number of faces along the axis, height, and base. Note that to create a mesh primitive with desired settings, first you need to set the values of the parameters and then create the mesh primitive.

The Preview Area

The **Preview** area displays the preview of the selected meshed primitive. The buttons in the **Preview** area are used to pan, zoom, or rotate the mesh primitive. To perform any of these operations, choose the respective button from the **Preview** area and then click in the preview screen. If you want the preview to be automatically updated according to the settings specified in the **Tessellation Divisions** rollout, select the **Auto-update** check box in the **Preview Update** area. Otherwise, choose the **Update** button to update the preview manually.

Note
*To change the visual style of the model in the **Preview** area, right-click and choose the required option from the **Visual Styles** menu.*

The **Preview's smoothness level** drop-down list in the **Preview** area is used to specify the level of smoothness of the mesh primitive. By default, Level 0 is set in the drop-down list, which means, the mesh primitives created will not be smooth by default. Figures 30-3 to 30-5 show a mesh box with different levels of smoothness. You can set the level of smoothness even after the primitive has been created. To do so, select the required primitive; the **Quick Properties** panel will be displayed. Select the level of smoothness from the **Smoothness** drop-down list in the **Quick Properties** panel; the mesh primitive will be smoothened.

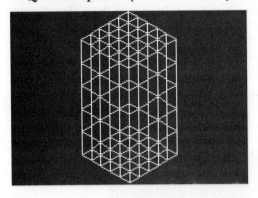

Figure 30-3 Mesh box with smoothness level=0

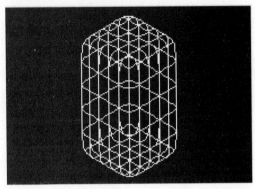

Figure 30-4 Mesh box with smoothness level=2

Like mesh box, the **MESH** command can also be used to create pointed or frustum mesh cones with circular or elliptical base, mesh cylinders with circular or elliptical bases, mesh pyramids, mesh spheres, mesh wedges, and mesh torus. You can set the corresponding parameters of the mesh primitive in the **Mesh Primitive Options** dialog box. The procedure to create primitives is the same as that of solids.

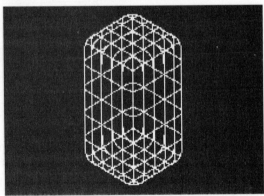

Figure 30-5 Mesh box with smoothness level=4

Note
*By default, the **Realistic** visual style is selected in the **Visual Styles** drop-down list of the **Visual Styles** panel in the **View** tab. For clarity of printing, the visual style is set to **Wireframe**.*

CREATING SURFACE MESHES

Apart from creating basic mesh primitives, you can create various types of surface meshes. Surface meshes are created by filling the space between objects consisting of straight or curved edges. There are various types of surface meshes such as revolved surface, edge surface, ruled surface, and tabulated surface mesh. These surface meshes as well as the commands used to create them are discussed next.

Creating Revolved Surface Meshes

Ribbon: Mesh > Primitives > Modeling, Meshes, Revolved Surface   **Command:** REVSURF

In AutoCAD, you can create a revolved surface by choosing the **Revolved Surface** tool from the **Primitives** panel. A revolved surface is formed by revolving a path curve about the axis of rotation. You can select an open or a closed entity as the path curve. The direction of the revolution depends upon the side from which you select the revolution axis. Since it is not possible to define the clockwise and counterclockwise directions in 3D space, therefore, AutoCAD uses the right-hand thumb rule (refer to Chapter 25 "Getting Started with 3D") to determine the direction of revolution. To create a revolved surface mesh, choose the **Modeling, Meshes, Revolved Surface** tool from the **Primitives** panel and follow the prompt sequence given next:

Command: _REVSURF
Current wire frame density: SURFTAB1=6 SURFTAB2=6
Select object to revolve: *Select the object to revolve around the axis*
Select object that defines the axis of revolution: *Select the axis of revolution*
Specify start angle <0>: Specify second point: *Specify the start angle to start the mesh at an offset from the profile curve*
Specify included angle (+=ccw, -=cw) <360>: *Specify the included angle that indicates the distance through which the profile curve will be swept*

Figure 30-6 shows a path curve and a revolution axis and Figure 30-7 shows the resultant revolved surface. You can specify the start angle and the included angle for the revolved surface as per your requirement, see Figures 30-8 and 30-9.

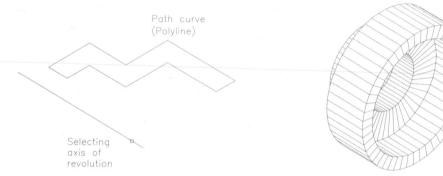

Figure 30-6 *The path curve and revolution axis*

Figure 30-7 *The resultant revolved surface*

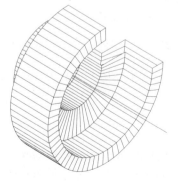

Figure 30-8 *Revolved surface created with the start angle **0** and included angle **270***

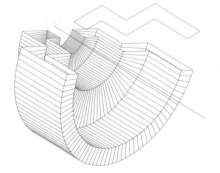

Figure 30-9 *Revolved surface created with the start angle **25** and included angle **180***

The smoothness of the revolved surface and the total number of lines that AutoCAD window will draw depend upon the value of the **SURFTAB1** system variable. AutoCAD draws tabulated lines in the direction of revolution. The number of tabulated lines is defined by the value of the **SURFTAB2** system variable (Figure 30-10).

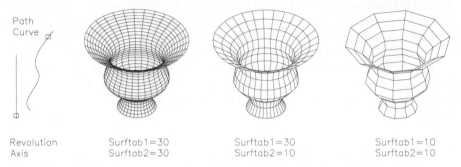

Path
Curve

Revolution   Surftab1=30   Surftab1=30   Surftab1=10
Axis   Surftab2=30   Surftab2=10   Surftab2=10

Figure 30-10 Revolved surface created with different values of the
SURFTAB1 and SURFTAB2 system variables

Creating Edge Surface Meshes

Ribbon: Mesh > Primitives > Modeling, Meshes, Edge Surface   **Command:** EDGESURF

The **Modeling, Meshes, Edge Surface** tool is used to create a 3D polygon mesh from the four adjoining edges that form a topologically closed loop. The entities that you select as edges can be lines, arcs, splines, open 2D polylines, or open 3D polylines. While creating an edge surface mesh, you can select the edges of the closed loop in any sequence. The **SURFTAB1** and **SURFTAB2** system variables control the appearance of the mesh and can be used to set the wireframe density. The edge that is selected first will define the M-direction (**SURFTAB1** direction). The other two edges that start from the endpoints of the first edge will define the N-direction (**SURFTAB2** direction), see Figures 30-11 and 30-12.

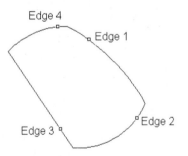

Edge 4

Edge 1

Edge 3

Edge 2

Figure 30-11 Edges to be selected   *Figure 30-12 An edge surface mesh created*

Creating Ruled Surface Meshes

Ribbon: Mesh > Primitives > Modeling, Meshes, Ruled Surface   **Command:** RULESURF

You can create a ruled surface between two defined entities by choosing the **Modeling, Meshes, Ruled Surface** tool from the **Primitives** panel. The defined entities can be lines, circles, ellipses, arcs, polylines, curves, or splines. You can create a ruled surface mesh between two closed entities, two open entities, a point and a closed entity, and a point and an open entity. However, you cannot create a ruled surface between a closed and an open entity. Also, two points do not define a surface and therefore, a ruled surface cannot be created between them.

If the two defining entities are closed, the selection point does not make any difference in the resultant surface. In case of circles, the start point of the surface is determined by the combination of direction of the *X* axis and the current value of the **SNAPANG** system variable. In case of closed polylines, the ruled surface starts from the last vertex of the polyline and proceeds backward along the polyline segment.

If two defined entities are open, the selection points do make a difference in the resultant surface. If you select open entities at the same end, a straight polygon mesh is created. If you select open entities at opposite ends, a self-intersecting polygon mesh is created, refer to Figures 30-13 and 30-14.

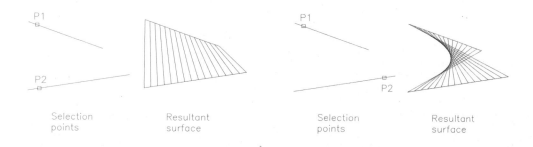

Selection points   Resultant surface   Selection points   Resultant surface

Figure 30-13 *Selecting objects at same ends and the resultant surface*   ***Figure 30-14*** *Selecting objects at opposite ends and the resultant surface*

The smoothness of the ruled surface can be increased using the **SURFTAB1** system variable, see Figures 30-15 and 30-16. The default value of this system variable is 6, but it can accept any integer between 2 and 32766 as its value. Note that this value must be increased before creating the surface.

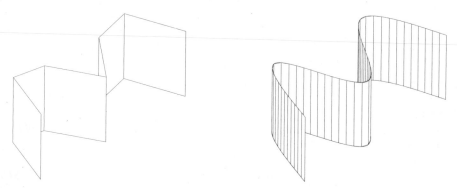

Figure 30-15 *The ruled surface created with 6 as the value of the **SURFTAB1** system variable*   ***Figure 30-16*** *The ruled surface created with 50 as the value of the **SURFTAB1** system variable*

Creating Tabulated Surface Meshes

Ribbon: Mesh > Primitives > Modeling, Meshes, Tabulated Surface   **Command:** TABSURF

You can create a tabulated surface mesh by choosing **Modeling, Meshes, Tabulated Surface** from the **Primitives** panel. The tabulated surface mesh is formed when a profile is swept along the direction defined by the direction vector. The profile curves can be lines, circles, arcs, splines, 2D polylines, 3D polylines, or ellipses. The length of the tabulated surface will be equal to the length of the direction vector, which in turn will be determined by the point selected for the direction vector, see Figures 30-17 and 30-18.

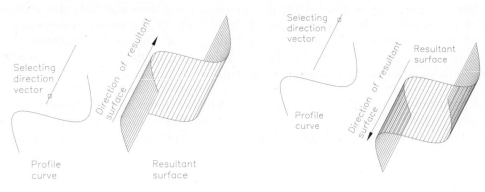

Figure 30-17 *Direction of the tabulated surface is upward*

Figure 30-18 *Direction of the tabulated surface is downward*

If there are more than one direction vectors, all intermediate points will be ignored and an imaginary vector will be drawn between the first and last points on the polyline. The imaginary vector will be considered as the direction vector and the tabulated surface will be created along it, as shown in Figure 30-19.

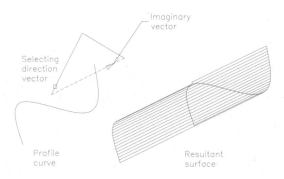

Figure 30-19 *Tabulated surface created along the imaginary vector*

Note
You cannot select a curved entity as the direction vector for defining the direction of the tabulated surface.

*You can increase the smoothness of the tabulated surface by increasing the value of the **SURFTAB1** system variable.*

*Apart from surface meshes, the **REVSURF**, **EDGESURF**, **RULESURF**, and **TABSURF** commands can be also used to create surface models.*

EXERCISE 1   *Surfaces*

In this exercise, you will create the table shown in Figure 30-20 by using different types of surfaces. Assume the dimensions for the table.

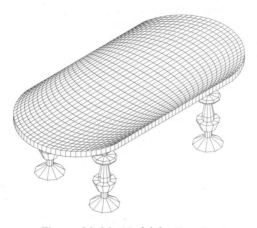

Figure 30-20 *Model for Exercise 1*

MODIFYING MESH OBJECTS

Mesh models can be modified to create freeform designs. This can be done by changing their smoothing levels, refining specific areas, adding creases to sharpen the edges, and so on. Various types of modifications that can be applied on meshes are discussed next.

Adding Smoothness to Meshes

Ribbon: Mesh > Mesh > Smooth Object	**Command:** MESHSMOOTH
Menu Bar: Modify > Mesh Editing > Smooth More/Less	**Toolbar:** Smooth Mesh

Adding smoothness to a mesh results in roundness of the sharp edges of the meshed model. To change the level of smoothness of a mesh, select it; the **Quick Properties** panel will be displayed. In this panel, select the required level of smoothness from the **Smoothness** drop-down list; the selected smoothness level will be applied to the mesh. Note that **Level 0** indicates minimum smoothness, whereas **Level 4** indicates maximum smoothness. You can switch between different levels of smoothness with the help of the **Quick Properties** panel. Remember that the higher level of smoothness makes the mesh complex and thus, decreases the system performance. Figure 30-21 shows a mesh with different levels of smoothness applied to it. It is recommended that you first perform the required editing of the mesh and then apply smoothness to it.

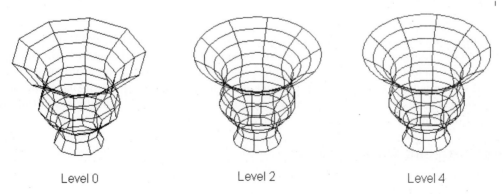

Level 0   Level 2   Level 4

Figure 30-21 *Mesh with different levels of smoothness*

Note
*You can also set the level of smoothness of the mesh from the **Properties** palette.*

If you need to apply smoothness to objects other than the mesh objects, first convert the objects into mesh models. The objects that can be converted into mesh models include 3D solids, 3D surfaces, 3D faces, polygon meshes, regions, and closed polylines.

To convert an object into a mesh object, choose the **Smooth Object** tool from the **Mesh** panel; you will be prompted to select the object to convert. Select an object and press ENTER; the 3D object will be converted into a mesh object. Now, you can perform the editing operations such as smoothening, refining, splitting, and so on on it. Note that the smoothening operation works best for primitive solids. But if you select objects other than the primitive solids, a message box will be displayed, as shown in Figure 30-22. Choose **Create mesh** from the message box; the selected objects will be converted into mesh models with default parameters.

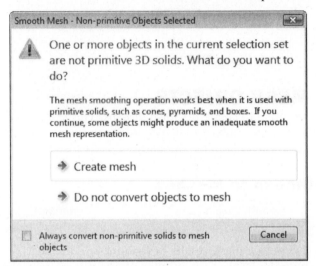

*Figure 30-22 The **Smooth Mesh - Non-primitive Objects Selected** message box*

However, you can modify the default parameters as per your requirement by invoking the **Mesh Tessellation Options** dialog box. To invoke this dialog box, click on the inclined arrow in the **Mesh** panel; the **Mesh Tessellation Options** dialog box will be displayed, as shown in Figure 30-23.

The options in this dialog box are discussed next.

Select objects to tessellate

This button enables you to select objects to tessellate while the **MESHOPTIONS** command is active. On choosing this button, the **Mesh Tessellation Options** dialog box will disappear allowing you to select objects. Select the required objects from the drawing area and press ENTER; the dialog box will reappear and the number of objects selected will be displayed below the **Select objects to tessellate** button. Now, you can specify the required parameters in the **Mesh Tessellation Options** dialog box.

Mesh Type and Tolerance Area

The options in this area are discussed next.

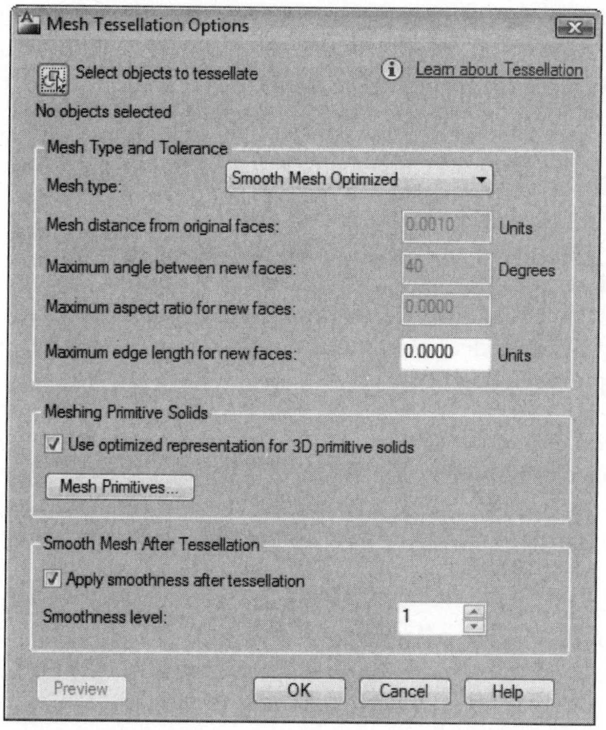

Figure 30-23 *The* **Mesh Tessellation Options** *dialog box*

Mesh type

This drop-down list is used to specify the mesh type that you want to create. This drop-down list has three options: **Smooth Mesh Optimized**, **Mostly Quads**, and **Triangle**. By default, the **Smooth Mesh Optimized** option is selected and therefore, the shape of the mesh face will be the optimized mesh shape that can adapt to the shape of the mesh object. If you select the **Mostly Quads** option, the shape of the mesh face will be quadrilateral to its maximum possible, and if you select the **Triangle** option, the shape of the mesh face will be triangular to its maximum possible.

Mesh distance from original faces

This edit box helps you specify how closely the converted mesh object will adhere to the original shape of solid or surface. Smaller the values specified in this edit box, lesser the deviation from the original faces, and therefore, greater the number of faces created in the mesh. This means small value results in more accurate mesh objects but with slower performance of system. However, this edit box is not enabled if the **Smooth Mesh Optimized** option is selected in the **Mesh type** drop-down list

Maximum angle between new faces

The **Maximum angle between new faces** edit box is not enabled if the **Smooth Mesh Optimized** option is selected in the **Mesh type** drop-down list. This edit box allows you to specify the maximum angle between the surface normal and continuous mesh faces. Larger the value specified in this edit box, more dense will be the mesh at higher curvatures and less dense at flat areas. This setting is used to refine the appearance of the curved features like holes or fillets.

Maximum aspect ratio for new faces

This edit box is not enabled in case the **Smooth Mesh Optimized** option is selected. This edit box allows you to set the maximum aspect ratio, which is, height/width ratio of new faces. This option is mostly used to avoid long and thin cylindrical faces. Note that smaller the value specified in this edit box, better will be the shape of the mesh.

Maximum edge length for new faces

This edit box is used to you to set the maximum length of the edges of the mesh objects that are created by converting solids and surfaces. Smaller the value specified in this edit box, better will be the shape/appearance of the mesh.

Meshing Primitive Solids

The options in this area are discussed next.

Use optimized representation for 3D primitive solids

Select this check box to apply the mesh settings specified in the **Mesh Primitive Options** dialog box and clear this check box to apply the settings specified in the **Mesh Tessellation Options** dialog box.

Mesh Primitives

The **Mesh Primitives** button will be enabled only when the **Use optimized representation for 3D primitive solids** check box is selected. On choosing the **Mesh Primitives** button, the **Mesh Primitive Options** dialog box will be displayed, where you can specify the required settings for the mesh object.

Smooth Mesh After Tessellation

This area will be active only when the **Smooth Mesh Optimized** option is selected in the **Mesh type** drop-down list. This is because the mesh objects converted on selecting any of the other two options from the drop-down list do not have any smoothness.

Apply smoothness after tessellation

Select this check box to apply smoothness to newly created mesh objects.

Smoothness level

This spinner will be available only when the **Apply smoothness after tessellation** check box is selected and you can set the level of smoothness of the mesh object in it.

Preview

This button is available when the objects required to be converted into meshes are selected. You can choose this button to preview the effect of the current settings in the drawing area. To go back to the dialog box, press ESC, or press ENTER to accept the mesh tessellation options.

Note

*To increase the smoothness of mesh by one level, choose the **Smooth More** button from the **Mesh** panel; the mesh will become smoother (rounded). Similarly, if you choose the **Smooth Less** button, the smoothness of the mesh will be decreased by one level.*

Refining the Meshes

Ribbon: Mesh > Mesh > Refine Mesh	**Command:** MESHREFINE	
Menu Bar: Modify > Mesh Editing > Refine Mesh	**Toolbar:** Smooth Mesh	

Refining the mesh means increasing the number of divisions in the mesh of an object, resulting in a dense mesh. You can refine the complete mesh object or a particular face of an object. Refining is mostly used for finer and detailed models, in which modification of small or refined divisions of the mesh is required. But remember that, denser the mesh, slower will be performance of the system. To refine the mesh, choose the **Refine Mesh** tool from the **Mesh** panel; you will be prompted to select the mesh object or faces of the mesh object to refine. Select the entity and press ENTER; the selected entity will get refined. Note that the **Refine Mesh** tool works only when the smoothness option of the selected mesh object is set to

Level 1 or higher. Also note that, as you refine the object, its level of smoothness becomes zero. To refine it again, you need to set the smoothness level to 1 or higher, and then follow the same procedure of refining. Figure 30-24 shows the mesh object before and after refinement.

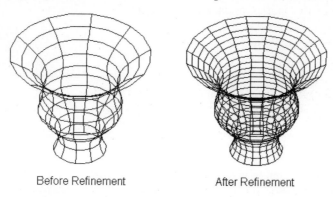

Before Refinement   After Refinement

Figure 30-24 *Object before and after refinement*

To refine the faces of the selected mesh object, first you need to set the criteria for selecting faces. To do so, right-click anywhere on the screen, and then choose **Subobject Selection Filter > Face** from the menu displayed. Alternatively, choose **Face** from the **Subobject Selection filter** drop-down in the **Subobject** panel. After the selection criteria have been set, invoke the **Refine Mesh** tool. Next, you need to select a face of the mesh object. To do so, press CTRL and click on the required mesh face, and then press ENTER; the specified face will be subdivided into four faces, as shown in Figure 30-25.

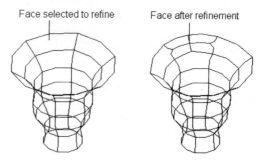

Face selected to refine   Face after refinement

Figure 30-25 *Face before and after refinement*

Note that refining an object resets the smoothness level to 0, whereas refining face/s does not affect the smoothness level.

Adding Crease to Meshes

Ribbon: Mesh > Mesh > Add Crease	**Command:** MESHCREASE
Menu Bar: Modify > Mesh Editing > Crease	**Toolbar:** Smooth Mesh

You can also modify a mesh object by adding creases to it. Creases are added to an object in order to sharpen the mesh sub-objects, thereby deforming the mesh object. You can sharpen the faces, edges, or vertices of a mesh object. To add crease to the selected mesh object, choose the **Add Crease** tool from the **Mesh** panel; you will be prompted to select the required mesh sub-objects. Select the mesh sub-objects and press ENTER; the **Specify crease value [Always] <Always>:** prompt will be displayed. By default, the crease is set to **Always**, which means the crease will retain its sharpness even if you smoothen or refine the mesh object. You can also specify a crease value at which the crease will start losing its sharpness. The crease value set to **-1** is the same as the crease value set to **Always**. If you specify **0** as the crease value, the existing crease will be removed. On sharpening a sub-object, the edges adjoining the selected sub-object are updated according to the modified model. Figures 30-26 through 30-31 show the effect of applying crease on various sub-objects.

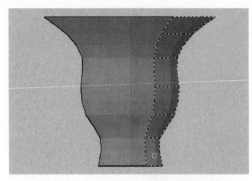

Figure 30-26 *Faces selected to be creased*

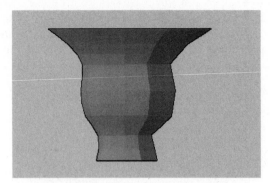

Figure 30-27 *Faces after adding crease*

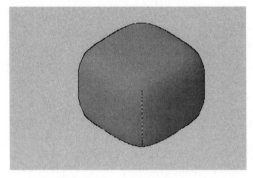

Figure 30-28 *Edge selected to be creased*

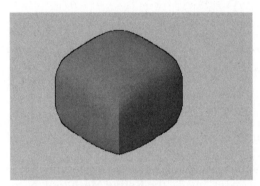

Figure 30-29 *Edge after adding crease*

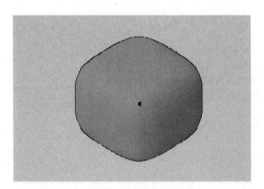

Figure 30-30 *Vertex selected to be creased*

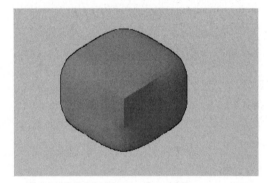

Figure 30-31 *Vertex after adding crease*

Note

*To select specific entities, right-click in the drawing area; a shortcut menu is displayed. Choose the desired sub-option of the **Subobject Selection Filter** option from this shortcut menu.*

The crease will not be visible when the level of smoothness is 0. So, increase the smoothness level to make it visible.

You can also remove the crease added to an object. To do so, choose the **Remove Crease** tool from the **Mesh** panel. Next, select the crease that you want to remove and press ENTER; the selected crease will be removed from the object and the corresponding faces will be updated. Alternatively, to remove the crease from the object, enter **MESHUNCREASE** at the Command prompt.

EDITING MESH FACES

You can edit the faces of a mesh object either by splitting or extruding its faces. In this process, the extruded or the split part merges with the original mesh object and act as a single entity. The splitting and extruding operations are useful for detailed and advanced modeling. To edit the faces of a mesh object, you can use the editing tools in the **Mesh Edit** panel. These tools are discussed next.

Splitting the Mesh Faces

Ribbon: Mesh > Mesh Edit > Split Face	**Command:** MESHSPLIT
Menu Bar: Modify > Mesh Editing > Split Mesh	

 In the process of Refining, the entire mesh model is split into various divisions, whereas in Splitting, a particular location of split is defined. As a result, splitting avoids larger deformations for smaller modifications. You can edit a mesh object by splitting it using the **Split Face** tool. This tool is used to divide the selected mesh face into two parts. To split the mesh, choose the **Split Face** tool from the **Mesh Edit** panel; you will be prompted to select the face of mesh to split. Select the face to split; a knife symbol will be displayed with the cursor and you will be prompted to specify the cut in the mesh. Now, specify the start and end points of the split line; the mesh will split along the line joining the specified points, and the adjoining faces will separate accordingly. Figure 30-32 shows the points selected for splitting the mesh face.

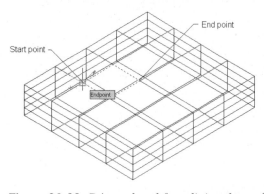

Figure 30-32 *Points selected for splitting the mesh*

 Tip
In case of detailed modeling, it is advisable to first split the mesh faces and then refine the mesh model.

Extruding the Mesh Faces

Ribbon: Mesh > Mesh Edit > Extrude Face	**Command:** MESHEXTRUDE

Extruding the mesh models is similar to extruding the 3D models. The only difference between the two processes is that in 3D modeling, the extruded part is created as a separate entity, whereas in mesh modeling, the extruded part merges with the mesh object and acts as a single entity. To extrude the mesh faces, choose the **Extrude Face** tool from the **Mesh Edit** panel; the Command sequence that will be displayed is given next.

Command: **MESHEXTRUDE**
Adjacent extruded faces set to: Join
Select mesh face(s) to extrude or [Setting]: *Select one or more faces to extrude*
Specify height of extrusion or [Direction/Path/Taper angle] <100>: *Specify the height of extrusion.*

You can control the extrusion style of multiple adjacent mesh faces. To do so, set the adjacent mesh faces to be extruded as a single unit or select each face separately. By default, the setting is **Join**. As a result, the adjacent faces are extruded as a single unit. If you want to change the setting to **Unjoin**, then press **S** when **Select mesh face(s) to extrude or [Setting]** prompt is displayed; you will be prompted to specify whether to join the adjacent mesh faces or not. Choose **No**; the extrusion style is set to **Unjoin**.

Chapter 30

You can extrude the mesh faces dynamically. Alternatively you can do so by specifying height, by specifying the direction of extrusion, by specifying the path of extrusion, or by specifying an angle of taper for extrusion. All these options work in the same way as in the 3D modeling. Figure 30-33 shows a mesh model after split and Figure 30-34 shows a mesh box with extruded split faces. The splitting and extruding operations on mesh models result in creation of new designs.

Tip
To select the faces that you want to extrude, click on the required faces of the mesh object keeping the CTRL key pressed. Continue till the selection is completed. To remove a face from the selection, press SHIFT + CTRL and click on the face to be deselected.

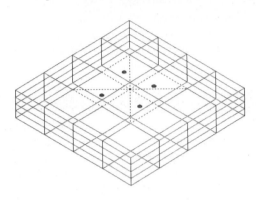

Figure 30-33 Mesh model after splitting

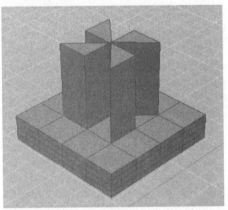

Figure 30-34 Mesh model with extruded split faces

Merging the Mesh Faces

| Ribbon: Mesh > Mesh Edit > Merge Face | Command: MESHMERGE |
| Menu Bar: Modify > Mesh Editing > Merge Face | |

Using **Merge Face** tool, you can merge one or more faces adjacent to a face into a single face. To merge the faces, invoke the **Merge Face** tool from the **Mesh Edit** panel; you will be prompted to select the adjacent faces of the specified face. Select the required number of faces and press ENTER; the selected faces will be merged.

Closing the Gaps

| Ribbon: Mesh > Mesh Edit > Close Hole | Command: MESHCAP |

Sometimes, some gaps or holes are created due to open edges on the mesh object. You can fill these holes or gaps by selecting the edges surrounding the hole by using the **Close Hole** tool. To do so, invoke the **Close Hole** tool from the **Mesh Edit** panel; you will be prompted to select the connecting edges to form a mesh face. Select the edges and press ENTER; a new mesh face will be formed between the connecting edges. Figure 30-35 shows the edges connecting the hole and Figure 30-36 shows a cap created on the hole.

You can also form gaps in the mesh objects. To do so, you can remove the faces, edges, or vertices of the mesh object by pressing the DELETE key from the keyboard. You can also use the **ERASE** command to delete faces and form gaps. Note that on removing a face, only the selected face will be removed. However, on removing an edge, the adjacent face will also be removed and on removing a vertex, all the faces shared by the vertex will also be removed.

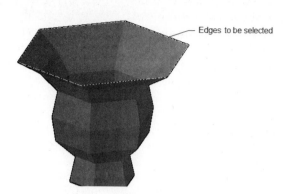

Edges to be selected

Figure 30-35 *Edges connecting the hole*

Figure 30-36 *Cap created*

Collapsing the Mesh Vertices

Ribbon: Mesh > Mesh Edit > Collapse Face or Edge   **Command:** MESHCOLLAPSE

The **Collapse Face or Edge** tool is used to converge the vertices of surrounding mesh faces to the center of the selected edge or face. To do so, invoke this tool from the **Mesh Edit** panel; you will be prompted to select the mesh face or edge to collapse. Select a mesh face or an edge; the vertices of the selected mesh face will be merged. Also, the adjacent faces will be updated based on the removed vertices. Figure 30-37 shows the model after collapsing the top face of the model shown in Figure 30-36.

Spinning the Edges of Triangular Faces

Ribbon: Mesh > Mesh Edit > Spin Triangle Face
Command: MESHSPIN

You can rotate the edge that is shared by two triangular mesh faces. To do so, invoke the **Spin Triangle Face** tool from the **Mesh Edit** panel; you will be prompted to select the triangular mesh faces adjacent to each other. Select the mesh faces; the edge shared by the two triangular mesh faces spins to connect the opposite vertices. For example, Figure 30-38 shows the model after spinning the triangular faces of the model created in Figure 30-33.

Figure 30-37 *The top face collapsed*

Sometimes, while selecting the subobjects of a mesh model, the hidden subobjects also get selected. You can control the selection of the subobjects hidden in the view by using the **Culling Object** button. This button is available in the **Selection** panel of the **Mesh** tab. When this button is active, only those subobjects that are normal to the view are selected or highlighted on rolling the mouse over them. However, if this button is not active, all subobjects including the hidden ones will be selected or highlighted when the mouse is rolled over them.

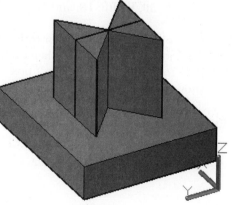

Figure 30-38 *Model after spinning*

EXAMPLE 1   *Mesh Edit Tools*

In this example, you will create a flash drive, as shown in Figure 30-39, with dimensions = 50 X 18 X 8.

Figure 30-39   *Flash drive*

Creating a Mesh Box

To create a mesh box, first you need to specify the parameters of the tessellation divisions on the mesh box.

1. Start a new file with the **Acad3d.dwt** template in the **3D Modeling** workspace.

2. Invoke the **Mesh Primitive Options** dialog box by clicking on the inclined arrow displayed at the lower right corner of the **Primitives** panel in the **Mesh** tab.

3. Select the **Box** option from the **Mesh Primitive** list box and set the following tessellation divisions: Length: **7**, Width: **7**, and Height: **7**. Choose the **OK** button from the **Mesh Primitive Options** dialog box to accept the settings and close the dialog box.

4. Choose the **Mesh Box** tool from **Mesh > Primitives > Mesh Primitives** drop-down; you are prompted to specify the start point.

5. Turn on the Ortho mode and specify the start point of the mesh box by clicking at the desired location in the drawing area; the **Specify other corner or [Cube/Length]** prompt is displayed.

6. Enter **L** at the command prompt and specify the following dimensions to create a mesh box:

   Length: **50**, Width: **18**, and Height: **8**

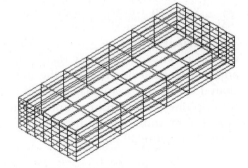

7. Click on the **Visual Style Controls** on the top left corner of drawing area and then change the visual style to Wireframe. The mesh box is created, as shown in Figure 30-40.

Figure 30-40   *The mesh box created*

Extruding the Selected Faces

Now, you need to extrude the inner faces at the left of the mesh box to create the port of the flash drive.

1. Press CTRL and select the inner faces at the left of the mesh box, as shown in Figure 30-41.

2. Choose the **Extrude Face** tool from the **Mesh Edit** panel of the **Mesh** tab; you are prompted to specify the height of extrusion.

3. Enter **8** as the value of extrusion in the **Dynamic Input**; the selected faces are extruded.

   The final extruded model after changing the visual style is shown in Figure 30-42.

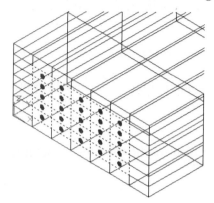

Figure 30-41 Faces selected to extrude   *Figure 30-42 The model after extruding*

Creating a Cut in the Port

To create a cut in the port, you need to extrude the inner faces of the port in negative direction.

1. Choose the **Extrude Face** tool from the **Mesh Edit** panel and then select the inner faces on the left of the port to create a cut, as shown in Figure 30-43.

2. Enter **-5** in the edit box displayed in the drawing area to specify the height of extrusion. Note that the negative value will extrude the inner faces inward, thereby creating a cut.

   Next, you need to create a cut on the top of the port. But before creating the cut, you need to split the top mesh faces to the size of the cut.

3. Choose the **Split Face** tool from the **Mesh Edit** panel and select the second mesh face on the top of the port; you are prompted to select the first split point on the selected face.

4. Specify the first split point on the mesh face at a distance of 2 units from the outer end by using the **From** snap. Then, specify the second split point using the **Perpendicular** snap; one split line is created on it.

5. Create the other split line at a distance of 2 units from the first split line, refer to Figure 30-44.

6. Create another split on the fourth face at the top of the port, refer to Figure 30-44.

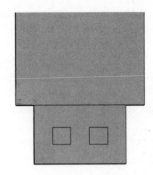

Figure 30-43 *Mesh faces selected to create a cut*   **Figure 30-44** *Model after splitting faces*

7. Next, you need to create a cut at the top. Select the new mesh faces created by splitting and extrude them downward upto the inner face so that the model looks like Figure 30-45.

Creating a Loop

Next, you need to create a loop of torus shape which can be used for hanging the flash drive.

1. Invoke the **Mesh Primitive Options** dialog box and change the tessellation division of torus as follows:

    Radius: **30**   Sweep path: **30**

2. To create a loop, you need to create a torus of radius 3 and tube radius 0.5.

3. Place the torus at the end of the flash drive, as shown in Figure 30-46.

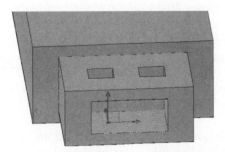

Figure 30-45 *Mesh faces selected to create a cut*   **Figure 30-46** *Model after adding a loop*

Merging Faces

Next, you need to merge faces and then create a cut on it to place the text.

1. Choose the **Merge Face** tool from the **Mesh Edit** panel of the **Mesh** tab; you are prompted to select the adjacent mesh faces on the model.

2. Select the inner faces of the mesh, as shown in Figure 30-47 and press ENTER; the selected faces are merged.

3. Select the merged face and extrude it to 0.5 unit along negative direction to create a cut, refer to Figure 30-48.

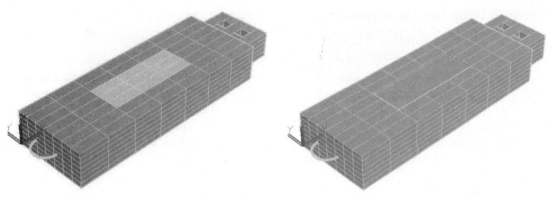

Figure 30-47 *Faces selected to be merged*   **Figure 30-48** *Model after extruding the merged face*

4. Choose **Smooth More** from the **Mesh** panel of the **Mesh** tab; you are prompted to select the objects to be smoothened.

5. Select the mesh object and press ENTER; the entire model gets smoothened.

   As the port and the cut created cannot be smoothened, therefore, you need to sharpen them.

6. Choose the **Add Crease** tool from the **Mesh** panel; you are prompted to select the mesh subobjects to add crease.

7. Select the mesh objects, as shown in Figure 30-49 and press ENTER; you are prompted to specify the crease value. Press ENTER to retain the sharpness of the selected mesh object; the crease is applied to the selected edges, refer to Figure 30-50.

8. Similarly, select the subobjects to be sharpened such that the final model created looks similar to Figure 30-50.

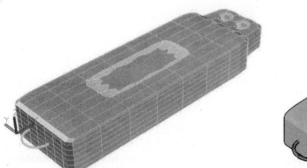

Figure 30-49 *Mesh faces selected to add crease*

Figure 30-50 *The final model after adding crease to the required entities*

CONVERTING MESH OBJECTS

You have already learned that 3D solids and surfaces can be converted into mesh objects using the **MESHSMOOTH** command so that they can be smoothened, refined, creased, and so on. Now, you will learn how to convert mesh objects into 3D solids or surfaces so that you can perform the solid modeling techniques such as union, subtraction, intersection, on them. Also, you can specify the settings such that the mesh objects converted into 3D solids or surfaces are smooth or faceted, or have merged faces.

Converting Mesh Objects into Solids

Ribbon: Mesh > Convert Mesh > Convert to Solid		**Command:** CONVTOSOLID
Menu Bar: Modify > 3D Operations > Convert to Solid		

 To convert a mesh object into solid, choose the **Convert to Solid** tool from the **Convert Mesh** panel; you will be prompted to select a mesh object. Select the required mesh object and press ENTER; the selected mesh object will be converted into a solid with default settings. To convert a mesh object into a solid with the user-defined settings, you need to specify the new settings before the conversion. You can also specify whether the converted solid will be smooth or faceted, or have merged faces.

Note

Other than the meshes, the closed polygons and circles with thickness, as well as the 3D surfaces can also be converted into solids. However, the objects like exploded solids, separate and continuous edges forming a surface, separate and continuous surfaces forming volumes, meshes with gaps between faces, and meshes with intersecting boundaries cannot be converted into solids.

To specify pre-defined settings for the solids that will be created from meshes, choose **Smooth, optimized** from the **Smooth Mesh Convert** drop-down, see Figure 30-51. Choose the desired pre-defined option from this drop-down. The different options in this drop-down are discussed next.

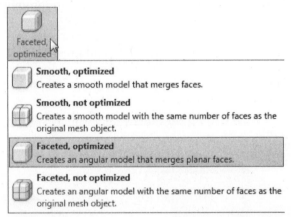

*Figure 30-51 The **Smooth Mesh Convert** drop-down*

Smooth, optimized

On choosing this option, all tessellations of the mesh in the same plane will be merged into a single face in the solid. Also, the edges of the solid model will be rounded. Figure 30-52 shows a mesh box and Figure 30-53 shows the solid created from the mesh box on choosing the **Smooth, optimized** option.

Smooth, not optimized

On choosing this option, the edges of the mesh will be rounded when the mesh is converted to solid, but the resulting faces will not be merged, and will remain the same as in the mesh object, as shown in Figure 30-54.

Faceted, optimized

On choosing this option, all tessellations of the mesh in the same plane will be merged into a single face in the solid. Also, the edges in the resulting solid will not be rounded, but creased, as shown in Figure 30-55.

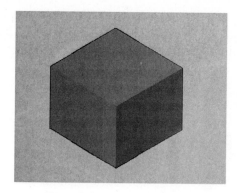

Figure 30-52 *A mesh box*

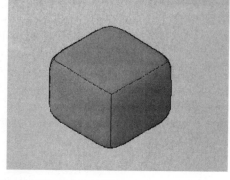

Figure 30-53 *The solid created on choosing the* **Smooth, optimized** *option*

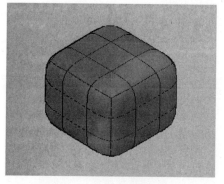

Figure 30-54 *The solid created on choosing the* **Smooth, not optimized** *option*

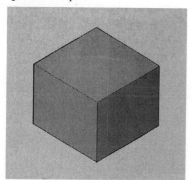

Figure 30-55 *The solid created on choosing the* **Faceted, optimized** *option*

Faceted, not optimized

On choosing this option, the faces of the converted solid will not be merged but the edges will be rounded, as shown in Figure 30-56.

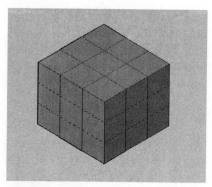

Figure 30-56 *The solid created on choosing the* **Faceted, not optimized** *option*

Converting Mesh Objects into Surfaces

| **Ribbon:** Mesh > Convert Mesh > Convert to Surface | **Command:** CONVTOSURFACE |
| **Menu Bar:** Modify > 3D Operations > Convert to Surface | |

 Like converting a mesh object into solid, you can also convert a mesh object into a surface. To do so, choose the **Convert to Surface** tool from the **Convert Mesh** panel; you will be prompted to select a mesh object. Select the required mesh object; the mesh object will be converted into a surface with default settings. You can change the default settings of the surface created as you did with solids. Figure 30-57 shows a mesh object and Figures 30-58 through 30-61 show the surfaces created on applying different settings on this object.

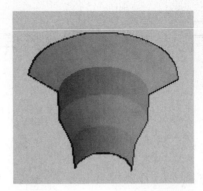

Figure 30-57 A mesh object

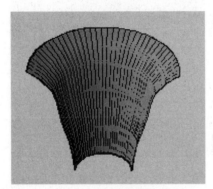

*Figure 30-58 The surface created on choosing the **Smooth, optimized** option*

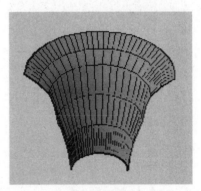

*Figure 30-59 The surface created on choosing the **Smooth, not optimized** option*

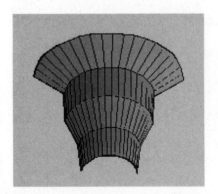

*Figure 30-60 The surface created on choosing the **Faceted, optimized** option*

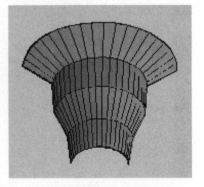

*Figure 30-61 The surface created on choosing the **Faceted, not optimized** option*

Note
Other than meshes, regions, lines and arcs with some thickness, the planar 3D faces, and 2D solids can also be converted into surfaces. Refer to Chapter 27 for details.

Tip
*You can create section views of the meshed models as you did for solid models in Chapter 26. The tools for creating section views are available in the **Section** panel of the **Home** tab or the **Mesh** tab.*

WORKING WITH GIZMOS

Ribbon: Mesh > Selection > Gizmo drop-down

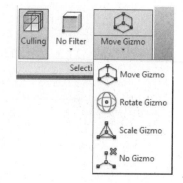

Gizmo is a tool that helps you move, rotate, or scale 3D objects or their sub-objects along a specified axis or plane. When you select objects or subobjects in any 3D visual style, a gizmo is displayed. A gizmo is of three types: move, rotate, and scale, as shown in Figure 30-62. To set the gizmo that you want to be displayed by default, choose the required gizmo from the **Selection** panel. Also, if you do not want any gizmo to be displayed, specify it in the **Selection** panel. Various types of gizmos are discussed next.

*Figure 30-62 The **Gizmo** drop-down*

Move Gizmo

When **Move Gizmo** is chosen from the **Selection** panel, the move gizmo is displayed on the first selected object or sub-object, see Figure 30-63. The move gizmo tool has three axis handles, three planes, and a base point. Using this gizmo, you can move the selected faces or edges of objects along any axis or plane, and thus, create freeform designs as per your requirement.

To move the selected object along any particular axis, click on the axis handle of the gizmo and drag along the axis; the selected object will move accordingly. To move the selected object anywhere along a plane, click on the rectangle formed in between the axes handles; the object will move along the specified plane. To relocate the base point of the gizmo, right-click on the gizmo, and then choose **Relocate Gizmo** from the shortcut menu displayed. You can also switch between various gizmos using this shortcut menu.

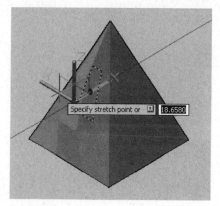

Figure 30-63 The move gizmo on the mesh object

Rotate Gizmo

The rotate gizmo is displayed on the mesh object when **Rotate Gizmo** is chosen from the **Selection** panel. The rotate gizmo is displayed as a sphere with three axes and a base point on it. Figure 30-64 shows the rotate gizmo on the mesh model. The **Rotate Gizmo** tool is used to rotate the selected objects about a particular axis. By default, the gizmo is displayed at the center of the selected object. As you move the cursor over the rotate gizmo, the nearest axis of the gizmo will be highlighted. Click when the required axis gets highlighted and specify the rotation angle. You can relocate the gizmo as well as switch between various gizmos using the shortcut menu that is displayed on right-clicking on the gizmo.

Scale Gizmo

The scale gizmo is displayed when **Scale Gizmo** is chosen from the **Selection** panel. This tool is used to scale the selected objects uniformly along an axis or a plane. As you move the cursor toward the gizmo, the axis of scale will be displayed. Click and drag the axis to scale the object in that direction. To scale the object uniformly, click on the triangle that is formed between the axes. Figure 30-65 shows the scale gizmo on the mesh model.

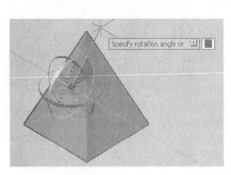

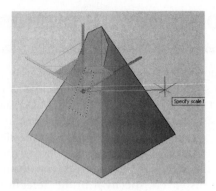

Figure 30-64 *The rotate gizmo on the mesh object*

Figure 30-65 *The scale gizmo on the mesh object*

To summarize, in this section, you learned that all the three gizmos can be used to create freeform designs from primitive meshes. The next example illustrates the use of these gizmos.

Note
*You can choose the **No Gizmo** option from the **Selection** panel, if you do not want any gizmo to be displayed on selecting the objects in the 3D visual style.*

EXAMPLE 2   *Gizmos*

In this example, first you will create an elliptical mesh cylinder with the following dimensions: major axis = 55, minor axis = 15, and height = 100. Then, you will modify it using various gizmos. The final model should look similar to Figure 30-66.

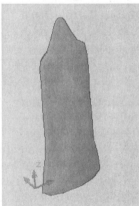

Figure 30-66 *The final model of an elliptical mesh cylinder*

Creating an Elliptical Cylinder

First, you need to start a new drawing file, specify the settings for the number of faces in the cylinder, and then create the cylinder.

1. Start a new file with the **Acad3D.dwt** template in the **3D Modeling** workspace.

2. Choose the inclined down-arrow displayed at the lower right corner of the **Primitives** panel of the **Mesh** tab; the **Mesh Primitive Options** dialog box is displayed.

3. In this dialog box, select **Cylinder** from the **Mesh Primitives** area and set the following parameters in the **Tessellation Divisions** rollout:

Axis: **10**   Height: **5**   Base: **3**

4. Choose **OK** to accept the settings specified in the rollout and exit the dialog box.

5. Turn on the **Ortho Mode** and invoke the **Mesh Cylinder** tool from **Mesh > Primitives > Primitives** drop-down. Then, follow the Command sequence given next to create the mesh cylinder.

   Command: _mesh
   Current smoothness level is set to: 0
   Enter an option [Box/Cone/CYlinder/Pyramid/Sphere/Wedge/Torus/SEttings]
   <Box>: _cylinder
   Specify center point of base or [3P/2P/Ttr/Elliptical]: *Enter **E** to create elliptical base*
   Specify endpoint of first axis or [Center]: *Enter **C** to specify the center of base*
   Specify center point: *Specify the center point for the base of the cylinder*
   Specify distance to first axis: *Drag the cursor along X axis and enter **55** in the Dynamic Input to specify a point on the first axis*
   Specify endpoint of second axis: *Drag the cursor along Y axis and enter **15** in the Dynamic Input to specify a point on the second axis*
   Specify height or [2Point/Axis endpoint]: *Enter **100** to specify the height of the cylinder*

6. Change the view to **SW Isometric** and the visual style to **Wireframe**. Figure 30-67 shows the mesh model.

Moving the Edges

To move the edges, you first need to select the Edge selection filter and then move them.

1. To filter the selection, choose the **Edge** option from **Mesh > Selection > Subobject Selection Filter** drop-down.

2. Change the visual style to **Realistic** from the **In-Canvas Visual Style Controls** at the top-right of the drawing area.

3. Press CTRL and then select the required edge, as shown in Figure 30-68; the move gizmo is displayed by default.

Note
*If the move gizmo is not displayed by default, choose the **Move Gizmo** option from the **Default Gizmo** drop-down list in the **Selection** panel.*

4. Click on the X handle of the move gizmo such that the base of the cylinder is elongated; the preview of the moved edge is displayed. Next, click on the screen at the required distance (here, it is **20**). Similarly, move the edge on the other side to the same distance.

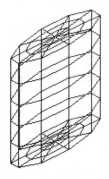

Edge selected

Figure 30-67 *The mesh model created*   ***Figure 30-68*** *Edge selected to be moved*

5. Now, you need to smoothen the mesh model. To do so, select it; the **Quick Properties** panel is displayed (if it does not appear, choose the **Quick Properties** button from the Status Bar). In this panel, select **Level 1** from the **Smoothness** drop-down list.

Rotating the Edges

You need to rotate the edges. Make sure the **Edge** option is selected in the **Subobject Selection Filter** drop-down list.

1. To invoke the rotate gizmo, choose the **Rotate Gizmo** option from **Mesh > Selection > Gizmo** drop-down.

2. Select the required edges from the mesh model, as shown in Figure 30-69; the rotate gizmo is displayed on the first selected edge.

3. Click on the red handle of the rotate gizmo such that the adjoining faces are rotated inward; the preview of rotation is displayed. Click when the desired rotation is achieved (here, it is 60°).

   Similarly, select the third set of edges in the same face and rotate the face. The final model, after rotating the edges, is shown in Figure 30-70.

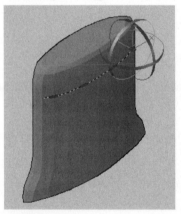 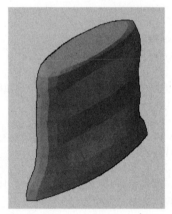

Figure 30-69 *Edges selected to be rotated*   *Figure 30-70* *Final model after rotating the edges*

Moving the Inner Faces of the Model

Now, you need to change the selection filter from edges to faces and then move faces.

1. To select faces, choose **Face** from **Mesh > Selection > Subobject Selection Filter** drop-down.

2. Choose the **Move Gizmo** tool from **Mesh > Selection > Gizmo** drop-down.

3. Select the faces in the inner circular region at the top of the mesh model; the move gizmo is displayed on the selected faces.

4. Click on the Z handle of the gizmo and drag it; the selected faces are elongated. Click when the desired height is achieved.

Scaling the Entire Model

Now, you can elongate the model using the scale gizmo.

1. Choose the **Scale Gizmo** tool from **Mesh > Selection > Gizmo** drop-down.

2. Select the model; the scale gizmo is displayed, as shown in Figure 30-71.

3. Click on the Z handle; the model gets elongated in the Z-direction. Click in the drawing area at the required distance. The final model is created, as shown in Figure 30-72.

Figure 30-71 *The scale gizmo displayed*

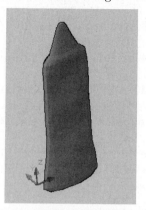

Figure 30-72 *The final model*

EXAMPLE 3   *Mesh Tools*

In this example, you will create a toy plane (see Figure 30-73) using various mesh tools.

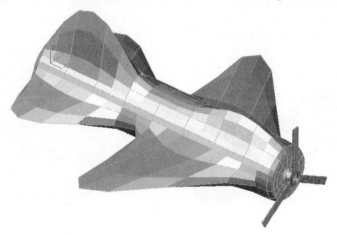

Figure 30-73 *Final model of the toy plane*

Creating the Cylinder

First, you need to create a cylinder facing the right view.

1. Start a new file by choosing the **Acad3d.dwt** template in the **3D Modeling** workspace.

2. Set UCS to WCS and switch to the **Right** viewport. Rotate the WCS around the Y-axis by -90°.

3. Select the **Wireframe** view style from the **In-Canvas View Controls**.

4. Invoke the **Mesh Primitive Options** dialog box and then choose **Cylinder** from the **Mesh Primitive** list.

5. Enter the following data in the **Tessellation Divisions** rollout:

   Axis = **16**   Height = **8**   Base = **5**

6. Choose **OK** to exit the dialog box.

7. Choose the **Mesh Cylinder** tool from **Mesh > Primitives > Mesh Primitives** drop-down; you are prompted to specify a centerpoint for the base of the cylinder.

8. Enter **0**; you are prompted to specify the base radius.

9. Enter **4** as radius; you are prompted to specify the height of the cylinder.

10. Enter **40** as the height; the cylinder is created.

    Now, you need to change the **UCS** to **WCS** and change the view to **Isometric**.

11. Choose the arrow besides the new UCS; a flyout is displayed. Choose **WCS** to change the view to WCS and click on the SW-Isometric hotspot, the mesh cylinder is created, as shown in Figure 30-74.

Extruding the Faces of the Cylinder

Now, you need to extrude the faces of the cylinder.

1. Select the inner faces on the right side of the cylinder, as shown in Figure 30-75.

2. Click on the red handle of the move gizmo to extrude; an edit box is displayed.

3. Enter **2** in the edit box; the faces are extruded.

Figure 30-74 Mesh cylinder created

Figure 30-75 Faces selected to be moved

Creating the blades

To create blades, you need to create a thin and long mesh box. Before creating the mesh box, you need to change the UCS to Right.

1. Set UCS to WCS and switch to the **Right** viewport. Rotate the WCS around the Y-axis by -90°.

2. Choose the **Mesh Box** tool from **Mesh > Primitives > Mesh Primitives** drop-down; you are prompted to specify the first point of the mesh box.

3. Specify the point anywhere in the drawing, but not on the mesh model and create a mesh box of dimensions = 1 x 7 x 0.1.

4. Move the blades to the point on the circumference of extruded face of the cylinder, see Figure 30-76.

5. Now, you need to create a polar array of the blade. To do so, select the blade and invoke the **Array** tool; you are prompted to select the array type.

6. Choose the **POlar** option from the dynamic input box and specify the center point of the cylinder as the center point of the array. You can also enter 0,0 as the coordinates of the centerpoint.

7. Enter **3** in the number of items to fill; you are now prompted to specify the angle to fill.

8. Press ENTER to accept the default angle which is 360°; the preview of the array is displayed and you are prompted to accept or modify the settings.

The model after creating the blades is shown in Figure 30-76.

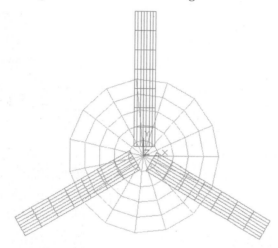

Figure 30-76 *Model created by combining various features*

Creating a Torus

You need to create a torus around the blades.

1. Choose the **Mesh Torus** tool from **Mesh > Primitives > Mesh Primitives** drop-down; you are prompted to specify a center point for the torus.

2. Specify the center point of the cylinder as the centerpoint of torus and create the torus covering the start points of the blades.

The model after creating the torus is shown in Figure 30-77.

Using Transform Gizmos and Stretching the Edge to Create a Plane

Now, you need to use various gizmos to get the shape of the plane. Before that set UCS to WCS and then switch to the Right viewport from the Viewport cube.

1. Select the **Edge** option from the **Subobject Selection Filter** drop-down list and the **Move Gizmo** from the **Default Gizmo** drop-down list in the **Selection** panel.

2. Select the edge that is fourth on the right and third from the starting point of the mesh cylinder, as shown in Figure 30-78, to transform.

3. Now, click on the green handle; an edit box is displayed. Enter **20** in it.

4. Next, click on the red handle, move the handle backward and enter **10** in the edit box displayed.

   The model after stretching the edge on the right is shown in Figure 30-79.

5. Similarly, stretch the edge on the other side of the model. The top view of the model after stretching edges at both sides is shown in Figure 30-80.

   Now, you need to stretch the edges on the bottom right side of the model.

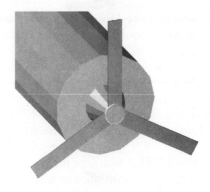

Figure 30-77 *Model after creating the torus*

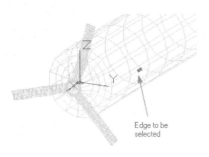

Figure 30-78 *Model created by combining various features*

Figure 30-79 *Model after stretching the edge*

6. Select the edge that is second from the bottom right end, as shown in Figure 30-81. For easy identification of the edge, you can use the **Back** viewport.

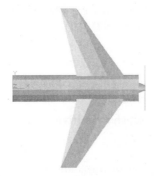

Figure 30-80 *Model after stretching the edge on the other side*

Figure 30-81 *Edge selected*

7. Drag the green handle and enter **10** in the edit box displayed.

8. Drag the red handle backward and enter **5** in the edit box displayed.

9. Similarly, modify the edge on the other side. The model after stretching the edges on both sides is shown in Figure 30-82.

   Now, you will create a bulge at the top of the toy plane.

Figure 30-82 *Model after stretching all the edges*

10. Select the second edge on the top of the plane, as shown in Figure 30-83.

11. Drag the blue handle and specify 2 as a distance of bulge.

    Now, you need to create a bulge on the top at the end of the model.

12. Select the edge second from the end and drag it toward the top to create the bulge, as shown in Figure 30-84.

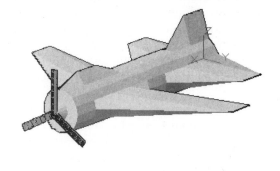

Figure 30-83 *Selecting an edge to create a bulge*

Figure 30-84 *Model after creating both bulges*

Smoothening the Mesh

1. Select the mesh model and choose the **Smooth More** tool from the **Mesh** panel of the **Mesh** tab.

    The final model of the toy plane after smoothening is shown in Figure 30-85.

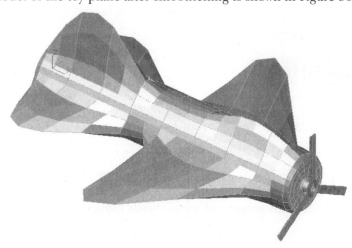

Figure 30-85 *Final model of the toy plane*

Self-Evaluation Test

Answer the following questions and then compare them to those given at the end of this chapter:

1. A mesh model is similar to a solid model without mass and volume properties. (T/F)

2. You can specify values for the number of faces in the primitive mesh in the _____ dialog box.

3. The _____ command is used to create mesh primitives.

4. The _____ command is used to create a mesh by adjoining four continuous edges or curves.

5. Which of the following objects can be converted into mesh objects?

   (a) 3D objects   (b) Closed Polygons
   (c) Regions   (d) All the above

6. Which of the following options is used to convert 3D objects into meshed objects?

   (a) **Refine Mesh**   (b) **Smooth More**
   (c) **Smooth Object**   (d) None of these

7. You can invoke the _____ tool to sharpen faces, edges, or vertices of mesh objects.

8. Which of the following options is used to specify the default settings for the solid created from the meshed object?

   (a) **Smooth, optimized**   (b) **Smooth, not optimized**
   (c) **Faceted, optimized**   (d) All the above

9. The mesh editing tools are available in the _____ panel of the **Mesh** tab of the **Ribbon**.

10. To retain the sharpness of the crease even when you smoothen or refine the mesh, you need to set the crease to _____.

Review Questions

Answer the following questions:

1. The _____ and _____ commands are used to set the wireframe density of primitives.

2. In smoothening, the complexity of a model is decreased. (T/F)

3. You can set the level of smoothness of mesh objects using the **Properties** palette. (T/F)

4. To create mesh cones with frustum, you need to choose the _____ option of the **Mesh Cone** tool.

5. The curves used to create ruled surfaces can be lines, arcs, polylines, curves, and splines. (T/F)

6. The **Mesh Tessellation Options** dialog box will be displayed on entering _____ at the Command prompt.

7. The **Refine Mesh** tool works only when the smoothening of the mesh object is set to **Level 1** or higher. (T/F)

8. Refining an object resets the smoothening level to 0, whereas refining a face does not reset the smoothing level. (T/F)

9. To remove creases from the model, you need to enter _____ at the Command prompt.

10. Which of the following is used to relocate a gizmo?

  (a) Mouse   (b) shortcut menu
  (c) **Ribbon**   (d) All the above

EXERCISE 2   *Mesh Surface*

Create the 3D model of a globe shown in Figure 30-86 arbitrarily, using the mesh modeling tools.

Figure 30-86 *Model of a globe*

EXERCISE 3   *Mesh Surface*

Create the drawing shown in Figure 30-87. First, create the base unit (transition) and then the dish and cone.

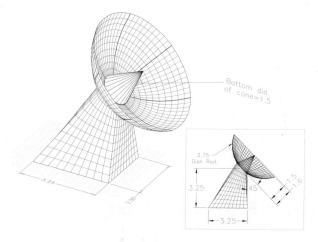

Figure 30-87 *Drawing for Exercise 3*

EXERCISE 4   *Mesh Surface*

Create the drawing shown in Figure 30-88. First, create one of the segments of the base and then use the **ARRAY** command to complete it.

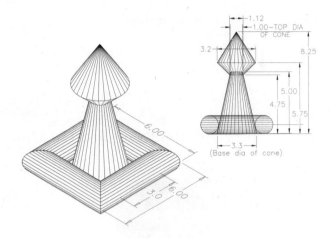

Figure 30-88 *Drawing for Exercise 4*

Answers to Self-Evaluation Test

1. T, **2. Mesh Primitive Options, 3. MESH, 4. EDGESURF,** 5. d, 6. c, **7. Add Crease,** 8. d, **9. Mesh Edit, 10. Always**

Chapter **31**

Rendering and Animating Designs

CHAPTER OBJECTIVES

In this chapter, you will learn:
- *To render objects and about the need to render them.*
- *To browse and manage material libraries.*
- *To assign materials to the drawing objects.*
- *To create new materials as well as edit existing materials.*
- *To map materials on objects.*
- *To convert old materials into new materials.*
- *To insert and modify light sources.*
- *To convert old lights into new lights.*
- *To select rendering types and render objects.*
- *To define and modify rendering settings.*
- *To replay and print renderings.*
- *To define and apply background and fog.*
- *To save the rendering objects.*
- *To configure, load, and unload AutoCAD Render.*
- *To work with cameras.*
- *To create animation*
- *To create animations by defining the motion path.*

KEY TERMS

- *Materials Browser*
- *Libraries*
- *Attach By Layer*
- *Materials Editor*
- *RENDER*
- *Material types*

- *Material properties*
- *Material Mapping*
- *Texture Editor*
- *Default Light*
- *Point Light*
- *Spot Light*

- *Distant Light*
- *Web Light*
- *Sun Light*
- *Set Location*
- *Render Presets*
- *FOG*

- *History pane*
- *Shade plot*
- *CAMERA*
- *Set Camera View*
- *Animation Settings*
- *Motion Path*

UNDERSTANDING THE CONCEPT OF RENDERING

A rendered image makes it easier to visualize the shape and size of a 3D object as compared to a wireframe image or a shaded image. A rendered object also makes it easier to express your design ideas to other people. For example, to make a presentation of your project or design, you do not need to build a prototype. You can use a rendered image to explain your design more clearly because you have a complete control over the shape, size, color, and surface material and lighting of the rendered image. Additionally, any required changes can be incorporated into the object, and it can be rendered to check or demonstrate the effect of these changes. Thus, rendering is a very effective tool for communicating ideas and demonstrating the shape of an object.

Generally, the process of adding a realistic effect to models (rendering) can be divided into the following four steps:

1. Selecting the material to be assigned to the design.
2. Assigning materials and textures to the design.
3. Applying additional effects to the model by assigning lights at different locations.
4. Using the **Render** tool to render the objects.

ASSIGNING MATERIALS*

After creating a model, you need to assign various types of materials to the objects in it to give it a realistic appearance. These materials create three types of effects. These effects are discussed next.

In the first type of material effect, the object is rendered without actually displaying the effect of the material applied. A global material is assigned to all objects. On rendering, the color of the rendered image depends on the color of the object. Rendering by using this type of material effect does not give a realistic effect to the model, see Figure 31-1.

In the second type of material effect, a material is assigned to the object. AutoCAD provides you with a number of predefined materials and material libraries from which you can select a material and apply to the model. The material libraries contain thousands of materials and you can browse them using the **Material Browser**. The **Material Browser** is discussed next. Once you have selected the material, you can assign it directly to the object, individual faces, blocks, or layers, and render the model. Figure 31-2 shows a model to which a material has been applied and then rendered.

Figure 31-1 *Model rendered after applying the default global material*

Figure 31-2 *Model rendered after applying the material*

The third type of material effect is photorealistic. In this type, additional parameters such as the mirror effect, reflection, and refraction can be defined for the material, see Figure 31-3. You can set these parameters in the **Material Editor**.

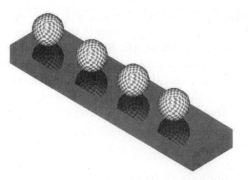

Figure 31-3 *Model rendered by adding effects such as mirror*

Material Browser

Ribbon: Render > Materials > Materials Browser	**Toolbar:** Render
Menu Bar: View > Render > Materials Browser	**Command:** MAT

You can browse materials, manage material libraries, and apply the selected material to the drawing objects using the **Materials Browser**. You can download the Autodesk material library available on the internet and install it, whenever required. When you choose the library to install, the library automatically gets installed in *Program Files\Common Files\Autodesk Shared\ Materials\2012* folder. Also, the materials get automatically arranged into sub-folders. When you invoke the **3D Modeling** workspace, the **Material Browser** is displayed, by default. If it is not displayed, you can invoke it by choosing the **Material Browser** button in the **Materials** panel. Figure 31-4 shows the **Material Browser**. For better utilization of the material library, materials are arranged in nested categories. The different areas of the **Material Browser** are discussed next.

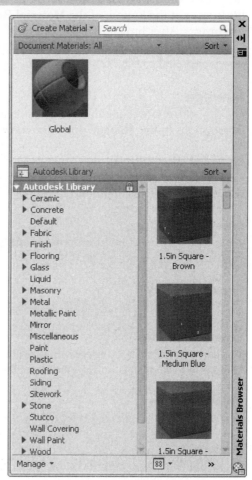

Figure 31-4 *The **Material Browser***

Search

You can search a material from all the available material libraries by entering its name or keyword in the **Search** edit box.

Create Material

You can create your own materials using the **Create Material** option available at the top of the browser. When you select this option, a drop-down is displayed with a list of material types. You can select a material type from the drop-down. On doing so, the **Material Editor** is displayed with the predefined settings of the selected type. Using the **Material Editor**, you can change the settings of the material to modify the material as per your requirement. The **Material Editor** is discussed later in this chapter.

Document Materials

The **Document Materials** drop-down acts as a filter to the materials used in the current drawing. The options in this drop-down are **Show All**, **Show Applied**, **Show Selected**, **Show Unused**, and **Purge All Unused**. You can filter materials display in the **Material Browser** to display either all materials in the browser, only the materials applied to the drawing, only the materials of the selected entities, or only the materials that are not used in the drawing by selecting the respective option from the **Document Materials** drop-down. You can also purge or delete the unused materials in the drawing.

Libraries

In this area, the Autodesk library with its pre-defined materials as well as other libraries with their user-defined materials are displayed. You can sort these materials by name, category, type, or material color using the **Sort** drop-down. On the right of the library list is the **Preview** area. In the **Preview** area, the preview swatches of the materials for the selected category/library are displayed. You can set the size of the preview swatches using the **Swatch** slider at the bottom of the **Material Browser**. You can also control the display of the preview using the **View** drop-down list located at the bottom of the **Material Browser**. You can set the display of swatches to icons, list, or text. After browsing to the required material, select the required material by clicking on it; the selected material will be added to the **Document Materials** list area. This material can now be applied to the drawing objects.

Manage

You can add, remove, or edit a library, or create or delete a category, using the **Manage** drop-down. The options in the **Manage** drop-down are discussed next.

Open Existing Library

On selecting this option, the **Add Library** dialog box will be displayed. Using this dialog box, you can open an existing library.

Create New Library

On selecting this option, the **Create Library** dialog box will be displayed. Using this dialog box, you can save the newly created library.

Remove Library

You can remove the selected library using this option.

Create Category

You can create a new category of materials in the library by using the **Create Category** option. You can add materials under a new category by dragging and dropping.

Delete Category

You can delete the selected category by using the **Delete Category** option.

Rename

To rename a category, double-click on it; an edit box is displayed. Enter the new name in it.

Assigning Selected Materials to Objects

Once you have selected the materials to be applied to the object, you need to assign them to the objects in the current drawing. To do so, first, invoke the **Material Browser** by choosing the **Materials Browser** button from **Materials** panel of the **Render** tab and then browse to the required material from the library. After selecting the required material, drag and drop it on the drawing object on which you want to apply the selected material. If the visual style is set to Realistic, the material will be displayed instantaneously on the objects you select. Note that if a material is already applied to an object and you apply a new material on it again, then the new material will be updated on the object automatically.

Apart from dragging and dropping, you can also assign materials to the selected entities. To assign materials to the selected entities, first select them and then click on the desired material from the browser; the material gets applied to the selected objects. Also, the material will be added to the **Document Materials** list automatically. Alternatively, select the entities, right-click on the selected material in the Browser and choose **Assign to Selection** from the shortcut menu.

If you need to reuse certain materials again and again then you can add those materials to the **My Materials** list, to the current drawing list, or to the active **Tool Palette**. To do so, right-click on the desired material; a shortcut menu will be displayed. Choose **Add to** from the shortcut menu; the **In Scene Materials**, **My Materials**, and **Active Tool Palette** sub-options are displayed. Using these sub-options, you can add the selected material to the required place.

Remove Materials from Selected Objects

To detach a material that has been applied to an object, invoke the **Remove Materials** tool from the **Materials** panel; the cursor will change into paintbrush cursor and you will be prompted to select the object from which you want to detach the material. Select the object; the material will be removed from them. By default, the global material will again be assigned to the object from which you have detached the material.

Attaching Material by Layers

Ribbon: Render > Materials > Attach By Layer   **Command:** MATERIALATTACH

You can assign a particular material to a layer by choosing the **Attach By Layer** tool from the **Materials** panel. This means that you can apply a specified material to all the objects that are placed on a layer. When you choose this tool, the **Material Attachment Options** dialog box is displayed. In this dialog box, all materials in the current drawing are placed under the **Material Name** column, the layers are placed in the **Layer** column, and the material attached to this layer under the **Material** column, refer to Figure 31-5. To assign a material to a layer, select the material from the **Material Name** column and drag it to the desired layer in the **Layer** column. All objects placed in the selected layer will now be assigned with the specified material. After attaching the material to a layer, the **Detach** button is automatically added on the right of the **Material** column. You can detach the material from the selected layer by choosing this **Detach** button.

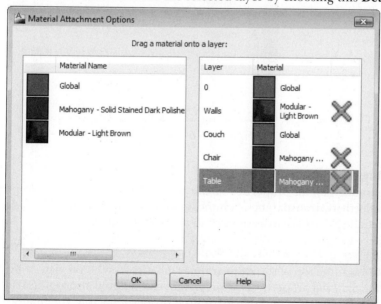

*Figure 31-5 The **Material Attachment Options** dialog box*

Creating and Editing Materials

Ribbon: Render > Materials > Materials Editor (Inclined arrow)   **Toolbar:** Render
Menu Bar: View > Render > Materials Editor

In AutoCAD, you can create a new material or an edit existing material according to your requirements. This is done using the **Materials Editor**. To edit an existing material, double-click on it in the **Materials Browser**; the **Materials Editor** will be displayed, as shown in Figure 31-6. Using the **Materials Editor**, you can change the color, adjust material properties, and add textures to the selected material. The **Materials Editor** has 2 tabs: **Appearance** and **Information** and these tabs are discussed next.

Note
*The **Material Editor** is also used to create new materials. In this case, you need not double-click on any material to invoke it. You can invoke it directly by choosing the inclined arrow on the title bar of the **Materials** panel of the **Render** tab. The procedure to create new materials is discussed later.*

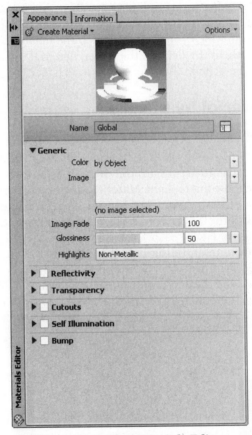

Figure 31-6 The Materials Editor

Appearance Tab
In the **Appearance** tab, all properties of the material are defined. These properties include glossiness, reflectivity, transparency, luminosity and so on. You can also add textures to the material, if required. Various areas in this tab are discussed next.

Preview
The preview area at the top of the **Appearances** tab displays the preview of the material. When you modify the material properties, the preview of the material also gets updated automatically. This helps you preview the materials with desired properties before you finally set the properties of the material and apply them to the model.

You can also select the shape of the swatch and render quality of the preview for better visualization of the material. To do so, choose the down arrow available in the lower right corner of the preview area; a list of preview types and render qualities is displayed. The preview types include **Sphere**, **Cube**, **Cylinder**, **Canvas**, **Plane**, **Object**, **Vase**, **Draped Fabric**, **Glass Curtain Wall**, **Walls**, **Pool of liquid**, and **Utility**. Select the required preview swatch. The render quality types include **Fastest Renderer**, **mental ray-Draft Quality**, **mental ray-Medium Quality**, or **mental ray-Production Quality**. Select the required render quality of the preview of the material. Note that higher the quality, slower the process of updating the materials.

Name

You can specify a name to an existing/new material by entering a new name in the **Name** edit box. On entering a new name; the name will be updated in the **Material Browser** also. In the **Description** edit box, enter the material category. For easy indexing in the Search option, enter the keywords in the Keywords edit box.

About

This area displays the type and version of material.

Texture Paths

This area displays the path of the edited material in the selected storage media.

> **Note**
> *You cannot rename the **Global** material.*

Displays the Material Browser

This button is available at bottom right of the **Material Editor**. This is a toggle button and is used to display and hide the **Material Browser**.

Information Tab

In this tab, you can view the file properties as well as specify the description and keywords for the material created. The **Material Editor** with the **Information** tab active is shown in Figure 31-7.

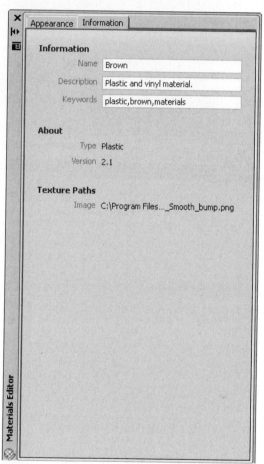

*Figure 31-7 The **Materials Editor** with the **Information** tab chosen*

The process of creating materials and adding texture maps is discussed later in this chapter.

BASIC RENDERING

Ribbon: Render > Render > Render drop-down > Render	**Toolbar:** Render
Menu Bar: View > Render > Render	**Command:** RENDER

In this section, you will learn about elementary or basic rendering. When you invoke the **Render** tool, the **Render** window will be displayed and the rendering process will be carried out. After the rendering has been completed, the rendered image will be displayed in the **Render** window, see Figure 31-8.

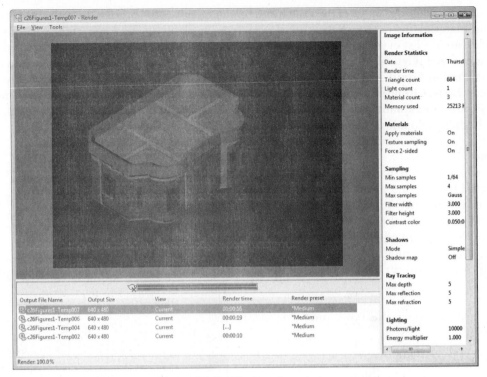

*Figure 31-8 The **Render** window*

EXAMPLE 1

In this example, you will draw the staircase shown in Figure 31-9. You can assume arbitrary dimensions. Assign materials to objects and then render the model. The materials that you have to assign to the model are given next.

Stairs	Flooring - Wood - Parquet - Brown
Balusters and Handrail	Metal Fabricated - Handrails and Railing - Painted White

1. Create the model shown in Figure 31-9 assuming arbitrary dimensions.

2. Invoke **Materials Browser** from the **Materials** panel of the **Render** tab.

3. Select the **Flooring - Wood** category from the **Autodesk Library** list; all materials in the category are displayed in the preview list. Browse to the **Parquet - Brown**, select it, and then drag and drop it on stairs; the material is applied to the stairs.

4. As explained in Step 3, select the **Metals Fabricated** category from the **Autodesk Library** list. Next, select **Handrails and Railways - Painted White** from the materials displayed and apply it to the balusters and handrails.

 Note
*You will notice that the materials selected in Step 3 and Step 4 are automatically attached to the current drawing and displayed in the **Document Materials** list in the **Materials Browser**.*

Figure 31-9 Model for Example 1

5. Choose the **Render** tool from the **Render** panel; the rendering process starts and the rendered image is displayed in the **Render** window. The rendered image should look similar to the one shown in Figure 31-10. The background of the image is black by default. However, in this figure, the background has been changed to white for the clarity in print.

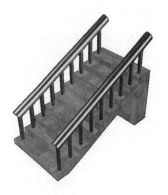

Figure 31-10 Rendered image of staircase

CREATING NEW MATERIALS

In AutoCAD, you can create new materials according to your requirement. To create a new material, click on the **Create Material** drop-down at the top of the **Material Browser**; a drop-down list with various material types will be displayed. Select the required material type; the **Materials Editor** with the properties of the sample material type will be displayed, refer to Figure 31-6.

You can create a new material directly from the **Materials Editor**. To do so, choose the **Create or duplicate Material** button available on the right of the **Name** edit box in the **Materials Editor**; a drop-down list will be displayed, see Figure 31-11. Select the required material type from the drop-down list; all material properties based on the selected material type will be grouped together. For example, if you select glass, then the properties such as reflectivity, refraction, and so on will be displayed. Whereas, if you select the **Wood** material; the properties displayed will be different. The preview of the selected material is displayed in the preview area. In the **Information** tab, you can rename the new material in the **Name** edit box. Also, you can give its description as well as specify the keywords to be used for searching it. This material will be automatically added in the

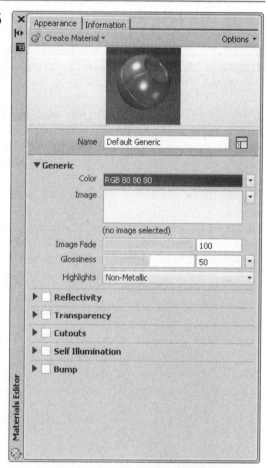

*Figure 31-11 The **Materials Editor***

Material Browser. Now, you can define the properties of the new material in the **Material Editor** by modifying the properties of the old material.

Generic

In the **Generic** material type, almost all material properties are displayed. These properties are discussed next.

Color

By default, the **Color by Object** option is selected in this area; and therefore, the color of the object is assigned to the material. To assign any other color to the material, click on the arrow displayed on the right of the **Color** option; a drop-down list will be displayed, as shown in Figure 31-12. Choose the **Color** option from the drop-down list; a color swatch will be displayed. Click on the swatch; the **Select Color** dialog box will be displayed. Select the required color from the dialog box; the selected color will be assigned to the material.

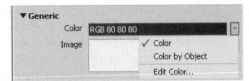

*Figure 31-12 The **Color** drop-down list*

Image

You can assign textures to the color of the material. To do so, choose the arrow on the right of the **Image** area; a drop-down list will be displayed with a list of texture maps, see Figure 31-13. There are nine types of texture maps: **Image**, **Checker**, **Gradient**, **Marble**, **Noise**, **Speckle**, **Tiles**, **Waves**, and **Wood**.

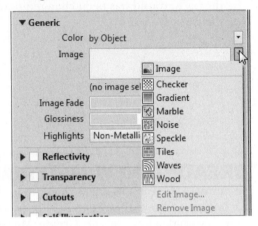

*Figure 31-13 The **Image** drop-down list*

If you select the **Image** option, the **Material Editor Open File** dialog box will be displayed, as shown in Figure 31-14. Browse to the required image and choose the **OK** button from the dialog box; the **Texture Editor** will be displayed. The **Texture Editor** displays the link for the texture image used, a slider to adjust the brightness of the image, a Transform node that enables you to set the position, scale, and repetition of the image. Similarly, you can select other texture maps such as checker or gradient and then edit the appearance properties of the image and transform it according to your requirement. If you want to change the selected image, then click in the preview area and browse for the required image map. After the image has been added, the **Edit** link is displayed below the swatch. If you click on the link, the **Texture Editor** is displayed allowing you to modify the image map settings.

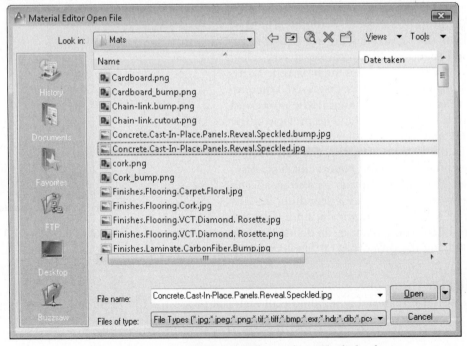

*Figure 31-14 The **Material Editor Open File** dialog box*

If you select the **Marble** option from the drop-down list, the material will be mapped to the image of the marble and the properties of the marble will be displayed in the **Texture Editor**, refer to Figure 31-15. Click on the **Stone color** swatch to change the color of the base stone and the **Vein color** swatch to change the color of the veins displayed on the base stone. Choose the **Swap Colors** option that is displayed on clicking on the **Stone/Vein** color swatch to interchange the color of the stone and vein. The **Vein spacing** and **Vein width** edit boxes are used to control the relative spacing between the veins and the relative size of the width of the vein, respectively.

If you select the **Wood** option from the drop-down list, the material will be mapped to the image of the wood. Also, the **Texture Editor** will be displayed. You can edit the image of an existing wood structure applied on the material. To do so, click in the preview swatch; the **Texture Editor** will be displayed with the properties of **Wood**, see Figure 31-16. Click on the required color buttons to change the color of the wood base and the pattern. Choose the **Swap Colors** option displayed that is displayed on right-clicking on the color swatch to interchange the wood base and pattern colors. The **Radial noise** and the **Axial noise** options are used to decide the relative occurrence of patches of the wood in the radial and axial directions, respectively. The **Grain thickness** option is used to specify the relative size of the grain patches in the material. All the values specified in this edit box are relative to the size of the object on which it will be applied.

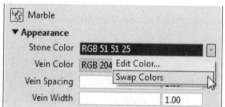

*Figure 31-15 Material properties for **Marble***

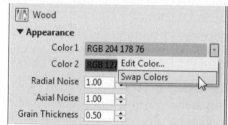

*Figure 31-16 Material properties for **Wood***

Image Fade

Use this slider bar to specify the extents of the effect of the texture map on the material. Figure 31-17 shows the Image Fade slider. At 0 value, there will be no effect of the texture map and only the material will be displayed. Also, as this value increases, the effect of the texture map becomes more dominant on the material and at the value of 100, only the texture map will be displayed on the object.

*Figure 31-17 The **Image Fade** slider*

Glossiness

This property is used to set the shininess of the texture. It is actually the ability of a surface to reflect the light. If the material is more glossy, it will reflect more light and the color of the portion of the object hit by the light will become whitish because white is the default color assigned to glossiness. However, you can assign any other color to it. You can set the shininess value as well as apply an image or texture to the gloss.

Highlight

This property is used to disperse light non-uniformly from the material. You can set metallic or non-metallic highlight for the texture. The **Metallic** option is used to highlight the color of the material and the **Non-Metallic** option is used to highlight the color of the light hitting the material.

Apart from the general properties discussed above, there are some other properties that are used to add extra effects to the material. These properties are discussed next.

Reflectivity

Reflectivity is the reflective property of the material. A hundred percent reflective material displays the surrounding objects on its surface. The **Reflectivity** area is used to specify the environment reflecting qualities of a material. Select the **Reflectivity** check box on the left of the node to apply the reflection effect. By default, sliders are displayed for both the **Direct** and **Oblique** options. To apply a reflection map to an object, click on the black arrow at the right of the **Direct/Oblique** slider; a list of options will be displayed, see Figure 31-18. Choose the required option from it. You can set an image file or some predefined textures to be reflected from the object as if the surrounding of the material is composed of this image/texture file. If you choose the **Image** option, the **Material Editor Open File** dialog box will be displayed. In this dialog box, you can select an image file to be reflected from the object. Similarly, you can select some predefined reflection maps from the drop-down list such as **Checker**, **Marble**, **Noise**, and so on. After selecting the map type, the **Texture Editor** will be displayed. Use the slider bar to set the brightness of the reflection map. The name of the texture file is displayed below the preview swatch and you can click on it to change the attached texture file. You can select or clear the **Reflectivity** check box to turn on or off its effect.

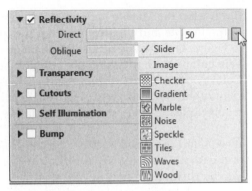

Figure 31-18 Options available in the ***Reflectivity*** *node*

Transparency

Transparency controls the opaqueness of the material. The material with no transparency is opaque. The **Transparency** area is used to define an imaginary transparency or opacity effect of the material. Select the **Transparency** check box to apply the opacity map settings on the object. You can specify the extent of transparency on the material by using the **Amount** slider. At 0 value, the material appears to be totally opaque and as this value increases, the transparency of the object also increases. You can set the image or texture for the transparency and adjust its color as discussed earlier. You can turn on or off the transparency effect by selecting or clearing this check box.

Translucency controls the amount of light transmitted and scattered within an object. At 0 translucency, all the light will be transmitted through the material without scattering. As the value of translucency increases, the

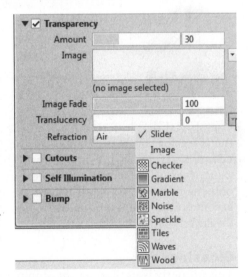

Figure 31-19 The opacity map settings available for the ***Translucency*** *option in the* ***Transparency*** *node*

amount of light scattered within the material also increases. You can set the image or texture for the transparency (Figure 31-19) and adjust its color as discussed earlier. Refractive index defines the extent by which the view of an object will distort if it is placed behind a translucent object. For a refractive index of 1 (Air), the view of the object will not distort at all. If this value is increased (1.5 for glass, 2.5 for diamond and so on), the extent of distortion will also increase. The maximum value of refractive index is 3.

Cutouts

Cutouts are used to create a perforation effect on the material and make the material partially transparent. When you select the **Cutouts** check box in the **Materials Editor**, the **Material**

Editor Open File dialog box is displayed. In this dialog box, you can select cutout maps. After selecting the texture to open, the **Texture Editor** is displayed. In the **Texture Editor**, you can set the color, softness, and size of the cutout map. Note that light colors appear opaque and dark colors appear transparent. Figure 31-20 shows a checker map without cutout effect and Figure 31-21 shows a checker map with cutout effect.

Figure 31-20 Checker map without cutout effect   *Figure 31-21 Checker map with cutout effect*

Self-Illumination

Self-illumination makes the material appear as if the light is coming from the material itself. To apply this effect, select the **Self-Illumination** check box. You can control the **Filter Color**, **Luminance**, and **Color Temperature** by using the options activated upon selecting this check box. The **Filter Color** is used to define a color on the illuminated material. Luminance controls the amount of light reflected from a surface. It is a measure of how bright or dark a surface will appear. By default, the **Dim Glow** option is set in this option. The other options in the **Luminance** drop-down list are shown in Figure 31-22. You can specify the value of luminance in the edit box in terms of candela/meter^2. You can also define the color of the illumination using the **Color Temperature** option. The options displayed in the **Color Temperature** drop-down list are shown in Figure 31-23.

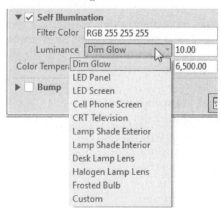

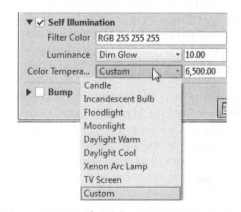

*Figure 31-22 The **Luminance** options*   *Figure 31-23 The **Color Temperature** options*

Bump

The **Bump** area is used to add a bumpy or irregular effect to the surface of the material. Select the **Bump** check box to apply the reflection map settings on the object. On doing so, the **Material Editor Open File** dialog box will be displayed. You can use this dialog box to attach the required bump map. The attached map type in this area will appear to be engraved on the surface of the material. The light colors in this area will be projected outward or inward and the dark colors will remain as they are. Use the slider bar to adjust the extent of the bump. The value of the bump ranges from -1000 to 1000. For negative values, the brighter colors will appear to be projected inwards and for positive values the brighter colors will appear to be projected outwards.

Note that the depth effect of a bump map is limited because it does not affect the profile of the object and cannot be self-shadowing. If you want extreme depth in a surface, you should use modeling techniques instead.

Note
The additional effects add realism to the object but at the same time increase the rendering time significantly.

*You can control the display of materials and textures from the **Material/Texture** drop-down in the **Materials** panel of the **Render** tab in the **Ribbon**. You can select the **Materials/Textures Off** or **Materials On/Textures Off**, or **Materials/Textures On** option from the **Material/Texture** drop-down. On doing this, you can control the rendering time as well.*

*You can create a material similar to the original AutoCAD material. To do so, select the original material from the **Material Browser** and double-click to invoke the **Material Editor**. Next, click on the **Create or duplicate material** swatch and select the **Duplicate** option; the duplicate of the selected material will be created.*

*The other options in the **Materials** window, the concept of mapping materials on an object, and the procedure of adjusting and modifying the maps will be discussed in later topics of this chapter.*

Tip
*The bitmap images in AutoCAD are stored in C:\Program Files\Common Files\Autodesk Shared\ Materials\2012\assetlibrary_base.fbm\Mats folder. These bitmap images are in the *.jpg format.*

MAPPING MATERIALS ON OBJECTS

Ribbon: Render > Materials > Material Mapping	**Command:** MATERIALMAP or SETUV
Menu Bar: View > Render > Mapping	**Toolbar:** Mapping or Render > Mapping

In AutoCAD, mapping is defined as the method used for adjusting the coordinates and the bitmap of the pattern of the material attached to a solid model. Mapping is required when the pattern of the material is not properly displayed on the object after rendering. Therefore, to display the material properly, you need to adjust the coordinates and the bitmap of the material properly.

To perform mapping on a model, choose the **Material Mapping** drop-down from the **Materials** panel; different types of mapping will be displayed, as shown in Figure 31-24. Choose the mapping type that you want to use on the object. The options for the mapping type are: **Planar**, **Box**, **Cylindrical**, and **Spherical**. First select the object; a material mapper grip tool will be displayed on the object. Use the grip tools on the object to adjust the display of the image on the model or on the face of the model. You can move or rotate the mapping about an axis by selecting the axis and right-clicking. Figure 31-25 shows a model with the **Planar** mapping and Figure 31-26 shows a model with the **Cylindrical** mapping.

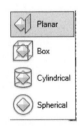

*Figure 31-24 The **Material Mapping** drop-down*

You can invoke the following options depending on your requirements.

Box

This option is used to map an image on a box such that the image is repeated on each side of the object. The prompt sequence that will be followed on selecting the **Box** option from the **Material Mapping** drop-down is given next.

Select an option [Box/Planar/Spherical/Cylindrical/copY mapping to/Reset mapping]<Box>: **_B**
Select faces or objects: *Select a face or object on which the image will be mapped.*
Select faces or objects: *Select more faces or objects on which the image will be mapped or* ⏎.

Figure 31-25 *Model with the **Planar** mapping*

Figure 31-26 *Model with the **Cylindrical** mapping*

Accept the mapping or [Move/Rotate/reseT/sWitch mapping mode]: *Accept the mapping as it is or select an option to modify it.*

The sub-options that you can select at this prompt are given next.

Move

On selecting this sub-option, the move gizmo will be displayed on the object. You can move and adjust the mapped image along any selected axis handle. Dynamic grips are also displayed on the mapping box, which can be dragged to resize the map.

Rotate

On selecting this sub-option, the rotate gizmo will be displayed on the object. You can rotate and adjust the mapped image along any selected axis handle. Dynamic grips are also displayed on the mapping box, which can be dragged to resize the map.

reseT

This sub-option is used to reset all adjustments made in the mapping coordinate to the default one.

sWitch mapping mode

This sub-option is used to redisplay the main prompt sequence on the screen to change the type of mapping.

Planar

This option is used to map the image on an object such that it seems being projected on that object. The projected image is scaled automatically according to the size of the object. The prompt sequence that will follow on selecting the **Planar** option from the **Material Mapping** drop-down is given next.

Select an option [Box/Planar/Spherical/Cylindrical/copY mapping to/Reset mapping]<Box>: **P**
Select faces or objects: *Select a face or object on which the image will be mapped*
Select faces or objects: *Select more faces or objects on which the image will be mapped or* Enter
Accept the mapping or [Move/Rotate/reseT/sWitch mapping mode]: *Accept the mapping as it is or select an option to modify it*

The sub-options available at this prompt are similar to the ones discussed in the **Box** option.

Spherical

This option is used to map the image by warping the image horizontally and vertically. In this type of mapping, the top edge of the image converges at the north pole of the object and the bottom edge converges at the south pole. The prompt sequence for this option is similar to that discussed in the **Box** option.

Cylindrical

This option is used to map the image by wrapping the image on a cylinder. The height of the image is scaled along the cylinder axis. The prompt sequence for this option is similar to that discussed in the **Box** option.

copY mapping to

You can also acquire the mapping from an existing object and copy it on the selected object. You can do so by choosing the **Copy Mapping Coordinates** tool from the expanded options of the **Materials** panel in the **Render** tab. The prompt sequence for this option is given next

> Select an option [Box/Planar/Spherical/Cylindrical/copY mapping to/Reset mapping]<Box>: **Y**
> Select faces or objects: *Select the source object to copy the mapping information*
> Select the objects to copy the mapping to: *Select the object to apply the copied mapping information*
> Select faces or objects: *Select more objects or* [Enter]

Reset mapping

This option is used to reset all the adjustments made in the mapping coordinates of the selected object to the default one produced by AutoCAD. To invoke this option, choose the **Reset Mapping Coordinates** tool from the expanded options of the **Materials** panel in the **Render** tab.

Adjusting the Bitmap Image on a Model

You can control the scale, tiling, and offset properties of a mapped image through the **Transforms** node of the **Texture Editor**.

To invoke the **Texture Editor**, click in the texture map swatch of the **Material Editor**. Figure 31-27 shows the **Texture Editor** displayed on clicking in the **Checker** map applied to the color of the object in the **Generic** node.

In the **Appearance** area, you can modify the color of the selected texture map. The options displayed depend on the material selected. When you expand the **Transforms** node in the **Texture Editor**, the **Position**, **Scale**, and **Repeat** sub-nodes are displayed to enable you to transform the map, see Figure 31-28. Note that these sub-nodes are available only for the **Image**, **Checker**, **Gradient**, and **Tiles** map types. In the **Position** sub-nodes, you can set the offset properties of the texture map. You can specify the start point of the map along the X and Y coordinates as well as rotate the map in this area. In the **Scale** area, you can specify the scale factor of the texture map in horizontal and vertical directions. You can also specify the method to be followed to create a pattern of the image. The options available in the **Repeat** area are used to set whether to tile the map in the Horizontal/Vertical directions or not.

Figures 31-29 and 31-30 show the models rendered with different **Width** and **Height** sizes in the **Scale** area of the **Texture Editor**.

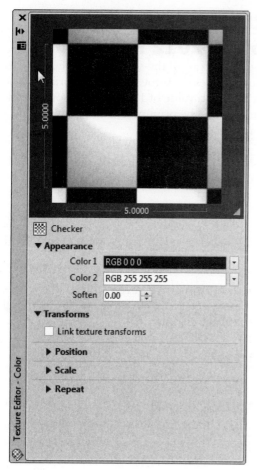

Figure 31-27 *The Texture Editor*

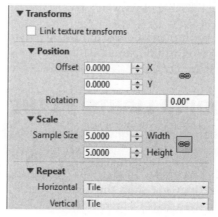

Figure 31-28 *The Transforms area of the Texture Editor*

Figure 31-29 *Model rendered with Width = 8, Height = 2*

Figure 31-30 *Model rendered with Width = 4, Height = 1*

CONVERTING MATERIALS CREATED IN AutoCAD PREVIOUS RELEASE TO AutoCAD 2012 FORMAT

The **3DCONVERSIONMODE** system variable is used to control the settings for the automatic conversion of the materials created in the previous releases of AutoCAD. If the variable is set to 0, no conversion of material will take place. If it is set to a value of 1, the conversion will take place automatically. If the value is set to 2, then the user will be prompted when materials need to be converted and will have the option to convert or not to convert.

You can convert the materials created in the previous releases of AutoCAD manually by the **CONVERTOLDMATERIALS** command. The prompt sequence is given next.

> Command: **CONVERTOLDMATERIALS**
> Loading materials...done.
> 4 material(s) created.
> 7 object(s) updated.

The prompt sequence indicates that the 4 materials applied to seven objects have been converted to the AutoCAD 2012 format. This prompt sequence also indicates the materials which are not converted to AutoCAD 2012 format, if there are any.

ADDING LIGHTS TO THE DESIGN

Ribbon: Render > Lights	**Toolbar:** Lights or Render > New Point Light
Command: LIGHT	**Menu Bar:** View > Render > Light

Lights are vital to rendering a realistic image of an object. Without proper lighting, the rendered image may not show the features the way you expect. The lights can be added to the design using the **LIGHT** command.

AutoCAD supports two types of lighting arrangements: Standard workflow and photometric workflow. The Photometric workflow is the default workflow in AutoCAD. The photometric workflow provides more accurate control over lighting through the photometric properties. The photometric workflow supports two types of lighting units: International (for example lux), and American (for example foot-candles). You can select the lighting workflow and its unit from the **Units for specifying the intensity of lighting** drop-down list in the **Lighting** area of the **Drawing Units** dialog box. To invoke this dialog box, choose **Drawing Utilities > Units** from the **Application Menu**. Alternatively, expand the **Lights** panel in the **Render** tab and specify the units from the **Lighting Units** drop-down.

Note
*The Standard workflow is enabled if **Generic lighting units** is selected as a unit in the **Lights** panel. Similarly, the Photometric workflow is enabled if **American lighting units** or **International lighting units** is selected as a unit in the **Lights** panel.*

All types of light sources supported by AutoCAD are discussed next.

Default Light

The default light is the natural light source that equally illuminates all surfaces of the objects, see Figure 31-31. The default light consists of two distance lights that do not have a source, and therefore, no location or direction. It is the default light that is automatically applied to the design. However, you can modify the brightness, contrast, and mid tone settings of the environment light by using the respective slider from the **Lights** panel in the **Render** tab, see Figure 31-32.

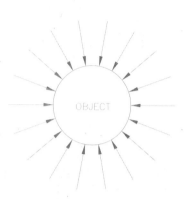

Figure 31-31 Default light emitting constant illumination

*Figure 31-32 The expanded **Lights** panel of the **Render** tab*

The **Brightness** and **Contrast** slider bars can be used to increase or decrease the brightness and contrast of the default light. By default, the color of the default light is white. Figures 31-33 and 31-34 show the same model rendered using different values of brightness and contrast of the default light.

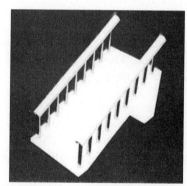

Figure 31-33 Model rendered with less brightness and contrast value of ambient light

Figure 31-34 Model rendered with more brightness and contrast value of ambient light

When you invoke any new light to create user-defined lights, the **Lighting - Viewport Lighting Mode** message will be displayed on the screen, see Figure 31-35. If you want to render the model using the user-defined lights only, then choose the **Turn off the default lighting (recommended)** option. But if you want to render the model using both the default and user defined lights, then choose the **Keep the default lighting turned on** option. However, if you will keep the default lights turned on, the effects of the user-defined lights will not be displayed properly, so you need to turn off the default lighting manually. If you invoke the **LIGHT** command again after turning off the default light, the **Lighting - Viewport Lighting Mode** message will not be displayed on the screen.

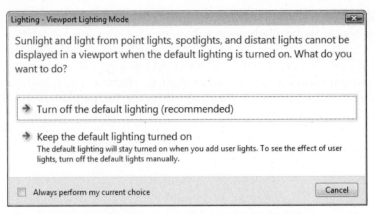

Figure 31-35 The **Lighting - Viewport Lighting Mode** *message box*

Point Light

Ribbon: Render > Lights > Create Light drop-down > Point	**Command:** POINTLIGHT
Toolbar: Lights > New Point Light	**Menu Bar:** View > Render > Light > New Point Light

A point light source emits light in all directions and the intensity of the emitted light is uniform. You can visualize an electric bulb as a point light source. In AutoCAD render, if you select to add a point light, you can set the options of casting the shadow of the objects in the design. Figure 31-36 shows a point light source that radiates light uniformly in all directions.

As this is not the default light source, therefore, you will have to add it manually. This light source can be added by choosing the **Point** light tool from the **Create Light** drop-down. The prompt sequence for the **Point** light tool is given next.

Figure 31-36 *Point light source emitting light uniformly in all directions*

> Command: _pointlight
> Specify source location <0,0,0>: *Specify the location where you want to place the point light*
> Enter an option to change [Name/Intensity factor/Status/Photometry/shadoW/Attenuation/filterColor/eXit] <eXit>: *Specify an option or press* ⏎

The following options can be selected from this prompt:

Name
This option is used to specify the name of the point light. AutoCAD, by default, specifies a name to the newly created light. You can change the name by using this option.

Intensity factor
This option is used to specify the intensity of the point light. The prompt sequence that will be followed is given next.

> Enter an option to change [Name/Intensity factor/Status/Photometry/shadoW/Attenuation/filterColor/eXit] <eXit>: **I** ⏎
> Enter intensity (0.00 - max float) <1.0000>: *Enter a value for the intensity of the light or press the* ⏎ *key to accept the default value*

Status

This option is used to turn the point light on and off.

Photometry

This property will be available only when you work in the photometric environment, which means the **American lighting units** or **International lighting units** is selected as a unit in the **Lights** panel. The **Photometry** option is used to control the intensity, unit of intensity, color, and type of point lights, so that you can define the lights accurately according to the real world requirement. Instead of specifying the photometric properties of the point light in the command itself, you can also specify or modify all the photometric properties later in the **Properties** palette of the point light, see Figure 31-37. The options in the **Photometric properties** rollout are discussed next.

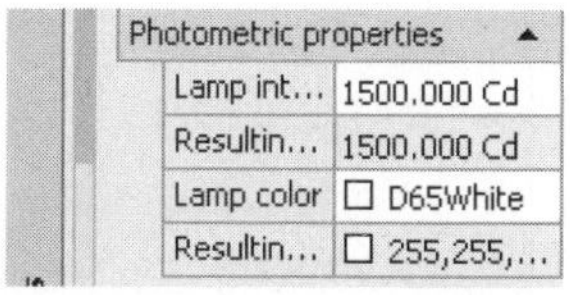

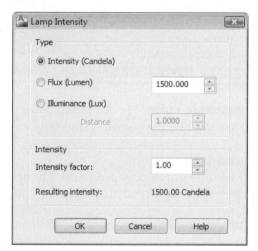

*Figure 31-37 The **Photometric properties** rollout of the **Properties** palette in the point light*

Lamp intensity. This property is used to specify the intensity, flux, or illuminance of the lamp. By default, the intensity of the light is specified in Candela. To change the properties, select the **Lamp intensity** edit box and choose the swatch displayed on the right; the **Lamp Intensity** dialog box will be displayed, see Figure 31-38. Select the **Intensity (Candela)** radio button to specify the intensity of the point light in Candela. The **Flux (Lumen)** radio button is used to specify the flux of light in Lumen. The flux is the rate of total energy leaving the light source. The spinner on the right of the above mentioned radio button is used to change the value for both the intensity and flux depending on the radio button selected. The **Illuminance (Lux)** radio button is used to specify the amount of energy received per unit area by an object. In this case, you have to specify the distance at which the object is placed from the light source because the photometric lights are divergent by nature and their intensities will reduce with the increase of distance between the light source and the object. You can also increase or decrease the resulting intensity, flux, or illuminance of the light source by specifying a multiplying factor in the **Intensity factor** spinner of the **Resulting intensity** area. The resulting intensity will be the product of intensity input and the intensity factor.

*Figure 31-38 The **Lamp Intensity** dialog box*

Resulting intensity. The **Resulting intensity** property displays the resulting final intensity of the light source. This information is read-only.

Lamp color. This property is used to specify the lamp color of the light source. To specify or modify the lamp color, select the **Lamp color** edit box and choose the swatch displayed on the right; the **Lamp Color** dialog box will be displayed, see Figure 31-39. Select the **Standard colors** radio button; the **Standard colors** drop-down list will be displayed, see Figure 31-40. Select the standard predefined colored lights listed in this drop-down list. The preview of the selected light will be displayed in the color swatch on the right of the **Standard colors** drop-down list. An ideal black body emits a particular color at a particular temperature. To specify the lamp color with respect to temperature, select the **Kelvin colors** radio button. You can specify the temperature in Kelvin in the spinner just below the **Kelvin colors** radio button. The **Filter color** drop-down list in the **Resulting color** area is used to mix some colors with the lamp color. The final resulting color of the lamp will be a mixture of the lamp color and filter color. The preview of the resulting lamp color will be displayed in the color swatch in the **Resulting color** area.

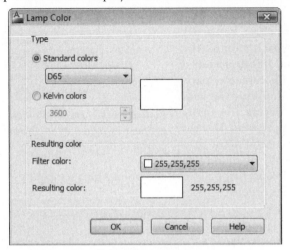

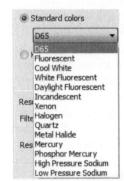

*Figure 31-39 The **Lamp Color** dialog box*

*Figure 31-40 The **Standard colors** drop-down list*

Resulting color. The **Resulting color** property displays the resulting final color of the light source. This information is read-only.

shadoW

The sub-options provided under this option are used to control the display of the shadows of the objects after rendering. The sub-options are given next.

Off. This option is used to turn off the display of the shadow in rendering.

Sharp. This option is used to display the shadow of the objects with sharp edges.

soFtmapped. This option is used to display the shadow with blurred edges. This gives a more realistic effect to the rendered image. If this sub-option is selected, you will have to specify the map size and softness of the shadow. The map size is calculated in terms of pixels. The greater the number of pixels, the better is the shadow. However, if the map size is large, the time taken to render the model will also increase. The shadow softness is used to specify the softness for the shadow. If the value of this option is increased, the shadow will be blurred. Figure 31-41 shows a model with a shadow map size=128 and softness=5, and Figure 31-42 shows a model with a shadow map size=1024 and softness=2.

Figure 31-41 Shadow map size =128 and softness = 5

Figure 31-42 Shadow map size =1024 and softness = 2

Tip
You can view the best shadow effect by keeping the value of the shadow map size 512 or 1024 and the shadow softness between 3 and 4.

softsAmpled. This option is used to specify the shape of the point light lamp. The shadow of the object will be displayed depending upon the shape of the point light lamp. The prompt sequence for this option is given next.

Enter an option to change [Name/Intensity factor/Status/Photometry/shadoW/ Attenuation/filterColor/eXit] <eXit>: **W** Enter
Enter [Off/Sharp/soFtmapped/softsAmpled] <Sharp>: **A** Enter
Enter an option to change [Shape/sAmples/Visible/eXit]<eXit>: **S** Enter
Enter shape [Linear/ Disk/ Rect/ Sphere/ Cylinder] <Sphere>: *Enter an option to specify the shape of the slit from which the point light will pass through.*

After specifying the shape of the slit, you will be prompted to specify the dimensions of the shape. For example, you can specify the radius of the sphere or the length and width of a rectangle. The prompt sequence for the **Cylinder** option is given next.

Enter shape [Linear/Disk/Rect/Sphere/Cylinder] <Sphere>: **C** Enter
Enter radius <0.0000>: *Specify the radius of the cylindrical slit.*
Enter Length <0.0000>: *Specify the height of the cylindrical slit.*
Enter an option to change [Shape/sAmples/Visible/eXit]<eXit>: *Specify an option or press* Enter

The **sAmples** option in the above prompt is used to specify the size of the shadow samples for light. The default value of the shadow sample is 16. The **Visible** option is used to specify whether to display the shape of the lamp in rendering or not. Figure 31-43 displays the rendered image of a cylindrical shape point light without the visibility of the point light shape. Figure 31-44 displays the same rendered image with the point light shape visible.

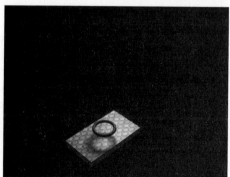

Figure 31-43 *Rendered image with the shape of cylindrical point light not visible*

Figure 31-44 *Rendered image with the shape of point light visible*

Note
The shape of the point light is also displayed in the drawing area close to the point light. When you move the cursor over the point light, the shape of the point light is highlighted in default set brown color.

Attenuation

The light intensity is defined as the amount of light falling per unit of area. The intensity of light is inversely proportional to the distance between the light source and the object. Therefore, the intensity decreases as the distance increases. This phenomenon is called Attenuation. In AutoCAD, it occurs only with spotlights and a point light.

In Figure 31-45, light is emitted by a point source. Assume the amount of light incident on Area1 is I. Therefore, the intensity of light on Area1 = I/Area. As light travels farther from the source, it covers a larger area. The amount of light falling on Area2 is the same as on Area1 but the area is larger. Therefore, the intensity of light for Area2 is smaller (Intensity of light for Area2 = I/Area). Area1 will be brighter than Area2 because of a higher light intensity. The prompt sequence that will be followed after selecting this option is given next.

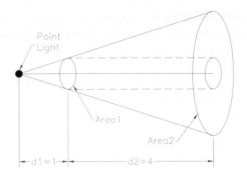

Figure 31-45 Example of attenuation

Enter an option to change
[attenuation Type/Use limits/
attenuation start
Limit/attenuation End limit/eXit] <eXit>: *Select a sub-option or press the* Enter *key.*

The sub-options that you can select at the above prompt are:

attenuation Type. AutoCAD render provides you with the following three types of attenuation.

None. If you select the **None** sub-option for the light falloff, the brightness of the objects will be independent of the distance. This means that the objects that are far away from the point light source will be as bright that are close to it.

Inverse linear. In this sub-option, the light falling on the object (brightness) is inversely proportional to the distance of the object from the light source (Brightness = 1/Distance). As the distance increases, the brightness decreases. For example, assume that the intensity of the light source is I and the object is located at a distance of 2 units from the light source. Now, the brightness or intensity = I/2. If the distance is 8 units, the intensity (light falling on the object per unit area) = I/8. The brightness is a linear function of the distance of the object from the light source.

inverse Squared. In this sub-option, the light falling on the object (brightness) is inversely proportional to the square of the distance between the object and the light source (Brightness = $1/Distance^2$). For example, assume that the intensity of the light source is I and the object is located at a distance of 2 units from the light source. Now, the brightness or intensity = $I/(2)^2$ = I/4. If the distance is 8 units, the intensity (light falling on the object per unit area) = $I/(8)^2$ = I/64.

Use limits. This option is used to specify whether you want to use the sub-options of **attenuation start Limit and attenuation End limit**. The effect of these limits can only be visualized in rendering, if the sub-option of **Use limits** is turned on.

attenuation start Limit. This option is used to specify the distance from the centre of the light to where the light will start.

attenuation End limit. This option is used to specify the distance from the center of the light at which the light will finish. The objects beyond this limit will not be illuminated by this point light.

filterColor

This option is used to specify the color for the point light. You can define a custom color by using the RGB combination. You can select a color by choosing the **Index color** sub-option. Select the sub-option **Hsl** to define the color by Hue, Saturation, and Luminance. You can also specify the color of the point light from the color book by choosing the **colorBook** sub-option. Next, enter the name of the color book from which you want to choose the color.

eXit

This option is used to finish the process of defining the point light properties and to exit the **POINTLIGHT** command.

Note
The lightings in the photometric workflow always follow the inverse squared attenuation type, and the attenuation properties cannot be modified later on for photometric lights.

Tip
*Instead of specifying all the above-mentioned properties of the point light in the command itself, you can also specify or modify all the properties later on from the **Properties** palette of the point light.*

EXAMPLE 2

In this example, you will apply the point light at the lintel of the window and at the lintel of the door of the model, as shown in Figures 31-46 and 31-47. Choose your own materials to make the model look realistic and then render it.

Figure 31-46 Model for Example 2

Figure 31-47 Model after rendering

1. Create the model shown in Figure 31-46.

2. Choose materials from **Material Browser** and attach them to the model as described in Example 1. (Recommended: Use **Masonry - Stone > Limestone - Ashlar Coursed** for Walls, **Ceramic - Tile > Mosaic Blue** for flooring, and **Flooring - Vinyl - Diamonds 1** for roof.)

3. Choose the **Point** tool from **Render > Lights > Create Light** drop-down to add a point light.

4. Choose the **Turn off the default lighting (recommended)** option from the **Lighting - Viewport Lighting Mode** message box. (This message box will be displayed when you create the light for the first time).

5. Now, follow the prompt sequence of the **Point** light tool given next.

Command: _pointlight
Specify source location <0,0,0>: *Specify the location of the point light at the lintel of the first window*
Enter an option to change [Name/Intensity factor/Status/Photometry/shadoW/Attenuation/filterColor/eXit] <eXit>: **N** [Enter]
Enter light name <Pointlight>: **1** [Enter]
Enter an option to change [Name/Intensity factor/Status/Photometry/shadoW/Attenuation/filterColor/eXit] <eXit>: **C** [Enter]
Enter true color (R,G,B) or enter an option [Index color/Hsl/colorBook]<255,255,255>: **I** [Enter]
Enter color name or number (1-255): **Yellow** [Enter]
Enter an option to change [Name/Intensity factor/Status/Photometry/shadoW/Attenuation/filterColor/eXit] <eXit>: **I** [Enter]
Enter intensity (0.00 - max float) <1.0000>: *Specify the intensity based on the size of the objects in the drawing and press the* [Enter] *key*
Enter an option to change [Name/Intensity factor/Status/Photometry/shadoW/Attenuation/filterColor/eXit] <eXit>: **S** [Enter]
Enter status [oN/oFf] <On>: [Enter]

6. Similarly, you can set the second light at the lintel of the door. Set the color of the light to green.

   Now, you need to set the background of the image from black to white. You can also set any image as the background of the image.

7. Invoke the **VIEW** command; the **View Manager** dialog box is displayed. Choose the **New** button from it; the **New View/Shot Properties** dialog box is displayed.

8. Select the **Solid** option from the **Background** drop-down list in the **Background** area in the **View** tab of the dialog box; the **Background** dialog box is displayed.

9. Click on the **Color** swatch and select the white color as the background color and exit the dialog box.

10. Assign a name to the view in the **New View/ Shot Properties** dialog box and then exit it .

11. Select the view from the **Views** list and choose the **Apply** button.

12. Choose the **Render** tool in the **Render** panel; the **Render** window is displayed and the rendering process starts. The rendered image is displayed in the **Render** window. The rendered image looks similar to the image shown in Figure 31-48.

Figure 31-48 *Rendered model of a house*

Tip
You can increase the smoothness of the rendered curved object by increasing the value of the **FACETRES** *system variable. The default value of this variable is 0.5. Set this value close to* **6***.*

Spotlight

Ribbon: Render > Lights > Create Light drop-down > Spot	**Toolbar:** Lights > New Spotlight
Menu Bar: View > Render > Light > New Spotlight	**Command:** SPOTLIGHT

A spotlight emits light in the defined direction with a cone-shaped light beam. This light has a focused beam that starts from a point and is targeted at another point. This type of light is generally used to illuminate a specific point, such as illuminating a person on a stage. The phenomenon of attenuation also applies to spotlights. This light source can be added by choosing the **Spot** light tool from the **Create Light** drop-down. When you choose this tool, you will be prompted to enter two values. As already mentioned, this type of light source has a start point and a target point. Therefore, you will first be prompted to specify the light location and then the light target. The prompt sequence displayed when the **Spot** light tool is invoked is given next.

> Command: _spotlight
> Specify source location <0,0,0>: *Specify the point where you want to place the spotlight*
> Specify target location <0,0,-10>: *Specify the location on the object where you want to focus the spotlight*
> Enter an option to change [Name/Intensity factor/Status/Photometry/Hotspot/Falloff/shadoW/Attenuation/filterColor/eXit] <eXit>: *Specify an option or press* Enter

Most of the options in the **Spot** light tool are similar to those in the **Point** tool. The options that are different are discussed next.

Hotspot/Falloff

As mentioned earlier, the spotlight has a focused beam of light that is targeted at a particular point. Therefore, there are two cones that comprise the spotlight.

The hotspot is the same as the cone that carries the highest intensity light beam. In this cone, the light beam is the most focused and is defined in terms of an angle, see Figure 31-49(a) and 31-49(b).

The prompt sequence for the **Hotspot** option is given next.

> Enter an option to change [Name/Intensityfactor/Status/Photometry/Hotspot/Falloff/
> shadoW/Attenuation/filterColor/eXit]<eXit>: **H** [Enter]
> Enter hotspot angle (0.00-160.00) <45.0000>: *Specify the interior angle of the cone in which the intensity of light will be high or press the* [Enter] *key to accept the default value*

The other cone is called a falloff and it specifies the full cone of light. It is the area around the hotspot where the intensity of the beam of light is not very high, as shown in Figures 31-49(a) and 31-49(b). It is defined in terms of an angle. The value of the hotspot and the falloff can vary from 0 to 160. The prompt sequence for the **Falloff** option is given next.

> Enter an option to change [Name/Intensity factor/Status/Photometry/Hotspot/Falloff/
> shadoW/Attenuation/filterColor/eXit]<eXit>: **F** [Enter]
> Enter falloff angle (0.00-160.00) <current>: *Specify the interior angle of the cone in which the intensity of the light will be high or press the* [Enter] *key to accept the default value*

Figure 31-49(a) *The hotspot and falloff cones*

Figure 31-49(b) *The hotspot and falloff after rendering*

Tip
The value of hotspot should always be less than the value of falloff. If you set the value of the hotspot more than the value of the falloff, then the value of the falloff will automatically change to one degree greater than the value of the hotspot.

Distant Light

Ribbon: Render > Lights > Create Light drop-down > Distant	**Toolbar:** Lights > New Distant Light
Menu Bar: View > Render > Light > New Distant Light	**Command:** DISTANTLIGHT

A distant light source emits a uniform parallel beam of light in a single direction only (Figure 31-50). The intensity of the light beam does not decrease with the distance. It remains constant. For example, the sunrays can be assumed to be a distant light source because the light rays are parallel. When you use a distant light source in a drawing, the location of the light source does not matter; only the direction is critical. Distant light is mostly used to light objects or a backdrop uniformly and for getting the effect of the sunlight.

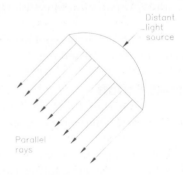

Figure 31-50 *Distant light source*

The distant lights are inaccurate and their use in photometric workflow is not preferred. Therefore, the **Distant** light tool is not available for the use in the photometric workflow. However, to use the distant light in photometric workflow, choose the **Distant** light tool from the **Lights** tab; the **Lighting - Viewport Lighting Mode** message box will be displayed on the screen. Choose the **Turn off the default lighting (recommended)** option so that the effect of the distant lights can be seen. On doing so, the **Lighting - Photometric Distant Lights** message box is displayed, see Figure 31-51. Choose the **Allow distant lights** button from this message box to invoke the **Distant** light tool. The prompt sequence displayed when the **Distant** light tool is invoked is given next.

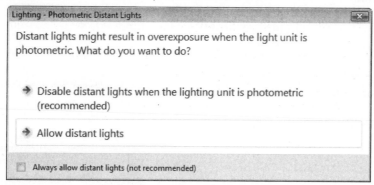

Figure 31-51 *The* ***Lighting - Photometric Distant Lights*** *message box*

Command: _distantlight
Specify light direction FROM <0,0,0> or [Vector]: *Specify the start point to represent the viewing direction of the distant light or type* **V** *and press the ENTER key.*
Specify light direction TO <1,1,1>: *Specify the endpoint to represent the viewing direction of the distant light.*
Enter an option to change [Name/Intensity factor/Status/Photometry/shadoW/filterColor/eXit] <eXit>: *Specify an option.*

If you select the **Vector** option at the **Specify light direction FROM <0,0,0> or [Vector]** prompt, then you will have to specify the vector direction in which the distant light will focus. The options in the **Distant** light tool are similar to those in the **Point** tool.

Web Light

Ribbon: Render > Lights > Create Light drop-down > Weblight
Command: WEBLIGHT

Web lights are used to create illuminated scenes that look more realistic than those created by using the spot or point lights. You can create a web light by invoking the **Weblight** tool from the **Lights** panel. On invoking this tool, you are prompted to specify the source and then the target points. Specify the source and then the target point. Next, specify the name, intensity, status, photometry, web file, shadows, etc to create the web light. The web lights offer a detailed 3D distribution of the intensity of a light in all directions. The distribution can be isotropic or anisotropic. The isotropic distribution represents a sphere-shaped web, and its effect will be the same as that of point light. However, the web light is generally used for anisotropic or non-uniform light distribution. The anisotropic distribution gives a unique washed effect on the surface it hits. The information regarding this light distribution is saved in an AutoCAD program file having extension *.ies*. These files are generally provided by manufactures of lights and fixtures. You can load the manufacturers' *.ies* files in the **Photometric Web** panel of the **Properties** palette. In this palette, you can set the other properties such as intensity, as well.

Sun Light

Ribbon: Render > Sun & Location (Inclined Arrow)	**Toolbar:** Light > Sun Properties
Menu Bar: View > Render > Light > Sun Properties	**Command:** SUNPROPERTIES

This light is similar to the actual sun light in which the rays are parallel and have the same intensity at any distance. The angle subtended by rays on an object depends on the location, date, and time specified. You can also change the intensity and color of the sunlight. Unlike distant light, sunlight can only be one in number. In the photometric workflow, you can also control the illumination effect when the sky comes between the sunlight and the object. This helps in adding more realistic lighting caused by the interaction between the sun and the sky.

To use sunlight, click on the inclined arrow in the **Sun & Location** panel of the **Render** tab; the **Sun Properties** palette will be displayed, as shown in Figure 31-52. The options in this palette are discussed next.

General

In this area, you can switch on or off the sunlight and change its intensity and color. You can also specify whether the object will cast a shadow due to the sunlight.

Sky Properties

These properties are only available while working in the photometric workflow. You can use the sky in between the model and the sun for the illumination of the model. The options available in the **Status** edit box are **Sky Off**, **Sky Background**, and **Sky Background and Illumination**. These options are also available in the **Sky Status** drop-down list of the **Render** tab.

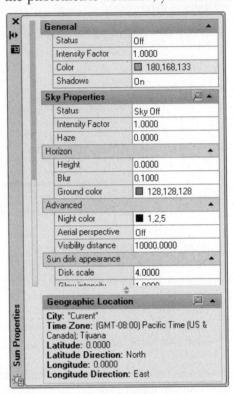

*Figure 31-52 The **Sun Properties** palette*

Figure 31-53 displays a model rendered by using the **Sky Off** option, Figure 31-54 displays a model rendered by using the **Sky Background** option, and Figure 31-55 displays the model rendered by using the **Sky Background and Illumination** option. The **Intensity Factor** edit box is used to magnify the brightness of the light reaching the object after passing through the sky. Figure 31-55 displays the rendered model by choosing the **Sky Background and Illumination** option and intensity factor set to 1 and Figure 31-56 displays the same model rendered by selecting the **Sky Background and Illumination** option and intensity factor set to **0.2**. The **Haze** edit box determines the amount of scattering in the atmosphere and the haze value can vary from 0 to 15. Figure 31-55 displays a rendered model with haze value set to **0** and Figure 31-57 displays the same model rendered with haze value set to **7**.

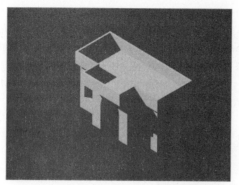

Figure 31-53 *Model rendered using the* **Sky Off** *option*

Figure 31-54 *Model rendered using the* **Sky Background** *option*

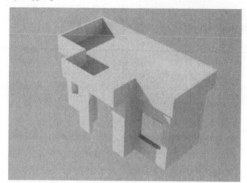

Figure 31-55 *Model rendered using the* **Sky Background and Illumination** *option with the intensity and haze values set to* **1** *and* **0**, *respectively*

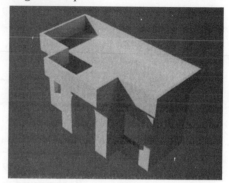

Figure 31-56 *Model rendered using the* **Sky Background and Illumination** *option with the intensity value set to* **0.2**

Horizon. The options in this area are used to control the location and appearance of the ground. Figure 31-58 displays a rendered model created by selecting the **Sun & Sky Background** option and setting that view as current. The ground color for the rendered image is set to **Cyan**.

Advanced. The properties under the **Advanced** rollout are used to produce more artistic effects in rendering such as the color of the night sky, whether to use the Aerial perspective view on or off, and so on.

Sun disk appearance. The properties under this rollout are used to control the appearance of the sun disk such as its size, intensity, and so on, when a view is saved with the **Sun & Sky Background**.

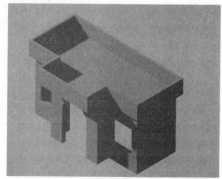

Figure 31-57 *Model rendered using the* **Sky Background and Illumination** *option with the intensity and haze values set to* **1** *and* **7**, *respectively*

Figure 31-58 *Model rendered with the view saved as* **Sun & Sky Background** *and ground color set to* **Cyan**

Note

*You can also specify or modify the **Sky Properties** from the **Adjust Sun & Sky Background** dialog box that will be displayed when you choose the **Sky Properties** tool displayed at the upper right corner of the **Sky Properties** title bar of the **Sun Properties** window, see Figure 31-59.*

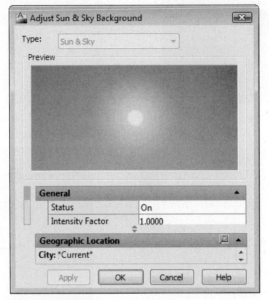

Figure 31-59 The Adjust Sun & Sky Background *dialog box*

Sun Angle Calculator

In this area, you can set the date and time in the **Date** and **Time** edit boxes, respectively. You can also switch on or off the effect of day light settings in the rendering. Apart from this, you can also view some read only information in this field such as the azimuth and altitude of the sun light direction. You can also view the default light source vector as the coordinates in the X, Y, and Z list boxes in the **Source Vector** area.

Rendered Shadow Details

This area gives you the option to control the shadow cast by the object in rendering. All the options in this field are similar to the **Point** tool.

Geographic Location

This area displays information about the geographic location of the place where you want to view the light effect. You can edit this information by choosing the **launch Geographic Location** button at the upper right corner of the **Geographic Location** area. Alternatively, choose the **Set Location** tool from the **Sun & Location** panel of the **Render** tab; the **Geographic Location- Define Geographic Location** window will be displayed with various options to define the location. To define the location, you can import the exact location data like longitude, latitude, and altitude from the .kml or .kmz files. Alternatively, you can pick your desired location from Google Earth, if it is already installed on your system or you can directly enter the location values. On choosing the **Enter the location values** button from the **Geographic Location- Define Geographic Location** window, the **Geographic Location** dialog box will be displayed, as shown in Figure 31-60. The options in this dialog box are given next.

You can set the latitude and longitude in the **Latitude** and **Longitude** edit boxes, respectively, and the directions as north, south, east, and west in the **North** and **West** drop-down lists in the dialog box. By default, the *Y* axis is assumed to be in the north direction in AutoCAD. You can also change this default direction in the **Angle** edit box.

You can select various locations by choosing the **Use Map** button available in the **Latitude & Longitude** area of the **Geographic Location** dialog box. On doing so, the **Location Picker** dialog box will be displayed, see Figure 31-61. You can select different locations from the map in this dialog box. On selecting the locations; a red colored cursor will be displayed on the selected locations. You can select the required map from the **Region** drop-down list. Also, all the nearest cities and the time zone, in which the selected city is located, will be listed in the **Nearest City** and **Time Zone** drop-down lists, respectively. Select the city and the time zone from these drop-down lists. You can also select a location in the World map. On doing so, the corresponding city and its latitude, longitude, and time zone are displayed in the respective drop-down lists. Choose the **OK** button; a **Geographic Location - Time Zone Updated** message box is displayed prompting you to either keep the updated time or select a different time zone. You can set the date and time in the **Date** and **Time** edit boxes of the **Sun & Location** panel of the **Render** tab.

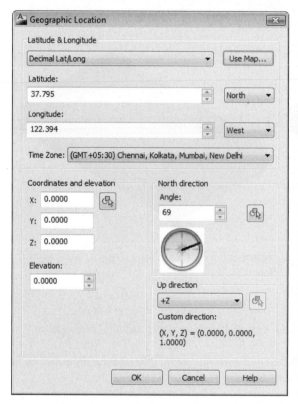

Figure 31-60 *The **Geographic Location** dialog box*

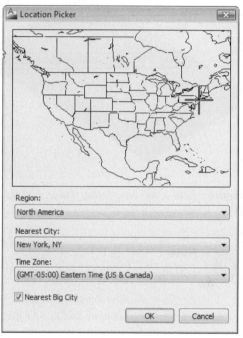

Figure 31-61 *The **Location Picker** dialog box*

Figures 31-62 through 31-67 show different positions of the Sun at different times and locations.

Figure 31-62 *City: Chicago, Time: 11:00, Date: 1st May*

Figure 31-63 *City: Chicago, Time: 16:00, Date: 1st May*

Figure 31-64 *City: Tokyo, Time: 11:00, Date: 1st May*

Figure 31-65 *City: Tokyo, Time: 16:00, Date: 1st May*

Figure 31-66 *City: New Delhi, Time: 11:00, Date: 1st May*

Figure 31-67 *City: New Delhi, Time: 16:00, Date: 1st May*

 Note

*1. If the **Sun Status** button is chosen in the **Sun & Location** panel, sunlight is turned on.*

2. Except for the geographic location, the rest of the settings for the sun and the sky are saved for a particular viewport, and not for the entire drawing.

You can also switch the sun light on and off, set the sky background illumination, change the date and time, open and close the **Sun Properties** palette and the **Geographic Location - Define Geographic Location** window directly from the **Render** tab of the **Ribbon**, see Figure 31-68.

Figure 31-68 *The **Render** tab of the **Ribbon***

EXAMPLE 3

In this example, you will create the model shown in Figures 31-69. Choose the materials of your choice. Set the sun light using the geographic location and render it, as shown in Figure 31-70. The geographic location and other parameters are given next.

City: New York, Date: 20 June, 2011   Time: 04:00 pm

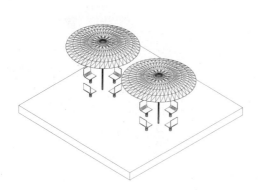

Figure 31-69 *Model for Example 3 before rendering*

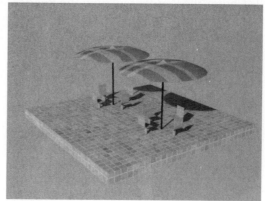

Figure 31-70 *Model for Example 3 after rendering*

1. Create the model shown in Figure 31-69.

2. Select the materials of your choice and apply them to the model, as explained earlier.

3. Select the **Sky Background and Illumination** option from **Render > Sun & Location > Sky** drop-down.

4. Click on the inclined arrow in the **Sun & Location** panel of the **Render** tab to invoke the **Sun Properties** palette and set the **Intensity Factor** value to **2.0** under the **Sky Properties** rollout.

5. Specify a color of your choice for the ground in the **Ground color** edit box under the **Horizon** sub-rollout and then exit the **Sun Properties** palette.

6. Choose the **Set Location** tool from the **Sun & Location** panel of the **Render** tab; the **Geographic Location - Define Geographic Location** window is displayed. Choose the **Enter the location values** button from the window; the **Geographic Location** dialog box is displayed.

7. Choose the **Use Map** button in this dialog box; the **Location Picker** dialog box is displayed.

8. Select **North America** from the **Region** drop-down list. Next, select **New York, NY** from the **Nearest City** drop-down list. You can also select a city by clicking on the map of **North America** displayed in the **Location Picker** dialog box. Choose **OK** to exit the dialog box; the **Geographic Location - Time Zone Updated** message box is displayed. Also, you are prompted to update the time zone or return to the previous time zone.

9. Choose **Accept updated time zone** from the message box; the updated locations are displayed in the **Geographic Location** dialog box. Choose the **OK** button to exit the dialog box.

10. Choose the **Sun Status** button from the **Sun & Location** panel of the **Render** tab; the **Lighting - Viewport Lighting Mode** message box is displayed. Choose the **Turn off the default lighting (recommended)** option.

11. Set the date and time as **6/20/2011** and **4:00 PM** in the **Date** and **Time** edit boxes, respectively in the **Sun & Location** panel of the **Render** tab.

12. Choose the **Render** tool from the **Render** tab to invoke the **Render** window. The rendering process will start and the rendered image will be displayed in the **Render** window, refer to Figure 31-70.

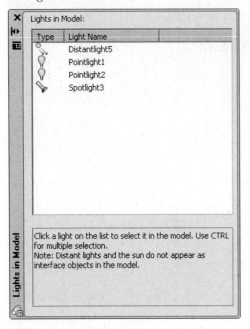

Figure 31-71 The **Lights in Model** palette

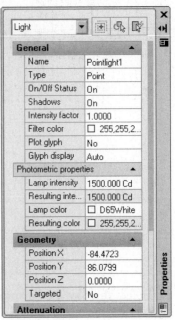

Figure 31-72 The **Properties** palette for distant light

CONVERTING LIGHTS CREATED IN AutoCAD'S PREVIOUS RELEASE TO AutoCAD 2012 FORMAT

For converting a model created in AutoCAD 2011 or earlier version of AutoCAD to AutoCAD 2012 format such that it supports both the standard workflow and the photometric workflow, set the value of **LIGHTINGUNITS** system variable to 1 or 2. The models and lights created prior to AutoCAD 2012 can be converted to the lights of AutoCAD 2012 by the **CONVERTOLDLIGHTS** command. The prompt sequence to do so is given next.

    Command: **CONVERTOLDLIGHTS**
    Loading materials...done.
    2 light(s) converted

In the prompt sequence, only 2 lights have been converted for the current drawing. Therefore, the prompt sequence displays **2 light(s) converted**. The prompt sequence will display the number of lights converted and also the lights that are not converted. Note that after converting the lights, their effect is not always the same as in their original versions. Therefore, you have to modify some parameters such as intensity, color, and so on.

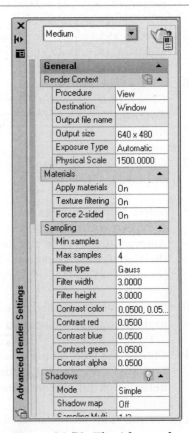

Figure 31-73 The **Advanced Render Settings** window

MODIFYING LIGHTS

As mentioned earlier, the lights and the lighting effects are important in rendering in order to create a realistic representation of an object. The sides of an object that face the light must appear brighter and the sides that are on the other side of the object must be darker. This smooth gradation of light produces a realistic image of the object. If the light intensity is uniform on the entire surface, the rendered object will not look realistic. For example, if you use the **SHADE** command to shade an object, it does not look realistic because the displayed model lacks any kind of gradation of light. Any number of lights can be added to the drawing. The color, location, and direction of all the lights can be specified individually. As mentioned earlier, you can specify an attenuation for the point lights and the spotlights. AutoCAD also allows you to change the color, position, and intensity of any light source. The point light and the spotlight can be interchanged. But you cannot change a distant light into a point light or a spotlight. All lights, except the default light and the sun light, attached to the current drawing are displayed in the light list. Choose the inclined arrow on the **Lights** panel of the **Render** tab of the **Ribbon**; the **Lights in Model** palette will be displayed, as shown in Figure 31-71. From this list box, select the light you want to modify, right-click, and then choose the **Properties** option from the shortcut menu displayed. Depending on the type of light selected, the **Properties** palette will be displayed. For example, if you select a distant light for modifications, the **Properties** palette with the distant light properties will be displayed. One such **Properties** palette is shown in Figure 31-72. All options in this window are similar to those under the **POINTLIGHT** command.

You can also modify the lights dynamically with the help of grips. This method provides a dynamic preview of the modifications made in lights. The symbols of the point light and the spot light, displaying their location in the drawing, can also be shown or hidden by choosing the **Light glyph display** button from the **Lights** panel of the **Render** tab.

UNDERSTANDING ADVANCED RENDERING

As mentioned earlier, rendering allows you to control the appearance of the objects in the design. This is done by defining the surface material and the reflective quality of the surface and by adding lights to get the desired effects. The basic rendering has already been discussed earlier in this chapter. You will now learn about the options of the **Advanced Render Settings** window, as shown in Figure 31-73.

You can also render the model using the **Advanced Render Settings** window by choosing the **Render** button displayed at the upper right corner. To invoke the **Advanced Render Settings** window, click on the inclined arrow in the **Render** panel of the **Render** tab.

Render Presets

The drop-down list at the top of the **Advanced Render Settings** window contains five standard qualities of rendering as presets. These are **Draft**, **Low**, **Medium**, **High**, and **Presentation**. The lowest quality of rendering is done in the **Draft** preset. The quality of rendering goes on increasing with the options and it is the best in the **Presentation** preset. However, you can also create your own render presets by selecting the **Manage Render Preset** option from the drop-down list; the **Render Presets Manager** dialog box is displayed, which will be explained in detail later in this chapter.

Managing Render Presets

The **Render Presets Manager** dialog box, as shown in Figure 31-74, is used to save any changes made to the render settings such that they can be used later as render presets. You can also select any standard preset render setting to get a preview of the settings and make them the current settings to be used for the rendering. But you cannot change the settings of the standard render preset from this dialog box. To open this dialog box, select the **Manage Render Presets** option from the **Render Presets** in the **Render** tab. You can also invoke this dialog box using the **RENDERPRESETS** command. The options available in the **Render Presets Manager** dialog box are given next.

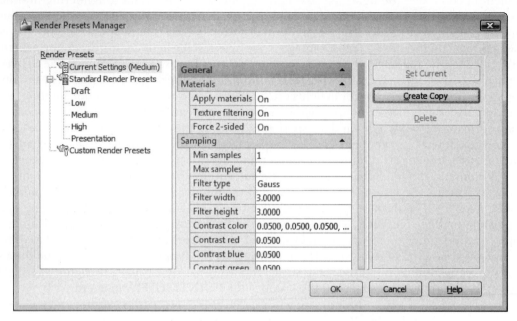

Figure 31-74 The **Render Presets Manager** *dialog box*

Render Presets List

This list displays all render presets saved in the current drawing and the render setting that is set as the current for rendering. The render presets are displayed in two categories, **Standard** and **Custom**. The **Standard Render Presets** displays the standard render presets provided in AutoCAD. The **Custom Render Presets** displays the user-defined render presets.

Properties Panel

This panel displays the settings of the render properties. You can also edit these properties for the custom presets, but the standard presets cannot be edited.

Set Current

This button is used to set current the selected render preset as current so that the renderer uses this preset.

Create Copy

This button is used to create a copy of the selected render preset or a new render preset based on the selected render preset. Choose this button; the **Copy Render Preset** dialog box will be displayed on the screen, see Figure 31-75. Enter a name and description about the render preset you are creating and choose the **OK** button. Now, change the render properties of the newly created render preset in the **Properties** panel.

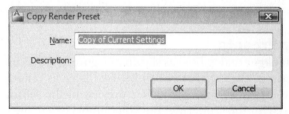

Figure 31-75 The **Copy Render Preset** *dialog box*

Delete

This button is used to delete the selected custom render preset. You cannot delete the standard render preset.

Thumbnail Viewer

This area displays the preview of the selected render settings if an image file with these render settings is selected in the **Thumbnail Image** option of the **Preset Info** area of the **Properties** panel.

General Rollout

The **General** rollout of the **Advanced Render Settings** window (Figure 31-73) has the following areas:

Render Context Area

The **Determines if file is written** button on the right of the **Render Context** area is used to specify if you want to save the rendered image as a file. The other options in the **Render Context** area are given next.

Procedure. This field is used to specify the portion of the model to be included in the rendering. When you click on this field, it turns into a drop-down list. The **View** option renders the current view of the model in the present window without opening the render window. The **Crop** option allows you to specify an area of the drawing to be included in the rendering. When you choose the **Render** button, you will be prompted to specify an area on the screen before rendering. The **Crop** option is available only if you select the **Viewport** option in the **Destination** drop-down list. This option can also be invoked by the **RENDERCROP** command. The **Selected** option is used to render only the selected objects.

Destination. This field is used to specify where to render and show the model. The **Window** option opens a separate **Render** window to render and display the rendered image. The **Viewport** option renders the model in the current viewport only and displays the rendered image in the drawing itself.

Output file name. This edit box is used to specify the file name, file type, and location to save the rendered image. This option will be available only when you choose the **Save File** button on the right of the **Render Context** area.

Output size. This field is used to select the standard output resolution of the rendered image file. However, you can also specify the desired value of width and height. To do so, click on this field and choose the **Specify Output Size** option; the **Output Size** dialog box will be displayed, see Figure 31-76.

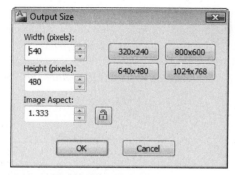

Exposure Type. This field is used to control the tone settings. The **Automatic** option changes the tone operator according to the current viewport tone operator settings. The **Logarithmic** option is used to apply the log exposure control.

*Figure 31-76 The **Output Size** dialog box*

Physical Scale. This field used to specify the physical scale.

Materials Area

This area is used to specify how the materials will be included in the rendering. The options in this area are given next.

Apply materials. This option specifies whether to render the model with or without the material attached to it by the user. If you render the model by selecting the **off** option from the **Apply materials** rollout, the global material will be attached to the model while rendering.

Texture filtering. This option is used to specify how the texture maps are being filtered.

Force 2-sided. This option is used to specify whether to render one side or both the sides of a surface. If this option is on, both sides of a face will be rendered.

Sampling Area

The images on the monitor are displayed as discrete pixels and so the slant and curved edges are displayed as distorted lines. Sampling is used to remove this distortion. The higher the value of Minimum Sample and Maximum Sample, the smoother will be the resulting edges in the rendering. But the rendering time increases considerably with the increase in the number of the samples. Figure 31-77 shows a model rendered with Min. sample=1 and Max. sample=16, and Figure 31-78 shows a model with Min. sample=1/64 and Max. sample=1/4. Compare the display of the slant and curved edges in the two figures.

Figure 31-77 *Model rendered keeping the Min. sample=1 and Max. sample=16*

Figure 31-78 *Model rendered keeping the Min. sample=1/64 and Max. sample=1/4*

 Note

*Some of the options in the **Advanced Render Settings** window that are used for rendering a cropped region, selecting render presets, specifying the sampling limits, saving the rendered image to a file, and specifying the resolution size of the rendered file, can be directly changed from the **Render** panel in the **Render** tab, see Figure 31-79.*

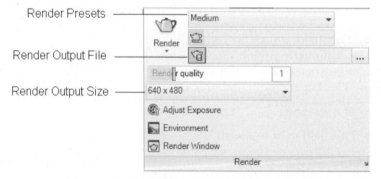

Figure 31-79 *The **Render** panel of the **Render** tab*

Shadows Area

This area is used to control the display of shadows in the rendered image. The **Bulb** button on the right of the **Shadows** area is used to specify whether to display shadows in rendering. The other options of the **Shadows** area will be highlighted only if this button is on. The options in this area are given next.

Mode. This field is used to specify the order in which the shadow is generated in the **Rendering** window. There are three modes to specify the order of the shadow generation: **Simple**, **Sorted**, and **Segment**. The **Simple** mode is the default mode of shadow generation.

Shadow map. This field is used to specify whether the shadow in rendering will be cast as a mapped image or as ray traced. If you switch on the option of **Shadow map**, the shadows generated will not be sharp. If you switch off the **Shadow map**, then the shadow will be generated as ray-traced. The shadows generated by this option are more accurate and sharp as compared to shadow map.

Sampling Multiplier. This field is used to specify the multiplying factor to control the overall shadow sampling for rendering.

Ray Tracing Rollout

This area controls the reflection and refraction caused by the model while ray-tracing. The **Bulb** button on the right of the **Ray Tracing** area is used to specify whether ray-tracing will be performed while calculating the reflection and refraction. All the other options of this area will be available only if this button is on. The other options in this area are discussed next.

Max reflections

This option controls the total number of times the ray will be reflected by the model, after which the ray-tracing will stop.

Max refractions

This option controls the total number of times the ray will be refracted by the model, after which the ray-tracing will stop.

Max depth

This option controls the total number of times, a ray reflects and refracts, after which the ray-tracing will stop. This is irrespective of whether the ray has attained the limits set using the **Max reflections** and **Min refractions** options.

Indirect Illumination Rollout

The **Indirect Illumination** rollout of the **Advanced Render Settings** window has the following areas:

Global Illumination Area

This area controls the effect of colors in the indirectly illuminated portions. For example, if a white object has dark blue objects surrounding it, then this white object will appear light blue in color. To calculate such indirect illumination effects, the global illumination uses a photon map. The quality and accuracy of the illumination is controlled by the **Photons/sample** and **Radius** options.

Photons/sample. The brightness of the indirect illumination depends on the number of photons. The larger the number of photons, the brighter will be the indirect illumination. But in this case, the sharpness of the indirect illumination will decrease.

Use radius. If this option is turned on, the photon radius is decided by the value specified in the **Radius** option. If this option is turned off, each photon radius becomes equal to 1/10 of the radius of the full scene.

Radius. Increasing the radius of the photon increases the amount of smoothing. This results into more natural indirect illumination.

Max depth/Max reflections/Max refractions. These options are similar to those explained in the **Ray Tracing** area. Here, these options are applicable for the photons used for indirect illumination.

Final Gather Area

Final gathering is used to improve global illumination. It increases the number of rays used to calculate global illumination. Because global illumination is calculated from a photon map, dark corners in the lighting may occur. By activating the final gathering, you can increase the number of rays used to calculate global illumination, which reduces these defects.

Mode. This field is used to turn the final gathering on, off, or set it to automatic mode so that the final gathering can be turned on, off, or function automatically, depending on the status of the skylight.

Rays. This option is used to specify the number of rays required for indirect illumination.

Radius mode. This option is used to select the radius mode for the final gathering calculation. If the radius mode is turned on, the radius specified in the **Max radius** option in world units is used for the final gathering. If the radius mode is turned off, the maximum radius is the default value of 10 percent of the maximum model radius, in world units. If the **View** option is selected in the **Radius mode** edit box, then the maximum radius will be considered in terms of pixels instead of world units.

Max radius. This option controls the maximum radius within which the final gathering will be calculated. Reducing this value increases the quality of rendering.

Use min. This option controls whether the minimum radius is to be included in the process of the final gathering.

Min radius. This option controls the minimum radius for the final gathering. Increasing this value improves the quality of rendering.

Light Properties Area

This area is used to define the properties of the lights used for indirect illumination. However, the properties defined here affect all the lights present in the drawing. The options in this area are given next.

Photons/light. This option controls the number of photons emitted per light for calculating the global illumination. Increasing this number, increases the quality of the rendered image.

Energy multiplier. This option multiplies the intensity of the indirect illumination with the factor specified in the **Energy multiplier** field.

Diagnostic Rollout

The **Diagnostic** rollout of the **Advanced Render Settings** window has the following areas:

Visual Area

The options in this area are used to control the visibility of visual aids such as grids, photon density, and so on during rendering. The options in this area are given next.

Grid. This option is used to render the model such that its coordinate spaces are displayed on the model. The coordinates can be displayed in three different ways: **Object**, **World**, and **Camera**. By default, it is set to **OFF**. In the **Object** option, the local coordinate spaces of each object are displayed on it. In the **World** option, the world coordinate space is displayed and it remains the same for all objects. In the **Camera** option, the camera coordinate spaces are displayed on object. The camera coordinates consist of rectangular coordinates superimposed on the model from the viewing direction, see Figure 31-80.

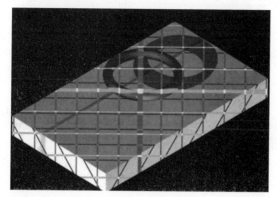

Figure 31-80 Rendered model with camera coordinates

Grid size. This option is used to specify the size of the grid to be displayed on the rendered model.

Photon. This option is used to display the distribution of photons on the rendered model. The global illumination should be **ON** to display the photon distribution. The **Density** option displays the distribution of the photon density. The location of the highest photon density is displayed in red and the location of the minimum photon density is displayed in light color, see Figure 31-81. The **Irradiance** option displays the photon based on its irradiance.

The maximum irradiance is displayed in red color and the minimum irradiance is displayed in light color.

Samples. By default, this option is turned off. Figure 31-82 shows the rendered model with the **Samples** option turned on.

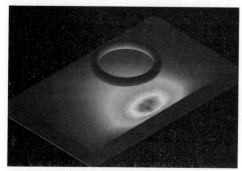

Figure 31-81 Rendered model displaying the photon density

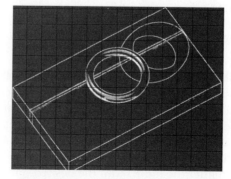

Figure 31-82 Rendered model with the Samples option turned on

BSP. This option renders a visualization of the parameters defined for the tree. The **Depth** option shows the depth of the tree with top faces in bright red and deep faces in increasingly light colors. The **Size** option shows the size of leaves in the tree with differently sized leaves indicated by different colors.

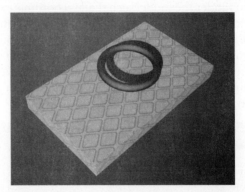

Figure 31-83 The rendered model with the gradient background

Processing Rollout

The **Processing** rollout of the **Advanced Render Settings** window has the following options:

Tile size

This option is used to specify the size of the tile that is displayed on the rendering screen while carrying out the rendering processes. The rendering time is inversely proportional to the size of the tile.

Tile order

This option is used to specify the order of appearance of the next tile on the render screen. Choose the tile order depending on the order in which you want the rendered image to appear on the screen. There are six orders in which the next tile appears on the screen: **Hilbert**, **Spiral**, **Left to right**, **Right to left**, **Top to bottom**, and **Bottom to top**. **Hilbert** is the default tile order.

Memory limit

This option controls the memory allocated to the rendering processes.

CONTROLLING THE RENDERING ENVIRONMENT

You can enhance the rendering by adding effects such as the surrounding environment of the model or by adding colors or images to the background of the model. The commands to do so are discussed next.

Rendering with a Background

You can render a model by changing its background. You can set a single color, combination of two colors, combination of three colors, an image, or sun and sky as the background of the model. To render the model with a background, you need to save a named view of the model with a background by using the **VIEW** command, as explained in the **View Manager** dialog box (refer to Chapter-6). Next, set the named view with a background as the current view and then render the model with the desired render settings. Figure 31-83 shows a rendered model with its named view saved with the gradient background, Figure 31-84 shows a rendered model with its named view saved with an image as the background, and Figure 31-85 shows a rendered model with its named view saved with the sun and sky as the background.

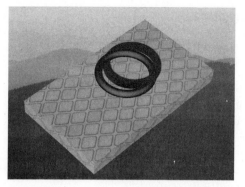

Figure 31-84 *The rendered model with* **Image** *as the background*

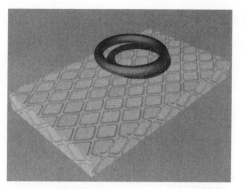

Figure 31-85 *The rendered model with the* **Sun & Sky** *as the background*

Rendering with the Fog Effect

Ribbon: Render > Render > Environment   **Toolbar:** Render > Render Environment
Menu Bar: View > Render > Render Environment   **Command:** FOG

You can use this option to add a misty effect to an object when it is rendered. You can also assign a color to the fog. When you choose the **Environment** tool from the **Render** tab, the **Render Environment** dialog box is displayed, as shown in Figure 31-86. This dialog box can also be invoked using the **FOG** command.

Click on the **Enable Fog** field and select **On** from the drop-down list to enable the fog. Enabling the fog does not affect the other settings of the rendering. Click on the color edit box to specify the color of the fog using various options specified in this drop-down list. Click on the **Fog Background** field and select **On** from the drop-down list to enable the Fog and Background options at the same time. The **Near Distance** and **Far Distance** options define where the fog begins and ends. The values specified in the **Near Distance** and **Far Distance** edit boxes are the percentages of the distances between the camera and the back clipping plane. The **Near Fog Percentage** and **Far Fog Percentage** edit boxes describe the percentage of the fog at the near and far distances. Figure 31-87 shows a rendered scene displaying the fog effect.

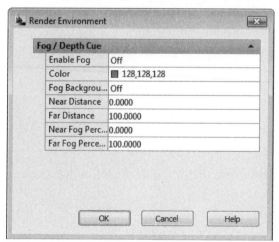

Figure 31-86 *The* **Render Environment** *dialog box*

Figure 31-87 *A design rendered with the fog*

Adjusting the Lighting Exposure to Rendered Image

Ribbon: Render > Render > Adjust Exposure
Command: RENDEREXPOSURE

 You can control the global lighting effect of a model dynamically even after it has been rendered. This saves a lot of time needed to render the drawing each time when you make some change in the render settings. The settings changed at this stage will only affect the most recent rendered output file. To change the settings after rendering, choose the **Adjust Exposure** tool from the **Render** panel in the **Render** tab; the **Adjust Rendered Exposure** dialog box will be displayed, with the thumbnail preview of the most recent rendered file, see Figure 31-88. You can use this dialog box to control the brightness, contrast, mid tones, and the exposure of exterior objects illuminated by the sun. You can also specify whether to include the background for exposure control or not.

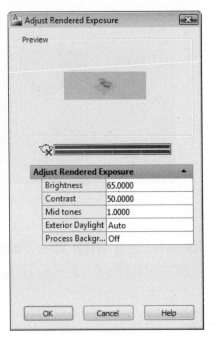

*Figure 31-88 The **Adjust Rendered Exposure** dialog box*

RENDERING A MODEL WITH DIFFERENT RENDER SETTINGS

The rendering depends on the view that is current and the lights that are defined in the drawing. Sometimes, the current view or the lighting setup may not be enough to show all features of an object. You might need different views with a certain light configuration to show different features of the object. But when you change the view or define the lights for a rendering, the previous setup is lost. To bypass this, you can use the rendering information in the history pane of the **Render** window. In the lower portion of the **Render** window, the history pane is located. In this pane, you can get the recent history of rendered images of the model in the drawing, see Figure 31-89. From this history pane, you can regain the previous rendering settings and can also save the rendered images. The data stored in the history pane includes: file name of the rendering, image size, view name (if no named view is used, the view is stored as current view), render time, and the name of the render preset used for the rendering.

Output File Name	Output Size	View	Render time	Render preset	
c26exm4-Temp033	1024 x 768	Current	00:03:02	Presentation	
c26exm4-Temp032	320 x 240	Current	00:00:28	High	
c26exm4-Temp031	640 x 480	Current	00:00:12	Medium	
c26exm4-Temp029	640 x 480	Current	00:00:44	Medium	

*Figure 31-89 The history pane in the **Render** window*

Right-clicking on a history entry displays a shortcut menu that contains the following options:

Render Again

This option restarts the renderer for the selected history entry with the same settings.

Save

This option displays the **Render Output File** dialog box where you can specify the name and format of the image file. Choose the **Save** button; the **BMP Image Options** dialog box will be displayed. You can choose the color quality of the image file from this dialog box, see Figure 31-90. Select the desired radio button and then choose the **OK** button to save the image file.

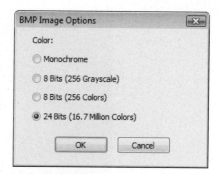

Save Copy

This option saves the image to a new location without affecting the location stored in the entry.

*Figure 31-90 The **BMP Image Options** dialog box*

Make Render Settings Current

This option loads the render settings associated with the selected history entry, if multiple history entries are present with different render presets.

Remove From The List

This option removes the entry from the history pane.

Delete Output File

This option removes the rendered image from the **Image** pane, but the entry remains in the history pane.

> **Note**
> *Any history entry that is not saved with the drawing only exists for the duration of the current drawing session. Once you close the file or exit the program, the unsaved history entries are lost.*

EXAMPLE 4

In this example, you will open the model used in Example 2 of this chapter and then create two rendering settings. Next, you will render the model with these settings and save the rendered image in a file. The first setting should display the effect of light 1 and the second setting should display the effect of light 2.

1. Open the drawing of Example 2.

2. Choose the **Light glyph display** tool from the **Lights** panel of the **Render** tab.

3. Select light 2 from the **Lights in Model** palette and choose the **Properties** option from the shortcut menu; the **Properties** palette is displayed. In this palette, select the **Off** option from the **On/Off Status** drop-down list in the **General** rollout.

4. Choose the **Render** tool from the **Render** panel in the **Render** tab to perform rendering. The rendered model is displayed in the **Render** window with the effect of light 1 only, as shown in Figure 31-91.

5. Next, switch off light 1 and switch on light 2, as explained in the previous steps, and then render the model. The rendered model is displayed in the **Render** window with the effect of light 2 only, as shown in Figure 31-92.

Figure 31-91 *Rendered model with the effect of light 1*

Figure 31-92 *Rendered model with the effect of light 2*

6. The history pane of the **Render** window displays both the render entries. Right click on one entry and choose the **Save** option from the shortcut menu. Now, specify the location, name, and format in which you want to save the image file in the **Render Output File** dialog box and choose the **Save** button.

7. Specify the **Color Mode** and choose the **OK** button in the **BMP Image Options** dialog box.

8. Similarly, save the second history entry, as explained in Steps 6 & 7.

Note
*In future, to render the model with the same rendering settings, select the history entry with that setting and choose the **Make Render Settings Current** option from the shortcut menu.*

OBTAINING RENDERING INFORMATION

The right side area of the **Render** window is known as the statistics pane. This portion displays the details about the rendering and render settings in effect when the image was created. The information stated in the statistics pane is derived from the settings made in the **Render Presets Manager** dialog box and the information that is generated at the time of the rendering.

SAVING A RENDERED IMAGE

A rendered image can be saved by rendering to a file or by rendering to the screen and then saving it. Redisplaying a saved rendering image requires very less time compared to the time involved in rendering. Various methods of saving the rendered image are discussed next.

Saving the Rendered Image to a File

You can save a rendered image directly to a file. One of the advantages of saving a rendered image to a file is that you can redisplay the rendered image in less time as compared to rendering the image again. Another advantage of saving a rendered image is that when you render a file in the viewport, the resolution of the image on rendering depends on the resolution of your current display. Whereas, if you render the image to a file, you can specify a higher resolution. Later, you can display this rendered image that has a higher resolution. The rendered images can be saved in different formats, such as BMP, PCX, TGA, TIF, JPEG, and PNG. The following steps explain the procedure of saving a rendered image to a file.

1. Choose the **Render Output File** tool from the **Render** panel of the **Render** tab to activate it.

2. Choose on the **Browse for File** button; the **Render Output File** dialog box is displayed.

3. Specify the file name and file type. Then, choose the **Save** button; the **BMP Image Options** dialog box is displayed.

4. Specify the Color Mode and choose **OK** in the **BMP Image Options** dialog box.

5. Choose the **Render** tool from the **Render** panel of the **Render** tab. The rendered image is saved at the specified location.

Saving the Viewport Rendering

In the **Advanced Render Settings** window, if you have selected the **Viewport** option in the **Destination** field, then the model will be rendered in the viewport. A rendered image in the viewport can be saved by using the **Save** option. To do so, choose the **Save** option from the **Tools > Display Image** in the menu bar; the **Render Output File** dialog box will be displayed. This dialog box is used to specify the file name and type. The valid output file formats are BMP, PCX, TGA, TIF, JPEG and PNG.

The steps given next explain the procedure of saving a viewport rendering.

1. Choose the **Save** option from **Tools > Display Image** in the menu bar to invoke the **Render Output File** dialog box.

2. Specify the file name and file type. Then, choose the **Save** button.

3. Specify the Color Mode and choose the **OK** button in the **BMP Image Options** dialog box.

4. Choose the Render tool from the Render panel in the Render tab. The rendered image is saved at the specified location.

Saving the Rendered Image from the Render Window

A rendered image in the **Render** window can be saved using the **Save** option from the **File** menu of the **Render** window. In this case, the rendered image can be saved in any of these formats: BMP, PCX, TGA, TIF, JPEG, and PNG. The following steps explain the procedure for saving a render-window image:

1. Choose the **Render** tool from the **Render** tab; the rendered image will be displayed in the **Render** window.

2. Choose the **Save** option from the shortcut menu displayed on right-clicking on the temporary image. Alternatively, use the **File** menu of the **Render** window to save the rendered image; the **Render Output File** dialog box is displayed.

3. Specify the file name and file type. Then, choose the **Save** button.

4. Specify the Color Mode and choose the **OK** button in the **BMP Image Options** dialog box.

PLOTTING RENDERED IMAGES

Plotting the rendered design was one of the biggest problems in the earlier releases of AutoCAD. You had to go through a number of steps to plot a rendered drawing. However, from AutoCAD 2007, plotting of rendered designs became extremely easy. To plot a rendered design, use the following steps:

1. Open the design to be plotted and then invoke the **Plot** dialog box. Remember that you do not need to first render the design and then invoke the **Plot** dialog box. You can also render the design while plotting. However, the render settings should be specified in advance.

2. Select the printer that supports rendered printing from the **Printer/plotter** area and select the other options.

3. Select the **Rendered** option from the **Shade plot** drop-down list in the **Shaded viewport options** area. If this area is not available by default, then choose **More Options** from the bottom right corner of the **Plot** dialog box.

4. Set the other options in this dialog box and then plot the design. The design will be first rendered and then plotted in the rendered form.

 Note
*If the **Rendered** option is used for a highly complex set of objects, the hardcopy output might contain only the viewport border.*

UNLOADING AutoCAD RENDER

When you invoke any AutoCAD **RENDER** command, AutoCAD Render is loaded automatically.

If you do not need AutoCAD Render, you can unload it by entering the **ARX** at the Command prompt. You can reload AutoCAD Render by invoking the **RENDER** command or any other command associated with rendering (such as **FOG**, **LIGHT**, and so on).

Command: **ARX**
Enter an option [?/Load/Unload/Commands/Options]: **U**
Enter ARX/DBX file name to unload: **ACRENDER**
ACRENDER successfully unloaded.

WORKING WITH CAMERAS

Cameras are used to define the 3d perspective view of a model. In AutoCAD, you can place a camera, edit camera settings, turn on/off the camera and save the perspective view. By default, the **Camera** panel is not displayed in the **Render** tab. To add this panel, right-click on the **Ribbon** area and choose **Panels > Camera** from the shortcut menu.

Create Camera

Ribbon: Render > Camera > Create Camera *(Customize to add)*
Command: CAMERA

To place a camera, choose the **Create Camera** tool from the **Render** panel; you will be prompted to specify the camera location. Specify the camera position; the preview of the camera focus will be displayed and you will be prompted to specify the target location. Click in the drawing area to specify the required target location; the **Enter an option [?/Name/LOcation/Height/Target/LEns/Clipping/View/eXit]<eXit>:** prompt will be displayed. The options in the prompt are discussed next.

?

This option lists all cameras created in the current drawing.

Name

This option is used to specify a name for the camera.

LOcation

This option is used to change the location of the camera.

Height

This option is used to specify the height of the camera.

Target

This option is used to modify the target of the camera.

LEns

This option is used to define the magnification length or zoom factor of camera lens. Note that greater the lens length, narrower will be the view.

Clipping

Clipping planes define the boundaries for a view. On choosing this option, you can specify the location of the front and back clipping planes. On defining the front and back clipping planes, everything between the camera and front clipping plane and the target and back clipping plane will be hidden.

If you choose this option, a message prompting you to enable the front clipping plane will be displayed. If you choose **Yes**, then you will be prompted to specify the offset distance of the clipping plane from the target plane. Specify the required distance and press ENTER; a message will be displayed prompting you to enable the back clipping plane. If you choose **Yes**, you will be prompted to specify the offset distance of the clipping plane from the target plane.

Figure 31-93 shows the model with the front and back clipping planes defined at 25 units from the target point. Figure 31-94 shows the preview of the resultant model.

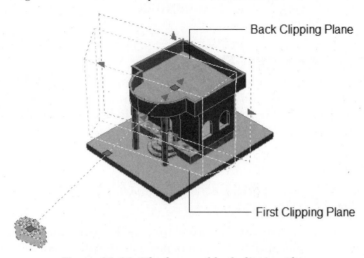

Figure 31-93 *The front and back clipping planes*

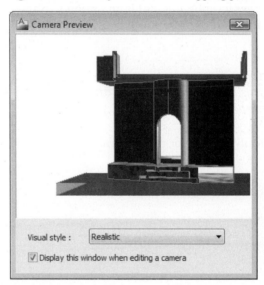

Figure 31-94 *The preview of the model*

You can turn off the display of front, back, or both clipping planes, display both the clipping planes, or adjust the offset values of the clipping planes using the **Properties** palette. To invoke the **Properties** palette, select the camera, right-click, and then choose the **Properties** option. You can also dynamically adjust the position of the clipping planes by using the grips displayed on selecting the camera.

View

This option is used to set the current view as the camera view.

eXit

This option is used to exit the tool.

Editing the Cameras

To edit a camera, select it from the drawing area; the **Camera Preview** dialog box will be displayed, as shown in Figure 31-95. Also, the grips will be displayed on the length of the camera. You can adjust the focus of the camera by dragging these grips. The move gizmo will also be displayed when you move your cursor toward the source or target grip. You can use this gizmo to move the camera in the desired direction. The corresponding changes can also be viewed in the **Camera Preview** dialog box. By default, the preview of the model will be displayed in the Wireframe mode, but we can change it to other visual styles. You can do so by using the **Visual style** drop-down in the **Camera Preview** dialog box.

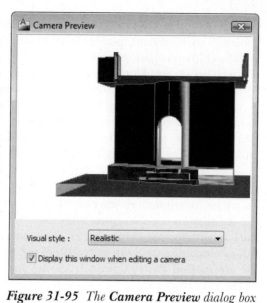

*Figure 31-95 The **Camera Preview** dialog box*

 Note

*The **Camera Preview** dialog box is displayed on selecting the camera because by default the*

***Display this window when editing a camera** check box is selected in the **Camera Preview** dialog box. If you do not want the dialog box to be displayed, you need to clear this check box. If you want to view it later, then select the camera, right-click, and then choose **View Camera Preview**; the **Camera Preview** dialog box will be displayed.*

Note that you can have more than one camera in one drawing. You can edit the camera properties by using the **Properties** palette. Using this palette, you can modify a camera's lens length, change its front and back clipping planes, change the name of the camera, and turn the display of all cameras on or off in a drawing.

You can set the camera view as the current view by selecting the camera from the **Views** drop-down list in the **View** tab. Then, you can save the view by using the **View Manager** dialog box. You can also set the camera view by selecting the camera and then choosing the **Set Camera View** option from the shortcut menu displayed on right-clicking.

 Note

*To toggle the display of camera glyph, choose the **Show Cameras** button from the **Camera** panel of the **Render** tab.*

You can set the camera template from the **Tool Palette**. To do so, invoke the **Tool Palette** and right-click on its title bar; a shortcut menu will be displayed. Select **Cameras** from the shortcut menu and then select the required option. The three camera options available in the menu are: **Normal Camera**, **Wide-angle Camera**, and **Extreme Wide-angle camera**.

CREATING ANIMATIONS

You can create a walk through or fly through animation by using the motion path animation. You can record and play back these presentations, whenever required. This helps you communicate your design intent dynamically and with more impact. The tools to create an animation are not displayed in the **Ribbon** by default. To display them, choose the **Render** tab and right-click on any one of the buttons; a shortcut menu will be displayed. Choose **Panels > Animations** from the shortcut menu; the **Animations** panel will be added to the **Ribbon**. The methods of creating animations are discussed next.

Creating Animation of 3D Navigations

You can create, record, playback, and save animations of any 3D navigation modes, such as **Walk** and **Fly** navigation modes. Before creating animations, you should specify the animation settings. To do so, choose the **Walk** or **Fly** button and right-click. Next, Choose **Animation Settings** from the shortcut menu; the **Animation Settings** dialog box will be displayed, see Figure 31-96. In this dialog box, you can specify the visual style in which you want to view the model in animation, resolution of the frames, number of frames to be captured in one second, and the format (AVI, MPG, or WMV) in which you want to save the file while creating the animation of the 3D navigation.

Recording Animation

To create an animation, start any 3D navigation command; the **Animation Record** button in the **Animations** panel in the **Tools** tab of the **Ribbon** will be enabled. Choose this button to start recording, see Figure 31-97. Navigate through the drawing model and all navigations will get recorded. You can also change the navigation mode while recording. To do so, choose the **Animation Pause** button from the **Animations** panel of the **Ribbon**. Next, right-click in the drawing area and choose an option from the **Other Navigation Modes**. Choose the **Animation Record** button again and the recording will start with the changed 3D navigation mode without any break in recording. If you are in the **Walk** or **Fly** navigation mode, you can also adjust the settings from the **POSITION LOCATOR** window, in the same way as explained above.

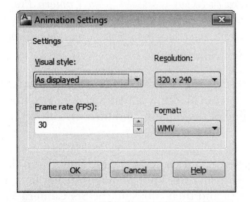

Figure 31-96 The Animation Settings dialog box

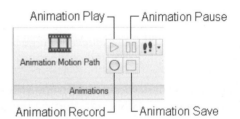

Figure 31-97 The Animations panel in the Render tab

Playing Animation

To preview a recorded animation, choose the **Animation Play** tool from the **Animations** panel in the **Render** tab; the recording will pause and the **Animation Preview** window will be displayed, see Figure 31-98. The **Animation Preview** window displays the animation. If you want to preview the animation in a different visual style, you can select another style from the drop-down list. The options in the visual style drop-down list are **Current**, **3D Hidden**, **3D Wireframe**, **Conceptual**, **Hidden**, **Realistic**, **Shaded**, **Shaded with edges**, **Shades of Gray**, **Sketchy**, **Wireframe**, and **X-Ray**. After playing the animation, you can also resume the recording by choosing the **Animation Record** button displayed in red color in the **Animation Preview** window.

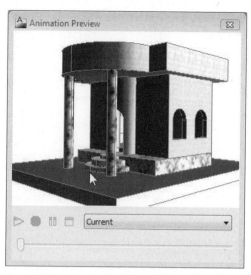

Figure 31-98 The Animation Preview window

Saving Animation

If you are satisfied with the animation playback, choose the **Save** button in the **Animation Preview** dialog box or the **Animations** panel; the **Save As** dialog box will be displayed on the screen. Select the file location, file name, type of file and then choose the **Save** button. In **Files of type** drop-down list of the **Save As** dialog box, you will get only the file type selected in the **Animation Settings** dialog box. To change the file type, choose the **Animation Settings** button from the **Save As** dialog box; the **Animation Settings** dialog box will be displayed. Select the desired file type from the **Format** drop-down list of the **Animation Settings** dialog box.

Creating Animation by Defining the Path of the Camera Movement

Ribbon: Render > Animations > Animation Motion Path   **Command:** ANIPATH
Menu Bar: View > Motion Path Animations

In this method of creating an animation, you can control the camera motion by linking the camera movement and its target to a point or a path. If you want the camera or target to remain still while creating animation, link the camera or target to a point. However, if you want the camera or target to move along a path while creating animation, link the camera or target to a path. To create a motion path animation, choose the **Animation Motion Path** tool from the **Animations** panel in the **Render** tab; the **Motion Path Animation** dialog box will be displayed, see Figure 31-99. Using this dialog box, specify the settings for the camera motion path and create the animation file. The options of **Motion Path Animation** dialog box are discussed next.

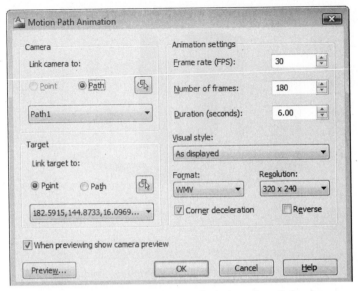

Figure 31-99 The **Motion Path Animation** *dialog box*

Camera Area

The options in this area are used to specify the camera movement. Select the **Point** or **Path** radio button, based on your requirement. Select the **Point** radio button, if you want the camera to remain stationary at the specified point in the drawing throughout the animation. Select the **Path** radio button, if you want the camera to move along the specified path in the drawing. Next, choose the **Pick Point** / **Select Path** button

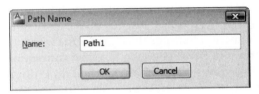

Figure 31-100 The **Path Name** *dialog box*

available next to the radio buttons to select the point/path in the drawing. Specify the point/ path in the drawing; the **Point Name** or **Path Name** dialog box will be displayed, depending on whether the **Point** or **Path** radio button is selected as the camera link, respectively, see Figure 31-100. Specify a name for the point/path and choose the **OK** button. You can also select a previously created point/path from the drop-down list that displays a list of named points or paths.

Target Area

The options in this area are used to specify the movement of the target with the movement of the camera. Select the **Point** or **Path** radio button, based on your requirement. To specify a target point, choose the **Pick Point** button and specify a point in the drawing. Enter a name for the point and choose the **OK** button in the **Point Name** dialog box. To specify a target path, click the **Select Path** button and specify a path in the drawing. Enter a name for the path and choose the **OK** button in the **Path Name** dialog box. To specify an existing target point or path, select it from the drop-down list.

Note
You cannot link both the camera and the target to a point.

To create a path, you can link a camera to a line, arc, elliptical arc, circle, polyline, 3D polyline, or spline.

Tip
You must create the path object before you create the motion path animation. The path that you create will not be visible in the animation.

Animation settings Area

The options in this area are used to control the output quality and display of the animation file. The options in this area are discussed next.

Frame rate (FPS). The frames per second is the speed at which the animation will be played. The value of the FPS can vary from 1 to 60. The default value for the FPS is 30.

Number of frames. This option is used to specify the total number of frames to be created in the animation. This value divided by the FPS determines the duration of the animation file.

Duration (seconds). This option specifies the duration of the animation in seconds. Any change in the duration automatically resets the **Number of frames** value.

Visual style. This option displays a drop-down list of the visual styles and the render presets that can be applied to an animation file.

Note
You cannot create an animation of 3D navigation with a rendered model, however, you can create it by using a motion path.

Format. This drop-down list is used to specify the file format for the animation. You can save an animation to an **AVI, MOV, MPG**, and **WMV** file format.

Note
*The **MOV** format is available only if QuickTime Player is installed. The **WMV** format is available only if Microsoft Windows Media Player 9 or later is installed. Otherwise, the **AVI** format is the default selection.*

Resolution. This drop-down list is used to specify the resolution of the resulting animation in terms of the width and height. The default value is 320 x 240.

Corner deceleration. This check box is used to specify the speed of the camera movement at the corners. If this check box is selected, the camera will move at a slower rate as it turns at a corner.

Reverse. This check box, if selected, will reverse the direction of the animation display.

When previewing show camera preview. This check box, if selected, will display the **Animation Preview** dialog box when you choose the **Preview** button so that you can preview the animation before you save it.

Preview. This button is used to get a preview of the animation in the **Animation Preview** dialog box, before saving it.

After you have finished adjusting the points, paths, and settings, choose **Preview** to view the animation, Next, choose the **OK** button to save the animation. The **Save As** dialog box will be displayed. Specify the location, file name, and file type of the animation to save and choose the **OK** button to save the animation.

EXAMPLE 5

In this example, you will create and animate the model shown in Figure 31-101 and then save this animation to a file.

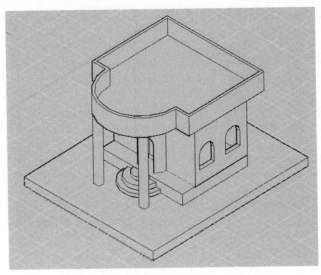

Figure 31-101 *Model for Example 5*

1. Create the model shown in Figure 31-101.

2. Select the materials of your choice and attach them to the model, as described in Example 1. The rendered image of the model is shown in Figure 31-102.

3. Change the view of the model to the top view and create the path to attach the camera, see Figure 31-103.

 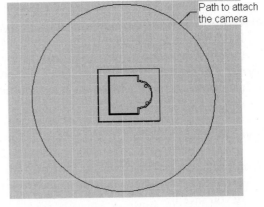

Figure 31-102 *Rendered model for Example 5*   **Figure 31-103** *Top view of the model displaying the path to attach the camera*

4. Move the path created in the previous step along the Z axis by a distance equal to half the height of the model.

5. Invoke the **Animation Motion Path** tool from the **Animations** panel in the **Render** tab; the **Motion Path Animation** dialog box is displayed (Figure 31-104).

6. Select the **Path** radio button and choose the **Select Path** button in the **Camera** area of the

Motion Path Animation dialog box. Next, select the path created in the previous step; the **Path Name** dialog box is displayed. Enter the name of the path as **Path1**.

7. Select the **Point** radio button and choose the **Select Point** button in the **Target** area of the **Motion Path Animation** dialog box. Next, specify a point in the middle of the model, which will act as the target of the camera. On doing so, the **Point Name** dialog box is displayed. In this dialog box, enter the name as **Point1**.

8. Change the values of the **Animation settings** area to the values shown in Figure 31-104.

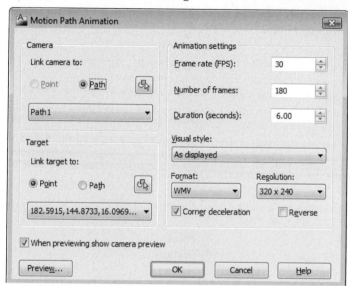

*Figure 31-104 The **Motion Path Animation** dialog box for Example 5*

9. Choose the **Preview** button to take a preview of the animation in the **Animation Preview** dialog box. Next, close the **Animation Preview** dialog box.

10. Choose the **OK** button to save the animation; the **Save As** dialog box is displayed. Specify the location, file name, and file type of the animation to save and then choose the **Save** button from the **Save As** dialog box.

Self-Evaluation Test

Answer the following questions and then compare them to those given at the end of this chapter:

1. A rendered image makes it easier to visualize the shape and size of a 3D object. (T/F)

2. You can unload AutoCAD Render by invoking the **ARX** command. (T/F)

3. The falloff occurs with a distant source of light. (T/F)

4. You can change the color of the ambient light. (T/F)

5. You cannot take the printout of a model without rendering it. (T/F)

6. The _____ command is used to attach material to a layer in a drawing.

7. The _____ source emits a focused beam of light in the defined direction.

8. The _____ and _____ commands are used to adjust the coordinates and the bitmap of the pattern of the material attached to the solid model.

9. The _____ tool is used to apply a fog effect to the rendering.

10. A _____ light source emits light in all directions, and the intensity of the emitted light is uniform.

Review Questions

Answer the following questions:

1. You can also import materials to the current drawing directly using the **RMAT** command. (T/F)

2. You can assign materials to an object using the AutoCAD Color Index. (T/F)

3. The phenomenon of attenuation occurs in case of direction lights. (T/F)

4. You can link both the camera and the target to a point. (T/F)

5. Which light does not have a source, and therefore, no location or direction?

    (a) **Default**   (b) **Point**
    (c) **Spot**   (d) **Distant**

6. The intensity of light decreases as the distance increases. This phenomenon is called:

    (a) **Attenuation**   (b) **Frequency**
    (c) **Light Effect**   (d) None of these

7. Which of the following lights allows you to define the geographic location?

    (a) **Sun**   (b) **Point**
    (c) **Spot**   (d) **Distant**

8. Which of the following commands can be used to define the background of an object while it is being rendered?

    (a) **FOG**   (b) **ARX**
    (c) **BACKGROUND**   (d) **VIEW**

9. Which of the following commands is used to create a motion path animation?

    (a) **ANIMAP**   (b) **ANITRAC**
    (c) **ANIPATH**   (d) **ANIMATION**

10. In the _____ source, the brightness of the incident light on an object is inversely proportional to the distance of the object from the light source.

11. The _____ command allows you to set the rendering preferences for the rendering.

12. A spotlight source emits a _____ beam of light in one direction only.

13. Redisplaying a saved rendered image takes _____ time as compared to the time involved in rendering.

14. The _____ command is used to render only the selected portion of a drawing.

15. Attenuation is defined as _____.

EXERCISE 1

Create the 3D drawings, shown in Figure 31-105. Next, render the drawing after applying materials and inserting lights at appropriate locations to get a realistic model. Assume the dimensions.

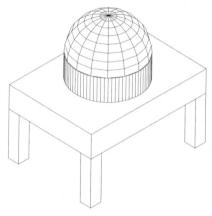

Figure 31-105 *Drawing for Exercise 1*

EXERCISE 2

Create the model shown in Figure 31-106 and apply materials of your choice. Apply the Point light and the Spotlight at the sill of the window and at bottom of the door. Also, attach a background image to it and then render the model. Assume the missing dimensions.

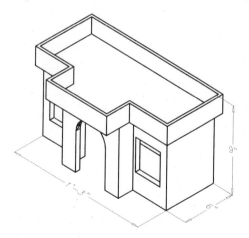

Figure 31-106 *Model for Exercise 2*

EXERCISE 3

Create the model as shown in Figure 31-107 and render with the effect of a fog. Choose your own materials.

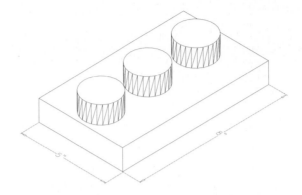

Figure 31-107 *Model for Exercise 3*

 Tip
*The display of fog is based on the trial and error method. Use different combinations of options available in the **Fog** dialog box. Use white color as the background to clearly visualize the fog.*

EXERCISE 4

Render the Tool Organizer created in Chapter 28. Assign a dark red wood material to the peg board and white oak to the remaining members, Figure 31-108.

Figure 31-108 *Model for Exercise 4*

EXERCISE 5

Render the Work Bench of Chapter 28. Assign white oak to all members. Adjust the mapping coordinates so that the rendered image appears as shown in Figure 31-109.

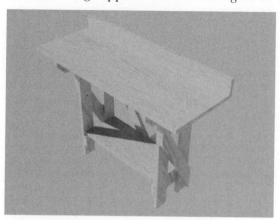

Figure 31-109 *Model for Exercise 5*

EXERCISE 6

Create the solid model shown in Figure 31-110. Also, assign brass material to the solid and render it.**

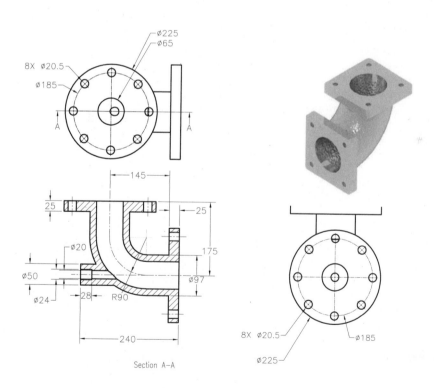

Figure 31-110 *Views and dimensions of the model for Exercise 6*

EXERCISE 7

Create the model shown in Figure 31-111. Also, assign materials and textures to different parts of the model and then render it. **

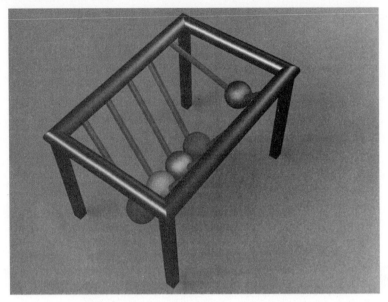

Figure 31-111 Rendered image of the model

Answers to Self-Evaluation Test

1. T, **2.** T, **3.** F, **4.** F, **5.** F, **6.** **MATERIALATTACH**, **7.** spotlight, **8.** **MATERIALMAP, SETUV,** **9. Environment, 10.** distant

Chapter 32

AutoCAD on Internet

CHAPTER OBJECTIVES

In this chapter, you will learn:
- *To launch a Web browser from AutoCAD.*
- *About the importance of the Uniform Resource Locator (URL).*
- *To open and save drawings to and from the Internet.*
- *To place hyperlinks in a drawing.*
- *About the Autodesk Express Viewer plug-in and the DWF file format.*
- *To view DWF files with a Web browser.*
- *To convert drawings into the DWF file format.*

KEY TERMS

- *Browser*
- *Hyperlink*
- *DWF*
- *URL*
- *i-Drop*
- *Export to DWF*

INTRODUCTION

The Internet has become the most important and the fastest way to exchange information in the world. AutoCAD allows you to interact with the Internet in several ways. It can open and save files located on the Internet; it can launch a Web browser; and it can create Drawing Web Format (DWF) files for viewing as drawings on Web pages. Before you can use the Internet features in AutoCAD 2012, some components of Microsoft Internet Explorer must be present in your computer. If Internet Explorer version 6.1 (or later) is already installed, then you have these required components. If not, then the components are installed automatically during AutoCAD 2012's setup when you select: (1) the **Full Install**; or (2) the **Internet Tools** option during the **Custom** installation.

This chapter introduces the following Web-related commands:

BROWSER

Launches a Web browser from within AutoCAD.

HYPERLINK

Attaches and removes a uniform resource locator (URL) to an object or an area in the drawing.

HYPERLINKFWD

Move to the next hyperlink (an undocumented command).

HYPERLINKBACK

Moves back to the previous hyperlink (an undocumented command).

HYPERLINKSTOP

Stops the hyperlink access action (an undocumented command).

PASTEASHYPERLINK

Attaches a URL to an object in the drawing from text stored in the Windows Clipboard (an undocumented command).

HYPERLINKBASE

A system variable for setting the path used for relative hyperlinks in the drawing.

In addition, all of AutoCAD 2012's file-related dialog boxes are "Web enabled."

Attached URLs from R14

In AutoCAD R14, attached URLs were not active until the drawing was converted to a DWF file; in AutoCAD 2012, URLs are active in the drawing.

If you attached URLs to drawings in Release 14 or later versions, they are converted to AutoCAD 2012-style hyperlinks the first time you save the drawings in the AutoCAD 2012 DWG format.

CHANGED INTERNET COMMANDS

The Internet-related commands of the previous releases of AutoCAD have been discontinued and replaced. The summary given next describes the status of the Internet commands used in the current release and the previous releases of AutoCAD:

BROWSER (launches a Web browser from within AutoCAD): continues to work in AutoCAD 2012; the default URL is now *http://www.autodesk.com*. However, you can type the URL without http. Also, you can specify the FTP location.

ATTACHURL (attaches a URL to an object or an area in the drawing): continues to work in AutoCAD 2012, but has been superceded by the **-HYPERLINK** command's **Insert** option.

GOTOURL (Opens the URL attached to the selected object): Select an object that has a hyperlink, the file or web page that is associated with the hyperlink will open.

DETACHURL (removes the URL from an object): continues to work in AutoCAD 2012, but has been superceded by the **-HYPERLINK** command's **Remove** option.

DWFOUT (exports the drawing and embedded URLs as a DWF file): continues to work in AutoCAD 2012, but has been superceded by the **Plot** dialog box's ePlot option and the **PUBLISH** command. You can also choose the **DWF** tool from **Output > Export to DWF/PDF > Export** drop-down or **Export > DWF** from the **Application Menu** to export the drawing to DWF.

INSERTURL (inserts a block from the Internet into the drawing): automatically executes the **INSERT** command and displays the **Insert** dialog box. Choose the **Browse** and then the **Search the Web** button and type a URL for the block name.

OPENURL (opens a drawing from the Internet): automatically executes the **OPEN** command and displays the **Select File** dialog box. You may type a URL for the file name.

SAVEURL (saves the drawing to the Internet): automatically executes the **SAVE** command and displays the **Save Drawing As** dialog box. You may type a URL for the file name.

You are probably already familiar with the best-known uses of the Internet: e-mail and the www (short for "World Wide Web"). E-mail allows the user to quickly exchange messages and data at very low cost. The Web brings together text, graphics, audio, and movies in an easy-to-use format. You can also use file transfer protocol (FTP) for effortless binary file transfer.

AutoCAD allows you to interact with the Internet in several ways. It can launch a Web browser from within AutoCAD with the **BROWSER** command. Hyperlinks can be inserted in drawings with the **HYPERLINK** command, which lets you link the drawing with the other documents on your computer and the Internet. With the **PLOT** command's ePlot option (short for "electronic plot), AutoCAD creates DWF files for viewing drawings in two-dimensional (2D) format on Web pages. AutoCAD can open, insert, and save drawings to and from the Internet through the **OPEN**, **INSERT**, and **SAVEAS** commands.

UNDERSTANDING URLS

The **Uniform Resource Locator**, known as URL, is the file naming system of the Internet. The URL system allows you to find any resource (a file) on the Internet. For example, resources include a text file, Web page, program file, an audio or movie clip, in short, anything you might also find on your own computer. The primary difference is that these resources are located on somebody else's computer. A typical URL looks like the following examples.

Example of URL	Meaning
http://www.autodesk.com	Autodesk Primary Web site
news://adesknews.autodesk.com	Autodesk News Server
ftp://ftp.autodesk.com	Autodesk FTP Server
http://www.autodeskpress.com	Autodesk Press Web Site

Note that the **http://** prefix is not required. Most of today's Web browsers automatically add it in the *routing* prefix, which saves you a few keystrokes.

URLs can access several different kinds of resources such as Web sites, e-mail, and news groups, but they always take the same general format as follows:

scheme://netloc

The scheme accesses the specific resource on the Internet including these.

Scheme	Meaning
file://	File is located on your computer's hard drive or local network
ftp://	File Transfer Protocol (used for downloading files)
http://	Hyper Text Transfer Protocol (the basis of Web sites)
mailto://	Electronic mail (e-mail)
news://	Usenet news (news groups)
telnet://	Telnet protocol
gopher://	Gopher protocol

The **://** characters indicate a network address. Autodesk recommends the following format for specifying URL-style file names with AutoCAD:

Resource	URL Format
Web Site	**http:**//*servername/pathname/filename*
FTP Site	**ftp:**//*servername/pathname/filename*
Local File	**file:**///*drive:/pathname/filename*
or	*drive:\pathname\filename*
or	**file:**///*drive|/pathname/filename*
or	**file://***localPC\pathname\filename*
or	**file:**////*localPC/pathname/filename*
Network File	**file:**//*localhost/drive:/pathname/filename*
or	*localhost\drive:\pathname\filename*
or	**file:**//*localhost/drive|/pathname/filename*

The terminology can be confusing. The following definitions will help you clarify these terms:

Term	Meaning
server name	The name or location of a computer on the Internet, for example: *www.autodesk.com*
path name	The same as a subdirectory or folder name
drive	The driver letter, such as C: or D:
local pc	A file located on your computer
local host	The name of the network host computer

If you are not sure of the name of the network host computer, use Windows Explorer to check the Network Neighborhood for the network names of computers.

Launching a Web Browser

The **BROWSER** command lets you start a Web browser from within AutoCAD. The commonly used Web browsers are Microsoft Internet Explorer, Google Chrome, and Mozilla Firefox.

By default, the **BROWSER** command uses whatever brand of Web browser program is registered in your computer's Windows operating system. AutoCAD prompts you for the URL, such as *http://www.autodeskpress.com*. The **BROWSER** command can be used in scripts, toolbars or menu macros, and AutoLISP routines to automatically access the Internet.

> Command: **BROWSER**
> Enter Web location (URL) <http://www.autodesk.com>: *Enter a URL.*

The default URL is an HTML file added to your computer during AutoCAD's installation. After you type the URL and press ENTER, AutoCAD launches the Web browser and contacts the Web site. Figure 32-1 shows the Internet Explorer with the Autodesk Web site.

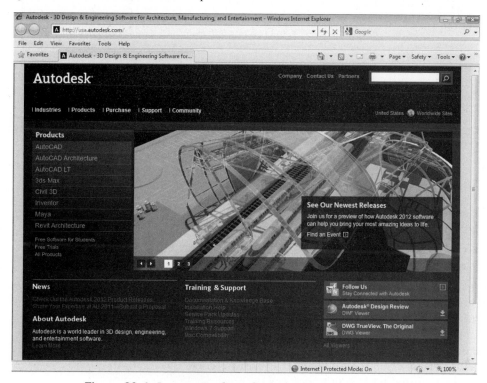

Figure 32-1 *Internet Explorer displaying the Autodesk website*

Changing the Default Website

To change the default Web page that your browser starts from within AutoCAD, and change the setting in the **INETLOCATION** system variable. The variable stores the URL used by the last executed **BROWSER** command and the **Browse the Web** dialog box. Make the change as follows.

> Command: **INETLOCATION**
> Enter new value for INETLOCATION <"http://www.autodesk.com">: *Type URL.*

DRAWINGS ON THE INTERNET

When a drawing is stored on the Internet, you can access it from within AutoCAD 2012 using the standard **OPEN**, **INSERT**, and **SAVE** commands. (In Release 14, these commands were known as **OPENURL**, **INSERTURL**, and **SAVEURL**.) Instead of specifying the file's location with the usual drive-subdirectory-file name format such as *c:\acad2012\filename.dwg*, use the URL format. (Recall that the URL is the universal file-naming system used by the Internet to access any file located on any computer hooked up to the Internet).

Opening Drawings from the Internet

Drawings from the Internet can easily be opened using the **Select File** dialog box, which can be displayed by invoking the **OPEN** command. To do so, choose the **Search the Web** button from the **Select File** dialog box, see Figure 32-2.

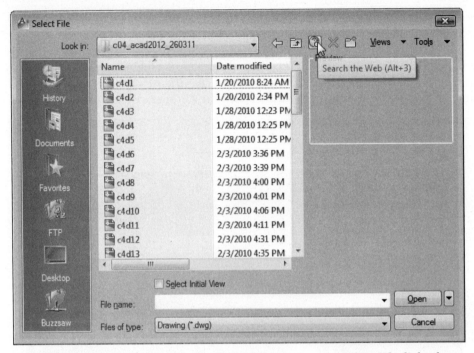

*Figure 32-2 Choosing the **Search the Web** button from the **Select File** dialog box*

When you choose the **Search the Web** button, AutoCAD opens the **Browse the Web** dialog box, see Figure 32-3. This dialog box is a simplified version of the Microsoft brand of Web browser. Its purpose is to allow you to browse the files in a Web site.

By default, the **Browse the Web** dialog box displays the contents of the URL stored in the **INETLOCATION** system variable. You can easily change this to another folder or website, by entering the required URL in the **Look in** edit box.

In top, the dialog box has six buttons. They are discussed next.

Back. Go back to the previous URL.

Forward. Go forward to the next URL.

Stop. Halt displaying the Web page (useful if the connection is slow or the page is very large).

Refresh. Redisplay the current Web page.

Home. Return to the location specified by the **INETLOCATION** system variable.

Favorites. List stored URLs (hyperlinks) or bookmarks. If you have previously used Internet Explorer, you will find all your favorites listed here. Favorites are stored in the *Windows\Favorites* folder on your computer.

The **Look in** field allows you to type the URL. Alternatively, click the down arrow to select a previous destination. If you have stored Web site addresses in the Favorites folder, then select a URL from that list.

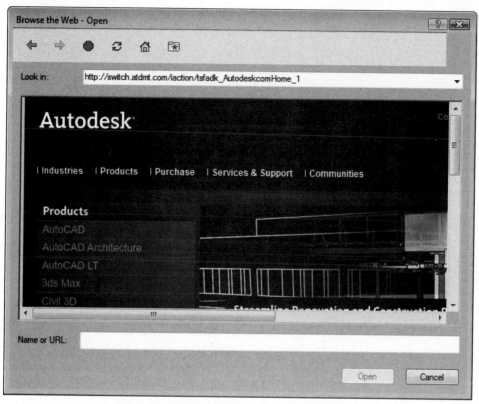

Figure 32-3 The **Browse the Web** *dialog box*

You can either double-click on the name of the file in the window, or type a URL in the **Name or URL** edit box. The following table shows the templates for typing the URL to open a drawing file:

Drawing Location	Template URL
Web or HTTP Site	*http://servername/pathname/filename.dwg*
	http://practicewrench.autodeskpress.com/wrench.dwg
FTP Site	*ftp://servername/pathname/filename.dwg*
	ftp://ftp.autodesk.com
Local File	*drive:\pathname\filename.dwg*
	c:\acad 2010\sample\tablet.dwg
Network File	*\\localhost\drive:\pathname\filename.dwg*
	\\upstairs\d:\install\sample.dwg

When you open a drawing from the Internet, it will probably take more time than opening a file on your computer. During the transfer of file, AutoCAD displays the **File Download** dialog box to report the progress, see Figure 32-4. If your computer uses a 29.8 Kbps modem, you should allow about 5 to 10 min/MB of a drawing file size. If your computer has access to a faster T1 connection to the Internet, you should expect a transfer speed of about 1 min/MB. It may be helpful to understand that the **OPEN** command does not copy the file from the Internet location directly into AutoCAD. Instead, it copies the file from the Internet to your computer's designated **Temporary** subdirectory such as *C:\Windows\Temp* (and then loads the drawing from the hard drive into AutoCAD). This is known as caching. It helps to speed up the processing of the drawing, since the drawing file is now located on your computer's fast hard drive, instead of the relatively slow Internet.

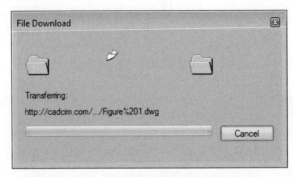

Figure 32-4 *The **File Download** dialog box*

Note that the **Locate** and **Find** options (in the **Select File** dialog box) do not locate files on the Internet.

EXAMPLE 1

The author's website has an area with a link that allows can be used to open a drawing in AutoCAD using the Internet. In this example, you will open a drawing file located at the author's website.

1. Start a new AutoCAD 2012 session.

2. Ensure that you have a live Internet connectivity. If you normally access the Internet via a telephone (modem) connection, dial your Internet service provider now.

3. Choose the **Open** button from the **Quick Access Toolbar**; the **Select File** dialog box is displayed.

4. Choose the **Search the Web** button to display the **Browse the Web** window.

5. In the **Look in** field, type **www.cadcim.com/faculty** and press ENTER; CADCIM Technologies's website is opened.

6. Click on the **AutoCAD 2012** link containing the link of **AutoCAD 2012: A Problem Solving Approach** textbook.

7. Next, click on this link to display the page containing the resources of this textbook.

8. Click on the **Figure1** option in the **Drawings for Practice of AutoCAD on Internet** area; the name of the corresponding file is displayed in the **Name or URL** edit box of the **Browse the Web** window.

9. Choose the **Open** button; AutoCAD begins transferring the file. Once the file has been downloaded completely, the drawing will be displayed in the drawing area, see Figure 32-5.

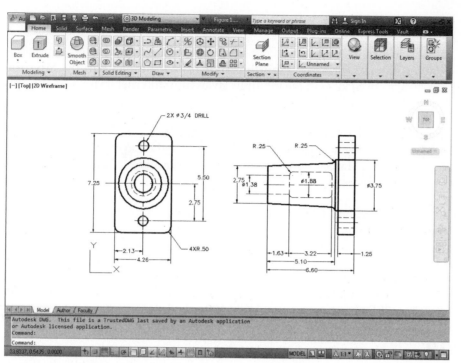

Figure 32-5 *Drawing opened in AutoCAD*

INSERTING A BLOCK FROM THE INTERNET

When a block (symbol) is stored on the Internet, you can access it from within AutoCAD using the **INSERT** command. When the **Insert** dialog box appears, choose the **Browse** button to display the **Select Drawing File** dialog box. This is identical to the dialog box discussed earlier.

After you select the file, AutoCAD downloads the file and continues with the prompt sequence of the **INSERT** command.

The process is identical for accessing external reference (xref) and raster image files. Other files that AutoCAD can access over the Internet include 3D Studio, SAT (ACIS solid modeling), DXB (drawing exchange binary), WMF (Windows metafile), and EPS (encapsulated PostScript). All of these options are found in the **Insert** menu on the menu bar.

ACCESSING OTHER FILES ON THE INTERNET

Most of the other file-related dialog boxes allow you to access files from the Internet or the Intranet. This allows your firm or agency to have a central location that stores drawing standards. If you need to use a linetype or a hatch pattern, you can access the LIN or PAT file over the Internet. More than likely, you will have the location of these files stored in the Favorites list. Some examples include the following:

Linetypes. Invoke the **Drafting & Annotation** workspace. Next, choose the **Other** option from **Home > Properties > Linetypes** drop-down list or **Format > Linetype** from the Menu Bar; the **Linetype Manager** dialog box will be displayed. Choose the **Load** button to invoke the **Load or Reload Linetypes** dialog box. Next, choose the **File** button to invoke the **Select Linetype File** dialog box. From this dialog box, choose the **Search the Web** button; the **Browse the Web** window is displayed. This window can be used to load the linetype files from Internet.

Hatch Patterns. Use the Web browser to copy *.pat* files from a remote location to your computer.

Multiline Styles. Choose **Format > Multiline Style** from the Menu Bar; the **Multiline Styles** dialog box is displayed. Next, load the multiline styles from Internet by following the procedure used for loading the linetype files.

Layer Name. Choose the **Layer Properties** button from the **Layers** panel to invoke the **Layer Properties Manager** dialog box. In this dialog box, choose the **Layer States Manager** button to display the **Layer States Manager** dialog box. In the **Layer States Manager** dialog box, choose the **Import** button to import the required layer state.

LISP and ARX Applications. Choose the **Load Application** tool from the **Applications** panel in the **Manage** tab.

Scripts. Choose the **Run Script** tool from the **Applications** panel in the **Manage** tab.

You cannot access text files, text fonts (SHX and TTF), color settings, lineweights, dimension styles, plot styles, OLE objects, or named UCSs from the Internet.

i-DROP

The i-drop plug-in is installed on your system while installing AutoCAD 2012. If your system has the i-drop installed, you can drag a drawing from a Web site and drop it in the AutoCAD main window. The drawing is displayed in AutoCAD. Using the i-drop, you can insert blocks, symbols, and so on, in your open drawing.

The i-drop can be used with the earlier releases of AutoCAD and it can be downloaded for free from *www.autodesk.com/idrop*. A sample drawing of a chair is provided on this Website, which you can drag and drop in the AutoCAD window.

Note
*The drawings that are inserted using the i-drop are inserted as a block and the **INSERT** command is executed automatically when you drop the drawing in the main window of AutoCAD.*

SAVING A DRAWING ON THE INTERNET

When you are finished with editing a drawing in AutoCAD, you can save it to a file server on the Internet with the **SAVE** command. If you inserted the drawing from the Internet (using **INSERT**) into the default *Drawing.dwg* drawing, AutoCAD insists you to first save the drawing on your computer's hard drive.

When a drawing of the same name already exists at that URL, AutoCAD warns you, just as it does when you use the **SAVEAS** command.

ONLINE RESOURCES

To access online resources, choose **Help > Additional Resources** from the menu bar; the cascading menu will be displayed. The options of online resources are **Support Knowledge Base**, **Online Training Resources**, **Online Developer Center**, **Developer Help**, and **Autodesk User Group International**. Their functions are discussed next.

Support Knowledge Base

When you choose this option, the AutoCAD Services and Support Web page is displayed in the Web browser. Use the search box to search the database.

Note

*To obtain more information on the related topics in **Product Support** and the other **Online Resources** option, it is recommended to connect to the Internet and then choose the **Online Resources** option.*

Online Training Resources

When you choose this option, the Web page of training is displayed in the Web browser. The information on the following can be obtained by selecting this option.

- Information on Autodesk authorized training centers.
- General training centers.
- Autodesk Certification.
- Discussion groups.
- Learning tools.

Online Developer Center

When you choose this option, the Web page is displayed providing the following information.

- Answers to the questions that were asked by the users.
- Sample applications.
- Documentation.

Note

*When you choose any one of the **Online Resources** options, the **Live Update Status** Web browser window is displayed.*

Developer Help

When you choose this button, the **AutoCAD 2012 Help** window will be displayed containing developers help. You need not to be connected to internet for accessing this option.

Autodesk User Group International

When you choose this window, **Autodesk User Group International** Web browser window is displayed. This Web browser gives information about the AutoCAD user groups.

USING HYPERLINKS WITH AutoCAD

AutoCAD 2012 allows you to employ URLs in two ways:

1. Directly within an AutoCAD drawing.
2. Indirectly in DWF files displayed by a Web browser.

URLs are also known as hyperlinks, the term that is used throughout this book. Hyperlinks are created, edited, and removed with the **HYPERLINK** command. You can also use the command line with the help of the **-HYPERLINK** command.

HYPERLINK Command

Ribbon: Insert > Data > Hyperlink	**Menu Bar:** Insert > Hyperlink
Command: HYPERLINK	

When you invoke this command, you will be prompted to select objects. Once the objects has been selected, the **Insert Hyperlink** dialog box will be displayed, see Figure 32-6.

If you select an object that has a hyperlink already attached to it, the **Edit Hyperlink** dialog box will be displayed, as shown in Figure 32-7. The **Remove Link** button available in the dialog box is used to remove the hyperlink from an object.

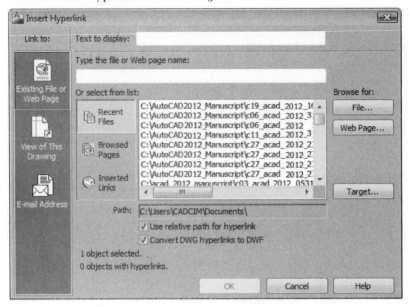

*Figure 32-6 The **Insert Hyperlink** dialog box*

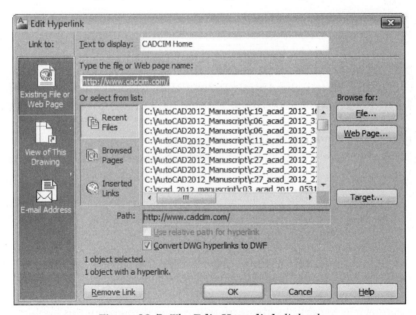

*Figure 32-7 The **Edit Hyperlink** dialog box*

If you use the Command line for inserting hyperlinks (**-HYPERLINK** command), you are also allowed to create hyperlink areas. A hyperlink area is a rectangular area that can be thought of as a 2D hyperlink (the dialog box-based **HYPERLINK** command does not create hyperlink areas). When you select the **Area** option from the Command line, a rectangular boundary object is placed automatically on the layer URLLAYER and turns red color, see Figure 32-8.

In the following sections, you will learn to apply and use hyperlinks in an AutoCAD drawing and in a Web browser through the dialog box-based **HYPERLINK** command.

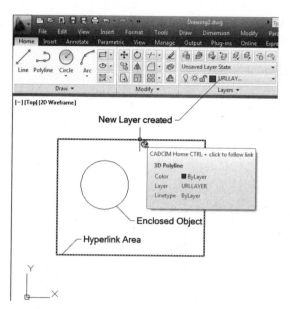

Figure 32-8 *A rectangular hyperlink area*

Hyperlinks Inside AutoCAD

As mentioned earlier, AutoCAD allows you to add a hyperlink to any object in a drawing. An object is permitted just a single hyperlink. On the other hand, a single hyperlink may be applied to a selected set of objects.

You can determine whether an object has a hyperlink or not by moving the cursor over it. The cursor displays the "linked Earth" icon, as well as a tooltip describing the link, see Figure 32-9.

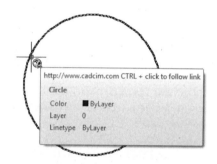

Figure 32-9 *The cursor displays a hyperlink*

If, for some reason, you do not want to see the hyperlink cursor, you can turn it off. To do so, invoke the **Options** dialog box, choose the **User Preferences** tab from it. The **Display hyperlink cursor, tooltip, and shortcut menu** check box in the **Hyperlink** area toggles the display of the hyperlink cursor, as well as the **Hyperlink** tooltip option in the shortcut menu. Select the check box as per requirement.

Attaching Hyperlinks

The following example explains attaching hyperlinks:

EXAMPLE 2

In this example, you are given the drawing of a floor plan to which you will add hyperlinks. These hyperlinks are another AutoCAD drawing. They must be attached to objects. For this reason, you will place some text in the drawing and then attach the hyperlinks to it.

1. Start the AutoCAD 2012 session.

2. Open the *db_samp.dwg* file from the location *C:\Program Files\Autodesk\AutoCAD 2012 - English\ Sample\Database Connectivity*. The drawing will open as a **Read-Only** file.

3. Save this file with the name *Example2.dwg*.

4. Invoke the **Single Line** tool from **Home > Annotation > Text** drop-down and write the texts, as shown in Figure 32-10.

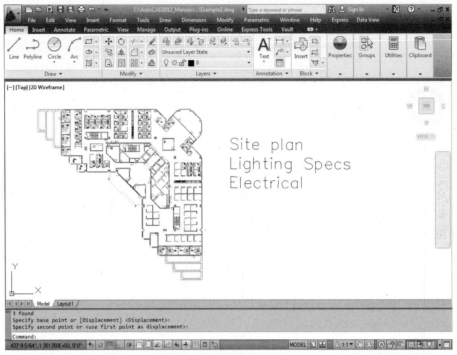

Figure 32-10 *Text placed in a drawing*

5. Choose the **Hyperlink** tool from the **Data** panel in the **Insert** tab; you are prompted to select objects. Select the text **Site Plan** and press ENTER to display the **Insert Hyperlink** dialog box.

6. Choose the **File** button from the **Insert Hyperlink** dialog box to display the **Browse the Web – Select Hyperlink** dialog box. Browse to *C:\Program Files\Autodesk\AutoCAD 2012 - English\ Sample\DesignCenter* folder, if it is not already opened, and select the *Home - Space Planner. dwg* file, see Figure 32-11.

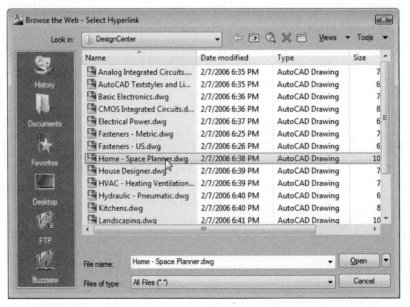

Figure 32-11 *The Browse the Web - Select Hyperlink dialog box*

7. Choose **Open** from the dialog box. AutoCAD does not open the drawing; rather, it copies the file's name to the **Insert Hyperlink** dialog box. The name of the drawing is displayed in the **Type the file or web page name** edit box. The same name and path are also displayed in the **Text to display** edit box. Using this edit box, retain only the file name from the path and remove the rest.

8. Choose the **OK** button to close the dialog box. Move the cursor over the **Site Plan** text. Notice the display of the "linked Earth" icon. A moment later, the tooltip displays the location of the drawing, see Figure 32-12.

Figure 32-12 *The Hyperlink cursor and tooltip*

9. Connect to the Internet and then connect the URL *www.autodesk.com* to **Lighting Specs** using the **Web Page** button of the **Insert Hyperlink** dialog box.

10. You can directly open the file attached as hyperlink to the object by using the linked object. To do so, select the object to which the file is linked and then right-click to display the shortcut menu. In the shortcut menu, choose **Hyperlink > Open "file path'**. In this case, the file name is *Home - Space Planner.dwg*. Therefore, right-click and then choose **Hyperlink > Open "Space planner.dwg"** from the shortcut menu, see Figure 32-13.

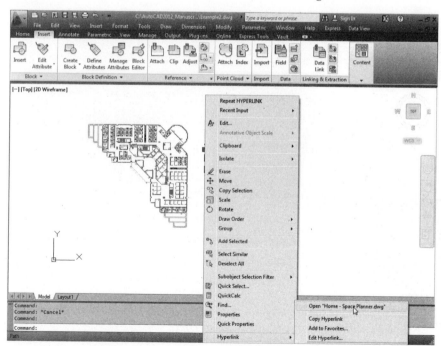

Figure 32-13 *Opening a hyperlink*

11. Choose the **Tile Vertically** button from the **Windows** panel in the **View** tab to view both the drawings together, see Figure 32-14.

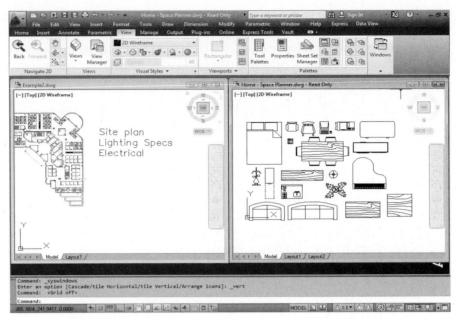

Figure 32-14 *Displaying both drawings in the drawing window*

12. Select the **Lighting Specs** text and then right-click on it. Choose **Hyperlink > Open "http:// www.autodesk.com"** from the shortcut menu. Notice that Windows starts your Web browser and opens the *www.autodesk.com* URL, see Figure 32-15. Make sure you are connected to the Internet before opening this URL.

Figure 32-15 *Web browser opens the URL that was hyperlinked to the selected text*

Note
If you have changed the default website using the **INTLOCATION** *system variable, then your web browser will be opened with the modified URL.*

13. Save the drawing.

PASTING AS HYPERLINK

AutoCAD 2012 has a shortcut method for pasting hyperlinks in the drawing. The hyperlink from one object can be copied and pasted on another object. This can be achieved by using the **PASTEASHYPERLINK** command.

1. In AutoCAD, select an object that has a hyperlink. Right-click to invoke the shortcut menu.

2. Choose **Hyperlink > Copy Hyperlink** from the shortcut menu.

3. Choose the **Paste as Hyperlink** tool from **Home > Clipboard > Paste** drop-down; you are prompted to select the object on which the hyperlink is to be pasted.

4. Select the object and press ENTER. The hyperlink is pasted to the new object.

Note

*The **MATCHPROP** command does not copy the hyperlinks from the source objects to the destination objects.*

EDITING HYPERLINKS

AutoCAD allows you to edit the hyperlinks attached to an object. To do so, select the hyperlinked object and right-click. Choose **Hyperlink > Edit Hyperlink** from the shortcut menu; the **Edit Hyperlink** dialog box appears, which looks identical to the **Insert Hyperlink** dialog box. Make the changes in the hyperlink and choose **OK**.

REMOVING HYPERLINKS FROM OBJECTS

To remove a URL from an object, use the **HYPERLINK** command on the object. When the **Edit Hyperlink** dialog box appears, choose the **Remove Link** button.

To remove a rectangular area hyperlink, you can simply use the **ERASE** tool; select the rectangle and AutoCAD erases the rectangle.

THE DRAWING WEB FORMAT

To display AutoCAD drawings on the Internet, Autodesk created a file format called drawing Web format (DWF). The DWF file has several benefits and some drawbacks over DWG files. The DWF file is compressed eight times smaller than the original DWG drawing file. Therefore, it takes less time to transmit these files over the Internet, particularly with the relatively slow telephone modem connections. The DWF format is more secure, since the original drawing is not being displayed; another user cannot tamper with the original DWG file.

However, the DWF format has some drawbacks, which are mentioned below:
- You must go through the extra step of translating from DWG to DWF.
- DWF files cannot display rendered or shaded drawings.
- DWF is a flat 2D-file format; therefore, it does not preserve 3D data, although you can export a 3D view.
- AutoCAD itself cannot display DWF files.
- DWF files cannot be converted back to DWG format without using file translation software from a third-party vendor.
- Earlier versions of DWF did not handle paper space objects (version 2.x and earlier), or line widths, lineweights, and non-rectangular viewports (version 3.x and earlier).

To view a DWF file on the Internet, your Web browser needs to have a *plug-in* software extension called **Autodesk DWF Viewer**. This viewer is installed on your system with AutoCAD 2012 installation. **Autodesk DWF Viewer** allows Internet Explorer 6.1 (or later) to handle a variety of file formats. Autodesk makes this DWF Viewer plug-in freely available from its Web site at *http://www.autodesk.com*. It is a good idea to regularly check for updates to the DWF plug-in, which is updated frequently.

CREATING A DWF FILE

You can use two methods to create DWF files from AutoCAD. The first method is to use the **PLOT** command and the second method is to use the **PUBLISH** command.

Creating a DWF File Using the PLOT Command

The following steps explain the procedure to create a DWF file using the **PLOT** command:

1. Open the file to be converted into a DWF file. Choose the **Plot** tool from the **Quick Access Toolbar**; the **Plot** dialog box is displayed.

2. Select **DWF6 ePlot.pc3** from the **Name** drop-down list in the **Printer/plotter** area, see Figure 32-16.

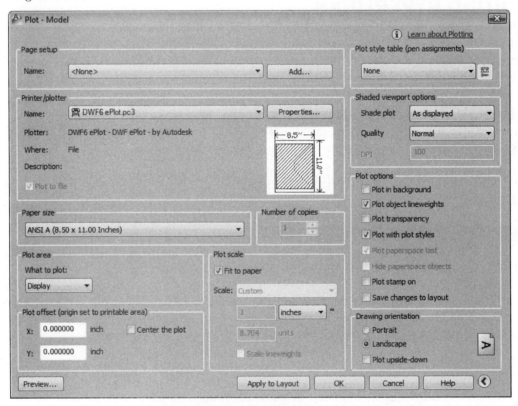

*Figure 32-16 The **Plot** dialog box*

3. Choose the **Properties** button to display the **Plotter Configuration Editor** dialog box, see Figure 32-17.

4. In the **Device and Documents Settings** tab of this dialog box, select **Custom Properties** in the tree view; the **Access Custom Dialog** area is displayed in the lower half of the dialog box.

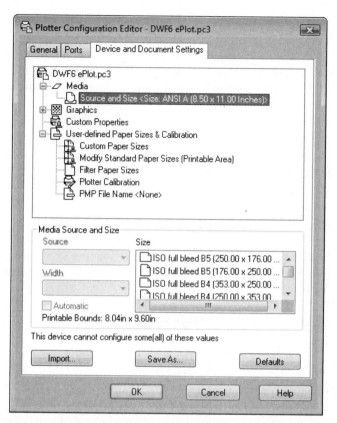

*Figure 32-17 The **Plotter Configuration Editor** dialog box*

5. Choose the **Custom Properties** button from the **Access Custom Dialog** area to display the **DWF6 ePlot Properties** dialog box, see Figure 32-18.

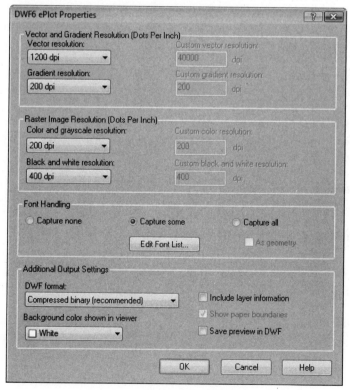

*Figure 32-18 The **DWF6 ePlot Properties** dialog box*

6. Set the required options in these dialog boxes and choose **OK** to exit the dialog box and return to the **Plot** dialog box.

7. In the **Plot** dialog box, use the required options to create the DWF file. Choose **OK** from the **Plot** dialog box; the **Browse for Plot File** dialog box will be displayed.

8. Specify the name of the DWF file in the **File name** edit box. If necessary, change the location where the file is to be stored.

DWF6 ePlot Properties Dialog Box Options

The options in this dialog box are discussed next.

Vector and Gradient Resolution (Dots Per Inch) Area

This area allows you to set the resolution for the vector information and gradients. Unlike AutoCAD DWG files that are based on real numbers, DWF files are saved using integer numbers. The higher the resolution, the better the quality of the DWF files. However, the size of the files also increases with the resolution.

Raster Image Resolution (Dots Per Inch) Area

This area allows you to set the resolution for the raster images in the DWF files.

Font Handling

This area of the **DWF6 ePlot Properties** dialog box is used to select the fonts that you need to include in the DWF file. You can choose the **Edit Font List** button from this area to display the list of the fonts. It should be noted that the fonts add to the size of the DWF file.

Additional Output Settings Area

The options available in this area are discussed next.

DWF format. This drop-down list provides the options for compressing or zipping while creating the DWF file.

Background color shown in viewer. This drop-down list is used to specify the background color of the resulting DWF file.

Include layer information. This check box is selected to include the layer information in the DWF file. If the layer information is included in the resulting DWF file, you can toggle layers off and on when the DWF file is being viewed.

Show paper boundaries. Selecting this check box results in a rectangular boundary at the drawing's extents, as displayed in the layouts.

Save preview in DWF. Saves the preview in DWF file.

Creating a DWF File Using the PUBLISH Command

The following is the procedure to create a DWF file using the **PUBLISH** command:

Choose the **Batch Plot** tool from the **Plot** panel in the **Output** tab; the **Publish** dialog box will be displayed, see Figure 32-19. The options in this dialog box are discussed next.

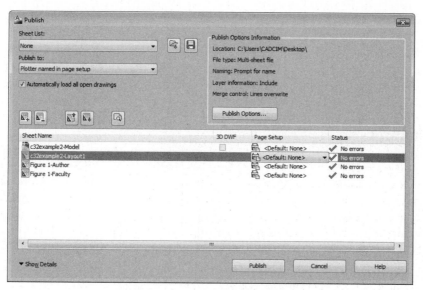

*Figure 32-19 The **Publish** dialog box*

The list box in the **Publish** dialog box lists the model and paper space layouts that will be included for publishing. Right-click on any of the sheets in the **Sheet Name** column to display a shortcut menu. The options in it can be used on the drawings listed in this area. You can change the page setup option of the sheets using the field in the **Page Setup/3D DWF** column. You can also import a page setup using the drop-down list that is displayed when you click on the field in this column. The remaining options in this area are discussed next.

Note

*If you have not saved the drawing before invoking the **PUBLISH** command, the sheet names in the list box will have **Unsaved** as prefix. Also, the status of the layouts that are not initialized will show **Layout not initialized**.*

Add Sheets. This button, when chosen, displays the **Select Drawings** dialog box. Using this dialog box, you can select the drawing files (*.dwg*). These files are then included in the creation of the DWF file.

Remove Sheets. This button is chosen to remove the sheet selected from the **Sheet Name** column.

Move Sheet Up. This button is chosen to move the selected sheet up by one position in the list.

Move Sheet Down. This button is chosen to move the selected sheet down by one position in the list.

Save Sheet List. This button, when chosen, displays the **Save List As** dialog box. This dialog box allows you to save the current list of drawing files as DSD files.

Load Sheet List. After saving a list, if you choose this button, the **Load List of Sheets** dialog box will be displayed with the saved list in the **Sheet List** drop-down list. You can select the *.dsd* or *.bp3* file format from this dialog box and select the list to load. After selecting the list, you can proceed to publish the selected sheet list. But if you want to publish a different list, select the required list from the **Load List of Sheets** dialog box. On doing so, the **Load Sheet List** message box will be displayed. Choose the required option from this message box to continue with the publishing list.

Preview. This button is chosen to preview the sheet selected from the **Sheet Name** column.

Publish to

This drop-down list provides the options of specifying whether the output of the **PUBLISH** command should be plotted on a sheet using the printer mentioned in the page setup of the selected sheet or as a DWF or DWFx or PDF file.

Publish Output

This area provides the additional options while publishing the sheets. The **Number of copies** spinner is used to specify the number of copies to be published. Note that if you want the output in the form of a DWF file, the number of copies can only be one. The button below the **Plot Stamp Settings** button is used to reverse the default order of publishing the drawings listed in the list box. Select the **Include plot stamp** check box to include a plot stamp while publishing. The **Plot Stamp Settings** button is chosen to invoke the **Plot Stamp** dialog box for configuring the settings of the plot stamp. The **Publish in background** check box is used to publish the selected drawings in background, while you can continue working with AutoCAD. Select the **Open in viewer when done** check box to open the published file in a new window.

Publish Options

When you choose this button, the **Publish Options** dialog box is displayed, as shown in Figure 32-20. It provides you with the options to publish the drawings. These options are discussed next.

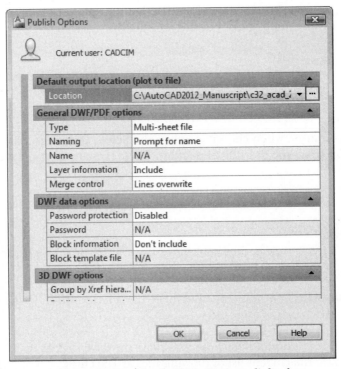

Figure 32-20 The **Publish Options** *dialog box*

Default output location (plot-to-file) rollout. The options in this rollout are used to specify the location of the output of the **PUBLISH** command. By default, the sheets are published locally on the hard drive of your computer. You can modify the default location of the file using the edit box or the swatch [...] button, which is displayed when you click on the field that shows the name and location of the file.

General DWF/PDF options. The options in this rollout are used to specify DWF/PDF files. Click the particular field and specify the appropriate options.

DWF data options. The options in this area are used to specify the password protection. You can also specify whether the block information should be included in the resulting DWF file or not.

3D DWF options. The options in this area are used to specify the data that you can include in the 3D DWF publishing. When only 2D DWF files are listed for publishing, all of the **3D DWF options** are set to **N/A** and cannot be modified.

This completes the options in the **Publish Options** dialog box. The remaining options of the **Publish** dialog box are discussed next.

Show Details

When you choose this button, the **Publish** dialog box expands and shows the details related to the sheet selected from the **Sheet Name** list box.

Publish

This button is used to start the process of publishing or creating the DWF file.

Note
*You may need to set the value of the **BACKGROUNDPLOT** system variable to **0** if AutoCAD gives an error while publishing.*

Export to DWF Files

You can export a drawing to a DWF file by choosing the **DWF** tool from **Output > Export to DWF/PDF > Export** drop-down; the **Save As DWF** dialog box will be displayed, as shown in Figure 32-21. In this dialog box, the current settings for the DWF file will be displayed in the **Current Settings** area. To edit the current settings, choose the **Options** button; the **Export to**

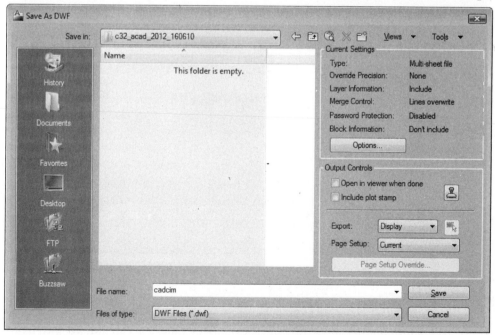

*Figure 32-21 The **Save As DWF** dialog box*

DWF/PDF Options dialog box will be displayed. In this dialog box, you can change the general information for the DWF file like file location, password protection, layer information, and so on. You can also invoke this dialog box by choosing the **Export to DWF/PDF Options** tool from the **Export to DWF/PDF** panel.

You can also select which portion of the drawing to be exported from the **Export** drop-down list in the **Save As DWF** dialog box. If you are in the **Model** tab, you can select the **Display** (objects currently displayed on the screen), **Extents** (drawing extents) or **Window** (selected area from the drawing) option and if you are in the **Layout** tab, you can select between the active layout or all layouts. These options are also available in the **What to Export** drop-down list of the **Export to DWF/PDF** panel in the **Output** tab of the **Ribbon**. You can also specify the page settings for the DWF file such as paper size, scale, and plot area. If you want to use the current settings of the **Page Setup Manager**, select the **Current** option from the **Page Setup** drop-down list in the **Save As DWF** dialog box or the **Export to DWF/PDF** panel. To override the current setting, select the **Override** button. If you want to view the DWF in the Autodesk Design Review, select the **Open in viewer when done** check box. Select the **Include plot stamp** check box to include drawing name, paper size, date and time, etc to the DWF file. After setting the parameters, choose **Save** from the **Save As DWF** dialog box; the drawing is exported.

Generating DWF Files for 3D Models

To create DWF files for 3D models, it is necessary to install 3D DWF Publish feature, while installing AutoCAD 2012. Choose the **3D DWF** tool from the **Export to DWF/PDF** panel in the **Output** panel to open the **Export 3D DWF** dialog box, as shown in Figure 32-22. Browse the location where you want to save the DWF file, and then specify the name of the file. Next, choose the **Save** button to publish the drawing.

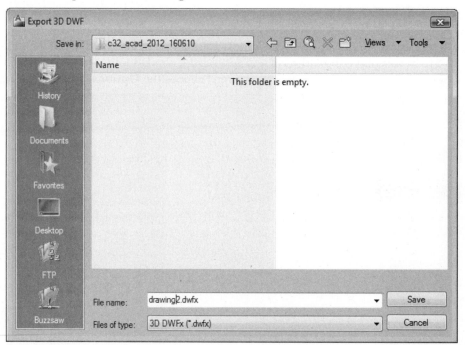

*Figure 32-22 The **Export 3D DWF** dialog box*

Viewing DWF Files

The Autodesk Design Review is automatically installed on your computer when you install AutoCAD 2012. This plug-in can be used to view the DWF files. You can also use a Web browser with a special plug-in that allows the browser to correctly interpret the file for viewing the DWF files. Remember that you cannot view a DWF file with AutoCAD.

You can use various buttons provided in the Autodesk Design Review to manipulate the view of a DWF file. However, remember that the original objects of a DWF file cannot be modified. You can also right-click on the display screen to display a shortcut menu. The shortcut menu provides you with the options such as **Cut**, **Copy**, **Paste**, **Pan**, **Zoom**, **Steering Wheels**, and so on, see Figure 32-23. Choose an option from this shortcut menu to invoke the corresponding operation.

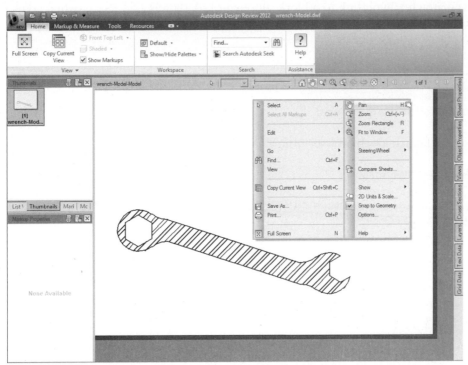

*Figure 32-23 The **Autodesk Design Review** with the shortcut menu*

Note
*To publish a sheet, the **DWF6 eplot.pc3** plotter should be selected from the **Page Setup** dialog box.*

Self-Evaluation Test

Answer the following questions and then compare them to those given at the end of this chapter:

1. To access online resources, choose **Help > Online Resources** from the title bar. (T/F)

2. You cannot find out whether an object has a hyperlink or not. (T/F)

3. By default, the **Browse the Web** dialog box displays the contents of the URL stored in the **INETLOCATION** system variable. (T/F)

4. URLs are also known as hyperlinks. (T/F)

5. You cannot select text in an AutoCAD drawing to paste it as hyperlink. (T/F)

6. Compression in a DWF file helps transmit the file in _____ time to Internet.

7. Rectangular area hyperlinks are stored on the _____ layer.

8. To see the location of hyperlinks in a drawing, use the _____ command.

9. The _____ is an HTML tag that is used to embed objects in a webpage.

10. If you attach a hyperlink to a block, the hyperlink data is _____ when you scale the block unevenly, stretch the block, or explode it.

Review Questions

Answer the following questions:

1. Can you launch a web browser from within AutoCAD?

2. What does DWF mean?

3. What is the purpose of DWF files?

4. Expand the acronym URL?

5. Which of the following URLs are valid?

   (a) *www.autodesk.com*
   (b) *http://www.autodesk.com*
   (c) Both (a) and (b).
   (d) None of the above.

6. Expand the acronym FTP?

7. What is a "local host"?

8. Are hyperlinks active in an AutoCAD 2012 drawing?

9. The purpose of URLs is to let you create _____ between files.

10. Can you attach a URL to any object?

Answers to Self-Evaluation Test

1. T, **2.** F, **3.** T, **4.** T, **5.** T, **6.** less, **7.** URLLAYER, **8. HYPERLINK**, **9.** <embed>, **10.** lost

AutoCAD
PART III

Author's Website

For Faculty: Please contact the author at **Stickoo@purduecal.edu** or **tickoo525@gmail.com** to access the website that contains the following:

1. PowerPoint presentations, program listings, and drawings used in this textbook.
2. Syllabus, chapter objectives and hints, and questions with answers for every chapter.

For Students: To download drawings, exercises, tutorials, and programs, please visit the author's website: *http://www.cadcim.com*. Alternatively, you can visit: *www.cengage.com/cad/autodeskpress*, choose the *Online Companions* option, and then click on the *AutoCAD 2012: A Problem Solving Approach* link to download student resources.

Chapter **33**

Accessing External Database

CHAPTER OBJECTIVES

After completing this chapter, you will learn:
* *About database and the database management system (DBMS).*
* *About AutoCAD database connectivity feature.*
* *To configure an external database.*
* *To access and edit a database using the dbConnect Manager.*
* *To create links with graphical objects.*
* *To create and display labels in a drawing.*
* *To understand AutoCAD SQL Environment (ASE).*
* *To create queries using Query Editor.*
* *To form selection sets using Link Select.*
* *To convert ASE links into AutoCAD 2012 format.*

KEY TERMS

* *Database*
* *Database Management System*
* *Database Connectivity*
* *Database Configuration*
* *dbConnect Manager*
* *Link Template*
* *Label Template*
* *SQL Environment*
* *Query Editor*
* *Quick Query*
* *Range Query*
* *Query Builder*
* *SQL Query*
* *Link Select*
* *Import Query Set*
* *Export Query Set*

UNDERSTANDING DATABASE

Database

Database is a collection of data arranged in a logical order. In a database, the columns are known as fields, individual rows are called records, and the entries in the database tables, that store data for a particular variable, are known as cells. For example, there are six computers in an office, and we want to keep a record of these computers on a sheet of paper. One of the ways of recording this information is to make a table with rows and columns, as shown in Figure 33-1. Each column can have a heading that specifies a certain feature of the computer, such as COMP_CFG, CPU, HDRIVE, or RAM. Once the columns are labeled, the computer data can be placed in the columns. By doing this, you have created a database on a sheet of paper that contains information about the computers. The same information stored on a computer is known as a computerized database.

COMPUTER

COMP_CFG	CPU	HDRIVE	RAM	GRAPHICS	INPT_DEV
1	PENTIUM350	4300MB	64MB	SUPER VGA	DIGITIZER
2	PENTIUM233	2100MB	32MB	SVGA	MOUSE
3	MACIIC	40MB	2MB	STANDARD	MOUSE
4	386SX/16	80MB	4MB	VGA	MOUSE
5	386/33	300MB	6MB	VGA	MOUSE
6	SPARC2	600MB	16MB	STANDARD	MOUSE

Figure 33-1 A table containing computer information

Most of the database systems are extremely flexible and any modifications to or additions of fields or records can be done easily. Database systems also allow you to define relationships between multiple tables so that if the data of one table is altered, the corresponding values of another table with predefined relationships changes automatically.

Database Management System

The database management system (DBMS) is a program or a collection of programs (software) used to manage the database. For example, DBASE, INFORMIX, and ORACLE are database management systems.

Components of a Table

A database table is a two-dimensional data structure that consists of rows and columns, as shown in Figures 33-2 and 33-3.

Row

The horizontal group of data is called a row. For example, Figure 33-2 shows three rows of a table. Each value in a row defines an attribute of the item. For example, in Figure 33-2, the attributes assigned to COMP_CFG (1) include PENTIUM350, 4300MB, 64MB, and so on. These attributes are arranged in the first row of the table.

Column

A vertical group of data (attribute) is called a column. (See Figure 33-3.) **HDRIVE** is the column heading that represents a feature of a computer, and the **HDRIVE** attributes of each computer are placed vertically in this column.

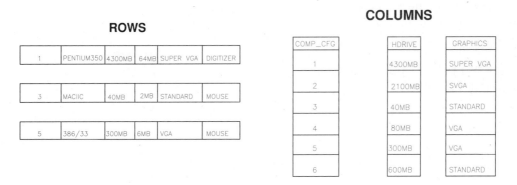

Figure 33-2 *Rows in a table (horizontal group)*   **Figure 33-3** *Columns in a table (vertical group)*

AutoCAD DATABASE CONNECTIVITY

AutoCAD can be effectively used in associating data contained in an external database table with the AutoCAD graphical objects by linking. The Links are the pointers to the database tables from which the data can be referred. AutoCAD can also be used to attach Labels that will display data from the selected tables as text objects. AutoCAD database connectivity offers the following facilities:

1. A **dbConnect Manager** that can be used to associate links, labels, and queries with AutoCAD drawings.
2. An **External Configuration Utility** that enables AutoCAD to access the data from a database system.
3. A **Data View** window that displays the records of a database table within the AutoCAD session.
4. A **Query Editor** that can be used to construct, store, and execute SQL queries. SQL is an acronym for Structured Query Language.
5. A **Migration Tool** that converts links and other display able attributes of files created by earlier releases to current release.
6. The Link Select Operation that creates iterative selection sets based on queries and graphical objects.

DATABASE CONFIGURATION

An external database can be accessed within AutoCAD only after configuring AutoCAD using Microsoft **ODBC** (Open Database Connectivity) and **OLE DB** (Object Linking and Embedding Database) programs. AutoCAD is capable of utilizing data from other applications, regardless of the format and the platform on which the file is stored. Configuration of a database involves creating a new **data source** that points to a collection of data and information about the required drivers to access it. A **data source** is an individual table or a collection of tables created and stored in an environment, catalog, or schema. Environments, catalogs, and schemes are the hierarchical database elements in most of the database management systems and they are analogous to Window-based directory structure in many ways. Schemes contain a collection of tables, while Catalogs contain subdirectories of schemes and Environment holds subdirectories of catalogs. The external applications supported by AutoCAD 2012 are **DBASE®**, **Oracle®**, **Microsoft Access®**, **Microsoft Visual FoxPro®**, **SQL Server**, and **Paradox®**. The configuration process varies slightly from one database system to other.

dbConnect Manager

Menu Bar: Tools > Palettes > dbConnect
Command: DBCONNECT

When you invoke this command, the **dbConnect Manager** will be displayed, as shown in Figure 33-4. Also, the **dbConnect** menu will be added to the menu bar. The **dbConnect Manager** enables you to access information from an external database more effectively. The **dbConnect Manager** is dockable as well as resizable. And, it contains a set of buttons and a tree view showing all the configured and available databases. You can use **dbConnect Manager** for associating various database objects with an AutoCAD drawing. It contains two nodes in the tree view. The Drawing node displays all the open drawings and the associated database objects with the drawing. The **Data Sources** node displays all the configured data on your system.

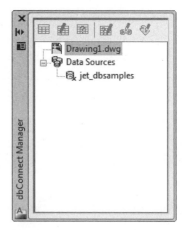

Figure 33-4 The dbConnect Manager

AutoCAD contains several Microsoft Access sample database tables and a direct driver (*jet_dbsamples.udl*). The following example shows the procedure to configure a database with a drawing using the **dbConnect Manager**.

Note
Users having 64-bit Operating System may face problem in using the Microsoft Access sample database table (jet_db-sample.udl) because of non-availability of appropriate drivers.

EXAMPLE 1   *Configuring Data Source with Drawing*

Configure a data source of Microsoft Access database with a drawing. Update and use the **jet_samples.udl** configuration file with new information.

1. Invoke the **DBCONNECT** command to display the **dbConnect Manager**. The **dbConnect** menu will be inserted between the **Parametric** and **Window** menus in the menu bar.

2. The **dbConnect Manager** will display **jet_dbsamples** under **Data Sources**. Right-click on **jet_dbsamples**; a shortcut menu is displayed.

3. In the shortcut menu, choose **Configure**, as shown in Figure 33-5; the **Data Link Properties** dialog box is displayed. Choose the **Connection** tab; the options in the dialog box is changed, as shown in Figure 33-6.

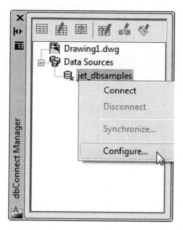

Figure 33-5 *Choosing the* **Configure** *option from the* **dbConnect Manager**

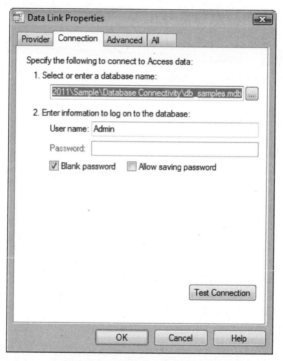

Figure 33-6 *The* **Data Link Properties** *dialog box*

4. In the **Data Link Properties** (**Connection** tab) dialog box, choose the [**...**] button adjacent to the **Select or enter a database name** text box; the **Select Access Database** dialog box is displayed, as shown in Figure 33-7. Select the *db_samples.mdb* file and choose the **Open** button.

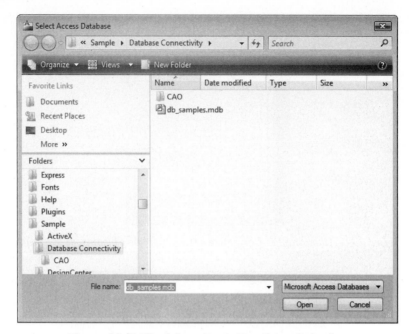

*Figure 33-7 The **Select Access Database** dialog box*

 Note
*If you configure a driver other than **Microsoft Jet** for database linking, you should refer the appropriate documentation for the configuring process.*

*The other tabs in the **Data Link Properties** dialog box are required for configuring various database providers supported by AutoCAD.*

5. Once the database name is selected, choose the **Test Connection** button from the **Data Link Properties** dialog box to ensure that the database source has been configured correctly. If it is configured correctly, the **Microsoft Data Link** message box is displayed, see Figure 33-8.

*Figure 33-8 The **Microsoft Data Link** message box*

6. Choose the **OK** button to end the message and then choose the **OK** button to complete the configuration process.

7. After configuring the data source, double-click on **jet_dbsamples** in **dbConnect Manager**; all the sample tables will be displayed in the tree view. You can connect any table and link its records to the entities in the drawing.

VIEWING AND EDITING TABLE DATA FROM AutoCAD

After configuring a data source using the **dbConnect Manager**, you can view as well as edit its tables within the AutoCAD session using the **Data View** window. You can open tables in **Read-only** mode to view their content. But you cannot edit their records in the **Read-only** mode. Some database systems may require a valid user-name and password before connecting to AutoCAD drawing files. The records of a database table can be edited by opening in the **Edit** mode. The procedure to open a table in various modes is given next.

Read-only Mode

The tables can be opened in the **Read-only** mode by selecting the table and then choosing the **View Table** button from the **dbConnect Manager**. You can also choose **View Table** from the shortcut menu displayed upon right-clicking on the selected table. If you choose the **View External Table** tool from **View Data > dbConnect** in the menu bar, the **Select Data Object** dialog box will be displayed. Select a table and choose the **OK** button. You will notice that the table that is displayed has a gray background and you cannot edit the entries in it.

Edit Mode

The tables can be opened in the **Edit** mode by choosing the **Edit Table** button from the **dbConnect Manager**. You can also choose **Edit Table** from the shortcut menu that is displayed upon right-clicking on the selected table. If you choose the **View External Table** tool from **View Data > dbConnect** in the menu bar, the **Select Data Object** dialog box will be displayed. Select a table and choose the **OK** button.

> **Note**
> The **db_samples.mdb** file is available in the \AutoCAD 2012\Sample\Database Connectivity directory.
>
> If you are using Vista Operating System, make sure that you are running AutoCAD as administrator.

EXAMPLE 2   *Editing Data Source*

In this example, you will select a **Computer** table from the **jet_dbsamples** data source to edit the rows in the table. Add a new computer entry, edit the existing entry, and save changes.

1. Select the **Computer** table from the **jet_dbsamples** in the **dbConnect Manager**, right-click to invoke the shortcut menu, and choose **Edit Table** (Figure 33-9). You can also do the same by double-clicking on the **Computer** table.

*Figure 33-9 Choosing **Edit Table** from the shortcut menu*

2. The **Data View** window is displayed with the **Computer** table in it. In the **Data View** window, you can resize, sort, hide, or freeze the columns according to your requirements.

3. To add a new item in the table, double-click on the last empty row. In the **Tag_Number** column, type **24675**. Then, type the following data into the columns:

Manufacturer:	**IBM**
Equipment_Description:	**PIV/450, 4500GL, NETX**
Item_Type:	**CPU**
Room:	**6035**

4. To edit the record for **Tag_Number 60298** in the **Data View** window (Figure 33-10), double-click on each cell and type the following.

Manufacturer:	**CREATIVE**
Equipment_Description:	**INFRA 6000, 40XR**
Item_Type:	**CD DRIVE**
Room:	**6996**

5. After making all changes in the database table, you have to save the changes for further use. To save the changes in the table, right-click on the **Data View grid header** (a triangle mark on the left of the **Tag_Number** column head). Choose **Commit** from the shortcut menu. The changes in the current table is saved. The new record that you have entered does not necessarily get added at the end of the table, so if you close the table and then reopen it, you might have to search for the new record in the table. Note that if you quit the **Data View** window without committing, all the changes you have made during the editing session are automatically added to the table.

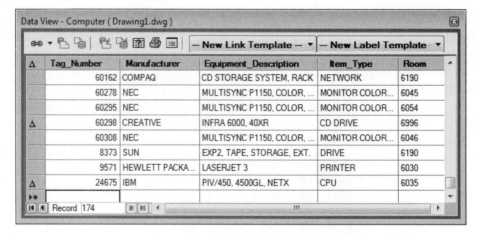

Figure 33-10 *The **Data View** window after editing*

Note

*After making changes, if you do not want to save the changes, choose **Restore** from the shortcut menu mentioned above.*

CREATING LINKS WITH GRAPHICAL OBJECTS

The main function of the database connectivity feature of AutoCAD is to associate data from external sources with its graphical objects. You can establish the association of the database table with the drawing objects by developing a link, which will make a reference to one or more records from the table. But you cannot link nongraphical objects such as layers or linetypes with the external database. Links are very closely related with graphical objects and change in the link will change the graphical objects.

To develop links between database tables and graphical objects, you must create a link template that identifies the fields of the tables with which the links are associated to share the template. For example, you can create a link template that uses the **Tag_Number** from the **COMPUTER** database table. The link template also acts as a shortcut that points to the associated database tables. You can associate multiple links to a single graphical object using different link templates. This is useful in associating multiple database tables with a single drawing object. The following example will describe the procedure of linking using the link template creation.

EXAMPLE 3   *Linking Data Source with Drawing*

Create a link template between the **Computer** database table from **jet_dbsamples** and your drawing, and then use the **Tag_Number** as the key field for linking.

1. Open the drawing that has to be linked with the **Computer** database table.

2. Invoke the **dbConnect Manager**. Select **Computer** from the **jet_dbsamples** data source and right-click on it to invoke the shortcut menu.

3. Choose **New Link Template** from the shortcut menu to invoke the **New Link Template** dialog box (Figure 33-11). You can also select the table in the tree view and choose the **New Link Template** button from the toolbar in the **dbConnect Manager**.

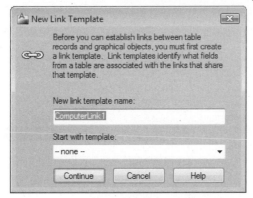

*Figure 33-11 The **New Link Template** dialog box*

4. Choose the **Continue** button to accept the default link template named **ComputerLink1**. On doing so, the **Link Template** dialog box is displayed. Select the check box adjacent to the **Tag_Number** to accept it as the **Key field** for associating the template with the block reference in the diagram. (See Figure 33-12.)

5. Choose the **OK** button to complete the new link template. The name of the link template is displayed under the **Drawing name** node of the tree view in the **dbConnect Manager**.

6. To link a record, double-click on the **Computer** table to invoke the **Data View** window (**Edit** mode). In the table, go to the **Tag_Number 24675** and highlight the record (row).

7. Right-click on the row header and choose **Link!** from the shortcut menu (Figure 33-13). Alternatively, Choose the down arrow on the **Link** button in the **Data View** window; a flyout will be displayed. Choose the **Create Links** option from the flyout.

Chapter 33

Figure 33-12 *The **Link Template** dialog box*

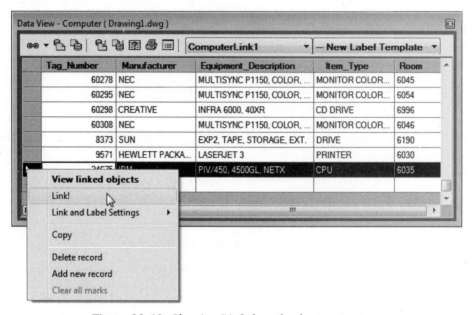

Figure 33-13 *Choosing **Link** from the shortcut menu*

8. On choosing the **Link!** option, you are prompted to select the objects to be linked. Select the required objects to link with the record and press the ENTER key. Repeat the process to link all the records to the corresponding block references in the diagram. By default, all the linked rows gets highlighted in yellow color.

9. After linking, you can view the linked objects by choosing the **View Linked Objects in Drawing** button in the **Data View** window, and the linked objects gets highlighted in the drawing area.

Additional Link View Settings

You can set a number of viewing options for linked graphical objects and linked records by using the **Data View and Query Options** dialog box (Figure 33-14). This dialog box can be invoked by choosing the **Data View and Query Options** button from the **Data View** window.

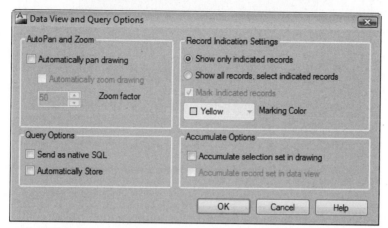

Figure 33-14 *The* **Data View and Query Options** *dialog box*

You can select the **Automatically pan drawing** check box from the **Data View and Query Options** dialog box, so that the drawing is panned automatically to display the objects linked with the current set of selected records in the **Data View** window. You can also select the **Automatically zoom drawing** check box and specify the value of zoom factor in the **Zoom factor** spinner. Moreover, you can change the record indication settings and the indication marking color using the other options available in this dialog box.

Editing Link Data

After linking data with the drawing objects, you may need to edit the data or update the **Key field** values. For example, you may need to reallocate the **Tag_Number** for the computer equipment or **Room** for each of the linked items. The **Link Manager** can be used for changing the key values. The next example describes the procedure of editing the linked data.

EXAMPLE 4   *Link Manager*

Use the **Link Manager** to edit the data linked to the **Computer** table and change the Key Values from **24675** to **24875**.

1. Open the diagram that was linked with the **Computer** table.

2. Select the linked object from the drawing area and right-click; a shortcut menu is displayed. Choose **Link > Link Manager** option from the shortcut menu displayed; the **Link Manager** dialog box for the **Computer** table is displayed, refer to Figure 33-15.

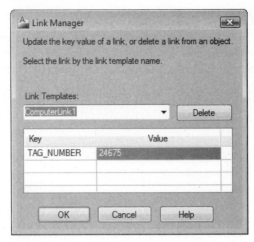

*Figure 33-15 The **Link Manager** dialog box*

3. Select the **24675** field and choose the [**...**] button in the **Value** column to invoke the **Column Values** dialog box.

4. Select **24857** from the list in the **Column Values** dialog box (Figure 33-16) and choose the **OK** button.

5. Again, choose the **OK** button in **Link Manager** to accept the changes.

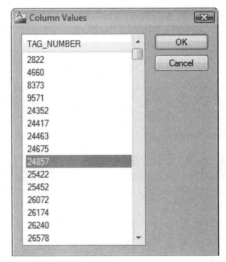

*Figure 33-16 The **Column Values** dialog box*

CREATING LABELS

You can use linking as a powerful mechanism to associate drawing objects with external database tables. You can directly access associated records in the database table by selecting linked objects in the drawing. But linking has some limitations. Consider that you want to include the associated external data with the drawing objects. Since during printing, the links are only the pointers to the external database table, they will not appear in the printed drawing. In such situations, a feature called Labels proves useful. The labels can be used for the visible representation of external data in the drawing.

Labels are the multiline text objects that display data from the selected fields in the AutoCAD drawing. The labels are of the following two types:

Freestanding labels

Freestanding labels exist in the AutoCAD drawing independent of the graphical objects. Their properties do not change with any change of graphical objects in the drawing.

Attached labels

The Attached labels are connected with the graphical objects they are associated with. If the graphical objects are moved, then the labels also move. If the objects are deleted, then the labels attached to them also get deleted.

Labels associated with the graphical objects in AutoCAD drawing are displayed with a leader. Labels are created and displayed by Label Templates and all their properties can be controlled by using the **Label Template** dialog box. The next example will demonstrate the complete procedure of creating and displaying labels in AutoCAD drawing using the label template.

EXAMPLE 5	*Creating New Label Template*

Create a new label template in the **Computer** database table and use the following specifications for the display of labels in the drawing.

a.   The labels include **Tag_Number**, **Manufacturer**, and, **Item_Type** fields.
b.   The fields in the label are having: **0.25** height, **Times New Roman** font, **Index color:** 18 color, and they are **Middle-Left** justified.
c.   The label offset starts with **Middle Center** justified and leader offset is **X=1.5** and **Y=1.5**.

1.   Open the drawing where you want to attach a label with the drawing objects.

2.   Choose the **dbConnect** option from **Tools > Palettes** menu bar to invoke the **dbConnect Manager**. Select the **Computer**.

3.   Right-click on **Computer** table from the **jet_dbsamples** data source and choose **New Label Template** from the shortcut menu to invoke the **New Label Template** dialog box (Figure 33-17). You can also invoke it by choosing the **New Label Template** button from the **dbConnect Manager**.

4.   In the **New Label Template** dialog box, choose the **Continue** button to accept the default label template name **Computer Label1**. The **Label Template** dialog box is displayed.

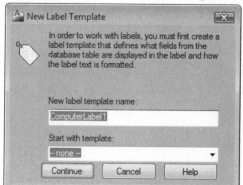

*Figure 33-17 The **New Label Template** dialog box*

5. In the **Label Template** dialog box, choose the **Label Fields** tab. From the **Field** drop-down list, add the **Tag_Number**, **Manufacturer**, and **Item_Type** fields one by one by using the **Add** button (Figure 33-18).

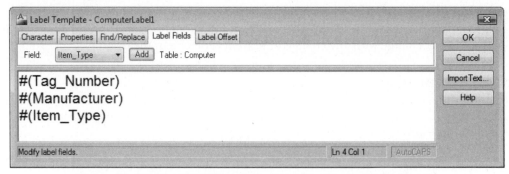

*Figure 33-18 The **Label Template** dialog box (the **Label Fields** tab)*

6. Choose the **Character** tab and highlight the field names by selecting them from the text area. Select the **Times New Roman** font from the **fonts** drop-down list. Next, set the font height to **0.25**. Set the **Index color** to **18**. Also, select the **Middle Left** justification in the **Properties** tab (Figure 33-19).

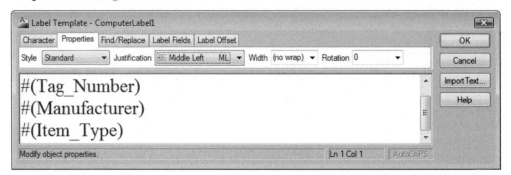

*Figure 33-19 The **Label Template** dialog box (the **Properties** tab)*

7. Choose the **Label Offset** tab and select **Middle Center** in the **Start** drop-down list. Set the **Leader offset** value to **X: 1.5** and **Y: 1.5** (Figure 33-20). Choose **OK**.

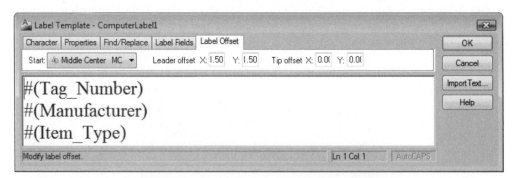

*Figure 33-20 The **Label Template** dialog box (the **Label Offset** tab)*

8. To display the label in the drawing, select the **Computer** table in the **dbConnect Manager**. Right-click on the table and choose **Edit Table** from the shortcut menu.

9. Select **ComputerLink1** (created in an earlier example) from the **Select a Link Template** drop-down list and **ComputerLabel1** from the **Select a Label Template** drop-down list in the **Data View** window.

10. Select the record (row) you want to use as a label. Then choose the down arrow button provided on the right side of the **Link** button. Choose **Create Attached Labels** from the shortcut menu (Figure 33-21). Then, choose the **Create Attached Labels** button that replaces the **Link** button. Select the drawing object that you want to label. The label, with the given specifications, is displayed in the drawing.

Similarly, you can create **Freestanding Labels** by choosing **Create Freestanding Labels** from the **Link and Label Settings** menu in the **Data View** window.

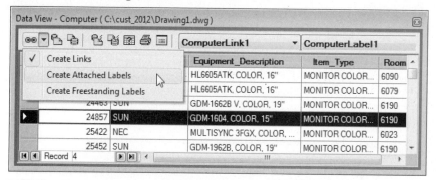

Figure 33-21 Attaching the label to the record

Updating Labels with New Database Values

You may have to change the data values in the database table after adding a label in the AutoCAD drawing. Therefore, you should update the labels in the drawing after making any alteration in the database table the drawing is linked with. The following is the procedure for updating all label values in the AutoCAD drawing:

1. After editing the database table, open the drawing that has to be updated. Select the diagram, right-click and choose **Labels > Reload** from the shortcut menu.

2. The details in the label will be modified automatically and the changes will be reflected in the label attached to the diagram.

Importing and Exporting Link and Label Templates

You may want to use the link and label templates that have been developed by some other AutoCAD users. This is very useful when developing a set of common tools to be shared by all the team members in a project. AutoCAD is capable of importing as well as exporting all the link and label templates that are associated with a drawing. The following is the procedure to export a set of templates from the current drawing:

1. Choose the **Export Template Set** tool from **dbConnect > Templates** in the menu bar to invoke the **Export Template Set** dialog box. From the **Save In** drop-down list of this dialog box, select the directory to save the template set.

2. Under **File Name**, specify a name for the template set, and then choose the **Save** button to save the template in the specified directory.

The following is the procedure to import a set of templates into the current drawing:

1. Choose the **Import Template Set** tool from **dbConnect > Templates** in the menu bar to invoke the **Import Template Set** dialog box, see Figure 33-22.

Chapter 33

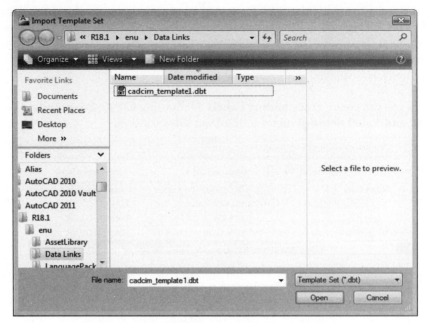

Figure 33-22 The Import Template Set dialog box

2. Select the template and choose **Open** to import the template set into the current drawing; AutoCAD displays an **Alert** box that can be used to provide a unique name for the template, if there is a link or label template with the same name associated with the current drawing.

AutoCAD SQL ENVIRONMENT (ASE)

SQL is an acronym for Structured Query Language. It is often referred to as Sequel. SQL is a format in computer programming that lets the user ask questions about a database according to specific rules. The AutoCAD SQL Environment (ASE) lets you access and manipulate the data that is stored in the external database table and link data from the database to objects in a drawing. Once you access the table, you can manipulate the data. The connection is made through a database management system (DBMS). The DBMS programs have their own methodology for working with the database. However, the ASE commands work the same way regardless of the database being used. This is made possible by the ASE drivers that come with AutoCAD software. For example, you want to prepare a report that lists the computer equipment that costs more than $25. AutoCAD Query Editor can be used to easily construct a query that returns a subset of records or linked graphical objects that follow the previously mentioned criterion.

AutoCAD QUERY EDITOR

The **Query Editor** consists of four tabs that can be used to create new queries. The tabs are arranged in order of increasing complexity. For example, if you are not familiar with **SQL** (Structured Query Language), you can start with Quick Query and Range Query initially to get familiar with the query syntax.

You can start developing a query in one tab and subsequently add and refine the query conditions in the following tabs. For example, if you have created a query in the **Quick Query** tab and then decide to add an additional query using the **Query Builder** tab, when you choose the **Query Builder** tab, all the values initially selected in the previous tabs are displayed in this tab and additional conditions can be added to the query. But it is not possible to go backwards through the tabs once you have created queries with one of the advanced tabs. The reason for this is that the additional functions are not available in the simpler tabs. A warning will be prompted indicating that the query will be reset to its default values if you attempt to move backward through the query tabs. The **Query Editor** has the following tabs for building queries:

Quick Query

This tab provides an environment where simple queries can be developed based on a single database field, single operator, and a single value. For example, you can find all records from the current table where the value of the '**Item_Type**' field equals '**CPU**'.

Range Query

This tab provides an environment where a query can be developed to return all records that fall within a given range of values. For example, you can find all records from the current table where the value of the '**Room**' field is greater than or equal to 6050 and less than or equal to 6150.

Query Builder

This tab provides an environment where more complicated queries can be developed based on a multiple search criteria. For example, you can find all records from the current table where the '**Item_Type**' equals '**CPU**' and '**Room**' number is greater than 6050.

SQL Query

This tab provides an environment where queries can be developed that confirm with the SQL 92 protocol. For example, you can select * from **Item type.Room. Tag_Number**.

> **Item_type = 'CPU'; Room >= 6050 and <= 6150**
> **and**
> **Tag_number > 26072**

The following example describes the complete procedure of creating a new query using all the tabs of the **Query Editor**:

EXAMPLE 6   *Query Editor*

Create a new Query for the **Computer** database table and use all tabs of the **Query Editor** to prepare a SQL query.

1. Right-click on **Computer** in the **dbConnect Manager** to display the shortcut menu. Choose **New Query** from it to invoke the **New Query** dialog box (Figure 33-23). You can also select **Computer** and choose the **New Query** button from the toolbar in the **dbConnect Manager** to invoke this dialog box.

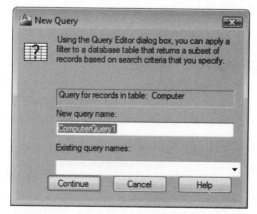

Figure 33-23 The New Query dialog box

2. In the **New Query** dialog box, choose the **Continue** button to accept the default query name **ComputerQuery1**; the **Query Editor** is displayed.

3. In the **Quick Query** tab of the **Query Editor**, select **Item_Type** from the **Field** list box, **=Equal** from the **Operator** drop-down list, and type **CPU** in the **Value** text box (Figure 33-24). You can also select **CPU** from the **Column Values** dialog box by choosing the **Look up values** button. Choose the **Store** button to save the query.

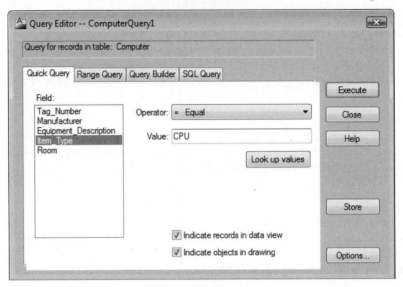

*Figure 33-24 The **Query Editor** (the **Quick Query** tab)*

Note
*To view the query, choose the **Execute** button from the **Query Editor**. However, you will not be able to continue with the example using the same **Query** dialog box.*

4. Choose the **Range Query** tab and select **Room** from the **Field** list box. Enter **6050** in the **From** edit box and **6150** in the **Through** edit box (Figure 33-25). You can also select the values from the **Column Values** dialog box by choosing the **Look up values** button. Choose the **Store** button to save the query.

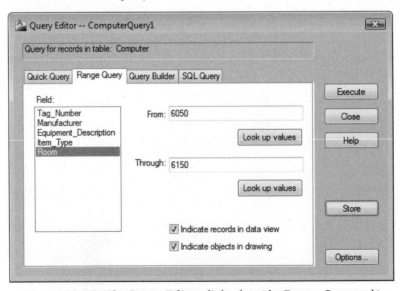

*Figure 33-25 The **Query Editor** dialog box (the **Range Query** tab)*

5. Choose the **Query Builder** tab and add **Item_Type** in the **Show fields** list box by selecting **Item_Type** from the **Fields in table** list box and choosing the **Add** button. In the table area, change the entries of **Field**, **Operator**, **Value**, **Logical**, and **Parenthetical** grouping criteria, as shown in Figure 33-26, by selecting each cell and selecting the values from the respective drop-down list or options lists. Choose the **Store** button to save the query.

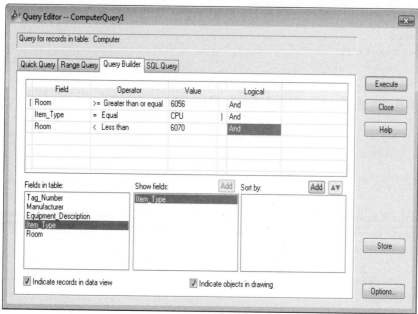

Figure 33-26 *The* **Query Editor** *(the* **Query Builder** *tab)*

6. Choose the **SQL Query** tab; the query conditions specified earlier will be carried to the tab automatically. Here you can create a query using multiple tables. Select **Computer** from the **Table** list box, **Tag_Number** from the **Fields** list box, **>= Greater than or equal** from the **Operator** drop-down list and enter **26072** in the **Values** edit box (Figure 33-27). You can also select any value from the available values by choosing the [**...**] button. Choose the **Store** button to save the query.

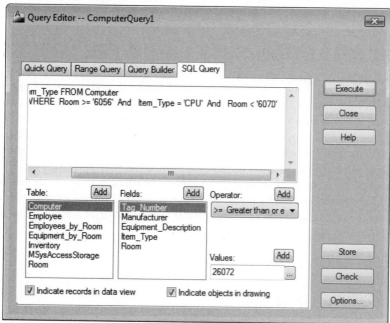

Figure 33-27 *The* **Query Editor** *dialog box (the* **SQL Query** *tab)*

7. To check whether the SQL syntax is correct, choose the **Check** button; AutoCAD will display the **Information** box to determine whether the syntax is correct. Choose **OK** to exit the dialog box.

8. Choose the **Execute** button to display the **Data View** window showing a subset of records matching the specified query criteria.

Note
You can view the subset of records conforming to your criterion with any of the four tabs during the creation of queries.

Importing and Exporting SQL Queries

You may occasionally be required to use the queries made by some other user in your drawing or vice-versa. AutoCAD allows you to import or export stored queries. Sharing queries is very useful when developing common tools used by all team members on a project.

The following is the procedure of exporting a set of queries from the current drawing:

1. Choose the **Export Query Set** tool from **dbConnect > Queries** in the menu bar to invoke the **Export Query Set** dialog box. In the dialog box, select the directory where you want to save the query set from the **Save In** drop-down list.

2. Under the **File name** box, specify a name for the query set, and then choose the **Save** button.

The following is the procedure of importing a set of queries into the current drawing:

1. Choose the **Import Query Set** tool from **dbConnect > Queries** in the menu bar to invoke the **Import Query Set** dialog box. In the dialog box, select the query set to be imported.

2. Choose the **Open** button to import the query set into the current drawing.

AutoCAD displays an alert box that you can use to provide a unique name for the query, if there is a query with the same name that is already associated with the current drawing.

FORMING SELECTION SETS USING THE LINK SELECT

It is possible to locate objects on the drawing on the basis of the linked nongraphic information. For example, you can locate the object that is linked to the first row of the **Computer** table or to the first and second rows of the **Computer** table. You can highlight specified objects or form a selection set of the selected objects. The **Link Select** is an advanced feature of the **Query Editor** that can be used to construct iterative selection sets of AutoCAD graphical objects and the database records. You can start constructing a query or selecting AutoCAD graphical objects for an iterative selection process. The initial selection set is referred to as set A. Now you can select an additional set of queries or graphical objects to further refine your selection set. The second selection set, is referred to as set B. To refine your final selection set, you must establish a relation between set A and set B. The following is the list of available relationships or set operations (Figure 33-28).

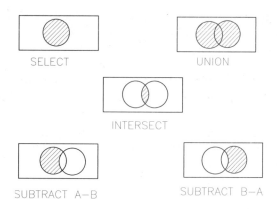

Figure 33-28 *Available relationships or Set operations*

Select

This creates an initial query or graphical objects selection set. This selection set can be further refined or modified using subsequent Link Select operations.

Union

This operation adds the outcome of a new selection set or query to the existing selection set. Union returns all the records that are members of set A or set B.

Intersect

This operation returns the intersection of the new selection set and the existing or running selection set. Intersection returns only the records that are common to both set A and set B.

Subtract A - B

This operation subtracts the result of the new selection set or query from the existing set.

Subtract B - A

This operation subtracts the result of the existing selection set or query from the new set.

After any of the Link Selection operation is executed, the result of the operation becomes the new running selection set and is assigned as set A. You can extend refining your selection set by creating a new set B and then continuing with the iterative process.

Following is the procedure to use **Link Select** for refining a selection set:

1. Choose the **Link Select** tool from **dbConnect > Links** in the menu bar to invoke the **Link Select** dialog box (Figure 33-29).

2. Select the **Select** option from the **Do** drop-down list for creating a new selection set. Also, select a link template from the **Using** drop-down list.

3. Choose either the **Use Query** or the **Select in Drawing <** radio button for creating a new query or drawing an object selection set.

4. Choose the **Execute** button to execute the refining operation or choose the **Select** operation from the **Do:** drop-down list to add your query or graphical object selection set.

5. Again, choose any **Link Selection** operation from the **Do:** drop-down list.

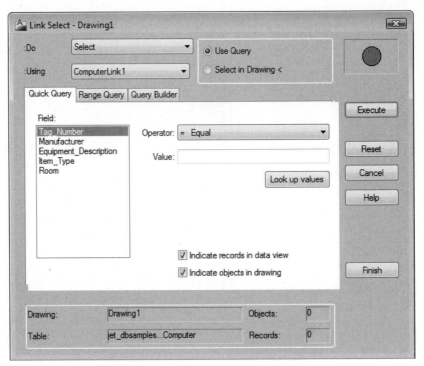

Figure 33-29 The **Link Select** dialog box

6. Repeat steps 2 through 4 for creating a set B for the **Link Select** operation and then choose the **Finish** button to complete the operation.

Note
*You can select the **Use Query** radio button to construct a new query or the **Select in Drawing <** radio button to select a graphical object from the drawing as a selection set.*

*If you select **Indicate records in data view**, then the **Link Select** operation result will be displayed in the **Data View** window and if you select **Indicate objects in drawing**, then AutoCAD displays a set of linked graphical objects in the drawing.*

Self-Evaluation Test

Answer the following questions and then compare them to those given at the end of this chapter:

1. A horizontal group of data is called a _____.

2. A vertical group of data (attribute) is called a _____.

3. You can connect to an external database table by using the _____ **Manager**.

4. You can edit an external database table within an AutoCAD session. (T/F)

5. You can resize, dock, and hide the **Data View** window. (T/F)

6. Once you define a link, AutoCAD does not store that information with the drawing. (T/F)

7. To display the associated records with the drawing objects, AutoCAD provides_____.

8. Labels are of two types, _____ and _____.

9. It is possible to import as well as export Link and label templates. (T/F)

10. ASE stands for _____.

Review Questions

Answer the following questions:

1. The SQL statements let you search through the database and retrieve the information as specified in the SQL statements. (T/F)

2. What are the various components of a table?

3. What is a database?

4. What is a database management system (DBMS)?

5. A row is also referred to a _____.

6. A column is sometimes called a _____.

7. The _____ acts like an identification tag for locating and linking a row.

8. Only one row can be manipulated at a time. This row is called the _____.

9. The **Query Editor** has four tabs, namely _____, _____, _____, and _____.

10. It is possible to go back to the previous query tab without resetting the values. (T/F)

11. You can execute a query after any tab in the **Query Editor**. (T/F)

12. Link select is an _____ implementation of **Query Editor**, which constructs selection sets of graphical objects or database records.

13. AutoCAD writes data source mapping information in the _____ file during the conversion process of links.

14. AutoCAD links _____ be converted into AutoCAD R12 format.

EXERCISE 1   *Query Editor*

In this exercise, select the table **Employee** from **jet_dbsamples** data source in **dbConnect Manager**, and then edit the sixth row of the table (EMP_ID=1006, Keyser). Also, add a new row to the table, set the new row as current, and then view it.

The row to be added should have the following values:

EMP_ID	**1064**
LAST_NAME	**Joel**
FIRST_NAME	**Billy**
Gender	**M**
TITLE	**Marketing Executive**
Department	**Marketing**
ROOM	**6071**

From the **Inventory** table in **jet_dbsamples** data source, build an SQL query step-by-step by using all tabs of the **Query Editor**. The conditions to be implemented are as follows:

1. Type of Item = **Furniture**
2. Range of Cost = **200 to 650**
3. Manufacturer = **Office master**

Answers to Self-Evaluation Test

1. row, **2.** column, **3. dbConnect**, **4.** T, **5.** T, **6.** F, **7.** Leaders, **8.** Freestanding, Attached, **9.** T, **10.** AutoCAD SQL Environment

Chapter 34

Script Files and Slide Shows

CHAPTER OBJECTIVES

After completing this chapter, you will learn:
- *To write script files and use the Run Script option to run script files.*
- *To use the RSCRIPT and DELAY commands in script files.*
- *To run script files while loading AutoCAD.*
- *To create a slide show.*
- *To preload slides when running a slide show.*

KEY TERMS

- *Script Files*
- *RESUME*
- *Run Script*

- *Switches*
- *Preloading Slide*
- *Slide Library*

- *RSCRIPT*
- *Slides*
- *Delay Time*

- *Slide Show*

WHAT ARE SCRIPT FILES?

AutoCAD has provided a facility called script files that allow you to combine different AutoCAD commands and execute them in a predetermined sequence. The commands can be written as a text file using any text editor like Notepad. These files, generally known as script files, have an extension *.scr* (example: *plot1.scr*). A script file is executed with the AutoCAD **SCRIPT** command.

Script files can be used to generate a slide show, do the initial drawing setup, or plot a drawing to a predefined specification. They can also be used to automate certain command sequences that are used frequently in generating, editing, or viewing drawings. Remember that the script files cannot access dialog boxes or menus. When commands that open dialog boxes are issued from a script file, AutoCAD runs the command line version of the command instead of opening the dialog box.

EXAMPLE 1	*Initial Setup of Drawing*

Write a script file that will perform the following initial setup for a drawing (file name *script1.scr*). It is assumed that the drawing will be plotted on 12x9 size paper (Scale factor for plotting = 4).

Ortho	On		Zoom	All
Grid	2.0		Text height	0.125
Snap	0.5		LTSCALE	4.0
Limits	0,0	48.0,36.0	DIMSCALE	4.0

Step 1: Understanding commands and prompt entries

Before writing a script file, you need to know the AutoCAD commands and the entries required in response to the Command prompts. To find out the sequence of the Command prompt entries, you can type the command and then respond to different Command prompts. The following is the list of AutoCAD commands and prompt entries for Example 1:

Command: **ORTHO**
Enter mode [ON/OFF] <OFF>: **ON**

Command: **GRID**
Specify grid spacing(X) or [ON/OFF/Snap/Major/aDaptive/Limits/Follow/Aspect] <0.5000>: **2.0**

Command: **SNAP**
Specify snap spacing or [ON/OFF/Aspect/Style/Type]<0.5000>: **0.5**

Command: **LIMITS**
Reset Model space limits:
Specify lower left corner or [ON/OFF] <0.0,0.0>: **0,0**
Specify upper right corner <12.0,9.0>: **48.0,36.0**

Command: **ZOOM**
Specify corner of window, enter a scale factor (nX or nXP), or
[All/Center/Dynamic/Extents/Previous/Scale/Window/Object] <real time>: **A**

Command: **TEXTSIZE**
Enter new value for TEXTSIZE <0.02>: **0.125**

Command: **LTSCALE**
Enter new linetype scale factor <1.0000>: **4.0**

Command: **DIMSCALE**
Enter new value for DIMSCALE <1.0000>: **4.0**

Step 2: Writing the script file

Once you know the commands and the required prompt entries, you can write a script file using any text editor such as the Notepad.

As you invoke the **NOTEPAD** command, AutoCAD prompts you to enter the file to be edited. Press ENTER in response to the prompt to display the **Notepad** editor. Write the script file in the **Notepad** editor. Given below is the listing of the script file for Example 1:

```
ORTHO
ON
GRID
2.0
SNAP
0.5
LIMITS
0,0
48.0,36.0
ZOOM
ALL
TEXTSIZE
0.125
LTSCALE
4.0
DIMSCALE 4.0
```
(Blank line for Return.)

Note that the commands and the prompt entries in this file are in the same sequence as mentioned earlier. You can also combine several statements in one line, as shown in the following list:

```
;This is my first script file, SCRIPT1.SCR
ORTHO ON
GRID 2.0
SNAP 0.5
LIMITS 0,0 48.0,36.0
ZOOM
ALL
TEXTSIZE 0.125
LTSCALE 4.0
DIMSCALE 4.0
```
(Blank line for Return.)

Save the script file as *SCRIPT1.scr* and exit the text editor. Remember that if you do not save the file in the *.scr* format, it will not work as a script file. Notice the space between the commands and the prompt entries. For example, between the **ORTHO** command and **ON** command, there is a space. Similarly, there is a space between **GRID** and **2.0**.

Note
In the script file, a space is used to terminate a command or a prompt entry. Therefore, spaces are very important in these files. Make sure there are no extra spaces, unless they are required to press ENTER more than once.

Chapter 34

*After you change the limits, it is a good practice to use the **ZOOM** command with the **All** option to increase the drawing display area.*

Keyboard shortcuts are not allowed in the script files. Therefore, make sure not to use them in the script files.

Tip
AutoCAD ignores and does not process any lines that begin with a semicolon (;). This allows you to put comments in the related file or line.

RUNNING SCRIPT FILES

The **SCRIPT** command allows you to run a script file while you are at the drawing editor. Choose the **Run Script** tool from the **Applications** panel of the **Manage** tab; the **Select Script File** dialog box will be invoked, as shown in Figure 34-1. You can enter the name of the script file or you can accept the default file name. The default script file name is the same as the drawing name. If you want to enter a new file name, type the name of the script file without the file extension (**.SCR**). (The file extension is assumed and need not be included with the file name.)

Step 3: Running the script file

To run the script file of Example 1, invoke the **SCRIPT** command, select the file **SCRIPT1**, and then choose the **Open** button in the **Select Script File** dialog box (Figure 34-1) . You will notice the changes taking place on the screen as the script file commands are executed.

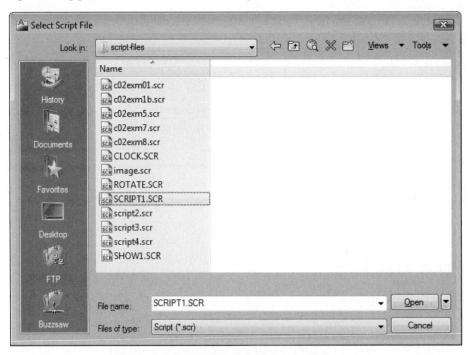

*Figure 34-1 The **Select Script File** dialog box*

You can also enter the name of the script file at the Command prompt by setting **FILEDIA**=0. The sequence for invoking the script using the Command line is given next.

Command: **FILEDIA**
Enter new value for FILEDIA <1>: **0**

Command: **SCRIPT**
Enter script file name <current>: *Specify the script file name.*

EXAMPLE 2   *Layers*

Write a script file that will set up the following layers with given colors and linetypes (file name *script2.scr*).

Layer Names	Color	Linetype	Line Weight
OBJECT	Red	Continuous	default
CENTER	Yellow	Center	default
HIDDEN	Blue	Hidden	default
DIMENSION	Green	Continuous	default
BORDER	Magenta	Continuous	default
HATCH	Cyan	Continuous	0.05

Step 1: Understanding commands and prompt entries

You need to know the AutoCAD commands and the required prompt entries before writing a script file. You need the following commands to create the layers with the given colors and linetypes.

Command: **-LAYER**
Enter an option
[?/Make/Set/New/Rename/ON/OFF/Color/Ltype/LWeight/TRansparency/MATerial/Plot/ Freeze/Thaw/LOck/Unlock/stAte/Description/rEconcile]: **N**
Enter name list for new layer(s): **OBJECT,CENTER,HIDDEN,DIMENSION,BORDER, HATCH**

Enter an option
[?/Make/Set/New/Rename/ON/OFF/Color/Ltype/LWeight/TRansparency/MATerial/Plot/ Freeze/Thaw/LOck/Unlock/stAte/Description/rEconcile]: **L**
Enter loaded linetype name or [?] <Continuous>: **CENTER**
Enter name list of layer(s) for linetype "CENTER" <0>: **CENTER**

Enter an option
[?/Make/Set/New/Rename/ON/OFF/Color/Ltype/LWeight/TRansparency/MATerial/Plot/ Freeze/Thaw/LOck/Unlock/stAte/Description/rEconcile]: **L**
Enter loaded linetype name or [?] <Continuous>: **HIDDEN**
Enter name list of layer(s) for linetype "HIDDEN" <0>: **HIDDEN**

Enter an option
[?/Make/Set/New/Rename/ON/OFF/Color/Ltype/LWeight/TRansparency/MATerial/Plot/ Freeze/Thaw/LOck/Unlock/stAte/Description/rEconcile]: **C**
New color [Truecolor/COlorbook]: **RED**
Enter name list of layer(s) for color 1 (red) <0>: **OBJECT**

Enter an option
[?/Make/Set/New/Rename/ON/OFF/Color/Ltype/LWeight/TRansparency/MATerial/Plot/ Freeze/Thaw/LOck/Unlock/stAte/Description/rEconcile]: **C**
New color [Truecolor/COlorbook]: **YELLOW**
Enter name list of layer(s) for color 2 (yellow) <0>: **CENTER**

Enter an option
[?/Make/Set/New/Rename/ON/OFF/Color/Ltype/LWeight/TRansparency/MATerial/Plot/

Freeze/Thaw/LOck/Unlock/stAte/Description/rEconcile]: **C**
New color [Truecolor/COlorbook]: **BLUE**
Enter name list of layer(s) for color 5 (blue)<0>: **HIDDEN**

Enter an option
[?/Make/Set/New/Rename/ON/OFF/Color/Ltype/LWeight/TRansparency/MATerial/Plot/
Freeze/Thaw/LOck/Unlock/stAte/Description/rEconcile]: **C**
New color [Truecolor/COlorbook]: **GREEN**
Enter name list of layer(s) for color 3 (green)<0>: **DIMENSION**

Enter an option
[?/Make/Set/New/Rename/ON/OFF/Color/Ltype/LWeight/TRansparency/MATerial/Plot/
Freeze/Thaw/LOck/Unlock/stAte/Description/rEconcile]: **C**
New color [Truecolor/COlorbook]: **MAGENTA**
Enter name list of layer(s) for color 6 (magenta)<0>: **BORDER**

Enter an option
[?/Make/Set/New/Rename/ON/OFF/Color/Ltype/LWeight/TRansparency/MATerial/Plot/
Freeze/Thaw/LOck/Unlock/stAte/Description/rEconcile]: **C**
New color [Truecolor/COlorbook]: **CYAN**
Enter name list of layer(s) for color 4 (cyan)<0>: **HATCH**

Enter an option
[?/Make/Set/New/Rename/ON/OFF/Color/Ltype/LWeight/TRansparency/MATerial/Plot/
Freeze/Thaw/LOck/Unlock/stAte/Description/rEconcile]: **LW**
Enter lineweight (0.0mm - 2.11mm): **0.05**
Enter name list of layers(s) for lineweight 0.05mm <0>: **HATCH**
[?/Make/Set/New/Rename/ON/OFF/Color/Ltype/LWeight/TRansparency/MATerial/Plot/
Freeze/Thaw/LOck/Unlock/stAte/Description/rEconcile]: Enter

Step 2: Writing the script file

The following file is a listing of the script file that creates different layers and assigns the given colors and linetypes to them:

```
;This script file will create new layers and
;assign different colors and linetypes to layers
LAYER
NEW
OBJECT,CENTER,HIDDEN,DIMENSION,BORDER,HATCH
L
CENTER
CENTER
L
HIDDEN
HIDDEN
C
RED
OBJECT
C
YELLOW
CENTER
C
BLUE
HIDDEN
```

C
GREEN
DIMENSION
C
MAGENTA
BORDER
C
CYAN
HATCH

(This is a blank line to terminate the **LAYER** *command. End of script file.)*

Save the script file as *script2.scr*.

Step 3: Running the script file

To run the script file of Example 2, choose the **Run Script** tool from the **Applications** panel of the **Manage** tab or enter **SCRIPT** at the Command prompt to invoke the **Select Script File** dialog box. Select **script2.scr** and then choose **Open**. You can also enter the **SCRIPT** command and the name of the script file at the Command prompt by setting **FILEDIA**=0.

EXAMPLE 3   *Rotating the Objects*

Write a script file that will rotate the line and the circle, as shown in Figure 34-2, around the lower endpoint of the line through 45° increments. The script file should be able to produce a continuous rotation of the given objects with a delay of two seconds after every 45° rotation (file name *script3.scr*). It is assumed that the line and circle are already drawn on the screen.

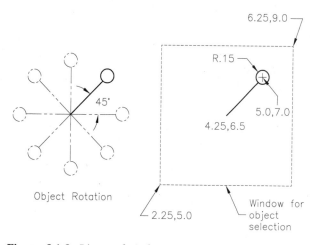

Figure 34-2 Line and circle rotated through 45° increments

Step 1: Understanding commands and prompt entries

Before writing the script file, enter the required commands and prompt entries. Write down the exact sequence of the entries in which they have been entered to perform the given operations. The following is the list of the AutoCAD command sequences needed to rotate the circle and the line around the lower endpoint of the line:

Command: **ROTATE**
Current positive angle in UCS: ANGDIR=counterclockwise ANGBASE=0
Select objects: **W**   *(Window option to select object)*
Specify first corner: **2.25, 5.0**

Specify opposite corner: **6.25, 9.0**
Select objects: Enter
Specify base point: **4.25,6.5**
Specify rotation angle or [Reference]: **45**

Step 2: Writing the script file

Once the AutoCAD commands, command options, and their sequences are known, you can write a script file. You can use any text editor to write a script file. The following is a listing of the script file that will create the required rotation of the circle and line of Example 3. **The line numbers on the right and the text written as '*(Blank line for Return)*' are not a part of the file; they are shown here for reference only.**

ROTATE	1
W	2
2.25,5.0	3
6.25,9.0	4
(Blank line for Return.)	5
4.25,6.5	6
45	7

Line 1
ROTATE
In this line, **ROTATE** is an AutoCAD command that rotates the objects.

Line 2
W
In this line, **W** is the **Window** option for selecting the objects that need to be edited.

Line 3
2.25,5.0
In this line, 2.25 defines the *X* coordinate and 5.0 defines the *Y* coordinate of the lower left corner of the object selection window.

Line 4
6.25,9.0
In this line, 6.25 defines the *X* coordinate and 9.0 defines the *Y* coordinate of the upper right corner of the object selection window.

Line 5
Line 5 is a blank line that terminates the object selection process.

Line 6
4.25,6.5
In this line, 4.25 defines the *X* coordinate and 6.5 defines the *Y* coordinate of the base point for rotation.

Line 7
45
In this line, 45 is the incremental angle of rotation.

Note
*One of the limitations of the script file is that all the information has to be contained within the file. These files do not let you enter information. For instance, in Example 3, if you want to use the **Window** option to select the objects, the **Window** option (W) and the two points that define*

this window must be contained within the script file. The same is true for the base point and all other information that goes in a script file. There is no way that a script file can prompt you to enter a particular piece of information and then resume the script file, unless you embed AutoLISP commands to prompt for user input.

Step 3: Saving the script file
Save the script file with the name *script3.scr*.

Step 4: Running the script file
Choose **Tools > Run Script** from the menu bar, or choose the **Run Script** tool from the **Applications** panel of the **Manage** tab, or enter **SCRIPT** at the Command prompt to invoke the **Select Script File** dialog box. Select **script3.scr** and then choose **Open**. You will notice that the line and circle that were drawn on the screen are rotated once through an angle of 45°. However, there will be no continuous rotation of the sketched entities. The next section (Repeating Script Files) explains how to continue the steps mentioned in the script file. You will also learn how to add a time delay between the continuous cycles in later sections of this chapter.

REPEATING SCRIPT FILES
The **RSCRIPT** command allows the user to execute the script file indefinitely until canceled. It is a very desirable feature when the user wants to run the same file continuously. For example, in the case of a slide show for a product demonstration, the **RSCRIPT** command can be used to run the script file repeatedly until it is terminated by pressing the ESC key. Similarly, in Example 3, the rotation command needs to be repeated indefinitely to create a continuous rotation of the objects. This can be accomplished by adding **RSCRIPT** at the end of the file, as shown in the following listing of the script file:

```
ROTATE
W
2.25,5.0
6.25,9.0
            (Blank line for Return.)
4.25,6.5
45
RSCRIPT
```

The **RSCRIPT** command in line 8 will repeat the commands from line 1 to line 7, and thus set the script file in an indefinite loop. If you run the *script3.scr* file now, you will notice that there is a continuous rotation of the line and circle around the specified base point. However, the speed at which the entities rotate makes it difficult to view the objects. As a result, you need to add time delay between every repetition. The script file can be stopped by pressing the ESC or the BACKSPACE key.

Note
You cannot provide conditional statements in a script file to terminate the file when a particular condition is satisfied unless you use the AutoLISP functions in the script file.

INTRODUCING TIME DELAY IN SCRIPT FILES
As mentioned earlier, some of the operations in the script files happen very quickly and make it difficult to see the operations taking place on the screen. It might be necessary to intentionally introduce a pause between certain operations in a script file. For example, in a slide show for a product demonstration, there must be a time delay between different slides so that the audience have enough time to see each slide. This is accomplished by using the **DELAY** command, which introduces a delay before the next command is executed. The general format of the **DELAY** command is given next.

Command: **DELAY Time**
Where **Command** -----AutoCAD Command prompt
DELAY ---------DELAY command
Time ------------Time in milliseconds

The **DELAY** command is to be followed by the delay time in milliseconds. For example, a delay of 2,000 milliseconds means that AutoCAD will pause for approximately two seconds before executing the next command. It is approximately two seconds because computer processing speeds vary. The maximum time delay you can enter is 32,767 milliseconds (about 33 seconds). In Example 3, a two-second delay can be introduced by inserting a **DELAY** command line between line 7 and line 8, as in the following file listing:

```
ROTATE
W
2.25,5.0
6.25,9.0
          (Blank line for Return.)
4.25,6.5
45
DELAY 2000
RSCRIPT
```

The first seven lines of this file rotate the objects through a 45° angle. Before the **RSCRIPT** command on line 8 is executed, there is a delay of 2,000 milliseconds (about two seconds). The **RSCRIPT** command will repeat the script file that rotates the objects through another 45° angle. Thus, a slide show is created with a time delay of two seconds after every 45° increment.

RESUMING SCRIPT FILES

If you cancel a script file and then want to resume it, you can use the **RESUME** command.

Command: **RESUME**

The **RESUME** command can also be used if the script file has encountered an error that causes it to be suspended. The **RESUME** command will skip the command that caused the error and continue with the rest of the script file. If the error occurs when the command is in progress, use a leading apostrophe with the **RESUME** command (**'RESUME**) to invoke the **RESUME** command in the transparent mode.

Command: **'RESUME**

COMMAND LINE SWITCHES

The command line switches can be used as arguments to the *acad.exe* file that launches AutoCAD. You can also use the **Options** dialog box to set the environment or by adding a set of environment variables in the *autoexec.bat* file. The command line switches and environment variables override the values set in the **Options** dialog box for the current session only. These switches do not alter the system registry. The following is the list of the command line switches:

Switch	Function
/c	Controls where AutoCAD stores and searches for the hardware configuration file. The default file is *acad 2012.cfg*
/s	Specifies which directories to search for support files if they are not in the current directory
/b	Designates a script to run after AutoCAD starts

/t	Specifies a template to use when creating a new drawing
/nologo	Starts AutoCAD without first displaying the logo screen
/v	Designates a particular view of the drawing to be displayed on start-up of AutoCAD
/r	Reconfigures AutoCAD with the default device configuration settings
/p	Specifies the profile to use on start-up

RUNNING A SCRIPT FILE WHILE LOADING AutoCAD

The script files can also be run while loading AutoCAD, before it is actually started. The following is the format of the command for running a script file while loading AutoCAD.

"Drive**Program Files\Autodesk\AutoCAD 2012\acad.exe**" [existing-drawing] [/t template] [/v view] /b Script-file

In the following example, AutoCAD will open the existing drawing (MYdwg1) and then run the script file (Setup) through the **Run** dialog box, as shown in Figure 34-3.

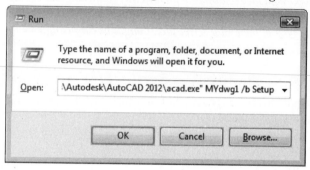

*Figure 34-3 Invoking the script file while loading AutoCAD using the **Run** dialog box*

Example:
"**C:\Program Files\Autodesk\AutoCAD 2012\acad.exe**" **MYdwg1 /b Setup**
      Where **AutoCAD 2012**   ---------AutoCAD 2012 subdirectory containing
                    ---------------------AutoCAD system files
             acad.exe --------ACAD command to start AutoCAD
             MYDwg1 -------Existing drawing file name
             Setup -----------Name of the script file

Note
Make sure that the existing drawing file that you want to open is saved in the C drive of your system. Also, you must have the administrative privileges to save and open the drawing files from the C drive.

In the following example, AutoCAD will start a new drawing with the default name (Drawing), using the template file temp1, and then run the script file (Setup).

Example:
"**C:\Program Files\Autodesk\AutoCAD 2012\acad.exe**" **/t temp1 /b Setup**
      Where **temp1** -----------Existing template file name
            Setup -----------Name of the script file

or

"**C:\ProgramFiles\Autodesk\AutoCAD 2012\acad.exe**"**/t temp1** "**C:\MyFolder**"**/b Setup**
      Where **C:\Program Files\AutoCAD 2012\acad.exe** Path name for acad.exe
        C:\MyFolder ---Path name for the Setup script file

In the following example, AutoCAD will start a new drawing with the default name (Drawing), and then run the script file (Setup).

Example:
"C:\Program Files\Autodesk\AutoCAD 2012\acad.exe" /b Setup
Where **Setup** ------------Name of the script file

Here, it is assumed that the AutoCAD system files are loaded in the AutoCAD 2012 directory.

Note
*To invoke a script file while loading AutoCAD, the drawing file or the template file specified in the command must exist in the search path. You cannot start a new drawing with a given name. You can also use any template drawing file that is found in the template directory to run a script file through the **Run** dialog box.*

Tip
You should avoid abbreviations to prevent any confusion. For example, C can be used as a close option when you are drawing lines. It can also be used as a command alias for drawing a circle. If you use both of them in a script file, it might be confusing.

EXAMPLE 4   *Running a Script File while Loading AutoCAD*

Write a script file that can be invoked while loading AutoCAD and create a drawing with the following setup (filename *script4.scr*).

Grid	3.0
Snap	0.5
Limits	0,0
	36.0,24.0
Zoom	All
Text height	0.25
LTSCALE	3.0
DIMSCALE	3.0

Layers

Name	Color	Linetype
OBJ	Red	Continuous
CEN	Yellow	Center
HID	Blue	Hidden
DIM	Green	Continuous

Step 1: Writing the script file

Write a script file and save the file under the name *script4.scr*. The following is the listing of the script file that performs the initial setup for a drawing:

```
GRID 3.0
SNAP 0.5
LIMITS 0,0 36.0,24.0 ZOOM ALL
TEXTSIZE 0.25
LTSCALE 3
DIMSCALE 3.0
LAYER NEW
OBJ,CEN,HID,DIM
L CENTER CEN
```

    L HIDDEN HID
    C RED OBJ
    C YELLOW CEN
    C BLUE HID
    C GREEN DIM
                    (Blank line for ENTER.)

Step 2: Loading the script file through the Run dialog box

After you have written and saved the file, quit the text editor. To run the script file, *script4*, select **Start > Run** and then enter the following command line.

"C:\Program Files\Autodesk\AutoCAD 2012\acad.exe" /t EX4 /b script4

> Where **acad.exe** --------ACAD to load AutoCAD
> **EX4** ------------Template file name
> **SCRIPT4** ------Name of the script file

Here, it is assumed that the template file (EX4) and the script file (script4) is on C drive. When you enter this line, AutoCAD is loaded and the file *ex4.dwt* is opened. The script file, script4, is then automatically loaded and the commands defined in the file are executed.

In the following example, AutoCAD will start a new drawing with the default name (Drawing), and then run the script file (script4) (Figure 34-4).

Example:
"C:\Program Files\Autodesk\AutoCAD 2012\acad.exe" /b script4
> Where **script4** ----------Name of the script file

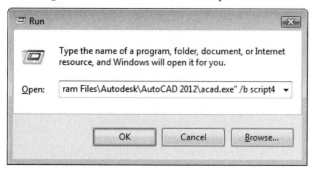

Figure 34-4 *Invoking the script file while loading AutoCAD using the* ***Run*** *dialog box*

Here, it is assumed that the AutoCAD system files are loaded in the AutoCAD 2012 directory.

EXAMPLE 5   *Plotting a Drawing*

Write a script file that will plot a 36" by 24" drawing to the maximum plot size on a 8.5" by 11" paper, using your system printer/plotter. Use the **Window** option to select the drawing to be plotted.

Step 1: Understanding commands and prompt entries

Before writing a script file to plot a drawing, find out the plotter specifications that must be entered in the script file to obtain the desired output. To determine the prompt entries and their sequence to set up the plotter specifications, enter the **-PLOT** command. Note the entries you make and their sequence (the entries for your printer or plotter will probably be different).

The following is a listing of the plotter specifications with the new entries:

> Command: **-PLOT**
> Detailed plot configuration? [Yes/No] <No>: **Y**
> Enter a layout name or [?] <Model>: Enter
> Enter an output device name or [?] <Default Printer name>: *Name of printer or plotter* Enter
> Enter paper size or [?] <Letter>: Enter
> Enter paper units [Inches/Millimeters] <Inches>: **I**
> Enter drawing orientation [Portrait/Landscape] <Landscape>: **L**
> Plot upside down? [Yes/No] <No>: **N**
> Enter plot area [Display/Extents/Limits/View/Window] <Display>: **W**
> Enter lower left corner of window <0.000000,0.000000>: **0,0**
> Enter upper right corner of window <0.000000,0.000000>: **36,24**
> Enter plot scale (Plotted Inches=Drawing Units) or [Fit] <Fit>: **F**
> Enter plot offset (x,y) or [Center] <0.00,0.00>: **0,0**
> Plot with plot styles? [Yes/No] <Yes>: **Yes**
> Enter plot style table name or [?] (enter . for none) <>: **.**
> Plot with lineweights? [Yes/No] <Yes>: **Y**
> Enter shade plot setting [As displayed/Wireframe/Hidden/Visual styles/Rendered] <As displayed>: Enter
> Write the plot to a file [Yes/No] <N>: **N**
> Save changes to page setup [Yes/No]? <N>: Enter
> Proceed with plot [Yes/No] <Y>: **Y**

Note that you need to have a printer installed on the system for successful execution of script.

Step 2: Writing the script file

Now you can write the script file by entering the responses of these prompts in the file. The following file is a listing of the script file that will plot a 36" by 24" drawing on 8.5" by 11" paper after making the necessary changes in the plot specifications. The comments on the right are not a part of the file.

> PLOT
> Y
>   *(Blank line for ENTER, selects default layout.)*
>   *(Blank line for ENTER, selects default printer.)*
>   *(Blank line for ENTER, selects the default paper size.)*
> I
> L
> N
> w
> 0,0
> 36,24
> F
> 0,0
> Y
> .   *(Enter . for none)*
> Y
>   *(Blank line for ENTER, plots as displayed.)*
> N
> N
> Y

The method of saving and running the script file for this example is the same as that described in the previous examples. You can use a blank line to accept the default value for a prompt. A blank line in the script file will cause a Return. However, you must not accept the default plot specifications because the file may have been altered by another user or by another script file. Therefore, always enter the actual values in the file so that when you run a script file, it does not take the default values.

EXAMPLE 6 *Animating a Clock*

Write a script file to animate a clock with continuous rotation of the second hand (longer needle) through 5° and the minutes hand (shorter needle) through 2° clockwise around the center of the clock, (Figure 34-5).

Figure 34-5 *Drawing for Example 6*

The specifications are given next.

Specifications for the rim made of donut.

Color of Donut	Blue
Inner diameter of Donut	8.0
Outer diameter of Donut	8.4
Center point of Donut	5,5

Specifications for the digit mark made of polyline.

Color of the digit mark	Green
Start point of Pline	5,8.5
Initial width of Pline	0.25
Final width of Pline	0.25
Height of Pline	0.25

Specifications for second hand (long needle) made of polyline.

Color of the second hand	Red
Start point of Pline	5,5
Initial width of Pline	0.5
Final width of Pline	0.0
Length of Pline	3.5
Rotation of the second hand	5 degree clockwise

Specifications for minute hand (shorter needle) made of polyline.

Color of the minute hand	Cyan
Start point of Pline	5,5
Initial width of Pline	0.35
Final width of Pline	0.0

Length of Pline	3.0
Rotation of the minute hand	2 degree clockwise

Step 1: Understanding the commands and prompt entries for creation of the clock

For this example, you can create two script files and then link them. The first script file will demonstrate the creation of the clock on the screen. The next script file will demonstrate the rotation of the needles of the clock.

First write a script file to create the clock as follows and save the file under the name *clock.scr*. The following is the listing of this file:

```
Command: -COLOR
Enter default object color [Truecolor/COlorbook] <BYLAYER>: Blue
Command: DONUT
Specify inside diameter of donut<0.5>: 8.0
Specify outside diameter of donut<0.5>: 8.4
Specify center of donut or <exit>: 5,5
Specify center of donut or <exit>: Enter
Command: -COLOR
Enter default object color [Truecolor/COlorbook] <BYLAYER>: Green
Command: PLINE
Specify start point: 5,8.5
Specify next point or [Arc/Halfwidth/Length/Undo/Width]: Width
Specify starting width<0.00>: 0.25
Specify ending width<0.25>: 0.25
Specify next point or [Arc/Halfwidth/Length/Undo/Width]: @0.25<270
Specify next point or [Arc/Halfwidth/Length/Undo/Width]: Enter
Command: -ARRAY
Select objects: Last
Select objects: Enter
Enter the type of array[Rectangular/PAth/Polar]<R>: Polar
Specify center point of array or [Base]: 5,5
Enter the number of items in the array: 12
Specify the angle to fill(+= ccw,  -=cw)<360>: 360
Rotate arrayed objects ? [Yes/No]<Y>: Y
Command: -COLOR
Enter default object color [Truecolor/COlorbook] <BYLAYER>: RED
Command: PLINE
Specify start point: 5,5
Specify next point or [Arc/Close/Halfwidth/Length/Undo/Width]: Width
Specify starting width<.25>: 0.5
Specify ending width<0.5>: 0
Specify next point or [Arc/Halfwidth/Length/Undo/Width]: @3.5<0
Specify next point or [Arc/Close/Halfwidth/Length/Undo/Width]: Enter
Command: -COLOR
Enter default object color [Truecolor/COlorbook] <BYLAYER>: Cyan
Command: PLINE
Specify start point: 5,5
Specify next point or [Arc/Halfwidth/Length/Undo/Width]: Width
Specify starting width<0.0000>: 0.35
Specify ending width<0.35>: 0
Specify next point or [Arc/Halfwidth/Length/Undo/Width]: @3<90
Specify next point or [Arc/Close/Halfwidth/Length/Undo/Width]: Enter
Command: SCRIPT
ROTATE.SCR
```

Now you will write the script file by entering the responses to these prompts in the file *clock.scr*.

Next, you will write the script to rotate the clock hands and save the file with the name *ROTATE. scr*. Remember that while entering the commands in the script files, you do not need to add a hyphen (-) as a prefix to the command name to execute them from the command line. When a command is entered using the script file, the dialog box is not displayed and it is executed using the command line. For example, in this script file, the **COLOR** and **ARRAY** commands will be executed using the command line. Listing of the script file is given next.

Color
Blue
Donut
8.0
8.4
5,5
                        (Blank line for ENTER.)
Color
Green
Pline
5,8.5
W
0.25
0.25
@0.25<270
                        (Blank line for ENTER.)
Array
L
                        (Blank line for ENTER.)
P
5,5
12
360
Y
Color
Red
Pline
5,5
W
0.5
0
@3.5<0
                        (Blank line for ENTER.)
Color
Cyan
Pline
5,5
w
0.35
0
@3<90
                        (Blank line for ENTER.)
Script
ROTATE.scr   (Name of the script file that will cause rotation)
                        (Blank line for ENTER.)

Chapter 34

Save this file as *clock.scr* in a directory that is specified in the AutoCAD support file search path. It is recommended that the *ROTATE.scr* file should also be saved in the same directory. Remember that if the files are not saved in the directory that is specified in the AutoCAD support file search path using the **Options** dialog box, the linked script file (*ROTATE.scr*) may not run.

Step 2: Understanding the commands and sequences for rotation of the needles

The last line in the above script file is *ROTATE.scr*. This is the name of the script file that will rotate the clock hands. Before writing the script file, enter the **ROTATE** command and respond to the command prompts that will cause the desired rotation. The following is the listing of the AutoCAD command sequences needed to rotate the objects:

Command: **ROTATE**
Select objects: **L**
Select objects: ⌨Enter
Specify base point: 5,5
Specify rotation angle or [Copy/Reference] <0>: -2
Command: **ROTATE**
Select objects: **C**
Specify first corner: 3,3
Specify other corner: 7,7
Select objects: **Remove**
Remove objects: **L**
Remove objects: ⌨Enter
Specify base point: 5,5
Specify rotation angle or [Copy/Reference]: -5

Now you can write the script file by entering the responses to these prompts in the file *ROTATE.scr*. The following is the listing of the script file that will rotate the clock hands:

Rotate
L
                              (Blank line for ENTER.)

5,5
-2
Rotate
c
3,3
7,7
R
L
                              (Blank line for ENTER.)

5,5
-5
Rscript
                              (Blank line for ENTER.)

Save the above script file as *ROTATE.scr*. Now, run the script file *clock.scr*. Since this file is linked with *ROTATE.scr*, it will automatically run *ROTATE.scr* after running *clock.scr*. Note that if the linked file is not saved in a directory specified in the AutoCAD support file search path, the last line of the *clock.scr* must include a fully-resolved path to *ROTATE.scr*, or AutoCAD would not be able to locate the file.

EXERCISE 1   *Plotting a Drawing*

Write a script file that will plot a 288' by 192' drawing on a 36" x 24" sheet of paper. The drawing scale is 1/8" = 1'. (The filename is *script9.scr*. In this exercise, assume that AutoCAD is configured for the HPGL plotter and the plotter description is HPGL-Plotter.)

WHAT IS A SLIDE SHOW?

AutoCAD provides a facility of using script files to combine the slides in a text file and display them in a predetermined sequence. In this way, you can generate a slide show for a slide presentation. You can also introduce a time delay in the display so that the viewer has enough time to view each slide.

A drawing or parts of a drawing can also be displayed using the AutoCAD display commands. For example, you can use **ZOOM**, **PAN**, or other commands to display the details you want to show. If the drawing is very complicated, it takes quite some time to display the desired information and it may not be possible to get the desired views in the right sequence. However, with slide shows you can arrange the slides in any order and present them in a definite sequence. In addition to saving time, this also helps to minimize the distraction that might be caused by constantly changing the drawing display. Also, some drawings are confidential in nature and you may not want to display some portions or views of them. You can send a slide show to a client without losing control of the drawings and the information that is contained in them.

WHAT ARE SLIDES?

A slide is the snapshot of a screen display; it is like taking a picture of a display with a camera. The slides do not contain any vector information like AutoCAD drawings, which means that the entities do not have any information associated with them. For example, the slides do not retain any information about the layers, colors, linetypes, start point, or endpoint of a line or viewpoint. Therefore, slides cannot be edited like drawings. If you want to make any changes in the slide, you need to edit the drawing and then make a new slide from the edited drawing.

CREATING SLIDES

In AutoCAD, slides are created using the **MSLIDE** command. If **FILEDIA** is set to 1, the **MSLIDE** command displays the **Create Slide File** dialog box (Figure 34-6) on the screen. You can enter the slide file name in this dialog box. If **FILEDIA** is set to 0, the command will prompt you to enter the slide file name.

> Command: **MSLIDE**
> Enter name of slide file to create <Default>: *Slide file name.*

Example:
Command: **MSLIDE**
Slide File: <Drawing1> **SLIDE1**
       Where   **Drawing1** ------Default slide file name
                 SLIDE1 --------Slide file name

In this example, AutoCAD will save the slide file as *slide1.sld*.

Note
*In model space, you can use the **MSLIDE** command to make a slide of the existing display in the current viewport. If you are in the paper space viewport, you can make a slide of the display in the paper space that includes all floating viewports created in it. When the viewports are not active, the **MSLIDE** command will make a slide of the current screen display.*

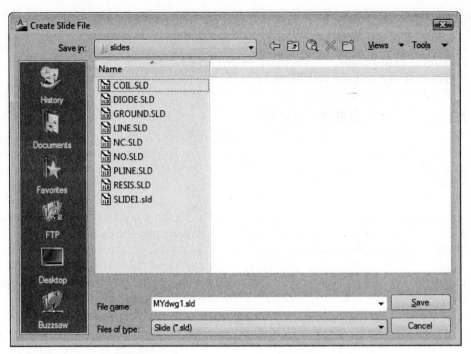

*Figure 34-6 The **Create Slide File** dialog box*

VIEWING SLIDES

To view a slide, use the **VSLIDE** command at the Command prompt; the **Select Slide File** dialog box will be displayed, as shown in the Figure 34-7. Choose the file that you want to view and then choose **OK**. The corresponding slide will be displayed on the screen. If **FILEDIA** is 0, the slide that you want to view can be directly entered at the Command prompt.

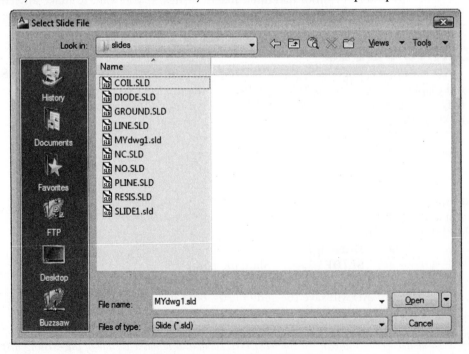

*Figure 34-7 The **Select Slide File** dialog box*

Command: **VSLIDE**
Enter name of slide file to view<Default>: *Name*.

Example:
Command: **VSLIDE**
Slide file <Drawing1>: SLIDE1
　　　　Where **Drawing1** ------Default slide file name
　　　　　　　SLIDE1 --------Name of slide file

Note
*After viewing a slide, you can use the **REDRAW** command, roll the wheel, or pan with a mouse to remove the slide display and return to the existing drawing on the screen.*

*Any command that is automatically followed by a **REDRAW** command will also display the existing drawing. For example, AutoCAD **GRID**, **ZOOM ALL**, and **REGEN** commands will automatically return to the existing drawing on the screen.*

You can view the slides on a high-resolution or a low-resolution monitor. Depending on the resolution of the monitor, AutoCAD automatically adjusts the image. However, if you are using a high-resolution monitor, it is better to make the slides using the same monitor to take full advantage of that monitor.

EXAMPLE 7　　　　　　　　　　　　　　　　　*Slide Show*

Write a script file that will create a slide show of the following slide files, with a time delay of 15 seconds after every slide (Figure 34-8).

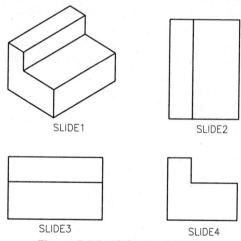

Figure 34-8 Slides for slide show

Step 1: Creating the slides

The first step in a slide show is to create the slides using the **MSLIDE** command. The **MSLIDE** command will invoke the **Create Slide File** dialog box. Enter the name of the slide as **SLIDE1** and choose the **Save** button to exit the dialog box. Similarly, other slides can be created and saved. Figure 34-8 shows the drawings that have been saved as slide files **SLIDE1**, **SLIDE2**, **SLIDE3**, and **SLIDE4**. The slides must be saved to a directory in AutoCAD's search path otherwise, the script would not find the slides.

Step 2: Writing the script file

The second step is to find out the sequence, in which you want these slides to be displayed, with the necessary time delay, if any, between slides. Then you can use any text editor or the AutoCAD **EDIT** command (provided the *acad.pgp* file is present and **EDIT** is defined in the file) to write the script file with the extension *.scr*.

The following is the listing of the script file that will create a slide show of the slides in Figure 34-8. The name of the script file is **SLDSHOW1.scr**.

```
VSLIDE SLIDE1
DELAY 15000
VSLIDE SLIDE2
DELAY 15000
VSLIDE SLIDE3
DELAY 15000
VSLIDE SLIDE4
DELAY 15000
```

Step 3: Running the script file

To run this script file, choose the **Run Script** tool from **Manage > Applications** or enter **SCRIPT** at the Command prompt to invoke the **Select Script File** dialog box. Choose **SLDSHOW1** and then choose the **Open** button from the **Select Script File** dialog box. You can see the changes taking place on the screen.

PRELOADING SLIDES

In the script file of Example 7, VSLIDE SLIDE1 in line 1 loads the slide file, **SLIDE1**, and displays it on the screen. After a pause of 15,000 milliseconds, it starts loading the second slide file, **SLIDE2**. Depending on the computer and the disk access time, you will notice that it takes some time to load the second slide file. The same is true for the other slides. To avoid the delay in loading the slide files, AutoCAD has provided a facility to preload a slide while viewing the previous slide. This is accomplished by placing an asterisk (*) in front of the slide file name.

VSLIDE SLIDE1	*(View slide, SLIDE1.)*
VSLIDE *SLIDE2	*(Preload slide, SLIDE2.)*
DELAY 15000	*(Delay of 15 seconds.)*
VSLIDE	*(Display slide, SLIDE2.)*
VSLIDE *SLIDE3	*(Preload slide, SLIDE3.)*
DELAY 15000	*(Delay of 15 seconds.)*
VSLIDE	*(Display slide, SLIDE3.)*
VSLIDE *SLIDE4	
DELAY 15000	
VSLIDE	
DELAY 15000	
RSCRIPT	*(Restart the script file.)*

EXAMPLE 8   *Preloading Slides*

Write a script file to generate a continuous slide show of the following slide files, with a time delay of two seconds between slides SLD1, SLD2, and SLD3.

The slide files are located in different subdirectories, as shown in Figure 34-9.

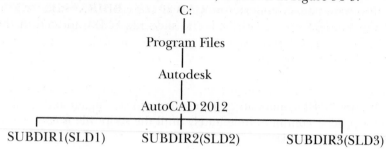

Figure 34-9 *Subdirectories of the C drive*

Where	**C:**	------------------*Root directory.*
	Program Files	-------------*Root directory.*
	Autodesk	-------------------*Root directory.*
	AutoCAD 2012	---------- *Subdirectory where the AutoCAD files are loaded.*
	SUBDIR1	------------------ *Drawing subdirectory.*
	SUBDIR2	----------------- *Drawing subdirectory.*
	SUBDIR3	----------------- *Drawing subdirectory.*
	SLD1	--------------------*Slide file in SUBDIR1 subdirectory.*
	SLD2	--------------------*Slide file in SUBDIR2 subdirectory.*
	SLD3	--------------------*Slide file in SUBDIR3 subdirectory.*

The following is the listing of the script files that will generate a slide show for the slides in Example 8:

```
VSLIDE "C:/Program Files/Autodesk/AutoCAD 2012/SUBDIR1/SLD1.SLD"    1
DELAY  2000                                                          2
VSLIDE "C:/Program Files/Autodesk/AutoCAD 2012/SUBDIR2/SLD2.SLD"    3
DELAY 2000                                                           4
VSLIDE "C:/Program Files/Autodesk/AutoCAD 2012/SUBDIR3/SLD3.SLD"    5
DELAY 2000                                                           6
RSCRIPT                                                              7
```

Line 1
VSLIDE "C:/Program Files/Autodesk/AutoCAD 2012/SUBDIR1/SLD1.SLD"
In this line, the AutoCAD command **VSLIDE** loads the slide file **SLD1**. The path name is mentioned along with the command **VSLIDE**. If the path name directory contains spaces, then the path name must be enclosed in quotes.

Line 2
DELAY 2000
This line uses the AutoCAD **DELAY** command to create a pause of approximately two seconds before the next slide is loaded.

Line 3
VSLIDE "C:/Program Files/Autodesk/AutoCAD 2012/SUBDIR2/SLD2.SLD"
In this line, the AutoCAD command **VSLIDE** loads the slide file **SLD2**, located in the subdirectory **SUBDIR2**. If the slide file is located in a different subdirectory, you need to define the path with the slide file.

Line 5
VSLIDE "C:/Program Files/Autodesk/AutoCAD 2012/SUBDIR3/SLD3.SLD"
In this line, the **VSLIDE** command loads the slide file SLD3, located in the subdirectory **SUBDIR3**.

Line 7
RSCRIPT
In this line, the **RSCRIPT** command executes the script file again and displays the slides on the screen. This process continues indefinitely until the script file is canceled by pressing the ESC key or the BACKSPACE key.

SLIDE LIBRARIES

AutoCAD provides a utility, SLIDELIB, which constructs a library of the slide files. The format of the **SLIDELIB** utility command is as follows:

> **SLIDELIB (Library filename) <(Slide list filename)**

Example:
SLIDELIB SLDLIB <SLDLIST
> Where **SLIDELIB** -----AutoCAD's SLIDELIB utility
> **SLDLIB** --------Slide library filename
> **SLDLIST** ------List of slide filenames

The **SLIDELIB** utility is supplied with the AutoCAD software package. You can find this utility (SLIDELIB.EXE) in the *C:\Program Files\Autodesk\AutoCAD 2012* subdirectory. The slide file list is a list of the slide file names that you want in a slide show. It is a text file that can be written by using any text editor like Notepad.

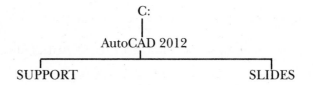

The slide file list can also be created by using the following command, if you have a DOS version 5.0 or above. You can use the make directory (md) or change directory (cd) commands in the DOS mode while making or changing directories.

> C:\AutoCAD 2012\SLIDES>**DIR *.SLD/B>SLDLIST**

In this example, assume that the name of the slide file list is **SLDLIST** and all slide files are in the SLIDES subdirectory. To use this command to create a slide file list, all slide files must be in the same directory.

When you use the SLIDELIB utility, it reads the slide file names from the file that is specified in the slide list and the file is then written to the file specified by the library. In Example 9, the SLIDELIB utility reads the slide filenames from the file SLDLIST and writes them to the library file SLDSHOW1:

> C:\AutoCAD 2012\SLIDES\ **SLIDELIB SLDSHOW1 <SLDLIST**

Note

You cannot edit a slide library file. If you want to change anything, you have to create a new list of the slide files and then use the SLIDELIB utility to create a new slide library.

*If you edit a slide while it is being displayed on the screen, the slide will not be edited. Instead, the current drawing that is behind the slide gets edited. Therefore, do not use any editing commands while you are viewing a slide. Use the **VSLIDE** and **DELAY** commands only while viewing a slide.*

The path name is not saved in the slide library. This is the reason if you have more than one slide with the same name, even though they are in different subdirectories, only one slide will be saved in the slide library.

EXAMPLE 9 *Slide Library*

Use AutoCAD's SLIDELIB utility to generate a continuous slide show of the following slide files with a time delay of 2.5 seconds between the slides. (The filenames are: SLDLIST for slide list file, SLDSHOW1 for slide library, SHOW1 for script file.)

front, top, rside, 3dview, isoview

The slide files are located in different subdirectories, as shown in Figure 34-10.

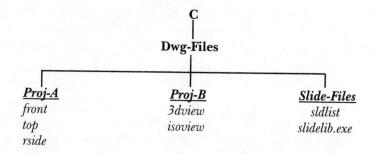

Figure 34-10 *Subdirectories to be created in C drive*

Where **C** ----------------- *(C drive.)*
        Dwg-Files ---- *(Subdirectory where drawing files are located)*
        Proj-A -------- *(Subdirectory where the slide files are located)*
        Proj-B --------- *(Subdirectory where the slide files are located)*
        Slide-Files --- *(Directory where Slidelib.exe and sldlist are copied)*

Step 1: Creating a list of the slide file names

The first step is to create a list of the slide file names with the drive and the directory information. Assume that you are in the **Slide-Files** subdirectory. You can use a text editor like Notepad to create a list of the slide files that you want to include in the slide show. After creating the list, save the text file and then remove its file extension. The following file is a listing of the file SLDLIST for Example 9:

    c:\Dwg-Files\Proj-A\front
    c:\Dwg-Files\Proj-A\top
    c:\Dwg-Files\Proj-A\rside
    c:\Dwg-Files\Proj-B\3dview
    c:\Dwg-Files\Proj-B\isoview

Chapter 34

Step 2: Copying the SLIDELIB utility

The **SLIDELIB** utility is supplied with the AutoCAD software package. You can find this utility (SLIDELIB.EXE) in the *C:\Program Files\Autodesk\AutoCAD 2012* subdirectory. Copy it to the **Slide-Files** folder.

Note

All related directories should be added in the AutoCAD's support files search path.

Step 3: Running the SLIDELIB utility

The third step is to use AutoCAD's SLIDELIB utility program to create the slide library. The name of the slide library is assumed to be **sldshow1** for this example. Before creating the slide library, copy the slide list file (SLDLIST) and the SLIDELIB utility from the support directory to the Slide-Files directory. This ensures that all the required files are in one directory. Enter **SHELL** command at AutoCAD Command prompt and then press the ENTER key at the OS Command prompt. The **AutoCAD Shell Active** Command window will be displayed on the screen, see Figure 34-11. You can also use Windows DOS box instead of **AutoCAD Shell Active** Command window by choosing **All Programs > Accessories > Command Prompt**. Make sure that the drawing file is saved before using the **SHELL** command.

Command: **SHELL**
OS Command: [Enter]

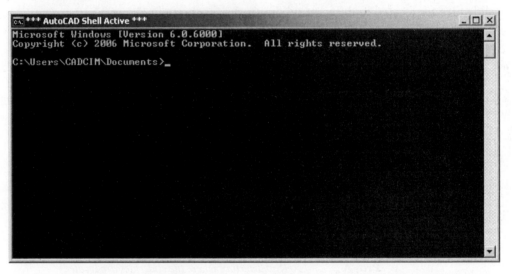

*Figure 34-11 The **AutoCAD Shell Active** Command window*

Now, enter the following command in the Command window to run the SLIDELIB utility and create the slide library. Here it is assumed that **Slide-Files** directory is the current directory. Use the **cd** command in the Command window to change the directory.

C:\Dwg-Files\Slide-Files>SLIDELIB sldshow1 <sldlist

Where **SLIDELIB** -----AutoCAD's SLIDELIB utility
sldshow1 -------Slide library
sldlist -----------Slide file list

Step 4: Writing the script file

Now, you can write a script file for the slide show that will use the slides in the slide library. The name of the script file for this example is assumed to be SHOW1.

VSLIDE C:\Dwg-Files\Slide-Files\sldshow1(front)
DELAY 2500
VSLIDE C:\Dwg-Files\Slide-Files\sldshow1(top)
DELAY 2500
VSLIDE C:\Dwg-Files\Slide-Files\sldshow1(rside)
DELAY 2500
VSLIDE C:\Dwg-Files\Slide-Files\sldshow1(3dview)
DELAY 2500
VSLIDE C:\Dwg-Files\Slide-Files\sldshow1(isoview)
DELAY 2500
RSCRIPT

Step 5: Running the script file

Invoke the **Select Script File** dialog box, as shown in Figure 34-12, by choosing the **Run Script** tool from **Manage > Applications** or enter the **SCRIPT** command at the Command prompt. You can also enter the **SCRIPT** command at the Command prompt after setting the system variable FILEDIA to 0.

Command: **SCRIPT**
Enter script file name<default>: **SHOW1**

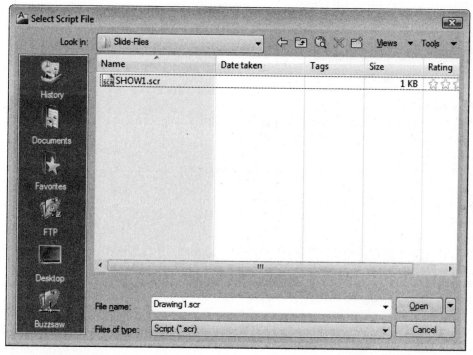

*Figure 34-12 Selecting the script file from the **Select Script File** dialog box*

Self-Evaluation Test

Answer the following questions and then compare them to those given at the end of this chapter:

SCRIPT FILES

1. AutoCAD has provided a facility of _____ that allows you to combine different AutoCAD commands and execute them in a predetermined sequence.

2. Before writing a script file, you need to know the AutoCAD _____ and the _____ required in response to the command prompts.

3. The AutoCAD _____ command is used to run a script file.

4. In a script file, the _____ is used to terminate a command or a prompt entry.

5. The **DELAY** command is to be followed by _____ in milliseconds.

SLIDE SHOW

6. Slides do not contain any _____ information, which means that the entities do not have any information associated with them.

7. Slides _____ edited like a drawing.

8. Slides can be created using the AutoCAD _____ command.

9. To view a slide, use the AutoCAD _____ command.

10. AutoCAD provides a utility that constructs a library of the slide files. This is done with AutoCAD's utility program called _____.

Review Questions

Answer the following questions:

SCRIPT FILES

1. The _____ files can be used to generate a slide show, perform the initial drawing setup, or plot a drawing to a predefined specification.

2. In a script file, you can _____ several statements in one line.

3. When you run a script file, the default script file name is the same as the _____ name.

4. Type the name of the script file without the _____ file, when you run a script file.

5. One of the limitations of script files is that all the information has to be contained _____ the file.

6. The AutoCAD _____ command allows you to re-execute a script file indefinitely until the command is canceled.

7. You cannot provide a _____ statement in a script file to terminate the file when a particular condition is satisfied.

8. The AutoCAD _____ command schedules a delay before the next command is executed.

9. If the script file is canceled and you want to resume the script file, you need to use the _____ command.

SLIDE SHOW

10. AutoCAD provides a facility through _____ files to combine the slides in a text file and display them in a predetermined sequence.

11. A _____ can also be introduced in the script file so that the viewer has enough time to view a slide.

12. Slides are the _____ of a screen display.

13. In model space, you can use the **MSLIDE** command to make a slide of the _____ display in the _____ viewport.

14. If you are in paper space viewport, you can make a slide of the display in paper space that _____ all floating viewports created in it.

15. If you want to make any changes in the slide, you need to _____ the drawing, then make a new slide from the edited drawing.

16. If the slide is in the slide library and you want to view it, the slide library name has to be _____ with the slide filename.

17. You cannot _____ a slide library file. If you want to change anything, you have to create a new list of the slide files and then use the _____ utility to create a new slide library.

18. The path name _____ be saved in the slide library. Therefore, if you have more than one slide with the same name, although with different subdirectories, only one slide will be saved in the slide library.

EXERCISE 2   *Script Files*

Write a script file that will perform the following initial setup for a drawing.

Grid	2.0
Snap	0.5
Limits	0,0
	18.0,12.0
Zoom	All
Text height	0.25
LTSCALE	2.0

Overall dimension scale factor is 2
Aligned dimension text with the dimension line
Dimension text above the dimension line
Size of the center mark is 0.75

EXERCISE 3   *Script Files*

Write a script file that will set up the following layers with the given colors and linetypes.

Layers

Name	**Color**	**Linetype**
Contour	Red	Continuous
SPipes	Yellow	Center
WPipes	Blue	Hidden
Power	Green	Continuous
Manholes	Magenta	Continuous
Trees	Cyan	Continuous

EXERCISE 4 — *Script Files*

Write a script file that will perform the following initial setup for a new drawing.

Limits	0,0 24,18
Grid	1.0
Snap	0.25
Ortho	On
Snap	On
Zoom	All
Pline width	0.02
PLine	0,0 24,0 24,18 0,18 0,0
Ltscale	1.5
Units	Decimal units
	Precision 0.00
	Decimal degrees
	Precision 0
	Base angle East (0.00)
	Angle measured counterclockwise

Layers

Name	Color	Linetype
Obj	Red	Continuous
Cen	Yellow	Center
Hid	Blue	Hidden
Dim	Green	Continuous

EXERCISE 5 — *Script Files*

Write a script file that will plot a given drawing according to the following specifications. (Use the plotter for which your system is configured and adjust the values accordingly.)

Plot, using the Window option
Window size (0,0 24,18)
Do not write the plot to file
Size in inch units
Plot origin (0.0,0.0)
Maximum plot size (8.5,11 or the smallest size available on your printer/plotter)
90° plot rotation
No removal of hidden lines
Plotting scale (Fit)

EXERCISE 6 — *Script Files*

Write a script file that will continuously rotate a line in 10° increments around its midpoint (Figure 34-13). The time delay between increments is one second.

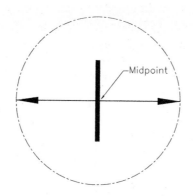

Figure 34-13 *Drawing for Exercise 6*

EXERCISE 7   *Script Files*

Write a script file that will continuously rotate the arrangement shown in Figure 34-14 as per the following instructions:

One set of two circles and one line should rotate clockwise, while the other set of two circles and the other line should rotate counterclockwise. Assume the rotation to be 5° around the intersection of the lines for both sets of arrangements.

Specifications are given below:

Start point of the horizontal line	2,4
End point of the horizontal line	8,4
Center point of circle at the start point of horizontal line	2,4
Diameter of the circle	1.0
Center point of circle at the end point of horizontal line	8,4
Diameter of circle	1.0
Start point of the vertical line	5,1
End point of the vertical line	5,7
Center point of circle at the start point of the vertical line	5,1
Diameter of the circle	1.0
Center point of circle at the end point of the vertical line	5,7
Diameter of the circle	1.0

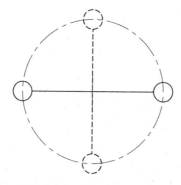

Figure 34-14 *Drawing for Exercise 7*

Hint: *Select one set of two circles and one line and then create one group. Similarly, select another set of two circles and one line and then create another group. Now, rotate one group clockwise and another group counterclockwise.*

EXERCISE 8   *Slide Show*

Make the slides shown in Figure 34-15 and write a script file for a continuous slide show. Provide a time delay of 5 seconds after every slide. (You are not restricted to use only the slides shown in Figure 34-15; you can use any slides of your choice.)

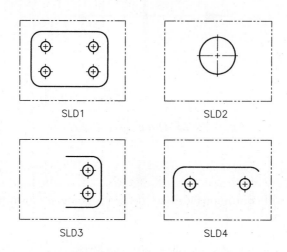

Figure 34-15 Slides for slide show

EXERCISE 9   *Slide Library*

List the slides used in Exercise 8 in a file SLDLIST2 and create a slide library file SLDLIB2. Then write a script file SHOW2 using the slide library with a time delay of 5 seconds after every slide.

Answers to Self-Evaluation Test

1. SCRIPT files, **2.** Commands, options, **3. SCRIPT**, **4.** blank space, **5.** time, **6.** vector, **7.** cannot be, **8. MSLIDE**, **9. VSLIDE**, **10. SLIDELIB**

Chapter 35

Creating Linetypes
and Hatch Patterns

CHAPTER OBJECTIVES

After completing this chapter, you will learn to:
Create Linetypes:
- *Write linetype definitions.*
- *Create different linetypes.*
- *Create linetype files.*
- *Determine LTSCALE for plotting the drawing to given specifications.*
- *Define alternate linetypes and modify existing linetypes.*
- *Create string and shape complex linetypes.*

Create Hatch Patterns:
- *Understand hatch pattern definition.*
- *Create new hatch patterns.*
- *Determine the effect of angle and scale factor on hatch.*
- *Create hatch patterns with multiple descriptors.*
- *Save hatch patterns in a separate file.*
- *Define custom hatch pattern file.*

KEY TERMS

- *Linetype*
- **.lin Files*
- *Header Line*
- *Pattern Line*
- *LTSCALE*
- *CELTSCALE*
- *String Complex Linetypes*
- *Shape Complex Linetypes*
- *Hatch Descriptors*
- **.pat Files*

STANDARD LINETYPES

The AutoCAD software package comes with a library of standard linetypes that has 38 different standard linetypes, including ISO linetypes and seven complex linetypes. These linetypes are saved in the *acad.lin* and *acadiso.lin* file. You can modify the existing linetypes or create new ones.

LINETYPE DEFINITIONS

All linetype definitions consist of two parts: **header line** and **pattern line**.

Header Line

The **header line** consists of an asterisk (*) followed by the name of the linetype and the linetype description. The name and the linetype description should be separated by a comma. If there is no description, the comma that separates the linetype name and the description is not required.

The format of the header line is:

* Linetype Name, Description

Example:
*HIDDENS,__ __ __ __ __ __

```
Where  *  -----------------Asterisk sign
       HIDDENS  -----Linetype name
       ,  ------------------Comma
       __ __ __ __  -----Linetype description
```

All linetype definitions require a linetype name. When you want to load a linetype or assign a linetype to an object, AutoCAD recognizes the linetype by the name you have assigned to the linetype definition. The names of the linetype definition should be selected to help the user recognize the linetype by its name. For example, the linetype name LINEFCX does not give the user any idea about the type of line. However, a linetype name like DASHDOT gives a better idea about the type of line that a user can expect.

The linetype description is a textual representation of the line. This representation can be generated by using dashes, dots, and spaces at the keyboard. The graphic given in the code is used for displaying preview of the linetypes. The linetype description cannot exceed 47 characters.

Pattern Line

The **pattern line** contains the definition of the line pattern consisting of the alignment field specification and the linetype specification, separated by a comma.

The format of the pattern line is:

Alignment Field Specification, Linetype Specification

Example:
A,.75,-.25,.75
```
       Where  A  -----------------Alignment field specification
              ,  ------------------Comma
              .75,-.25,.75  ----Linetype specification
```

The letter used for alignment field specification is A. This is the only alignment field supported by AutoCAD; therefore, the pattern line will always start with the letter A. The linetype specification defines the configuration of the dash-dot pattern to generate a line. The maximum number for dash length specification in the linetype is 12, provided the linetype pattern definition fits on one 80-character line.

ELEMENTS OF LINETYPE SPECIFICATION

All linetypes are created by combining the basic elements in a desired configuration. There are three basic elements that can be used to define a linetype specification.

Dash	(Pen down)
Dot	(Pen down, 0 length)
Space	(Pen up)

Example:

_____ . _____ . _____ . _____

Where . ------------------Dot (pen down with 0 length)
Blank space ----------------Space (pen up)
_____ ------------------------------Dash (pen down with specified length)

The dashes are generated by defining a positive number. For example, .5 will generate a dash 0.5 units long. Similarly, spaces are generated by defining a negative number. For example, -.2 will generate a space 0.2 units long. The dot is generated by defining 0 length.

Example:
A,.5,-.2,0,-.2,.5
Where **0** ------------------Dot (zero length)
-.2 ----------------Length of space (pen up)
.5 ----------------Length of dash (pen down)

CREATING LINETYPES

Before creating a linetype, you need to decide the type of line that you want to generate. Draw the line on a piece of paper and measure the length of each element that constitutes the line. You need to define only one segment of the line because the pattern is repeated when you draw a line. Linetypes can be created or modified by any one of the following methods:

1. Using a text editor like Notepad
2. Adding a new linetype in the *acad.lin* or *acadiso.lin* file
3. Using the **-LINETYPE** command

The following example explains how to create a new linetype using the three methods mentioned above.

EXAMPLE 1   *Linetype*

Create a linetype DASH3DOT (Figure 35-1) with the following specifications:

Length of the first dash 0.5
Blank space 0.125
Dot
Blank space 0.125
Dot

Blank space 0.125
Dot
Blank space 0.125

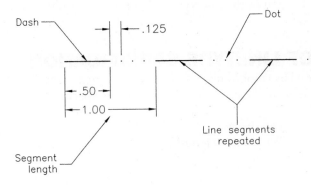

Figure 35-1 Linetype specifications of DASH3DOT

Using a Text Editor

Step 1: Writing definition of linetype
You can start a new linetype file and then add the line definitions to this file. To do this, use any text editor like Notepad to start a new file (*newlt.lin*) and then add the linetype definition of the DASH3DOT linetype. The name and the description must be separated by a comma (,). The description is optional. If you decide not to give one, omit the comma after the linetype name DASH3DOT.

> ***DASH3DOT,___ . . . ___ . . . ___**
> **A,.5,-.125,0,-.125,0,-.125,0,-.125**

Save it as *newlt.lin* in AutoCAD's Support directory.

Step 2: Loading the linetype
To load this linetype, choose the **Other** option from **Properties > Linetype** drop-down list in the **Drafting & Annotation** workspace to display the **Linetype Manager** dialog box. Next, choose the **Load** button in the **Linetype Manager** dialog box to display the **Load or Reload Linetypes** dialog box. Choose the **File** button in the **Load or Reload Linetypes** dialog box to display the **Select Linetype File** dialog box, as shown in Figure 35-2. Select the *newlt.lin* file from the location where you have saved it and then choose **Open**. Again, the **Load or Reload Linetypes** dialog box is displayed. Choose the **DASH3DOT** linetype in the **Available Linetypes** area and then choose **OK**. The **Linetype Manager** dialog box is displayed. Select the **DASH3DOT** linetype and then choose the **Current** button to make the selected linetype as the current linetype. Then, choose **OK**. Now, if you draw any object, then the new linetype will be used to draw that object.

Adding a New Linetype in the *acad.lin* File

Step 1: Adding a new linetype in the *acad.lin*
Browse to the *acad.lin* file and double-click on it; the *acad.lin* file opens in a Notepad. In this text editor, you can insert the line definition that defines the new linetype. Given next is the partial listing of the *acad.lin* file after adding a new linetype to it:

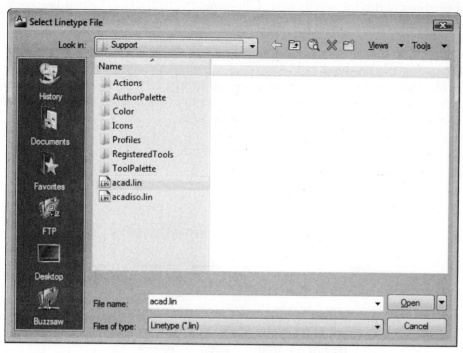

Figure 35-2 *The **Select Linetype File** dialog box*

```
*BORDER,Border __ __ . __ __ . __ __ . __ __ . __ __ .
A,.5,-.25,.5,-.25,0,-.25
*BORDER2,Border (.5x) __._.__._.__._.__._.__._.__._.__.
A,.25,-.125,.25,-.125,0,-.125
*BORDERX2,Border (2x) ____ ____ . ____ ____ . ___
A,1.0,-.5,1.0,-.5,0,-.5


*CENTER,Center ____ _ ____ _ ____ _ ____ _ ____ _ ___
A,1.25,-.25,.25,-.25
*CENTER2,Center (.5x) __ _ __ _ __ _ __ _ __ _ __ __
A,.75,-.125,.125,-.125
*CENTERX2,Center (2x) _____ __ _____ __ ____
A,2.5,-.5,.5,-.5


*DASHDOT,Dash dot __ . __ . __ . __ . __ . __ . __ . __
A,.5,-.25,0,-.25
*DASHDOT2,Dash dot (.5x) _._._._._._._._._._._._._.
A,.25,-.125,0,-.125
*DASHDOTX2,Dash dot (2x) ____ . ____ . ____ . ___
A,1.0,-.50,-.5
|
|
*GAS_LINE,Gas line ----GAS----GAS----GAS----GAS----GAS----GAS--
A,.5,-.2,["GAS",STANDARD,S=.1,R=0.0,X=-0.1,Y=-.05],-.25
*ZIGZAG,Zig zag /\/\/\/\/\/\/\/\/\/\/\/\/\/\
A,.0001,-.2,[ZIG,ltypeshp.shx,x=-.2,s=.2],-.4,[ZIG,ltypeshp.shx,r=180,x=.2,s=.2],-.2
*DASH3DOT,____ . . . ____ . . . ____
A,.5,-.125,0,-.125,0,-.125,0,-.125
```

The last two lines of this file define the new linetype, DASH3DOT. The first line contains the name DASH3DOT and the description of the line (__ . . .__). The second line contains the alignment and the pattern definition.

Step 2: Loading the linetype
Save the file and then load the linetype using the **LINETYPE** command. The procedure of loading the linetype is the same as described earlier in this example. The lines and polylines that this linetype will generate are shown in Figure 35-3.

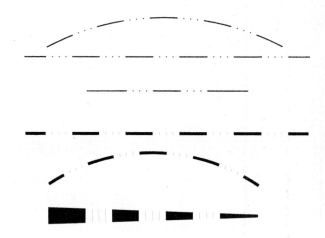

Figure 35-3 Lines created by the linetype DASH3DOT

Note

If you change the LTSCALE factor, all lines in the drawing are affected by the new ratio.

Using the LINETYPE Command

Step 1: Creating a linetype
To create a linetype using the **LINETYPE** command, first make sure that you are in the drawing editor. Then enter the **-LINETYPE** command and choose the **Create** option to create a linetype.

> Command: **-LINETYPE**
> Enter an option [?/Create/Load/Set]: **C**

Enter the name of the linetype and the name of the library file in which you want to store the definition of the new linetype.

> Enter name of linetype to create: **DASH3DOT**

If **FILEDIA**=1, the **Create or Append Linetype File** dialog box (Figure 35-4) will appear on the screen. If **FILEDIA**=0, you are prompted to enter the name of the file.

> Enter linetype file name for new linetype definition <default>: **Acad**

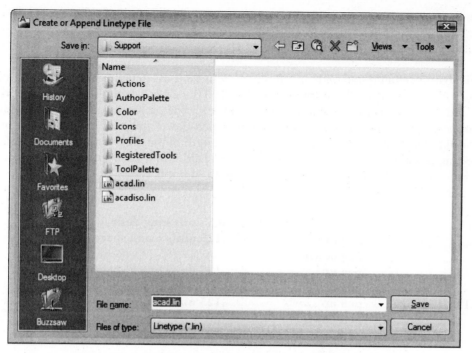

Figure 35-4 *The **Create or Append Linetype File** dialog box*

If the linetype already exists, the following message will be displayed on the screen:

> Wait, checking if linetype already defined...
> "Linetype" already exists in this file. Current definition is:
> alignment, dash-1, dash-2, _____.
> Overwrite?<N>

If you want to redefine the existing line style, enter **Y**. Otherwise, type **N** or press ENTER to choose the default value of N. You can then repeat the process with a different name of the linetype. After entering the name of the linetype and the library file name, you are prompted to enter the descriptive text and the pattern of the line.

> Descriptive text: ***DASH3DOT,___ . . . ___ . . . ___**
> Enter linetype pattern (on next line):
> **A,.5,-.125,0,-.125,0,-.125,0,-.125**

Descriptive Text

> ***DASH3DOT,___ . . . ___ . . . ___**

For the descriptive text, you have to type an asterisk (*) followed by the name of the linetype. For Example 1, the name of the linetype is DASH3DOT. The name *DASH3DOT can be followed by the description of the linetype; the length of this description cannot exceed 47 characters. In this example, the description is dashes and dots ___ . . . ___. It could be any text or alphanumeric string. The description is displayed on the screen when you list the linetypes.

Pattern

A,.5,-.125,0,-.125,0,-.125,0,-.125

The line pattern should start with an alignment definition. By default, AutoCAD supports only one type of alignment—A. Therefore, it is displayed on the screen when you select the **LINETYPE** command with the **Create** option. After entering **A** for the pattern alignment, define the pen position. A positive number (.5 or 0.5) indicates a "pen-down" position, and a negative number (-.25 or -0.25) indicates a "pen-up" position. The length of the dash or the space is designated by the magnitude of the number. For example, 0.5 will draw a dash 0.5 units long, and -0.25 will leave a blank space of 0.25 units. A dash length of 0 will draw a dot (.). The following are the pattern definition elements for Example 1:

.5	pen down	**0.5 units long dash**
-.125	pen up	**.125 units blank space**
0	pen down	**dot**
-.125	pen up	**.125 units blank space**
0	pen down	**dot**
-.125	pen up	**.125 units blank space**
0	pen down	**dot**
-.125	pen up	**.125 units blank space**

After you enter the pattern definition, the linetype (DASH3DOT) is automatically saved in the *acad.lin* file.

Step 2: Loading the linetype

Choose the **Other** option from **Home>Properties > Linetype** drop-down list to load the linetype or use the **LINETYPE** command. The linetype (DASH3DOT) can also be loaded using the **-LINETYPE** command and selecting the **Load** option.

ALIGNMENT SPECIFICATION

As the name suggests, the alignment specifies the pattern alignment at the start and the end of the line, circle, or arc. In other words, the line always starts and ends with the dash (___). The alignment definition "A" requires the first element to be a dash or dot (pen down), followed by a negative (pen up) segment. The minimum number of dash segments for alignment A is two. If there is not enough space for the line, a continuous line is drawn.

For example, in the linetype DASH3DOT of Example 1, the length of each line segment is 1.0 (.5 + .125 + .125 + .125 + .125 = 1.0). If the length of the line drawn is less than 1.00, a single line is drawn that looks like a continuous line, see Figure 35-5. If the length of the line is 1.00 or greater, the line will be drawn according to DASH3DOT linetype. AutoCAD automatically adjusts the length of the dashes and the line always starts and ends with a dash. The length of the starting and ending dashes is at least half the length of the dash as specified in the file. If the length of the dash as specified in the file is 0.5, the length of the starting and ending dashes is at least 0.25. To fit a line that starts and ends with a dash, the length of these dashes can also increase, as shown in Figure 35-5.

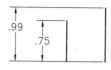

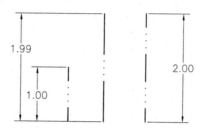

Figure 35-5 *Alignment of linetype DASH3DOT*

LTSCALE COMMAND

As mentioned earlier, the length of each line segment in the DASH3DOT linetype is 1.0 (.5 + .125 + .125 + .125 + .125 = 1.0). If you draw a line that is less than 1.0 units long, a single dash is drawn that looks like a continuous line, see Figure 35-6. This problem can be rectified by changing the linetype scale factor variable **LTSCALE** to a smaller value. This can be accomplished using the **LTSCALE** command.

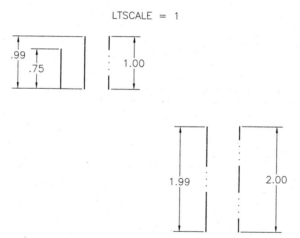

Figure 35-6 *Alignment when LTSCALE = 1*

Command: **LTSCALE**
Enter new linetype scale factor <default>: *New value.*

The default value of the **LTSCALE** variable is 1.0. If the LTSCALE is changed to 0.75, the length of each segment is reduced by 0.75 (1.0 x 0.75 = 0.75). Then, if you draw a line 0.75 units or longer, it will be drawn according to the definition of DASH3DOT (___ . . . ___) (Figures 35-7 and 35-8).

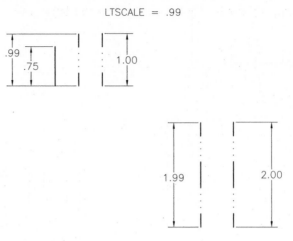

Figure 35-7 *Alignment when LTSCALE = 0.99*

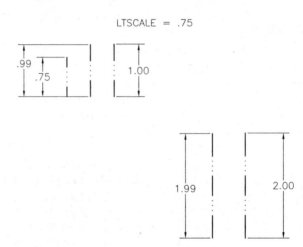

Figure 35-8 *Alignment when LTSCALE = 0.75*

The appearance of the lines is also affected by the limits of the drawing. Most of the AutoCAD linetypes work fine for drawings that have the limits 12,9. Figure 35-9 shows a line of linetype DASH3DOT that is four units long and the limits of the drawing are 12,9. If you increase the limits to 48,36 the lines will appear as continuous lines. If you want the line to appear the same as before on the screen, the LTSCALE needs to be changed. Since the limits of the drawing have increased four times, the LTSCALE should also be increased by the same amount. If you change the scale factor to four, the line segments will also increase by a factor of four. As shown in Figure 35-9, the length of the starting and the ending dash has increased to one unit.

In general, the approximate LTSCALE factor for screen display can be obtained by dividing the X -limit of the drawing by the default X -limit (12.00). However, it is recommended that the linetype scale must be set according to plot scale discussed in the next section.

LTSCALE factor for SCREEN DISPLAY = X -limits of the drawing/12.00

Example:
Drawing limits are 48,36
LTSCALE factor for screen display= 48/12 = 4

Drawing sheet size is 36,24 and scale is 1/4" = 1'
LTSCALE factor for screen display = 12 x 4 x (36 / 12) = 144

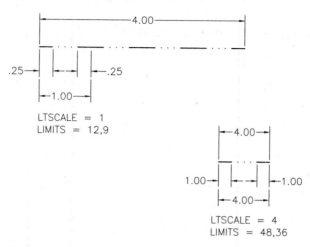

Figure 35-9 *Linetype DASH3DOT before and after changing the LTSCALE factor*

LTSCALE FACTOR FOR PLOTTING

The LTSCALE factor for plotting depends on the size of the sheet used to plot the drawing. For example, if the limits are 48 by 36, the drawing scale is 1:1, and you want to plot the drawing on a 48" by 36" size sheet, the LTSCALE factor is 1. If you check the specification of a hidden line in the *acad.lin* file, the length of each dash is 0.25. Therefore, when you plot a drawing with 1:1 scale, the length of each dash in a hidden line is 0.25.

However, if the drawing scale is 1/8" = 1' and you want to plot the drawing on a 48" by 36" paper, the LTSCALE factor must be 96 (8 x 12 = 96). If you increase the LTSCALE factor to 96, the length of each dash in the hidden line will increase by a factor of 96. As a result, the length of each dash will be 24 units (0.25 x 96 = 24). At the time of plotting, the scale factor must be 1:96 to plot the 384' by 288' drawing on a 48" by 36" size paper. Each dash of the hidden line that was 24" long on the drawing will be 0.25 (24/96 = 0.25) inch long when plotted. Similarly, if the desired text size on the paper is 1/8", the text height in the drawing must be 12" (1/8 x 96 = 12").

LTSCALE Factor for PLOTTING = Drawing Scale

Sometimes your plotter may not be able to plot a 48" by 36" drawing or you may like to decrease the size of the plot so that the drawing fits within the specified area. To get the correct dash lengths for hidden, center, or other lines, you must adjust the LTSCALE factor. For example, if you want to plot the previously mentioned drawing in a 45" by 34" area, the correction factor is:

Correction factor	= 48/45
	= 1.0666
New LTSCALE factor	= LTSCALE factor x Correction factor
	= 96 x 1.0666
	= 102.4

New LTSCALE Factor for PLOTTING = Drawing Scale x Correction Factor

Note

If you change the LTSCALE factor, all lines in the drawing are affected by the new ratio.

CURRENT LINETYPE SCALING (CELTSCALE)

Like **LTSCALE**, the **CELTSCALE** system variable controls the linetype scaling. The difference is that **CELTSCALE** determines the current linetype scaling. For example, if you set the **CELTSCALE** to 0.5, all lines drawn after setting the new value for **CELTSCALE** will have the linetype scaling factor of 0.5. The value is retained in the **CELTSCALE** system variable. The first line (a) in Figure 35-10 is drawn with the **CELTSCALE** factor of 1 and the second line (b) is drawn with the **CELTSCALE** factor of 0.5. The length of the dashes is reduced by a factor of 0.5 when the **CELTSCALE** is 0.5.

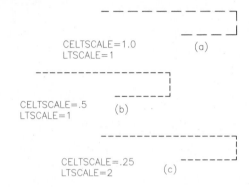

*Figure 35-10 Using **CELTSCALE** to control current linetype scaling*

The **LTSCALE** system variable controls the global scale factor. For example, if **LTSCALE** is set to 2, all lines in the drawing will be affected by a factor of 2. The net scale factor is equal to the product of **CELTSCALE** and **LTSCALE**. Figure 35-10(c) shows a line that is drawn with **LTSCALE** of 2 and **CELTSCALE** of 0.25. The net scale factor is = **LTSCALE** x **CELTSCALE** = 2 x 0.25 = 0.5.

Note

*You can change the current linetype scale factor of a line by using the **Properties** palette that can be invoked by choosing the **Properties** tool from the **Quick Access** Toolbar (Customize to add). You can also use the **CHANGE** command and then choose the **ltScale** option.*

ALTERNATE LINETYPES

One of the problems with the LTSCALE factor is that it affects all the lines in the drawing. As shown in Figure 35-11(a), the length of each segment in all DASH3DOT type lines is approximately equal, no matter how long the lines are. You may want to have a small **segment** length if the lines are small and a longer segment length if the lines are long. You can accomplish this by using CELTSCALE (discussed later in this chapter) or by defining an alternate linetype with a different segment length. For example, you can define a linetype DASH3DOT and DASH3DOTX with different line pattern specifications.

> *DASH3DOT,____ . . . ____ . . . ____ . . . ____
> A,0.5,-.125,0,-.125,0,-.125,0,-.125
> *DASH3DOTX,_____ . . . _____
> A,1.0,-.25,0,-.25,0,-.25,0,-.25

In the DASH3DOT linetype, the segment length is one unit; whereas, in the DASH3DOTX linetype the segment length is two units. You can have several alternate linetypes to produce the lines with different segment lengths. Figure 35-11(b) shows the lines generated by DASH3DOT and DASH3DOTX.

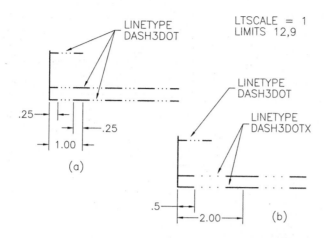

Figure 35-11 *Linetypes generated by DASH3DOT and DASH3DOTX*

Note
Although you may have used various linetypes with different segment lengths, the lines will be affected equally when you change the LTSCALE factor. For example, if the LTSCALE factor is 0.5, the segment length of DASH3DOT line will be 0.5 unit and the segment length of DASH3DOTX will be 1.0 unit.

MODIFYING LINETYPES

You can also modify the linetypes that are defined in the *acad.lin* or *acadiso.lin* file. You must save a copy of the original *acad.lin* or *acadiso.lin* file before making any changes to it. You need a text editor, such as Notepad, to modify the linetype. You can also use the **EDIT** function of DOS, or the **EDIT** command (provided the *acad.pgp* file is present and **EDIT** is defined in the file). For example, if you want to change the dash length of the border linetype from 0.5 to 0.75, load the file, and then edit the pattern line of the border linetype. The following file is a partial listing of the *acad.lin* file after changing the BORDER and CENTERX2 linetypes.

```
;;  AutoCAD Linetype Definition file
;;  Version 3.0
;;  Copyright (C) 1991-2010 by Autodesk, Inc.  All Rights Reserved.


*BORDER,Border __ __ . __ __ . __ __ . __ __ . __ __ .
A,.5,-.25,.5,-.25,0,-.25
*BORDER2,Border (.5x) __._._._._._._._._._._._.
A,.25,-.125,.25,-.125,0,-.125
*BORDERX2,Border (2x) ___ ___ . ___ ___ . ___
A,1.0,-.5,1.0,-.5,0,-.5

*CENTER,Center ____ _ ____ _ ____ _ ____ _ ____
A,1.25,-.25,.25,-.25
*CENTER2,Center (.5x) ___ _ ___ _ ___ _ ___ _ ___
A,.75,-.125,.125,-.125
```

*CENTERX2,Center (2x) _____ __ _____ __ ____
A,2.5,-.5,.5,-.5

*DASHDOT,Dash dot __ . __ . __ . __ . __ . __ . __
A,.5,-.25,0,-.25
*DASHDOT2,Dash dot (.5x) _._._._._._._._._._._._.
A,.25,-.125,0,-.125
*DASHDOTX2,Dash dot (2x) ____ . ____ . ____ . __
A,1.0,-.5,0,-.5

*DASHED,Dashed __ __ __ __ __ __ __ __ __ __
A,.5,-.25
*DASHED2,Dashed (.5x) _ _ _ _ _ _ _ _ _ _ _ _ _
A,.25,-.125
*DASHEDX2,Dashed (2x) ____ ____ ____ ____ ____ __
A,1.0,-.5

*DIVIDE,Divide ____ .. ____ .. ____ .. ____ .. ____
A,.5,-.25,0,-.25,0,-.25
*DIVIDE2,Divide (.5x) _.._.._.._.._.._.._.._.._.
A,.25,-.125,0,-.125,0,-.125
*DIVIDEX2,Divide (2x) _____ . . _____ . . _
A,1.0,-.5,0,-.5,0,-.5

*DOT,Dot . . . .
A,0,-.25
*DOT2,Dot (.5x) . . . .
A,0,-.125
*DOTX2,Dot (2x) . . . .
A,0,-.5

*HIDDEN,Hidden __ __ __ __ __ __ __ __ __ __ __ __
A,.25,-.125
*HIDDEN2,Hidden (.5x) _ _ _ _ _ _ _ _ _ _ _ _ _ _
A,.125,-.0625
*HIDDENX2,Hidden (2x) ____ ____ ____ ____ ____ ____
A,.5,-.25

*PHANTOM,Phantom _____ __ __ _____ __ __ _____
A,1.25,-.25,.25,-.25,.25,-.25
*PHANTOM2,Phantom (.5x) ___ __ __ ___ __ __ ___ __ __
A,.625,-.125,.125,-.125,.125,-.125
*PHANTOMX2,Phantom (2x) _____ ____ ____ _
A,2.5,-.5,.5,-.5,.5,-.5

;;
;; ISO 128 (ISO/DIS 12011) linetypes
;;
;; The size of the line segments for each defined ISO line is
;; defined for an usage with a pen width of 1 mm. To use them with
;; the other ISO predefined pen widths, the line has to be scaled
;; with the appropriate value (e.g. pen width 0,5 mm -> ltscale 0.5).
;;
*ACAD_ISO02W100,ISO dash __ __ __ __ __ __ __ __ __ __ __

A,12,-3
|
|
*ACAD_ISO15W100,ISO double-dash triple-dot __ __ . . . __ __ . .
A,12,-3,12,-3,0,-3,0,-3,0,-3

;; Complex linetypes
;; Complex linetypes have been added to this file.
;; These linetypes were defined in LTYPESHP.LIN in
;; Release 13, and are incorporated in ACAD.LIN in
;; Release 14.
;;
;; These linetype definitions use LTYPESHP.SHX.
;;
*FENCELINE1,Fenceline circle ----0-----0----0-----0----0-----0--
A,.25,-.1,[CIRC1,ltypeshp.shx,x=-.1,s=.1],-.1,1
*FENCELINE2,Fenceline square ----[]-----[]----[]-----[]----[]---
A,.25,-.1,[BOX,ltypeshp.shx,x=-.1,s=.1],-.1,1
*TRACKS,Tracks -|-|-|-|-|-|-|-|-|-|-|-|-|-|-|-|-|-|-|-|-|-
A,.15,[TRACK1,ltypeshp.shx,s=.25],.15
*BATTING,Batting SSSSSSSSSSSSSSSSSSSSSSSSSSSSSSSSSSSSSSSSSSSSSSSSSS
A,.0001,-.1,[BAT,ltypeshp.shx,x=-.1,s=.1],-.2,[BAT,ltypeshp.shx,r=180,x=.1,s=.1],-.1
*HOT_WATER_SUPPLY,Hot water supply ---- HW ---- HW ---- HW ----
A,.5,-.2,["HW",STANDARD,S=.1,R=0.0,X=-0.1,Y=.05],-.2
*GAS_LINE,Gas line ----GAS----GAS----GAS----GAS----GAS----GAS--
A,.5,-.2,["GAS",STANDARD,S=.1,R=0.0,X=-0.1,Y=-.05],-.25
*ZIGZAG,Zig zag /\/\/\/\/\/\/\/\/\/\/\/\/\/\/\
A,.0001,-.2,[ZIG,ltypeshp.shx,x=-.2,s=.2],-.4,[ZIG,ltypeshp.shx,r=180,x=.2,s=.2],-.2

EXAMPLE 2   *Linetype*

Create a new file, *newlint.lin*, and define a linetype VARDASH with the following specifications:

Length of first dash 1.0
Blank space 0.25
Length of second dash 0.75
Blank space 0.25
Length of third dash 0.5
Blank space 0.25
Dot
Blank space 0.25
Length of next dash 0.5
Blank space 0.25
Length of next dash 0.75

Step 1: Writing definition of linetype
Use a text editor and insert the following lines to define the new linetype **VARDASH**.

*VARDASH,————— ——— —— . —— ——— ————
A,1,-.25,.75,-.25,.5,-.25,0,-.25,.5,-.25,.75,-.25

Now, save this file with the name *newlint.lin*.

Step 2: Loading the linetype

You can use the **LINETYPE** command or choose **Home > Properties > Linetype > Other** from the **Ribbon** to load the linetype. The types of lines that this linetype will generate are shown in Figure 35-12.

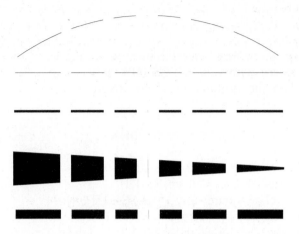

Figure 35-12 *Lines generated by linetype VARDASH*

COMPLEX LINETYPES

AutoCAD has provided a facility to create complex linetypes. The complex linetypes can be classified into two groups: string complex linetype and shape complex linetype. The difference between the two is that the string complex linetype has a text string inserted in the line, whereas the shape complex linetype has a shape inserted in the line. The facility of creating complex linetypes increases the functionality of lines. For example, if you want to draw a line around a building that indicates the fence line, you can do it by defining a complex linetype that will automatically give you the desired line with the text string (Fence). Similarly, you can define a complex linetype that will insert a shape (symbol) at predefined distances along the line.

CREATING A STRING COMPLEX LINETYPE

While writing the definition of a string complex linetype, the actual text and its attributes must be included in the linetype definition, refer to Figure 35-13. The format of the string complex linetype is:

["String", Text Style, Text Height, Rotation, X-Offset, Y-Offset]

String. It is the actual text that you want to insert along the line. The text string must be enclosed in quotation marks (" ").

Text Style. This is the name of the text style file that you want to use for generating the text string. The text style must be predefined.

Text Height. This is the actual height of the text, if the text height defined in the text style is 0. Otherwise, it acts as a scale factor for the text height specified in the text style. In Figure 35-13, the height of the text is 0.1 units.

Rotation. The rotation can be specified as an absolute or relative angle. In the absolute rotation, the angle is always measured with respect to the positive X axis, no matter what AutoCAD's direction setting is. The absolute angle is represented by the letter "a".

In relative rotation, the angle is always measured with respect to orientation of dashes in the linetype. The relative angle is represented by the letter "r". The angle can be specified in radians (r), grads (g), or degrees (d). The default is degrees.

X-Offset. This is the distance of the lower left corner of the text string from the endpoint of the line segment measured along the line. If the line is horizontal, then the X-Offset distance is measured along the X axis. In Figure 35-13, the X-Offset distance is 0.05.

Y-Offset. This is the distance of the lower left corner of the text string from the endpoint of the line segment measured perpendicular to the line. If the line is horizontal, then the Y-Offset distance is measured along the Y axis. In Figure 35-13, the Y-Offset distance is -0.05. The distance is negative because the start point of the text string is 0.05 units below the endpoint of the first line segment.

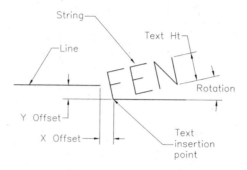

Figure 35-13 *Attributes of a string complex linetype*

EXAMPLE 3 — *String Complex Linetype*

In the following example, you will write the definition of a string complex linetype that consists of the text string "Fence" and line segments. The length of each line segment is 0.75. The height of the text string is 0.1 units, and the space between the end of the text string and the following line segment is 0.05, see Figure 35-14.

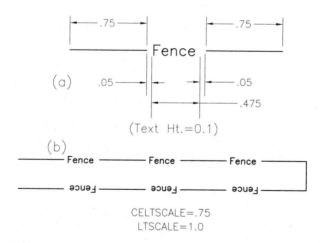

Figure 35-14 *The attributes of a string complex linetype and line specifications for Example 3*

Step 1: Determining the line specifications

Before writing the definition of a new linetype, it is important to determine the line specification. One of the ways to do this is to actually draw the lines and the text the way you want them to appear in the drawing. Once you have drawn the line and the text to your satisfaction, measure the distances needed to define the string complex linetype. The values are given as follows:

Text string	=Fence
Text style	=Standard
Text height	=0.1
Text rotation	=0
X-Offset	=0.05
Y-Offset	=-0.05
Length of the first line segment	=0.75
Distance between the line segments	=0.575

Step 2: Writing the definition of string complex linetype

Use a text editor to write the definition of the string complex linetype. You can add the definition to the *acad.lin* or *acadiso.lin* file or create a separate file. The extension of the file must be *.lin*. The following file is the listing of the *fence.lin* file for Example 3. The name of the linetype is NEWFence1.

```
*NEWFence1,New fence boundary line
A,0.75,["Fence",Standard,S=0.1,A=0,X=0.05,Y=-0.05],-0.575
or
A,0.75,-0.05,["Fence",Standard,S=0.1,A=0,X=0,Y=-0.05],-0.525
```

Step 3: Loading the linetype

Choose the **Other** option from **Properties > Linetype** drop-down list to load the linetype or use the **LINETYPE** command. Now, draw a line or any object to check if the line is drawn as per the given specifications, as shown in Figure 35-15. Notice that the text is always drawn along the *X* axis. Also, when you draw a line at an angle, polyline, circle, or spline, the text string does not align with the object (Figure 35-15).

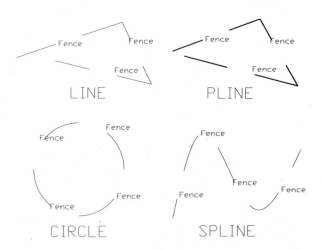

Figure 35-15 *Using string complex linetype with angle A=0*

Step 4: Aligning the text with the line

In the NEWFence linetype definition, the specified angle is 0° (Absolute angle A = 0). Therefore, when you use the NEWFence linetype to draw a line, circle, polyline, or spline, the text string (Fence) will be at zero degrees. If you want the text string (Fence) to align with the polyline

(Figure 35-16), spline, or circle, specify the angle as relative angle (R = 0) in the NEWFence linetype definition. The following is the linetype definition for NEWFence linetype with relative angle R = 0:

*NEWFence2,New fence boundary line
A,0.75,["Fence",Standard,S=0.1,R=0,X=0.05,Y=-0.05],-0.575

Note
You have to load the linetype again to get the updated changes.

Step 5: Aligning the midpoint of text with the line

In Figure 35-16, you will notice that the text string is not properly aligned with the circumference of the circle. This is because AutoCAD draws the text string in a direction that is tangent to the circle at the text insertion point.

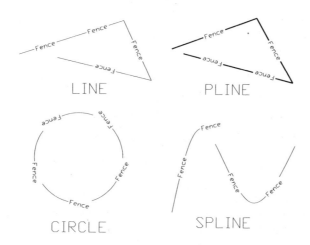

Figure 35-16 *Using a string complex linetype with angle R = 0*

To resolve this problem, you must define the middle point of the text string as the insertion point. Also, the line specifications should be measured accordingly. Figure 35-17 gives the measurements of the NEWFence linetype with the middle point of the text as the insertion point and Figure 35-18 shows the entities sketched with the selected linetype.

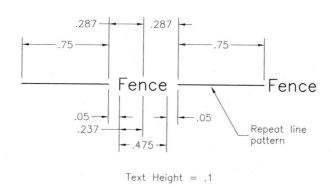

Figure 35-17 *Specifications of a string complex linetype with the middle point of the text string as the text insertion point*

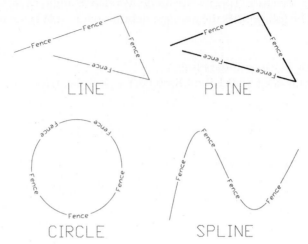

Figure 35-18 *Using a string complex linetype with the middle point of the text string as the text insertion point*

The following is the linetype definition for NEWFence linetype:

*NEWFence3,New fence boundary line
A,0.75,-0.287,["Fence",Standard,S=0.1,X=-0.237,Y=-0.05],-0.287

Note

If no angle is defined in the line definition, it acquires the default value of angle R = 0. Also, the text does not automatically insert to its midpoint like the regular text with MID justification.

Creating a Shape Complex Linetype

As with the string complex linetype, when you write the definition of a shape complex linetype, the name of the shape, the name of the shape file, and other shape attributes, like rotation, scale, X-Offset, and Y-Offset, must be included in the linetype definition. The format of the shape complex linetype is:

[Shape Name, Shape File, Scale, Rotation, X-Offset, Y-Offset]

The following is the description of the attributes of Shape Complex Linetype (Figure 35-19).

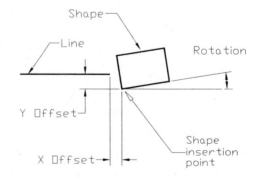

Figure 35-19 *The attributes of a shape complex linetype*

Shape Name. This is the name of the shape that you want to insert along the line. The shape name must exist; otherwise, no shape will be generated along the line.

Shape File. This is the name of the compiled shape file (*.shx*) that contains the definition of the shape being inserted in the line. The name of the subdirectory where the shape file is located must be in the AutoCAD search path. The shape files (*.shp*) must be compiled before using the **SHAPE** command to load the shape.

Scale. This is the scale factor by which the defined shape size is to be scaled. If the scale is 1, the size of the shape will be the same as defined in the shape definition (*.shp* file).

Rotation. The rotation can be specified as an absolute or relative angle. In absolute rotation, the angle is always measured with respect to the positive *X* axis, no matter what AutoCAD's direction setting. The absolute angle is represented by letter "a." In relative rotation, the angle is always measured with respect to the orientation of dashes in the linetype. The relative angle is represented by the letter "r." The angle can be specified in radians (r), grads (g), or degrees (d). The default is degrees.

X-Offset. This is the distance of the shape insertion point from the endpoint of the line segment measured along the line. If the line is horizontal, then the X-Offset distance is measured along the *X* axis. In Figure 35-20, the X-Offset distance is 0.2.

Y-Offset. This is the distance of the shape insertion point from the endpoint of the line segment measured perpendicular to the line. If the line is horizontal, then the Y-Offset distance is measured along the *Y* axis. In Figure 35-20, the Y-Offset distance is 0.

EXAMPLE 4   *Shape Complex Linetype*

Write the definition of a shape complex linetype that consists of a shape (Manhole; the name of the shape is MH) and a line. The scale of the shape is 0.1, the length of each line segment is 0.75, and the space between line segments is 0.2.

Step 1: Determining the line specifications

Before writing the definition of a new linetype, it is important to determine the line specifications. One of the ways to do this is to actually draw the lines and the shape the way you want them to appear in the drawing (Figure 35-20). Once you have drawn the line and the shape to your satisfaction, measure the distances needed to define the shape complex linetype. In this example, the values are as follows:

Shape name	=MH
Shape file name file.)	=*mhole.shx* (Name of the compiled shape
Scale	=0.1
Rotation	=0
X-Offset	=0.2
Y-Offset	=0
Length of the first line segment	= 0.75
Distance between the line segments	= 0.2

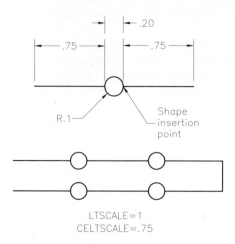

Figure 35-20 *The attributes of the shape complex linetype and line specifications for Example 4*

Step 2: Writing the definition of the shape

Use a text editor to write the definition of the shape file. The extension of the file must be *.shp*. The following file is the listing of the *mhole.shp* file for Example 4. The name of the shape is MH.

```
*215,9,MH
001,10,(1,007),
001,10,(1,071),0 [Enter]
```

Step 3: Compiling the shape

Use the **COMPILE** command to compile the shape file (*.shp* file). Remember that the path of the folder in which you save the shape file should be specified in the AutoCAD support file search path. When you use this command, the **Select Shape or Font File** dialog box will be displayed (Figure 35-21). If **FILEDIA** = 0, this command will be executed using the command line. The following is the command sequence for compiling the shape file:

Command: **COMPILE**
Enter shape (.SHP) or PostScript font (.PFB) file name: **MHOLE**

Step 4: Writing the definition of the shape complex linetype

Use a text editor to write the definition of the shape complex linetype. You can add the definition to the *acad.lin* file or create a separate file. The extension of the file must be *.lin*. The following file is the listing of the *mhole.lin* file for Example 4. The name of the linetype is MHOLE.

```
*MHOLE,Line with Manholes
A,0.75,[MH,MHOLE.SHX,S=0.10,X=0.2,Y=0],-0.2 [Enter]
```

Step 5: Loading the linetype

To test the linetype, load the linetype using the **LINETYPE** command. Assign the linetype to a layer. Draw a line or any object to check if the line is drawn to the given specifications. The shape is drawn upside down when you draw a line from right to left. Figure 35-22 shows the execution of the linetype *mhole.lin* using line, pline, and spline. Figure 35-23 displays the region hatched using the string complex and shape complex line types respectively.

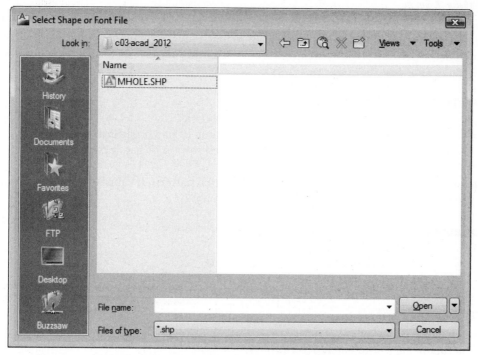

Figure 35-21 *The **Select Shape or Font File** dialog box*

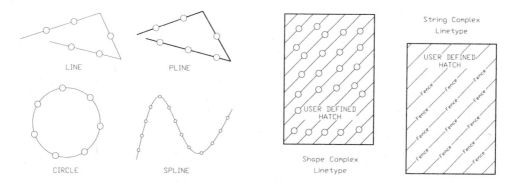

Figure 35-22 *Using a shape complex linetype*

Figure 35-23 *Using shape and string complex linetypes to create custom hatch*

HATCH PATTERN DEFINITION

AutoCAD has a default hatch pattern library file, *acad.pat* and *acadiso.pat*, that contains 72 hatch patterns. Generally, you can hatch all the drawings using these default hatch patterns. However, if you need a different hatch pattern, AutoCAD lets you create your own hatch patterns. There is no limit to the number of hatch patterns you can define.

The hatch patterns you define can be added to the hatch pattern library file, *acad.pat or acadiso. pat*. You can also create a new hatch pattern library file, provided the file contains only one hatch pattern definition, and the name of the hatch is the same as the name of the file.

The hatch pattern definition consists of the following two parts: **Header Line** and **Hatch Descriptors**.

Header Line

The **header line** consists of an asterisk (*) followed by the name of the hatch pattern. The hatch name is the name used in the hatch command to hatch an area. The hatch name is followed by the hatch description. Both are separated from each other by a comma (,). The general format of the header line is:

*HATCH Name [, Hatch Description]
            Where * ----------------------------Asterisk
                        HATCH Name -----------Name of hatch pattern
                        Hatch Description   --Description of hatch pattern

The description can be any text that describes the hatch pattern. It can also be omitted, in which case, a comma should not follow the hatch pattern name.

Example:
*DASH45, Dashed lines at 45°
            Where DASH45 --------Hatch name
                        Dashed lines at 45°   --Hatch description

Hatch Descriptors

The **hatch descriptors** consist of one or more lines that contain the definition of the hatch lines. The general format of the hatch descriptor is:

Angle, X-origin, Y-origin, D1, D2 [,Dash Length.....]
            Where Angle -----------Angle of hatch lines
                        X-origin --------X coordinate of hatch line
                        Y-origin --------Y coordinate of hatch line
                        D1 --------------Displacement of second line (Delta-X)
                        D2 --------------Distance between hatch lines (Delta-Y)
                        Length   Length of dashes and spaces (Pattern line definition)
Example:
45,0,0,0,0.5,0.5,-0.125,0,-0.125
            Where 45 ----------------Angle of hatch line
                        0 ------------------X-Origin
                        0 ------------------Y-Origin
                        0 ------------------Delta-X
                        0.5 ---------------Delta-Y
                        0.5 --------------Dash (pen down)
                        -0.125 ----------Space (pen up)
                        0 ------------------Dot (pen down)
                        -0.125 ----------Space (pen up)
                        0.5,-0.125,0,-0.125 Pattern line definition

Hatch Angle

X-origin and Y-origin. The hatch angle is the angle that the hatch lines make with the positive X axis. The angle is positive if measured counterclockwise (Figure 35-24), and negative if the angle is measured clockwise. When you draw a hatch pattern, the first hatch line starts from the point defined by X-origin and Y-origin.

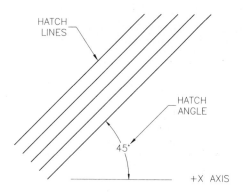

Figure 35-24 *Hatch angle*

The remaining lines are generated by offsetting the first hatch line by a distance specified by delta-X and delta-Y. In Figure 35-25(a), the first hatch line starts from the point with the coordinates X = 0 and Y = 0. In Figure 35-25(b), the first line of hatch starts from a point with the coordinates X = 0 and Y = 0.25.

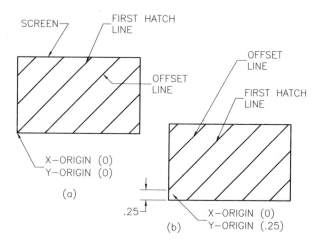

Figure 35-25 *X-origin and Y-origin of hatch lines*

Delta-X and Delta-Y. Delta-X is the displacement of the offset line in the direction in which the hatch lines are generated. For example, if the lines are drawn at a 0° angle and delta-X = 0.5, the offset line will be displaced by a distance delta-X (0.5) along the 0-angle direction. Similarly, if the hatch lines are drawn at a 45° angle, the offset line will be displaced by a distance delta-X (0.5) along a 45° direction (Figure 35-26). Delta-Y is the displacement of the offset lines measured perpendicular to the hatch lines. For example, if delta-Y = 1.0, the space between any two hatch lines will be 1.0 (Figure 35-26).

HOW HATCH WORKS?

When you hatch an area, infinite number of hatch lines of infinite length are generated. The first hatch line always passes through the point specified by the X-origin and Y-origin. The remaining lines are generated by offsetting the first hatch line in both directions. The offset distance is determined by delta-X and delta-Y, as shown in Figure 35-26. All selected entities that form the boundary of the hatch area are then checked for intersection with these lines. Any hatch lines found within the defined hatch boundaries are turned on, and the hatch lines outside the hatch boundary are turned off, as shown in Figure 35-27. Since the hatch lines are generated by offsetting, the hatch lines in all areas of drawing are automatically aligned relative to the drawing's snap origin. Figure 35-27(a) shows the hatch lines as computed by AutoCAD. These lines are not drawn on the screen; they are shown here for illustration only. Figure 35-27(b) shows the hatch lines generated in the circle that was defined as the hatch boundary.

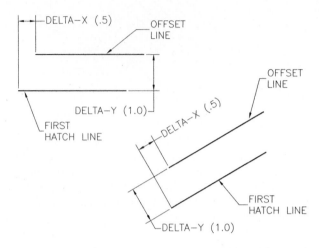

Figure 35-26 Delta-X and Delta-Y of hatch lines

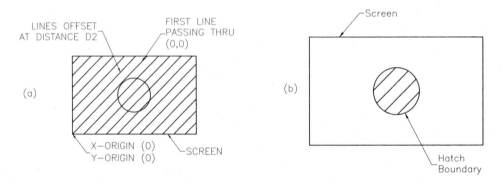

Figure 35-27 Hatch lines outside the hatch boundary are turned off

SIMPLE HATCH PATTERN

It is good practice to develop the hatch pattern specification before writing a hatch pattern definition. For simple hatch patterns it may not be that important, but for more complicated hatch patterns you should know the detailed specifications. Example 5 illustrates the procedure for developing a simple hatch pattern.

EXAMPLE 5   *Hatch Pattern*

Write a hatch pattern definition for the hatch pattern shown in Figure 35-28, with the following specifications:

Name of the hatch pattern	=HATCH1
X-Origin	=0
Y-Origin	=0
Distance between hatch lines	=0.5
Displacement of hatch lines	=0
Hatch line pattern	=Continuous

Step 1: Creating the hatch pattern file

This hatch pattern definition can be added to the existing *acad.pat* hatch file. You can use any text editor (like Notepad) to write the file. Load the *acad.pat* file that is located in the *C:\Users\User\AppData\Roaming\Autodesk\AutoCAD 2012\R18.2\enu\Support* directory and insert the following two lines at the end of the file.

*HATCH1,Hatch Pattern for Example 5
45,0,0,0,.5

Where **45** ----------------Hatch angle
       0 ------------------X-origin
       0 ------------------Y-origin
       0 ------------------Displacement of second hatch line
      .5 ----------------Distance between hatch lines

The first field of hatch descriptors contains the angle of the hatch lines and the value of this angle in this particular case is 45° with respect to the positive X axis. The second and third fields describe the X and Y coordinates of the first hatch line origin. The first line of the hatch pattern will pass through this point. If the values of the X-origin and Y-origin were 0.5 and 1.0, respectively, then the first line would pass through the point with the X coordinate of 0.5 and the Y coordinate of 1.0, with respect to the drawing origin 0,0. The remaining lines are generated by offsetting the first line, as shown in Figure 35-28.

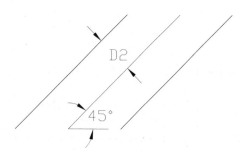

***Figure 35-28** Hatch pattern angle and offset distance*

Step 2: Applying the hatch pattern

Choose the **Hatch** tool from the **Draw** panel; the **Hatch Creation** context tab will get added to the **Ribbon**. Select the **HATCH1** pattern from **Pattern > Hatch Creation** context tab. If needed, change the **Scale** and **Angle** from the **Properties** panel of the **Hatch Creation** tab. Next, hatch a circle to test the hatch pattern.

EFFECT OF ANGLE AND SCALE FACTOR ON HATCH

When you hatch an area, you can alter the angle and displacement of hatch lines you have specified in the hatch pattern definition to get the desired hatch spacing. You can do this by entering an appropriate value for angle and scale factor in the **HATCH** command.

To understand how the angle and the displacement can be changed, hatch an area with the hatch pattern HATCH1 in Example 5. You will notice that the hatch lines have been generated according to the definition of hatch pattern HATCH1. Notice the effect of hatch angle and scale factor on the hatch. Figure 35-29(a) shows a hatch with a 0° angle and a scale factor of 1.0. If the angle is 0, the hatch will be generated with the same angle as defined in the hatch pattern definition (45° in Example 5). Similarly, if the scale factor is 1.0, the distance between the hatch lines will be the same as defined in the hatch pattern definition. Figure 35-29(b) shows a hatch that is generated when the hatch scale factor is 0.5. If you measure the distance between the successive hatch lines, it will be 0.5 x 0.5 = 0.25. Figures 3-29(c) and (d) show the hatch when the angle is 45° and the scale factors are 1.0 and 0.5, respectively.

Scale and Angle can also be set by entering **-HATCH** at the Command prompt.

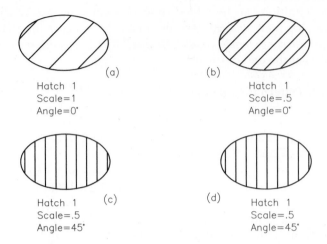

Figure 35-29 *Delta-X and Delta-Y of hatch lines*

HATCH PATTERN WITH DASHES AND DOTS

The lines that you can use in a hatch pattern definition are not restricted to continuous lines. You can define any line pattern to generate a hatch pattern. The lines can be a combination of dashes, dots, and spaces in any configuration. However, the maximum number of dashes you can specify in the line pattern definition of a hatch pattern is six. Example 6 uses a dash-dot linetype to create a hatch pattern.

EXAMPLE 6	Hatch Pattern

Write a hatch pattern definition for the hatch pattern shown in Figure 35-30, with the following specifications. Define a new path say *C:\Program Files\Hatch1* and save the hatch pattern in that path.

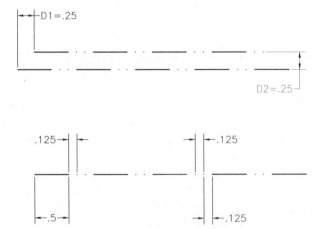

Figure 35-30 *Hatch lines made of dashes and dots*

Name of the hatch pattern =	HATCH2
Hatch angle =	0
X-origin =	0
Y-origin =	0

Displacement of lines (D1) =	0.25
Distance between lines (D2) =	0.25
Length of each dash =	0.5
Space between dashes and dots =	0.125
Space between dots =	0.125

Writing the definition of a hatch pattern

You can use any text editor (Notepad) to edit the *acad.pat* file. The general format of the header line and the hatch descriptors is:

*HATCH NAME, Hatch Description
Angle, X-Origin, Y-Origin, D1, D2 [,Dash Length.....]

Substitute the values from Example 6 in the corresponding fields of the header line and field descriptor:

> *HATCH2,Hatch with dashes and dots
> **0,0,0,0.25,0.25,0.5,-0.125,0,-0.125,0,-0.125**
>
> Where **0** ------------------Angle
> **0** ------------------X-origin
> **0** -----------------Y-origin
> **0.25** --------------Delta-X
> **0.25** --------------Delta-Y
> **0.5** ---------------Length of dash
> **-0.125** -----------Space (pen up)
> **0** -----------------Dot (pen down)
> **-0.125** -----------Space (pen up)
> **0** -----------------Dot
> **-0.125** -----------Space

Specifying a New Path for Hatch Pattern Files

When you enter a hatch pattern name for hatching, AutoCAD looks for that file name in the **Support** directory or the directory paths specified in the support file search path. You can specify a new path and directory to store your hatch files.

Create a new folder *Hatch1* in *C* drive under the *Program Files* folder. Save the *acad.pat* file with hatch pattern **HATCH2** definition in the same subdirectory, Hatch1. Right-click in the drawing area to activate the shortcut menu. Choose **Options** from the shortcut menu to display the **Options** dialog box. The **Options** dialog box can also be invoked by choosing the **Options** tool from the **Application Menu** or by directly entering **OPTIONS** at the Command prompt. Choose the **Files** tab in the **Options** dialog box to display the **Search paths, file names and file locations** area. Click on the **+** sign of the **Support File Search Path** to display the different subdirectories of the **Support File Search Path**, as shown in Figure 35-31. Now, choose the **Add** button to display the space to add a new subdirectory. Enter the location of the new subdirectory, **C:\Program Files\Hatch1** or click on the **Browse** button to specify the path. Choose the **Apply** button and then choose **OK** to exit the **Options** dialog box. You have created a subdirectory and specified the search path for the hatch files.

Follow the procedure as described in Example 5 to activate the hatch pattern. The hatch thus generated is shown in Figure 35-32. Figure 35-32(a) shows the hatch with a 0° angle and a scale factor of 1.0. Figure 35-32(b) shows the hatch with a 45° angle and a scale factor of 0.5.

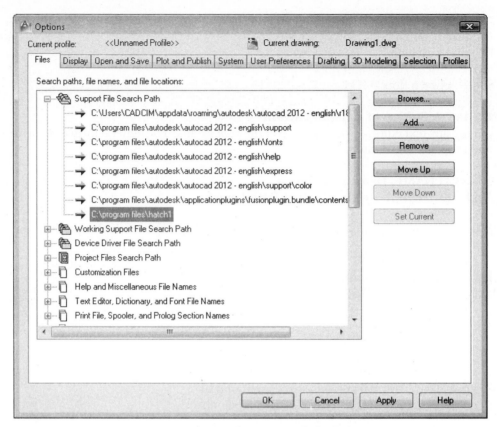

*Figure 35-31 The **Options** dialog box*

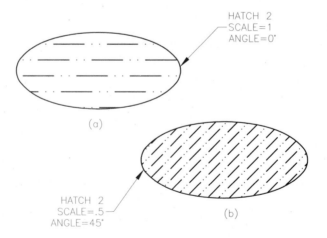

Figure 35-32 Hatch pattern at different angles and scales

HATCH WITH MULTIPLE DESCRIPTORS

Some hatch patterns require multiple lines to generate a shape. For example, if you want to create a hatch pattern of a brick wall, you need a hatch pattern that has four hatch descriptors to generate a rectangular shape. You can have any number of hatch descriptor lines in a hatch pattern definition. It is up to the user to combine them in any conceivable order. However, there are some shapes you cannot generate. A shape that has a nonlinear element, such as an arc, cannot be generated by hatch pattern definition. However, you can simulate an arc by defining short line segments because you can use only straight lines to generate a hatch pattern. Example 7 uses three lines to define a triangular hatch pattern.

EXAMPLE 7 *Hatch Pattern*

Write a hatch pattern definition for the hatch pattern shown in Figure 35-33, with the following specifications:

Name of the hatch pattern	=HATCH3
Vertical height of the triangle	=0.5
Horizontal length of the triangle	=0.5
Vertical distance between the triangles	=0.5
Horizontal distance between the triangles	=0.5

Each triangle in this hatch pattern consists of the following three elements: a vertical line, a horizontal line, and a line inclined at 45°.

Step 1: Defining specifications for vertical line

For the vertical line (see figure 35-34) the specifications are:

Hatch angle	=90°
X-origin	=0
Y-origin	=0
Delta-X (D1)	=0
Delta-Y (D2)	=1.0
Dash length	=0.5
Space	=0.5

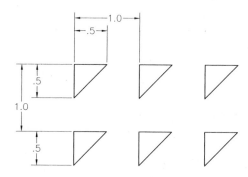

Figure 35-33 *Triangle hatch pattern*

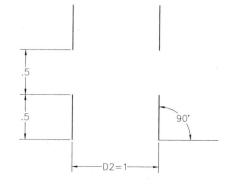

Figure 35-34 *Vertical line*

Substitute the values from the vertical line specification in various fields of the hatch descriptor to get the following line:

90,0,0,0,1,.5,-.5

Where	**90**	----------------Hatch angle
	0	------------------X-origin
	0	------------------Y-origin
	0	------------------Delta-X
	1	------------------Delta-Y
	.5	-----------------Dash (pen down)
	-.5	---------------Space (pen up)

Step 2: Defining specifications of horizontal line

For the horizontal line (Figure 35-35), the specifications are:

Hatch angle	=0°
X-origin	=0
Y-origin	=0.5
Delta-X (D1)	=0
Delta-Y (D2)	=1.0
Dash length	=0.5
Space	=0.5

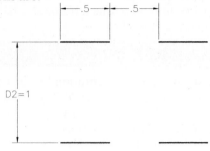

Figure 35-35 Horizontal line

The only difference between the vertical line and the horizontal line is the angle. For the horizontal line, the angle is 0°, whereas for the vertical line, the angle is 90°. Substitute the values from the vertical line specification to obtain the following line:

0,0,0.5,0,1,.5,-.5

Where	**0**	------------------Hatch angle
	0	------------------X-origin
	0.5	---------------Y-origin
	0	------------------Delta-X
	1	------------------Delta-Y
	.5	-----------------Dash (pen down)
	-.5	----------------Space (pen up)

Step 3: Defining specifications of the inclined line

This line is at an angle; therefore, you need to calculate the distances delta-X (D1) and delta-Y (D2), the length of the dashed line, and the length of the blank space. Figure 35-36 shows the calculations to find these values.

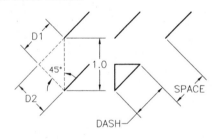

$$D1 = 1.0 \times COS\ 45 \qquad D2 = 1.0 \times SIN\ 45$$
$$D1 = .7071 \qquad\qquad D2 = .7071$$

$$DASH = SQRT(0.5^2 + 0.5^2)$$
$$= .7071$$
$$SPACE = DASH = .7071$$

Figure 35-36 Line inclined at 45°

Hatch angle	=45°
X-Origin	=0
Y-Origin	=0
Delta-X (D1)	=0.7071
Delta-Y (D2)	=0.7071
Dash length	=0.7071
Space	=0.7071

After substituting the values in the general format of the hatch descriptor, you will obtain the following line:

45,0,0,.7071,.7071,.7071,-.7071

> Where **45** ----------------Hatch angle
> **0** ------------------X-origin
> **0** ------------------Y-origin
> **.7071** ------------Delta-X
> **.7071** ------------Delta-Y
> **.7071** ------------Dash (pen down)
> **-.7071** ----------Space (pen up)

Step 4: Loading the hatch pattern

Now, you can combine three lines and insert them at the end of the *acad.pat* file or you can enter the values in a separate hatch file and save it. You can also use any text editor and insert the line definition.

The following file is a partial listing of the *acad.pat* file, after adding the hatch pattern definitions from Examples 5, 6, and 7.

```
*SOLID, Solid fill
45, 0,0, 0,.125
*Angle,Angle steel
0, 0,0, 0,.275, .2,-.075
90, 0,0, 0,.275, .2,-.075
*Ansi31,ANSI Iron, Brick, Stone masonry
45, 0,0, 0,.125
*Ansi32,ANSI Steel
45, 0,0, 0,.375
45, .176776695,0, 0,.375
*Ansi33,ANSI Bronze, Brass, Copper
45, 0,0, 0,.25
45, .176776695,0, 0,.25, .125,-.0625
*Ansi34,ANSI Plastic, Rubber
45, 0,0, 0,.75
45, .176776695,0, 0,.75
45, .353553391,0, 0,.75
45, .530330086,0, 0,.75
*Ansi35,ANSI Fire brick, Refractory material
45, 0,0, 0,.25
45, .176776695,0, 0,.25, .3125,-.0625,0,-.0625
*Ansi36,ANSI Marble, Slate, Glass
45, 0,0, .21875,.125, .3125,-.0625,0,-.0625

*Swamp,Swampy area
0, 0,0, .5,.866025403, .125,-.875
90, .0625,0, .866025403,.5, .0625,-1.669550806
90, .078125,0, .866025403,.5, .05,-1.682050806
90, .046875,0, .66025403,.5, .05,-1.682050806
|
|
0, 0,0, .125,.125, .125,-.125
90, .125,0, .125,.125, .125,-.125
*HATCH1,Hatch at 45 Degree Angle
```

45,0,0,0,.5
***HATCH2,Hatch with Dashes & Dots:**
0,0,0,.25,.25,0.5,-.125,0,-.125,0,-.125
***HATCH3,Triangle Hatch:**
90,0,0,0,1,.5,-.5
0,0,0.5,0,1,.5,-.5
45,0,0,.7071,.7071,.7071,-.7071

Load the hatch pattern as described in Example 5 and test the hatch. Figure 35-37 shows the hatch pattern that will be generated by this hatch pattern (HATCH3). In Figure 35-37(a) the hatch pattern is at a 0° angle and the scale factor is 0.5. In Figure 35-37(b) the hatch pattern is at a -45° angle and the scale factor is 0.5.

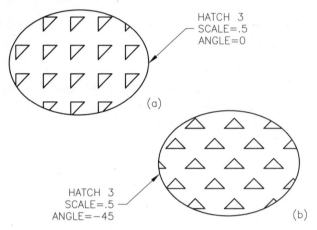

Figure 35-37 *Hatch generated by HATCH3 pattern*

SAVING HATCH PATTERNS IN A SEPARATE FILE

When you load a certain hatch pattern, AutoCAD looks for that definition in the *acad.pat* or *acadiso.pat* file. This is the reason the hatch pattern definitions must be in that file. However, you can add the new pattern definition to a different file and then copy that file to *acad.pat* or *acadiso.pat*. Be sure to make a copy of the original *acad.pat* or *acadiso.pat* file so that you can copy that file back when needed. Assume the name of the file that contains your custom hatch pattern definitions is *customh.pat*.

1. Copy *acad.pat* or *acadiso.pat* file to *acadorg.pat*
2. Copy *customh.pat* to *acad.pat* or *acadiso.pat*

If you want to use the original hatch pattern file, copy the *acadorg.pat* file to *acad.pat* or *acadiso.pat*.

CUSTOM HATCH PATTERN FILE

As mentioned earlier, you can add the new hatch pattern definitions to the *acad.pat* or *acadiso.pat* file. There is no limit to the number of hatch pattern definitions you can add to this file. However, if you have only one hatch pattern definition, you can define a separate file. It has the following requirements:

1. The name of the file has to be the same as the hatch pattern name.
2. The file can contain only one hatch pattern definition.
3. The hatch pattern name and the hatch file name should be unique.
4. If you use the hatch patterns saved on the removable drive quite often to hatch the drawings,

you can add removable drive to the AutoCAD search path using the **Options** dialog box. AutoCAD will automatically search the file on the removable drive and will display it in the **Hatch Creation** tab of the **Ribbon**.

*HATCH3,Triangle Hatch:
90,0,0,0,1,.5,-.5
0,0,0.5,0,1,.5,-.5
45,0,0,.7071,.7071,.7071,-.7071

Note
*The hatch lines can be edited after exploding the hatch with the **EXPLODE** command. After exploding, each hatch line becomes a separate object.*

However, it is recommended not to explode a hatch because it increases the size of the drawing database. For example, if a hatch consists of 100 lines, save it as a single object. However, after you explode the hatch, every line becomes a separate object and you have 99 additional objects in the drawing.

Keep the hatch lines in a separate layer to facilitate editing of the hatch lines.

Assign a unique color to hatch lines so that you can control the width of the hatch lines at the time of plotting.

Tip
*1. The file or the subdirectory in which hatch patterns have been saved must be defined in the **Support File Search Path** in the **File** tab of the **Options** dialog box.*

*2. The hatch patterns that you create automatically get added to AutoCAD's slide library as an integral part of AutoCAD and are displayed in the **Preview Area** in the **Hatch Pattern Palette** dialog box under the **Hatch Creation** tab of the **Ribbon**. Hence, there is no need to create a slide library.*

Self-Evaluation Test

Answer the following questions and then compare them to those given at the end of this chapter:

CREATING LINETYPES

1. The _____ command can be used to change the linetype scale factor.

2. The linetype description should not be more than _____ characters long.

3. A positive number denotes a pen _____ segment.

4. The segment length _____ generates a dot.

5. A negative number denotes a pen _____ segment.

6. The _____ option of the **LINETYPE** Command is used to generate a new linetype.

7. The description in the case of header line is _____. (optional/necessary)

8. The standard linetypes are stored in the _____ file.

9. The_____ determines the current linetype scaling.

CREATING HATCH PATTERNS

10. The header line consists of an asterisk, the pattern name, and _____.

11. The *acad.pat* or *acadiso.pat* file contains _____ number of hatch pattern definitions.

12. The standard hatch patterns are stored in the _____ file.

13. The first hatch line passes through a point whose coordinates are specified by _____ and _____.

Review Questions

Answer the following questions:

CREATING LINETYPES

1. The _____ command can be used to create a new linetype.

2. The _____ command can be used to load a linetype.

3. In AutoCAD, the linetypes are saved in the _____ file.

4. AutoCAD supports only _____ alignment field specification.

5. A line pattern definition always starts with _____.

6. A header line definition always starts with _____.

CREATING HATCH PATTERNS

7. The perpendicular distance between the hatch lines in a hatch pattern definition is specified by _____.

8. The displacement of the second hatch line in a hatch pattern definition is specified by _____.

9. The maximum number of dash lengths that can be specified in the line pattern definition of a hatch pattern is _____.

10. The hatch lines in different areas of the drawing will automatically _____ because the hatch lines are generated by offsetting.

11. The hatch angle as defined in the hatch pattern definition can be changed further when you use the AutoCAD _____ command.

12. When you load a hatch pattern, AutoCAD looks for that hatch pattern in the _____ file.

13. The hatch lines can be edited after _____ the hatch by using the _____ command.

EXERCISE 1   *Linetype*

Using the **LINETYPE** command, create a new linetype "DASH3DASH" with the following specifications:

Length of the first dash 0.75
Blank space 0.125
Dash length 0.25
Blank space 0.125
Dash length 0.25
Blank space 0.125
Dash length 0.25
Blank space 0.125

EXERCISE 2 — *Linetype*

Use a text editor to create a new file, *newlt2.lin*, and a new linetype, DASH2DASH, with the following specifications:

Length of the first dash 0.5
Blank space 0.1
Dash length 0.2
Blank space 0.1
Dash length 0.2
Blank space 0.1

EXERCISE 3 — *String Complex Linetype*

a. Write the definition of a string complex linetype (hot water line), as shown in Figure 35-38(a). To determine the length of the HW text string, you should first draw the text (HW) using any text command and then measure its length.

b. Write the definition of a string complex linetype (gas line), as shown in Figure 35-38(b). Determine the length of the text string as mentioned in part a.

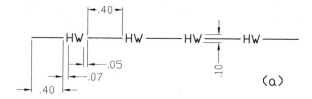

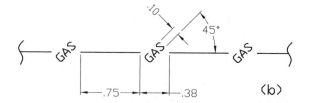

Figure 35-38 *Specifications for a string complex linetype*

EXERCISE 4 — *Hatch Pattern*

Determine the hatch pattern specifications and then write a hatch pattern definition for the hatch pattern shown in Figure 35-39.

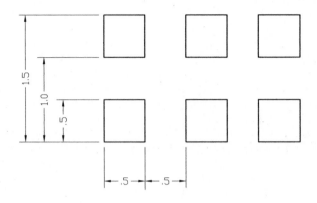

Figure 35-39 Drawing for Exercise 4

EXERCISE 5   *Hatch Pattern*

Determine the hatch pattern specifications and then write a hatch pattern definition for the hatch pattern shown in Figure 35-40.

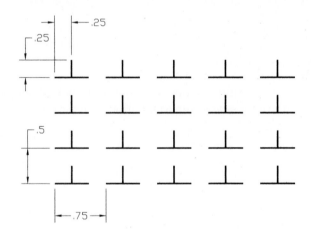

Figure 35-40 Hatch pattern for Exercise 5

Answers to Self-Evaluation Test

1. LTSCALE, **2.** 47, **3.** down, **4.** zero, **5.** up, **6.** Create, **7.** optional, **8.** *acad.lin* or *acadiso.lin*, **9.** CELTSCALE, **10.** Pattern Description, **11.** 67, **12.** *acad.pat* or *acadiso.pat*, **13.** X-origin, Y-origin.

Chapter **36**

Customizing the acad.pgp File

CHAPTER OBJECTIVES

After completing this chapter, you will be able to:
* *Customize the acad.pgp file.*
* *Edit different sections of the acad.pgp file.*
* *Abbreviate commands by defining command aliases.*
* *Use the REINIT command to reinitialize the PGP file.*

KEY TERMS

* *acad.pgp file*
* *OS Command Name*
* *Command Aliases*
* *Command Name*
* *Bit Flag*

WHAT IS THE acad.pgp FILE?

AutoCAD software comes with the program parameters file *acad.pgp*, which defines aliases for the operating system commands and some of the AutoCAD commands. When you install AutoCAD on a computer that runs in Windows 2000, XP, Vista, or Windows 7.0 operating system, this file is automatically copied on the *C:\Users\User\AppData\Roaming\Autodesk\AutoCAD 2012\R18.2\ enu\Support* subdirectory of the hard drive. The *acad.pgp* file lets you access the commands of operating system from the drawing editor. For example, if you want to delete a file, all you need to do is enter **DEL** at the Command prompt (Command: **DEL**), and AutoCAD will prompt you to enter the name of the file that you want to delete.

The file also contains command aliases of some frequently used AutoCAD commands. For example, the command alias for the **LINE** command is L. This is the reason if you enter **L** at the Command prompt (**Command: L**), AutoCAD will treat it as the **LINE** command. The *acad.pgp* file also contains comment lines that give you some information about different sections of the file. The *acad.pgp* file can be opened by choosing the **Edit Aliases** tool from the **Customization** panel in the **Manage** tab. You can also open the *acad.pgp* file by choosing the **Edit Program Parameters (acad.pgp)** tool from **Customize > Tools** in the menu bar. The following file is a listing of the standard *acad.pgp* file. Some of the lines have been deleted to make the file shorter.

```
;
;
; Program Parameters File For AutoCAD 2012
; External Command and Command Alias Definitions

; Copyright (C) 1997-2011 by Autodesk, Inc.  All Rights Reserved.

; Each time you open a new or existing drawing, AutoCAD searches
; the support path and reads the first acad.pgp file that it finds.

; -- External Commands --
; While AutoCAD is running, you can invoke other programs or utilities
; such Windows system commands, utilities, and applications.
; You define external commands by specifying a command name to be used
; from the AutoCAD command prompt and an executable command string
; that is passed to the operating system.

; -- Command Aliases --
; The Command Aliases section of this file provides default settings for
; AutoCAD command shortcuts.  Note: It is not recommended that  you directly
; modify this section of the PGP file., as any changes you make to this section of the
; file will not migrate successfully if you upgrade your AutoCAD to a
; newer version.  Instead, make changes to the new
; User Defined Command Aliases
; section towards the end of this file.

; -- User Defined Command Aliases --
; You can abbreviate frequently used AutoCAD commands by defining
; aliases for them in the User Defined Command Aliases section of acad.pgp.
; You can create a command alias for any AutoCAD command,
; device driver command, or external command.

; Recommendation: back up this file before editing it.  To ensure that
```

; any changes you make to PGP settings can successfully be migrated
; when you upgrade to the next version of AutoCAD, it is suggested that
; you make any changes to the default settings in the User Defined Command
; Aliases section at the end of this file.

; External command format:
; <Command name>,[<Shell request>],<Bit flag>,[*]<Prompt>,

; The bits of the bit flag have the following meanings:
; Bit 1: if set, don't wait for the application to finish
; Bit 2: if set, run the application minimized
; Bit 4: if set, run the application "hidden"
; Bit 8: if set, put the argument string in quotes
;
; Fill the "bit flag" field with the sum of the desired bits.
; Bits 2 and 4 are mutually exclusive; if both are specified, only
; the 2 bit is used. The most useful values are likely to be 0
; (start the application and wait for it to finish), 1 (start the
; application and don't wait), 3 (minimize and don't wait), and 5
; (hide and don't wait). Values of 2 and 4 should normally be avoided,
; as they make AutoCAD unavailable until the application has completed.
;
; Bit 8 allows commands like DEL to work properly with filenames that
; have spaces such as "long filename.dwg". Note that this will interfere
; with passing space delimited lists of file names to these same commands.
; If you prefer multiplefile support to using long file names, turn off
; the "8" bit in those commands.

; Examples of external commands for command windows

DEL,   DEL,   8,File to delete: ,
DIR,   DIR,   8,File specification: ,
SH,   ,   1,*OS Command: ,
SHELL,   ,   1,*OS Command: ,
START,   START,   1,*Application to start: ,
TYPE,   TYPE,   8,File to list: ,

(External commands area)

; Examples of external commands for Windows
; See also the (STARTAPP) AutoLISP function for an alternative method.

EXPLORER, START EXPLORER, 1,,
NOTEPAD,   START NOTEPAD, 1,*File to edit: ,
PBRUSH,   START PBRUSH,   1,,

; Command alias format:
;   <Alias>,*<Full command name>

; The following are guidelines for creating new command aliases.
; 1. An alias should reduce a command by at least two characters.
;   Commands with a control key equivalent, status bar button,
;   or function key do not require a command alias.
;   Examples: Control N, O, P, and S for New, Open, Print, Save.

; 2. Try the first character of the command, then try the first two,
;   then the first three.
; 3. Once an alias is defined, add suffixes for related aliases:
;   Examples: R for Redraw, RA for Redrawall, L for Line, LT for
;   Linetype.
; 4. Use a hyphen to differentiate between command line and dialog
;   box commands.
;   Example: B for Block, -B for -Block.
;
; Exceptions to the rules include AA for Area, T for Mtext, X for Explode.

; -- Sample aliases for AutoCAD commands --
; These examples include most frequently used commands. NOTE: It is recommended
; that you not make any changes to this section of the PGP file to ensure the
; proper migration of your customizations when you upgrade to the next version of
; AutoCAD. The aliases listed in this section are repeated in the User Custom
; Settings section at the end of this file, which can safely be edited while
; ensuring your changes will successfully migrate.

3A,	*3DARRAY
3DMIRROR,	*MIRROR3D
3DNavigate,	*3DWALK
3DO,	*3DORBIT
3DP,	*3DPRINT
3DPLOT,	*3DPRINT
3DW,	*3DWALK
3F,	*3DFACE
3M,	*3DMOVE
3P,	*3DPOLY
3R,	*3DROTATE
3S,	*3DSCALE
A,	*ARC
AC,	*BACTION
ADC,	*ADCENTER
AECTOACAD,	*-ExportToAutoCAD
AA,	*AREA
AL,	*ALIGN
3AL,	*3DALIGN
AP,	*APPLOAD
APLAY,	*ALLPLAY
AR,	*ARRAY
-AR,	*-ARRAY
ARR,	*ACTRECORD
ARM,	*ACTUSERMESSAGE
-ARM,	*-ACTUSERMESSAGE
ARU,	*ACTUSERINPUT
ARS,	*ACTSTOP
-ARS,	*-ACTSTOP
ATI,	*ATTIPEDIT
ATT,	*ATTDEF
-ATT,	*-ATTDEF

External commands area

ATE,	*ATTEDIT
-ATE,	*-ATTEDIT
ATTE,	*-ATTEDIT
B,	*BLOCK
|	
|	
|	
|	
HE,	*HATCHEDIT
HB,	*HATCHTOBACK
HI,	*HIDE
I,	*INSERT
-I,	*-INSERT
IAD,	*IMAGEADJUST
IAT,	*IMAGEATTACH
ICL,	*IMAGECLIP
IM,	*IMAGE
-IM,	*-IMAGE
IMP,	*IMPORT
IN,	*INTERSECT
INSERTCONTROLPOINT,	*CVADD
INF,	*INTERFERE
IO,	*INSERTOBJ
ISOLATE,	*ISOLATEOBJECTS
QVD,	*QVDRAWING
QVDC,	*QVDRAWINGCLOSE
QVL,	*QVLAYOUT
QVLC,	*QVLAYOUTCLOSE
J,	*JOIN
L,	*LINE
LA,	*LAYER
-LA,	*-LAYER
LAS,	*LAYERSTATE
LE,	*QLEADER
LEN,	*LENGTHEN
LESS,	*MESHSMOOTHLESS
LI,	*LIST
LINEWEIGHT,	*LWEIGHT
LMAN,	*LAYERSTATE
LO,	*-LAYOUT
LS,	*LIST
LT,	*LINETYPE
-LT,	*-LINETYPE
LTYPE,	*LINETYPE
-LTYPE,	*-LINETYPE
LTS,	*LTSCALE
LW,	*LWEIGHT
M,	*MOVE
MA,	*MATCHPROP
MAT,	*MATERIALS
ME,	*MEASURE

```
MEA,                          *MEASUREGEOM
MI,                           *MIRROR
ML,                           *MLINE
MLA,                          *MLEADERALIGN
MLC,                          *MLEADERCOLLECT
MLD,                          *MLEADER
|
|
|
R,                            *REDRAW
RA,                           *REDRAWALL
RC,                           *RENDERCROP
RE,                           *REGEN
REA,                          *REGENALL
REBUILD,                      *CVREBUILD
REC,                          *RECTANG
REFINE,                       *MESHREFINE
REG,                          *REGION
REMOVECONTROLPOINT,           *CVREMOVE
REN,                          *RENAME
-REN,                         *-RENAME
REV,                          *REVOLVE
RO,                           *ROTATE
|
|
|
; The following are alternative aliases and aliases as supplied
;  in AutoCAD Release 13.

AV,                           *DSVIEWER
CP,                           *COPY
DIMALI,                       *DIMALIGNED
DIMANG,                       *DIMANGULAR
DIMBASE,                      *DIMBASELINE
DIMCONT,                      *DIMCONTINUE
DIMDIA,                       *DIMDIAMETER
DIMED,                        *DIMEDIT
DIMTED,                       *DIMTEDIT
DIMLIN,                       *DIMLINEAR
DIMORD,                       *DIMORDINATE
DIMRAD,                       *DIMRADIUS
DIMSTY,                       *DIMSTYLE
DIMOVER,                      *DIMOVERRIDE
LEAD,                         *LEADER
TM,                           *TILEMODE

; Aliases for Hyperlink/URL Release 14 compatibility
SAVEURL,                      *SAVE
OPENURL,                      *OPEN
INSERTURL,                    *INSERT
```

```
; Aliases for commands discontinued in AutoCAD 2000:
AAD,                    *DBCONNECT
AEX,                    *DBCONNECT
ALI,                    *DBCONNECT
ASQ,                    *DBCONNECT
ARO,                    *DBCONNECT
ASE,                    *DBCONNECT
DDATTDEF,               *ATTDEF
DDATTEXT,               *ATTEXT
DDCHPROP,               *PROPERTIES
DDCOLOR,                *COLOR
DDLMODES,               *LAYER
DDLTYPE,                *LINETYPE
DDMODIFY,               *PROPERTIES
DDOSNAP,                *OSNAP
DDUCS,                  *UCS

; Aliases for commands discontinued in AutoCAD 2004:
ACADBLOCKDIALOG,        *BLOCK
ACADWBLOCKDIALOG,       *WBLOCK
ADCENTER,               *ADCENTER
BMAKE,                  *BLOCK
BMOD,                   *BLOCK
BPOLY,                  *BOUNDARY
|
|
|
PAINTER,                *MATCHPROP
PREFERENCES,            *OPTIONS
RECTANGLE,              *RECTANG
SHADE,                  *SHADEMODE
VIEWPORTS,              *VPORTS

; Aliases for commands discontinued in AutoCAD 2007:
RMAT,                   *MATERIALS
FOG,                    *RENDERENVIRONMENT
FINISH,                 *MATERIALS
SETUV,                  *MATERIALMAP
SHOWMAT,                *LIST
RFILEOPT,               *RENDERPRESETS
RENDSCR,                *RENDERWIN

; Aliases for commands discontinued in AutoCAD 2009:
DASHBOARD,              *RIBBON
DASHBOARDCLOSE,         *RIBBONCLOSE

;  -- User Defined Command Aliases --
; Make any changes or additions to the default AutoCAD command aliases in
; this section to ensure successful migration of these settings when you
; upgrade to the next version of AutoCAD.  If a command alias appears more
; than once in this file, items in the User Defined Command Alias take
; precedence over duplicates that appear earlier in the file.
; **********---------********** ; No xlate ; DO NOT REMOVE
```

SECTIONS OF THE acad.pgp FILE

The contents of the AutoCAD program parameters file (*acad.pgp*) can be categorized into three sections based on the information that is defined in the *acad.pgp* file. They do not appear in any definite order in the file, and have no section headings. For example, the comment lines can be entered anywhere in the file; the same is true with external commands and AutoCAD command aliases. The *acad.pgp* file can be divided into three sections: comments, external commands, and command aliases.

Comments

The comments of *acad.pgp* file can contain any number of comment lines and can occur anywhere in the file. Every comment line must start with a semicolon (;) (This is a comment line). Any line that is preceded by a semicolon is ignored by AutoCAD. You should use the comment line to give some relevant information about the file that will help other AutoCAD users to understand, edit, or update the file.

External Command

In the external command section, you can define any valid external command that is supported by your system. The information must be entered in the following format:

<Command name>, [OS Command name],<Bit flag>, [*]<Command prompt>,

Command Name. This is the name you want to use to activate the external command from the AutoCAD drawing editor. For example, you can use **GOWORD** as a command name to load the Word program (Command: **GOWORD**). The command name must not be an AutoCAD command name or an AutoCAD system variable name. If the name is an AutoCAD command name, the command name in the **PGP** file will be ignored. Also, if the name is an AutoCAD system variable name, the system variable will be ignored. You should use the command names that reflect the expected result of the external commands. (For example, **HELLO** is not a good command name for a directory file.) The command names can be in uppercase or lowercase.

OS Command Name. The OS Command name is the name of a valid system command that is supported by an operating system. For example, in DOS, the command to delete files is DEL and therefore, the OS Command name used in the *acad.pgp* file must be DEL. The following is a list of the type of commands that can be used in the PGP file:

OS Commands (DEL, DIR, TYPE, COPY, RENAME, EDLIN, etc.)
Commands for starting a word processor, or a text editor (WORD, SHELL, etc.)
Name of the user-defined programs and batch files

Bit Flag. This field must contain a number, preferably 8 or 1. The following are the bit flag values and their meaning:

Bit flag set to	Meaning
0	Starts the application and wait for it to finish
1	Do not waits for the application to finish
2	Runs the application minimized
4	Runs the application hidden
8	Puts the argument string in quotes

Command Prompt. The Command prompt field of the command line contains the prompt you want to display on the screen. It is an optional field that must be replaced by a comma if there is no prompt. If the operating system (OS) command that you want to use contains spaces, the prompt must be preceded by an asterisk (*). For example, the DOS command **EDIT NEW.PGP** contains a space between EDIT and NEW; therefore, the prompt used in this command line must be preceded by an asterisk. The command can be terminated by pressing ENTER. If the OS command consists of a single word (DIR, DEL, TYPE), the preceding asterisk must be omitted. In this case, you can terminate the command by pressing the SPACEBAR or the ENTER key.

Command Aliases

It is time-consuming to enter AutoCAD commands at the keyboard because it requires typing the complete command name before pressing ENTER. AutoCAD provides a facility that can be used to abbreviate the commands by defining aliases for the commands. This is made possible by the AutoCAD program parameters file (*acad.pgp* file). Each command alias line consists of two fields (**L, *LINE**). The first field (**L**) defines the alias of the command; the second field (***LINE**) consists of the AutoCAD command. The command must be preceded by an asterisk for AutoCAD to recognize the command line as a command alias. The two fields must be separated by a comma. The blank lines and the spaces between the two fields are ignored. In addition to AutoCAD commands, you can also use aliases for AutoLISP command names, provided the programs that contain the definition of these commands are loaded.

EXAMPLE 1 — Adding New Command Aliases

Add the following external commands and AutoCAD command aliases to the AutoCAD program parameters file (*acad.pgp*).

External Commands

Abbreviation	Command Description
GOWORD	This command loads the word processor (Winword) program from *C:\Program Files\Winword*
RN	This command executes the rename command of DOS.
COP	This command executes the copy command of DOS.

Command Aliases Section

Abbreviation	Command	Abbreviation	Command
EL	Ellipse	T	Trim
CO	Copy	CH	Chamfer
O	Offset	ST	Stretch
S	Scale	MI	Mirror

The *acad.pgp* file is an ASCII text file. To edit this file, choose the **Edit Aliases** tool from the **Customization** panel in the **Manage** tab. You can also use text editor like notepad to edit the *acad.pgp* file. The following is a partial listing of the *acad.pgp* file after the insertion of the lines for the command aliases of Example 1. **The line numbers on the right are not a part of the file; they are given here for reference only.** The lines that have been added to the file are highlighted in bold face.

```
DEL,DEL,            8,File to delete: ,              1
DIR,DIR,            8,File specification ,           2
EDIT, START EDIT,   8,File to edit: ,                3
SH,                 1,*OS Command: ,                 4
SHELL,              1,*OS Command: ,                 5
START,START,        1,Application to start: ,        6
TYPE,TYPE,          8,File to list: ,                7
                                                     8
```

GOWORD, START WINWORD,1,,		**9**
RN, RENAME,8,File to rename:,		**10**
COP,COPY,8,File to copy:,		**11**
DIMLIN	*DIMLINEAR	12
DIMORD,	*DIMORDINATE	13
DIMRAD,	*DIMRADIUS	14
DIMSTY,	*DIMSTYLE	15
DIMOVER,	*DIMOVERRIDE	16
LEAD,	*LEADER	17
TM,	*TILEMODE	18
EL,	***ELLIPSE**	**19**
CO,	***COPY**	**20**
O,	***OFFSET**	**21**
S,	***SCALE**	**22**
MI,	***MIRROR**	**23**
ST,	***STRETCH**	**24**

Explanation

Line 9

GOWORD, START WINWORD,1,,

In line 9, **GOWORD** loads the word processor program **(WINWORD)**. The **winword.exe** program is located in the winword directory under Program Files.

Lines 10 and 11

RN, RENAME,8,File to rename:,

COP,COPY,8,File to copy:,

Line 10 defines the alias for the DOS command **RENAME,** and the next line defines the alias for the DOS command **COPY**. The **8** is a bit flag, and the Command prompt **File to rename and File to copy** is automatically displayed to let you know the format and the type of information that is expected.

Lines 19 and 20

EL, *ELLIPSE

CO, *COPY

Line 19 defines the alias **(EL)** for the **ELLIPSE** command, and the next line defines the alias **(CO)** for the **COPY** command. The commands must be preceded by an asterisk. You can put any number of spaces between the alias abbreviation and the command.

Note

*If a command alias definition duplicates an existing one then the one that is lower down in the file is given preference and is allowed to work. For example, in a standard file, if you add S, *SCALE to the end of the file then your definition works and the one higher up in the file is ignored.*

REINITIALIZING THE acad.pgp FILE

When you make any changes in the *acad.pgp* file, there are two ways to reinitialize the *acad.pgp* file. One is to quit AutoCAD and then restart the AutoCAD program. When you start AutoCAD, the *acad.pgp* file is automatically loaded. You can also reinitialize the *acad.pgp* file by using the **REINIT** command. The **REINIT** command lets you reinitialize the I/O ports, digitizer, and AutoCAD program parameters file, *acad.pgp*. When you enter the **REINIT** command, AutoCAD will display the **Re-initialization** dialog box (Figure 36-1). To reinitialize the *acad.pgp* file, select the corresponding check box, and then choose **OK**. AutoCAD will reinitialize the program parameters file (*acad.pgp*), and then you can use the command aliases defined in the file.

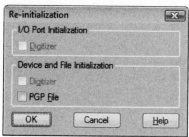

Figure 36-1 The Re-initialization dialog box

Tip
Copy and save the original acad.pgp file at some other location, so that after you have made changes in the acad.pgp file and used it, you can restore the default settings by copying and pasting the original acad.pgp file. This lets other users use the original, unedited file.

Self-Evaluation Test

Answer the following questions and then compare them to those given at the end of this chapter:

1. One way of reinitializing the *acad.pgp* file is to _____ AutoCAD and then _____ it.

2. The command used to reinitialize the *acad.pgp* file is _____ .

3. In the command alias section, the AutoCAD command must be preceded by an_____ symbol.

4. Every comment line must start with a _____ .

5. The command alias and the AutoCAD command must be separated by a_____ .

Review Questions

Indicate whether the following statements are true or false.

1. The comment section can contain any number of lines. (T/F)

2. AutoCAD ignores any line that is preceded by a semicolon. (T/F)

3. The command alias must not be an AutoCAD command. (T/F)

4. The bit flag field must be set to 8. (T/F)

5. In the command alias section, the command alias must be preceded by a semicolon. (T/F)

6. You cannot use aliases for AutoLISP commands. (T/F)

7. The *acad.pgp* file does not come with AutoCAD software. (T/F)

8. The *acad.pgp* file is an ASCII file. (T/F)

EXERCISE 1   *Adding New Command Aliases*

Add the following external commands and AutoCAD command aliases to the AutoCAD program parameters file (*acad.pgp*):

External Command Section

Abbreviation	Command Description
MYWORDPAD	This command loads the WORDPAD program that resides in the **Program Files\Accessories** directory.
MYEXCEL	This command loads the EXCEL program that resides in the **Program Fies\Microsoft Office** directory.
CD	This command executes the CHKDSK command of DOS.
FORMAT	This command executes the FORMAT command of DOS.

Command Aliases Section

Abbreviation	Command	Abbreviation	Command
BL	**BLOCK**	LTS	**LTSCALE**
INS	**INSERT**	EXP	**EXPLODE**
DIS	**DISTANCE**	GR	**GRID**
TE	**TIME**		

Answers to Self-Evaluation Test

1. quit, restart, **2. REINIT**, **3.** asterisk, **4.** semicolon, **5.** comma.

Index

Symbols

A